THE HUMAN AUDITORY SYSTEM

HANDBOOK OF CLINICAL NEUROLOGY

Series Editors

MICHAEL J. AMINOFF, FRANÇOIS BOLLER, AND DICK F. SWAAB

VOLUME 129

ELSEVIER

EDINBURGH LONDON NEW YORK OXFORD PHILADELPHIA
ST LOUIS SYDNEY TORONTO 2015

THE HUMAN AUDITORY SYSTEM: FUNDAMENTAL ORGANIZATION AND CLINICAL DISORDERS

Series Editors

MICHAEL J. AMINOFF, FRANÇOIS BOLLER, AND DICK F. SWAAB

Volume Editors

GASTONE G. CELESIA AND GREGORY HICKOK

VOLUME 129

3rd Series

ELSEVIER

EDINBURGH LONDON NEW YORK OXFORD PHILADELPHIA
ST LOUIS SYDNEY TORONTO 2015

ELSEVIER B.V.
Radarweg 29, PO Box 211, 1000 AE Amsterdam, Netherlands
The Boulevard, Langford Lane, Kidlington, Oxford OX5 1GB, UK
225 Wyman Street, Waltham, MA 02451, USA

ISBN: 9780444626301

British Library Cataloguing in Publication Data
A catalogue record for this book is available from the British Library

Library of Congress Cataloging in Publication Data
A catalog record for this book is available from the Library of Congress

Notices
Knowledge and best practice in this field are constantly changing. As new research and experience broaden our understanding, changes in research methods, professional practices, or medical treatment may become necessary.

Practitioners and researchers must always rely on their own experience and knowledge in evaluating and using any information, methods, compounds, or experiments described herein. In using such information or methods they should be mindful of their own safety and the safety of others, including parties for whom they have a professional responsibility.

With respect to any drug or pharmaceutical products identified, readers are advised to check the most current information provided (i) on procedures featured or (ii) by the manufacturer of each product to be administered, to verify the recommended dose or formula, the method and duration of administration, and contraindications. It is the responsibility of practitioners, relying on their own experience and knowledge of their patients, to make diagnoses, to determine dosages and the best treatment for each individual patient, and to take all appropriate safety precautions.

To the fullest extent of the law, neither the Publisher nor the authors, contributors, or editors, assume any liability for any injury and/or damage to persons or property as a matter of products liability, negligence or otherwise, or from any use or operation of any methods, products, instructions, or ideas contained in the material herein.

The Publisher

Commissioning Editor: Mica Haley
Development Editor: Michael Parkinson
Project Manager: Sujatha Thirugnana Sambandam
Designer/Design Direction: Alan Studholme

The Publisher's policy is to use **paper manufactured from sustainable forests**

Handbook of Clinical Neurology 3rd Series

Available titles

Vol. 79, The human hypothalamus: basic and clinical aspects, Part I, D.F. Swaab ISBN 9780444513571
Vol. 80, The human hypothalamus: basic and clinical aspects, Part II, D.F. Swaab ISBN 9780444514905
Vol. 81, Pain, F. Cervero and T.S. Jensen, eds. ISBN 9780444519016
Vol. 82, Motor neurone disorders and related diseases, A.A. Eisen and P.J. Shaw, eds. ISBN 9780444518941
Vol. 83, Parkinson's disease and related disorders, Part I, W.C. Koller and E. Melamed, eds. ISBN 9780444519009
Vol. 84, Parkinson's disease and related disorders, Part II, W.C. Koller and E. Melamed, eds. ISBN 9780444528933
Vol. 85, HIV/AIDS and the nervous system, P. Portegies and J. Berger, eds. ISBN 9780444520104
Vol. 86, Myopathies, F.L. Mastaglia and D. Hilton Jones, eds. ISBN 9780444518996
Vol. 87, Malformations of the nervous system, H.B. Sarnat and P. Curatolo, eds. ISBN 9780444518965
Vol. 88, Neuropsychology and behavioural neurology, G. Goldenberg and B.C. Miller, eds. ISBN 9780444518972
Vol. 89, Dementias, C. Duyckaerts and I. Litvan, eds. ISBN 9780444518989
Vol. 90, Disorders of consciousness, G.B. Young and E.F.M. Wijdicks, eds. ISBN 9780444518958
Vol. 91, Neuromuscular junction disorders, A.G. Engel, ed. ISBN 9780444520081
Vol. 92, Stroke – Part I: Basic and epidemiological aspects, M. Fisher, ed. ISBN 9780444520036
Vol. 93, Stroke – Part II: Clinical manifestations and pathogenesis, M. Fisher, ed. ISBN 9780444520043
Vol. 94, Stroke – Part III: Investigations and management, M. Fisher, ed. ISBN 9780444520050
Vol. 95, History of neurology, S. Finger, F. Boller and K.L. Tyler, eds. ISBN 9780444520081
Vol. 96, Bacterial infections of the central nervous system, K.L. Roos and A.R. Tunkel, eds. ISBN 9780444520159
Vol. 97, Headache, G. Nappi and M.A. Moskowitz, eds. ISBN 9780444521392
Vol. 98, Sleep disorders Part I, P. Montagna and S. Chokroverty, eds. ISBN 9780444520067
Vol. 99, Sleep disorders Part II, P. Montagna and S. Chokroverty, eds. ISBN 9780444520074
Vol. 100, Hyperkinetic movement disorders, W.J. Weiner and E. Tolosa, eds. ISBN 9780444520142
Vol. 101, Muscular dystrophies, A. Amato and R.C. Griggs, eds. ISBN 9780080450315
Vol. 102, Neuro-ophthalmology, C. Kennard and R.J. Leigh, eds. ISBN 9780444529039
Vol. 103, Ataxic disorders, S.H. Subramony and A. Durr, eds. ISBN 9780444518927
Vol. 104, Neuro-oncology Part I, W. Grisold and R. Sofietti, eds. ISBN 9780444521385
Vol. 105, Neuro-oncology Part II, W. Grisold and R. Sofietti, eds. ISBN 9780444535023
Vol. 106, Neurobiology of psychiatric disorders, T. Schlaepfer and C.B. Nemeroff, eds. ISBN 9780444520029
Vol. 107, Epilepsy Part I, H. Stefan and W.H. Theodore, eds. ISBN 9780444528988
Vol. 108, Epilepsy Part II, H. Stefan and W.H. Theodore, eds. ISBN 9780444528995
Vol. 109, Spinal cord injury, J. Verhaagen and J.W. McDonald III, eds. ISBN 9780444521378
Vol. 110, Neurological rehabilitation, M. Barnes and D.C. Good, eds. ISBN 9780444529015
Vol. 111, Pediatric neurology Part I, O. Dulac, M. Lassonde and H.B. Sarnat, eds. ISBN 9780444528919
Vol. 112, Pediatric neurology Part II, O. Dulac, M. Lassonde and H.B. Sarnat, eds. ISBN 9780444529107
Vol. 113, Pediatric neurology Part III, O. Dulac, M. Lassonde and H.B. Sarnat, eds. ISBN 9780444595652
Vol. 114, Neuroparasitology and tropical neurology, H.H. Garcia, H.B. Tanowitz and O.H. Del Brutto, eds. ISBN 9780444534903
Vol. 115, Peripheral nerve disorders, G. Said and C. Krarup, eds. ISBN 9780444529022
Vol. 116, Brain stimulation, A.M. Lozano and M. Hallett, eds. ISBN 9780444534972
Vol. 117, Autonomic nervous system, R.M. Buijs and D.F. Swaab, eds. ISBN 9780444534910
Vol. 118, Ethical and legal issues in neurology, J.L. Bernat and H.R. Beresford, eds. ISBN 9780444535016
Vol. 119, Neurologic aspects of systemic disease Part I, J. Biller and J.M. Ferro, eds. ISBN 9780702040863
Vol. 120, Neurologic aspects of systemic disease Part II, J. Biller and J.M. Ferro, eds. ISBN 9780702040870
Vol. 121, Neurologic aspects of systemic disease Part III, J. Biller and J.M. Ferro, eds. ISBN 9780702040887
Vol. 122, Multiple sclerosis and related disorders, D.S. Goodin, ed. ISBN 9780444520012
Vol. 123, Neurovirology, A.C. Tselis and J. Booss, eds. ISBN 9780444534880

Vol. 124, Clinical neuroendocrinology, E. Fliers, M. Korbonits and J.A. Romijn, eds. ISBN 9780444596024
Vol. 125, Alcohol and the nervous system, E.V. Sullivan and A. Pfefferbaum, eds. ISBN 9780444626196
Vol. 126, Diabetes and the nervous system, D.W. Zochodne and R.A. Malik, eds. ISBN 9780444534804
Vol. 127, Traumatic brain injury Part I, J. Grafman and A.M. Salazar, eds. ISBN 9780444528926
Vol. 128, Traumatic brain injury Part II, J. Grafman and A.M. Salazar, eds. ISBN 9780444635211

Foreword

It is curious how little attention most neurologists direct at hearing, despite the fact that its impairment disrupts communication and can lead to social withdrawal and isolation. Our ability to hear sounds and clothe them with meaning forms the basis of language, music, learning, and culture. In other words, it helps to define humankind.

We have long felt the need for a volume on the human auditory system in the *Handbook of Clinical Neurology* series. Such a volume could summarize the basic aspects of the auditory system, detail the methods that have been used to establish our current knowledge and further our understanding of how the system functions, review the various clinical disorders that result from gross or subtle abnormalities in the auditory system, and perhaps predict future developments in the field. It is therefore with great pleasure that we welcome this new volume to the series, edited by Gastone G. Celesia, former Professor and Chair of the Department of Neurology, Loyola University, Stritch School of Medicine, Chicago, and past Editor-in-Chief of the journal *Electroencephalography and Clinical Neurophysiology*, who has long had an interest in sensory systems and their integration centrally, and by Gregory Hickok, Professor and Director of the Center for Language Sciences at the University of California at Irvine, and former President of the Society for the Neurobiology of Language. Both are world authorities in their respective fields and together have made an outstanding team to develop and edit this authoritative volume.

The volume contains 38 chapters divided into three sections. The first section, which contains 12 chapters, discusses the anatomy and physiology of the human auditory system and also includes chapters on speech perception, production, and processing and on the perception of music. The second section contains six chapters on methodologies and techniques for investigating the auditory system. In a final section, 20 chapters discuss various disorders involving the auditory system, from such common problems as tinnitus or deafness from cochlear and auditory nerve disorders to less commonly encountered abnormalities such as congenital and acquired amusias and musicogenic epilepsy. The volume editors have added a final chapter in this section, dedicated to future advances in the field and thus providing insight into what might be expected in the years ahead.

Professors Celesia and Hickok have gathered about them a group of international experts to contribute to the present compendium, which they have crafted to ensure that the various contributions fit seamlessly together to produce a truly valuable appraisal and synthesis of the literature. We are grateful to them and to all the contributors. We have read and commented on each of the chapters in our capacity as series editors and believe that both clinicians and basic scientists will find much to appeal to them in this volume. Furthermore, the availability of the volume electronically on Elsevier's Science Direct site should ensure its ready accessibility and facilitate searches for specific information.

As always, it is a pleasure to thank Elsevier, our publishers – and in particular Mica Haley, Michael Parkinson, and Kristi Anderson – for their assistance in the development and production of this volume.

Michael J. Aminoff
François Boller
Dick F. Swaab

Preface

Nature has given man one tongue, but two ears, that we may hear twice as much as we speak.
Epictetus (c. AD *50–138),* Fragment VI

At the beginning of recorded history, philosopher/physicians began to observe and record the human ability to hear, see, smell, and taste. Alcmaeon of Croton (sixth–fifth century BC) was the first to write about hearing, as reported by Theophrastus (371–287 BC) in *De Sensibus*:

Hearing is by means of the ears, he [Alcmaeon] says, because within them is an empty space, and this empty space resounds. A kind of noise is produced by the cavity, and the internal air re-echoes this sound.

The first historical document that reported impaired hearing is found in a set of tablets excavated in Epidaurus dating to the fourth century BC. Sigerist in his *History of Medicine* states that two cases of deafness were listed among the 70 cases found. Around 385 BC Plato stated that deaf Athenians used a "sign language" to communicate, although at that time it was believed that deaf people could not be educated. It was only 2000 years later, during the Renaissance, that systematic studies of the anatomy of hearing began and the stigma of deafness started to change. Pedro Ponce de Leon (1510?–1584), a Benedictine monk, is recognized as the first teacher interested in helping deaf children. Progress in knowledge of the auditory system has been slow but steady; we have now come a long way. In the last 50 years, the literature on the anatomy and physiology of hearing and its disorders has reached encyclopedic proportion. Yet most of the knowledge of the human auditory systems is scattered among diverse publications and from different disciplines: from neurophysiology to audiology, from otolaryngology to neuropsychology, and from neurology to psychiatry. It is the task of this volume of the *Handbook of Clinical Neurology* to assemble current knowledge into a single volume with the collaboration of world-renowned experts from the basic and clinical sciences. The reader will have the opportunity to obtain information on the auditory pathways and auditory cortex, on the relationship between sound, music, and language, and on how the human auditory system can be tested. The various disorders of audition from the cochlea to the cortex are detailed, providing a comprehensive description of their clinical features, diagnosis, and therapy.

Section 1 of the volume describes the anatomy and physiology of the human auditory system. It includes descriptions of auditory pathways and the organization of the auditory cortex as well as an account of the development, and of learning and memory, of the auditory system. The representation of pitch, loudness, sound localization, and temporal coding is emphasized and based on human data. Speech perception, production, music perception, and the relationship of music and language are described.

Section 2 reviews the methodology and techniques used to study the system. It includes direct recording of the human auditory cortex as well as non-invasive electromagnetic and electrophysiologic testing of the system. Psychophysical and behavioral auditory testing is included. Neuroimaging techniques of the auditory cortices and the white-matter pathways are described.

Section 3 describes disorders of the auditory system. This section includes neurocognitive development in deaf children, aging of the system, central auditory processing disorders, and altered perceptions of sound from hyperacusis, diplacousis, tinnitus, auditory synesthesias, hallucinations, and palinacousis. Musicogenic epilepsy, auditory neglect, agnosia, and word deafness are described. Deafness in cochlear and auditory nerve disorders is covered as well as auditory neuropathy, hearing disorders in brainstem lesions, congenital and acquired amusias. Hearing difficulties in strokes, multiple sclerosis, and dementia are described. Lastly, the subject of future advances is addressed, with the knowledge that they are difficult to predict, and only educated guesses are possible.

We hope to have been successful in our endeavors and that this handbook will serve as the benchmark summarizing present knowledge and inspiring future research.

Gastone G. Celesia
Gregory Hickok

Contributors

P. Afra
Department of Neurology, School of Medicine, University of Utah, Salt Lake City, UT, USA

P. Albouy
Auditory Cognition and Psychoacoustics Team and Brain Dynamics and Cognition Team, Lyon Neuroscience Research Center, and University Lyon 1, Lyon, France

L.H. Arnal
Department of Neurosciences, University Medical Centre, Geneva, Switzerland and Department of Psychology, New York University, New York, NY, USA

D.E. Bamiou
Ear Institute, University College London and Neuro-otology Department, National Hospital for Neurology and Neurosurgery, London, UK

J.D. Bellis
Auditory Neuroscience Laboratory, Department of Communication Sciences and Disorders, University of South Dakota, Vermillion, SD, USA

T.J. Bellis
Department of Communication Sciences and Disorders, USD Speech-Language-Hearing Clinics and Division of Basic Biomedical Sciences, University of South Dakota, Vermillion, SD, USA

J.D. Blom
Parnassia Psychiatric Institute, The Hague and University of Groningen, Groningen, The Netherlands

A. Caclin
Brain Dynamics and Cognition Team, Lyon Neuroscience Research Center and University Lyon 1, Lyon, France

M. Catani
NatBrainLab, Institute of Psychiatry, King's College London, London, UK

G.G. Celesia
Department of Neurology, Loyola University of Chicago; Chicago Council for Science and Technology, Chicago, IL, USA

G.D. Chermak
Department of Speech and Hearing Sciences, Washington State University Spokane, Spokane, WA, USA

M.L. Chow
School of Medicine, University of California, San Francisco, CA, USA

C.N. Clark
Dementia Research Centre, UCL Institute of Neurology, University College London, Queen Square, London, United Kingdom

A. Dykstra
Department of Neurology, Ruprecht-Karls-Universität Heidelberg, Heidelberg, Germany

M.C. Fields
Department of Neurology, Icahn School of Medicine at Mount Sinai, Mount Sinai Hospital, New York, NY, USA

E. Fitzpatrick
Faculty of Health Sciences, University of Ottawa and Children's Hospital of Eastern Ontario Research Institute, Otttawa, ON, Canada

A.D. Friederici
Department of Neuropsychology, Max Planck Institute for Human Cognitive and Brain Sciences, Leipzig, Germany

M. Furst
School of Electrical Engineering, Tel Aviv University, Tel Aviv, Israel

A-L. Giraud
Department of Neurosciences, University Medical Centre, Geneva, Switzerland

H.L. Golden
Dementia Research Centre, UCL Institute of Neurology, University College London, Queen Square, London, United Kingdom

F.H. Guenther
Departments of Speech, Language, and Hearing Sciences and Biomedical Engineering, Boston University, Boston, MA, USA

A. Gutschalk
Department of Neurology, Ruprecht-Karls-Universität Heidelberg, Heidelberg, Germany

T.A. Hackett
Department of Hearing and Speech Sciences, Vanderbilt University School of Medicine and Department of Psychology, Vanderbilt University, Nashville, TN, USA

D.A. Hall
National Institute for Health Research, Nottingham Hearing Biomedical Research Unit, University of Nottingham, Nottingham, UK

G. Hickok
Department of Cognitive Sciences, Center for Language Science and Center for Cognitive Neuroscience, University of California, Irvine, CA, USA

K. Hopkins
School of Psychological Sciences, University of Manchester, Manchester, UK

M.A. Howard III
Department of Neurosurgery, University of Iowa, Iowa City, IA, USA

P. Janata
Center for Mind and Brain and Department of Psychology, University of California, Davis, CA, USA

M.M. Jastreboff
JHDF, Inc., Ellicott City, MD, USA

P.J. Jastreboff
Department of Otolaryngology, Emory University School of Medicine, Atlanta, GA, USA

J.K. Johnson
Institute for Health and Aging, University of California, San Francisco, CA, USA

N. Kraus
Auditory Neuroscience Laboratory, Department of Communication Sciences, Department of Neurobiology and Physiology and Department of Otolaryngology, Northwestern University, Evanston, IL, USA

A.D. Legatt
Departments of Neurology and Neuroscience, Montefiore Medical Center and Albert Einstein College of Medicine, Bronx, NY, USA

R.A. Levine
Department of Ear, Nose and Throat and Head and Neck Surgery, Tel Aviv Sourasky Medical Center, Tel Aviv, Israel

R. Litovsky
Binaural Hearing and Speech Laboratory, Waisman Center, University of Wisconsin-Madison, Madison, WI, USA

C. Maffei
NatBrainLab, Institute of Psychiatry, King's College London, London, UK

B.J. Malone
Center for Integrative Neuroscience and Coleman Memorial Laboratory, Department of Otolaryngology – Head and Neck Surgery, University of California San Francisco, San Francisco, CA, USA

L.V. Marcuse
Department of Neurology, Icahn School of Medicine at Mount Sinai, Mount Sinai Hospital, New York, NY, USA

J.C. Middlebrooks
Departments of Otolaryngology, Neurobiology and Behavior, Cognitive Sciences, and Biomedical Engineering, University of California at Irvine, Irvine, CA, USA

F.E. Musiek
Department of Communication Sciences and Department of Surgery (Otolaryngology), School of Medicine, University of Connecticut, Farmington, CT, USA

K.V. Nourski
Department of Neurosurgery, University of Iowa, Iowa City, IA, USA

Y. Oron
Department of Otolaryngology, Head and Neck Surgery, E. Wolfson Medical Centre, Holon, Israel

J.O. Pickles
Department of Physiology and Pharmacology, School of Biomedical Sciences, University of Queensland, St. Lucia, Queensland, Australia

D. Poeppel
Department of Psychology, New York University, New York, NY, USA

A. Prats-Galino
Laboratory of Surgical NeuroAnatomy, Human Anatomy and Embryology Unit, Faculty of Medicine, Universitat de Barcelona, Barcelona, Spain

G. Rance
School of Audiology, University of Melbourne, Melbourne, Australia

T.N. Roth
Department of Otorhinolaryngology, Head and Neck Surgery, University Hospital of Zurich, Zurich, Switzerland

C.E. Schreiner
Center for Integrative Neuroscience and Coleman Memorial Laboratory, Department of Otolaryngology – Head and Neck Surgery, University of California San Francisco, San Francisco, CA, USA

A.R. Shell
Department of Psychology, University of Maryland, College Park, MD, USA

J. Slater
Auditory Neuroscience Laboratory and Department of Communication Sciences, Northwestern University, Evanston, IL, USA

L.R. Slevc
Department of Psychology, University of Maryland, College Park, MD, USA

G. Soria
Experimental 7T MRI Unit, Institut d'Investigacions Biomediques August Pi i Sunyer, Barcelona, Spain

A. Starr
Departments of Neurology and Neurobiology, University of California, Irvine, CA, USA

J. Stern
Department of Neurology, Geffen School of Medicine, University of California, Los Angeles, CA, USA

K. Susi
Division of Psychology, School of Social Sciences, Nottingham Trent University, Nottingham, UK

B. Tillmann
Auditory Cognition and Psychoacoustics Team, Lyon Neuroscience Research Center and University Lyon 1, Lyon, France

J.D. Warren
Dementia Research Centre, UCL Institute of Neurology, University College London, Queen Square, London, United Kingdom

N.M. Weinberger
Center for the Neurobiology of Learning and Memory and Department of Neurobiology and Behavior, University of California, Irvine, CA, USA

Contents

Foreword vii
Preface ix
Contributors xi

SECTION 1 Anatomy and physiology of the human auditory system

1. **Auditory pathways: anatomy and physiology**
J.O. Pickles (St. Lucia, Australia) **3**

2. **Anatomic organization of the auditory cortex**
T.A. Hackett (Nashville, USA) **27**

3. **Development of the auditory system**
R. Litovsky (Madison, USA) **55**

4. **Representation of loudness in the auditory cortex**
C.E. Schreiner and B.J. Malone (San Francisco, USA) **73**

5. **Temporal coding in the auditory cortex**
L.H. Arnal, D. Poeppel, and A-L. Giraud (Geneva, Switzerland and New York, USA) **85**

6. **Sound localization**
J.C. Middlebrooks (Irvine, USA) **99**

7. **New perspectives on the auditory cortex: learning and memory**
N.M. Weinberger (Irvine, USA) **117**

8. **Neural basis of speech perception**
G. Hickok and D. Poeppel (Irvine and New York, USA) **149**

9. **Role of the auditory system in speech production**
F.H. Guenther and G. Hickok (Boston and Irvine, USA) **161**

10. **White-matter pathways for speech and language processing**
A.D. Friederici (Leipzig, Germany) **177**

11. **Neural basis of music perception**
P. Janata (Davis, USA) **187**

12. **Music and language: relations and disconnections**
N. Kraus and J. Slater (Evanston, USA) **207**

SECTION 2 Methodology and techniques

13. **Invasive recordings in the human auditory cortex**
K.V. Nourski and M.A. Howard III (Iowa City, USA) **225**

14. **Electromagnetic recording of the auditory system**
D. Poeppel and G. Hickok (New York and Irvine, USA) **245**

15. **Hemodynamic imaging of the auditory cortex**
D.A. Hall and K. Susi (Nottingham, UK) **257**

16. **Imaging white-matter pathways of the auditory system with diffusion imaging tractography**
C. Maffei, G. Soria, A. Prats-Galino, and M. Catani (London, UK and Barcelona, Spain) **277**

17. **Electrophysiologic auditory tests**
A.D. Legatt (Bronx, USA) **289**

18. **Psychophysical and behavioral peripheral and central auditory tests**
F.E. Musiek and G.D. Chermak (Farmington and Spokane, USA) **313**

SECTION 3 Disorders of the auditory system

19. **Neurocognitive development in congenitally deaf children**
E. Fitzpatrick (Ottawa, Canada) **335**

20. **Aging of the auditory system**
T.N. Roth (Zurich, Switzerland) **357**

21. **Decreased sound tolerance: hyperacusis, misophonia, diplacousis, and polyacousis**
P.J. Jastreboff and M.M. Jastreboff (Atlanta and Ellicott City, USA) **375**

22. **Auditory synesthesias**
P. Afra (Salt Lake City, USA) **389**

23. **Tinnitus**
R.A. Levine and Y. Oron (Tel Aviv and Holon, Israel) **409**

24. **Auditory hallucinations**
J.D. Blom (Groningen, The Netherlands) **433**

25. **Palinacousis**
M.C. Fields and L.V. Marcuse (New York, USA) **457**

26. **Musicogenic epilepsy**
J. Stern (Los Angeles, USA) **469**

27. **Deafness in cochlear and auditory nerve disorders**
K. Hopkins (Manchester, UK) **479**

28. **Auditory neuropathy**
A. Starr and G. Rance (Irvine, USA and Melbourne, Australia) **495**

29. Hearing disorders in brainstem lesions
G.G. Celesia (Chicago, USA) **509**

30. Central auditory processing disorders in children and adults
T.J. Bellis and J.D. Bellis (Vermillion, USA) **537**

31. Auditory neglect and related disorders
A. Gutschalk and A. Dykstra (Heidelberg, Germany) **557**

32. Auditory agnosia
L.R. Slevc and A.R. Shell (College Park, USA) **573**

33. Congenital amusias
B. Tillmann, P. Albouy, and A. Caclin (Lyon, France) **589**

34. Acquired amusia
C.N. Clark, H.L. Golden, and J.D. Warren (London, UK) **607**

35. Hearing disorders in stroke
D.E. Bamiou (London, UK) **633**

36. Hearing disorders in multiple sclerosis
M. Furst and R.A. Levine (Tel Aviv, Israel) **649**

37. Hearing and music in dementia
J.K. Johnson and M.L. Chow (San Francisco, USA) **667**

38. Future advances
G.G. Celesia and G. Hickok (Chicago and Irvine, USA) **689**

Index **693**

Section 1

Anatomy and physiology of the human auditory system

Handbook of Clinical Neurology, Vol. 129 (3rd series)
The Human Auditory System
G.G. Celesia and G. Hickok, Editors

Chapter 1

Auditory pathways: anatomy and physiology

JAMES O. PICKLES*

Department of Physiology and Pharmacology, School of Biomedical Sciences, University of Queensland, St. Lucia, Queensland, Australia

INTRODUCTION AND OVERVIEW

The auditory brainstem, midbrain, and cortex have a multiplicity of parallel and overlapping pathways, which have parallel but overlapping and interrelated functions. In addition, the stages of analysis of the auditory signal are not as clearly separated or as clearly comprehensible as in for instance the visual system. It is difficult therefore to use a simple functional framework to help understand the anatomic and physiologic results. It is hoped that this chapter will provide a scheme by which the auditory system can be more easily approached and understood.

The auditory signal is a time-dependent variation in sound pressure. From the one-dimensional stimuli as received by each ear, the whole multifeatured auditory world is constructed. Therefore the auditory system accomplishes an outstanding feat of both analysis and synthesis. This chapter will describe some of the anatomy and physiology that underlies this. The issues described in this chapter are also described in more detail in *An Introduction to the Physiology of Hearing*, to which the reader is referred for further information (Pickles, 2012).

THE OUTER AND MIDDLE EARS

The input impedance of the cochlea (defined as the pressure required to produce a unit displacement of the oval window) is some 200 times greater than that of free air (Nakajima et al., 2009). If the sound vibrations met the oval window directly, we can calculate that most of the energy would be reflected, with only 2% of the energy being transmitted. However, the outer and middle ears increase this transmission substantially. The increase in transmission is accomplished at two stages.

Firstly, the outer ear acts as a directionally sensitive ear trumpet, collecting sound pressures over the area of the pinna, and by a set of resonances, increasing the sound pressure at the rather smaller tympanic membrane. The frequency peaks of the major resonances are complementary, so that the pressure at the eardrum is raised relatively uniformly, by 15–20 dB, over the frequency range from 2 to 8 kHz, with transmission being similarly raised.

Secondly, there is an impedance transformer in the middle ear; this stage makes the major contribution. The middle-ear transformer has two components. Firstly, the largest factor arises from the ratio of the area of the tympanic membrane to the area of the footplate of the stapes in the oval window. The two areas are 60 mm^2 and 3.2 mm^2 respectively. The pressure on the oval window, and hence the pressure/displacement ratio, is therefore increased $60/3.2 = 18.75$ times. The second factor is the lever action: the arm of the malleus (i.e., the umbo) is 2.1 times longer than the arm of the stapes. Therefore the force at the round window, and hence the pressure, is increased 2.1 times, while the displacement is decreased 2.1 times. The impedance ratio, being pressure/displacement, is therefore increased $2.1^2 = 4.4$ times. The overall impedance change of the driving stimulus is therefore increased by 18.75×4.4 times $= 82.5$ times. The overall effect of the outer and middle ears is to increase the transmission efficiency, at its optimum frequency of 1 kHz, to 35% (Rosowski, 1991). Most of the losses in transmission are due to friction in the middle ear.

The absolute threshold and relation to outer- and middle-ear transmission

Over a wide range of frequencies, and in a variety of mammals, including human beings, the auditory absolute threshold corresponds to a power of the order of 10^{-18} W absorbed by the inner ear (Rosowski, 1991). Since at threshold we integrate energy for approximately

*Correspondence to: James O. Pickles, Department of Physiology and Pharmacology, School of Biomedical Sciences, University of Queensland, St. Lucia, Queensland 4072, Qld, Australia. E-mail: j.pickles@uq.edu.au

300 ms to make a decision, this equals an energy detection threshold of 3×10^{-19} J. This corresponds to the energy in a single quantum of red light. Therefore the fundamental energy sensitivities of the eye and the ear are comparable.

In young human beings with good hearing, the threshold of 10^{-18} W delivered to the cochlea applies over the range of 450 Hz to at least 10 kHz. Therefore, within this range the shape of the audiogram, i.e., the variation in the threshold of hearing as a function of frequency, can be described by the variation in the efficiency of transmission through the outer and middle ears to the inner ear. At higher and lower frequencies, however, other factors come into play. The upper frequency limit of hearing in human beings is commonly taken as 20 kHz in young children and 15 kHz in young adults (e.g., Dadson and King, 1952). The upper frequency limit of hearing arises because the cochlea itself becomes unresponsive to stimuli of higher frequency (Ruggero and Temchin, 2002). At low frequencies (<1 kHz), the threshold rises gradually as the stimulus frequency is lowered, so there is no clear lower frequency limit of hearing. The gradually increasing threshold arises because below 1 kHz there is reduced transmission of power through the middle ear, and at still lower frequencies (<450 Hz), the traveling wave reaches the apex of the cochlea. In this case, some of the power is shunted through the helicotrema, an opening between the scala vestibuli and the scala tympani at the extreme apex of the cochlea, and is not able to activate the hair cells.

THE COCHLEA

Overall anatomy

The cochlea is a spiral fluid-filled tube, with (in human beings) 2.5 turns, an overall width of 1 cm, and standing 5 mm high. It is unfortunate that, following the well-known illustrations of Netter (1948), many common illustrations show the cochlea at about three times this size. The fluid-filled tube has three divisions or scalae, which spiral together around the central core, the modiolus, containing the auditory nerve and many of the blood vessels. One membrane between the scalae, the basilar membrane, contains the organ of Corti, the site of the receptor cells (Fig. 1.1). The spiral allows a long (35 mm long) basilar membrane and organ of Corti to be packed into a small overall dimension.

Anatomy in relation to function

The cochlea performs an amazing feat of sound detection and analysis. The absolute threshold mentioned above is many times lower than the thermal noise expected in the detector. Moreover, the cochlea performs a frequency analysis with a high degree of resolution, but without the long temporal "ringing" that would normally accompany such a high degree of frequency resolution.

The input vibrations of the cochlea produce the well-known traveling wave on the basilar membrane, which peaks more basally for stimuli of higher frequencies, and more apically for stimuli of lower frequencies (Fig. 1.2A). Therefore stimulus frequency is mapped on to place of stimulation in the cochlea. As originally measured by Békésy in human cadavers (see Békésy, 1960), the traveling wave was relatively small and broad, and showed poor frequency selectivity as the frequency was changed. It is now recognized that when the cochlea is in good physiologic condition, the traveling wave is much larger and more sharply peaked, showing a much greater frequency selectivity (Fig. 1.2B; for review see Robles and Ruggero, 2001).

The sharp tuning and great sensitivity arise because the organ of Corti acts as a regenerative mechanical amplifier, with some of the mechanical energy detected by the hair cells being amplified by the hair cells and fed back as mechanical energy into the traveling wave.

While the basilar papillae of non-mammals probably also contain some form of mechanical feedback, the cochleae of mammals far surpass them in performance, achieving much lower absolute thresholds and also much faster temporal responses, which give much higher upper frequency limits of detection. When the hair cells are stimulated, the changes in electric potential within the outer hair cells drive a mechanical response in the outer hair cells so that they change length. This changes the thickness of the organ of Corti, changing the fluid flow within the cochlea in such a way that it mechanically enhances the traveling wave. The amplitude of the traveling wave is in this way amplified in magnitude by approximately 40–60 dB (100–1000 times). A further consequence is that, because the traveling wave is progressively amplified by mechanical feedback as it travels up the cochlea, it grows exponentially in amplitude as a function of distance. Then it reaches a point at which the traveling wave is not mechanically possible for the particular frequency of stimulation. At this point, the amplitude of the wave falls abruptly. The sharp steep lower, and particularly steep upper, slopes of the traveling wave envelope give rise to the high-frequency resolution of the cochlea. However, because the sharp tuning is not produced by a simple mechanical resonance (as, e.g., a mass-and-spring system), the peak of the traveling wave can be relatively rounded, giving relatively good temporal resolution in the responses.

The active process, otherwise known as reverse transduction, is driven by the protein prestin in the walls of the outer hair cells, which causes the cells to change length when the transmembrane voltage varies. Prestin is a modified anion transporter. It acts as an anion

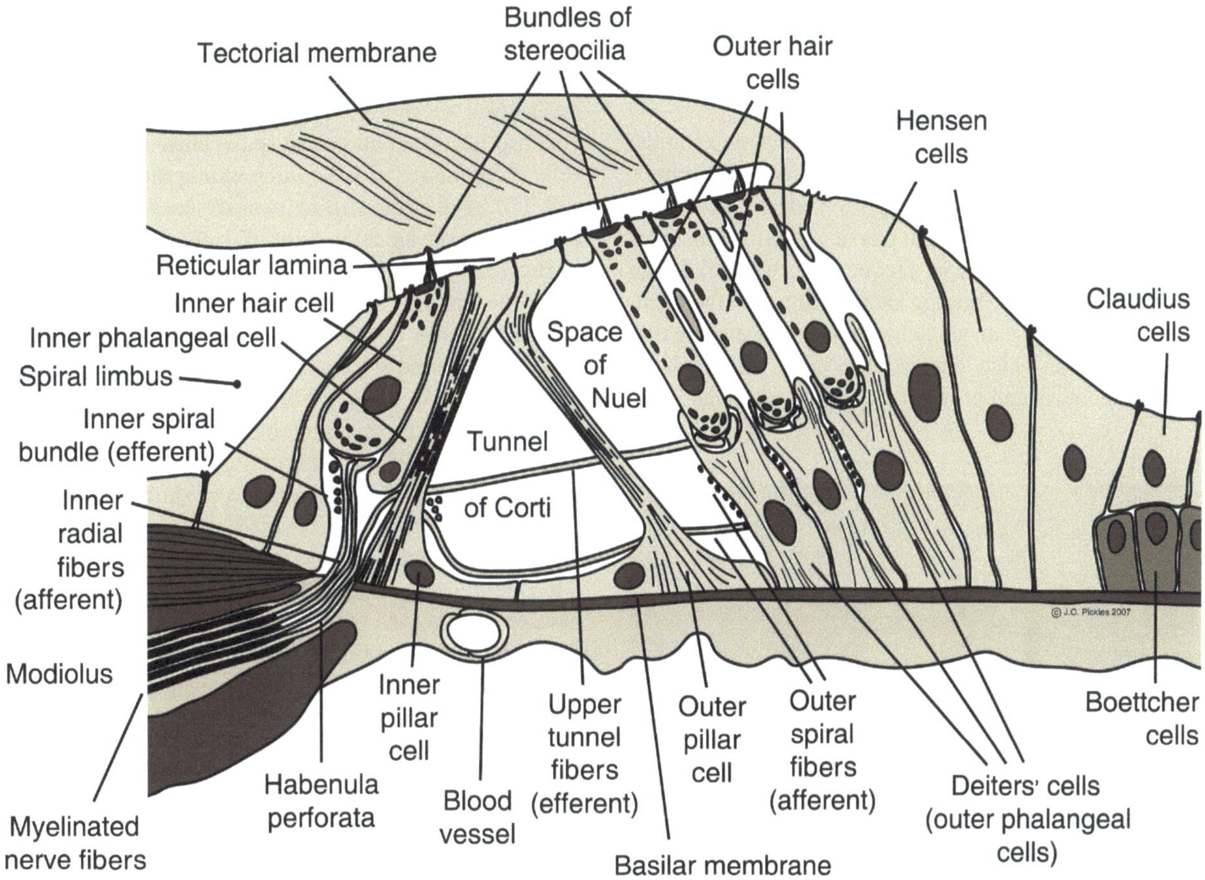

Fig. 1.1. Cross-section of the organ of Corti, as it appears in the basal turn, showing the hair cells in the reticular lamina, with their bundles of stereocilia running to the tectorial membrane. Deiters' cells send extensions (phalanges) up to the reticular lamina, running in the spaces around the outer hair cells, although the phalanges are incompletely shown in this particular cross-section. The modiolus is to the left of the figure. (Reproduced rom Pickles, 2013.)

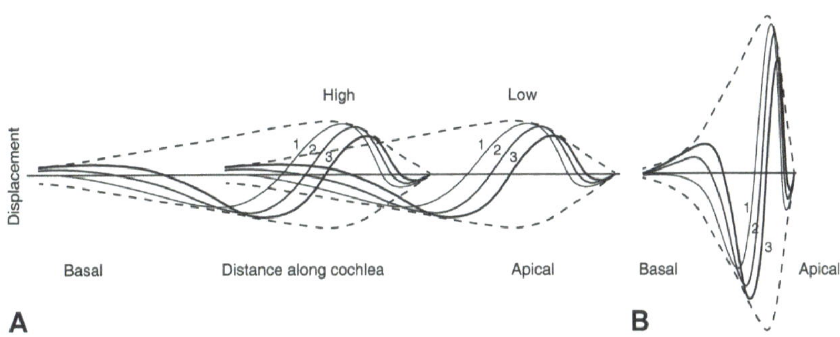

Fig. 1.2. (**A**) The traveling wave of the basilar membrane, as first demonstrated by Békésy (see Békésy, 1960). Traveling waves to stimuli of lower frequency peak more apically in the cochlea. Waves at successive instants of time (1, 2, 3) are shown; the waves move apically within an envelope (dashed line) that stays constant for tones of constant frequency.

(**B**) Current experiments show that, with the cochlea in optimal condition, the traveling wave is much larger and peaks much more sharply than shown by Békésy. The traveling wave, if it was drawn on the same scales as the waves in (**A**), would be five times narrower and 30 times higher than shown in (**B**). (Reproduced from Pickles, 2013.)

transporter in zebrafish, chicken, lizards, and frogs and does not have any motile function in these species. However in mammals prestin has acquired a motile function while losing its ability as an anion transporter. Motile prestin is found in monotremes (platypus) as well as in eutherian mammals, showing that this form of the active process developed early in mammalian evolution (Tang et al., 2013). The structure of the organ of Corti, with an open lattice work of outer hair cells and supporting cells running at an angle to each other, allows the organ of

Corti to change thickness as the outer hair cells change length; this open structure is seen in monotremes as well as in eutherian mammals (Fig. 1.3; Ladhams and Pickles, 1996).

An active process arising from length changes in the outer hair cells means that the sensitivity of mammals far surpasses that of any other vertebrate orders. It allows mammals to detect sounds of very low intensities over a very wide range of frequencies, while the high upper frequencies of hearing allow accurate sound localization in animals with small heads. It is conceivable that these abilities, which allowed the early mammals to detect predators and prey before they themselves could be detected, helped to permit the early expansion of the mammals.

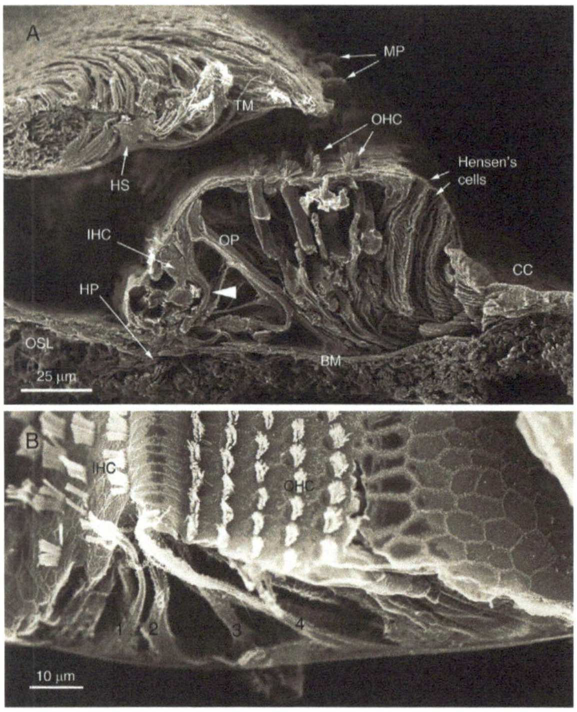

Fig. 1.3. (**A**) Human organ of Corti: scanning electron micrograph of a fractured cross-section from the mid-turn of the cochlea (the 500-Hz place). In this specimen, the inner pillar cell (arrowhead) has partly collapsed, and the tectorial membrane has shrunk away from the reticular lamina. There are four rows of outer hair cells (OHC). (Reproduced from Glueckert et al., 2005.)

(**B**) Scanning micrograph of a fractured cross-section of a monotreme (echidna) organ of Corti, showing four to five irregular rows of inner hair cells, and five rows of outer hair cells. There are also four rows of pillar cells (1, 2, 3, 4), compared with just one row of inner and one row of outer pillar cells in eutherian mammals. (Reproduced rom Ladhams and Pickles, 1996.)

BM, basilar membrane; CC, Claudius cell; HP, habenula perforata; HS, Hensen's stripe; IHC, inner hair cell; MP, marginal pillars (of tectorial membrane); OP, outer pillar cell; OSL, osseous spiral lamina; TM, tectorial membrane.

Forward transduction refers to the transformation of mechanical vibration into an electrochemical change. Forward transduction occurs in the inner and outer hair cells in fundamentally the same way in the two types of cell. The large number of hairs or stereocilia (60–120 per hair cell) on the apical surface of each hair cell are interconnected by fine fibers called tip links, which emerge from the tips of the shorter stereocilia on the bundle, to run upwards to join the adjacent, taller stereocilia. Deflection of the stereocilia in the excitatory direction stretches the tip links, pulling the mechanotransducer channels open by a direct mechanical action (Corey and Hudspeth, 1983; Pickles et al., 1984). The mechanotransducer channels are thought to be situated at the tips of the shorter stereocilia on each bundle, and so would be associated with the bottom ends of each tip link (Hudspeth, 1982; Beurg et al., 2009). Outer hair cells have specializations for their particular mechanical role in reverse transduction, with bundles of stereocilia that are relatively stiff, and with motile cell bodies. Inner hair cells have a more direct function, acting as simple sensory receptor cells. Here, the changes in intracellular potential caused by opening and closing of the mechanotransducer channels modify the release of neurotransmitter at the cells' basal ends, to activate the fibers of the auditory nerve.

The output of the cochlea

Auditory nerve fibers innervate inner hair cells which are activated by the local vibration of the organ of Corti and tectorial membrane. Inner hair cells are simple receptors without any motile function. The responses of auditory nerve fibers therefore follow the vibration in a relatively straightforward way. Their responses underlie the responses of all later stages of the auditory system.

The frequency selectivity of individual nerve fibers can be demonstrated by their tuning curves, or frequency–threshold curves, which show the sound intensity necessary to produce a certain threshold increase in firing rate, as a function of frequency (Fig. 1.4). Fibers innervating the high-frequency base of the cochlea are tuned to high frequencies, while those innervating the apex are tuned to low frequencies. As well as the sensitive, sharply tuned tip, arising from the peak of the traveling wave, the curves have a long, high-threshold tail stretching towards low frequencies, reflecting an insensitive, shallow, low-frequency slope of the traveling wave.

For low-frequency sounds, auditory nerve fibers are preferentially activated in one phase of the stimulus, reflecting displacement of the stereocilia of the inner hair cells in the effective direction. Therefore the responses of the auditory nerve fibers are phase-locked, though in a probabilistic way, to the individual cycles of the stimulating waveform (Fig. 1.5). As the stimulus frequency is raised (to 1 kHz and above), the phase locking becomes gradually weaker, so that by 5 kHz auditory nerve fibers fire with nearly equal probability at all phases of the waveform. Both the identity (i.e., site of origin) of the nerve fiber, which determines its characteristic frequency, and the timing of its action potentials can convey information about the frequency of the stimulus. The extent to which these two cues are actually used in judgements of frequency is still a matter of controversy.

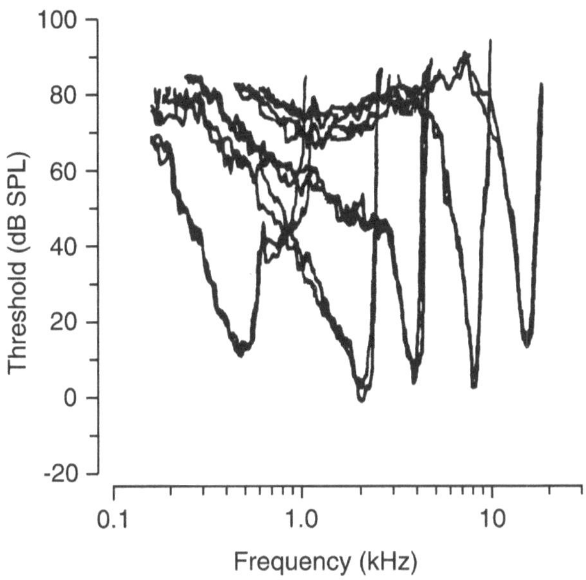

Fig. 1.4. Tuning curves (frequency–threshold curves) recorded from five sites in the cat auditory nerve. Two fibers of similar characteristic frequency and threshold have been recorded at each site, showing the constancy of tuning under these circumstances. The tuning curves have a sharp low-threshold tip at their characteristic frequency, and a long, high-threshold tail stretching to lower frequencies. SPL, sound pressure level. Redrawn with original curves from Liberman and Kiang (1978). (Reproduced rom Pickles, 2013.)

THE AUDITORY CENTRAL NERVOUS SYSTEM: INTRODUCTION TO CENTRAL PROCESSING

The central auditory system has multiple parallel and overlapping pathways, which diverge and converge throughout the system. Any evolutionary pressure for simplicity has clearly been constrained by the demands of the types of analysis being undertaken by the system. The auditory system, unlike the visual system, deals with very rapid temporal fluctuations and preserves timing of a few tens of microseconds in the overall population response. However, many neuronal circuits needed for stimulus extraction, such as those undertaking lateral inhibition to pick out dominant spectral features from a background, introduce much longer time uncertainties, on the order of milliseconds. I suggest that the complexity of the auditory system arises because of the incompatibility of these two types of analysis. The result is that there are multiple parallel systems, some preserving temporal information with a high degree of fidelity, others dedicated to the analysis of patterns of activity measured in the mean response rates over populations of neurons. Pathways of the two types are separate at the very first stage of the central auditory nervous system, in the input from the auditory

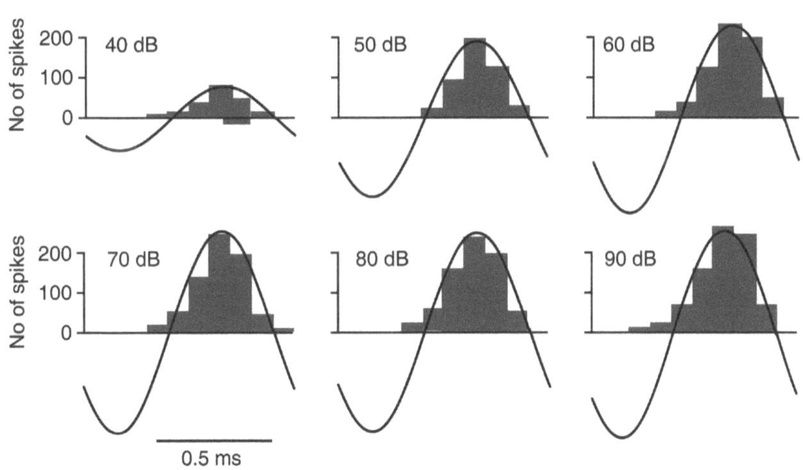

Fig. 1.5. Period histograms show that auditory nerve fibers tend to be activated by only one-half of the auditory stimulating waveform (waveform shown by superimposed sine wave). Although the number of action potentials does not increase above about 70 dB sound pressure level, the histograms still preserve the shape of the stimulating waveform, with no tendency to square. Based on original data of Rose et al. (1971). (Reproduced from Pickles, 2013.)

nerve to the cochlear nucleus. At the inferior colliculus (IC) and later, the results of the analyses are progressively combined and are further refined, so that neural responses become driven by combinations of both types of feature. These neurons eventually produce the neural representation of the auditory objects in the external world.

The division into the two pathways first occurs in the auditory nerve. Each auditory nerve fiber branches on entering the brainstem, to give rise to the anterior and posterior branches of the nerve. The anterior branch innervates the anterior cochlear nucleus, and gives rise to the ventral, sound-localizing, auditory stream of the brainstem. The posterior branch innervates the dorsal cochlear nucleus (DCN), and, in passing, the posteroventral cochlear nucleus (PVCN), to give rise to the dorsal auditory stream of the brainstem, involved in complex stimulus analysis (Fig. 1.6). There are therefore three divisions of the cochlear nucleus, one, the anteroventral cochlear nucleus (AVCN), being part of the ventral stream of the brainstem, and the other two, the PVCN and DCN, part of the dorsal stream of the brainstem. The PVCN also makes some contribution to the ventral auditory stream.

The auditory nerve fibers entering the cochlear nucleus preserve the spatial arrangements arising from their sites of origin in the cochlea. The incoming fibers are therefore tonotopically arranged, that is, their best or characteristic frequencies are spatially ordered. The regular pattern of innervation means that each of the separate divisions of the cochlear nucleus are also tonotopically arranged, each with its own separate progression of characteristic frequencies. This tonotopic organization is then preserved in the higher central auditory structures.

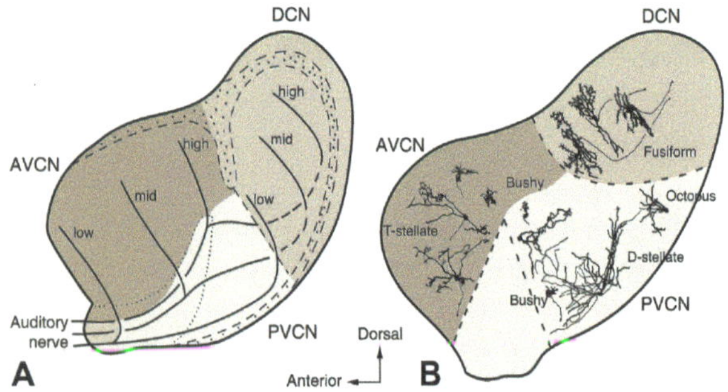

Fig. 1.6. (**A**) A sagittal section of the cat cochlear nucleus shows the three divisions of the nucleus. The paths of branching auditory nerve fibers of high, mid and low frequency are illustrated. (**B**) Principal cell types in each division of the nucleus. AVCN, anteroventral cochlear nucleus; DCN, dorsal cochlear nucleus; PVCN, posteroventral cochlear nucleus. (Reproduced from Osen and Roth, 1969, and Oertel, 1991.)

THE VENTRAL AUDITORY STREAM OF THE BRAINSTEM; SOUND LOCALIZATION BY COMPARING RESPONSES AT THE TWO EARS

The anteroventral cochlear nucleus

The AVCN preserves and further enhances the precision of temporal information in the neural firing. It feeds via the ventral auditory stream of the brainstem into the superior olivary complex. Here, sounds are localized by comparing the responses to the stimuli, including their relative intensities and relative timings, at the two ears.

The AVCN contains two types of cell, known as spherical and globular bushy cells. They both receive direct inputs from the auditory nerve. Spherical bushy cells receive inputs from one to four individual auditory nerve fibers, via giant synaptic terminals called end-bulbs of Held, which wrap around the cell bodies. Globular bushy cells receive their inputs from four to 40 auditory nerve fibers, via smaller modified end-bulbs of Held. Their frequency–response areas are similar to those found in the auditory nerve (type I, Fig. 1.7A).

Activation of a single end-bulb of Held is able to trigger an action potential in the spherical bushy cell. Therefore the firing in the spherical bushy cells can follow the firing in the input fibers in a temporally accurate 1:1 fashion. However, as the stimulus intensity is raised, inhibitory inputs from elsewhere in the nucleus reduce the sensitivity of the cell, so that a larger number of inputs arriving at the same moment are needed to trigger an action potential (Kopp-Scheinpflug et al., 2002). The resulting temporal averaging over a larger number of inputs increases the temporal accuracy of the action potentials in the cell. They therefore more accurately represent the timing of the mean firing in the auditory nerve and so of the stimulating auditory waveform. The averaging also occurs and is more powerful in globular bushy cells because of their larger number of synaptic inputs. The temporal accuracy of the output of the AVCN is thereby increased, and is much greater than in the individual input auditory nerve fibers (Fig. 1.8; Joris et al., 1994). The inhibitory areas, which may overlap only the excitatory areas, or which may extend over a greater frequency range to give lateral inhibition, also serve to preserve spectral contrasts in the stimuli over a wide range of stimulus intensities.

The spherical bushy cells project via the ventral acoustic stria to the medial superior olivary nuclei (MSO) of both sides (Fig. 1.9). The MSOs contribute to sound localization by comparing the timing of the stimuli at the two ears. Clearly, accurate preservation of timing is necessary for this. The spherical bushy cells project also to the ipsilateral lateral superior olive (LSO), where the intensities of the stimuli at the two ears are compared.

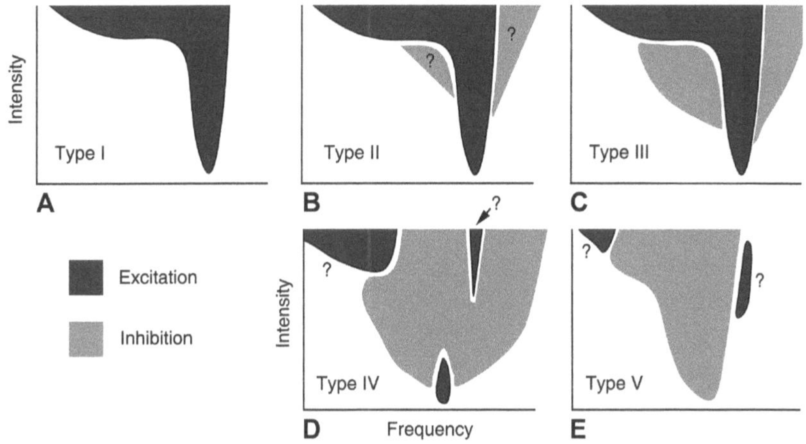

Fig. 1.7. Tuning curves of excitation and inhibition in the cat cochlear nucleus are shown in order of increasing amounts of inhibition (**A–E**). Purely excitatory responses (**A**: type I) are predominant in the anteroventral cochlear nucleus. Greater amounts of inhibition are found towards the dorsal cochlear nucleus (especially in the cat; e.g., **C–E**, types III–V). Question marks show variable features. (Reproduced from Pickles, 2012.)

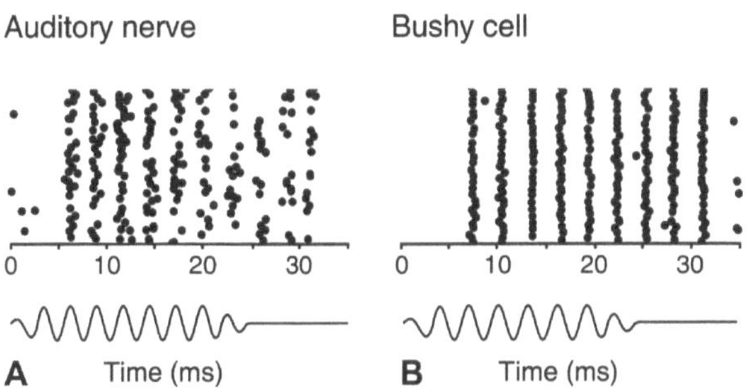

Fig. 1.8. Raster diagram of firing to tone burst in an auditory nerve fiber (**A**) and in a bushy cell (**B**: type not described). Note greater precision of firing in (**B**) than in (**A**). Redrawn using data from Joris (1998). (Reproduced from Pickles, 2012.)

The globular bushy cells project primarily to the contralateral medial nucleus of the trapezoid body (MNTB), a stage in the pathway to the contralateral LSO, and therefore also contribute to localization by comparison of the intensities of the stimuli at the two ears.

The medial superior olive

The MSO has a flat, disk-like structure, and receives direct inputs from the AVCNs of both sides. The constituent cells are bipolar; contralateral inputs tend to contact the medial part of their adjacent dendritic arbor, while ipsilateral inputs tend to contact the distal parts of their adjacent dendritic arbors. There are also inhibitory synapses on the soma and proximal dendrites. The cells respond optimally when the ipsilateral and contralateral excitatory drives to the cells coincide. The combination of the differences in delay due to differences in neural

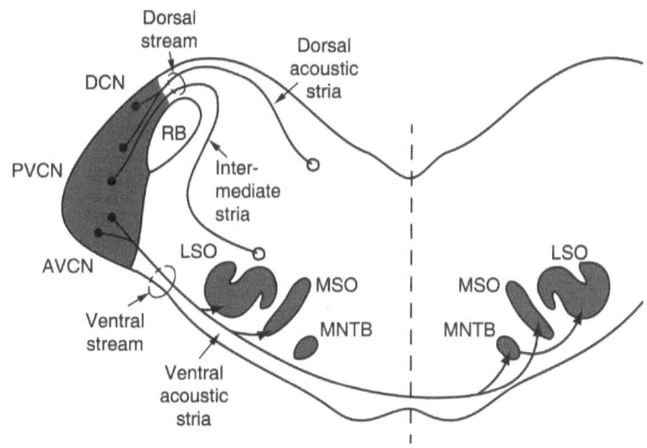

Fig. 1.9. The major outflows from the cochlear nucleus are shown on a transverse section of the cat brainstem. The ventral stream (binaural sound localization stream) arises in the anteroventral cochlear nucleus (AVCN) with a contribution from the posteroventral cochlear nucleus (PVCN). It runs in the ventral acoustic stria to the superior olivary nuclei of both sides (see Fig. 1.10 for further details). The dorsal stream (mainly sound identification stream) runs in the dorsal and intermediate acoustic striae and passes dorsally over the restiform body (RB) or inferior cerebellar peduncle. It arises in the dorsal cochlear nucleus (DCN) and PVCN. Most fibers of the dorsal stream leave the plane of the section (shown by small circles) to run to higher levels of the brainstem, while others (not shown) run to the cells surrounding the lateral superior olive (LSO). MSO, medial superior olivary nucleus; MNTB, medial nucleus of the trapezoid body. Other fibers (not shown) run between the cochlear nuclei of both sides. The fibers are represented diagrammatically and do not necessarily branch or join as indicated. For this diagram, anterior sections containing the AVCN and more posterior sections containing the PVCN and DCN have been superimposed. (Reproduced from Pickles, 2012.)

path lengths from the ipsilateral and contralateral AVCNs, differences in speed of effects from the MSO cells' distal and proximal dendrites, and the effect of the inhibition mean that the cells respond optimally when the auditory stimulus strikes the contralateral ear first (for reviews, see Grothe et al., 2010; Golding and Oertel, 2012). This means that cells in the MSO preferentially respond to sound sources on the contralateral side of the head, a selectivity based on the ability to detect precise differences in timing.

The lateral superior olive

The LSO is a flat sheet of cells, folded in a complex way which is species-dependent, but which in the cat give it a characteristic S-shape in transverse sections, giving rise to the name the "S-segment." The LSO receives two major inputs. It receives a direct excitatory input from the ipsilateral AVCN, predominantly from the spherical bushy cells. The second major input is inhibitory, and arrives via a synapse in the adjacent MNTB, from globular bushy cells of the contralateral AVCN. Therefore the great majority of cells in the LSO are excited by ipsilateral stimuli and inhibited by contralateral ones. Sounds on, say, the left side of the head will be more intense in the left ear. Therefore the excitatory–inhibitory interactions mean that neurons in the LSO preferentially respond to sounds on the same side of the head, a selectivity that is based on intensity cues. Since the thresholds and tunings for the ipsilateral and contralateral effects are approximately the same, the cells of the LSO are able to lateralize sounds in a manner that is frequency-specific.

Outputs of the ventral auditory stream of the brainstem

The main output of the MSO is uncrossed, and projects to the ipsilateral IC. Since the MSO preferentially responds to sounds on the opposite side of the head, this means that its target cells in the IC also preferentially respond to sounds on the opposite side of the head (Fig. 1.10).

The main output of the LSO, preferentially driven by sounds on the ipsilateral side of the head, is however crossed, and so goes to the IC of the opposite side. This means that cells in the IC are driven by sounds on the opposite side of the head as a result of input from the LSO, as well as from the MSO. Therefore the timing-based localization signals from the MSO and the

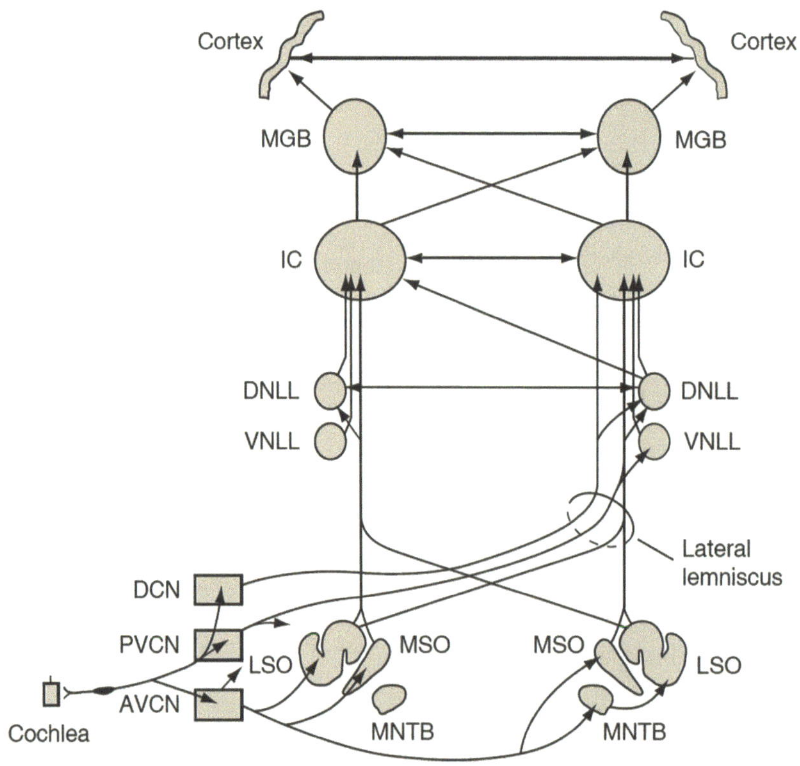

Fig. 1.10. The main ascending pathways of the brainstem. Many minor pathways, including many inhibitory pathways, are not shown. DNLL, dorsal nucleus of the lateral lemniscus; IC, inferior colliculus; MGB, medial geniculate body; VNLL, ventral nucleus of the lateral lemniscus. For other abbreviations, see caption to Figure 1.9. The branching and joining of arrows do not necessarily mean that fibers branch or join. (Reproduced from Pickles, 2012.)

intensity-based localization signals from the LSO summate in their effects at the IC.

The fibers to the IC travel in the tract known as the lateral lemniscus. The tract contains two nuclei, called the dorsal and ventral nuclei of the lateral lemniscus (DNLL and VNLL; the VNLL will be discussed below). The DNLL serves to increase the accuracy, contrast, and dynamic range of the localization information, compared with that found in the superior olive (Pecka et al., 2010). It receives its inputs from the ipsilateral MSO, the LSOs of both sides, and the contralateral cochlear nucleus. The DNLL sends a substantially inhibitory projection to the ICs of both sides as well as to the contralateral DNLL. These crossed inhibitory inputs increase the neural contrast between the responses to sound sources localized in different directions in space.

THE DORSAL AUDITORY STREAM OF THE BRAINSTEM: COMPLEX STIMULUS ANALYSIS

In comparison with the relatively straightforward functions of the ventral stream, the dorsal auditory stream of the brainstem performs a multiplicity of functions, from signal extraction based on complex spectral analysis and sound localization in the vertical plane, to the preservation of the accurate timing in broadband stimuli. The stream feeds to later stages of the auditory system, and in particular, to the IC, where the results of the analyses are combined with the spatial information signaled from the superior olive.

The dorsal cochlear nucleus

The DCN extracts complex patterns in the auditory stimulus, in particular in its spectrum. It contains a multiplicity of cell types and has many interneurons. As well as many neurons with central excitatory response areas reminiscent of those of the auditory nerve, the DCN contains many neurons with large inhibitory bands. Figure 1.7 shows many of the different excitatory–inhibitory response types found in the different divisions of the cochlear nuclei. In some cells of the DCN, the inhibition is so great that the response areas contains only a few excitatory islands among otherwise entirely inhibitory responses (classed as types IV and V: Fig. 1.7D and 1.E). The inhibition arises from interneurons within the nucleus. Since the interneurons may themselves receive inhibitory inputs, clearly there can be very complex patterns of response to spectrally complex stimuli.

One pattern of complex analysis has been shown to be useful in cats. The pinna produces notches in the spectrum of broadband stimuli, which vary with the elevation of the sound source; certain cells of the dorsal nucleus have response areas that appear specialized for detecting these notches, and therefore can code the elevation of sound sources. Cats with lesions of the dorsal nucleus are unable to make reflexive responses to sound elevations, though they can still learn to make responses with training (Sutherland et al., 1998). The failure with reflexive, but not learned, responses suggests that the dorsal nucleus is the beginning of certain auditory reflexive pathways. This is consistent with its multimodal inputs from vestibular, somatosensory, and pontine nuclei (Oertel and Young, 2004), and its possible role in the generation of tinnitus (Kaltenbach, 2011).

These results also suggests that the dorsal nucleus contributes to the start of yet another functional division of the auditory system, which is known as the "non-lemniscal" auditory system. The non-lemniscal auditory system can be classed as a separate functional division of the whole ascending auditory system. Its pathways lie separate from, but anatomically adjacent to, those of the "lemniscal" auditory system. The lemniscal auditory pathways (originally defined as ascending via the lateral lemniscus) predominantly deal with unimodal auditory processing. The non-lemniscal pathways primarily deal with complex multimodal integration and reflexes. Some of the nuclei beyond the cochlear nuclei (the IC, the medial geniculate) have separate divisions which appear to belong primarily to the non-lemniscal rather than to the lemniscal auditory system. This division is continued into the auditory cortex, the non-lemniscal pathways having a stronger input into the areas surrounding the core auditory areas.

The posteroventral cochlear nucleus

The PVCN undertakes both timing and spectral analysis. It contributes mainly to the dorsal auditory stream of the brainstem (Fig. 1.9). One area of the PVCN contains only one type of cell, the octopus cell. Octopus cells receive a very large number (>60) of auditory nerve afferents, and are driven only when a large number of the relatively small excitatory postsynaptic potentials arrive together within a short time period and summate (McGinley and Oertel, 2006). The averaging in the response means that they fire with great accuracy, in synchrony with rapid fluctuations in the auditory stimulus which they can follow at high rates, greater than 800 repetitions/second (Golding and Oertel, 2012). They also respond to tones over a very wide range of stimulus frequencies. They are therefore specialized for extracting the temporal fluctuations in complex broadband stimuli such as speech sounds (Cao and Oertel, 2010). Octopus cells project mainly to the ventral nucleus of the lateral lemniscus of the opposite side.

The PVCN also contains stellate (multipolar) cells of two types. They have narrow excitatory areas and wider inhibitory areas, and send collaterals, some of which are inhibitory, widely within the cochlear nuclei and to the contralateral VNLL, as well as to the contralateral IC and areas around the LSO.

The ventral nucleus of the lateral lemniscus

The VNLL, situated within the lemniscal tract between the superior olive and IC, relates to monaural pattern identification, i.e., is functionally part of the dorsal auditory stream of the brainstem. It receives its input from all cell types of the contralateral ventral cochlear nucleus as well as from the ipsilateral MNTB (Kelly et al., 2009). It does not receive an input from the major binaurally innervated nuclei of the superior olive (the MSO and LSO) and therefore does not appear to be involved in binaural sound localization. For this reason it is here classed as part of the dorsal auditory stream of the brainstem, in spite of having an input from the MNTB, which is part of the ventral stream. Many neurons in the VNLL have complex, multipeaked, tuning curves, covering a wide frequency range, while some have precise temporal responses, appropriate for processing temporal information. Therefore, the neurons appear to be specialized for extracting the temporal patterns in complex sounds, with many possibilities for wide cross-frequency interaction, which they then transmit to the ipsilateral IC (Malmierca et al., 1998).

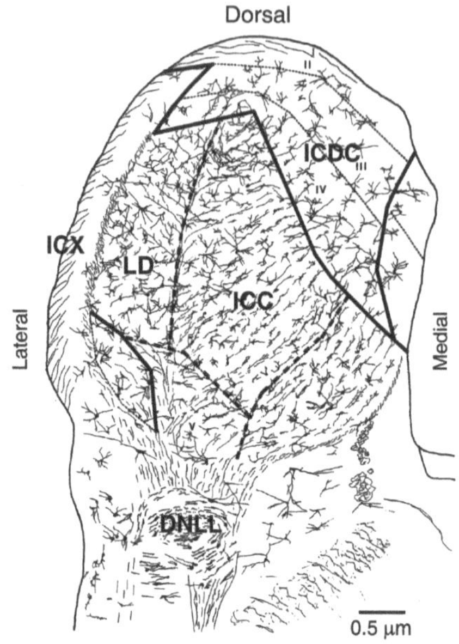

Fig. 1.11. The inferior colliculus as seen in Golgi-impregnated material, showing divisions of the nucleus according to Morest and Oliver (1984), and the lamination in part of the central nucleus (central inferior colliculus (ICC)). I–IV, layers of dorsal cortex. ICDC, dorsal cortex of inferior colliculus; ICX, external nucleus of inferior colliculus; DNLL, dorsal nucleus of the lateral lemniscus; LD, lateral division of ICC. (Reproduced from Morest and Oliver, 1984.)

THE INFERIOR COLLICULUS

The IC is the primary site of convergence of the dorsal and ventral auditory streams of the brainstem, i.e., of the sound pattern identification and sound localization streams. The IC shows a jump in the complexity of responses: the integration of the sound identification and the sound localization streams, and the response to temporally complex stimuli, suggest that this is a critical stage in the transformation from responses dominated by the simple acoustic characteristics to those which integrate acoustic properties in a way that begins to define specific "auditory objects."

The inputs to the IC arise from a great diversity of sources, including the DNLL, the VNLL, the LSO, the MSO, and the cochlear nuclei, and include both excitation and inhibition, and innervate the specific central nucleus as well as the multisensory (non-lemniscal) external nucleus and the dorsal cortex of the IC (ICDC in Fig. 1.11; see Pickles, 2012, for further details).

The central nucleus of the inferior colliculus

The largest of the constituent nuclei, the central nucleus (the central inferior colliculus (ICC)), is part of the specific ("lemniscal") auditory pathway. The nucleus is tonotopically organized and so fibers from different sources but of the same characteristic frequency manage to meet in the same isofrequency plane. The ICC has a laminar anatomic structure with jumps in neurons' characteristic frequency as an exploring electrode moves from lamina to lamina, suggesting that each lamina covers a single frequency range. A high proportion of neurons in the ICC are classed as disk-shaped neurons. Their dendrites arborize in a flat plane, along the plane of the laminations, facilitating frequency-specific interactions within one lamina. The remaining neurons, the stellate cells, have more spherical arborizations and facilitate cross-laminar interactions.

In anesthetized animals, the majority of neurons in the ICC have sharp tuning curves, while a minority have long, shallow, low-frequency tails to the curves. In unanesthetized animals, some neurons have more complex response areas, with more than one excitatory area and multiple areas of inhibition. In some neurons, flanking inhibitory areas stop the excitatory area widening at higher stimulus intensities, so that sharp tuning is maintained at all intensities. In some cases, this is so extreme that the excitatory response area disappears at the highest intensities, giving a so-called closed excitatory response area, while in others there may be an inhibitory island within the excitatory response curve (Alkhatib et al., 2006). The inhibition arises from both extrinsic and intrinsic synaptic inputs (LeBeau et al., 2001).

As might be expected from their inputs from the spatially directionally sensitive LSO and MSO, many cells in the ICC show a spatial directional selectivity. The majority of the cells have a broad directional selectivity towards the contralateral spatial hemifield, with the responses being dominated by inputs from the contralateral ear, but refined and enhanced by input from the ipsilateral ear (Kuwada et al., 2011). There is patchiness in the types of directional cues that dominate. There are some areas where timing cues dominate, others where intensity differences dominate, and others where the complex spectral analysis of high-frequency notches as undertaken in the DCN dominate. In many cases, these can be correlated with the detailed anatomic patterns of inputs from the related lower-brainstem nuclei (Loftus et al., 2004).

The different inputs on to a binaurally sensitive cell in the ICC can arrive at slightly different times, and the different inputs can adapt at different rates, so that in many cells the pattern of excitation and inhibition from the two ears changes over time (Zhang and Kelly, 2010). An implication of this is that some cells in the ICC are sensitive to the direction of movement of sound sources, in ways that are not predictable from their responses to static sound sources. This further enables animals to detect and localize a single moving sound source among other sound sources (for review, see Pollak, 2012).

Neurons in the ICC also show multiple phases of excitation and inhibition to stimulus onsets, likely to arise from the different times of arrival of the inputs from the different projecting nuclei, as well as from the different temporal properties of the various ion channels on the neuron. We expect these effects to enhance the response to temporal modulations in the stimuli. Amplitude-modulated stimuli give bandpass rate functions in a high proportion (50%) of cells. In these cells, the mean firing rate shows a peak for certain frequencies of amplitude modulation, with suppression of the mean rate for higher and lower frequencies of modulation (Langner and Schreiner, 1988). It is possible that these neurons have a role in periodicity detection, as discussed in other chapters of this volume.

Neurons in the ICC can also show preferential responses to frequency-modulated stimuli. In the study of Poon et al. (1992) in the rat, 75% of cells showed an enhanced response to frequency-modulated stimuli, in a way that could not be simply predicted from their ongoing responses to continuous tones (see also Escabi and Schreiner, 2002). Some cells were driven particularly well by fast sweeps and also were selective to the direction of the sweep. Intracellular labeling and later morphologic analysis of the latter cells showed that they were stellate cells, that is, the cells with large spheric or ovoid dendritic trees which cross the isofrequency laminae and which therefore would be in a position to detect cross-frequency information.

The external nucleus and dorsal cortex of the inferior colliculus

The external nucleus of the IC receives inputs from the contralateral cochlear nucleus, including the contralateral DCN, and from the ICC. It also receives a somatosensory input from the dorsal columns and the trigeminal nuclei. The dorsal cortex of the IC, on the other hand, does not receive a direct ascending auditory projection, but receives inputs from the contralateral IC. It also receives descending inputs from the auditory cortex (for review, see Irvine, 1986).

Neurons in the external nucleus and dorsal cortex of the IC tend to be broadly tuned and to habituate rapidly. Many neurons in the external nucleus are bimodal, most of those being excited by auditory stimuli and inhibited by somatosensory ones. In the dorsal cortex of the colliculus, neurons show strong responses at the onset of stimuli. They also can adapt to ongoing stimuli, but show a strong response to changes, a phenomenon known as stimulus-specific adaptation (SSA). This suggests that the nucleus might be important for detecting novel sounds (Lumani and Zhang, 2010).

The external nucleus is likely to be an auditory and somatosensory integrative area, governing, among other things, spatial reflex responses to sound (e.g., Jain and Shore, 2006). Both the external nucleus and dorsal cortex of the IC are likely to form part of what has been called the "diffuse," "belt," non-specific or extralemniscal auditory system, often with multisensory interactions, which starts in the DCN and which surrounds the specific "core" or lemniscal auditory system. This division into specific and diffuse auditory systems is carried up through the medial geniculate body (MGB) to the auditory cortex (see Lee and Sherman, 2011, for further analysis of the distinction between the two types of pathway).

THE MEDIAL GENICULATE BODY

The MGB is the specific thalamic relay of the auditory system, receiving afferents from the IC, and projecting to the cerebral cortex. It also has heavy reciprocal interactions with the cortex, so that the different structures are grouped together as a single functional unit. It continues the integration of the spatial, spectral, and temporal features that have been extracted separately and partially reintegrated within the nuclei of the lower levels of the auditory system, to refine the neural coding of specific "auditory objects" which is then presented to the auditory cortex.

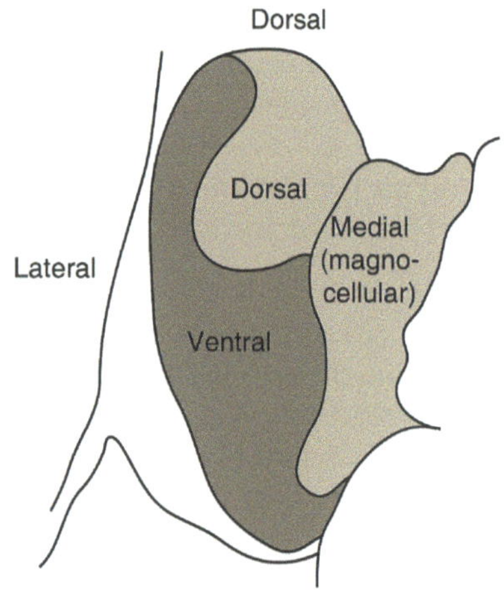

Fig. 1.12. The ventral, dorsal, and medial (magnocellular) divisions of the medial geniculate complex of the macaque, shown in a coronal section. Redrawn using data from Jones (2003). (Reproduced from Pickles, 2013.)

Overall anatomy and inputs

Figure 1.12 shows the divisions of the MGB that are currently recognized. There is a ventral, principal nucleus, a dorsal nucleus (divided in primates into anterodorsal and posterodorsal nuclei), and a medial or magnocellular nucleus (Jones, 2003).

The ventral division is the specific auditory relay in all species and is classed as part of the lemniscal or specific auditory pathway. It receives its afferents predominantly from the ipsilateral central nucleus of the IC, and projects to the ipsilateral tontopically organized core areas of the auditory cortex, that is, to the primary auditory cortex (AI), rostral (R), and rostrotemporal (RT) cortical fields (as defined in primates; see section on the auditory cortex, below, and Chapter 2, this volume, and Hackett, 2011, for reviews).

In contrast, the medial (i.e., magnocellular) division has a multisensory and integrative role. As well as receiving inputs from the central nucleus of the IC, it receives inputs from the external nucleus and the dorsal cortex of the IC, from the lateral tegmental system running just medial to the brachium of the IC, from the superior colliculus, the vestibular system, and from the spinal cord. The medial division projects widely to the different auditory cortices, as well as to the core. In addition, it has connections with the lateral nucleus of the amygdala and is therefore engaged in emotional conditioning.

The dorsal division receives afferents primarily from the dorsal cortex of the IC and also from the region medial to the brachium and from the somatosensory system. It projects to the non-tonotopic cortical belt areas which surround the core auditory cortex (i.e., the belt areas known in the cat as secondary auditory cortex (AII), posterior ectosylvian gyrus (Ep), and insulotemporal (I-T)), as well to anterior auditory field (AAF) within the core (Winer, 1985 for the cat). The medial and dorsal divisions, classed as part of the extralemniscal auditory pathway, therefore can be viewed as part of the "diffuse" or non-specific auditory system that surrounds the specific auditory system, that is engaged in more complex cross-sensory, and in particular in auditory-somatosensory, interactions. The different patterns of inputs, and the different patterns of projections, suggest that the three divisions of the MGB are parts of three separate and parallel projection pathways to the auditory cortex (reviewed by Smith and Spirou, 2002; Winer et al., 2005).

The ventral nucleus

Anatomy and frequency organization

The ventral nucleus or division has a laminar structure, formed by the interdigitating layers of afferent axons and intrinsic neurons. There are two types of intrinsic neuron, namely tufted neurons and short-axon interneurons. The tufted neurons have large dendritic fields (approximately 450 μm across) with axons that project out of the nucleus. The short-axon interneurons (Golgi type II cells) may have large or small dendritic trees depending on species and are inhibitory. The tufted cells receive a multiplicity of inputs, including excitatory and inhibitory inputs from the IC, excitatory inputs from the auditory cortex, inhibitory inputs from the short-axon interneurons, and inhibitory inputs from the reticular thalamic nucleus.

The laminae are almost certainly parallel to the isofrequency planes in the nucleus, with low frequencies represented dorsally in the nucleus and high frequencies ventrally. However, as an electrode passes through the nucleus in a dorsal to ventral direction, the best frequency does not increase steadily, but in jumps of around 0.8 octave, every 300 μm or so (Cetas et al., 2001). The measurement 300 μm is greater than the thickness of the individual anatomic laminae, and this suggests that the individual anatomic laminae are organized into what have been called functional "slabs," with a common frequency representation within each slab. The thickness of a slab roughly corresponds to the size of the dendritic fields of the tufted neurons, suggesting the possibility of strong interactions within each slab. However all frequencies must eventually be represented somewhere within the ventral nucleus, so the finding of the large-frequency jumps suggests that there must be a further level of organization within the nucleus. This suggests that the intermediate frequencies must be represented in other parts of any two-dimensional slab, and that within each slab patches of neurons may well be joined into functional groups which together cover a small range of frequencies.

There is a further organization orthogonal to the isofrequency planes, since the neurons with different classes of binaural response (e.g., the balance of excitation and inhibition from ipsilateral and contralateral ears) are segregated in different patches within the isofrequency planes. This is correlated with a patchiness in the projections to the auditory cortex and a patchiness in the different cells' binaural response types within the cortex (Velenovsky et al., 2003).

Responses to sound

Neurons in the ventral division, the specific auditory nucleus, have relatively sharp tuning curves (Bartlett and Wang, 2011). A great variety of temporal response patterns have been found, with sustained responses, either sustained excitation or sustained inhibition, more common in unanesthetized animals (Aitkin and Prain, 1974). In general, neurons with complex temporal properties have complex frequency response areas, as would be expected for neurons with a multiplicity of excitatory and inhibitory inputs.

Some investigators have shown that the tips of the tuning curves are much sharper than those of the auditory nerve fibers. Some of the sharpening is contributed by local inhibition, since it can be reversed by the local addition of inhibitory blockers (e.g., Suga et al., 1997). Very sharp tuning has also been found in a high proportion of neurons in awake animals (the marmoset: Bartlett et al., 2011). Some of these aspects of MGB responses may reflect cortical processing, as there are substantial corticofugal influences on the nucleus, and cortical manipulations can change the frequency resolution of neurons in the MGB (Zhang and Suga, 2000). The close bi-directional interaction between the MGB and the auditory cortex means that the two must be considered as one unit. This close interaction enhances the difficulty of analyzing the stimulus transformations that are specific to the medial geniculate itself.

The medial and dorsal nuclei of the MGB

The medial division is part of the "diffuse" or extralemniscal auditory system which projects more generally to the auditory cortex. It has sometimes been grouped with a number of adjacent nuclei which together are known as the paralaminar nuclei as well as with the posterior thalamic nucleus which is adjacent to the MGB rostrally.

In the medial division, many neurons in unanesthetized animals show many very wide, multipeaked, and complex response areas (Aitkin, 1973). Three-quarters of the neurons are binaural and are therefore expected to show spatial selectivity. Many neurons have short first-spike latencies, and show highly reliable firing at the onsets of stimuli (Anderson and Linden, 2011). Neurons also show stimulus-specific adaptation, in which the response to an ongoing stimulus declines with time but returns on presentation of a novel stimulus (Bäuerle et al., 2011). These findings suggest a particular role in the response to novel stimuli arriving from specific directions in space.

The responses in the medial nucleus are affected by associative learning. They change as a result of a conditioned fear response, in which sounds have been associated with a negative stimulus, such as a puff of air to the cornea (McEchron et al., 1996). The amygdalae are involved in learning fear in response to negative stimuli. The medial division of the MGB and the posterior intralaminar nucleus, both of which receive an input from the IC, have particularly close anatomic associations with the amygdalae. Disruption of the former nuclei interrupts conditioned responses to sound stimuli (Halverson et al., 2008). In turn, outputs from the amygdalae modify the responses of neurons in the medial MGB, as shown by pairing the auditory stimulus with a fear-inducing stimulus such as an electric shock to the foot (Duvel et al., 2001).

The dorsal division projects to the auditory areas surrounding AI (Smith and Spirou, 2002). Neurons in the deep division of the dorsal nucleus have short latency and sharply tuned responses to sound in anesthetized cats; those situated more dorsally and medially respond more weakly and with much habituation (Anderson and Linden, 2011). The responses show strong stimulus-specific adaptation (Antunes et al., 2010), and may also be modulated by training paradigms.

The medial and dorsal divisions of the MGB therefore show multisensory responses that are modifiable depending on the behavioral environment of the animal. By projecting widely to many areas of the auditory cortex, they would be able to prepare the cortex for responding to stimuli which are of particular significance for the organism.

THE AUDITORY CORTEX

The auditory cortex is discussed in detail elsewhere in this volume, and so the anatomy will be discussed relatively briefly, with more emphasis on the functional aspects. However a brief description of the anatomy is a necessary prerequisite.

While current research is directed more strongly to the primate cortex than previously, many of the pioneering concepts, as well as some of the current important functional observations, have been made in the cat. Therefore a brief description of the cat as well as the primate cortex is required.

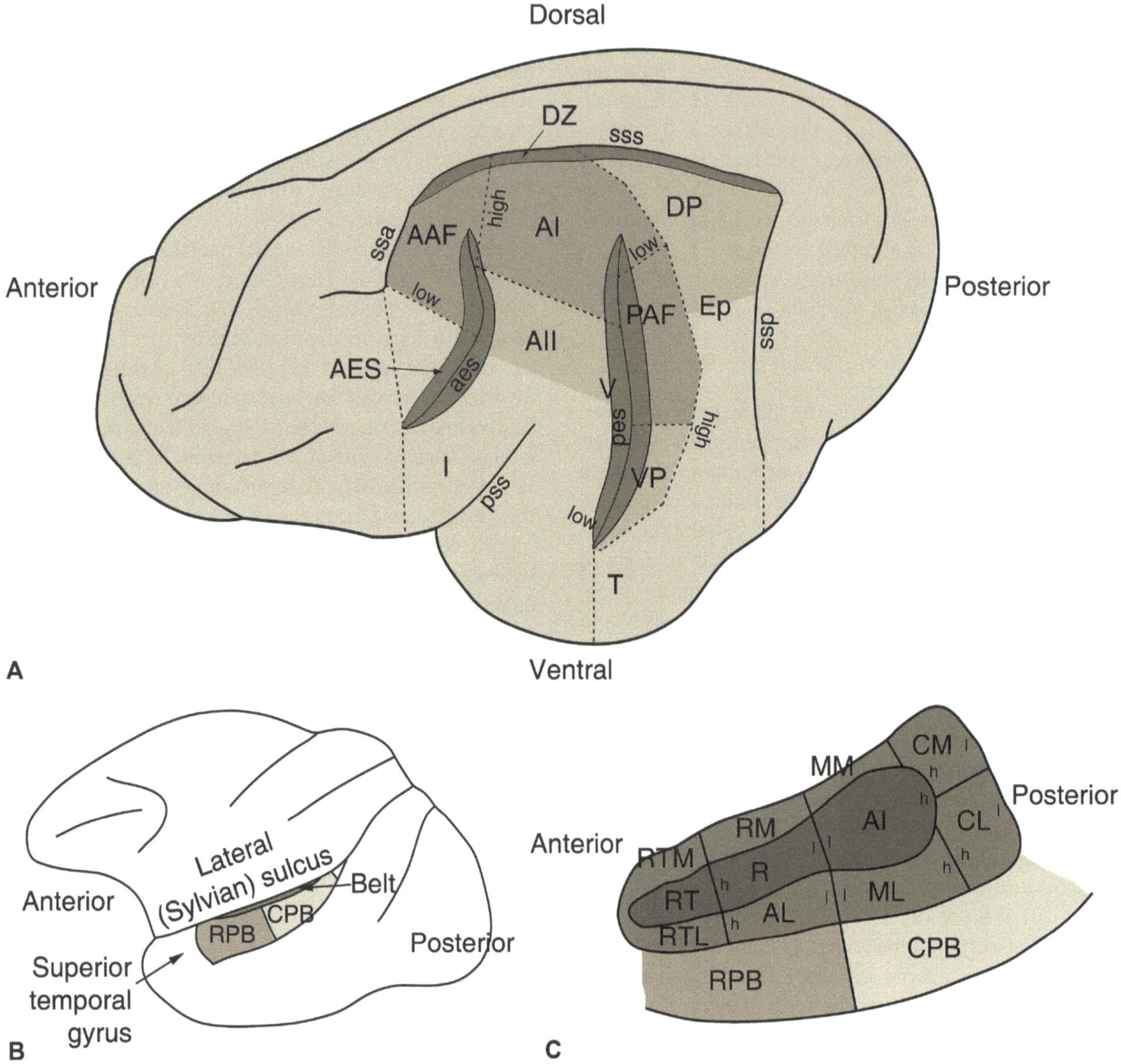

Fig. 1.13. (**A**) Auditory areas recognized in the cat cortex. Core areas: AI, AAF, and PAF. Other areas are belt, surrounded by parabelt. Where the fields are tonotopically organized (AI, AAF, PAF, and VP), the representation of highest frequencies (high) and lowest frequencies (low) is marked. Areas shaded darker are hidden in the sulci, which have been opened slightly to show the fields within the sulci. AI, primary auditory cortex; AII, secondary auditory cortex; AAF, anterior auditory field; AES, field of anterior ectosylvian sulcus (buried in sulcus); DP, dorsal posterior area; DZ, dorsal zone, buried on the ventral (lower) surface of the suprasylvian sulcus; Ep, posterior ectosylvian gyrus; I, insula; PAF, posterior auditory field; Sulci, aes, and pes, anterior and posterior ectosylvian sulci; pss, pseudosylvian sulcus; ssa and ssp, anterior and posterior suprasylvian sulci; sss, suprasylvian sulcus; T, temporal area; V, ventral field; VP, ventral posterior field. Adapted from Reale and Imig (1980), Fig. 1, including data from Clarey and Irvine (1990).

(**B**) Auditory areas from an external lateral view of the macaque cortex, showing the caudal and rostral parabelt areas (CPB, RPB) on the superior temporal gyrus. The other cortical areas are not visible and are buried in the lateral sulcus. (Modified from Pickles, 2012.)

(**C**) Auditory cortical areas in the macaque, shown on a projection of the upper surface of the superior temporal lobe, according to Hackett et al. (1998). Where the area is tonotopically organized, the representations of the high (h) and low (l) frequencies are marked. Core areas: AI, primary auditory cortex; R, rostral area; RT, rostrotemporal area. Belt areas: AL, anterolateral area; CL, caudolateral area; CM, caudomedian area; ML, middle lateral area; MM, middle medial area; RM, rostromedial area; RTL, lateral rostrotemporal area; RTM, medial rostrotemporal area. Parabelt areas: CPB, caudal parabelt; RPB, rostral parabelt. (Redrawn from Hackett et al., 1998.) (Reproduced from Pickles, 2012).

Anatomic introduction to the auditory cortex

Figure 1.13A shows the main areas of the cat auditory cortex. The primary area AI together with the anterior and posterior auditory fields (AAF and PAF) are classed as core areas. Core areas are defined by a number of histologic criteria, and have substantial direct inputs from the specific auditory division of the MGB, that is, from its ventral or lemniscal division. In addition, each of the areas within the core has its own separate tonotopic progression of best or characteristic frequencies, which is the criterion by which the different core areas can be distinguished from each other. The core areas are heavily interconnected by reciprocal connections, which is a further criterion by which they are grouped together.

Figure 1.13B and C shows the auditory cortical areas currently recognized in the macaque. In the macaque, the core areas are the AI, R, and RT areas.

The core areas are surrounded by the belt areas, and the belt areas by the parabelt. The belt areas are defined by having major connections with the dorsal or medial divisions of the medial geniculate, and no or only minor connections with the ventral division of the medial geniculate, while having recordable auditory responses. Each belt area receives inputs from multiple core areas, though with a heavier input from the nearest core area. Many of the belt areas have their own separate representation of the cochlea and their own tonotopic organization (Fig. 1.13C).

The parabelt areas have interconnections with the belt, but not with the core areas (Hackett et al., 1998). In the macaque, the belt is divided into caudal and rostral divisions, on the basis of its interconnections. The parabelt is also connected to the frontal lobes, including the frontal eye fields which are concerned with eye movements.

In human beings, it has been more difficult to define areas with certainty. Functional imaging and myeloarchitectonic mapping in human beings suggest that there are three areas in the core, likely to correspond to the macaque AI, R, and RT fields, situated on and sometimes around the posterior-medial part of Heschl's gyrus, within Brodmann's area 41 (Barton et al., 2012; Dick et al., 2012). As in other primates, the primary cortex is surrounded by further areas, with functional imaging suggesting that, as in the macaque, there are a further eight belt areas situated around the core (Barton et al., 2012). Cytologic studies suggest that these areas are further surrounded by areas with the properties of parabelt (see Chapter 2, Figures 2.1, 2.9, and 2.10).

Tonotopic organization

If the cortical surface of AI is sampled with a microelectrode, and the best or characteristic frequencies of the cells plotted as a function of distance across the cortex, a progression of characteristic frequency with position is found. On the two-dimensional cortical surface, it is possible to draw lines that connect points of similar characteristic frequency, giving iso-frequency lines (Fig. 1.14). Note however that the lines joining points of equal-frequency representation are irregular, likely to be a result of local variation in the characteristic frequencies, combined with relatively sparse sampling of the individual cells or multiunit clusters. For this reason, the term "frequency-band strips" rather than "isofrequency lines" is often preferred.

Adjacent cortical fields have reversed tonotopic maps; that is, if an electrode sampling across the surface finds a progression of best frequencies going from high to low and then to high, the reversal in progression is taken as a sign that a field boundary has been crossed and that the area in fact contains two separate fields.

In the cat, tonotopic maps have been found in the three fields of the core (AI, AAF, and PAF) and one field of the belt (VP, the ventral posterior field). In the macaque, the core fields (AI, R, RT) are tonotopically organized, as are four of the fields of the belt (caudomedian (CM), caudolateral (CL), middle lateral (ML), and anterolateral (AL); Petkov et al., 2006). In human beings, current data suggest that the three fields of the core (likely to be homologues of the AI, R, and RT found in other primates) are also tonotopically organized (Barton et al., 2012). Further frequency-selective representations have been found outside these areas, showing

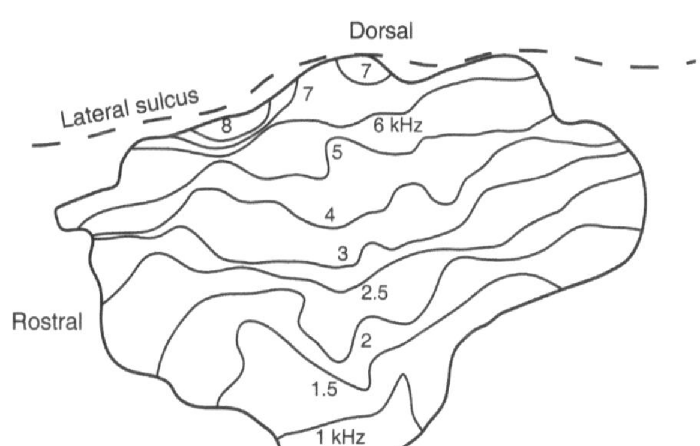

Fig. 1.14. Tonotopic organization within the primary auditory cortex of the squirrel monkey. The characteristic frequencies of single units or multiunit clusters were sampled densely across the cortical surface. Lines have been drawn to separate recording points with different characteristic frequencies (marked on curves in kHz). The irregularity of the lines stems from the variability of characteristic frequencies at closely adjacent points in the cortex. Data from Cheung et al. (2001). (Reproduced from Pickles, 2013.)

the existence of further tonotopically organized fields. However, variability between subjects, and the difficulty of defining the position of frequency reversals, means that the details of the fields, which vary between investigators, must be seen at the moment as provisional (e.g., Barton et al., 2012; Moerel et al., 2012).

Organization along frequency-band strips

Driven by a search for evidence on the functional significance of the auditory cortex, further patterns of organization have been sought within the tonotopic maps. In the cat, sound localization seems to be one such factor. Cells excited by sounds in one ear but inhibited by sounds in the other are separated from cells excited by both ears. They are segregated within broad strips in the cortex, the strips running at right angles to the frequency-band strips, suggesting an important role for the auditory cortex in sound localization (Imig and Adrian, 1977). Further spatially organized variations in breadth of tuning, threshold, latency, periodicity, and sensitivity to frequency modulation have been found in many primate and non-primate species. In some cases, the progressions of properties defined in these ways also run at right angles to the tonotopic progression; however in others they bear no clear relation to the tonotopic map (see e.g., Cheung et al., 2001; Read et al., 2002; Bizley et al., 2009; Barton et al., 2012; Camalier et al., 2012).

Responses of single neurons: responses in the core

Anesthesia, and particularly barbiturate anesthesia, suppresses cortical responses, increasing the difficulty of analyzing auditory cortex. In unanesthetized animals, a great variety of responses are seen in AI, with many cells having sustained and transient responses, in addition to cells with only on, off, or on-off responses, and with the most preferred stimuli tending to give the more sustained responses (Abeles and Goldstein, 1972; Wang, 2007).

Most responsive neurons in AI have sharp tuning curves with a single frequency region of maximum sensitivity; the sharp tuning is particularly found in the sustained, rather than the onset, part of the response (Bartlett et al., 2011). In addition, many neurons have two or more regions of maximum sensitivity, giving what are known as multipeaked responses. Other neurons have very broad tuning curves, covering several octaves. In the cat, as in primates, broadly tuned neurons are spatially segregated from those showing sharp tuning. Many neurons also have inhibitory sidebands. In contrast to lower centers, where inhibitory sidebands tend to immediately abut the excitatory area in frequency, cortical inhibition can often arise from one or more discrete frequency bands which are remote from the central excitatory or inhibitory area (Kadia and Wang, 2003). Cells with multiple and widely spaced excitatory or inhibitory bands would appear to be specialized for extracting sounds with complex spectral features spread over a wide bandwidth.

The cat AAF and PAF are also classed as part of the core. Compared with AI, neurons in cat AAF are more broadly tuned, have shorter latencies of response, and are particularly responsive to tones with rapid-frequency sweeps, in many cases being selective for the direction of the sweep (Tian and Rauschecker, 1994). No multipeaked neurons have been reported in AAF as they have been in AI. These results imply that the AAF is involved in faster, higher-frequency processing than AI. In the PAF, neural excitatory response areas have a wide variety of shapes, to include some multipeaked and some very broadly tuned neurons (Loftus and Sutter, 2001). Neurons in PAF are also sensitive to the location of sound sources, more so than neurons in AI, and are also particularly responsive to frequency modulation (Stecker et al., 2003). These results imply that the PAF is involved in analyzing the more complex aspects of the auditory stimulus, including sound localization (see section on cortical processing of sound location, below).

Relatively few comparisons of responses in the different core areas have been published in primates. Neurons in the R core area of unanesthetized and behaving macaque monkeys appear to be more sharply tuned to frequency, and more non-monotonic, than in AI. Similarly, spatial tuning is broader in AI, and other core areas, than in the caudal belt (for review, see Recanzone et al., 2011).

Responses of single neurons: responses in the belt

There are multiple separate areas in the belt, not all of which may be tonotopically organized. These areas have been incompletely investigated, reflecting the large numbers of areas and the many analyses possible. In general, compared with the core, the responses suggest that the belt areas have a particular role in processing of the more complex aspects of the stimulus, such as decoding communication calls.

In cat AII, responses from single cells or small clusters of cells show some poor tonotopic organization, with higher neuronal thresholds than in AI (Schreiner and Cynader, 1984). Compared with AI, in AII the spatial patterns of binaural interaction are more patchy and more variable from animal to animal.

This pattern may be compared with the cat dorsal zone (DZ), which borders AI dorsally on the ventral surface of the suprasylvian sulcus. DZ may in fact be part of AI; that is, part of the core, rather than the belt. Cells in DZ have complex frequency tuning with multiple excitatory and inhibitory domains, together with a high proportion of binaurally selective neurons, and sharp selectivity to direction in the horizontal plane. These results suggest that DZ might have a role in the spatial representation of sounds.

Further areas that have been investigated are thought to be involved in sound localization and in the processing of complex stimuli. In the marmoset, neurons in the medial belt area CM have been found to be just as responsive to tones as neurons in AI, though with lower thresholds, shorter response latencies, and broader tuning curves (Kajikawa et al., 2005). Neurons in the lateral belt areas (AL, ML, and CL) are much less responsive to pure tones than are neurons in AI. However, they seem particularly responsive to frequency sweeps and are selective to both the rate and direction of the sweeps (rhesus monkey: Tian and Rauschecker, 2004). The maximal sensitivity of AL neurons was in the range of communication sounds, and it was suggested that AL was specialized for the decoding communication, while CL was specialized for localization.

Cortical processing of sound location

The importance of the auditory cortex for sound localization was shown in the early finding that, after large bilateral lesions, cats were unable to localize sounds in space (e.g., Neff, 1968). Many later experiments over the years showed that unilateral lesions interfered with the localization of sounds in the contralateral hemifield of space. This suggests that, as expected, each cortex preferentially processes stimuli on the contralateral side.

More recently, Malhotra, Lomber and colleagues placed cooling probes over different cortical areas in the cat. Deficits in sound localization were found after unilateral cooling of AI, and a further area of the core, the PAF, or a field in the belt, the field of the anterior ectosylvian sulcus (AES). Cooling of any of these fields (AI/DZ, PAF, or AES) on their own disrupted localization, indicating that they all need to be operating together for effective sound localization. On the other hand, cooling of the remaining area in the core, AAF, or any of the other fields in the belt, left sound localization unaffected (Lomber and Malhotra, 2008). In conjunction with the known anatomic connections, the results can be interpreted to suggest that sound localization is first processed in AI and then in DZ and PAF (see also Lee and Middlebrooks, 2013). The information is then passed to AES, which transmits the information to the superior colliculus, where lesions have a yet more profound effect on orientation to auditory stimuli (Lomber et al., 2007).

It has been difficult to obtain a clear picture in human beings because of the variability of the effects between patients. Spierer et al. (2009), who analyzed patients with a variety of lesions in the auditory cortex, found that lesions in the right hemisphere affected the localization of both contralateral and ipsilateral sound sources. Interaural timing and level cues tended to be affected together, suggesting that this hemisphere had an integrative role in representing sound location. The deficits were more profound and severe than those arising from left-hemisphere lesions, which affected the localization of only contralateral sound sources (see Chapter 32 of this volume).

Electrophysiologic evidence in cats shows that the majority of cells in AI are indeed responsive to spatial location, based on differences in either interaural intensity or timing, or a combination of the two. When responses to external sound sources are measured in the intact animal, some cells are found to be broadly responsive to sounds all over the contralateral hemifield, while others have more spatially circumscribed response areas (e.g., Brugge et al., 1996). In these latter experiments, the responses will be affected by the directional selectivity of the pinna, as well as by the between-ear comparisons which are made at every binaural stage of the auditory system and which are further refined in the cortex. Cells in the cat DZ, AES, and PAF, which behavioural experiments had shown were also involved in sound localization, also show spatial selectivity (Lee and Middlebrooks, 2013).

In the macaque, Miller and Recanzone (2009) found that neurons in the CL field signaled direction most accurately, and more so than neurons in AI. Their results suggested a possible hierarchic refinement of information during the transfer from AI to CL, and a particular role for CL in sound localization.

Cortical processing in relation to stimulus complexity

The auditory cortex is necessary both for the simple detection of sound and for the discrimination of frequency. While the earliest experiments showed no changes after bilateral cortical lesions, more recent experiments have shown deficits in these basic functions. For instance, temporary bilateral inactivation of AI in rats by the gamma-aminobutyric acid agonist muscimol raises rats' auditory detection thresholds for several hours, and raises frequency discrimination thresholds for a further period after that (Talwar et al., 2001). Moreover, after lesions of the auditory cortex, animals seem to have particular difficulty with very short stimuli, suggesting that the area is important for registering the trace left by short stimuli (e.g., Cranford, 1979).

In human patients, Tramo et al. (2005) reported that lesions of the auditory cortex interfered with detection of the discrimination of frequency changes, with a particularly critical role for the right hemisphere.

Cortical lesions also affect more complex auditory tasks. Lomber and Malhotra (2008) showed in cats that, after bilateral cooling of AAF, i.e., an anterior field in the core, the discrimination of auditory temporal patterns was significantly disrupted. Discrimination was also lost after large lesions of the auditory cortex, which included AI, AII, Ep, and I-T, but survived lesions of AI alone, suggesting a particular role for the core outside AI and for the belt and parabelt in this task (Diamond and Neff, 1957).

Further complex tasks affected by auditory cortical lesions include the discrimination of species-specific vocalizations in non-human primates, and complex auditory perceptual tasks, including the perception of speech, in human beings. Given that the auditory cortex seems necessary for performance of complex auditory tasks, it is therefore not surprising that many cells in the auditory cortex respond selectively to complex aspects of the stimulus.

Many neurons in AI have multipeaked tuning curves. Responses to stimuli presented in the different peaks can be facilitatory, that is, the neuron responds non-linearly such that response to both of the stimuli together can be greater than the arithmetic sum of the responses to the two stimuli separately (Kadia and Wang, 2003). Commonly, the constituent peaks are harmonically related; clearly, such neurons would be specialized for representing stimulus complexes where the stimulus components are harmonically related. Such stimuli include the vocalizations of the species concerned (the marmoset, in the above study) which are rich in harmonics. Other patterns of response area have been found where inhibitory bands would preferentially select against the detection of sounds where the components are harmonically related (Kadia and Wang, 2003).

Outside AI, complex stimuli seem particularly effective at driving neurons, with frequency modulation being one critical cue. In the cat, neurons in AAF and PAF (non-AI parts of the core) are particularly responsive to frequency sweeps, with many neurons being selectively responsive to the direction of the sweep (Tian and Rauschecker, 1998). Neurons in PAF are also particularly responsive to changes in the carrier frequency of species-specific vocalizations (Gourevitch and Eggermont, 2007). In cat PAF, unlike in AI and AAF, the response to frequency modulation cannot be simply predicted from the response to steady tones, suggesting that the responses reflect processing at a more complex level (Tian and Rauschecker, 1998). In the macaque lateral belt areas, neurons are also particularly strongly driven by frequency-modulated stimuli. Neurons in CL have been found to be responsive to the fast sweep rates, while in AL, neurons responded to slow sweep rates, in the range found in species-specific vocalizations. Those in ML responded to the whole range of sweep rates (Tian and Rauschecker, 2004).

In responding to complex stimuli, the auditory cortex seems to be responding to specific combinations of fundamental stimuli. Such responses could be determined by coincidence networks in the cortex. The combinations that are effective probably depend both on genetic programming, and on the stimuli to which the animal has been previously been exposed.

Analysis of speech will be dealt with only briefly here, since it is dealt with in detail in other chapters of this volume. Functional magnetic resonance imaging studies in human beings and other primates have shown that stimuli with complex spectral features preferentially activate the more lateral areas of the auditory cortex (for summary, see Evans et al., 2014). Moreover, vowel sounds and sounds that evoke a strong sensation of pitch produce strong responses in anterolateral Heschl's gyrus, or in the anterior planum temporale, suggesting that they might be centers for the extraction of pitch, an important component of speech (e.g., Barker et al., 2011). It has been suggested that there is a hierarchic analysis of information, running from Heschl's gyrus to the planum temporale and from the planum temporale to the superior temporal sulcus (Kumar et al., 2007). Multiple other areas, including temporal and posterior areas, are heavily involved in the response to speech as well (e.g., Evans et al., 2014). Speech first appears to be treated differently from non-speech stimuli in an area lateral and inferior to Heschl's gyrus, and is then distributed along the superior temporal gyrus.

The involvement of anterior and lateral areas in processing communication calls is consistent with the notion of a ventral cortical stream primarily concerned with object identification (the "what" cortical stream). It may be contrasted with a second, more dorsal and posterior stream across the cortex, which is activated where a sensorimotor response is required.

Finally, it should be remarked that the responsiveness and connectivity of the auditory cortex are continually and dynamically adjusted in response to the behavioral demands at the time (e.g., Chait et al., 2012). Cortical responses to acoustic stimuli, together with the synchronization of neuronal firing, are enhanced when the stimuli are associated with other stimuli of strong significance for the animal (e.g., Headley and Weinberger, 2013). The cortex is therefore likely to be able to devote larger networks of neurons to stimuli which are of particular current significance for the organism.

THE CENTRIFUGAL SYSTEM

As well as the centripetal system, i.e., the system which conveys information from the auditory periphery to the central structures, the auditory system contains a rich set of pathways that run in the reverse direction, forming the centrifugal system. Figure 1.15 shows some of the known pathways.

One particularly rich network of pathways starts in the auditory cortices, and runs to the MGB and the IC. Electric stimulation of the cortex can modify the responses of the latter nuclei, enhancing their responses to frequencies represented in the areas of the cortex being stimulated (for review, see Suga and Ma, 2003). The cortex therefore appears able to enhance the responses of stimuli of significance, not only in the cortex, but at other stages of the auditory system. These enhanced responses will then in turn be able to further enhance the responses within the cortex (see Bajo and King, 2013, for a further review of corticofugal pathways to the IC).

The IC sends centrifugal fibers to the superior olivary complex and the cochlear nucleus, while other centrifugal pathways run from the superior olivary complex to the cochlear nucleus. A system that has generated particular interest is the olivocochlear system, where fibers run from the superior olivary nuclei to end within the organ of Corti of the cochlea.

One component of the olivocochlear bundle, the medial olivocochlear system, runs from the superior olive, predominantly contralaterally though with a small ipsilateral component, to end on the outer hair cells ("m" in Fig. 1.15). It inhibits the responses of the cochlea by reducing the degree of active mechanical amplification of the mechanical traveling wave (for details, see Guinan, 2010; Pickles, 2012). A second component, the lateral olivocochlear system ("lat" in Fig. 1.15), runs predominantly ipsilaterally, though with a small contralateral component as well, to end on or near the synaptic terminals of the afferent auditory nerve fibers on the inner hair cells. It also inhibits auditory responses, this

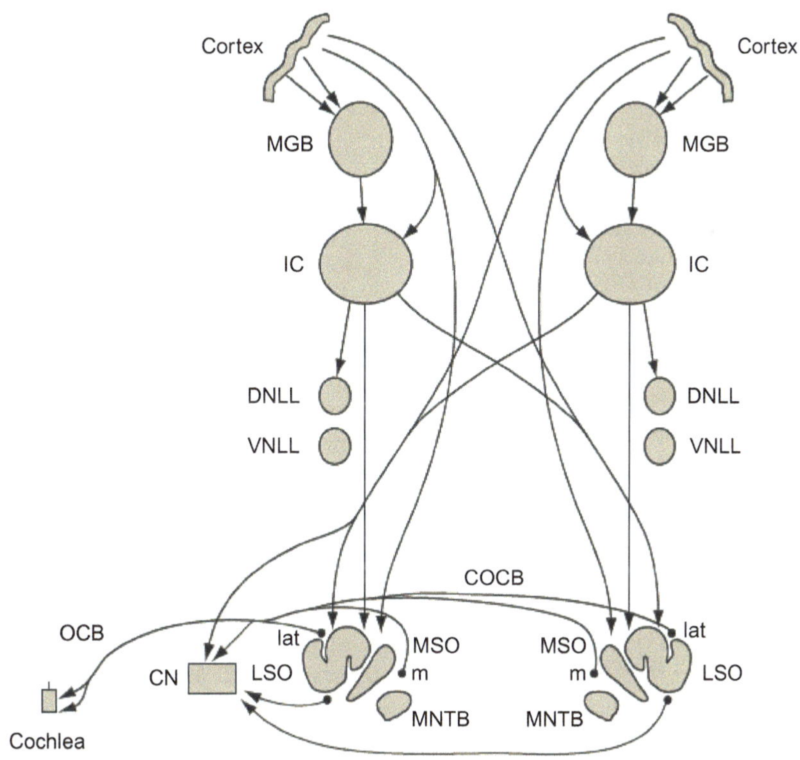

Fig. 1.15. Major centrifugal or descending pathways in the auditory system (compare with the centripetal or ascending pathways in Fig. 1.10). Emphasis is given to pathways that have been investigated physiologically; other, less characterized pathways also exist. In general, the descending pathways end on neurons in areas surrounding, and less commonly within, the nuclei associated with the ascending system. The branching and joining of arrows do not necessarily mean that fibers branch or join. CN, cochlear nucleus; COCB, crossed fibers of the olivocochlear bundle; DNLL, dorsal nucleus of the lateral lemniscus; IC, inferior colliculus; lat, lateral component of the olivocochlear bundle; LSO, lateral superior olive; m, medial component of the olivocochlear bundle; MGB, medial geniculate body; MNTB, medial nucleus of the trapezoid body; MSO, medial superior olive; OCB, olivocochlear bundle; VNLL, ventral nucleus of the lateral lemniscus, (Reproduced from Pickles, 2013).)

time at the neural level. Animal experiments have shown that the olivocochlear fibers, which can be driven by sound in the contralateral ear, can have a powerful effect in suppressing noise-induced hearing loss, via their effect on the mechanical traveling wave and also probably by reducing excitotoxicity in the afferent nerve fibers (e.g., Kujawa and Liberman, 1997). Other experiments in animals have suggested that activation of the olivocochlear bundle could help increase the perceptual contrast between a signal such as speech and a broadband background. However, observations in human patients with unilateral surgical section of the peripheral vestibular nerve, which carries the olivocochlear fibers, have in general been disappointing for these conclusions, the effects being either equivocal or contradictory (Giraud et al., 1997; Scharf et al., 1997). In view of this, the strong effect on acoustic trauma as found in animals cannot be viewed as verified in human beings. Further tests have however shown a very small, though statistically significant effect, with selective attention to stimuli of expected versus unexpected frequency (Scharf et al., 1997).

References

Abeles M, Goldstein Jr MH (1972). Responses of single units in the primary auditory cortex of the cat to tones and to tone pairs. Brain Res 42: 337–352.

Aitkin LM (1973). Medial geniculate body of the cat: responses to tonal stimuli of neurons in medial division. J Neurophysiol 36: 275–283.

Aitkin LM, Prain SM (1974). Medial geniculate body: unit responses in the awake cat. J Neurophysiol 37: 512–521.

Alkhatib A, Biebel UW, Smolders JW (2006). Inhibitory and excitatory response areas of neurons in the central nucleus of the inferior colliculus in unanesthetized chinchillas. Exp Brain Res 174: 124–143.

Anderson LA, Linden JF (2011). Physiological differences between histologically defined subdivisions in the mouse auditory thalamus. Hear Res 274: 48–60.

Antunes FM, Nelken I, Covey E et al. (2010). Stimulus-specific adaptation in the auditory thalamus of the anesthetized rat. PLoS One 5: e14071.

Bajo VM, King AJ (2013). Cortical modulation of auditory processing in the midbrain. Front Neural Circuits 6: 114.

Barker D, Plack CJ, Hall DA (2011). Human auditory cortical responses to pitch and to pitch strength. Neuroreport 22: 111–115.

Bartlett EL, Wang X (2011). Correlation of neural response properties with auditory thalamus subdivisions in the awake marmoset. J Neurophysiol 105: 2647–2667.

Bartlett EL, Sadagopan S, Wang X (2011). Fine frequency tuning in monkey auditory cortex and thalamus. J Neurophysiol 106: 849–859.

Barton B, Venezia JH, Saberi K et al. (2012). Orthogonal acoustic dimensions define auditory field maps in human cortex. Proc Natl Acad Sci U S A 109: 20738–20743.

Bäuerle P, von der Behrens W, Kossl M et al. (2011). Stimulus-specific adaptation in the gerbil primary auditory thalamus is the result of a fast frequency-specific habituation and is regulated by the corticofugal system. J Neurosci 31: 9708–9722.

Békésy G (1960). Experiments in Hearing. McGraw-Hill, New York.

Beurg M, Fettiplace R, Nam JH et al. (2009). Localization of inner hair cell mechanotransducer channels using high-speed calcium imaging. Nat Neurosci 12: 553–558.

Bizley JK, Walker KM, Silverman BW et al. (2009). Interdependent encoding of pitch, timbre, and spatial location in auditory cortex. J Neurosci 29: 2064–2075.

Brugge JF, Reale RA, Hind JE (1996). The structure of spatial receptive fields of neurons in primary auditory cortex of the cat. J Neurosci 16: 4420–4437.

Camalier CR, D'Angelo WR, Sterbing-D'Angelo SJ et al. (2012). Neural latencies across auditory cortex of macaque support a dorsal stream supramodal timing advantage in primates. Proc Natl Acad Sci U S A 109: 18168–18173.

Cao XJ, Oertel D (2010). Auditory nerve fibers excite targets through synapses that vary in convergence, strength, and short-term plasticity. J Neurophysiol 104: 2308–2320.

Cetas JS, Price RO, Velenovsky DS et al. (2001). Frequency organization and cellular lamination in the medial geniculate body of the rabbit. Hear Res 155: 113–123.

Chait M, Ruff CC, Griffiths TD et al. (2012). Cortical responses to changes in acoustic regularity are differentially modulated by attentional load. Neuroimage 59: 1932–1941.

Cheung SW, Bedenbaugh PH, Nagarajan SS et al. (2001). Functional organization of squirrel monkey primary auditory cortex: responses to pure tones. J Neurophysiol 85: 1732–1749.

Clarey JC, Irvine DR (1990). The anterior ectosylvian sulcal auditory field in the cat: I. An electrophysiological study of its relationship to surrounding auditory cortical fields. J Comp Neurol 301: 289–303.

Corey DP, Hudspeth AJ (1983). Kinetics of the receptor current in bullfrog saccular hair cells. J Neurosci 3: 962–976.

Cranford JL (1979). Detection versus discrimination of brief tones by cats with auditory cortex lesions. J Acoust Soc Am 65: 1573–1575.

Dadson RS, King M (1952). A determination of the normal threshold of hearing and its relation to the standardization of audiometers. J Laryngol Otol 66: 366–378.

Diamond IT, Neff WD (1957). Ablation of temporal cortex and discrimination of auditory patterns. J Neurophysiol 20: 300–315.

Dick F, Tierney AT, Lutti A et al. (2012). In vivo functional and myeloarchitectonic mapping of human primary auditory areas. J Neurosci 32: 16095–16105.

Duvel AD, Smith DM, Talk A et al. (2001). Medial geniculate, amygdalar and cingulate cortical training-induced neuronal activity during discriminative avoidance learning in rabbits with auditory cortical lesions. J Neurosci 21: 3271–3281.

Escabi MA, Schreiner CE (2002). Nonlinear spectrotemporal sound analysis by neurons in the auditory midbrain. J Neurosci 22: 4114–4131.

Evans S, Kyong JS, Rosen S et al. (2014). The pathways for intelligible speech: multivariate and univariate perspectives. Cereb Cortex 24: 2350–2361.

Giraud AL, Garnier S, Micheyl C et al. (1997). Auditory efferents involved in speech-in-noise intelligibility. Neuroreport 8: 1779–1783.

Glueckert R, Pfaller K, Kinnefors A et al. (2005). High resolution scanning electron microscopy of the human organ of Corti. A study using freshly fixed surgical specimens. Hear Res 199: 40–56.

Golding NL, Oertel D (2012). Synaptic integration in dendrites: exceptional need for speed. J Physiol 590: 5563–5569.

Gourevitch B, Eggermont JJ (2007). Spatial representation of neural responses to natural and altered conspecific vocalizations in cat auditory cortex. J Neurophysiol 97: 144–158.

Grothe B, Pecka M, McAlpine D (2010). Mechanisms of sound localization in mammals. Physiol Rev 90: 983–1012.

Guinan Jr JJ (2010). Cochlear efferent innervation and function. Curr Opin Otolaryngol Head Neck Surg 18: 447–453.

Hackett TA (2011). Information flow in the auditory cortical network. Hear Res 271: 133–146.

Hackett TA, Stepniewska I, Kaas JH (1998). Subdivisions of auditory cortex and ipsilateral cortical connections of the parabelt auditory cortex in macaque monkeys. J Comp Neurol 394: 475–495.

Halverson HE, Poremba A, Freeman JH (2008). Medial auditory thalamus inactivation prevents acquisition and retention of eyeblink conditioning. Learn Mem 15: 532–538.

Headley DB, Weinberger NM (2013). Fear conditioning enhances gamma oscillations and their entrainment of neurons representing the conditioned stimulus. J Neurosci 33: 5705–5717.

Hudspeth AJ (1982). Extracellular current flow and the site of transduction by vertebrate hair cells. J Neurosci 2: 1–10.

Imig TJ, Adrian HO (1977). Binaural columns in the primary field (A1) of cat auditory cortex. Brain Res 138: 241–257.

Irvine DRF (1986). The auditory brainstem. Progress in Sensory Physiology 7: 1–279.

Jain R, Shore S (2006). External inferior colliculus integrates trigeminal and acoustic information: unit responses to trigeminal nucleus and acoustic stimulation in the guinea pig. Neurosci Lett 395: 71–75.

Jones EG (2003). Chemically defined parallel pathways in the monkey auditory system. Ann N Y Acad Sci 999: 218–233.

Joris PX (1998). Response classes in the dorsal cochlear nucleus and its output tract in the chloralose-anesthetized cat. J Neurosci 18: 3955–3966.

Joris PX, Carney LH, Smith PH et al. (1994). Enhancement of neural synchronization in the anteroventral cochlear nucleus. I. Responses to tones at the characteristic frequency. J Neurophysiol 71: 1022–1036.

Kadia SC, Wang X (2003). Spectral integration in A1 of awake primates: neurons with single- and multipeaked tuning characteristics. J Neurophysiol 89: 1603–1622.

Kajikawa Y, de La Mothe L, Blumell S et al. (2005). A comparison of neuron response properties in areas A1 and CM of the marmoset monkey auditory cortex: tones and broadband noise. J Neurophysiol 93: 22–34.

Kaltenbach JA (2011). Tinnitus: Models and mechanisms. Hear Res 276: 52–60.

Kelly JB, van Adel BA, Ito M (2009). Anatomical projections of the nuclei of the lateral lemniscus in the albino rat (*Rattus norvegicus*). J Comp Neurol 512: 573–593.

Kopp-Scheinpflug C, Dehmel S, Dorrscheidt GJ et al. (2002). Interaction of excitation and inhibition in anteroventral cochlear nucleus neurons that receive large endbulb synaptic endings. J Neurosci 22: 11004–11018.

Kujawa SG, Liberman MC (1997). Conditioning-related protection from acoustic injury: effects of chronic deefferentation and sham surgery. J Neurophysiol 78: 3095–3106.

Kumar S, Stephan KE, Warren JD et al. (2007). Hierarchical processing of auditory objects in humans. PLoS Comput Biol 3: e100.

Kuwada S, Bishop B, Alex C et al. (2011). Spatial tuning to sound-source azimuth in the inferior colliculus of unanesthetized rabbit. J Neurophysiol 106: 2698–2708.

Ladhams A, Pickles JO (1996). Morphology of the monotreme organ of Corti and macula lagena. J Comp Neurol 366: 335–347.

Langner G, Schreiner CE (1988). Periodicity coding in the inferior colliculus of the cat. I. Neuronal mechanisms. J Neurophysiol 60: 1799–1822.

LeBeau FE, Malmierca MS, Rees A (2001). Iontophoresis in vivo demonstrates a key role for GABA(A) and glycinergic inhibition in shaping frequency response areas in the inferior colliculus of guinea pig. J Neurosci 21: 7303–7312.

Lee CC, Middlebrooks JC (2013). Specialization for sound localization in fields A1, DZ, and PAF of cat auditory cortex. J Assoc Res Otolaryngol 14: 61–82.

Lee CC, Sherman SM (2011). On the classification of pathways in the auditory midbrain, thalamus, and cortex. Hear Res 276: 79–87.

Liberman MC, Kiang NY (1978). Acoustic trauma in cats. Cochlear pathology and auditory-nerve activity. Acta Otolaryngol Suppl 358: 1–63.

Loftus WC, Sutter ML (2001). Spectrotemporal organization of excitatory and inhibitory receptive fields of cat posterior auditory field neurons. J Neurophysiol 86: 475–491.

Loftus WC, Bishop DC, Saint Marie RL et al. (2004). Organization of binaural excitatory and inhibitory inputs to the inferior colliculus from the superior olive. J Comp Neurol 472: 330–344.

Lomber SG, Malhotra S (2008). Double dissociation of 'what' and 'where' processing in auditory cortex. Nat Neurosci 11: 609–616.

Lomber SG, Malhotra S, Hall AJ (2007). Functional specialization in non-primary auditory cortex of the cat: Areal and laminar contributions to sound localization. Hear Res 229: 31–45.

Lumani A, Zhang H (2010). Responses of neurons in the rat's dorsal cortex of the inferior colliculus to monaural tone bursts. Brain Res 1351: 115–129.

Malmierca MS, Leergaard TB, Bajo VM et al. (1998). Anatomic evidence of a three-dimensional mosaic pattern of tonotopic organization in the ventral complex of the lateral lemniscus in cat. J Neurosci 18: 10603–10618.

McEchron MD, Green EJ, Winters RW et al. (1996). Changes of synaptic efficacy in the medial geniculate nucleus as a result of auditory classical conditioning. J Neurosci 16: 1273–1283.

McGinley MJ, Oertel D (2006). Rate thresholds determine the precision of temporal integration in principal cells of the ventral cochlear nucleus. Hear Res 216–217: 52–63.

Miller LM, Recanzone GH (2009). Populations of auditory cortical neurons can accurately encode acoustic space across stimulus intensity. Proc Natl Acad Sci U S A 106: 5931–5935.

Moerel M, De Martino F, Formisano E (2012). Processing of natural sounds in human auditory cortex: tonotopy, spectral tuning, and relation to voice sensitivity. J Neurosci 32: 14205–14216.

Morest DK, Oliver DL (1984). The neuronal architecture of the inferior colliculus in the cat: defining the functional anatomy of the auditory midbrain. J Comp Neurol 222: 209–236.

Nakajima HH, Dong W, Olson ES et al. (2009). Differential intracochlear sound pressure measurements in normal human temporal bones. J Assoc Res Otolaryngol 10: 23–36.

Neff WD (1968). Localization and lateralization of sound in space. In: AVS de Reuck, J Knight (Eds.), Hearing Mechanisms in Vertebrates, Churchill, London, pp. 207–231.

Netter FH (1948). The CIBA Collection of Medical Illustrations, Vol. 1, Nervous System, CIBA Foundation, Summit, NJ.

Oertel D (1991). The role of intrinsic neuronal properties in the encoding of auditory information in the cochlear nuclei. Curr Opin Neurobiol 1: 221–228.

Oertel D, Young ED (2004). What's a cerebellar circuit doing in the auditory system? Trends Neurosci 27: 104–110.

Osen KK, Roth K (1969). Histochemical localization of cholinesterases in the cochlear nuclei of the cat, with notes on the origin of acetylcholinesterase-positive afferents and the superior olive. Brain Res 16: 165–185.

Pecka M, Siveke I, Grothe B et al. (2010). Enhancement of ITD coding within the initial stages of the auditory pathway. J Neurophysiol 103: 38–46.

Petkov CI, Kayser C, Augath M et al. (2006). Functional imaging reveals numerous fields in the monkey auditory cortex. PLoS Biol 4: e215.

Pickles JO (2012). An introduction to the physiology of hearing. 4th edn. Brill, Leiden.

Pickles JO (2013). Illustrations of the inner ear, 2nd edn. Graceville Press, Graceville.

Pickles JO, Comis SD, Osborne MP (1984). Cross-links between stereocilia in the guinea pig organ of Corti, and their possible relation to sensory transduction. Hear Res 15: 103–112.

Pollak GD (2012). Circuits for processing dynamic interaural intensity disparities in the inferior colliculus. Hear Res 288: 47–57.

Poon PW, Chen X, Cheung YM (1992). Differences in FM response correlate with morphology of neurons in the rat inferior colliculus. Exp Brain Res 91: 94–104.

Read HL, Winer JA, Schreiner CE (2002). Functional architecture of auditory cortex. Curr Opin Neurobiol 12: 433–440.

Reale RA, Imig TJ (1980). Tonotopic organization in auditory cortex of the cat. J Comp Neurol 192: 265–291.

Recanzone GH, Engle JR, Juarez-Salinas DL (2011). Spatial and temporal processing of single auditory cortical neurons and populations of neurons in the macaque monkey. Hear Res 271: 115–122.

Robles L, Ruggero MA (2001). Mechanics of the mammalian cochlea. Physiol Rev 81: 1305–1352.

Rose JE, Hind JE, Anderson DJ et al. (1971). Some effects of stimulus intensity on response of auditory nerve fibers in the squirrel monkey. J Neurophysiol 34: 685–699.

Rosowski JJ (1991). The effects of external- and middle-ear filtering on auditory threshold and noise-induced hearing loss. J Acoust Soc Am 90: 124–135.

Ruggero MA, Temchin AN (2002). The roles of the external, middle, and inner ears in determining the bandwidth of hearing. Proc Natl Acad Sci U S A 99: 13206–13210.

Scharf B, Magnan J, Chays A (1997). On the role of the olivocochlear bundle in hearing: 16 case studies. Hear Res 103: 101–122.

Schreiner CE, Cynader MS (1984). Basic functional organization of second auditory cortical field (AII) of the cat. J Neurophysiol 51: 1284–1305.

Smith PH, Spirou GA (2002). From the cochlea to cortex and back. In: D Oertel, RR Fay, AN Popper (Eds.), Integrative Functions in the Mammalian Auditory Pathway, Springer Handbook of Auditory Research, Vol. 15. Springer, New York, pp. 6–71.

Spierer L, Bellmann-Thiran A, Maeder P et al. (2009). Hemispheric competence for auditory spatial representation. Brain 132: 1953–1966.

Stecker GC, Mickey BJ, Macpherson EA et al. (2003). Spatial sensitivity in field PAF of cat auditory cortex. J Neurophysiol 89: 2889–2903.

Suga N, Ma X (2003). Multiparametric corticofugal modulation and plasticity in the auditory system. Nat Rev Neurosci 4: 783–794.

Suga N, Zhang Y, Yan J (1997). Sharpening of frequency tuning by inhibition in the thalamic auditory nucleus of the mustached bat. J Neurophysiol 77: 2098–2114.

Sutherland DP, Masterton RB, Glendenning KK (1998). Role of acoustic striae in hearing: reflexive responses to elevated sound-sources. Behav Brain Res 97: 1–12.

Talwar SK, Musial PG, Gerstein GL (2001). Role of mammalian auditory cortex in the perception of elementary sound properties. J Neurophysiol 85: 2350–2358.

Tang J, Pecka JL, Fritzsch B et al. (2013). Lizard and frog prestin: evolutionary insight into functional changes. PLoS One 8: e54388.

Tian B, Rauschecker JP (1994). Processing of frequency-modulated sounds in the cat's anterior auditory field. J Neurophysiol 71: 1959–1975.

Tian B, Rauschecker JP (1998). Processing of frequency-modulated sounds in the cat's posterior auditory field. J Neurophysiol 79: 2629–2642.

Tian B, Rauschecker JP (2004). Processing of frequency-modulated sounds in the lateral auditory belt cortex of the rhesus monkey. J Neurophysiol 92: 2993–3013.

Tramo MJ, Cariani PA, Koh CK et al. (2005). Neurophysiology and neuroanatomy of pitch perception: auditory cortex. Ann N Y Acad Sci 1060: 148–174.

Velenovsky DS, Cetas JS, Price RO et al. (2003). Functional subregions in primary auditory cortex defined by thalamocortical terminal arbors: an electrophysiological and anterograde labeling study. J Neurosci 23: 308–316.

Wang X (2007). Neural coding strategies in auditory cortex. Hear Res 229: 81–93.

Winer JA (1985). The medial geniculate body of the cat. Adv Anat Embryol Cell Biol 86: 1–97.

Winer JA, Miller LM, Lee CC et al. (2005). Auditory thalamocortical transformation: structure and function. Trends Neurosci 28: 255–263.

Zhang H, Kelly JB (2010). Time dependence of binaural responses in the rat's central nucleus of the inferior colliculus. Hear Res 268: 271–280.

Zhang Y, Suga N (2000). Modulation of responses and frequency tuning of thalamic and collicular neurons by cortical activation in mustached bats. J Neurophysiol 84: 325–333.

Handbook of Clinical Neurology, Vol. 129 (3rd series)
The Human Auditory System
G.G. Celesia and G. Hickok, Editors

Chapter 2

Anatomic organization of the auditory cortex

TROY A. HACKETT*
Department of Hearing and Speech Sciences, Vanderbilt University School of Medicine and Department of Psychology, Vanderbilt University, Nashville, TN, USA

ABBREVIATIONS

A1 auditory area 1 (core)
AAF anterior auditory field
AChE acetylcholinesterase
AIP anterior intraparietal cortex
AL anterolateral area (belt)
AS arcuate sulcus
CB calbindin
Cis circular sulcus
CL caudolateral area (belt)
CM caudomedial area (belt)
CO cytochrome oxidase
CPB caudal parabelt area (parabelt)
CS central sulcus
DTI diffusion tensor imaging
fMRI functional magnetic resonance imaging
HG Heschl's gyrus
Ig granular insula
Ins insula
IPS intraparietal sulcus
ITG inferior temporal gyrus
LA lateral nucleus of the amygdala
Lim limitans
LIP lateral intraparietal cortex
LS lateral sulcus
LuS lunate sulcus
MGad medial geniculate complex, anterodorsal division
MGB medial geniculate body
MGC medial geniculate complex
MGd medial geniculate complex, dorsal division
MGm medial geniculate complex, magnocellular division
MGpd medial geniculate complex, posterodorsal division
MGv medial geniculate complex, ventral division
ML middle lateral area (belt)
MM middle medial
MT middle temporal area
PM medial pulvinar
Po posterior
PP peripeduncular
Pro proisocortical area
proA prokoniocortex area
PS principal sulcus
PT planum temporale
PV parvalbumin
R rostral area (core)
Ri retroinsular area
RM rostromedial area (belt)
RPB rostral parabelt area (parabelt)
RT rostrotemporal area (core)
RTL rostrotemporal lateral area (belt)
RTM rostrotemporal medial area (belt)
S2 somatosensory area 2
Sg suprageniculate
STG superior temporal gyrus
STGr rostral STG
STP superior temporal polysensory region
STS superior temporal sulcus
TPO temporal parietal occipital area
Tpt temporal parietotemporal area
TRN thalamic reticular nucleus
Ts1 superior temporal area 1
Ts2 superior temporal area 2
TTG transverse temporal gyrus
VIP ventral intraparietal cortex
VGlut2 vesicular glutamate transporter 2

*Correspondence to: Troy A. Hackett, Vanderbilt University, 465 21st Avenue South, MRB-3 Suite 7114, Nashville, TN 37232, USA. E-mail: troy.a.hackett@vanderbilt.edu

WHAT IS AUDITORY CORTEX?

In every lobe of the brain are regions that contain neurons responsive to sound. Are all of these regions part of the auditory cortex? The answer depends on how auditory cortex is defined. No single definition is entirely satisfactory, but to facilitate the present discussion, it will be helpful to establish some boundaries. For example, a functional definition could be adopted such that any portion of the cerebral cortex containing neurons responsive to sound is part of auditory cortex. While that definition appears straightforward, auditory-responsive neurons in several brain areas are intermingled with neurons responsive to tactile or visual stimulation. Indeed, it has been suggested that neuronal activity in most of the cerebral cortex may be influenced by the inputs of more than one sensory modality, including areas typically considered to be unisensory (Ghazanfar and Schroeder, 2006). The notion that auditory cortex is widely distributed throughout the brain and overlaps with other sensory fields is unnecessarily complicated and subject to repeated modifications over time as new "auditory" fields are discovered. To avoid these difficulties, we have adopted an anatomic definition, based on subcortical inputs to the cerebral cortex. By this definition, auditory cortex includes any portion of the cerebral cortex that receives its main thalamic input from one or more divisions of the medial geniculate body (MGB) in the thalamus. The bases for this definition are that the MGB is the principal target of the ascending brainstem auditory pathways, and that the MGB is the primary source of auditory input to cortex. By this definition, the auditory cortex of most mammals is restricted to an assembly of adjacent areas located in a portion of the temporal lobe (or region) (Fig. 2.1). In contrast, cortical areas that receive inputs from auditory cortex and/or minor inputs from the MGB are referred to as auditory-related. Such areas are widely distributed throughout the brain and typically receive additional inputs from other sensory or motor modalities. The principles of auditory cortical organization discussed in this chapter are built around these definitions.

PRINCIPLES OF AUDITORY CORTICAL ORGANIZATION

As detailed in Chapter 1, auditory information from the cochlea is processed in five major nuclear complexes in the brainstem and thalamus before it reaches the auditory cortex. In cortex, that information is distributed among a

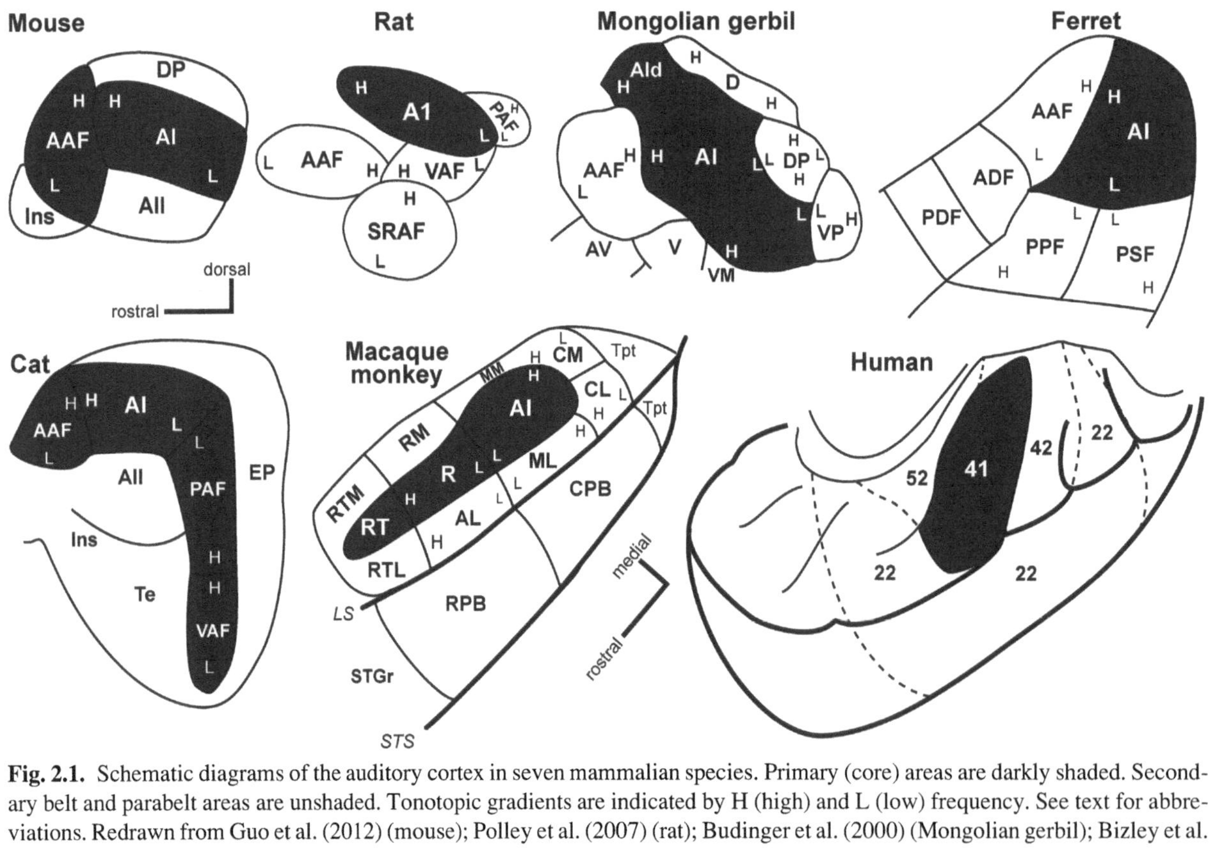

Fig. 2.1. Schematic diagrams of the auditory cortex in seven mammalian species. Primary (core) areas are darkly shaded. Secondary belt and parabelt areas are unshaded. Tonotopic gradients are indicated by H (high) and L (low) frequency. See text for abbreviations. Redrawn from Guo et al. (2012) (mouse); Polley et al. (2007) (rat); Budinger et al. (2000) (Mongolian gerbil); Bizley et al. (2005) (ferret); Lee et al. (2004) (cat); Hackett et al. (1998a) (macaque monkey); Brodmann (1909) (human).

network of auditory and auditory-related areas that presumably mediate the conscious awareness and perception of sound. Over decades of research, several principles of organization have been derived from the major anatomic and physiologic features that serve to govern information flow in the network of auditory areas in the brain. While these principles apply generally to all mammalian species, the discussion below focuses primarily on auditory cortex of non-human primates and humans.

Principle 1: Auditory cortex can be divided into regions

The first principle of organization is that the auditory cortex can be divided into regions. A cortical region may be further subdivided, but is distinguished from adjacent regions by a unique set of anatomic or physiologic properties. For example, sensory regions of cortex have commonly been referred to by terms such as primary, secondary, or association cortex. The dark- and medium-shaded areas of auditory cortex in the species illustrated in Figure 2.1 would be referred to as primary and secondary regions of auditory cortex. In non-human primates, the major regions of auditory cortex have come to be known as the core, belt, and parabelt (Fig. 2.2). These three regions occupy the lower bank of the lateral sulcus (LS) and superior temporal gyrus (STG), forming parallel bands elongated along the rostrocaudal axis of the temporal lobe. In primates with larger brains (e.g., macaque monkeys), most of the core and belt regions (Fig. 2.2C) are completely covered by the parietal operculum, so that only the parabelt and most lateral portion of the belt are visible on the surface of the brain (Fig. 2.2A, B). In smaller-brained primates (e.g., marmosets, squirrel monkeys), the lateral sulcus is shallower and some portion of the core region extends on to the STG surface. In humans, where there is not yet an established layout, it seems fairly certain that there is a central core region surrounded by a belt region and likely at least one other region. The core–belt–parabelt schema has been applied to humans in some studies (e.g., Sweet et al., 2005), but none of the current models has been widely adopted and this effort remains an ongoing area of research (see below).

Each region is characterized by a unique set of architectonic features and connection patterns, which

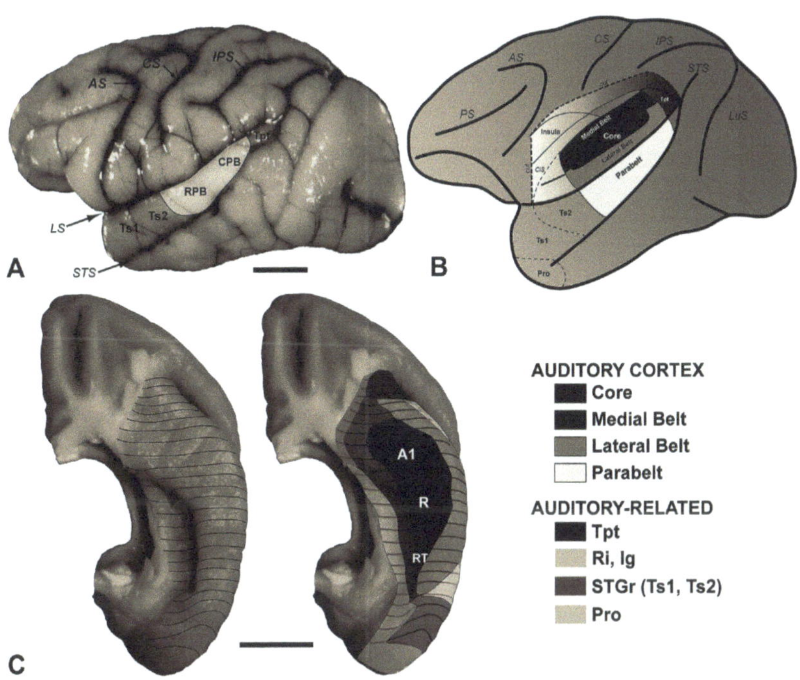

Fig. 2.2. Location of auditory cortex and adjacent auditory-related areas in superior temporal cortex of the macaque monkey. (**A**) Lateral view of the left hemisphere. Location of the parabelt areas (CPB, caudal; RPB, rostral) on the superior temporal gyrus is highlighted. AS, arcuate sulcus; CS, central sulcus; IPS, intraparietal sulcus; LS, lateral sulcus; STS, superior temporal sulcus; Ts1, Ts2, Tpt, temporal parietotemporal area. (**B**) Schematic diagram showing the location of the core, belt, and parabelt regions after removal of the overlying parietal and frontal opercula (cut). The estimated locations of the rostral superior temporal gyrus (STGr) areas Ts2, Ts1, and temporal pole (Pro) are also drawn (Galaburda and Pandya, 1983). (**C**) Dorsal views of the superior temporal plane showing locations of each region. The approximate location of areas primary auditory cortex (A1), rostral (R), and rostrotemporal (RT) are indicated within the core region (darkest shading). Caudal is up, rostral is down. Black contour lines run across the brain surface from medial (left) to lateral (right). Ri, retroinsular area; Ig, granular insula. Scale bars: (**A–C**) 10 mm.

distinguish it from other regions. Connections refer to axonal connections between auditory cortex and nuclei in the thalamus and other areas of cortex. Architecture, or architectonics, refers to the numerous structural features that can be identified following various histologic treatments of the tissue. The main categories include cytoarchitecture (arrangement and types of neurons), myeloarchitecture (orientation and density of myelinated axons), and chemoarchitecture (distribution and expression of proteins, enzymes, and other substances within neurons and their branches).

In the auditory cortex of primates, systematic transitions in the architecture occur along two anatomic axes (medial–lateral and caudal–rostral), which contribute to its division into regions. At low magnification, robust differences between regions can be seen in preparations that reveal expression of various chemoarchitectonic markers, as illustrated in Figure 2.3. Transitions that distinguish the core, belt, and parabelt occur along the medial–lateral axis. The core region, for example, is characterized by the colocalization of several protein markers (e.g., cytochrome oxidase, acetylcholinesterase, parvalbumin, vesicular glutamate transporter 2), which are concentrated in a prominent horizontal band encompassing layer IV and the lower part of layer III (Fig. 2.3, asterisk). This band is also the main site of thalamocortical inputs to auditory cortex. Moving from the core into the belt (medial and lateral) and parabelt regions, note the stepwise reductions in the density of these markers. Their density in the layer III/IV band is highest in core, intermediate in the belt, and lowest in the parabelt. Additional details about these architectonic features are available elsewhere, and are regularly updated as the expression of other proteins in auditory cortex is characterized (Pandya and Sanides, 1973; Galaburda and Pandya, 1983; Morel and Kaas, 1992; Morel et al., 1993; Jones et al., 1995; Kosaki et al., 1997; Hackett et al., 1998a; de la Mothe et al., 2006a). Connection patterns and physiologic features tend to be correlated with these architectonic features, and contribute to the profile of each region.

Principle 2: Regions of auditory cortex are subdivided into areas

The second principle of organization is that each region of auditory cortex may contain more than one subdivision, or area. The subdivision of regions into areas reflects heterogeneity within a region based on structural and functional gradients, such as changes in architecture, connections and physiologic response properties (Merzenich and Brugge, 1973; Pandya and Sanides, 1973; Imig et al., 1977; Galaburda and Pandya, 1983; Morel and Kaas, 1992; Morel et al., 1993; Jones et al.,

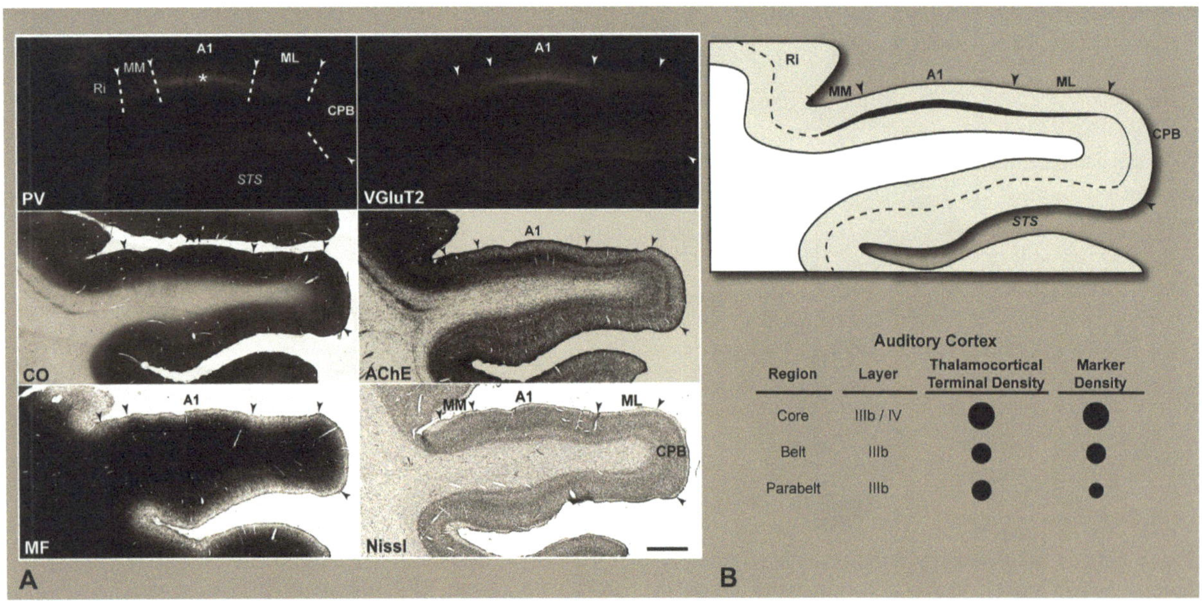

Fig. 2.3. Architectonic gradients in the macaque monkey auditory cortex. (**A**) Examples of protein marker expression in coronal sections through primary auditory cortex (A1) of the macaque monkey. From the top left, parvalbumin (PV); vesicular glutamate transporter 2 (VGlut2); cytochrome oxidase (CO); acetylcholinesterase (AChE); myelinated fibers (MF); and thionin stain (Nissl). The layer III/IV band in A1 is indicated by a white asterisk in top left panel. Note systematic changes in expression in the layer III/IV band across the medial to lateral axis: density is highest in the core (A1), intermediate in the medial and lateral belt areas (MM, ML), and lowest in the parabelt (CPB). Ri, retroinsular area. (**B**) Summary of architectonic gradients in auditory cortex. Stepwise transitions in density of the layer III/IV band between regions are noted along the medial to lateral axis. STS, superior temporal sulcus.

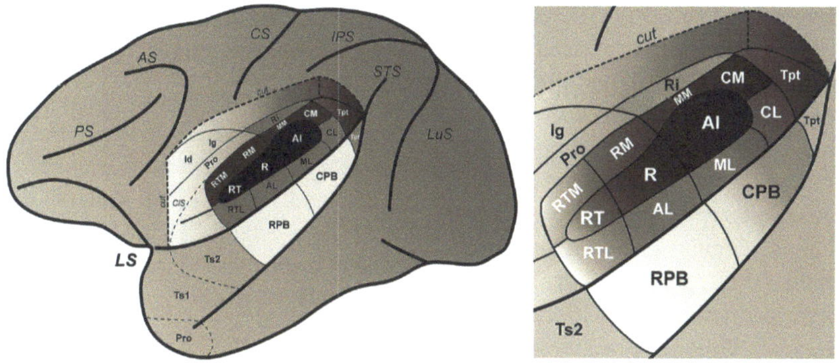

Fig. 2.4. Schematic diagrams of the macaque monkey auditory cortex in the left hemisphere showing the individual areas of the core, belt, and parabelt regions and adjoining areas of the insula and rostral superior temporal gyrus. See text for abbreviations.

1995; Hackett et al., 1998a; de la Mothe et al., 2006a; Smiley et al., 2007; Hackett and de la Mothe, 2009). The current version of our primate model (Fig. 2.4) retains most of the areas originally identified in macaques (Merzenich and Brugge, 1973; Pandya and Sanides, 1973) and owl monkeys (Imig et al., 1977), along with the subsequent addition of the parabelt (Morel and Kaas, 1992; Hackett et al., 1998a) and other elaborations. The core region has three areas (auditory area 1 (A1), rostral (R), rostrotemporal (RT)). The belt region has at least seven or eight (medial belt: caudomedial (CM)/middle medial, rostromedial (RM), rostrotemporal medial (RTM); lateral belt, caudolateral (CL), middle lateral (ML), anterolateral (AL), rostrotemporal lateral (RTL)). The parabelt region has at least two areas (caudal parabelt (CPB), rostral parabelt (RPB)). As relevant data have accumulated, uncertainty over the precise number of areas has diminished, but some remains. For example, characterization of the rostral and medial areas of the belt remains rather meager to date, and significant modifications are still possible, whereas the caudal and lateral belt areas have become better established by more intensive study. As more is learned, it is possible that some of the areas included in current models may be modified. Thus, it is important to be mindful that the figures contained in this review are representative of the available data, and are subject to change.

Principle 3: Individual areas of auditory cortex are tonotopically organized

The third principle of organization is the topographic organization of acoustic frequency, known as tonotopy or cochleotopy. Beginning with the cochlear nucleus and continuing throughout the auditory pathway, a representation, or map, of the cochlea is maintained at each level. The highest frequencies are represented at one end of each structure, the lowest frequencies at the other, with a continuous gradient in between. In the auditory cortex of primates and most other studied mammals, a complete or partial representation of the cochlea has been identified for most of the areas that have been studied. In monkeys, microelectrode recordings have revealed a complete or partial representation of the cochlea in all three core areas and several belt areas, as indicated by the designation of high- (H) and low- (L) frequency gradients in Figure 2.5 (Merzenich and Brugge, 1973; Imig et al., 1977; Luethke et al., 1989; Morel and Kaas, 1992; Morel et al., 1993; Rauschecker et al., 1995; Kosaki et al., 1997; Rauschecker and Tian, 2004; Bendor and Wang, 2005, 2008a; Kajikawa et al., 2005). Frequency gradients tend to reverse between adjacent areas along the rostral–caudal axis (e.g., A1–R–RT), but are usually parallel between adjacent areas along the medial–lateral axis (e.g., A1–ML, R–AL). Areas that remain to be mapped in this way include several rostral and medial belt areas, and all of the parabelt, although functional magnetic resonance imaging (fMRI) studies in the macaque revealed tonotopic

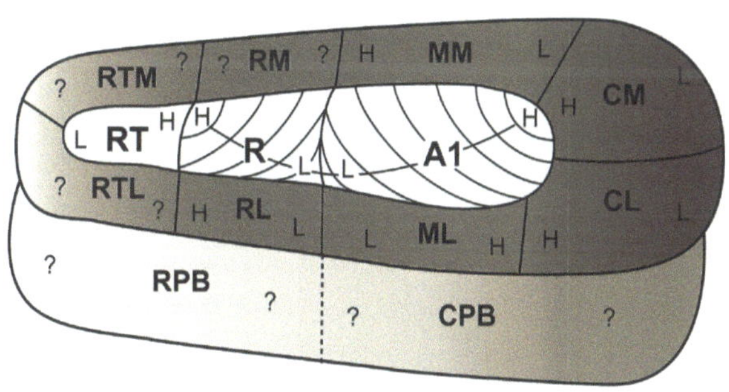

Fig. 2.5. Tonotopic gradients of areas within auditory cortex as established by microelectrode recordings and functional magnetic resonance imaging studies (see text). High best frequency (H); low best frequency (L). Areas in which tonotopic organization is not known or uncertain are labeled by question marks (?). See text for abbreviations.

gradients in all of the core and belt areas (Petkov et al., 2006). Functionally, the representation of frequency is the most robust topographically distributed physiologic feature known in auditory cortex. The presence of these maps indicates that auditory processing within each area is spectrally distributed, such that the interactions between auditory areas appear to be shaped by tonotopic organization. For example, connections between matching isofrequency domains in different areas are stronger than connections between unmatched locations (Morel and Kaas, 1992; Morel et al., 1993; Lee et al., 2004). Other feature maps and neurophysiologic gradients have also been observed in auditory cortex of primates and other species, commonly in orientations that are orthogonal or tangential to the map of frequency (e.g., latency, intensity, bandwidth, periodicity, binaural properties) (Schreiner and Sutter, 1992; Schreiner et al., 1992; Recanzone et al., 1999; Schulze et al., 2002; Schreiner and Winer, 2007; Nelken et al., 2008; Langner et al., 2009; Barton et al., 2012; Herdener et al., 2013).

Principle 4: Thalamic inputs to auditory cortex vary by region and layer

The thalamus is the primary source of sensory information to the cerebral cortex. Inputs to the auditory and auditory-related areas from the thalamus include the MGB and several other nuclei in the posterior thalamus that have established multisensory features (Fig. 2.6). The main thalamic inputs to auditory cortex arise from among three major divisions of the MGB (Jones, 1998, 2003, 2007; Lee and Winer, 2008): ventral (MGv), dorsal (MGd), and magnocellular or medial (MGm). In primates, the MGd is subdivided into two divisions: anterodorsal (MGad) and posterodorsal (MGpd). Additional smaller divisions have been identified in other species, such as the cat and rat (Lee and Winer, 2008). These subdivisions may also be present in primates, but the detailed anatomic studies needed to validate those details and reconcile apparent species differences have not yet been completed. As a result, there are differences

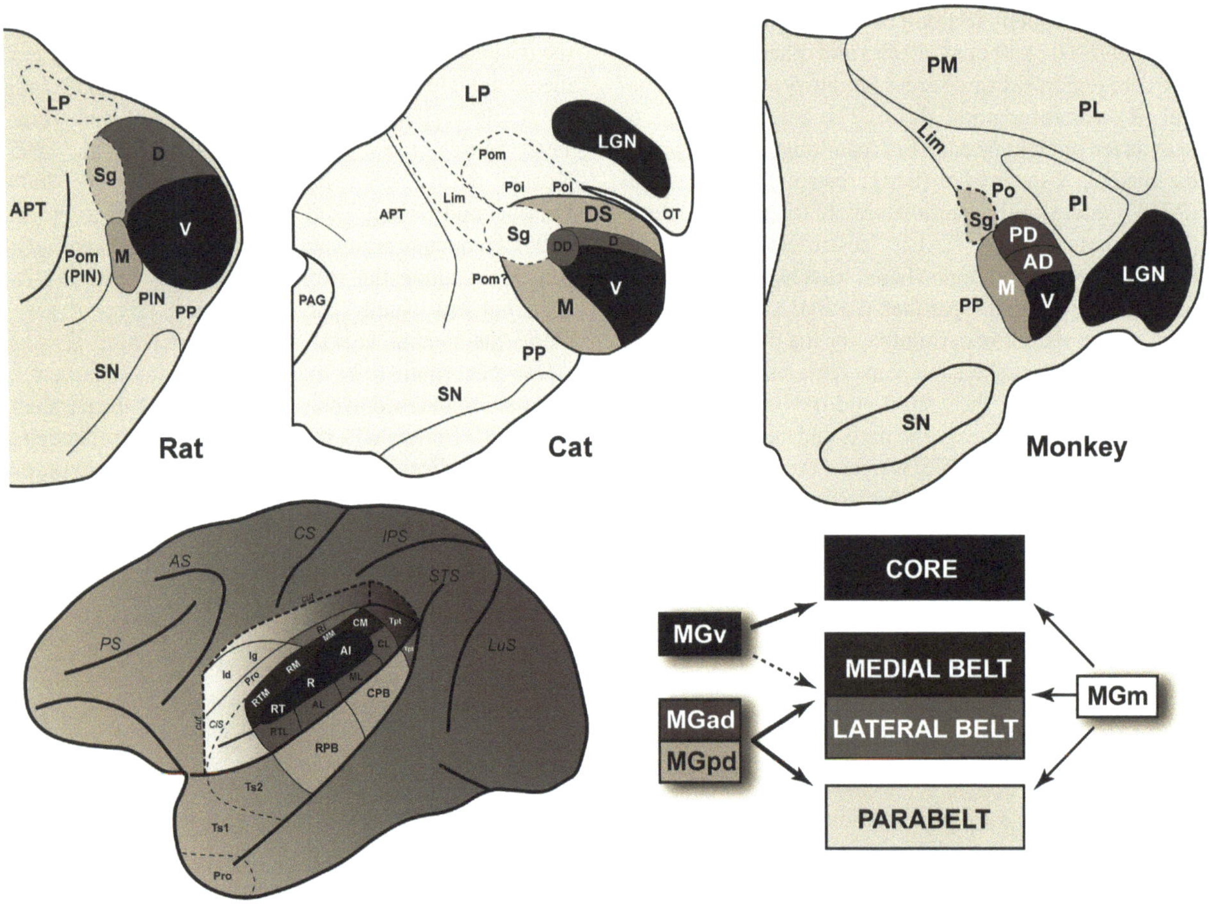

Fig. 2.6. Top, schematic diagrams of the medial geniculate complex (MGC) in three species. Abbreviations of nuclei are given in the list of abbreviations. Bottom, summary of major thalamocortical projections to auditory cortex. The core region receives inputs from the ventral division of the MGC (MGv). The belt and parabelt regions receive inputs from the anterodorsal (MGad) and posterodorsal (MGpd) divisions of the MGC. The magnocellular division (MGm) of the MGC projects to all areas of auditory cortex.

in the layout and nomenclature of the MGB and adjacent divisions between species (Fig. 2.6, top). In addition to the MGB, inputs to auditory cortex arise from several multisensory nuclei in the thalamus, which represent smaller fractions of the total input to auditory cortex, and favor the non-primary fields. These nuclei include the suprageniculate (Sg), limitans (Lim), posterior (Po), medial pulvinar (PM), and peripeduncular (PP) nuclei, among others. Some of the auditory-related areas in the temporal lobe (e.g., proisocortical (Pro), Ts1, Ts2, temporal parietotemporal (Tpt)) also receive inputs from the MGB and multisensory nuclei, but the MGB accounts for a smaller proportion of the total connections, in accordance with our definition of auditory cortex (Markowitsch et al., 1985; Kosmal et al., 1997). The general trend is that the MGB projections to the auditory-related areas become progressively weaker with distance from auditory cortex. To some extent, this reduction is offset by a relative increase in inputs from the multisensory nuclei.

Connections of the major MGB divisions

Each MGB division is unique with respect to its inputs, outputs, and cellular architecture, and can therefore be considered to provide a distinct set of projections to cortex (Fig. 2.6, bottom). The thalamocortical terminations of the MGB are concentrated in, though not limited to, the layer III/IV band in all of their cortical targets. The relative strength of each projection varies by region and by area. Specifically, principal thalamic inputs to the core region come from the ventral (MGv) division of the MGB, whereas the belt and parabelt regions mainly receive inputs from the dorsal division (MGd). The magnocellular division (MGm) projects to all three regions. These patterns are briefly described in more detail below.

MGv

The projection from the ventral division is the main source of thalamic inputs to the core region (Fig. 2.6). Weaker inputs to areas outside the core have been noted in some of the belt areas (Morel and Kaas, 1992; Morel et al., 1993; de la Mothe et al., 2006b, 2012; Hackett et al., 2007b), but not to the parabelt (Hackett et al., 1998b). The MGv and its projection to cortex are tonotopically organized, reflecting inputs from the central nucleus of the inferior colliculus and the primary auditory pathways in the brainstem.

MGd

Inputs from the MGd are concentrated in the belt and parabelt regions. In primates, two divisions are recognized: anterodorsal (MGad) and posterodorsal (MGpd) (Fig. 2.6). The anterodorsal division is largely located in the anterior (rostral) pole of the MGB. This nucleus is most strongly connected with the caudal belt areas in primates (de la Mothe et al., 2006b; Hackett et al., 2007a). The MGad and its projection may be tonotopically organized (Gross et al., 1974), as are corresponding nuclei located in the rostral pole of the MGB in other mammals (Lee et al., 2004; de la Mothe et al., 2006b; Hackett et al., 2007a; Lee and Winer, 2008). The projection from the posterodorsal division (MGpd) is the principal source of inputs to most of the rostral belt and parabelt areas. In contrast to the MGad, it does not appear to exhibit a tonotopic pattern of frequency organization.

MGm

The magnocellular, or medial, division projects broadly to areas in all three regions of auditory cortex, and also to auditory-related areas such as Tpt (Fig. 2.6). On the basis of studies in other mammals, it is also thought that non-auditory (somatosensory, vestibular, visual) inputs are integrated by at least some MGm neurons, but this not been clearly demonstrated in primates (reviewed in Jones, 2007; Hackett, 2011).

In addition to inputs from the MGB to cortex, there is a dense system of descending (corticothalamic) projections originating from neurons in layers V and VI of all areas of auditory cortex and targeting all divisions of the MGB. The descending projections target the MGB via a direct system involving both small and giant axon terminals, and an indirect system that sends collateral inputs to the posterior (auditory) aspect of the thalamic reticular nucleus (TRN) (Bajo et al., 1995; Crabtree, 1998; Rouiller and Welker, 2000; He, 2003; Rouiller and Durif, 2004; Hackett, 2007b; Zhang et al., 2008). Both systems project topographically to the MGB, where their effects on activity vary by division. Descending projections from many areas of auditory cortex also target the dorsomedial and ventromedial divisions of the inferior colliculus (FitzPatrick and Imig, 1978; Luethke et al., 1989; Morel and Kaas, 1992; Winer et al., 2002; de la Mothe et al., 2006b), but since these projections have not been systematically studied for all areas, it is not yet clear whether they vary topographically.

Principle 5: The connections of auditory cortex have serial and parallel features

A fifth principle of organization is that the circuits formed by the inputs and outputs of auditory cortical areas have both serial (hierarchical) and parallel properties (Fig. 2.7). These properties largely derive from regularities in the patterns of thalamic and cortical connections, which provide an indirect means of

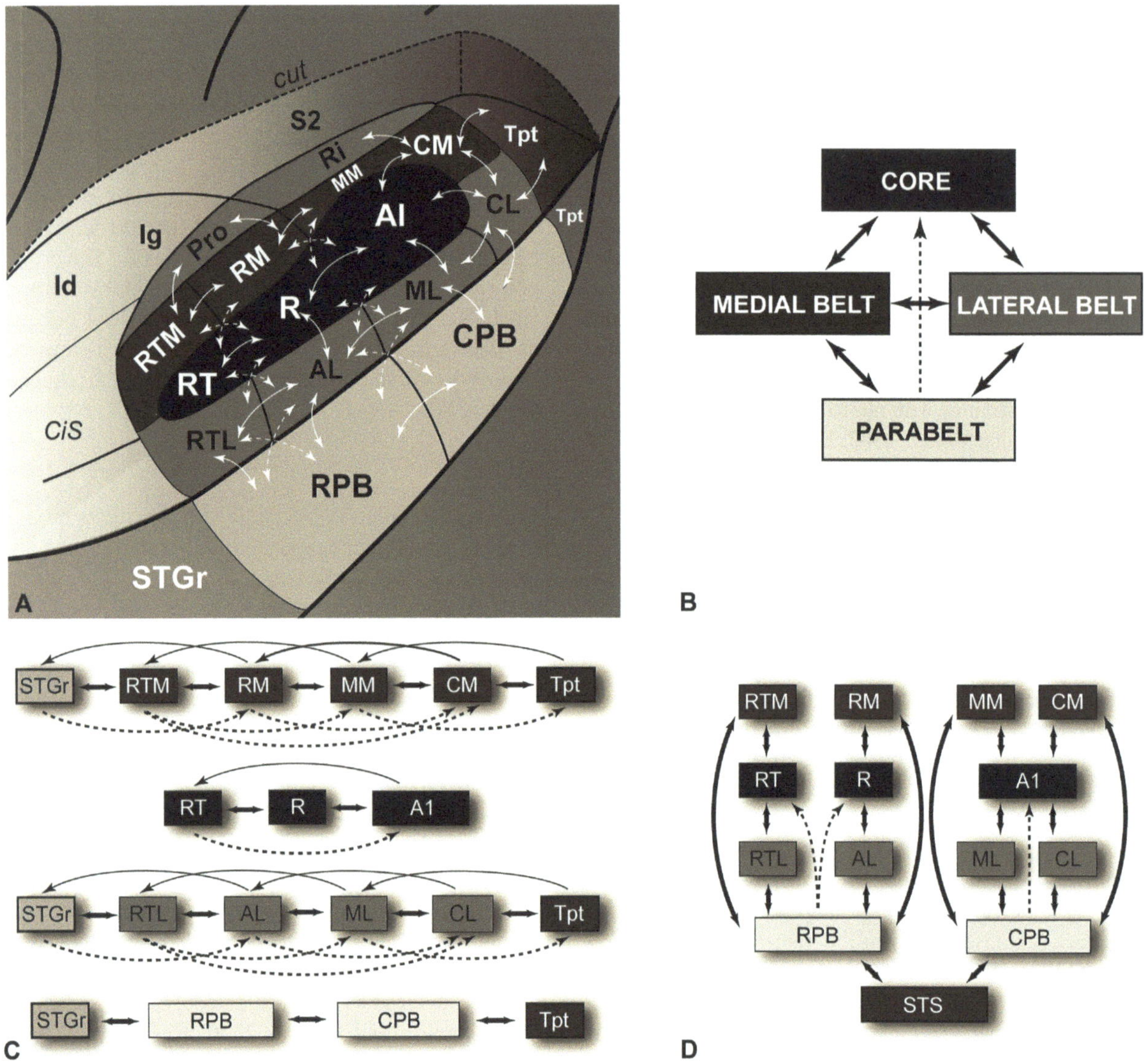

Fig. 2.7. Patterns in the connections between areas of the monkey auditory cortex. Only major connections are shown to improve clarity. (**A**) Schematic of local reciprocal connections (arrows) between areas. Dashed lines denote weaker connection. (**B**) Summary of major connections between regions. Dashed line denotes a one-way feedback projection from the parabelt to the core. (**C, D**) Schematics showing typical gradients in the connections between auditory areas along two axes: (**C**) connections along the rostrocaudal axis; (**D**) connections between core, belt, and parabelt regions. Dashed lines indicate feedback projections from infragranular layers. Thin lines denote weaker connections.

determining the flow of information into, within, and out of auditory cortex.

Serial connections and hierarchic relationships

Within auditory cortex of primates, connection patterns indicate that information flow mainly occurs along two axes: core–belt–parabelt and caudal–rostral. Although the laminar relationships between areas have not been fully determined, there is evidence of processing hierarchies along both axes in the connection patterns currently available.

Core–belt–parabelt axis

The core–belt–parabelt hierarchy is based on two related sets of observations (Kaas and Hackett, 1998; Rauschecker, 1998b). First, the belt areas receive dense inputs from the core areas, and the parabelt areas receive strong inputs from the belt areas. The parabelt does not receive significant input from the core (Fig. 2.7B, D) (Hackett et al., 1998a), although we now have some evidence of a direct projection back to the core from the infragranular layers of the parabelt (de la Mothe et al., 2006a). Second, the belt and especially parabelt areas are densely connected with auditory-related areas along the upper bank and fundus of the superior temporal sulcus (STS). These patterns suggest that some information is processed in a hierarchy beginning in the core, passed on to the belt, and then on to the parabelt and STS. Note that the flow of information along this axis is also constrained by topographic relationships along the caudal–rostral axis. Thus, the caudal areas of the core, belt, and parabelt are the most strongly connected and have weaker connections with the rostral areas, and vice versa. The medial belt area, RM, is a notable exception, as it is broadly interconnected with the rostral and caudal divisions of the parabelt (Hackett et al., 1998a; de la Mothe et al., 2006a). This is not reflected in Figure 2.7A.

Caudal–rostral axis

Information along the caudal–rostral axis appears to be dominated by a feedforward flow of projections in the rostral direction, although some feedback connections are oriented in the opposite direction (Fig. 2.7C). The rostral flow extends beyond auditory cortex to auditory-related areas of the rostral STG (STGr: Ts2, Ts1, Pro) (Galaburda and Pandya, 1983; Morel and Kaas, 1992; Morel et al., 1993; Jones et al., 1995; de la Mothe et al., 2006a; Smiley et al., 2007; Hackett et al., 2014).

Parallel connections

Parallel processing of multiple, but not necessarily identical, inputs characterizes all areas of auditory cortex, as each area receives inputs from multiple nuclei in the thalamus and other areas of cortex. With respect to the thalamocortical circuits, each MGB division projects to more than one area of auditory cortex, and each cortical area receives a unique blend of inputs from more than one division of the MGB and other thalamic nuclei (Fig. 2.6) (Molinari et al., 1995; de la Mothe et al., 2006b; Hackett et al., 2007a). Added to these are the corticocortical inputs, whereby each area of auditory cortex projects to and receives inputs from several areas of auditory and/or auditory-related cortex (Fig. 2.7C, D). In these two panels, the major connections of single areas of the core, belt, and parabelt regions are illustrated. Adjacent areas are the most densely interconnected, but each area is connected with several others. Thus, each area integrates information from the MGB and more than one cortical field. It is of additional interest that only a small percentage of single neurons in one thalamic nucleus or cortical area projects in parallel to more than one area of auditory cortex (Lee et al., 2004; de la Mothe et al., 2006a, b; Hackett et al., 2007a). For example, ML and CL both receive inputs from A1 and MGpd, but most of those inputs arise from different neuronal populations within each nucleus.

Functional considerations

It is expected that the connection patterns linking neurons between areas of auditory cortex will be reflected in their neurophysiologic response properties. A reasonable prediction is that these response properties such as latency, spectrotemporal tuning, and stimulus specificity will increase from one stage or level to the next (Rauschecker, 1998a). Thus, neurons in the core region could be expected to respond to sound at shorter latencies, with higher temporal precision, and be more sharply tuned to frequency, as compared to neurons in the belt or parabelt. Similarly, due to gradients along the caudal–rostral axis, neurons in the rostral areas would be expected to have longer latencies, reduced temporal precision, and greater specificity compared to neurons in caudal areas.

Neurophysiologic validation of these predictions has been slow in coming, but there is a wide range of supportive findings. With respect to the core–belt–parabelt hierarchy, neurons in the lateral belt areas do appear to have longer response latencies, broader tuning, and reduced temporal precision compared to neurons in the adjacent core regions (Rauschecker et al., 1995, 1997; Kosaki et al., 1997; Recanzone, 2000a; Recanzone et al., 2000a; Rauschecker and Tian, 2004; Kajikawa et al., 2005; Lakatos et al., 2005a; Camalier et al., 2012). Along the caudal–rostral axis, several studies indicate that selectivity for simple and complex auditory stimuli, including vocalizations, is greater in the rostral auditory and auditory-related areas (Rauschecker et al., 1995; Tian et al., 2001; Poremba et al., 2004; Bendor and Wang, 2008b; Petkov et al., 2008; Kusmierek and Rauschecker, 2009; Kikuchi et al., 2010; Camalier et al., 2012; Kusmierek et al., 2012).

These examples suggest that there is a fairly consistent relationship between the anatomic and physiologic gradients in auditory cortex. It is expected that a better understanding of those relationships will emerge over time, especially from studies that combine anatomic and physiologic techniques. As discussed in the next two sections, the complexity of those interactions becomes even greater when the influence of auditory-related and multisensory connections is considered, yet some informative trends in the topographic relationships between areas are now becoming much clearer.

Principle 6: The auditory-related connections of auditory cortex are topographically organized

The sixth principle of organization is that the connections between auditory and auditory-related areas throughout the brain are topographically organized.

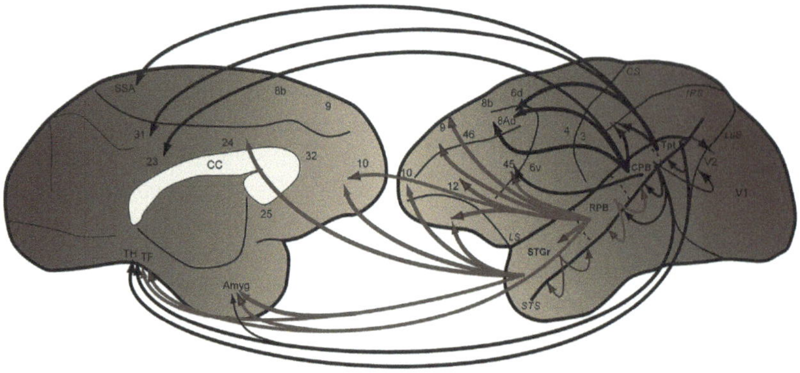

Fig. 2.8. Long-range connections between auditory and auditory-related areas, showing rostrocaudal topography between areas on the medial (left) and lateral (right) surfaces of the left hemisphere. Most connections are reciprocal, except as noted in text. Also note that each arrow represents connections of the belt and parabelt areas at that rostrocaudal level. For simplicity, the belt areas are not shown. See text for abbreviations.

These connections primarily involve the belt and parabelt regions, since the core region has few connections beyond the belt. For the most part, the connection patterns are an extension of the topographic relationships among areas within auditory cortex, especially along the mediolateral and rostrocaudal axes (e.g., Fig. 2.7). The summary diagram in Figure 2.8 reveals the rather-striking topography that links auditory cortex with auditory-related areas widely distributed in the temporal, prefrontal, posterior parietal, and occipital cortex, as well as the amygdala and striatum.

The topographic patterns suggest that multiple streams of information arise from auditory cortex to influence activity in areas downstream. Because these areas are functionally distinct, it can be assumed that the projections from auditory cortex relay information that is relevant to the function of each. In that sense, these pathways can be regarded as functionally distinct processing streams. An important question is whether the functional significance of a given stream is determined by the properties of its target or its source, or both. Completed and ongoing investigations suggest the latter. Because of the expansive nature and functional importance of these circuits, substantial space has been devoted to their description in this section of the review.

Superior temporal cortex

Within the temporal lobe, auditory cortex is topographically interconnected with adjoining regions on its caudal, rostral, medial, and lateral borders. Connections with these auditory-related areas are largely confined to superior temporal fields, including the STG, insula, and portions of the dorsal STS. With few exceptions, connections with the inferior and medial temporal lobe are lacking. These are discussed in subsequent sections of this review.

Caudally, the belt and parabelt regions are bordered by the Tpt, which is mostly known for its multisensory features (Leinonen et al., 1980). Recent studies involving Tpt have added new insights into its anatomic and functional organization. Tracer injections of Tpt and adjacent belt areas (CM, CL) reveal that its strongest cortical connections include the caudal belt and parabelt regions, while connections with the core are sparse (Smiley et al., 2007). The principal thalamic inputs to Tpt include the MGm and multisensory nuclei of the posterior thalamus (i.e., Sg, Lim, Po, PM), whereas inputs from the MGpd or MGad are sparse and variable. (Hackett et al., 2007a). Therefore, Tpt is best described as an auditory-related field that is strongly influenced by the caudal belt and parabelt. Although systematic neurophysiologic studies of Tpt have not been conducted, an auditory role for Tpt has been known for some time (Leinonen et al., 1980).

Rostrally, auditory cortex is bordered by a region (STGr) that includes several areas: Ts2, Ts1, and Pro (retaining the nomenclature of the original studies) (Pandya and Sanides, 1973; Galaburda and Pandya, 1983). The field adjacent to the rostral belt and parabelt corresponds to area Ts2. Ts2 receives inputs from the rostral auditory cortex and MGB, but connections with the MGB are reduced compared to auditory cortex and thalamic inputs are dominated by the multisensory nuclei (e.g., Sg, Lim, PM) (Trojanowski and Jacobson, 1975; Galaburda and Pandya, 1983; Markowitsch et al., 1985; Pandya and Rosene, 1993; Pandya et al., 1994; Kosmal et al., 1997; Hackett et al., 1998a, b). Moving further rostral along the STGr are areas Ts1 and the temporal pole (Pro). Compared with Ts2, the density of inputs from the auditory cortex and MGB is even weaker in Ts1, and reaches a minimum in the temporal pole.

Medially, auditory cortex is bounded by secondary somatosensory areas (e.g., retroinsular (Ri), retroinsular temporal area), the insula, and poorly defined areas of the circular sulcus, such as Pro, part of which extends to the temporal pole. Connections between auditory cortex and these fields are not well understood, but some connections between these areas and the medial belt area of auditory cortex have been identified (de la Mothe et al., 2006a; Smiley et al., 2007).

Laterally, the principal connections of the auditory (belt, parabelt) and auditory-related areas of the STG are with areas in the fundus and upper bank of the STS, known as the superior temporal polysensory region (STP). The STP region corresponds to the temporal parietal occipital area (TPO) and temporal area A. The connections with auditory cortex do not extend significantly beyond the fundus of the STS to its ventral bank or to the inferior temporal gyrus (ITG), nor do the belt and parabelt have any clear connections with the middle temporal (MT) complex, or V5, which is a visual region known to be involved in visual motion processing (Hackett et al., 1998a; de la Mothe et al., 2006a). Otherwise, the connections between auditory cortex and the STS are rather dense and exhibit some degree of rostral–caudal topography. However, there is also substantial overlap of rostral and caudal areas of the belt and parabelt along this axis, and the connections with other cortical and sensory regions can be patchy, overlapping, or interdigitating (Seltzer and Pandya, 1978, 1994; Galaburda and Pandya, 1983; Cipolloni and Pandya, 1985; Barnes and Pandya, 1992; Cusick et al., 1995; Seltzer et al., 1996; Hackett et al., 1998a, 2014). These inputs appear to provide the anatomic basis for multisensory interactions that characterize this part of the STS. Finally, other than Tpt, the temporal pole, and TPO, there is currently no evidence of direct connections between auditory cortex and the visual areas in inferior temporal lobe.

Prefrontal and cingulate cortex

Intensive studies of the superior temporal connections with prefrontal and cingulate cortex in the last decade have led to a much more complete rendering of their topographic patterns, summarized in Figure 2.8 (Pandya et al., 1969; Hackett et al., 1999; Romanski et al., 1999a, b; Cavada et al., 2000; Petrides and Pandya, 2002, 2007; Morecraft et al., 2004; Barbas et al., 2005; Barbas, 2007; Roberts et al., 2007; Saleem et al., 2008). In brief, the rostral areas of the STG (i.e., rostral belt and parabelt, Ts1, Ts2, temporal pole) have reciprocal connections favoring the rostral, ventrolateral, rostromedial, and orbital areas of prefrontal cortex, whereas connections of the caudal belt, CPB, and Tpt favor caudal and dorsal prefrontal areas. A comparable topographic gradient links caudal STG to caudal and mid-level cingulate cortex, and rostral STG to rostral cingulate (Vogt and Pandya, 1987; Romanski et al., 1999a; Morecraft et al., 2004). The most widely separated auditory areas along this axis (i.e., Tpt and temporal pole) have the fewest connections in common. In between, segregation is less strict, suggesting that auditory-related connections with prefrontal and cingulate cortex form a gradient along the rostral–caudal axis of the temporal lobe (Kaas and Hackett, 1999).

Posterior parietal cortex

The intraparietal cortex contains several functionally distinct areas that contribute to a multimodal representation of space that can used by motor structures to guide movements (Andersen, 1997). Compared to prefrontal cortex, much less is known about connections between auditory and posterior parietal cortex. The data available at the present time suggest that there are at least modest connections linking the lateral (LIP) and especially ventral (VIP) areas with Tpt and the caudal belt (Fig. 2.8), but not rostral auditory areas (Pandya and Kuypers, 1969; Divac et al., 1977; Hyvarinen, 1982; Lewis and Van Essen, 2000; Smiley et al., 2007). Auditory-related activity in LIP and VIP has been frequently described, and may be related to those connections (Mazzoni et al., 1996; Stricanne et al., 1996; Grunewald et al., 1999; Linden et al., 1999; Cohen et al., 2002, 2004a, 2005; Gifford and Cohen, 2004, 2005; Mullette-Gillman et al., 2005; Schlack et al., 2005). Connections with area 7a (not illustrated) are relatively sparse, as is auditory activity in this region (Rozzi et al., 2006; Smiley et al., 2007), and there appear to be no connections with area 7b (Pandya and Seltzer, 1982; Lewis and Van Essen, 2000; Smiley et al., 2007) or the anterior (AIP) field (Borra et al., 2008).

Occipital cortex

Connections between auditory and visual areas have been uncovered in a number of studies (Fig. 2.8). The connections include projections to the peripheral representations in V1 and dorsal V2, found after injections of caudal belt and parabelt areas of auditory cortex (Falchier et al., 2002, 2010; Rockland and Ojima, 2003; Cappe and Barone, 2005). These projections arose mainly from infragranular neurons and target laminae I and VI. From our study of the caudal superior temporal plane of macaques (Smiley et al., 2007), labeled cells (infragranular layers) were found in the peripheral representation of V2 and area prostriata after injections of the caudal belt, parabelt, and Tpt, but not A1 (Falchier et al., 2010). Area prostriata is adjacent to V1 in the upper bank of the anterior calcarine sulcus (Allman and Kaas,

1971; Ding et al., 2003), and broadly connected with the peripheral visual field representations of other visual areas, including V1, V2, and MT (Sousa et al., 1991; Rosa et al., 1993; Palmer and Rosa, 2006), orbitofrontal Barbas, 1993; Cavada et al., 2000) and cingulate cortex (Morecraft et al., 2004). The laminar patterns of connections between the auditory and visual areas across studies are concentrated in the superficial and infragranular layers, consistent with a feedback relationship between the regions. Although little is known about the functional impact of these connections at present, the anatomic details suggest that secondary auditory and visual areas respectively influence the processing of unimodal information in each modality. These connections may contribute to the results of electrophysiologic and imaging studies in which visual (Schroeder and Foxe, 2002; Ghazanfar et al., 2005; Lakatos et al., 2005b; Kayser et al., 2007, 2008) and visual-motor (Werner-Reiss et al., 2003; Fu et al., 2004) interactions were observed in the core and belt areas of monkeys (Bulkin and Groh, 2006).

Anterior parietal cortex

Connections between any part of auditory and somatosensory areas in the anterior parietal cortex have been sporadically reported in studies of either modality (Jones and Powell, 1969; Friedman et al., 1986; Krubitzer and Kaas, 1990; Burton and Fabri, 1995; Lewis and Van Essen, 2000; Qi et al., 2002; Disbrow et al., 2003; Cappe and Barone, 2005). Generally, when connections were found, they were not especially robust, and often limited to a single case or injection. Recent studies of areas in the caudal superior temporal plane of marmoset and macaque monkeys (de la Mothe et al., 2006a; Smiley et al., 2007) in which tracer injections were placed in the caudal belt areas consistently revealed reciprocal connections with the posterior granular insula (Ig) and Ri, which is a somatotopically organized area in the fundus of the lateral sulcus (Leinonen, 1980; Robinson and Burton, 1980a) (Fig. 2.2). Areas Ri and Ig are strongly interconnected with somatosensory areas (3a, 3b, 1, 2, S2) (Friedman and Murray, 1986; Friedman et al., 1986; Cusick et al., 1989; Guldin et al., 1992; Burton et al., 1995; Krubitzer et al., 1995; Qi et al., 2002; Disbrow et al., 2003). Some neurons in Ri may also have vestibular function (Grusser et al., 1990a, b, Guldin et al., 1992; Akbarian et al., 1994). Accordingly, the connections between auditory cortex and Ri and Ig may account for the auditory–somatosensory interactions observed in several neurophysiologic studies of the caudal auditory belt (Robinson and Burton, 1980a, b; Schroeder et al., 2001; Schroeder and Foxe, 2002; Fu et al., 2003; Brosch et al., 2005; Kayser et al., 2005; Lakatos et al., 2007) and Tpt (Leinonen et al., 1980). Compared with the caudal belt areas, we have not yet found significant connections between the rostral belt areas and either Ri or posterior Ig in our studies of primates. These results suggest that audiotactile interactions in auditory cortex could vary topographically along the rostrocaudal axis (i.e., becoming less robust rostrally). We expect that a more complete and accurate understanding of the relationship between the auditory and somatosensory systems will develop over time as additional studies focus on the connections between the modalities.

Striatum

Most of the auditory and auditory-related areas of the STG project to the striatum, with the exception of the core region (Selemon and Goldman-Rakic, 1985; Yeterian and Pandya, 1998; Borgmann and Jurgens, 1999; Jung and Hong, 2003; de la Mothe et al., 2006a). The connections are not reciprocal, as neurons in the striatum do not project back to the STG. This unidirectional input pattern is typical of cortical projections to the striatum. There is also some evidence of projections from the MGd, MGm, and adjacent thalamic nuclei to the caudal parts of the caudate and putamen (LeDoux et al., 1991; Shinonaga et al., 1994).

Inputs from the STG are concentrated in parts of the caudate and putamen alongside inputs from numerous other cortical regions. Some topography is evident in these projections. Along the rostrocaudal axis, rostral auditory and auditory-related areas project to rostral and ventromedial portions of the caudate and ventromedial putamen. Moving caudally along the STG, projections shift caudally and dorsally in the caudate, but remain concentrated ventromedially in the putamen. In contrast, the projections of caudal and rostral areas tend to overlap in the medial part of the tail of the caudate nucleus. Thus, there is topography in the STG projection to the main part of the caudate, whereas inputs to the tail of the caudate and putamen are more overlapping.

Along the core–belt–parabelt axis, two observations are notable. First, except for sparse connections in one report (Yeterian and Pandya, 1998), tracer injections into either the STG or striatum do not reveal connections with the core region. Second, inspection of the available data suggests that projections from the belt areas are not as dense or widespread as projections from the parabelt and rostral STG. These patterns suggest that the higher levels of auditory processing may have the greatest impact on activity in the striatum.

Lastly, expanding our review of striatal connections to include regions beyond the STG, two important observations emerge (Yeterian and Van Hoesen, 1978;

Alexander et al., 1986; Yeterian and Pandya, 1991; Nakano, 2000; Nakano et al., 2000). First, functionally distinct regions of cortex (i.e., motor, somatosensory, visual, prefrontal, posterior partietal) have a distinctive pattern of inputs to the striatum. The projections of these different regions tend to occupy segregated zones, although the segregation is not complete, as there are some zones in which overlap occurs. These anatomic relationships are consistent with physiologic studies in primates and other species that have found zones of response segregation in which there is little evidence of interactions between modalities, as well as zones of overlap in which multisensory or sensorimotor interactions are common (Rolls, 1994; Aosaki et al., 1995; Kolomiets et al., 2001; Glynn and Ahmad, 2002). Second, the topography of projections to the striatum is similar among areas of cortex that are also interconnected. For example, Tpt and the caudal belt/parabelt are strongly interconnected with the dorsal part of area 8, and the caudal portion of dorsal area 46. The striatal projections of these auditory and prefrontal fields are concentrated in dorsal portions of the caudate and putamen. By comparison, the temporal pole and rostral areas of the STG are most strongly interconnected with ventrolateral and orbital prefrontal areas. The striatal projections of these temporal and prefrontal areas are shifted to ventral portions of the caudate and putamen. Thus, we see in these patterns additional evidence of topographic clustering in the circuits involving the rostral and caudal areas of auditory cortex.

Amygdala

The amygdala plays an important role in mediating emotional and other behavioral responses to sensory stimuli of all modalities. Accordingly, it is widely connected with sensory, limbic, and executive regions in the forebrain (McDonald, 1998). With respect to its connections with auditory structures, anatomic studies in rodents, carnivores, and primates have revealed fairly consistent patterns of inputs to the nuclear complex of the amygdala from auditory and auditory-related areas (Heath and Jones, 1971; Herzog and Van Hoesen, 1976; Aggleton et al., 1980; Turner et al., 1980; Reale and Imig, 1983; Mascagni et al., 1993; Romanski and LeDoux, 1993; Kosmal et al., 1997; Shi and Cassell, 1997; Ghashghaei and Barbas, 2002; Yukie, 2002). Projections from temporal cortex are concentrated in the lateral nucleus of the amygdala (LA), with additional inputs to lateral parts of the basal and central nuclei. Studies in cats and rats indicate that these projections overlap with inputs from the MGm, MGd and adjacent nuclei in the posterior thalamus (e.g., peripeduncular, paralaminar, suprageniculate), some of which also project to the caudal striatum (Jones et al., 1976; Aggleton et al., 1980; LeDoux et al., 1990, 1991; Shinonaga et al., 1994). Thus, there is a network of connections linking the amygdala and striatum with the medial geniculate complex (MGC) and auditory cortex. By comparison, there is little overlap between the auditory projections to the amygdala and those from prefrontal cortex or other sensory areas in the LA (McDonald, 1998). Overlap among projections from prefrontal and sensory cortices mainly occurs in the central lateral nucleus, whereas projections from multisensory areas project broadly across the various sensory domains.

Projections to the amygdala from the STG exhibit clear topographic patterns (Fig. 2.8). Along the core–belt–parabelt axis, projections originate in the belt and parabelt regions, but not the core. The absence of input from primary areas has been consistently reported in non-primates, as well. Along the rostrocaudal axis of the STG, tracer injections into its lateral surface indicate that the caudal areas of auditory cortex and Tpt lack significant projections to the amygdala. Rostrally, however, projection density increases and reaches a maximum in TS1 and the temporal pole (Kosmal et al., 1997; Ghashghaei and Barbas, 2002; Yukie, 2002). These topographic patterns are also true of the medial belt, which has projections from RM, but not CM or the core (de la Mothe et al., 2006a). Thus, there is a distinct rostral bias in the projections of auditory and auditory-related superior temporal areas to the amygdala. This appears to mirror the pattern of projections from the visual cortex to the amygdala, which are limited to the inferior temporal areas of the ventral stream and exclude the occipital areas. Similarly, projections from prefrontal cortex to the amygdala primarily arise from medial and orbital areas which are most densely interconnected with the rostral temporal lobe (Fig. 2.8).

Overall, the cortical input patterns suggest that the higher levels of the auditory-processing hierarchy have the greatest influence on activity in the amygdala. Since those inputs are limited to the rostral areas, the patterns also indicate that the type of information processed by these areas (e.g., stimulus identity) is fundamentally important to the function of the amygdala with respect to sound. This seems likely, given the role of the amygdala in generating responses to sensory stimuli that are behaviorally relevant.

Functional considerations

The topographic patterns of connections within auditory cortex and their extensions to auditory-related areas have provided strong support for the dual-streams hypothesis, whereby rostral areas of auditory cortex contribute to a ventral stream involving the rostral STG and its auditory-related targets, while the caudal areas of auditory cortex contribute to a largely segregated dorsal stream (Colombo et al., 1996; Kaas and Hackett, 1999; Rauschecker, 1998a, b, 2011; Romanski et al., 1999b, 2000; Rauschecker and Tian, 2000). According to this model, the ventral stream is preferentially engaged in non-spatial auditory processing, such as auditory object recognition, whereas the dorsal stream is more involved in processing of spatial and sensorimotor information (Hickok et al., 2000; Hickok and Poeppel, 2000; Wise et al., 2001; Rauschecker and Scott, 2009; Rauschecker, 2011). This theory has spawned numerous physiologic studies of the auditory and auditory-related areas distributed along these axes.

Within auditory cortex, for example, Tian et al. (2001) reported that neurons in AL of the lateral belt were more selective for the type of call, whereas neurons in CL were consistently more selective for sound location. Neurons in the intermediate area, ML, showed mixed selectivity. In a subsequent study by these investigators, Tian and Rauschecker (2004) added that neurons in CL were more selective for rapid changes in frequency modulation compared to units in AL, which preferred slower rates of frequency modulation. The authors suggested that slow rate preferences found in AL are consistent with higher selectivity for vocalizations, whereas higher rate preferences are supportive of sound localization. In addition to these examples, several neurophysiologic studies of rostral and caudal belt areas have provided data supporting such functional specialization in the auditory cortical areas that contribute to the ventral and dorsal streams, respectively (Recanzone, 2000b, 2001; Recanzone et al., 2000b; Woods et al., 2006; Kusmierek and Rauschecker, 2009; Miller and Recanzone, 2009; Kikuchi et al., 2010; Tsunada et al., 2011; Kusmierek et al., 2012).

These data are complemented by functional imaging studies of auditory and auditory-related areas within the STG in which activation gradients to various types of sounds were observed along its rostrocaudal axis. For example, in a positron emission tomography imaging of passively listening macaques, Poremba et al. (2004) found that temporal-lobe activation to five classes of sounds was greatest in areas corresponding to auditory cortex, but reduced in the auditory-related areas caudally (~Tpt) and rostrally (Ts1, Ts2), reaching a minimum in the temporal pole (Pro). This result implies that activity generated by auditory stimulation, at least under passive listening conditions, may be greater in areas with heavier MGB inputs (i.e., auditory cortex), compared to auditory-related areas that have weaker MGB inputs and likely depend on inputs from auditory cortex (Fig. 2.6). Using functional magnetic resonance imaging (fMRI), Petkov et al. (2008) reported preferential activation of the rostral superior temporal plane of macaques, corresponding to dorsal parts of Ts2 and Ts1, for species-specific vocalizations. The activated area was also selective for the identity of the caller, suggesting that this rostral field has a role in processing these features. Finally, in the caudal STG of macaques, Gil-da-Costa et al. (2006) found that fMRI activation of Tpt was significantly stronger for macaque vocalizations compared to core areas (A1, R), which were more strongly activated by non-biologic stimuli with comparable acoustic features. These findings point to several ways in which auditory processing in the STGr and Tpt is distinct from auditory cortex. More specifically, activation within auditory cortex across sound class is generally higher in magnitude and tends to be less selective for sound class, compared to the STGr or Tpt, where selectivity appears to be higher. Together, these findings are consistent with the anatomic gradients and support the general conclusion that these auditory-related areas operate at a relatively higher level of auditory processing (Kaas and Hackett, 1998; Rauschecker, 1998a; Rauschecker et al., 1995; Kaas et al., 1999).

To be complete, it is important to note here that exceptions to the topographic segregation of function have been reported, suggesting that neuronal activity in the rostral and caudal areas is not always cleanly segregated into spatial and non-spatial domains (for review, see Recanzone and Cohen, 2010). There are significant anatomic interactions between the rostral and caudal areas of the superior temporal region, as well as some overlap in their projections to auditory-related areas elsewhere in the brain (Kaas and Hackett, 1999), suggesting that neurons in these areas could be sensitive to both spatial and non-spatial features. This is supported by neurophysiologic data showing that elements of both types of processing can be found in both rostral and caudal areas of auditory cortex, respectively, depending on experimental conditions (Wollberg and Sela, 1980; Watanabe, 1992; Tian et al., 2001; Cohen et al., 2004b; Ghazanfar et al., 2005; Maier and Ghazanfar, 2007; Kajikawa et al., 2008; Recanzone, 2008). As one example, in a series of studies by Ghazanfar and colleagues, neuronal activity in ML reflected the integration of facial expression and conspecific vocalizations as well as sensitivity to intensity changes associated with looming sounds (Ghazanfar et al., 2005; Maier and Ghazanfar, 2007). Thus, while it is quite clear that the caudal and rostral areas make unique contributions to auditory processing – many of which support the notion of dual streams – there is much more to be learned about the functional roles of these pathways (Recanzone and Cohen, 2010).

Auditory processing in areas beyond the temporal lobe is being intensively studied in humans and non-human primates using a wide range of techniques.

A complete review of these studies is beyond the scope of this chapter, but a glimpse into some of these active areas of research is provided here as these data relate specifically to the topographic organization of the projections from the rostral and caudal portions of auditory cortex.

In prefrontal cortex, evidence of processing related to object recognition and vocalizations has been observed in several studies in a ventral region that receives inputs from rostral auditory and auditory-related areas in the STG. The ventrolateral prefrontal cortex (~area 12vl) contains a zone in which neurons are responsive to auditory, visual, and auditory-visual stimuli, including complex sounds, vocalizations, and faces, as originally noted by Romanski (Romanski and Goldman-Rakic, 2002; Romanski et al., 2005; Sugihara et al., 2006; Romanski, 2007). In other studies of the ventrolateral field, Cohen and colleagues found that neurons in this area can discriminate between different classes of vocalizations, forming representations of auditory objects, and object categorization, but not reliably activated by stimuli in a spatial task (Gifford et al., 2005; Cohen et al., 2006, 2007, 2009; Lee et al., 2009). Rolls et al. (2006) found that certain neurons in the orbitofrontal cortex can be responsive to vocalizations amidst cells selective for faces. Note that the ventrolateral and orbitofrontal areas are within the projection zones of rostral auditory and auditory-related areas, indicating that a network of areas in the rostral temporal lobe is tied to the processing of this type of information.

In the dorsolateral prefrontal cortex, which contains targets of the caudal auditory cortex, neurons exhibit behaviorally relevant cross-modal (auditory-visual) associations, are active in tasks invoking visuospatial and audiospatial working memory, and most are selective for spatial location (Ito, 1982; Azuma and Suzuki, 1984; Vaadia et al., 1986; Vaadia, 1989; Russo and Bruce, 1994; Bodner et al., 1996; Fuster et al., 2000; Kikuchi-Yorioka and Sawaguchi, 2000; Artchakov et al., 2007). These cells are located in the dorsal periprincipal region (areas 46d and 8d), which has dense connections with the caudal belt, parabelt, and Tpt (Fig. 2.8). The studies in monkeys are consistent with a variety of evidence that points to involvement of dorsal stream areas in the processing of space and motion (Warren et al., 2002; Brunetti et al., 2005; Krumbholz et al., 2005; Zimmer and Macaluso, 2005; Deouell et al., 2007). Yet, it has also been noted that dorsal stream areas are active in various non-spatial processes (Hickok et al., 2000; Hickok and Poeppel, 2004, 2007). Neurons in posterior parietal cortex of monkeys contribute to a range of goal-directed functions with both spatial and non-spatial components (Cohen et al., 2005; Nieder et al., 2006). In addition, rapidly growing evidence from studies of humans and non-human primates suggests that connections linking premotor, parietal, and posterior superior temporal cortex contribute not only to spatial processing, but also to sensorimotor integration related to the processing of sound sequences, and the planning and control of speech production (Hashimoto and Sakai, 2003; Hickok et al., 2003, 2009; Hickok and Poeppel, 2004, 2007; Buchsbaum et al., 2005; Fu et al., 2006; Christoffels et al., 2007; Pa and Hickok, 2008; Tourville et al., 2008; Cohen, 2009; Okada and Hickok, 2009; Zheng, 2009; Leaver et al., 2009; Leaver and Rauschecker, 2010; Rauschecker, 2011; Isenberg et al., 2012).

CORRESPONDENCE OF HUMAN AND NON-HUMAN PRIMATE AUDITORY CORTEX

A thorough discussion of this subject is beyond the scope of this review, but it is important to at least briefly mention here that some degree of correspondence is beginning to emerge from ongoing studies of the human and non-human primate auditory cortex. While it is not likely that the cortical organization of any two species is identical, especially in the greatly expanded cortex of humans, the basic principles of organization described above are likely common to all and could therefore provide a foundation for systematic comparison. Some of this information is reviewed in greater detail elsewhere (Hackett, 2003, 2007a).

Where is auditory cortex in the human brain?

The portion of the cerebral cortex that can be defined as auditory cortex is not yet known. By classic anatomic and functional criteria, we can be rather certain that the auditory cortex is largely confined to the superior temporal gyrus, as it is in other primates. However, the precise size and extent have not been defined. In part, the uncertainties could be cleared up by knowledge of the thalamocortical projections from the MGB. Assuming some homology with monkeys, most of the caudal two-thirds of the superior temporal plane and gyrus should receive significant MGB input. Unfortunately, most of the available anatomic literature on the auditory radiation from the MGB to the cortex emphasizes the projection to the transverse temporal gyrus (TTG), or Heschl's gyrus (HG) (Poljak, 1932; Walker, 1937; Woollard and Harpman, 1939; Ades and Felder, 1942). There are few data on projections to the putative belt or parabelt areas surrounding the HG, which presumably receive inputs from the secondary pathways through the MGd and MGm. So far, studies utilizing diffusion tensor imaging (DTI) have been able to resolve the main auditory radiation from the MGB to HG, but there is much less information on connections outside the HG

(Lutz et al., 2007; Berman et al., 2013). In one recent study, the connections of areas in the superior and middle temporal gyri activated by voices with emotional content were examined with DTI (Ethofer et al., 2012). These investigators noted that a connection between the STG (area 22) and the MGB was present. This suggests the possibility that DTI or a similar technology could be used to show the distribution of thalamic inputs to the superior temporal region and define the boundaries of the auditory cortex, as we have defined it. Until then, we largely rely on descriptions of the architectonic features of the tissue, which combine classic and modern approaches to reveal the structural properties of different brain areas.

Regions and areas

Both classic (Wernicke, 1874; Beck, 1928, 1929; Broca, 1865; Brodmann, 1905, 1909; Campbell, 1905; von Economo and Koskinas, 1925; von Economo and Horn, 1930) and modern studies (Galaburda and Sanides,

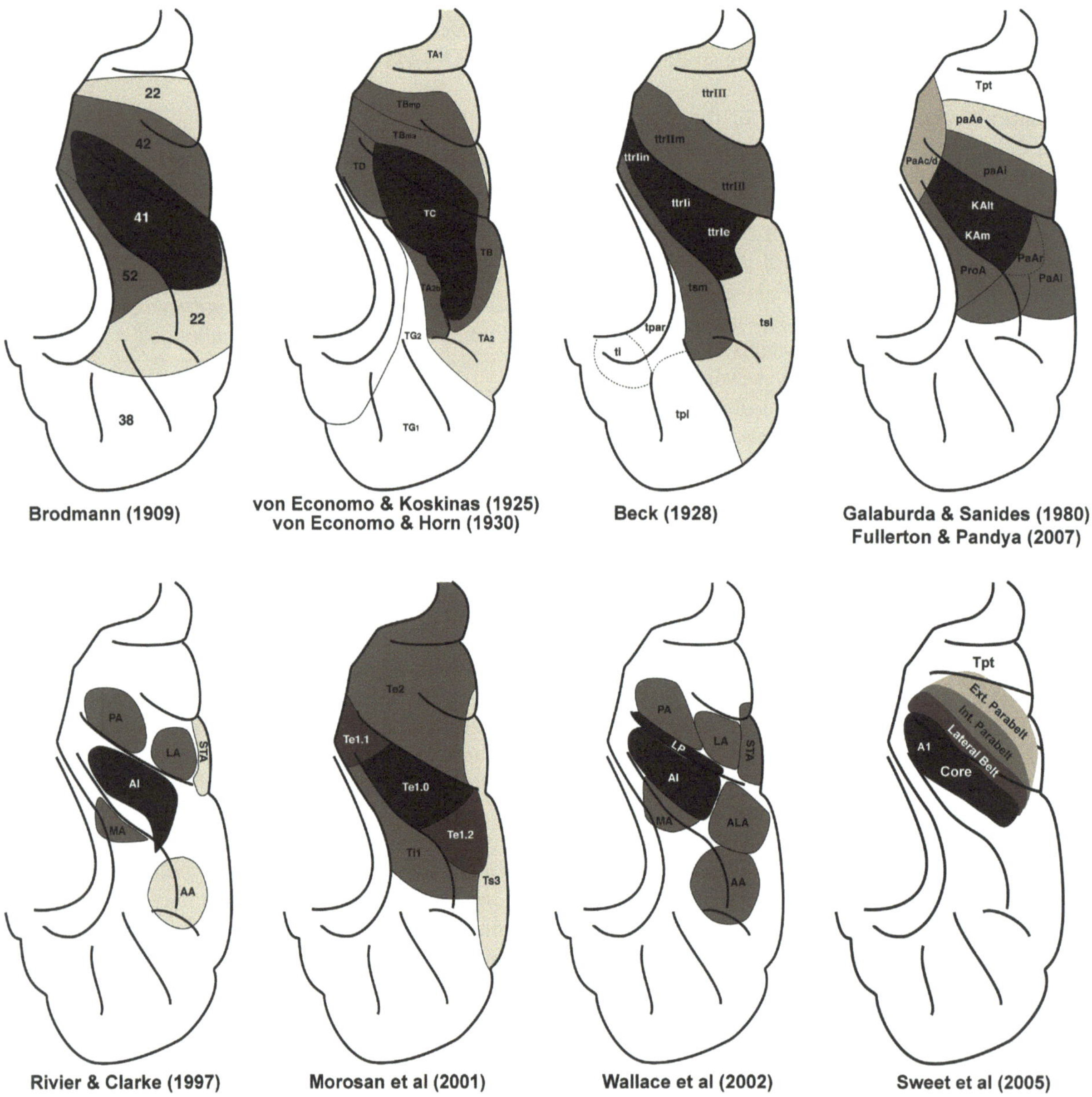

Fig. 2.9. Architectonic parcellations of the human superior temporal region by different investigators. In each panel, the locations of major auditory cortical regions and selected subdivisions are drawn on to a standardized schematic of the superior temporal plane (dorsal view). The superior temporal gyrus is not visible. Author impressions of regional identities are indicated by shading. Dark shading, core region; medium shading, belt region; light shading, parabelt and possibly other regions. Posterior is up, lateral is right. (Modified from Hackett, 2003, 2007a.)

1980; Rademacher et al., 1993, 2001; Rivier and Clarke, 1997; Clarke and Rivier, 1998; Nakahara et al., 2000; Hackett et al., 2001; Morosan et al., 2001, 2005; Wallace et al., 2002; Bailey et al., 2007; Fullerton and Pandya, 2007) have produced evidence consistent with the division of human auditory cortex into distinct regions, each of which is further subdivided (Fig. 2.9). Although the precise number of subdivisions identified ranges from four to over 30, a consistent pattern can be discerned, as follows. There is a centrally located core region (~area 41), confined to the caudal two-thirds of HG. The medial two-thirds of the TTG has architectonic characteristics matching that of the core in apes and monkeys (Pandya, 1995; Hackett et al., 2001; Sweet et al., 2005; Fullerton and Pandya, 2007) and more than one tonotopic representation (discussed below) (Fig. 2.9). The core is flanked by bands of areas that extend anteriorly on to the planum polare and posteriorly on the planum temporale (PT). A simplistic interpretation is that the anterior band (~area 52) corresponds to the medial belt, while the posterior bands correspond to the lateral belt and parabelt regions (~area 42 and part of posterior 22) (Fig. 2.10). Thus, the most caudal areas of monkey auditory cortex would correspond to the posteromedial portions of the TTG and PT in humans, while the rostral auditory areas of monkeys correspond to the anterolateral TTG and PT. Depending on the extent to which the belt and parabelt regions extend on to the surface of the STG, this accounting leaves variable portions of area 22 unaccounted for. Whether this reflects the tremendous expansion of the human auditory cortex by the addition or modification of areas remains to be determined.

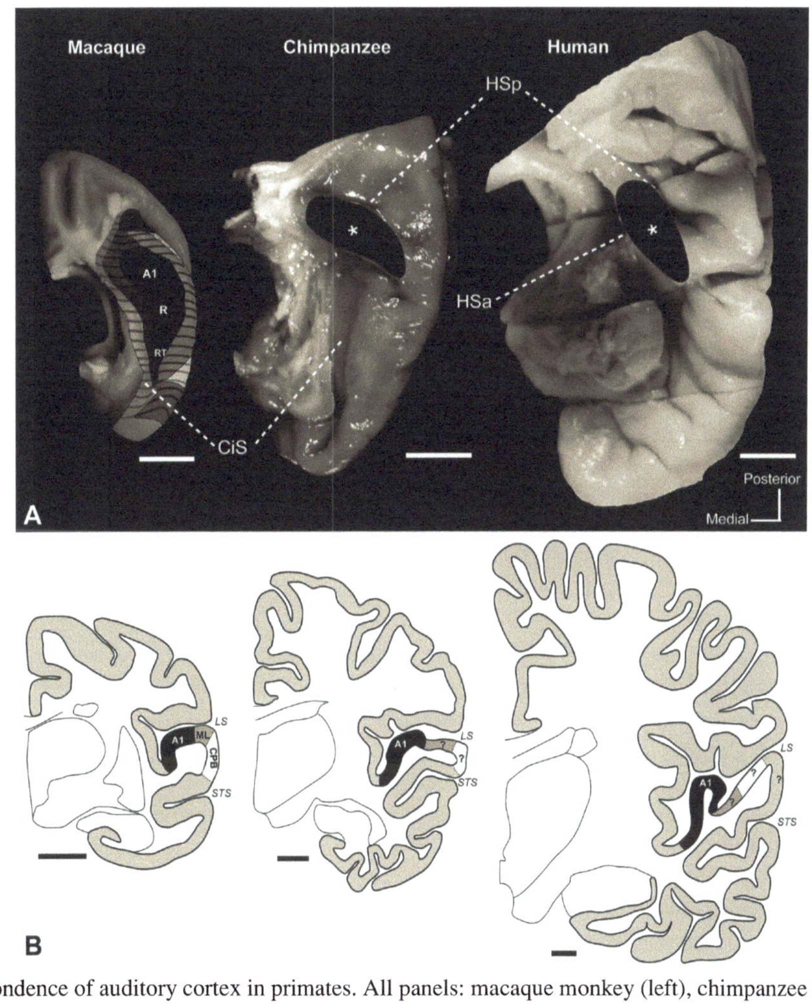

Fig. 2.10. Correspondence of auditory cortex in primates. All panels: macaque monkey (left), chimpanzee (middle), and human (right). (**A**) Location of the core region (black shading). Asterisk denotes location of transverse temporal gyrus of Heschl in chimpanzee and human. HSa and HSp, anterior and posterior sulci of Heschl; CiS, circular sulcus. (**B**) Drawings of coronal sections showing locations of core, belt, and parabelt areas in the macaque (left). Putative locations of corresponding areas in the chimpanzee (middle) and human (right). Question marks denote uncertainty about correspondence with the regions identified in monkeys. Scale bars: (**A**) 10 mm; (**B**) 5 mm.

This schema is admittedly speculative and oversimplified, but represents a working hypothesis that can be and is being tested in ongoing architectonic and imaging studies. As such, it is a useful guide for future research. At present, there is a broad range of imaging and electrophysiologic support for a model of human auditory cortex organization that approximates various aspects of this schema and is governed by the principles of organization identified in non-human primates, some of which is discussed in later chapters in this volume (Hickok and Poeppel, 2000, 2004, 2007; Scott et al., 2000; Wessinger et al., 2001; Griffiths and Warren, 2002; Hart et al., 2002, 2004; Seifritz et al., 2002, 2006; Davis and Johnsrude, 2003; Hall et al., 2003; Scott and Johnsrude, 2003; Sigalovsky et al., 2006; Griffiths et al., 2007; Langers et al., 2007; Obleser et al., 2007; Patterson and Johnsrude, 2008; Okada et al., 2010; Smith et al., 2010; Woods et al., 2010; Barton et al., 2012; Dick et al., 2012; Isenberg et al., 2012; Harinen and Rinne, 2013).

One line of research has focused on identifying tonotopic gradients in different areas of the human auditory cortex. In practice, identification of these gradients has been challenging. Until recently, the non-invasive imaging and electrophysiologic techniques that are used to study activity in the human brain have not been able to clearly resolve tonotopic patterns, even in the HG, where such patterns should be the most robust. The stimulus levels needed to reliably generate recordable signals have been too high to estimate frequency specificity in voxel space, and factors such as high intersubject variability have also been a confound that has led to poor agreement between studies. To some extent, the expected reversals in the frequency gradient along the long axis of the HG have been observed in a majority of studies, supporting the existence of at least two core areas (Talavage and Edmister, 2004; Upadhyay et al., 2007; Ozaki and Hashimoto, 2007; Woods et al., 2010). Recently, however, several studies produced tonotopic maps of HG that are in fairly close agreement with each other, but show frequency reversals that tend to run orthogonal to the HG (Humphries et al., 2010; Da Costa et al., 2011; Barton et al., 2012; Dick et al., 2012; Langers and van Dijk, 2012). Barton et al. (2012) added that periodotopic organization gradients are orthogonal to tonotopic gradients, which enabled the authors to improve correspondence between current human and monkey models of auditory field organization. While consensus has not been reached on the representation of frequency or other physiologic features in these areas, the data do suggest the presence of more than one tonotopically organized field on the HG, consistent with available anatomic data and expectations from the primate model, in general.

In addition to tonotopic gradients, evidence of functional distinctions between regions and areas continues to grow. A few of these studies are highlighted briefly here. In one recent study, Okada et al. (2010) found that the human core region was sensitive to low-level acoustic features of speech sounds, whereas downstream areas in the STS were more sensitive to whether the speech sounds were intelligible. These findings are evidence of increasing integration along a hierarchically organized network of areas between the core and STS.

Several imaging studies have focused on functionally distinct domains in the PT. Division of the PT into functionally distinct areas would be consistent with detailed anatomic descriptions of the region (Griffiths and Warren, 2002; Warren et al., 2002; Hickok et al., 2003, 2009; Isenberg et al., 2012). Isenberg et al. (2012) found that two areas on the PT were differentially activated in tasks involving sound. One area in the posterior PT was activated by a sensorimotor task involving the shadowing of pseudowords. The other area, located on the anterior-medial PT, was preferentially activated by a spatial task involving noise in motion. These distinctions are consistent with prior studies cited just above that revealed spatial and sensorimotor engagement by discrete areas or domains of the PT.

Areas on the posterior STG, corresponding in general to area 22, have been the least studied, but show consistent responsiveness to auditory stimulation, as well as regional variations in response to various classes of stimuli. Data from laboratories equipped to obtain surface recordings from the human brain have revealed robust auditory activity to a wide range of simple and complex stimuli, including evidence of functional topography for sound class, suggesting that division of the region into multiple fields is in order (Howard et al., 2000; Edwards et al., 2005; Chang et al., 2010; Leaver and Rauschecker, 2010; Besle et al., 2011; Steinschneider and Fishman, 2011; Nourski et al., 2013, 2014; Pasley et al., 2012; Greenlee et al., 2013). Like much of the functional imaging studies, these datasets are running far ahead of anatomic surveys of the posterior STG, where distinct divisions have not yet been identified.

CONCLUDING REMARKS

The purpose of this chapter was to provide a summary of the major organizational features of the primate auditory cortex. We began by making an operational distinction between auditory and auditory-related areas of cortex. These definitions provided the basis for a subsequent overview of six principles of organization that have emerged from over 40 years of research by scores of investigators. The bulk of the chapter was devoted to descriptions of these principles, with an emphasis on the most relevant functional discoveries in neurophysiology and functional imaging. This was followed by a brief discussion of the organization of the human auditory cortex, and its correspondence with the non-human primate model. Many important questions concerning the organization and function of the auditory cortex remain. Hopefully, interest in these areas will increase over the next decade as our field moves toward a more comprehensive description of auditory cortical processing. Given recent methodologic advances and continued support for research, significant progress toward this goal can be, and is being, made.

ACKNOWLEDGMENTS

The authors gratefully acknowledge the support of NIH/NIDCD grant R01 DC04318 to T.A.H.

References

Ades HW, Felder R (1942). The acoustic area of the monkey (*Macaca mulatta*). J Neurophysiol 5: 49–59.

Aggleton JP, Burton MJ, Passingham RE (1980). Cortical and subcortical afferents to the amygdala of the rhesus monkey (*Macaca mulatta*). Brain Res 190: 347–368.

Akbarian S, Grusser OJ, Guldin WO (1994). Corticofugal connections between the cerebral cortex and brainstem vestibular nuclei in the macaque monkey. J Comp Neurol 339: 421–437.

Alexander GE, Delong MR, Strick PL (1986). Parallel organization of functionally segregated circuits linking basal ganglia and cortex. Annu Rev Neurosci 9: 357–381.

Allman JM, Kaas JH (1971). Representation of the visual field in striate and adjoining cortex of the owl monkey (*Aotus trivirgatus*). Brain Res 35: 89–106.

Andersen RA (1997). Multimodal integration for the representation of space in the posterior parietal cortex. Philos Trans R Soc Lond B Biol Sci 352: 1421–1428.

Aosaki T, Kimura M, Graybiel AM (1995). Temporal and spatial characteristics of tonically active neurons of the primate's striatum. J Neurophysiol 73: 1234–1252.

Artchakov D, Tikhonravov D, Vuontela V et al. (2007). Processing of auditory and visual location information in the monkey prefrontal cortex. Exp Brain Res 180: 469–479.

Azuma M, Suzuki H (1984). Properties and distribution of auditory neurons in the dorsolateral prefrontal cortex of the alert monkey. Brain Res 298: 343–346.

Bailey L, Abolmaesumi P, Taml J et al. (2007). Customised cytoarchitectonic probability maps using deformable registration: primary auditory cortex. Med Image Comput Comput Assist Interv Int Conf Med Image Comput Comput Assist Interv 10: 760–768.

Bajo VM, Rouiller EM, Welker E et al. (1995). Morphology and spatial distribution of corticothalamic terminals originating from the cat auditory cortex. Hear Res 83: 161–174.

Barbas H (1993). Organization of cortical afferent input to orbitofrontal areas in the rhesus monkey. Neuroscience 56: 841–864.

Barbas H (2007). Specialized elements of orbitofrontal cortex in primates. Ann N Y Acad Sci 1121: 10–32.

Barbas H, Medalla M, Alade O et al. (2005). Relationship of prefrontal connections to inhibitory systems in superior temporal areas in the rhesus monkey. Cereb Cortex 15: 1356–1370.

Barnes CL, Pandya DN (1992). Efferent cortical connections of multimodal cortex of the superior temporal sulcus in the rhesus monkey. J Comp Neurol 318: 222–244.

Barton B, Venezia JH, Saberi K et al. (2012). Orthogonal acoustic dimensions define auditory field maps in human cortex. Proc Natl Acad Sci U S A 109: 20738–20743.

Beck E (1928). Die myeloarchitektonische Felderung es in der Sylvischen Furche gelegenen Teiles des menschlichen Schläfenlappens. J Psych Neurol 36: 1–21.

Beck E (1929). Der myeloarchitektonische Bau des in der Sylvischen Furche gelegenen Teiles des Schläfenlappens beim Schimpansen (*Troglodytes niger*). J Psych Neurol 38: 309–428.

Bendor D, Wang X (2005). The neuronal representation of pitch in primate auditory cortex. Nature 436: 1161–1165.

Bendor D, Wang X (2008a). Neural response properties of core fields AI, R, and RT in the auditory cortex of marmoset monkeys. J Neurophysiol 100: 888–906.

Bendor D, Wang X (2008b). Neural response properties of primary, rostral, and rostrotemporal core fields in the auditory cortex of marmoset monkeys. J Neurophysiol 100: 888–906.

Berman JI, Lanza MR, Blaskey L et al. (2013). High angular resolution diffusion imaging probabilistic tractography of the auditory radiation. AJNR Am J Neuroradiol 34: 1573–1578.

Besle J, Schevon CA, Mehta AD et al. (2011). Tuning of the human neocortex to the temporal dynamics of attended events. J Neurosci 31: 3176–3185.

Bizley JK, Nodal FR, Nelken I et al. (2005). Functional organization of ferret auditory cortex. Cereb Cortex 15: 1637–1653.

Bodner M, Kroger J, Fuster JM (1996). Auditory memory cells in dorsolateral prefrontal cortex. Neuroreport 7: 1905–1908.

Borgmann S, Jurgens U (1999). Lack of cortico-striatal projections from the primary auditory cortex in the squirrel monkey. Brain Res 836: 225–228.

Borra E, Belmalih A, Calzavara R et al. (2008). Cortical connections of the macaque anterior intraparietal (AIP) area. Cereb Cortex 18: 1094–1111.

Broca P (1865). Remarques sur le siège de la faculté du langage articulé. Bull Soc Anthropol 6: 377–393.

Brodmann K (1905). Beitrage zur histologischen Lokalisation der Grosshirnrinde. Dritte Mitteilung: Die Rinderfelder der niederen Affen. J Psych Neurol 4: 177–226.

Brodmann K (1909). Vergleichende Lokalisationslehre der Grosshirnrinde. Barth, Leipzig.

Brosch M, Selezneva E, Scheich H (2005). Nonauditory events of a behavioral procedure activate auditory cortex of highly trained monkeys. J Neurosci 25: 6797–6806.

Brunetti M, Belardinelli P, Caulo M et al. (2005). Human brain activation during passive listening to sounds from different locations: an fMRI and MEG study. Hum Brain Mapp 26: 251–261.

Buchsbaum BR, Olsen RK, Koch P et al. (2005). Human dorsal and ventral auditory streams subserve rehearsal-based and echoic processes during Verbal working memory. Neuron 48: 687–697.

Budinger E, Heil P, Scheich H (2000). Functional organization of auditory cortex in the Mongolian gerbil (*Meriones unguiculatus*). III. Anatomical subdivisions and cortico-cortical connections. Eur J Neurosci 12: 2425–2451.

Bulkin DA, Groh JM (2006). Seeing sounds: visual and auditory interactions in the brain. Curr Opin Neurobiol 16: 415–419.

Burton H, Fabri M (1995). Ipsilateral intracortical connections of physiologically defined cutaneous representations in areas 3b and 1 of macaque monkeys: projections in the vicinity of the central sulcus. J Comp Neurol 355: 508–538.

Burton H, Fabri M, Alloway K (1995). Cortical areas within the lateral sulcus connected to cutaneous representations in areas 3b and 1: a revised interpretation of the second somatosensory area in macaque monkeys. J Comp Neurol 355: 539–562.

Camalier CR, D'Angelo WR, Sterbing-D'Angelo SJ et al. (2012). Neural latencies across auditory cortex of macaque support a dorsal stream supramodal timing advantage in primates. Proc Natl Acad Sci U S A 109: 18168–18173.

Campbell HW (1905). Histological Studies on the Localization of Cerebral Function, Cambridge University Press, Cambridge, UK.

Cappe C, Barone P (2005). Heteromodal connections supporting multisensory integration at low levels of cortical processing in the monkey. Eur J Neurosci 22: 2886–2902.

Cavada C, Company T, Tejedor J et al. (2000). The anatomical connections of the macaque monkey orbitofrontal cortex. A review. Cereb Cortex 10: 220–242.

Chang EF, Rieger JW, Johnson K et al. (2010). Categorical speech representation in human superior temporal gyrus. Nat Neurosci 13: 1428–1432.

Christoffels IK, Formisano E, Schiller NO (2007). Neural correlates of verbal feedback processing: an fMRI study employing overt speech. Hum Brain Mapp 28: 868–879.

Cipolloni PB, Pandya DN (1985). Topography and trajectories of commissural fibers of the superior temporal region in the rhesus monkey. Exp Brain Res 57: 381–389.

Clarke S, Rivier F (1998). Compartments within human primary auditory cortex: evidence from cytochrome oxidase and acetylcholinesterase staining. Eur J Neurosci 10: 741–745.

Cohen YE (2009). Multimodal activity in the parietal cortex. Hear Res 258: 100–105.

Cohen YE, Batista AP, Andersen RA (2002). Comparison of neural activity preceding reaches to auditory and visual stimuli in the parietal reach region. Neuroreport 13: 891–894.

Cohen YE, Cohen IS, Gifford 3rd GW (2004a). Modulation of LIP activity by predictive auditory and visual cues. Cereb Cortex 14: 1287–1301.

Cohen YE, Russ BE, Gifford 3rd GW et al. (2004b). Selectivity for the spatial and nonspatial attributes of auditory stimuli in the ventrolateral prefrontal cortex. J Neurosci 24: 11307–11316.

Cohen YE, Russ BE, Gifford 3rd GW (2005). Auditory processing in the posterior parietal cortex. Behav Cogn Neurosci Rev 4: 218–231.

Cohen YE, Hauser MD, Russ BE (2006). Spontaneous processing of abstract categorical information in the ventrolateral prefrontal cortex. Biol Lett 2: 261–265.

Cohen YE, Theunissen F, Russ BE et al. (2007). Acoustic features of rhesus vocalizations and their representation in the ventrolateral prefrontal cortex. J Neurophysiol 97: 1470–1484.

Cohen YE, Russ BE, Davis SJ et al. (2009). A functional role for the ventrolateral prefrontal cortex in non-spatial auditory cognition. Proc Natl Acad Sci U S A 106: 20045–20050.

Colombo M, Rodman HR, Gross CG (1996). The effects of superior temporal cortex lesions on the processing and retention of auditory information in monkeys (*Cebus apella*). J Neurosci 16: 4501–4517.

Crabtree JW (1998). Organization in the auditory sector of the cat's thalamic reticular nucleus. J Comp Neurol 390: 167–182.

Cusick CG, Wall JT, Felleman DJ et al. (1989). Somatotopic organization of the lateral sulcus of owl monkeys: area 3b, S-II, and a ventral somatosensory area. J Comp Neurol 282: 169–190.

Cusick CG, Seltzer B, Cola M et al. (1995). Chemoarchitectonics and corticocortical terminations within the superior temporal sulcus of the rhesus monkey: evidence for subdivisions of superior temporal polysensory cortex. J Comp Neurol 360: 513–535.

Da Costa S, Van Der Zwaag W, Marques JP et al. (2011). Human primary auditory cortex follows the shape of Heschl's gyrus. J Neurosci 31: 14067–14075.

Davis MH, Johnsrude IS (2003). Hierarchical processing in spoken language comprehension. J Neurosci 23: 3423–3431.

De La Mothe LA, Blumell S, Kajikawa Y et al. (2006a). Cortical connections of auditory cortex in marmoset monkeys: core and medial belt regions. J Comp Neurol 496: 27–71.

De La Mothe LA, Blumell S, Kajikawa Y et al. (2006b). Thalamic connections of auditory cortex in marmoset monkeys: core and medial belt regions. J Comp Neurol 496: 72–96.

De La Mothe LA, Blumell S, Kajikawa Y et al. (2012). Thalamic connections of auditory cortex in marmoset monkeys: lateral belt and parabelt regions. Anat Rec (Hoboken) 295: 822–836.

Deouell LY, Heller AS, Malach R et al. (2007). Cerebral responses to change in spatial location of unattended sounds. Neuron 55: 985–996.

Dick F, Tierney AT, Lutti A et al. (2012). In vivo functional and myeloarchitectonic mapping of human primary auditory areas. J Neurosci 32: 16095–16105.

Ding SL, Morecraft RJ, Van Hoesen GW (2003). Topography, cytoarchitecture, and cellular phenotypes of cortical areas that form the cingulo-parahippocampal isthmus and adjoining retrocalcarine areas in the monkey. J Comp Neurol 456: 184–201.

Disbrow E, Litinas E, Recanzone GH et al. (2003). Cortical connections of the second somatosensory area and the parietal ventral area in macaque monkeys. J Comp Neurol 462: 382–399.

Divac I, Lavail JH, Rakic P (1977). Heterogeneous afferents to the inferior parietal lobule of the rhesus monkey revealed by the retrograde transport method. Brain Res 123: 197–207.

Edwards E, Soltani M, Deouell LY et al. (2005). High gamma activity in response to deviant auditory stimuli recorded directly from human cortex. J Neurophysiol 94: 4269–4280.

Ethofer T, Bretscher J, Gschwind M et al. (2012). Emotional voice areas: anatomic location, functional properties, and structural connections revealed by combined fMRI/DTI. Cereb Cortex 22: 191–200.

Falchier A, Clavagnier S, Barone P et al. (2002). Anatomical evidence of multimodal integration in primate striate cortex. J Neurosci 22: 5749–5759.

Falchier A, Schroeder CE, Hackett TA et al. (2010). Projection from visual areas V2 and prostriata to caudal auditory cortex in the monkey. Cereb Cortex 20: 1529–1538.

Fitzpatrick KA, Imig TJ (1978). Projections of auditory cortex upon the thalamus and midbrain in the owl monkey. J Comp Neurol 177: 537–555.

Friedman DP, Murray EA (1986). Thalamic connectivity of the second somatosensory area and neighboring somatosensory fields of the lateral sulcus of the macaque. J Comp Neurol 252: 348–373.

Friedman DP, Murray EA, O'Neill JB et al. (1986). Cortical connections of the somatosensory fields of the lateral sulcus of macaques: evidence for a corticolimbic pathway for touch. J Comp Neurol 252: 323–347.

Fu KM, Johnston TA, Shah AS et al. (2003). Auditory cortical neurons respond to somatosensory stimulation. J Neurosci 23: 7510–7515.

Fu KM, Shah AS, Mn OC et al. (2004). Timing and laminar profile of eye position effects on auditory responses in primate auditory cortex. J Neurophysiol 92: 3522–3531.

Fu CH, Mcintosh AR, Kim J et al. (2006). Modulation of effective connectivity by cognitive demand in phonological verbal fluency. Neuroimage 30: 266–271.

Fullerton BC, Pandya DN (2007). Architectonic analysis of the auditory-related areas of the superior temporal region in human brain. J Comp Neurol 504: 470–498.

Fuster JM, Bodner M, Kroger JK (2000). Cross-modal and cross-temporal association in neurons of frontal cortex. Nature 405: 347–351.

Galaburda AM, Pandya DN (1983). The intrinsic architectonic and connectional organization of the superior temporal region of the rhesus monkey. J Comp Neurol 221: 169–184.

Galaburda A, Sanides F (1980). Cytoarchitectonic organization of the human auditory cortex. J Comp Neurol 190: 597–610.

Ghashghaei HT, Barbas H (2002). Pathways for emotion: interactions of prefrontal and anterior temporal pathways in the amygdala of the rhesus monkey. Neuroscience 115: 1261–1279.

Ghazanfar AA, Schroeder CE (2006). Is neocortex essentially multisensory? Trends Cogn Sci 10: 278–285.

Ghazanfar AA, Maier JX, Hoffman KL et al. (2005). Multisensory integration of dynamic faces and voices in rhesus monkey auditory cortex. J Neurosci 25: 5004–5012.

Gifford 3rd GW, Cohen YE (2004). Effect of a central fixation light on auditory spatial responses in area LIP. J Neurophysiol 91: 2929–2933.

Gifford 3rd GW, Cohen YE (2005). Spatial and non-spatial auditory processing in the lateral intraparietal area. Exp Brain Res 162: 509–512.

Gifford 3rd GW, Maclean KA, Hauser MD et al. (2005). The neurophysiology of functionally meaningful categories: macaque ventrolateral prefrontal cortex plays a critical role in spontaneous categorization of species-specific vocalizations. J Cogn Neurosci 17: 1471–1482.

Gil-Da-Costa R, Martin A, Lopes MA et al. (2006). Species-specific calls activate homologs of Broca's and Wernicke's areas in the macaque. Nat Neurosci 9: 1064–1070.

Glynn G, Ahmad SO (2002). Three-dimensional electrophysiological topography of the rat corticostriatal system. J Comp Physiol A Neuroethol Sens Neural Behav Physiol 188: 695–703.

Greenlee JD, Behroozmand R, Larson CR et al. (2013). Sensory-motor interactions for vocal pitch monitoring in non-primary human auditory cortex. PLoS One 8: e60783.

Griffiths TD, Warren JD (2002). The planum temporale as a computational hub. Trends Neurosci 25: 348–353.

Griffiths TD, Kumar S, Warren JD et al. (2007). Approaches to the cortical analysis of auditory objects. Hear Res 229: 46–53.

Gross NB, Lifschitz WS, Anderson DJ (1974). The tonotopic organization of the auditory thalamus of the squirrel monkey (*Saimiri sciureus*). Brain Res 65: 323–332.

Grunewald A, Linden JF, Andersen RA (1999). Responses to auditory stimuli in macaque lateral intraparietal area. I. Effects of training. J Neurophysiol 82: 330–342.

Grusser OJ, Pause M, Schreiter U (1990a). Localization and responses of neurones in the parieto-insular vestibular cortex of awake monkeys (*Macaca fascicularis*). J Physiol 430: 537–557.

Grusser OJ, Pause M, Schreiter U (1990b). Vestibular neurones in the parieto-insular cortex of monkeys (*Macaca fascicularis*): visual and neck receptor responses. J Physiol 430: 559–583.

Guldin WO, Akbarian S, Grusser OJ (1992). Cortico-cortical connections and cytoarchitectonics of the primate vestibular cortex: a study in squirrel monkeys (*Saimiri sciureus*). J Comp Neurol 326: 375–401.

Guo W, Chambers AR, Darrow KN et al. (2012). Robustness of cortical topography across fields, laminae, anesthetic states, and neurophysiological signal types. J Neurosci 32: 9159–9172.

Hackett T (2003). The comparative anatomy of the primate auditory cortex. In: A Ghazanfar (Ed.), Primate Audition: Behavior and Neurobiology, CRC Press, Boca Raton.

Hackett TA (2007a). Organization and correspondence of the auditory cortex of humans and nonhuman primates. In: J Kaas (Ed.), Evolution of the Nervous System, Elsevier, Oxford.

Hackett TA (2007b). Organization of the thalamocortical auditory pathways in primates. In: RF Burkard, M Don, JJ Eggermont (Eds.), Auditory Evoked Potentials: Basic Principles and Clinical Application, Lippincott Williams & Wilkins, Baltimore.

Hackett TA (2011). Information flow in the auditory cortical network. Hear Res 271(1–2): 133–146.

Hackett TA, De La Mothe LA (2009). Regional and laminar distribution of the vesicular glutamate transporter, VGluT2, in the macaque monkey auditory cortex. J Chem Neuroanat 38: 106–116.

Hackett TA, de la Mothe LA, Camalier CR et al. (2014). Feedforward and feedback projections of caudal belt and parabelt areas of auditory cortex: refining the hierarchical model. Front Neurosci 8: 72.

Hackett TA, Stepniewska I, Kaas JH (1998a). Subdivisions of auditory cortex and ipsilateral cortical connections of the parabelt auditory cortex in macaque monkeys. J Comp Neurol 394: 475–495.

Hackett TA, Stepniewska I, Kaas JH (1998b). Thalamocortical connections of the parabelt auditory cortex in macaque monkeys. J Comp Neurol 400: 271–286.

Hackett TA, Stepniewska I, Kaas JH (1999). Prefrontal connections of the parabelt auditory cortex in macaque monkeys. Brain Res 817: 45–58.

Hackett TA, Preuss TM, Kaas JH (2001). Architectonic identification of the core region in auditory cortex of macaques, chimpanzees, and humans. J Comp Neurol 441: 197–222.

Hackett TA, De La Mothe LA, Ulbert I et al. (2007a). Multisensory convergence in auditory cortex. II. Thalamocortical connections of the caudal superior temporal plane. J Comp Neurol 502: 924–952.

Hackett TA, Smiley JF, Ulbert I et al. (2007b). Sources of somatosensory input to the caudal belt areas of auditory cortex. Perception 36: 1419–1430.

Hall DA, Hart HC, Johnsrude IS (2003). Relationships between human auditory cortical structure and function. Audiol Neurootol 8: 1–18.

Harinen K, Rinne T (2013). Activations of human auditory cortex to phonemic and nonphonemic vowels during discrimination and memory tasks. Neuroimage 77: 279–287.

Hart HC, Palmer AR, Hall DA (2002). Heschl's gyrus is more sensitive to tone level than non-primary auditory cortex. Hear Res 171: 177–190.

Hart HC, Palmer AR, Hall DA (2004). Different areas of human non-primary auditory cortex are activated by sounds with spatial and nonspatial properties. Hum Brain Mapp 21: 178–190.

Hashimoto Y, Sakai KL (2003). Brain activations during conscious self-monitoring of speech production with delayed auditory feedback: an fMRI study. Hum Brain Mapp 20: 22–28.

He J (2003). Corticofugal modulation on both ON and OFF responses in the nonlemniscal auditory thalamus of the guinea pig. J Neurophysiol 89: 367–381.

Heath CJ, Jones EG (1971). The anatomical organization of the suprasylvian gyrus of the cat. Ergeb Anat Entwicklungsgesch 45: 3–64.

Herdener M, Esposito F, Scheffler K et al. (2013). Spatial representations of temporal and spectral sound cues in human auditory cortex. Cortex 49: 2822–2833.

Herzog AG, Van Hoesen GW (1976). Temporal neocortical afferent connections to the amygdala in the rhesus monkey. Brain Res 115: 57–69.

Hickok G, Poeppel D (2000). Towards a functional neuroanatomy of speech perception. Trends Cogn Sci 4: 131–138.

Hickok G, Poeppel D (2004). Dorsal and ventral streams: a framework for understanding aspects of the functional anatomy of language. Cognition 92: 67–99.

Hickok G, Poeppel D (2007). The cortical organization of speech processing. Nat Rev Neurosci 8: 393–402.

Hickok G, Erhard P, Kassubek J et al. (2000). A functional magnetic resonance imaging study of the role of left posterior superior temporal gyrus in speech production: implications for the explanation of conduction aphasia. Neurosci Lett 287: 156–160.

Hickok G, Buchsbaum B, Humphries C et al. (2003). Auditory-motor interaction revealed by fMRI: speech, music, and working memory in area Spt. J Cogn Neurosci 15: 673–682.

Hickok G, Okada K, Serences JT (2009). Area Spt in the human planum temporale supports sensory-motor integration for speech processing. J Neurophysiol 101: 2725–2732.

Howard MA, Volkov IO, Mirsky R et al. (2000). Auditory cortex on the human posterior superior temporal gyrus. J Comp Neurol 416: 79–92.

Humphries C, Liebenthal E, Binder JR (2010). Tonotopic organization of human auditory cortex. Neuroimage 50: 1202–1211.

Hyvarinen J (1982). Posterior parietal lobe of the primate brain. Physiol Rev 62: 1060–1129.

Imig TJ, Ruggero MA, Kitzes LM et al. (1977). Organization of auditory cortex in the owl monkey (Aotus trivirgatus). J Comp Neurol 171: 111–128.

Isenberg AL, Vaden Jr KI, Saberi K et al. (2012). Functionally distinct regions for spatial processing and sensory motor integration in the planum temporale. Hum Brain Mapp 33: 2453–2463.

Ito SI (1982). Prefrontal unit activity of macaque monkeys during auditory and visual reaction time tasks. Brain Res 247: 39–47.

Jones EG (1998). Viewpoint: the core and matrix of thalamic organization. Neuroscience 85: 331–345.

Jones EG (2003). Chemically defined parallel pathways in the monkey auditory system. Ann N Y Acad Sci 999: 218–233.

Jones EG (2007). The Thalamus, Cambridge University Press, Cambridge.

Jones EG, Powell TP (1969). Connexions of the somatic sensory cortex of the rhesus monkey. I. Ipsilateral cortical connexions. Brain 92: 477–502.

Jones EG, Burton H, Saper CB et al. (1976). Midbrain, diencephalic and cortical relationships of the basal nucleus of Meynert and associated structures in primates. J Comp Neurol 167: 385–419.

Jones EG, Dell'anna ME, Molinari M et al. (1995). Subdivisions of macaque monkey auditory cortex revealed by calcium binding protein immunoreactivity. J Comp Neurol 362: 153–170.

Jung Y, Hong S (2003). Corticostriatal connections of the superior temporal regions in the macaque monkey. Korean J Biol Sci 7: 317–325.

Kaas JH, Hackett TA (1998). Subdivisions of auditory cortex and levels of processing in primates. Audiol Neurootol 3: 73–85.

Kaas JH, Hackett TA (1999). 'What' and 'where' processing in auditory cortex. Nat Neurosci 2: 1045–1047.

Kaas JH, Hackett TA, Tramo MJ (1999). Auditory processing in primate cerebral cortex. Curr Opin Neurobiol 9: 164–170.

Kajikawa Y, De La Mothe LA, Blumell S et al. (2005). A comparison of neuron response properties in areas A1 and CM of the marmoset monkey auditory cortex: tones and broad band noise. J Neurophysiol 93: 22–34.

Kajikawa Y, De La Mothe LA, Blumell S et al. (2008). Coding of FM sweep trains and twitter calls in area CM of marmoset auditory cortex. Hear Res 239: 107–125.

Kayser C, Petkov CI, Augath M et al. (2005). Integration of touch and sound in auditory cortex. Neuron 48: 373–384.

Kayser C, Petkov CI, Augath M et al. (2007). Functional imaging reveals visual modulation of specific fields in auditory cortex. J Neurosci 27: 1824–1835.

Kayser C, Petkov CI, Logothetis NK (2008). Visual modulation of neurons in auditory cortex. Cereb Cortex 18: 1560–1574.

Kikuchi Y, Horwitz B, Mishkin M (2010). Hierarchical auditory processing directed rostrally along the monkey's supratemporal plane. J Neurosci 30: 13021–13030.

Kikuchi-Yorioka Y, Sawaguchi T (2000). Parallel visuospatial and audiospatial working memory processes in the monkey dorsolateral prefrontal cortex. Nat Neurosci 3: 1075–1076.

Kolomiets BP, Deniau JM, Mailly P et al. (2001). Segregation and convergence of information flow through the cortico-subthalamic pathways. J Neurosci 21: 5764–5772.

Kosaki H, Hashikawa T, He J et al. (1997). Tonotopic organization of auditory cortical fields delineated by parvalbumin immunoreactivity in macaque monkeys. J Comp Neurol 386: 304–316.

Kosmal A, Malinowska M, Kowalska DM (1997). Thalamic and amygdaloid connections of the auditory association cortex of the superior temporal gyrus in rhesus monkey (*Macaca mulatta*). Acta Neurobiol Exp (Wars) 57: 165–188.

Krubitzer LA, Kaas JH (1990). The organization and connections of somatosensory cortex in marmosets. J Neurosci 10: 952–974.

Krubitzer L, Clarey J, Tweedale R et al. (1995). A redefinition of somatosensory areas in the lateral sulcus of macaque monkeys. J Neurosci 15: 3821–3839.

Krumbholz K, Schonwiesner M, Von Cramon DY et al. (2005). Representation of interaural temporal information from left and right auditory space in the human planum temporale and inferior parietal lobe. Cereb Cortex 15: 317–324.

Kusmierek P, Rauschecker JP (2009). Functional specialization of medial auditory belt cortex in the alert rhesus monkey. J Neurophysiol 102: 1606–1622.

Kusmierek P, Ortiz M, Rauschecker JP (2012). Sound-identity processing in early areas of the auditory ventral stream in the macaque. J Neurophysiol 107: 1123–1141.

Lakatos P, Pincze Z, Fu KG et al. (2005a). Timing of pure tone and noise-evoked responses in macaque auditory cortex. Neuroreport 16: 933–937.

Lakatos P, Shah AS, Knuth KH et al. (2005b). An oscillatory hierarchy controlling neuronal excitability and stimulus processing in the auditory cortex. J Neurophysiol 94: 1904–1911.

Lakatos P, Chen CM, O'Connell MN et al. (2007). Neuronal oscillations and multisensory interaction in primary auditory cortex. Neuron 53: 279–292.

Langers DR, Van Dijk P (2012). Mapping the tonotopic organization in human auditory cortex with minimally salient acoustic stimulation. Cereb Cortex 22: 2024–2038.

Langers DRM, Backes WH, Van Dijk P (2007). Representation of lateralization and tonotopy in primary versus secondary human auditory cortex. Neuroimage 34: 264–273.

Langner G, Dinse HR, Godde B (2009). A map of periodicity orthogonal to frequency representation in the cat auditory cortex. Front Integr Neurosci 3: 27.

Leaver AM, Rauschecker JP (2010). Cortical representation of natural complex sounds: effects of acoustic features and auditory object category. J Neurosci 30: 7604–7612.

Leaver AM, Van Lare J, Zielinski B et al. (2009). Brain activation during anticipation of sound sequences. J Neurosci 29: 2477–2485.

Ledoux JE, Farb C, Ruggiero DA (1990). Topographic organization of neurons in the acoustic thalamus that project to the amygdala. J Neurosci 10: 1043–1054.

Ledoux JE, Farb CR, Romanski LM (1991). Overlapping projections to the amygdala and striatum from auditory processing areas of the thalamus and cortex. Neurosci Lett 134: 139–144.

Lee CC, Winer JA (2008). Connections of cat auditory cortex: I. Thalamocortical system. J Comp Neurol 507: 1879–1900.

Lee CC, Schreiner CE, Imaizumi K et al. (2004). Tonotopic and heterotopic projection systems in physiologically defined auditory cortex. Neuroscience 128: 871–887.

Lee JH, Russ BE, Orr LE et al. (2009). Prefrontal activity predicts monkeys' decisions during an auditory category task. Front Integr Neurosci 3: 16.

Leinonen L (1980). Functional properties of neurones in the parietal retroinsular cortex in awake monkey. Acta Physiol Scand 108: 381–384.

Leinonen L, Hyvarinen J, Sovijarvi AR (1980). Functional properties of neurons in the temporo-parietal association cortex of awake monkey. Exp Brain Res 39: 203–215.

Lewis JW, Van Essen DC (2000). Corticocortical connections of visual, sensorimotor, and multimodal processing areas in the parietal lobe of the macaque monkey. J Comp Neurol 428: 112–137.

Linden JF, Grunewald A, Andersen RA (1999). Responses to auditory stimuli in macaque lateral intraparietal area. II. Behavioral modulation. J Neurophysiol 82: 343–358.

Luethke LE, Krubitzer LA, Kaas JH (1989). Connections of primary auditory cortex in the New World monkey, Saguinus. J Comp Neurol 285: 487–513.

Lutz J, Hemminger F, Stahl R et al. (2007). Evidence of subcortical and cortical aging of the acoustic pathway: a diffusion tensor imaging (DTI) study. Acad Radiol 14: 692–700.

Maier JX, Ghazanfar AA (2007). Looming biases in monkey auditory cortex. J Neurosci 27: 4093–4100.

Markowitsch HJ, Emmans D, Irle E et al. (1985). Cortical and subcortical afferent connections of the primate's temporal pole: a study of rhesus monkeys, squirrel monkeys, and marmosets. J Comp Neurol 242: 425–458.

Mascagni F, McDonald AJ, Coleman JR (1993). Corticoamygdaloid and corticocortical projections of the rat temporal cortex: a *Phaseolus vulgaris* leucoagglutinin study. Neuroscience 57: 697–715.

Mazzoni P, Bracewell RM, Barash S et al. (1996). Spatially tuned auditory responses in area LIP of macaques performing delayed memory saccades to acoustic targets. J Neurophysiol 75: 1233–1241.

McDonald AJ (1998). Cortical pathways to the mammalian amygdala. Prog Neurobiol 55: 257–332.

Merzenich MM, Brugge JF (1973). Representation of the cochlear partition of the superior temporal plane of the macaque monkey. Brain Res 50: 275–296.

Miller LM, Recanzone GH (2009). Populations of auditory cortical neurons can accurately encode acoustic space across stimulus intensity. Proc Natl Acad Sci U S A 106: 5931–5935.

Molinari M, Dell'anna ME, Rausell E et al. (1995). Auditory thalamocortical pathways defined in monkeys by calcium-binding protein immunoreactivity. J Comp Neurol 362: 171–194.

Morecraft RJ, Cipolloni PB, Stilwell-Morecraft KS et al. (2004). Cytoarchitecture and cortical connections of the posterior cingulate and adjacent somatosensory fields in the rhesus monkey. J Comp Neurol 469: 37–69.

Morel A, Kaas JH (1992). Subdivisions and connections of auditory cortex in owl monkeys. J Comp Neurol 318: 27–63.

Morel A, Garraghty PE, Kaas JH (1993). Tonotopic organization, architectonic fields, and connections of auditory cortex in macaque monkeys. J Comp Neurol 335: 437–459.

Morosan P, Rademacher J, Schleicher A et al. (2001). Human primary auditory cortex: cytoarchitectonic subdivisions and mapping into a spatial reference system. Neuroimage 13: 684–701.

Morosan P, Schleicher A, Amunts K et al. (2005). Multimodal architectonic mapping of human superior temporal gyrus. Anat Embryol (Berl) 210: 401–406.

Mullette-Gillman OA, Cohen YE, Groh JM (2005). Eye-centered, head-centered, and complex coding of visual and auditory targets in the intraparietal sulcus. J Neurophysiol 94: 2331–2352.

Nakahara H, Yamada S, Mizutani T et al. (2000). Identification of the primary auditory field in archival human brain tissue via immunocytochemistry of parvalbumin. Neurosci Lett 286: 29–32.

Nakano K (2000). Neural circuits and topographic organization of the basal ganglia and related regions. Brain Dev 22 (Suppl 1): S5–S16.

Nakano K, Kayahara T, Tsutsumi T et al. (2000). Neural circuits and functional organization of the striatum. J Neurol 247 (Suppl 5): V1–V15.

Nelken I, Bizley JK, Nodal FR et al. (2008). Responses of auditory cortex to complex stimuli: functional organization revealed using intrinsic optical signals. J Neurophysiol 99: 1928–1941.

Nieder A, Diester I, Tudusciuc O (2006). Temporal and spatial enumeration processes in the primate parietal cortex. Science 313: 1431–1435.

Nourski KV, Brugge JF, Reale RA et al. (2013). Coding of repetitive transients by auditory cortex on posterolateral superior temporal gyrus in humans: an intracranial electrophysiology study. J Neurophysiol 109: 1283–1295.

Nourski KV, Steinschneider M, Oya H et al. (2014). Spectral organization of the human lateral superior temporal gyrus revealed by intracranial recordings. Cereb Cortex 24: 340–352.

Obleser J, Zimmermann J, Van Meter J et al. (2007). Multiple stages of auditory speech perception reflected in event-related FMRI. Cereb Cortex 17: 2251–2257.

Okada K, Hickok G (2009). Two cortical mechanisms support the integration of visual and auditory speech: a hypothesis and preliminary data. Neurosci Lett 452: 219–223.

Okada K, Rong F, Venezia J et al. (2010). Hierarchical organization of human auditory cortex: evidence from acoustic invariance in the response to intelligible speech. Cereb Cortex 20: 2486–2495.

Ozaki I, Hashimoto I (2007). Human tonotopic maps and their rapid task-related changes studied by magnetic source imaging. Can J Neurol Sci 34: 146–153.

Pa J, Hickok G (2008). A parietal-temporal sensory-motor integration area for the human vocal tract: evidence from an fMRI study of skilled musicians. Neuropsychologia 46: 362–368.

Palmer SM, Rosa MG (2006). A distinct anatomical network of cortical areas for analysis of motion in far peripheral vision. Eur J Neurosci 24: 2389–2405.

Pandya DN (1995). Anatomy of the auditory cortex. Rev Neurol (Paris) 151: 486–494.

Pandya DN, Kuypers HG (1969). Cortico-cortical connections in the rhesus monkey. Brain Res 13: 13–36.

Pandya DN, Rosene DL (1993). Laminar termination patterns of thalamic, callosal, and association afferents in the primary auditory area of the rhesus monkey. Exp Neurol 119: 220–234.

Pandya DN, Sanides F (1973). Architectonic parcellation of the temporal operculum in rhesus monkey and its projection pattern. Z Anat Entwicklungsgesch 139: 127–161.

Pandya DN, Seltzer B (1982). Intrinsic connections and architectonics of posterior parietal cortex in the rhesus monkey. J Comp Neurol 204: 196–210.

Pandya DN, Hallett M, Mukherjee SK (1969). Intra- and interhemispheric connections of the neocortical auditory system in the rhesus monkey. Brain Res 14: 49–65.

Pandya DN, Rosene DL, Doolittle AM (1994). Corticothalamic connections of auditory-related areas of the temporal lobe in the rhesus monkey. J Comp Neurol 345: 447–471.

Pasley BN, David SV, Mesgarani N et al. (2012). Reconstructing speech from human auditory cortex. PLoS Biol 10: e1001251.

Patterson RD, Johnsrude IS (2008). Functional imaging of the auditory processing applied to speech sounds. Philos Trans R Soc Lond B Biol Sci 363: 1023–1035.

Petkov CI, Kayser C, Augath M et al. (2006). Functional imaging reveals numerous fields in the monkey auditory cortex. PLoS Biol 4: 1–14.

Petkov CI, Kayser C, Steudel T et al. (2008). A voice region in the monkey brain. Nat Neurosci 11: 367–374.

Petrides M, Pandya DN (2002). Comparative cytoarchitectonic analysis of the human and the macaque ventrolateral prefrontal cortex and corticocortical connection patterns in the monkey. Eur J Neurosci 16: 291–310.

Petrides M, Pandya DN (2007). Efferent association pathways from the rostral prefrontal cortex in the macaque monkey. J Neurosci 27: 11573–11586.

Poljak S (1932). Origin, course, termination and internal organization of the auditory radiation. In: HM Evans, IM Thompson (Eds.), The Main Afferent Fiber Systems of the Cerebral Cortex in Primates, University of California Press, Berkeley.

Polley DB, Read HL, Storace DA et al. (2007). Multiparametric auditory receptive field organization across five cortical fields in the albino rat. J Neurophysiol 97: 3621–3638.

Poremba A, Malloy M, Saunders RC et al. (2004). Species-specific calls evoke asymmetric activity in the monkey's temporal poles. Nature 427: 448–451.

Qi HX, Lyon DC, Kaas JH (2002). Cortical and thalamic connections of the parietal ventral somatosensory area in marmoset monkeys (*Callithrix jacchus*). J Comp Neurol 443: 168–182.

Rademacher J, Caviness VJ, Steinmetz H et al. (1993). Topographical variation of the human primary cortices: implications for neuroimaging, brain mapping, and neurobiology. Cereb Cortex 3: 313–329.

Rademacher J, Morosan P, Schormann T et al. (2001). Probabilistic mapping and volume measurement of human primary auditory cortex. Neuroimage 13: 669–683.

Rauschecker JP (1998a). Cortical processing of complex sounds. Curr Opin Neurobiol 8: 516–521.

Rauschecker JP (1998b). Parallel processing in the auditory cortex of primates. Audiol Neurootol 3: 86–103.

Rauschecker JP (2011). An expanded role for the dorsal auditory pathway in sensorimotor control and integration. Hear Res 271: 16–25.

Rauschecker JP, Scott SK (2009). Maps and streams in the auditory cortex: nonhuman primates illuminate human speech processing. Nat Neurosci 12: 718–724.

Rauschecker JP, Tian B (2000). Mechanisms and streams for processing of "what" and "where" in auditory cortex. Proc Natl Acad Sci U S A 97: 11800–11806.

Rauschecker JP, Tian B (2004). Processing of band-passed noise in the lateral auditory belt cortex of the rhesus monkey. J Neurophysiol 91: 2578–2589.

Rauschecker JP, Tian B, Hauser M (1995). Processing of complex sounds in the macaque nonprimary auditory cortex. Science 268: 111–114.

Rauschecker JP, Tian B, Pons T et al. (1997). Serial and parallel processing in rhesus monkey auditory cortex. J Comp Neurol 382: 89–103.

Reale RA, Imig TJ (1983). Auditory cortical field projections to the basal ganglia of the cat. Neuroscience 8: 67–86.

Recanzone GH (2000a). Response profiles of auditory cortical neurons to tones and noise in behaving macaque monkeys. Hear Res 150: 104–118.

Recanzone GH (2000b). Spatial processing in the auditory cortex of the macaque monkey. Proc Natl Acad Sci U S A 97: 11829–11835.

Recanzone GH (2001). Spatial processing in the primate auditory cortex. Audiol Neurootol 6: 178–181.

Recanzone GH (2008). Representation of con-specific vocalizations in the core and belt areas of the auditory cortex in the alert macaque monkey. J Neurosci 28 (49): 13184–13193.

Recanzone GH, Cohen YE (2010). Serial and parallel processing in the primate auditory cortex revisited. Behav Brain Res 206: 1–7.

Recanzone GH, Schreiner CE, Sutter ML et al. (1999). Functional organization of spectral receptive fields in the primary auditory cortex of the owl monkey. J Comp Neurol 415: 460–481.

Recanzone GH, Guard DC, Phan ML (2000a). Frequency and intensity response properties of single neurons in the auditory cortex of the behaving macaque monkey. J Neurophysiol 83: 2315–2331.

Recanzone GH, Guard DC, Phan ML et al. (2000b). Correlation between the activity of single auditory cortical neurons and sound-localization behavior in the macaque monkey. J Neurophysiol 83: 2723–2739.

Rivier F, Clarke S (1997). Cytochrome oxidase, acetylcholinesterase, and NADPH-diaphorase staining in human supratemporal and insular cortex: evidence for multiple auditory areas. Neuroimage 6: 288–304.

Roberts AC, Tomic DL, Parkinson et al. (2007). Forebrain connectivity of the prefrontal cortex in the marmoset monkey (*Callithrix jacchus*): an anterograde and retrograde tract-tracing study. J Comp Neurol 502: 86–112.

Robinson CJ, Burton H (1980a). Organization of somatosensory receptive fields in cortical areas 7b, retroinsula, postauditory and granular insula of *M. fascicularis*. J Comp Neurol 192: 69–92.

Robinson CJ, Burton H (1980b). Somatic submodality distribution within the second somatosensory (SII), 7b, retroinsular, postauditory, and granular insular cortical areas of *M. fascicularis*. J Comp Neurol 192: 93–108.

Rockland KS, Ojima H (2003). Multisensory convergence in calcarine visual areas in macaque monkey. Int J Psychophysiol 50: 19–26.

Rolls ET (1994). Neurophysiology and cognitive functions of the striatum. Rev Neurol (Paris) 150: 648–660.

Rolls ET, Critchley HD, Browning AS et al. (2006). Face-selective and auditory neurons in the primate orbitofrontal cortex. Exp Brain Res 170: 74–87.

Romanski LM (2007). Representation and integration of auditory and visual stimuli in the primate ventral lateral prefrontal cortex. Cereb Cortex 17 (Suppl 1): i61–i69.

Romanski LM, Goldman-Rakic PS (2002). An auditory domain in primate prefrontal cortex. Nat Neurosci 5: 15–16.

Romanski LM, Ledoux JE (1993). Information cascade from primary auditory cortex to the amygdala: corticocortical and corticoamygdaloid projections of temporal cortex in the rat. Cereb Cortex 3: 515–532.

Romanski LM, Bates JF, Goldman-Rakic PS (1999a). Auditory belt and parabelt projections to the prefrontal cortex in the rhesus monkey. J Comp Neurol 403: 141–157.

Romanski LM, Tian B, Fritz J et al. (1999b). Dual streams of auditory afferents target multiple domains in the primate prefrontal cortex. Nat Neurosci 2: 1131–1136.

Romanski LM, Tian B, Fritz JB et al. (2000). Reply to 'What', 'where' and 'how' in auditory cortex'. Nat Neurosci 3: 966.

Romanski LM, Averbeck BB, Diltz M (2005). Neural representation of vocalizations in the primate ventrolateral prefrontal cortex. J Neurophysiol 93: 734–747.

Rosa MG, Soares JG, Fiorani Jr M et al. (1993). Cortical afferents of visual area MT in the Cebus monkey: possible homologies between New and Old World monkeys. Vis Neurosci 10: 827–855.

Rouiller EM, Durif C (2004). The dual pattern of corticothalamic projection of the primary auditory cortex in macaque monkey. Neurosci Lett 358: 49–52.

Rouiller EM, Welker E (2000). A comparative analysis of the morphology of corticothalamic projections in mammals. Brain Res Bull 53: 727–741.

Rozzi S, Calzavara R, Belmalih A et al. (2006). Cortical connections of the inferior parietal cortical convexity of the macaque monkey. Cereb Cortex 16: 1389–1417.

Russo GS, Bruce CJ (1994). Frontal eye field activity preceding aurally guided saccades. J Neurophysiol 71: 1250–1253.

Saleem KS, Kondo H, Price JL (2008). Complementary circuits connecting the orbital and medial prefrontal networks with the temporal, insular, and opercular cortex in the macaque monkey. J Comp Neurol 506: 659–693.

Schlack A, Sterbing-D'Angelo SJ, Hartung K et al. (2005). Multisensory space representations in the macaque ventral intraparietal area. J Neurosci 25: 4616–4625.

Schreiner CE, Sutter ML (1992). Topography of excitatory bandwidth in cat primary auditory cortex: single-neuron versus multiple-neuron recordings. J Neurophysiol 68: 1487–1502.

Schreiner CE, Winer JA (2007). Auditory cortex mapmaking: principles, projections, and plasticity. Neuron 56: 356–365.

Schreiner CE, Mendelson JR, Sutter ML (1992). Functional topography of cat primary auditory cortex: representation of tone intensity. Exp Brain Res 92: 105–122.

Schroeder CE, Foxe JJ (2002). The timing and laminar profile of converging inputs to multisensory areas of the macaque neocortex. Brain Res Cogn Brain Res 14: 187–198.

Schroeder CE, Lindsley RW, Specht C et al. (2001). Somatosensory input to auditory association cortex in the macaque monkey. J Neurophysiol 85: 1322–1327.

Schulze H, Hess A, Ohl FW et al. (2002). Superposition of horseshoe-like periodicity and linear tonotopic maps in auditory cortex of the Mongolian gerbil. Eur J Neurosci 15: 1077–1084.

Scott SK, Johnsrude IS (2003). The neuroanatomical and functional organization of speech perception. Trends Neurosci 26: 100–107.

Scott SK, Blank CC, Rosen S et al. (2000). Identification of a pathway for intelligible speech in the left temporal lobe. Brain 123 (Pt 12): 2400–2406.

Seifritz E, Esposito F, Hennel F et al. (2002). Spatiotemporal pattern of neural processing in the human auditory cortex. Science 297: 1706–1708.

Seifritz E, Di Salle F, Esposito F et al. (2006). Enhancing BOLD response in the auditory system by neurophysiologically tuned fMRI sequence. Neuroimage 29: 1013–1022.

Selemon LD, Goldman-Rakic PS (1985). Longitudinal topography and interdigitation of corticostriatal projections in the rhesus monkey. J Neurosci 5: 776–794.

Seltzer B, Pandya DN (1978). Afferent cortical connections and architectonics of the superior temporal sulcus and surrounding cortex in the rhesus monkey. Brain Res 149: 1–24.

Seltzer B, Pandya DN (1994). Parietal, temporal, and occipital projections to cortex of the superior temporal sulcus in the rhesus monkey: a retrograde tracer study. J Comp Neurol 343: 445–463.

Seltzer B, Cola MG, Gutierrez C et al. (1996). Overlapping and nonoverlapping cortical projections to cortex of the superior temporal sulcus in the rhesus monkey: double anterograde tracer studies. J Comp Neurol 370: 173–190.

Shi CJ, Cassell MD (1997). Cortical, thalamic, and amygdaloid projections of rat temporal cortex. J Comp Neurol 382: 153–175.

Shinonaga Y, Takada M, Mizuno N (1994). Direct projections from the non-laminated divisions of the medial geniculate nucleus to the temporal polar cortex and amygdala in the cat. J Comp Neurol 340: 405–426.

Sigalovsky IS, Fischl B, Melcher JR (2006). Mapping an intrinsic MR property of gray matter in auditory cortex of living humans: a possible marker for primary cortex and hemispheric differences. Neuroimage 32: 1524–1537.

Smiley JF, Hackett TA, Ulbert I et al. (2007). Multisensory convergence in auditory cortex, I. Cortical connections of the caudal superior temporal plane in macaque monkeys. J Comp Neurol 502: 894–923.

Smith KR, Hsieh IH, Saberi K et al. (2010). Auditory spatial and object processing in the human planum temporale: no evidence for selectivity. J Cogn Neurosci 22: 632–639.

Sousa AP, Pinon MC, Gattass R et al. (1991). Topographic organization of cortical input to striate cortex in the Cebus monkey: a fluorescent tracer study. J Comp Neurol 308: 665–682.

Steinschneider M, Fishman YI (2011). Enhanced physiologic discriminability of stop consonants with prolonged formant transitions in awake monkeys based on the tonotopic organization of primary auditory cortex. Hear Res 271: 103–114.

Stricanne B, Andersen RA, Mazzoni P (1996). Eye-centered, head-centered, and intermediate coding of remembered sound locations in area LIP. J Neurophysiol 76: 2071–2076.
Sugihara T, Diltz M, Averbeck B et al. (2006). Integration of auditory and visual communication information in the primate ventrolateral prefrontal cortex. J Neurosci 26: 11138–11147.
Sweet RA, Dorph-Petersen KA, Lewis DA (2005). Mapping auditory core, lateral belt, and parabelt cortices in the human superior temporal gyrus. J Comp Neurol 491: 270–289.
Talavage TM, Edmister WB (2004). Nonlinearity of FMRI responses in human auditory cortex. Hum Brain Mapp 22: 216–228.
Tian B, Rauschecker JP (2004). Processing of frequency-modulated sounds in the lateral auditory belt cortex of the rhesus monkey. J Neurophysiol 92: 2993–3013.
Tian B, Reser D, Durham A et al. (2001). Functional specialization in rhesus monkey auditory cortex. Science 292: 290–293.
Tourville JA, Reilly KJ, Guenther FH (2008). Neural mechanisms underlying auditory feedback control of speech. Neuroimage 39: 1429–1443.
Trojanowski JQ, Jacobson S (1975). A combined horseradish peroxidase-autoradiographic investigation of reciprocal connections between superior temporal gyrus and pulvinar in squirrel monkey. Brain Res 85: 347–353.
Tsunada J, Lee JH, Cohen YE (2011). Representation of speech categories in the primate auditory cortex. J Neurophysiol 105: 2634–2646.
Turner BH, Mishkin M, Knapp M (1980). Organization of the amydalopetal projections from modality-specific cortical association areas in the monkey. J Comp Neurol 191: 515–543.
Upadhyay J, Ducros M, Knaus T et al. (2007). Function and connectivity in human primary auditory cortex: a combined fMRI and DTI study at 3 Tesla. Cereb Cortex 17: 2420–2432.
Vaadia E (1989). Single-unit activity related to active localization of acoustic and visual stimuli in the frontal cortex of the rhesus monkey. Brain Behav Evol 33: 127–131.
Vaadia E, Benson DA, Hienz RD et al. (1986). Unit study of monkey frontal cortex: active localization of auditory and of visual stimuli. J Neurophysiol 56: 934–952.
Vogt BA, Pandya DN (1987). Cingulate cortex of the rhesus monkey: II. Cortical afferents. J Comp Neurol 262: 271–289.
Von Economo C, Horn L (1930). Uber Windungsrelief, Masse und Rindenarchitektonik der Supratemporalflache, ihre individuellen und ihre Seitenunterschiede. Z Neurol Psychiatr 130: 678–757.
Von Economo C, Koskinas G (1925). Die Cytoarchitectonik der Hirnrinde des erwachsenen menschen. Julius-Springer, Berlin.
Walker AE (1937). The projections of the medial geniculate body to the cerebral cortex in the macaque monkey. J Anat (London) 71: 319–331.
Wallace MN, Johnston PW, Palmer AR (2002). Histochemical identification of cortical areas in the auditory region of the human brain. Exp Brain Res 143: 499–508.
Warren JD, Zielinski BA, Green GGR, Rauschecker JP et al. (2002). Perception of sound source motion by the human brain. Neuron 34: 139–148.
Watanabe M (1992). Frontal units of the monkey coding the associative significance of visual and auditory stimuli. Exp Brain Res 89: 233–247.
Werner-Reiss U, Kelly KA, Trause AS et al. (2003). Eye position affects activity in primary auditory cortex of primates. Curr Biol 13: 554–562.
Wernicke C (1874). Der aphasische Symptomenkomplex, Cohn and Weigert, Breslau.
Wessinger CM, Vanmeter J, Tian B et al. (2001). Hierarchical organization of the human auditory cortex revealed by functional magnetic resonance imaging. J Cogn Neurosci 13: 1–7.
Winer JA, Chernock ML, Larue DT et al. (2002). Descending projections to the inferior colliculus from the posterior thalamus and the auditory cortex in rat, cat, and monkey. Hear Res 168: 181–195.
Wise RJ, Scott SK, Blank SC et al. (2001). Separate neural subsystems within 'Wernicke's area'. Brain 124: 83–95.
Wollberg Z, Sela J (1980). Frontal cortex of the awake squirrel monkey: responses of single cells to visual and auditory stimuli. Brain Res 198: 216–220.
Woods TM, Lopez SE, Long JH et al. (2006). Effects of stimulus azimuth and intensity on the single-neuron activity in the auditory cortex of the alert macaque monkey. J Neurophysiol 96: 3323–3337.
Woods DL, Herron TJ, Cate AD et al. (2010). Functional properties of human auditory cortical fields. Front Syst Neurosci 4: 155.
Woollard H, Harpman A (1939). The cortical projection of the medial geniculate body. J Neurol Psych 2: 35–44.
Yeterian EH, Pandya DN (1991). Prefrontostriatal connections in relation to cortical architectonic organization in rhesus monkeys. J Comp Neurol 312: 43–67.
Yeterian EH, Pandya DN (1998). Corticostriatal connections of the superior temporal region in rhesus monkeys. J Comp Neurol 399: 384–402.
Yeterian EH, Van Hoesen GW (1978). Cortico-striate projections in the rhesus monkey: the organization of certain cortico-caudate connections. Brain Res 139: 43–63.
Yukie M (2002). Connections between the amygdala and auditory cortical areas in the macaque monkey. Neurosci Res 42: 219–229.
Zhang Z, Liu CH, Yu YQ et al. (2008). Corticofugal projection inhibits the auditory thalamus through the thalamic reticular nucleus. J Neurophysiol 99: 2938–2945.
Zheng ZZ (2009). The functional specialization of the planum temporale. J Neurophysiol 102: 3079–3081.
Zimmer U, Macaluso E (2005). High binaural coherence determines successful sound localization and increased activity in posterior auditory areas. Neuron 47 (6): 893–905.

Handbook of Clinical Neurology, Vol. 129 (3rd series)
The Human Auditory System
G.G. Celesia and G. Hickok, Editors

Chapter 3

Development of the auditory system

RUTH LITOVSKY*

Binaural Hearing and Speech Laboratory, Waisman Center, University of Wisconsin-Madison, Madison, WI, USA

INTRODUCTION

When considering auditory development, we may ask ourselves: what is the purpose of auditory perception and processing of auditory signals? Generally, the auditory system serves the role of constructing a perceptual space that extracts information from objects (animate and inanimate), groups together some objects, and segregates sounds from one another. In other words, the auditory system engages in an analysis of the auditory world, so that the listener can accomplish the goal of communication and learning. While the auditory system regularly interacts with other sensory modalities, much of what we understand about auditory perception and development is rooted in studies that focus the auditory signal *per se*. Auditory processing occurs in a series of analysis stages, beginning with peripheral auditory mechanisms devoted to encoding sound, and proceeding to more complex stages at which sound processing leads to perception and object recognition. An important feature of the staging is the combination of inputs from the right and left ears, which takes place several synapses after the initial impact of sound on the auditory sensory organ, the cochlea.

It is in this context that we can begin to consider the goal of auditory development, the conditions under which changes in auditory mechanisms generally result in beneficial outcomes for the listener, and, conversely, conditions under which changes bring about negative consequences. This all depends on whether these changes occur because the organism is experiencing a "normal" or "typical" trajectory of exposure to sound during development, or whether the changes are occurring due to unanticipated events, such as disease or trauma, in the auditory pathway or in other brain areas that interact with the processing of incoming signals or the production of sound by the individual.

Auditory development is a broadly defined term, which refers to the fact that perception is influenced by a combination of innate, genetically programmed changes in anatomy and physiology, combined with auditory experience. Two primary questions resonate with research to date. First, how does experience with and exposure to sound during early stages in development impact the ability to process sound? Second, how does lack of exposure to sound influence the anatomic organization and physiologic functioning of the auditory pathways? One way to address these questions, which is the approach taken in this chapter, is to consider developmental changes in the auditory pathway as one progresses from peripheral to central mechanisms. The peripheral system generally encodes stimuli based on temporal, spectral, and intensity cues. The subsequent extraction of complex combinations of those cues, assignment of auditory features to meaningful stimuli such as speech and music, and the role of non-auditory processing occur at more central levels of the auditory pathway. Finally, this chapter focuses attention on the development of the human auditory system; however, much is to be learned also from research done with animal models, which will be referred to as well. As such, from a methodologic perspective, there are a number of approaches taken towards describing and understanding human auditory development. Structural changes are generally studied using histologic approaches by reconstructing connectivity between neurons. In humans this would be done either by using neuroimaging techniques or with postpartum labeling of neural connections in cadavers. A second approach, to explore functional maturation, also relies on imaging, primarily functional

*Correspondence to: Ruth Litovsky, Binaural Hearing and Speech Laboratory, Waisman Center, University of Wisconsin-Madison, 1500 Highland Ave, #521, Madison WI 53705, USA. Tel: +1-608-262-5045, E-mail: litovsky@waisman.wisc.edu

neuroimaging techniques. Third, behavioral studies, which depend on extracting information from listeners who are performing psychoacoustic tasks, provide a window into the extent to which underlying structure and function support perception.

DEVELOPMENT OF THE EAR

The ear is responsible for the initial encoding of acoustic input. The functional status of the ear has been studied using various investigational and clinical tools. Anatomic studies suggest than numerous changes occur after full-term birth. For example, the anatomy of the human ear canal continues to mature; in infants the canal is shorter and straighter than it is in adults (Northern and Downs, 1984). Another example is the increase in ear canal diameter and length during the first 2 years of life (Keefe et al., 1993). Finally, even more prolonged is the development of the middle-ear cavity volume, which extends into the late teenage to years, and is likely to influence the mechanics involved in middle-ear function (Eby and Nadol, 1986). For a full review of these and other changes, see Abdala and Keefe (2011). The postnatal maturation of the external and middle ear is important to note, because they can have substantial effects on how sounds are absorbed, processed, filtered, and transmitted to the auditory system.

Cochlear development in mammals generally proceeds from the basal region to the apex of the cochlea, that is, on a gradient that follows high-to-low-frequency organization (Pujol and Hilding, 1973). The mammalian cochlea contains two types of sensory cells (the outer and inner hair cells). The innervation from the eighth nerve consists of primarily two types of ganglion neurons (type I and type II).

In utero, hair cells are fairly mature from a morphologic standpoint some time in the third trimester, although synaptic connections with the auditory nerve likely continue to mature after birth. Although not fully mature, the developing auditory system enables response to sound *in utero*, as has been shown by studies that measure fetal movement or heart rate response to sound at around 25–27 weeks' gestation. Another estimate of the onset of hearing is 27–28 weeks, obtained with the ubiquitous electrophysiologic measures of functionally, the auditory brainstem response, which shows synchronous eighth-nerve activity and average brainstem responses (Galambos and Hecox, 1978). Little is known about maturation of the basilar membrane, a structure within the inner ear that serves as the anatomic base upon which the sensory hair cells are positioned. Functionally, the basilar membrane changes in mass and elasticity along its basal-to-apical distribution, resulting in a "place–frequency" mapping resulting from resonance dependence response to sound vibrations at different frequencies.

Peripheral maturation of the auditory system can be studied using a non-invasive and clinically relevant technique whereby otoacoustic emissions (OAEs) are recorded with a small sensitive microphone inserted in the entrance to the ear canal. OAEs reflect normal physiology of the outer hair cells, which act as a cochlear amplifier. OAEs are generally present at birth; however, numerous types of OAEs have been described, which vary by the region in the cochlea and mechanisms responsible for their functionality. OAEs have become a highly utilized tool for investigating maturation of cochlear function, and thus also for identifying immature and abnormal peripheral auditory function. At birth, in contrast with the immature outer and middle ear, the inner ear seems to be more mature, as characterized by OAEs. Studies generally suggest that newborn cochleae produce robust responses that are adult-like in many ways. Nuanced research into this area further suggests that maturational changes occur particularly in the apical region of the cochlea, where low-frequency information is generally processed. How these maturational changes influence perceptual outcomes is yet to be understood (Keefe and Abdala, 2011).

BEHAVIORAL TESTING AND PSYCHOACOUSTICS

Research reviewed in this chapter focuses on how the human auditory system develops, what aspects of sound are perceived by infants, children and non-human animals, and the ways in which those sounds are interpreted at higher levels of neural encoding. It is thus prudent to begin by understanding how perception can be measured reliably. In humans, over the lifespan, depending on the age of the infant/toddler/child, the testing methods that can be used vary dramatically. The testing methodologies used over the years were developed with specific goals in mind, focusing on the need for accurate reliable and replicable measurement of perception. Another issue is the need for bias-free assessment, within a reasonable testing period, given the short attention span of most young listeners. In the late 20th century, behavioral habituation and dishabituation methods were developed, whereby infants are exposed to a stimulus, and once they habituate to it they stop responding. Subsequent responses to novel stimuli represent a demonstrable and measurable indication that the infant has detected a change in sensory stimulation. This technique may ultimately underestimate sensitivity, because conditioned reflexes/responses subside after repeated presentation of novel stimuli. Nonetheless, basic responses have been measured, and are rather useful in underscoring development of auditory sensitivity in infancy. For example, by 35 weeks' gestation fetuses can apparently discriminate between 250-Hz and 500-Hz tones (Shahidullah and Hepper, 1994). Another approach, that of visual reinforcement audiometry (Moore et al., 1975; Trehub et al., 1981), leads to prolonged responses and may produce better and more reliable threshold estimates with repeated stimulation, because the infant has a motivation to respond. However, the reinforcement method only works once an infant is mature enough to engage in that task (~5 months of age).

In the 1980s Olsho and colleagues (1987) developed the observer-based psychophysical method (OBPM), to test infants younger than 5 months as well as older infants. The OBPM relies on bias-free ways of determining whether an infant hears or does not hear a given sound; similarly, whether an infant has heard a difference between two sounds. A researcher, i.e., "observer," is "blind" as to the stimuli being presented, and is trained to watch the infant during specific testing intervals. The observer determines whether, based on changes in the infant's behavior, a change in stimulus has occurred. The behaviors that are reasonable to include, in addition to overt head turn and eye gaze, are more subtle behaviors such as eye widening and brow movement. Although not often used with special populations such as persons with cognitive and/or physical disability, this approach has the strength in its objectivity, such that it could theoretically be adapted to any population. With normal-hearing infants, the OBPM can be used to determine thresholds for hearing sounds of various frequencies and other characteristics as early as 2–4 months' age postnatal (Olsho et al., 1987, 1988), and has been applied to testing of toddlers (Grieco-Calub et al., 2008; Grieco-Calub and Litovsky, 2010). However, in newborn infants the method is less reliable (Tharpe and Ashmead, 2001).

Finally, by the time that children reach 3 years of age or older, they can typically engage in tasks involving computerized, interactive platforms. Many of the studies in the past ~20 years with ages 3+ years have utilized these approaches to study auditory maturation in children. An obvious user-friendly approach is one whereby the child is given options for responding, known as the "forced-choice" methods. For example, if one is measuring the child's ability to perceive a difference between two tones, a two-alternative forced-choice method would provide the child with two types of stimuli and two response options. Experimenters vary the parameters that change and the size of change (large change, easier to hear vs small change, harder to hear). Litovsky (2005) used a four-alternative forced-choice procedure to test speech intelligibility, whereby four pictures of known words appeared on the computer screen and the child's task was to identify the picture that matched the word that was presented from a loudspeaker. By varying the sound intensity at which the word was presented, speech reception thresholds were then estimated, in quiet and in noise. Similar approaches have been used to study the development of sensitivity to important features of sound. These include detection of sounds at the lowest intensity possible, discrimination for stimuli varying in intensity and loudness, frequency and pitch, as well as sensitivity to duration of a stimulus and detection of a gap in a stimulus. Finally, studies focus on effects of masking on threshold for detecting stimuli, on how subjects group together sounds that share characteristics, and on sound localization.

Estimation of sensitivity and threshold in the world of psychoacoustics takes two primary approaches. The first is to design experiments that measure psychometric functions, whereby responses to many stimulus values are collected and percent correct or percent of trials with detection is computed, and from those functions, threshold is defined as the point on the function that meets some agreed-upon criteria that is different from chance (50%, 71%, or higher, depending on the experimental procedure). This approach provides abundant data at all stimulus values (e.g., low to high levels, or small to large differences between stimulus features being discriminated). However, this method can be quite laborious and may not be feasible with young listeners. An alternative, more efficient method involves an adaptive tracking algorithm, whereby the stimulus feature of interest changes following the response of the listener, and is adapted to "track" performance. Thus, when the stimulus values presented result in success, changes are implemented to bring the values to a more challenging value. In contrast, failure to hear a sound or to discriminate between two values of interest results in values that are less challenging. Much work has been done on identifying more efficient and revealing algorithms for achieving adaptive tracking goals. For further readings on this topic please refer to Macmillan and Creelman (2004).

In non-human species behavioral testing is somewhat similar to testing very young human listeners. The animal must be trained to respond to sounds and motivated to do so efficiently and effectively. Typically, studies with non-human animals involve situations in which the animal is deprived of food and/or water outside of the experimental setup, and the animal "earns" its nourishment by producing behavioral responses to stimulation. These studies are typically conducted with fully mature animals, rather than developing ones, because even the mature animal is often very difficult to train. In fact, in most cases, animals are trained for weeks, if not months, before the experimenter is convinced of their ability to perform on the task reliably.

It is fairly accurate to note that mature non-human animals behave much like very young humans. The common characteristic is that they do not share a spoken language with the experimenter through which instructions can be given. Thus behavioral training is an essential component of successful experimentation, and reliable results are ultimately dependent on the integrity with which the listener is trained to understand the task and is motivated to perform. To date, many animals have been studied behaviorally and a comprehensive review of these is outside the scope of this chapter. Finally, as is discussed below, auditory abilities emerge at various times during postnatal maturation. The estimate of the age at which these abilities reach adult level maturity is determined by the behavioral testing done. Table 3.1 summarizes these by indicating the age range during which development reaches adult-like maturity, for aspects of these developmental milestones.

Table 3.1

Summary of auditory development findings

	Infants <6 months	Infants <12 months	Toddlers <3 years	Preschool age	Early School age	Middle School age	Teenage years
High Frequency Detection	↔						
Low Frequency Detection					↔		
Intensity Discrimination					↔		
Frequency Discrimination		↔					
Loudness Perception		⇠⇢	⇠⇢	⇠⇢			
Simultaneous Masking					⇠⇢	⇠⇢	
Auditory Scene : Streaming						↔	↔
Co-modulation Masking Release				⇠⇢	⇠⇢	⇠⇢	
Informational Masking				⇠⇢	⇠⇢	⇠⇢	⇠⇢
Backward Masking					↔	↔	↔
Spatial Release from Masking			↔				
Contralateral Un-Masking				↔	↔	↔	↔
Minimum Audible Angle (MAA)					↔		
MAA with echo (precedence)							↔

Continuous arrows, age range during which perceptual ability is known to be mature; dashed arrows, age range during which perceptual ability is somewhat mature but additional research may be required.

CODING OF AUDITORY FEATURES

While behavioral responses to sound can be seen in infants as early as 28 weeks' gestation (Birnholz and Benacerraf, 1983), the ability of humans to perform on behavioral auditory tasks matures into the late teenage years (Maxon and Hochberg, 1982; Fischer and Hartnegg, 2004). As discussed below in greater detail, the auditory system undergoes protracted maturation, which is one of the mitigating factors in auditory development. However, non-sensory factors, including attention, cognition, and memory, are known to play a crucial role in maturation of auditory perception, and in the individual variability that is seen within age.

Detection of sound

Much of the early work on the stimulus level required to detect stimuli was done on auditory sensitivity in quiet, that is, in absence of noise or other masking stimuli such as speech. For example, it was well documented that thresholds for detecting stimuli in quiet improve between infancy and early school age (Olsho et al., 1988; Trehub et al., 1988). Infants can have thresholds up to 25 dB worse than adults; rapid improvement occurs before the age of 6 months (Tharpe and Ashmead, 2001), with a difference narrowing down to a 10–15 dB gap by the time that children reach 5 years of age. Differences in detecting sounds seem to also depend on the frequency, whereby sensitivity is near adult-like by age 6 months for high-frequency sounds of around 4000 Hz, but not for low-frequency sounds around 500 Hz (Olsho et al., 1988; Trehub et al., 1988). The physiologic basis for this difference has not been identified, and does not appear to result from maturation in conductive properties of the peripheral auditory system or in the sensory development that takes place during infancy and childhood. More likely is the fact that the tasks used to measure perception do not separate out effects of cognitive ability, motivation, memory and variability in neural representation of the stimuli, otherwise known as internal noise.

Frequency and intensity discrimination

Frequency discrimination, which refers to the ability to perceive a change in the frequency of tonal stimuli, undergoes considerable maturation between 3 and 6 months of age. Testing for this ability is done by presenting a train of tones, and switching to a tone of different frequency midstream, then noting whether the infants' behavior changed. While discrimination ability is poorer for 4000-Hz tones than 500-Hz tones at 3 months, at 6 months the pattern is reversed and infants are better at discriminating changes imposed on 4000-Hz tones than 500-Hz tones (Olsho et al., 1987). Remarkably, adult-like performance of discriminating ~1% change in the frequency is reached between 6 and 12 months of age for 4000-Hz tones; infants can discriminate changes of 2% in the frequency. At 4000 Hz that would mean that infants can perceive a change of 8 Hz (4000 Hz–4008 Hz). Interestingly, measures of frequency discrimination in children suggest that thresholds are higher than those seen in infants, but the tasks used with children are notably different, and likely place a heavier memory and attentional load on the child than the infant task (Moore et al., 2008). Thus, an important conclusion is that basic auditory abilities may appear to be highly developed in infancy, if tested using simple tasks that select for measuring sensation rather than attention or memory.

Intensity discrimination refers to the ability to detect a change in the level (in dB sound pressure level) at which a sound source is presented, and this ability is typically measured separately for narrow ranges of frequencies (tones or narrowband stimuli). Intensity processing is thought to reflect a change in the rate of firing of auditory neurons and possible "recruitment" of auditory nerve fibers that are anatomically neighboring those at which the stimulus is presented. Similar to frequency sensitivity measures, intensity discrimination can be tested by measuring the ability of a listener to detect a change from a background stimulus. Adult listeners can hear differences in intensity as small as 1–2 dB (Sinnott and Aslin, 1985), and, in contrast, infants between the ages of 5 and 7 months need approximately 6 dB difference. As far as stimuli are concerned, infants are particularly worse than adults for low-frequency stimuli (~400 Hz) than high-frequency stimuli (4000 Hz), and for stimuli that have a standard (sound to which changes are compared) that is low in intensity to begin with. There are clear methodologic issues here related to how exactly sensitivity is measured, and whether the task is difficult or much more subtle for infants. This issue is not exclusive to intensity discrimination, but may be appear to be more relevant because there are not many demonstrations of mature ability for intensity discrimination in infancy. In fact, it may be the case that maturation of perception for changes in intensity continues into childhood (Buss and Hall, 2009). It is possible that cognitive abilities, including memory, play an important role in the tasks used to date; however, tasks that lead to equal performance between very young listeners and adults have not been identified, suggesting that intensity discrimination is more immature than frequency discrimination early in life.

Loudness

Another aspect of auditory perception that relates to sound level or intensity is that of perceived loudness of a stimulus. Whereas discrimination, whereby we measure whether a listener was correct when s/he reported whether a stimulus changed or did not change, is objective, loudness perception is subjective. Any subjective perception is difficult to measure in listeners, including not only infants and children, but adults as well. Loudness is an attribute for a sound that places perception on a scale ranging from inaudible/quiet to loud/uncomfortable, in response to change in sound pressure level (intensity). Because there is no correct answer, there is some challenge in knowing when and how to reinforce a child and how to train the child to respond. Nonetheless, it appears that, while some children have difficulty learning the task, others can perform similarly to adults (Serpanos and Gravel, 2000). Because loudness growth is abnormal in people with hearing loss, such that loudness grows rapidly over a small range of intensities, emphasis on understanding the importance of loudness perception maturation may come from the audiologic literature, with hearing-impaired children. Evaluations of hearing-aid fittings and perceived loudness of speech signals after amplification become clinically crucial, so that the speech signal is heard, understood, and comfortably presented (e.g., Scollie et al., 2010; Ching and Dillon, 2013). Future work on basic psychophysics may be important in order to capture the perception of young infants and children with typical hearing, and benchmark their abilities, so that expectations are appropriate for children who are fitted with hearing aids.

MASKING AND AUDITORY SEGREGATION

Background on grouping and segregation

Natural auditory environments such as classrooms, home environments, entertainment areas, and other public space are typically complex in terms of sensory stimulation. In the auditory domain the listener is faced with a cacophony of information that varies in content, location, identity of the source, frequency, intensity, and other; these features are also typically dynamic rather than static, thus they are likely to change over time. It is the infant's or child's job to learn how to cope with these environments, to organize the onslaught of stimulation into perceptible nuggets of information, so that the child can learn about his or her environment, attain language capacity, and master numerous abilities that rely on auditory-based communication.

At the initial stages of auditory perception, as described earlier in this chapter, the auditory system masters the ability to detect and discriminate between sounds. A further stage of analysis requires that decisions be made regarding which sounds "belong together" and ought to be "grouped," and which sounds "do not belong together" and ought to therefore be "segregated." This issue is not so much about categorization, but rather this is ultimately about identification of sounds and extraction of meaning from sound sources. Categorization might be relevant when determining which sounds are emitted by humans vs nonhumans, or deciding whether sounds arise from musical instruments vs natural environmental sounds (a stream or the sound of the wind rustling leaves on a tree). In contrast, identification and information extraction require that the listeners be able to segregate a sound source from other sounds in the environment. For example, a child who is attempting to listen to her mother's voice in a crowded room must separate that voice of interest from other people's voices and other sounds in the environment. In some sense, the "other sounds" can be thought of as competitors or interferers; in the classic psychoacoustic literature they are also referred to as "maskers." The basic process through which the neural mechanisms involved in perception are able to parse out various sound sources and assign meaning to appropriate sound sources has been termed auditory scene analysis (ASA: Bregman, 1990). A more specific example of ASA that occurs for speech sounds is the cocktail-party effect (Cherry, 1953). Bregman's notion was consistent with two stages – one that is more "primitive" and does not require much learning or experience, and a second stage that is "schema-based" and is organized through each listener's experiences and learned abilities during a lifetime of listening. Most of the work done in the field in general has been with adult subjects. Developmental work in this area has primarily attempted to understand the "primitive" case for ASA, because the lack of dependence on experience means that we can begin to understand mechanisms involved in the organization of incoming stimuli that are existent in young listeners.

To the extent that these mechanisms exist at a young age they can potentially serve as the scaffolding upon which more complex schema-based organization of information can occur with experience. This issue of course lends itself to fascinating questions about ways in which deprivation of stimulation during development results in disruption of the primitive organizational structures and/or in ways in which schemas can in fact develop properly. Very little is known about ASA and effects of deprivation during development on the ability of listeners to organize incoming information once the auditory system has been reactivated.

One reason that little is known about this issue is that there are no examples of cases in which human listeners have undergone auditory deprivation early in life and have had natural hearing restored henceforth. The closest that we can get to a "natural" experiment of this sort is to study effects of auditory deprivation in people who are born deaf and who are able to receive auditory stimulation later in life through artificial means. Today's technologic advances offer the cochlear implant (CI), which provides hearing through electric stimulation of the auditory nerve (Wilson and Dorman, 2008). An important caveat is that the CI does not restore normal hearing to recipients; rather, the auditory signal is compromised by spectral degradation, poor specificity of stimulation along the tonotopic (frequency-dependent) mapping of the auditory system, and neural death following deprivation, to name a few limitations (Shepherd and Hardie, 2001; see Litovsky et al., 2012). Thus, interpretation of data on how deprivation might affect ASA are unavoidably compromised by the inability to know whether poor performance following auditory stimulation with a CI is attributable to the CI-related issues or to factors related to the deprivation *per se*.

Energetic masking

The concept of energetic masking was introduced when another form of masking, known as informational masking, became prominent in the literature (see below). Energetic masking is thought to be the interference produced by noise or other stimuli that can vary in characteristic, but that share features with the target sound, and can be accounted for by considering the peripheral auditory system. A simple example of energetic masking is when a noise burst occurs simultaneously with a tone, and the two-frequency bandwidth of the noise is centered around the frequency of the target zone. In energetic masking, one sound interferes with our ability to detect or otherwise hear another sound because the two sounds excite similar auditory neurons in the periphery, thereby limiting the extent to which information about the target can be perceived in the presence of the masker (Fletcher, 1940). Several studies have examined the ability of infants and children to detect a signal in background noise, either when the sounds are presented simultaneously, or when they are temporally offset. For simultaneous masking detection, thresholds of infants aged 6–24 months are higher than those of adults by 15–25 dB (e.g., Bull et al., 1981; Bargones et al., 1995), suggesting that their ability to extract a simple signal in background noise is worse than that of adults. Similar to the thresholds discussed above for detecting signals in quiet, there appears to be a more rapid maturation for hearing signals in noise, with high-frequency thresholds reaching adult-like performance sooner than low-frequency thresholds (e.g., He et al., 2010).

Auditory streaming

The field of perception has broadly defined relevant terms that address the fascinating problem of ASA. For example, auditory streaming refers to the notion that when listeners are presented with sounds that share some dimensions and vary along other dimensions, they perceive them either as one coherent sound, or as two distinct sounds. A classic example of auditory streaming is when two sequences of tones that vary in frequency are presented; the more similar the frequencies, the more likely it is that listeners will perceive a single, coherent auditory image. Auditory stream formation appears to depend on acoustic parameters, including frequency, spatial locations, sex of the talker, spectrum, and common temporal onsets/offsets (Bregman, 1990; Yost, 1997). Using a method whereby infants are habituated to a signal, then dishabituated to a novel stimulus, it has been shown that by the age of $\sim$4 months infants can use similar acoustic cues to those used by adults to perceive two stimuli as either belonging to a single stream or to two streams (McAdams and Bertoncini, 1997). It is important however to note that the sensitivity to these cues was not measured with these studies, thus it remains to be understood whether infants need the same amounts of cue similarity or difference in order for the effect to be operative.

In contrast to studies with infants, by the time that children are 4–5 years of age, methods for testing auditory perception can be fairly similar to methods used in adults. When children and adults were tested on their ability to hear one or two tone sequences, developmental changes have been observed between children who are 8 years old or younger and children ages 9–11 years. Older children, like adults, required small differences between the frequencies of the two tone sequences in order to perceptually segregate them into two "streams" whereas younger children required larger frequency differences in order to hear the tone sequences as segregated (Sussman et al., 2007). These findings suggest that ASA, as measured with auditory streaming, develops well into school-age years, as will be discussed below for a number of other auditory abilities.

Co-modulation masking release

Auditory grouping can also be studied using a completely different paradigm, known as co-modulation masking release (CMR), which was originally described by Hall et al. (1984). This fascinating perceptual phenomenon can be elicited under conditions when a target signal is masked by another sound that has intensity fluctuations known as amplitude modulations. When a masker at a frequency that is different from that of the target signal is added, and has the same temporally varying modulations in intensity, i.e., when the target signal and added masker are "co-modulated" in time, the two sounds are perceptually grouped and perceived as a single auditory image. This grouping leads to benefits in signal detection, i.e., reduction in listeners' thresholds for hearing the target signal. Studies on CMR in children suggest that, although children experience more masking in the presence of noise compared with adults, when the added co-modulated noise is introduced, they show similar amounts of benefit as adults. Thus, CMR is adult-like under certain conditions, suggesting that mechanisms involved in grouping under noisy conditions, and the ability to benefit from the temporally co-modulated noise, are developed by 4–5 years of age (Veloso et al., 1990; Grose et al, 1993). Other studies have shown that CMR can also undergo more protracted maturation, for example, when the target signal and noise are both modulated, but with a temporal asynchrony between them. In that scenario, CMR in children was eliminated, whereas in adults the effect was reduced relative to the condition with temporal synchrony (Hall et al., 1997).

Informational masking

An important concept in the context of how it is that we extract information in complex auditory scenes is that of informational masking. In contrast to energetic masking (see above), informational masking refers to masking that cannot be accounted for by peripheral mechanisms, for example when a tone is difficult to hear despite the fact that the masking noise is presented at a frequency region that is remote and that stimulates a different population of auditory neurons. A simple example of informational masking is when a series of "off-frequency" tones change in frequency through time in a manner that is unpredictable each time they are presented. Because the listener does not know which aspects of the spectral, temporal, or other features of the masker to ignore, uncertainty occurs, which results in increased threshold for the target tone. In other words, informational masking might simply indicate that the listener is confused about which features of the stimulus belong to the target and which to the masker (Durlach et al., 2003).

Studies on informational masking began to get attention 30–40 years ago (Watson et al., 1976; Neff and Green, 1987), when several paradigms were implemented using tone bursts to study the ability of adults with normal hearing to hear tones in the presence of multitonal background maskers, or "interferers," whose content was varied in an unpredictable manner. By randomizing the frequencies of the tones from presentation to presentation, researchers introduced uncertainty about what listeners should ignore. This was quite unlike, then traditional, studies with energetic masking, where the maskers are consistent across presentations, and listeners know what sound to ignore and which sound to attend to. Informational masking was defined then as the amount of masking observed, above and beyond that which was observed with energetic masking conditions. This paradigm resulted in demonstrating vast individual differences (Oh and Lutfi, 1998). In addition, there is an interesting non-monotonic relationship between the amount of masking and the number of tones, such that, with a small number of tones, as more are added masking increases, suggesting that the confusability or uncertainty may be related to attention or memory load. However, at some stage the increase in number of tones results in the tone complex being perceived as a noise with little information, hence the amount of masking begins to drop (e.g., Neff and Green, 1987; Oh and Lutfi, 1998).

This paradigm has been successfully applied to studies with infants and children and has led to important findings regarding auditory development. In general, studies suggest that infants and children are more vulnerable to effects of informational masking, that is, the amount of masking they experience is larger than that seen in adults (e.g., Oh et al., 2001; Leibold and Werner, 2006). Lutfi et al. (2003) investigated informational masking using tone bursts in 38 children aged 4–16 years, and focused on identifying potential sources of variability to account for different strategies that children might use to perform on this task. The authors concluded that, under easy-listening conditions, children generally utilize what is in the world of psychoacoustics known as an "optimal" decision strategy to solve this problem. However, in the informational-masking paradigm, differences between children existed at all ages, as is commonly found in other studies on auditory development. Moreover, these were thought to reflect the extent to which uncertainty in the masker affects the ability of the child to "focus in" on the signal. However, the authors did not discuss the possibility of utilizing this tool for ultimately measuring auditory-based, but not speech- or language-based, attention and memory in children.

Leibold and colleagues have also examined effects of uncertainty when masker spectra do not vary, and found developmental differences, suggesting that infants and children are perhaps generally poorer at segregating the target from masker, even when there is uncertainty regarding "what to ignore" (Leibold and Werner, 2006; Leibold and Neff, 2007). This clearly suggests that the conclusion regarding developmental effects depends on exactly how the experiments are set up and what the stimuli and tasks consist of. Nonetheless, the fact that developmental findings are clearly found suggests that there are reasons to be concerned about the ability of children to listen in complex, realistic acoustic environments, where sound sources vary in many features, including location, number, time, spectrum, and amplitude. More emphasis on effects of location and reflections is placed on sections below, where spatial hearing is discussed. In general, when it comes to speech sounds and the effects of informational vs energetic masking, again the issue revolves around the extent to which the masker is similar to the target, and thus the extent to which the two can be confused with one another. For example, the voice of one person can be easily confused with that of another if the two are similar, and if the content overlaps. In contrast, different voices, such as those of a young child (high pitch) uttering a nursery rhyme vs a male with low fundamental frequency talking about the stock market, are less likely to be confused.

Furthermore, there is the issue of whether content can be extracted from the maskers vs whether it sounds like noise. Newman (2005) found that infants can recognize known words in the presence of multitalker babble, which is more similar to noise and has less discernable content than speech, and in contrast they failed to do so in the presence of a single "masking" talker, perhaps because its content was more distracting and thus confusing. Finally, there is the added issue of whether the sounds are familiar to the infants. Barker and Newman (2004) found that when the target voice was their own mother (thus familiar), infants were better able to understand speech in the presence of background talkers compared with a situation in which the voice was not familiar. Performance is also measured by obtaining the signal-to-noise ratio (level in decibels of the target re: the masker) that is needed for the target to be understood. Compared with adults, infants generally require a higher signal-to-noise ratio than adults in order to recognize their name in the presence of a noise masker (Newman and Jusczyk, 1996). These level differences are generally higher for children than for adults (Elliott et al., 1979; Nittrouer and Boothroyd, 1990; Hall et al., 2002; Litovsky, 2005).

These studies are also related to the issues described above in relation to the cocktail-party effect, and raise the issue of whether children are simply poorer at selectively attending to the target stimulus in the presence of the masker. Some of the more interesting studies in this area are done with event-related potentials, which measure physiologic responses in the central auditory pathways. For example, when adults and children were each asked to attend to one of two passages read to them, the brain activity as measured with the N1 wave was not adult-like at age 6–8 years (Coch et al., 2005). While somewhat remote from neural mechanisms involved in the task, one might argue that lacking characteristics that are adult-like in brain waves may reflect the immaturity that is observed in behavioral studies. It is not surprising that neural maturity for central processing of complex information continues to mature beyond ages 6–8 years. Cortical maturation is known to extend into the late teenage years, and effects of experience are certainly bound to have significant effects on these maturational changes.

Backward masking and auditory maturation

As discussed below, one way to study masking is to present signal and noise at the same time and to vary few parameters. Thus, the interference from the noise can be easily accounted for by peripheral mechanisms. In contrast, backward masking is studied by presenting the signal before the noise; hence, a more central mechanism that weights information arriving from the signal and noise must be considered. In fact, it appears that, unlike simultaneous masking, which is adult-like by age 6–10 years, backward masking results in significantly worse performance in children than adults. It has been suggested that temporal resolution is involved in backward masking and is an ability that continues to refine during childhood (Hartley et al., 2000). The physiologic mechanisms underlying these developments are not well understood, but it has been suggested that peripheral frequency resolution cannot explain these findings (Spetner and Olsho, 1990).

Backward-masking paradigms have been ubiquitous in studies with populations of children who demonstrate language delays and reading disorders. It appears that maturational delays occur when the brain processes temporal information. For example, there appear to be delays in maturation on tasks that measure backward masking and on other masking paradigms in which there are delays between the signal and masker. It has been argued that language deficits may result from limitations in auditory processing (Wright et al., 2000), which may have implications for diagnosis and/or remediation in children with language impairment or other disorders with a related underlying mechanism. Wright et al. (1997) argued that brains of individuals with language delays and reading disorders develop at a slower rate than brains of their unaffected peers. They honed in on the age of $\approx$ 10 years, when brain changes associated with onset of puberty could affect perceptual development. Numerous other studies have identified interactions between delayed brain maturation and perceptual and cognitive abilities and suggest that puberty may be an important factor contributing to learning problems (Doupe and Kuhl, 1999; Bourgeois et al., 2000).

Finally, the field of rehabilitation, training, and remediation for those affected with learning disorders is robust and the treatment options are beyond the scope of this chapter. One interesting point, however, is whether effects of treatment or training can be captured in the neural representation of cues that are important for perception, including speech cues. There appears to be some evidence for short-term training effects on improving neural representations of auditory cues that are known to be important for understanding speech in noise (i.e., masked speech). This evidence comes from measurements of biologic responses at the level of the brainstem (Song et al., 2012).

SPATIAL AND BINAURAL HEARING

Some of the most robust effects from listening in noise, and the ability to overcome challenges that arise in noise, are observed when listeners have the capacity to utilize input arriving at both ears. Thus the cocktail-party effect has been widely studied in relation to what are known as binaural and bilateral benefits. Related to these benefits is the ability to locate where sounds are coming from, which also depends greatly on acoustic cues known as binaural cues. Finally, as discussed below, the attainment of a normally developed sound localization system depends on the integrity of the auditory system's ability to extract spatial cues and present them to the auditory system with fidelity.

Binaural cues

In order to consider the development of spatial and binaural hearing, one must first consider the acoustic cues that arise when sound sources occur from various locations in the environment. Sources arriving from the side will reach the two ears with differences in time of arrival and intensity. For example, a sound at 90° to the right will produce a difference of ~0.7 ms in interaural timing, otherwise known as the ITD. This cue is robust at low frequencies, typically below 1500 Hz. At high frequencies the head creates an acoustic "shadow," so that the near ear receives a greater intensity than the far ear; sounds are thus received at the two ears with interaural level differences, otherwise known as ILDs, and these can be as large as 20 dB. For sounds that are amplitude-modulated, such as speech, ITD cues are also available from differences in the timing of the envelopes (slowly varying amplitude) of the stimuli. Figure 3.1 provides a schematic of the directionally dependent cues

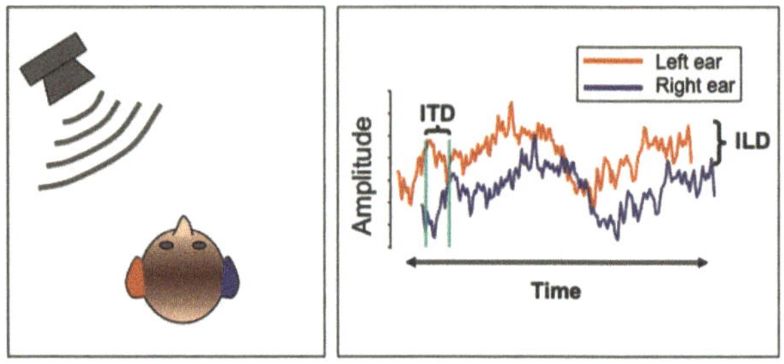

Fig. 3.1. Schematic of the directionally dependent cues that would be potentially available to listeners in the horizontal plane for a broadband sound. The left panel shows the sound emitted from a loudspeaker and arriving at the left ear first and with greater intensity. The right panel shows the measurements at the ear canal of the two ears; both interaural time difference (ITD) and interaural level difference (ILD) cues are present. (From "Bilateral Cochlear Implants in children" by Ruth Y. Litovsky and Jane Madell in *Clinical Management of children with cochlear implants* (P. 61) by Laurie S. Eisenberg. Copyright © 2009 Plural Publishing Inc. All rights reserved. Used with permission).

that would be potentially available to listeners in the horizontal plane for a brief signal such a click.

Measuring localization

Sound localization tasks reflect the extent to which an organism has a well-developed spatial-hearing "map," a perceptual representation of where sounds are located relative to the head. This ability has been studied in adult listeners for decades, with a focus on the extent to which listeners can either identify the exact position of a sound or discriminate changes in the location of sounds. In the former case, localization accuracy has been measured by asking listeners to point to source locations, either by using a pointer object, directing their nose or head in that direction, or labeling source locations using numeric values that refer to angles in the horizontal and/or vertical dimension. Accuracy depends on factors such as the task, instructions to listeners, stimulus frequency content and duration, and response options. Error rates are often quantified by the root-mean-square (RMS) error, an estimate of the deviation of responses from the true source location. Results from some studies suggest that RMS errors can be as small as a few degrees (Middlebrooks and Green, 1991).

With children, responses regarding perceived source location are challenging to obtain, and few studies on this topic exist. However, similarly to studies on other aspects of auditory signals described above, localization prediction has been measured in infants and children using the discrimination paradigm. A common measure is the minimum audible angle (MAA), the smallest change in the angular position of a sound source that can be reliably discriminated (Mills, 1958; Litovsky and

Macmillan, 1994). It is important to note that the ability of a listener to discriminate small changes in source location does not automatically generalize to having acute sound localization abilities (Hartmann and Rakerd, 1989; Grieco-Calub and Litovsky, 2010).

Developmental findings on localization

Newborn infants orient towards the direction of auditory stimuli within hours after birth. This head-orienting response is unconditioned, or reflexive, and seems to enable the infant to bring visual targets into view and to integrate auditory and visual information. At approximately 3–5 months of age the conditioned head-turn behavior emerges, whereby the infant's response can be shaped such that a reinforcing stimulus is associated with the behavior (Muir et al., 1989). Figure 3.2 shows the developmental trend in right–left discrimination ability, as measured with head-orienting responses. Data represent MAA thresholds from several studies in which the sound intensity was held constant from trial to trial. From these studies, one can conclude that the largest decrease in MAA occurs between 2 months of age and 2 years of age, with continued improvement through 5 years of age, when children's thresholds are similar to those of adults (see Litovsky, 1997).

Children between the ages of 4 and 10 years can localize sounds with average error rates ranging from <5° to >30°. Grieco-Calub and Litovsky (2010) found that

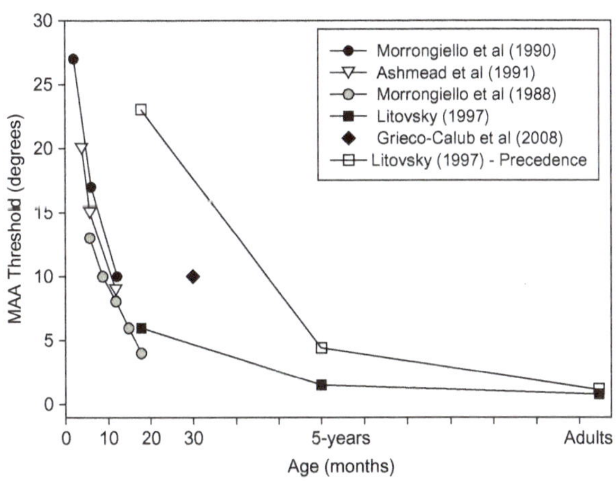

Fig. 3.2. Developmental trend in right–left discrimination ability, as measured with head-orienting responses. Data represent minimum audible angle (MAA) thresholds from several studies in which the sound intensity was held constant from trial to trial. From these studies, one can conclude that the largest decrease in MAA occurs between 2 months and 2 years of age, with continued improvement through 5 years of age, when children's thresholds are similar to those of adults. All studies by one used stimulus levels that were fixed. The diamond at 30 months represents data collected with stimulus levels roving an 8 dB range. (From Litovsky, 2011, with permission from Springer Science and Business Media).

RMS errors ranged from 9° to 29° (average of 18.3° ± 6.9° SD). RMS errors were generally smaller in two other studies. Litovsky and Godar (2010) reported RMS errors ranging from 1.4 to 38° (average of 10.2° ± 10.72° SD), which overlapped with that observed with adults. Van Deun et al. (2009) found that RMS errors averaged 10°, 6°, and 4° for children aged 4, 5, and 6 years, respectively. The question is whether sound localization using this type of source location identification paradigm can also be measured in very young children. Recent work by Litovsky and colleagues suggests that 2-year-old children can respond to sounds arriving from one of nine locations (Ehlers et al., 2013). Interestingly, most children identify locations correctly on >95% of trials (RMS errors <10°), while a small number of children identify incorrectly more often (RMS errors nearly 30°). These findings are summarized in Figure 3.3.

Across these studies it is also clear that children vary in their ability to perform tasks of absolute localization. While some children's performance is well developed by age 4–5, fairly similar to that of adults, other children find the task to be difficult and display large errors in location identification. The extent to which this finding reflects individual differences in sensory as opposed to non-sensory factors, as discussed above for other findings in perceptual development, needs to be further explored. It is unlikely that children would differ in the maturation of the sensory apparatus. Non-sensory factors are more likely to be responsible for individual differences or age-related differences. Although the studies have not been designed to look at this, children are likely to have moments of inattention that add variance to the data.

There is evidence from the animal literature to suggest that the auditory cortex plays a key role in determining the ability of an animal to localize, and to learn to localize or to relearn novel maps of space. Studies in which the cortex was ablated, conducted over 40 years ago, led to fairly impaired performance, but the damage to the cortex was difficult to control (Whitfield et al., 1972). More recently the ferret has been used as an

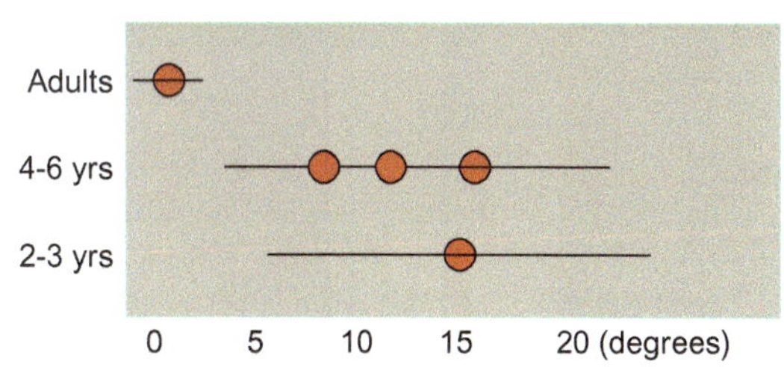

Fig. 3.3. Root-mean-square (RMS) error values collected in various studies are shown here, with age on the *y* axis and mean (±range) of RMS values measured. This is intentionally vague to represent the range of abilities tested, because the measures depend greatly on the exact stimulus used, loudspeaker configuration and task.

animal model of the effects of unilateral occlusion on spatial hearing. Most notable are two findings. First is that lesions of the auditory cortex (the primary auditory cortex (A1) in particular) lead to a reduction in the ability of animals to relearn spatial hearing maps (Nodal et al., 2010, 2012). Second, their behavior improves most dramatically with training and feedback (Kacelnik et al., 2006), suggesting again that non-sensory factors are likely to be involved in the emergence and preservation of spatial hearing maps.

Binaural unmasking

In this section, the issues related to complex auditory environments and ASA are reintroduced. However, the focus here becomes the availability of spatial and binaural cues for segregating target speech from competing speech and/or noise. Normal-hearing listeners rely on a complex set of auditory computations that involve both monaural and binaural processes and that depend on features of the competing sounds (Hawley et al., 1999, 2004; Bronkhorst, 2000; Culling et al., 2004). Spatial cues play a key role in facilitating source segregation: speech intelligibility improves when the target speech and competing sounds are spatially separated, resulting in spatial release from masking (SRM) (Plomp and Mimpen, 1981; Hawley et al., 1999, 2004; Arbogast et al., 2002; Bronkhorst, 2000). Average data from several studies are shown in Figure 3.4; these values can be quite large (10–12 dB) or relatively small SRM (3–5 dB), depending on the stimulus of choice, type of masker, and task. The larger effects seem to occur when the competing sound, or "masker," and target can be easily confused, and when listeners are unsure as to what aspects of the masker to ignore (i.e., informational masking: see above). The magnitude of SRM with speech-on-speech masking can be quite large relative to noise maskers (Durlach et al., 2003; Jones and Litovsky, 2008). The magnitude of SRM can also be divided into binaural and monaural components (Hawley et al., 2004). Binaural-dependent SRM occurs when target and maskers differ in ITDs and/or ILDs, whereas SRM can also occur under conditions in which the monaural "head shadow" occurs.

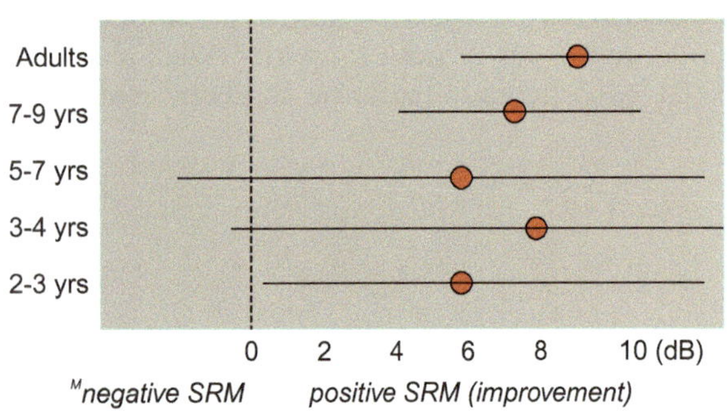

Fig. 3.4. Range of spatial release from masking (SRM) data collected from numerous studies is shown. Age on the *y*-axis and mean (±range) of SRM values measured with numerous studies is shown. This is intentionally vague to represent the range of findings, which depend greatly on the stimulus used, choice of words, and sentences, and whether informational or energetic masking occurred.

Litovsky (2005) first demonstrated SRM in children aged 4–7. Target stimuli were presented from the front at 0°. Maskers were presented from locations that were either co-located with the target, or spatially separated from the target. For children, a four-alternative forced-choice task is implemented in order to enable children to indicate what they hear rapidly and reliably. In the co-located condition children experience masking that is greater than that seen in the separated condition. The difference between the two conditions, or SRM, represents the ability of the listener to segregate sources based on location. SRM reported by Litovsky (2005) was 5–7 dB improvement in speech reception threshold. This test is an excellent tool for delineating children who are unable to utilize spatial cues for source segregation, as demonstrated by Misurelli and Litovsky (2012) with CI users. Garadat and Litovsky (2007) were able to show the effects in 3-year-old children, and more recently, Hess (2013) found SRM that was fairly mature in 2-year-old children. One key factor is to select target words that are within the receptive language and vocabulary of the children being tested.

Finally, as was discussed above in regard to informational masking, SRM is largest when target masker stimuli are more easily confusable with one another, such as if they comprise the same sex vs different sex talkers (Johnstone, 2006; Misurelli et al., 2013). Source segregation has also been studied over headphones, with a focus on non-speech stimuli. Thus, a related measure of the binaural masking level difference (BMLD) provides vital information regarding separation of a target signal such as a tone or narrow band from masking noise. BMLD can be measured, for example, by comparing threshold for tone detection when: both the noise and tone are in phase at the two ears – the N_0S_0 condition – and when the noise is in phase at the two ears but the tone signal is out of phase at the two ears – the N_0S_π condition. Presumably, the tone and noise are perceived as co-located intracranially in the N_0S_0 condition, while they are perceived as spatially separated in the N_0S_π..

The difference in threshold between these conditions ranges from 8 to 30 dB, depending on the specific condition (Zurek and Durlach, 1987; van de Par and Kohlrausch, 1999). Nozza et al. (1988) reported BMLDs for speech stimuli of 5, 8, and 10 dB for infants, 4-year-olds and adults, respectively, and suggested that there are developmental effects due to changes over time in absolute sensitivity to signals presented in masked conditions. Other studies, with children between the ages of 4 and 12 years, have shown that BMLDs are similar in children and adults for wideband noise bursts. However, there are age effects when narrowband noise is used, such that children ~4 years old are generally slightly worse than older children or adults (~7–12 dB depending on the noise bandwidth; Grose et al., 1993; Van Deun et al., 2009). It is possible that BMLDs are related to the ability of the listener to integrate information through time, and that this temporal window develops with age. Thus, adults might be more adept than children at utilizing smaller and more optimal temporal windows during which they can listen and optimize the ability to extract information from relevant stimuli. In contrast, children may be less able to focus their listening efforts over a small time interval, which effectively leads to integration of information from stimuli that fall into these larger time windows (Hall et al., 2007).

The precedence effect

In reverberant environments sound arrives at the listener's ears through a direct path, which is the most rapid and least disturbed path. Reflections of the sound from nearby surfaces, including walls and various objects, reach the ears with a time delay, and offer their own set of localization cues (Fig. 3.5A). In reverberant rooms, although listeners are aware of the presence of reflections, localization cues carried by reflections are de-emphasized relative to the cues carried by the source. This phenomenon is commonly attributed to auditory mechanisms that assign greater weight to the localization cues belonging to the preceding or first-arriving sound, henceforth referred to as the precedence effect (PE) (Blauert, 1997; Litovsky et al., 1999). While the PE is not a direct measure of people's ability to function in realistic acoustic environments, it provides an excellent tool for exploring how the auditory system assigns greater weight to the location associated with the source.

As is illustrated in Figure 3.5B, stimuli in PE studies typically contain a simulated, single source and a simulated single "echo" or reflection, known as the lagging sound. At short delays the lead and lag perceptually fuse into a single auditory percept; when the delay is between 0 and 1 ms, summing localization occurs, whereby both lead and lag contribute to the perceived location of the fused image. As the delay increases to 1 ms and beyond, the location of the lead dominates the perceived location of the fused auditory image, a phenomenon that has become known as localization dominance (Litovsky et al., 1999). The delay at which the lead and lag break apart into two auditory events is known as fusion echo threshold. Another way of quantifying the extent to which the directional cues from the lag are available to the listener is to measure discrimination suppression, whereby the listener discriminates changes in the location, or interaural parameters related to the lag. As delays increase, the ability of the listener to extract directional cues from the lag improves, indicating that discrimination suppression diminishes with delay in a similar fashion to reduced fusion.

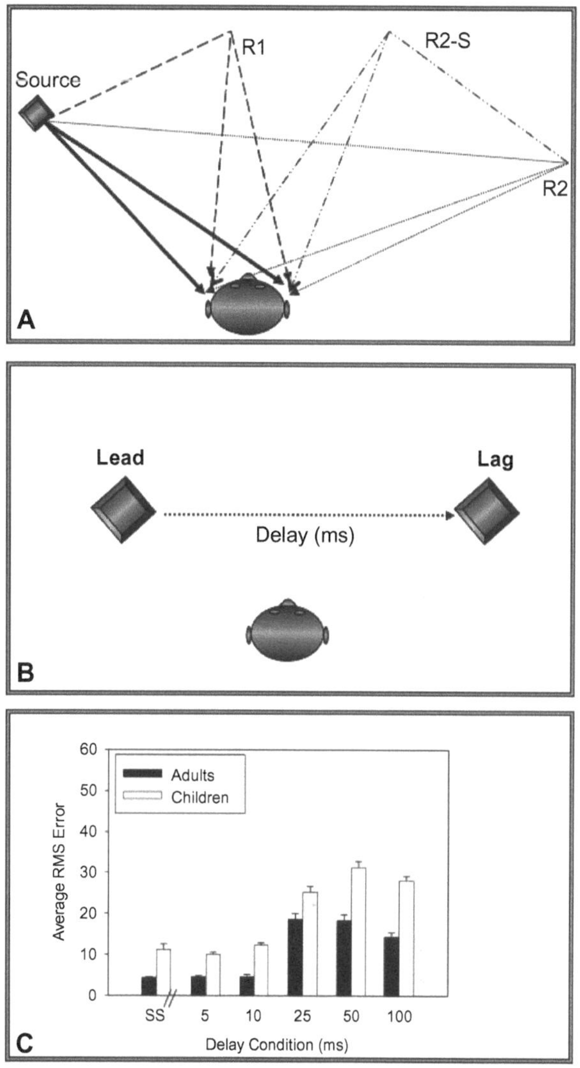

Fig. 3.5. (**A**) Rendition of the source and reflections that a listener might encounter in a room. Sounds reach the ears through a direct path and through reflections from the walls. (**B**) Stimuli used in precedence effect experiments, which typically contain a simulated source (lead) and a single "echo" or reflection (lag), thus a much more simplistic version than the ones shown in (**A**). (**C**) RMS, root mean square. Data from Litovsky and Godar (2010). (Reprinted with permission from Litovsky RY and Godar SP. Difference in precedence effect between children and adults signifies development of sound localization abilities in complex listening tasks. J Acoust Soc Amer 128 (4) 1979–1991. Copyright (2010), Acoustic Society of America.)

In newborns, the use of the reflexive unconditioned head turns has been used, with the lead stimulus to one side and the lag towards the other side; a listener who has functional PE would hear the image on the side containing the lead – "localization dominance." As reported by Clifton et al. (1981), infants do not orient their heads towards the location of the leading source, suggesting that lagging sounds or "unheard echoes" can exert a strong influence on directional hearing in young listeners. At age 4–5 months localization dominance seems to have developed (Muir et al., 1989); however, the strength of echo suppression continues to mature into middle to late childhood. Using MAA to measure localization acuity in young children, Litovsky (1997) found that the presence of a lagging sound at delays below echo threshold (i.e., when it is not heard as a separate sound) degraded localization acuity significantly. This can be seen in Figure 3.2, where the data from Litovsky (1997) in the PE condition are plotted along those from single-source measurements. The fact that MAA thresholds in PE conditions exceed those of single-source conditions, for the children but not adults, is an indication of the maturation of neural circuits involved in mediating perceptual weighting of sources and their reflections. Thus, there appears to be an immature ability in young children to suppress directional information from a delayed sound that is not perceived as a separate auditory event. In absence of the echo, MAA acuity is mature by age 5 years. Spatial hearing acuity in children is therefore vulnerable to degradation by virtue of the fact that the auditory system is unable to dismiss directional cues carried by the delayed stimulus.

More recently, Litovsky and Godar (2010) studied localization dominance (the extent to which the leading source dominates sound localization) in children aged 4–5 years and adults. Figure 3.5C shows localization data with RMS errors for localizing the lead, or lag. RMS errors for lead localization were fairly similar to errors observed with single-source stimuli, when the lead lag delays were short, suggesting that the presence of an "unheard echo" does result in worse sound localization. As delays increased, RMS errors for the lead increased. That is, lead localization worsened, for both adults and children, suggesting that a "heard echo" disrupts localization. In addition, RMS errors were consistently higher in children than adults, suggesting that their ability to suppress or ignore directional cues from the lagging source was not as fully developed as that of adults. The implication of this finding is that in reverberant environments children may have more difficulty than adults in perceptual weighting of spatial cues arriving from sources and their reflections. What factors could account for immature PE in children and infants? Sensitivity to simple (i.e., single-source) configurations can be modeled by invoking fairly low-level processing in the binaural pathway (Xia et al., 2010), whereas localization of PE stimuli requires higher-level (central) processes. Neural inputs that are initially involved in establishing the PE probably exist at peripheral (Hartung and Trahiotis, 2001) and brainstem (Litovsky and Yin, 1998; Litovsky and Delgutte, 2002) levels. However, more central (cortical) mechanisms seem to be necessary for mediating behavioral components associated with the PE (Litovsky, 1998). Finally, it is important to keep in mind the fact that simple stimulus scenarios described here and used in studies of the PE do not represent the "real world" because here only a single reflection is simulated. In addition, in PE studies the reflection has a similar intensity to that of the original sound source. However, the disruption to localization by a single simulated echo is indicative of the potential difficulties that children might experience in realistic, reverberant listening situations.

RELATIONSHIP BETWEEN AGE OF DEVELOPMENT AND DESIRABLE AGE OF INTERVENTION IN DEAF CHILDREN

The primary purpose of providing hearing aids and/or CIs to young children is to habilitate language acquisition, to provide the signal processing necessary for hearing speech in noise, and to provide each individual with the skills needed to function in a mainstream environment. The poignant question thus becomes: "what is the age at which auditory stimulation can minimize loss of auditory function and maximize the ability of the individual to access sound?" In other words, the concern is whether there exists a "critical period" for providing sound, beyond which the child will not be able to maximize his or her potential. On the other hand, even with significant hearing loss and lack of recovery of auditory function, children can function very well in an auditory environment. The answer is complex, because the outcomes in each individual can be quite different, and depend on factors that are not always well understood. The acclimatization strategies vary, and include multimodal use of speech cues, such as lip reading; top-down processing that enables hearing-impaired people to fill in the gaps where speech information is blurred; and numerous other possible strategies.

It is well known that the auditory system can undergo significant changes due to lack of sensory input, resulting in loss of function. This degeneration due to lack of stimulation occurs both peripherally and centrally (Shepherd and McCreery, 2006). At fairly peripheral levels in the auditory system there is known to be degradation in size and function of neural ganglion cells following a prolonged period of auditory deprivation (Leake et al., 1999). Profound deafness in the early developmental period seems to result in loss of normal tonotopic organization of primary auditory cortex, although there is some reversal following reactivation of afferent input (e.g., Kral et al., 2009). In addition to neural degeneration there is reassignment following deafness (Kawano et al., 1998). There is some evidence to suggest that outcomes in deaf individuals are maximized if stimulation is provided at an early stage in language acquisition, i.e., by approximately 18 months of age (e.g., Niparko et al., 2010).

Regarding spatial hearing mechanisms, there is important application in the area of single-sided deafness, since unilateral hearing loss is prevalent amongst children. In bilaterally deaf children, the issues raised above regarding bilateral implantation become relevant as clinicians determine whether children should receive a second CI at a young age or hold off until the child is older. From research in animal models (e.g., Kral et al., 2013) and humans (e.g., Gordon et al., 2013), there is evidence that early auditory deprivation in one ear, and prolonged stimulation in the other ear, results in aural preferences that are perhaps irreversible. Auditory plasticity though has two important ingredients. First, the level within the auditory system at which studies are conducted is crucial. Peripheral mechanisms are in general more susceptible to deprivation, and less likely to recover function. In contrast, central auditory mechanisms are susceptible to deprivation, likely to show dysfunction due to hearing loss, but also more amenable to plasticity and some recovery of function. Thus, while auditory deprivation can lead to damage in the structures responsible for early stimulus analysis, cortical development follows more iterative interactions and depends on stimulus-driven learning. This means that there is tremendous potential for recovery of some function with training and habilitation. However, it is important to bear in mind the fact that early access to sound does not guarantee good auditory function, because the brain mechanisms must be stimulated and utilized in the growth and learning process (Flexer, 2011).

Finally, windows of developmental plasticity, and ages beyond which the nervous system loses robust responses, are much easier to study with control in non-human species. In humans, studies are conducted retrospectively, by examining children after they have been treated clinically. Thus, combined efforts across the non-human and human approaches can be instrumental in determining best outcomes for children in the long run.

CONCLUSIONS

In this chapter we considered auditory development in the context of the need of humans to achieve auditory perception with fidelity, and interact and learn in a complex world, surrounded by multiple sounds that arrive from different locations. The auditory system reconstructs the perceptual space, groups together some objects, and segregates sounds from one another. In order to engage in an analysis of the auditory world, features of sounds are extracted, and the development of these abilities is staggered throughout development. Auditory development is a broadly defined term, and many approaches have been taken to study this fascinating topic. Here we focused on behavioral studies, which depend on extracting information from listeners who are performing psychoacoustic tasks. These approaches provide a window into the extent to which underlying structure and function support perception. To a large extent, the discussion here focuses on the fact that both sensory and non-sensory, cognitive or listening strategy-driven factors account for developmental changes, and differences between children and adults. Thus interpreting developmental "delays" must be done with caution, and must allow for separation of sensory and non-sensory factors.

ACKNOWLEDGMENT

The author is very grateful to students, postdoctoral fellows, and collaborators whose participation in the cited studies was important to the success of this chapter. The author received support for her work from the National Institutes of Health (R01 DC 003083 and 5R01 DC 008365). The research in the Litovsky lab is also supported in part by a Core Grant to the Waisman center from the NIH-NICHD (P30 HD 003352).

References

Abdala C, Keefe DH (2011). Morphological and functional ear development. In: LA Werner, RR Fay, AN Popper (Eds.), Springer Handbook of Auditory Research, Springer-Verlag, New York, pp. 19–57.

Arbogast TL, Mason CR, Kidd Jr G (2002). The effect of spatial separation on informational and energetic masking of speech. J Acoust Soc Am 112 (5 Pt 1): 2086–2098.

Bargones JY, Werner LA, Marean GC (1995). Infant psychometric functions for detection: mechanisms of immature sensitivity. J Acoust Soc Am 98 (1): 99–111.

Barker BA, Newman RS (2004). Listen to your mother! The role of talker familiarity in infant streaming. Cognition 94 (2): B45–B53.

Birnholz JC, Benacerraf BR (1983). The development of human fetal hearing. Science 222 (4623): 516–518.

Blauert J (1997). Spatial Hearing: The Psychophysics of Human Sound Localization, revised edn. MIT Press, Cambridge, Massachusetts.

Bourgeois JP, Goldman-Rakic PS, Rakic P (2000).In: MS Gazzaniga (Ed.), The New Cognitive Neurosciences, 2nd edn. MIT Press, Cambridge, MA, pp. 45–53.

Bregman AS (1990). Auditory scene analysis: the perceptual organization of sound. MIT Press, Cambridge, MA.

Bronkhorst AW (2000). The cocktail party phenomenon: a review of research on speech intelligibility in multiple-talker conditions. Acustica 86: 117–128.

Bull D, Schneider BA, Trehub SE (1981). The masking of octave-band noise by broad-spectrum noise: a comparison of infant and adult thresholds. Percept Psychophys 30 (2): 101–106.

Buss E, Hall 3rd JW (2009). Effects of masker envelope coherence on intensity discrimination. J Acoust Soc Am 126 (5): 2467–2478.

Cherry EC (1953). Some experiments on the recognition of speech, with one and with two ears. J Acoust Soc Am 25: 975–979.

Ching TY, Dillon H (2013). Major findings of the LOCHI study on children at 3 years of age and implications for audiological management. Int J Audiol 2013 (52 Suppl 2): S65–S68.

Clifton RK, Morrongiello BA, Kulig JW et al. (1981). Newborns' orientation toward sound: Possible implications for cortical development. Child Dev 52: 833–838.

Coch D, Sanders LD, Neville HJ (2005). An event-related potential study of selective auditory attention in children and adults. J Cogn Neurosci 17 (4): 605–622.

Culling JF, Hawley ML, Litovsky RY (2004). The role of head-induced interaural time and level differences in the speech reception threshold for multiple interfering sound sources. J Acoust Soc Am 116: 1057–1065.

Doupe AJ, Kuhl PK (1999). Annu Rev Neurosci 22: 567–631.

Durlach NI, Mason CR, Shinn-Cunningham BG et al. (2003). Informational masking: counteracting the effects of stimulus uncertainty by decreasing target-masker similarity. J Acoust Soc Am 114 (1): 368–379.

Eby TL, Nadol Jr JB (1986). Postnatal growth of the human temporal bone. Implications for cochlear implants in children. Ann Otol Rhinol Laryngol 95 (4 Pt 1): 356–364.

Ehlers E, Hess C, Litovsky R (2013). Spatial hearing abilities in toddlers with bilateral cochlear implants. Presented at the 12th Annual Early Hearing Detection and Intervention Meeting, Glendale, AZ.

Elliott LL, Connors S, Kille E et al. (1979). Children's understanding of monosyllabic nouns in quiet and in noise. J Acoust Soc Am 66 (1): 12–21.

Fischer B, Hartnegg K (2004). On the development of low-level auditory discrimination and deficits in dyslexia. Dyslexia 10 (2): 105–118.

Fletcher H (1940). Auditory patterns. Rev Mod Phys 12: 47–65.

Flexer C (2011). Cochlear implants and neuroplasticity: linking auditory exposure and practice. Cochlear Implants Int 12 (Suppl 1): S19–S21.

Galambos R, Hecox KE (1978). Clinical applications of the auditory brain stem response. Otolaryngol Clin North Am 11 (3): 709–722.

Garadat SN, Litovsky RY (2007). Speech intelligibility in free field: spatial unmasking in preschool children. J Acoust Soc Am 121 (2): 1047–1055.

Gordon KA, Wong DD, Papsin BC (2013). Bilateral input protects the cortex from unilaterally-driven reorganization in children who are deaf. Brain 136 (Pt 5): 1609–1625.

Grieco-Calub T, Litovsky RY (2010). Sound localization skills in children who use bilateral cochlear implants and in children with normal acoustic hearing. Ear Hearing 31 (5): 645–656.

Grieco-Calub TM, Litovsky RY, Werner LA (2008). Using the observer-based psychophysical procedure to assess localization acuity in toddlers who use bilateral cochlear implants. Otology Neurotology 29 (2): 235–239.

Grose JH, Hall 3rd JW, Gibbs C (1993). Temporal analysis in children. J Speech Hear Res 36 (2): 351–356.

Hall JW, Haggard MP, Fernandes MA (1984). Detection in noise by spectro-temporal pattern analysis. J Acoust Soc Am 76 (1): 50–56.

Hall 3rd JW, Grose JH, Dev MB (1997). Auditory development in complex tasks of comodulation masking release. J Speech Lang Hear Res 40 (4): 946–954.

Hall 3rd JW, Grose JH, Buss E et al. (2002). Spondee recognition in a two-talker masker and a speech-shaped noise masker in adults and children. Ear Hear 23 (2): 159–165.

Hall 3rd JW, Buss E, Grose JH (2007). The binaural temporal window in adults and children. J Acoust Soc Am 121 (1): 401–410.

Hartley DE, Wright BA, Hogan SC et al. (2000). Age-related improvements in auditory backward and simultaneous masking in 6- to 10-year-old children. J Speech Lang Hear Res 43 (6): 1402–1415.

Hartmann WM, Rakerd B (1989). On the minimum audible angle – a decision theory approach. J Acoust Soc Am 85: 2031–2041.

Hartung K, Trahiotis C (2001). Peripheral auditory processing and investigations of the "precedence effect" which utilize successive transient stimuli. J Acoust Soc Am 110: 1505–1513.

Hawley ML, Litovsky RY, Colburn HS (1999). Speech intelligibility and localization in complex environments. J Acoust Soc Am 105: 3436–3448.

Hawley ML, Litovsky RY, Culling JF (2004). The benefit of binaural hearing in a cocktail party: effect of location and type of interferer. J Acoust Soc Am 115: 833–843.

He S, Buss E, Hall 3rd JW (2010). Monaural temporal integration and temporally selective listening in children and adults. J Acoust Soc Am 127 (6): 3643–3653.

Hess CL (2013). Speech discrimination and spatial hearing in toddlers with bilateral cochlar implants. PhD Dissertation, University of Wisconsin-Madison.

Johnstone PM (2006). Informational Masking and Spatial Asymmetry in a "Cocktail Party" Environment. University of Wisonsin-Madison, Unpublished Dissertation.

Jones GL, Litovsky RY (2008). Effects of uncertainty in a cocktail party environment in adults. J Acoust Soc Am 124: 3818–3830.

Kacelnik O, Nodal FR, Parsons CH et al. (2006). Training-induced plasticity of auditory localization in adult mammals. PLoS Biol 4 (4): e71.

Kawano A, Seldon HL, Clark GM et al. (1998). Intracochlear factors contributing to psychophysical percepts following cochlear implantation. Acta Otolaryngol 118: 313–326.

Keefe DH, Abdala C (2011). Distortion-product otoacoustic-emission suppression tuning in human infants and adults using absorbed sound power. J Acoust Soc Am 129 (4): EL108–EL113.

Keefe DH, Bulen JC, Arehart KH et al. (1993). Ear-canal impedance and reflection coefficient in human infants and adults. J Acoust Soc Am 94 (5): 2617–2638.

Kral A, Tillein J, Hubka P et al. (2009). Spatiotemporal patterns of cortical activity with bilateral cochlear implants in congenital deafness. J Neurosci 29 (3): 811–827.

Kral A, Heid S, Hubka P et al. (2013). Unilateral hearing during development: hemispheric specificity in plastic reorganizations. Front Syst Neurosci 7: 93.

Leake PA, Hradek GT, Snyder RL (1999). Chronic electrical stimulation by a cochlear implant promotes survival of spiral ganglion neurons after neonatal deafness. J Comp Neurol 412 (4): 543–562.

Leibold LJ, Neff DL (2007). Effects of masker-spectral variability and masker fringes in children and adults. J Acoust Soc Am 121 (6): 3666–3676.

Leibold LJ, Werner LA (2006). Effect of masker-frequency variability on the detection performance of infants and adults. J Acoust Soc Am 119 (6): 3960–3970.

Litovsky R (1997). Developmental changes in the precedence effect: estimates of minimal audible angle. J Acoust Soc Am 102: 1739–1745.

Litovsky RY (1998). Physiological studies of the precedence effect in the inferior colliculus of the kitten. J Acoust Soc Am 103: 3139–3152.

Litovsky RY (2005). Speech intelligibility and spatial release from masking in young children. J Acoust Soc Am 117 (5): 3091–3099.

Litovsky RY (2011). Development of binaural and spatial hearing. Springer Handbook of Auditory Research. Springer-Verlag, New York, pp. 163–195.

Litovsky RY, Delgutte B (2002). Neural correlates of the precedence effect in the inferior colliculus: effect of localization cues. J Neurophysiol 87: 976–994.

Litovsky RY, Godar SP (2010). Difference in precedence effect between children and adults signifies development of sound localization abilities in complex listening tasks. J Acoust Soc Am 128 (4): 1979–1991.

Litovsky R, Macmillan N (1994). Minimum auditory angle for clicks with simulated echoes: effects of azimuth and standard. J Acoust Soc Am 96 (2): 752–758.

Litovsky RY, Madell J (2009). Bilateral cochlear implants in children. In: L Eisenberg (Ed.), Clinical Management of Children with Cochlear Implants, Plural Publishing, San Diego, CA.

Litovsky RY, Yin TC (1998). Physiological studies of the precedence effect in the inferior colliculus of the cat. I. Correlates of psychophysics. J Neurophysiol 80 (3): 1285–1301.

Litovsky RY, Colburn HS, Yost WA et al. (1999). The precedence effect. J Acoust Soc Am 106: 1633–1654.

Litovsky RY, Goupell MJ, Godar S et al. (2012). Studies on bilateral cochlear implants at the University of Wisconsin's binaural hearing and speech lab. J Am Acad Audiol 23 (6): 474–494.

Lutfi RA, Kistler DJ, Oh EL et al. (2003). One factor underlies individual differences in auditory informational masking within and across age groups. Percept Psychophys 65 (3): 396–406.

Macmillan NA, Creelman CD (2004). Detection Theory: A User's Guide. Lawrence Erlbaum, Hillsdale, NJ.

Maxon AB, Hochberg I (1982). Development of psychoacoustic behavior: sensitivity and discrimination. Ear Hear 3 (6): 301–308.

McAdams S, Bertoncini J (1997). Organization and discrimination of repeating sound sequences by newborn infants. J Acoust Soc Am 102 (5 Pt 1): 2945–2953.

Middlebrooks JC, Green DM (1991). Sound localization by human listeners. Annu Rev Psychol 42: 135–159.

Mills A (1958). On the minimum audible angle. J Acoust Soc Am 30: 237–246.

Misurelli SM, Litovsky RY (2012). Spatial release from masking in children with normal hearing and with bilateral cochlear implants: effect of interferer asymmetry. J Acoust Soc Am 132 (1): 380–391.

Misurelli SM, Bernstein SR, Johnsrude IS et al. (2013). How do cognitive factors interact with speech-in-noise segregation in normal hearing children and adults? J Assoc Res Otolaryngol, Abstracts Feb 16–23.

Moore JM, Thompson G, Thompson M (1975). Auditory localization of infants as a function of reinforcement conditions. J Speech Hear Disord 40: 29–34.

Moore DR, Ferguson MA, Halliday LF et al. (2008). Frequency discrimination in children: perception, learning and attention. Hear Res 238 (1–2): 147–154.

Muir DW, Clifton RK, Clarkson MG (1989). The development of a human auditory localization response: a U-shaped function. Can J Psychol 43 (2): 199–216.

Neff DL, Green DM (1987). Masking produced by spectral uncertainty with multicomponent maskers. Percept Psychophys 41 (5): 409–415.

Newman RS (2005). The cocktail party effect in infants revisited: listening to one's name in noise. Dev Psychol 41 (2): 352–362.

Newman RS, Jusczyk PW (1996). The cocktail party effect in infants. Percept Psychophys 58 (8): 1145–1156.

Niparko JK, Tobey EA, Thal DJ et al. (2010). JAMA 303 (15): 1498–1506.

Nittrouer S, Boothroyd A (1990). Context effects in phoneme and word recognition by young children and older adults. J Acoust Soc Am 87 (6): 2705–2715.

Nodal FR, Kacelnik O, Bajo VM et al. (2010). Lesions of the auditory cortex impair azimuthal sound localization and its recalibration in ferrets. J Neurophysiol 103 (3): 1209–1225.

Nodal FR, Bajo VM, King AJ (2012). Plasticity of spatial hearing: behavioural effects of cortical inactivation. J Physiol 590 (Pt 16): 3965–3986.

Northern JL, Downs MP (1984). Hearing in children, Williams & Wilkins, Baltimore.

Nozza RJ, Wagner EF, Crandell MA (1988). Binaural release from masking for a speech sound in infants, preschoolers, and adults. J Speech Hear Res 31 (2): 212–218.

Oh EL, Lutfi RA (1998). Nonmonotonicity Of informational masking. J Acoust Soc Am 104 (6): 3489–3499.

Oh EL, Wightman F, Lutfi RA (2001). Children's detection of pure-tone signals with random multitone maskers. J Acoust Soc Am 109 (6): 2888–2895.

Olsho LW, Koch EG, Halpin CF (1987). Level and age effects in infant frequency discrimination. J Acoust Soc Am 82 (2): 454–464.

Olsho LW, Koch EG, Carter EA et al. (1988). Pure-tone sensitivity of human infants. J Acoust Soc Am 84 (4): 1316–1324.

Plomp R, Mimpen AM (1981). Effect of the orientation of the speaker's head and the azimuth of a noise source on the speech reception threshold for sentences. Acustica 48: 325–328.

Pujol R, Hilding D (1973). Early synapse formation in vestibular system of fetal guinea pig. Acta Otolaryngol 76 (1): 1–10.

Scollie SD, Ching TY, Seewald RC et al. (2010). Children's speech perception and loudness ratings when fitted with hearing aids using the DSL v.4.1 and the NAL-NL1 prescriptions. Int J Audiol 49 (Suppl 1): S26–S34.

Serpanos YC, Gravel JS (2000). Assessing growth of loudness in children by cross-modality matching. J Am Acad Audiol 11 (4): 190–202.

Shahidullah S, Hepper PG (1994). Frequency discrimination by the fetus. Early Hum Dev 36 (1): 13–26.

Shepherd RK, Hardie NA (2001). Deafness-induced changes in the auditory pathway: implications for cochlear implants. Audiol Neurootol 6 (6): 305–318.

Shepherd RK, McCreery DB (2006). Basis of electrical stimulation of the cochlea and the cochlear nucleus. Adv Otorhinolaryngol 64: 186–205.

Sinnott JM, Aslin RN (1985). Frequency and intensity discrimination in human infants and adults. J Acoust Soc Am 78 (6): 1986–1992.

Song JH, Skoe E, Banai K et al. (2012). Training to improve hearing speech in noise: biological mechanisms. Cereb Cortex 22 (5): 1180–1190.

Spetner NB, Olsho LW (1990). Auditory frequency resolution in human infancy. Child Dev 61 (3): 632–652.

Sussman E, Wong R, Horváth J et al. (2007). The development of the perceptual organization of sound by frequency separation in 5-11-year-old children. Hear Res 225 (1–2): 117–127.

Tharpe AM, Ashmead DH (2001). A longitudinal investigation of infant auditory sensitivity. Am J Audiol 10 (2): 104–112.

Trehub SE, Bull D, Schneider BA (1981). Infants' detection of speech in noise. J Speech Hear Res 24 (2): 202–206.

Trehub SE, Schneider BA, Morrongiello BA et al. (1988). Auditory sensitivity in school-age children. J Exp Child Psychol 46 (2): 273–285.

van de Par S, Kohlrausch A (1999). Dependence of binaural masking level difference on center frequency, masker bandwidth, and interaural parameters. J Acoust Soc Am 106: 1940–1947.

Van Deun L, van Wieringen A, Van den Bogaert T et al. (2009). Sound localization, sound lateralization, and binaural masking level differences in young children with normal hearing. Ear Hear 30: 178–190.

Veloso K, Hall JW, Grose JH (1990). Frequency selectivity and comodulation masking release in adults and in 6-year-old children. J Speech Hear Res 33: 96–102.

Watson CS, Kelly WJ, Wroton HW (1976). Factors in the discrimination of tonal patterns. II. Selective attention and learning under various levels of stimulus uncertainty. J Acoust Soc Am 60 (5): 1176–1186.

Whitfield IC, Cranford J, Ravizza R et al. (1972). Effects of unilateral ablation of auditory cortex in cat on complex sound localization. J Neurophysiol 35 (5): 718–731.

Wilson BS, Dorman MF (2008). Cochlear implants: current designs and future possibilities. J Rehabil Res Dev 45 (5): 695–730.

Wright BA, Bowen RW, Zecker SG (2000). Nonlinguistic perceptual deficits associated with reading and language disorders. Curr Opin Neurobiol 10 (4): 482–486.

Wright BA, Lombardino LJ, King WM et al. (1997). Deficits in auditory temporal and spectral resolution in language-impaired children. Nature 387 (6629): 176–178.

Xia J, Brughera A, Colburn HS et al. (2010). Physiological and Psychophysical Modeling of the Precedence Effect. J Assoc Res Otolaryngol 11 (3): 495–513.

Yost WA (1997). The cocktail party problem: Forty years later. In: RH Gilkey, TR Anderson (Eds.), Binaural and Spatial Hearing in Real and Virtual Environments, Lawrence Erlbaum, Hillsdale, NJ, pp. 329–348.

Zurek PM, Durlach NI (1987). Masker-bandwidth dependence in homophasic and antiphasic tone detection. J Acoust Soc Am 81: 459–464.

Handbook of Clinical Neurology, Vol. 129 (3rd series)
The Human Auditory System
G.G. Celesia and G. Hickok, Editors

Chapter 4

Representation of loudness in the auditory cortex

CHRISTOPH E. SCHREINER* AND BRIAN J. MALONE
Center for Integrative Neuroscience and Coleman Memorial Laboratory, Department of Otolaryngology – Head and Neck Surgery, University of California San Francisco, San Francisco, CA, USA

INTRODUCTION

Studies of sensory processing in humans have the distinct advantage that the subjects can report directly on the perception that a specific stimulus can elicit. Human psychophysics has quantified that relationship in some detail. By contrast, understanding the relationship between acoustic stimuli and neural processing has been largely limited to animal studies, due to the often invasive nature of probing the various stations and their elements for their role in stimulus encoding. Animal studies are, however, much more limiting with regard to assessing perceptual attributes, as opposed to perceptual performance, i.e., signal detection or discrimination. Non-invasive approaches (e.g., electroencephalography, magnetoencephalography, positron emission tomography (PET) and functional magnetic resonance imaging (fMRI)) to assess human brain processes involved in sensory processing and the generation of perception have created the opportunity to study the relationship between physical stimulus properties, stimulus representation in the brain, and their connection with perceptual attributes more directly. Many auditory perceptual attributes, such as pitch, timbre, and sound location, depend on relatively complex spectral and temporal aspects of sounds, and so result in correspondingly multidimensional percepts. An argument can be made that perceived loudness is most closely related to the underlying physical quantity, signal intensity. It is hoped that characterizing a largely one-dimensional relationship may inform the question of how stimulus, neural, and perceptual properties are related more generally. In the following review, we focus on how loudness may be reflected in brain processes, especially in the auditory cortex, in an attempt to provide a useful perspective on how the neural bases of other, more complex, perceptual attributes can be explored.

The auditory cortex processes sound information, as it relates to perception, communication, emotive behavior, and cognition. It transforms the ascending information in appropriate ways to support these global tasks within many widely distributed cerebral networks and provides feedback to many subcortical stations, including the ascending auditory stations from the cochlea to the thalamus (Winer and Schreiner, 2010). Auditory cortex consists of a number of distinct areas or fields, acting in parallel or hierarchically, that process and represent acoustic as well as perceptual attributes of sounds (Fig. 4.1). It is likely comprised of functionally distinct streams that may have evolved interdependently and interact cooperatively (Rauschecker and Tian, 2000; Schreiner and Winer, 2007). A ventral stream is postulated to comprise areas more involved in the decoding of spectrally complex sounds, while a dorsal stream is thought to play a role in sensorimotor integration and control (Hickok and Poeppel, 2007; Hickok et al., 2011; Rauschecker, 2012; and see Chapter 9, this volume). The multitude of specific connections within auditory cortex and with many extra-auditory regions is a reflection of the many tasks carried out within its confines. However, the nature and precise purpose of the feedforward and recurrent processing within and between those streams and different fields remain unexplained. In humans, for example, it is not clear how the spectrotemporal distribution of energy present in speech sounds is being preprocessed in the auditory cortex before reaching the various regions dedicated to speech and language analysis. Since cortex is considered to be necessary for awareness and perception (e.g., Crick and Koch, 1995), a fruitful approach would be to consider first the transformation of more basic physical sound properties into neural representations and then the potential relationship between representational properties and perceptual attributes.

*Correspondence to: Christoph E. Schreiner, Coleman Memorial Laboratory, Department of Otolaryngology – Head and Neck Surgery, University of California San Francisco, 675 Nelson Rising Lane, San Francisco, CA 94143-0444, USA. Tel: +1-415 502 7375, E-mail: chris@phy.ucsf.edu

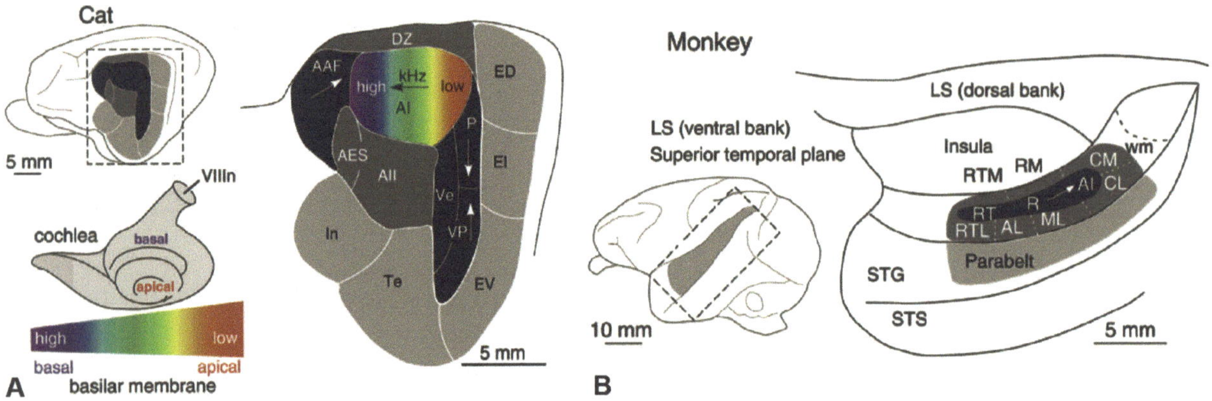

Fig. 4.1. Organization of auditory cortical areas in cat and rhesus monkey. (**A**) Cat auditory cortex has at least 13 areas: five are tonotopic (black or color gradient), three are non-tonotopic (dark gray), and five are multimodal and/or limbic-related (light gray). The color gradient indicates the frequency map along the cochlea and in the primary auditory cortex (AI). Arrows: low-to high-frequency gradients in five tonotopic fields. (**B**) Rhesus monkey auditory cortex is divided into tonotopic core fields (R, AI, etc.), belt (AL, ML, etc.), and less well-defined parabelt regions along the superior temporal plane. (Reproduced from Schreiner and Winer, 2007; with permission). AAF, anterior auditory field; AES, anterior ectosylvian area; aes, anterior ectosylvian sulcus; AI, primary auditory cortex; AII, second auditory field; AL, anterolateral belt; CL, caudolateral belt; CM, caudomedial auditory belt; DZ, dorsal auditory zone; ED, posterior ectosylvian gyrus, dorsal part; EI, posterior ectosylvian gyrus, intermediate part; EV, posterior ectosylvian gyrus, ventral part; In, insular cortex; pes, posterior ectosylvian sulcus; LS, lateral sulcus; P, posterior auditory field; R, rostral auditory field; RM, rostromedial region; RT, rostrotemporal area; RTL, rostrotemporal cortex, lateral area; RTM, rostrotemporal medial auditory belt; STG, superior temporal gyrus; sss, suprasylvian sulcus; STS, superior temporal sulcus; Te, temporal cortex; Ve, ventral auditory area; VIIIn, eighth nerve; VP, ventroposterior area; wm, white matter.

Psychologists and psychophysicists have identified various perceptual dimensions that our brains generate in an effort to assess the state of our environment and inform our interactions with it. In audition, basic perceptual dimensions of an ongoing sound include loudness, pitch, timbre, and the location of the sound source (e.g., Zwicker and Fastl, 1999; Moore, 2012). These attributes usually describe perceptual aspects that are largely, but not completely, independent from each other. For example, our perception of the duration of a sound is usually independent of its pitch or loudness. However, some interactions between different perceptual dimensions do occur, e.g., the pitch of some tones may be slightly affected by loudness (Morgan et al., 1951) and loudness does in fact depend on the duration of the sound (e.g., Small et al., 1962).

Loudness is the psychophysical correlate of the strength (amplitude) of a sound wave. The relationships between physical stimulus characteristics and the sensations and perceptual attributes they elicit have been established in some detail (see Zwicker and Fastl, 1999; Moore, 2012). The psychophysical study of loudness over many decades has resulted in the development of loudness models which include features of the physical signals as well as psychologic factors (Florentine et al., 2011) since perceived loudness can vary significantly from one listener to another (Marks and Florentine, 2011). The mapping of physical sound characteristics on to perceptual scales is usually not straightforward. For example, doubling/halving the intensity of a tone generally does not lead to a doubling/halving of the perceived loudness (Zwicker and Fastl, 1999; Florentine et al., 2011; Moore, 2012). This non-linear transformation of physical scales to perceptual ones should help in distinguishing the neural codes that give rise to our perceptual experiences as opposed to those that are more directly reflective of features of the physical stimuli themselves. While significant correlations in the relationship between neural stimulus encoding and the representation of perceptual dimensions have been established (Zwicker and Fastl, 1999; Florentine et al., 2011; Moore, 2012), significant gaps regarding the precise link remain. This is partially due to the range of potential codes available to the brain to represent auditory information, such as location within a spatial parameter representation (i.e., a map), firing rate, spike timing, spiking patterns, and the relationships among spiking patterns across multiple neurons, to name a few. In addition, auditory cortex contains a large number of structurally, connectionally, and functionally distinguishable regions that function both hierarchically and in parallel (see Winer and Schreiner, 2010). The contributions of individual, functionally distinguishable cortical fields to the generation of a percept has not been determined, nor have the rules for integrating information across multiple cortical fields. While temporarily disabling a specific cortical field through reversible cooling can result in specific electrophysiologic, perceptual, or behavioral deficits (Lomber et al., 1999), it does not necessarily follow that the disabled field is solely responsible for a specific perceptual attribute given the complex hierarchical structure of cortex and the many recurrent pathways and functional interactions between the fields (Winer and Schreiner, 2010).

Despite these challenges, clever experimental work in both human and animal subjects hints at future progress in relating neural representations to perceptual experience. The goal of the following sections is to review prevailing notions of how a basic auditory perceptual dimension – loudness – may be encoded in the auditory cortex of animals, and how this may inform the interpretation of invasive or non-invasive cortical studies in humans.

PSYCHOPHYSICS

Loudness is the perceived or subjective intensity of a sound (Scharf, 1978) (for details, see Florentine et al., 2011). For clarity, we will use the term "perceived loudness." Psychophysically, it is defined as "that attribute of auditory sensation in terms of which sounds can be ordered on a scale extending from quiet to loud" (American National Standards Institute, 1973). It is a psychologic correlate (measured in sones) of the physical stimulus amplitude (measured as sound pressure, sound pressure level (SPL) (20 times the logarithm of the ratio of a sound pressure to a reference sound pressure of 20 μPascal: dB SPL), or sound intensity).

The perceptual scale of loudness (L) determined through magnitude estimation results in a ratio scale that relates to the physical dimension of sound intensity (I) roughly following a power law: $L = kI^{0.3}$ (Fig. 4.2A; k is a constant that is subject-specific). A simple way to understand the implications of this function is that increasing the physical intensity of the sound by a certain amount (e.g., 5 dB SPL) will produce a greater increase in perceived loudness at lower intensities (e.g., 30 dB SPL) than at higher intensities (e.g., 70 dB SPL). Loudness can be affected by parameters other than sound pressure, including stimulus frequency, spectral bandwidth, and duration. For example, the sound pressure to create the same loudness for different frequencies (the equal-loudness contours) varies at low intensities similar to changes of the absolute detection threshold but becomes fairly flat at higher intensities (Fig. 4.2B).

Psychophysical experiments have determined that loudness can be accounted for by two basic integration mechanisms, both related to the fundamental transformation of sound energy along the auditory receptor surface, the basilar membrane in the cochlea. Vibrations on the

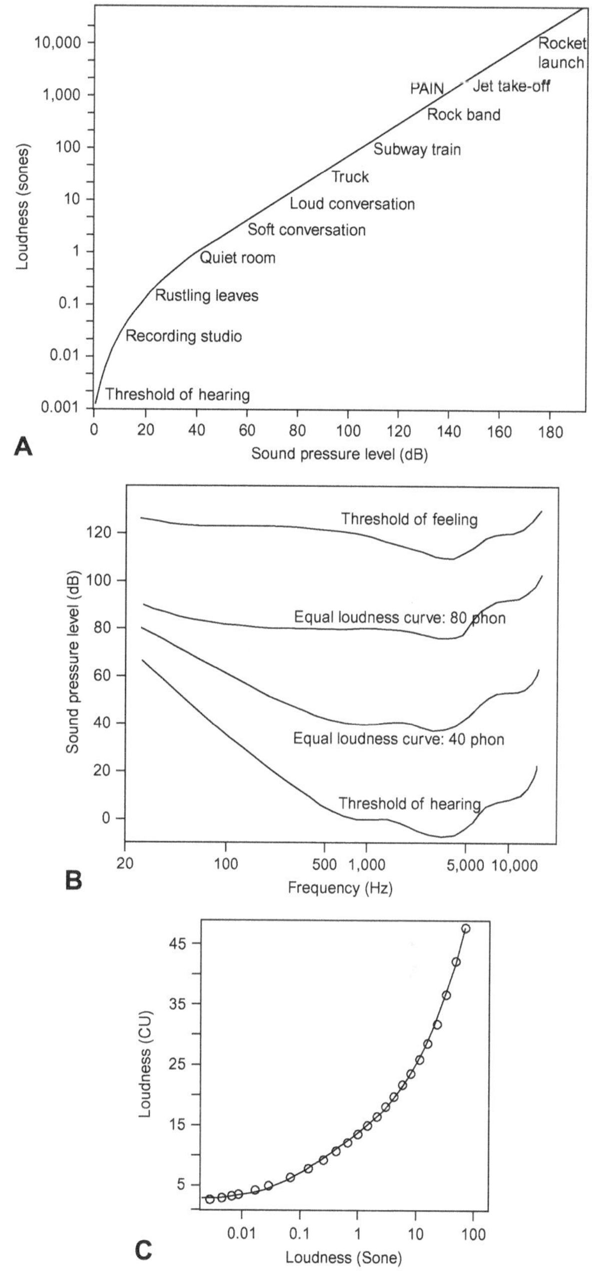

Fig. 4.2. Relationships between sound intensity and perceptual measures of loudness. (**A**) Loudness in sones as a function of sound pressure level. The approximate loudness of familiar sound sources is indicated. (**B**) Loudness is frequency-dependent, as indicated by the curvature of the equal-loudness contours in a plot of sound pressure level as a function of stimulus frequency. The Phon is a unit of loudness equal to the loudness level of a 1 kHz tone expressed in sound pressure level. (**C**) Relationship between categorical loudness units (CU) and loudness expressed in sones. A cubic fit (solid line) to the data (circles represent median subject data at 1000 Hz) relates loudness in sones to CUs at 1 kHz. (Adapted from Heeren et al., 2013, with permission.)

eardrum are communicated to the fluid-filled cochlea, causing ripples in the basilar membrane that propagate in what is called the "travelling wave" (Robles and Ruggero, 2001). These vibrations in the basilar membrane are then transduced into neural activity in the auditory nerve. Due to resonance properties of the cochlea, the dispersion of sound energy along the basilar membrane is frequency-specific, resulting in a precise and systematic spatial segregation of different input frequencies. The resulting cochleotopic or tonotopic mapping of the receptor surface is approximately proportional to the logarithm of the input frequency. While the vibration pattern has a sharp frequency-specific amplitude peak, the overall pattern is spread out across a wider range of the basilar membrane in a roughly triangular envelope shape. This results in a fairly broad neural excitation pattern across the array of auditory nerve fibers that transmit the outcome of this initial transformation of sound from a physical to a neural entity to more central structures in the auditory pathway. As a consequence, increasing the sound intensity has two main effects on the neural activity emerging from the cochlea: First, neural elements at the location of the stimulus frequency increase their firing rate since the peak amplitude of the vibration amplitude increases. Second, the vibrational pattern spreads – especially to locations associated with higher frequencies – with the consequence that a wider range of auditory nerve fibers are excited for more intense stimuli.

How is this excitation pattern related to the perceived loudness of a sound? First, it has been psychophysically determined that sound energy falling within about one-third octave of the stimulus frequency is integrated following a power law (see above) to generate the perceived loudness of the sound. This filter band has been named the "critical band" (Fig. 4.3) and it roughly corresponds to a fixed segment along the basilar membrane whose length is largely independent of the stimulus frequency (in humans, this segment is $\sim$1.3 mm long). Since the frequency map along the basilar membrane nearly follows a logarithmic scale, the bandwidth of the critical band is frequency-dependent (Fig. 4.3B for cats). In addition to the critical band at the stimulus frequency, directly neighboring critical bands also integrate energy following the power law, according to the energy spread along the receptor surface (Fig. 4.3C). The total loudness estimate, then, corresponds to the sum of the output of each critical band (the so-called specific loudness) that receives input from the excitation pattern elicited by the stimulus (Zwicker et al., 1957). Although it is not yet clear where and how this integration process is actually realized in the auditory system, psychophysical models based on critical bands have proven quite accurate in predicting the perceived loudness of various sounds (Glasberg and Moore, 2002).

The unit of perceived loudness is the sone. One sone is defined as the loudness of a tone at 1 kHz at 40 dB SPL. Doubling of perceived loudness (e.g., from 1 to 2 sones) corresponds approximately to a 10-dB SPL increase of the stimulus level. This perceptual loudness measure is based on a continuous-value ratio scale that is relatively cumbersome to assess for each individual. Other measures have been introduced that allow a more rapid but perhaps coarser estimate of the perceived loudness, such as an ordinal or categoric scale. In that case, subjects are asked to assign the loudness of a given stimulus to a named categoric loudness unit (e.g., very soft, soft, medium, loud, very loud) or unnamed intermediate units. In contrast to a ratio scale, the perceptual difference between neighboring categories may not be equal. Thus, the two measures are not equivalent, though they have been shown to be related via a cubic function (Fig. 4.2C; Heeren et al., 2013). Use of the categoric methods provides investigators with a more convenient procedure to establish psychophysically rigorous estimates of loudness.

ANIMAL PHYSIOLOGY

Physiologic studies can reveal neural mechanisms of the encoding and decoding of stimulus information at various stations along the auditory pathway (Heinz, 2012). Various electrophysiologic measures of neural activity have been used, from intracellular potentials and extracellular action potentials of individual neurons to summed potentials of small and large groups of neural elements, including contributions from neuronal cell bodies, axons, and dendrites, as well as from glial cells. While it is fairly straightforward to relate any of these measures to the physical parameters of the stimulus, it is much more difficult to relate these measures to units of perceptual attributes. Establishing perceptual magnitudes in animal models is difficult and therefore rarely done. Instead, it is often assumed that the relationships between stimulus and perceived attributes measured in humans are sufficiently similar in animals to forgo a direct perceptual measurement. However, this uncertainty greatly diminishes the interpretive power of physiologic experiments with regard to relating stimuli to perceptual attributes.

Two fundamental response properties of auditory neurons are related to stimulus intensity coding: each responsive neuron has a minimum sound intensity at which it will increase the rate at which it fires action potentials, known as the response threshold (Fig. 4.4A–C, arrows). The values of the response threshold can vary widely across neurons, from values near or even slightly below the animal's minimum behavioral detection threshold to values 60–80 dB above it. With increasing sound intensity above

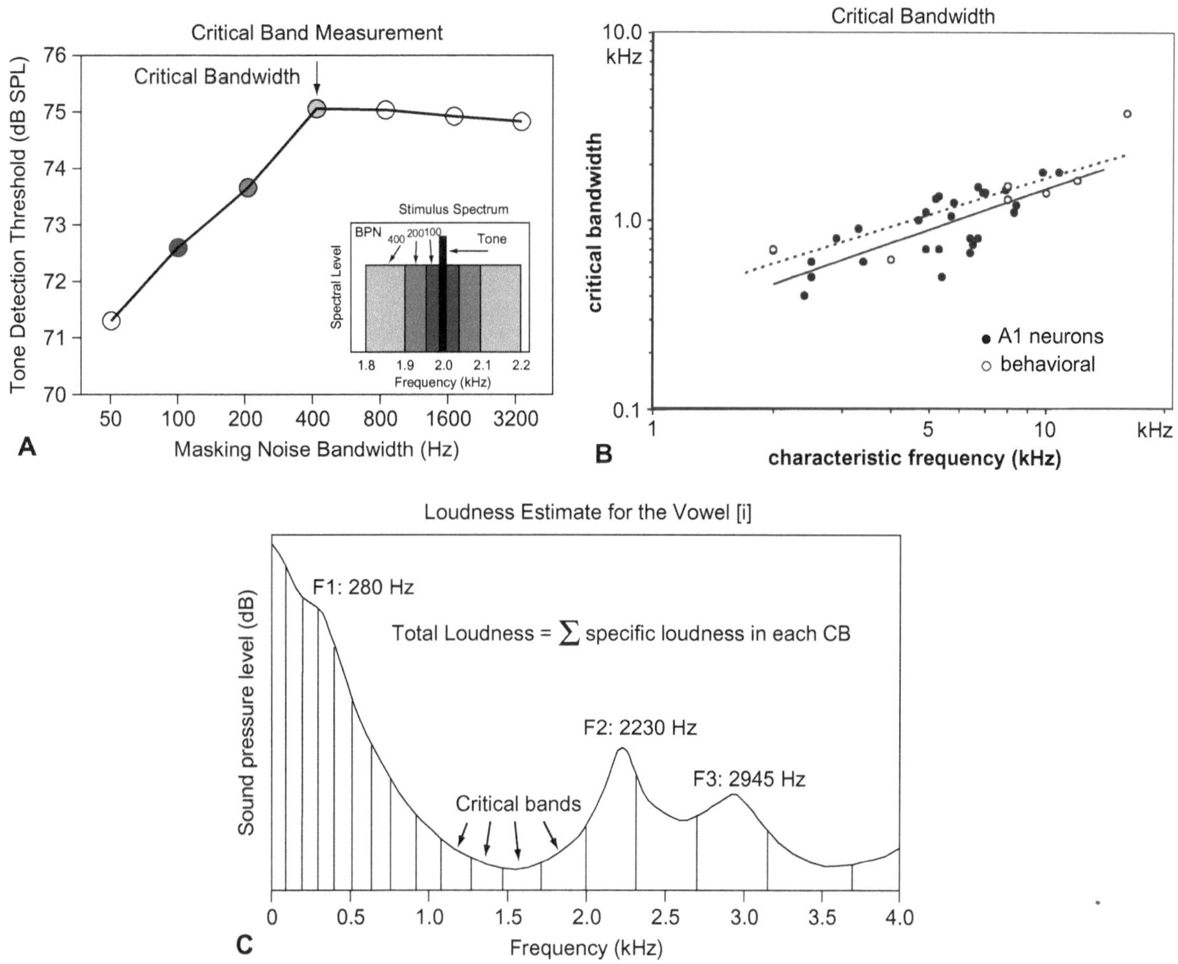

Fig. 4.3. Critical bands (CBs) and loudness formation. (**A**) Measurement of critical bandwidth in a tone detection experiment. A tone at 2 kHz is presented in bandpass noise (BPN) of different bandwidths. As the noise bandwidth increases, the detection threshold for the tone increases because the CB centered at the tone frequency accumulates more stimulus energy, which requires the tone level to be raised for successful detection. For bandwidths larger than 400 Hz (light gray dot) the detection threshold no longer increases because additional sound energy outside the CB nearest the tone frequency no longer contributes to masking of the tone. (Adapted from Moore, 2012, with permission.) (**B**) Critical bandwidth is frequency-dependent. Behaviorally determined critical bandwidths in the cat (open circles, dashed line) broaden with increases in frequency. Some auditory cortical neurons reveal the same relationship (filled circles, solid line). (Reproduced from Ehret and Schreiner, 1997; with permission.) (**C**) Simplified depiction of loudness formation for the vowel [i]. The spectral envelope of the vowel [i] with its main three formant frequencies (F1–F3) is depicted. The edges of CBs from 0 and 4 kHz are indicated by the vertical lines. The overall loudness (in sones) of the vowel can be estimated by calculating the specific loudness of each CB (proportional to the stimulus intensity in the critical band to the power of 0.3) then summing the specific loudness from all CBs (Zwicker and Fastl, 1999).

threshold, neural firing rates may increase until a maximum rate is reached (i.e., saturation) or the firing rate may even decline at the highest stimulus intensities (Fig. 4.4A–C). When firing rates increase or saturate with increasing intensity the response is said to be monotonic. When rates decrease with increasing intensities the response is described as non-monotonic (Fig. 4.4C).

An important concern when relating firing rates to the total range of perceptible intensities is the relatively narrow range – called the dynamic range – over which a neuron's firing rate increases as a function of sound intensity. For the majority of neurons, the dynamic range is less than 50 dB and often as small as 10–20 dB, compared to a total behavioral intensity range of more than 120 dB. This means that no single neuron can encode the full range of environmental stimulus intensities in terms of average firing rates. However, the response thresholds of different neurons also vary over at least 50 dB. Given appropriate distributions of response thresholds and dynamic ranges across neurons, then, the full population could encode the full range of behaviorally relevant intensities via changes in firing rates. It is

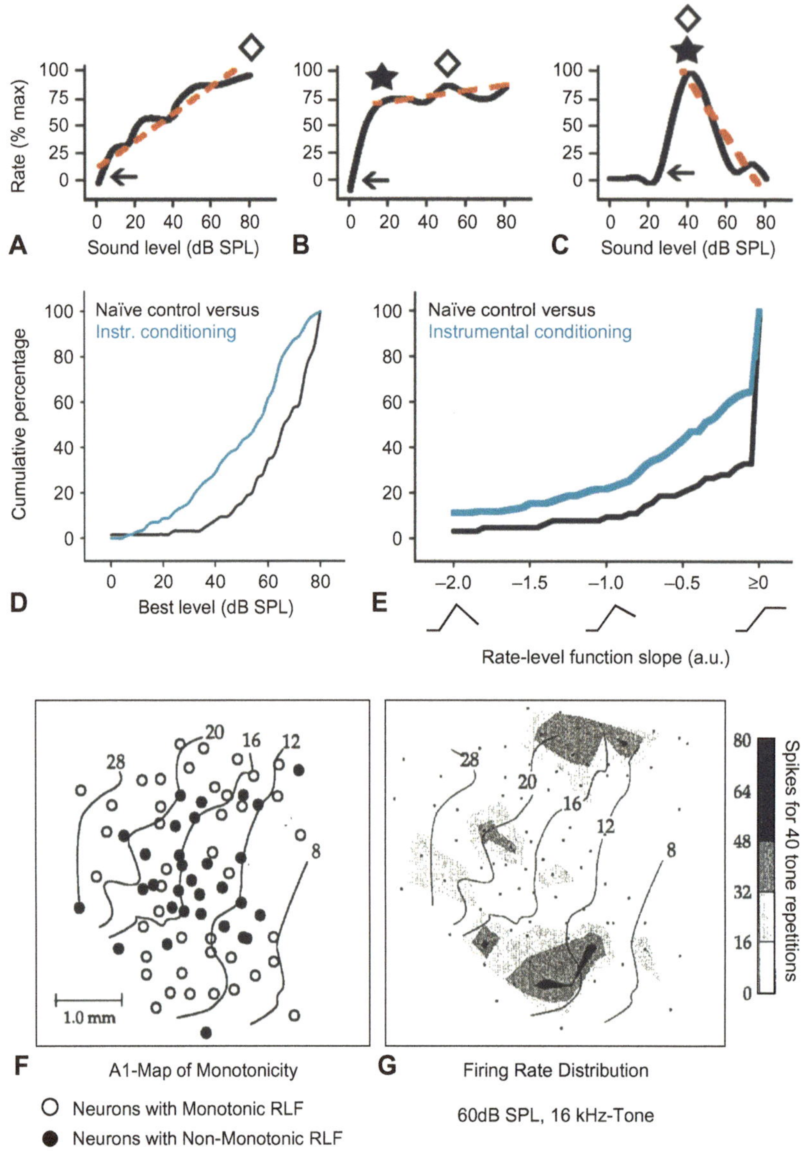

Fig. 4.4. Relationships between neural response strength and spatial cortical response distribution as a function of stimulus level. (**A–C**) Firing rates in response to tones with increasing sound level typically increase (**A**), increase and saturate (**B**), or increase and decrease (**C**). Type C is called a non-monotonic rate-level function (RLF), in contrast to the monotonic growth functions shown in A and B. (Diamond, highest firing rate = "best level"; star, inflection point for saturating and non-monotonic RLFs; arrow, response threshold). (**D**) Training an animal on identifying low-intensity sounds increases the proportion of neurons with low best levels, as evidenced by the shift of the cumulative function to the left. (**E**) Training an animal on identifying low-intensity sounds increases the proportion of neurons with non-monotonic RLFs. (**A–E**, Reproduced from Polley et al., 2004, with permission.) (**F**) Neurons with non-monotonic RLFs form spatial clusters in cat primary auditory cortex (lines correspond to locations of neurons with similar best frequency (in kHz)). (**G**) Firing rate distribution in cat primary auditory cortex in response to a 16-kHz, 60-dB sound pressure level (SPL) tone (data from the same A1 as shown in **F**). Spatial clusters with high firing rates correspond to locations with monotonic RLFs (see **F**). Areas with non-monotonic RLFs (see **F**) respond only weakly, indicating that the best level for neurons at those locations was below 60 dB SPL. (**F**, **G**, Reproduced from Phillips et al., 1994, with permission.)

also possible that aspects of neural firing patterns other than average firing rate could extend the effective dynamic range (Malone et al., 2010).

An important question about neural intensity coding is whether there is a spatial map for intensity analogous to that observed for frequency. In fact, some evidence of spatial maps of response thresholds in auditory cortex has already been obtained. Response thresholds along the isofrequency domain in primary auditory cortex systematically change from higher thresholds near the edge of the isofrequency domain, to lower thresholds in the middle, and again to higher thresholds at the other end of the isofrequency domain (Schreiner et al., 1992; Heil et al., 1994; Phillips et al., 1994; Sutter and Schreiner, 1995). This would suggest that low- and high-intensity stimuli activate different regions in the primary auditory cortex, corresponding to a place map. In addition, a complementary coding strategy may be provided by neurons with non-monotonic intensity tuning functions. These neurons fire at their maximum rate for a specific stimulus intensity – their "best intensity." Local spatial clustering of neurons with similar monotonic or non-monotonic properties and/or best intensities have been observed that suggest more granular spatial organization of intensity coding-related neural properties (Fig. 4.4 F and G). Low- and high-intensity stimuli could be represented by complementary distributions of activity alternating along the horizontal extent of isofrequency strips due to differences in the distribution of threshold and best intensity (Schreiner et al., 1992; Heil et al., 1994; Phillips et al., 1994; Sutter and Schreiner, 1995). However, these "maps" are noisier and less salient than the basic tonotopic organization. By contrast, significant similarity of intensity-based response parameters throughout the cortical depth has been demonstrated in primary auditory cortex of awake primates (Scott et al., 2011). That study showed that the apparent clustering of level-tuning properties within the same electrode penetration is greater than would be expected by chance, suggesting a locally clustered organization of sound intensity independently of a particular subtype of neurons. Overall, diversity in minimum response thresholds, dynamic ranges, and best intensities, combined with non-homogeneous horizontal spatial distributions of differently tuned neurons as well as columnar distribution of similarly tuned neurons may ensure adequate coverage of the range of stimulus intensities in the acoustic environment.

Evidence for a neural code based on preferred intensities has been obtained in the auditory cortex of rodents (Polley et al., 2004). Animals were trained to find a place on an experimental platform that was associated with a specific sound intensity and rewarded for reaching that location when a sound of the appropriate intensity was presented. After several weeks of training, neural responses in the primary auditory cortex of those animals was examined for evidence for intensity tuning. Whereas untrained animals showed a wide distribution of response thresholds and increasing or saturating firing rates for increasing intensities, trained animals exhibited best intensities at or near the rewarded training intensity and a preponderance of neurons with non-monotonic intensity-tuning functions (Fig. 4.4D and E). Whole-cell recordings have demonstrated that intensity tuning can be inherited from subcortical stations but also created within auditory cortex itself through appropriate adjustment of the relative magnitude and timing of local excitatory and inhibitory inputs and their interactions (Tan et al., 2007; Zhou et al., 2012). Plastic changes in cortical intensity coding indicate a significant link between cortical stimulus representation and behavioral stimulus relevance.

Neuronal tuning to a specific, behaviorally relevant stimulus magnitude does not necessarily clarify the relationship between the neural representation and perceived loudness. Indeed, it has not been unequivocally demonstrated that increasing the behavioral relevance of a given sound intensity alters its apparent loudness. Training might instead alter the attention to or awareness of the presence of the sound. Behavioral training (Polley et al., 2004) or pairing acetylcholine release with a specific stimulus configuration induces plastic increases in the representation of the paired stimulus in auditory cortex (Bakin and Weinberger, 1996; Kilgard and Merzenich, 1998; see Chapter 8, this volume). Behavioral testing of the animal's detection ability of soft sounds before and after pairing them with the release of the neurotransmitter acetylcholine in auditory cortex via stimulation of the nucleus basalis resulted in an increased probability of detection after pairing (Froemke et al., 2013). While behavioral training or neuromodulatory intervention may increase the behavioral relevance and salience of a stimulus, it has not been shown – and it appears unlikely – that the involved changes also affect the perceptual attribute, in this case perceived loudness, of the stimulus. This suggests that encoding of perceptual attributes and of behavioral relevance/salience may involve distinctly different aspects of neural activity.

As we noted above, perceived loudness is affected by the spectral distribution of sound, not simply the total acoustic energy comprising it. Correlates of the psychophysically defined critical bandwidth, a potential key contributor to the formation of loudness, have been observed in physiologic experiments. For example, the frequency spacing of the laminae of the concentrically organized central nucleus of the inferior colliculus is closely related to the critical bandwidth (Schreiner and Langner, 1997). Many neurons in the inferior colliculus (Ehret and Merzenich, 1985) and some in the primary auditory cortex (Ehret and Schreiner, 1997) have filter properties that correspond to psychophysical critical bands (Fig. 4.3B). These include spectral integration bandwidths and the relative independence of those bandwidths from overall stimulus intensity. However, not all neurons conform, particularly in the auditory cortex, where most neurons do not show the filter and integration behavior predicted from the psychophysical critical band model. For example, many neurons in primary and non-primary auditory cortical fields show a much broader spectral integration behavior (>2 octaves) than predicted for a critical band behavior (~1/3 octave) (Ehret and Schreiner, 1997). These observations do not rule out, however, the validity of the psychophysically well-supported critical band model *per se*, since it is possible that an appropriate decoding model downstream could instantiate critical bands not evident in the tuning functions of individual neurons located in the inferior colliculus or the primary auditory cortex.

As the foregoing paragraphs suggest, physiologic studies have identified a number of features of the cortical representation of sound that are related to sound intensity, or to mechanisms that might contribute to the formation of the perceptual attribute of loudness. However, one problem is that intensity and loudness are correlated, but related non-linearly (Fig. 4.2A), and we generally do not have access to perceived loudness in experiments involving animal subjects. As a result, no conclusive model has been developed that can account for the means by which the perceptual attribute loudness is related to those local encoding features (but see Micheyl et al., 2013, for perceptual performance).

It should be emphasized that very few physiologic studies in animals have considered intensity coding outside primary auditory cortical fields and no systematic effort has been undertaken to elucidate how the physiologic representation of sound intensity relates to the perceptual attribute loudness. Further, we have not yet fully determined how the physical intensity of sound is encoded in the auditory system. Experiments involving fMRI in awake monkeys have demonstrated significant and complex interactions between signal intensity and the distribution of activity among auditory cortical fields, including the lateral spread of activation to belt regions at higher intensities, and even frequency- and intensity-specific deactivation of particular regions (Tanji et al., 2010). We have focused here on average firing rates, but Malone et al. (2010) demonstrated that cortical spiking patterns contained more information about stimulus intensity than average firing rates, even when those patterns were normalized to eliminate differences in the firing rates evoked by sounds at different intensities. Thus, it is possible that a more complex encoding model will be necessary for a complete description of the neural representation of intensity and perceived loudness.

HUMAN CORTICAL STUDIES

How then do studies of stimulus intensity coding in human auditory cortex relate to the observations in animal models? Multiple studies of human auditory cortex, predominantly with fMRI approaches, but also with PET and event-related potential approaches, have confirmed that many regions do show clear response changes with increasing stimulus intensity (Fig. 4.5A). Studies reliant on fMRI signals are limited by the temporal resolution of the technique, which is orders of magnitude slower than the temporal precision seen in the patterns of action potentials generated by auditory neurons. In addition, the spatial resolution of the method is limited, though it may in some cases resolve single cortical columns or layers. Local cortical circuits or groups of neurons still cannot be resolved by these non-invasive approaches. Therefore, interpretation of fMRI data is largely based on relating changes in secondary, indirect measures of activity, such as blood oxygen level-dependent (BOLD) response magnitude and spatial spread (e.g., cortical volume) to changes in the stimulus intensity.

Increased activation with increasing level has been observed in the cochlear nucleus, superior olive, inferior colliculus, medial geniculate body, and auditory cortical areas (Sigalovsky and Melcher, 2006). In auditory cortex, the involvement of a number of different areas or subareas has been demonstrated. Primary auditory cortex seems particularly affected by increasing intensities (Hart et al., 2002; Lasota et al., 2003; Ernst et al., 2008). The planum temporale and anterior lateral areas of Heschl's gyrus also show response changes with sound intensity, but less so than in primary auditory cortex (Brechmann et al., 2002; Hart et al., 2002; Ernst et al., 2008). Contralateral-dominant activation effects in response magnitude and volume have also been reported (Lasota et al., 2003). In the non-primary cortex of the left hemisphere, stimulus intensity was correlated with activated volume but not with response magnitude (Brechmann et al., 2002).

The growth pattern of activity depends on the spectrotemporal content of the stimulus, not unlike psychophysical and neurophysiologic observations. The growth function for low-frequency tones in Heschl's gyrus was found to be flat at low intensities and steeply increasing at higher intensities. Higher frequencies tend to show a more steady growth function (Hart et al., 2003). In the superior temporal gyrus, increasing intensity resulted in an increase in activation volume for single-frequency tones but not harmonic, multitone complexes, although both stimuli increased response magnitude (Hall et al., 2001). The fact that harmonic-complex tones produced greater activation than the single tone suggests that bandwidth had a greater effect on the pattern of auditory activation than sound level (Hall et al., 2001).

There is some evidence to suggest the existence of a spatial, "ampliotopic" representation of stimulus intensity perhaps similar to such an organization in cat auditory cortex (Fig. 4.4 F and G; Schreiner et al., 1992; Heil

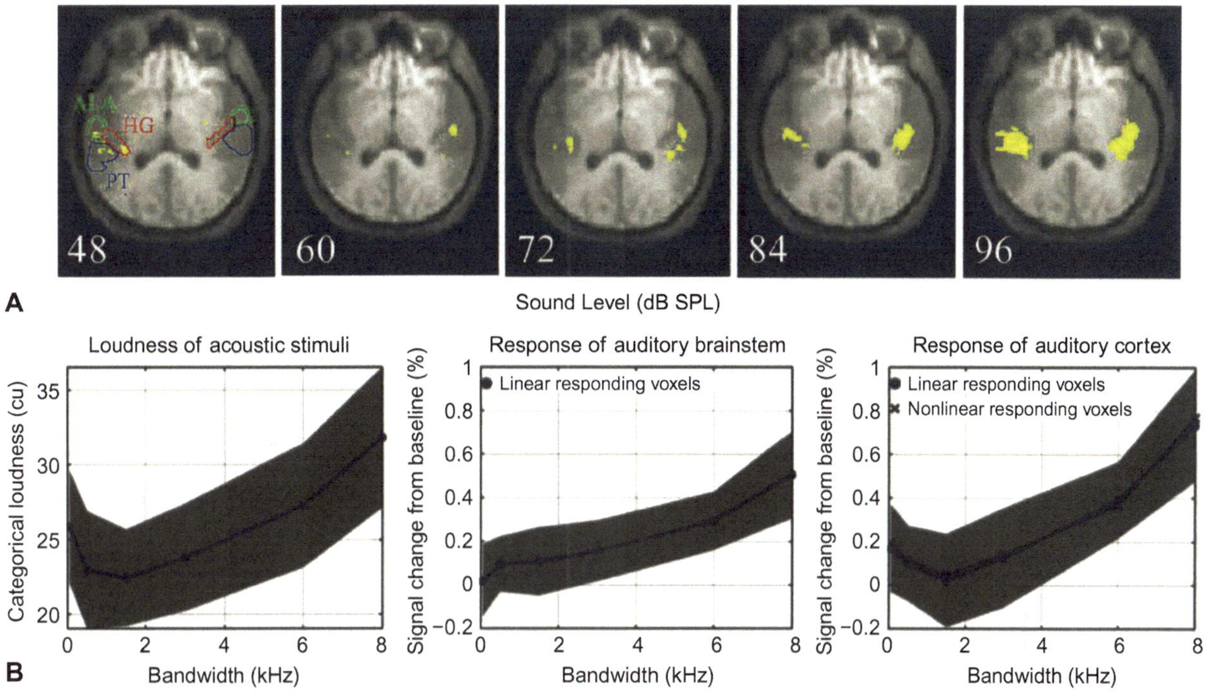

Fig. 4.5. Functional magnetic resonance imaging response changes in human auditory cortex for increasing stimulus intensity or bandwidth. (**A**) Group activation for five different sound levels (indicated at bottom left of each panel in dB sound pressure level (SPL)), analyzed relative to the silent baseline. Significantly activated voxels, shown in yellow, are overlaid on to a structural image. The distribution of significantly activated voxels generally expands with increases in sound level. Approximate locations of Heschl's gyrus (HG, red), anterior lateral area (ALA, green), and planum temporale (PT, blue) are shown in the left panel. (**B**) Comparisons between loudness and cortical activation of band passed pink noise centered at 4 kHz presented at 70 dB SPL as a function of bandwidth. (Loudness increases with increasing bandwidth, although the overall stimulus level remains constant. The broader spectra activate more critical bands than the narrower stimuli, resulting in a greater loudness value.) Left panel: Perceived loudness for different stimulus bandwidths in categorical loudness units (gray area corresponds to standard deviation across the subjects). Middle panel: Percent blood oxygen level-dependent (BOLD) signal change from baseline of voxels in the auditory brainstem. Right panel: Percent BOLD signal change for primary auditory cortex. (Linear and non-linear (linear plus quadratic) fits are shown.) The response in auditory cortex reflects the perceived loudness more closely than the response in the brainstem. (Reproduced from Röhl et al., 2011, with permission.)

et al., 1994; Phillips et al., 1994). Some studies describe a shift of the spatial distribution of activated clusters under different sound intensities across the temporal plane of Heschl's gyrus (Lockwood et al., 1999; Bilecen et al., 2002). However, this issue has not been completely resolved since other studies showed no clear signs of such an organization (Hart et al., 2002; Sigalovsky and Melcher, 2006).

Sound intensity effects on cortex have also been investigated for some time with event-related potential. Direct comparison of the results from the two methodologies suggests a close relationship between the fMRI signal and event-related potential activity (Mulert et al., 2005).

The physiologic interpretation of intensity-related response growth in human cortex is largely based on observations obtained from animal studies and the coding principles derived from them. Increases in signal strength with increasing intensity may reflect increases in the firing rates of single neurons or local networks, as observed in many cortical neurons. The spread of excitation along spatially mapped frequency gradients, already present in the cochlea, might account for the increased activated cortical volume observed at higher stimulus intensities. Importantly, however, activation measures cannot clearly distinguish between excitatory or inhibitory components since both increase energy consumption. For example, non-monotonic intensity tuning may involve activation of both excitatory and inhibitory inputs. Thus, BOLD magnitude may not be an accurate reflection of the relevant output activity in a given structure. Neurons with non-monotonic intensity coding behavior may contribute substantially to the level dependence of cortical fMRI activation for tones (Polley et al., 2004; Sigalovsky and Melcher, 2006). Such intensity coding may shift with increasing stimulus level to subpopulations of higher thresholds and less saturation as well as to neurons with different frequency spread (Lasota et al., 2003). However, the relationship of the neural activation pattern and their changes with stimulus intensity to the perceived loudness remains unresolved.

Thus, a major goal would be to determine the relationship of cortical activation to the perceptual attribute of signal intensity, i.e., loudness. The complicated relationship between intensity and perceived loudness suggests that a transformation in the neural representation of intensity occurs within the ascending auditory pathway, which changes the more straightforward relationship between intensity and response measures in the cochlea. Several studies have begun to address this subject in order to establish a link between cortical activity patterns, even the relatively crude ones provided by the non-invasive recording methods available today. As mentioned above, the growth of perceptual loudness is non-linearly related to the growth of stimulus intensity and may help distinguish intensity encoding from perceptual magnitude.

The relationship between loudness growth versus stimulus growth can be captured with a categoric loudness scale (see above). This specific type of loudness estimate has been shown to correlate with fMRI signal changes in auditory cortex (Röhl and Uppenkamp, 2012). The correlation between the slopes of loudness function and response magnitude growth was significant in auditory cortex but not in subcortical stations (Fig. 4.5B). Furthermore, the variance in perceived, categorized loudness in individual subjects correlated with response change from quiet to loud stimuli. In other words, those subjects who perceived the sound as being louder also exhibited a greater measured activity. This suggests that aspects of the neural code in auditory cortex captured by the fMRI method do reflect perceptual measures.

Another way to assess the relationship between activity in auditory cortex and loudness growth employs the idea of spectral loudness summation. Increasing the bandwidth of the signal while keeping the overall sound intensity constant results in a non-linear loudness growth function due to the partial loudness summation across different critical bands. This is a consequence of the intensity-dependent frequency spread of cochlear activation (see section on psychophysics, above). The perceived loudness of stimuli with the same overall intensity is greater when the total acoustic energy is distributed over several critical bands than when it is concentrated within a single critical band. In the brainstem, a linear relationship between bandwidth and fMRI magnitude change was found, in contrast to the psychophysical categoric loudness function, which shows an initial decrease with increasing bandwidth before beginning to increase again for bandwidths just under 2 kHz. In primary auditory cortex, however, the growth of response magnitude was compatible with the behavior of the bandwidth-based loudness function (Fig. 4.5B; Röhl et al., 2011). This also indicates that the perceptual attributes of sound intensity may be related to activation of the auditory cortex.

Loudness estimates in patients with an elevated hearing threshold have shown that, for relatively low sound intensities, loudness grows very rapidly, whereas at high intensities the growth is less steep and matches that of normal-hearing subjects. This phenomenon is called loudness recruitment (e.g., Moore, 2012). In hearing-impaired subjects, the abnormally steep loudness growth function above the elevated response threshold corresponded with the growth function of fMRI signals obtained in such patients. Therefore, the observation of different cortical activity growth functions in people with normal hearing and hearing loss supports the idea that mass activity in auditory cortex is more closely linked to the perceived loudness than to the intensity of the physical signal (Langers et al., 2007). It should be noted, however, that there is evidence that specific cell types in the ventral cochlear nucleus also exhibit growth curves compatible with recruitment (Cai et al., 2009; Joris, 2009), suggesting an important subcortical contribution to the phenomenon.

Overall, however, these studies suggest that auditory cortical activation – but not subcortical activation – assessed with fMRI is more closely related to the perceived loudness attribute of the stimulus than its physical characteristics.

Nevertheless, the foregoing studies do not reveal the precise neural mechanisms responsible for the perceptual correlate of stimulus intensity. Further studies are required to establish detailed links between the activation rules, neural coding principles, contributions of multiple primary and non-primary areas, and the emergence and further propagation of the neural correlate of the perceptual attribute loudness. For example, it is not currently known how information across cortical cells is combined to form the basis of intensity discrimination, although models have been proposed (Micheyl et al., 2013). Such models will likely be important in interpreting results from methods that measure the massed activation of cortical neurons. Using imaging or evoked potential methods in human cortex to identify the neural foundation of perceptual attributes is challenging. However, improvements in the temporal and, especially, spatial resolution of these methods hold great promise if and only if physical and perceptual attributes of sound are properly disambiguated. These studies could facilitate the development of therapies for clinically important phenomena such as tinnitus, where the neural correlates of the loudness of such "phantom" sensations are currently being investigated (Balkenhol et al., 2013).

ACKNOWLEDGMENTS

The research of the authors is supported by grants from the National Institutes of Health DC002260 (C.E.S.) and DC011843 (B.J.M.).

References

American National Standards Institute (1973). American national psychoacoustical terminology. S3.20. Anonymous, New York.

Bakin JS, Weinberger NM (1996). Induction of a physiological memory in the cerebral cortex by stimulation of the nucleus basalis. Proc Natl Acad Sci U S A 93: 11219–11224.

Balkenhol T, Wallhäusser-Franke E, Delb W (2013). Psychoacoustic tinnitus loudness and tinnitus-related distress show different associations with oscillatory brain activity. PLoS One 10: 1371.

Bilecen D, Seifritz E, Scheffler K et al. (2002). Amplitopicity of the human auditory cortex: an fMRI study. Neuroimage 17: 710–718.

Brechmann A, Baumgart F, Scheich H (2002). Sound-level-dependent representation of frequency modulations in human auditory cortex: a low-noise fMRI study. J Neurophysiol 87: 423–433.

Cai S, Ma WL, Young ED (2009). Encoding intensity in ventral cochlear nucleus following acoustic trauma: implications for loudness recruitment. J Assoc Res Otolaryngol 10: 5–22.

Crick F, Koch C (1995). Are we aware of neural activity in primary visual cortex? Nature 375: 121–123.

Ehret G, Merzenich MM (1985). Auditory midbrain responses parallel spectral integration phenomena. Science 227: 1245–1247.

Ehret G, Schreiner CE (1997). Frequency resolution and spectral integration (critical band analysis) in single units of the cat primary auditory cortex. J Comp Physiol A 181: 635–650.

Ernst SMA, Verhey JL, Uppenkamp S (2008). Spatial dissociation of changes of level and signal-to-noise ratio in auditory cortex for tones in noise. Neuroimage 43: 321–328.

Florentine M, Popper A, Fay RR (2011). Loudness. Springer Handbook of Auditory Research. Springer, New York.

Froemke RC, Carcea I, Barker AJ et al. (2013). Long-term modification of cortical synapses improves sensory perception. Nat Neurosci 16: 79–88.

Glasberg BR, Moore BCJ (2002). A model of loudness applicable to time-varying sounds. J Audio Eng Soc 50: 331–342.

Hall DA, Haggard MP, Summerfield AQ et al. (2001). Functional magnetic resonance measurements of sound-level encoding in the absence of background scanner noise. J Acoust Soc Am 109: 1559–1570.

Hart HC, Palmer AR, Hall DA (2002). Heschl's gyrus is more sensitive to tone level than non-primary auditory cortex. Hear Res 171: 177–190.

Hart HC, Hall DA, Palmer AR (2003). The sound-level-dependent growth in the extent of fMRI activation in Heschl's gyrus is different for low- and high-frequency tones. Hear Res 179: 104–112.

Heeren W, Hohmann V, Appell JE et al. (2013). Relation between loudness in categorical units and loudness in phons and sones. J Acoust Soc Am 133: 314–319.

Heil P, Rajan R, Irvine DRF (1994). Topographic representation of tone intensity along the isofrequency axis of cat primary auditory cortex. Hear Res 76: 188–202.

Heinz MG (2012). Intensity coding throughout the auditory system. In: KL Tremblay, RF Burkard (Eds.), Translational Perspectives in Auditory Neuroscience: Normal Aspects of Hearing, Plural Publishing, San Diego, CA, pp. 349–386.

Hickok G, Poeppel D (2007). The cortical organization of speech processing. Nat Rev Neurosci 8: 393–402.

Hickok G, Houde J, Rong F (2011). Sensorimotor integration in speech processing: computational basis and neural organization. Neuron 69: 407–422.

Joris PX (2009). Recruitment of neurons and loudness. Commentary. J Assoc Res Otolaryngol 10: 1–4.

Kilgard MP, Merzenich MM (1998). Cortical map reorganization enabled by nucleus basalis activity. Science 279: 1714–1718.

Langers DRM, van Dijk P, Schoenmaker ES (2007). fMRI activation in relation to sound intensity and loudness. Neuroimage 35: 709–718.

Lasota KJ, Ulmer JL, Firszt JB et al. (2003). Intensity-dependent activation of the primary auditory cortex in functional magnetic resonance imaging. J Comput Assist Tomogr 27: 213–218.

Lockwood AH, Salvi RJ, Coad ML et al. (1999). The functional anatomy of the normal human auditory system: responses to 0.5 and 4.0 kHz tones at varied intensities. Cereb Cortex 9: 65–76.

Lomber SG, Bertram R, Payne BP et al. (1999). The cryoloop: an adaptable reversible cooling deactivation method for behavioral or electrophysiological assessment of neural function. J Neurosci Meth 86: 179–194.

Malone BJ, Scott BH, Semple MN (2010). Temporal codes for amplitude contrast in auditory cortex. J Neurosci 30: 767–784.

Marks LE, Florentine M (2011). Measurement of Loudness, Part I: Methods, Problems, and Pitfalls. In: M Florentine, AN Popper, RR Fay (Eds.), Springer Handbook of Auditory Research Vol. 37, Loudness. New York, Springer, pp. 17–56.

Micheyl C, Schrater PR, Oxenham AJ (2013). Auditory frequency and intensity discrimination explained using a cortical population rate code. PLoS Comput Biol 9: e1003336.

Moore BCJ (2012). An Introduction to Hearing, 4th edn. Academic Press, London.

Morgan CT, Garner WR, Galambos R (1951). Pitch and intensity. J Acoust Soc Am 23: 658–663.

Mulert C, Jäger L, Propp S et al. (2005). Sound level dependence of the primary auditory cortex: Simultaneous measurement with 61-channel EEG and fMRI. Neuroimage 28: 49–58.

Phillips DP, Semple MN, Calford MB et al. (1994). Level-dependent representation of stimulus frequency in cat primary auditory cortex. Exp Brain Res 102: 210–226.

Polley DP, Heiser MA, Blake DT et al. (2004). Associative learning shapes the neural code for stimulus magnitude in primary auditory cortex. Proc Natl Acad Sci U S A 101: 16351–16356.

Rauschecker JP (2012). Ventral and dorsal streams in the evolution of speech and language. Front Evol Neurosci 15: 7.

Rauschecker JP, Tian B (2000). Mechanisms and streams for processing of "what" and "where" in auditory cortex. Proc Natl Acad Sci U S A 97: 11800–11806.

Robles L, Ruggero MA (2001). Mechanics of the mammalian cochlea. Physiol Rev 81: 1305–1352.

Röhl M, Uppenkamp S (2012). Neural coding of sound intensity and loudness in the human auditory system. J Assoc Res Otolaryngol 13: 369–379.

Röhl M, Kollmeier B, Uppenkamp S (2011). Spectral loudness summation takes place in the primary auditory cortex. Hum Brain Mapp 32: 1483–1496.

Scharf B (1978). Comparison of normal and impaired hearing. I. Loudness, localization. Scand Audiol Suppl 6: 49–79.

Schreiner CE, Langner G (1997). Laminar fine structure of frequency organization in auditory midbrain. Nature 388: 383–386.

Schreiner CE, Winer JA (2007). Auditory cortex mapmaking: principles, projections, and plasticity. Neuron 56: 356–365.

Schreiner CE, Mendelson JR, Sutter ML (1992). Functional topography of cat primary auditory cortex: representation of tone intensity. Exp Brain Res 92: 105–122.

Scott BH, Malone BJ, Semple MN (2011). Transformation of temporal processing across auditory cortex of awake macaques. J Neurophysiol 105: 712–730.

Sigalovsky IS, Melcher JR (2006). Effects of sound level on fMRI activation in human brainstem, thalamic and cortical centers. Hear Res 215: 67–76.

Small Jr AM, Brandt JF, Cox PG (1962). Loudness as a function of signal duration. J Acoust Soc Am 34: 513–514.

Sutter ML, Schreiner CE (1995). Topography of intensity tuning in cat primary auditory cortex: single-neuron versus multiple-neuron recordings. J Neurophysiol 73: 190–204.

Tan AY, Atencio CA, Polley DB et al. (2007). Unbalanced synaptic inhibition can create intensity-tuned auditory cortex neurons. Neuroscience 146: 449–462.

Tanji K, Leopold DA, Ye FQ et al. (2010). Effect of sound intensity on tonotopic fMRI maps in the unanesthetized monkey. Neuroimage 49: 150–157.

Winer JA, Schreiner CE (2010). The auditory cortex, Springer, New York.

Zhou M, Tao HW, Zhang LI (2012). Generation of intensity selectivity by differential synaptic tuning: fast-saturating excitation but slow-saturating inhibition. J Neurosci 32: 18068–18078.

Zwicker E, Fastl H (1999). Psychoacoustics — facts and models. Springer Series on Information Science, 2nd edn. Springer, Berlin.

Zwicker E, Flottorp G, Stevens SS (1957). Critical bandwidth in loudness summation. J Acoust Soc Am 29: 548–557.

Handbook of Clinical Neurology, Vol. 129 (3rd series)
The Human Auditory System
G.G. Celesia and G. Hickok, Editors

Chapter 5

Temporal coding in the auditory cortex

LUC H. ARNAL[1,2], DAVID POEPPEL[2], AND ANNE-LISE GIRAUD[1*]
[1]*Department of Neurosciences, University Medical Centre, Geneva, Switzerland*
[2]*Department of Psychology, New York University, New York, NY, USA*

OVERVIEW

Of all the signals the human auditory system has to process, the one with the most compelling relevance to the listener is arguably speech. Speech perception is learned and executed with automaticity and great ease, even by very young children, but is handled surprisingly poorly by even the most sophisticated automatic devices. Parsing and decoding speech can hence be considered one of the main challenges of the auditory system. This chapter focuses on how the human auditory cortex uses the temporal structure of the acoustic signal to extract phonemes and syllables, the two major types of events that need to be identified in connected speech.

Speech is a complex "multiplexed" acoustic signal exhibiting a quasiperiodic behavior at several timescales. The neural signals recorded from the auditory cortex using electroencephalography (EEG) or magnetoencephalography (MEG) also show a quasiperiodic structure, whether in response to speech or not. In this chapter we review recent neurophysiologic models of speech perception, grounded on the assumption that the quasiperiodic structure of collective neural activity in auditory cortex represents the ideal mechanical infrastructure to solve the speech demultiplexing problem, i.e., the fractioning of speech into linguistic constituents of variable size. The theoretic models presented here remain largely hypothetic. That being said, we believe that they constitute exciting new hypotheses, and should lead to new research questions and incremental progress on this foundational question about human perception.

The chapter proceeds as follows. First, some of the essential features of natural and speech auditory stimuli are outlined. Next, the properties of auditory cortex that reflect its sensitivity to these features are reviewed, and finally current ideas about the neurophysiologic mechanisms underpinning the processing of connected speech are discussed.

Timescales in auditory perception

Sounds are audible over a broad frequency range between 20 and 20 000 Hz. They enter the outer ear and travel through the middle ear to the inner ear, where they provoke the basilar membrane to vibrate at a specific location, depending on the sound frequency. Low and high frequencies induce vibrations of the apex and base of the basilar membrane, respectively. The deformation of the membrane upon acoustic stimulation provokes the deflection of inner hair cell ciliae, and the emission of a neural signal to cochlear neurons, subsequently transmitted to neurons of the cochlear nucleus in the brainstem. Each cochlear neuron is sensitive to a specific range of acoustic frequencies between 20 Hz and 20 kHz. Owing to their regular position along the basilar membrane, the cochlear neurons ensure the place coding of acoustic frequencies, also called "tonotopy," which is preserved up to the cortex (Moerel et al., 2013; Saenz and Langers, 2014).

Acoustic fluctuations below 20 Hz are not audible. They do not elicit place-specific response in the cochlea. Low frequencies <300 Hz are present in complex sounds as temporal fluctuations of audible frequencies, and are encoded through the discharge rate of cochlear neurons (Zeng, 2002). There is hence a range of frequencies from 20 to 300 Hz that are both place- and rate-coded at the auditory periphery. Temporal modulation of sounds in these frequencies typically elicits a sensation of pitch. Figure 5.1A and B summarizes the correspondence between categories of perceptual attributes and the sound modulation frequency (see also Nourski and Brugge, 2011 for a review). When sounds are modulated at very slow

*Correspondence to: Anne-Lise Giraud, Department of Neuroscience, University Medical Centre (CMU), 1, rue Michel-Servet, 1211 Geneva, Switzerland. Tel: +41-(0)223795547, Fax: +41-(0)223795452, E-mail: Anne-Lise.Giraud@unige.ch

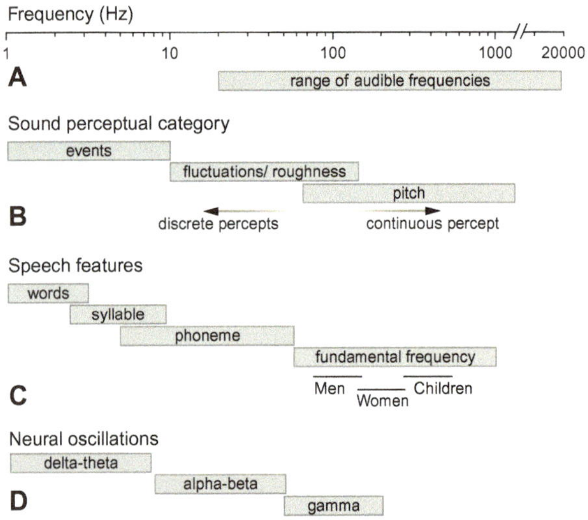

Fig. 5.1. (**A**) Scale of perceived temporal modulation. (**B**) Relevant psychophysical parameters (perceptual changes) of the spectrogram reflect the temporal constraints that superimpose on the structure of linguistic signals. (**C**) Temporal structure of linguistic features. (**D**) The length of linguistic features remarkably matches the frequency of oscillations that are observed at rest in the brain. Note that the frequency ranges at which auditory percepts switch from discrete (flutter) to continuous (pitch) roughly match the upper limit at which gamma rhythms can be entrained by the stimulus (≈200 Hz). (Modified from Joris et al., 2004, with permission from the American Physiological Society, and from Nourski and Brugge, 2011).

rates <10 Hz, a sequence of distinct events is perceived. When modulations accelerate from 10 to about 100 Hz distinct events merge into a single auditory stream, and the sensation evolves from fluctuating magnitude to a sensation of acoustic roughness (Fig. 5.1B).

Speech sounds are complex acoustic signals that involve only the lower part of audible frequencies (20–8000 Hz). They are "complex" in the sense that both their frequency distribution and their magnitude vary strongly and quickly over time. In natural speech, amplitude modulations (AMs) at slow (<20 Hz) and fast (>100 Hz) timescales are coupled (Fig. 5.2) and slower temporal fluctuations modulate the amplitude of spectral fluctuations. Current views suggest that slow modulations (<5 Hz) signal word and syllable boundaries (Hyafil et al., 2012), which are hence perceived as a sequence of distinct events, whereas phonemes (speech sounds) are signaled by fast spectrotemporal modulations (<30 Hz). They can be perceived as distinct events only when being discriminated from each other. Faster modulations, such as those imposed by the glottal pulse (100–300 Hz), indicate the voice pitch (Fig. 5.1C). Figure 5.1D shows how these perceptual events relate to the different frequency ranges of the EEG.

The temporal structure of speech sounds

Figure 5.2 illustrates two useful ways to visualize speech signals: as a waveform (A) and as a spectrogram (B). The waveform represents energy variation over time – the input that the ear actually receives. The outlined "envelope" (thick line) reflects that there is a temporal regularity in the signal at relatively low modulation frequencies. These modulations of signal energy (in reality, spread out across the "cochlear" filterbank) are below 20 Hz and peak roughly at a rate of 4–6 Hz (Steeneken and Houtgast, 1980; Elliott and Theunissen, 2009). From the perspective of what auditory cortex receives as input, namely the modulations at the output of each frequency

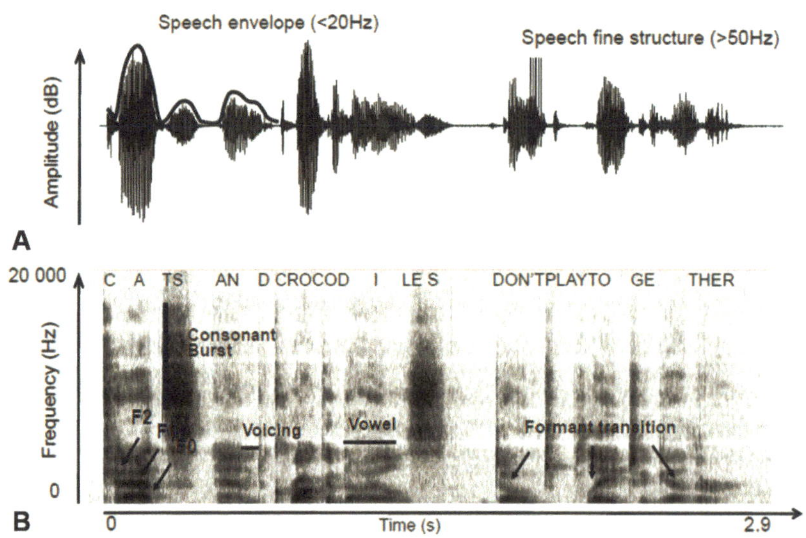

Fig. 5.2. (**A**) Waveform and (**B**) spectrogram of the same sentence uttered by a male speaker. Some of the key acoustic cues in speech comprehension are highlighted in black. (From Giraud and Poeppel, 2012b, with permission from Springer Science and Business Media.)

channel of the cochlear filterbank, these energy fluctuations can be characterized by the modulation spectrum (Kingsbury et al., 1998; Kanedera et al., 1999). At the shortest timescale (below 1 ms or equivalently above 1 kHz), the very fast temporal fluctuations are transformed into a spectral representation at the cochlea and the neural processing of these features is generally known as spectral processing. At an intermediate timescale (~70 Hz – 1 kHz), the temporal fluctuations are usually referred to as the temporal fine structure. The temporal fine structure is critical to the perception of pitch and interaural time differences that are important cues for sound source localization (Plack et al., 2005; Grothe et al., 2010). Temporal fluctuations on an even longer timescale (~1–10 Hz) are heard as a sequence of discrete events. Acoustic events occurring on this timescale include syllables and words in speech and notes and beats in music. Of course, there are no clear boundaries between these timescales; they are divided here based on human auditory perception.

The second analytic representation, the spectrogram, decomposes the acoustic signal in frequency, time, and amplitude domains (Fig. 5.2B). Although the human auditory system captures frequency information between 20 Hz and 20 kHz (as in Fig. 5.2), most of the information that is extracted for effective recognition lies below 8 kHz. It is worth remembering that speech transmitted over telephone landlines contains an even narrower bandwidth (200–3600 Hz) and is comfortably understood by normal listeners.

A number of critical acoustic features can be identified in the speech spectrogram. The faintly visible vertical stripes represent the glottal pulse, which reflects the speaker's fundamental frequency, F0. This can range from approximately 100 Hz (male adult) to 300 Hz (child; Fig. 5.1D). The horizontal bands of energy show where in the frequency space a particular speech sound is carried. The spectral structure thus reflects the articulator configuration. These bands of energy include the formants (F1, F2, etc.), definitional of vowel identity; high-frequency bursts associated, for example, with frication in certain consonants (e.g., /s/, /f/); and formant transitions that signal the change from a consonant to a vowel or vice versa.

Notwithstanding the perceptual importance of the spectral fine structure, there is a big caveat: speech can be understood, in the sense of being intelligible in psychophysical experiments, when the spectral structure is replaced by noise and only the envelope is preserved. For speech to remain intelligible, this manipulation should be done in separate bands across the spectrum. With training, it is still possible to grasp the speech content when the speech envelope, that is, temporal modulations of speech at relatively slow rates, is applied to only four separate frequency bands (e.g., Shannon et al., 1995). Such speech signals, containing only envelope but no fine structure information, are called vocoded speech (Faulkner et al., 2000). Compelling demonstrations, exemplified by this type of signal decomposition, illustrate that the speech signal can undergo radical alterations and distortions and yet remain intelligible (Shannon et al., 1995; Smith et al., 2002). Such findings have led to the idea that the temporal envelope is sufficient to yield speech comprehension (Rosen, 1992; Drullman et al., 1994a, b; Shannon et al., 1995; Giraud et al., 2004; Scott et al., 2006; Loebach and Wickesberg, 2008; Souza and Rosen, 2009). When using stimuli in which the fine structure is compromised or not available at all, envelope modulations below 16 Hz appear to suffice for adequate intelligibility. The remarkable comprehension level reached by most patients with cochlear implants, in whom about 15–20 electrodes replace 3000 hair cells, remains the best empiric demonstration that the spectral content of speech can be degraded with tolerable alteration of speech perception (Roberts et al., 2011).

A related demonstration showing the resilience of speech comprehension in the face of radical signal impoverishment is provided by sine-wave speech (Remez et al., 1981). In these stimuli both envelope and spectral content are degraded but enough information is preserved to permit intelligibility. Typically, sine-wave speech preserves the modulations of the three first formants, which are replaced by sine-waves centered on F0, F1, and F2. In sum, dramatically impoverished stimuli remain intelligible insofar as enough information in the spectrum is available to convey temporal modulations at appropriate rates.

Based on this brief and selective summary, two concepts merit emphasis: first, the extended speech signal contains critical information that is modulated at rates below 20 Hz, with the modulation peaking around 5 Hz (Edwards and Chang, 2013). This low-frequency information correlates closely with the syllabic structure of connected speech (Hyafil et al., 2012). Second, the speech signal contains critical information at modulation rates higher than, say, 50 Hz. This rapidly changing information is associated with fine spectral changes that carry information about the speaker's gender or identity and other relevant speech attributes (Elliott and Theunissen, 2009). Thus, there exist two surprisingly different timescales concurrently at play in the speech signal. This important issue is taken up in the text that follows. In this chapter, we discuss the timescales longer than 5 ms (<200 Hz) with a focus on the timescale between 100 ms and 1 second (1–10 Hz). Temporal features that contribute to the spatial localization of sounds will not be discussed (see e.g. Grothe et al., 2010 for a review).

ENCODING OF SPECTROTEMPORAL FEATURES IN THE AUDITORY CORTEX

Sensitivity to temporal modulations in the primary auditory cortex

Speech temporal variations in signal frequency and magnitude are at the basis of articulated speech and require specific neural encoding and decoding properties. A critical property is the sensitivity of auditory neurons to temporal modulations. Auditory neurons in subcortical nuclei (cochlear nucleus, inferior colliculus, and medial geniculate nucleus of the thalamus) phase-lock very well with fast trains of stimuli presented at rates over 200 Hz (Fishman et al., 2000; Brugge et al., 2009; see also Joris et al., 2004 and Nourski and Brugge, 2011 for reviews on the topic). This property to adjust the response rate to the rate of acoustic modulations gradually diminishes along the auditory hierarchy (see Sharpee et al., 2011; for a review). In the primary auditory cortex, the firing rate of a large number of neurons is phase-locked to slow temporal modulations below 30 Hz (Liang et al., 2002; Malone et al., 2010; Yin et al., 2011). Progressive slowing of phase-locking properties across the auditory hierarchy has been observed across species, e.g., monkeys, cats, and ferrets, for both anesthetized and awake animals. The degree of neural phase locking to temporal AM is often characterized as a function of the modulation rate, using temporal modulation transfer functions.

In the auditory cortex, the encoding of temporal features is implemented using at least two complementary strategies. Some neurons are sensitive to slow modulation rates, and the degree of phase locking decreases with increasing rates. Other neurons are tuned to a preferred modulation rate and the degree of phase locking decreases when the modulation rate deviates from the preferred one. Some neurons even show multiple preferred modulation rates (Malone et al., 2010; Yin et al., 2011). Single-unit recordings from the primary auditory cortex of awake marmoset suggest that AM is encoded by different neural codes and/or different neurons (Wang, 2007; Wang et al., 2008). One population of neurons, called the synchronized population, encodes AM by spikes phase-locked to the stimulus envelope. A second neural population, called the non-synchronized population, encodes AM by the mean firing rate rather than the timing of spikes. The two neural populations in general do not spatially overlap. While the synchronized population encodes slow temporal modulations (mostly below 50 Hz), the non-synchronized population encodes faster temporal modulations (mostly above 50 Hz). Studies in awake macaque monkey, however, suggested that neurons might actually encode both slow temporal modulations by phase-locked activity and fast temporal modulations by the firing rate of non-phase-locked activity (Malone et al., 2010). In other words, the same neuron might encode slow and fast modulations, suggesting that primary auditory cortex neurons may carry multiplexed temporal and rate codes.

Although single auditory cortical neurons generally cannot phase lock to temporal modulations above 100 Hz (Wang, 2007), larger-scale recordings such as local field potentials and human MEG/EEG show phase locking to the stimulus modulations beyond 100 Hz. This suggests that macro-scale measurements that spatially integrate neuronal responses at the population level reflect the ability of auditory cortical regions to maintain a – spatially distributed – representation of higher temporal frequencies. In humans, intracranial recordings from the core auditory cortex and the lateral surface of posterolateral superior temporal gyrus show that phase-locked neural activity is most prominent below 50 Hz but remains measurable up to ~200 Hz (Brugge et al., 2009). Similarly, the human MEG response also shows phase locking up to 100 Hz (Lehongre et al., 2011; Miyazaki et al., 2013). That neural oscillations phase lock to exogenous entrainment up to this frequency possibly determines the transition between discrete and continuous auditory percepts (Fig. 5.1B).

Sensitivity to spectrotemporal modulations

Speech signals are characterized by modulations in both spectral and temporal domains (frequency modulations (FM) and AM). Two separate possible codes to represent complex stimuli such as speech have been implicated in the preceding text, a place code for FMs and a temporal code for AMs. Whether the encoding of frequency and AMs is implemented by a single or by distinct mechanisms is a long-lasting question, whether at the periphery or at the cortical level. The idea of a single code for spectrotemporal modulations is supported by the presence of neurons that respond to FMs but not AMs (Gaese and Ostwald, 1995) and by complex responses to spectrotemporal modulations (Pienkowski and Eggermont, 2009; Schonwiesner and Zatorre, 2009). Luo et al. (2006, 2007) and Ding and Simon (2009) tested, based on MEG recordings in human listeners, whether FM and AM used the same coding principles (Fig. 5.3). The authors argue that if coding equivalence (or similarity) is the case, cortical responses as assessed by MEG should be the same when the carrier of slow AM is rapidly frequency-modulated, or when a slowly changing carrier sound is amplitude-modulated at fast rate (AM–FM comodulation experiments). Yet, they observed that only the phase of fast AM auditory responses (auditory steady-state responses at 40 Hz) is modulated by slow FM, while both the phase and the amplitude of fast

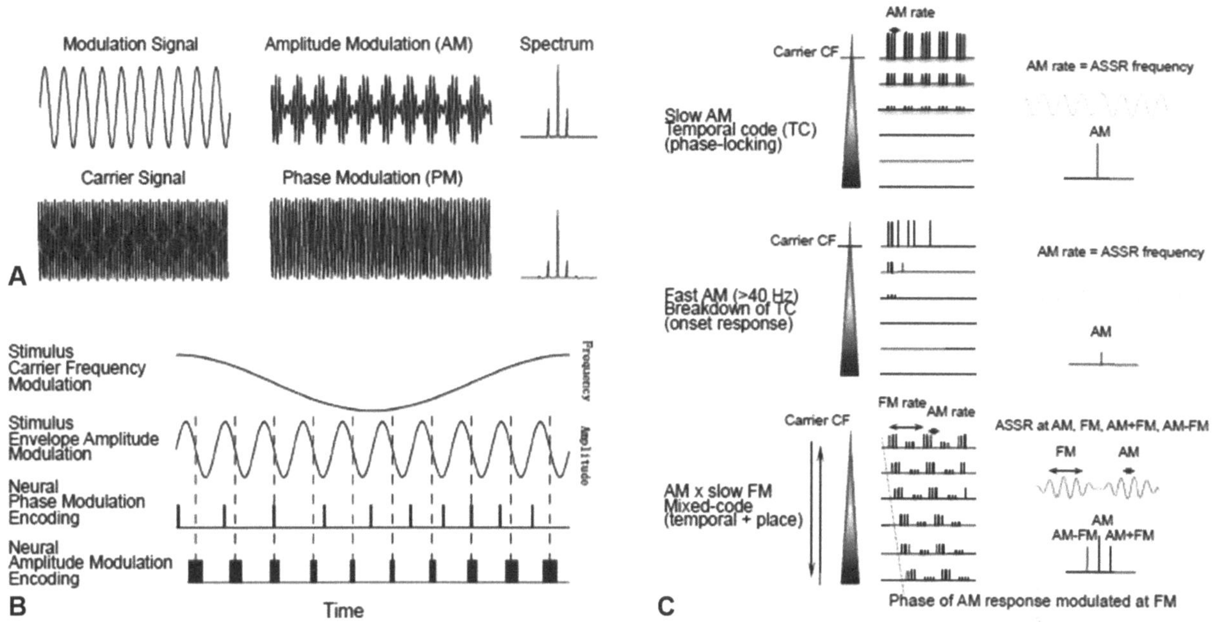

Fig. 5.3. Principles of amplitude and frequency modulation encoding in auditory cortex. (**A**) In radio engineering, modulation is used to encode acoustic stimuli, which can be either amplitude-modulated (AM; upper row) or phase-modulated (PM; second row). (**B**) Proposals for neural AM and PM encoding. A stimulus is made of a frequency-varying signal (upper row) and an amplitude modulation (second row). Using a PM encoding (third row), a neuron fires one spike per stimulus envelope cycle (dotted line) and the spikes' precise timing (phase) depends on the carrier frequency. Alternatively, using AM encoding (last row), a neuron changes its firing rate according to the instantaneous frequency of the carrier, while keeping constant the firing phase. (**C**) AM coding is illustrated in more details in three different conditions: slow AM (upper row), fast AM (second row), and when AM and PM covary (last row). CF, characteristic frequency. ASSR, auditory steady-state responses. (From Luo et al., 2006, with permission from the American Physiological Society.)

FM auditory responses (auditory steady-state responses at 40 Hz) are modulated by slow AM. That AM and FM interact non-linearly is beyond doubt. However, the mere fact that the spectral place coding, present in several auditory territories, plays a more important role in FM processing than in AM processing could account for the asymmetry in the results (Barton et al., 2012). Whereas FM, by hypothesis, is encoded by a combination of place and temporal coding, AM is mostly encoded by temporal coding. As a consequence, critical features of speech signals may plausibly be encoded based on processing units that have a tonotopic axis and incorporate distinct thresholds for temporal stimulus modulations.

The asymmetric response pattern to fast and slow AM/FM might also depend on coding differences for fast and slow modulations. Whereas very slow FMs are perceived as pitch variations, fast modulations are perceived as varying loudness. On the other hand, slow-amplitude modulations are perceived as variations of loudness, whereas fast modulations are perceived as roughness, or flutter, or pitch (Fig. 5.1). These sharp perceptual transitions could be underpinned by both the size and the place of the population recruited by each of these stimulus types. Whereas slow FM presumably allows for both a temporal and spatial segregation of cortical responses, entailing distinct percepts varying in pitch, fast FM presumably phase locks together the entire population stimulated by the varying carrier. In a similar way, fast AM is possibly no longer perceived as variations of loudness when the ability of neurons to phase lock is overridden (beyond 40 Hz). Flutter (and then pitch sensations) for AM higher than 40 Hz superimposed on the primary spectral content of the modulated sound might reflect the additional excitation of (pitch) neurons with very low characteristic frequency. The spectral place code, the transition from phase locking to rate coding for higher stimulus rates, and ensemble neuronal behavior, that is, the size of the population targeted by a stimulus, provides enough representational complexity to account for non-linear neuronal responses to spectrotemporal acoustic modulations without invoking a specific AM/FM code.

CORTICAL PROCESSING OF CONTINUOUS SOUND STREAMS

The discretization problem

Most of the aforementioned experimental settings involved artificial sinusoidally modulated stimuli. Yet, natural sounds, speech in particular, present temporal modulations that are not strictly periodic. On the other hand, experimental research on speech has focused on the processing of individually presented speech sounds, such as vowels, syllables, or single words. The related findings underpin most current models of speech perception. However, in natural connected speech, speech information is embedded in a continuous acoustic flow, and sentences are not "pre-segmented" in perceptual units of analysis. Recent work on sentence-level stimuli (i.e., materials with a duration exceeding 1–2 seconds), using experimental tasks as intelligibility, demonstrate the fundamental importance of long-term temporal parameters of the acoustic signal. Online segmentation remains a major challenge to contemporary models of speech perception as well as automatic speech recognition.

Interestingly, a large body of psychophysical work studied speech perception and intelligibility using phrasal or sentential stimuli (see, e.g., Miller, 1951 for a summary of many experiments and Allen, 2005 for a review of the influential work of Fletcher and others). Fascinating findings emerged from that work, emphasizing the role of signal-to-noise ratio in speech comprehension, but perhaps the most interesting feature is that connected speech has principled and useful temporal properties that may play a key role in the problem of speech parsing and decoding. Natural speech usually comes to the listener as a continuous stream and needs to be analyzed online and decoded by mechanisms that are unlikely to be continuous (Giraud and Poeppel, 2012a). The parsing mechanism corresponds to the discretization of the continuous input signal into subsegments of speech information that are read out, to a certain extent, independently from each other. The notion that perception is discrete has been extensively discussed and generalized in numerous sensory modalities and contexts (Pöppel, 1988; VanRullen and Koch, 2003; VanRullen et al., 2014). Here we discuss the hypothesis that neural oscillations constitute a possible mechanism for discretizing temporally complex sounds such as speech (Giraud and Poeppel, 2012a).

Analysis at multiple timescales

Speech is a multiplexed signal, that is, it interlinks several levels of complexity, and organizational principles and perceptual units of analysis exist at distinct timescales. Using data from linguistics, psychophysics, and physiology, Poeppel and colleagues proposed that speech is analyzed in parallel at multiple timescales (Poeppel, 2001, 2003; Boemio et al., 2005; Poeppel et al., 2008). The central idea is that both local-to-global and global-to-local types of analyses are carried out concurrently (multitime resolution processing). This assumption adds to the notion of reverse hierarchy (Hochstein and Ahissar, 2002; Nahum et al., 2008) and other hierarchic models in perception, which propose that the hierarchic complexification of sensory information (e.g., the temporal hierarchy) maps on to the anatomofunctional hierarchy of the brain (Giraud et al., 2000; Kiebel et al., 2008). The motivations for extending such a hypothesis are twofold. First, a single, short temporal window that forms the basis for hierarchic processing, that is, increasingly larger temporal analysis units as one ascends the processing system, fails to account for the spectral and temporal sensitivity of the speech-processing system and is hard to reconcile with behavioral performance. Second, the computational strategy of analyzing information on multiple scales is widely used in engineering and biologic systems, and the neuronal infrastructure exists to support multiscale computation (Canolty and Knight, 2010). According to the view summarized here, speech is chunked into segments of roughly featural or phonemic length, and then integrated into larger units, as segments, diphones, syllables, words. In parallel, there is a fast global analysis that yields coarse inferences about speech (akin to Stevens' "landmarks" hypothesis: Stevens, 2002), and that subsequently refines segmental analysis. Here, we propose that segmental and suprasegmental analyses could be carried out concurrently and "packaged" for parsing and decoding by neuronal oscillations at different rates.

The notion that speech analysis occurs in parallel at multiple timescales justifies moving away from strictly hierarchic models of speech perception (e.g., Giraud and Price, 2001). Accordingly, the simultaneous extraction of different acoustic cues permits simultaneous high-order processing of different information from a unique input signal. That speech should be analyzed in parallel at different timescales derives, among other reasons, from the observation that articulatory–phonetic phenomena occur at different timescales.

It was noted previously (Fig. 5.2) that the speech signal contains events of different durations: short energy bursts and formant transitions occur within a 20–80-ms timescale, whereas syllabic information occurs over 150–300 ms. The processing of both types of events could be accounted for either by a hierarchic model in which smaller acoustic units (segments) are concatenated into larger units (syllables) or by a parallel model in which both temporal units are extracted independently, and then combined. A certain degree of independence in the processing of long (slow modulation) and short (fast modulation) units is observed at the behavioral level. For instance, speech can be understood well when it is first segmented into units up to 60 ms and when these local units are temporally reversed (Saberi and Perrott, 1999; Greenberg and Arai, 2001). Because the correct extraction of short units is not a prerequisite for comprehension, this rules out the notion that speech processing relies solely on hierarchic processing of short and then larger units. Overall, there appears to be a grouping of psychophysical phenomena such that some cluster at thresholds of approximately 50 ms and below and others cluster at approximately 200 ms and above (a similar clustering is observed for temporal properties in vision; Holcombe, 2009).

Importantly, non-speech signals are subject to similar thresholds. For example, 15–20 ms is the minimal stimulus duration required for correctly identifying upward versus downward FM sweeps (Luo et al., 2007). By comparison, 200 ms stimulus duration underlies loudness judgments. In sum, physiologic events at related scales form the basis for processing at that level. Therefore, the neuronal oscillatory machinery (together with motor constraints related to speech production; Morillon et al., 2010) presumably imposed strong temporal constraints that might have shaped the size of acoustic features selected to carry speech information. This is consistent with the notion that perception is discrete and that the exogenous recruitment of neuronal populations is followed by refractory periods that temporarily reduce the ability to optimally extract sensory information (Ghitza and Greenberg, 2009; Ghitza, 2011). According to this hypothesis, the temporally limited capacity of gamma oscillations to integrate information over time possibly imposes a lower limit to the phoneme length. This also suggests that oscillatory constraints in the delta-theta range possibly constrained the size of syllables to be roughly the size of a delta-theta cycle. Considering that the average length of phoneme and syllable is about 25–80 ms and 150–300 ms respectively (Figs. 5.1 and 5.2), the dual timescale segmentation requires two parallel sampling mechanisms, one at about 40 Hz (or, more broadly, in the low gamma range) and one at about 4 Hz (or in the theta range).

Neural oscillations as endogenous temporal constraints

Neural oscillations correspond to synchronous activity of neuronal assemblies that are both intrinsically coupled and coupled by a common input. It was proposed that these oscillations reflect modulations of neuronal excitability that temporally constrain the sampling of sensory information (Schroeder and Lakatos, 2009a). The intriguing correspondence between the size of certain speech temporal units and the frequency of oscillations in certain frequency bands (Fig. 5.1) has elicited the intuition that they might play a functional role in sensory sampling (see below). Oscillations are evidenced by means of a spectrotemporal analysis of electrophysiologic recordings (see Wang, 2010, for a review). The requirements for measuring oscillations and spiking activity are different. The presentation of an exogenous stimulus typically results in an increase of spiking activity in those brain areas that are functionally sensitive to such inputs. Neural oscillations, on the other hand, can be observed in local field potential recordings in the absence of any external stimulation. Exogenous stimulation however typically modulates oscillatory activity, resulting either in a reset of their phase and/or a change (increase or decrease) in the magnitude of these oscillations (Howard and Poeppel, 2012).

Cortical oscillations are proposed to shape spike-timing dynamics and to impose phases of high and low neuronal excitability (Britvina and Eggermont, 2007; Schroeder and Lakatos, 2009a, b; Panzeri et al., 2010). The assumption that it is oscillations that cause spiking to be temporally clustered derives from the observation that spiking tends to occur in specific phases (i.e., the trough) of oscillatory activity (Womelsdorf et al., 2007). It is also assumed that spiking and oscillations do not reflect the same aspect of information processing. Whereas spiking reflects axonal activity, oscillations are said to reflect mostly dendritic synaptic activity (Wang, 2010). While both measures are relevant to address how sensory information is encoded in the brain, we believe that the ability of neural oscillations to temporally organize spiking activity supports the functional relevance of neural oscillations to solve the discretization problem and to permit the integration of complex sensory signals across time.

Neuronal oscillations are ubiquitous in the brain, but they vary in strength and frequency depending on their location and the exact nature of their neuronal generators (Mantini et al., 2007; Hyafil et al., 2012). The notion that neural oscillations shape the way the brain processes sensory information is supported by a wealth of electrophysiologic findings in humans and animals. On the one hand, stimuli that occur in the ideal excitability phase of slow oscillations ($<$12 Hz) are processed faster and with a higher accuracy (Lakatos et al., 2008; Busch et al., 2009; Henry and Obleser, 2012; Ng et al., 2012; Wyart et al., 2012). On the other hand, gamma-band 40-Hz activity (low gamma band) can be observed at rest in both monkey (Fukushima et al., 2012) and human auditory cortex. In humans, it can be measured using EEG, MEG, and with a more precise localization with concurrent EEG and functional magnetic resonance imaging (Morillon et al., 2010) and intracranial electroencephalographic recordings (stereotactic EEG (sEEG), Electrocorticography (EcoG)) in patients. Neural oscillations in this range are endogenous in the sense that one can observe a spontaneous spike clustering at approximately 40 Hz even in the absence of external stimulation. This gamma activity is thought to be generated by a "ping-pong" interaction between pyramidal cells and inhibitory interneurons (Borgers et al., 2005, 2008), or even just among interneurons that are located in superficial cortical layers (Tiesinga and Sejnowski, 2009). Exogenous inputs usually increase gamma-band activity in sensory areas, presumably clustering spiking activity that is propagated to higher hierarchic processing stages (Arnal et al., 2011; Arnal and Giraud, 2012; Bastos et al., 2012). By analogy with the proposal of Elhilali et al. (2004) that slow responses gate faster ones, it is interesting to envisage this periodic modulation of spiking by oscillatory activity as an endogenous mechanism to optimize the extraction of relevant sensory input in time. Such integration could occur under the patterning of slower oscillations in the delta-theta range.

Alignment of neuronal excitability with speech timescales

Experimental exploration of how speech parsing and encoding is carried out by the brain is non-trivial. One approach has been to explore how neural responses can discriminate different sentences, assuming that the features of neural signals that are sensitive to such differences (e.g., frequency band, amplitude, phase) should reveal the features that are key to sentence decoding. Using this approach, it was shown that the phase of theta-band neural activity reliably discriminates different sentences (Luo and Poeppel, 2007). Specifically, when one sentence is repeatedly presented to listeners, the phase of ongoing theta-band activity follows a consistent phase sequence. When different sentences are played, however, different phase sequences are observed. Since theta-band (4–8 Hz) falls around the mean syllabic rate of speech (~5 Hz), the phase of theta-band activity likely tracks syllabic-level features of speech (Giraud and Poeppel, 2012a; Hyafil et al., 2012; Edwards and Chang, 2013). These findings support the notion that the syllabic timescale has adapted to a pre-existing cortical preference for temporal information in this frequency range. At this point, however, it is not clear whether the phase locking between the speech input and neural oscillations is necessary for speech intelligibility. On the one hand, sentences played backward (and therefore unintelligible) can similarly be discriminated on the basis of their phase course, which tempers the interpretation that these oscillations play a causal role in speech perception (Howard and Poeppel, 2011). On the other hand, two recent studies using distinct ways of acoustically degrading speech intelligibility demonstrate that the temporal alignment between the stimulus and delta-theta band responses is higher when the stimulus is intelligible (Peelle et al., 2013; Doelling et al., 2014). This, again, supports the notion that those neural oscillations that match the slow (syllabic) speech timescales are useful (if not necessary) for the extraction of relevant speech information.

Neural oscillatory responses can also be entrained at much higher rates in the middle to high (40—200 Hz) gamma band (Fishman et al., 2000; Brugge et al., 2009). This could suggest that faster speech segments such as phonemic transitions could be extracted using the same encoding principle. High gamma responses in early auditory regions (Ahissar et al., 2001; Nourski et al., 2009; Mesgarani and Chang, 2012; Morillon et al., 2012) reflect the fast temporal fluctuations in the speech envelope. A recent EcoG study succeeded at reconstructing the original speech input using a combination of linear and non-linear methods to decode neural responses from high gamma activity in auditory cortical regions (Pasley et al., 2012). Therefore, the decoding of auditory activity on a large spatial scale (at the population level) demonstrates that the auditory cortex maintains a high-fidelity representation of temporal modulations up to 200 Hz. However, according to psychophysiologic findings described earlier, speech intelligibility mostly relies on the preservation of the low-frequency (<50 Hz) temporal fluctuations rather than on higher-frequency information. Therefore, whether it is necessary to maintain a representation of such acoustic features to correctly perceive speech remains unclear. The following section aims at clarifying the putative neural mechanisms underpinning the segmentation and the integration of auditory speech signals into an intelligible percept.

Parallel processing at multiple timescales

Schroeder and Lakatos (2009a, b) have argued that oscillations correspond to the alternation of phases of high and low neuronal excitability, which temporally constrain sensory processing. This means that gamma oscillations, which have a period of approximately 25 ms, provide a 10–15-ms window for integrating spectrotemporal information (low spiking rate) followed by a 10–15-ms window for propagating the output (high spiking rate; see, for illustration, Fig. 5.4A). However, because the average length of a phoneme is about 50 ms, a 10–15-ms window might be too short for integrating this information. Using a computational model of gamma oscillations generated by a pyramidal interneuron network (PING model: Borgers et al., 2005; Shamir et al., 2009) shows that the shape of a sawtooth input signal designed to have the typical duration and AM of a diphone (~50 ms; typically a consonant–vowel or vowel–consonant transition) can correctly be represented by three gamma cycles, which act as a three-bit code. Such a code has the capacity required to distinguish different shapes of the stimulus and is therefore a plausible means

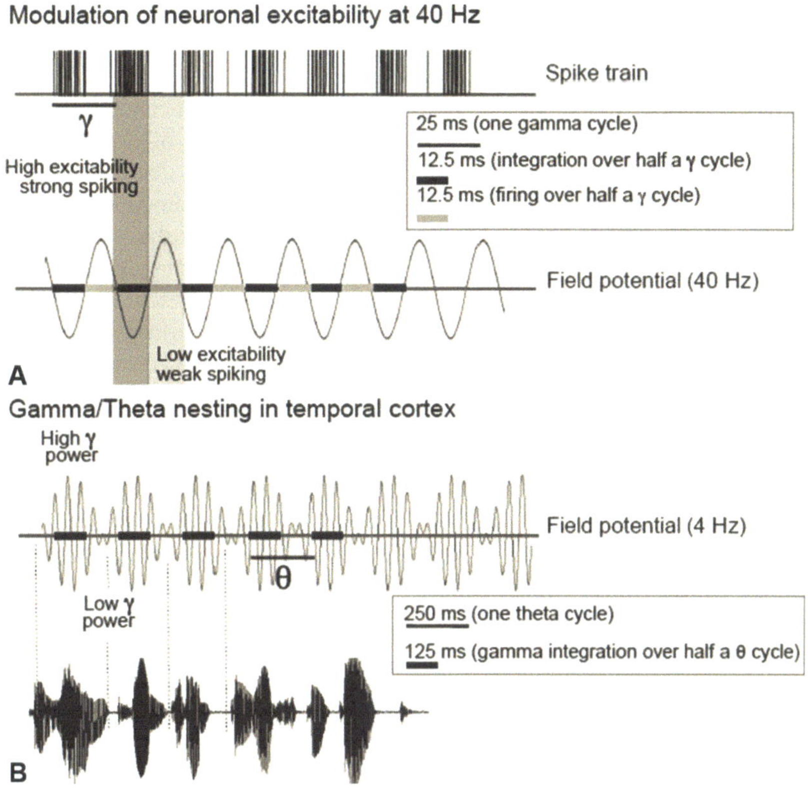

Fig. 5.4. The temporal relationship between speech and brain oscillations. (**A**) Gamma oscillations periodically modulate neuronal excitability and spiking. The hypothesized mechanism is that neurons fire for about 12.5 ms and integrate for the rest of the 25-ms time window. Note that these values are approximate, as we consider the relevant gamma range for speech to lie between 28 and 40 Hz. (**B**) Gamma power is modulated by the phase of the theta rhythm (about 4 Hz). Theta rhythm is reset by speech, resulting in maintaining the alignment between brain rhythms and speech bursts.

to distinguish between phonemes. That 50-ms diphones could be correctly discriminated with three gamma cycles suggests that phonemes could be sampled with one/two gamma cycles. This issue is critical, as the frequency of neural oscillations in the auditory cortex might constitute a strong biophysical determinant with respect to the size of the minimal acoustic unit that can be manipulated for linguistic purposes.

In a recent extension of this model, the parsing and encoding capacity of coupled theta and gamma oscillating modules was studied (Hyafil et al., 2012). In combination, these modules succeed in signaling syllable boundaries and to orchestrate spiking within syllabic windows, so that online speech decoding becomes possible with a similar accuracy as experimental findings using intracortical recordings in monkeys (Kayser et al., 2012).

An important requirement of the computational model mentioned previously (Shamir et al., 2009) is that ongoing gamma oscillations are phase-reset, for example, by a population of onset excitatory neurons. In the absence of this onset signal the performance of the model drops. Ongoing intrinsic oscillations appear to be effective as a segmenting tool only if they align with the stimulus. Schroeder and colleagues suggest that gamma and theta rhythms work together, and that the phase of theta oscillations determines the power and possibly also the phase of gamma oscillations (Fig. 5.4B; Schroeder et al., 2008). This cross-frequency relationship is referred to as "nesting." Electrophysiologic recordings suggest that theta oscillations can be phase-reset by several means, through multimodal corticocortical pathways (Lakatos et al., 2007; Arnal et al., 2009; Thorne et al., 2011) or through predictive top-down modulations, but most probably by the stimulus onset itself (Fig. 5.4B). This phase reset would align the speech signal and the cortical theta rhythm, the proposed instrument of speech segmentation into syllable/word units. As speech is strongly amplitude-modulated at the theta rate, this would result in aligning neuronal excitability with those parts of the speech signals that are most informative in terms of energy and spectrotemporal content (Fig. 5.4B). There remain critical computational issues, such as the means to get strong gamma activity at the moment of theta reset. Recent psychophysical research emphasizes the importance of aligning the acoustic speech signal with the brain's oscillatory/quasi-rhythmic activity. Ghitza and Greenberg (2009) demonstrated that, while comprehension was drastically reduced by time-compressing speech signals by a factor of 3, comprehension was restored by artificially inserting periods of silence. The mere fact of restoring "syllabicity" by adding silent periods to speech improves performance, even though the speech segments that remained available are still compressed. Optimal performance is obtained when 80-ms silent periods alternate with 40-ms time-compressed speech signals. These time constants allowed the authors to propose a phenomenologic model involving three nested rhythms in the theta (5 Hz), beta, or low gamma (20–40 Hz) and gamma (80 Hz) domains (for an extended discussion, see Ghitza, 2011).

Parallel processing in bilateral auditory cortices

There is emerging consensus, based on neuropsychologic and imaging data, that speech perception is mediated bilaterally. Poeppel (2003) attempted to integrate and reconcile several of the strands of evidence: first, speech signals contain information on at least two critical timescales, correlating with segmental and syllabic information; second, many non-speech auditory psychophysical phenomena fall in two groups, with integration constants of approximately 25–50 ms and 200–300 ms; third, both patient and imaging data reveal cortical asymmetries such that both sides participate in auditory analysis but are optimized for different types of processing in left versus right; and fourth, crucially for the present chapter, neuronal oscillations might relate in a principled way to temporal integration constants of different sizes. Poeppel (2003) proposed that there are asymmetric distributions of neuronal ensembles between hemispheres with preferred shorter versus longer integration constants; these cell groups "sample" the input with different sampling integration constants (Fig. 5.5A). Specifically, left auditory cortex has a relatively higher proportion of short-term (gamma) integrating cell groups, whereas right auditory cortex has a larger long-term (theta) integrating proportion (Fig. 5.5B). As a consequence, left-hemisphere auditory cortex is likely better

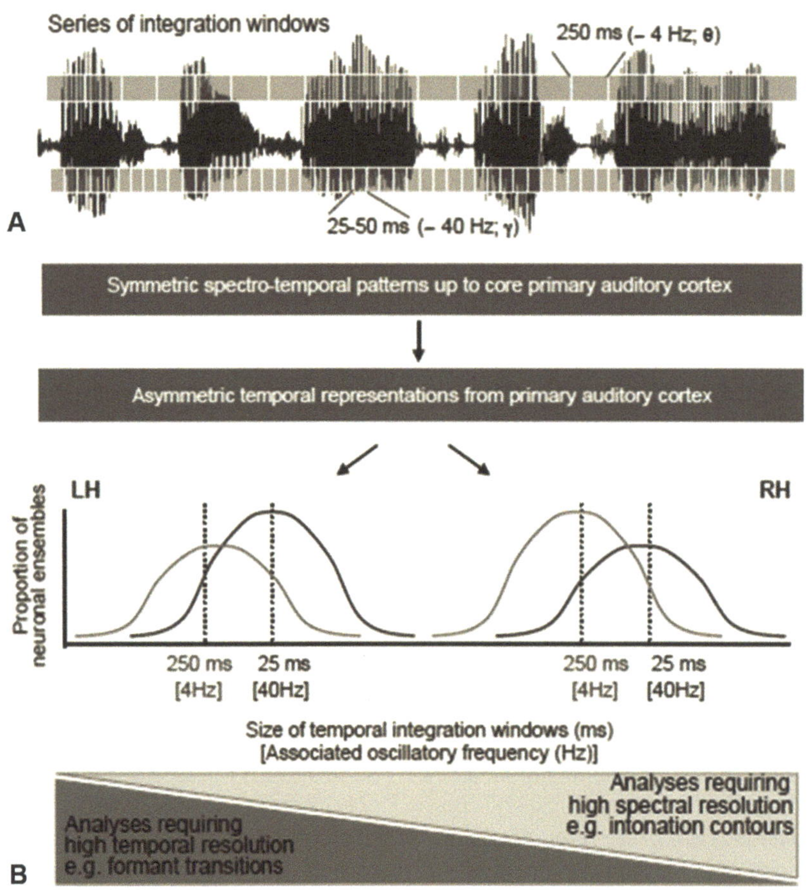

Fig. 5.5. The asymmetric sampling in time hypothesis. (**A**) Temporal relationship between the speech waveform and the two proposed integration timescales (in ms) and associated brain rhythms (in Hz). (**B**) Proposed mechanisms for asymmetric speech parsing: left auditory cortex (LH) contains a larger proportion of neurons able to oscillate at gamma frequency than the right one (RH). (From Giraud and Poeppel, 2012b, with permission from Springer Science and Business Media.)

equipped for parsing speech at the segmental scale, and right auditory cortex for parsing speech at the syllabic timescale. This hypothesis, referred to as the asymmetric sampling in time (AST) theory, is summarized in Figure 5.5. It accounts for a variety of psychophysical and functional neuroimaging results that show that left temporal cortex responds better to many aspects of rapidly modulated speech content, while right temporal cortex responds better to slowly modulated signals, including music, voices, and other sounds (Zatorre et al., 2002; Warrier et al., 2009). A difference in the size of the basic integration window between left and right auditory cortices would explain speech functional asymmetry by a better sensitivity of left auditory cortex to information carried in fast temporal modulations that convey, for example, phonetic cues. A specialization of right auditory cortex to slower modulations would grant it a better sensitivity to slower and stationary cues such as harmonicity and periodicity (Rosen, 1992) that are important to identify vowels, syllables, and thereby speaker identity. The AST theory is very close, in kind, to the spectrotemporal asymmetry hypothesis promoted by Zatorre (e.g., Zatorre et al., 2002; Zatorre and Gandour, 2008), which originally proposed that, whereas the left auditory cortex is better suited to process temporal information, the right auditory cortex is better at processing spectral information. While many psychophysics and neurophysiologic experiments seem to support this idea (see Poeppel, 2003; Poeppel et al., 2008 and Giraud and Poeppel, 2012a for reviews on the topic) there is a lot of work in progress regarding this unresolved question.

Dysfunctional oscillatory sampling

Additional evidence to support the notion that neural oscillations play an instrumental role in speech processing would be to show that dysfunctional oscillatory mechanisms result in speech-processing impairments. Dyslexia, which is a phonologic deficit, i.e., a deficit in processing speech sounds, presumably constitutes a good candidate to test this hypothesis. Temporal sampling mediated by cortical oscillations has recently been proposed to be a central mechanism in several aspects of dyslexia (Goswami, 2011). This proposal suggests that a deficit involving theta oscillations might impair the tracking of low temporal modulations in the syllabic range. In a complementary way, it was proposed recently that gamma oscillations might play a role in yielding an auditory phonemic deficit.

Interestingly, at around 30 Hz, the left-dominant phase-locking profile of auditory responses in MEG (auditory steady-state responses) was only present in subjects with normal reading ability (Lehongre et al., 2011). Because this response is absent in dyslexic participants, the authors suggested that the ability of their left auditory cortex to parse speech at the appropriate phonemic rate was altered. Those with dyslexia had a strong response at this frequency in right auditory cortex and therefore presented an abnormal asymmetry between left and right auditory cortices. Importantly, the magnitude of the anomalous asymmetry correlated with behavioral measures in phonology (such as non-word repetition and rapid automatic naming). Finally it was also shown that dyslexic readers had a stronger resonance than controls in both left and right auditory cortices at frequencies between 50 and 80 Hz. This supports the notion that these subjects had a tendency to oversample information in the phonemic range, this latter effect being positively correlated with a phonologic memory deficit. As a consequence, if dyslexia induces speech parsing at a wrong frequency, phonemic units would be sampled erratically, without necessarily inducing major perceptual deficits (Ramus and Szenkovits, 2008; Ziegler et al., 2009). As a consequence, the phonologic impairment could take different forms, with a stronger impact on the acoustic side for undersampling (insufficient acoustic detail per time unit) and on the memory side for oversampling (too many frames to be integrated per time unit).

Although important, the observation that oscillatory anomalies co-occur with atypical phonologic representations remains insufficient to establish a causal role of dysfunctional oscillatory sampling. Causal evidence that auditory sampling is determined by cortical columnar organization could be obtained from knockout animal models comparing neuronal activity to continuous auditory stimuli in sites with various degrees of columnar disorganization. However, such animal work can only indirectly address a specific relation to speech processing.

CONCLUSION

Time is an essential feature of speech perception. No speech sound can be identified without integrating the acoustic input over time, and the temporal scale at which such integration operates determines whether we are hearing phonemes, syllables, or words. The central idea of this chapter is that, unlike subcortical processing that faithfully encodes speech sounds in their precise spectrotemporal structure, processing in primary and association auditory cortices results in the discretization of spectrotemporal patterns, using variable temporal integration scales. By analogy with Heisenberg's uncertainty principle (Ozawa, 2003), speech representations cannot be precise in both time and space. The limited phase-locking capacity of the auditory cortex thus appears a likely counterpart to its spatial integration properties (across cortical layers and functional regions). Speech processing through and across cortical columns containing complex recurrent circuits bears a cost on the temporal precision of speech representations, and integration at gamma scale could be a direct consequence of processing at the cortical column scale. In this chapter we argue that the auditory cortex uses gamma oscillations to integrate the speech auditory stream at the phonemic timescale, and theta oscillations to signal syllable boundaries and orchestrate gamma activity. Although the generation mechanisms are less well known for theta than for gamma oscillations, at present we see no alternative computational solution to the online speech segmentation and integration problem than invoking coupled theta and gamma activity. More research is needed to evaluate the detailed neural operations that are necessary to transform the acoustic input into linguistic representations, and it might turn out that non-oscillatory mechanisms succeed more efficiently in achieving these transformations.

References

Ahissar E, Nagarajan S, Ahissar M et al. (2001). Speech comprehension is correlated with temporal response patterns recorded from auditory cortex. Proc Natl Acad Sci U S A 98: 13367–13372.

Allen JB (2005). Articulation and intelligibility. Synthesis Lectures on Speech and Audio Processing 1 (1): 1–124.

Arnal LH, Giraud AL (2012). Cortical oscillations and sensory predictions. Trends Cogn Sci 16: 390–398.

Arnal LH, Morillon B, Kell CA et al. (2009). Dual neural routing of visual facilitation in speech processing. J Neurosci 29: 13445–13453.

Arnal LH, Wyart V, Giraud AL (2011). Transitions in neural oscillations reflect prediction errors generated in audiovisual speech. Nat Neurosci 14: 797–801.

Barton B, Venezia JH, Saberi K et al. (2012). Orthogonal acoustic dimensions define auditory field maps in human cortex. Proc Natl Acad Sci U S A 109: 20738–20743.

Bastos AM, Usrey WM, Adams RA et al. (2012). Canonical microcircuits for predictive coding. Neuron 76: 695–711.

Boemio A, Fromm S, Braun A et al. (2005). Hierarchical and asymmetric temporal sensitivity in human auditory cortices. Nat Neurosci 8: 389–395.

Borgers C, Epstein S, Kopell NJ (2005). Background gamma rhythmicity and attention in cortical local circuits: a computational study. Proc Natl Acad Sci U S A 102: 7002–7007.

Borgers C, Epstein S, Kopell NJ (2008). Gamma oscillations mediate stimulus competition and attentional selection in a cortical network model. Proc Natl Acad Sci U S A 105: 18023–18028.

Britvina T, Eggermont JJ (2007). A Markov model for inter-spike interval distributions of auditory cortical neurons that do not show periodic firings. Biol Cybern 96: 245–264.

Brugge JF, Nourski KV, Oya H et al. (2009). Coding of repetitive transients by auditory cortex on Heschl's gyrus. J Neurophysiol 102: 2358–2374.

Busch NA, Dubois J, VanRullen R (2009). The phase of ongoing EEG oscillations predicts visual perception. J Neurosci 29: 7869–7876.

Canolty RT, Knight RT (2010). The functional role of cross-frequency coupling. Trends Cogn Sci 14: 506–515.

Ding N, Simon JZ (2009). Neural representations of complex temporal modulations in the human auditory cortex. J Neurophysiol 102: 2731–2743.

Doelling KB, Arnal LH, Ghitza O et al. (2014). Acoustic landmarks drive delta-theta oscillations to enable speech comprehension by facilitating perceptual parsing. Neuroimage 85: 761–768.

Drullman R, Festen JM, Plomp R (1994a). Effect of reducing slow temporal modulations on speech reception. J Acoust Soc Am 95: 2670–2680.

Drullman R, Festen JM, Plomp R (1994b). Effect of temporal envelope smearing on speech reception. J Acoust Soc Am 95: 1053–1064.

Edwards E, Chang EF (2013). Syllabic (approximately 2–5 Hz) and fluctuation (approximately 1–10 Hz) ranges in speech and auditory processing. Hear Res 305: 113–134.

Elhilali M, Fritz JB, Klein DJ et al. (2004). Dynamics of precise spike timing in primary auditory cortex. J Neurosci 24: 1159–1172.

Elliott TM, Theunissen FE (2009). The modulation transfer function for speech intelligibility. PLoS Comput Biol 5: e1000302.

Faulkner A, Rosen S, Smith C (2000). Effects of the salience of pitch and periodicity information on the intelligibility of four-channel vocoded speech: implications for cochlear implants. J Acoust Soc Am 108: 1877–1887.

Fishman YI, Reser DH, Arezzo JC et al. (2000). Complex tone processing in primary auditory cortex of the awake monkey. II. Pitch versus critical band representation. J Acoust Soc Am 108: 247–262.

Fukushima M, Saunders RC, Leopold DA et al. (2012). Spontaneous high-gamma band activity reflects functional organization of auditory cortex in the awake macaque. Neuron 74: 899–910.

Gaese BH, Ostwald J (1995). Temporal coding of amplitude and frequency modulation in the rat auditory cortex. Eur J Neurosci 7: 438–450.

Ghitza O (2011). Linking speech perception and neurophysiology: speech decoding guided by cascaded oscillators locked to the input rhythm. Front Psychol 2: 130.

Ghitza O, Greenberg S (2009). On the possible role of brain rhythms in speech perception: intelligibility of time-compressed speech with periodic and aperiodic insertions of silence. Phonetica 66: 113–126.

Giraud AL, Poeppel D (2012a). Cortical oscillations and speech processing: emerging computational principles. Nat Neurosci 15: 511–517.

Giraud AL, Poeppel D (2012b). Speech perception from a neurophysiological perspective. In: D Poeppel, T Overath, AN Popper et al. (Eds.), The Human Auditory Cortex, Springer, New York, pp. 225–260.

Giraud AL, Price CJ (2001). The constraints functional neuroimaging places on classical models of auditory word processing. J Cogn Neurosci 13: 754–765.

Giraud AL, Lorenzi C, Ashburner J et al. (2000). Representation of the temporal envelope of sounds in the human brain. J Neurophysiol 84: 1588–1598.

Giraud AL, Kell C, Thierfelder C et al. (2004). Contributions of sensory input, auditory search and verbal comprehension to cortical activity during speech processing. Cereb Cortex 14: 247–255.

Goswami U (2011). A temporal sampling framework for developmental dyslexia. Trends Cogn Sci 15: 3–10.

Greenberg S, Arai T (2001). The relation between speech intelligibility and the complex modulation spectrum. In: Proceedings of the 7th Eurospeech Conference on Speech Communication and Technology (Eurospeech-2001), pp. 473–476.

Grothe B, Pecka M, McAlpine D (2010). Mechanisms of sound localization in mammals. Physiol Rev 90: 983–1012.

Henry MJ, Obleser J (2012). Frequency modulation entrains slow neural oscillations and optimizes human listening behavior. Proc Natl Acad Sci U S A 109: 20095–20100.

Hochstein S, Ahissar M (2002). View from the top: hierarchies and reverse hierarchies in the visual system. Neuron 36: 791–804.

Holcombe AO (2009). Seeing slow and seeing fast: two limits on perception. Trends Cogn Sci 13: 216–221.

Howard MF, Poeppel D (2011). Discrimination of speech stimuli based on neuronal response phase patterns depends on acoustics but not comprehension. J Neurophysiol 104: 2500–2511.

Howard MF, Poeppel D (2012). The neuromagnetic response to spoken sentences: Co-modulation of theta band amplitude and phase. Neuroimage 60: 2118–2127.

Hyafil A, Fontolan L, Gutkin B et al. (2012). Theoretical exploration of speech/neural oscillation alignment for speech parsing. FENS Abstract 6, S47.04.

Joris PX, Schreiner CE, Rees A (2004). Neural processing of amplitude-modulated sounds. Physiol Rev 84: 541–577.

Kanedera N, Arai T, Hermansky H et al. (1999). On the relative importance of various components of the modulation spectrum for automatic speech recognition. Speech Comm 28: 43–55.

Kayser C, Ince RA, Panzeri S (2012). Analysis of slow (theta) oscillations as a potential temporal reference frame for information coding in sensory cortices. PLoS Comput Biol 8: e1002717.

Kiebel SJ, Daunizeau J, Friston KJ (2008). A hierarchy of time-scales and the brain. PLoS Comput Biol 4: e1000209.

Kingsbury BED, Morgan N, Greenberg S (1998). Robust speech recognition using the modulation spectrogram. Speech Comm 25: 117–132.

Lakatos P, Chen CM, O'Connell MN et al. (2007). Neuronal oscillations and multisensory interaction in primary auditory cortex. Neuron 53: 279–292.

Lakatos P, Karmos G, Mehta AD et al. (2008). Entrainment of neuronal oscillations as a mechanism of attentional selection. Science 320: 110–113.

Lehongre K, Ramus F, Villiermet N et al. (2011). Altered low-gamma sampling in auditory cortex accounts for the three main facets of dyslexia. Neuron 72: 1080–1090.

Liang L, Lu T, Wang X (2002). Neural representations of sinusoidal amplitude and frequency modulations in the primary auditory cortex of awake primates. J Neurophysiol 87: 2237–2261.

Loebach JL, Wickesberg RE (2008). The psychoacoustics of noise vocoded speech: a physiological means to a perceptual end. Hear Res 241: 87–96.

Luo H, Poeppel D (2007). Phase patterns of neuronal responses reliably discriminate speech in human auditory cortex. Neuron 54: 1001–1010.

Luo H, Wang Y, Poeppel D et al. (2006). Concurrent encoding of frequency and amplitude modulation in human auditory cortex: MEG evidence. J Neurophysiol 96: 2712–2723.

Luo H, Boemio A, Gordon M et al. (2007). The perception of FM sweeps by Chinese and English listeners. Hear Res 224: 75–83.

Malone BJ, Scott BH, Semple MN (2010). Temporal codes for amplitude contrast in auditory cortex. J Neurosci 30: 767–784.

Mantini D, Perrucci MG, Del Gratta C et al. (2007). Electrophysiological signatures of resting state networks in the human brain. Proc Natl Acad Sci U S A 104: 13170–13175.

Mesgarani N, Chang EF (2012). Selective cortical representation of attended speaker in multi-talker speech perception. Nature 485: 233–236.

Miller GA (1951). Language and communication, McGraw-Hill, New York.

Miyazaki T, Thompson J, Fujioka T et al. (2013). Sound envelope encoding in the auditory cortex revealed by neuromagnetic responses in the theta to gamma frequency bands. Brain Res 1506: 64–75.

Moerel M, De Martino F, Santoro R et al. (2013). Processing of natural sounds: characterization of multipeak spectral tuning in human auditory cortex. J Neurosci 33: 11888–11898.

Morillon B, Lehongre K, Frackowiak RS et al. (2010). Neurophysiological origin of human brain asymmetry for speech and language. Proc Natl Acad Sci U S A 107: 18688–18693.

Morillon B, Liegeois-Chauvel C, Arnal LH et al. (2012). Asymmetric function of theta and gamma activity in syllable processing: an intra-cortical study. Front Psychol 3: 248.

Nahum M, Nelken I, Ahissar M (2008). Low-level information and high-level perception: the case of speech in noise. PLoS Biol 6: e126.

Ng BS, Schroeder T, Kayser C (2012). A precluding but not ensuring role of entrained low-frequency oscillations for auditory perception. J Neurosci 32: 12268–12276.

Nourski KV, Brugge JF (2011). Representation of temporal sound features in the human auditory cortex. Rev Neurosci 22: 187–203.

Nourski KV, Reale RA, Oya H et al. (2009). Temporal envelope of time-compressed speech represented in the human auditory cortex. J Neurosci 29: 15564–15574.

Ozawa M (2003). Universally valid reformulation of the Heisenberg uncertainty principle on noise and disturbance in measurement. Phys Rev A 67: 042105.

Panzeri S, Brunel N, Logothetis NK et al. (2010). Sensory neural codes using multiplexed temporal scales. Trends Neurosci 33: 111–120.

Pasley BN, David SV, Mesgarani N et al. (2012). Reconstructing speech from human auditory cortex. PLoS Biol. http://dx.doi.org/10.1371/journal.pbio.1001251.

Peelle JE, Gross J, Davis MH (2013). Phase-locked responses to speech in human auditory cortex are enhanced during comprehension. Cereb Cortex 23: 1378–1387.

Pienkowski M, Eggermont JJ (2009). Nonlinear cross-frequency interactions in primary auditory cortex spectro-temporal receptive fields: a Wiener-Volterra analysis. J Comput Neurosci 28: 285–303.

Plack CJ, Oxenham AJ, Fay RR et al. (2005). Pitch: neural coding and perception, Springer, New York.

Poeppel D (2001). New approaches to the neural basis of speech sound processing: introduction to special section on brain and speech. Cognit Sci 25: 659–661.

Poeppel D (2003). The analysis of speech in different temporal integration windows: cerebral lateralization as 'asymmetric sampling in time'. Speech Comm 41: 245–255.

Poeppel D, Idsardi WJ, van Wassenhove V (2008). Speech perception at the interface of neurobiology and linguistics. Philos Trans R Soc Lond B Biol Sci 363: 1071–1086.

Pöppel E (1988). Mindworks: Time and conscious experience, Harcourt Brace Jovanovich, Boston.

Ramus F, Szenkovits G (2008). What phonological deficit? Q J Exp Psychol 61: 129–141.

Remez RE, Rubin PE, Pisoni DB et al. (1981). Speech perception without traditional speech cues. Science 212: 947–949.

Roberts B, Summers RJ, Bailey PJ (2011). The intelligibility of noise-vocoded speech: spectral information available from across-channel comparison of amplitude envelopes. Proc Biol Sci 278: 1595–1600.

Rosen S (1992). Temporal information in speech: acoustic, auditory and linguistic aspects. Philos Trans R Soc Lond B Biol Sci 336: 367–373.

Saberi K, Perrott DR (1999). Cognitive restoration of reversed speech. Nature 398: 760.

Saenz M, Langers DR (2014). Tonotopic mapping of human auditory cortex. Hear Res 307: 42–52.

Schonwiesner M, Zatorre RJ (2009). Spectro-temporal modulation transfer function of single voxels in the human auditory cortex measured with high-resolution fMRI. Proc Natl Acad Sci U S A 106: 14611–14616.

Schroeder CE, Lakatos P (2009a). Low-frequency neuronal oscillations as instruments of sensory selection. Trends Neurosci 32: 9–18.

Schroeder CE, Lakatos P (2009b). The gamma oscillation: master or slave? Brain Topogr 22: 24–26.

Schroeder CE, Lakatos P, Kajikawa Y et al. (2008). Neuronal oscillations and visual amplification of speech. Trends Cogn Sci 12: 106–113.

Scott SK, Rosen S, Lang H et al. (2006). Neural correlates of intelligibility in speech investigated with noise vocoded speech – a positron emission tomography study. J Acoust Soc Am 120: 1075–1083.

Shamir M, Ghitza O, Epstein S et al. (2009). Representation of time-varying stimuli by a network exhibiting oscillations on a faster time scale. PLoS Comput Biol 5: e1000370.

Shannon RV, Zeng FG, Kamath V et al. (1995). Speech recognition with primarily temporal cues. Science 270: 303–304.

Sharpee TO, Atencio CA, Schreiner CE (2011). Hierarchical representations in the auditory cortex. Curr Opin Neurobiol 21: 761–767.

Smith ZM, Delgutte B, Oxenham AJ (2002). Chimaeric sounds reveal dichotomies in auditory perception. Nature 416: 87–90.

Souza P, Rosen S (2009). Effects of envelope bandwidth on the intelligibility of sine- and noise-vocoded speech. J Acoust Soc Am 126: 792–805.

Steeneken HJ, Houtgast T (1980). A physical method for measuring speech-transmission quality. J Acoust Soc Am 67: 318–326.

Stevens KN (2002). Toward a model for lexical access based on acoustic landmarks and distinctive features. J Acoust Soc Am 111: 1872–1891.

Thorne JD, De Vos M, Viola FC et al. (2011). Cross-modal phase reset predicts auditory task performance in humans. J Neurosci 31: 3853–3861.

Tiesinga P, Sejnowski TJ (2009). Cortical enlightenment: are attentional gamma oscillations driven by ING or PING? Neuron 63: 727–732.

VanRullen R, Koch C (2003). Is perception discrete or continuous? Trends Cogn Sci 7: 207–213.

VanRullen R, Zoefel B, Ilhan B (2014). On the cyclic nature of perception in vision versus audition. Philos Trans R Soc Lond B Biol Sci 369: 20130214.

Wang X (2007). Neural coding strategies in auditory cortex. Hear Res 229: 81–93.

Wang XJ (2010). Neurophysiological and computational principles of cortical rhythms in cognition. Physiol Rev 90: 1195–1268.

Wang X, Lu T, Bendor D et al. (2008). Neural coding of temporal information in auditory thalamus and cortex. Neuroscience 157: 484–494.

Warrier C, Wong P, Penhune V et al. (2009). Relating structure to function: Heschl's gyrus and acoustic processing. J Neurosci 29: 61–69.

Womelsdorf T, Schoffelen JM, Oostenveld R et al. (2007). Modulation of neuronal interactions through neuronal synchronization. Science 316: 1609–1612.

Wyart V, de Gardelle V, Scholl J et al. (2012). Rhythmic fluctuations in evidence accumulation during decision making in the human brain. Neuron 76: 847–858.

Yin P, Johnson JS, O'Connor KN et al. (2011). Coding of amplitude modulation in primary auditory cortex. J Neurophysiol 105: 582–600.

Zatorre RJ, Gandour JT (2008). Neural specializations for speech and pitch: moving beyond the dichotomies. Philos Trans R Soc Lond B Biol Sci 363: 1087–1104.

Zatorre RJ, Belin P, Penhune VB (2002). Structure and function of auditory cortex: music and speech. Trends Cogn Sci 6: 37–46.

Zeng FG (2002). Temporal pitch in electric hearing. Hear Res 174: 101–106.

Ziegler JC, Pech-Georgel C, George F et al. (2009). Speech-perception-in-noise deficits in dyslexia. Dev Sci 12: 732–745.

Handbook of Clinical Neurology, Vol. 129 (3rd series)
The Human Auditory System
G.G. Celesia and G. Hickok, Editors

Chapter 6

Sound localization

JOHN C. MIDDLEBROOKS*
Departments of Otolaryngology, Neurobiology and Behavior, Cognitive Sciences, and Biomedical Engineering, University of California at Irvine, Irvine, CA, USA

INTRODUCTION

Spatial hearing permits a listener to identify the locations of sound sources and it aids in detection and recognition of sounds in the presence of other competing sounds. Of these functions, sound localization *per se* probably is the best understood, and that will be the main topic of this chapter. In most natural listening conditions, normal human listeners can localize sounds within a few degrees of accuracy in both horizontal and vertical dimensions (Stevens and Newman, 1936; Makous and Middlebrooks, 1990; Carlile et al., 1997; Yost et al., 2013). Figure 6.1 shows the performance of a listener in localizing brief broadband sounds.

Localization is a special challenge for the auditory system. In the visual and somatosensory systems, stimulus location is mapped directly on the sensory surfaces, i.e., the retina and the skin, respectively. In the auditory system, it is frequency that is mapped on the cochlear sensory surface, not location. Sound location must be inferred from spatial acoustic cues that arise from the physical interactions of sounds with the head and external ears. Those cues must be analyzed within the central auditory system and, in some way, integrated to create a neural representation of auditory space. We will see that this is not possible in every stimulus condition: examples are failures of horizontal localization of tones with frequencies around 3 kHz and failures of vertical or front/back localization of any sound that is restricted in bandwidth.

In this chapter, we will consider the acoustic cues on which listeners rely for source localization, the brainstem pathways in which those cues are analyzed, and the representations of sound locations in the midbrain and cerebral cortex. We treat the horizontal (lateral) dimension separately from vertical and front/back dimensions because it turns out that localization in those dimensions relies on different acoustic cues and, necessarily, different brainstem pathways. We will consider the somewhat special problems of localization in distance and perception of sound motion. We will address some of the mechanisms that permit a listener to defeat spatial ambiguity potentially caused by the sound reflections in typical acoustic environments. Finally, we will consider present-day understanding of the representation of sound locations within the auditory cortex.

SOME TERMS AND TECHNIQUES

We begin by briefly defining some terms and describing some experimental techniques that have been used in spatial hearing research. The most direct way of testing sound localization seemingly would be to present sounds at varying locations remote from the listener and to have the listener make some judgment of those locations. Experimental conditions in which the sound source is at a distance from the listener are known as "open-field" conditions. Most often, investigators will strive for the particular open-field condition known as a "free field" in which the sound field is free of obstacles and in which there are no reflections from room surfaces. Although free-field conditions are possible in natural environments (e.g., a grassy or snowy field), such experiments most often are conducted in special anechoic chambers in which the walls, floor, and ceiling are treated to absorb sound. Generally, free-field conditions also imply that sound sources are presented in the "far field," more than about 1 m from the center of the listener's head. "Near-field" conditions, discussed below in the context of distance perception, introduce changes in localization cues that require special attention. Study of sensitivity to

*Correspondence to: John C. Middlebrooks, Ph.D., 116 Medical Science Building E, University of California at Irvine, Irvine, CA 92697-5310, USA. Tel: +1-949-824-0119, E-mail: j.midd@uci.edu

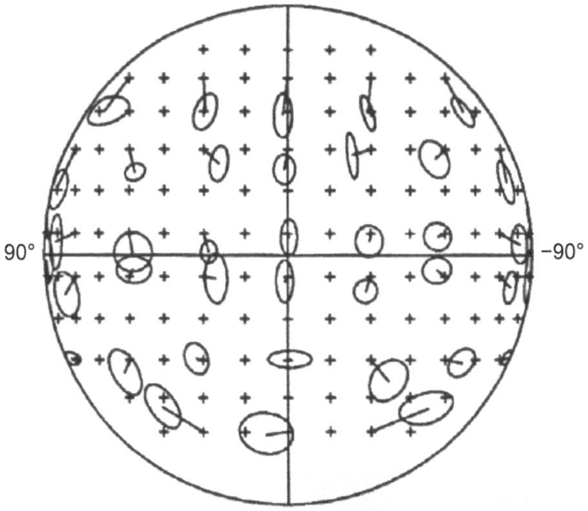

Fig. 6.1. Sound localization by a human listener. The coordinate system is drawn as if looking toward the listener, with 90° at the listener's right side. Stimuli were 150-ms broadband noise bursts presented from the locations indicated with plus signs, 1.2 m from the center of the listener's head. Listeners reported the locations by orienting to face the source while the head orientation was monitored electronically. Source locations were spaced in 10° intervals, although for clarity, responses are shown only in 20° intervals. The line segments connect each source location to the mean of five response locations. Horizontal and vertical axes of the ellipses indicate ±1 SD of responses in horizontal and vertical dimensions. (Reproduced from Makous and Middlebrooks, 1990.)

particular acoustical cues to sound location might be conducted in "closed-field" conditions, meaning that sounds are presented through headphones. Typically in such experiments, a listener will attempt to distinguish a "dichotic" presentation, in which there is a difference in sound level or timing between the two ears, from a "diotic" presentation, in which sounds are identical at the two ears.

Finally, we consider "virtual auditory space" conditions, in which an investigator will attempt to present through headphones sounds containing all the natural localization cues, thereby eliciting a sensation of a realistic sound source at a well-defined location at some distance from the listener (Wightman and Kistler, 1989a, b). The procedure requires detailed measurements of filter characteristics of the listener's ears, typically derived by recording sounds from the two ear canals while a broadband probe sound is presented from systematically varying locations; the filters are referred to as "head-related transfer functions" (HRTFs). Arbitrary sounds then are processed with the HRTFs from the two ears and are presented through headphones, with or without some experimental manipulation. As we will see in a later section, head and ear acoustics vary among listeners. For that reason, optimal virtual audio is achieved with use of HRTFs recorded from a listener's own ears, although use of "generic" HRTFs is acceptable in many situations. Indeed, many studies have employed HRTFs recorded from an acoustic manikin, such as the Knowles electronics manikin for auditory research (KEMAR). Virtual audio techniques have become refined and more widely accessible in the past decade and have provided many useful insights.

HORIZONTAL LOCALIZATION

Duplex theory

"Horizontal localization" primarily refers to the localization of a sound source to the left or right relative to the midline plane. Discrimination of front from back might be thought of as a component of horizontal localization, but the mechanisms of front/back localization are more like those of vertical localization and will be considered in that section. The foundation of our understanding of horizontal localization can be traced to the work of the English physicist and baron, Lord Rayleigh (Strutt, 1907). He worked largely at the end of the 19th century, often with Lady Rayleigh as the experimental subject, often using tuning forks as the sound sources. Rayleigh observed that left-versus-right locations of tuning forks could be discriminated across a broad range of frequencies. He appreciated that, at relatively high frequencies, the head is larger than the wavelength of the sound and casts an acoustic shadow. For that reason, the sound level at the ear nearer to a sound source would be greater than that at the farther ear, thereby providing a cue to the sound-source location. He reported: "When the pitch is pretty high, there is no doubt that this [intensity theory] explanation is adequate." That explanation was more doubtful for sounds of lower frequency, at which the sound wavelength can be several times longer than the diameter of the human head. In that condition, the sound wave "bends" around the head and the difference in sound level at the two ears is negligible. After a series of ingenious experiments using pairs of slightly mistuned tuning forks or sounds directed independently to the ears with "India-rubber" tubes, Rayleigh determined that "the only alternative to the intensity theory is to suppose that the judgment is founded upon the difference of *phases* at the two ears." That is, the difference in path lengths from a source to the two ears introduces differential delays, resulting in an interaural phase difference. The notion of a listener being sensitive to interaural phase differences was somewhat difficult to accept, because it implies that the auditory system is sensitive to interaural time differences (ITDs) of less than a tenth of a millisecond near the threshold of location discrimination. Nevertheless, such sensitivity has been confirmed with modern psychophysical (i.e., quantitative perceptual) procedures and has been demonstrated at the level of brainstem neurons in animal physiology experiments.

Rayleigh's view of horizontal sound localization has come to be known as the "duplex" theory: that sound localization at low frequencies relies on differences in phase at the two ears, or ITDs, and that high-frequency localization relies on interaural differences in sound level (ILDs). Early free-field studies of localization of tonal stimuli (Stevens and Newman, 1936; Sandel et al., 1955; Mills, 1958) confirmed that sound location and location discrimination were accurate for sound frequencies below ~1.5 kHz, consistent with listeners' use of ITDs, and for sound frequencies above ~4 kHz, in the range at which sound wavelengths shorter than the diameter of the head result in substantial ILDs. There was an intermediate range of frequencies, roughly 2–4 kHz, at which neither ITD nor ILD provided a reliable cue and at which localization errors were greatest. Fortunately, this range of sounds that cannot be localized well has little impact on everyday sound localization because one is rarely asked to localize sounds limited to the 2–4-kHz frequency band.

The basic elements of the duplex theory have been quantified with modern psychophysical measures, and their anatomic and physiologic correlates are largely understood. Rayleigh's work was conducted using more or less pure-tone stimuli. The demonstrations that ITDs and ILDs are used for localization of low- and high-frequency sounds, respectively, were extended to broader-band sounds by Wightman and Kistler (1992), who used virtual auditory-space technology to deliver sounds in which ITDs and ILDs signaled opposing directions. Those authors demonstrated that, when sounds had broad frequency spectra containing low frequencies (below ~2 kHz), listeners' localization judgments were dominated by ITD cues irrespective of conflicting ILDs. When low frequencies were eliminated by high-pass filtering, the ITD dominance was lost. Macpherson and Middlebrooks (2002) extended those observations by showing that localization of high-pass (4–16 kHz) sounds is dominated by ILD cues, even in the presence of conflicting ITD cues.

Interaural time-difference cues

ITDs result from differences in the travel time from a sound source to the ears closer to and farther from the source. Woodworth (1938) modeled the ears and head as two points on a rigid sphere. Based on the geometry, he calculated ITD as:

$$\text{ITD} = a/c\,[\theta_{rad} + \sin(\theta_{rad})]$$

with a equal to the radius of the sphere, c the speed of sound (~344 m/s at 20° C), and θ_{rad} equal to the angle in radians of the sound source relative to the midline; $\theta_{rad} - \theta_{deg} \times \pi/180$, where θ_{deg} is the angle in degrees. Woodworth used $a = 8.75$ cm as the effective radius of the sphere, whereas measures based on the best fits to actual ITD measurements in adult humans range from 8.75 to 10.57 cm (Kuhn, 1977; Middlebrooks, 1999a). The Woodworth model fits the data well for steady-state tones at frequencies ≥2 kHz. Kuhn (1977) showed that the effective radius of the head increases with frequency decreasing below ~2 kHz, as the wavelength of sound approaches the dimensions of the head, such that the effective head radius is larger by a factor of 1.5 at 0.5 kHz. Similarly, maximum values of ITD increase with frequency decreasing from 2 to 0.5 kHz.

The thresholds for detection of ITDs in pure-tone stimuli by well-practiced human listeners are around 10–20 μs (Klumpp and Eady, 1956; Zwislocki and Feldman, 1956; Brughera et al., 2013). That is remarkable acuity when one considers that it is accomplished with neural machinery having time constants of several hundreds of microseconds or longer. Thresholds are minimal (i.e., sensitivity is greatest) for frequencies from 0.7 to 1.0 kHz, they increase slowly for frequencies <0.7 kHz, and they increase dramatically above 1.0 kHz such that listeners are essentially insensitive to ITDs at frequencies 1.4 kHz and higher (Brughera et al., 2013). These thresholds refer to ITDs in the cycle-by-cycle fine structure of the waveforms of low-frequency sounds having periods on a time scale of ~0.7–3 ms; it is fine-structure ITDs that dominate localization judgments. Listeners also are sensitive to ITDs in the envelopes of sounds, where "envelope" refers to variation in the amplitude of complex sounds on a time scale longer than ~10 ms. Thresholds for envelope ITDs tend to be 2–10 times larger (i.e., less sensitive) than for low-frequency fine structure (Henning, 1974; McFadden and Pasanen, 1976; Nuetzel and Hafter, 1981; Blauert, 1982; Bernstein and Trahiotis, 1994, 2002). Envelope ITDs are rather ineffective in eliciting a sensation of a sound displaced away from the midline (Trahiotis and Bernstein, 1986), and listeners give relatively little weight to envelope ITDs when judging locations of high-frequency sounds (Macpherson and Middlebrooks, 2002).

Our understanding of the brainstem mechanisms of ITD sensitivity is shaped by the influential model of a "place" mechanism for ITD representation proposed by Jeffress (1948). In the Jeffress model (Fig. 6.2) pathways originate at the two ears, and signals travel with varying delays to converge on an array of "tertiary" binaural comparator neurons. The delay lines would compensate for differences in times of arrival of sounds at the two ears such that, for a given ITD, signals from

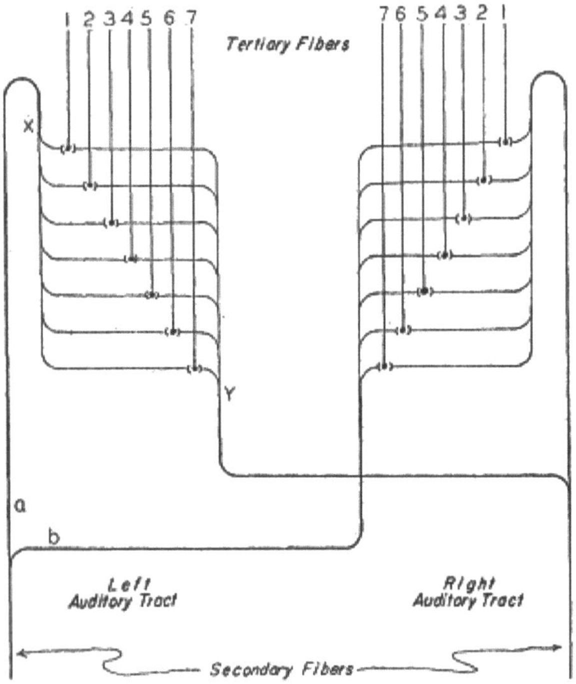

Fig. 6.2. A hypothetical mechanism for detection of interaural time differences. (Reproduced from Jeffress, 1948.)

the two ears would arrive simultaneously at certain tertiary fibers, eliciting an action potential. In contrast, mistimed signals traveling on different delay lines would not elicit responses. The Jeffress model supposed that there would be comparator neurons for every realistic ITD. The bank of delay lines and comparators presumably would be replicated for each frequency band.

The Jeffress ITD-versus-place model appears to correspond very well to the functional organization of the avian nucleus laminaris (Carr and Konishi, 1990; Overholt et al., 1992; Koppl and Carr, 2008), which is a structure homologous to the medial superior olive (MSO) in mammals. In mammals, ITD-sensitive neurons much like those implied by Jeffress's hypothetic tertiary neurons have been identified in the MSO (Goldberg and Brown, 1969; Yin and Chan, 1990). The pathways to the MSO neurons begin with auditory nerve fibers firing action potentials phase-locked to the fine structure of low-frequency sounds. The auditory nerve fibers synapse on the spherical bushy cells of the anterior ventral cochlear nucleus (AVCN) with high-temporal-fidelity end-bulbs of Held (Ryugo and Seishiro, 1991). Each principal neuron in the MSO receives excitatory projections from both the ipsilateral AVCN and, via the trapezoid body, the contralateral AVCN. The MSO neurons are activated by synchronized inputs from the ipsi- and contralateral inputs. The MSO also receives inhibition input, from the medial and lateral nuclei of the trapezoid body (MNTB and LNTB). The role of that inhibition in ITD sensitivity presently is a topic of debate (Pecka et al., 2008; Roberts et al., 2013; van der Hiejden et al., 2013).

The MSO principal cells exhibit robust sensitivity to ITD, and for many years there was a general acceptance that something like the Jeffress model lived in the mammalian MSO. Recent studies, however, have cast some doubt on that notion. Specifically, the Jeffress model predicts that there must be neurons tuned to a range of ITDs at each frequency. In contrast, at least in small-headed mammals like gerbils, preferred ITDs of MSO neurons tend to vary with neurons' characteristic frequencies (CFs), and low-CF neurons tend to prefer interaural delays longer than the animal ever would experience in nature (Brand et al., 2002). These neurons tuned to unrealistic ITDs seemingly would be worthless in a strict place model of ITD because they would never receive their optimal stimulus. Conversely, there would be a dearth of neurons tuned to realistic ITDs. Nevertheless, a neuron that responds maximally to large, even unrealistic, ITDs will necessarily have a spike-rate-versus-ITD function that varies most steeply across ITDs near zero, corresponding to the range of sound-source locations at which spatial hearing is most acute. A model is gaining favor in which locations of low-frequency sounds are coded by the relative spike rates of neurons tuned to large ipsi- and contralateral-leading ITDs (McAlpine et al., 2001). Whether such a rate difference model applies to large-headed animals like humans presently is controversial (Joris and Yin, 2007). Relevant to that controversy is a recent study that attempted to model human detection of ITDs, tested at narrowly spaced pure-tone frequencies (Brughera et al., 2013). Across the frequency range in which ITD thresholds could be measured, 0.25 to near 1.4 kHz, the data were fit best by a hybrid model consisting of a Jeffress-like place model at the higher frequencies and a rate difference model at lower frequencies.

Interaural level-difference cues

The head casts an acoustic shadow for sounds having wavelengths shorter than the dimensions of the head, such that the sound level at the ear closer to the sound source is greater than that at the far, shadowed, ear. The resulting ILD potentially is a cue to the sound-source location. Conversely, sounds at frequencies lower than ~1 kHz have wavelengths longer than the diameter of the head, and ILDs at those frequencies are negligible. The magnitudes and spatial dependence of ILDs can be measured by recording from the ear canals with miniature microphones while presenting probe sounds in the free field (Shaw, 1974; Mehrgardt and Mellert, 1977). Given a sound source around 90° from the midline, ILDs are around 20 dB at 4 kHz, increasing to ~35 dB at

10 kHz (Middlebrooks et al., 1989). At any given frequency, ILDs are roughly proportional to the sine of the angle of the sound source relative to the midline. The threshold for detection of an ILD in a high-frequency tone is ~1 dB (Mills, 1960; Hafter et al., 1977).

The lowest-level brainstem nucleus containing neurons sensitive to ILD is the lateral superior olive (LSO) (Boudreau and Tsuchitani, 1968). Principal cells in the LSO receive excitatory inputs from spherical bushy cells in the ipsilateral AVCN. The contralateral pathway is inhibitory, involving an excitatory projection from globular bushy cells in the contralateral AVCN, crossing the midline in the trapezoid body, and terminating in the MNTB. Neurons in the MNTB send inhibitory glycinergic projections to the LSO. The ipsilateral excitation and contralateral inhibition that converge on single LSO neurons result in sigmoid spike-rate-versus-ILD in the LSO: activity is low for ITDs favoring the contralateral ear, increases most steeply for ILDs around 0 dB, and is high for ipsilaterally favoring ILDs.

Monaural conditions

Loss of hearing in one ear disrupts interaural difference cues for horizontal localization. If one ear of a normal-hearing listener is plugged, the listener will tend to localize all sounds to the side of the open ear. It is thought that most individuals who have acquired hearing loss in one ear after some stage of maturity are unable to localize sounds. A study of monaural localization included five individuals who had congenital deafness in one ear and normal hearing in the other (Slattery and Middlebrooks, 1994). Two of those unilaterally deaf patients localized all sounds to the side of the hearing ear, resembling normal-hearing listeners wearing an earplug. The other three unilaterally deaf individuals localized reasonably well. They could localize sources on both sides of head with errors within a factor of 2 of those of normal-hearing controls and with accuracy far better than that of normal-hearing controls wearing an earplug. The interpretation is that the successful unilaterally deaf localizers had learned to make use of spectral-shape cues for horizontal localization in a manner analogous to the use of such cues by normal-hearing listeners for vertical and front/back localization, as described in the next section.

VERTICAL AND FRONT/BACK LOCALIZATION

Interaural differences in time and level are effective cues for the angle of displacement of a sound source relative to the midline plane. Knowledge of that lateral displacement, however, narrows the range of possible sound sources only to a "cone of confusion" centered on the interaural axis (Woodworth, 1938) – a given ITD or ILD could be produced by a source at any vertical or front/back location on such a cone. If asked to localize a pure tone, a listener can respond fairly accurately if the possible sources are restricted to the front half of the horizontal plane. If the source location can vary throughout the horizontal plane, however, the listener will make front/back confusions. If the source also is free to vary in elevation, pure-tone localization is pretty hopeless.

Vertical and front/back localization can improve dramatically if the sound source has a relatively broad, flat spectrum, like that of many natural sounds; the important spectral range is ~4 kHz and higher. Direction-dependent filtering of such a spectrum by the head and pinnae (i.e., the external flaps of the outer ears) provides spectral-shape cues that are the principal cues for vertical and front/back source location. Specifically, the various convolutions of the pinnae introduce resonances and antiresonances that result in spectral peaks and notches in the spectrum of sound reaching the tympanic membranes. Those spectral features vary in frequency according to the angle of incidence of sound relative to the pinna. Batteau (1967) first considered the filtering action of the pinna in the time domain, but most other authors have emphasized the effects of the pinna on spectral amplitude. The importance of pinna acoustics for localization has been demonstrated by filling or otherwise disrupting the cavities of the pinna, which results in prominent disruptions of vertical localization (Musicant and Butler, 1984; Humanski and Butler, 1988). Spectral-shape cues often are referred to as "monaural spectral cues" – monaural because the necessary spectral analysis can occur at each ear individually. Spectral cues are available to both ears of a normal-hearing listener, but when a sound source is located to one side or the other, the greatest weight in localization judgments is given to spectral cues from the pinna on the same side as the source (Macpherson and Sabin, 2007).

Successful use of spectral-shape cues for localization requires that listeners be familiar with the filter properties of their own ears and that the sound source is relatively broad and flat in spectrum. Listeners appear to hold to the broad, flat spectrum assumption even when that assumption obviously is invalid. When asked to localize 1/6- or 1/3-octave band-passed sounds, listeners report erroneous vertical and front/back locations that vary with center frequency of the stimulus, irrespective of the actual source locations (Blauert, 1983; Butler, 1987; Middlebrooks, 1992). In such narrowband vertical-localization experiments, comparison of the patterns of localization judgments with the spectral transformations by listeners' pinnae (quantified by their HRTFs) suggests that listeners tended to misinterpret the narrowband sounds as broadband sounds that had been filtered by their pinnae to introduce a spectral peak, even though it was clear to the listener that the source spectrum was narrowband.

A quantitative model predicted vertical localization judgments successfully by assuming that listeners correlate the spectra of sounds in their ear canals with internal libraries of spectral templates, corresponding to their HRTFs for various sound directions, and report the elevation corresponding to the best-fitting template (Middlebrooks, 1992). That model, and several subsequent template-matching schemes (Hofman and Van Opstal, 1998; Chung et al., 2000; Langendijk and Bronkhorst, 2002; Bremen et al., 2010), did not offer any specific physiologic mechanism for template matching. Visual inspection of HRTFs reveals prominent spectral notches that vary in frequency according to sound direction, and a number of studies have seized on notches as the key vertical localization cue (Hebrank and Wright, 1974; Bloom, 1977; Musicant et al., 1990), but one is not aware of a conclusive demonstration that spectral notches signal vertical sound location exclusive of other spectral features.

The sizes and detailed shapes of heads and ears vary among listeners and, not surprisingly, their HRTFs also vary. A perceptual consequence of that interlistener variability can be demonstrated by recording HRTFs from one individual and processing sounds through those HRTFs to present to a different listener. Such an ear swap can result in prominent vertical localization errors, especially as an increase in the incidence of front/back confusions (Wenzel et al., 1993; Middlebrooks, 1999b). The differences among ears is not haphazard. Basic features of HRTFs are conserved among listeners but tend to scale in frequency roughly in proportion to listeners' physical sizes (Middlebrooks, 1999a). Hofman and colleagues (1998) have demonstrated that vertical localization can adapt to physical changes in the acoustics of a person's pinnae over the course of ~6 weeks. All these results are consistent with the view that the aspects of head and ear shape that are relevant for vertical localization are similar among most individuals, and that corresponding spectral templates are fine-tuned based on listeners' experience.

It is difficult to point to a single brainstem nucleus that could accomplish the broadband spectral template matching that is needed for vertical localization. Nevertheless, there is good evidence that type IV cells in the DCN are sensitive to spectral notches like those found in HRTFs (Young et al., 1992; Imig et al., 2000). Those cells are excited by broadband sounds, and that excitatory response is inhibited by the introduction of a spectral notch close to the cell's best frequency. Neurons in the inferior colliculus that appear to receive a direct projection from the contralateral DCN show excitatory responses that are selective to the center frequencies of notches in broadband sounds. These observations demonstrate that neurons in the DCN can code at least one element of the spectral cues for vertical localization. A necessary role of the DCN in vertical localization behavior has been shown by demonstrations that lesions of dorsal acoustic stria, which carries the output of the DCN, can result in deficits in unconditioned (Sutherland et al., 1998) or conditioned (May, 2000) orienting responses to elevated sounds.

The spectral features of HRTFs that vary as a function of vertical source location also vary systematically with horizontal location (Shaw, 1974). For that reason, one might expect spectral-shape cues to contribute to horizontal localization. Indeed, as mentioned in a previous section, there is some evidence of use of spectral cues for horizontal localization by some unilaterally deaf individuals (Slattery and Middlebrooks, 1994). In listeners with normal binaural hearing, however, the interaural difference cues (i.e., ITD and ILD) appear to dominate horizontal judgments, with little or no influence of spectral-shape cues (Macpherson and Middlebrooks, 2002). Interaural difference cues are less vulnerable to peculiarities in the source spectrum than are spectral-shape cues. It seems as if, when confronted with conflicting cues, the auditory system favors the cues that are the most reliable.

DISTANCE LOCALIZATION

Sound localization in the third dimension, distance, is not nearly as accurate as that in the horizontal and vertical dimensions. In optimal conditions (e.g., distance <2 m, broadband noise bursts, reflective room acoustics, target located to one side), reported distances can correlate closely with actual distances; correlation coefficients were >0.8 (Kopco and Shinn-Cunningham, 2011). In studies using unfavorable conditions (e.g., far field, recorded sounds presented at randomized levels, anechoic environment, target in front), distance judgments show random-chance performance levels (Coleman, 1962; Gardner, 1969; Holt and Thurlow, 1969). Listeners tend to underestimate distances that are greater than ~1.5 m and tend to overestimate distances that are <1.5 m (Zahorik, 2002; Zahorik et al., 2005). Distance judgments generally are more accurate for lateral sounds than for sounds in the midline plane, both for far and for nearby sources (Holt and Thurlow, 1969; Brungart et al., 1999; Zahorik, 2002; Kopco and Shinn-Cunningham, 2011). Distance judgments are influenced by familiarity of the sound source, source sound pressure level, source spectrum, source radiation pattern, azimuth of the source relative to the listener, acoustics of the surroundings, and near-versus-far range of distances. Visual cues also influence sound distance judgments, but they are beyond the scope of this review.

The principal acoustic cues for distance perception are intensity (i.e., sound level arriving at the listener's ears), direct-to-reverberant (D/R) energy ratio, and ILD. The relative importance of these cues varies widely across conditions. The intensity cue arises from the physical attenuation of a sound with distance. Given a point sound source in anechoic conditions, sound intensity arriving at the listener will decay by 6 dB with every doubling of the distance; the rate of decay is lower in reflective surroundings or if the source is directional. High and low intensities arriving at the listener signal near and far source distances, respectively. Of course, use of the intensity cue is confounded by uncertainty about source levels. Familiarity with the source can improve use of the intensity cue. For instance, a quiet word from a nearby speaker or a distant shout might arrive at the listener with equal intensities, but the listener could recognize the vocal effort and attribute the low intensity of the shout to attenuation across a greater distance (Gardner, 1969). Listeners give particular importance to intensity cues for the distances of sources in the midline plane, where D/R cues are weak and ILD cues are negligible (Zahorik, 2002).

D/R energy ratios are probably the most reliable distance cues in reverberant conditions as, for instance, in most office- to classroom-sized rooms. In such a room, a listener will receive sound directly from the source and also indirectly from reflections from multiple room surfaces. As discussed, the direct energy tends to decay with increasing distance, whereas the reverberant energy reaching the listener is largely independent of the source-to-listener distance. For those reasons, the ratio of D/R energy tends to decay with increasing source distance. The D/R ratio at a particular ear tends to be maximum for sound sources straight to the side, declining for midline source locations, and declining further for contralateral locations (Zahorik, 2002; Kopco and Shinn-Cunningham, 2011). In a study that simulated the spatial cues available in a small classroom, Kopco and Shinn-Cunningham (2011) demonstrated that distance judgments for simulated sources in the range of 15–170 cm correlated with D/R cues more than with any other potential cue.

ILDs are essentially independent of distances >1 m, but ILDs can increase markedly as lateral sources approach closer than ~ 1 m (Brungart and Rabinowitz, 1999); ILDs are negligible for midline sources regardless of distance. Near-field ILDs are present even at low frequencies at which they are absent in the far field. Use of ILD for distance judgments potentially could be confounded with uncertainty about the lateral location of the source. That confound can be resolved, however, by identification of lateral location on the basis of ITD cues; ITD cues give a reliable signal for lateral location, largely independent of distance (Brungart and Rabinowitz, 1999). Brungart (1999) has argued that ILDs are the dominant cues for judgments of the distances of nearby sources, and that is almost certainly true for anechoic conditions. Results from reverberant conditions, however, suggest that D/R cues dominate distance judgments when room reflections are available (Kopco and Shinn-Cunningham, 2011).

No brainstem structure has been identified that is sensitive specifically to distance cues aside from the LSO, which was discussed above as the likely site of ILD analysis.

MOTION PERCEPTION

One's everyday experience affirms the perceptual salience of a moving sound source. A sound source that moves in distance relative to a listener produces both Doppler shifts in frequency and modulation in loudness. We will focus here, however, on the problem of motion in azimuth or elevation, i.e., change in direction. It is a matter of some controversy whether such directional motion perception is a product of motion-specific brain mechanisms, or whether it is simply another aspect of localization. There certainly is no evidence in hearing for the degree of specialization for motion detection that is well known in the visual system, in which neurons are specific for the direction and velocity of visual motion (Hubel and Wiesel, 1962) and in which a motion-specific cortical area has been characterized (i.e., area MT of primates) (Mikami et al., 1986). In psychophysical studies, humans can discriminate velocities of visual targets differing by $<10\%$, whereas discrimination of auditory target velocities require 20–70% differences (Carlile and Best, 2002).

A number of early studies represented auditory motion sensitivity by the minimum audible movement angle (MAMA), which was the smallest detectable change in location of a moving stimulus. The MAMA may be compared with the minimum audible angle (MAA), which is the smallest detectable difference in location of two static sounds presented successively. In the motion studies considered in an early review (Middlebrooks and Green, 1991), MAMAs all were substantially larger than MAAs, indicating that motion did not improve (and generally impaired) detection of a change in location. Based on such data, a number of investigators have concluded that auditory motion detection does not reflect sensitivity to specific dynamic cues such as changing ITD or ILD. Instead it might be that the auditory system simply takes two or more "snapshots" of source locations at successive times and interprets changes in location as evidence for motion (Grantham, 1986). At the time of that 1991 review, we concluded that "As yet there is no compelling evidence for motion-sensitivity systems in the auditory system." Several more recent observations tip the balance somewhat in favor of motion-sensitive systems, although it is not yet clear that the sum of evidence reaches the threshold for "compelling." The newer observations to be considered include those of motion aftereffects, neurons sensitive to dynamic ITDs and ILDs, and cortical areas in humans activated by moving sounds.

"Motion aftereffects" refer to phenomena in which prolonged exposure to a moving sound influences perception of the motion of a following probe sound; this is analogous to the "waterfall illusion" in vision. Grantham (1986) found that exposure to a moving noise source could result in a bias of the judgment of the direction of movement of a probe. The auditory aftereffect was considerably less robust than is seen in vision in that the magnitude of the auditory aftereffect was smaller and its duration was only a few seconds; the visual-motion aftereffect can last for a minute or more. Grantham speculated that the auditory aftereffect that he observed was due to some combination of response bias (a tendency to report probe motion contrary to that of the adaptor) and a sensory effect consisting of a loss of sensitivity to probe velocity. Grantham (1992) later demonstrated that exposure to a moving adaptor could decrease motion sensitivity as indicated by enlarged MAMAs. The notion of generalized suppression of motion sensitivity by a moving adaptor is supported by a study of auditory evoked potentials (Magezi et al., 2013). The auditory event potentials elicited by a stationary probe stimulus differed depending on the characteristics of a preceding auditory adaptor. There was no significant difference in the responses conditioned by leftward compared to rightward adaptor motion, suggesting a lack of directionality in a putative motion system. There was, however, a significant difference between effects of a bidirectional-moving adaptor compared to a stationary adaptor, consistent with the notion of generalized suppression of motion sensitivity.

Studies of ITD sensitivity neurons in the inferior colliculus have demonstrated that the response to a particular ITD can vary according to the context of a varying ITD (Spitzer and Semple, 1991, 1998). That is, the best ITD of a neuron can be influenced by the range of ITDs that is traversed in a sweep of varying ITD. Similarly, a study of ILD sensitivity in the inferior colliculus (Sanes et al., 1998) demonstrated that the ILD sensitivity of neurons depended on their history of prior stimulation by a dynamic stimulus. One might regard these demonstrations of sensitivity to dynamic ITDs and ILDs as a substrate for motion perception. Alternatively, as pointed out by Spitzer and Semple (1998), these phenomena might be "non-adaptive by-products" of neural mechanisms that might account for the psychophysical observations of degraded spatial acuity for moving sounds (i.e., that MAMAs are wider than MAAs).

Several human cortical imaging studies using positron emission tomography or functional magnetic resonance imaging have identified cortical areas that show enhanced activity during presentation of sounds that simulate sound-source motion compared to control conditions of static locations (Griffiths et al., 1994, 1998; Baumgart et al., 1999; Lewis et al., 2000; Parvani et al., 2002; Warren et al., 2002; Krumbholz et al., 2005). Involved areas generally are located in the right hemisphere, including the insula, planum temporale, and posterior parietal cortex. Although one could readily accept those results as evidence of motion-specific brain areas, one must also entertain the alternative interpretation that those cortical areas simply are sensitive to static sound-source location and that they respond more strongly to the broad range of locations provided by the moving source than to a single fixed source. Indeed, studies that have attempted to control for the number of stimulus locations have failed to show motion selectivity (Smith et al., 2004, 2007). Similarly, there are multiple interpretations possible in a case study in which a right-sided lesion involving the insula, the planum temporale, and the posterior parietal cortex resulted in a pronounced deficit in sensitivity to simulated moving sounds (Griffiths et al., 1996). Although the loss of sensitivity to dynamic sounds was clear, that patient also exhibited a deficit in "fixed lateralization," including elevated thresholds for detection of static ITD and ILD. In support of the argument that responses to changes in location can appear as motion sensitivity, Getzmann and Lewald (2012) recorded very similar "motion-onset responses" (i.e., auditory evoked potentials recorded from the human scalp, elicited by sound movement) in response to simulated-motion sounds that moved continuously, moved abruptly from one sound hemifield to the other, or moved randomly among a range of static locations.

In summary, listeners are sensitive to motion of sound sources, and their judgments of the movement of a probe sound can be influenced by a recent history of motion stimulation. Similarly, the sensitivity of neurons to spatial cues can be modulated by prior stimulation. Clearly, there are brain regions that respond well to moving sounds, but it is difficult to distinguish responses to continuous motion from responses to multiple static stimuli. Although the presence of motion-sensitivity systems in the auditory system is an attractive notion, evidence for such systems still is less than compelling.

LOCALIZATION IN REVERBERANT SPACES: THE PRECEDENCE EFFECT

Most of our listening occurs in reverberant environments in which sounds reflect from walls, floor, ceiling, and other structures. That means that when a sound is emitted, the listener first is exposed to the sound traveling on a direct path from the source, followed by a few early reflections, followed by the reverberation due to the interaction of all the reflections. We already have considered how the ratio of D/R energy can aid estimation of source distance, but one might expect the reverberations to confound estimation of source direction. Our everyday experience, as well as a large body of research, indicates that the auditory system is quite effective in limiting localization judgments to the sound that arrives first, via the direct path, and disregarding the directions of sound reflections. The group of auditory phenomena encompassing the factors that facilitate localization in the presence of reflections is known as the "precedence effect"; one of several good reviews is by Litovsky and colleagues (1999).

The precedence effect has most often been studied using brief sounds presented from two free-field locations or, under headphones, from two locations simulated using pairs of ITDs. A range of perceptions is reported depending on the delay between the leading and lagging sound. When the delay is less than ~5 ms, the two sounds tend to fuse and the listener reports hearing only a single sound. At 0 ms delay, the listener tends to localize the sound midway between the two sources. As the delay increases from 0 to ~1 ms, the perceived source location gradually drifts toward that of the leading source. This is the range of "summing localization," which is exploited for the purpose of stereophonic sound reproduction. At further increasing delays, "localization dominance" occurs in which the perceived location corresponds to that of the leading source. At a delay >5 ms, known as the "echo threshold," the listener begins to report hearing two sounds. Even though two sounds are heard in this condition, the localization judgment is still dominated by the leading source and discrimination of multiple lagging-source locations is impaired (this is "discrimination suppression"). Only after delays of >10 ms is a listener able to judge accurately the location of the lagging-sound source. The various delays given here are minimum values, appropriate for brief sources. The respective delays generally all are longer under conditions of longer-duration sounds.

The physiologic mechanisms of the precedence effect involve some combination of the binaural neurons of the superior olivary complex plus suppressive mechanisms at midbrain and cortical levels. In anesthetized cats, correlates of fusion are present at the level of the inferior colliculus in the sense that single neurons respond reliably to the first of two sounds and respond to the lagging sound only after some delay (Yin, 1994). The delay at which the neural response to the lagging sound recovers to half that of the leading sound varies widely across neurons but has a median of 20 ms, not much longer than that observed in human psychophysics. In unanesthetized cats, locations of sound sources can be estimated from the spike patterns of auditory cortical neurons. In the primary auditory cortex, neurons reliably indicated the locations of leading sounds. Responses of those neurons to lagging sounds were suppressed, and accurate estimates of the locations of lagging sounds could be obtained from only a small subset of neurons (Mickey and Middlebrooks, 2005).

The notion that the precedence effect requires considerable central integration is supported by studies of its development in humans (Muir et al., 1989; Litovsky et al., 1999). Infants can distinguish between two sound sources within hours of birth, but localization dominance appears first at 4–5 months of age, maturing fully by about 5 years.

CENTRAL REPRESENTATION OF SOUND-SOURCE LOCATIONS

We have considered the various cues for sound-source locations and described brainstem pathways that process those cues. Now we address the question of how and where in the auditory system information from the spatial cues might be integrated to form neural representations of the locations of sources in auditory space. Existing evidence suggests that there are at least two very different modes of central spatial representation: a topographic representation in the superior colliculus (SC) and distributed representations in the auditory cortex.

Spatial topography in the superior colliculus

The SC is a sensorimotor integrative structure that directs orienting movements of the head and eyes toward sources of visual, tactile, and auditory stimuli. Studies conducted under anesthetized conditions in a number of non-primate species have demonstrated topographic maps of auditory space in the intermediate and deep layers of the SC: cat (Gordon, 1973; Middlebrooks and Knudsen, 1984); guinea pig (Palmer and King, 1982); ferret (King and Hutchings, 1987); rat (Gaese and Johnen, 2000); a map also is seen in a homologous structure, the optic tectum, in barn owls (Knudsen, 1982). Here, "topographic" means that neurons in the SC respond maximally to sounds presented within a more-or-less restricted region of space (often called the "best area") and that best areas vary systematically in space according to locations of neurons within the SC. Most SC neurons that respond to sounds also respond to visual stimuli. The auditory map lies in approximate spatial register with visual maps in the SC. This is only "approximate" because auditory spatial sensitivity extends further lateral than do visual responses, consistent with the 360° spatial sensitivity of hearing compared with the more restricted visual fields. Auditory and visual spatial tuning for near-midline locations tends to show fairly close mutual alignment whereas SC neurons with more eccentric receptive fields tend to show auditory best areas shifted systematically lateral to the centers of visual receptive fields.

The static auditory space maps that have been described in the SC in non-primate species arguably reflect the use of general anesthesia in studying those species. Under those conditions, neurons show sensitivity to source locations relative to the positions of head and external ears, i.e., in head-centered coordinates. In non-human primates, study of the SC generally has been conducted in awake conditions in which the animal is required to make eye movements to visual and/or auditory targets (Jay and Sparks, 1987). Under such conditions, the spatial sensitivity is interpreted more easily as a motor error signal reflecting the gaze shift needed to acquire the sensory target, and the auditory topography appears integrated within a map of motor error. Auditory spatial signals appear to transform within the SC from head-centered to eye-centered coordinates (Lee and Groh, 2012).

Distributed spatial representation in the auditory cortex

Normal auditory spatial perception and behavior requires intact auditory cortex. Experimental lesions or reversible inactivation of the auditory cortex in laboratory animals (Thompson and Cortez, 1983; Jenkins and Merzenich, 1984; Kavanagh and Kelly, 1987; Malhotra et al., 2004) and clinical lesions of auditory cortex in humans (Sanchez-Longo and Forster, 1958; Klingon and Bontecou, 1966; Zatorre and Penhune, 2001) result in deficits in the ability to report the location of a sound source, either by walking to the source (by animals) or by some other means of report by humans (see Chapter 32).

Given the apparent role of auditory cortex in sound localization, it may come as a surprise that no topographic representation of auditory space has been identified in the cortex, despite the efforts of several research groups to find such a map (Middlebrooks and Pettigrew, 1981; Imig et al., 1990; Rajan et al., 1990). In the primary auditory cortex (A1), a large proportion of neurons responds with more than half of neurons' maximum responses to sounds from all tested locations; such "omnidirectional" neurons constitute around half the sample in some species and some experimental conditions (Harrington et al., 2008; Werner-Reiss and Groh, 2008; Zhou and Wang, 2012). Among the neurons that exhibit sharper spatial sensitivity, many have "contralateral hemifield" spatial tuning, meaning that they respond strongly to sound sources located throughout the contralateral half of space, with their responses cutting off near the midline. A smaller proportion of units responds most strongly to near-midline or ipsilateral sources. In area A1 of the cat, study in awake and in awake-behaving conditions (Mickey and Middlebrooks, 2003; Lee and Middlebrooks, 2013) reveals spatial tuning that is somewhat more restricted and biased toward the frontal midline than is seen in anesthetized conditions (Harrington et al., 2008). Nevertheless, it would be difficult to argue that spatial sensitivity in awake conditions is qualitatively different from that observed with anesthesia. Contrary to the requirement for a topographic representation, there has been no indication of a systematic progression of best area as a function of cortical area in either awake or anesthetized conditions in any species. Instead, the cortical distribution of spatial tuning tends to be modular in that neurons within a restricted cortical region tend to show one spatial preference while neighboring regions can show quite different preferences (Middlebrooks and Pettigrew, 1981; Rajan et al., 1990; Middlebrooks and Bremen, 2013). There is some indication of a systematic progression of spatial receptive field cutoffs in area A1 in the pallid bat (Razak, 2011), but no such organization has been seen in any of the other species that have been examined.

Examination of the temporal response patterns of auditory cortical neurons has suggested a form of spatial representation that is qualitatively different from the topographic place codes that are familiar in visual and somatosensory systems. In many cases, response patterns vary systematically with sound-source location such that the locations of stimulus sound sources can be identified with some accuracy for source locations throughout 360° of space (Middlebrooks et al., 1994, 1998). Figure 6.3 shows an example of one neuron in the second auditory area (A2) of an anesthetized cat. The dot raster plot in Figure 6.3A shows how the temporal response pattern varied with sound-source location, and Figure 6.3B shows estimates of source location based on computer recognition of those response patterns. The property of single neurons signaling source locations throughout a broad range of locations has been referred to as "panoramic" location coding (Middlebrooks et al., 1994). That individual neurons can signal locations panoramically and that such neurons are distributed throughout the auditory cortex indicate that the representation of any particular source location is distributed throughout the auditory cortex.

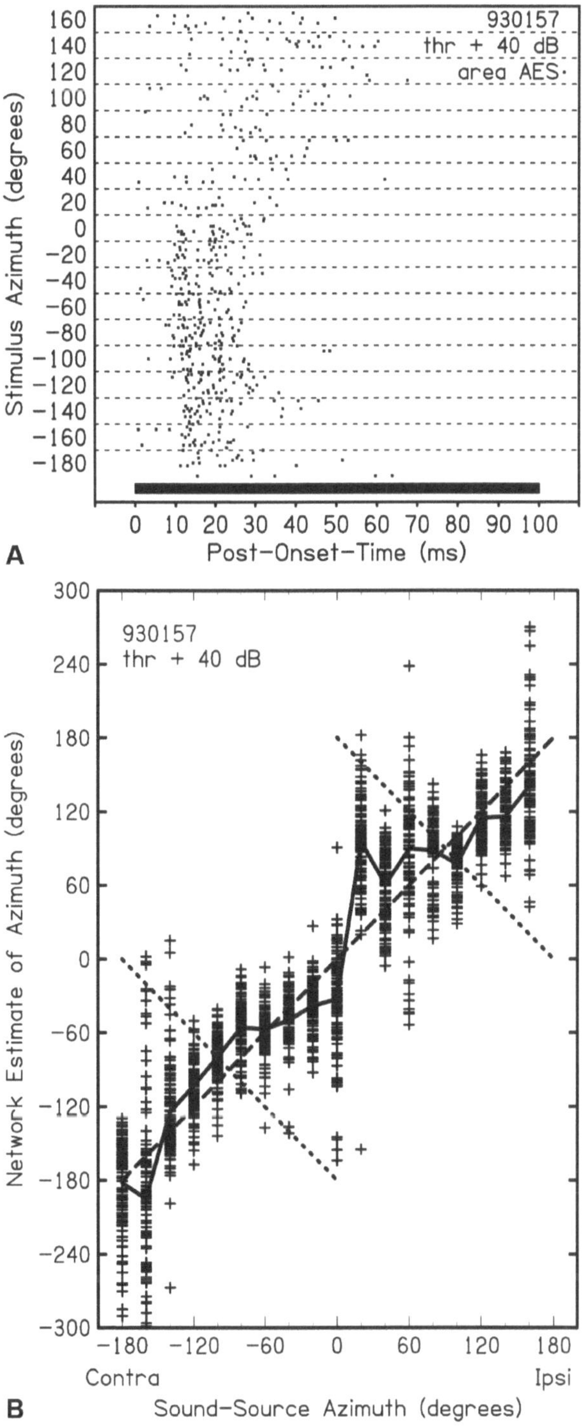

Fig. 6.3. Spatial sensitivity of a cortical neuron in area A2 of an anesthetized cat. (**A**) In this dot-raster plot, each row of dots represents the pattern of action potentials elicited by a 100-ms broadband noise burst presented from the indicated location in azimuth. Responses to eight such bursts are shown for each azimuth. (**B**) Estimates of source locations based on spike patterns from the neuron in (**A**). Each symbol represents one estimate based on a bootstrapped response pattern based on responses to eight stimulus presentation. The dashed line indicates loci of perfect performance, and the solid curve connects the mean values of estimates for each source location. The estimates involved recognition of spike patterns using an artificial neural network. AES, anterior ectosylvian sulcus. (Reproduced from Middlebrooks et al., 1998.)

Cortical neurons transmit information about sound-source locations both with their spike counts and with the timing of spikes. Several studies have demonstrated that the accuracy of location estimation based on cortical spike patterns is degraded when temporal information is degraded or eliminated (Middlebrooks et al., 1994; Furukawa and Middlebrooks, 2002; Mickey and Middlebrooks, 2003). Indeed, in some conditions first-spike latencies transmit more location-related information than do spike counts (Furukawa and Middlebrooks, 2002). The accuracy of panoramic coding by cortical spike patterns improves when information is combined across multiple neurons (Furukawa et al., 2000; Miller and Recanzone, 2009; Lee and Middlebrooks, 2013). In one study, information from ensembles of 16 of the most informative neurons signaled source locations throughout space with accuracy comparable to the performance of cats in a psychophysical localization task (Furukawa et al., 2000).

The sensitivity of cortical neurons for locations of sound sources varies considerably depending on anesthetic and behavioral conditions and on the presence of other sounds. As noted above, the spatial sensitivity of neurons in cat area A1 tends to be sharper and the distribution of best areas shifted closer to the frontal midline in awake compared to anesthetized conditions (Mickey and Middlebrooks, 2003; Lee and Middlebrooks, 2013). In awake cats, spatial sensitivity of neurons in A1 sharpens on individual trials on which the animal performs a listening task compared to trials on which it is idle, and many neurons show additional sharpening when the task requires localization compared to a non-spatial task (Lee and Middlebrooks, 2011). A recent study of cortical responses to temporally interleaved sequences of sounds from spatially separated sources demonstrated that the selectivity of neurons for the location of a target source sharpens dramatically in the presence of a competing sound from another location (Middlebrooks and Bremen, 2013). These observations all support the picture of a dynamic, highly distributed cortical representation of auditory space rather than a rigid point-to-point topography.

Despite their panoramic spatial sensitivity, in most cases the spatial acuity of cortical neurons is greatest for sound sources located near the frontal midline (Stecker et al., 2005). This accords with human psychophysical results showing highest acuity (Mills, 1958) and greatest localization accuracy (Stevens and Newman, 1936; Makous and Middlebrooks, 1990) for near-midline targets. Several investigators have considered the possibility that sound locations are represented by the opponent activity of a small number (i.e., possibly two or three) of neural channels, each tuned to a rather broad region of space; channels tuned to the right or left half of space might be one example. Opponent-channel models draw support from human psychophysical studies of gap detection, spatial release from masking, and selective adaptation in humans (Boehnke and Phillips, 1999; Phillips et al., 2003; Dingle et al., 2012, 2013), from cortical single-neuron recordings in animals (Stecker et al., 2005), and from far-field magnetic or electric recordings in humans (Salminen et al., 2009; Magezi and Krumbholz, 2010; Briley et al., 2012). Models that assume summed activity across large populations of similarly tuned neurons capture the generally contralateral bias of neural spatial tuning in each hemisphere and can account well for results of low-resolution scalp recordings. Nevertheless, the variety of spatial sensitivity that can be recorded at the level of single cortical neurons and that is not captured by limited-channel models suggests that two- or three-channel models should be regarded as only first-order approximations of the details of cortical spatial processing.

Cortical areas specialized for sound localization

In the previous section, we emphasized the primary auditory cortex. Here, we consider additional auditory cortex areas that do or do not show specialization for localization, we consider the issue of laterality of spatial representation in humans and non-humans, and we consider hypothetic "what" and "where" processing streams in the human and non-human primate cortex.

Single-neuron recordings in cats have demonstrated at least some spatial sensitivity in each of six distinct cortical areas in which such sensitivity was evaluated (Middlebrooks et al., 1998; Harrington et al., 2008). Nevertheless, the sharpness of spatial tuning and the sound-level invariance of spatial tuning varies considerably among areas, with the sharpest spatial tuning seen in the posterior auditory field (PAF) and the dorsal zone (DZ), intermediate sharpness of tuning in A1, A2, and the anterior ectosylvian area, and the weakest tuning seen in the anterior auditory field (AAF). That ranking of the sharpness of spatial tuning parallels the results of behavioral experiments involving reversible inactivation of restricted cortical areas. Inactivation of PAF or DZ disrupted trained cats' ability to walk to the location of a sound source (Malhotra et al., 2004). In a related study, inactivation of AAF disrupted performance of a task requiring discrimination of temporal patterns while sparing localization performance (Lomber and Malhotra, 2008). In the same cats, inactivation of PAF disrupted localization but spared the temporal discrimination. An interpretation is that intact PAF and DZ are necessary for sound localization, whereas the spatial sensitivity in AAF might support some other aspect of spatial hearing, such as spatial release from masking (Middlebrooks et al., 2002).

Studies in cats and monkeys suggest specialization of cortical areas for localization in particular regions of space. In awake cats (Lee and Middlebrooks, 2013), best areas of neurons in area DZ show a bias toward representation of frontal space, with a large majority of spatially tuned neurons showing best areas centered within 45° of the frontal midline. In contrast, area PAF in awake cats shows a more uniform representation of space, with about equal numbers of spatially tuned neurons showing best areas centered in the frontal or contralateral quadrant of space. Analysis of signaling of sound-source location by responses of ensembles of 16 neurons shows most accurate localization of near-midline sources by DZ neurons and of more peripheral sources by PAF neurons (Lee and Middlebrooks, 2013). In the awake macaque (Woods et al., 2006), distributions of best azimuths in most of the tested cortical areas show large proportions of neurons showing far-peripheral best areas. In contrast, the largest proportion of neurons showing best areas around the frontal midline was in the core area (R). Miller and Recanzone (2009) tested identification of sound-source locations based on responses of ensembles of cortical neurons in the macaque data set (Woods et al., 2006). Across all source locations, localization was most accurate based on responses of neurons in the caudolateral area (CL). Examination of the figures, however, shows that accurate localization by CL neurons was seen only for far-contralateral targets. Localization of sources around the frontal midline was most accurate based on responses of neurons in R; note that the frontal midline is where psychophysical spatial acuity is highest. It is difficult to draw homologies between particular cortical areas in cats and macaques, but both species show analogous areas that apparently are specialized for frontal (areas DZ and R) and more peripheral (PAF and CL) localization.

Rauschecker and Tian (2000) proposed a hypothesis that has provided a very influential framework for study of cortical spatial processing in human and non-human primates. The hypothesis draws an analogy with the visual system, in which ventrally and dorsally directed hierarchical "streams" are specialized for recognizing, respectively, "what" an object is or "where" it is (Ungerleider and Mishkin, 1982). In the auditory system, the "where" pathway was proposed to originate in the caudal part of the superior temporal gyrus (encompassing the planum temporale in humans and auditory areas CL and caudomedial (CM) in monkeys) and to project to the dorsolateral prefrontal cortex, both directly and by way of the parietal cortex (Romanski et al., 1999; Rauschecker and Tian, 2000). Those parietal and prefrontal areas also participate in visual spatial processing. In contrast, the "what" pathway was proposed to originate from anterior lateral belt areas and to project toward the temporal pole. Human functional imaging studies support the "where" part of the hypothesis in that auditory spatial tasks tend to activate parietal and prefrontal areas that also are activated during visual spatial tasks (Rauschecker and Scott, 2009; Recanzone and Cohen, 2010). Functional imaging offers less consistent support for the hypothetic special role of the caudal auditory cortex as the origin of the spatial-processing stream. Several studies that used simulated moving sounds or sounds that varied in their spatial distribution failed to demonstrate specific activation of posterior auditory areas (Griffiths et al., 1994, 1998; Zatorre et al., 2002). Other studies have demonstrated specific activation of the planum temporale, but only when the stimuli incorporated perceived movement or complex spectral changes (Baumgart et al., 1999; Warren et al., 2002; Krumbholz et al., 2005; Ahveninen et al., 2006).

Physiologic studies of the auditory cortex in non-human primates do not consistently support the notion that caudal areas are specialized for spatial processing. The prediction would be that caudal belt areas would show sharper spatial tuning or a higher percentage of sharply tuned neurons than do R or A1 core areas or middle or R belt areas. In marmoset monkeys, overall distributions of best areas and modulation of responses by source location were "markedly similar" between area A1 and the CL and CM fields (Zhou and Wang, 2012). One study in macaques demonstrated that neurons in CL tended to have narrower spatial tuning for monkey calls than did neurons in anterolateral or middle lateral (Tian et al., 2001). Another study in macaques is often cited in support of the what/where hypothesis (Recanzone et al., 2000). That study, however, found no significant difference in the proportion of spatially sensitive neurons in cortical areas A1 and CM; no R belt areas were studied. Another macaque study (Woods et al., 2006) found that a middle belt area consistently had a higher proportion of spatially sensitive neurons than did CL and that at moderate-to-high sound levels area A1 had a higher proportion of spatially sensitive neurons than did either CL or CM; again, no R belt area was sampled. As noted above, computational analysis of those data demonstrated that neurons in area CL signaled far-contralateral locations accurately but that signaling of near-midline regions by CL neurons was poor compared to that by neurons in area R (Miller and Recanzone, 2009). Inasmuch as psychophysical localization is most accurate for near-midline targets, it is difficult to take those physiologic results as evidence that CL is the principal auditory cortical area driving sound localization.

The cortical representation of auditory space appears to be strictly contralateral in carnivores and non-human primates. Restricted auditory cortical lesions or inactivation disrupt the trained ability to walk to the source of a sound contralateral to the lesion or inactivation (Thompson and Cortez, 1983; Jenkins and Merzenich, 1984; Kavanagh and Kelly, 1987; Malhotra et al., 2004). The contralaterality of auditory spatial representation in humans is less clear. Early clinical studies suggested that contralesional deficits were most common (Sanchez-Longo and Forster, 1958; Klingon and Bontecou, 1966). In conflict with a notion of contralateral dominance, however, functional imaging studies tend to show enhanced activity in right parietal areas during auditory spatial tasks, regardless of the laterality of the auditory stimulus (Griffiths et al., 1998; Bushara et al., 1999; Zatorre et al., 1999). A study of patients who underwent partial temporal-lobe resections for relief of epilepsy demonstrated a substantial right-sided dominance of auditory spatial representation (Zatorre and Penhune, 2001). In that study, most patients having right-sided resections demonstrated symmetric bilateral sound localization deficits, whereas most patients having left-sided resections showed no localization deficit. Substantial deficits were observed even in cases in which the resection was entirely rostral to Heschl's gyrus, involving hypothetic "what" pathways but sparing the primary auditory cortex and the putative "where" area on the planum temporale.

BEYOND LOCALIZATION

Sound localization can aid a cat in catching a mouse and can help a person orient to a talker of interest in a crowded party. Beyond just localization, however, normal-hearing listeners can exploit spatial hearing mechanisms to detect and recognize sounds of interest in the presence of other competing sounds. Conversely, people with mild-to-moderate hearing loss report that among their greatest disabilities are the inability to segregate multiple talkers and to understand speech in noisy environments (Gatehouse and Nobel, 2004). Spatial hearing has long been recognized as contributing to so-called "auditory scene analysis" (Cherry, 1953; Bregman, 1990), and there has been growing interest in spatial release from masking, which is the improvement of sound detection and recognition when the signal is separated in space from maskers (Edmonds and

Culling, 2005; Ihlefeld and Shinn-Cunningham, 2008; Marrone et al., 2008). Spatial release from masking shares many properties with localization – the dominance of low-frequency ITD cues (Kidd et al., 2010; Middlebrooks and Onsan, 2012; Bremen and Middlebrooks, 2013) is one important example. Understanding of relevant spatial cues and low-level mechanisms gained from ongoing research will lead to enhanced processing schemes for hearing aids and cochlear implants that will benefit both localization and spatial release. There is evidence, however, that localization and spatial release from masking might employ discrete cortical structures. Specifically, cortical lesions can disrupt localization while sparing spatial release from masking or (less commonly) can disrupt spatial release from masking while sparing localization (Thiran and Clarke, 2003; Duffour-Nikolov et al., 2012). One is hopeful that enhanced non-invasive imaging and encephalographic techniques in humans and invasive cortical recording techniques in behaving animals will yield new understanding of these relevant cortical mechanisms.

References

Ahveninen J, Jääskeläinen IP, Raij T et al. (2006). Task-modulated "what" and "where" pathways in human auditory cortex. Proc Natl Acad Sci U S A 103: 14608–14613.

Batteau DW (1967). The role of the pinna in human localization. Proc R Soc Lond B Biol Sci 168: 158–180.

Baumgart F, Gaschler-Markefski B, Woldorff MG et al. (1999). A movement sensitive area in auditory cortex. Nature 4000: 724–726.

Bernstein LR, Trahiotis C (1994). Detection of interaural delay in high-frequency sinusoidally amplitude-modulated tones, two-tone complexes, and bands of noise. J Acoust Soc Am 95: 3561–3567.

Bernstein LR, Trahiotis C (2002). Enhancing sensitivity to interaural delays at high frequencies by using "transposed stimuli". J Acoust Soc Am 112: 1026–1036.

Blauert J (1982). Binaural localization: Multiple images and application in room- and electroacoustics. In: RW Gatehouse (Ed.), Localization of Sound: Theory and Application. Amphora, Groton, CT, pp. 65–84.

Blauert J (1983). Spatial Hearing. MIT Press, Cambridge, Mass.

Bloom PJ (1977). Creating source elevation illusions by spectral manipulation. J Audio Eng Soc 25: 560–565.

Boehnke SE, Phillips DP (1999). Azimuthal tuning of human perceptual channels for sound location. J Acoust Soc Am 106: 1948–1955.

Boudreau JC, Tsuchitani C (1968). Binaural interaction in the cat superior olive S segment. J Neurophysiol 31: 442–454.

Brand A, Behrend O, Marquardt T et al. (2002). Precise inhibition is essential for microsecond interaural time difference coding. Nature 417: 543–547.

Bregman AS (1990). Auditory Scene Analysis: The Perceptual Organization of Sound. MIT Press, Cambridge, MA.

Bremen P, Middlebrooks JC (2013). Weighting of spatial and spectro-temporal cues for auditory scene analysis by human listeners. PLoS One 8: e59815.

Bremen P, van Wanrooij MM, van Opstal AJ (2010). Pinna cues determine orienting response modes to synchronous sounds in elevation. J Neurosci 30: 194–204.

Briley PM, Kitterick PT, Summerfield AQ (2012). Evidence for opponent process analysis of sound source location in humans. J Assoc Res Otolaryngol 14: 83–101.

Brughera A, Dunai L, Hartmann WM (2013). Human interaural time difference thresholds for sine tones: The high-frequency limit. J Acoust Soc Am 133: 2839–2855.

Brungart DS (1999). Auditory localization of nearby sources. III. Stimulus effects. J Acoust Soc Am 106: 3589–3602.

Brungart DS, Rabinowitz WM (1999). Auditory localization of nearby sources. Head-related transfer functions. J Acoust Soc Am 106: 1465–1479.

Brungart DS, Durlach NI, Rabinowitz WM (1999). Auditory localization in nearby sources. II. Localization of a broadband source. J Acoust Soc Am 106: 1956–1968.

Bushara KO, Weeks RA, Ishii K et al. (1999). Modality-specific frontal and parietal areas for auditory and visual spatial localization in humans. Nat Neurosci 2: 759–766.

Butler RA (1987). An analysis of the monaural displacement of sound in space. Percept Psychophys 41: 1–7.

Carlile S, Best V (2002). Discrimination of sound source velocity in human listeners. J Acoust Soc Am 111: 1026–1035.

Carlile S, Leong P, Hyams S (1997). The nature and distribution of errors in sound localization by human listeners. Hear Res 114: 179–196.

Carr CE, Konishi M (1990). A circuit for detection of interaural time differences in the brain stem of the barn owl. J Neurosci 10: 3227–3246.

Cherry CE (1953). Some experiments on the recognition of speech, with one and two ears. J Acoust Soc Am 25: 975–979.

Chung W, Carlile S, Leong P (2000). A performance adequate computational model for auditory localization. J Acoust Soc Am 107: 432–445.

Coleman PD (1962). Failure to localize the source distance of an unfamiliar sound. J Acoust Soc Am 34: 345–346.

Dingle R, Hall S, Phillips D (2012). The three-channel model of sound localization mechanisms: interaural level differences. J Acoust Soc Am 131: 4023–4029.

Dingle RN, Hall SE, Phillips DP (2013). The three-channel model of sound localization mechanisms: Interaural time differences. J Acoust Soc Am 133: 417–424.

Duffour-Nikolov C, Tardif E, Maeder P et al. (2012). Auditory spatial deficits following hemispheric lesions: Dissociation of explicit and implicit processing. Neuropsychol Rehabil 22: 674–696.

Edmonds BA, Culling JF (2005). The spatial unmasking of speech: evidence for within-channel processing of interaural time delay. J Acoust Soc Am 117: 3069–3078.

Furukawa S, Middlebrooks JC (2002). Cortical representation of auditory space: Information bearing features of spike patterns. J Neurophysiol 87: 1749–1762.

Furukawa S, Xu L, Middlebrooks JC (2000). Coding of sound-source location by ensembles of cortical neurons. J Neurosci 20: 1216–1228.

Gaese BH, Johnen A (2000). Coding for auditory space in the superior colliculus of the rat. Eur J Neurosci 12: 1739–1752.

Gardner MB (1969). Distance estimation of 0 deg. or apparent 0 deg.-oriented speech signals in anechoic space. J Acoust Soc Am 45: 47–53.

Gatehouse S, Nobel W (2004). The speech, spatial, and qualities of hearing scale (SSQ). Int J Audiol 43 (85): 99.

Getzmann S, Lewald J (2012). Cortical processing of change in sound location: Smooth motion versus discontinuous displacement. Brain Res 1466: 119–127.

Goldberg JM, Brown PB (1969). Response of binaural neurons of dog superior olivary complex to dichotic tonal stimuli: Some physiological mechanisms of sound localization. J Neurophysiol 32: 613–636.

Gordon B (1973). Receptive fields in deep layers of cat superior colliculus. J Neurophysiol 36: 157–178.

Grantham DW (1986). Detection and discrimination of simulated motion of auditory targets in the horizontal plane. J Acoust Soc Am 79: 1939–1949.

Grantham D (1992). Adaptation to auditory motion in the horizontal plane: Effect of prior exposure to motion on motion detectability. Percept Psychophys 52: 144–150.

Griffiths TD, Bench CJ, Frackowiak RSJ (1994). Human cortical areas selectively activated by apparent sound movement. Curr Biol 4: 892–895.

Griffiths TD, Rees A, Witton C et al. (1996). Evidence for a sound movement area in the human cerebral cortex. Nature 383: 425–427.

Griffiths T, Rees G, Rees A et al. (1998). Right parietal cortex is involved in the perception of sound movement in humans. Nat Neurosci 1: 74–78.

Hafter ER, Dye RH, Nuetzel JM et al. (1977). Difference thresholds for interaural intensity. J Acoust Soc Am 61: 829–834.

Harrington IA, Stecker GC, Macpherson EA et al. (2008). Spatial sensitivity of neurons in the anterior, posterior, and primary fields of cat auditory cortex. Hear Res 240: 22–41.

Hebrank J, Wright D (1974). Spectral cues used in the localization of sound sources on the median plane. J Acoust Soc Am 56: 1829–1834.

Henning GB (1974). Detectability of interaural delay in high-frequency complex waveforms. J Acoust Soc Am 55: 84–90.

Hofman PM, Van Opstal JA (1998). Spectro-temporal factors in two-dimensional human sound localization. J Acoust Soc Am 103: 2634–2648.

Hofman PM, Van Riswick JGA, Van Opstal JA (1998). Relearning sound localization with new ears. Nat Neurosci 1: 417–421.

Holt RE, Thurlow Wr (1969). Subject orientation and judgment of distance of a sound source. J Acoust Soc Am 46: 1584–1585.

Hubel DH, Wiesel TN (1962). Receptive fields, binocular interaction and functional architecture in the cat's visual cortex. J Physiol 160: 106–154.

Humanski RA, Butler RA (1988). The contribution of the near and far ear toward localization of sound in the sagittal plane. J Acoust Soc Am 83: 2300–2310.

Ihlefeld A, Shinn-Cunningham B (2008). Spatial release from energetic and informational masking in a selective speech identification task. J Acoust Soc Am 123: 4369–4379.

Imig TJ, Irons WA, Samson FR (1990). Single-unit selectivity to azimuthal direction and sound pressure level of noise bursts in cat high-frequency primary auditory cortex. J Neurophysiol 63: 1448–1466.

Imig TJ, Bibikov NG, Poirier P et al. (2000). Directionality derived from pinna-cue spectral notches in cat dorsal cochlear nucleus. J Neurophysiol 83: 907–925.

Jay MF, Sparks DL (1987). Sensorimotor integration in the primate superior colliculus. II. Coordinates of auditory signals. J Neurophysiol 57: 35–55.

Jeffress LA (1948). A place theory of sound localization. J Comp Physiol Psychol 41: 35–39.

Jenkins WM, Merzenich MM (1984). Role of cat primary auditory cortex for sound-localization behavior. J Neurophysiol 52: 819–847.

Joris PH, Yin TCT (2007). A matter of time: internal delays in binaural processing. Trends Neurosci 30: 70–78.

Kavanagh GL, Kelly JB (1987). Contribution of auditory cortex to sound localization by the ferret (*Mustela putorius*). J Neurophysiol 57: 1746–1766.

Kidd Jr G, Mason CR, Best V et al. (2010). Stimulus factors influencing spatial release from speech-on-speech masking. J Acoust Soc Am 128: 1965–1978.

King AJ, Hutchings ME (1987). Spatial response properties of acoustically responsive neurons in the superior colliculus of the ferret: A map of auditory space. J Neurophysiol 57: 596–624.

Klingon GH, Bontecou DC (1966). Localization in auditory space. Neurol 16: 879–886.

Klumpp RG, Eady HR (1956). Some measurements of interaural time difference thresholds. J Acoust Soc Am 28: 859–860.

Knudsen EI (1982). Auditory and visual maps of space in the optic tectum of the owl. J Neurosci 2: 1177–1194.

Kopco N, Shinn-Cunningham B (2011). Effect of stimulus spectrum on distance perception for nearby sources. J Acoust Soc Am 130: 1530–1541.

Koppl C, Carr CE (2008). Maps of interaural time difference in the chicken's brainstem nucleus laminaris. Biol Cybern 98: 541–559.

Krumbholz K, Schönwiesner M, Rübsamen R et al. (2005). Hierarchical processing of sound location and motion in the human brainstem and planum temporale. Eur J Neurosci 21: 230–238.

Kuhn GF (1977). Model for the interaural time differences in the azimuthal plane. J Acoust Soc Am 62: 157–167.

Langendijk EHA, Bronkhorst AW (2002). Contribution of spectral cues to human sound localization. J Acoust Soc Am 112: 1583–1596.

Lee J, Groh JM (2012). Auditory signals evolve from hybrid- to eye-centered coordinates in the primate superior colliculus. J Neurophysiol 108: 227–242.

Lee C-C, Middlebrooks J (2011). Auditory cortex spatial sensitivity sharpens during task performance. Nat Neurosci 14: 108–114.

Lee C-C, Middlebrooks J (2013). Specialization for sound localization in fields A1, DZ, and PAF of cat auditory cortex. J Assoc Res Otolaryngol 14: 61–82.

Lewis JW, Meauchamp MS, DeYoe EA (2000). A comparison of visual and auditory motion processing in human cerebral cortex. Cereb Cortex 10: 873–888.

Litovsky RY, Colburn HS, Yost WA et al. (1999). The precedence effect. J Acoust Soc Am 106: 1633–1654.

Lomber S, Malhotra S (2008). Double dissociation of 'what' and 'where' processing in auditory cortex. Nat Neurosci 11: 609–616.

Macpherson EA, Middlebrooks JC (2002). Listener weighting of cues for lateral angle: the duplex theory of sound localization revisited. J Acoust Soc Am 111: 2219–2236.

Macpherson EA, Sabin AT (2007). Binaural weighting of monaural cues for sound localization. J Acoust Soc Am 121: 3677–3688.

Magezi DA, Krumbholz K (2010). Evidence for opponent-channel coding of interaural time differences in human auditory cortex. J Neurophysiol 104: 1997–2007.

Magezi D, Buetler K, Chouiter L et al. (2013). Electrical neuroimaging during auditory motion aftereffects reveals that auditory motion processing is motion sensitive but not direction selective. J Neurophysiol 109: 321–331.

Makous JC, Middlebrooks JC (1990). Two-dimensional sound localization by human listeners. J Acoust Soc Am 87: 2188–2200.

Malhotra S, Hall AJ, Lomber SG (2004). Cortical control of sound localization in the cat: Unilateral cooling deactivation of 19 cerebral areas. J Neurophysiol 92: 1625–1643.

Marrone N, Mason CR, Kidd Jr G (2008). Tuning in the spatial dimension: Evidence from a masked speech identification task. J Acoust Soc Am 124: 1146–1158.

May BJ (2000). Role of the dorsal cochlear nucleus in the sound localization behavior of cats. Hear Res 148: 74–87.

McAlpine D, Jiang D, Palmer AR (2001). A neural code for low-frequency sound localization in mammals. Nat Neurosci 4: 396–401.

McFadden D, Pasanen EG (1976). Lateralization at high frequencies based on interaural time differences. J Acoust Soc Am 59: 634–639.

Mehrgardt S, Mellert V (1977). Transformation characteristics of the external human ear. J Acoust Soc Am 61: 1567–1576.

Mickey BJ, Middlebrooks JC (2003). Representation of auditory space by cortical neurons in awake cats. J Neurosci 23: 8649–8663.

Mickey BJ, Middlebrooks JC (2005). Sensitivity of auditory cortical neurons to the locations of leading and lagging sounds. J Neurophysiol 94: 979–989.

Middlebrooks JC (1992). Narrow-band sound localization related to external ear acoustics. J Acoust Soc Am 92: 2607–2624.

Middlebrooks JC (1999a). Individual differences in external-ear transfer functions reduced by scaling in frequency. J Acoust Soc Am 106: 1480–1492.

Middlebrooks JC (1999b). Virtual localization improved by scaling nonindividualized external-ear transfer functions in frequency. J Acoust Soc Am 106: 1493–1510.

Middlebrooks JC, Bremen P (2013). Spatial stream segregation by auditory cortical neurons. J Neurosci 33: 10986–11001.

Middlebrooks JC, Green DM (1991). Sound localization by human listeners. Annu Rev Psychol 42: 135–159.

Middlebrooks JC, Knudsen EI (1984). A neural code for auditory space in the cat's superior colliculus. J Neurosci 4: 2621–2634.

Middlebrooks JC, Onsan ZA (2012). Stream segregation with high spatial acuity. J Acoust Soc Am 132: 3896–3911.

Middlebrooks JC, Pettigrew JD (1981). Functional classes of neurons in primary auditory cortex of the cat distinguished by sensitivity to sound location. J Neurosci 1: 107–120.

Middlebrooks JC, Makous JC, Green DM (1989). Directional sensitivity of sound-pressure levels in the human ear canal. J Acoust Soc Am 86: 89–108.

Middlebrooks JC, Clock AE, Xu L et al. (1994). A panoramic code for sound location by cortical neurons. Science 264: 842–844.

Middlebrooks JC, Xu L, Eddins AC et al. (1998). Codes for sound-source location in nontonotopic auditory cortex. J Neurophysiol 80: 863–881.

Middlebrooks JC, Xu L, Furukawa S et al. (2002). Cortical neurons that localize sounds. Neuroscientist 8: 73–83.

Mikami A, Newsome W, Wurtz RH (1986). Motion selectivity in macaque visual cortex. I. Mechanisms of direction and speed selectivity in extrastriate area MT. J Neurophysiol 55: 1308–1327.

Miller LM, Recanzone GH (2009). Populations of auditory cortical neurons can accurately encode acoustic space across stimulus intensity. Proc Natl Acad Sci U S A 106: 5931–5935.

Mills AW (1958). On the minimum audible angle. J Acoust Soc Am 30: 237–246.

Mills AW (1960). Lateralization of high-frequency tones. J Acoust Soc Am 32: 132–134.

Muir D, Clifton R, Clarkson M (1989). The development of a human auditory localization response: A U-shaped function. Can J Psychol 43: 199–216.

Musicant AD, Butler RA (1984). The influence of pinnae-based spectral cues on sound localization. J Acoust Soc Am 75: 1195–1200.

Musicant AD, Chan JCK, Hind JE (1990). Direction-dependent spectral properties of cat external ear: New data and cross-species comparisons. J Acoust Soc Am 87: 757–781.

Nuetzel JM, Hafter ER (1981). Discrimination of interaural delays in complex waveforms: Spectral effects. J Acoust Soc Am 69: 1112–1118.

Overholt EM, Rubel EW, Hyson RL (1992). A circuit for coding interaural time differences in the chick brainstem. J Neurosci 12: 1698–1708.

Palmer AR, King AJ (1982). The representation of auditory space in the mammalian superior colliculus. Nature 299: 248–249.

Parvani F, Macaluso E, Warren J et al. (2002). A common cortical substrate activated by horizontal and vertical sound movement in the human brain. Curr Biol 12: 1584–1590.

Pecka M, Brand A, Behrend O et al. (2008). Interaural time difference processing in the mammalian medial superior olive: The role of glycinergic inhibition. J Neurosci 28: 6914–6925.

Phillips DP, Vigneault-MacLean BK, Boehnke SE et al. (2003). Acoustic hemifields in the spatial release from masking of speech by noise. J Am Acad Audiol 14: 518–524.

Rajan R, Aitkin LM, Irvine DRF (1990). Azimuthal sensitivity of neurons in primary auditory cortex of cats. II. Organization along frequency-band strips. J Neurophysiol 64: 888–902.

Rauschecker JP, Scott SK (2009). Maps and streams in the auditory cortex: nonhuman primates illuminate human speech processing. Nat Neurosci 12: 718–724.

Rauschecker JP, Tian B (2000). Mechanisms and streams for processing of "what" and "where" in auditory cortex. Proc Natl Acad Sci U S A 97: 11800–11806.

Razak KA (2011). Systematic representation of sound locations in the primary auditory cortex. J Neurosci 31: 13848–13859.

Recanzone GH, Cohen YE (2010). Serial and parallel processing in the primate auditory cortex revisited. Behav Brain Res 206: 1–7.

Recanzone GH, Guard DC, Phan ML et al. (2000). Correlation between the activity of single auditory cortical neurons and sound-localization behaviour in the macaque monkey. J Neurophysiol 83: 2723–2739.

Roberts MT, Seeman SC, Golding NL (2013). A mechanistic understanding of the role of feedforward inhibition in the mammalian sound localization circuitry. Neuron 78: 923–935.

Romanski LM, Tian B, Fritz J et al. (1999). Dual streams of auditory afferents target multiple domains in the primate prefrontal cortex. Nat Neurosci 2: 1131–1136.

Ryugo DK, Seishiro S (1991). Synaptic connections of the auditory nerve in cats: relationship between endbulbs of held and spherical bushy cells. J Comp Neur 305: 35–48.

Salminen N, May P, Alku P et al. (2009). A population rate code of auditory space in the human cortex. PLoS One 4: e7600.

Sanchez-Longo LP, Forster FM (1958). Clinical significance of impairment of sound localization. Neurol 8: 119–125.

Sandel TT, Teas DC, Feddersen WE et al. (1955). Localization of sound from single and paired sources. J Acoust Soc Am 27: 842–852.

Sanes DH, Malone BJ, Semple MN (1998). Role of synaptic inhibition in processing of dynamic binaural level stimuli. J Neurosci 18: 794–803.

Shaw EAG (1974). Transformation of sound pressure level from the free field to the eardrum in the horizontal plane. J Acoust Soc Am 56: 1848–1861.

Slattery III WH, Middlebrooks JC (1994). Monaural sound localization: Acute versus chronic impairment. Hear Res 75: 38–46.

Smith KR, Okada K, Saberi K et al. (2004). Human cortical auditory motion areas are not motion selective. Neuroreport 15: 1523–1526.

Smith KR, Saberi K, Hickok G (2007). An event-related fMRI study of auditory motion perception: no evidence for a specialized cortical system. Brain Res 1150: 94–99.

Spitzer MW, Semple N (1991). Interaural phase coding in auditory midbrain: Influence of dynamic stimulus features. Reports 254: 721–724.

Spitzer M, Semple M (1998). Transformation of binaural response properties in the ascending auditory patheway: Influence of time-varying interaural phase disparity. J Neurophysiol 80: 3062–3076.

Stecker GC, Harrington IA, Middlebrooks JC (2005). Location coding by opponent neural populations in the auditory cortex. PLoS Biol 3: 520–528.

Stevens SS, Newman EB (1936). The localization of actual sources of sound. Am J Psychol 48: 297–306.

Strutt JW (1907). On our perception of sound direction. Philos Mag 13: 214–232.

Sutherland DP, Masterton RB, Glendenning KK (1998). Role of acoustic striae in hearing: reflexive responses to elevated sound sources. Behav Brain Res 97: 1–12.

Thiran AB, Clarke S (2003). Preserved use of spatial cues for sound segregation in a case of spatial deafness. Neuropsychologia 41: 1254–1261.

Thompson GC, Cortez AM (1983). The inability of squirrel monkeys to localize sound after unilateral ablation of auditory cortex. Behav Brain Res 8: 211–216.

Tian B, Reser D, Durham A et al. (2001). Functional specialization in rhesus monkey auditory cortex. Science 292: 290–293.

Trahiotis C, Bernstein LR (1986). Lateralization of bands of noise and sinusoidally amplitude-modulated tones: Effects of spectral locus and bandwidth. J Acoust Soc Am 79: 1950–1957.

Ungerleider LG, Mishkin M (1982). Two cortical visual systems. In: DJ Ingle, MA Goodale, RJWOe Mansfields (Eds.), Analysis of Visual Behaviour, MIT Press, Cambridge, pp. 549–586.

van der Hiejden M, Lorteije JAM, Plaushka A et al. (2013). Directional hearing by linear summation of binaural inputs at the medial superior olive. Neuron 78: 936–948.

Warren JD, Zielinski BA, Green GGR et al. (2002). Perception of sound-source motion by the human brain. Neuron 34: 139–148.

Wenzel EM, Arruda M, Kistler DJ et al. (1993). Localization using nonindividualized head-related transfer functions. J Acoust Soc Am 94: 111–123.

Werner-Reiss U, Groh J (2008). A rate code for sound azimuth in monkey auditory cortex: implications for human neuroimaging studies. J Neurosci 28: 3747–3758.

Wightman FL, Kistler DJ (1989a). Headphone simulation of free field listening. I: Stimulus synthesis. J Acoust Soc Am 85: 858–867.

Wightman FL, Kistler DJ (1989b). Headphone simulation of free-field listening. II: Psychophysical validation. J Acoust Soc Am 85: 868–878.

Wightman FL, Kistler DJ (1992). The dominant role of low-frequency interaural time differences in sound localization. J Acoust Soc Am 91: 1648–1661.

Woods TM, Lopez SE, Long JH et al. (2006). Effects of stimulus azimuth and intensity on the single-neuron activity in the auditory cortex of the alert macaque monkey. J Neurophysiol 96: 3323–3337.

Woodworth RS (1938). Experimental Psychology. Holt, Rinehart, and Winston, New York.

Yin TCT (1994). Physiological correlates of the precedence effect and summing localization in the interior colliculus of the cat. J Neurosci 14: 5170–5186.

Yin TCT, Chan JCK (1990). Interaural time sensitivity in medial superior olive of cat. J Neurophysiol 64: 465–488.

Yost WA, Loiselle L, Dorman M et al. (2013). Sound source localization of filtered noises by listeners with normal hearing: A statistical analysis. J Acoust Soc Am 133: 2876–2882.

Young ED, Spirou GA, Rice JJ et al. (1992). Neural organization and responses to complex stimuli in the dorsal cochlear nucleus. Philos Trans R Soc Lond B Biol Sci 336: 407–413.

Zahorik P (2002). Assessing auditory distance perception using virtual acoustics. J Acoust Soc Am 111: 1832–1846.

Zahorik P, Brungart DS, Bronkhorst AW (2005). Auditory distance perception in humans: A summary of past and present research. Acta Acustica 91: 409–420.

Zatorre RJ, Penhune VB (2001). Spatial localization after excision of human auditory cortex. J Neurosci 21: 6321–6328.

Zatorre RJ, Mondor T, Evans AC (1999). Functional activation of right parietal and frontal cortex during auditory attention of space and frequency. Neuroimage 10: 544–554.

Zatorre RJ, Bouffard PA, Belin P (2002). Where is 'where' in the human auditory cortex? Nat Neurosci 5: 905–909.

Zhou Y, Wang X (2012). Level dependence of spatial processing in the primate auditory cortex. J Neurophysiol 108: 810–826.

Zwislocki J, Feldman RS (1956). Just noticeable differences in dichotic phase. J Acoust Soc Am 28: 860–864.

Handbook of Clinical Neurology, Vol. 129 (3rd series)
The Human Auditory System
G.G. Celesia and G. Hickok, Editors

Chapter 7

New perspectives on the auditory cortex: learning and memory

NORMAN M. WEINBERGER*
Center for the Neurobiology of Learning and Memory and Department of Neurobiology and Behavior, University of California, Irvine, CA, USA

INTRODUCTION

This chapter concerns a paradigm shift in conceptions of the primary auditory cortex (A1). Alone among the multitudinous cortical fields, the primary sensory cortices have had the dubious honor of having their functions "known" before any physiologic data had been obtained. This function has been assumed to be "pure sensory analysis," specifically excluding any psychologic or cognitive functions. Campbell crystallized this view in his highly influential 1905 monograph, *Histological Studies on the Localisation of Cerebral Function*. An insightful perspective on the matter has been provided by Irving Diamond, late professor of neuroscience at Duke University: "Campbell's influential monograph was largely responsible for cleaving learning and memory from sensory processes" (Diamond, 1979). It might seem odd that Campbell's aged work should be cited here, but the conception that he championed has become so woven into the fabric of neuroscience as to constitute an invisible core assumption that still guides much research and thought.

Nonetheless, the position that the primary sensory cortices are exclusively "sensory analytic," and proscribed from learning, memory, and a variety of other cognitive processes has been eroded to the point of collapse by two lines of research. First, Galambos and colleagues (1956) discovered that training a cat with a click followed by an air puff to its eye both produced learning to predict this annoyance and also greatly increased the amplitude of the click-evoked potential in A1. Thus, the responses of the auditory cortex to sound were shown to be altered when it became behaviorally important. This finding, using classic (pavlovian) associative conditioning, was soon pursued in many laboratories internationally to include controls for general arousal and sensitization, and extend the phenomenon to various species, including humans, different types of auditory cues, multisound discrimination training, and more complex tasks such as instrumental/operant conditioning (reviewed in Weinberger and Diamond, 1987).

Second, studies that combined auditory neurophysiology with associative conditioning discovered representational plasticity (RP) in the primary auditory cortex. Beyond "physiological plasticity" (hereafter "plasticity"), which refers to most types of non-transient changes in the nervous system, RP refers to a systematic change in the encoding of sensory stimuli along a dimension, e.g., acoustic frequency, level, duration tuning, frequency-modulated (FM) envelope, repetition rate, and localization in space. In the best-studied example, the coding of acoustic frequency is modified so that neuronal receptive fields ("tuning functions") are systematically altered in learning and memory. More specifically, the frequency tuning of neurons in A1 shifts toward and to the frequency of the signal tone ("receptive field shift") (Diamond and Weinberger, 1986; Bakin and Weinberger, 1990; Edeline and Weinberger, 1993).

The implications of RP are great. Thus, simple "plasticity" in the form of an increased cortical response to an important sound may be limited to that sound. In stark contrast, RP is not limited to a particular experienced stimulus, but rather biases an entire sensory dimension to emphasize behaviorally important sensory stimuli. Therefore, the processing of future and hitherto non-experienced stimuli in that dimension will be altered. Such learning-based predispositions to classes of sounds or types of acoustic parameters are likely to have major consequences for each individual, including modifying the effects of remedial auditory training in ways not yet appreciated.

*Correspondence to: Norman M. Weinberger, PhD, Center for the Neurobiology of Learning and Memory, 309 Qureshey Research Laboratory, University of California, Irvine, CA 92697-3800, USA. Tel: +1-949-824-5512, Fax: +1-949-824-4576, E-mail: nmweinbe@uci.edu

These types of finding have at least three major implications. First, they provide a foundation for understanding how the contents of auditory experience are acquired, represented, stored, and ultimately employed in the service of adaptive behavior. Second, they force a reconceptualization of the auditory cortex (and other "early" sensory cortices). Third, they demonstrate the need for a new way of thinking about overall cortical organization. Insofar as the primary auditory cortex is clearly deeply involved in learning, memory, and other cognitive processes (as documented below), it seems likely that all of the cerebral cortex is "cognitive," but various regions conduct different and somewhat specialized computations. Much less research on this new approach has been conducted in the primary somatosensory (S1) and visual (V1) cortices. But the findings to date appear to be fundamentally the same as for A1 (e.g., S1: Wiest et al., 2010; V1: Miller et al., 2008). In short, "pure sensory cortex" now appears to be a misconception. This no longer tenable belief about primary sensory cortices remains largely unexamined and dominant in neuroscience, perhaps because it seems to be supported by personal experience. Thus, "introspection" may suggest that sensation and perception precede and are separate from all cognition. But how the brain creates experience is not bound by folk psychology. It is past time to drop this long-outdated concept, which is not only demonstrably false but also shackles thought and impedes progress both in the laboratory and in the clinic (see also Masterton, 1993).

What follows is an account of the current state of the involvement of the primary auditory cortex in auditory associative learning and memory, as these are dominant processes in knowledge acquisition and behavioral adaptation. Implications for the treatment of certain human disorders are included. Emphasis is placed on animal studies because they have been most revealing, including permitting manipulations unavailable in humans. However, the recording and imaging studies reported to date for the human brain appear to be consistent with the findings from animal studies. Rather than provide an exhaustive review of the literature, the choice of examples is based on explaining types of findings and illustrating specific issues. Readers who wish to pursue various other aspects of this topic may consult prior reviews (e.g., Palmer et al., 1998; Rauschecker, 1999; Kilgard et al., 2002; Edeline, 2003; Suga and Ma, 2003; Weinberger, 2004 (analysis of experimental designs), 2007 (intersections of learning/memory and sensory physiology), 2008 (cognitive functions of auditory cortex); Ohl and Scheich, 2005; Scheich et al., 2011). This article draws from Weinberger (2011b) for early neurobehavioral research.

BASIC CONSIDERATIONS AND EARLY FINDINGS

Introduction

For present purposes, we use broad definitions of learning and memory to give them widespread applicability. "Learning" is the acquisition of information and "memory" is the storage of information for periods as short as many seconds (e.g., memorizing a telephone number to order a pizza) to lifetime (one's own name). Two important points should be kept in mind. First, neither learning nor memory is directly observable. They must always be inferred from behavior. Therefore, simply placing a subject in a learning situation, such as repeated presentation of a sound that is followed by a reward, cannot justify the conclusion that learning has taken place. Rather, learning must be inferred from changes in behavior. The necessity of behavioral validation means that neural plasticity cannot be used to verify that learning has occurred; rather, it is a presumptive substrate of learning and memory. Second, because learning and memory are inferred from behavior, all causes of a behavioral change, other than the acquisition and storage of information, must be ruled out. These include general changes in a subject's state of arousal or excitability, drug condition, sensory abilities, and capability for motor action.

Much learning-brain research concerns the relationship between two or more stimuli, experiences, or events, i.e., the learning of associations. For example, whenever there are regularities in experience, such that one sound predicts another event, an association can form so that the sound predicts (and may bring to mind or elicit an image of) the second event. A common example is the expectation of seeing a car after hearing a car horn. Such predictive associations occur continually and in myriad ways, thus enabling people and other animals to predict the consequences of events or of their own behavior, rather than simply be passive recipients of stimuli. In the laboratory, daily associations are modeled generally by classic (pavlovian) conditioning, in which one stimulus regularly precedes another experience, and instrumental/operant conditioning, in which one's own behavior is regularly followed by an event, e.g., a reward.

Background findings

This section provides a summary of the approximately 30-year period (1956–1984) that formed the foundations for contemporary research. During this era, learning-induced plasticity in A1 was discovered, validated, and extended to many learning tasks. The issues focused on the fate of sensory signals when they predicted an important event, such as reward or negative reinforcement. The main question, often implicit, was: "When a stimulus gains behavioral relevance through experience, do the neural bases of learning (i.e., plasticity) begin during or after analysis of that signal?" This issue was part of the larger program of seeking the neural circuitry underlying learning and memory. Naturally, researchers simply recorded the responses of the auditory cortex to signal sounds during training trials.

Galambos and his colleagues (1956) published the first western study on learning and plasticity in primary sensory cortex. Cats received an auditory (click) conditioned stimulus (CS) paired with an immediately following puff of air to the face (unconditioned stimulus, US). As a result of this classic conditioning procedure, CS-elicited local field potentials (LFPs) in A1 increased in magnitude. The authors validated learning by noting the development of eyeblink conditioned responses. Furthermore, they also controlled for CS constancy at the ear by obtaining the same findings with subjects under neuromuscular blockade, thus preventing head and pinna movements and possible contractions of the middle-ear muscles. Interestingly, the authors failed to include a non-associative control for sensitization or general excitability, such as a group that received the CS and US unpaired. However, subsequent investigators did include proper controls, confirming the associative nature of increased response magnitude of LFPs in A1 (Marsh et al., 1961; Majkowski and Sobieszek, 1975; Molnár et al., 1988).

LFP research was extended to different conditioning tasks and pavlovian processes, with essentially the same findings of enhanced responses to sounds that became behaviorally important (reviewed in Weinberger and Diamond, 1987). Similar results were obtained for instrumental avoidance learning. In fact, systematic increases in CS-evoked LFPs could be obtained in the same dogs that were successively switched between classic conditioning and instrumental avoidance conditioning with the same CS and US, despite the learning of different responses for the two types of conditioning (Cassady et al., 1973).

In addition to the use of LFPs, workers began recording cellular discharges in A1. Most studies recorded from clusters of cells ("multiple-unit" activity). The results generally were the same, i.e., increased discharges to an acoustic CS that predicted reward or punishment, as animals formed behavioral conditioned responses indicating successful associations. This increase in response was fairly consistent, both in simple (one-tone) conditioning and in two-tone discrimination training (reinforced CS+, non-reinforced CS–) (Buchwald et al., 1966; Saunders and Chabora, 1969). Reversal was also obtained when the CS+and CS– acoustic stimuli were interchanged after initial learning. Moreover, acoustic CS+stimuli acquired the ability to elicit responses in the sensory cortex of the shock US, i.e., in primary somatosensory cortex (Oleson et al., 1975).

Such "cluster" recordings have the advantage over single-unit recordings of yielding consistent data over days. They have the disadvantage of being unable to determine if single cells develop different directions of plasticity, i.e., either increased or decreased responses. That is, if associative processes produce both increased and decreased responses to the CS, but increased responses dominate, then the decreased responses would not be detected.

Studies of single units in auditory cortex during learning also found plasticity. However, as suspected, despite the detection of many cells that developed increased responses to the CS, a substantial number of cells developed decreased responses, and yet others exhibited no change (Gasanov and Galashina, 1976; Woody et al., 1976; Dumenko and Sachenko, 1978, 1979; Weinberger et al., 1984). Similar heterogeneity of unit discharge plasticity was also found in auditory field A2 (Diamond and Weinberger, 1984). Such divergent results were not attributable to inadvertent changes in effective acoustic stimulus level at the auditory periphery, undetected movements, muscle contractions, or feedback from muscle spindles, because the same results were obtained when animals were trained while under neuromuscular blockade (Weinberger et al., 1984).

Although studies of associative learning had firmly established that the neural bases of learning start with sensory analysis of acoustic signals, and that single-unit plasticity in A1 was associative, the findings of opposite sign made no functional sense. Thus, while recording in A1 during training trials had provided foundational information, this approach appeared to have reached an impasse. A new direction was needed.

CONTEMPORARY APPROACHES: REPRESENTATIONAL PLASTICITY

A synthesis of two disciplines

Paradigm shifts in both concept and method developed once the issue changed from whether the auditory cortex was involved in learning and memory to that of how A1 was involved. Thus, while one could easily determine that plasticity had developed during training trials, it was impossible to know whether changes were restricted to, e.g., the signal tone, or were perhaps even greater to other frequencies that had not been tested. The situation was similar for behavior; detection that learning had happened merely required observing behavioral changes during trials, but it was impossible thereby to know what had been learned. Thus, while the subject might have learned that a particular acoustic frequency predicted or provided the opportunity to press a button to get reward, she might have merely learned that sound predicted reward, or simply that a sudden stimulus predicted reward. In fact, subjects can learn about any single stimulus parameter or any combination of parameters of a given sound (e.g., frequency, level, duration, location, pattern) because they coexist and therefore are all predictive of reinforcement. In short, research began to address the specificity of both cortical plasticity and memory.

The solution to solving the specificity problem for both plasticity and memory was to invent a hybrid approach that was based on a synthesis of two previously non-intersecting disciplines: auditory neurophysiology and the neurobiology of learning and memory. Thus, to determine if learning alters neuronal coding of sound in the auditory cortex, researchers began to obtain auditory cortical receptive fields before and after training subjects in any desired learning task that used sounds. For example, receptive fields for acoustic frequency were obtained by giving many tones to obtain tuning functions before training, then training with one of the tones, and again obtaining tuning functions after training. To determine if learning specifically alters tuning, one merely subtracts the pretraining from the posttraining receptive fields.

The same logic and general procedure apply to behavior. Subjects are trained with one tone but tested after learning with many tones. The resultant behavioral data generate a "behavioral stimulus generalization gradient," the peak of which reflects the stimulus value that had gained the greatest memory value due to training. In most tasks, pretraining gradients cannot be obtained because no behavioral response is available. In these cases, a control group can be trained with a signal and reinforcement not paired, and the posttraining gradient from this group is subtracted from that of the paired (associative) group. However, investigators have also made use of recording autonomic behavioral responses that respond to sound before training (changes in heart rate or respiration) (McLin et al., 2002), and so have been able to obtain behavioral frequency gradients before and after training in the same subjects (Miasnikov et al., 2006).

Representational plasticity reveals specificity of auditory cortical dynamics

The "contemporary era" may be considered to have been initiated in the 1980s when a new question was posed that focused on specificity: "Does learning cause a 'retuning' of the primary auditory cortex?" Initially, studies concentrated on neural recording of the frequency-receptive fields of neurons in A1 before and after training. Later, behavioral frequency generalization gradients were also obtained. In so doing, new dimensions of learning and plasticity were uncovered, including unsuspected mnemonic functions and the role of the cholinergic system.

Gonzalez-Lima and Scheich (1984, 1986) found specificity of A1 plasticity by determining the spatially restricted increase in metabolic activity (2-deoxyglucose uptake) in the gerbil after classic conditioning with tone and aversive brain stimulation. Posttraining analysis of patterns of 2-deoxyglucose uptake in A1 revealed a CS-specific increase in metabolic activity for the cortical area that represented the CS frequency. However, most other research has employed neurophysiologic recording, primarily because of its greater spatial and temporal specificity and ability to repeatedly obtain data from the same subjects.

Receptive field analysis was first applied to classic auditory (tone) fear conditioning in the cat. Single-unit discharges were recorded in two non-primary auditory fields, "secondary" (A2) and ventral ectosylvian (VE) cortices (Diamond and Weinberger, 1986, 1989). Neural tuning became biased toward the tonal CS in a paired group but not in unpaired controls, demonstrating specific associative plasticity. Extinction (additional CS presentation without the shock US) produced loss of the receptive field plasticity. The findings received little notice, perhaps because these "higher" auditory fields were both expected to have plasticity and also were not well understood compared to A1.

Similar studies were then undertaken in A1 of the guinea pig with behavioral validation of associative learning, viz., conditioned bradycardia. Auditory fear conditioning produced increased discharges to the CS frequency with simultaneous decreased responses to many other frequencies. These opposing changes produced CS-directed shifts of tuning, so that the signal frequency could become the new "best frequency (BF)," i.e., peak of the tuning function (Bakin and Weinberger, 1990) (Fig. 7.1). Thus, associative learning can retune cells in the primary auditory cortex to increase the representation of sounds that become behaviorally important.

Representational plasticity across species

CS-specific associative tuning shifts develop in the primary auditory cortex of all species studied to date: guinea pig (Bakin and Weinberger, 1990), big brown bat (Gao and Suga, 2000), cat (Diamond and Weinberger, 1986), and rat (Kisley and Gerstein, 2001).

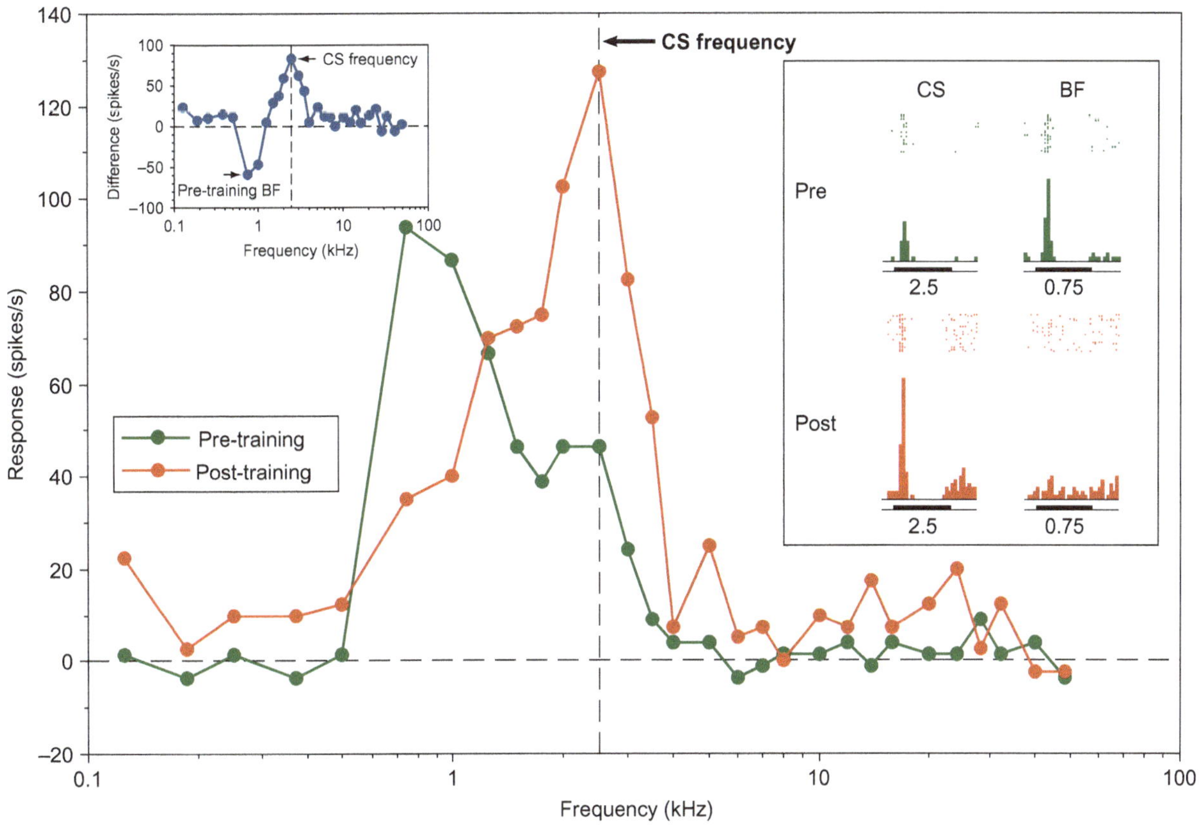

Fig. 7.1. Basic associative learning (classical conditioning) produces conditioned stimulus (CS)-specific facilitation and tuning shifts. An example of a complete shift of frequency tuning of a single cell in A1 of the guinea pig, from a pretraining best frequency (BF) of 0.75 kHz to the CS frequency of 2.5 kHz after 30 trials of auditory fear conditioning (tone–shock pairing). Right inset shows pre- and posttraining poststimulus time histograms (PSTHs) for the pretraining BF and the CS frequencies. Note the marked increase in response to the tone signal (2.5 kHz) and the concomitant decrease in response to the pretraining peak of the tuning function (0.75 kHz). The overall changes (post- minus pretuning) are shown in the upper left inset, revealing the high level of specificity of retuning to emphasize the signal frequency, caused by associative learning.

Furthermore, CS-specific expanded representations in the tonotopic map of A1, which are predicted from CS-directed frequency tuning shifts, develop in the owl monkey (Recanzone et al., 1993) and the rat (Rutkowski and Weinberger, 2005).

Associative plasticity also develops in the auditory cortex of humans during auditory learning. For example, Molchan and coworkers (1994) paired tone with air puffs to the right eye, causing the development of a conditioned eyeblink; this was reversed when the air puff was eliminated (i.e., "experimental extinction"). Compared to control conditions (tone and air puff not paired), they found an increase in regional cerebral blood flow (rCBF) in the left primary auditory cortex. These authors replicated and extended this finding, applying air puff to the left eye; they found associative increased rCBF in the right A1 (Schreurs et al., 1997). Thus, the auditory cortex contralateral to the locus of the noxious air puff developed associative plasticity.

Specificity within the tonotopic map also has been reported. Morris and associates (1998) conducted a two-tone discrimination study in which one tone (CS+) predicted a noxious sound (100 dB white noise) while the other (CS–) was presented without noise; the two types of trial were randomly intermixed. All subjects developed discriminative conditioned associations, detected as a change in skin conductance (galvanic skin response) only to the CS+tone. Positron emission tomography imaging revealed changes in rCBF within the tonotopic representation that were specific to the tones: responses to the CS+changed, whereas responses to the CS– did not change.

Thus, RP appears to be a ubiquitous aspect of auditory cortical functioning, in which associative learning specifically modifies the coding of acoustic stimuli, favoring the processing of behaviorally important stimuli at the expense of less relevant sounds. It is expected that the context of learning is part of a broad network that includes the representation of sounds, so that auditory cortical processing is sufficiently flexible to enable stimuli that are not important in one situation to become behaviorally relevant in another situation. However, research on contextual aspects of RP has not yet been actively pursued.

ISN'T ALL AUDITORY LEARNING ACTUALLY "PERCEPTUAL LEARNING"?

The early studies of RP in the auditory cortex had established that learning alters the coding of sounds. While this constituted a "sea change" in conceptions of auditory cortical function, it did not have much influence on assumptions about cortical organization in general. Rather, Campbell's (1905) exclusion of learning, memory, and cognition from the primary auditory cortex underwent a slight modification. Instead of having the function of pure sensory analysis, A1 is often now viewed as an "adaptive auditory analyzer." It simply adjusts its analysis of the physical parameters of sound to emphasize those that become better perceived. That is, a widespread assumption is now that all auditory learning is simply "perceptual learning" and unrelated to associative learning and genuine behavioral memory.

Perceptual learning refers to an increase in sensory acuity within a stimulus dimension, due usually to extensive practice. Everyone is familiar with the highly tuned sensitivities of "noses" (perfumers), oenophiles, and musicians. Perceptual learning typically is the result of extensive and increasingly difficult discrimination training (Kellman, 2002). Such learning may require thousands of trials, over many days or weeks: e.g., frequency discrimination learning of 4000–5000 trials (Irvine et al., 2000); pitch discrimination of more than 10 000 trials (Demany and Semal, 2002); melodic patterns trained extensively with 1200 different stimuli (Tervaniemi et al., 2001). There are reports of perceptual learning in considerably fewer trials (Hawkey et al., 2004), but even so, associative learning can develop in five or fewer trials (Lennartz and Weinberger, 1992; Edeline et al., 1993). Associative learning can include discrimination learning but, unlike much perceptual learning, associations can be formed in the absence of discrimination training. However, that is not the most distinctive difference.

We can best appreciate the distinction by asking, "After an episode of perceptual learning, what is changed in the auditory cortex?" The answer would seem to be that the "machinery" of the cortex has been altered to enable greater acuity. This is certainly an interesting aspect of learning. However, in contrast to associative learning, perceptual learning does not seem to include "perceptual memory." This absence might reflect the fact that most investigators of perceptual learning are more concerned with sensory/perceptual processes than with mnemonic processes. But of greater import, subjects apparently do not actually remember the specific "contents" of their experience, that is, the particular stimuli or stimulus values given during certain of their multitude of trials. Thus, while increased acuity is an important aspect of auditory perception, it appears to be a process that is distinct from the storage of sounds. Therefore, perceptual learning might be thought of as controlling the gateway to memory; sounds that can be remembered first have to be distinguished and increased acuity provides increased acoustic detail of sounds that may enter memory, i.e., become the contents of experience.

An argument can be made that perceptual learning is a subclass of associative learning because subjects generally first have to learn an association between a sound and whatever comes next, usually a reward. Next, the discriminations become increasingly difficult. Because basic associative learning and its correlated cortical plasticity develop rapidly, they may help elucidate mechanisms of subsequent perceptual learning. In short, associative learning and perceptual learning are not the same, although both can affect primary sensory cortices in general, and the auditory cortex in particular (Weinberger, 2008).

DOES THE PRIMARY AUDITORY CORTEX HOLD SPECIFIC MEMORY TRACES?

The preceding review indicates that the coding of acoustic frequency in the primary auditory cortex is systematically modified by associative learning, to remodel this dimension to those spectra that become behaviorally important are emphasized at the expense of unimportant sounds. Moreover, the findings cannot be explained simply as only subserving perceptual learning. These considerations suggest that RP may be a basis for specific auditory memory. We evaluate this possibility in the following sections.

Forms of representational plasticity

To this point, we have focused on shifts in frequency tuning toward or to the training frequency. This has been justified by the fact that this was the first form of RP discovered (Diamond and Weinberger, 1986; Bakin and Weinberger, 1990). However, the possible domain of forms of plasticity is very large. It encompasses all parameters of acoustic response. There are many ways to organize and quantify neuronal responses to sound. Each of them might develop RP, i.e., exhibit bias toward sounds that gain behavioral significance, generally accompanied by relative loss of emphasis on some other stimulus values within the domain of the modified acoustic dimension (but see the section on three classes of auditory tasks and representational plasticity, below).

One of the most extensively used methods of characterizing response to the acoustic spectrum is to randomly and repeatedly present brief pure tone pips across acoustic frequency and stimulus level ("intensity"), yielding a frequency response area while recording unit or multiunit discharges from microelectrodes in the animal brain. This approach delimits the spectro-level space within which neurons respond. The "characteristic frequency" (CF) is a cardinal parameter of response: the frequency to which the cells are most sensitive; the level at which the CF occurs is the threshold. Response selectivity can be determined by measuring the bandwidth (BW) of response, usually from 10–30 dB above threshold. The BW is simply the frequency range over which the neurons respond at each of the levels specified; it is usually given in octaves. In addition, researchers often designate the BF, which is the frequency–level combination that elicits the greatest magnitude of neuronal response (usually given in spikes/second above pretone background activity rate). Figure 7.2A presents a typical frequency response area.

In addition, the primary auditory cortex contains a "tonotopic map," in which the CFs of neighboring cells are likely to be similar, but not identical, and the organization of CFs across A1 is across the spectrum (low–high), very roughly like the layout of a piano keyboard, although far less precise (Fig. 7.2B). The amount of area representing different frequencies (or fairly narrow frequency bands) is determined by interpolation from the "maps" of CFs, which usually contain 60–80 spaced recording sites, and require many hours to obtain by moving microelectrodes across the cortex in anesthetized subjects.

In summary, three major forms of RP for acoustic frequency are available: (1) specific decrease in threshold and BW near threshold; (2) tuning shift toward or to a signal frequency above threshold; and (3) specific increase of signal area in the tonotopic map of A1. These forms may be linked along the dimension of magnitude of effect. Thus, decreased threshold and BW may be the minimal expression of RP. Greater effects may be manifest as actual changes in frequency tuning above threshold, and the largest expression of plasticity seems to be a frank increase in representational area. The relationship between tuning shifts and map expansions is most clearly understood, because the area of representation is simply the distribution of frequency tuning across the entire primary auditory cortex. Therefore, if many loci shift their tuning to favor the signal frequency, then an increase in representational area for that frequency should result. However, research on the relationships of different forms of RP is still in its earliest stages and so no definitive conclusions can be drawn at this time.

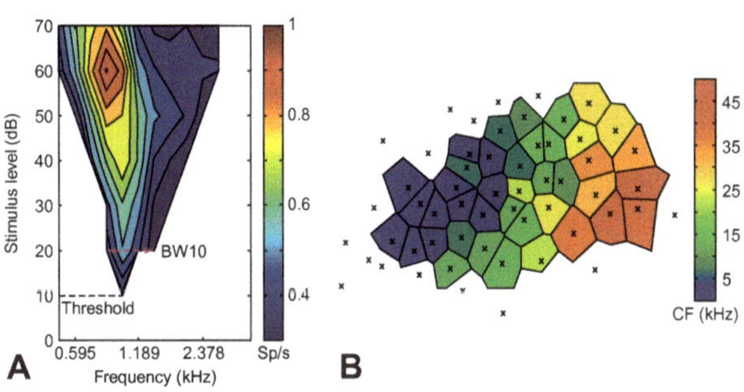

Fig. 7.2. (**A**) Example of a frequency response area (FRA), which provides the response of neurons in the auditory cortex to tonal frequencies (kHz) over a wide range of stimulus levels (intensity, dB). The frequency tuning of cells at threshold (characteristic frequency, CF) and the selectivity of tuning (measured at 10 dB above threshold) are indicated. The CF in this case is ~1.0 kHz. The colors indicate the magnitude of response, which in this case is at 60 dB. (**B**) Example of a tonotopic map in the primary auditory cortex of the rats. The CFs of all recordings are represented in the standard measure as interpolated polygons based on Voronoi tessellation methods. This enables determination of the area of the cortex that represents each acoustic frequency band. In this case, frequencies below ~5.0 kHz are overrepresented because the animal was trained with a low-frequency tone. The increase in representational area is attributable to tuning many shifts toward and to the signal tone frequency in the auditory cortex.

Basic considerations of specific memory traces

If RP is part of the substrate of specific auditory memory, then it might be expected to possess the major characteristics of behavioral associative memory itself. This would impose a set of criteria that have not previously been demanded of any other neurophysiologic correlates of learning and memory. The apparent "double standard" does not reflect a prejudice against sensory cortical bases of memory, but rather that studies of non-sensory regions of the brain (perhaps 99% of literature) are generally not concerned with memory content, i.e., the neural bases of specific representations of stimuli, but rather focus on the processes that enable memory, such as the time-limited role of the medial temporal lobe in long-term memories stored in the cerebral cortex (Squire et al., 2004).

What are these major characteristics of associative memory? In addition to being associative, specific memory traces (SMTs) should also exhibit specificity, develop fairly rapidly, exhibit long-term retention, and even show consolidation, i.e., continued strengthening over time after training in the absence of additional reinforcement.

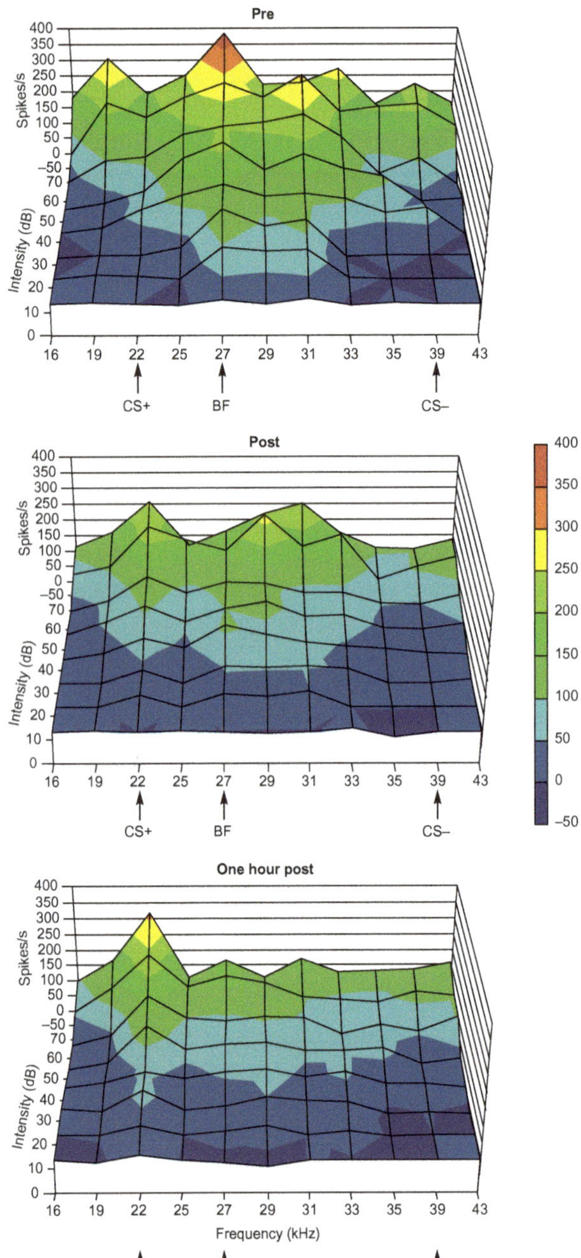

Fig. 7.3. Retuning and neural consolidation of the auditory cortex in a two-tone discrimination task. Representation of neuronal responses in A1 before (Pre), immediately after (Post), and 1 hour after two-tone discrimination training (30 × each: conditioned stimulus (CS)+(22.0 kHz) and CS– (39.0 kHz), intermixed trials). Displayed are rates of discharge (*x*-axis) as a function of tonal frequency (*y*-axis) and level of testing stimuli (*z*-axis) (10–70 dB). Note that associative learning produced a massive remodeling of the auditory cortex, i.e., the "topography" of neuronal response across frequency and stimulus level. The pretraining best frequency (BF) of 27.0 kHz suffered a reduction in response, as did the CS– frequency. In contrast, responses to the CS + frequency increased. Note consolidation, in the form of a continued development of these changes; after 1 hour of silence, the only excitatory response is at the CS+frequency. (Based on Edeline and Weinberger, 1993.)

In addition to having the major features of memory itself, SMTs should be found for whatever the types of CS or signal stimuli are used in training, i.e., should not be limited to plasticity of frequency representation. Another feature of memory is that it can be formed in a wide variety of tasks, rather than being confined to, e.g., classic conditioning. A further key feature of memory is that it transcends a particular type of motivation, but develops in both appetitive and aversive situations. Finally, as in the case of memory, SMTs should be biologically conserved, that is, developing across diverse taxa. This last criterion has been met, as reviewed above (see the section on representational plasticity across species, above). As described below, the findings from various laboratories support the conclusion that SMTs develop in the primary auditory cortex.

Cardinal characteristics of representational plasticity

How well do the major characteristics of RP match those of behavioral associative memory? The most direct test of associativity is to train separate groups, one receiving a tone paired with reinforcement (reward or punishment), the other with the same sound and reinforcement but not paired. In such cases, only the paired group develops tuning shifts (e.g., Diamond and Weinberger, 1986; Bakin and Weinberger, 1990; Bakin et al., 1992). Another test is discrimination training, in which one tone is followed by reinforcement (CS+) while another tone is not (CS–). The development of behavioral responses only to the CS+demonstrates associativity rather than enhanced excitability to all (or both) sounds. A selective enhancement of responses to the CS+in A1 does develop in discrimination learning (Edeline and Weinberger, 1993) (Fig. 7.3). This finding demonstrates not only associativity but also the highly selective nature of RP. Like memory, RP can be established rapidly, within five training trials (Edeline et al., 1993) (Fig. 7.4), and exhibits long-term retention, enduring for the longest periods tested, up to 8 weeks after a single 30-trial conditioning session (Weinberger et al., 1993; see also Edeline et al., 1990) (Fig. 7.5). Finally, RP consolidates, i.e., continues to develop increased responses to the frequency of the CS vs decreased responses to other frequencies in the absence of further training, over hours (Edeline and Weinberger, 1993) and days after a single training session (Galván and Weinberger, 2002) (Fig. 7.6). In summary, RP satisfies the criterion of having the same cardinal characteristics as behavioral associative memory.

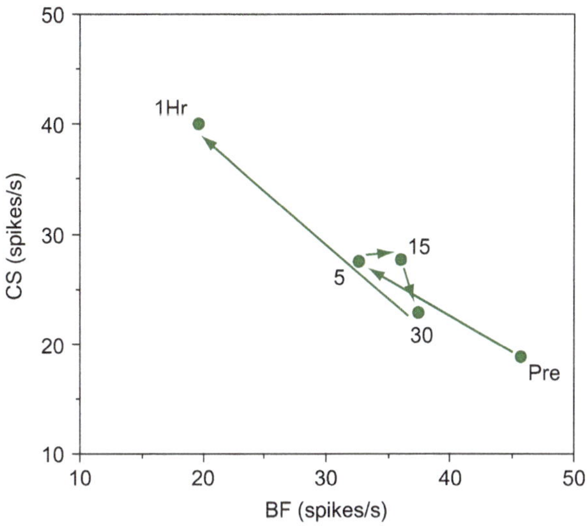

Fig. 7.4. Rapid induction of receptive field (RF) plasticity shown as vector diagrams of changes in response to the pretraining best frequency (BF) and the conditioned stimulus (CS) frequency during auditory fear learning. After five trials, responses to the BF had decreased, while those to the CS increased, changes that were maintained after 15 and 30 trials. However neural consolidation (continued development of neural substrates of memory) was detected 1 hour after training, at which time the CS frequency became the new BF. These findings illustrate that retuning of the primary auditory cortex can develop very rapidly (e.g., after only 5 minutes of training) and that the full extent of this plasticity cannot be known immediately after training but must be evaluated after the potential for neural consolidation has been tested. (Reproduced from Edeline et al., 1993.)

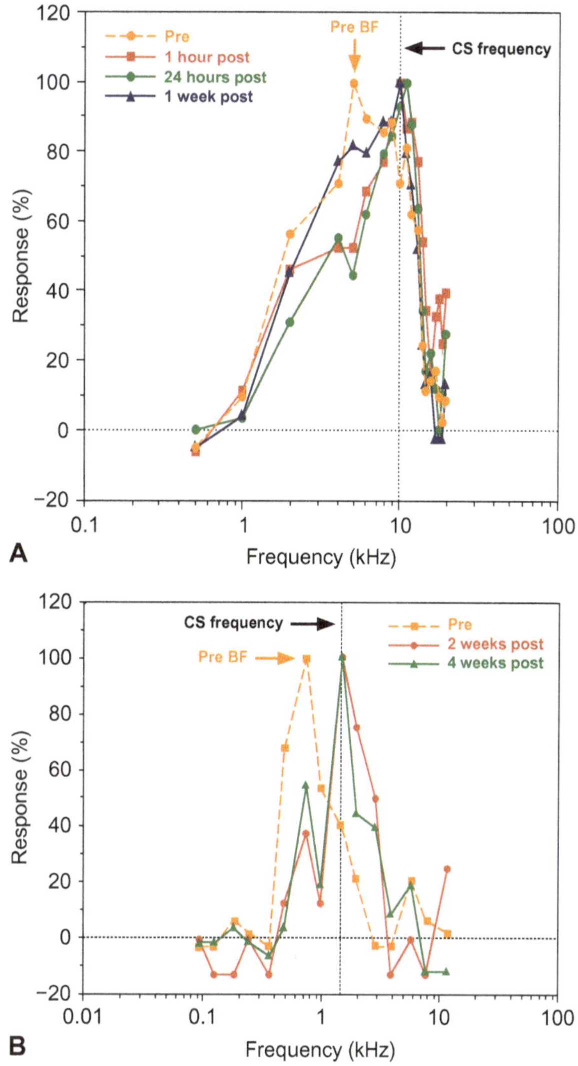

Fig. 7.5. Long-term retention of associative, specific receptive field plasticity in the primary auditory cortex. In both examples, conditioning induced a tuning shift to the frequency of the conditioned stimulus (CS). (**A**) An example of a CS-specific tuning shift over a period of 1 week. The best frequency (BF) shifted from 4.9 kHz to the CS frequency of 10.0 kHz, detected 1 hour after completion of training (tone–shock pairing: 30 trials). This shift was maintained at 24 hours and 1 week posttraining. (**B**) An example of a CS-specific tuning shift over a period of 4 weeks. The peak of tuning shifted from the pretraining BF of 0.69 kHz to the CS frequency of 1.45 kHz. Data depict tuning at 2 and 4 weeks posttraining (tone–shock pairing: 30 trials). (Based on Weinberger et al., 1993.)

Generality across types of acoustic stimulus parameters

Although most research has concerned acoustic frequency, RP has been found for all other stimulus parameters examined. For example, when sound level ("loudness") is the relevant cue, cells in A1 become selectively responsive to those intensities (Polley et al., 2004, 2006). Comparable findings apply to the repetition rate of noise pulses (Bao et al., 2004), the envelope frequency of a sinusoidally modulated tone (Beitel et al., 2003), sequences of particular sounds (Zhou et al., 2010), the size of step change in frequency (Selezneva et al., 2006), and the direction of frequency modulation (ascending vs descending) (Ohl et al., 2001). Moreover, RP linked to hemispheric specialization has been found in humans. Using functional magnetic resonance imaging (fMRI), Brechmann and Scheich (2005) engaged subjects in different tasks to identical stimuli, which were FM glides of various frequencies and durations. When their task was to categorize these sounds according to their direction of change (ascending or descending), the blood oxygen level-dependent (BOLD) signal increased in the right auditory cortex. In contrast, during judging of the duration of the tones, activation was found only in the left auditory cortex. Thus, the responses of auditory cortex can depend not merely on the involvement in a task, but rather on the nature of the task itself. Overall, the generality of RP includes

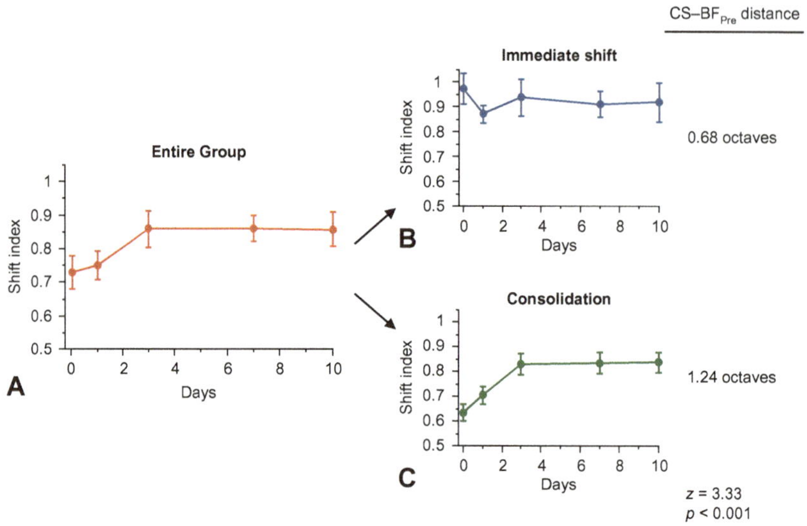

Fig. 7.6. Neural consolidation of representational plasticity over a period of 10 days posttraining. The magnitudes of tuning shifts (y-axes) are shown as a function of the time (1 hour to 10 days) after a single training session (auditory fear conditioning). An index of 1.0 would indicate a complete shift to the training frequency. (**A**) Group data show tuning shifts had developed by 1 hour post-training, increasing over 3 days to asymptote. A breakdown according to the frequency distance of the training tone from the pretraining best frequency (BF) (cells' baseline tuning preference) further reveals the temporal dynamics of neural consolidation. (**B**) Cells tuned near the frequency of the signal tone (±0.68 octaves) shifted greatly (~90% of the distance to the training tone) immediately and maintained their new tuning for at least 10 days. (**C**) Cells tuned farther away (±1.24 octaves) required 3 days to complete their tuning shift (viz., 72 hours of neural consolidation). Overall, the findings reveal the dynamics by which neurons are "recruited" to increase the representation of behaviorally important sounds, with completion of retuning (neural consolidation) requiring 3 days for neurons originally tuned distantly from the training tone. Thus, the full effects of learning on cortical representations of behaviorally important stimuli cannot be determined until days after a learning experience. CS, conditioned stimulus. (Reproduced from Galván and Weinberger, 2002.)

both the physical parameters of sound and types of tasks. No limit on either one has yet been reached.

Generality across motivational states and tasks

RP develops in all tasks reported to date, and for both positive (reward) and negative (fear) motivations: e.g., simple classic conditioning of fear (Diamond and Weinberger, 1986; Bakin and Weinberger, 1990; Gao and Suga, 2000) and reward (Kisley and Gerstein, 2001; Blake et al., 2006), discriminative classic conditioning of fear (Edeline et al., 1990; Edeline and Weinberger, 1993; Ohl and Scheich, 1996, 1997) and reward (Blake et al., 2002; Polley et al., 2004, 2006), simple and two-tone discriminative instrumental avoidance (fear) association (Bakin et al., 1996), operant reward training (Hui et al., 2009), and categoric discrimination for reward (Wetzel et al., 1998; Ohl et al., 2001; Selezneva et al., 2006).

Therefore, RP is not confined to particular tasks or types of motivation, but appears to develop regardless of task or motivation. It thus satisfies additional criteria for matching major features of associative learning.

Multiple rule tasks and the role of learning strategy

Most of the findings reviewed above have concerned learning situations in which subjects confront a relatively well-defined task like learning that a sound predicts a reward or the threat of nociception, or learning to produce a particular behavioral response to achieve reward or to avoid discomfort. To what extent does RP develop when subjects are in a more complex situation, such as having to solve a problem that has several rules? One particular type of such problem has been well studied.

Rats were trained to bar-press for water reward during occasional 10-s tone presentations. However, bar-presses made after the tone ended, during silent intertrial periods, were subjected to two rules: if made within 2 seconds of tone offset, they were neither rewarded nor punished ("grace period"); if made later, they were punished by presentation of a flashing-light error signal and a time-out penalty that lengthened the current intertrial interval. This task can be solved in two ways. First, animals could use the strategy of starting to respond at tone onset and stopping at tone offset ("tone duration" strategy: T-Dur). Second, they could start at onset and continue to respond after offset until receiving an error signal ("tone onset to error" strategy: TOTE). A second group was trained the same way except that they had no grace period and tended to use the T-Dur strategy. In contrast, rats trained with the grace period were inclined to use the TOTE strategy; that is, they paid attention to tone onset but not to tone offset. While both groups achieved the same level of correct performance, only the animals that used the TOTE strategy developed RP. In this case, the signal tone's threshold and BW decreased, increasing its sensitivity and selectivity (Fig. 7.7) (Berlau and Weinberger, 2008). The plasticity based on tone onset may reflect a proclivity of A1 to process onset transients (Phillips and Hall, 1990; Phillips et al., 2002), while other auditory fields may be more computationally adept for acoustic steady state (plateaus) and offsets. But whatever the explanation, these findings demonstrate that learning strategy, i.e., how an auditory problem is solved, can determine the development of RP.

The form of RP depends on the amount of use of the TOTE strategy: more extensive use (viz., on a greater proportion of training trials) produced a frank increase in the area of the cortex representing the training frequency band. Moreover, the amount of area gain is proportional to the amount of TOTE strategy use (Bieszczad and Weinberger, 2010b) (Fig. 7.8). Surprisingly, learning strategy is more critical in determining whether or not RP would develop in A1 than the level of motivation. In animal studies, motivational level can be determined by the hedonic value of a reward and validated by increased behavioral responses as a function of increasing value. Thus, animals will make more responses (e.g., lever presses) to attain a more desirable reward provided by the experimenter. Highly motivated subjects that do not use the TOTE strategy do not develop RP, despite making more responses, whereas moderately motivated animals that do use TOTE also develop RP (Bieszczad and Weinberger, 2010a).

Reversal and loss of representational gain in A1

Representational plasticity is subject to reversal when a tone no longer signals reward, as revealed in experimental extinction. Importantly, the amount of area loss is correlated with the amount of behavioral extinction (Bieszczad and Weinberger, 2012). These findings indicate that A1 "tracks" the current level of behavioral importance of sound and also that a loss of representational area is an active rather than a passive process.

Dissipation of area gain has also been reported in a very different circumstance, with undoubtedly different mechanisms. Rats were trained in a discrimination task to a high level of performance, and then overtrained for several thousand trials. At the end of overtraining, the previous gain in area was no longer evident; this has been referred to as "renormalization" (Reed et al., 2011). It has long been known that overtraining promotes a shift from cognitive memory, which is thought to be cortically dependent, to more habit-like response strategies, considered to be more dependent on subcortical striatal and cerebellar circuits (Mishkin and Petrie, 1984; Yin and Knowlton, 2006). In fact, since the dawn of experimental psychology, a distinction has been made between "memory" and "habit," to the extent that they were considered as separate chapters in *The Principles of Psychology* (James, 1893). So it is possible that the loss of area gain is due to overtraining.

To test this possibility, two groups of rats were trained in an appetitive tonal discrimination task; one was then mapped, while the other was overtrained for 2 weeks before being mapped. Both groups learned the task, but only the first group developed tone-specific area expansions; cortical representations in the second overtrained group did not did not differ from naïve subjects. During overtraining, the second group had shifted its learning strategy away from use of the TOTE strategy. Thus, "renormalization" appears to be attributable to a shift in how the task was solved, caused by overtraining (Elias, 2014). Two important questions remain to be answered: (1) Is the behavioral change a shift from cognitive memory to habit-like behavior? (2) Is there a neural shift of the critical locus from the primary auditory cortex to subcortical structures, such as the striatum?

Individual vs group analysis

The vast majority of findings in neuroscience are based on group analyses. In the case of the relationships between RP and learning, they have yielded many important results. However, individualistic analyses can be even more revealing. This advantage is evident in a study of gamma activity and auditory learning. It is well known that the magnitude of gamma-band oscillations (~40–120 Hz) in the electroencephalogram (EEG) is related to cognitive processes in humans and animals (see section on mechanisms of representational plasticity and specific auditory memory, below). For example, an increase in tone-induced gamma-band activity in A1 of the rat is correlated with the magnitude of subsequent behavioral conditioned responses to the signal tone. However, an analysis of posttraining responses across the tonal spectrum revealed that many animals did not exhibit maximum behavioral responses to the training frequency. This is not unusual (although often overlooked), because subjects are not *tabula rasa* but have different prior auditory experiences that can interact with new learning. A reanalysis of the data based on

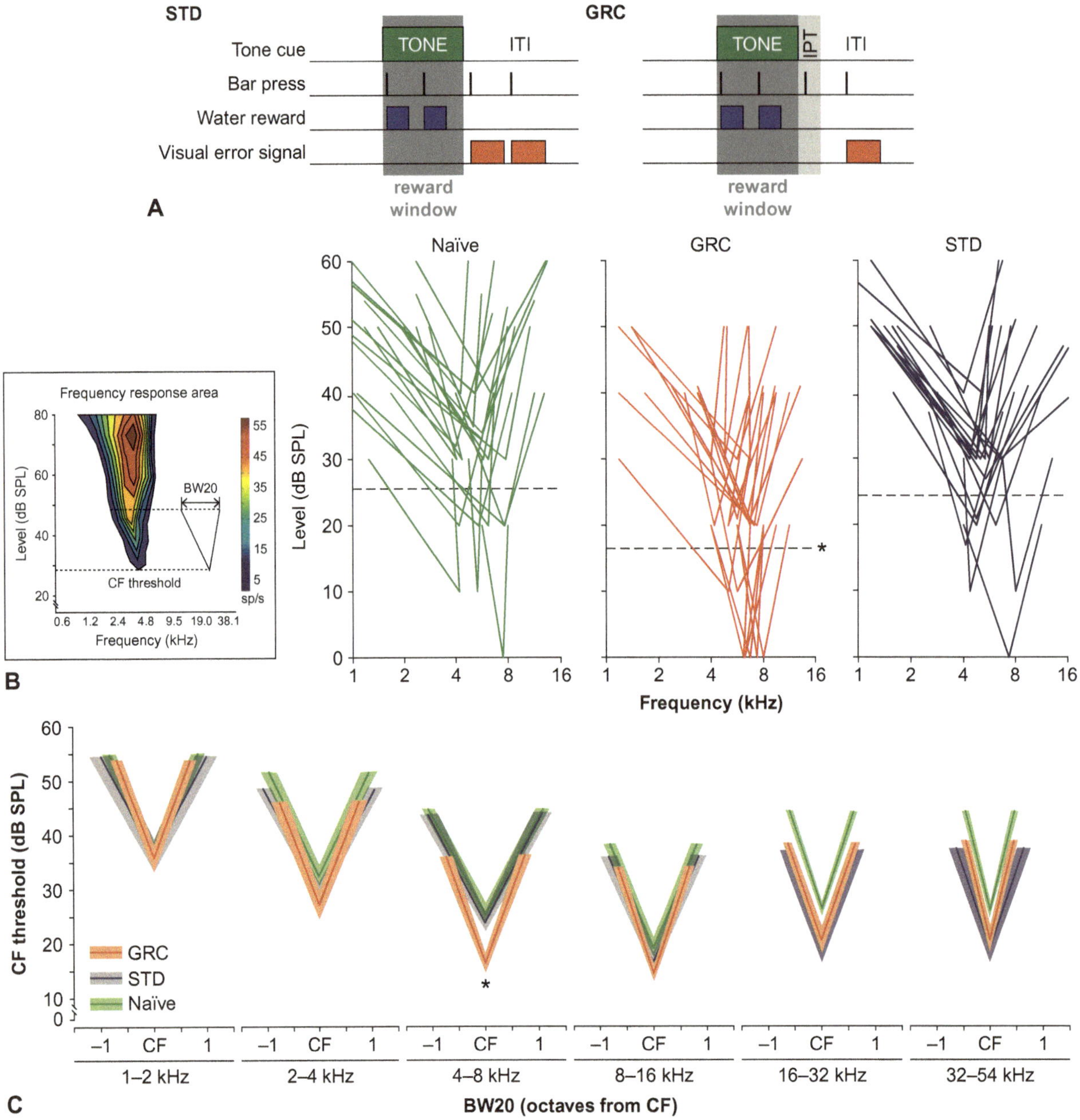

Fig. 7.7. Learning strategy determines auditory cortical plasticity. Rats learned to bar-press (BP) during a 5.0-kHz (10 seconds, 70 dB) tone for water reward. BPs made during intertrial interval silences resulted in a flashing-light error signal and a time-out penalty. (**A**) For the standard (STD) group all BPs during silence were signaled as errors. For the grace period (GRC) group, BPs made within 2 seconds of tone offset were neither rewarded nor signaled as an error, thus directing attention away from tone offset. Both groups learned the task to the same high level of performance, but used different strategies. STD behavior was based on attention to both tone onset and tone offset (tone duration (T-DUR) strategy). In contrast, GRC behavior was based on attention to tone onset and responding until receiving an error signal for BPs made after the grace period (tone onset to error (TOTE) strategy). (**B**) Posttraining frequency response areas revealed that only the GRC group developed specific representational plasticity (RP). Inset shows typical frequency response area (FRA) with characteristic frequency (CF) tuning and bandwidth (BW) indicated. Use of the TOTE strategy produced reduced threshold (increased sensitivity) and bandwidth (increased selectivity) in the signal-band tone. Shown are FRA "tuning tips" in groups STD, GRC, and a naïve group. Each V shape delineates the CF threshold, and BW20 of a recorded FRA (inset) that had a CF within the signal-tone frequency band (4–8 kHz) in each respective group. For sake of clarity, subsets of the total population of FRAs are depicted. Dashed lines represent the mean CF threshold for the entire population of each group. The asterisk shows that the GRC group had significantly lower CF thresholds than either naïve or STD groups, which are not different from each other. SPL, sound pressure level. (**C**) Plasticity in threshold and bandwidth in the

(Continued)

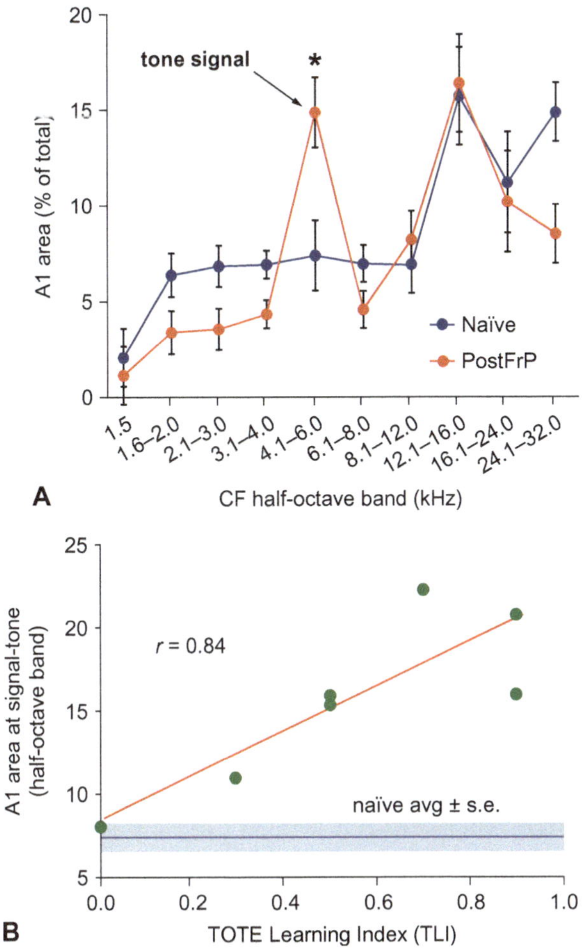

Fig. 7.8. The degree to which animals use the tone-onset-to-error (TOTE) strategy determines the magnitude of gain in representational area of the signal tone. (**A**) The magnitude of representational area for a group that was trained with a strategy ("PostFrP") that promoted use of the TOTE strategy, compared to an untrained control group. Note the significant increase in area of the trained group for the frequency band (4.0–6.0 kHz) centered on the signal tone (5.0 kHz). There were no other differences from the controls, showing specificity of representational plasticity. (**B**) The relationship between the degree of use of the TOTE strategy and the area representing the signal frequency band. Within the trained group, individual variability revealed a significant positive relationship: the greater the use of TOTE (i.e., the more trials on which it was used), the greater the gain in representational area. Thus, learning strategy, i.e., how a problem is solved, dictates the nature and amount of plasticity in the primary auditory cortex. CF, characteristic frequency. (Reproduced from Bieszczad and Weinberger, 2010b.)

the maximum response (peak of the behavioral frequency gradient) for each animal yielded a significantly stronger relationship between gamma activity and behavior. Thus, gamma-band plasticity was revealed as a signature of the frequency memory that gained the most strength due to training, not the memory of the training stimulus (Weinberger et al., 2013) (Fig. 7.9).

Therefore, studies of learning, whether pertaining to the auditory cortex or to other brain sites, can benefit from determining what each subject has actually learned or remembered. This is easily accomplished by training with one stimulus and performing posttraining assessment by testing with many stimuli (i.e., yielding "generalization gradients"). The peak of the gradient shows the stimulus that is most strongly linked to the learning. Even when group analyses yield significant effects, they may not reveal the full extent of relationships. Furthermore, lacking knowledge of the actual contents of individual learning, negative findings based on group treatments may mask significant relationships that exist at the level of the individual subjects. Insofar as individualized analyses (based on within-group variability) can be performed without compromising group-based evaluations, they ought to be encouraged.

THREE CLASSES OF AUDITORY TASKS AND REPRESENTATIONAL PLASTICITY

An important distinction has been made that classifies auditory associative learning tasks into three types: (1) detection; (2) discrimination; and (3) categorization (Scheich et al., 2011). The first refers to single-stimulus situations, e.g., in which a sound becomes associated with a reward or opportunity to avoid an aversive stimulus. The second refers to distinguishing between two or more sounds, only one of which is associated with reinforcement. The third refers to having two or more sounds that are discriminated from each other, not on the basis of having different physical properties (e.g., frequencies) that are differential-reinforced, but as a member of a differentially reinforced class of sounds. An example is differentiating between ascending and descending frequency sweeps (frequency modulations) irrespective of the particular acoustic frequencies they contain. Thus categorization requires abstracting a common property among sounds, although their particular physical features differ.

Fig. 7.7.—Cont'd GRC group is specific to the frequency band of the signal tone (asterisk). Both threshold and BW20 decreased only in the signal-tone frequency band. CF threshold and BW20 values are not significantly different from naïves in any frequency band in the STD group. Solid lines surrounded by shaded areas represent group means ± SE, respectively. Thus, specific RP in primary auditory cortex is determined by learning strategy, not merely whether or not learning occurred. The findings support the theory that different cortical fields are somewhat specialized to compute the relationship of different stimulus parameters to the acquired behavioral meaning of stimuli. A1 is known to be particularly sensitive to onset transients of sound and the GRC group attended only to tone onset, not to tone offset. (Reproduced from Berlau and Weinberger, 2008.)

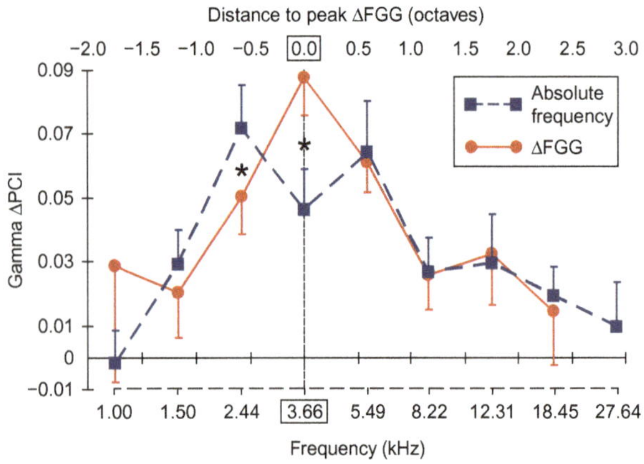

Fig. 7.9. Individual analyses based on what each rat learned (associative conditioning with signal frequency = 3.66 kHz) revealed a stronger relationship between the magnitude of tone-induced gamma activity in A1 (*y*-axis) and behavioral conditioned responses. There are two *x*-axes: bottom is standard, i.e., for absolute acoustic frequency; top is for octave distance from the peak of the individual behavioral frequency generalization gradients (FGG). Although all animals were trained at 3.66 kHz, the peaks of individual gradients varied widely (not shown), reflecting normal differential acoustic experiences outside of the lab that interact with the training frequency. The blue line shows the standard group analysis reflecting the amount of increase in gamma activity as a function of absolute acoustic frequency. Note that the peak increase in gamma is at 2.44 kHz, not at the training frequency. However, when the individual gradients are expressed as (octave) distance from their own behavioral generalization peaks (red line), then the greatest mean increase in gamma activity is precisely at the behavioral peak, i.e., the largest increase in gamma oscillations in A1 is at the frequency that was strongest in the memory of the animals. In short, increased gamma activity is a signature of the most behaviorally important memory, not the memory of the training frequency. The difference between standard group and individualized analyses is statistically significant (asterisks). PCI, power change index. (Reproduced from Weinberger et al., 2013).

According to this schema, three different forms of plasticity accompany the three types of tasks. Detection is accompanied by tuning shifts and expanded representation. This is consistent with the findings reviewed above. However, discrimination is said to be not accompanied by enhanced representation of the signal stimulus, but by increased responses to adjacent frequencies, often with a decrease of response to the CS+, producing an overall increase in local contrast (Ohl and Scheich, 1996, 1997). These findings are based on the use of one reinforced (shock) CS+tone and numerous (12–30) intermixed non-reinforced CS– tones. Insofar as two-tone discrimination studies had found increased responses to the CS+and tuning shifts rather than decreased responses to the CS+and local contrast change (Fig. 7.3), this formulation may not hold for simple discriminations. Categorization of the direction of FM sweep revealed a different form of RP; similar sounds (e.g., upward sweeps) became identified with a particular spatiotemporal activity pattern of LFPs in A1. This pattern emerged at the time that behavior switched from discrimination based on sound frequencies to categoric behavior (Ohl et al., 2001). Another form of plasticity was reported for a more complex variation of direction of FM in the monkey (Brosch et al., 2005; Selezneva et al., 2006). In all of these classes of auditory learning tasks, studies have shown that the primary auditory cortex is part of a multimodal matrix, in which its neurons are not merely modulated by other sensory stimuli but actually respond thereto (Scheich et al., 2011). No doubt future research will elaborate on the properties and mechanisms of complex discriminations and concept formation.

FUNCTIONS OF REPRESENTATIONAL PLASTICITY IN THE PRIMARY AUDITORY CORTEX

Given findings that support the establishment of SMTs in the primary auditory cortex, what can be said about their functions? While research is still in early stages, several studies have delineated various functions, all of which support auditory memory.

Encoding the behavioral importance of a sound

One possibility is that the area of representation of a tone encodes the degree of its behavioral importance. To test this hypothesis, rats were trained at different motivational levels (amounts of water restriction) to bar-press for reward contingent on the presence of a tone. They achieved corresponding levels of correct performance. Posttraining mapping of A1 revealed that subjects developed expanded representation for the tone's frequency; the higher their level of correct performance (presumably reflecting the tone's importance), the greater the area of tonal representation (Rutkowski and Weinberger, 2005) (Fig. 7.10).

Reinforcement prediction

Another potential function of RP is that of predicting reinforcement outcome, which is clearly related to the representation of stimulus importance. If expanded representation serves to predict reinforcement, then it should be reversed by extinction because reward is withdrawn. However, if it is a sign of any relevant change in auditory predictions of the state of affairs, then extinction should not eliminate cortical expansion because the

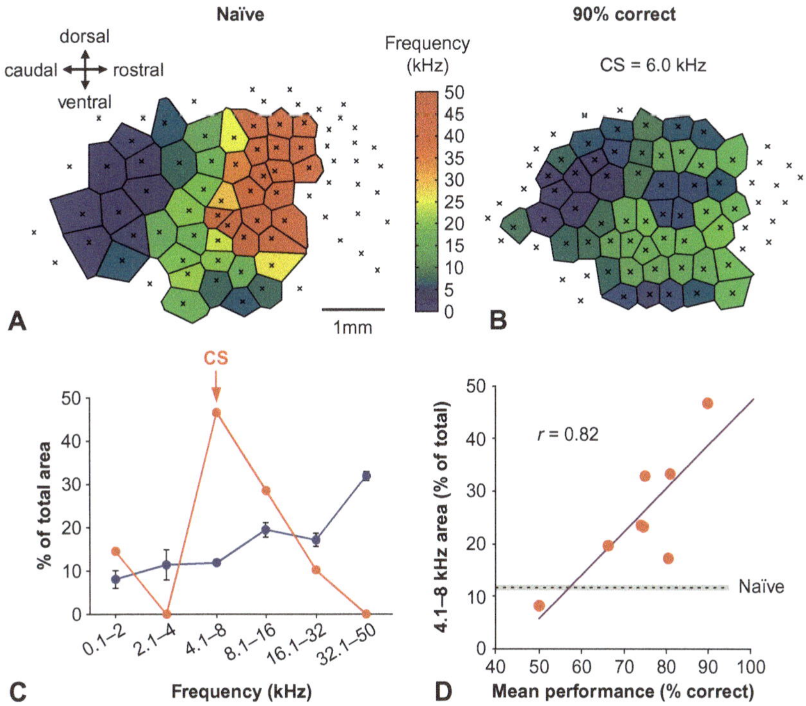

Fig. 7.10. The degree of acquired importance of a tone is correlated directly with the amount of area of frequency representation in the tonotopic map of A1. Trained rats received water reward for bar-presses in the presence of a 6.0 kHz tone. (**A, B**) Tonotopic maps for a naïve rat and a rat that attained over 90% correct performance, respectively. (**C**) Quantification of areas of representation across all frequency bands. Note that training greatly increased the area of representation for the frequency band containing the 6.0-kHz tone signal compared to the naïve animal. (**D**) Evidence of a "memory code" for the acquired behavioral importance of sound. The area of representation of the frequency band containing the 6.0-kHz tone signal increases as a function of the level of behavioral importance of the tone, as operationally indexed by the level of correct performance. CS, conditioned stimulus. (Reproduced from Rutkowski and Weinberger, 2005; Weinberger, 2007.)

absence of expected reward following response to an acoustic signal is itself a behaviorally significant event. Animals were trained in the same bar-pressing task described previously (see section on multiple rule tasks and the role of learning strategy, above). Following extinction, signal-specific area gains were reversed, supporting the "reinforcement prediction" hypothesis (Bieszczad and Weinberger, 2012). This is consistent with evidence of reward feedback in primary auditory cortex of primates (Brosch et al., 2011).

Substrate of memory strength

A third potential function of RP is as a neural substrate for the strength of memory, for which mechanisms are currently unknown. Rats trained in the same task later underwent experimental extinction. Memory strength is considered inversely proportional to the rate of extinction: the longer animals persist in bar-pressing, the slower the extinction and the stronger the memory. Terminal mapping revealed a significant positive relationship between the amount of area representing the signal tone and the strength of memory. Thus, the number of neurons that become tuned to a sound, the stronger the memory for that sound (Bieszczad and Weinberger, 2010c, 2012) (Fig. 7.11).

ARGUMENTS TO THE CONTRARY

Representational plasticity in A1 has the major characteristics of associative memory, as indexed in the forms of specific decreases in threshold and BW, tuning shifts and expansion of cortical area. Further, it exhibits generality across tasks, motivational valences, types of acoustic stimuli and species. While evidence continues to accumulate in this very active area of research, the overall picture emerging from research over approximately the past 25 years is one of consistency and replicability. On the other hand, there have been challenges to the conception of A1 as being deeply involved in learning and memory, particularly in storing SMTs.

It is entirely appropriate and necessary for new ideas and findings to be scrutinized deeply. In particular, the idea that the primary auditory cortex is a site of specific auditory memory storage is at variance with traditional views of A1 as exclusively an acoustic analyzer.

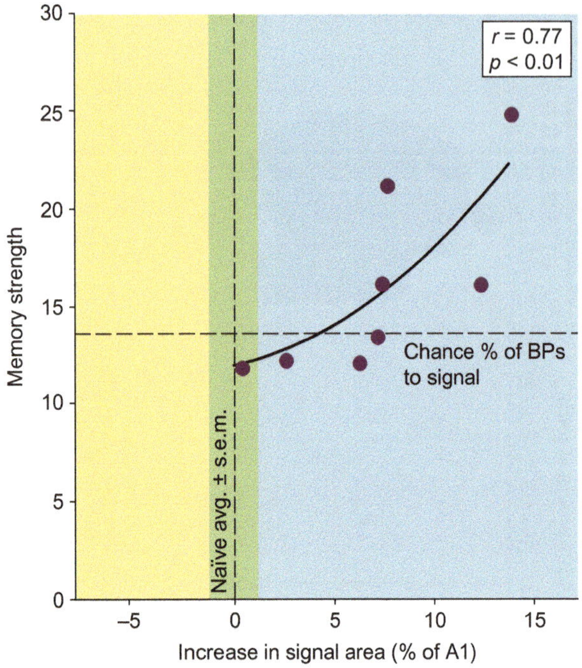

Fig. 7.11. Memory strength correlates with the amount of signal-specific area gains in the primary auditory cortex (A1). The *x*-axis shows the amount of relative area in A1 for the signal frequency (half-octave band). The *y*-axis shows the strength of memory for the signal tone (5.0 kHz) determined by its resistance to extinction relative to non-signal test frequencies. Trained animals showed increases in area relative to the naïve group (vertical dashed line, SE marked by shaded area). Increased memory strength for the signal tone is significantly positively correlated with the amount of area gain in A1 (best-fit regression, curvilinear: $r = 0.77$, $p < 0.01$). BP, bar-press. (Reproduced from Bieszczad and Weinberger, 2010c.)

Additionally, it is not yet clear how the modulation and plasticity of auditory neurons by the acquired behavioral meaning of sounds can be reconciled with the traditional and dominant concept that cellular discharges always encode the same specific physical parameters of sound. We will consider some major criticisms of the idea of, and findings supporting, SMTs in the primary auditory cortex.

Is representational plasticity caused by fear or increased arousal?

It has been argued that confounded state changes, not learning or memory, produce plasticity in A1. For example, the state of fear during the use of negative reinforcement (Hall and Mark, 1967; Mark and Hall, 1967) or increased arousal with any reinforcement (Kitzes et al., 1978) are claimed to explain facilitated responses to signal stimuli. Such state effects might either be elicited by an acoustic signal, be present as background throughout training, or both. However, these possibilities can be ruled out. First, the latency of most signal-evoked plasticity in A1 (~10–100 ms) is less than sound-induced fear or arousal, which requires at least 100 ms. Second, potential confounds by changes in background state due to the presence of shock have been ruled out by the use of sensitization control groups (i.e., tone and shock randomly presented) and the use of two-tone discrimination protocols have shown for fear conditioning that pairing of tone and shock is necessary for RP (Edeline and Weinberger, 1993). Simply creating a state of fear or heightened arousal is insufficient. Third, and perhaps most importantly, receptive field analysis has revealed that such tonic state changes do not cause specific tuning shifts but rather produce increased responses non-selectively to all acoustic frequencies (Bakin et al., 1992). In contrast, associations are far more specific, as revealed by increased responses to the CS (and closely related frequencies), while there is a simultaneous decline in responses to other frequencies (Weinberger and Diamond, 1988; Bakin and Weinberger, 1990). Contemporary research based on the determination of specificity of plasticity resolves all such state issues, but any studies that fail to control for non-associative effects and fail to demonstrate specificity run the risk of similar criticisms.

Attention is often confused with increased arousal because they sometimes occur together. However, they are fundamentally different. Attention refers to preferential processing of particular stimuli, and is usually highly selective. For example, responses in A1 to specific acoustic parameters can be facilitated by selective changes in frequency tuning involving both augmentation of processing a target tone while attenuating responses to irrelevant frequencies (e.g., Fritz et al., 2003, 2005). Selective attention can take place without any change in the overall state of arousal or behavioral excitability, and can be switched between visual targets without any change in cortical state or even any eye movements. Similarly, auditory attention can be redirected rapidly and repeatedly, as when tracking one or another voice, without any movement of the head or change in arousal level. In most cases selective attention is a product of previously learning which stimuli or items in the environment are or may become behaviorally important in a given situation. Therefore, while selective attention can modify processing in A1 and other cortical fields, it is neither independent of, nor in any way in opposition to, prior learning processes.

Are tuning shifts spontaneous, not induced by learning?

Kisley and Gerstein (1999) questioned whether tuning shifts were due to learning or spontaneous changes, because they observed tuning changes over days in the absence of training. This concern is understandable in the abstract. However, it had already been established that tuning shifts during learning experiments are toward, not also away from, the CS frequency, are associative, as they develop only in animals receiving paired CS and US, and are discriminative, i.e., toward the CS+, not the CS– (Bakin and Weinberger, 1990; Bakin et al., 1992; Edeline and Weinberger, 1993). Spontaneous shifts of tuning could not account for any of these attributes of learning-induced RP.

Furthermore, the "spontaneous tuning shifts" that they reported are readily explained as an artifact of their data analysis. This is best understood in a subsequent experiment in which these authors studied classic conditioning, pairing a tone with reward (Kisley and Gerstein, 2001). They did find that conditioning produces shifts directed toward or even to the CS. Also in agreement with prior studies, this receptive field plasticity was associative because it required CS–US pairing. The learning effects were above and beyond any spontaneous changes in tuning. However, what was the evidence for spontaneous drifts of tuning? The authors reported that the entire tuning curves from the same recording site became less correlated over days in the absence of conditioning, but they gave equal weighting to each of the many test tones used, including those that did not excite cells. So such random activity added noise to the analysis. A more direct test of their hypothesis is to directly determine the BF tuning over days. This has been accomplished for a period of 14–21 days. BF did not drift over days, neither did thresholds nor BWs (Galván et al., 2001; Galván and Weinberger, 2002).

"But, cortical lesions don't prevent memory"

Lesions of the auditory cortex can spare the acquisition of auditory fear memory (Romanski and LeDoux, 1992a). Some authors have concluded that, as the auditory cortex is apparently not necessary for the acquisition of auditory memory, RP cannot be a substrate of SMTs (e.g., Ohl and Scheich, 2004; but see Weinberger, 2004, 2007). This conclusion seems to reflect the assumption that memory substrates are localized. In so doing, it supposes that the claim that A1 contains SMTs seems to imply that all of the relevant traces are localized there. Following this logic, then lesions of A1 should prevent all auditory learning.

However, contemporary formulations hold that memory is ordinarily multimodal and multidimensional, and so its substrates involve the distributed representation of stored experience. Therefore, lesions that fail to prevent acquisition do not necessarily imply that the destroyed tissue held no SMTs, but only that the tissue was not essential for the behavioral measure that was used to assess learning. It should be emphasized that, as associative learning can be expressed through many effector systems, impairment of one behavioral measure should be interpreted cautiously. For example, lesions of the amygdala impair freezing (inhibition of all movements except respiration) to sounds or environmental contexts that induce fear due to prior association with nociceptive stimulation. This change in behavior is assumed to demonstrate that memory traces are stored in the amygdala (LeDoux, 1995). However, basolateral amygdala lesions actually impair unconditioned freezing (Vazdarjanova et al., 2001), so the conclusion that amygdala lesions destroy learned fear is questionable. In fact, rats with amygdala lesions that are unable to freeze can still express conditioned fear by avoiding the place of shock administration if given the physical context to do, like a simple maze rather than an inescapable standard test box (Vazdarjanova and McGaugh, 1998). Therefore, it appears that the amygdala is necessary for the behavioral expression of fear as measured by freezing and changes in autonomic behavior such as heart rate and blood pressure.

Furthermore, auditory memory traces for simple tone-shock learning can be stored in parallel, at the levels of both the medial geniculate (magnocellular (MGm) and posterior intralaminar (PIN) nuclei) and the auditory cortex. Thus lesions of either level spare the acquisition of simple one-tone auditory fear conditioning as measured by freezing (Romanski and LeDoux, 1992b). In short, although A1 may not be essential for learning to fear a single, simple tone, it is nonetheless "interested"; thus it could use its memory traces for an unknown future, or for the present if the task becomes more complicated and also in the unlikely event of lesions to the MGm/PIN (Weinberger, 2011a). The transactions of A1 with other cortical and subcortical targets, even in a task as simple as tone–shock pairing, are likely to proceed regardless of the involvement of the thalamoamygdaloid circuit. However, they are likely to remain unknown to the experimenter unless the animal is appropriately interrogated. For example, while the development of simple one-tone auditory conditioning does not require an intact A1 (DiCara et al., 1970; Berntson et al., 1983; Romanski and LeDoux, 1992a, b), as soon as two-tone discrimination is demanded, A1 is required (Teich et al., 1988). An intact A1 is also obligatory to achieve experimental extinction in the future (Teich et al., 1989), so memory traces in the cortex appear necessary for behavioral inhibition to the tone when it no longer predicts shock. Of course, information stored in the auditory cortex has interactions with far more widespread inputs and outputs than the subcortical auditory system, and can handle far more complex problems while still solving simple problems.

Finally, it is important to distinguish acquisition processes from memory storage. Thus, even if cortical lesions do not prevent acquisition of simple associations (due to subcortical capabilities), it is nonetheless the case that destruction of the primary auditory cortex after learning can abolish auditory memory (Boatman and Kim, 2006). In summary, the failure of a lesion of A1 to prevent behavioral evidence of acquisition cannot itself demonstrate that it holds no SMTs.

MECHANISMS OF REPRESENTATIONAL PLASTICITY AND SPECIFIC AUDITORY MEMORY

To date, most studies of the mechanism underlying RP have concerned neuromodulators. Recent work has begun to address the role of gamma-band oscillations and neural synchrony as well as synaptic mechanisms. Considerations of the contributions of the thalamic auditory system are beyond the scope of this article and have been reviewed elsewhere (Weinberger, 1995). However, it can be said that RP in the cortex is not simply projected from the medial geniculate nucleus. Thus, while the ventral medial geniculate possesses the requisite frequency specificity in its own tonotopic organization and does develop specific tuning shifts, they are very restricted and not maintained even for 1 hour (Edeline and Weinberger, 1991).

Neuromodulators in representational plasticity

The following neuromodulators have been studied most in RP: acetylcholine (ACh), dopamine (DA), norepinephrine, (NE) and serotonin (5HT).

Acetylcholine

ACh is the only neuromodulator that has been implicated in both RP and in the induction of specific behavioral memory by activation of the cholinergic nucleus basalis (NB), producing "implanted memory."

ACh and representational plasticity

ACh, particularly acting at muscarinic receptors, has long been implicated as important for memory (Deutsch, 1971; Vanderwolf, 1987; van der Zee and Luiten, 1999). The NB supplies ACh to the cortex (Mesulam et al., 1983) and copious evidence implicates it in auditory cortical function. For example, NB ACh cells project directly to A1 (Bigl et al., 1982), and NB stimulation releases ACh (Casamenti et al., 1986) that produces cortical EEG activation, especially increased gamma-band power (~40–120 Hz) (Cape and Jones, 1998, 2000), which depends on the engagement of muscarinic cholinergic receptors (mAChRs) in A1 (Metherate et al., 1992). Conversely, NB lesions reduce cortical ACh (Arendash et al., 1987) and gamma activity (Berntson et al., 2002). NB stimulation modifies A1 responses to sensory volleys (Metherate and Ashe, 1991; Hars et al., 1993) and facilitates field potentials, cellular discharges, and excitatory postsynaptic potentials via mAChRs in A1 (Metherate and Ashe, 1991, 1993). NB cells can differentially respond to specific acoustic frequencies (Chernyshev and Weinberger, 1998). NB responses to tones that signal appetitive or aversive reinforcement increase during learning (Richardson and DeLong, 1986) and NB cells develop increased discharges to signal tones during auditory fear conditioning (Whalen et al., 1994) preceding neuronal plasticity in A1 (Maho et al., 1995). An important, but underappreciated, finding is that NB cells projecting to A1 selectively increase transcription of the gene for ACh's synthetic enzyme, choline acetyltransferase, during auditory fear conditioning (Oh et al., 1996). Finally, but of no less importance, ACh is selectively released in A1 as animals learn that a sound signals forthcoming reward (Butt et al., 2009).

Given the implication of ACh in learning and memory, coupled with the fact that RP is formed during learning, one might expect ACh to have an important role in RP. Indeed, this is the case. Pairing a tone with NB stimulation actually induces associative RP in A1, in the form of receptive field tuning shifts (Bakin and Weinberger, 1990). Moreover, RP induced by paired NB stimulation has the same characteristics as learning-induced RP: in addition to being associative, it is specific, rapidly acquired, consolidates, and is retained (Bakin and Weinberger, 1996; Bjordahl et al., 1998; Dimyan and Weinberger, 1999). Given the fact that gains in representational area in A1 must be based on many individual shifts in tuning, it is not surprising that tone paired with NB stimulation produces RP in the form of specific area gains (Kilgard and Merzenich, 1998). Also, as with learning-induced RP, stimulation of the NB can specifically modulate the representation of parameters other than acoustic frequency (Kilgard et al., 2001a, b). RP can be induced by pairing a tone with ACh applied directly to A1 and NB-induced RP is dependent on the engagement of mAChRs in A1 (Ji et al., 2001; Miasnikov et al., 2001; Ji and Suga, 2003; Yan and Zhang, 2005).

Finally, the NB is almost certainly activated by other structures that hold information about the current need to store information. For example, the basolateral amygdala is known to modulate the strength of memory that is stored elsewhere (McGaugh, 2004). Tone paired with stimulation of the basolateral amygdala induces the same specific tuning shifts as develops in natural learning and when tone is paired with stimulation of the NB (Chavez et al., 2009). This form of RP is highly specific, consolidates (becomes stronger over days) and lasts indefinitely, as tracked to 3 weeks (Chavez et al., 2013). These findings suggest that memory modulation by the basolateral amygdala is accomplished by activating the NB. Of course, as this nucleus receives input from numerous brain regions, the relevant area for a given learning situation may initiate release of ACh in the cortex.

ACh and the implantation of specific associative behavioral memory

Remarkably, a sound paired with NB stimulation also implants specific behavioral auditory associative memory (McLin et al., 2002). After pairing, many tones across the acoustic spectrum were presented and changes in heart rate and respiration were measured. A control group received tone and NB stimulation unpaired. The results showed that the paired tone alone elicited larger behavioral responses after pairing (Fig. 7.12). The behavioral frequency gradients are indistinguishable from those generated by natural memory. Moreover, like natural memory, implanted memory is associative and specific (McLin et al., 2002), rapidly acquired (Miasnikov et al., 2006), consolidates (Weinberger et al., 2009; Miasnikov et al., 2011), is long-lasting (Miasnikov et al., 2011) and extinguishes due to the removal of "reinforcement" (NB stimulation) (Miasnikov et al., 2009). Furthermore, implanted memory depends on intrinsic ACh (Miasnikov et al., 2008b).

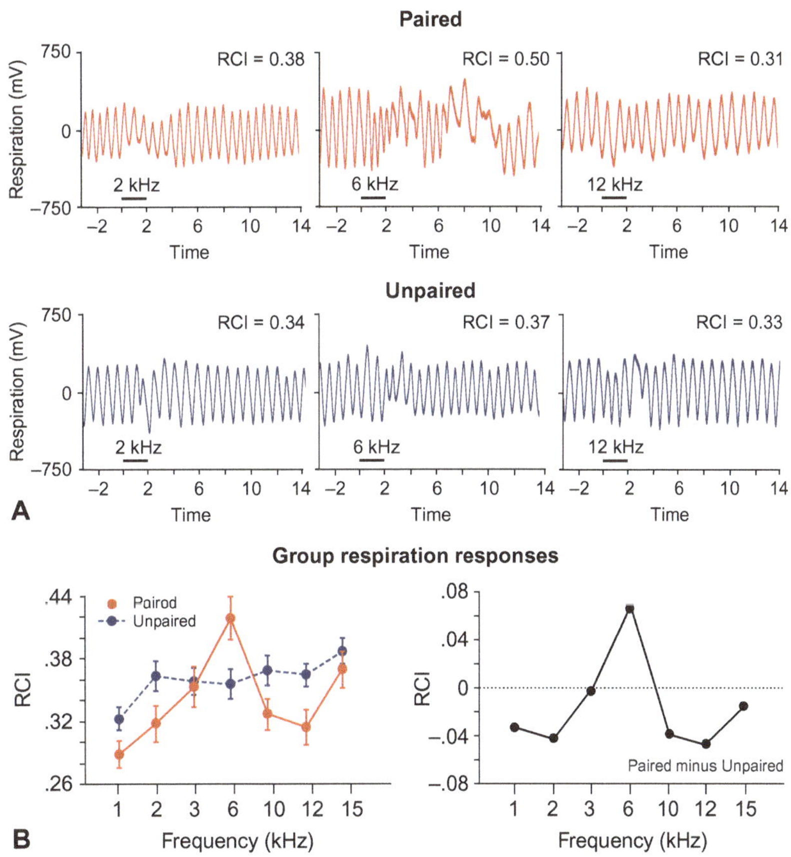

Fig. 7.12. Implantation of specific associative behavioral memory by pairing a tone with activation of the cholinergic nucleus basalis. Rats heard a 6.0-kHz tone followed by microstimulation of the nucleus basalis, which released acetylcholine into the cerebral cortex. After many pairing trials, they were tested for the implantation of specific memory of 6.0 kHz by being tested with many different frequencies, including the training signal. Interruption of ongoing respiration was used as a sensitive behavioral index of associative and specific memory. (**A**) Examples of individual respiration records to three frequencies (2.0, 6.0, and 12.0 kHz) for one animal each from the paired group and an unpaired control group. Note that the largest response was at 6.0 kHz for the paired animal. The control subject exhibited little response. Horizontal bars indicate tone duration. (**B**) Group mean change in respiration to all tones for both groups. Left shows that the maximal response was at 6.0 kHz for the paired group but not for the unpaired group. The group difference function (paired minus unpaired) (right) shows the high degree of specificity of associative respiratory responses to the 6.0-kHz training tone; it was the only stimulus to elicit a significant behavioral response. Implanted memory that the training tone was behaviorally important depends on the engagement of muscarinic cholinergic receptors in the primary auditory cortex. RCI, respiration change index. (Reproduced from McLin et al., 2002.)

In short, animals behave as though they had an experience that never happened; they have a false memory. What is likely to be the content of such implanted memory? It should not have any emotional content because the NB is "downstream" of motivational centers such as the amygdala and stimulation of this cholinergic nucleus is neither rewarding nor aversive (Miasnikov et al., 2008a). It seems that animals with implanted memory "know" that the signal tone is now important, but don't know why. This may not be as strange as it at first might seem. People often remember that an event was significant long after they have forgotten the particular details of how it became so.

The use of NB stimulation can yield information about cortical plasticity and memory that is not easily obtainable using natural memory tasks. For example, it permits control of the amount of specific detail in auditory memory, by varying the level of current; weak stimulation produces a general meaning that sound is important, but without any specificity for the training tone, whereas moderate stimulation produces highly specific meaning in memory (Weinberger et al., 2006). This finding suggests that the degree of detail in auditory memory may be determined in part or entirely by the amount of ACh released by an experience that engages mAChRs in the auditory cortex.

Overall, the attributes of natural associative learning, RP induced during natural learning, RP induced by tone paired with NB stimulation, and memory implanted by NB stimulation are the same (Table 7.1).

In toto, the ACh findings suggest that the direct induction of RP in A1 might be sufficient for the formation of specific auditory memory. However, no link had been demonstrated between artificial increase in RP for a tone and the implantation of memory.

To test this possibility, rats received a 3.66-kHz tone paired with NB stimulation and were tested for the implantation of specific behavioral memory. Next, their cortical tonotopic maps were determined. These revealed that pairing had induced gains in representational area. However, the areas of maximum increase were not exactly at the 3.66-kHz training frequency, but distributed across various frequencies. As noted previously (see section on individual vs group analysis, above), individuals are not functional or behavioral clones as each comes to a common training experience with different prior experiences; these often interact with new experiences, the result being that they often remember somewhat different things, in this case, exactly what frequency had been paired with NB stimulation. Nonetheless, there was a match on an individual basis between the frequency whose area increased the most and the behavioral frequency memory that was strongest (Bieszczad et al., 2013). This finding constitutes the first demonstration of the artificial induction of specific memory by direct remodeling of the cerebral cortex.

Naturally, such invasive research cannot be conducted with human subjects. However, other types of evidence show that the cholinergic system is involved in associative learning and auditory cortical plasticity in humans (reviewed in Thiel, 2003). For example, fMRI studies have revealed that scopolamine blocks discriminative conditioned changes in auditory cortical responses (enhanced BOLD signal) to tones that were paired with electric shock (CS+) compared to tones without shock (CS–) (Thiel et al., 2002).

Dopamine

There is increased release of homovanillic acid (a metabolite of DA) from the primary auditory cortex of gerbils during the initial stages of avoidance learning. Insofar as performance was maintained on succeeding days without the same elevation of release, DA appears to be involved in the acquisition of information rather than on retention of memory (Stark and Scheich, 1997). Increasing DA release by self-stimulation of the ventral tegmental area facilitates avoidance learning, potentially

Table 7.1

Comparison of the attributes of natural associative memory with learning-induced receptive field plasticity, nucleus basalis (NB)-induced plasticity, and specific behavioral memory implanted by NB stimulation

Characteristic	Associative memory	Receptive field plasticity	NB-induced receptive field plasticity	NB-induced memory
Associative?	Yes	Yes	Yes	Yes
Highly specific?	Yes	Yes	Yes	Yes
Rapidly acquired?	Yes	Yes	Yes	Yes
Consolidation?	Yes	Yes	Yes	Yes
Long-lasting?	Yes	Yes	Yes	Yes

acting in A1. Consistent with this, direct application of D1/D5 receptor agonists into the auditory cortex facilitates the consolidation of discrimination of auditory stimuli during avoidance learning, while D1/D5 receptor antagonists suppress this effect (Schicknick et al., 2008, 2012).

Norepinephrine

NE (noradrenaline: NA) is known to increase the frequency selectivity of neurons in the primary auditory cortex of animals (Edeline, 1995), apparently by reducing both spontaneous activity and receptive field size for frequency tuning (Manunta and Edeline, 1997, 1999). Most relevant for present purposes are the effects of pairing a tone with iontophoretic application of NE/NA to the primary auditory cortex. The increase in frequency selectivity by reduction of receptive field size was found to be specific to the paired tone, the effects extending only to 0.25 octaves. This plasticity was unaffected by propranolol but blocked by phentolamine, indicating the involvement of alpha receptors (Manunta and Edeline, 2004).

Serotonin

Very little is known about the role of 5HT in associative RP in the auditory cortex. A microdialysis study revealed no increase in release of its metabolite 5-hydroxyindoleacetic acid during avoidance learning. Levels of release appeared to be more closely related to the animals' stress levels (Stark and Scheich, 1997).

Neural synchrony and gamma-band oscillations

Cortical gamma-band oscillations (~40–120 Hz) reflect synchronous neuronal activity involving a systematic interplay between excitatory and inhibitory cells (Buzsáki and Wang, 2012). Gamma's time scale is appropriate for synaptic integration (Salinas and Sejnowski, 2001) and spike timing-dependent plasticity (Bi and Poo, 1998; Wespatat et al., 2004), which can enable synaptic strengthening both within auditory cortex and between auditory cortex and other regions. Indeed, synchronous activities do occur at different spatial scales, i.e., within a sensory field (Gray et al., 1989) and between cortical areas (Gregoriou et al., 2009). Within auditory cortex, synchronous activity representing a sound should be more effective in communicating with its targets than a less coherent representational population (Börgers et al., 2005, 2008). If synchrony increases after initial processing of sounds that become behaviorally significant, they could gain greater control both within A1 and over more distant targets. These effects could result in expanded representation locally and enhanced influence distally, promoting the formation of distributed networks for auditory memories and ultimately finding expression in behavior.

Gamma can easily be recorded in humans with scalp electrodes and increases during visual associative learning (Miltner et al., 1999). Animal studies that can reveal the relevant mechanisms have extended research in this relatively neglected area. Thus, increased gamma activity in A1 has been implicated in auditory learning and memory for both natural memory (Jeschke et al., 2008) and for memory implanted by stimulation of the NB (McLin et al., 2003). More recently, it has been found that both gamma and synchronous unit firing are enhanced in A1 during auditory fear conditioning when frequency tuning is shifted toward the signal tone and cells representing the signal tone increasingly discharge together, likely strengthening its effectiveness. Importantly, increased gamma activity predicts behavioral memory 24 hours later, greatly strengthening a causal role for increased neural synchrony in auditory learning and memory (Headley and Weinberger, 2011, 2013) (Fig. 7.13). Further, pairing a tone with stimulation of the basolateral amygdala, which can strengthen memories, also produces specific tuning shifts and increased gamma to the CS+tone (Chavez et al., 2013). *In toto*, these gamma findings bridge the gap between cholinergic mechanisms in RP and behavioral memory because gamma is itself induced by NB stimulation acting at muscarinic receptors in A1 (Metherate et al., 1992).

Circuit and synaptic processes

Research on circuit and synaptic processes in RP and learning is at a very early stage, but the findings appear promising. For example, given that associative learning produces tuning shifts and area gain for the signal frequency, one might suspect a "lateral" spread or recruitment of neighboring cells. Evidence for this mechanism has been recently reported. Rats were trained to detect a 5.0-kHz target to obtain food. After training, current-source density profiles obtained to a wide spectrum of test tones revealed that current sinks were enhanced in response to the training tone in layers 2–3 but not thalamo-recipient layer 4. Enhancement was limited to 1.0 octaves from the 5.0-kHz training frequency; sites more spectrally distant showed no change (Guo et al., 2013). The findings provide strong evidence that auditory learning increases the frequency domain of behaviorally important sounds via intracortical pathways.

Synaptic changes linked to behavior have been reported. Animals received extensive training of tone

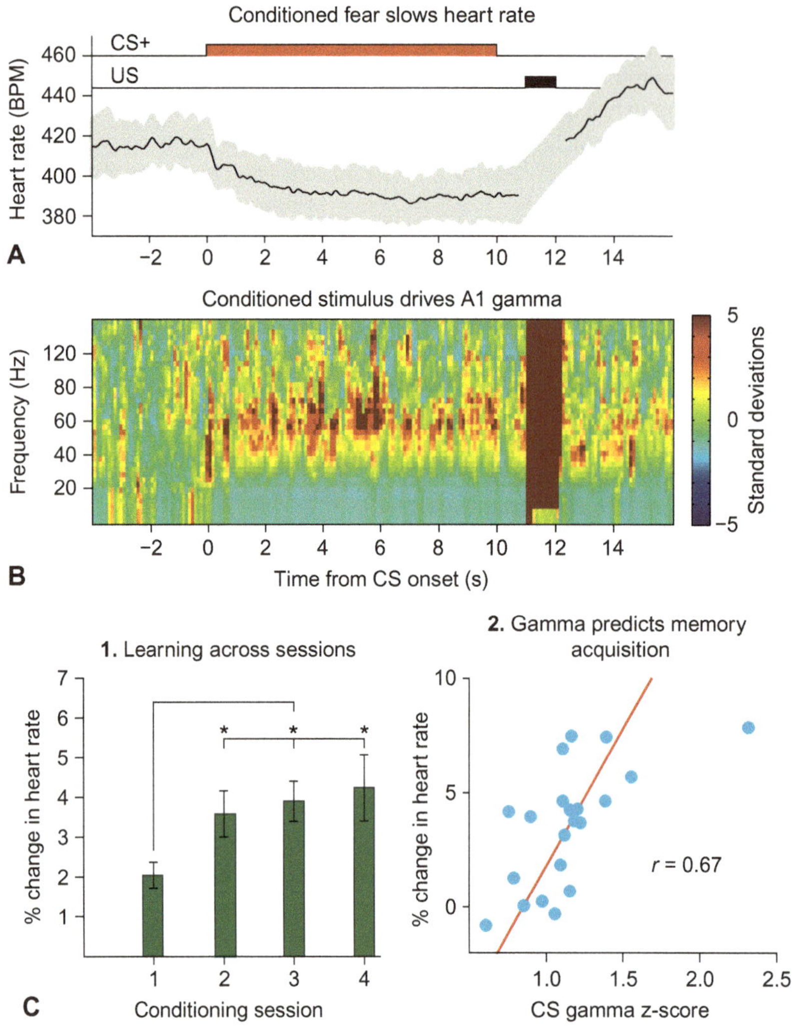

Fig. 7.13. Increased gamma-band activity in the primary auditory cortex predicts the formation of associative memory. Rats received a tone conditioned stimulus (CS) paired with shock (auditory fear conditioning). Conditioning cardiac slowing was used as the index of behavioral change. (**A**) Conditioned bradycardia (±SE) for an exemplar subject. Note the sustained reduction in heart rate during the 10-second conditioned stimulus. (**B**) The CS tone induces conditioned gamma oscillations (~40–120 Hz) in A1 after having been paired with shock; thick red line is shock artifact. (**C**) Group findings. (1) The magnitude of the conditioned response: asymptotic behavior was attained on session #2 and maintained thereafter. (2) The amount of tone-induced increased gamma activity (during tone minus pre-tone baseline) on session #1 predicts the magnitude of conditioned behavioral response on session #2. US, unconditioned stimulus. (Reproduced from Headley and Weinberger, 2011.)

paired with stimulation of the NB. This produced synaptic modifications in A1, the main effect of which was to decrease discharge variability and thereby increase neural signal-to-noise ratios. Neuronal sensitivity increased, e.g., thresholds were lowered and subjects were able to detect signals that were weak or previously undetectable (Froemke et al., 2013). Insofar as pairing a tone with NB stimulation implants a memory that the tone is behaviorally important (see section on ACh and the implantation of specific associative behavioral memory, above), it is very likely that the animals in this study also had memory implantation. Thus, the increased sensitivity to tones that had been paired with NB stimulation may have been a consequence of associative learning that was not evaluated in this study.

IMPLICATIONS FOR TREATMENT OF AUDITORY AND RELATED DISORDERS

The results of research on learning, plasticity, and the primary auditory cortex have implications for the clinic. Four of these are discussed below: (1) memory strength and posttraumatic stress disorder (PTSD); (2) gamma-band oscillations in assessment and treatment; (3) individualized analyses of neural plasticity, learning, and memory; and (4) cholinergic implantation of specific memory.

Memory strength and posttraumatic stress disorder

The area of cortical representation of a behaviorally important sound in A1 is increased; the greater the area, the stronger the memory (see section on substrate of memory strength, above). Less is known about representations for more complex important sounds (including speech) in other auditory cortical fields of both animals and humans. At this early stage of research, there is no reason to assume that RP is confined to A1; this is simply a convenient auditory field because its tonotopic organization provides for neural signatures of RP. On the contrary, although little studied, non-primary cortical fields also develop RP (Diamond and Weinberger, 1986; rat: Puckett et al., 2007). PTSD, with its intrusive "flashbacks" triggered by a stimulus related to the original traumatic experience, can be considered as an instance of abnormally strong memory. A testable hypothesis is that PTSD involves abnormally large representation in the cortex of all relevant sensory fields and modalities, as well as involved emotional neural representations. Such an extensive network of relevant neurons would likely be triggered by any stimulus whose similarity to a component of the traumatic memory is sufficient to activate part of the memory representation, and produce recall or flashback. Such flashbacks could even be triggered spontaneously, given the extremely large representational network, any of whose components might be activated in the normal course of daily events or dream sequences. Therefore, clinical studies should be directed to determining the size of involved representational networks in all cortical auditory fields, i.e., primary, belt and parabelt (Kaas and Hackett, 1998; Woods and Alain, 2009). If the hypothesis is confirmed, then treatments aimed at reducing the size of the representational area should be explored (see section on reversal and loss of representational gain in A1, above).

Gamma-band oscillations in assessment and treatment

Gamma-band oscillations index neural synchrony and such coordinated activity is likely to be more effective in communication with and influence on their targets, both locally and distally. Therefore sounds with greater gamma may have larger areas of tuning and promote associations. Indeed, tone signals do develop increased gamma and predict behavioral associations detected 24 hours later (see section on neural synchrony and gamma-band oscillations, above; Headley and Weinberger, 2011, 2013). Naturally, acoustic stimuli are used in treatments for auditory-processing disorders and other hearing problems. Such treatments constitute learning opportunities. Sounds that induce high levels of gamma prior to training may be more effective in forming new associations than sounds that induce lower levels of gamma. As gamma oscillations are routinely recorded from humans, we suggest that sounds planned for remedial training first be screened for their effectiveness in the induction of gamma oscillations, and the most effective relevant sounds be employed. Double-blind studies would reveal if this approach has hypothesized advantages.

Cholinergic implantation of specific memory

The importance of a given sound can be implanted by pairing it with stimulation of the NB, given the engagement of mAChRs in the primary auditory cortex (see section on acetylcholine, above). Similar effects may occur in other auditory fields when the sounds are complex or linguistic. It might be possible to induce or facilitate specific memory formation in humans by pairing a desired sound (simple, complex, linguistic) with the release of ACh in auditory cortical primary, "belt" and "parabelt" fields. It is already known that RP develops in non-primary fields even with tones (Diamond and Weinberger, 1986; Puckett et al., 2007). Indeed, RP probably develops in all auditory cortical fields simultaneously, if their constituent neurons respond reasonably strongly to the particular sounds that are used as a signal in associative learning.

Although direct stimulation of the NB is not possible in humans, RP can be induced by pairing a tone with stimulation of the vagus nerve, which already has been approved as therapeutic treatment for non-auditory problems (Engineer et al., 2011, 2013). Vagus stimulation is known to facilitate the memory of preceding stimuli in humans (Clark et al., 1999) as well as in the rat (Clark et al., 1998). Its effects depend upon activation of the basolateral amygdala, which has been established to modulate memory strength when activated shortly after an experience, via a cholinergic link (McGaugh, 2004). Vagus afferents appear to strengthen memory by detecting increased release of epinephrine in the periphery caused by an exciting event (Chen and Williams, 2012). The vagus then synapses in the nucleus of the solitary tract, which projects to the locus coeruleus, which releases NE that in turn activates the basolateral amygdala to facilitate the formation of memory traces in the cortex (McIntyre et al., 2012). Supporting this account, direct stimulation of the basolateral amygdala does induce long-lasting specific receptive field shifts in A1, possibly via the NB (Chavez et al., 2013).

Therefore, it should now be possible to determine if specific memory implantation or augmentation can be accomplished therapeutically for auditory-processing disorders, and perhaps facilitation of language comprehension in cochlear implant patients. An important consideration is that cholinergic effects depend on the relevance of a sound or visual stimulus; agonists enhance neural and behavioral measures only when the current stimuli are task-relevant, but reduce or have no effect when sensory stimuli are given in passive or task-irrelevant situations (Bentley et al., 2011). As signal sounds are task-relevant in auditory associative learning, it would probably be more efficacious to use pro-cholinergic agents therapeutically, with appropriate timing, during learning tasks rather than in passive situations. We recommend that human protocols be based on the animal protocols now in use for more than a decade (McLin et al., 2002), including within-subject evaluation of the difference between pre- and posttraining behavioral gradients in heart rate, respiration, or other convenient physiologic measure (Miasnikov et al., 2006). This approach could be used as an adjunct or sole treatment for many problems, including perception, learning, memory, attention, and a variety of cognitive disorders.

Individualized analyses of neural plasticity, learning, and memory

The ability to induce specific memory by directed reorganization of auditory cortical representations has translational implications. Thus, training with a given auditory stimulus does not necessarily produce the best learning about that particular training sound. Posttraining generalization gradients can reveal which sound gained the greatest strength through training. In fact, when memory is implanted by tone paired with stimulation of the NB, the area of greatest gain in A1 matches the stimulus at the peak of the frequency gradient, rather than matching the training tone (Bieszczad et al., 2013). Therefore, it is important to actually determine what was learned by each subject or patient, rather than assume that the learning was to the training signal. As noted above, stimulation-based treatments have already shown promise for cortical remodeling to alter perceptual functions (Reed et al., 2011), including the potential to reverse maladaptive effects of tinnitus (Engineer et al., 2011, 2013). However, the findings reviewed above reveal that individual differences must be factored into both conceptions and empirical aspects of research (see Three classes of auditory tasks and representational plasticity, above). Reduction of potentially harmful, e.g., phantom-limb-like, effects of pain or aggravation that may result if map expansion develops "off target" (i.e., for an unintended stimulus) (Lozano, 2011) may be achieved by taking advantage of individualized approaches that promote intended therapeutic functional benefits.

CONCLUSIONS AND GENERAL IMPLICATIONS

Primary auditory cortex, learning, and plasticity

The primary auditory cortex can no longer be considered as only a sensory analyzer. Likewise, it can no longer be excluded from being deeply involved in associative learning and memory. Three major implications of the paradigm-changing findings reviewed above are considered: (1) the attribution of meaning to sound; (2) reconceptualization of the primary auditory cortex and, by extension, (3) the need for a new model for the entire cerebral cortex.

Beyond perception: The acquisition of meaning to sound

Auditory comprehension is essential to hearing because it gives meaning to sounds, whether of natural or human origin. Yet, remarkably, the amount of research devoted to this central problem of how sounds gain behavioral meaning has been vanishingly small compared to the wealth of effort expended to understand the detection and perception of sound. However, without foundational research on the neural bases of auditory perception, progress on auditory meaning would not have been possible.

This problem is also of fundamental importance to the field of the neurobiology of learning and memory because the meaning of sensory events is acquired through experience. This article has concerned the results of combining the two disciplines of auditory and learning/memory neuroscience to solve this common problem.

An answer has begun to emerge. First, associative learning systematically modifies the representation of acoustic stimuli in A1. For the vast majority of studies to date, the processing of meaningful sounds is augmented, including learning that a sound predicts reinforcement, discriminating a reinforced from a non-reinforced stimulus, solving instrumental problems that require behavioral acts to be performed or withheld under a wide variety of simple to complex task structures and demands. Such RP develops across species, types of motivation, and acoustic stimulus parameters. This selective facilitation can take several forms, including increased sensitivity and selectivity (decreased threshold and BW, respectively), tuning shifts directed to the signal in question, and increased area of representation in the tonotopic map.

Second, RP is formed when sounds are paired with stimulation of the cholinergic NB. It appears that activation of mAChRs in the auditory cortex is a mandatory step in this process, although more research is needed. Third, specific behavioral memory is implanted by this same pairing of sound with NB stimulation. Fourth, directly increasing the area of representation artificially by pairing a tone with NB stimulation actually implants *de novo* specific behavioral auditory memory (Bieszczad et al., 2013).

The RP formed and the behavioral associative memory implanted by activation of the NB are indistinguishable from those that develop during natural learning. Overall, the findings satisfy cardinal criteria for a neural substrate of learning and memory (Martin et al., 2000).

The physiologic mode by which ACh acting at mAChRs may produce RP also has been identified, i.e., increasing neuronal synchrony, as detected in gamma oscillations and the covariance of discharge of separate neurons.

Future studies need not only to expand upon these findings but also to uncover the neural circuitry that normally activates the NB. One candidate is the basolateral amygdala, which projects to the NB and whose activation induces RP. However, numerous cortical structures project to the NB, including the frontal cortex, which probably is concerned with top-down control of ACh release.

Reconceptualizing the primary auditory cortex

Insofar as A1 is not simply an acoustic analyzer, what is it? We have summarized evidence for several mnemonic functions: (1) encoding the behavioral importance of sound; (2) reinforcement prediction; and (3) as a substrate of the strength of memory. Given this start, future research can be directed at the discovery of other potential functions, with the goal of achieving a comprehensive understanding of the functions of A1, and also of all other auditory cortical fields, and their modes of interaction. New conceptions of A1 have been advanced by two laboratories, which have approached the problem from different perspectives. Primary auditory cortex is viewed as a "semantic processor" (Scheich et al., 2011) or as an "acoustic problem solver" (Weinberger, 2011b). Both conceptions have converged on the similar formulation of A1 as a cortical field that is concerned with assigning meaning to sound, including all multimodal components of any auditory learning situation, be they other sensory, motivational, emotional, or the planning of behavioral acts. Regardless of the outcome, there can be no doubt it is essential to achieve a reconceptualization of the primary auditory cortex.

Toward a new model of the cerebral cortex

The cerebral cortex can no longer be considered to process information from the purely "sensory analytic" fields of A1, S1, and V1 because they are not merely sensory analyzers. It is speculative to consider that S1 and V1 have mnemonic and cognitive properties like those of A1, but not unreasonable given many similar reports for these two less-studied "early" sensory cortices (e.g., S1: Wiest et al, 2010; Kattoor et al., 2013; V1: Miller et al, 2008; Gavornik et al., 2009). Extensive research on the primary auditory cortex over the past 30 years or so, and particularly during the past decade, is incompatible with the view that its function is limited to the analysis of acoustic stimuli independent of their acquired behavioral significance or meaning. In short, responses of A1 neurons reflect both the physical and psychologic properties of sound. Therefore, primary auditory cortex should no longer be assumed to be merely an acoustic feature detector.

As A1 is influenced by and responds to multimodal inputs, and is involved in cognitive processes even beyond associative learning and memory, it would seem to perform highly complex, integrative computations. There is no reason to assume that A1 is the only cortical field so endowed. We suggest that the broad scope of A1, while specializing in auditory afferentation, is not fundamentally different from other cortical fields. Every field can be presumed to specialize in the processing of its input, the storage of somewhat specialized information, and the synthesis of both, to provide its special contribution to behavioral adaptation in concert with other fields.

While presently hypothetical, this concept suggests that the brain is more egalitarian and less authoritarian than assumed. This formulation does not eliminate hierarchic relationships but it does argue against the characterization of cortical fields as being isomorphic with common-sense or folk psychology ideas. This is the simplistic approach exemplified by Campbell's (1905) assumption that primary sensory fields are sensory analytic while "higher" adjacent cortical fields are cognitive ("psychic" in the terminology of his day). While specialized functions are maintained in this proposed model, they simply are not the functions that have been assumed. The challenge is not only to reconceptualize primary sensory cortices, but also to arrive at a far better model of a more dynamic, interactive cerebral cortex. We do not yet know how the computations of cortical fields should best be conceived. But that challenge has been with us for more than two decades. It is past time to refocus our thinking.

ACKNOWLEDGMENTS

I sincerely thank Gabriel K. Hui for preparation of this manuscript and figures, and Jacquie Weinberger for bibliographic and administrative assistance. It is a pleasure to recognize recent and current members of my laboratory whose research has contributed so much to the subject of this article: Kasia Bieszczad, Gabriel Elias, Drew Headley, and Alexandre Miasnikov. This work has been supported by research grants from the National Institutes of Health/National Institute on Deafness and Other Communication Disorders (NIDCD), DC-02938 and DC-010013 to N.M.W.

References

Arendash GW, Millard WJ, Dunn AJ et al. (1987). Long-term neuropathological and neurochemical effects of nucleus basalis lesions in the rat. Science 238 (4829): 952–956.

Bakin JS, Weinberger NM (1990). Classical conditioning induces CS-specific receptive field plasticity in the auditory cortex of the guinea pig. Brain Res 536 (1–2): 271–286.

Bakin JS, Weinberger NM (1996). Induction of a physiological memory in the cerebral cortex by stimulation of the nucleus basalis. Proc Natl Acad Sci U S A 93 (20): 11219–11224.

Bakin JS, Lepan B, Weinberger NM (1992). Sensitization induced receptive field plasticity in the auditory cortex is independent of CS-modality. Brain Res 577 (2): 226–235.

Bakin JS, South DA, Weinberger NM (1996). Induction of receptive field plasticity in the auditory cortex of the guinea pig during instrumental avoidance conditioning. Behav Neurosci 110 (5): 905–913.

Bao S, Chang EF, Woods J et al. (2004). Temporal plasticity in the primary auditory cortex induced by operant perceptual learning. Nat Neurosci 7 (9): 974–981.

Beitel RE, Schreiner CE, Cheung SW et al. (2003). Reward-dependent plasticity in the primary auditory cortex of adult monkeys trained to discriminate temporally modulated signals. Proc Natl Acad Sci U S A 100 (19): 11070–11075.

Bentley P, Driver J, Dolan RJ (2011). Cholinergic modulation of cognition: insights from human pharmacological functional neuroimaging. Prog Neurobiol 94: 360–388.

Berlau KM, Weinberger NM (2008). Learning strategy determines auditory cortical plasticity. Neurobiol Learn Mem 89 (2): 153–166.

Berntson GG, Tuber DS, Ronca AE et al. (1983). The decerebrate human: associative learning. Exp Neurol 81 (1): 77–88.

Berntson GG, Shafi R, Sarter M (2002). Specific contributions of the basal forebrain corticopetal cholinergic system to electroencephalographic activity and sleep/waking behaviour. Eur J Neurosci 16 (12): 2453–2461.

Bi GQ, Poo MM (1998). Synaptic modifications in cultured hippocampal neurons: dependence on spike timing, synaptic strength, and postsynaptic cell type. J Neurosci 18 (24): 10464–10472.

Bieszczad KM, Weinberger NM (2010a). Learning strategy trumps motivational level in determining learning-induced auditory cortical plasticity. Neurobiol Learn Mem 93 (2): 229–239.

Bieszczad KM, Weinberger NM (2010b). Remodeling the cortex in memory: increased use of a learning strategy increases the representational area of relevant acoustic cues. Neurobiol Learn Mem 94 (2): 127–144.

Bieszczad KM, Weinberger NM (2010c). Representational gain in cortical area underlies increase of memory strength. Proc Natl Acad Sci U S A 107 (8): 3793–3798.

Bieszczad KM, Weinberger NM (2012). Extinction reveals that primary sensory cortex predicts reinforcement outcome. Eur J Neurosci 35 (4): 598–613.

Bieszczad KM, Miasnikov AA, Weinberger NM (2013). Remodeling sensory cortical maps implants specific behavioral memory. Neuroscience 246: 40–51.

Bigl V, Woolf N, Butcher LL (1982). Cholinergic projections from the basal forebrain to frontal, parietal, temporal, occipital, and cingulate cortices: a combined fluorescent tracer and acetylcholinesterase analysis. Brain Res Bull 8 (6): 727–749.

Bjordahl TS, Dimyan MA, Weinberger NM (1998). Induction of long-term receptive field plasticity in the auditory cortex of the waking guinea pig by stimulation of the nucleus basalis. Behav Neurosci 112 (3): 467–479.

Blake DT, Strata F, Churchland AK et al. (2002). Neural correlates of instrumental learning in primary auditory cortex. Proc Natl Acad Sci U S A 99 (15): 10114–10119.

Blake DT, Heiser MA, Caywood M et al. (2006). Experience-dependent adult cortical plasticity requires cognitive association between sensation and reward. Neuron 52 (2): 371–381.

Boatman JA, Kim JJ (2006). A thalamo-cortico-amygdala pathway mediates auditory fear conditioning in the intact brain. Eur J Neurosci 24 (3): 894–900.

Börgers C, Epstein S, Kopell NJ (2005). Background gamma rhythmicity and attention in cortical local circuits: a computational study. Proc Natl Acad Sci U S A 102 (19): 7002–7007.

Börgers C, Epstein S, Kopell NJ (2008). Gamma oscillations mediate stimulus competition and attentional selection in a cortical network model. Proc Natl Acad Sci U S A 105 (46): 18023–18028.

Brechmann A, Scheich H (2005). Hemispheric shifts of sound representation in auditory cortex with conceptual listening. Cereb Cortex 15 (5): 578–587.

Brosch M, Selezneva E, Scheich H (2005). Nonauditory events of a behavioral procedure activate auditory cortex of highly trained monkeys. J Neurosci 25 (29): 6797–6806.

Brosch M, Selezneva E, Scheich H (2011). Representation of reward feedback in primate auditory cortex. Front Syst Neurosci 5: 5.

Buchwald JS, Halas ES, Schramm S (1966). Changes in cortical and subcortical unit activity during behavioral conditioning. Physiol Behav 1 (1): 11–22.

Butt AE, Chavez CM, Flesher MM et al. (2009). Association learning-dependent increases in acetylcholine release in the rat auditory cortex during auditory classical conditioning. Neurobiol Learn Mem 92 (3): 400–409.

Buzsáki G, Wang XJ (2012). Mechanisms of gamma oscillations. Annu Rev Neurosci 35: 203–225.

Campbell AW (1905). Histological Studies on the Localisation of Cerebral Function, Cambridge University Press, Cambridge UK.

Cape EG, Jones BE (1998). Differential modulation of high-frequency gamma-electroencephalogram activity and sleep–wake state by noradrenaline and serotonin microinjections into the region of cholinergic basalis neurons. J Neurosci 18 (7): 2653–2666.

Cape EG, Jones BE (2000). Effects of glutamate agonist versus procaine microinjections into the basal forebrain cholinergic cell area upon gamma and theta EEG activity and sleep–wake state. Eur J Neurosci 12 (6): 2166–2184.

Casamenti F, Deffenu G, Abbamondi AL et al. (1986). Changes in cortical acetylcholine output induced by modulation of the nucleus basalis. Brain Res Bull 16 (5): 689–695.

Cassady JM, Cole M, Thompson RF et al. (1973). Neural correlates of asymptotic avoidance and classical conditioned leg flexion. Exp Neurol 40 (1): 207–215.

Chavez CM, McGaugh JL, Weinberger NM (2009). The basolateral amygdala modulates specific sensory memory representations in the cerebral cortex. Neurobiol Learn Mem 91 (4): 382–392.

Chavez CM, McGaugh JL, Weinberger NM (2013). Activation of the basolateral amygdala induces long-term enhancement of specific memory representations in the cerebral cortex. Neurobiol Learn Mem 101: 8–18.

Chen CC, Williams CL (2012). Interactions between epinephrine, ascending vagal fibers, and central noradrenergic systems in modulating memory for emotionally arousing events. Front Behav Neurosci 6 (35): 1–20.

Chernyshev BV, Weinberger NM (1998). Acoustic frequency tuning of neurons in the basal forebrain of the waking guinea pig. Brain Res 793 (1–2): 79–94.

Clark KB, Smith DC, Hassert DL et al. (1998). Neurobiol Learn Mem 70 (3): 364–373.

Clark KB, Naritoku DK, Smith DC et al. (1999). Enhanced recognition memory following vagus nerve stimulation in human subjects. Nat Neurosci 2 (1): 94–98.

Demany L, Semal C (2002). Learning to perceive pitch differences. J Acoust Soc Am 111 (3): 1377–1388.

Deutsch JA (1971). The cholinergic synapse and the site of memory. Science 174 (4011): 788–794.

Diamond IT (1979). The subdivisions of the neocortex: A proposal to revise the traditional view of sensory, motor, and association areas. In: JM Sprague, AN Epstein (Eds.), Progress in Psychobiology and Physiological Psychology, Vol. 8, Chap. 1, Academic Press, New York, pp. 1–43.

Diamond DM, Weinberger NM (1984). Physiological plasticity of single neurons in auditory cortex of the cat during acquisition of the pupillary conditioned response: II. Secondary field (AII). Behav Neurosci 98 (2): 189–210.

Diamond DM, Weinberger NM (1986). Classical conditioning rapidly induces specific changes in frequency receptive fields of single neurons in secondary and ventral ectosylvian auditory cortical fields. Brain Res 372 (2): 357–360.

Diamond DM, Weinberger NM (1989). Role of context in the expression of learning-induced plasticity of single neurons in auditory cortex. Behav Neurosci 103 (3): 471–494.

DiCara LV, Braun JJ, Pappas BA (1970). Classical conditioning and instrumental learning of cardiac and gastrointestinal responses following removal of neocortex in the rat. J Comp Physiol Psychol 73 (2): 208–216.

Dimyan MA, Weinberger NM (1999). Basal forebrain stimulation induces discriminative receptive field plasticity in the auditory cortex. Behav Neurosci 113 (4): 691–702.

Dumenko VN, Sachenko VV (1978). Spontaneous activity of auditory cortical neurons of waking cats at rest and during defensive conditioning. Neurophysiology 10 (3): 159–168.

Dumenko VN, Sachenko VV (1979). Unit responses of the auditory cortex of waking cats at rest and after defensive conditioning. Neurophysiology 11 (1): 17–23.

Edeline J-M (1995). The alpha 2-adrenergic antagonist idazoxan enhances the frequency selectivity and increases the threshold of auditory cortex neurons. Exp Brain Res 107 (2): 221–240.

Edeline J-M (2003). The thalamo-cortical auditory receptive fields: regulation by the states of vigilance, learning and the neuromodulatory systems. Exp Brain Res 153 (4): 554–572.

Edeline J-M, Weinberger NM (1991). Thalamic short-term plasticity in the auditory system: associative returning of receptive fields in the ventral medial geniculate body. Behav Neurosci 105 (5): 618–639.

Edeline J-M, Weinberger NM (1993). Receptive field plasticity in the auditory cortex during frequency discrimination training: selective retuning independent of task difficulty. Behav Neurosci 107 (1): 82–103.

Edeline J-M, Neuenschwander-El Massioui N, Dutrieux G (1990). Discriminative long-term retention of rapidly induced multiunit changes in the hippocampus, medial geniculate and auditory cortex. Behav Brain Res 39 (2): 145–155.

Edeline J-M, Pham P, Weinberger NM (1993). Rapid development of learning-induced receptive field plasticity in the auditory cortex. Behav Neurosci 107 (4): 539–551.

Elias GA (2014). Associative Representational Plasticity in Primary Auditory Cortex: importance of Learning Strategies in Acquisition and Maintenance, Unpublished doctoral dissertation. University of California Irvine.

Engineer ND, Riley JR, Seale JD et al. (2011). Reversing pathological neural activity using targeted plasticity. Nature 470 (7332): 101–104.

Engineer ND, Møller AR, Kilgard MP (2013). Directing neural plasticity to understand and treat tinnitus. Hear Res 295: 58–66.

Fritz J, Shamma S, Elhilali M et al. (2003). Rapid task-related plasticity of spectrotemporal receptive fields in primary auditory cortex. Nat Neurosci 6: 1216–1223.

Fritz J, Elhilali M, Shamma S (2005). Active listening: task-dependent plasticity of spectrotemporal receptive fields in primary auditory cortex. Hear Res 206: 159–176.

Froemke RC, Carcea I, Barker AJ et al. (2013). Long-term modification of cortical synapses improves sensory perception. Nat Neurosci 16 (1): 79–88.

Galambos R, Sheatz G, Vernier VG (1956). Electrophysiological correlates of a conditioned response in cats. Science 123 (3192): 376–377.

Galván VV, Weinberger NM (2002). Long-term consolidation and retention of learning-induced tuning plasticity in the auditory cortex of the guinea pig. Neurobiol Learn Mem 77 (1): 78–108.

Galván VV, Chen J, Weinberger NM (2001). Long-term frequency tuning of local field potentials in the auditory cortex of the waking guinea pig. J Assoc Res Otolaryngol 2 (3): 199–215.

Gao E, Suga N (2000). Experience-dependent plasticity in the auditory cortex and the inferior colliculus of bats: role of the corticofugal system. Proc Natl Acad Sci U S A 97 (14): 8081–8086.

Gasanov UG, Galashina AG (1976). Study of plastic changes in cortical interneuronal connections. Pavlov J High Nerv Act 26: 820–827.

Gavornik JP, Shuler MG, Loewenstein Y et al. (2009). Learning reward timing in cortex through reward dependent expression of synaptic plasticity. Proc Natl Acad Sci U S A 106 (16): 6826–6831.

Gonzalez-Lima F, Scheich H (1984). Neural substrates for tone-conditioned bradycardia demonstrated with 2-deoxyglucose. I Activation of auditory nuclei. Behav Brain Res 14 (3): 213–233.

Gonzalez-Lima F, Scheich H (1986). Neural substrates for tone-conditioned bradycardia demonstrated with 2-deoxyglucose. II Auditory cortex plasticity. Behav Brain Res 20 (3): 281–293.

Gray CM, König P, Engel AK et al. (1989). Oscillatory responses in cat visual cortex exhibit inter-columnar synchronization which reflects global stimulus properties. Nature 338 (6213): 334–337.

Gregoriou GG, Gotts SJ, Zhou H et al. (2009). Long-range neural coupling through synchronization with attention. Prog Brain Res 176: 35–45.

Guo F, Intskirveli I, Blake DT et al. (2013). Tone-detection training enhances spectral integration mediated by intracortical pathways in primary auditory cortex. Neurobiol Learn Mem 101: 75–84.

Hall RD, Mark RG (1967). Fear and the modification of acoustically evoked potentials during conditioning. J Neurophysiol 30 (4): 893–910.

Hars B, Maho C, Edeline J-M et al. (1993). Basal forebrain stimulation facilitates tone-evoked responses in the auditory cortex of awake rat. Neuroscience 56 (1): 61–74.

Hawkey DJ, Amitay S, Moore DR (2004). Early and rapid perceptual learning. Nat Neurosci 7 (10): 1055–1056.

Headley DB, Weinberger NM (2011). Gamma-band activation predicts both associative memory and cortical plasticity. J Neurosci 31 (36): 12748–12758.

Headley DB, Weinberger NM (2013). Fear conditioning enhances gamma oscillations and their entrainment of neurons representing the conditioned stimulus. J Neurosci 33 (13): 5705–5717.

Hui GK, Wong KL, Chavez CM et al. (2009). Conditioned tone control of brain reward behavior produces highly specific representational gain in the primary auditory cortex. Neurobiol Learn Mem 92 (1): 27–34.

Irvine DR, Martin RL, Klimkeit E et al. (2000). Specificity of perceptual learning in a frequency discrimination task. J Acoust Soc Am 108 (6): 2964–2968.

James W (1893). The Principles of Psychology, H Holt, New York.

Jeschke M, Lenz D, Budinger E et al. (2008). Gamma oscillations in gerbil auditory cortex during a target-discrimination task reflect matches with short-term memory. Brain Res 1220: 70–80.

Ji W, Suga N (2003). Development of reorganization of the auditory cortex caused by fear conditioning: effect of atropine. J Neurophysiol 90 (3): 1904–1909.

Ji W, Gao E, Suga N (2001). Effects of acetylcholine and atropine on plasticity of central auditory neurons caused by conditioning in bats. J Neurophysiol 86 (1): 211–225.

Kaas JH, Hackett TA (1998). Subdivisions of auditory cortex and levels of processing in primates. Audiol Neurootol 3 (2–3): 73–85.

Kattoor J, Gizewski ER, Kotsis V et al. (2013). Fear conditioning in an abdominal pain model: Neural responses during associative learning and extinction in healthy subjects. PLoS One 8 (2): e51149.

Kellman PJ (2002). Perceptual learning. In: H Pashler, R Gallistel (Eds.), Learning, Motivation and Emotion, 3rd ed. John Wiley, Hoboken NJ, pp. 259–299.

Kilgard MP, Merzenich MM (1998). Cortical map reorganization enabled by nucleus basalis activity. Science 279 (5357): 1714–1718.

Kilgard MP, Pandya PK, Vazquez J et al. (2001a). Sensory input directs spatial and temporal plasticity in primary auditory cortex. J Neurophysiol 86 (1): 326–338.

Kilgard MP, Pandya PK, Vazquez JL et al. (2001b). Spectral features control temporal plasticity in auditory cortex. Audiol Neurootol 6 (4): 196–202.

Kilgard MP, Pandya PK, Engineer ND et al. (2002). Cortical network reorganization guided by sensory input features. Biol Cybern 87 (5–6): 333–343.

Kisley MA, Gerstein GL (1999). Trial-to-trial variability and state-dependent modulation of auditory-evoked responses in cortex. J Neurosci 19 (23): 10451–10460.

Kisley MA, Gerstein GL (2001). Daily variation and appetitive conditioning-induced plasticity of auditory cortex receptive fields. Eur J Neurosci 13 (10): 1993–2003.

Kitzes LM, Farley GR, Starr A (1978). Modulation of auditory cortex unit activity during the performance of a conditioned response. Exp Neurol 62 (3): 678–697.

LeDoux JE (1995). Emotion: clues from the brain. Annu Rev Psychol 46: 209–235.

Lennartz RC, Weinberger NM (1992). Analysis of response systems in Pavlovian conditioning reveals rapidly versus slowly acquired conditioned responses: support for two factors, implications for behavior and neurobiology. Psychobiology 20 (2): 93–119.

Lozano AM (2011). Harnessing plasticity to reset dysfunctional neurons. N Engl J Med 364 (14): 1367–1368.

Maho C, Hars B, Edeline J-M et al. (1995). Conditioned changes in the basal forebrain: relations with learning-induced cortical plasticity. Psychobiology 23 (1): 10–25.

Majkowski J, Sobieszek A (1975). Evolution of average evoked potentials in cats during conditioning before and after tegmental lesions. Physiol Behav 14 (2): 123–131.

Manunta Y, Edeline J-M (1997). Effects of noradrenaline on frequency tuning of rat auditory cortex neurons. Eur J Neurosci 9 (4): 833–847.

Manunta Y, Edeline J-M (1999). Effects of noradrenaline on frequency tuning of auditory cortex neurons during wakefulness and slow-wave sleep. Eur J Neurosci 11 (6): 2134–2150.

Manunta Y, Edeline J-M (2004). Noradrenergic induction of selective plasticity in the frequency tuning of auditory cortex neurons. J Neurophysiol 92 (3): 1445–1463.

Mark RG, Hall RD (1967). Acoustically evoked potentials in the rat during conditioning. J Neurophysiol 30 (4): 875–892.

Marsh JT, McCarthy DA, Sheatz G et al. (1961). Amplitude changes in evoked auditory potentials during habituation and conditioning. Electroencephalogr Clin Neurophysiol 13 (2): 224–234.

Martin SJ, Grimwood PD, Morris RG (2000). Synaptic plasticity and memory: an evaluation of the hypothesis. Annu Rev Neurosci 23: 649–711.

Masterton RB (1993). Central auditory system. ORL J Otorhinolaryngol Relat Spec 55 (3): 159–163.

McGaugh JL (2004). The amygdala modulates the consolidation of memories of emotionally arousing experiences. Annu Rev Neurosci 27: 1–28.

McIntyre CK, McGaugh JL, Williams CL (2012). Interacting brain systems modulate memory consolidation. Neurosci Biobehav Rev 36 (7): 1750–1762.

McLin 3rd DE, Miasnikov AA, Weinberger NM (2002). Induction of behavioral associative memory by stimulation of the nucleus basalis. Proc Natl Acad Sci U S A 99 (6): 4002–4007.

McLin 3rd DE, Miasnikov AA, Weinberger NM (2003). CS-specific gamma, theta, and alpha EEG activity detected in stimulus generalization following induction of behavioral memory by stimulation of the nucleus basalis. Neurobiol Learn Mem 79 (2): 152–176.

Mesulam M-M, Mufson EJ, Levey AI et al. (1983). Cholinergic innervation of cortex by the basal forebrain: cytochemistry and cortical connections of the septal area, diagonal band nuclei, nucleus basalis (substantia innominata), and hypothalamus in the rhesus monkey. J Comp Neurol 214 (2): 170–197.

Metherate R, Ashe JH (1991). Basal forebrain stimulation modifies auditory cortex responsiveness by an action at muscarinic receptors. Brain Res 559 (1): 163–167.

Metherate R, Ashe JH (1993). Nucleus basalis stimulation facilitates thalamocortical synaptic transmission in the rat auditory cortex. Synapse 14 (2): 132–143.

Metherate R, Cox CL, Ashe JH (1992). Cellular bases of neocortical activation: modulation of neural oscillations by the nucleus basalis and endogenous acetylcholine. J Neurosci 12 (12): 4701–4711.

Miasnikov AA, McLin 3rd D, Weinberger NM (2001). Muscarinic dependence of nucleus basalis induced conditioned receptive field plasticity. Neuroreport 12 (7): 1537–1542.

Miasnikov AA, Chen JC, Weinberger NM (2006). Rapid induction of specific associative behavioral memory by stimulation of the nucleus basalis in the rat. Neurobiol Learn Mem 86 (1): 47–65.

Miasnikov AA, Chen JC, Gross N et al. (2008a). Motivationally neutral stimulation of the nucleus basalis induces specific behavioral memory. Neurobiol Learn Mem 90 (1): 125–137.

Miasnikov AA, Chen JC, Weinberger NM (2008b). Specific auditory memory induced by nucleus basalis stimulation depends on intrinsic acetylcholine. Neurobiol Learn Mem 90 (2): 443–454.

Miasnikov AA, Chen JC, Weinberger NM (2009). Behavioral memory induced by stimulation of the nucleus basalis: effects of contingency reversal. Neurobiol Learn Mem 91 (3): 298–309.

Miasnikov AA, Chen JC, Weinberger NM (2011). Consolidation and long-term retention of an implanted behavioral memory. Neurobiol Learn Mem 95 (3): 286–295.

Miller MJ, Weiss C, Song X et al. (2008). Functional magnetic resonance imaging of delay and trace eyeblink conditioning in the primary visual cortex of the rabbit. J Neurosci 28 (19): 4974–4981.

Miltner WH, Braun C, Arnold M et al. (1999). Coherence of gamma-band EEG activity as a basis for associative learning. Nature 397 (6718): 434–436.

Mishkin M, Petrie HL (1984). Memories and habits: some implications for the analysis of learning and retention. In: LR Squire, N Butters (Eds.), The Neuropsychology of Memory, Guilford, New York, pp. 287–296.

Molchan SE, Sunderland T, McIntosh AR et al. (1994). A functional anatomical study of associative learning in humans. Proc Natl Acad Sci U S A 91 (17): 8122–8126.

Molnár M, Karmos G, Csépe V et al. (1988). Intracortical auditory evoked potentials during classical aversive conditioning in cats. Biol Psychol 26 (1–3): 339–350.

Morris JS, Friston KJ, Dolan RJ (1998). Experience-dependent modulation of tonotopic neural responses in human auditory cortex. Philos Trans R Soc Lond B Biol Sci 265 (1397): 649–657.

Oh JD, Edwards RH, Woolf NJ (1996). Choline acetyltransferase mRNA plasticity with Pavlovian conditioning. Exp Neurol 140 (1): 95–99.

Ohl FW, Scheich H (1996). Differential frequency conditioning enhances spectral contrast sensitivity of units in auditory cortex (field AI) of the alert Mongolian gerbil. Eur J Neurosci 8 (5): 1001–1017.

Ohl FW, Scheich H (1997). Learning-induced dynamic receptive field changes in primary auditory cortex of the unanaesthetized Mongolian gerbil. J Comp Physiol A 181 (6): 685–696.

Ohl FW, Scheich H (2004). Fallacies in behavioural interpretation of auditory cortex plasticity. Nat Rev Neurosci 5 (4). Published online November 20, 2004 at, http://www.nature.com/nrn/journal/v5/n4/full/nrn1366-c1.html.

Ohl FW, Scheich H (2005). Learning-induced plasticity in animal and human auditory cortex. Curr Opin Neurobiol 15 (4): 470–477.

Ohl FW, Scheich H, Freeman WJ (2001). Change in pattern of ongoing cortical activity with auditory category learning. Nature 412 (6848): 733–736.

Oleson TD, Ashe JH, Weinberger NM (1975). Modification of auditory and somatosensory system activity during pupillary conditioning in the paralyzed cat. J Neurophysiol 38 (5): 1114–1139.

Palmer CV, Nelson CT, Lindley 4th GA (1998). The functionally and physiologically plastic adult auditory system. J Acoust Soc Am 103 (4): 1705–1721.
Phillips DP, Hall SE (1990). Response timing constraints on the cortical representation of sound time structure. J Acoust Soc Am 88 (3): 1403–1411.
Phillips DP, Hall SE, Boehnke SE (2002). Central auditory onset responses, and temporal asymmetries in auditory perception. Hear Res 167 (1–2): 192–205.
Polley DB, Heiser MA, Blake DT et al. (2004). Associative learning shapes the neural code for stimulus magnitude in primary auditory cortex. Proc Natl Acad Sci U S A 101 (46): 16351–16356.
Polley DB, Steinberg EE, Merzenich MM (2006). Perceptual learning directs auditory cortical map reorganization through top-down influences. J Neurosci 26 (18): 4970–4982.
Puckett AC, Pandya DB, Moucha R et al. (2007). Plasticity in the rat posterior auditory field following nucleus basalis stimulation. J Neurophysiol 98 (1): 253–265.
Rauschecker JP (1999). Auditory cortical plasticity: a comparison with other sensory systems. Trends Neurosci 22 (2): 74–80.
Recanzone GH, Schreiner CE, Merzenich MM (1993). Plasticity in the frequency representation of primary auditory cortex following discrimination training in adult owl monkeys. J Neurosci 13 (1): 87–103.
Reed A, Riley J, Carraway R et al. (2011). Cortical map plasticity improves learning but is not necessary for improved performance. Neuron 70 (1): 121–131.
Richardson RT, DeLong MR (1986). Nucleus basalis of Meynert neuronal activity during a delayed response task in monkey. Brain Res 399 (2): 364–368.
Romanski LM, LeDoux JE (1992a). Bilateral destruction of neocortical and perirhinal projection targets of the acoustic thalamus does not disrupt auditory fear conditioning. Neurosci Lett 142 (2): 228–232.
Romanski LM, LeDoux JE (1992b). Equipotentiality of thalamo-amygdala and thalamo-cortico-amygdala circuits in auditory fear conditioning. J Neurosci 12 (11): 4501–4509.
Rutkowski RG, Weinberger NM (2005). Encoding of learned importance of sound by magnitude of representational area in primary auditory cortex. Proc Natl Acad Sci U S A 102 (38): 13664–13669.
Salinas E, Sejnowski TJ (2001). Correlated neuronal activity and the flow of neural information. Nat Rev Neurosci 2 (8): 539–550.
Saunders JC, Chabora JT (1969). Effects of appetitive drive on evoked potentials in cochlear nucleus and auditory cortex in cats. J Comp Physiol Psychol 69 (2): 355–361.
Scheich H, Brechmann A, Brosch M et al. (2011). Behavioral semantics of learning and crossmodal processing in auditory cortex: the semantic processor concept. Hear Res 271 (1–2): 3–15.
Schicknick H, Schott BH, Budinger E et al. (2008). Dopaminergic modulation of auditory cortex-dependent memory consolidation through mTOR. Cereb Cortex 18 (11): 2646–2658.
Schicknick H, Reichenbach N, Smalla KH et al. (2012). Dopamine modulates memory consolidation of discrimination learning in the auditory cortex. Eur J Neurosci 35 (5): 763–774.
Schreurs BG, McIntosh AR, Bahro M et al. (1997). Lateralization and behavioral correlation of changes in regional cerebral blood flow with classical conditioning of the human eyeblink response. J Neurophysiol 77 (4): 2153–2163.
Selezneva E, Scheich H, Brosch M (2006). Dual time scales for categorical decision making in auditory cortex. Curr Biol 16 (24): 2428–2433.
Squire LR, Stark CE, Clark RE (2004). The medial temporal lobe. Annu Rev Neurosci 27: 279–306.
Stark H, Scheich H (1997). Dopaminergic and serotonergic neurotransmission systems are differentially involved in auditory cortex learning: a long-term microdialysis study of metabolites. J Neurochem 68 (2): 691–697.
Suga N, Ma X (2003). Multiparametric corticofugal modulation and plasticity in the auditory system. Nat Rev Neurosci 4 (10): 783–794.
Teich AH, McCabe PM, Gentile CG et al. (1988). Role of auditory cortex in the acquisition of differential heart rate conditioning. Physiol Behav 44 (3): 405–412.
Teich AH, McCabe PM, Gentile CG et al. (1989). Auditory cortex lesions prevent the extinction of Pavlovian differential heart rate conditioning to tonal stimuli in rabbits. Brain Res 480 (1–2): 210–218.
Tervaniemi M, Rytkönen M, Schröger E et al. (2001). Superior formation of cortical memory traces for melodic patterns in musicians. Learn Mem 8 (5): 295–300.
Thiel CM (2003). Cholinergic modulation of learning and memory in the human brain as detected with functional neuroimaging. Neurobiol Learn Mem 80: 234–244.
Thiel CM, Friston JK, Dolan J (2002). Cholinergic modulation of experience-dependent plasticity in human auditory cortex. Neuron 35: 567–574.
van der Zee EA, Luiten PG (1999). Muscarinic acetylcholine receptors in the hippocampus, neocortex and amygdala: a review of immunocytochemical localization in relation to learning and memory. Prog Neurobiol 58 (5): 409–471.
Vanderwolf CH (1987). Near-total loss of 'learning' and 'memory' as a result of combined cholinergic and serotonergic blockade in the rat. Behav Brain Res 23 (1): 43–57.
Vazdarjanova A, McGaugh JL (1998). Basolateral amygdala is not critical for cognitive memory of contextual fear conditioning. Proc Natl Acad Sci U S A 95 (25): 15003–15007.
Vazdarjanova A, Cahill L, McGaugh JL (2001). Disrupting basolateral amygdala function impairs unconditioned freezing and avoidance in rats. Eur J Neurosci 14 (4): 709–718.
Weinberger NM (1995). Dynamic regulation of receptive fields and maps in the adult sensory cortex. Annu Rev Neurosci 18: 129–158.
Weinberger NM (2004). Experience-dependent response plasticity in the auditory cortex: issues, characteristics, mechanisms, and functions. In: TN Parks, EW Rubel, RR Fay et al. (Eds.), Plasticity of the Auditory System, Chap. 5.

Springer Handbook of Auditory Research, Vol. 23, Springer, New York, pp. 173–227.

Weinberger NM (2007). Associative representational plasticity in the auditory cortex: a synthesis of two disciplines. Learn Mem 14 (1–2): 1–16.

Weinberger NM (2008). The cognitive auditory cortex. In: AR Palmer, A Rees (Eds.), The Oxford Handbook of Auditory Science: The Auditory Brain, Oxford University Press, Oxford UK.

Weinberger NM (2011a). The medial geniculate, not the amygdala, as the root of auditory fear conditioning. Hear Res 274 (1–2): 61–74.

Weinberger NM (2011b). Reconceptualizing the primary auditory cortex: learning, memory and specific plasticity. In: JA Winer, CE Schreiner (Eds.), The Auditory Cortex, Chap. 22, Springer, New York, pp. 465–491.

Weinberger NM, Diamond DM (1987). Physiological plasticity in auditory cortex: rapid induction by learning. Prog Neurobiol 29 (1): 1–55.

Weinberger NM, Diamond DM (1988). Dynamic modulation of the auditory system by associative learning. In: GM Edelman, WE Gall, WM Cowan (Eds.), Auditory Function: Neurobiological Bases of Hearing, Chap. 16, John Wiley, New York, pp. 485–512.

Weinberger NM, Hopkins W, Diamond DM (1984). Physiological plasticity of single neurons in auditory cortex of the cat during acquisition of the pupillary conditioned response: I. Primary field (AI). Behav Neurosci 98 (2): 171–188.

Weinberger NM, Javid R, Lepan B (1993). Long-term retention of learning-induced receptive-field plasticity in the auditory cortex. Proc Natl Acad Sci U S A 90 (6): 2394–2398.

Weinberger NM, Miasnikov AA, Chen JC (2006). The level of cholinergic nucleus basalis activation controls the specificity of auditory associative memory. Neurobiol Learn Mem 86 (3): 270–285.

Weinberger NM, Miasnikov AA, Chen JC (2009). Sensory memory consolidation observed: increased specificity of detail over days. Neurobiol Learn Mem 91 (3): 273–286.

Weinberger NM, Miasnikov AA, Bieszczad KM et al. (2013). Gamma band plasticity in sensory cortex is a signature of the strongest memory rather than memory of the training stimulus. Neurobiol Learn Mem 104: 49–63.

Wespatat V, Tennigkeit F, Singer W (2004). Phase sensitivity of synaptic modifications in oscillating cells of rat visual cortex. J Neurosci 24 (41): 9067–9075.

Wetzel W, Wagner T, Ohl FW et al. (1998). Categorical discrimination of direction in frequency-modulated tones by Mongolian gerbils. Behav Brain Res 91 (1–2): 29–39.

Whalen PJ, Kapp BS, Pascoe JP (1994). Neuronal activity within the nucleus basalis and conditioned neocortical electroencephalographic activation. J Neurosci 14 (3 Pt 2): 1623–1633.

Wiest MC, Thomson E, Pantoja J et al. (2010). Changes in S1 neural responses during tactile discrimination learning. J Neurophysiol 104: 300–312.

Woods DL, Alain C (2009). Functional imaging of human auditory cortex. Curr Opin Otolaryngol Head Neck Surg 17 (5): 406–411.

Woody CD, Knispel JD, Crow TJ et al. (1976). Activity and excitability to electrical current of cortical auditory receptive neurons of awake cats as affected by stimulus association. J Neurophysiol 39 (5): 1045–1061.

Yan J, Zhang Y (2005). Sound-guided shaping of the receptive field in the mouse auditory cortex by basal forebrain activation. Eur J Neurosci 21 (2): 563–576.

Yin HH, Knowlton BJ (2006). The role of the basal ganglia in habit formation. Nat Rev Neurosci 7 (6): 464–476.

Zhou X, de Villers-Sidani E, Panizzutti R et al. (2010). Successive-signal biasing for a learned sound sequence. Proc Natl Acad Sci U S A 107 (33): 14839–14844.

Handbook of Clinical Neurology, Vol. 129 (3rd series)
The Human Auditory System
G.G. Celesia and G. Hickok, Editors

Chapter 8

Neural basis of speech perception

GREGORY HICKOK[1]* AND DAVID POEPPEL[2]

[1]*Department of Cognitive Sciences, Center for Language Science and Center for Cognitive Neuroscience, University of California, Irvine, CA, USA*

[2]*Department of Psychology, New York University, New York, NY, USA*

INTRODUCTION

The mind/brain must figure out at least two things when faced with the task of learning a natural language. One is how to transform the sound patterns of speech into a representation of the meaning of an utterance. The other is how to reproduce those sound patterns with the vocal tract (or, in the case of signed languages, with manual and facial gestures). Put differently, speech information must be processed along two different routes, an auditory-conceptual route and an auditory-motor route. These two processing streams involve partially segregated circuits in the brain and form the basis of the dual-route model of speech processing (Hickok and Poeppel, 2000, 2004, 2007), which traces its routes to the classic model of Wernicke (1874/1977), and parallels analogous proposals in the visual (Milner and Goodale, 1995) and somatosensory (Dijkerman and de Haan, 2007) systems. Thus, the division of labor proposed in dual-route models, wherein one route is sensory-conceptual and the other sensory-motor, appears to be a general organizational property of the cerebral cortex.

This chapter outlines the dual-route model as a foundation for understanding the functional anatomy of speech and language processing.

THE DUAL-ROUTE MODEL OF SPEECH PROCESSING

The dual-route model (Fig. 8.1) holds that a ventral stream, which involves structures in the superior and middle portions of the temporal lobe, is involved in processing speech signals for comprehension. A dorsal stream, which involves structures in the posterior planum temporale region (at the parietal-temporal junction) and the posterior frontal lobe, is involved in translating acoustic-based representations of speech signals into articulatory representations essential for speech production. In contrast to the canonical view that speech processing is mainly left-hemisphere-dependent, a wide range of evidence suggests that the ventral stream is bilaterally organized (although with important computational differences between the two hemispheres). The compelling extent to which neuroimaging data implicate both hemispheres has recently been reviewed (Turkeltaub and Coslett, 2010; Price, 2012; Schirmer et al., 2012). The dorsal stream, on the other hand, is traditionally, and in the model outlined here, held to be strongly left-dominant.

Ventral stream: mapping from sound to meaning

BILATERAL ORGANIZATION AND PARALLEL COMPUTATION

The ventral stream is bilaterally organized, although not computationally redundant in the two hemispheres. This may not be obvious based on a cursory evaluation of the clinical data. After all, left-hemisphere damage yields language deficits of a variety of sorts, including comprehension impairment, while, in most cases, right-hemisphere damage has little effect on phonologic, lexical, or sentence-level language abilities. A closer look tells a different story. In particular, research in the 1980s showed that auditory comprehension deficits in aphasia (caused by unilateral left-hemisphere lesions) were not caused primarily by impairment in the ability to perceive speech sounds, as Wernicke and later Luria proposed (Wernicke, 1874/1969; Luria, 1970). For example, when

*Correspondence to: Gregory Hickok, Department of Cognitive Sciences, University of California, Irvine, CA 92697-5100, USA. E-mail: greg.hickok@uci.edu

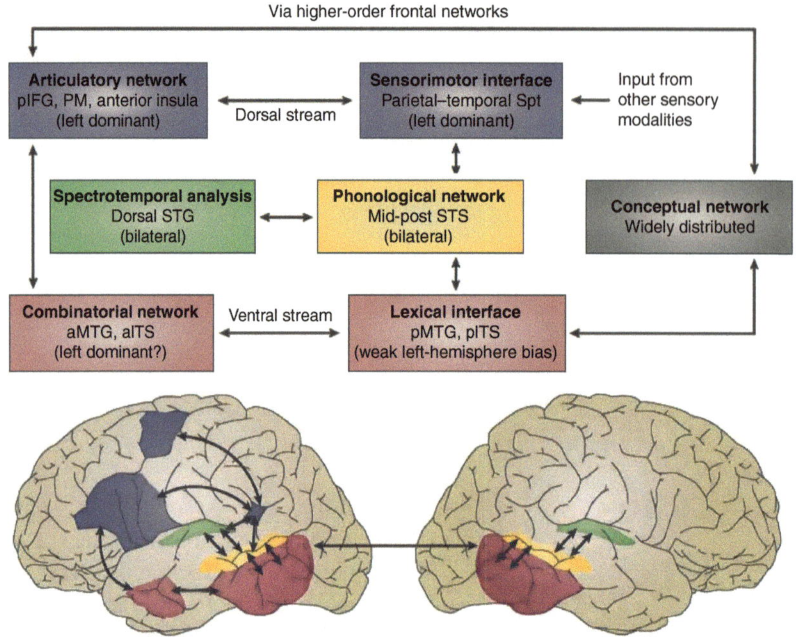

Fig. 8.1. Dual-stream model of speech processing. The dual-stream model (Hickok and Poeppel, 2000, 2004, 2007) holds that early stages of speech processing occur bilaterally in auditory regions on the dorsal superior temporal gyrus (STG) (spectrotemporal analysis: green) and superior temporal sulcus (STS) (phonologic access/representation: yellow), and then diverges into two broad streams: a temporal-lobe ventral stream supports speech comprehension (lexical access and combinatorial processes: pink) whereas a strongly left-dominant dorsal stream supports sensory-motor integration and involves structures at the Sylvian parietal–temporal (Spt) junction and frontal lobe. The conceptual network (gray box) is assumed to be widely distributed throughout cortex. IFG, inferior frontal gyrus; ITS, inferior temporal sulcus; MTG, middle temporal gyrus; PM, premotor; p, posterior; a, anterior. (Reproduced from Hickok and Poeppel, 2007, with permission.)

Wernicke's aphasics are asked to match pictures to auditorily presented words, their overall performance is well above chance, and when they do make errors they tend to confuse the correct answer with semantically similar alternatives more often than with phonemically similar foils (Miceli et al., 1980; Baker et al., 1981; Rogalsky et al., 2008, 2011). A similar pattern of performance has been observed following acute deactivation of the entire left hemisphere in Wada procedures (Fig. 8.2) (Hickok et al., 2008). "Speech perception" deficits can be identified in left-injured patients, but only on metalinguistic tasks, such as syllable discrimination, that involve some level of conscious attention to phonemic structure and working memory; the involvement of the left hemisphere in these tasks likely follows from the relation between working memory and speech articulation (Hickok and Poeppel, 2000, 2004, 2007). In contrast to the (minimal) effects of unilateral lesions on the processing of phoneme-level information during auditory comprehension, bilateral lesions involving the superior temporal lobe can have a devastating effect, as cases of word deafness attest (see Chapter 32) (Buchman et al., 1986; Poeppel, 2001).

Data from neuroimaging have been more controversial. One consistent and uncontroversial finding is that, when contrasted with a resting baseline, listening to speech activates the superior temporal gyrus (STG) bilaterally, including the dorsal STG and superior temporal sulcus (STS). However, when listening to connected, intelligible speech is contrasted with various acoustic baselines, some studies have reported left-dominant activation patterns (Scott et al., 2000; Narain et al., 2003), leading some authors to argue for a fully left-lateralized network for speech perception (Scott et al., 2000; Rauschecker and Scott, 2009). Other studies report bilateral activation even when acoustic controls are subtracted out of the activation pattern (Okada et al., 2010; for a review, see Hickok and Poeppel, 2007). The issue is still being actively debated within the functional imaging literature, although recent reviews and meta-analyses support the conjecture of bilateral STG/STS involvement (Turkeltaub and Coslett, 2010; Price, 2012; Schirmer et al., 2012).

Computational asymmetries

The hypothesis that sublexical-level processes in speech recognition are bilaterally organized does not imply that the two hemispheres are computationally identical. In fact there is strong evidence for hemispheric differences

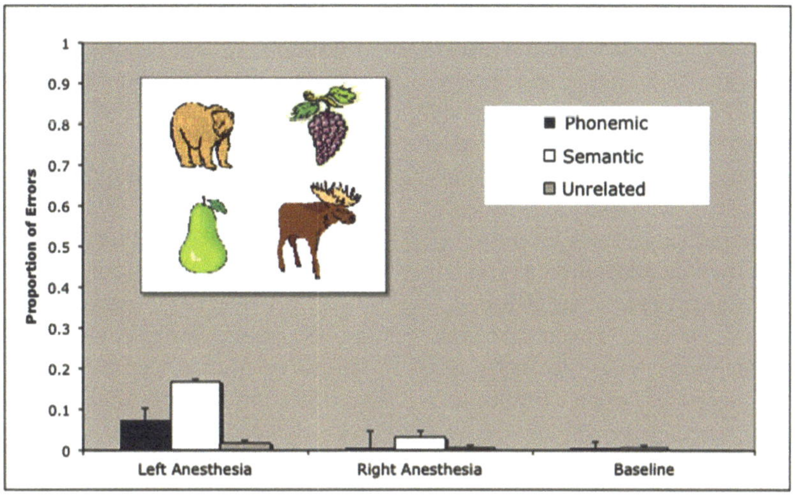

Fig. 8.2. Auditory comprehension performance during Wada procedure. Data show mean error rate on a four-alternative forced-choice auditory comprehension task with phonemic, semantic, and unrelated foils (inset shows sample stimulus card for the spoken target word, bear) in 20 patients undergoing clinically indicated Wada procedures. Error rate is shown as a function error type and amytal condition: left-hemisphere injection, right-hemisphere injection, and baseline. Note overall low error rate, even with left-hemisphere injection, and the dominance of semantic misselections when errors occur. (Modified from Hickok et al., 2008.)

in the processing of acoustic/speech information (Zatorre et al., 2002; Boemio et al., 2005; Giraud et al., 2007; Hickok and Poeppel, 2007; Abrams et al., 2008). The basis of these differences is currently being debated. One view, arguing for a domain-general perspective for all sounds, is that the difference turns on the selectivity for temporal (left-hemisphere) versus spectral (right-hemisphere) resolution (Zatorre et al., 2002; Obleser et al., 2008). That is, the left hemisphere may be particularly well suited for resolving rapid acoustic change (such as a formant transition), while the right hemisphere may have an advantage in resolving spectral frequency information. A closely related proposal is that the two hemispheres differ in terms of their preferred "sampling rate," with some left auditory cortical regions incorporating at bias for faster-rate (25–50 Hz) sampling and the right hemisphere for slower-rate sampling (4–8 Hz) (Poeppel, 2003). These two proposals are not incompatible as there is a relation between sampling rate and spectral vs temporal resolution: rapid sampling allows the system to detect changes that occur over short timescales, but sacrifices spectral resolution, and vice versa (Zatorre et al., 2002).

Further research is needed to address these hypotheses. For present purposes, the central point is that this asymmetry of function indicates that spoken word recognition involves parallel pathways – at least one in each hemisphere – in the mapping from sound to lexical meaning (Hickok and Poeppel, 2007), similar to well-accepted dual-route models of reading (phoneme-to-grapheme conversion and whole-word routes) (Coltheart et al., 1993). Although the parallel-pathway view differs from standard models of speech recognition (McClelland and Elman, 1986; Marslen-Wilson, 1987; Luce and Pisoni, 1998), wherein the processor proceeds from small to larger units in serial stages, it is consistent with the fact that speech contains redundant cues to phonemic information (e.g., in the speech envelope and fine spectral structure cues) and with behavioral evidence suggesting that the speech system can take advantage of these different cues (Remez et al., 1981; Shannon et al., 1995). It is worth bearing in mind that such computational asymmetries apply to all sounds that the auditory system analyzes. They reflect properties of neuronal ensembles that are like filters acting on any incoming signal. Specialization is likely to occur at the next stage at which signals are translated into a format suitable for lexical access.

Phonologic processing and the superior temporal sulcus

Beyond the earliest stages of speech recognition there is accumulating evidence that portions of the STS are important for representing and/or processing phonologic information (Price et al., 1996; Binder et al., 2000; Hickok and Poeppel, 2004, 2007; Indefrey and Levelt, 2004; Liebenthal et al., 2005). The STS is activated by language tasks that require access to phonologic information, including both the perception and production of speech (Indefrey and Levelt, 2004), and during active maintenance of phonemic information (Buchsbaum et al., 2001; Hickok et al., 2003). Portions of the STS seem to be relatively selective for acoustic signals that contain phonemic information when compared to complex non-speech signals (yellow shaded portion of Fig. 8.1) (Narain et al., 2003; Liebenthal et al., 2005; Hickok and Poeppel, 2007; Okada et al., 2010). STS activation can be modulated by the manipulation of psycholinguistic variables that tap phonologic networks (Okada and Hickok, 2006), such as phonologic neighborhood density (the number of words that sound similar to a target word), and this region shows neural adaptation effects to phonologic-level information (Vaden et al., 2010).

One currently unresolved question concerns the relative contribution of anterior versus posterior STS regions in phonologic processing. Lesion evidence indicates that damage to posterior temporal-lobe areas is most predictive of auditory comprehension deficits (Bates et al., 2003) and a majority of functional imaging studies targeting phonologic processing in perception have identified regions in the posterior half of the STS (Hickok and Poeppel, 2007). Other studies, however, have reported anterior STS activation in perceptual speech tasks (Mazoyer et al., 1993; Scott et al., 2000; Narain et al., 2003; Spitsyna et al., 2006). These studies typically involved sentence-level stimuli, raising the possibility that anterior STS regions may be responding to some other aspect of the stimulus, such as its syntactic or prosodic organization (Friederici et al., 2000; Humphries et al., 2001, 2005, 2006; Vandenberghe et al., 2002). Recent electrophysiologic work supports the hypothesis that the left anterior temporal lobe (ATL) is critical to elementary structure building (Bemis and Pylkkanen, 2011) in line with the view that intelligibility tasks tap into additional operations beyond speech recognition. It will, in any case, be important in future work to understand the role of various portions of the STS in auditory speech perception and language processing.

Lexical-semantic access

During auditory comprehension, the goal of speech processing is to use phonologic information to access words and conceptual-semantic representations that are critical to comprehension. The dual-stream model holds that conceptual-semantic representations are widely distributed throughout the cortex. However, a more focal system serves as a computational interface that maps between phonologic-level representations of words or morphologic roots and distributed conceptual representations (Hickok and Poeppel, 2000, 2004, 2007; Lau et al., 2008). This interface is not the site for storage of conceptual information. Instead, it is hypothesized to store information regarding the relation (or correspondences) between phonologic information on the one hand and conceptual information on the other. Most authors agree that the temporal lobes play a critical role in this process, but there is disagreement regarding the role of anterior versus posterior regions. The evidence for both of these viewpoints is briefly presented below.

Damage to posterior temporal-lobe regions, particularly along the middle temporal gyrus, has long been associated with auditory comprehension deficits (Damasio, 1991; Dronkers et al., 2000; Bates et al., 2003), an effect confirmed in a large-scale study involving 101 patients (Bates et al., 2003). We infer that these deficits are primarily postphonemic in nature, as phonemic deficits following unilateral lesions to this area are mild (Hickok and Poeppel, 2004). Data from direct cortical stimulation studies corroborate the involvement of the middle temporal gyrus in auditory comprehension, but also indicate the involvement of a much broader network involving most of the superior temporal lobe (including anterior portions) and the inferior frontal lobe (Miglioretti and Boatman, 2003). Functional imaging studies have also implicated posterior middle temporal regions in lexical-semantic processing (Binder et al., 1997; Rissman et al., 2003; Rodd et al., 2005). These findings do not preclude the involvement of more anterior regions in lexical-semantic access, but they do make a strong case for significant involvement of posterior regions. Electrophysiologic studies have successfully used paradigms building on the N400 response to study lexical-semantic processing. This response is very sensitive to a range of variables known to implicate lexical-level properties. A review of that literature (including source localization studies of the N400) also suggests that the posterior middle temporal gyrus plays a key role, although embedded in a network of anterior temporal, parietal, and inferior frontal regions (Lau et al., 2008).

ATL regions have been implicated both in lexical-semantic and sentence-level processing (syntactic and/or semantic integration processes). Patients with semantic dementia, who have been used to argue for a lexical-semantic function (Scott et al., 2000; Spitsyna et al., 2006), have atrophy involving the ATL bilaterally, along with deficits on lexical tasks, such as naming, semantic association, and single-word comprehension (Gorno-Tempini et al., 2004). However, these deficits are not specific to the mapping between phonologic and conceptual representations and indeed appear to involve more general semantic integration (Patterson et al., 2007). Further, given that atrophy in semantic dementia involves a number of regions in addition to the lateral ATL, including bilateral inferior and medial temporal lobe, bilateral caudate nucleus, and right posterior thalamus, among others (Gorno-Tempini et al., 2004), linking the deficits specifically to the ATL is difficult.

Higher-level syntactic and compositional semantic processing might involve the ATL. Functional imaging studies have found portions of the ATL to be more active while subjects listen to or read sentences rather than unstructured lists of words or sounds (Mazoyer et al., 1993; Friederici et al., 2000; Humphries et al., 2001, 2005; Vandenberghe et al., 2002). This structured-versus-unstructured effect is independent of the semantic content of the stimuli, although semantic manipulations can modulate the ATL response somewhat (Vandenberghe et al., 2002). Recent electrophysiologic data (e.g. Brennan and Pylkkanen, 2012; Bemis and Pylkkanen, 2013) also implicate the left ATL in elementary structure building. Damage to the ATL has also been linked to deficits in comprehending complex syntactic structures (Dronkers et al., 2004). However, data from semantic dementia are contradictory, as these patients are reported to have good sentence-level comprehension (Gorno-Tempini et al., 2004).

In summary, there is strong evidence that lexical-semantic access from auditory input involves the posterior lateral temporal lobe. In terms of syntactic and compositional semantic operations, neuroimaging evidence is converging on the ATL as an important component of the computational network (Vandenberghe et al., 2002; Humphries et al., 2005, 2006); however, the neuropsychologic evidence remains equivocal.

Dorsal stream: mapping from sound to action

The earliest proposals regarding the dorsal auditory stream argued that this system was involved in spatial hearing, a "where" function (Rauschecker, 1998), similar to the dorsal "where" stream proposal in the cortical visual system (Ungerleider and Mishkin, 1982). More recently, there has been some convergence on the idea that the dorsal stream supports auditory-motor integration (Hickok and Poeppel, 2000, 2004, 2007; Wise et al., 2001; Scott and Wise, 2004; Rauschecker and Scott, 2009; Rauschecker, 2011). Specifically, the idea is that the auditory dorsal stream supports an interface between auditory and motor representations of speech, a proposal similar to the claim that the dorsal visual stream has a sensory-motor integration function (Milner and Goodale, 1995; Andersen, 1997).

The need for auditory-motor integration

The idea of auditory-motor interaction in speech is not new. Wernicke's classic model of the neural circuitry of language incorporated a direct link between sensory and motor representations of words and argued explicitly that sensory systems participated in speech production (Wernicke, 1874/1969). More recently, research on motor control has revealed why this sensory-motor link is critical. Motor acts aim to hit sensory targets. In the visual-manual domain, we identify the location and shape of a cup visually (the sensory target) and generate a motor command that allows us to move our limb toward that location and shape the hand to match the shape of the object. In the speech domain, the targets are not external objects but internal representations of the sound pattern (phonologic form) of a word. We know that the targets are auditory in nature because manipulating a speaker's auditory feedback during speech production results in compensatory changes in motor speech acts (Houde and Jordan, 1998; Larson et al., 2001; Purcell and Munhall, 2006). For example, if a subject is asked to produce one vowel and the feedback that she hears is manipulated so that it sounds like another vowel, the subject will change the vocal tract configuration so that the feedback sounds like the original vowel. In other words, talkers will readily modify their motor articulations to hit an auditory target, indicating that the goal of speech production is not a particular motor configuration but rather a speech sound (Guenther et al., 1998). The role of auditory input is nowhere more apparent than in development, where the child must use acoustic information in the linguistic environment to shape vocal tract movements that must reproduce those sounds.

A great deal of progress has been made in mapping the neural organization of sensorimotor integration for speech. Early functional imaging studies identified an auditory-related area in the left planum temporale region as involved in speech production (Hickok et al., 2000; Wise et al., 2001). Subsequent studies showed that this left-dominant region, dubbed Spt for its location in the Sylvian fissure at the parietal–temporal boundary (Fig. 8.3A) (Hickok et al., 2003), exhibited a number of properties characteristic of sensorimotor integration areas such as those found in macaque parietal cortex (Andersen, 1997; Colby and Goldberg, 1999). Most fundamentally, Spt exhibits sensorimotor response properties, activating both during the passive perception of speech and during covert (subvocal) speech articulation (Buchsbaum et al., 2001, 2005; Hickok et al., 2003), and further that different subregional patterns of activity are apparent during the sensory and motor phases of the task (Hickok et al., 2009), likely reflecting the activation of different neuronal subpopulations (Dahl et al., 2009), some sensory- and others motor-weighted. Figure 8.3B–D shows examples of the sensory-motor response properties of Spt and the patchy organization of this region for sensory- versus motor-weighted voxels (Fig. 8.3C, inset).

Spt is not speech-specific; its sensorimotor responses are equally robust when the sensory stimulus consists of

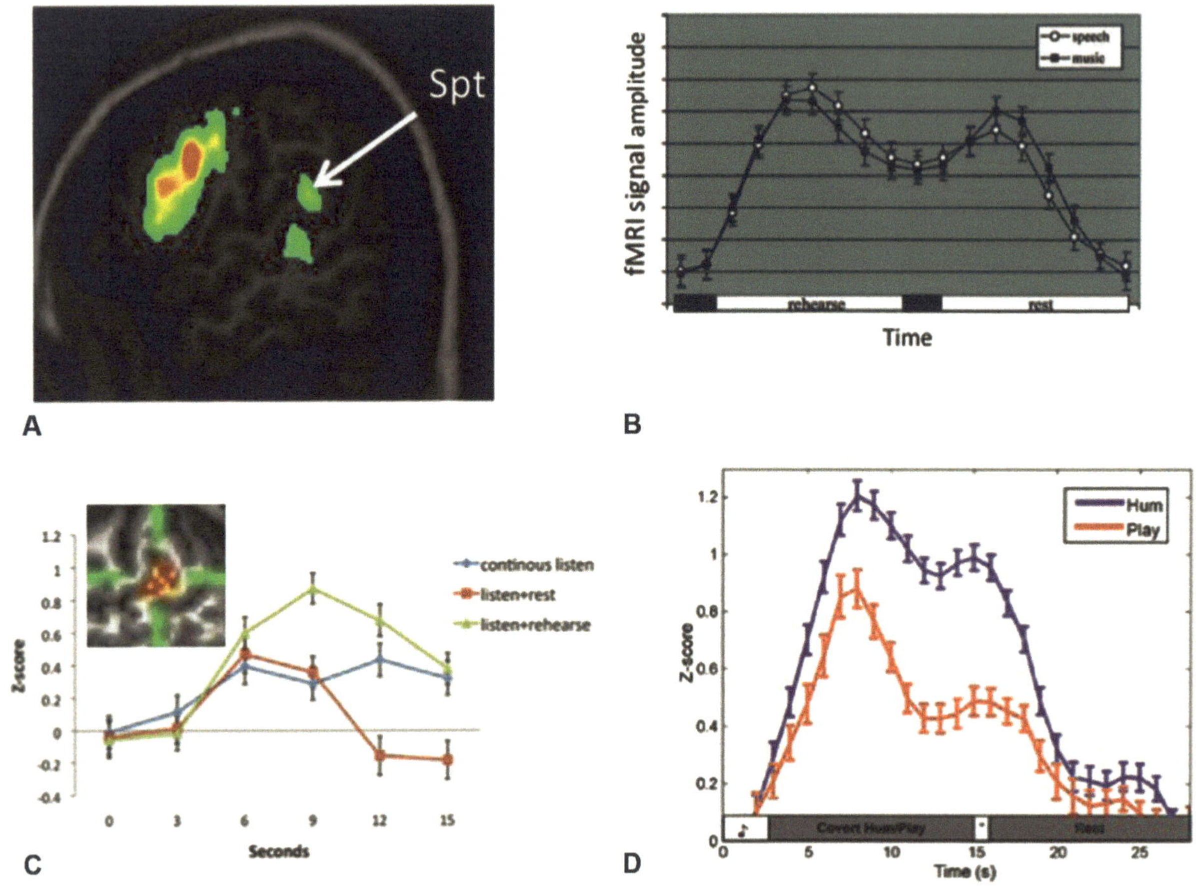

Fig. 8.3. Location and functional properties of area Sylvian parietal–temporal (Spt) junction. (**A**) Activation map for covert speech articulation (rehearsal of a set of non-words). (**B**) Activation timecourse (functional magnetic resonance imaging (fMRI) signal amplitude) in Spt during a sensorimotor task for speech and music. A trial is composed of 3 seconds of auditory stimulation followed by 15 seconds of covert rehearsal/humming of the heard stimulus, followed by 3 seconds of auditory stimulation, followed by 15 seconds of rest. The two humps represent the sensory responses, the valley between the humps is the motor (covert rehearsal) response, and the baseline values at the onset and offset of the trial reflect resting activity levels. Note similar response to both speech and music. (Adapted from Hickok et al., 2003.) (**C**) Activation timecourse in Spt in three conditions: continuous speech (15 seconds, blue curve), listen + rest (3 seconds speech, 12 seconds rest, red curve), and listen + covert rehearse (3 seconds speech, 12 seconds rehearse, green curve). The pattern of activity within Spt (inset) was found to be different for listening to speech compared to rehearsing speech assessed at the end of the continuous listen versus listen + rehearse conditions despite the lack of a significant signal amplitude difference at that time point. (Adapted from Hickok et al., 2009.) (**D**) Activation timecourse in Spt in skilled pianists performing a sensorimotor task involving listening to novel melodies and then covertly humming them (blue curve) vs listening to novel melodies and imagining playing them on a keyboard (red curve). This indicates that Spt is relatively selective for vocal tract actions. (Reproduced with permission from Hickok, 2009.) (Reproduced from Hickok and Buchsbaum, 2003.)

tonal melodies and (covert) humming is the motor task (see the two curves in Fig. 8.3B) (Hickok et al., 2003). Activity in Spt is highly correlated with activity in the pars opercularis (Buchsbaum et al., 2001, 2005), which is the posterior sector of Broca's region. White-matter tracts identified via diffusion tensor imaging suggest that Spt and the pars opercularis are densely connected anatomically (for review, see Friederici, 2009; Rogalsky and Hickok, 2011). Finally, consistent with some sensorimotor integration areas in the monkey parietal lobe (Andersen, 1997; Colby and Goldberg, 1999), Spt appears to be motor-effector-selective, responding more robustly when the motor task involves the vocal tract than the manual effectors (Fig. 8.2D) (Pa and Hickok, 2008). More broadly, Spt is situated in the middle of a network of auditory (STS) and motor (pars opercularis, premotor cortex) regions (Buchsbaum et al., 2001, 2005; Hickok et al., 2003), perfectly positioned both functionally and anatomically to support sensorimotor integration for speech and related vocal tract functions.

Lesion evidence is consistent with the functional imaging data implicating Spt as part of a sensorimotor integration circuit. Damage to auditory-related regions in the left hemisphere often results in speech production deficits (H. Damasio, 1991; A. R. Damasio, 1992), demonstrating that sensory systems participate in motor speech. More specifically, damage to the left temporal–parietal junction is associated with conduction aphasia, a syndrome that is characterized by good comprehension, but frequent phonemic errors in speech production (Damasio and Damasio, 1980; Goodglass, 1992; Baldo et al., 2008), and the lesion distribution overlaps with the location of functional area Spt (Fig. 8.4) (Buchsbaum et al., 2011). Conduction aphasia has classically been considered to be a disconnection syndrome involving damage to the arcuate fasciculus. However, there is now good evidence that this syndrome results from cortical dysfunction (Anderson et al., 1999; Hickok et al., 2000). The production deficit is load-sensitive: errors are more likely on longer, lower-frequency words, and verbatim repetition of strings of speech with little semantic constraint (Goodglass, 1992, 1993). In the context of the above discussion, the effects of such lesions can be understood as an interruption of the system that serves at the interface between auditory target and the motor speech actions that can achieve them (Hickok and Poeppel, 2000, 2004, 2007).

Recent theoretic work has clarified the computational details underlying auditory-motor integration in the dorsal stream. Drawing on advances in understanding motor control generally, speech researchers have emphasized the role of internal forward models in speech motor control (Golfinopoulos et al., 2010; Hickok et al., 2011; Houde and Nagarajan, 2011). The basic idea is that the nervous system makes forward predictions about the future state of the motor articulators and the sensory consequences of the predicted actions to control action. The predictions are assumed to be generated by an internal model that receives copies of motor commands and integrates them with information about the current state of the system and past experience (learning) of the relation between particular motor commands and their sensory consequences. This internal model affords a mechanism for detecting and correcting motor errors, i.e., motor actions that fail to hit their sensory targets.

Several models have been proposed with similar basic assumptions, but slightly different architectures (Guenther et al., 1998; Golfinopoulos et al., 2010; Hickok et al., 2011; Houde and Nagarajan, 2011). One such model is shown in Figure 8.5 (Hickok et al., 2011). Input to the system comes from a lexical-conceptual network as assumed by psycholinguistic models of speech production (Dell et al., 1997; Levelt et al., 1999). In between the input/output system is a phonologic system that is split into two components, corresponding to sensory input and motor output subsystems and mediated by a sensorimotor translation system, which corresponds to area Spt (Buchsbaum et al., 2001; Hickok et al., 2003, 2009). Parallel inputs to sensory and motor systems are needed to explain neuropsychologic observations (Jacquemot et al., 2007), such as conduction aphasia, as we will see below. Inputs to the auditory-phonologic network define the auditory targets of speech acts. As a motor speech unit (ensemble) begins to be activated, its predicted auditory consequences can be checked against the auditory target. If they match, then that unit will continue to be activated, resulting in an articulation that will hit the target. If there is a mismatch, then a correction signal can be generated to activate the correct motor unit.

This model provides a natural explanation of conduction aphasia. A lesion to Spt would disrupt the ability to generate forward predictions in auditory cortex and

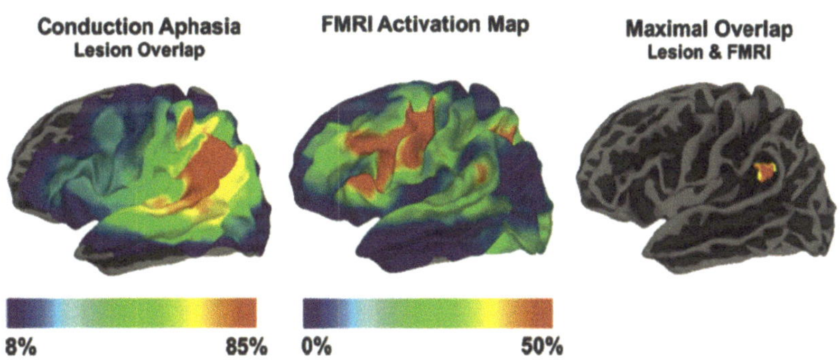

Fig. 8.4. Relation between lesions associated with conduction aphasia and the cortical auditory-motor network. A comparison of conduction aphasia, an auditory-motor task (listening to and then repeating back speech) in functional magnetic resonance imaging (fMRI), and their overlap. The uninflated surface in the left panel shows the regional distribution lesion overlap in patients with conduction aphasia (maximum is 12/14 or 85% overlap). The middle panel shows the auditory-motor network in the fMRI analysis. The right panel shows the area of maximal overlap between the lesion and fMRI surfaces (lesion >85% overlap and significant fMRI activity). (Modified from Buchsbaum et al., 2011.)

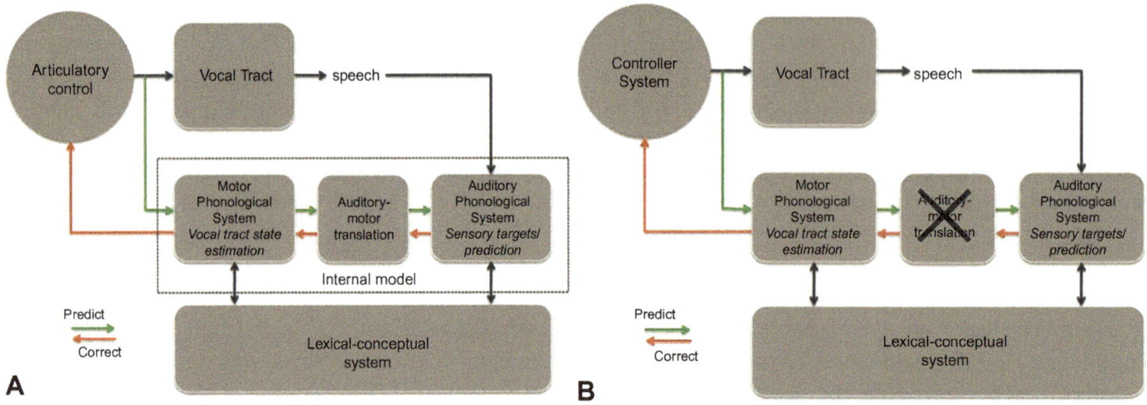

Fig. 8.5. An integrated state feedback control (SFC) model of speech production. (**A**) Speech models derived from the feedback control, psycholinguistic, and neurolinguistic literatures are integrated into one framework, presented here. The architecture is fundamentally that of an SFC system with a controller, or set of controllers (Haruno et al., 2001), localized to primary motor cortex, which generates motor commands to the vocal tract and sends a corollary discharge to an internal model, which makes forward predictions about both the dynamic state of the vocal tract and about the sensory consequences of those states. Deviations between predicted auditory states and the intended targets or actual sensory feedback generate an error signal that is used to correct and update the internal model of the vocal tract. The internal model of the vocal tract is instantiated as a "motor phonologic system," which corresponds to the neurolinguistically elucidated phonologic output lexicon, and is localized to premotor cortex. Auditory targets and forward predictions of sensory consequences are encoded in the same network, namely the "auditory phonologic system," which corresponds to the neurolinguistically elucidated phonologic input lexicon, and is localized to the superior temporal gyrus/superior temporal sulcus. Motor and auditory phonologic systems are linked via an auditory-motor translation system, localized to the Sylvian parietal–temporal junction. The system is activated via parallel inputs from the lexical-conceptual system to the motor and auditory phonologic systems. (**B**) Proposed source of the deficit in conduction aphasia: damage to the auditory-motor translation system. Input from the lexical-conceptual system to motor and auditory phonologic systems is unaffected, allowing for fluent output and accurate activation of sensory targets. However, internal forward sensory predictions are not possible, leading to an increase in error rate. Further, errors detected as a consequence of mismatches between sensory targets and actual sensory feedback cannot be used to correct motor commands. (Reproduced with permission from Hickok and Poeppel, 2004; Hickok et al., 2011.)

thereby the ability to perform internal feedback monitoring, making errors more frequent than in an unimpaired system (Fig. 8.5B). However, this would not disrupt the activation of auditory targets via the lexical-semantic system, thus leaving patients capable of detecting errors in their own speech, a characteristic of conduction aphasia. Once an error is detected however, the correction signal will not be accurately translated to the internal model of the vocal tract due to disruption of Spt. The ability to detect but not accurately correct speech errors should result in repeated unsuccessful self-correction attempts, again a characteristic of conduction aphasia.

CLINICAL CORRELATES OF THE DUAL-STREAM MODEL

The dual-stream model, like the classic Wernicke–Lichtheim model, provides an account of the major clinical aphasia syndromes (Hickok and Poeppel, 2004). Within the dual-stream model, Broca's aphasia and conduction aphasia are considered to be dorsal stream-related syndromes, while Wernicke's aphasia, word deafness, and transcortical sensory aphasia are considered ventral-stream syndromes. We have already noted that conduction aphasia can be conceptualized as a disruption of auditory-motor integration resulting from damage to area Spt. Broca's aphasia can be viewed as a disruption to representations that code for speech-related actions at multiple levels, from coding low-level phonetic features, to sequences of syllables, to sequences of words in structured sentences. Although Broca's area and Broca's aphasia are widely considered to be associated with deficits in receptive syntactic processing (Caramazza and Zurif, 1976; Grodzinsky, 2000), this issue is now being seriously questioned and remains debatable (Rogalsky and Hickok, 2011).

Word deafness is the "lowest-level" ventral-stream syndrome, according to the dual-stream model, affecting the processing of phonemic information during speech recognition. This differs from classic interpretations of word deafness as a disconnection syndrome (Geschwind, 1965). Due to the key role that auditory systems play in speech production, as discussed above, we should expect that disruption to auditory speech systems, as in word deafness, will impact production as well. Although the canonic description of word deafness is a syndrome in which speech production is preserved, the majority of case descriptions that provide information on the speech output of word-deaf patients report the presence of paraphasic errors (Buchman et al., 1986).

Wernicke's aphasia is explained in terms of damage to multiple ventral-stream processing levels in the dual-stream model. Given the rather extensive posterior lesions that are typically required to yield a chronic Wernicke's aphasia (Dronkers and Baldo, 2009), it is likely that this syndrome results from damage to auditory-motor area Spt, left-hemisphere auditory areas, and posterior middle temporal lexical-semantic interface systems. Such damage can explain the symptom complex: relatively good phonologic-level speech recognition (due to the bilateral organization, as described above), poor comprehension at the higher semantic level (due to damage to lexical-semantic interface systems), fluent speech (due to preserved motor-speech systems), poor repetition (due to disruption of auditory-motor interface network), and paraphasic errors (due to disruption of auditory motor interface network).

Transcortical sensory aphasia, which is similar to Wernicke's aphasia but with preserved repetition, is conceptualized as a functionally more focal deficit involving the lexical-semantic interface network but sparing the auditory-motor network. Damage to the lexical-semantic interface explains the poor comprehension, while sparing of the auditory-motor interface explains the preserved repetition.

SEX DIFFERENCES IN LANGUAGE ORGANIZATION

Substantial evidence exists for sexual dimorphism in the brain (Cahill, 2006), which raises the question of whether there are sex differences in organization within the dorsal and/or ventral speech streams. This issue has not been thoroughly investigated, in part because existing evidence for sex differences in language-related brain function has not yielded consistent results (Wallentin, 2009). More work is needed to address this question.

SUMMARY

Dual-stream models of cortical organization have proven useful in understanding both language and visual-related systems and indeed have been a recurrent theme in neural models stretching back more than a century (Wernicke, 1874/1977). Thus, the general concept underlying the model – that the brain must interface sensory information with two different systems, conceptual and motor – is not only intuitively appealing but has a proven track record across domains. In the language domain, the dual-stream model provides an explanation of classic language disorders (Hickok and Poeppel, 2004; Hickok et al., 2011) and provides a framework for integrating and unifying research across psycholinguistic, neurolinguistic, and neurophysiologic traditions. Recent work has shown that still further integration with motor control models is possible (Hickok et al., 2011). All of this suggests that the dual-stream framework is on the right track as a model of language organization and provides a rich context for guiding future research.

REFERENCES

Abrams DA, Nicol T, Zecker S et al. (2008). Right-hemisphere auditory cortex is dominant for coding syllable patterns in speech. J Neurosci 28 (15): 3958–3965.

Andersen R (1997). Multimodal integration for the representation of space in the posterior parietal cortex. Phil Trans Roy Soc Lond B Bio Sci 352: 1421–1428.

Anderson JM, Gilmore R, Roper S et al. (1999). Conduction aphasia and the arcuate fasciculus: A reexamination of the Wernicke-Geschwind model. Brain Lang 70: 1–12.

Baker E, Blumsteim SE, Goodglass H (1981). Interaction between phonological and semantic factors in auditory comprehension. Neuropsychologia 19: 1–15.

Baldo JV, Klostermann EC, Dronkers NF (2008). It's either a cook or a baker: patients with conduction aphasia get the gist but lose the trace. Brain Lang 105 (2): 134–140.

Bates E, Wilson SM, Saygin AP et al. (2003). Voxel-based lesion-symptom mapping. Nat Neurosci 6 (5): 448–450.

Bemis DK, Pylkkanen L (2011). Simple composition: a magnetoencephalography investigation into the comprehension of minimal linguistic phrases. J Neurosci 31 (8): 2801–2814.

Bemis DK, Pylkkanen L (2013). Basic linguistic composition recruits the left anterior temporal lobe and left angular gyrus during both listening and reading. Cereb Cortex 23 (8): 1859–1873.

Binder JR, Frost JA, Hammeke TA et al. (1997). Human brain language areas identified by functional magnetic resonance imaging. J Neurosci 17: 353–362.

Binder JR, Frost JA, Hammeke TA et al. (2000). Human temporal lobe activation by speech and nonspeech sounds. Cereb Cortex 10: 512–528.

Boemio A, Fromm S, Braun A et al. (2005). Hierarchical and asymmetric temporal sensitivity in human auditory cortices. Nat Neurosci 8 (3): 389–395.

Brennan J, Pylkkanen L (2012). The time-course and spatial distribution of brain activity associated with sentence processing. Neuroimage 60 (2): 1139–1148.

Buchman AS, Garron DC, Trost-Cardamone JE et al. (1986). Word deafness: One hundred years later. J Neurol Neurosurg Psychiatr 49: 489–499.

Buchsbaum B, Hickok G, Humphries C (2001). Role of left posterior superior temporal gyrus in phonological processing for speech perception and production. Cognit Sci 25: 663–678.

Buchsbaum BR, Olsen RK, Koch P et al. (2005). Human dorsal and ventral auditory streams subserve rehearsal-based and echoic processes during verbal working memory. Neuron 48 (4): 687–697.

Buchsbaum BR, Baldo J, Okada K et al. (2011). Conduction aphasia, sensory-motor integration, and phonological short-term memory – an aggregate analysis of lesion and fMRI data. Brain Lang 119 (3): 119–128.

Cahill L (2006). Why sex matters for neuroscience. Nat Rev Neurosci 7 (6): 477–484.

Caramazza A, Zurif EB (1976). Dissociation of algorithmic and heuristic processes in sentence comprehension: Evidence from aphasia. Brain Lang 3: 572–582.

Colby CL, Goldberg ME (1999). Space and attention in parietal cortex. Annu Rev Neurosci 22: 319–349.

Coltheart M, Curtis B, Atkins P et al. (1993). Models of reading aloud: Dual-route and parallel-distributed-processing approaches. Psychol Rev 100: 589–608.

Dahl CD, Logothetis NK, Kayser C (2009). Spatial organization of multisensory responses in temporal association cortex. J Neurosci 29 (38): 11924–11932.

Damasio H (1991). Neuroanatomical correlates of the aphasias. In: M Sarno (Ed.), Acquired aphasia, 2nd edn. Academic Press, San Diego, pp. 45–71.

Damasio AR (1992). Aphasia. New England Journal of Medicine 326: 531–539.

Damasio H, Damasio AR (1980). The anatomical basis of conduction aphasia. Brain 103: 337–350.

Dell GS, Schwartz MF, Martin N et al. (1997). Lexical access in aphasic and nonaphasic speakers. Psychol Rev 104: 801–838.

Dijkerman HC, de Haan EH (2007). Somatosensory processes subserving perception and action. Behav Brain Sci 30 (2): 189–201, discussion 201–139.

Dronkers N, Baldo J (2009). Language: Aphasia. In: LR Squire (Ed.), Encyclopedia of Neuroscience, Vol. 5. Academic Press, Oxford, pp. 343–348.

Dronkers NF, Redfern BB, Knight RT (2000). The neural architecture of language disorders. In: MS Gazzaniga (Ed.), The new cognitive neurosciences, MIT Press, Cambridge, MA, pp. 949–958.

Dronkers NF, Wilkins DP, Van Valin Jr RD et al. (2004). Lesion analysis of brain regions involved in language comprehension. Cognition 92: 145–177.

Friederici AD (2009). Pathways to language: fiber tracts in the human brain. Trends Cogn Sci 13 (4): 175–181.

Friederici AD, Meyer M, von Cramon DY (2000). Auditory languge comprehension: An event-related fMRI study on the processing of syntactic and lexical information. Brain Lang 74: 289–300.

Geschwind N (1965). Disconnexion syndromes in animals and man. Brain 88 (237–294): 585–644.

Giraud AL, Kleinschmidt A, Poeppel D et al. (2007). Endogenous cortical rhythms determine cerebral specialization for speech perception and production. Neuron 56 (6): 1127–1134.

Golfinopoulos E, Tourville JA, Guenther FH (2010). The integration of large-scale neural network modeling and functional brain imaging in speech motor control. Neuroimage 52 (3): 862–874.

Goodglass H (1992). Diagnosis of conduction aphasia. In: SE Kohn (Ed.), Conduction aphasia, Lawrence Erlbaum Associates, Hillsdale, NJ, pp. 39–49.

Goodglass H (1993). Understanding aphasia, Academic Press, San Diego.

Gorno-Tempini ML, Dronkers NF, Rankin KP et al. (2004). Cognition and anatomy in three variants of primary progressive aphasia. Ann Neurol 55 (3): 335–346.

Grodzinsky Y (2000). The neurology of syntax: Language use without Broca's area. Behav Brain Sci 23: 1–21.

Guenther FH, Hampson M, Johnson D (1998). A theoretical investigation of reference frames for the planning of speech movements. Psychol Rev 105: 611–633.

Haruno M, Wolpert DM, Kawato M (2001). Mosaic model for sensorimotor learning and control. Neural Comput 13 (10): 2201–2220.

Hickok G (2009). The functional neuroanatomy of language. Phys Life Rev 6: 121–143.

Hickok G, Buchsbaum B (2003). Temporal lobe speech perception systems are part of the verbal working memory circuit: evidence from two recent fMRI studies. Behav Brain Sci 26: 740–741.

Hickok G, Poeppel D (2000). Towards a functional neuroanatomy of speech perception. Trends Cogn Sci 4: 131–138.

Hickok G, Poeppel D (2004). Dorsal and ventral streams: A framework for understanding aspects of the functional anatomy of language. Cognition 92: 67–99.

Hickok G, Poeppel D (2007). The cortical organization of speech processing. Nat Rev Neurosci 8 (5): 393–402.

Hickok G, Erhard P, Kassubek J et al. (2000). A functional magnetic resonance imaging study of the role of left posterior superior temporal gyrus in speech production: implications for the explanation of conduction aphasia. Neurosci Lett 287: 156–160.

Hickok G, Buchsbaum B, Humphries C et al. (2003). Auditory-motor interaction revealed by fMRI: Speech, music, and working memory in area Spt. J Cogn Neurosci 15: 673–682.

Hickok G, Okada K, Barr W et al. (2008). Bilateral capacity for speech sound processing in auditory comprehension: evidence from Wada procedures. Brain Lang 107 (3): 179–184.

Hickok G, Okada K, Serences JT (2009). Area Spt in the human planum temporale supports sensory-motor integration for speech processing. J Neurophysiol 101 (5): 2725–2732.

Hickok G, Houde J, Rong F (2011). Sensorimotor integration in speech processing: computational basis and neural organization. Neuron 69 (3): 407–422.

Houde JF, Jordan MI (1998). Sensorimotor adaptation in speech production. Science 279: 1213–1216.

Houde JF, Nagarajan SS (2011). Speech production as state feedback control. Front Hum Neurosci 5: 82.

Humphries C, Willard K, Buchsbaum B et al. (2001). Role of anterior temporal cortex in auditory sentence comprehension: An fMRI study. Neuroreport 12: 1749–1752.

Humphries C, Love T, Swinney D et al. (2005). Response of anterior temporal cortex to syntactic and prosodic manipulations during sentence processing. Hum Brain Mapp 26: 128–138.

Humphries C, Binder JR, Medler DA et al. (2006). Syntactic and semantic modulation of neural activity during auditory sentence comprehension. J Cogn Neurosci 18 (4): 665–679.

Indefrey P, Levelt WJ (2004). The spatial and temporal signatures of word production components. Cognition 92 (1–2): 101–144.

Jacquemot C, Dupoux E, Bachoud-Levi AC (2007). Breaking the mirror: asymmetrical disconnection between the phonological input and output codes. Cogn Neuropsychol 24 (1): 3–22.

Larson CR, Burnett TA, Bauer JJ et al. (2001). Comparison of voice F0 responses to pitch-shift onset and offset conditions. J Acoust Soc Am 110 (6): 2845–2848.

Lau EF, Phillips C, Poeppel D (2008). A cortical network for semantics: (de)constructing the N400. Nat Rev Neurosci 9 (12): 920–933.

Levelt WJM, Roelofs A, Meyer AS (1999). A theory of lexical access in speech production. Behavioral and Brain Sciences 22 (1): 1–75.

Liebenthal E, Binder JR, Spitzer SM et al. (2005). Neural substrates of phonemic perception. Cereb Cortex 15 (10): 1621–1631.

Luce PA, Pisoni DB (1998). Recognizing spoken words: the neighborhood activation model. Ear Hear 19: 1–36.

Luria AR (1970). Traumatic aphasia, Mouton, The Hague.

Marslen-Wilson WD (1987). Functional parallelism in spoken word-recognition. Cognition 25: 71–102.

Mazoyer BM, Tzourio N, Frak V et al. (1993). The cortical representation of speech. J Cogn Neurosci 5: 467–479.

McClelland JL, Elman JL (1986). The TRACE model of speech perception. Cogn Psychol 18: 1–86.

Miceli G, Gainotti G, Caltagirone C et al. (1980). Some aspects of phonological impairment in aphasia. Brain Lang 11: 159–169.

Miglioretti DL, Boatman D (2003). Modeling variability in cortical representations of human complex sound perception. Exp Brain Res 153 (3): 382–387.

Milner AD, Goodale MA (1995). The visual brain in action, Oxford University Press, Oxford.

Narain C, Scott SK, Wise RJ et al. (2003). Defining a left-lateralized response specific to intelligible speech using fMRI. Cereb Cortex 13 (12): 1362–1368.

Obleser J, Eisner F, Kotz SA (2008). Bilateral speech comprehension reflects differential sensitivity to spectral and temporal features. J Neurosci 28 (32): 8116–8123.

Okada K, Hickok G (2006). Identification of lexical-phonological networks in the superior temporal sulcus using fMRI. Neuroreport 17: 1293–1296.

Okada K, Rong F, Venezia J et al. (2010). Hierarchical organization of human auditory cortex: evidence from acoustic invariance in the response to intelligible speech. Cereb Cortex 20 (10): 2486–2495.

Pa J, Hickok G (2008). A parietal-temporal sensory-motor integration area for the human vocal tract: Evidence from an fMRI study of skilled musicians. Neuropsychologia 46: 362–368.

Patterson K, Nestor PJ, Rogers TT (2007). Where do you know what you know? The representation of semantic knowledge in the human brain. Nat Rev Neurosci 8 (12): 976–987.

Poeppel D (2001). Pure word deafness and the bilateral processing of the speech code. Cognit Sci 25: 679–693.

Poeppel D (2003). The analysis of speech in different temporal integration windows: cerebral lateralization as "asymmetric sampling in time." Speech Communication 41: 245–255.

Price CJ (2012). A review and synthesis of the first 20 years of PET and fMRI studies of heard speech, spoken language and reading. Neuroimage 62 (2): 816–847.

Price CJ, Wise RJS, Warburton EA et al. (1996). Hearing and saying: the functional neuro-anatomy of auditory word processing. Brain 119: 919–931.

Purcell DW, Munhall KG (2006). Compensation following real-time manipulation of formants in isolated vowels. J Acoust Soc Am 119 (4): 2288–2297.

Rauschecker JP (1998). Cortical processing of complex sounds. Curr Opin Neurobiol 8 (4): 516–521.

Rauschecker JP (2011). An expanded role for the dorsal auditory pathway in sensorimotor control and integration. Hear Res 271 (1–2): 16–25.

Rauschecker JP, Scott SK (2009). Maps and streams in the auditory cortex: nonhuman primates illuminate human speech processing. Nat Neurosci 12 (6): 718–724.

Remez RE, Rubin PE, Pisoni DB et al. (1981). Speech perception without traditional speech cues. Science 212: 947–950.

Rissman J, Eliassen JC, Blumstein SE (2003). An event-related FMRI investigation of implicit semantic priming. J Cogn Neurosci 15 (8): 1160–1175.

Rodd JM, Davis MH, Johnsrude IS (2005). The neural mechanisms of speech comprehension: fMRI studeis of semantic ambiguity. Cereb Cortex 15: 1261–1269.

Rogalsky C, Hickok G (2011). The role of Broca's area in sentence comprehension. J Cogn Neurosci 23: 1664–1680.

Rogalsky C, Pitz E, Hillis AE et al. (2008). Auditory word comprehension impairment in acute stroke: relative contribution of phonemic versus semantic factors. Brain Lang 107 (2): 167–169.

Rogalsky C, Love T, Driscoll D et al. (2011). Are mirror neurons the basis of speech perception? Evidence from five cases with damage to the purported human mirror system. Neurocase 17 (2): 178–187.

Schirmer A, Fox PM, Grandjean D (2012). On the spatial organization of sound processing in the human temporal lobe: a meta-analysis. Neuroimage 63 (1): 137–147.

Scott SK, Wise RJ (2004). The functional neuroanatomy of prelexical processing in speech perception. Cognition 92 (1–2): 13–45.

Scott SK, Blank CC, Rosen S et al. (2000). Identification of a pathway for intelligible speech in the left temporal lobe. Brain 123: 2400–2406.

Shannon RV, Zeng F-G, Kamath V et al. (1995). Speech recognition with primarily temporal cues. Science 270: 303–304.

Spitsyna G, Warren JE, Scott SK et al. (2006). Converging language streams in the human temporal lobe. J Neurosci 26 (28): 7328–7336.

Turkeltaub PE, Coslett HB (2010). Localization of sublexical speech perception components. Brain Lang 114 (1): 1–15.

Ungerleider LG, Mishkin M (1982). Two cortical visual systems. In: DJ Ingle, MA Goodale, RJW Mansfield (Eds.), Analysis of visual behavior, MIT Press, Cambridge, MA, pp. 549–586.

Vaden Jr KI, Muftuler LT, Hickok G (2010). Phonological repetition-suppression in bilateral superior temporal sulci. Neuroimage 49: 1018–1023.

Vandenberghe R, Nobre AC, Price CJ (2002). The response of left temporal cortex to sentences. J Cogn Neurosci 14 (4): 550–560.

Wallentin M (2009). Putative sex differences in verbal abilities and language cortex: a critical review. Brain Lang 108 (3): 175–183.

Wernicke C (1874/1969). The symptom complex of aphasia: A psychological study on an anatomical basis. In: RS Cohen, MW Wartofsky (Eds.), Boston studies in the philosophy of science. D. Reidel, Dordrecht, pp. 34–97.

Wernicke C (1874/1977). Der aphasische Symptomencomplex: Eine psychologische Studie auf anatomischer Basis. In: GH Eggert (Ed.), Wernicke's works on aphasia: A sourcebook and review. Mouton, The Hague, pp. 91–145.

Wise RJS, Scott SK, Blank SC et al. (2001). Separate neural sub-systems within "Wernicke's area". Brain 124: 83–95.

Zatorre RJ, Belin P, Penhune VB (2002). Structure and function of auditory cortex: music and speech. Trends Cogn Sci 6: 37–46.

Handbook of Clinical Neurology, Vol. 129 (3rd series)
The Human Auditory System
G.G. Celesia and G. Hickok, Editors

Chapter 9

Role of the auditory system in speech production

FRANK H. GUENTHER[1]* AND GREGORY HICKOK[2]

[1]*Departments of Speech, Language, and Hearing Sciences and Biomedical Engineering, Boston University, Boston, MA, USA*

[2]*Department of Cognitive Sciences, Center for Language Science and Center for Cognitive Neuroscience, University of California, Irvine, CA, USA*

INTRODUCTION

Speech production is a highly complex motor act, involving the finely coordinated activation of approximately 100 muscles in the respiratory, laryngeal, and oral motor systems. To achieve this task, speakers utilize a large network of brain regions. This network includes regions involved in other motor tasks, such as the motor and somatosensory cortical areas, cerebellum, basal ganglia, and thalamus, as well as regions that are more specialized for speech and language, including inferior and middle prefrontal cortex and superior and middle temporal cortex. Our goal in this chapter is to describe the critical role of the auditory system in speech production. We will first discuss the role of sensory systems in motor control broadly and summarize the long history of ideas and research on the interaction between auditory and motor systems for speech. We then describe current research on speech planning, which strongly implicates the auditory system in this process. Two large-scale neurocomputational models of speech production are then discussed. Finally, we will highlight some future directions for research on speech production.

Movement is absolutely dependent on sensory information. We know where and how to reach for an object because we see its location and shape; we know how much force to exert while we are holding the object because we feel the pressure of the object on our hand and the weight on our limb; and we know how to initiate any of these movements because our sensory systems tell us where our limb is in relation to our body and the object. British neurologist and physiologist Henry Charlton Bastian (1837–1915) wrote on the topic of movement control in 1887, stating, "It may be regarded as a physiological axiom, that all purposive movements of animals are guided by sensations or by afferent impressions of some kind" (Bastian, 1887, p. 1). Experimental work over the decades backs up these claims. This work has found, for example, that blocking somatosensory feedback from a monkey's limb (while leaving motor fibers intact) causes the limb to go dead. With training the monkey can learn to reuse it clumsily, but only with visual feedback; blindfold the animal and motor control degrades dramatically (Sanes et al., 1984). Similar symptomology can be found in humans suffering from large-fiber sensory neuropathy, which deafferents the body sense while leaving motor fibers intact (Sanes et al., 1984).

Speech is no different. Without the auditory system, as in prelingual-onset peripheral deafness, normal speech development cannot occur. Importantly, it is not just during development that auditory information is critical. Experimental or naturally caused manipulations of acoustic input can have dramatic effects on speech production. For example, delayed auditory feedback induces non-fluency (Yates, 1963), altering feedback in the form of pitch or the formant frequency structure results in automatic and largely unconscious compensation in speech articulation (Burnett et al., 1998; Houde and Jordan, 1998; Larson et al., 2001), and exposure to a different linguistic environment can induce changes in the listener–speaker's articulation (picking up accents; Sancier and Fowler, 1997). Furthermore, although individuals who become deaf as adults can remain intelligible for years after they lose hearing, they show some speech output impairments immediately (including impaired ability to adjust pitch loudness in different listening conditions), and over time their phonetic contrasts become reduced (e.g., Perkell et al., 2000) and they exhibit articulatory decline (Waldstein, 1989).

*Correspondence to: Frank H. Guenther, Departments of Speech, Language, and Hearing Sciences and Biomedical Engineering, Boston University, 677 Beacon Street, Boston, MA 02215, USA. Tel: +1-617-353-5765, Fax: +1-617-353-7755, E-mail: guenther@bu.edu

The speech research literature contains numerous theoretic proposals that strongly link speech perception and speech production. Notable examples include the motor theory of speech perception (Liberman et al., 1967; Liberman and Mattingly, 1985), which posits that speech perception involves translating acoustic signals into the motor gestures that produce them, as well as acoustic theories of speech production (e.g., Fant, 1960; Stevens, 1998), which highlight the importance of acoustic or auditory targets in the speech production process. In the following sections we elaborate on the roles of auditory information in recent neural models of speech planning and execution.

THE PLANNING OF SPEECH MOVEMENTS

Although it may seem obvious that auditory information is in some way involved in speech production – the lack of normal speech development in the absence of auditory feedback provides incontrovertible evidence of this – there has been much debate in the speech motor control literature over exactly what role(s) auditory feedback plays. Much of this debate revolves around the following central question: what exactly are the goals, or targets, of the speech production planning process?

To produce a goal-directed movement, the nervous system must generate a complex muscle activation pattern that satisfies the goals of the movement. For example, reaching to a target position in three-dimensional (3-D) space requires coordinated activation of muscles in the arm, wrist, and hand that move the hand in space to the target position. Various explanations have been put forth regarding how this complex task may be solved by the nervous system, ranging from a motoric planning extreme to a task space planning extreme. In a motoric planning solution, the nervous system translates the desired spatial position of the hand into a corresponding target in motoric terms, such as a set of joint angles or muscle lengths that place the hand at the appropriate position in 3-D space. Movement planning is then carried out simply by interpolating between the current set of muscle lengths and the target set (e.g., Bullock and Grossberg, 1988; Rosenbaum et al., 1993, 1995). In a task space planning scheme, the movement is planned in a reference frame that is much further removed from the musculature. In reaching, for example, a 3-D spatial reference frame may be used to plan a hand path to the target. Muscle activation patterns must still be generated to move the arm and hand, and different proposals for this transformation have been put forth. In one solution, the spatial movement direction needed to keep the hand moving along the planned spatial trajectory is transformed into appropriate muscle velocities during the reach (Guenther, 1992; Bullock et al., 1993). An important consequence of this form of planning is that it does not involve a specific target configuration of the arm for a given spatial position of the hand.

To understand the planning of speech movements, the relationship between articulator configuration and acoustic signal must be understood. At any point in time, the 3-D shape of the vocal tract (i.e., the shape of the "air tube" that exists from the vocal folds to the lips) is determined by the locations of speech articulators such as the tongue, lips, jaw, larynx, and the soft palate (velum). This shape manifests itself in the speech signal as a set of peaks in the envelope of the frequency spectrum of the acoustic signal; these peaks are referred to as formants. Roughly speaking, the first formant frequency, or F1, corresponds to the degree of constriction formed by the tongue in the vocal tract, which in turn is closely related to tongue height. The second formant frequency, F2, is related to the location of the tongue constriction along the length of the vocal tract, with higher F2 values corresponding to constrictions closer to the lips. These characterizations are rather crude "rules of thumb" that don't apply to all vocal tract configurations and leave out many important aspects of the relationship between speech articulation and acoustics. The relationships between the third and higher formant frequencies and vocal tract shape become increasingly complex and are not easily related to articulator positions.

Evidence from studies of speech movements indicate that speech sounds (e.g., phonemes or syllables) are not encoded as muscle length or articulator configuration targets; i.e., the speech motor system does not use a purely motoric planning strategy, as defined above. Instead, a large amount of variability is seen in the vocal tract configurations used to realize a particular phoneme. A stark example is the American English phoneme /r/. This phoneme is characterized acoustically by a deep dip in the third formant frequency; this formant frequency transition is highly consistent across phonetic contexts (Boyce and Espy-Wilson, 1997). Although the acoustic signal characterizing /r/ remains nearly constant, completely different articulatory gestures are used in the different contexts (DeLattre and Freeman, 1968; Guenther et al., 1999; Nieto-Castanon et al., 2005). Figure 9.1 illustrates this articulatory variability. The top row shows tongue configurations at the start of /r/ production (dashed lines) and at the acoustic "center" of the /r/ (bold lines) when spoken by an American English participant in three different phonetic contexts (/ar/, /dr/, and /gr). In the /ar/ context, the participant raises the tongue tip to produce the /r/. In the /dr/ context, the speaker uses a backward movement of the tongue. In the /gr/ context, the speaker uses a downward movement of the tongue.

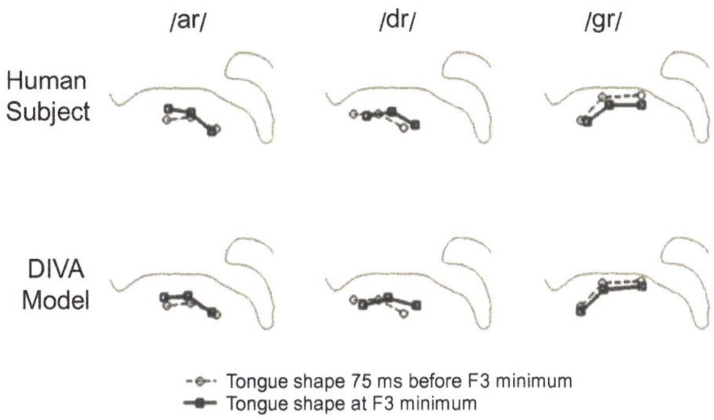

Fig. 9.1. Articulatory variability during American English /r/ production. Measured (top) and simulated (bottom) tongue movements during production of the phoneme /r/ when preceded by the phoneme /a/ (left column), /d/ (center), and /g/ (right), shown in the mid sagittal plane. Outline of the palate is shown for reference. Each connected line (bold or dashed) represents a tongue configuration; each circular or square data point corresponds to the position of a movement transducer on the tongue (tongue tip on the left; dorsum in the center; back on the right). The gesture used to produce /r/ is represented by the transition between the dashed and bold tongue configurations. Entirely different tongue gestures are used by both the human subject and the model in the three contexts. Simulations of the Directions Into Velocities of Articulators (DIVA) model utilizing a common auditory target for the three contexts capture this articulatory variability. (Adapted from Nieto-Castanon et al., 2005.)

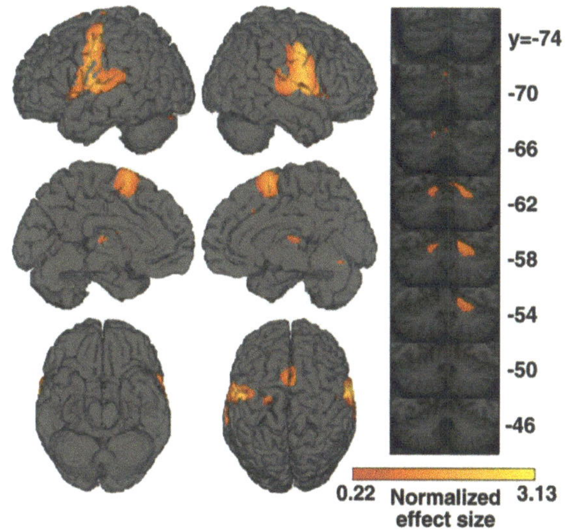

Fig. 9.2. Neural activity in the cerebral cortex and cerebellum during production of monosyllabic words contrasted with a silent baseline. In the lateral views of the cerebral cortex (top panels), bilateral activity can be seen in the precentral gyrus of the frontal lobe, the postcentral gyrus of the parietal lobe, and the superior temporal gyrus of the temporal lobe. Medial surfaces (middle panels) reveal activity in the supplementary motor area. Slices through the cerebellum (right column) indicate activity in the superior paravermal cerebellum, primarily in lobules V and VI. Additional activity in the basal ganglia and thalamus is not visible in this figure. (Adapted from Tourville et al. 2008).

The articulatory variability seen for /r/ can be accounted for in a straightforward manner by a task space planning scheme in which the targets of speech are auditory rather than articulatory. The bottom row of Figure 9.1 shows the results of computer simulations of the Directions Into Velocities of Articulators (DIVA) model of speech motor control (Guenther, 1995a; Guenther et al., 1998, 2006; Tourville and Guenther, 2011; Guenther and Vladusich, 2012; described further below) when producing /r/ in the three phonetic contexts. The same auditory target is used for /r/ in the three contexts, and articulator movements are planned in auditory space (see Nieto-Castanon et al., 2005 for details). This simple planning scheme accounts for the disparate tongue gestures seen in the three phonetic contexts, providing strong evidence for the use of an auditory planning space.

BRAIN REGIONS INVOLVED IN SPEECH ARTICULATION

Figure 9.2 illustrates neural activity in the cerebral cortex and cerebellum during a simple speech task (monosyllabic word production) contrasted with a baseline task of silently viewing letters. As in most speech production neuroimaging studies, activity is seen in the ventral precentral gyrus (motor and premotor cortex), ventral postcentral gyrus (somatosensory cortex), superior temporal gyrus (auditory cortex), supplementary motor area (SMA), and superior paravermal cerebellum (primarily lobules V and VI). Though not visible in Figure 9.2, the basal ganglia and thalamus are also active during speech. In the following sections we will discuss the roles of these regions within the context of neurocomputational models of speech production.

NEUROCOMPUTATIONAL MODELS OF SPEECH PRODUCTION

The extensive library of results from neuroimaging studies provides important insight into the roles of a large number of cortical and subcortical areas in speech production. In isolation, however, these results do not provide an integrated, mechanistic view of how the neural circuits engaged by speech tasks interact to produce fluent speech. To this end, computational models that both suggest the neural computations performed within specific modules and across pathways linking modules and propose specific neural correlates for these computational elements bridge this critical gap. Furthermore, these neurocomputational models lead to specific hypotheses that may be tested in both imaging and behavioral experiments, leading to a continuous cycle of model refinement that provides a unifying account of a wide range of experimental findings. Below we describe the DIVA model of speech motor control, which has been developed, tested, and refined by Guenther and colleagues over the last two decades, as well as an extension to that model, termed GODIVA, which begins to account for higher-level phonologic processes involved in speech planning. Our discussion of neurocomputational models ends with a treatment of the Hierarchical State Feedback Control (HSFC) model of Hickok (2012), which addresses data from studies of conduction aphasia and internal correction of upcoming phonemic errors before they are articulated.

The Directions Into Velocities of Articulators model

The DIVA model, schematized in Figure 9.3, provides the most detailed and thoroughly tested account of the neural processes underlying speech motor control. Each box in the diagram corresponds to a set of neurons, or map, and arrows correspond to synaptic projections that transform one type of neural representation into another. The model learns to control movements of a simulated vocal tract that produces an acoustic signal (Maeda, 1990). The results of numerous computer simulations of the model have been shown to account for a wide range of findings regarding the brain activity, sound output, and movements produced by human speakers, as detailed elsewhere (e.g., Guenther, 1995b; Guenther et al., 1998, 1999; Callan et al., 2000; Perkell et al., 2004a, b; Nieto-Castanon et al., 2005; Ghosh et al., 2008; Tourville et al., 2008; Peeva et al., 2010).

The production of a speech sound in the DIVA model starts with activation of neurons associated with that sound in the model's speech sound map. A "speech sound" in this map is defined as a segment of speech that has its own motor program, with the syllable being the most typical unit represented. The term "sound map" is used to highlight the fact that speech movement planning appears to aim at auditory targets rather than

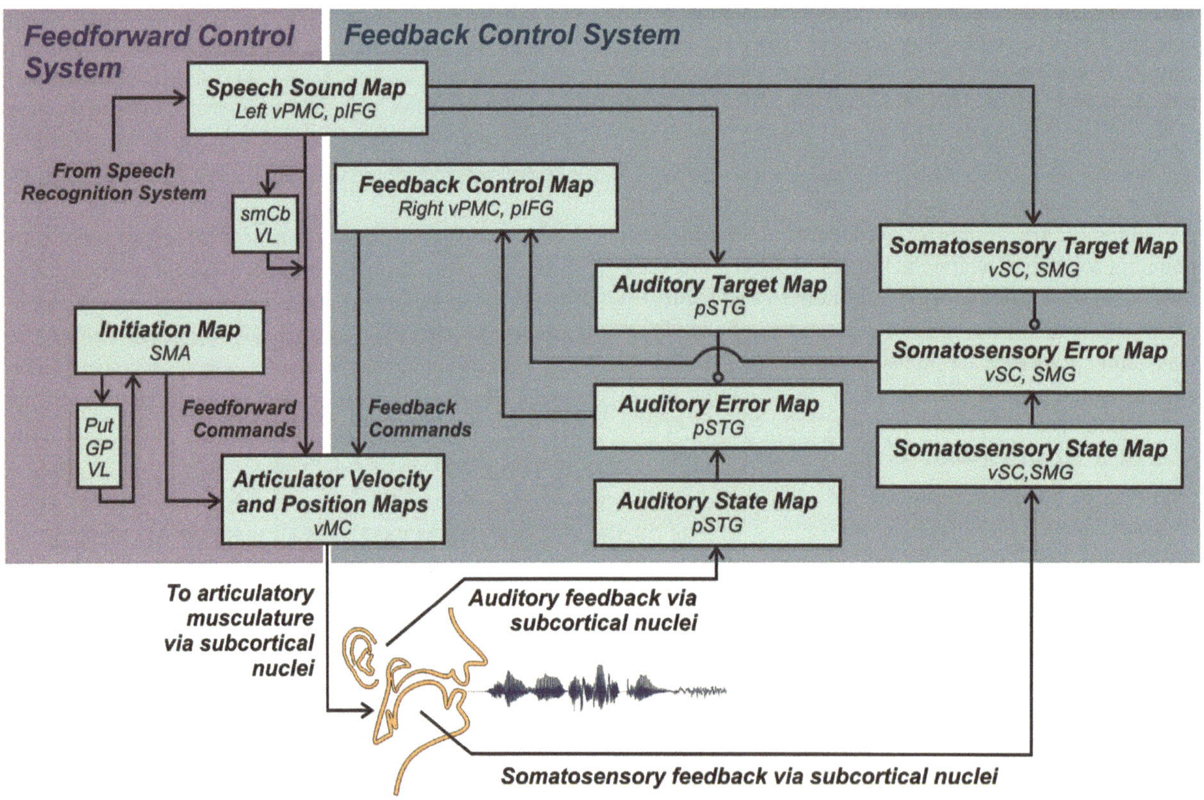

Fig. 9.3. Schematic overview of the Directions Into Velocities of Articulators (DIVA) model of speech production. Each box in the diagram corresponds to a group of neurons; arrows represent excitatory axonal projections and empty circles represent inhibitory projections. The model breaks the control of speech production into a feedforward control system and a feedback control system, with the latter further broken into auditory and somatosensory feedback control subsystems. See text for details. GP, globus pallidus; Put, putamen; pIFG, posterior inferior frontal gyrus; pSTG, posterior superior temporal gyrus; smCb, superior medial cerebellum; SMA, supplementary motor area; SMG, supramarginal gyrus; VL, ventral lateral nucleus of the thalamus; vMC, ventral motor cortex; vPMC, ventral premotor cortex; vSC, ventral somatosensory cortex.

articulatory targets, as described in the preceding section. In the computer implementation of the model, each neuron in the map represents a different speech sound. When a new sound is encountered, a new neuron that codes for that sound is added to the speech sound map.

Activation of the speech sound map leads to articulator movement commands that arrive in primary motor cortex via two control subsystems. The feedforward control system projects directly, as well as indirectly via a side loop through the cerebellum, from the speech soundmap to primary motor cortex. The feedback control system, including both an auditory feedback control subsystem and a somatosensory feedback control subsystem, involves indirect projections passing through sensory brain areas.

The DIVA model provides a description of how an infant learns to produce speech sounds through babbling and imitation processes. According to the model, a combination of motor, auditory, and somatosensory information generated during early random and reduplicated babbling is used to tune the synaptic projections of the feedback control system. The sensorimotor transformations learned during babbling allow detected sensory errors to be mapped into corrective motor commands during the subsequent imitation stage, when feedforward commands are learned for the speech sounds of the infant's native language.

The imitation stage describes how syllable-specific learning occurs when an infant is presented with a new speech sound to learn, corresponding to an infant learning a new syllable from his/her parent, for example. Detection of a novel sound leads to activation of a new speech sound map neuron that will represent that sound. This is presumed to occur via projections from the speech recognition system in the auditory cortical areas to the left ventral premotor cortex. Though not specified in the model, these projections may be part of the dorsal auditory stream (involving the superior longitudinal fasciculus or arcuate fasciculus) or the ventral stream (involving the uncinate fasciculus).

The model first learns an auditory target for the new sound from sound samples it is presented. This auditory target is encoded in the synaptic projections from the speech sound map to the auditory target map in Figure 9.3. The model then repeatedly attempts to produce the sound, improving the feedforward command for the sound with each iteration. The functions of the feedforward and feedback control systems are further detailed in the following sections.

Since its inception (Guenther, 1992), the DIVA model has hypothesized the existence of a speech sound map in premotor cortex that is activated both during speech perception (when it is necessary for learning sensory targets for speech sounds) and speech production (when it is necessary for driving speech articulator movements via the feedforward and feedback control systems). Cells in premotor cortical areas that are active during both action perception and production of a motor act have since been identified in numerous electrophysiologic studies in animals (see Rizzolatti and Craighero, 2004, for a review) and functional magnetic resonance imaging (fMRI) studies showing activity while listening to speech in ventral premotor cortical areas involved in speech production (Buchsbaum et al., 2001; Hickok et al., 2003; Wilson et al., 2004).

Auditory feedback control

As described in the introduction, auditory feedback plays an important role in tuning the speech motor control system. According to the DIVA model, axonal projections from speech sound map cells in the left inferior frontal gyrus/ventral premotor cortex to the higher-order auditory cortical areas (via the dorsal auditory stream; cf. Hickok and Poeppel, 2004) embody the auditory target region for the speech sound currently being produced. That is, they represent the auditory feedback that should arise when the speaker hears himself/herself producing the sound. This representation of expected auditory consequences is referred to as the auditory target map, and the model's auditory target consists of a temporally varying region of auditory space that encodes acceptable auditory variation for the current sound. The use of target regions, rather than points, is an important aspect of the DIVA model that provides a unified explanation for a wide range of speech production phenomena (see Guenther, 1995b for details). Projections such as those between the speech sound map and auditory state map in Figure 9.3, which predict the sensory state resulting from a movement, are often referred to as forward models (e.g., Guenther, 1995a; Miall and Wolpert, 1996; Guenther et al., 1998; Kawato, 1999; Desmurget and Grafton, 2000; Davidson and Wolpert, 2005), and they are closely related to the much older concepts of efference copy (originally proposed by von Helmholtz in the mid 19th century; see von Helmholtz, 1925) and corollary discharge (a term coined by Sperry, 1950).

The model's auditory state map represents the incoming auditory signal of one's own speech. This map is hypothesized to include a combination of relatively low-level auditory features such as pitch represented in the primary auditory cortex (cf. Bendor and Wang, 2005; Kumar and Schönwiesner, 2012), located in Heschl's gyrus, as well as more complex auditory representations such as a speaker-normalized formant representation hypothesized to reside in higher-order auditory cortical areas in the planum temporale and posterior superior temporal gyrus (including the superior bank of the superior temporal sulcus).

The auditory feedback control system detects speech errors, represented in the auditory error map, by comparing the auditory target map and auditory state map. Specifically, the auditory target map inhibits auditory error map cells, whereas the auditory state map excites auditory error map cells. If the incoming auditory signal is within the target region, the inhibition from the auditory target cancels the excitatory effects of the auditory state map. If the incoming auditory signal is outside the target region, the inhibitory target region will not completely cancel the excitatory input from the auditory periphery, resulting in activation of auditory error map cells. Evidence of inhibition in auditory cortical areas in the superior temporal gyrus during one's own speech comes from several different sources, including recorded neural responses during open-brain surgery (Creutzfeldt et al., 1989), magnetoencephalography (MEG) measurements (Numminen and Curio, 1999; Numminen et al., 1999; Houde et al., 2002), and positron emission tomography (PET) measurements (Wise et al., 1999). The auditory error map is hypothesized to reside in the same cortical areas as the auditory state and target maps; these regions include the Sylvian parietal temporal (Spt) area, which has been shown to become active during both speech perception and production (Buchsbaum et al., 2001; Hickock and Poeppel, 2004).

Activation of the auditory error map during speech leads to corrective motor signals in the model via projections to the feedback control map in right ventral premotor cortex. This map is responsible for transforming auditory errors into movements that correct these errors; such a transformation is often referred to as an inverse model (e.g., Kawato, 1999). The DIVA model derives its name from the particular form of inverse model used: a mapping from Directions (in sensory space) Into Velocities of Articulators. Such a mapping, which approximates the pseudoinverse of the Jacobian matrix relating changes in the auditory signal to the velocities of the speech articulators, results in many desirable properties, including motor equivalence (Guenther et al., 1998, 1999, 2006; see simulations of /r/ production in previous section) and efficient movement trajectories (Guenther et al., 1998).

Once the model has learned appropriate feedforward commands for a speech sound (described further below), it can successfully produce the sound using just those feedforward commands. That is, no auditory error will arise during production, and thus the auditory feedback control subsystem will not be activated. However, if an externally imposed perturbation occurs, such as a real-time "warping" of the subject's auditory feedback so that he hears himself producing the wrong sound (cf. Houde and Jordan, 1998), the auditory error cells will become active and attempt to correct for the perturbation through projections to the feedback control map in the right inferior frontal cortex. These predictions were verified by Tourville et al. (2008) in an fMRI study involving unexpected perturbations of auditory feedback during speech – specifically, upward or downward shifts of F1 by 30% on randomly dispersed trials. As illustrated in Figure 9.4, perturbed speech led to increased activity in the posterior superior temporal gyrus

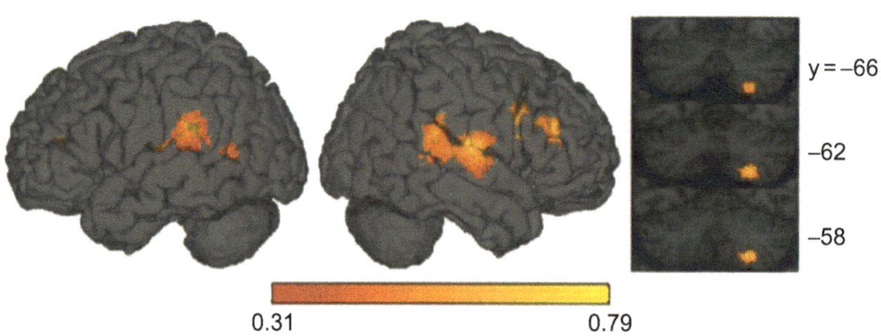

Fig. 9.4. Auditory feedback control network for speech production. Areas of increased brain activity for auditorily perturbed speech compared to normal speech are indicated on the left and right lateral cortical hemispheres and coronal slices through the cerebellum (see Tourville et al., 2008, for details). The auditory perturbation consisted of a 30% increase or decrease of the first formant frequency (F1), which has the effect of making the vowel in a word like "bet" sound more like "bat" or "bit." This perturbation occurred in real time, i.e., the perturbed speech signal was delivered to the subject over headphone within 16 ms of the actual acoustic signal from his/her own voice, and it was applied on a random subset of trials. Subjects partially compensated for this perturbation by shifting their produced F1 in the direction opposite the perturbation. Increased activity was found bilaterally in the auditory cortical areas of the posterior superior temporal gyrus during perturbed speech, supporting the Directions Into Velocities of Articulators (DIVA) model prediction of auditory error neurons in these areas. Increased activity was also seen in inferior frontal regions, with a right hemisphere bias; according to the DIVA model these regions are involved in generating corrective motor commands.

bilaterally and in the right frontal cortex. These regions correspond to the auditory feedback control system in the schematic of the DIVA model (Fig. 9.3).

Somatosensory feedback control

Like auditory information, somatosensory information has long been known to be important for speech production (e.g., Lindblom et al., 1979). The DIVA model posits a somatosensory feedback control subsystem operating in parallel with the auditory feedback control subsystem. The model's somatosensory state map corresponds to the representation of tactile and proprioceptive information from the speech articulators in primary and higher-order somatosensory cortical areas in the postcentral gyrus and supramarginal gyrus. The model's somatosensory target map and somatosensory error map, which play roles analogous to their auditory counterparts, are hypothesized to reside primarily in the supramarginal gyrus, a region that has been implicated in phonologic processing for speech perception (e.g., Caplan et al., 1995; Celsis et al., 1999) and speech production (e.g., Geschwind, 1965; Damasio and Damasio, 1980). This hypothesized role for inferior parietal cortex in the integration of motor commands and sensory feedback during speech production is analogous to the visual-motor integration role associated with more dorsal parietal regions during limb movements (Andersen, 1997; Rizzolatti et al., 1997).

According to the model, cells in the somatosensory error map become active during speech if the speaker's tactile and proprioceptive feedback from the vocal tract deviates from the somatosensory target region for the sound being produced; this prediction is supported by brain activity measured with fMRI during unexpected jaw perturbations, as illustrated in Figure 9.5. This study found increased activity bilaterally in the supramarginal gyrus and postcentral gyrus when contrasting the perturbed and unperturbed speaking conditions. The output of the somatosensory error map then propagates to the feedback control map in right ventral premotor cortex, which transforms somatosensory errors into motor commands that correct those errors in a manner analogous to the auditory feedback control system, i.e., using an approximation to the pseudoinverse of the Jacobian matrix relating changes in somatosensory state to velocities of the speech articulators.

Although the auditory and somatosensory feedback subsystems are implemented separately in computer simulations of the DIVA model, the existence of multisensory neurons that encode both auditory and somatosensory information during speech has been hypothesized (e.g., Guenther et al., 2006). Interactions between different sensory modalities involved in speech

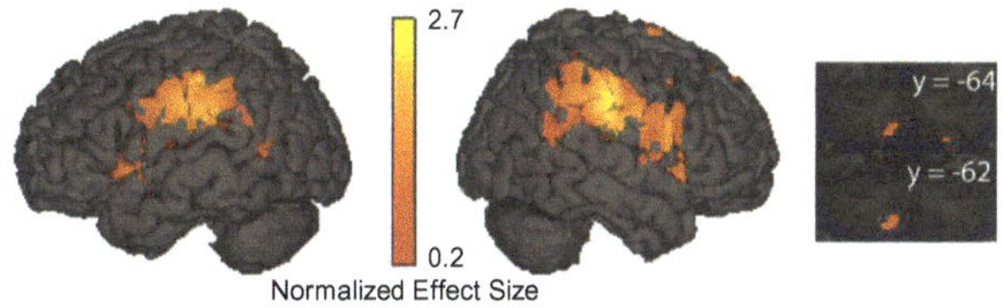

Fig. 9.5. Somatosensory feedback control network for speech production. Areas of increased brain activity for speech produced with an unexpected blockage of the jaw contrasted with normal speech are indicated on the left and right lateral cortical hemispheres and coronal slices through the cerebellum. The unexpected jaw constraint caused bilateral activity in the postcentral gyrus and supramarginal gyrus, supporting the Directions Into Velocities of Articulators (DIVA) model prediction of a somatosensory error map in these regions. Increased activity was also seen in inferior frontal regions, with a right hemisphere bias; according to the DIVA model these regions are involved in generating corrective motor commands. (Adapted from Golfinopoulos et al., 2011.)

have been noted in a number of studies; for example, it has been shown that somatosensory stimulation can modulate auditory perceptual responses under some conditions (Ito et al., 2009), auditory stimulation can alter skin sensation (Ito and Ostry, 2012), and viewing speech articulation can alter auditory perception, a phenomenon referred to as the McGurk effect (McGurk and MacDonald, 1976).

Feedforward control

According to the DIVA model, projections from left ventral premotor cortex to primary motor cortex, supplemented by a cerebellar sideloop between these two regions involving the superior paravermal region of the cerebellar cortex, constitute the feedforward motor commands for speech production. The primary motor and premotor cortices are well known to be strongly interconnected (e.g., Passingham, 1993; Krakauer and Ghez, 1999). Furthermore, the cerebellum is known to receive input via the pontine nuclei from premotor cortical areas, as well as higher-order auditory and somatosensory areas that can provide state information important for choosing motor commands (Schmahmann and Pandya, 1997), and it projects heavily to the primary motor cortex via the thalamus (e.g., Middleton and Strick, 1997).

The DIVA model's feedforward commands constitute a form of motor program, and as such they are closely related to the concepts of a gestural score (Browman and Goldstein, 1989) and mental syllabary (Levelt and Wheeldon, 1994). A gestural score is a stored representation of properly timed vocal tract gestures (defined primarily as constrictions in the vocal tract) needed to produce a phonologic message. The mental syllabary is a repository of motor programs (or gestural scores) for the most frequently used syllables in a language. The DIVA model's conception of feedforward commands differs from these characterizations primarily in the choice of representation for the motor programs (in the DIVA model these programs command articulator movements to hit auditory targets rather than constriction targets) and size of the motor programs (the DIVA model allows for phonemic motor programs as well as syllabic and multisyllabic motor programs for frequently used phoneme sequences).

Prior to the development of speech, infants do not possess fine-tuned motor programs for producing the sounds of their language; in other words, their feedforward control system is not yet tuned. Only by attempting to produce the sounds of their language can accurate feedforward commands be learned. In the DIVA model, feedforward commands for a syllable are tuned on each production attempt. On the first attempt to produce a new sound, the model relies very heavily on auditory feedback control to produce the sound since its feedforward commands are inaccurate, thus resulting in auditory errors which are detected by comparing the stored auditory target for the sound with the auditory feedback generated by the production attempt. These auditory error signals activate the feedback control subsystem, which leads to corrective motor commands via the feedback control map. The corrective commands issued by the auditory feedback control subsystem during the current attempt to produce the sound become stored in the feedforward command for use on the next attempt. The superior paravermal region of the cerebellum is hypothesized to be involved in this process (see Guenther et al., 2006 for details). Each subsequent attempt to produce the sound results in a better feedforward command and less auditory error, until the feedforward command is capable of producing the sound fluently, i.e., without producing any auditory error, at which point the auditory feedback subsystem no longer contributes to production unless speech is perturbed in some way or the sizes or shapes of the articulators change. As the speech articulators get larger with growth, the auditory feedback control subsystem continues to provide corrective commands that are subsumed into the feedforward controller, thus allowing the feedforward controller to stay properly tuned despite dramatic changes in the sizes and shapes of the speech articulators over the course of a lifetime.

A second component of the feedforward control system schematized in Figure 9.3 is a basal ganglia-thalamo-cortical loop involving an initiation map in the SMA. The SMA is strongly interconnected with lateral motor and premotor cortex and the basal ganglia (Jurgens, 1984; Luppino et al., 1993; Lehericy et al., 2004; Matsumoto et al., 2007), and a number of investigators have proposed that the SMA plays a critical role in controlling the initiation of speech motor commands (e.g., Jonas, 1987; Ziegler et al., 1997; Alario et al., 2006; Bohland and Guenther, 2006). Each speech sound is represented by a different neuron in the model's initiation map, and activation of a sound's initiation map cell leads to readout of the feedforward commands for that sound via projections to ventral premotor and primary motor cortex (Tourville and Guenther, 2011). The basal ganglia, which receive input from a wide range of sensory, motor, and association areas of the cerebral cortex (not shown in Fig. 9.3), are ideally suited for determining the proper sensorimotor and behavioral context for launching the feedforward commands for the next sound to be produced. This is hypothesized to occur via the basal ganglia-thalamo-cortical loop with SMA in Figure 9.3.

The feedforward control system of the DIVA model provides a mechanistic account for three disorders of speech motor control: apraxia of speech, ataxic dysarthria, and stuttering. Apraxia of speech is hypothesized to occur with damage to the speech sound map and/or projections from this map to primary motor cortex; such damage would impair the readout of speech motor programs, resulting in speech errors such as groping and syllable reduction that are associated with apraxia of speech (Ames, 2009). This account is consistent with the finding that damage to the brain regions thought to contain the speech sound map, left inferior frontal gyrus and ventral premotor cortex, is strongly associated with apraxia of speech (Hillis et al., 2004). Damage to the superior paravermal region of the cerebellar cortex results in ataxic dysarthria, a motor speech disorder characterized by slurred, poorly coordinated speech (Ackermann et al., 1992). This finding is in accord with the model's inclusion of this region for providing precisely timed feedforward commands necessary for fluent speech. Stuttering is hypothesized to arise from impairment of the basal ganglia-thalamo-cortical loop involving the initiation map in SMA (Civier et al., 2013). Alm (2004) provides a detailed treatment of the wide range of evidence implicating the basal ganglia in stuttering.

Interestingly, a number of auditory feedback manipulations have been shown to improve fluency, at least temporarily, in people who stutter. These include noise-masked auditory feedback (MacCulloch et al., 1970), delayed auditory feedback (Van Riper, 1982), and frequency-altered feedback (Hargrave et al., 1994). Alm (2004) proposes that such manipulations decrease the automaticity of speech and thereby reduce dependency on the impaired basal ganglia circuit, and Civier et al. (2010) propose that such alterations prevent the speaker from hearing his/her own auditory errors, which when perceived may prohibit the basal ganglia from recognizing the proper sensory state for launching the next sound's feedforward commands.

The GODIVA model of speech sound sequencing

The DIVA model accounts for production of individual speech motor programs, each corresponding to a different speech sound, with the typical unit of motor programming being the syllable. Additional brain areas, particularly in the left prefrontal cortex, become involved for longer speech utterances, which require the generation of an appropriately timed sequence of phonemes and syllables corresponding to a phrase or sentence. Here we briefly treat some of the key areas involved in speech sound sequencing and introduce an extension to the DIVA model, called gradient order DIVA (GODIVA), that addresses the neural computations underlying speech sound sequencing.

The pre-SMA, like the SMA that lies immediately posterior to it in the medial frontal cortex, has been implicated in movement sequencing (e.g., Clower and Alexander, 1998; Shima and Tanji, 2000), including the sequencing of speech movements (Bohland and Guenther, 2006). Shima and Tanji (2000) found cells in pre-SMA that code for an entire sequence to be produced, and other cells in pre-SMA that fired for the *n*th movement (e.g., the third movement) of a sequence regardless of the type of movement. Bohland and Guenther (2006) found increased activity in pre-SMA when the number of unique syllables in an utterance was increased, as well as when the complexity of the individual syllables was increased.

The lateral prefrontal cortex has been implicated in many studies of language and working memory (Fiez et al., 1996; D'Esposito et al., 1998; Gabrieli et al., 1998; Kerns et al., 2004) and in serial order processing (Petrides, 1991; Averbeck et al., 2002, 2003). Bohland and Guenther (2006) noted increased activity in lateral prefrontal cortex, specifically the region around the inferior frontal sulcus (IFS), with increased syllable sequence complexity of a spoken utterance.

The basal ganglia, particularly interactions between the cortex and the basal ganglia, are organized into multiple loop circuits (Alexander et al., 1986; Alexander and Crutcher, 1990; Middleton and Strick, 2000), including the sensorimotor loop described above. The basal ganglia are known to be involved in sequencing motor acts (e.g., Harrington and Haaland, 1998), and abnormalities in the basal ganglia and/or associated circuitry can impact speech production (Kent, 2000; Murdoch, 2001), with some patients having particular difficulty fluently sequencing articulatory movements (Ho et al., 1998; Pickett et al., 1998).

The GODIVA model (Bohland et al., 2010) provides an account of how the pre-SMA, IFS, and basal ganglia interact with the speech motor control network described by the DIVA model in order to produce utterances containing multiple syllables. Briefly, left IFS acts as a phonologic working memory that buffers the phonemes in an upcoming utterance, organized by their location within the syllable. Projections from left IFS to ventral premotor cortex are responsible for choosing the speech sound map representation for the next syllable. Left pre-SMA is responsible for the sequencing of items within a syllable; specifically, pre-SMA represents the syllabic frame structure of the upcoming syllables (cf. MacNeilage, 1998) without regard for the particular phonemes involved. Projections from pre-SMA to SMA activate, in the proper sequence, the appropriate neurons in the initiation map, which in turn leads to the readout of the feedforward commands for the next syllable via projections to the speech sound map. Loops through the basal ganglia are proposed to play central roles in the activation of pre-SMA, IFS, SMA, and ventral premotor cortex neurons.

The Hierarchical State Feedback Control model

DIVA is by far the most detailed and explicit model of speech motor control as well as the most influential to date. As noted, DIVA uses a feedback control architecture to detect and correct overtly produced errors. However, there is evidence in the motor control literature generally (Shadmehr and Mussa-Ivaldi, 1994; Wolpert et al., 1995; Tian and Poeppel, 2010) and in the speech production literature more specifically for internal feedback control, that is, detecting and correcting internal coding errors prior to overt speech production (see below). We describe one such model here, the HSFC model, illustrated in Figure 9.6.

The empiric motivation for an internal feedback loop in the speech domain comes from three sources. One is simply that we can imagine speaking and hear ourselves in our "mind's ear." Experimental research on such inner speech has shown that imagined speech can contain inadvertent errors that are internally detected. Further, the types and distribution of such errors are similar to what is observed in overt speech; e.g., phonemic errors show a "lexical bias" (Oppenheim and Dell, 2008; Nozari et al., 2011). A second source is that talkers correct partially articulated speech errors faster than they should be able to if they were relying on overt auditory feedback alone (Nozari et al., 2011); it is an open question whether somatosensory feedback may explain this phenomenon.

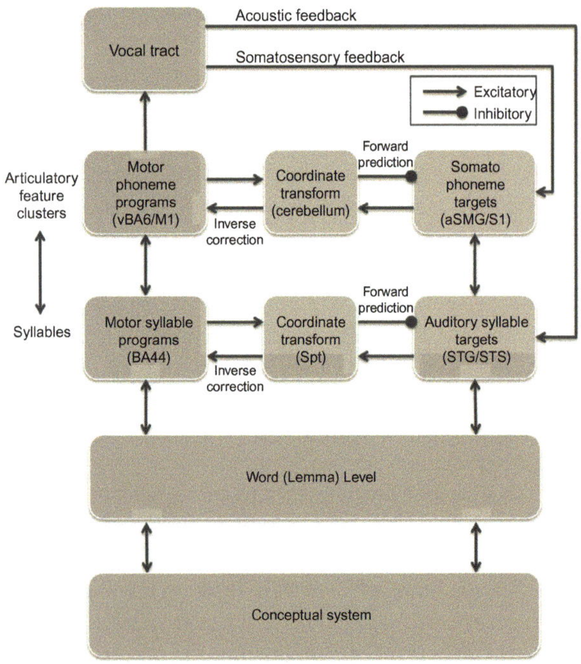

Fig. 9.6. The hierarchical state feedback control (HSFC) model. The HSFC model includes two hierarchical levels of feedback control, each with its own internal and external sensory feedback loops. As in psycholinguistic models, the input to the HSFC model starts with the activation of a conceptual representation that in turn excites a corresponding word (lemma) representation. The word level projects in parallel to sensory and motor sides of the highest, fully cortical level of feedback control, the auditory–Spt–BA44 loop. This higher-level loop in turn projects, also in parallel, to the lower-level somatosensory–cerebellum–motor cortex loop. Direct connections between the word level and the lower-level circuit may also exist, although they are not depicted here. The HSFC model differs from the state feedback control (SFC) model in two main respects. First, "phonologic" processing is distributed over two hierarchically organized levels, implicating a higher-level cortical auditory–motor circuit and a lower-level somatosensory–motor circuit, which roughly map on to syllabic and phonemic levels of analysis, respectively. Second, a true efference copy signal is not a component of the model. Instead, the function served by an efference copy is integrated into the motor planning process. BA, Brodmann area; M1, primary motor cortex; S1, primary somatosensory area; aSMG, anterior supramarginal gyrus; STG, superior temporal gyrus; STS, superior temporal sulcus; vBA6, ventral BA6. The HSFC model is squarely within the tradition of the Directions Into Velocities of Articulators (DIVA) model in that it assumes that the targets of speech gestures are coded in auditory space and that feedback control is a key computational operation of the network. HSFC differs from DIVA in three respects: (1) it assumes an internal as well as an external feedback detection/correction mechanism; (2) it situates auditory and somatosensory feedback loops in a hierarchical arrangement (auditory loop being higher-level and somatosensory loop being lower-level); and (3) it assumes a modified computational architecture for the feedback loops.

The third source is conduction aphasia, a syndrome characterized by fluent speech output and intact speech perception abilities but with a higher than normal rate of phonemic errors in production. Crucially, affected individuals can readily detect their own errors once spoken, but have trouble correcting them, even after they have been overtly detected (Goodglass, 1992). This pattern of speech behavior can be explained by a damaged internal error detection and correction loop (leading to the higher error rate) with an intact external feedback loop (allowing for detection of overtly produced errors). More details on conduction aphasia are provided below.

The HSFC model, like DIVA, assumes that a basic planning unit in auditory space is roughly at the syllable level. The somatosensory circuit, on the other hand, is hypothesized to code sensory targets at a lower level. The basic idea is that speech production involves a cyclic opening and closing of the vocal tract (roughly corresponding to vowels and consonants respectively) and that the somatosensory system defines the targets of these opening or closing gestures (Gracco, 1994; Gracco and Lofqvist, 1994) similar to the somatosensory target map in the DIVA model but involving targets for individual phonemes rather than a single target for a whole syllable. The HSFC model holds that the internal auditory loop comprises a fully cortical network including auditory regions in the superior temporal gyrus, motor regions in the inferior frontal gyrus, and an auditory–motor interface network, Spt, in the posterior planum temporale region (of course, the external feedback loop involves non-cortical structures). The somatosensory loop comprises somatosensory regions in the inferior parietal lobe, lower-level motor regions in primary motor cortex and/or Brodmann area 6, and a somatosensory–motor interface in the cerebellum. The hypothesis that the cerebellum is part of the lower-level sensorimotor circuit is motivated by the nature of speech deficits following cerebellar damage, which tend to be fairly low-level dysarthrias compared to the higher-level phonologic deficits found in conduction aphasics with cortical damage (Kohn, 1984; Ackermann et al., 2007; Baldo et al., 2008; Buchsbaum et al., 2011).

Architecturally and computationally the HSFC (Fig. 9.6) differs somewhat from DIVA (Fig. 9.3). In the HSFC there are two sensorimotor feedback loops, both of which involve three components: sensory target representations, motor codes for "hitting" those targets (learned via external feedback), and a sensorimotor coordinate transform network. The latter is assumed to compute the relation between sensory and motor representations of speech units. While in DIVA speech production begins with the activation of a speech sound map in the frontal lobe, in HSFC production begins with the activation of both auditory and motor units corresponding to the intended word. The auditory units comprise the target for motor unit selection in the same sense that a visually perceived object might comprise the target for activating motor units to execute a reaching action. The difference with speech is that the sensory (auditory) target is not physically present in the environment but is instead a (re-)activated mental representation of the sensory target, i.e., a sound pattern. During motor activation, the accuracy of motor unit selection is checked, internally, prior to speech articulation. If motor and sensory units match, then articulation proceeds. If there is a mismatch, a correction signal can be generated prior to articulation. Computationally, this internal "checking" mechanism is instantiated via excitatory connections from auditory target units to their previously learned corresponding motor units (via the interface network) and via inhibitory feedback connections from motor units to their corresponding auditory units. When the motor and auditory units match, the motor units will inhibit the auditory target units and carry on with their activation pattern. When there is a mismatch, motor units will inhibit non-target units in the auditory network, allowing the target auditory units to persist in exciting the correct motor units; this is the "error signal." Although a full-scale implementation of this architecture has not yet been demonstrated, a small-scale computational simulation has shown the feasibility of this architecture for internal error detection and correction (Hickok, 2012). A similar mechanism is assumed to hold at the various levels of the sensorimotor control hierarchy.

This architecture explains conduction aphasia as damage to the cortical interface network. Production is fluent because motor units are fully intact and can be activated directly from higher-level word representations. But because of damage to the auditory–motor interface, motor unit activation cannot be checked against their auditory targets and an increase in error rates is observed. Once the conduction aphasic overtly produces an error, it is immediately detected via external feedback because the auditory target network is intact and the appropriate units are activated. Subsequent attempts to correct such errors often fail, however, again because of the damage to the auditory–motor interface. Analysis of the relation between the lesions that typically cause conduction aphasia and the anatomic location of Spt, the auditory–motor interface, has shown good correspondence, lending further support for the proposed model (Buchsbaum et al., 2011).

The advantage of the HSFC model is that it incorporates an internal feedback loop that has some explanatory power regarding a speech disorder that has proven difficult to explain. It also includes a computational architecture that integrates auditory target activation, error detection, and error correction into a single process, which has some appeal from a parsimony standpoint. However, the model is far less developed than DIVA.

Although the models described herein can account for a wide range of experimental phenomena, it is important to note that these models are incomplete in their characterization of the vastly complex neural processes involved in speech production. An iterative process of: (1) generating testable predictions from a neurocomputational model; (2) experimentally testing that prediction; and (3) modifying the model as necessary to account for the experimental findings will lead to increasingly accurate accounts of the neural computations underlying speech.

FUTURE DIRECTIONS

The advent of non-invasive functional neuroimaging techniques such as PET and fMRI in the late 20th century has led to greatly accelerated progress towards understanding the neural mechanisms underlying speech production and perception. In more recent years, investigators have used multiple neuroimaging techniques in the same subject pool to overcome limitations in any given technique. For example, fMRI data, which has high spatial resolution but low temporal resolution, can be combined with either electroencephalograph (EEG) or MEG, which both have high temporal resolution but relatively low spatial resolution. New structural neuroimaging techniques such as diffusion-weighted imaging provide another dimension, namely the ability to correlate differences in brain function with differences in brain anatomy. We anticipate a continued increase in the use of multiple neuroimaging modalities as these neuroimaging technologies become more widely available and tools for analyzing the combined datasets improve.

The recent literature has seen a rapid rise in the number of publications that utilize electrocorticography (ECoG) to study the functioning human brain. ECoG involves electrodes placed on the surface of the cerebral cortex (either subdural or epidural); this technique is currently limited almost exclusively to epilepsy patients undergoing presurgical functional mapping to guide the surgical intervention. Despite the relatively limited opportunities for ECoG use, ECoG during speech production has already been used to investigate temporal-lobe activity in response to one's own voice (Creutzfeldt et al., 1989), to compare spoken and sign language (Crone et al., 2001), to identify the time course of activity across the speech network during overt (Pei et al., 2011; Leuthardt et al., 2012) and covert (Pei et al., 2011) word repetition, and to identify the spatio-temporal pattern of activity in articulatory representations in sensorimotor cortex (Bouchard et al., 2013).

We expect this trend to continue due to ECoG's unique combination of high spatial and temporal resolution.

Early EEG studies identified correlations between mental state or behavior and brain rhythms (e.g., alpha rhythm, delta rhythm), which are quasi-oscillatory neural activity patterns that can be identified by peaks in the frequency spectrum of the recorded signals. The higher spatial resolution afforded by MEG and ECoG compared to EEG has led to the development of computational methods aimed at analyzing interactions between large-scale brain rhythms in different cortical areas, or in different frequency bands within the same cortical area. Amongst other things, these techniques can identify functional connectivity between different brain regions, or functional coupling across frequency bands within a brain region. An increasing number of studies are now reporting relationships between brain rhythms and speech processing (e.g., Giraud et al., 2007; Luo and Poeppel, 2007; Leuthardt et al., 2011). We expect this trend to continue as MEG and ECoG data become more widely available.

The development of speech prostheses involving brain–computer interfaces is another promising area of future research in speech production. Guenther et al. (2009) utilized intracranial electrodes in the speech motor cortex to produce the first real-time speech prosthesis, a system that allowed an individual suffering from locked-in syndrome to produce vowels with a speech synthesizer using auditory feedback from the synthesizer to guide his speech "movements." In this study, the DIVA model prediction of a spatiotemporal formant frequency trajectory in the speech premotor/motor cortex was first verified from electrophysiologic recordings during attempted speech. A decoder was then built to translate brain activity into predicted formant frequencies, which were then synthesized into a speech signal using a formant synthesizer (Klatt, 1980). The delay between neural firing and corresponding sound output was 50 ms, which is approximately the delay from motor cortical activity to sound output in neurologically normal individuals. This real-time audio feedback allowed the user to improve his performance substantially with a small amount of practice. ECoG has also shown preliminary promise for brain–computer interfaces; for example, Leuthardt et al. (2011) developed an ECoG-based brain–computer interface that used activity in the speech network to control one-dimensional movement of a cursor.

ACKNOWLEDGMENTS

This study was supported by the National Institute on Deafness and other Communication Disorders grants R01 DC007683 (FHG), R01 DC002852 (FHG), R01 DC03681 (GH), and R01 DC009659 (GH). We thank Barbara Holland and Jason Tourville for assistance with manuscript preparation.

REFERENCES

Ackermann H, Vogel M, Petersen D et al. (1992). Speech deficits in ischaemic cerebellar lesions. J Neurol 239: 223–227.

Ackermann H, Mathiak K, Riecker A (2007). The contribution of the cerebellum to speech production and speech perception: clinical and functional imaging data. Cerebellum 6: 202–213.

Alario FX, Chainay H, Lehericy S et al. (2006). The role of the supplementary motor area (SMA) in word production. Brain Res 1076 (1): 129–143.

Alexander GE, Crutcher MD (1990). Functional architecture of basal ganglia circuits: neural substrates of parallel processing. Trends Neurosci 13: 266–271.

Alexander GE, DeLong MR, Strick KL (1986). Parallel organization of functionally segregated circuits linking basal ganglia and cortex. Annu Rev Neurosci 9: 357–381.

Alm P (2004). Stuttering and the basal ganglia circuits: a critical review of possible relations. J Commun Disord 37: 325–369.

Ames H (2009). Neural dynamics of speech perception and production: From speaker normalization to apraxia of speech, Boston University PhD Dissertation.

Andersen RA (1997). Multimodal integration for the representation of space in the posterior parietal cortex. Philos Trans R Soc Lond B Biol Sci 352 (1360): 1421–1428.

Averbeck BE, Chafee MV, Crowe DA et al. (2002). Parallel processing of serial movements in prefrontal cortex. Proc Natl Acad Sci 99 (20): 13172–13177.

Averbeck BB, Chafee MV, Crowe DA et al. (2003). Neural activity in prefrontal cortex during copying geometrical shapes: I. Single cells encode shape, sequence, and metric parameters. Exp Brain Res 150 (2): 127–141.

Baldo JV, Klostermann EC, Dronkers NF (2008). It's either a cook or a baker: patients with conduction aphasia get the gist but lose the trace. Brain Lang 105: 134–140.

Bastian HC (1887). The "muscular sense": its nature and cortical localisation. Brain 10: 1–89.

Bendor D, Wang X (2005). The neuronal representation of pitch in primate auditory cortex. Nature 436 (7054): 1161–1165.

Bohland JW, Guenther FH (2006). An fMRI investigation of syllable sequence production. Neuroimage 32: 821–841.

Bohland JW, Bullock D, Guenther FH (2010). Neural representations and mechanisms for the performance of simple speech sequences. J Cogn Neurosci 22: 1504–1529.

Bouchard KE, Mesgarani N, Johnson K et al. (2013). Functional organization of human sensorimotor cortex for speech articulation. Nature 495: 327–332.

Boyce S, Espy-Wilson CY (1997). Coarticulatory stability in American English /r/. J Acoust Soc Am 101: 3741–3753.

Browman CP, Goldstein L (1989). Articulatory gestures as phonological units. Phonology 6: 201–251.

Buchsbaum BR, Hickok G, Humphries C (2001). Role of left posterior superior temporal gyrus in phonological processing for speech perception and production. Cognitive Sci 25: 663–678.

Buchsbaum BR, Baldo J, Okada K et al. (2011). Conduction aphasia, sensory-motor integration, and phonological short-term memory - an aggregate analysis of lesion and fMRI data. Brain Lang 119: 119–128.

Bullock D, Grossberg S (1988). Neural dynamics of planned arm movements: emergent invariants and speed-accuracy properties during trajectory formation. Psychol Rev 95: 49–90.

Bullock D, Grossberg S, Guenther FH (1993). A self-organizing neural network model for redundant sensory-motor control, motor equivalence, and tool use. J Cogn Neurosci 5: 408–435.

Burnett TA, Freedland MB, Larson CR et al. (1998). Voice F0 responses to manipulations in pitch feedback. J Acoust Soc Am 103: 3153–3161.

Callan DE, Kent RD, Guenther FH et al. (2000). An auditory-feedback-based neural network model of speech production that is robust to developmental changes in the size and shape of the articulatory system. J Speech Lang Hear Res 43: 721–736.

Caplan D, Gow D, Makris N (1995). Analysis of lesions by MRI in stroke patients with acoustic-phonetic processing deficits. Neurology 45: 293–298.

Celsis P, Boulanouar K, Doyon B et al. (1999). Differential fMRI responses in the left posterior superior temporal gyrus and left supramarginal gyrus to habituation and change detection in syllables and tones. Neuroimage 9: 135–144.

Civier O, Tasko SM, Guenther FH (2010). Overreliance on auditory feedback may lead to sound/syllable repetitions: simulations of stuttering and fluency-inducing conditions with a neural model of speech production. J Fluency Disord 35: 246–279.

Civier O, Bullock D, Max L et al. (2013). Computational modeling of stuttering caused by impairments in a basal ganglia thalamo-cortical circuit involved in syllable selection and initiation. Brain Lang 126: 263–278.

Clower WT, Alexander GE (1998). Movement sequence-related activity reflecting numerical order of components in supplementary and presupplementary motor areas. J Neurophysiol 80: 1562–1566.

Creutzfeldt O, Ojemann G, Lettich E (1989). Neuronal-activity in the human lateral temporal-lobe.2. Responses to the subjects own voice. Exp Brain Res 77: 476–489.

Crone NE, Hao L, Hart Jr J et al. (2001). Electrocorticographic gamma activity during word production in spoken and sign language. Neurology 57 (11): 2045–2053.

D'Esposito M, Aguirre GK, Zarahn E et al. (1998). Functional MRI studies of spatial and nonspatial working memory. Cogn Brain Res 7 (1): 1–13.

Damasio H, Damasio AR (1980). The anatomical basis of conduction aphasia. Brain 103: 337–350.

Davidson PR, Wolpert DM (2005). Widespread access to predictive models in the motor system: a short review. J Neural Eng 2 (3): S313–S319.

Delattre P, Freeman D (1968). A dialect study of American R's by X-ray motion picture. Linguistics 44: 29–68.

Desmurget M, Grafton S (2000). Forward modeling allows feedback control for fast reaching movements. Trends Cogn Sci 4 (11): 423–431.

Fant G (1960). Acoustic theory of speech production, Mouton, The Hague.

Fiez JA, Raife EA, Balota DA et al. (1996). A positron emission tomography study of the short-term maintenance of verbal information. J Neurosci 16 (2): 808–822.

Gabrieli JDE, Poldrack RA, Desmond JE (1998). The role of left prefrontal cortex in language and memory. Proc Natl Acad Sci 95: 906–913.

Geschwind N (1965). Disconnexion syndromes in animals and man II. Brain 88: 585–644.

Ghosh SS, Tourville JA, Guenther FH (2008). A neuroimaging study of premotor lateralization and cerebellar involvement in the production of phonemes and syllables. J Speech Lang Hear Res 51: 1183–1202.

Giraud AL, Kleinschmidt A, Poeppel D et al. (2007). Endogenous cortical rhythms determine cerebral specialization for speech perception and production. Neuron 56: 1127–1134.

Golfinopoulos E, Tourville JA, Bohland JW et al. (2011). fMRI investigation of unexpected somatosensory feedback perturbation during speech. Neuroimage 55: 1324–1338.

Goodglass H (1992). Diagnosis of conduction aphasia. In: SE Kohn (Ed.), Conduction aphasia: 39-49, Lawrence Erlbaum, Hillsdale, NJ, pp. 39–49.

Gracco VL (1994). Some organizational characteristics of speech movement control. J Speech Hear Res 37: 4–27.

Gracco VL, Lofqvist A (1994). Speech motor coordination and control: evidence from lip, jaw, and laryngeal movements. J Neurosci 14: 6585–6597.

Guenther FH (1992). Neural models of adaptive sensory-motor control for flexible reaching and speaking, Boston University Ph.D. Dissertation.

Guenther FH (1995a). Speech sound acquisition, coarticulation, and rate effects in a neural network model of speech production. Psychol Rev 102: 594–621.

Guenther FH (1995b). A modeling framework for speech motor development and kinematic articulator control. In: K Elenius, P Branderud (Eds.), Proceedings of the XIIIth International Congress of Phonetic Sciences, Stockholm, Sweden, 13–19 August, 1995, KTH and Stockholm University, Stockholm, Sweden, Vol. 2: 92–99.

Guenther FH, Vladusich T (2012). A neural theory of speech acquisition and production. J Neuroling 25: 408–422.

Guenther FH, Hampson M, Johnson D (1998). A theoretical investigation of reference frames for the planning of speech movements. Psychol Rev 105: 611–633.

Guenther FH, Espy-Wilson CY, Boyce SE et al. (1999). Articulatory tradeoffs reduce acoustic variability during American English /r/ production. J Acoust Soc Am 105: 2854–2865.

Guenther FH, Ghosh SS, Tourville JA (2006). Neural modeling and imaging of the cortical interactions underlying syllable production. Brain Lang 96: 280–301.

Guenther FH, Brumberg JS, Wright EJ et al. (2009). A wireless brain–machine interface for real-time speech synthesis. PLoS One 4 (12), e8218+.

Hargrave S, Kalinowski JS, Stuart A et al. (1994). Effect of frequency-altered feedback on stuttering frequency at normal and fast speech rates. J Speech Hear Res 37: 1313–1319.

Harrington DL, Haaland KY (1998). Sequencing and timing operations of the basal ganglia. In: DA Rosenbaum, CE Collyer (Eds.), Timing of behavior: neural, psychological, and computational perspectives. MIT Press, Cambridge, MA, pp. 35–61.

Hickok G (2012). Computational neuroanatomy of speech production. Nat Rev Neurosci 13: 135–145.

Hickok G, Poeppel D (2004). Dorsal and ventral streams: a framework for understanding aspects of the functional anatomy of language. Cognition 92: 67–99.

Hickok G, Buchsbaum B, Humphries C et al. (2003). Auditory-motor interaction revealed by fMRI: speech, music, and working memory in area Spt. J Cogn Neurosci 15: 673–682.

Hillis AE, Work M, Barker PB et al. (2004). Reexamining the brain regions crucial for orchestrating speech articulation. Brain 127 (7): 1479–1487.

Ho AK, Bradshaw JL, Cunnington R et al. (1998). Sequence heterogeneity in Parkinsonian speech. Brain Lang 64: 122–145.

Houde JF, Jordan MI (1998). Sensorimotor adaptation in speech production. Science 279 (5354): 1213–1216.

Houde JF, Nagarajan SS, Sekihara K et al. (2002). Modulation of the auditory cortex during speech: an MEG study. J Cogn Neurosci 14: 1125–1138.

Ito T, Ostry DJ (2012). Speech sounds alter facial skin sensation. J Neurophysiol 107 (1): 442–447.

Ito T, Tiede M, Ostry DJ (2009). Somatosensory function in speech perception. Proc Natl Acad Sci U S A 106 (4): 1245–1248.

Jonas S (1987). The supplementary motor region and speech. In: E Perecman (Ed.), The frontal lobes revisited, IRBN Press, New York, pp. 241–250.

Jurgens U (1984). The efferent and efferent connections of the supplementary motor area. Brain Res 300: 63–81.

Kawato M (1999). Internal models for motor control and trajectory planning. Curr Opin Neurobiol 9 (6): 718–727.

Kent RD (2000). Research on speech motor control and its disorders: a review and prospective. J Commun Disord 33: 391–427.

Kerns JG, Cohen JD, Stenger VA et al. (2004). Prefrontal cortex guides context-appropriate responding during language production. Neuron 43: 283–291.

Klatt DH (1980). Software for a cascade/parallel formant synthesizer. J Acoust Soc Am 67: 971–995.

Kohn SE (1984). The nature of the phonological disorder in conduction aphasia. Brain Lang 23: 97–115.

Krakauer J, Ghez C (1999). Voluntary movement. In: ER Kandel, JH Schwartz, TM Jessell (Eds.), Principles of neural science, 4th edn. McGraw Hill, New York, pp. 756–781.

Kumar S, Schönwiesner M (2012). Mapping human pitch representation in a distributed system using depth-electrode recordings and modeling. J Neurosci 32: 13348–13351.

Larson CR, Burnett TA, Bauer JJ et al. (2001). Comparison of voice F0 responses to pitch-shift onset and offset conditions. J Acoust Soc Am 110: 2845–2848.

Lehericy S, Ducros M, Krainik A et al. (2004). 3-D diffusion tensor axonal tracking shows distinct SMA and pre-SMA projections to the human striatum. Cereb Cortex 14 (12): 1302–1309.

Leuthardt EC, Gaona C, Sharma M et al. (2011). Using the electrocorticographic speech network to control a brain–computer interface in humans. J Neural Eng 8: 036004.

Leuthardt EC, Pei XM, Breshears J et al. (2012). Temporal evolution of gamma activity in human cortex during an overt and covert word repetition task. Front Hum Neurosci 6: 99.

Levelt WJ, Wheeldon L (1994). Do speakers have access to a mental syllabary? Cognition 50: 239–269.

Liberman AM, Mattingly IG (1985). The motor theory of speech perception revised. Cognition 21: 1–36.

Liberman AM, Cooper FS, Shankweiler DP et al. (1967). Perception of the speech code. Psychol Rev 74 (6): 431–461.

Lindblom B, Lubker J, Gay T (1979). Formant frequencies of some fixed-mandible vowels and a model of speech motor programming by predictive simulation. J Phon 7: 147–161.

Luo H, Poeppel D (2007). Phase patterns of neuronal responses reliably discriminate speech in human auditory cortex. Neuron 54: 1001–1010.

Luppino G, Matelli M, Camarda R et al. (1993). Corticocortical connections of area F3 (SMA-proper) and area F6 (pre-SMA) in the macaque monkey. J Comp Neurol 338 (1): 114–140.

MacCulloch MJ, Eaton R, Long E (1970). The long term effect of auditory masking on young stutterers. Br J Disord Commun 5: 165–173.

MacNeilage PF (1998). The frame/content theory of evolution of speech production. Behav Brain Sci 21: 499–511.

Maeda S (1990). Compensatory articulation during speech: evidence from the analysis and synthesis of vocal tract shapes using an articulatory model. In: WJ Hardcastle, A Marchal (Eds.), Speech production and speech modeling. Kluwer Academic, Boston, pp. 131–149.

Matsumoto R, Nair DR, LaPresto E et al. (2007). Functional connectivity in human cortical motor system: a cortico-cortical evoked potential study. Brain 130 (1): 181–197.

McGurk H, MacDonald J (1976). Hearing lips and seeing voices. Nature 264 (5588): 746–748.

Miall RC, Wolpert DM (1996). Forward models for physiological motor control. Neural Netw 9 (8): 1265–1279.

Middleton FA, Strick PL (1997). Cerebellar output channels. Int Rev Neurobiol 41: 61–82.

Middleton FA, Strick PL (2000). Basal ganglia and cerebellar loops: motor and cognitive circuits. Brain Res Rev 31: 236–250.

Murdoch BE (2001). Subcortical brain mechanisms in speech and language. Folia Phoniatr Logop 53: 233–251.

Nieto-Castanon A, Guenther FH, Perkell JS et al. (2005). A modeling investigation of articulatory variability and acoustic stability during American English /r/ production. J Acoust Soc Am 117: 3196–3212.

Nozari N, Dell GS, Schwartz MF (2011). Is comprehension necessary for error detection? A conflict-based account of monitoring in speech production. Cognit Psychol 63: 1–33.

Numminen J, Curio G (1999). Differential effects of overt, covert and replayed speech on vowel-evoked responses of the human auditory cortex. Neurosci Lett 272: 29–32.

Numminen J, Salmelin R, Hari R (1999). Subject's own speech reduces reactivity of the human auditory cortex. Neurosci Lett 265: 119–122.

Oppenheim GM, Dell GS (2008). Inner speech slips exhibit lexical bias, but not the phonemic similarity effect. Cognition 106: 528–537.

Passingham RE (1993). The frontal lobes and voluntary action, Oxford University Press, Oxford.

Peeva MG, Guenther FH, Tourville JA et al. (2010). Distinct representations of phonemes, syllables, and supra-syllabic sequences in the speech production network. Neuroimage 50: 626–638.

Pei X, Leuthardt EC, Gaona CM et al. (2011). Spatiotemporal dynamics of electrocorticographic high gamma activity during overt and covert word repetition. Neuroimage 54: 2960–2972.

Perkell JS, Guenther FH, Lane H et al. (2000). A theory of speech motor control and supporting data from speakers with normal hearing and profound hearing loss. J Phon 28: 233–272.

Perkell JS, Guenther FH, Lane H et al. (2004a). The distinctness of speakers' productions of vowel contrasts is related to their discrimination of the contrasts. J Acoust Soc Am 116 (4): 2338–2344.

Perkell JS, Matthies ML, Tiede M et al. (2004b). The distinctness of speakers' /s-sh/ contrast is related to their auditory discrimination and use of an articulatory saturation effect. J Speech Lang Hear Res 47: 1259–1269.

Petrides M (1991). Functional specialization within the dorsolateral frontal cortex for serial order memory. Proc R Soc Lond B Biol Sci 246 (1317): 299–306.

Pickett ER, Kuniholm E, Protopapas A et al. (1998). Selective speech motor, syntax and cognitive deficits associated with bilateral damage to the putamen and the head of the caudate nucleus: a case study. Neuropsychologia 36: 173–188.

Rizzolatti G, Craighero L (2004). The mirror-neuron system. Annu Rev Neurosci 27: 169–192.

Rizzolatti G, Fogassi L, Gallese V (1997). Parietal cortex: from sight to action. Curr Opin Neurobiol 7 (4): 562–567.

Rosenbaum DA, Engelbrecht SE, Bushe MM et al. (1993). Knowledge model for selecting and producing reaching movements. J Mot Behav 25: 217–227.

Rosenbaum DA, Loukopoulos LD, Meulenbroek RGJ et al. (1995). Planning reaches by evaluating stored postures. Psychol Rev 192: 28–67.

Sancier ML, Fowler CA (1997). Gestural drift in a bilingual speaker of Brazilian Portuguese and English. J Phon 25: 421–436.

Sanes JN, Mauritz KH, Evarts EV et al. (1984). Motor deficits in patients with large-fiber sensory neuropathy. Proc Natl Acad Sci U S A 81: 979–982.

Schmahmann JD, Pandya DN (1997). The cerebrocerebellar system. Int Rev Neurobiol 41: 31–60.

Shadmehr R, Mussa-Ivaldi FA (1994). Adaptive representation of dynamics during learning of a motor task. J Neurosci 14: 3208–3224.

Shima K, Tanji J (2000). Neuronal activity in the supplementary and presupplementary motor areas for temporal organization of multiple movements. J Neurophysiol 84: 2148–2160.

Sperry RW (1950). Neural basis of the spontaneous optokinetic response produced by visual inversion. J Comp Physiol Psychol 43 (6): 482–489.

Stevens K (1998). Acoustic phonetics. MIT Press, Cambridge, MA.

Tian X, Poeppel D (2010). Mental imagery of speech and movement implicates the dynamics of internal forward models. Front Psychol 1: 166.

Tourville JA, Guenther FH (2011). The DIVA model: a neural theory of speech acquisition and production. Lang Cog Proc 26: 952–981.

Tourville JA, Reilly KJ, Guenther FH (2008). Neural mechanisms underlying auditory feedback control of speech. Neuroimage 39: 1429–1443.

Van Riper C (1982). The nature of stuttering, 2nd edn. Prentice-Hall, Englewood Cliffs, NJ.

von Helmholtz H (1925). Helmholtz's Treatise on Physiological Optics, 3rd Edition, 1910, translated by J.P.C. Southall, Optical Society of America, New York.

Waldstein RS (1989). Effects of postlingual deafness on speech production: implications for the role of auditory feedback. J Acoust Soc Am 88: 2099–2144.

Wilson SM, Saygin AEP, Sereno MI et al. (2004). Listening to speech activates motor areas involved in speech production. Nat Neurosci 7 (7): 701–702.

Wise RJ, Greene J, Buchel C et al. (1999). Brain regions involved in articulation. Lancet 353: 1057–1061.

Wolpert DM, Ghahramani Z, Jordan MI (1995). An internal model for sensorimotor integration. Science 269: 1880–1882.

Yates AJ (1963). Delayed auditory feedback. Psychol Bull 60: 213–251.

Ziegler W, Kilian B, Deger K (1997). The role of the left mesial frontal cortex in fluent speech: evidence from a case of left supplementary motor area hemorrhage. Neuropsychologia 35 (9): 1197–1208.

Handbook of Clinical Neurology, Vol. 129 (3rd series)
The Human Auditory System
G.G. Celesia and G. Hickok, Editors

Chapter 10

White-matter pathways for speech and language processing

ANGELA D. FRIEDERICI*

Department of Neuropsychology, Max Planck Institute for Human Cognitive and Brain Sciences, Leipzig, Germany

INTRODUCTION

The white-matter fiber bundles in the brain play a crucial role in the transmission of information between different regions. These fiber bundles are composed of millions of axons that are surrounded by myelin, which is essential for fast transmission of electrical impulses (Wake et al., 2011).

The relevance of white-matter fiber bundles for speech perception and speech production was first recognized in the late 19th century, when they were proposed to form possible connections between the different language centers (Wernicke, 1874). In his model, Wernicke (1874) proposed a speech production center (based on the work of Broca, 1861), a sensory language center, and a connection between the centers supporting their interaction. Broca's area in the inferior frontal cortex and Wernicke's area in the superior temporal cortex as the classical language areas are subparts of temporofrontal language network that also support higher-language function processes such as grammatical assignment and attribution of meaning (Vigneau et al., 2006; Hickok and Poeppel, 2007; Friederici, 2011).

The connections between the classic language regions that have been identified over the past decades are multifold. Initially, two broad processing streams connecting temporal and frontal areas were proposed: a ventral stream supporting sound-to-meaning mapping and a dorsal stream subserving sound-to-motor mapping (Hickok and Poeppel, 2000; Rauschecker and Scott, 2009).

In recent years, new imaging techniques, such as diffusion-weighted magnetic resonance imaging (dMRI), also called diffusion tensor imaging (DTI), allow white-matter fiber bundles to be tracked *in vivo* in the human brain (Mori and Zijl, 2002). DTI can provide information about the internal fibrous structure based on the measure of water diffusion. Since water will diffuse more rapidly in the direction aligned with the internal structure, the principal direction of the diffusion tensor can be used to infer white-matter connectivity in the brain. Often DTI is used to identify the different fiber tracts in the human brain (Behrens et al., 2003; Catani and Thiebaut de Schotten, 2008; Berthier et al., 2012). This approach is primarily structure-based and independent from function, but once identified, the role of those tracts that connect brain areas known to be involved in particular processes, e.g., language, are discussed (Catani et al., 2005; Makris and Pandya, 2009). This approach only allows very indirect conclusions about a particular fiber tract's function.

A second approach is called a function-based approach. It combines functional MRI (fMRI) and dMRI, and thereby allows specification of a given fiber tract's function, indirectly, in a two-step approach. In a first step, particular brain regions relevant for a specific language task are identified in a functional experiment by fMRI, and then in a second step these regions are used as seed regions from which a tractogram is calculated (Friederici et al., 2006; Saur et al., 2008). The resulting fiber tract is taken to transmit information relevant for the particular function investigated by the fMRI, although it is not always certain whether the tracing originates only from the seed region or also from adjacent parts. Within this function-based tractography approach, two methodologies can be applied: probabilistic and deterministic tracking. The probabilistic approach takes one seed point, which is neurofunctionally defined, as the starting point of tractography (Friederici et al., 2006). The deterministic approach takes two regions that are activated simultaneously as a function of a particular task and calculates the fiber tract between these (Saur et al., 2008).

The most direct test of a particular fiber tract's function, however, is to investigate language performance in

*Correspondence to: Angela D. Friederici, Max Planck Institute for Human Cognitive and Brain Sciences, Stephanstraße 1A, 04103 Leipzig, Germany. Tel: +49-341-99-40-112, Fax: +49-341-99-40-113, E-mail: friederici@cbs.mpg.de

patients with lesions in the white-matter fiber tracts (Galantucci et al., 2011; Papoutsi et al., 2011; Turken and Dronkers, 2011; Wilson et al., 2011), or to correlate language performance with the degree of myelination of particular fiber tracts in the developing brain (Brauer et al., 2011).

Based on these approaches different white-matter pathways that are relevant for auditory language processing have been identified. Most generally the long-range fiber bundles which connect the frontal cortex (including the classic language region of Broca's area and premotor cortex (PMC)) and the temporal cortex (including the classic language region of Wernicke's area and the auditory cortex) can be classified into two anatomic pathways, i.e., the dorsal and the ventral pathway. Each of the pathways consists of more than one fiber bundle. Below we will specify the different subparts of the dorsal and ventral pathways, both structurally and functionally. Note that, in addition to these long-range connections between the frontal and the temporal cortex, there are a number of short-range connections within the frontal cortex and the temporal cortex. Here we will only describe those that have been considered as relevant for speech and language.

THE LANGUAGE-RELEVANT BRAIN REGIONS

White-matter pathways that are relevant for speech and language can be defined neuroanatomically by their different target regions. Interestingly, these target regions often differ in their function (Friederici, 2002; Vigneau et al., 2006; Price, 2010), but moreover, with respect to their cytoarchitectonic structure (Brodmann, 1909; Amunts and Zilles, 2012) and their receptorarchitectonic structure (Zilles and Amunts, 2009; Amunts et al., 2010). Moreover, the different regions can be parcellated on the basis of their particular structural connectivity to other regions (Johansen-Berg et al., 2004; Anwander et al., 2007). Thus, target regions can be identified on the basis of different structural and functional aspects.

Left frontal cortex

The language-relevant areas in the frontal cortex have been cytoarchitectonically defined to span along a posterior-to-anterior gradient from the PMC and Brodmann's area 6 (BA 6) in the precentral gyrus to BA 44 (the pars opercularis) and BA 45 (the pars triangularis), which together constitute Broca's area in the inferior frontal gyrus (IFG), and then they continue to the more anterior located BA 47 (the pars orbitalis) (for a schematic view, see Fig. 10.1). Spanning more medial and ventrally to BA 44 and BA 45, there is the frontal operculum (FOP), which cytoarchitectonically subdivides into different subregions (op8 and op9; Amunts et al., 2010). Recent receptorarchitectonic analyses even suggest that area 44 can be further divided into a dorsal (44d) and a ventral (44v) portion, and area

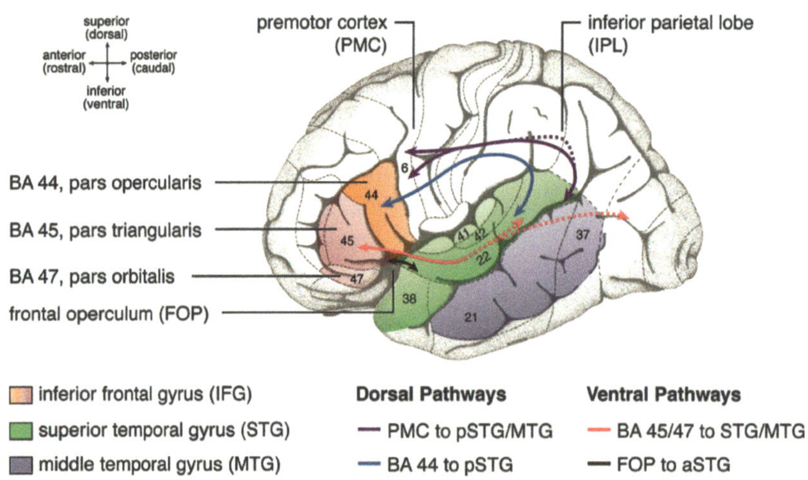

Fig. 10.1. Language-relevant brain regions and schematic fiber tracts (displayed in the left hemisphere). Numbers represent cytoarchitectonically defined Brodmann areas (BAs). BA 44 and BA 45 together constitute Broca's area. BA 22 constitutes Wernicke's area. Different pathways are color-coded according to the color legend in the figure. The connection from the prefrontal motor cortex (PMC) to the posterior superior temporal gyrus (pSTG)/middle temporal gyrus (MTG) involves the superior longitudinal fascicle (SLF). The connection from BA 44 to the pSTG involves the arcuate fascicle (AF). The connection from BA 45/47 to the superior temporal gyrus (STG)/MTG involves the extreme capsule fiber system (ECFS)/longitudinal inferior-fronto-occipital fascicle (IFOF). The connection from the frontal operculum to the anterior superior temporal gyrus (aSTG) involves the uncinate fascicle (UF). pMTG, posterior middle temporal gyrus. (Adapted from Friederici and Gierhan, 2013.)

45 can be separated into an anterior (45a) and a posterior (45p) portion (Amunts et al., 2010).

Functionally, these areas have been argued to support different aspects of language processing: BA 6 is involved in phonology and articulation; BA 44, together with area 45p, is involved in syntactic structure building; and area 45a, together with BA 47, supports semantic processes (Bookheimer, 2002; Friederici, 2002; Uddén and Bahlmann, 2012). Note that there is still some debate about the role of Broca's area in processing syntactic information (Musso et al., 2003; Hagoort, 2005; Novick et al., 2005; Grodzinsky and Amunts, 2006) and which part of Broca's area (BA 44/45) is involved: is it BA 44 proper or BA 45/BA 44 (Friederici, 2002; Hagoort, 2005; Grodzinsky and Santi, 2008; Tyler and Marslen-Wilson, 2008; Rogalsky and Hickok, 2011). This may be partly due to the fact that different fMRI studies have used different stimulus material: either natural sentences containing semantic and syntactic information, or so-called "jabberwocky" sentences that only contain syntactic information (function words) and no semantic information (pseudowords), or random word lists containing function words and pseudowords. Moreover, they have employed different presentation modes (written vs auditory) and different tasks (such as sentence comprehension, grammaticality judgement, plausibility judgement, or probe identification) (Caplan et al., 2008).

Left temporal cortex

The temporal cortex neuroanatomically subdivides into the inferior, middle, and superior temporal gyri (STGs). Cytoarchitectonically, the STG can be differentiated into BA 41/42, as the primary and secondary auditory cortex (roughly "core" and "belt" in animal research terminology), and BA 22 (very roughly, "parabelt"), which constitutes the large region of Wernicke's area (Brodmann, 1909; Morosan et al., 2005). Functional connectivity-based analyses suggest a subdivision of the STG into three parts: (1) the middle part of the STG (mSTG) located lateral to Heschl's gyrus (which contains BA 41/core auditory areas) and the primary auditory cortex; (2) a portion located anterior to the Heschl's gyrus (aSTG); and (3) a portion located posterior to Heschl's gyrus (pSTG) (Upadhyay et al., 2008) (Fig. 10.2).

This subdivision of the STG into subparts is generally in line with proposed functional differentiation of the STG and superior temporal sulcus (STS). An early functional imaging study already proposed a subdivision of the STS/STG into an anterior, middle, and posterior portion, with the middle portion being involved in sound and speech processing and both the anterior and posterior portions being involved only in speech (Giraud and Price, 2001). A recent meta-analysis over 115 studies claims a posterior-to-anterior gradient with increasing complexity of the auditory stimulus, going from phoneme processing (mSTG) to the processing of words and phrases (aSTG) (DeWitt and Rauschecker, 2012) (Fig. 10.3).

The middle temporal gyrus (MTG) is also considered to be language-relevant (Turken and Dronkers, 2011), subserving lexical-semantic and conceptual semantic aspects in particular (Patterson et al., 2007; Lau et al., 2008).

Left parietal cortex

In addition to the left frontal and temporal cortices, the left parietal cortex, in particular the inferior parietal lobe (IPL), is also seen to be involved in language processing whenever language tasks that require phonologic working memory have to be performed. Initially, the left inferior and posterior parietal cortices, in particular, were found to support aspects of phonologic working memory and storage (Jonides et al., 1998; Bushara et al., 1999; Gruber and von Cramon, 2001). Although this region is not considered as one of the classic language areas, it appears to come into play in language tasks that require certain memory capacities, such as in a verbal working memory task or a repetition task in which auditorily perceived words or sentences have to be recalled and/or actively reproduced (Jonides et al., 1998; Ravizza et al., 2011). Moreover, this region is activated during sentence comprehension tasks that require the binding of two distant elements across a number of intervening elements in a sentence (Grossman et al., 2002; Meyer et al., 2012). This was demonstrated in an fMRI sentence comprehension study varying systematically the working memory demands (long vs short distance of related elements) and the reordering demands of elements (subject-first structure vs object-first structure). The effect of reordering was found in BA 44 and the effect of storage in temporoparietal regions (Meyer et al., 2012) (Fig. 10.4). Note that temporoparietal regions have been previously claimed to be involved in the processing of working memory and language performance (Hickok et al., 2003; Buchsbaum and D'Esposito, 2008; Buchsbaum et al., 2011).

THE LANGUAGE PATHWAYS

Language regions in the frontal cortex and in the temporal cortex are neuroanatomically connected by both dorsally located and ventrally located long-range fiber tracts known as the dorsal and the ventral pathways.

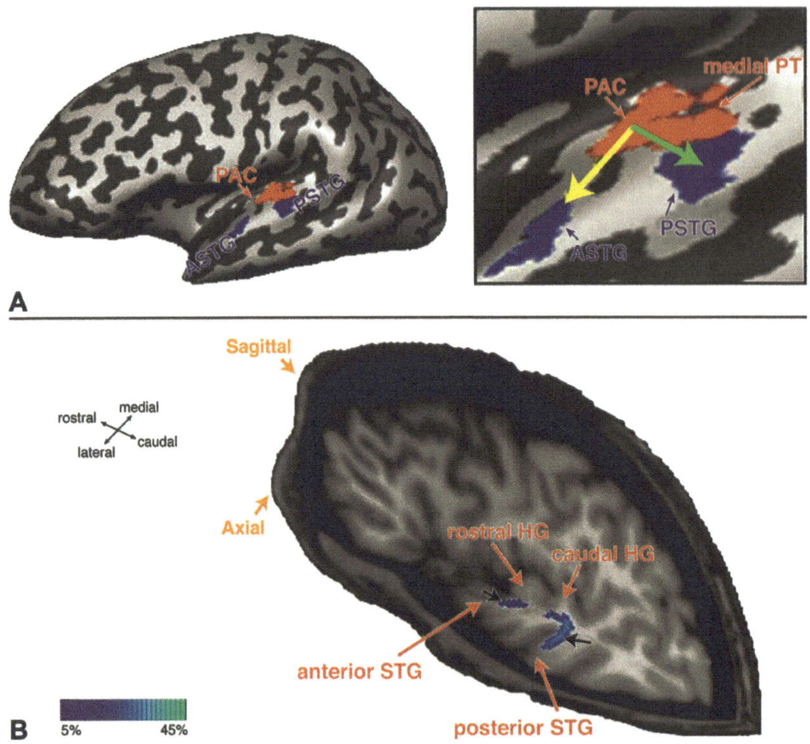

Fig. 10.2. Connectivities in the auditory cortex. (**A**) Effective connectivity in auditory cortex. An effective connectivity map in the auditory cortex is shown for a single subject. The red region corresponds to the reference region in the primary auditory cortex (PAC) as well as the medial regions of the planum temporale (PT), which showed functional connectivity with PAC. Blue regions depict the rostrolateral (lateral planum polare + anterior superior temporal gyrus, referred to as ASTG) and caudolateral (lateral PT + posterior superior temporal gyrus, referred to as PSTG) auditory cortex regions identified as being targets of influence of PAC. (**inset**) An enlarged view of the effective connectivity map is shown on the inflated brain surface. The green arrow depicts a caudal pathway as defined by granger causality mapping (GCM) analysis, and the yellow arrow represents a possible rostral pathway. (**B**) Rostral and caudal fiber pathways in auditory cortex. Two-dimensional probabilistic maps of the rostral and caudal fiber pathways (marked by the black arrows) are shown on sagittal and axial T1-weighted slices. Whereas the caudal pathway projects from more caudal regions of the rostral Heschl's gyrus, the rostral pathway projects from the rostral segments of the same gyrus. The two pathways also have distinct fiber densities or probabilities, as defined by probability values. High fiber density is indicated by light blue-green voxels; low fiber density is indicated by dark blue voxels. (Adapted from Upadhyay et al., 2008.)

Dorsal pathway

The distinction between a dorsal and a ventral pathway connecting the temporal cortex and the frontal cortex is based on the neuroanatomic differentiation in the monkey between the "where" stream, which goes from the auditory cortex to the dorsolateral prefrontal cortex and the ventrally located "what" stream, originally defined by Rauschecker and Tian (2000). The dorsal pathway was initially described as emanating from the planum temporale in the posterior superior temporal region running through the IPL to the dorsal PMC (BA 6) and was seen as responsible for sensory-to-motor mapping in speech processing (Hickok and Poeppel, 2000, 2007; Rauschecker and Scott, 2009).

Structurally, however, the dorsal pathway appears to consist of more than one white-matter fiber bundle (Catani et al., 2005; for a recent review, see Gierhan, 2013). Here, we will discuss two major fiber bundles, which can be differentiated by their termination regions in the frontal cortex and possibly by their particular language subfunctions as well. One dorsal fiber bundle connects the temporal cortex with the PMC through the mediation of the IPL, whereas the other bundle connects the temporal cortex to Broca's area, in particular BA 44 (Perani et al., 2011). Interestingly, this fiber bundle is not yet myelinized in newborns (Fig. 10.5).

Functionally, the dorsal stream has been discussed as being involved in speech repetition, that is, the production and articulation of perceived speech sounds (Hickok and Poeppel, 2000, 2007). A function-based DTI study has shown that the fiber tract terminating in the dorsal PMC can indeed be allocated to sensory-to-motor mapping functions (Saur et al., 2008) (Fig. 10.6).

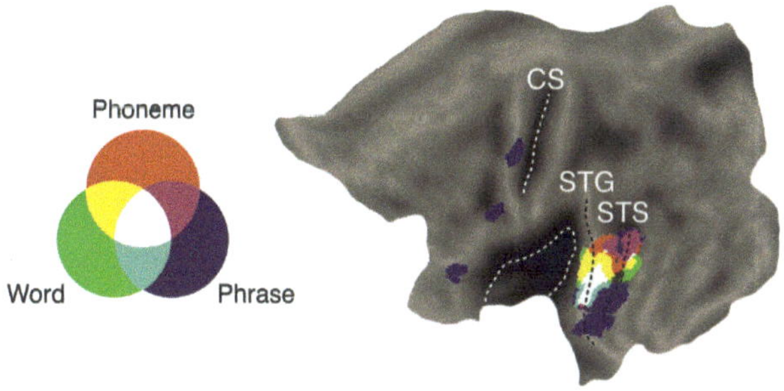

Fig. 10.3. Flat-map presentation of activation likelihood estimation for more than 115 studies on speech perception. Activations for phonemes (red), words (green), and phrases (blue), as well as the overlap (white). CS, central sulcus; STG, superior temporal gyrus; STS, superior temporal sulcus. (Adapted from DeWitt and Rauschecker, 2012.)

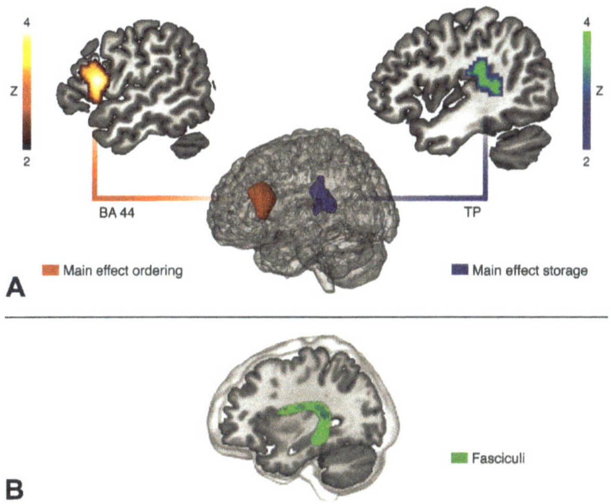

Fig. 10.4. Functional activations (FAs) and connectivity pattern. (**A**) Brain activations for the ordering effect (red cluster) and the storage effect (blue cluster). For the ordering factor, 64.8% of the obtained cluster is in the left BA 44 with a peak at $x=-54$, $y=10$, $z=18$ ($z=3.76$). For the storage factor, the activation is found in the medial left temporoparietal (TP) region, peaking at $x=-42$, $y=-40$, $z=10$ ($z=3.66$). (**B**) Correlation of FA values with signal change of the storage effect in left TP region (blue cluster in **A**) as compared to the ordering effect in the BA 44 (red cluster in **A**). The regression analysis resulting from the averaging of the individual fiber tracts was carried out in a volume of 1561 voxels of the arcuate fascicle/superior longitudinal fascicle (green). Clusters were thresholded at pb0.05 and a minimum cluster size of seven suprathreshold voxels to control for family-wise error at pb0.005. (Reproduced from Meyer et al., 2012.)

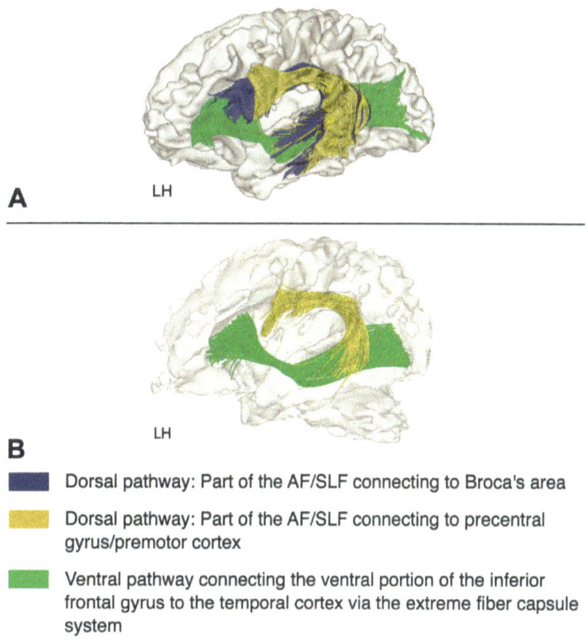

Fig. 10.5. Structural connectivity results. (**A**) Fiber tracking of diffusion tensor imaging data for adults for speech-relevant regions with seed in Broca's area and seed in the precentral gyrus/premotor cortex. Two dorsal pathways are present in adults – one connecting the temporal cortex via the arcuate fascicle (AF) and the superior longitudinal fascicle (SLF) to the inferior frontal gyrus, i.e., Broca's area (blue), and one connecting the temporal cortex via the AF and SLF to the precentral gyrus, i.e., premotor cortex (yellow). The ventral pathway connects the ventral inferior frontal gyrus via the extreme capsule fiber system to the temporal cortex (green). Left hemisphere (LH). (**B**) Fiber tracking for newborns. Dorsally only one pathway can be detected, i.e., the pathway connecting the temporal cortex to the precentral gyrus, i.e. premotor cortex (yellow). The ventral pathway (green) is present in newborns. (Adapted from Perani et al., 2011.)

The second major dorsal fiber tract terminates in BA 44, i.e., in the posterior portion of Broca's area, and connects this area directly with the posterior temporal cortex (Fig. 10.6). This fiber bundle, which connects the language areas Broca's area and Wernicke's area, has long been described as relevant for language processing. It is classically labeled as the arcuate fascicle (AF). Catani et al. (2005) tracked the AF using dMRI and already suggested two dorsal connections, one connecting Broca's region and Wernicke's region directly, and one connecting these regions indirectly. The figure in their publication, however, suggests the latter fiber bundle terminates in the PMC rather than in Broca's area (Fig. 10.7). Nevertheless, this early dMRI work provided some indication of there being two major dorsal connections, a direct one and an indirect one mediated by the IPL.

The neuroanatomic description and the labeling of the dorsally located fiber bundles in the literature are unfortunately somewhat heterogeneous and have even changed over the past decades. Today, the fiber bundles that connect the IPL to the PMC are most commonly named the superior longitudinal fascicle (SLF), which has three subcomponents: SLF I, SLF II, and SLF III,

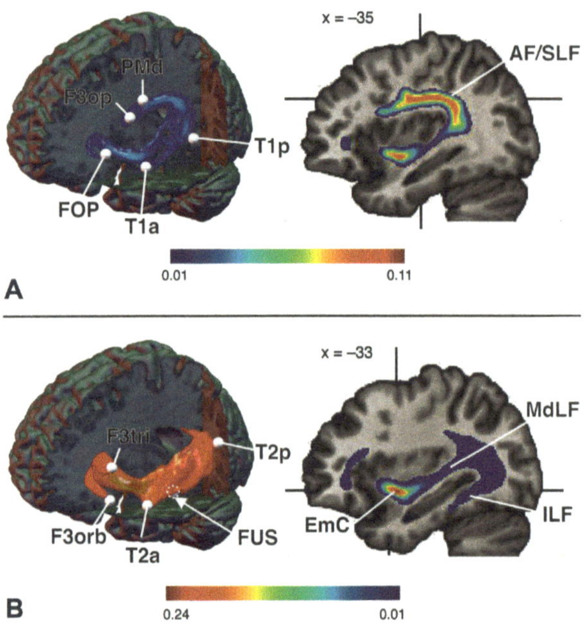

Fig. 10.6. Fiber tracking results. (**A**) Composite fiber networks subserving repetition and (**B**) comprehension computed by averaging the pairwise connections of 33 subjects defined in the repetition and comprehension experiment, respectively. Three-dimensional tractography renderings visualize the spatial orientation of both networks. Crosshairs on sagittal sections indicate the orientation of the coronal and axial sections. Maximum PIBI (probability index-forming part of the bundle of interest) values are given at the top of the color bar. T1a/p, anterior/posterior superior temporal gyrus; T2a/p, anterior/posterior middle temporal gyrus; FUS, fusiform gyrus; F3orb/tri/op, pars orbitalis/triangularis and opercularis of the inferior frontal gyrus; FOP, deep frontal operculum; PMd, dorsal premotor cortex; EmC, extreme capsule; AF/SLF, arcuate and superior longitudinal fascicle; MdLF/ILF, middle and inferior longitudinal fascicle. (Adapted from Saur et al., 2008. Copyright (2008) National Academy of Sciences, U.S.A.)

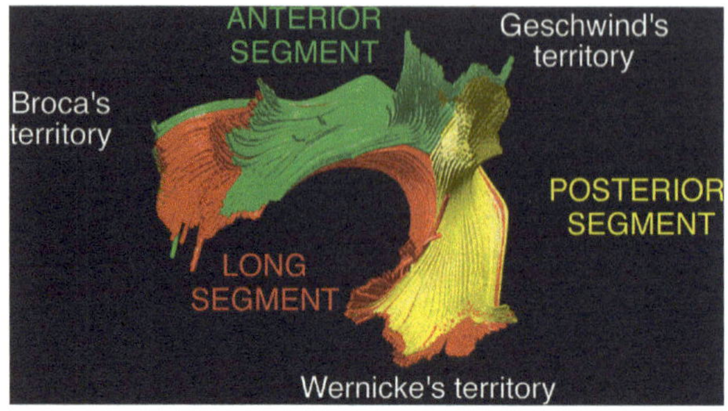

Fig. 10.7. Tractography reconstruction of the arcuate fascicle. Broca's and Wernicke's territories are connected through direct and indirect pathways in the average brain. The direct pathway (long segment shown in red) runs medially and corresponds to classical descriptions of the arcuate fascicle. The indirect pathway runs laterally and is composed of an anterior segment (green) connecting the inferior parietal cortex (Geschwind's territory) and Broca's territory and a posterior segment (yellow) connecting Geschwind's and Wernicke's territories. (Reproduced from Catani et al., 2005.)

initially identified in the monkey (Petrides and Pandya, 1984; Schmahmann et al., 2007). SLF I is the most dorsally located bundle connecting the superior parietal cortex with the superior frontal cortex, while SLF II and III are located more inferiorly. SLF III connects the PMC to the IPL. The nomenclature of SLF I–III has been adapted to the human brain, with SLF II and III connecting the PMC to the IPL being highlighted as relevant for language (Makris et al., 2005; Frey et al., 2008; Thiebaut de Schotten et al., 2012). The fiber bundle connecting the IPL to the temporal cortex has been named SLF-tp. The AF, which connects the prefrontal cortex to the posterior STG, runs closely in parallel with SLF II from prefrontal to parietal regions, but differs from the SLF in particular in its posterior portion curving into the temporal posterior cortex. But since the SLF and the AF run in parallel between the prefrontal and parietal cortex, some researchers refer to this fiber tract as AF/SLF.

The specific function of the AF connecting Broca's area to the posterior temporal cortex is still under debate. One proposed function is that it is particularly involved in processing syntactically complex sequences and sentences. This evidence comes from a number of studies using different approaches to identify the function of this fiber tract in language. First, a function-based dMRI study using probabilistic fiber tracking with seeds in BA 44 argued that it was involved in the processing of syntactically complex sequences and was structurally connected by a dorsal fiber tract to the temporal cortex (Friederici et al., 2006) (Fig. 10.8A).

This suggested that this fiber tract, the AF/SLF, is crucially involved in processing complex syntax. Second, non-fluent primary progressive aphasic patients, who had suffered from lesions in this fiber bundle, were found to be specifically deficient in processing syntactically complex sentences (Wilson et al., 2011) (Fig. 10.9). Third, a study of 7-year-old children found that their AF/SLFs had not yet fully matured and they had problems in processing syntactically complex sentences (Brauer et al., 2011). It is an open question how this pathway supports complex sentence processing, either by computing syntactic relations or via some form of working memory-related mechanism.

Ventral pathway

The language regions in the frontal cortex and in the temporal cortex are also connected to each other by at least two ventrally located pathways. First, the extreme

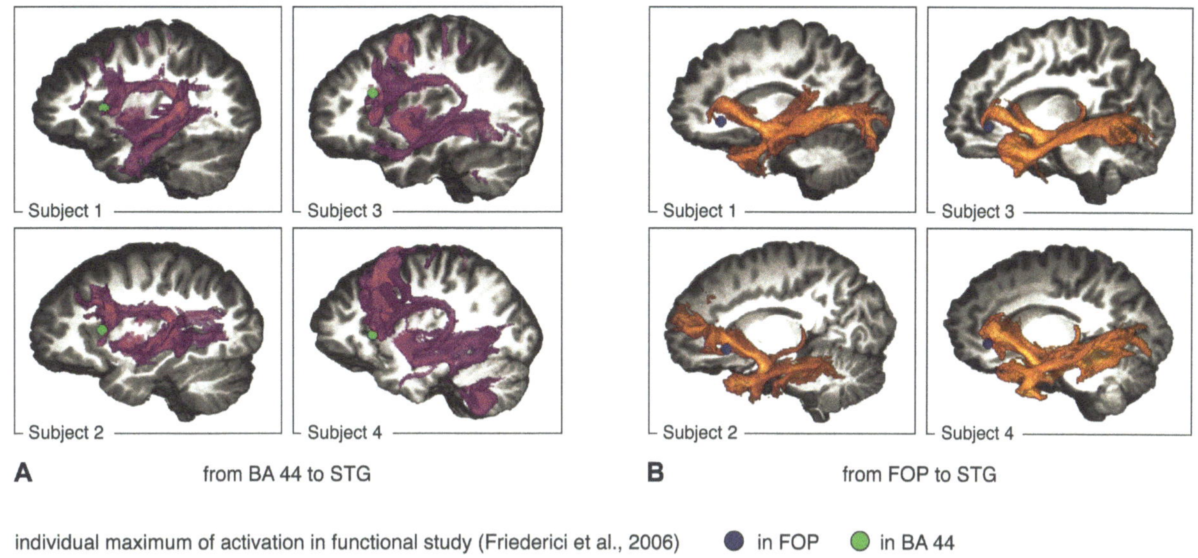

Fig. 10.8. Tractograms for two brain regions: Broca's area and frontal operculum (FOP). Three-dimensional rendering of the distribution of the connectivity values of two start regions with all voxels in the brain volume (purple, tractograms from Broca's area; orange, tractograms from FOP). (**A**) Four representative subjects of the group processing a phrase structure grammar with their individual activation maxima in Broca's area (green) in the critical contrast incorrect vs correct sequences ($P > 0.005$). For all subjects, the tractography detected connections from Broca's area to the posterior and middle portion of the superior temporal region via the fasciculus longitudinalis superior. (**B**) Four representative subjects of the group processing a finite-state grammar with their individual activation maxima in the FOP (blue) in the critical contrast incorrect vs correct sequences ($P > 0.005$). The individual peaks of the functional activation were taken as starting points for the tractography. For all subjects, connections to the anterior temporal lobe via the fasciculus uncinatus were detected. BA, Brodmann's area; STG, superior temporal gyrus. (Adapted from Friederici et al., 2006.)

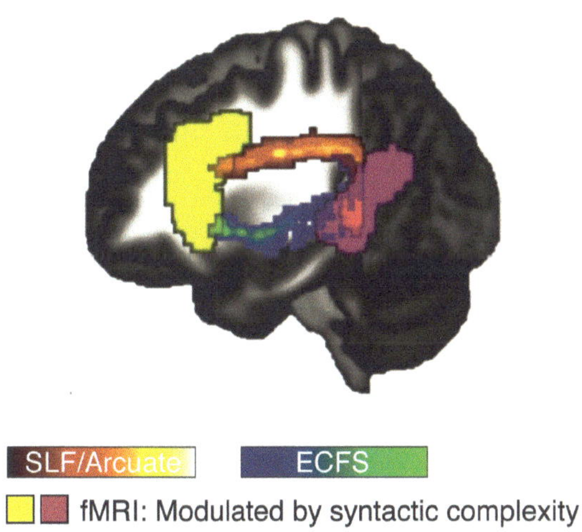

Fig. 10.9. Voxel-based morphometry and functional magnetic resonance imaging (fMRI). The left superior longitudinal fascicle (SLF)/arcuate fascicle (red/orange) and extreme capsule fiber system (ECFS) (blue/green) were constrained to connect the anterior language region (yellow) and posterior language region (magenta) that were modulated by syntactic complexity in normal controls in a previous fMRI study (Wilson et al., 2010a). Functional anisotropy (FA) values in the constrained SLF/arcuate fascicle were associated with syntactic deficits, but FA values in the constrained ECFS were not. (Adapted from Wilson et al., 2011.)

capsule fiber system (ECFS) connects BA 45 and BA 47, which are known to be involved in semantic processes, to the STG/MTG, the MTG, and the occipital cortex. Second, the uncinate fascicle (UF) connects the FOP to the anterior temporal cortex.

The ventral pathway connecting the frontal and temporal cortex ventrally was originally viewed as a unifunctional tract subserving sound-to-meaning mapping (Hickok and Poeppel, 2004, 2007; Rauschecker and Scott, 2009), but later different functions were assigned to subparts of the ventral pathway, specifically the ECFS, also named the inferior-fronto-occipital fascicle (IFOF) (Saur et al., 2008, 2010; Turken and Dronkers, 2011; Sarubbo et al., 2013) and the UF. From these subparts the ECFS/IFOF, with frontal terminations in BA 45, BA 47, and the orbitofrontal cortex and temporal terminations in the MTG, the STG, and the temporoparietal cortex, are seen as the major pathway supporting semantic processes (Saur et al., 2008) (Fig. 10.6).

The specific function of the UF, which connects the FOP and orbitofrontal cortex to the aSTG (Anwander et al., 2007; Hua et al., 2009; Thiebaut de Schotten et al., 2012), is still being debated. Early work proposed that the UF is relevant for language processing in general

(Wise, 2003; Grossman et al., 2004; Catani and Mesulam, 2008; Matsuo et al., 2008). However, Friederici et al. (2006), using function-based dMRI, suggested that the UF connecting the FOP and the aSTG may also be involved in syntactic processing such as the building up of local syntactic phrases through which adjacent elements are combined syntactically. Note that, for syntactic processes, a differentiation is made between local phrase structure building processes which recruit a ventral fiber tract (Fig. 10.8B) and the process of non-local syntactic dependencies in artificial grammar sequences and sentences with non-canonical word order which involve a dorsal pathway (Fig. 10.8A) (all the fMRI studies investigating such complex structures are discussed in a recent review: Friederici, 2011).

Both the proposal of allocating semantic processes to the ventral pathway and the proposal of allocating local phrase structure building to the ventral pathway may be valid since, functionally, the anterior STS has been found to be relevant for the intelligibility of speech in general (Scott et al., 2000; Crinion et al., 2003; Narain et al., 2003; Obleser et al., 2007; Friederici et al., 2010; Obleser and Kotz, 2010) and the aSTG as well as the anterior temporal lobe have been reported to be activated as a function of semantic and syntactic aspects (for a review of the functional studies, see Mazoyer et al., 1993; Stowe et al., 1998; Friederici et al., 2000, 2003; Vandenberghe et al., 2002; Humphries et al., 2005, 2006; Snijders et al., 2009; Friederici, 2011). Thus, both semantic and syntactic combinatorial processes that are elementary for language may involve the anterior temporal cortex.

The combined data suggest that the ventral pathway, with its different subparts, supports elementary semantic processes and local phrase structure building processes. The intraoperative stimulation of the two ventral fiber tracts suggests that the ECFS is more relevant for language than the UF because verbal fluency (naming as many words as possible in a particular category) and performance on a grammatical gender task (matching the article to the noun) were found not to be interrupted when stimulating the UF, but only when stimulating the ECFS (Duffau et al., 2008). However, these tasks may have challenged the semantic system more than the combinatorial system and, thus, do not rule out the possible involvement of the UF in combinatorial processes. It is undisputed that the ECFS/IFOF is language-relevant.

Thus, both the ventral and the dorsal pathway are involved in syntactic processes (Griffiths et al., 2013); however, the dorsal pathway (AF/SLF) is needed for the processing of sentences when they are syntactically complex (Brauer et al., 2011; Wilson et al., 2011).

CONCLUSIONS

The white-matter fiber tracts connecting the language-relevant regions in the frontal and temporal cortices can be classified into dorsal and ventral pathways. Each pathway consists of more than one major fiber tract. Within the ventral pathway one tract connects BA 45/47 with the STG and MTG and the other tract connects the orbitofrontal cortex with the anterior temporal cortex. The former ventral tract supports semantic processes, while the function of the latter is still debated. Within the dorsal pathway, one tract connects the posterior temporal cortex to the PMC and another one connects it to the posterior portion of Broca's area. While the former dorsal pathway is taken to support sensory-to-motor mapping, the latter appears to be relevant for the performance of complex syntactic tasks.

REFERENCES

Amunts K, Zilles K (2012). Architecture and organizational principles of Broca's region. Trends Cognit Sci 16: 418–426.

Amunts K, Lenzen M, Friederici AD et al. (2010). Broca's region: novel organizational principles and multiple receptor mapping. PLoS Biol 8: e1000489.

Anwander A, Tittgemeyer M, von Cramon DY et al. (2007). Connectivity-based parcellation of Broca's area. Cereb Cortex 17: 816–825.

Behrens TEJ, Johansen-Berg H, Woolrich MW et al. (2003). Non-invasive mapping of connections between human thalamus and cortex using diffusion imaging. Nat Neurosci 6: 750–757.

Berthier ML, Ralph MAL, Pujol J et al. (2012). Arcuate fasciculus variability and repetition: the left sometimes can be right. Cortex 48: 133–143.

Bookheimer S (2002). Functional MRI of language: new approaches to understanding the cortical organization of semantic processing. Annu Rev Neurosci 25: 151–188.

Brauer J, Anwander A, Friederici AD (2011). Neuroanatomical prerequisites for language functions in the maturing brain. Cereb Cortex 21: 459–466.

Broca P (1861). Remarques sur le siège de la faculté du langage articulé, suivies d'une observation d'aphémie (parte de la parole). Bull Soc Anat Paris 6: 330–357.

Brodmann K (1909). Beiträge zur histologischen Lokalisation der Grosshirnrinde. VI Die Cortexgliederung des Menschen. J Psychol Neurol 10: 231–246.

Buchsbaum BR, D'Esposito M (2008). The search for the phonological store: from loop to convolution. J Cogn Neurosci 20: 762–778.

Buchsbaum BR, Baldo J, Okada K et al. (2011). Conduction aphasia, sensory-motor integration, and phonological short-term memory – an aggregate analysis of lesion and fMRI data. Brain Lang 119: 119–128.

Bushara KO, Weeks RA, Ishii K et al. (1999). Modality-specific frontal and parietal areas for auditory and visual spatial localization in humans. Nat Neurosci 2: 759–766.

Caplan D, Chen E, Waters G (2008). Task-dependent and task-independent neurovascular responses to syntactic processing. Cortex 44: 257–275.

Catani M, Mesulam MM (2008). The arcuate fascicle and the disconnection theme in language and aphasia: history and current state. Cortex 44: 953–961.

Catani M, Thiebaut de Schotten M (2008). A diffusion tensor imaging tractography atlas for virtual in vivo dissections. Cortex 44: 1105–1132.

Catani M, Jones DK, Ffytche DH (2005). Perisylvian language networks of the human brain. Ann Neurol 57: 8–16.

Crinion JT, Lambon-Ralph MA, Warburton EA et al. (2003). Temporal lobe regions engaged during normal speech comprehension. Brain 126: 1193–1201.

DeWitt I, Rauschecker JP (2012). Phoneme and word recognition in the auditory ventral stream. Proc Natl Acad Sci U S A 109: E505–E514.

Duffau H, Gatignol P, Moritz-Gasser S et al. (2008). Dissociating the human language pathways with high angular resolution diffusion fiber tractography. J Neurosci 28: 11435–11444.

Frey S, Campbell JSW, Pike GB et al. (2008). Dissociating the human language pathways with high angular resolution diffusion fiber tractography. J Neurosci 28: 11435–11444.

Friederici AD (2002). Towards a neural basis of auditory sentence processing. Trends Cogn Sci 6: 78–84.

Friederici AD (2011). The brain basis of language processing: From structure to function. Physiol Rev 91: 1357–1392.

Friederici AD, Gierhan SME (2013). The language network. Curr Opin Neurobiol 23 (2): 250–254.

Friederici AD, Meyer M, von Cramon DY (2000). Auditory language comprehension: an event-related fMRI study on the processing of syntactic and lexical information. Brain Lang 74: 289–300.

Friederici AD, Rüschemeyer SA, Hahne A et al. (2003). The role of left inferior frontal and superior temporal cortex in sentence comprehension: localizing syntactic and semantic processes. Cereb Cortex 13: 170–177.

Friederici AD, Bahlmann J, Heim S et al. (2006). The brain differentiates human and non-human grammars: Functional localization and structural connectivity. Proc Natl Acad Sci U S A 103: 2458–2463.

Friederici AD, Kotz SA, Scott SK et al. (2010). Disentangling syntax and intelligibility in auditory language comprehension. Hum Brain Mapp 31: 448–457.

Galantucci S, Tartaglia MC, Wilson SM et al. (2011). White matter damage in primary progressive aphasias: a diffusion tensor tractography study. Brain 134: 3011–3029.

Gierhan SME (2013). Connections for auditory language in the human brain. Brain Lang 127: 205–221.

Giraud AL, Price CJ (2001). The constraints functional neuroimaging places on classical models of auditory word processing. J Cogn Neurosci 13: 754–765.

Griffiths JD, Marslen-Wilson WD, Stamatakis EA et al. (2013). Functional organization of the neural language system: dorsal and ventral pathways are critical for syntax. Cereb Cortex 23: 139–147.

Grodzinsky Y, Amunts K (Eds.), (2006). Broca's Region, Oxford University Press, New York.

Grodzinsky Y, Santi A (2008). The battle for Broca's region. Trends Cogn Sci 12: 474–480.

Grossman M, Cooke A, DeVita C et al. (2002). Age-related changes in working memory during sentence comprehension: an fMRI study. Neuroimage 15: 302–317.

Grossman M, McMillan C, Moore P et al. (2004). What's in a name: Voxel-based morphometric analyses of MRI and naming difficulty in Alzheimer's disease, frontotemporal dementia and corticobasal degeneration. Brain 127: 628–649.

Gruber O, von Cramon DY (2001). Domain-specific distribution of working memory processes along human prefrontal and parietal cortices: a functional magnetic resonance imaging study. Neurosci Lett 297: 29–32.

Hagoort P (2005). On Broca, brain, and binding: a new framework. Trends Cogn Sci 9: 416–423.

Hickok G, Poeppel D (2000). Towards a functional neuroanatomy of speech perception. Trends Cogn Sci 4: 131–138.

Hickok G, Poeppel D (2004). Dorsal and ventral streams: a framework for understanding aspects of the functional anatomy of language. Cognition 92: 67–99.

Hickok G, Poeppel D (2007). The cortical organization of speech perception. Nat Rev Neurosci 8: 393–402.

Hickok G, Buchsbaum B, Humphries C et al. (2003). Auditory-motor interaction revealed by fMRI: Speech, music, and working memory in area Spt. J Cogn Neurosci 15: 673–682.

Hua K, Oishi K, Zhang J et al. (2009). Mapping of functional areas in the human cortex based on connectivity through association fibers. Cereb Cortex 19: 1889–1895.

Humphries C, Love T, Swinney D et al. (2005). Response of anterior temporal cortex to prosodic and syntactic manipulations during sentence processing. Hum Brain Mapp 26: 128–138.

Humphries C, Binder JR, Medler DA et al. (2006). Syntactic and semantic modulation of neural activity during auditory sentence comprehension. J Cogn Neurosci 18: 665–679.

Johansen-Berg H, Behrens TEJ, Robson MD et al. (2004). Changes in connectivity profiles define functionally distinct regions in human medial frontal cortex. Proc Natl Acad Sci U S A 101: 13335–13340.

Jonides J, Schumacher EH, Smith EE et al. (1998). The role of parietal cortex in verbal working memory. J Neurosci 18: 5026–5034.

Lau EF, Phillips C, Poeppel D (2008). A cortical network for semantics: (de)constructing the N400. Nat Rev Neurosci 9: 920–933.

Makris N, Pandya DN (2009). The extreme capsule in humans and rethinking of the language circuitry. Brain Struct Funct 213: 343–358.

Makris N, Kennedy DN, McInerney S et al. (2005). Segmentation of subcomponents within the superior longitudinal fascicle in humans: a quantitative, in vivo, DT-MRI study. Cereb Cortex 15: 854–869.

Matsuo K, Mizuno T, Yamada K et al. (2008). Cerebral white matter damage in frontotemporal dementia assessed by diffusion tensor tractography. Neuroradiology 50: 605–611.

Mazoyer BM, Tzourio N, Frak V et al. (1993). The cortical representation of speech. J Cogn Neurosci 5: 467–479.

Meyer L, Obleser J, Anwander A et al. (2012). Linking ordering in Broca's area to storage in left temporo-parietal regions: The case of sentence processing. Neuroimage 62: 1987–1998.

Mori S, Zijl PCM (2002). Fiber tracking: principles and strategies – a technical review. NMR Biomed 15: 468–480.

Morosan P, Schleicher A, Amunts K et al. (2005). Multimodal architectonic mapping of human superior temporal gyrus. Anat Embryol (Berl) 210: 401–406.

Musso M, Moro A, Glauche V et al. (2003). Broca's area and the language instinct. Nat Neurosci 6: 774–781.

Narain C, Scott SK, Wise RJS et al. (2003). Defining a left-lateralized response specific to intelligible speech using fMRI. Cereb Cortex 13: 1362–1368.

Novick JM, Trueswell JC, Thompson-Schill SL (2005). Cognitive control and parsing: re-examining the role of Broca's area in sentence comprehension. Cogn Affect Behav Neurosci 5: 263–281.

Obleser J, Kotz SA (2010). Expectancy constraints in degraded speech modulate the language comprehension network. Cereb Cortex 20: 633–640.

Obleser J, Zimmermann J, Van Meter J et al. (2007). Multiple stages of auditory speech perception reflected in event-related FMRI. Cereb Cortex 17: 2251–2257.

Papoutsi M, Stamatakis EA, Griffiths JD et al. (2011). Is left fronto-temporal connectivity essential for syntax? Effective connectivity, tractography and performance in left-hemisphere damaged patients. Neuroimage 58: 656–664.

Patterson K, Nestor PJ, Rogers TT (2007). Where do you know what you know? The representation of semantic knowledge in the human brain. Nat Rev Neurosci 8: 976–987.

Perani D, Saccuman MC, Scifo P et al. (2011). The neural language networks at birth. Proc Natl Acad Sci U S A 108: 16056–16061.

Petrides M, Pandya DN (1984). Projections to the frontal cortex from the posterior parietal region in the rhesus monkey. J Comp Neurol 228: 105–116.

Price CJ (2010). The anatomy of language: a review of 100 fMRI studies published in 2009. Ann New York Acad Sci 1191: 62–88.

Rauschecker JP, Scott SK (2009). Maps and streams in the auditory cortex: nonhuman primates illuminate human speech processing. Nat Neurosci 12: 718–724.

Rauschecker JP, Tian B (2000). Mechanisms and streams for processing of "what" and "where" in auditory cortex. Proc Natl Acad Sci U S A 97: 11800–11806.

Ravizza SM, Hazeltine E, Ruiz S et al. (2011). Left TPJ activity in verbal working memory: implications for storage- and sensory-specific models of short term memory. Neuroimage 55: 1836–1846.

Rogalsky C, Hickok G (2011). The role of Broca's area in sentence comprehension. J Cogn Neurosci 23: 1664–1680.

Sarubbo S, De Benedictis A, Maldonado IL et al. (2013). Frontal terminations for the inferior fronto-occipital fascicle: anatomical dissection, DTI study and functional considerations on a multi-component bundle. Brain Struct Funct 218: 21–37.

Saur D, Kreher BW, Schnell S et al. (2008). Ventral and dorsal pathways for language. Proc Natl Acad Sci U S A 105: 18035–18040.

Saur D, Schelter B, Schnell S et al. (2010). Combining functional and anatomical connectivity reveals brain networks for auditor language comprehension. Neuroimage 49: 3187–3197.

Schmahmann JD, Pandya DN, Wang R et al. (2007). Association fibre pathways of the brain: Parallel observations from diffusion spectrum imaging and autoradiography. Brain 130: 630–653.

Scott SK, Blank CC, Rosen S et al. (2000). Identification of a pathway for intelligible speech in the left temporal lobe. Brain 123: 2400–2406.

Snijders TM, Vosse T, Kempen G et al. (2009). Retrieval and unification of syntactic structure in sentence comprehension: an fMRI study using word-category ambiguity. Cereb Cortex 19: 1493–1503.

Stowe LA, Broere CAJ, Paans AMJ et al. (1998). Localizing components of a complex task: sentence processing and working memory. Neuroreport 9: 2995–2999.

Thiebaut de Schotten M, Dell'Acqua F, Valabregue R et al. (2012). Monkey to human comparative anatomy of the frontal lobe association tracts. Cortex 48: 82–96.

Turken AU, Dronkers NF (2011). The neural architecture of the language comprehension network: converging evidence from lesion and connectivity analyses. Front Syst Neurosci 5: 1.

Tyler LK, Marslen-Wilson W (2008). Fronto-temporal brain systems supporting spoken language comprehension. Philos Trans R Soc Lond B Biol Sci 363: 1037–1054.

Uddén J, Bahlmann J (2012). A rostro-caudal gradient of structured sequence processing in the left inferior frontal gyrus. Philos Trans R Soc Lond B Biol Sci 367: 2023–2032.

Upadhyay J, Silver A, Knaus TA et al. (2008). Effective and structural connectivity in the human auditory cortex. J Neurosci 28: 3341–3349.

Vandenberghe R, Nobre AC, Price CJ (2002). The response of left temporal cortex to sentences. J Cogn Neurosci 14: 550–560.

Vigneau M, Beaucousin V, Herve PY et al. (2006). Meta-analyzing left hemisphere language areas: phonology, semantics, and sentence processing. Neuroimage 30: 1414–1432.

Wake H, Lee PR, Fields D (2011). Control of local protein synthesis and initial events in myelination by action potentials. Science 333: 1647–1651.

Wernicke C (1874). Der aphasische Symptomencomplex, Springer-Verlag, Berlin.

Wilson SM, Dronkers NF, Ogar JM et al. (2010a). Neural correlates of syntactic processing in the nonfluent variant of primary progressive aphasia. J Neurosci 30: 16845–16854.

Wilson SM, Galantucci S, Tartaglia MCRK et al. (2011). Syntactic processing depends on dorsal language tracts. Neuron 72: 397–403.

Wise RJS (2003). Language systems in normal and aphasic subjects: functional imaging studies and inferences from animal studies. Br Med Bull 65: 95–119.

Zilles K, Amunts K (2009). Receptor mapping: architecture of the human cerebral cortex. Curr Opin Neurol 22: 331–339.

Handbook of Clinical Neurology, Vol. 129 (3rd series)
The Human Auditory System
G.G. Celesia and G. Hickok, Editors

Chapter 11

Neural basis of music perception

PETR JANATA*
Center for Mind and Brain and Department of Psychology, University of California, Davis, CA, USA

INTRODUCTION

Music is a multifaceted phenomenon that engages humans of all ages and from all cultures. Music, like language, fundamentally exists as patterns of sound. By contrast to language, in which the morphology and sequencing of sound patterns must be constrained suitably in order to meet precise communicative goals in any given language, the palette of sounds and the possibilities for how they are arranged in time in music are vast. The organization of musical sound is constrained only insofar as it matters that it can be produced and also fall on appreciative ears.

A full accounting of the neuroscience of music would have to take into account the brain systems underlying action, emotion, social cognition, and mechanisms of memory and attention. The scope of this chapter is necessarily more limited. Here, the goal is to describe basic organizing principles of music and how these manifest themselves in the brain, primarily within the auditory cortex, so that music can be considered in relationship to more general aspects of auditory perception, covered in other chapters of this volume.

Music and the auditory scene

In order to link discussions of music to general conceptions of auditory experience and the role that the auditory cortex plays in supporting auditory experiences, it is useful to regard music as an auditory scene. Accordingly, music can be thought of as the organization of auditory objects along three primary dimensions: temporal, tonal, and timbral. The temporal dimension encompasses the organization of auditory objects in time, and is the basis for meter and rhythm. The tonal dimension pertains to the organization of discrete pitches into intervals, melodies, chords, and chord progressions. Tonal perception relies heavily on memory for specific collections of pitches (scales, modes, and keys) that guide perception. The timbral dimension provides the identity of auditory objects, be they sounds that are easily recognized as orchestral instruments or human voices, or more abstract sounds made possible by electronic synthesis and filtering techniques.

The phenomenon of pitch is central to most discussions of music, but it is not treated in any detail here, given other thorough treatments in this volume (see Chapter 4) and elsewhere (Zatorre and Zarate, 2012). In this chapter, aside from brief discussions of the chroma and timbre of isolated auditory objects, I focus mainly on combinations of pitches and musical events.

Summarizing the literature

The past 20 years have witnessed a dramatic increase in the number of studies aimed at understanding the representation and manipulation of musical material by the brain. Here, the functional anatomic organization of music processing in the auditory system, in particular in the auditory cortex, is discussed in the context of whether there is localization of function for representing musical relationships among auditory objects. This question has been addressed using the spectrum of cognitive neuroscience methods, including studies of patients with lesions, and structural and functional neuroimaging techniques such as volumetric measurements of gray matter, positron emission tomography (PET), and functional magnetic resonance imaging (fMRI) which provide indirect measures of regional neural activity, or electroencephalography (EEG) and magnetoencephalography (MEG) which are measures of mass activity of large populations of neurons.

Given research in other sensory modalities that has revealed orderly maps and principles of hierarchical

*Correspondence to: Petr Janata, Center for Mind and Brain, 267 Cousteau Pl., Davis, CA 95618, USA. Tel: +1-530-297-4471, E-mail: pjanata@ucdavis.edu

processing, coupled with the reasonable spatial resolution – on the order of several cubic millimeters – that PET and fMRI techniques provide, there is a temptation to assign the processing of specific musical features to specific cortical areas based on results of neuroimaging studies. In order to determine whether there is an emerging consensus across neuroimaging studies about the loci of processing core musical features described below – melody, harmony, rhythm, and timbre – I undertook a survey and quantitative meta-analysis of the pertinent neuroimaging literature to serve as a basis for this chapter.

A meta-analysis of the neuroimaging literature

Over 150 PET and fMRI studies, dating back to the early 1990s, were identified as pertaining to music, of which 42 had a primary focus on perception in one or more of the following categories: melody (19 studies), harmony (11 studies), rhythm (12 studies), and timbre (eight studies), and met the criterion that they reported anatomic coordinates of activation loci identified by analyses of the data across a group of subjects. Thus, several studies from the collection of 42 were excluded from the meta-analysis for failing to meet this criterion, but are mentioned below in the relevant sections.

Given that individual studies utilize different methodologies and seek to answer somewhat different questions, the specificity of any given meta-analysis is determined by the combined scope of the specific task or stimulus comparisons (contrasts) that enter into the analysis. Table 11.1 provides a listing of the studies and contrasts reported in each study that were entered into the meta-analysis. In summary, 255 activation loci (x, y, z, locations within the Montreal Neurological Institute brain template) from 28 contrasts entered the meta-analysis of melody, 341 loci from 17 contrasts were used for harmony, 324 loci from 23 contrasts were used for rhythm, and 100 loci from 10 contrasts were used for timbre.

Activation likelihood estimation (ALE) was used for the meta-analysis, as implemented in GingerALE using the Turkeltaub non-additive algorithm (Turkeltaub et al., 2012). Briefly, ALE utilizes the coordinates of activation loci that correspond to statistically significant effects of stimulus or task manipulations. After compiling the activation results reported for a related set of manipulations or comparisons (contrasts) across some number of papers, it is possible to use an ALE algorithm to identify brain areas that are observed to be activated consistently across the set of contrasts. The algorithm used here incorporates corrections for the number of contrasts associated with any given paper to minimize undue influence of papers reporting many different, but related, contrasts. Clusters of activation, aggregated across contrasts, are tested for statistical significance using a method that combines a magnitude threshold with a voxel cluster size threshold and permutation testing to estimate the likelihood of obtaining clusters of a given size and activation likelihood magnitude given the observed data.

Figure 11.1 shows the results of the meta-analyses performed for studies of melody, harmony, and rhythm. Figure 11.2 compares the results of the meta-analyses for tonality (melody or harmony), rhythm, and timbre. Because the auditory cortex is the focus of this chapter, the brain slices in the figures have been chosen to encompass the temporal lobes and thereby exclude some brain areas, including midline cortical areas such as the supplementary motor area (SMA), pre-SMA, and medial prefrontal cortex that are involved in musical behaviors. The images are presented as conjunctions of images so that brain areas engaged by more than one musical domain are represented by a color that is the combination of the primary colors used to display above-threshold voxels for each domain. In addition, the extent of auditory areas TE1 (primary auditory cortex) and TE3 along the superior temporal plane determined by probabilistic cytoarchitectonic mapping (Morosan et al., 2001) are displayed using black and light gray contour lines, respectively. Details of the figures and studies that contributed to them are described in the following sections organized by each of the musical domains.

MUSICAL PROPERTIES OF ISOLATED AUDITORY OBJECTS

Pitch chroma

Musical notes are referred to by note names, e.g. C or C-sharp. The same note name is used to refer to pitches that are in octave relationships to each other, underscoring a particular perceptual equivalence, commonly referred to as octave equivalence. Thus, a pair of pitches in which the fundamental frequency of the higher pitch is an integer multiple of the fundamental frequency of the lower pitch is said to have the same chroma or belong to the same pitch class, even though their pitch height differs (Shepard, 1964, 1982). Warren et al. (2003) investigated the neural underpinnings of the abstraction of pitch class, cleverly dissociating changes in pitch chroma from changes in pitch height using stimuli that were closely matched acoustically. They found that changes in both pitch chroma and height activated a large portion of the lateral aspects of Heschl's gyrus, but the preferential processing of changes in pitch height was displaced more posteriorly to the planum temporale, whereas preferential processing of changes in pitch chroma was found anterior to Heschl's gyrus in the planum polare.

Table 11.1

Neuroimaging studies incorporated into the meta-analyses of tonal, rhythmic, and timbral processing by the brain

Meta-analysis	Study	Contrast	# subjects
Melody			
	Angenstein et al. (2012)	Semitone vs non-semitone sequences	16
	Angenstein et al. (2012)	Interval size task vs pitch direction and passive listening	16
	Bangert et al. (2006)	Acoustic task	14
	Brown and Martinez (2007)	Melody discrimination vs button-pressing control	11
	Brown and Martinez (2007)	Melody listening vs rest	6
	Brown et al. (2013)	Linear response decrease with pitch repetition during "listen" trials	14
	Chen et al. (2012)	Listening to melodies (pre-training)	16
	Ellis et al. (2012)	Melodic discrimination vs rhythm discrimination	84
	Engel and Keller (2011)	Perception of improvised and imitated melodies	22
	Escoffier et al. (2013)	Music vs voice	16
	Griffiths et al. (1998)	Iteration × melody interaction	10
	Groussard et al. (2010a)	Music vs verbal (in familiarity experiment)	20
	Groussard et al. (2010a)	Effect of familiarity	20
	Groussard et al. (2010b)	Musical semantic vs rest	11
	Groussard et al. (2010b)	Musical reference vs rest	11
	Groussard et al. (2010b)	Musical semantic vs musical reference	11
	Groussard et al. (2010b)	Musical semantic	11
	Habermeyer et al. (2009)	Deviant stimuli	16
	Habermeyer et al. (2009)	Standard stimuli	16
	Herholz et al. (2012)	Listen vs baseline (encoding phase)	10
	Janata et al. (2002b)	Main effect of task	8
	Jerde et al. (2011)	Passive listening to melodies vs eyes-closed rest	7
	Jerde et al. (2011)	Working memory for melody vs passive listening to melodies	7
	Lee et al. (2011)	Melodic contour classifier	7
	Schulze et al. (2011a)	Tonal melody encoding vs atonal melody encoding	33
	Schulze et al. (2011b)	Tonal working memory (musicians)	16
	Schulze et al. (2011b)	Tonal working memory (non-musicians)	17
	Warren et al. (2003)	Change in chroma	10
Harmony			
	Alluri et al. (2012)	Key clarity	11
	Brown and Martinez (2007)	Harmony discrimination vs button-pressing control	11
	Brown et al. (2004a)	Harmonization vs rest	10
	Fujisawa and Cook (2011)	Harmony vs white noise	12
	Janata et al. (2002b)	Main effect of task	8
	Janata (2009)	Tonality-tracking clusters	13
	Klein and Zatorre (2011)	Adapt1	10
	Klein and Zatorre (2011)	Adapt2	10
	Klein and Zatorre (2011)	Disc3	10
	Koelsch et al. (2002)	Modulations	10
	Koelsch et al. (2005)	Neapolitan vs tonic chords (adults)	20
	Tillmann et al. (2003)	Harmonic context vs rest	15
	Tillmann et al. (2003)	Related consonant target	15
	Tillmann et al. (2003)	Related dissonant target	15
	Tillmann et al. (2003)	Unrelated consonant target	15
	Tillmann et al. (2003)	Unrelated dissonant target	15
	Tillmann et al. (2006)	Less related vs related	20

Continued

Table 11.1

Continued

Meta-analysis	Study	Contrast	# subjects
Rhythm			
	Alluri et al. (2012)	Increasing pulse clarity	11
	Alluri et al. (2012)	Decreasing pulse clarity	11
	Brown et al. (2013)	Linear response decrease with timing repetition during "listen" trials	14
	Chapin et al. (2010a)	Auditory attend 1 vs rest	13
	Chapin et al. (2010a)	Auditory attend 2 vs rest	13
	Chapin et al. (2010b)	Expressive vs mechanical (main effect)	14
	Chapin et al. (2010b)	Tempo	14
	Chen et al. (2008)	Listening with anticipation (experiment 1)	12
	Chen et al. (2008)	Passive listening (experiment 2)	12
	Chen et al. (2008)	Listening with anticipation (experiment 2)	12
	Engel and Keller (2011)	Entropy of timing in improvizations and imitation	22
	Grahn and Rowe (2009)	Beat vs no beat (experiment 1)	36
	Grahn and Rowe (2009)	Beat vs no beat (experiment 2)	36
	Grahn and Brett (2007)	Rhythms vs rest	27
	Grahn and McAuley (2009)	All stimuli vs rest	35
	Jerde et al. (2011)	Passive listening to rhythms vs eyes-closed rest	7
	Jerde et al. (2011)	Working memory for rhythms vs passive listening to rhythms	7
	Sakai et al. (1999)	1:2:4 rhythm vs control (experiment 1)	6
	Sakai et al. (1999)	1:2:3 rhythm vs control (experiment 1)	6
	Sakai et al. (1999)	1:2:4 rhythm vs control (experiment 2)	6
	Sakai et al. (1999)	1:2.5:3.5 rhythm vs control (experiment 2)	6
	Teki et al. (2011)	Absolute, duration-based timing	18
	Teki et al. (2011)	Relative, beat-based timing	18
Timbre			
	Alluri et al. (2012)	Brightness	11
	Alluri et al. (2012)	Timbral complexity	11
	Deike et al. (2004)	Double stream vs single stream	18
	Halpern et al. (2004)	Perception vs visual control	10
	Koelsch et al. (2002)	Deviant instruments	10
	Leaver and Rauschecker (2010)	Musical instrument category vs human speech, songbirds, other animal categories	15
	Margulis et al. (2009)	Instrument of expertise vs non-expertise	16
	Menon et al. (2002)	Difference between two timbres	10
	Warren et al. (2005)	Abstraction of spectral envelope	14
	Warren et al. (2005)	Changing spectral envelope vs constant spectral envelope	14

In the context of discussing pitch chroma, it is appropriate to mention that western tonal music utilizes only a subset of all possible pitches. Given a tuning reference, such as the note A with a fundamental frequency of 440 Hz, the remaining pitches that constitute the chromatic scale can be determined by subdividing each octave, either below or above the A440 reference, into 12 steps that are equally spaced on a logarithmic scale. This is what is known as "equal-temperament" tuning. The step between two adjacent pitches in the chromatic scale, e.g., G and G-sharp, is referred to as a semitone.

Timbre

Timbre refers to the spectral characteristics of an auditory object, apart from pitch and overall loudness, and effectively provides the auditory object with its identity. In music, timbre is what distinguishes different instruments from each other, but construed more broadly, timbre provides the distinctions among musical instruments, voices, and other natural and unnatural sounds. The timbre of a sound is described by its spectrotemporal properties, that is, the changes in the amplitudes of the sound's constituent frequencies over time. With such a definition, the overall

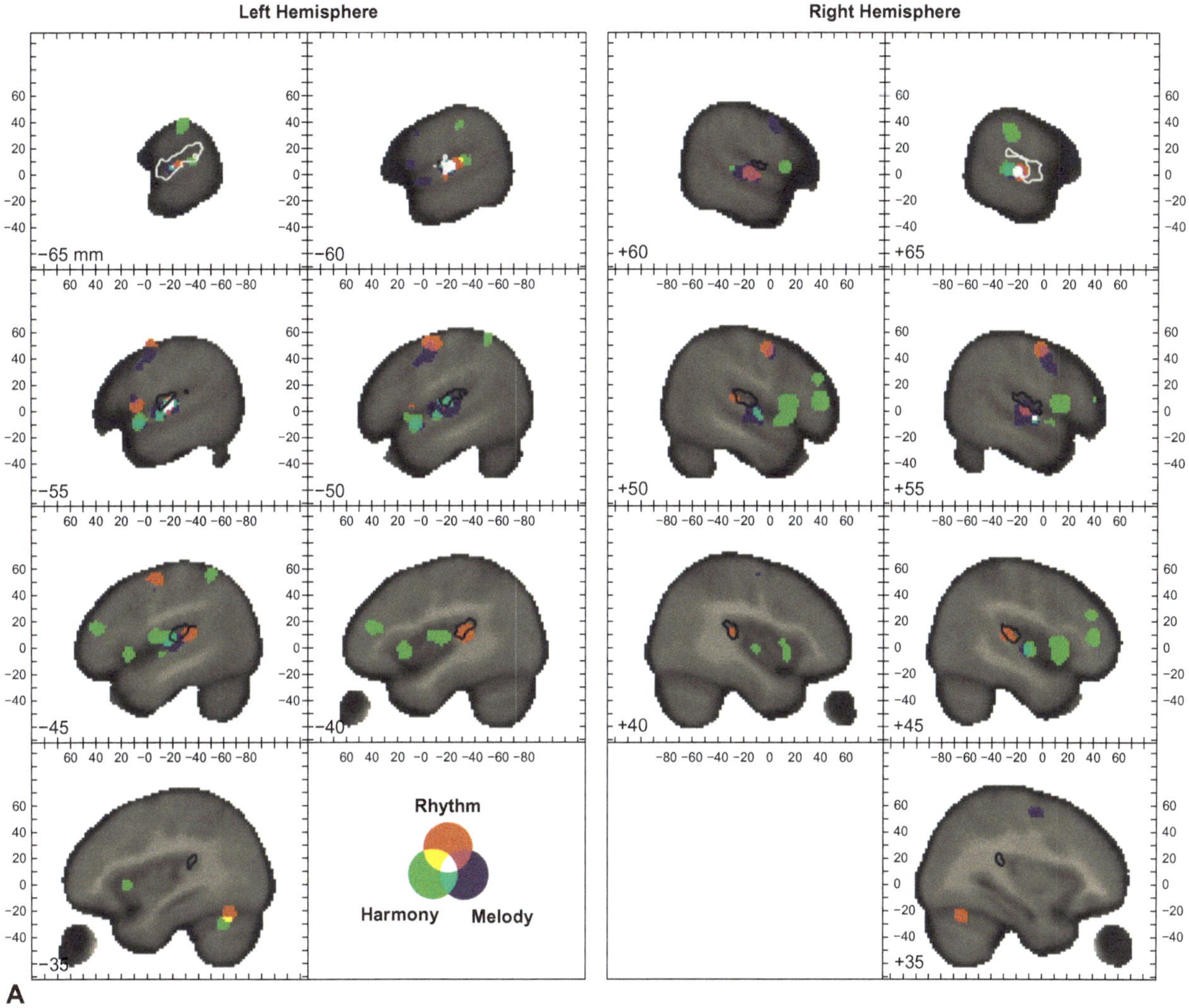

Fig. 11.1. (**A, B**) Results of meta-analyses of activation loci from experiments that manipulated melodic (blue), harmonic (green), and rhythmic (red) dimensions. See Table 11.1 for a listing of experiments included in the analysis. The underlying anatomic image is the Montreal Neurological Institute (MNI) average of 251 T1-weighted images. Analyses were performed and reported in the MNI coordinate space.

(Continued)

space of timbral possibilities is vast and relatively unconstrained when compared with phenomena of pitch chroma and pitch height. Accordingly, determining how timbral representations are organized in the brain has proven challenging.

Neuroimaging and electrophysiologic studies of timbral representations have made use of the fact that particular types of spectrotemporal features are perceptually salient and form a basis for making similarity judgments between pairs of heard instrument sounds. Multidimensional scaling, a technique that positions instrument sounds relative to each other in perceptual space based on similarity or distance judgments, has revealed that attack, the slope of the increase in energy following the onset of a sound, and spectral centroid or brightness, the center of mass along the frequency dimension, are two qualities of timbre that can be perceived readily, and for which a mapping to underlying acoustic dimensions can be discerned (Grey, 1977; McAdams et al., 1995; McAdams, 1999). A third dimension, labeled spectral flux, is generally identified, but because this refers to all of the spectrotemporal variation that is not captured by the attack or spectral centroid characteristics of the sound, the range of possibilities by which this spectrotemporal variation can be structured is extensive and poorly understood. It should be noted that somewhat different approaches to decomposing musical sounds identify perceptual dimensions of timbre that have combined spectrotemporal characteristics (Elliott et al., 2013).

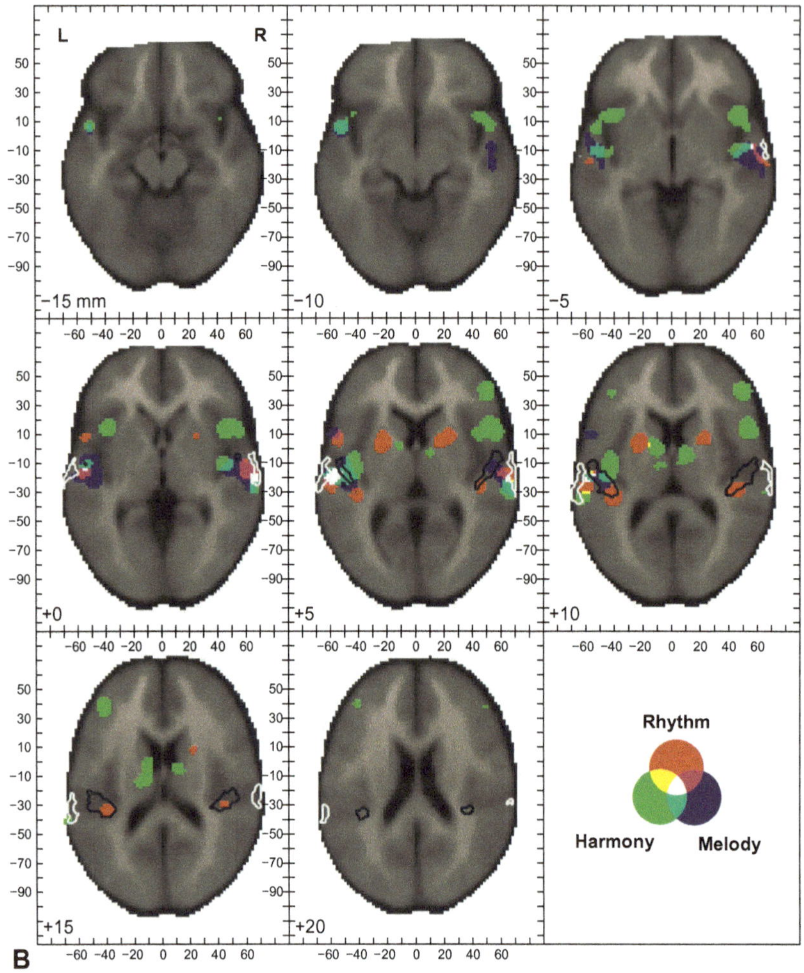

Fig. 11.1—Cont'd

Neuroimaging studies included in the timbre meta-analysis characterized or manipulated stimuli: (1) along the attack and spectral centroid dimensions (Menon et al., 2002; Halpern et al., 2004; Alluri et al., 2012); (2) in terms of change in spectral envelope that is more extensive than the change in spectral envelope that is used to effect a shift in spectral centroid (Warren et al., 2005); or (3) in terms of an aggregate measure of timbral complexity (Alluri et al., 2012). Other studies that were included in the meta-analysis involved discriminations among sounds in broader timbral categories such as musical instrument sounds, speech sounds, and other animal vocalizations (Leaver and Rauschecker, 2010), the detection of timbral deviants (Koelsch et al., 2002), or the perception of a sound sequence as two separate auditory streams as a function of the timbral identity of each stream (Deike et al., 2004). The activations from a study that examined the interaction of timbral representation with instrument-specific musical training were also included (Margulis et al., 2009).

The meta-analysis found extensive activations of the superior temporal gyrus (STG) in both hemispheres, with all eight of the included studies showing activation in the left hemisphere and six of the eight showing activation in the right hemisphere. The caudolateral extent of the activation clusters was greater for timbral manipulations than for either tonal or temporal manipulations, encompassing much of TE1 and extending into the planum temporale (Fig. 11.2).

COMPLEX MUSICAL OBJECTS AND THEIR COMBINATION

Although sounds played in isolation (auditory objects) may convey extensive semantic or emotional information, the real richness and motivation to engage with music have to do with temporally extended patterns of musical sounds. Rhythms, melodies, and harmonies all rely on combinations of musical sounds. Thus the question arises how and where increasingly complex

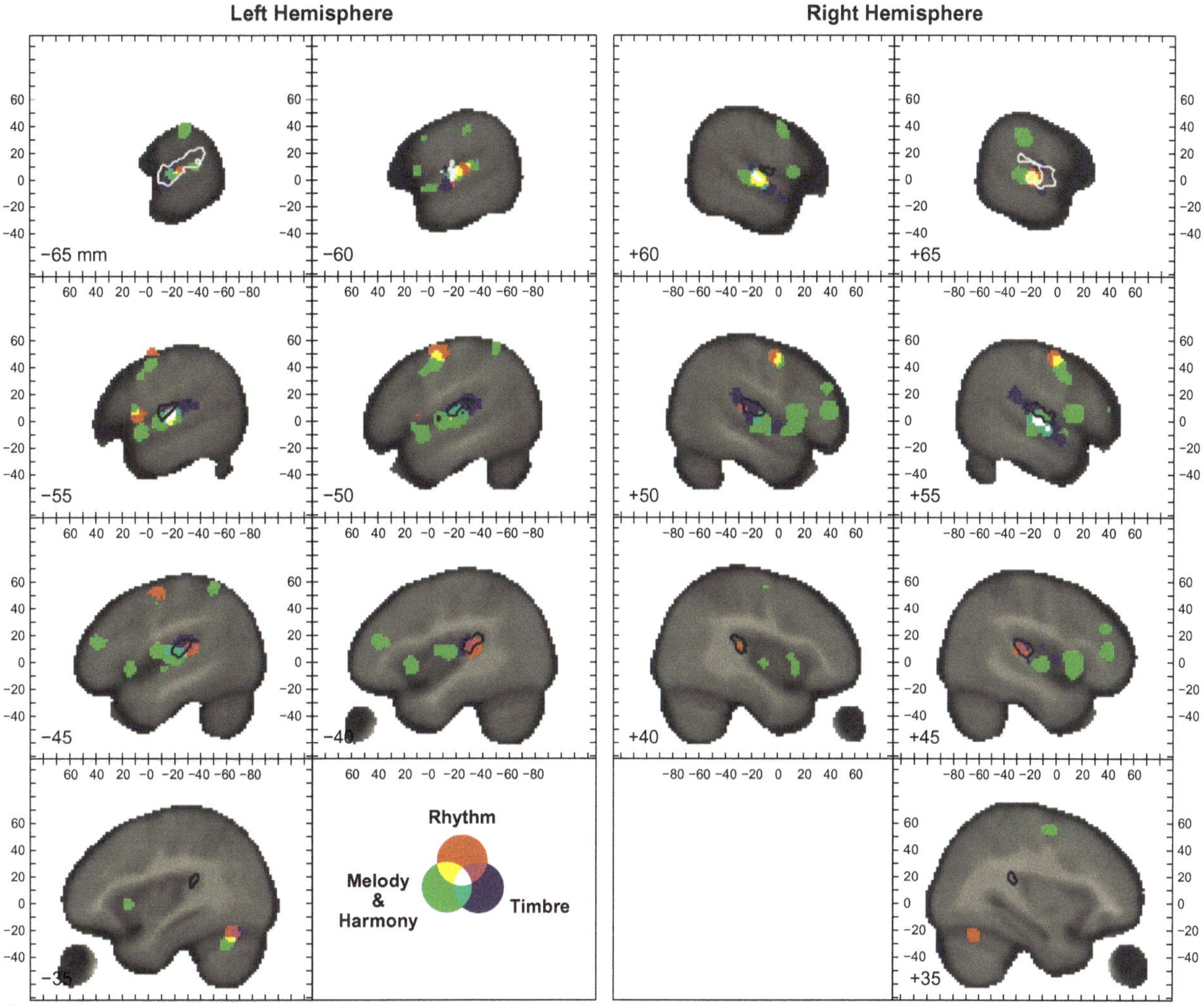

A

Fig. 11.2. (A, B) Results of meta-analyses of activation loci from experiments that manipulated tonal (green; the union of melodic and harmonic meta-analyses), timbral (blue), and rhythmic (red) dimensions. See Figure 11.1 for further information.

(Continued)

combinations of the underlying auditory objects are processed. Are there processing hierarchies akin to those found in the visual system in which simpler features are represented by neurons in the primary sensory area and combined into more complex feature and object representations in secondary sensory and polymodal association areas? Are there separate pathways for representing temporal patterns (rhythms) and pitch patterns (melodies) that arise from an apparent predisposition for the auditory cortices in the left and right hemispheres to process fine temporal and pitch differences, respectively (Zatorre et al., 2002)?

Interval, contour, and melody

The combination of two pitches is referred to as an interval. Two pitches played together constitute a harmonic interval, whereas two pitches played sequentially constitute a melodic interval. Harmonic intervals are associated with percepts of consonance and dissonance, and combinations of harmonic intervals give rise to chords. Consonance and dissonance arise from interactions of multiple pitches. Using the frequency-following response recording technique optimized to record brainstem responses in humans, Bidelman and Krishnan (2009) provided an elegant demonstration that the brainstem encodes the pitch of complex harmonic tones (tones consisting of a fundamental frequency and some number of frequencies that are integer multiples of the fundamental frequency) by synchronizing neural activity in a temporal pattern such that the period between the peaks in the neural activity corresponds to the fundamental frequency of the stimulus. The quality of the synchronization is used to estimate neural pitch salience,

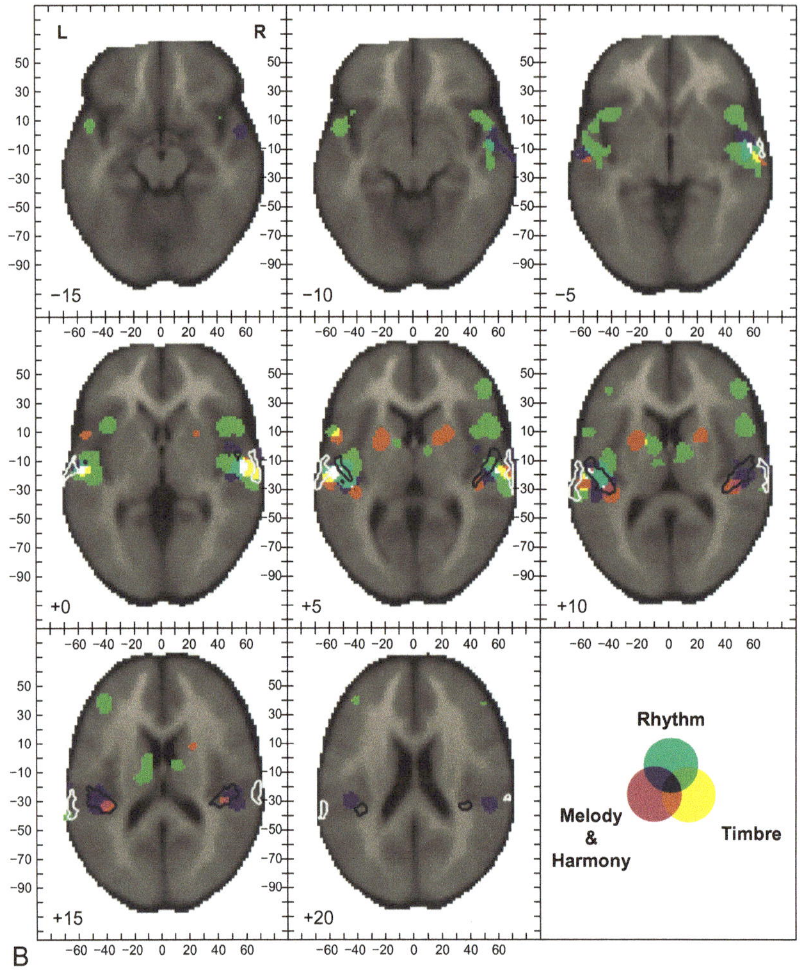

Fig. 11.2—Cont'd

and it has been shown that the neural pitch salience of the response to harmonic intervals used in western tonal music is strongly correlated with the perceptual consonance ratings for those intervals (Bidelman and Krishnan, 2009).

Melodic intervals are the building blocks of melodies (Fig. 11.3A). Melodies can be defined as sequences of intervals (changes in pitch). Each change is defined both by its magnitude (the size of the interval) and direction (an ascending or descending interval). The ability to perceive the direction of pitch change is fundamental to our ability to appreciate the contour of a melody, that is, the pattern of ups and downs in pitch. Moreover, the ability to perceive the direction of change is distinct from the ability to perceive a difference between two notes. In a study that compared patient groups with left- or right-hemisphere temporal-lobe resections with matched controls, Johnsrude et al. (2000) found that patients with right temporal-lobe lesions were impaired in determining the direction of pitch change, and the degree of impairment was greater when the lesion encroached on Heschl's gyrus. Remarkably, however, when judging pairs of tones as same or different, the patients' frequency discrimination thresholds were no larger than those of either normal controls or left-hemisphere patients. Such results suggest that the rudiments of melodic processing are supported preferentially by the right temporal lobe.

Although a melody can be construed simply as a sequence of changes in pitch, the details of the selection and ordering of specific pitches (and their timing) determine whether a melody sounds pleasing, coherent, or well formed. The principles and constraints underlying the composition of melodies are many, and are a steady topic of discussion among music theorists, musicologists, and psychologists. From a neuroscientific perspective, an understanding of how the brain processes melodies necessarily spans multiple levels, from the mechanisms that identify successive changes in pitch to mechanisms that associate an irresistible jingle with

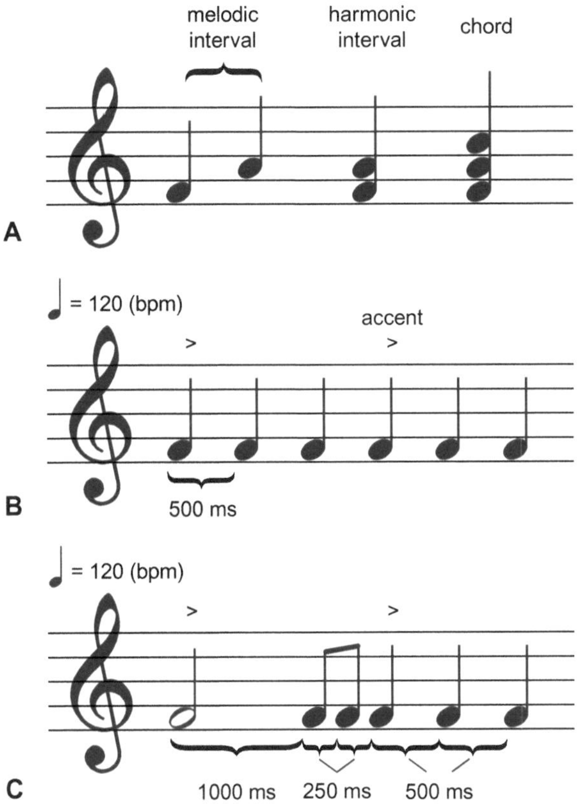

Fig. 11.3. Basic terminology illustrated on a musical staff. (**A**) A melodic interval consists of two notes played sequentially. Here the notes F and A are shown, which constitute an interval of a major third. When played together, these notes constitute a harmonic interval. A chord comprises three or more notes that are played simultaneously. If the chord shown here were played in isolation, it would most likely be interpreted as the tonic triad in the key of F-major. (**B**) Absolute and relative durations are indicated by a combination of a tempo marking, open and closed note symbols, dots, beams, and other markings connecting notes. The notes shown here represent quarter notes whose onsets are separated by 500 ms, corresponding to a tempo of 120 beats per minute (bpm). The accent mark above every third note indicates that it should be played more loudly. Marking such an isochronous sequence of events with regular accents imparts a sense of meter, in this case a ternary meter. (**C**) Rhythm arises from patterns of relative note durations. Illustrated here are events that span three different metric levels: the open half note (1000 ms), the quarter note (500 ms), and the two eighth notes (250 ms each) that are connected by a beam. The notes are written here such that their horizontal spacing accurately reflects their durations, though this is rarely the case in written music. In this example, the quarter note metric level would carry the tactus or beat.

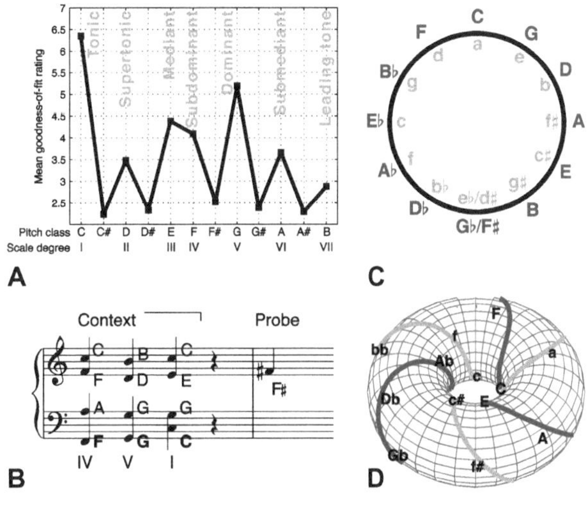

Fig. 11.4. Pitch classes can be further organized, giving rise to tonal space. (**A**) Of the 12 pitch classes, seven belong to any given key. Within a key there exists a tonal hierarchy whereby the first and fifth scale degrees are judged to fit better with a preceding context that has been used to establish the key. Melodies used in experiment settings are typically composed of notes falling on different scale degrees. Those pitch classes not labeled with a scale degree are notes from outside the key and are typically perceived as "wrong" notes. (**B**) The notes from a key combine to form chords and harmonic sequences, thus establishing tonal contexts (keys). (**C**) In music theory, the relationships among keys are commonly shown on the "circle of fifths," a name that arises from the fact that the fifth scale degree (dominant) of one key becomes the first scale degree (tonic) of its neighbor in a clockwise direction. (**D**) Behavioral studies have shown that the relationships between major and minor keys are fully captured when their respective circle of fifths is projected on to the surface of a torus. By strategically selecting within-key and out-of-key notes to create harmonic motion in tonal space, composers are able to create changing patterns of activation on the toroidal surface. (Reproduced from Collins et al., 2014), published by American Psychological Association. Reprinted with permission.)

personal memories. The collection of studies and task contrasts included in the meta-analysis of melody processing is therefore an amalgam of stimulus types and task conditions, although the preponderance of studies to date has focused on rudimentary aspects of melodic processing. Thus many studies use unfamiliar melodies that are as short as 4–5 notes in length, and stimuli typically last 4–8 seconds. The studies considered here all used melodies in which the pitches were drawn from the chromatic scale (Fig. 11.4A). Although melodies that vary in pitch steps other than the intervals formed by combining notes from the chromatic scale are encountered in certain musical cultures, e.g., Turkish makams, such melodies are not considered here, except to the extent that they serve as control sequences (e.g., Angenstein et al., 2012). Experimental tasks tend to fall into two broad categories: those that do not have a strong memory component to them (passive listening, or target/deviance detection tasks) and those that do (same/different discrimination and working-memory tasks).

Patterson et al. (2002) specifically examined where melodic information is processed in relation to auditory cortex situated on Heschl's gyrus. Although they found considerable interindividual variability, activation by melodic sequences (relative to sequences of repeated pitches) was biased toward the right hemisphere, extending from lateral Heschl's gyrus in a rostral direction into the planum polare of the STG. Activations within the superior temporal sulcus (STS) were also commonly observed. With the exception of the bias toward the right hemisphere, these observations are corroborated by the meta-analysis, in which the Patterson et al. (2002) study was not included. Within the meta-analysis, the extent of activation is best described as encompassing the STG immediately ventral and rostral to the borders of TE1 (Fig. 11.1). The restriction of melody-related foci in the meta-analysis to the posterior half of the lateral temporal lobe is consistent with the results of a study of 65 patients who had undergone either left or right temporal cortical resections, in which the greatest deficits in melody discrimination were found in patients with resections of approximately the posterior half of the lateral temporal lobes (Liegeois-Chauvel et al., 1998).

Aside from passive listening, a common experimental task context in which participants listen to melodic material involves discrimination tasks in which two versions of a melody, typically differing by only a single note, are presented and participants must make a same/different judgment (Brown and Martinez, 2007; Ellis et al., 2012). Such tasks require the encoding and maintenance of the melodies into short-term memory, and are thus related to tasks that explicitly focus on tonal working memory (Jerde et al., 2011; Schulze et al., 2011a, b). Within the meta-analysis, studies that involved an active memory component were observed to activate frontal-lobe premotor areas (Brodamann area (BA) 6 and BA 44), likely reflecting the motor component of a rehearsal mechanism.

Whereas studies of melodic processing have typically contrasted activations during conditions of pitch variation (melodies) with conditions in which pitch variation is severely restricted, e.g., sequences of repeated pitches (Patterson et al., 2002), more recent experiments have attempted to identify brain areas that are involved in the identification and categorization of short (~5-note) melodic sequences (melodic contours or "gestalts") (Lee et al., 2011; Schindler et al., 2013). Using multivoxel pattern analysis (MVPA) – an analysis technique that discriminates categories of stimuli based on differential patterns of voxel activations within small volumes – Lee et al. (2011) found that the region of the right STG/STS ventral to TE1 exhibited this type of discrimination activity across a variety of ascending and descending melodic contours. Very similar anatomic localization, albeit in both hemispheres, was obtained by Schindler et al. (2013; not included in the meta-analysis), who used MVPA to identify regions that could discriminate among specific four-note melodies, despite the presentation of the melodies using different instruments or in different keys.

Further evidence that the right hemisphere plays a prominent role in the processing of melodic patterns comes from a study that compared individuals with either left or right anterior temporal-lobe excisions with normal controls during performance of a task that required judging whether two notes played by different instruments have the same pitch (Warrier and Zatorre, 2004). Pitch constancy judgments are facilitated when a melodic context is provided (Warrier and Zatorre, 2002), but individuals with anterior temporal-lobe damage in the right hemisphere do not benefit from the tonal context provided by the melody (Warrier and Zatorre, 2004).

A role of the auditory cortex in the perception of melodies (tonal patterns) has also been established through studies that use the mismatch negativity (MMN) paradigm. The MMN is an event-related brain potential that manifests itself as a negative potential at frontocentral sites on the scalp in response to deviant auditory stimuli under conditions when attention is not explicitly oriented toward the stimuli (Naatanen et al., 2007). Accordingly, an MMN response to a deviant stimulus is commonly interpreted as evidence for automatic processing of those features in which there was a deviation. Importantly, the MMN is elicited not only by deviations in the absolute properties of a stimulus, e.g., a response to infrequent 420-Hz tones embedded in a sequence of frequent 400-Hz tones, but also to deviations in the abstract relationships among tones comprising patterns (Saarinen et al., 1992; Tervaniemi et al., 1994). As noted above, melodies are defined in terms of both their contour and the intervals between successive notes. The fact that recognition of melodies is preserved when they are shifted in pitch (i.e., transposed to different keys) (Dowling, 1978) highlights that relative patterns of intervallic and contour information, rather than absolute pitch values, are the defining features of melodies. MMN responses to both interval and contour deviants reinforce the idea that the auditory cortex is involved in the construction of melodic patterns (Trainor et al., 2002).

The distinction between interval and contour processing in melody perception has also been considered in terms of a local/global feature processing distinction, such that discriminating between melodies that differ in their contour requires a global process, whereas discriminating between melodies in which the global contour is preserved but the size of an interval between two notes in the melody has changed requires a local process. Processing of contour and interval has also been construed in hierarchic terms, such that more precise interval processing follows coarser and faster contour processing (Ouimet et al., 2012). An fMRI study aimed at testing this hierarchic model, and possible left/right hemispheric lateralization of local/global processing, respectively, found support for the hierarchic processing view insofar as detection of local interval differences elicited greater activation of the right posterior STS and the right planum temporale than did detection of changes in global contour (Stewart et al., 2008). However, the neuroimaging study did not reveal the same mapping of local/global processes to hemispheric differences that were suggested by earlier studies of patients in whom unilateral lesions of the left hemisphere were selectively associated with deficits in discrimination of interval discrimination, whereas right-hemisphere lesions resulted in impairments of both contour and interval processing (Peretz, 1990; Liegeois-Chauvel et al., 1998).

Familiar melodies

Although it is clearly possible to talk about melodies in the abstract as patterns of pitch change, and to present participants with melodic fragments or novel melodies in order to search for parts of the brain that are responsible for processing such pitch variations, the processing of melodies must also be considered in terms of higher cognitive functions. Melodies become familiar when they are repeatedly heard, and multi-instrumental music and music that is rich in harmonic variation is nonetheless commonly identified by characteristic melodic themes. Thus the question arises whether regions of the superior temporal lobe outside of the immediate vicinity of Heschl's gyrus play a role in the processing of familiar melodies above and beyond general melody processing. Given that voxel-based morphometry and lesion studies suggest a role of the anterior temporal lobes in processing familiar musical material (Samson and Zatorre, 1991; Johnson et al., 2011), fMRI studies that require processing associated with familiar melodies might be expected to show activation in these regions. Indeed, listening to, and imagining, familiar melodies activates rostral regions of the STG (Herholz et al., 2012), as does making familiarity judgments about brief (5-seconds) melodic excerpts (Platel et al., 2003). The role of the rostral temporal lobe in the associating of melodic information with semantic information that commonly accompanies musical knowledge, e.g., song titles, is treated in greater detail in other chapters of this volume (see Chapters 34 and 37).

Melody and song

Of particular relevance to melodies are lyrics. It is likely that music and language have been bound together in song since the earliest days of music in humans (Mithen, 2005). Several studies have sought to identify where in the brain melodies and lyrics might be bound, with an emphasis on the binding of pitches and syllables, and a particular focus on the role of the auditory cortex. A recent meta-analysis of these studies (Janata and Parsons, 2013) found that the perception and production of sung material recruited a somewhat broader extent of the STG, from the primary auditory areas in a rostral direction, than was observed for the melodic and harmonic material in the present meta-analysis. The more rostrally situated regions of the STG and STS may be involved in the binding of melodies and lyrics, as evidenced by larger responses to sung trisyllabic words than either the melodic fragments or spoken words alone (Schon et al., 2010), and adaptation of responses to repeated melodies and lyrics (Sammler et al., 2010). A direct comparison of speaking and singing the words to a familiar song found greater activation of anterior temporal areas in the right hemisphere when the words were sung (Jeffries et al., 2003).

Harmony and key

Perhaps the strongest constraint placed on the selection of notes for a melody is the notion of tonality or key. Just as the set of all possible pitches is constrained to the 12 pitch classes that constitute the chromatic scale, a further constraint divides the chromatic scale into seven diatonic and five non-diatonic notes (Fig. 11.4). The diatonic notes make up the scale that defines a key. For example, the key of C major has the pitch class C as the first scale degree and D, E, F, G, A, and B as the remaining scale degrees that are part of the set of diatonic notes. These are all the white keys on a piano keyboard. The remaining pitch classes (in this case the black keys on a piano keyboard) are the non-diatonic notes. Although non-diatonic notes are routinely encountered in a melody, they tend to be followed by (resolve to) diatonic notes in a process known as melodic anchoring (Bharucha, 1996). Among the diatonic tones, there is also a hierarchy in terms of perceived stability, with the first scale degree, called the tonic, as the most stable and defining note of the key. The next most stable note is the fifth scale degree, or dominant (i.e., the note G in the key of C major), which in turn is the tonic of a closely related key. Krumhansl (1990) and Krumhansl and Cuddy (2010) provide excellent reviews of the cognitive psychology of tonal structure.

As noted above, two notes played simultaneously form a harmonic interval, and multiple intervals combine to form chords. As in the case of tonal hierarchies, chords associated with a key differ in their stability within the key, i.e., the degree to which they belong to a particular key. For example, the chord built of the first, third, and fifth scale degrees is called the tonic triad and is the most stable and representative chord of a key. This is the harmony that would be most expected and provide a sense of closure at the end of a piece of western tonal music. The second most stable harmony within a key is built on the fifth scale degree.

A prominent distinction in western tonal music is between major and minor keys. The difference between major and minor keys lies in the sets of diatonic tones. Although there are different forms of minor keys, e.g., natural, melodic, and harmonic, the primary distinctions are third and sixth scale degrees that are one semitone lower than their major-key counterparts. One study that specifically studied the categoric distinction between major and minor keys found that an area of the mid-STS in the right hemisphere was sensitive to this distinction (Klein and Zatorre, 2011). Although additional varieties of chords, such as diminished chords, dominant-seventh chords, suspensions, ninth chords, are commonly heard in music, whether these musical objects are categorically represented in the auditory cortex is not known. When listeners have the task of determining the direction of change in the pitch of one of the constituent notes of a chord (Fujisawa and Cook, 2011), or whether one chord in the harmonic accompaniment to a melody changed between presentations of a melody (Brown and Martinez, 2007), activation is observed along the superior temporal lobe, extending through BA 21 to the temporal pole (BA 38). Activation of BA 38 is also observed during singing of melodies and harmonizing to melodies, with greater activation during harmonizing (Brown et al., 2004b).

Taken together, results arising from the collection of studies pertaining to the perception of melodic intervals, brief melodic patterns, chord pairs, or notes that are bound to sung syllables or lyrics suggest that regions of the superior temporal lobe (both the STG and the STS) rostral to the primary auditory cortex are responsible for the construction of basic tonal relationships and musical objects.

Harmonic sequences

Harmonic perception in music is shaped not only by the perceived stability of individual chords as a function of the scale degrees on which they are built, but also crucially on their sequential ordering. A chord that is either unlikely within the context of the instantiated key, or is unlikely as a successor to the preceding chords, creates a harmonic expectancy violation. Studying responses to harmonic expectancy violations using a variety of paradigms and measurement techniques has led to the understanding that knowledge of tonal structure is assimilated by most listeners, not only individuals with musical training. However, individuals with musical training commonly show stronger responses to contextual violations than do non-musicians (e.g., Steinbeis et al., 2006).

Given the apparent role that the areas of the auditory cortex play in processing the relational information in tonal patterns and melodies, the question arises whether these areas play a similar role in evaluating the relationship between incoming tonal information, whether melodic or harmonic, and any tonal context that has been established by a preceding sequence of notes or chords. The preponderance of EEG, MEG, and fMRI evidence from experiments that utilized, primarily, expectancy violation paradigms, suggests that regions of ventrolateral prefrontal cortex (VLPFC) rather than auditory cortical areas respond to harmonic expectancy violations. Most commonly, responses to harmonic expectancy violations are observed in the ventral aspects of the inferior frontal gyrus (BA 44) and the anterior insula (Patel et al., 1998; Koelsch et al., 2000, 2005; Maess et al., 2001; Tillmann et al., 2003, 2006; Loui et al., 2005; Poulin-Charronnat et al., 2006; Leino et al., 2007) (Fig. 11.1), though the anterior STG has been found to respond to the degree of expectancy violation in cadences (a cadence is the terminal sequence of chords in a musical phrase) (Seger et al., 2013). A further locus in the lateral prefrontal cortex that is commonly activated in tasks in which the evaluation of stimuli depends on the instantiated tonal context is situated slightly more anteriorly, in BA 46 (Janata et al., 2002b; Tillmann et al., 2003; Koelsch et al., 2005; Brown and Martinez, 2007; Fujisawa and Cook, 2011).

Like the MMN, the lateral prefrontal response to harmonic expectancy violations, termed the early-right anterior negativity (ERAN) in the event-related potential literature, is also elicited in the absence of overt attention to the musical stimulus, though its amplitude is modulated by attention (Loui et al., 2005; Maidhof and Koelsch, 2011). However, the MMN and the ERAN are dissociable processes in terms of their function and localization (Koelsch et al., 2001; Koelsch, 2009). Although MMN-like responses have been obtained in response to out-of-key notes in melodies, deviants created by mistuning in-key notes elicited larger responses (Brattico et al., 2006). Thus, with respect to music perception, the auditory cortex appears to be specialized for the organization of individual auditory objects, binding of auditory objects, e.g., pitches and syllables, and construction and maintenance of patterns of those objects, whereas contextual processing, such as that required for establishing syntactic relationships, appears to depend largely on regions outside the auditory cortex.

Rhythm and meter

The patterning of musical events in time is integral to any music we hear, and some musical traditions focus primarily on this dimension of music (Pressing, 2002).

Temporal patterns arise from variation in the durations between onsets of musical objects, be they the onsets of notes in a melody or the onsets of percussive timbres that need not be associated with a clear pitch percept. Figure 11.3B and C illustrates some of the basic principles and terminology that underlie rhythm and meter. The simplest type of sequence of inter-onset intervals is an isochronous sequence in which the time period between all onsets is equal. In such sequences, the inter-onset interval determines the tempo of the sequence, typically expressed in beats per minute (bpm).

The simplest elaboration of an isochronous sequence is to divide (or multiply) the time interval between events by a simple integer such as 2 or 3. If all of the events in the sequence are identical in terms of loudness, pitch, and timbre, the resulting percept will simply be that of a doubled or tripled tempo. However, if one out of every two or three events is accented, thus giving rise to percept of a more slowly occurring cycle, the more complex percept of meter arises. This is the percept that is associated with designations of musical forms, such as a march (a binary subdivision of the cycle) or a waltz (a ternary subdivision of a cycle). The division and multiplication of the duration between events that constitute a reference level, commonly called the beat or tactus, give rise to the levels of a metric hierarchy. Sequences or groups of events drawn from different metric levels, and therefore different durations, give rise to rhythmic patterns.

Perhaps the most striking difference in the results of the meta-analysis for rhythm and the meta-analyses for melody and harmony is the greater degree to which the STG focus for rhythm falls within the boundaries of the primary auditory cortex (TE1) in both hemispheres (Fig. 11.1). The focus for rhythm is caudomedial in relation to the more anterolateral focus for tonal relationships. This is similarly true in relation to timbre, though there is considerable overlap with the meta-analysis focus for timbre, which itself encompasses much of primary auditory cortex (Fig. 11.2). Aside from a region in both hemispheres between TE1 and TE3 that all dimensions of music appear to activate, no other lateral temporal-lobe foci were identified in the meta-analysis for rhythm. By contrast, foci were found throughout the motor system, including the cerebellum, putamen, SMA, and lateral premotor areas. Of these foci, the cerebellum is associated with the timing of absolute durations, whereas the remaining areas participate in a beat-based timing mechanism in which time intervals are judged relative to a discernable underlying beat (Teki et al., 2011).

Of note is the fact that many of the comparisons of conditions (contrasts) that were entered into the rhythm meta-analysis were rather unrestricted low-level contrasts that compared listening to rhythmic stimuli with resting-state activity (Grahn and Brett, 2007; Grahn and McAuley, 2009; Chapin et al., 2010b; Jerde et al., 2011), or attentive listening in anticipation of subsequent tapping with the heard rhythm (Chen et al., 2008). However, contrasts that are more descriptive of temporal variation, such as the tempo fluctuations in an expressively timed rendition of a piano composition (Chapin et al., 2010a) or the entropy in the timing of inter-onset intervals in imitated and improvised piano performances (Engel and Keller, 2011), also activate motor regions. Consistent with the idea that these areas are driven by temporal variation is the observation that a repeating temporal pattern causes an adaptation (decrease) of activity in these areas (Brown et al., 2013). Motor areas also follow fluctuations in the clarity of the pulse in a complex piece of real-world music (Alluri et al., 2012) and are engaged by a beat-based timing discrimination task (Teki et al., 2011).

MUSICAL EXPERTISE AND PLASTICITY IN THE AUDITORY CORTEX

Because developing expertise with music might be regarded as optional for survival, at least in comparison with language, the natural variability in the amounts of musical training across the human population affords the possibility to study the correlates, if not the direct consequences, of musical training on the functional neuroanatomy that underlies the perception and production of music (Munte et al., 2002; Pantev and Herholz, 2011). For example, the volume of gray matter in the anterior portion of Heschl's gyrus is positively correlated with musical training, musical aptitude, and the magnitude of early (<30 ms latency) auditory evoked magnetic fields in response to amplitude-modulated tones (Schneider et al., 2002, 2005). Relatedly, the tendency to perceive changes in pitch direction in pairs of complex harmonic tones in terms of the change in fundamental frequency or change in spectral centroid is reflected in asymmetries in the size of left and right lateral Heschl's gyri, respectively, with similarly asymmetric evoked response magnitudes (Fig. 11.5; Schneider et al., 2005). Such anatomic changes may arise as a consequence of musical training relatively early in development. In a sample of ~6–7-year-old children, 15 months of weekly 30-minute keyboard lessons resulted in greater relative voxel sizes in a number of brain areas, including motor areas and the right primary auditory cortex, relative to a group of similarly aged children who engaged in weekly music activities but did not have focused instrumental training (Hyde et al., 2009).

Given the structure/function correlations in Heschl's gyrus described in the preceding paragraph, it is perhaps

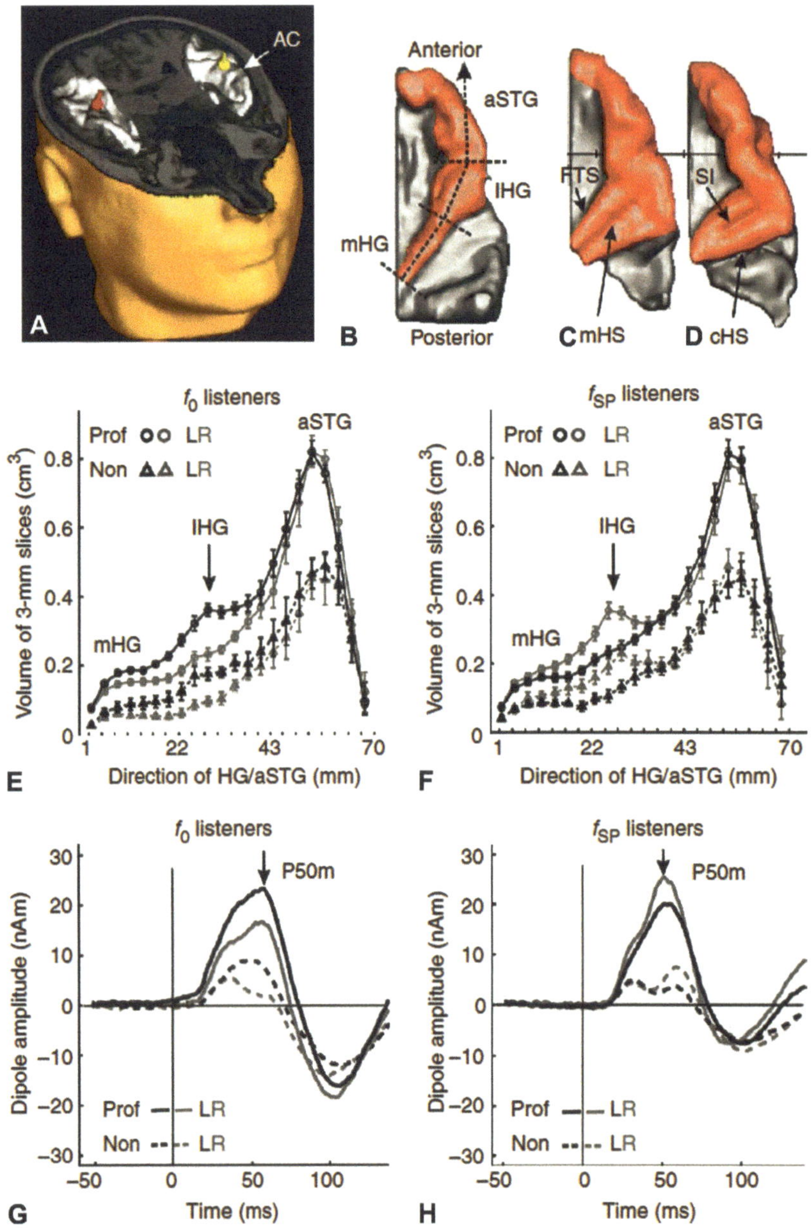

Fig. 11.5. An illustration of training- and perception-related morphologic differences in the auditory cortex. Here, the perceptual differences pertain to biases toward either hearing complex harmonic sounds in terms of their fundamental frequency (f_0 listeners) or spectral distribution (f_{SP} listeners). (**A**) Three-dimensional (3D) reconstruction of an individual auditory cortex and auditory evoked activity (blue and red dipoles). (**B-D**) The top view of three individual 3D surface reconstructions of right auditory cortex shows the pronounced gyral entity including Heschl's gyrus (HG) and the anterior superior temporal gyrus (colored red), bordered anteriorly by the first transverse sulcus (FTS) and posteriorly by the first complete Heschl's sulcus (cHS). HG may include a sulcus intermedius (SI), indenting locally the crown of HG, or a medial Heschl's sulcus (mHS), not reaching the lateral end. Gray-matter volume was successively calculated in cross-sectional slices along the medial–lateral–anterior progression. (**E, F**) Fundamental pitch listeners demonstrated a pronounced leftward asymmetry of gray-matter volume and spectral pitch listeners a rightward asymmetry, peaking within the lateral one-third of HG. (**G, H**) Functional asymmetry of the auditory evoked P50m source activity of lHG in response to harmonic complex tones. P50m magnitude was relatively larger in the left hemisphere for f_0 listeners and vice versa for f_{SP} listeners, irrespective of musical aptitude. Prof, professional musician; non, non-musician. (Reproduced from Schneider et al., 2005.)

no surprise that larger MMN responses have been found in musicians than non-musicians (the non/musician distinction is usually based on the amount of practice singing or playing an instrument) when presented with short (e.g., five-note) melodic patterns in which the contour (Tervaniemi et al., 1997, 2001) or grouping pattern (van Zuijen et al., 2004) is changed. Similar effects of training on the MMN are seen in responses to slightly mistuned chords embedded in sequences of major triads (Koelsch et al., 1999). The development of MMN responses to abstract musical patterns can be quite rapid and dependent on accurate behavior during an attentive deviance detection task. For example, Tervaniemi et al. (2001) found an MMN in response to contour deviants in five tone patterns to develop over the course of several hundred trials, but only in those subjects who accurately detected the deviants during interleaved blocks of trials in which it was their task to detect the deviants (Fig. 11.6A). Of note is the fact that, while musically trained individuals exhibit larger MMN responses to interval, contour, or key deviants in short melodies than do individuals without musical training (Fujioka et al., 2004, 2005; Fig. 11.6B), both types of individuals show comparable MMN response to deviants in sequences of isolated tones (Fujioka et al., 2004). (A variety of studies suggest, however, that responses to individual tones will be enhanced if they are played on an instrument on which the listener is trained (Shahin et al., 2008; Margulis et al., 2009; Pantev and Herholz, 2011).) A longitudinal study of children with and without musical training, across ages 9, 11, and 13 years, examined MMN responses to a variety of musical deviance types and found significantly larger MMN responses to mistuned notes, timbral deviants, contour deviants, and rhythmic deviants in musically trained children by ages 11–13 years (Putkinen et al., 2014). For further examples of the interactions between musical training, brain development, and auditory abilities as they pertain to

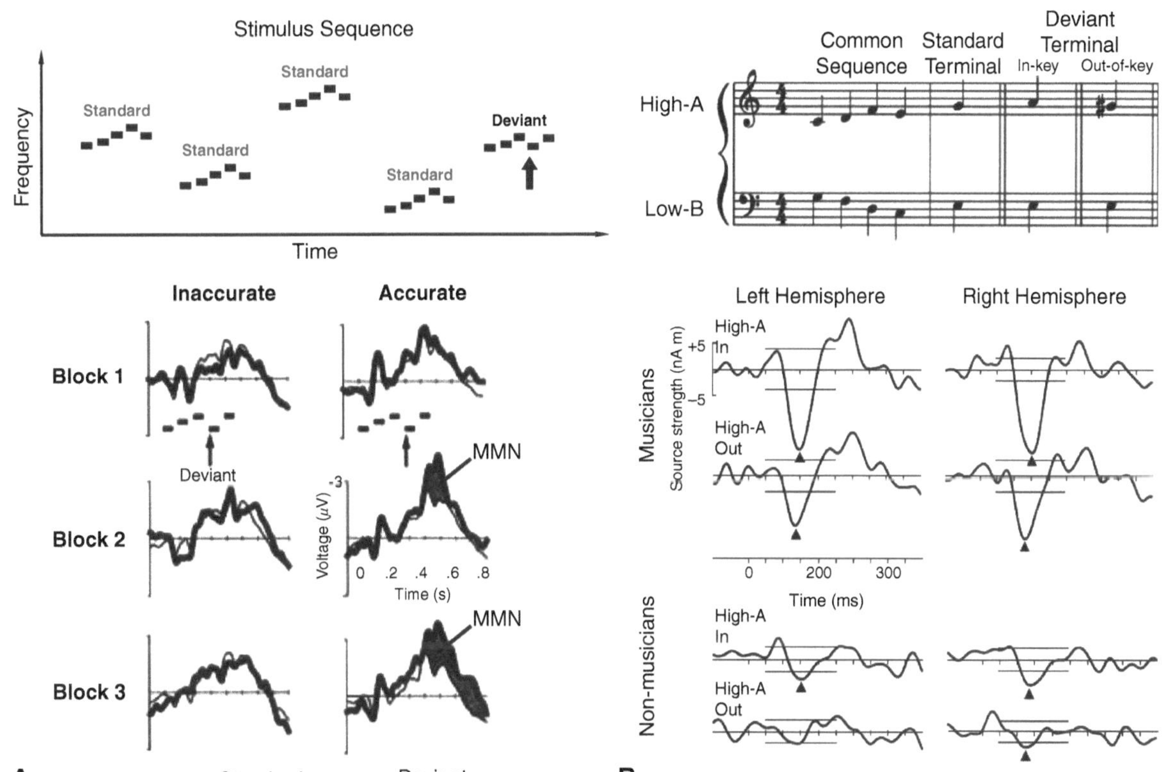

Fig. 11.6. Examples of the effects of short-term and long-term musical training on the amplitude of the mismatch negativity (MMN) brain potential. (**A**) Experimental conditions in MMN experiments typically consist of hundreds of repetitions of "standard" stimuli interspersed 10% of the time with "deviant" stimuli. In this case the stimuli are brief tonal patterns consisting of five 50-ms tones, and the deviant pattern constitutes a change in the contour of the pattern. Here, the MMN was assessed in three blocks of trials, each block immediately preceded by a block of trials in which subjects had to detect the occurrence of the deviant patterns. The MMN grew in size with increased training, but only in those subjects who accurately detected the deviant contours. (Adapted from Tervaniemi et al., 2001.) (**B**) In more complex musical situations, such as when two melodies are played simultaneously (High-A, Low-B), the MMN (assessed here using magnetoencephalography) to endings that are either in-key (interval) or out-of-key deviants is larger for musically trained than untrained subjects. (Adapted from Fujioka et al., 2005.)

speech and music, please see reviews by Kraus et al. (2009) and Pantev and Herholz (2011).

MUSIC BEYOND THE AUDITORY CORTEX

As the meta-analyses shown in Figures 11.1 and 11.2 suggest, a full accounting of how the brain perceives, responds to, and creates music must necessarily extend beyond consideration of the role that the auditory cortex plays in transforming auditory information into musically meaningful streams of information. Outside of the auditory cortex, areas of the frontal lobes have garnered the most attention. These include the VLPFC, activations of which in BA 44 have been observed during the detection or violations of tonal context as cited above, or segmentation of phrases or longer musical spans (Sridharan et al., 2007; Nan et al., 2008) in BA 47. The perception of coherent musical compositions versus scrambled versions of those compositions also engages the VLPFC (Levitin and Menon, 2003; Abrams et al., 2011), as does the time-varying structure of musical movement in tonal space (Janata et al., 2002b; Janata, 2009). As the studies of rhythm utilized in the meta-analysis indicate, premotor areas are commonly activated when rhythmic patterns of varying complexity are perceived with varying degrees of attention or performance intention (Janata and Grafton, 2003).

Studies that have focused on tasks that require maintenance of musical information in working memory (Zatorre et al., 1994; Koelsch et al., 2009), attention (Janata et al., 2002a), or mental manipulation of auditory images (Zatorre et al., 1996, 2010; Halpern and Zatorre, 1999; Herholz et al., 2012) have typically found combinations of dorsolateral prefrontal, medial and lateral premotor, and parietal areas to be active.

Perhaps most saliently, from the experiential perspective of an individual, music is tied to emotion, giving rise to a variety of emotions, ranging from moderately pleasing or displeasing experiencing of the music (Brown et al., 2004a; Menon and Levitin, 2005; Koelsch et al., 2006) to focal experiencing of intensely pleasurable "chills" (Blood et al., 1999; Blood and Zatorre, 2001; Salimpoor et al., 2011), or nostalgic remembering (Janata, 2009; Trost et al., 2012). Not surprisingly, experiments that have elicited such experiences have found activation of some combination of the amygdala, the striatum, and the medial prefrontal cortex. These areas increase in functional connectivity with the auditory cortex as a person makes affective valuations of novel pieces of music (Salimpoor et al., 2013), suggesting that there might be a direct emotional appraisal of the information, e.g., the auditory scene, represented in the auditory cortex.

ACKNOWLEDGMENTS

The author thanks Genie Pak and Natalia Karvelis for assistance with the meta-analyses.

References

Abrams DA, Bhatara A, Ryali S et al. (2011). Decoding temporal structure in music and speech relies on shared brain resources but elicits different fine-scale spatial patterns. Cereb Cortex 21: 1507–1518.

Alluri V, Toiviainen P, Jääskeläinen IP et al. (2012). Large-scale brain networks emerge from dynamic processing of musical timbre, key and rhythm. Neuroimage 59: 3677–3689.

Angenstein N, Scheich H, Brechmann A (2012). Interaction between bottom-up and top-down effects during the processing of pitch intervals in sequences of spoken and sung syllables. Neuroimage 61: 715–722.

Bangert M, Peschel T, Schlaug G et al. (2006). Shared networks for auditory and motor processing in professional pianists: evidence from fMRI conjunction. Neuroimage 30 (3): 917–926.

Bharucha JJ (1996). Melodic anchoring. Music Percept 13: 383–400.

Bidelman GM, Krishnan A (2009). Neural correlates of consonance, dissonance, and the hierarchy of musical pitch in the human brainstem. J Neurosci 29: 13165–13171.

Blood AJ, Zatorre RJ (2001). Intensely pleasurable responses to music correlate with activity in brain regions implicated in reward and emotion. Proc Natl Acad Sci U S A 98: 11818–11823.

Blood AJ, Zatorre RJ, Bermudez P et al. (1999). Emotional responses to pleasant and unpleasant music correlate with activity in paralimbic brain regions. Nat Neurosci 2: 382–387.

Brattico E, Tervaniemi M, Naatanen R et al. (2006). Musical scale properties are automatically processed in the human auditory cortex. Brain Res 1117: 162–174.

Brown S, Martinez MJ (2007). Activation of premotor vocal areas during musical discrimination. Brain Cogn 63: 59–69.

Brown S, Martinez MJ, Parsons LM (2004a). Passive music listening spontaneously engages limbic and paralimbic systems. Neuroreport 15: 2033–2037.

Brown S, Martinez MJ, Hodges DA et al. (2004b). The song system of the human brain. Cogn Brain Res 20: 363–375.

Brown RM, Chen JL, Hollinger A et al. (2013). Repetition suppression in auditory-motor regions to pitch and temporal structure in music. J Cogn Neurosci 25: 313–328.

Chapin H, Jantzen K, Kelso JAS et al. (2010a). Dynamic emotional and neural responses to music depend on performance expression and listener experience. PLoS One 5.

Chapin HL, Zanto T, Jantzen KJ et al. (2010b). Neural responses to complex auditory rhythms: the role of attending. Front Psychol 2: 12.

Chen JL, Penhune VB, Zatorre RJ (2008). Listening to musical rhythms recruits motor regions of the brain. Cereb Cortex 18: 2844–2854.

Chen JL, Rae C, Watkins KE (2012). Learning to play a melody: an fMRI study examining the formation of auditory-motor associations. Neuroimage 59 (2): 1200–1208.

Collins T, Tillmann B, Barrett FS et al. (2014). A combined model of sensory and cognitive representations underlying tonal expectations in music: From audio signals to behavior. Psychol Rev 121 (1): 33–65.

Deike S, Gaschler-Markefski B, Brechmann A et al. (2004). Auditory stream segregation relying on timbre involves left auditory cortex. Neuroreport 15: 1511–1514.

Dowling WJ (1978). Scale and contour: two components of a theory of memory for melodies. Psychol Rev 85: 341–354.

Elliott TM, Hamilton LS, Theunissen FE (2013). Acoustic structure of the five perceptual dimensions of timbre in orchestral instrument tones. J Acoust Soc Am 133: 389–404.

Ellis RJ, Norton AC, Overy K et al. (2012). Differentiating maturational and training influences on fMRI activation during music processing. Neuroimage 60: 1902–1912.

Engel A, Keller PE (2011). The perception of musical spontaneity in improvised and imitated jazz performances. Front Psychol 2: 83.

Escoffier N, Zhong J, Schirmer A et al. (2013). Emotional expressions in voice and music: same code, same effect? Hum Brain Mapp 34 (8): 1796–1810.

Fujioka T, Trainor LJ, Ross B et al. (2004). Musical training enhances automatic encoding of melodic contour and interval structure. J Cogn Neurosci 16: 1010–1021.

Fujioka T, Trainor LJ, Ross B et al. (2005). Automatic encoding of polyphonic melodies in musicians and nonmusicians. J Cogn Neurosci 17: 1578–1592.

Fujisawa TX, Cook ND (2011). The perception of harmonic triads: an fMRI study. Brain Imaging Behav 5: 109–125.

Grahn JA, Brett M (2007). Rhythm and beat perception in motor areas of the brain. J Cogn Neurosci 19: 893–906.

Grahn JA, McAuley JD (2009). Neural bases of individual differences in beat perception. Neuroimage 47: 1894–1903.

Grahn JA, Rowe JB (2009). Feeling the beat: premotor and striatal interactions in musicians and nonmusicians during beat perception. J Neurosci 29 (23): 7540–7548.

Grey JM (1977). Multidimensional perceptual scaling of musical timbres. J Acoust Soc Am 61: 1270–1277.

Griffiths TD, Buchel C, Frackowiak RSJ et al. (1998). Analysis of temporal structure in sound by the human brain. Nat Neurosci 1 (5): 422–427.

Groussard M, La Joie R, Rauchs G et al. (2010a). When music and long-term memory interact: effects of musical expertise on functional and structural plasticity in the hippocampus. PLoS One 5 (10).

Groussard M, Rauchs G, Landeau B et al. (2010b). The neural substrates of musical memory revealed by fMRI and two semantic tasks. Neuroimage 53 (4): 1301–1309.

Habermeyer B, Herdener M, Esposito F et al. (2009). Neural correlates of pre-attentive processing of pattern deviance in professional musicians. Hum Brain Mapp 30 (11): 3736–3747.

Halpern AR, Zatorre RJ (1999). When that tune runs through your head: a PET investigation of auditory imagery for familiar melodies. Cereb Cortex 9: 697–704.

Halpern AR, Zatorre RJ, Bouffard M et al. (2004). Behavioral and neural correlates of perceived and imagined musical timbre. Neuropsychologia 42: 1281–1292.

Herholz SC, Halpern AR, Zatorre RJ (2012). Neuronal correlates of perception, imagery, and memory for familiar tunes. J Cogn Neurosci 24: 1382–1397.

Hyde KL, Lerch J, Norton A et al. (2009). Musical training shapes structural brain development. J Neurosci 29: 3019–3025.

Janata P (2009). The neural architecture of music-evoked autobiographical memories. Cereb Cortex 19: 2579–2594.

Janata P, Grafton ST (2003). Swinging in the brain: shared neural substrates for behaviors related to sequencing and music. Nat Neurosci 6: 682–687.

Janata P, Parsons L (2013). Neural mechanisms of music, singing and dancing. In: MA Arbib (Ed.), Language, Music, and the Brain: A Mysterious Relationship, MIT Press, Cambridge, MA.

Janata P, Tillmann B, Bharucha JJ (2002a). Listening to polyphonic music recruits domain-general attention and working memory circuits. Cogn Affect Behav Neurosci 2: 121–140.

Janata P, Birk JL, Van Horn JD et al. (2002b). The cortical topography of tonal structures underlying Western music. Science 298: 2167–2170.

Jeffries KJ, Fritz JB, Braun AR (2003). Words in melody: an H-2 O-15 PET study of brain activation during singing and speaking. Neuroreport 14: 749–754.

Jerde TA, Childs SK, Handy ST et al. (2011). Dissociable systems of working memory for rhythm and melody. Neuroimage 57: 1572–1579.

Johnson JK, Chang CC, Brambati SM et al. (2011). Music recognition in frontotemporal lobar degeneration and Alzheimer disease. Cogn Behav Neurol 24: 85–92.

Johnsrude IS, Penhune VB, Zatorre RJ (2000). Functional specificity in the right human auditory cortex for perceiving pitch direction. Brain 123: 155–163.

Klein ME, Zatorre RJ (2011). A role for the right superior temporal sulcus in categorical perception of musical chords. Neuropsychologia 49: 878–887.

Koelsch S (2009). Music-syntactic processing and auditory memory: similarities and differences between ERAN and MMN. Psychophysiology 46: 179–190.

Koelsch S, Schroger E, Tervaniemi M (1999). Superior pre-attentive auditory processing in musicians. Neuroreport 10: 1309–1313.

Koelsch S, Gunter TC, Friederici AD et al. (2000). Brain indices of music processing: "Nonmusicians" are musical. J Cogn Neurosci 12: 520–541.

Koelsch S, Gunter TC, Schroger E et al. (2001). Differentiating ERAN and MMN: an ERP study. Neuroreport 12: 1385–1389.

Koelsch S, Gunter TC, v Cramon DY et al. (2002). Bach speaks: a cortical "language-network" serves the processing of music. Neuroimage 17: 956–966.

Koelsch S, Fritz T, Schulze K et al. (2005). Adults and children processing music: an fMRI study. Neuroimage 25: 1068–1076.

Koelsch S, Fritz T, v Cramon DY et al. (2006). Investigating emotion with music: an fMRI study. Hum Brain Mapp 27: 239s–250s.

Koelsch S, Schulze K, Sammler D et al. (2009). Functional architecture of verbal and tonal working memory: an fMRI study. Hum Brain Mapp 30: 859–873.

Kraus N, Skoe E, Parbery-Clark A et al. (2009). Experience-induced malleability in neural encoding of pitch, timbre, and timing. Ann N Y Acad Sci 1169: 543–557.

Krumhansl CL (1990). Cognitive foundations of musical pitch, Oxford University Press, New York.

Krumhansl CL, Cuddy LL (2010). A theory of tonal hierarchies in music. In: M Riess Jones, RR Fay, AN Popper (Eds.), Music Perception, Springer, New York, pp. 51–87.

Leaver AM, Rauschecker JP (2010). Cortical representation of natural complex sounds: effects of acoustic features and auditory object category. J Neurosci 30: 7604–7612.

Lee Y-S, Janata P, Frost C et al. (2011). Investigation of melodic contour processing in the brain using multivariate pattern-based fMRI. Neuroimage 57: 293–300.

Leino S, Brattico E, Tervaniemi M et al. (2007). Representation of harmony rules in the human brain: further evidence from event-related potentials. Brain Res 1142: 169–177.

Levitin DJ, Menon V (2003). Musical structure is processed in "language" areas of the brain: a possible role for Brodmann area 47 in temporal coherence. Neuroimage 20: 2142–2152.

Liegeois-Chauvel C, Peretz I, Babai M et al. (1998). Contribution of different cortical areas in the temporal lobes to music processing. Brain 121: 1853–1867.

Loui P, Grent-'t-Jong T, Torpey D et al. (2005). Effects of attention on the neural processing of harmonic syntax in Western music. Cognit Brain Res 25: 678–687.

Maess B, Koelsch S, Gunter TC et al. (2001). Musical syntax is processed in Broca's area: an MEG study. Nat Neurosci 4: 540–545.

Maidhof C, Koelsch S (2011). Effects of selective attention on syntax processing in music and language. J Cogn Neurosci 23: 2252–2267.

Margulis EH, Mlsna LM, Uppunda AK et al. (2009). Selective neurophysiologic responses to music in instrumentalists with different listening biographies. Hum Brain Mapp 30: 267–275.

McAdams S (1999). Perspectives on the contribution of timbre to musical structure. Comput Music J 23: 85–102.

McAdams S, Winsberg S, Donnadieu S et al. (1995). Perceptual scaling of synthesized musical timbres – common dimensions, specificities, and latent subject classes. Psychol Res-Psychol Forsch 58: 177–192.

Menon V, Levitin DJ (2005). The rewards of music listening: response and physiological connectivity of the mesolimbic system. Neuroimage 28: 175–184.

Menon V, Levitin DJ, Smith BK et al. (2002). Neural correlates of timbre change in harmonic sounds. Neuroimage 17: 1742–1754.

Mithen S (2005). The Singing Neanderthals, Weidenfeld & Nicolson, London.

Morosan P, Rademacher J, Schleicher A et al. (2001). Human primary auditory cortex: cytoarchitectonic subdivisions and mapping into a spatial reference system. Neuroimage 13: 684–701.

Munte TF, Altenmuller E, Jancke L (2002). The musician's brain as a model of neuroplasticity. Nat Rev Neurosci 3: 473–478.

Naatanen R, Paavilainen P, Rinne T et al. (2007). The mismatch negativity (MMN) in basic research of central auditory processing: a review. Clin Neurophysiol 118: 2544–2590.

Nan Y, Knosche TR, Zysset S et al. (2008). Cross-cultural music phrase processing: an fMRI study. Hum Brain Mapp 29: 312–328.

Ouimet T, Foster NEV, Hyde KL (2012). Auditory global-local processing: effects of attention and musical experience. J Acoust Soc Am 132: 2536–2544.

Pantev C, Herholz SC (2011). Plasticity of the human auditory cortex related to musical training. Neurosci Biobehav Rev 35: 2140–2154.

Patel AD, Gibson E, Ratner J et al. (1998). Processing syntactic relations in language and music: an event-related potential study. J Cogn Neurosci 10: 717–733.

Patterson RD, Uppenkamp S, Johnsrude IS et al. (2002). The processing of temporal pitch and melody information in auditory cortex. Neuron 36: 767–776.

Peretz I (1990). Processing of local and global musical information by unilateral brain-damaged patients. Brain 113: 1185–1205.

Platel H, Baron JC, Desgranges B et al. (2003). Semantic and episodic memory of music are subserved by distinct neural networks. Neuroimage 20: 244–256.

Poulin-Charronnat B, Bigand E, Koelsch S (2006). Processing of musical syntax tonic versus subdominant: an event-related potential study. J Cogn Neurosci 18: 1545–1554.

Pressing J (2002). Black Atlantic rhythm: its computational and transcultural foundations. Music Percept 19: 285–310.

Putkinen V, Tervaniemi M, Saarikivi K et al. (2014). Investigating the effects of musical training on functional brain development with a novel melodic MMN paradigm. Neurobiol Learn Mem 110: 8–15.

Saarinen J, Paavilainen P, Schöger E et al. (1992). Representation of abstract attributes of auditory stimuli in the human brain. Neuroreport 3: 1149–1151.

Sakai K, Hikosaka O, Miyauchi S et al. (1999). Neural representation of a rhythm depends on its interval ratio. J Neurosci 19 (22): 10074–10081.

Salimpoor VN, Benovoy M, Larcher K et al. (2011). Anatomically distinct dopamine release during anticipation and experience of peak emotion to music. Nat Neurosci 14: 257, U355.

Salimpoor VN, van den Bosch I, Kovacevic N et al. (2013). Interactions between the nucleus accumbens and auditory cortices predict music reward value. Science 340: 216–219.

Sammler D, Baird A, Valabregue R et al. (2010). The relationship of lyrics and tunes in the processing of unfamiliar songs: a functional magnetic resonance adaptation study. J Neurosci 30: 3572–3578.

Samson S, Zatorre RJ (1991). Recognition memory for text and melody of songs after unilateral temporal lobe lesion – evidence for dual encoding. J Exp Psychol Learn Mem Cognit 17: 793–804.

Schindler A, Herdener M, Bartels A (2013). Coding of melodic gestalt in human auditory cortex. Cereb Cortex 23: 2987–2993.

Schneider P, Scherg M, Dosch HG et al. (2002). Morphology of Heschl's gyrus reflects enhanced activation in the auditory cortex of musicians. Nat Neurosci 5: 688–694.

Schneider P, Sluming V, Roberts N et al. (2005). Structural and functional asymmetry of lateral Heschl's gyrus reflects pitch perception preference. Nat Neurosci 8: 1241–1247.

Schon D, Gordon R, Campagne A et al. (2010). Similar cerebral networks in language, music and song perception. Neuroimage 51: 450–461.

Schulze K, Mueller K, Koelsch S (2011a). Neural correlates of strategy use during auditory working memory in musicians and non-musicians. Eur J Neurosci 33: 189–196.

Schulze K, Zysset S, Mueller K et al. (2011b). Neuroarchitecture of verbal and tonal working memory in nonmusicians and musicians. Hum Brain Mapp 32: 771–783.

Seger CA, Spiering BJ, Sares AG et al. (2013). Corticostriatal contributions to musical expectancy perception. J Cogn Neurosci 25: 1062–1077.

Shahin AJ, Roberts LE, Chau W et al. (2008). Music training leads to the development of timbre-specific gamma band activity. Neuroimage 41: 113–122.

Shepard RN (1964). Circularity in judgments of relative pitch. J Acoust Soc Am 36: 2346.

Shepard RN (1982). Geometrical approximations to the structure of musical pitch. Psychol Rev 89: 305–333.

Sridharan D, Levitin DJ, Chafe CH et al. (2007). Neural dynamics of event segmentation in music: converging evidence for dissociable ventral and dorsal networks. Neuron 55: 521–532.

Steinbeis N, Koelsch S, Sloboda JA (2006). The role of harmonic expectancy violations in musical emotions: evidence from subjective, physiological, and neural responses. J Cogn Neurosci 18: 1380–1393.

Stewart L, Overath T, Warren JD et al. (2008). fMRI evidence for a cortical hierarchy of pitch pattern processing. PLoS One 3.

Teki S, Grube M, Kumar S et al. (2011). Distinct neural substrates of duration-based and beat-based auditory timing. J Neurosci 31: 3805–3812.

Tervaniemi M, Maury S, Näätänen R (1994). Neural representations of abstract stimulus features in the human brain as reflected by the mismatch negativity. Neuroreport 5: 844–846.

Tervaniemi M, Ilvonen T, Karma K, Alho K, Näätänen R (1997). The musical brain: Brain waves reveal the neurophysiological basis of musicality in human subjects. Neurosci Lett 226: 1–4.

Tervaniemi M, Rytkönen M, Schröger E et al. (2001). Superior formation of cortical memory traces for melodic patterns in musicians. Learn Mem 8: 295–300.

Tillmann B, Janata P, Bharucha JJ (2003). Activation of the inferior frontal cortex in musical priming. Cognit Brain Res 16: 145–161.

Tillmann B, Koelsch S, Escoffier N et al. (2006). Cognitive priming in sung and instrumental music: activation of inferior frontal cortex. Neuroimage 31: 1771–1782.

Trainor LJ, McDonald KL, Alain C (2002). Automatic and controlled processing of melodic contour and interval information measured by electrical brain activity. J Cogn Neurosci 14: 430–442.

Trost W, Ethofer T, Zentner M et al. (2012). Mapping aesthetic musical emotions in the brain. Cereb Cortex 22: 2769–2783.

Turkeltaub PE, Eickhoff SB, Laird AR et al. (2012). Minimizing within-experiment and within-group effects in activation likelihood estimation meta-analyses. Hum Brain Mapp 33: 1–13.

van Zuijen TL, Sussman E, Winkler I et al. (2004). Grouping of sequential sounds—an event-related potential study comparing musicians and nonmusicians. J Cogn Neurosci 16: 331–338.

Warren JD, Uppenkamp S, Patterson RD et al. (2003). Separating pitch chroma and pitch height in the human brain. Proc Natl Acad Sci U S A 100: 10038–10042.

Warren JD, Jennings AR, Griffiths TD (2005). Analysis of the spectral envelope of sounds by the human brain. Neuroimage 24: 1052–1057.

Warrier CM, Zatorre RJ (2002). Influence of tonal context and timbral variation on perception of pitch. Percept Psychophys 64: 198–207.

Warrier CM, Zatorre RJ (2004). Right temporal cortex is critical for utilization of melodic contextual cues in a pitch constancy task. Brain 127: 1616–1625.

Zatorre RJ, Zarate JM (2012). Cortical Processing of Music. In: D Poeppel, T Overath, AN Popper, RR Fay (Eds.), The Human Auditory Cortex, Springer, New York, pp. 261–294.

Zatorre RJ, Evans AC, Meyer E (1994). Neural mechanisms underlying melodic perception and memory for pitch. J Neurosci 14: 1908–1919.

Zatorre RJ, Halpern AR, Perry DW et al. (1996). Hearing in the mind's ear: a PET investigation of musical imagery and perception. J Cogn Neurosci 8: 29–46.

Zatorre RJ, Belin P, Penhune VB (2002). Structure and function of auditory cortex: music and speech. Trends Cogn Sci 6: 37–46.

Zatorre RJ, Halpern AR, Bouffard M (2010). Mental reversal of imagined melodies: a role for the posterior parietal cortex. J Cogn Neurosci 22: 775–789.

Handbook of Clinical Neurology, Vol. 129 (3rd series)
The Human Auditory System
G.G. Celesia and G. Hickok, Editors

Chapter 12

Music and language: relations and disconnections

NINA KRAUS[1,2,3,4*,†] AND JESSICA SLATER[1,2]
[1]*Auditory Neuroscience Laboratory, Northwestern University, Evanston, IL, USA*
[2]*Department of Communication Sciences, Northwestern University, Evanston, IL, USA*
[3]*Department of Neurobiology and Physiology, Northwestern University, Evanston, IL, USA*
[4]*Department of Otolaryngology, Northwestern University, Evanston, IL, USA*

INTRODUCTION

Above all, music and language are forms of human communication. Since sensory function is ultimately shaped by what is biologically important to the organism, the human urge to communicate has been a powerful driving force in both the evolution of auditory function and the ways in which it can be changed by experience within an individual lifetime. The ability to extract meaning from sound requires that all aspects of the auditory system work together in concert: by considering human auditory function in the context of communication, we hope to emphasize the highly interactive nature of the auditory system as well as the depth of its integration with other sensory and cognitive systems. Through consideration of relations and dissociations between music and language we will explore key themes in auditory function, learning and plasticity.

First we will look to the origins of music and language: we shall not attempt to resolve the ongoing debate regarding their evolutionary status, but we will consider what their structural characteristics and functional origins reveal about the human relationship with sound. Then we will focus on the role of temporal structure in music and language, and how temporal patterns provide a framework for information processing in the brain. Next we will consider how music and language are learned, emphasizing interactions between the auditory system, cognition, and emotion and reward centers in the brain. The final section of this chapter will focus on the biology of sound processing and how experience with music and language influences underlying neural function. We will show that auditory expertise strengthens some of the very same aspects of sound encoding that are deficient in impaired populations and discuss the implications for clinical and social contexts.

MUSIC AND LANGUAGE: STRUCTURAL AND FUNCTIONAL ORIGINS

Structure rooted in sound

Both music and language are complex communication systems, in which basic components are combined into higher-order structures in accordance with rules. Whether music was an evolutionary precursor to language (Mithen, 2005) or merely a byproduct of cognitive faculties that developed to support language (Pinker, 1999), music is pervasive across human cultures and throughout history, with evidence of prehistoric bone and ivory flutes dating back over 42,000 years (Higham et al., 2012). The capacity of music to convey emotion is driven not only by cultural conventions but by the underlying psychophysical properties of sound and cochlear mechanics (Balkwill and Thompson, 1999; Tramo et al., 2001; Gomez and Danuser, 2007). For example, the perception of consonance and dissonance arises at least in part from the different vibration patterns generated by combinations of frequencies and their harmonics on the basilar membrane, where consonant intervals comprise simple frequency ratios and generate less interference than dissonant intervals (Von Helmholtz, 1912). Consonant intervals are typically considered to be pleasing to the ear (Trainor, 2008), which may reflect a relationship between ease of perceptual processing and positive emotional state (Reber et al.,

*Correspondence to: Nina Kraus, Ph.D., Northwestern University, 2240 Campus Drive, Evanston, IL 60208, USA. Tel: +1-847-491-3181, Fax: +1-847-491-2523, E-mail: nkraus@northwestern.edu
†Auditory neuroscience lab website: www.brainvolts.northwestern.edu

1998, 2004). Children's innate receptivity to music from a very young age, including their ability to perceive consonance and dissonance (Trainor and Heinmiller, 1998; Zentner and Kagan, 1998; Trehub, 2003), reinforces the idea that musical structure is rooted in fundamental characteristics of sound and sensory perception.

An important distinction between music and language is in their differing degrees of semantic specificity. While language can be used to convey precise semantic content between individuals, precision can also lead to disagreement; music lacks semantic specificity but plays an important role in promoting social bonding (Cross, 1999; Trehub, 2003), as demonstrated experimentally by an increase in cooperative behavior in young children following joint music making (Kirschner and Tomasello, 2010). In this sense, language enables articulation of what is within us, whereas music strengthens what is shared between us.

Communication in context

While we are largely focused on music and language as forms of auditory communication, it is important to acknowledge that the auditory system does not function in isolation, especially within the context of communication. Other sensory modalities, such as vision, help the brain to construct a percept of sound object (Schutz and Lipscomb, 2007; Musacchia et al., 2008) and can influence speech perception. This is demonstrated by the McGurk effect, in which a video presenting repeated utterances of the syllable [ba] dubbed on to the lip movements for [ga] result in normal-hearing adults perceiving a hybrid percept, the syllable [da] (McGurk and MacDonald, 1976). In the case of music and language, this cross-modal integration is important in helping to determine the communicative intent which produced a sound, and thereby determining its meaning, thus: "Speech is processed both as a sound and as an action" (Scott and Johnsrude, 2003). Similarly, it has been shown that the perception of musical performances can be significantly influenced by non-auditory cues, such as body movements (Tsay, 2013).

Music and language are likely to have evolved in conjunction with important shifts in the emergence of human cognition, and while the exact sequences and interdependence of these evolutionary trajectories are debated (Cross, 1999; Mithen, 2005; Fitch, 2006), it is clear that processing of music and language relies upon cognitive capacities, such as working memory. There is evidence that expertise such as musical training is associated with greater auditory working-memory capacity across the lifespan (Bugos et al., 2007; Kraus et al., 2012), which also relates to better performance on everyday listening tasks such as perceiving speech in a noisy background (Parbery-Clark et al., 2009b; Kraus and Chandrasekaran, 2010; Anderson et al., 2013). Taken together, these outcomes emphasize that the ability to make sense of sound in a communication context relies heavily upon integration between the auditory system and other cognitive and sensory systems. We will provide further discussion of how these relationships are mediated by experience with music and language in the final section of this chapter.

TEMPORAL PROCESSING IN MUSIC AND LANGUAGE

The auditory system plays a unique role as the temporal processing "expert" in the nervous system, with microsecond precision required for the localization of sounds (Mauk and Buonomano, 2004). The ability to lock on to temporal patterns is a fundamental aspect of auditory processing, and is especially important for the perception of communication sounds: meaningful information unfolds simultaneously over multiple timescales in both speech and music, from overarching rhythms and stress patterns, to the fine-grained timing differences which differentiate consonants and characterize the distinctive timbre of a voice or musical instrument.

The timing precision of the auditory system is also important for sensorimotor coordination, with a study of 16 normal adults showing that synchronization of finger tapping to an auditory stimulus is more accurate than to a visual stimulus (Repp and Penel, 2002), and a study from our laboratory with 124 typically developing adolescents showing that the ability to synchronize with a beat tracks with more consistent subcortical neural timing in the auditory system in response to speech sounds (Tierney and Kraus, 2013a).

The integration between auditory and motor systems is especially strong in vocal learning species, with co-activation of comparable motor and auditory brain regions in both humans and songbirds during vocal learning tasks (Brown et al., 2004). Rhythm perception involves activation of motor regions of the brain (Zatorre et al., 2007; Chen et al., 2008), and it has been proposed that the ability to synchronize to a beat may rely upon the same auditory-motor neural infrastructure that evolved to support vocal learning (Patel, 2006; Patel et al., 2008, 2009; Patel and Iversen, 2014). This ability has been observed in humans and other vocal learning species but is otherwise very rare in non-human animals, however, individual cases of beat-keeping abilities in non-vocal-learning species such as the sea lion (Cook et al., 2013) leave this a matter of continuing research.

The motor system is not only involved in the production of linguistic and musical sounds (Zatorre et al., 2007; Nasir and Ostry, 2009); motor areas are also activated during the perception of speech and music, when no explicit movement is involved (Zatorre et al., 2007; Chen et al., 2008). In particular, motor regions of the brain such as the cerebellum and basal ganglia are activated during the perception of rhythm (Grahn and Brett, 2007; Chen et al., 2008), which is unsurprising given the direct origins of rhythm in physical movement. Together, these findings emphasize that auditory processing does not happen in isolation: the neural networks involved in the acquisition and production of communication sounds also play a role in their perception. Further, the involvement of other brain areas can be influenced by an individual's previous experience with specific sounds, as demonstrated by a study with 15 adult pianists in which experimenters observed greater excitability in motor regions of the brain while listening to a rehearsed piece of music compared with an unfamiliar piece (D'Ausilio et al., 2006).

Rhythm as a "temporal map"

In the perception of both music and language, rhythm provides a "temporal map" with signposts to the most likely locations of meaningful input. The ability to anticipate based on prior experience can help to streamline attention and processing resources; for example, timing regularities in speech may help a listener develop temporal expectations about when important phonetic information is likely to occur (Pitt and Samuel, 1990; Quene and Port, 2005; Schmidt-Kassow and Kotz, 2008, 2009; Roncaglia-Denissen et al., 2013). Sensitivity to temporal patterns is critical to language acquisition, since it assists in the detection of word boundaries in a continuous speech stream (Nakatani and Schaffer, 1978; Cutler and Butterfield, 1992). These parsing mechanisms are critical to developing accurate representations of meaningful sound elements which can then be mapped to written language, and the ability to make use of rhythmic cues when perceiving speech has been linked to reading ability (Wood and Terrell, 1998; Whalley and Hansen, 2006; Gutiérrez-Palma and Palma Reyes, 2007; Holliman et al., 2008, 2010). One study found that performance on a metric perception task strongly predicted reading ability and phonologic awareness across a population of normal and dyslexic subjects (Huss et al., 2011) and a longitudinal study in a sample of 695 French children showed that rhythm reproduction skills in kindergarten were predictive of reading ability in second grade (Dellatolas et al., 2009). A recent study with preschoolers revealed that those who were able to synchronize with a beat performed better on tests of pre-reading skills, including phonologic awareness, and demonstrated more precise neural encoding of the temporal modulations in speech than those who were unable to synchronize (Woodruff Carr et al., 2014). Impaired rhythmic abilities have been associated with language disorders such as dyslexia (Overy et al., 2003; Thomson and Goswami, 2008; Corriveau and Goswami, 2009; Huss et al., 2011) and there has been some success in using rhythm-based interventions to address reading difficulties in children (Bhide et al., 2013). Music-based therapies, such as melodic intonation therapy, have been used to facilitate the rehabilitation of speech function in aphasic patients by emphasizing "musical" features of speech and thereby engaging neural networks normally involved in the processing of music, to compensate for damaged language-processing regions (Schlaug et al., 2008). Interestingly, a recent study with 17 aphasic patients revealed that rhythmic speech improved articulatory quality in patients whereas melodic intonation alone did not, suggesting that rhythm may play a critical role in the effectiveness of this therapy (Stahl et al., 2011).

In the musical realm, a 2010 study looked at how rhythm influences attention during the process of stream segregation while listening to interleaved melodies, showing that subjects more effectively suppressed a distractor melody when the distractor was rhythmically regular (Devergie et al., 2010). The authors highlight this as evidence for the importance of rhythmic attention in auditory scene analysis, within the framework of the dynamic attending theory (Large and Jones, 1999; Jones et al., 2002). Another study demonstrated that musical training improved performance in a lexical stress-processing task (Kolinsky et al., 2009), suggesting a transfer of rhythmic expertise to speech perception. These studies support the idea that, despite its natural irregularities, metric structure is an important organizing principle for language as well as music, that our perceptual systems are tuned to take advantage of this, and that attention plays a significant role in that tuning process (Kotz et al., 2009; Schmidt-Kassow and Kotz, 2009).

Temporal predictability also serves an important function in communication by facilitating interpersonal synchrony (Phillips-Silver et al., 2010): from the exchange between a mother and her newborn (Censullo et al., 1985) to the exquisitely choreographed complexities of conversational turn taking (Auer et al., 1999) and the interplay of improvisational jazz (Berkowitz, 2010), the ability to coordinate human behavior in time not only streamlines the flow of information, but also engenders greater affiliation between individuals (Hove and Risen, 2009; Cirelli et al., 2012; Launay et al., 2013).

Regularity and variability in temporal structure

One notable difference between music and language is that music typically contains a regular beat, whereas the temporal patterns of natural speech are less regular and more free-flowing (Patel, 2006, 2010). It has been proposed that much of the emotional power of music stems from the creation (and violation) of temporal expectancies within a metrically predictable framework (Huron, 2006). It is also important to note that, although music is formulated around a regular pulse, live music performances contain significant timing fluctuations and exhibit more complex temporal patterns, closer to the free-flowing rhythms of speech (Repp, 1992; Palmer, 1997; Large et al., 2002; Ashley, 2002; Levitin et al., 2012).

Despite the tempo fluctuations of live performances, listeners have no difficulty perceiving the underlying pulse of music (Large and Snyder, 2009; Rankin et al., 2009), indicating that the ability to extract a beat does not rely upon rigid predictability but rather can be conveyed through more subtle timing cues. Analyses of musical excerpts have revealed that the timing variations in musical performance exhibit fractal characteristics (Rankin et al., 2009; Levitin et al., 2012) that are found in many other naturally generated time series, from heart beats (West and Shlesinger, 1989) to the dynamic oscillations of neuronal populations (Engel et al., 2001; Linkenkaer-Hansen et al., 2001; Large and Snyder, 2009). Human sensory systems have an inherent preference for signals that exhibit these fractal characteristics (Gilden et al., 1995; Yu et al., 2005), and naturally generated movement sequences, such as gait (Hausdorff, 2007) and tapping to a beat (Torre and Delignières, 2008) vary over time according to the same underlying mathematic relations.

Neural basis of temporal processing: oscillatory rhythms

In recent years there has been increased interest in the role of oscillatory brain rhythms in the coordination of activity across cortical regions and in establishing temporal binding and prediction mechanisms in the brain (Engel and Singer, 2001; Engel et al., 2001; Buzsaki and Draguhn, 2004; Sauseng and Klimesch, 2008; Buzsaki, 2009; Uhlhaas et al., 2009). It is argued that predictable variations in the balance of ongoing neuronal activity increase the efficiency of information processing (Buzsaki, 2009) and that some degree of noise in the nervous system may, in fact, confer processing benefits (Faisal et al., 2008). The dynamic attending theory mentioned above proposes that attention can be modulated over time based on temporal expectancies that are tracked by neuronal oscillatory activity (Large and Jones, 1999). Recent research into the neural bases of rhythm perception indicates that neural oscillatory activity also plays an important role in encoding musical rhythm (Fujioka et al., 2009, 2012; Large and Snyder, 2009; Levitin, 2009; Grahn, 2012; Nozaradan et al., 2012).

The brain's ability to make use of temporal patterns in extracting meaning from speech may also be driven by underlying oscillatory activity, which allows the brain to "entrain" to various aspects of temporal structure (Luo and Poeppel, 2007; Schmidt-Kassow and Kotz, 2008; Ghitza and Greenberg, 2009; Giraud and Poeppel, 2012), with different frequency bands of brain activity concurrently synchronizing to different components. It has further been suggested that poor reading skills and language-based learning impairment may result from deficient oscillatory mechanisms (Abrams et al., 2009; Goswami, 2011; Kraus, 2012). If the brain cannot lock on to underlying temporal information patterns, the efficiency of information processing is greatly reduced. This causes disruption in attention and memory resources and impedes coordination between brain regions, as is evidenced in impaired multisensory integration common in dyslexics (Facoetti et al., 2010; Goswami, 2011). The cerebral hemispheres show specialization in their encoding of different temporal integration windows based on distinct frequency bands of dominant oscillatory activity, with the left hemisphere specialized for fast acoustic processing and the right hemisphere specialized for slower temporal processing (Belin et al., 1998). Stronger reading abilities have been linked to stronger lateralization of relevant timing rates within the speech signal, such as right-hemisphere specialization for the slower syllable rate of speech (Poeppel, 2003; Abrams et al., 2009; Goswami, 2011). Oscillatory mechanisms may therefore play an important role not only in integrating activity across disparate regions of the brain, but in melding the activity within more specialized local networks.

As with the rhythmicities of speech and music, neuronal entrainment is happening simultaneously on multiple timescales. This increases the information-processing capacity by using pattern extraction and prediction to allocate neural resources as efficiently and effectively as possible (Fig. 12.1). We have seen that deficiencies in oscillatory mechanisms may result in language-related difficulties; further, musical training has the potential to strengthen oscillatory function (Trainor et al., 2009). This is an exciting area of research in which there is much work still to be done, with great potential for informing the use of music as a means of strengthening language abilities.

In summary, despite greater variability in the rhythms of natural language compared with the perceived musical beat, both exhibit complex temporal structure. Neural oscillations across different frequency bands provide an important mechanism by which the brain is able to lock

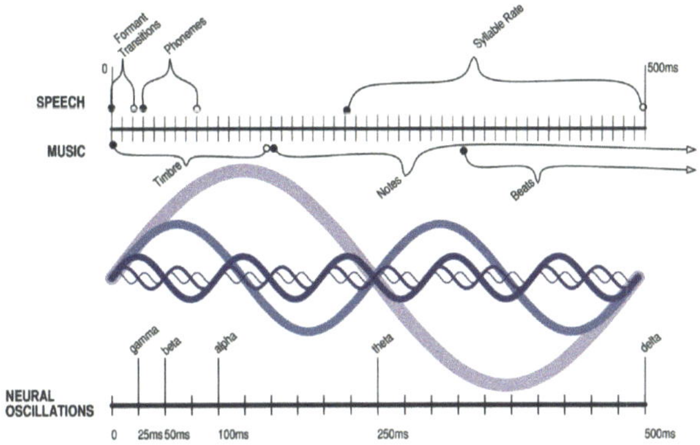

Fig. 12.1. Timescales of speech and music characteristics: different frequencies of neural oscillatory activity lock on to meaningful temporal patterns in speech and music.

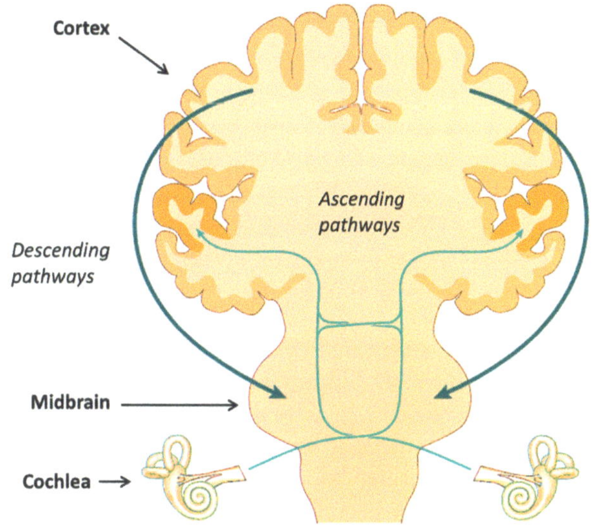

Fig. 12.2. Patterns and predictability are important for processing both speech and music. Neural responses to speech sounds are enhanced under predictable conditions in good readers, but not in poor readers, and this enhancement tracks with rhythm skills.

on to underlying temporal patterns in both music and speech. Taken together, these findings show that the communicative systems of music and language are exquisitely in tune (or in time!) with the temporal processing mechanisms of the human brain.

MUSIC AND LANGUAGE: MODELS OF LEARNING AND PLASTICITY

Rule-based learning

Mastery of both music and language depends upon rule-based learning, which involves an ongoing process of perception, prediction, feedback, and correction. A comparison of grammar acquisition in human infants and non-human primates suggests that human learning mechanisms are distinctively tuned to complex predictive patterns (Saffran et al., 2008), and while there is some evidence of recursive sequence generation in non-human communication such as birdsong (Gentner et al., 2006), the degree of complexity of grammatic structure in human communication may be unique. As infants acquire language, they are able to implicitly extract statistical regularities to help guide the segmentation of a continuous acoustic stream into syllables and words, thereby constructing a framework from which they can begin to map sounds to meaning (Saffran et al., 1996, 2001; Conway and Pisoni, 2008; Conway et al., 2010; Romberg and Saffran, 2010).

Motivation plays an important role in learning, and activation of reward centers in the brain has been associated with neural plasticity (Wise, 2004; Day et al., 2007). As the brain starts to build predictions based on experience, the reward pathways are activated not only by the reward itself, but in anticipation of likely reward (Knutson et al., 2001), with the degree of activation reflecting the statistical probabilities derived from prior experience (Morris et al., 2004). Release of the neurotransmitter dopamine provides a chemical mechanism by which patterns, predictions, and feedback are tracked within the brain (Salimpoor and Zatorre, 2013), and the same chemical signal is involved in the implicit learning functions of the basal ganglia, which are implicated in the acquisition of linguistic and musical grammar (Ullman, 2001, 2005; Conway and Pisoni, 2008; Ettlinger et al., 2011).

Evidence of relationships between rapid auditory processing skills and neural indices of implicit learning (McNealy et al., 2006) suggest that basic temporal processing abilities may provide the foundation for more complex, rule-based learning. There is ongoing debate as to whether the same neural processes underlie learning in both music and language (Rogalsky et al., 2011), but there is evidence of some overlap in the neural resources involved in the processing of linguistic and musical syntax (see Patel, 2003, for a review), and that musical expertise results in improved implicit learning abilities for both musical and linguistic structures (Ettlinger et al., 2011; Francois and Schön, 2011; François et al., 2013; Skoe et al., 2013).

Further, there is evidence that the statistical context of a sound influences neural encoding in the auditory brainstem (Skoe et al., 2013), and that the extent of this neural enhancement relates to language-related skills such as speech-in-noise perception (Chandrasekaran et al., 2009; Parbery-Clark et al., 2011), as well as reading abilities and rhythm-related skills (Strait et al., 2011). These outcomes suggest that common underlying pattern detection abilities are involved in the extraction of meaning from sound in both music and language (Fig. 12.2).

Learning trajectories in music and language

While the developmental trajectory of native language acquisition is relatively consistent, at least within a given culture, there is much greater diversity in the extent of musical participation and style of musical training: from none at all, to a lifetime of intensive instruction and practice. Given evidence for some degree of commonality between music and language learning, the greater diversity in musical experience can provide invaluable insight into the underlying mechanisms of learning and in illuminating how training and experience interact with development. In particular, while it is widely acknowledged that acquiring a second language or learning to play a musical instrument is much easier at a young age, there is ongoing debate regarding the existence of so-called "sensitive periods" and what they reflect about the underlying neural circuitry of learning. It has been shown that, while infants are able to differentiate phonetic categories in a non-native language, this ability diminishes within the first year of life (Werker and Tees, 1984). Similar constraints are evident in the acquisition of musical structure; for example, infants are able to differentiate metrical categories from another musical culture while adults only respond to those which exist in their own culture (Hannon and Trehub, 2005). However these constraints may not be set in stone, as there is evidence to suggest that enriched experience such as musical training can heighten the nervous system's receptivity to learning during these sensitive periods (Skoe and Kraus, 2013).

Children are natural improvisers: it is notable that young children will often create their own songs before they can reproduce learned songs (Trehub, 2003), and research into childhood play behavior suggests that high-quality play is important for the development of abstract thinking, as well as social and linguistic competence (Bergen, 2002). Adult second-language instruction provides an interesting contrast with native-language learning since typically it involves greater emphasis on explicit learning of rules and memorization of phrases, with less experimentation and play than occurs in childhood language acquisition. There is evidence that the balance of learning style may shift over development, from implicit procedural-based learning in childhood (Kuhl, 2004) to more explicit, declarative learning styles in adulthood, and that this reflects a shift in the underlying memory systems involved in learning between childhood and adulthood (Ullman, 2005).

There is much still to be discovered about how we learn to communicate through music and language, and the wide range of musical learning styles and experiences present within the normal population presents a rich opportunity for further investigation. This area of research is of particular interest to educators and clinicians given that musical experience can strengthen aspects of brain function which also support language-related skills (Patel, 2011; Strait and Kraus, 2014), and may thereby offer a framework for the remediation of language difficulties. Since music making integrates brain activity across sensory, motor, and reward regions, engagement in musical activities strengthens neural networks that benefit other non-musical activities and promotes neural plasticity throughout the lifespan (Wan and Schlaug, 2010; Herholz and Zatorre, 2012); in other words, music may help prime the nervous system for learning, in part by "exercising" the reward systems which are critical to plasticity (Salimpoor and Zatorre, 2013). In the case of expert musicians, the drudgery of daily practice may not offer immediate positive reward but still represents emotional engagement in the task, and the longer-term goal of skilled communication and expression through music provides a motivational framework that is critical to the attainment of expertise (Ericsson et al., 1993).

NEURAL PLASTICITY: THE INTERACTIVE AUDITORY SYSTEM

In this section we will look at relations between music and language from the perspective of how sound is represented by the nervous system. Many aspects of the neural encoding of speech and music have been covered in previous chapters, with the first chapter focused on ascending auditory pathways: it is perhaps fitting that this chapter will close the section by emphasizing the great importance of the descending pathways. In particular, we will examine how experience can fine-tune the auditory system's ability to extract meaning from sound through the interaction of sensory and cognitive processing (Suga and Ma, 2003; Bajo et al., 2009) (see Figs 1.10 and 1.15 in Chapter 1).

There has been extensive and important debate regarding the degree to which music and language functions are distinct within the brain (Patel and Peretz, 1997; Peretz and Hyde, 2003; Patel, 2010). Rather than reviewing the cortical areas and neural networks that are involved in specialized aspects of sound processing, we will focus on the underlying subcortical representation of sound, where sensory and cognitive influences converge. We shall summarize evidence for neural plasticity resulting from music and language experience, as well as showing that many of the same aspects of sound processing that are selectively enhanced through expertise are deficient in populations with language-based impairments such as dyslexia. We will end the section with a discussion of the clinical implications of these findings for the remediation of language and learning disorders.

The auditory brainstem: hub of auditory information processing

The brainstem auditory nuclei are highly specialized for very precise neural signaling, which underlies the complex temporal processing necessary for biologically critical functions such as sound localization (Oertel, 1999). Importantly, the auditory system is not a one-way street: the extensive network of efferent pathways enables dynamic modulation of signal processing, as well as long-term tuning of sensory function with experience (Suga and Ma, 2003; Bajo et al., 2009) (see Fig. 1.15 in Chapter 1). The inferior colliculus in the brainstem is a critical information-processing hub in which top-down cortical influences converge with multiple bottom-up sensory inputs to compile an integrated and comprehensive representation of the auditory scene. On this basis, the role of the inferior colliculus has been described as the auditory analog to the primary visual cortex in the visual system, generating a "whiteboard" of all available information from which higher processing pathways diverge (Nelken, 2008). There are also direct connections between the inferior colliculus and brain regions important for motor control and coordination, including output to the cerebellum and input from the basal ganglia (Casseday et al., 2002), which also play an important role in pattern learning, as discussed above (see Fig. 1.15 in Chapter 1).

In our lab, investigation of the evoked auditory brainstem response to complex sounds (of which the inferior colliculus is a primary generator) has proved an extremely fruitful approach in exploring mechanisms of experience-based plasticity (Kraus and Chandrasekaran, 2010; Skoe and Kraus, 2010). The evoked brainstem response preserves fundamental characteristics of the incoming signal such as its timing, fundamental frequency, and harmonics, allowing direct comparison between the response and its evoking stimulus. By comparing stimulus and response characteristics in both time and frequency domains, it is possible to assess the neural representation of spectral characteristics important for the perception of pitch and timbre, as well as neural response timing and consistency across trials. The evoked responses has also been shown to be sensitive to experience (Krishnan et al., 2005; Tzounopoulos and Kraus, 2009; Krizman et al., 2012; Kraus and Nicol, 2014) and therefore offers an objective biologic snapshot of how sound is processed in an individual. We have used this approach to explore how music and language are processed by "auditory experts" such as bilinguals and musicians, as well as in individuals with language impairments (Fig. 12.3).

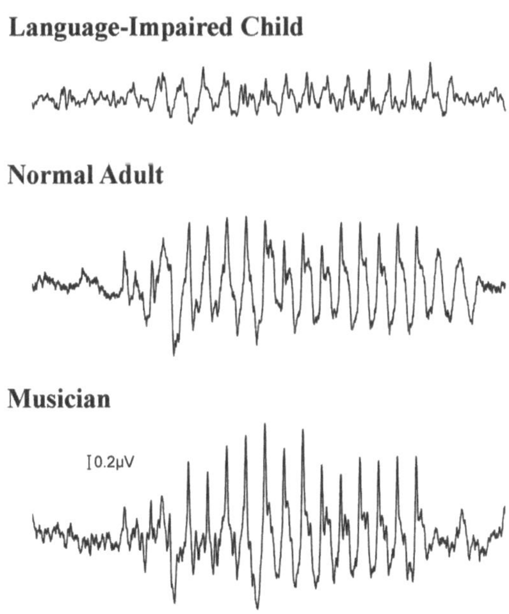

Fig. 12.3. Representative waveforms showing individual variation in brainstem responses to speech. Auditory expertise such as musical training has been shown to strengthen many of the same aspects of the neural encoding of sound that are deficient in individuals with language impairment.

Selective enhancement: neural signatures of auditory expertise

Musical expertise

Experience does not shape auditory function with a simple volume knob effect: rather, the effect is more like a mixing board, boosting important components, attenuating irrelevant inputs, and exploiting patterns in the incoming signal to reduce redundant processing. In combination these modulations help to bring out what is meaningful with the greatest possible clarity and efficiency. Of course, what is meaningful to any given individual will depend upon their specific type of experience, resulting in distinct "neural signatures" of expertise (Kraus and Nicol, 2014). Musicians' neural responses can be enhanced specifically for the sound of their own instrument: pianists demonstrate more robust subcortical representation of a piano sound than of a bassoon or cello tone (Strait et al., 2012a), and there are timbre-specific effects on cortical evoked responses (Pantev et al., 2001) as well as neural oscillatory activity (Shahin et al., 2008) based on the instrument of practice. The style of music played may also influence plasticity effects, with jazz musicians demonstrating greater sensitivity to acoustic features in their preattentive brain responses to sound (Vuust et al., 2012).

In addition to the selective enhancement of the sounds or patterns that are meaningful to an individual, extended experience also shapes the automatic processing of sound, and thereby influences perception in auditory domains beyond the specific area of expertise. Musicians demonstrate more robust brainstem encoding

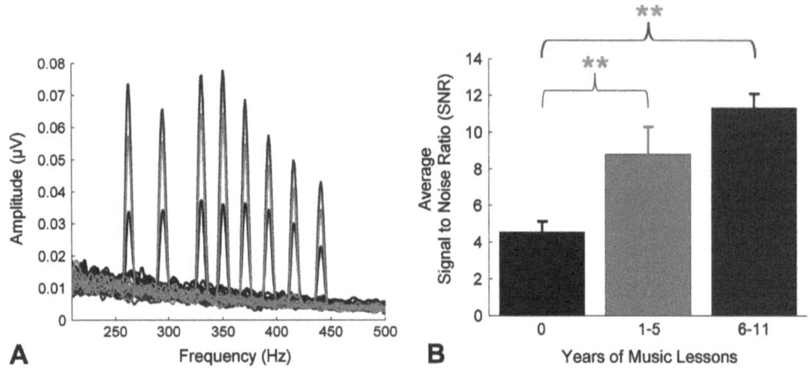

Fig. 12.4. Musical practice during childhood is associated with more robust neural responses to sound in adulthood, years after training has stopped. Adults with no past musical experience (in black) have smaller-amplitude brainstem responses (smaller signal-to-noise ratios (SNRs)) relative to adults who played a musical instrument from around the age of 9, and continued to play for either 1–5 (light blue) or 6–11 (dark blue) years. (**A**) Frequency-following responses to eight sounds of varying frequency. (**B**) Average SNR across the eight sounds for each group (mean 1 SEM). $^{**}P < 0.01$. (Adapted from Skoe and Kraus, 2012.)

of speech sounds in noise (Parbery-Clark et al., 2009a; Strait et al., 2012b, 2013b) and less degradative effects of reverberation on neural encoding of the harmonic information in speech sounds (Bidelman and Krishnan, 2010) compared with non-musician counterparts. Musicians also demonstrate enhanced neural representation of the emotion-carrying portion of a baby's cry (Strait et al., 2009), more accurate brainstem encoding of linguistic pitch patterns in an unfamiliar tonal language (Wong et al., 2008), and a greater benefit of audiovisual integration in their encoding of music and language sounds (Musacchia et al., 2007).

Neural representation of meaningful timing characteristics is also more precise in musicians; for example, they show greater neural differentiation of contrastive consonant sounds than non-musicians across the lifespan (Parbery-Clark et al., 2012c; Strait et al., 2013a; Kraus et al., 2014a). There is evidence that musical experience offsets age-related delays in neural timing in older adults and increases the consistency of neural responses (Parbery-Clark et al., 2012a, b). Further, musicians' subcortical encoding of sound is influenced by harmonic context (Bidelman and Krishnan, 2009; Lee et al., 2009; Marmel et al., 2011) as well as by statistical predictability (Parbery-Clark et al., 2011; Skoe et al., 2013), suggesting that musical experience may strengthen the ability of the nervous system to benefit from underlying patterns in sound in both musical and linguistic contexts.

Taken together, these findings indicate that auditory expertise such as musical experience tunes the auditory system so that it can more effectively pull out a meaningful signal from a complex soundscape. Recent studies from our lab demonstrate that just a few years of musical training in childhood can influence the neural encoding of sound in adulthood, years after the training has ceased (Skoe and Kraus, 2012; White-Schwoch et al., 2013) (Fig. 12.4), emphasizing that attention to sound in the past can result in more efficient automatic processing of sound in the present, even many years later.

There is much work still to be done in further understanding the parameters of plasticity across the lifespan. It seems plausible that similar overall "information processing" benefits could result from very different types of musical expertise, each with their own underlying neural signatures: a vocalist may tune into pitch and a drummer may tune into timing, but both may be better at "tuning in" to meaningful sound in general (Slater et al., 2014). This is reflected in behavior, with musicians showing enhanced speech-in-noise perception across the lifespan (see Strait and Kraus, 2014 for a review). Much of the work so far in the field of musical experience-based plasticity has focused on the comparison between musicians and non-musicians; further investigation into the distinct effects of different types of musical activity on plasticity and learning presents a promising area for future research and would advance understanding of how selective neural enhancements relate with more general perceptual benefits.

Bilingualism

In the language domain, bilinguals represent another example of auditory expertise: there is converging evidence that bilingual experience engenders functional and structural changes in cortical language networks (Kim et al., 1997; Crinion et al., 2006). Bilingual experience also strengthens cognitive function (Bialystok et al., 2012) and may counteract age-related cognitive decline, with bilinguals demonstrating later onset of dementia than monolingual peers matched for potentially confounding factors such as education and socioeconomic background (Bialystok et al., 2007; Alladi et al., 2013).

Recent work from our lab reveals more robust subcortical representation of a speech syllable in bilingual adolescents compared with age-matched monolinguals (Krizman et al., 2012, 2014). Specifically, bilinguals show a selective enhancement in the encoding of the fundamental frequency, which has particular salience for bilinguals in helping to identify a language switch (Altenberg and Ferrand, 2006), and this enhanced subcortical representation is related to stronger attention skills (Fig. 12.5).

Both music and language expertise have been shown to strengthen cognitive function, such as auditory attention, working memory, and executive control (Bialystok, 2012; Bugos et al., 2007; Kraus et al., 2012). It is proposed that the observed enhancements in the neural encoding of sound may be driven in a top-down manner by this strengthened cognitive function (Nelken and Ulanovsky, 2007; Ahissar et al., 2009; Conway et al., 2009; Kraus and Chandrasekaran, 2010; Kraus et al., 2012) by more effectively homing in on relevant aspects of the signal, as well as by coordinating underlying neural activity, thereby promoting more consistent and robust neural responses (Krizman et al., 2012; Parbery-Clark et al., 2012b; Skoe and Kraus, 2013).

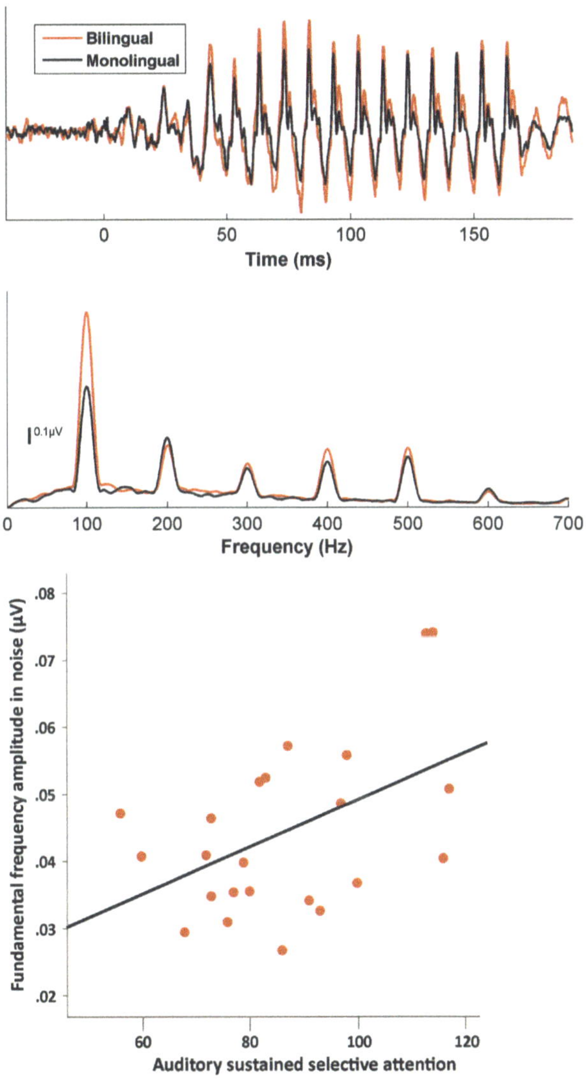

Fig. 12.5. Bilinguals (red) show a larger neural response to the fundamental frequency (100 Hz) of a speech sound than monolinguals (black). Within the bilingual group only, sustained selective attention ability correlated with the amplitude of the fundamental frequency encoding ($r = 0.483$, $P = 0.02$). (Adapted from Krizman et al., 2012.)

Neural underpinnings of language ability and impairment

In this section we will consider the relationship between reading and the neural encoding of sound in both typically developing and impaired populations. While it is not always intuitive to think of reading as an auditory skill, the first internal representations of linguistic meaning in a typically developing child will come from exposure to speech; the accurate mapping of orthographic representations to meaning is therefore dependent upon these sound-based representations, and auditory processing skills in infants have been shown to predict later reading performance (Benasich and Tallal, 2002; Boets et al., 2007, 2008, 2011; Corriveau et al., 2010). An important characteristic of speech is that the most meaningful parts of the signal are carried by the fastest-changing components, namely the consonants. The ability to encode these subtle timing differences is critical to developing accurate phonologic representations and therefore essential for the development of strong reading skills, with more precise subcortical timing linked to better reading abilities (Fig. 12.6) (Banai et al., 2009; Hornickel et al., 2009),

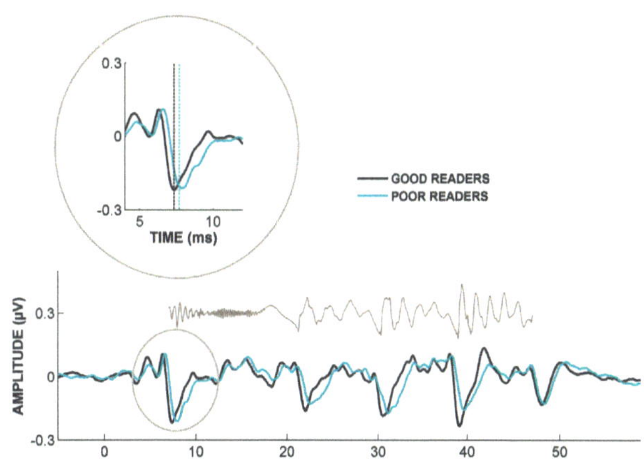

Fig. 12.6. Subcortical timing in response to speech is delayed in poor (blue) readers compared with good (black) readers. Grand average waveforms of responses to the syllable /da/. The stimulus is presented in gray, shifted by 8 ms to align stimulus features with the response. Magnified peak is shown. Standard errors surrounding the mean latency of each group are denoted by the dashed lines. (Modified from Banai et al., 2009.)

and reading difficulties associated with less consistent timing in neural responses to speech (Hornickel and Kraus, 2013).

As discussed above, oscillatory activity plays an important role in coordinating neural synchrony within different temporal frameworks which are important for speech perception, and there is evidence of links between abnormal oscillatory activity and reading impairments such as dyslexia (Nagarajan et al., 1999; Goswami, 2011; Heim et al., 2011; Kraus, 2012). The left-hemispheric specialization which supports rapid acoustic processing is predicted by the precision of brainstem encoding of speech sounds, suggesting that accurate timing in the brainstem is an important foundation for higher-level processing of the fast acoustic properties of speech which are critical to the development of reading skills (Abrams et al., 2006; Tierney and Kraus, 2013b).

Clinical implications

We have shown that auditory expertise with music and language can strengthen many of the same aspects of neural sound encoding that are found to be deficient in populations with language-related disorders (see Tierney and Kraus, 2013b for review) (Fig. 12.7). A number of studies have demonstrated that deficient neural processing of speech can be strengthened by short-term training in children with language-based learning problems (see Kraus and Hornickel, 2012 for review), such as a study of children with clinical diagnoses of language-based learning difficulties (e.g., dyslexia), in which 9 children underwent 3 months of computer-based training, while a control group of 10 children did not participate in any remediation program. The children who received training showed a significant improvement in the fidelity of their neural response to the evoking stimulus, while the control children showed no such improvement (Russo et al., 2005). Another study with 38 dyslexic children showed that use of an assistive-listening device (classroom-based FM system) for 1 year increased the consistency of neural responses to speech and that these neural changes were associated with improvements in phonologic awareness and reading skills (Hornickel et al., 2012). Taken together, the evidence for strengthened neural processing in musicians and the potential for improved speech processing following short-term training suggest that music-based interventions may also be effective in the treatment of language disorders, and this is an important area for future research.

Recent longitudinal assessments of existing music education programs provides support for the educational merits of musical training in fostering the development of critical language and learning skills (Slater et al., in press; Kraus et al., 2014b; Tierney et al., 2013c). Given that music can play an important role in promoting social development, interpersonal skills, and community building, the integration of music-based programs offers particular value in at-risk settings. Real world education programs, in combination with laboratory-based research, there is still a great need for longitudinal assessments to begin differentiating which aspects of musical training are driving language-related benefits. Certainly the accumulating evidence for shared temporal processing mechanisms underlying musical rhythm and language skills offers a promising avenue for future research.

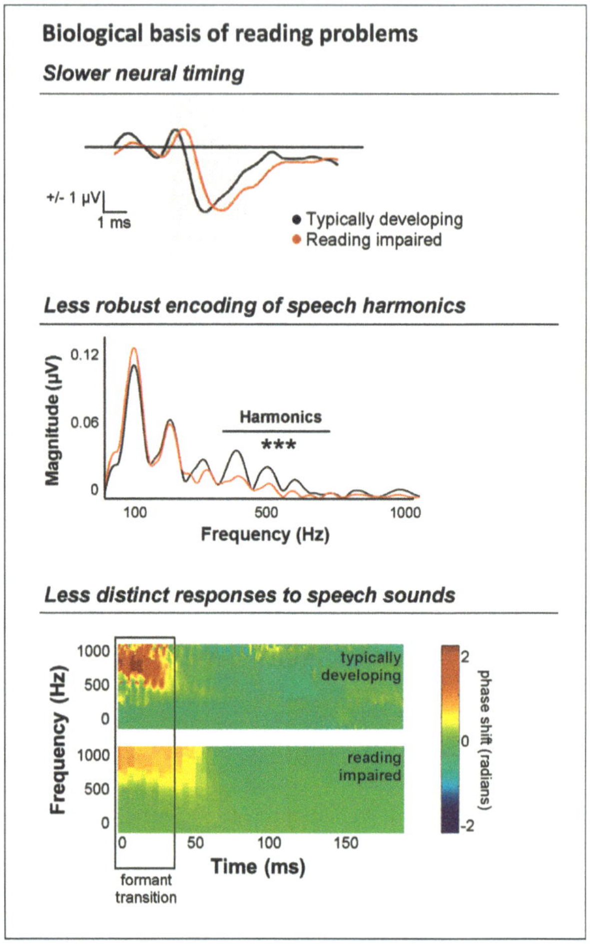

Fig. 12.7. Reading problems are associated with deficits in the neural encoding of sound. Many of the same aspects of sound processing that are deficient in populations with language-based impairments are strengthened in auditory experts such as musicians. (For review, see Tierney and Kraus, 2013b.)

CONCLUSIONS

Music and language are two sides of the human communication coin: while language is effective for semantically precise communication, the great strength of music lies in its facilitation of social bonding and shared emotion. Both systems of communication are derived from the fundamental building blocks of sound, its inherent harmonic properties, and its temporal patterns. In many senses music and language are sewn from the same cloth, but their complementary strengths may have played distinct and important roles in the emergence of human cognition and learning.

A dominant theme in this chapter has been that of temporal processing: from the temporal structure of music and language, to the role of anticipation, prediction, and reward in learning, to the oscillatory rhythms which orchestrate perception and action. The exquisite temporal precision of the auditory system provides the biologic basis for sophisticated information processing, enabling the extraction of meaning from a complex soundscape. Through the convergence of sensory and cognitive influences, auditory function is not only shaped by the immediate demands of the environment but fine-tuned by experience over the course of a lifetime. We have demonstrated the significant impact of experience with language and music on the neural encoding of sound, and highlighted the potential for auditory training to support the development of language and listening skills, and in the remediation of language deficits.

While many ingredients of sound processing are common across species, examination of music and language reveals a degree of sophistication in both communication and underlying cognition which may be uniquely human. These rich connections illuminate our understanding of human auditory function and present many exciting avenues for future research in further disentangling the myriad rhythms of music, language and the brain.

ACKNOWLEDGMENTS

Research supported by NSF BCS-1057556, NSF BCS-0921275, NSF SMA-1015614, Mathers Foundation, National Association of Music Merchants (NAMM), and the Hugh Knowles Hearing Center. We would like to thank Erika Skoe, Dana Strait, Adam Tierney, and Trent Nicol for comments on a previous version of this manuscript and Adam Max Trefonides for contributions to figures.

Conflict of interest: The authors declare no competing financial interests.

References

Abrams DA, Nicol T, Zecker SG et al. (2006). Auditory brainstem timing predicts cerebral asymmetry for speech. J Neurosci 26: 11131–11137.

Abrams DA, Nicol T, Zecker S et al. (2009). Abnormal cortical processing of the syllable rate of speech in poor readers. J Neurosci 29: 7686–7693.

Ahissar M, Nahum M, Nelken I et al. (2009). Reverse hierarchies and sensory learning. Phil Trans Biol Sci 364: 285–299.

Alladi S, Bak TH, Duggirala V et al. (2013). Bilingualism delays age at onset of dementia, independent of education and immigration status. Neurology 81: 1938–1944.

Altenberg EP, Ferrand CT (2006). Fundamental frequency in monolingual English, bilingual English/Russian, and bilingual English/Cantonese young adult women. J Voice 20: 89–96.

Anderson S, White-Schwoch T, Parbery-Clark A et al. (2013). A dynamic auditory-cognitive system supports speech-in-noise perception in older adults. Hear Res 300: 18–32.

Ashley R (2002). Do [n't] change a hair for me: the art of jazz rubato. Music Percept 19: 311–332.

Auer P, Couper-Kuhlen E, Müller F (1999). Language in time: The rhythm and tempo of spoken interaction. Oxford University Press, Oxford.

Bajo VM, Nodal FR, Moore DR et al. (2009). The descending corticocollicular pathway mediates learning-induced auditory plasticity. Nat Neurosci 13: 253–260.

Balkwill L-L, Thompson WF (1999). A cross-cultural investigation of the perception of emotion in music: Psychophysical and cultural cues. Music Percept 43–64.

Banai K, Hornickel J, Skoe E et al. (2009). Reading and subcortical auditory function. Cereb Cortex 19: 2699–2707.

Belin P, Zilbovicius M, Crozier S et al. (1998). Lateralization of speech and auditory temporal processing. J Cogn Neurosci 10: 536–540.

Benasich AA, Tallal P (2002). Infant discrimination of rapid auditory cues predicts later language impairment. Behav Brain Res 136: 31–49.

Bergen D (2002). The role of pretend play in children's cognitive development. Early Childhood Research and Development 4 (1): Available online at, http:/ecrp.uiuc.edu/v4n1/bergen.html.

Berkowitz A (2010). The improvising mind: Cognition and creativity in the musical moment. Oxford University Press, Oxford.

Bhide A, Power A, Goswami U (2013). A rhythmic musical intervention for poor readers: a comparison of efficacy with a letter-based intervention. Mind Brain Educ 7: 113–123.

Bialystok E, Craik FI, Freedman M (2007). Bilingualism as a protection against the onset of symptoms of dementia. Neuropsychologia 45: 459–464.

Bialystok E, Craik FI, Luk G (2012). Bilingualism: consequences for mind and brain. Trends Cogn Sci 16: 240–250.

Bidelman GM, Krishnan A (2009). Neural correlates of consonance, dissonance, and the hierarchy of musical pitch in the human brainstem. J Neurosci 29: 13165–13171.

Bidelman GM, Krishnan A (2010). Effects of reverberation on brainstem representation of speech in musicians and non-musicians. Brain Res 1355: 112–125.

Boets B, Wouters J, Van Wieringen A et al. (2007). Auditory processing, speech perception and phonological ability in pre-school children at high-risk for dyslexia: a longitudinal study of the auditory temporal processing theory. Neuropsychologia 45: 1608–1620.

Boets B, Wouters J, Van Wieringen A et al. (2008). Modelling relations between sensory processing, speech perception, orthographic and phonological ability, and literacy achievement. Brain Lang 106: 29–40.

Boets B, Vandermosten M, Poelmans H et al. (2011). Preschool impairments in auditory processing and speech perception uniquely predict future reading problems. Res Dev Disabil 32: 560–570.

Brown S, Martinez MJ, Hodges DA et al. (2004). The song system of the human brain. Cogn Brain Res 20: 363–375.

Bugos J, Perlstein W, McCrae C et al. (2007). Individualized piano instruction enhances executive functioning and working memory in older adults. Aging Ment Health 11: 464–471.

Buzsaki G (2009). Rhythms of the Brain. Oxford University Press, Oxford.

Buzsaki G, Draguhn A (2004). Neuronal oscillations in cortical networks. Science 304: 1926–1929.

Casseday JH, Fremouw T, Covey E (2002). The inferior colliculus: a hub for the central auditory system. Integrative functions in the mammalian auditory pathway. Springer, New York.

Censullo M, Lester B, Hoffman J (1985). Rhythmic patterning in mother-newborn interaction. Nurs Res 34: 342.

Chandrasekaran B, Hornickel J, Skoe E et al. (2009). Context-dependent encoding in the human auditory brainstem relates to hearing speech in noise: Implications for developmental dyslexia. Neuron 64: 311–319.

Chen JL, Penhune VB, Zatorre RJ (2008). Listening to musical rhythms recruits motor regions of the brain. Cereb Cortex 18: 2844–2854.

Cirelli LK, Einarson K, Trainor LJ (2012). Bouncing babies to the beat: Music and helping behaviour in infancy. In: 12th International Conference on Music Perception and Cognition, Thessaloniki, Greece.

Conway CM, Pisoni DB (2008). Neurocognitive basis of implicit learning of sequential structure and its relation to language processing. Ann N Y Acad Sci 1145: 113–131.

Conway CM, Pisoni DB, Kronenberger WG (2009). The importance of sound for cognitive sequencing abilities: the auditory scaffolding hypothesis. Curr Dir Psychol Sci 18: 275–279.

Conway CM, Bauernschmidt A, Huang SS et al. (2010). Implicit statistical learning in language processing: word predictability is the key. Cognition 114: 356–371.

Cook P, Rouse A, Wilson M et al. (2013). A California sea lion (*Zalophus californianus*) can keep the beat: motor entrainment to rhythmic auditory stimuli in a non vocal mimic. J Comp Psychol 127: 412–427.

Corriveau KH, Goswami U (2009). Rhythmic motor entrainment in children with speech and language impairments: tapping to the beat. Cortex 45: 119–130.

Corriveau KH, Goswami U, Thomson JM (2010). Auditory processing and early literacy skills in a preschool and kindergarten population. J Learn Disabil 43: 369–382.

Crinion J, Turner R, Grogan A et al. (2006). Language control in the bilingual brain. Science 312: 1537–1540.

Cross I (1999). Is music the most important thing we ever did? Music, development and evolution. Music, Mind and Science 10–39.

Cutler A, Butterfield S (1992). Rhythmic cues to speech segmentation: evidence from juncture misperception. J Mem Lang 31: 218–236.

D'Ausilio A, Altenmüller E, Olivetti Belardinelli M et al. (2006). Cross-modal plasticity of the motor cortex while listening to a rehearsed musical piece. Eur J Neurosci 24: 955–958.

Day JJ, Roitman MF, Wightman RM et al. (2007). Associative learning mediates dynamic shifts in dopamine signaling in the nucleus accumbens. Nat Neurosci 10: 1020–1028.

Dellatolas G, Watier L, Le Normand MT et al. (2009). Rhythm reproduction in kindergarten, reading performance at second grade, and developmental dyslexia theories. Arch Clin Neuropsychol 24: 555–563.

Devergie A, Grimault N, Tillmann B et al. (2010). Effect of rhythmic attention on the segregation of interleaved melodies. J Acoust Soc Am 128: EL1–EL7.

Engel AK, Singer W (2001). Temporal binding and the neural correlates of sensory awareness. Trends Cogn Sci 5: 16–25.

Engel AK, Fries P, Singer W (2001). Dynamic predictions: Oscillations and synchrony in top-down processing. Nat Rev Neurosci 2: 704–716.

Ericsson KA, Krampe RT, Tesch-Römer C (1993). The role of deliberate practice in the acquisition of expert performance. Psychol Rev 100: 363.

Ettlinger M, Margulis EH, Wong PC (2011). Implicit memory in music and language. Front Psychol 2: 211.

Facoetti A, Trussardi AN, Ruffino M et al. (2010). Multisensory spatial attention deficits are predictive of phonological decoding skills in developmental dyslexia. J Cogn Neurosci 22: 1011–1025.

Faisal AA, Selen LP, Wolpert DM (2008). Noise in the nervous system. Nat Rev Neurosci 9: 292–303.

Fitch WT (2006). The biology and evolution of music: a comparative perspective. Cognition 100: 173–215.

Francois C, Schön D (2011). Musical expertise boosts implicit learning of both musical and linguistic structures. Cereb Cortex 21: 2357–2365.

François C, Chobert J, Besson M et al. (2013). Music training for the development of speech segmentation. Cereb Cortex 23: 2038–2043.

Fujioka T, Trainor LJ, Large EW et al. (2009). Beta and gamma rhythms in human auditory cortex during musical beat processing. Ann N Y Acad Sci 1169: 89–92.

Fujioka T, Trainor LJ, Large EW et al. (2012). Internalized timing of isochronous sounds is represented in neuromagnetic beta oscillations. J Neurosci 32: 1791–1802.

Gentner TQ, Fenn KM, Margoliash D et al. (2006). Recursive syntactic pattern learning by songbirds. Nature 440: 1204–1207.

Ghitza O, Greenberg S (2009). On the possible role of brain rhythms in speech perception: intelligibility of time-compressed speech with periodic and aperiodic insertions of silence. Phonetica 66: 113–126.

Gilden D, Thornton T, Mallon M (1995). 1/f noise in human cognition. Science 267: 1837–1839.

Giraud A-L, Poeppel D (2012). Cortical oscillations and speech processing: emerging computational principles and operations. Nat Neurosci 15: 511–517.

Gomez P, Danuser B (2007). Relationships between musical structure and psychophysiological measures of emotion. Emotion 7: 377–387.

Goswami U (2011). A temporal sampling framework for developmental dyslexia. Trends Cogn Sci 15: 3–10.

Grahn JA (2012). Neural mechanisms of rhythm perception: current findings and future perspectives. Top Cogn Sci 4: 585–606.

Grahn JA, Brett M (2007). Rhythm and beat perception in motor areas of the brain. J Cogn Neurosci 19: 893–906.

Gutiérrez-Palma N, Palma Reyes A (2007). Stress sensitivity and reading performance in Spanish: a study with children. J Res Read 30: 157–168.

Hannon EE, Trehub SE (2005). Metrical categories in infancy and adulthood. Psychol Sci 16: 48–55.

Hausdorff JM (2007). Gait dynamics, fractals and falls: finding meaning in the stride-to-stride fluctuations of human walking. Hum Mov Sci 26: 555–589.

Heim S, Friedman JT, Keil A et al. (2011). Reduced sensory oscillatory activity during rapid auditory processing as a correlate of language-learning impairment. J Neurolinguist 24: 538–555.

Herholz SC, Zatorre RJ (2012). Musical training as a framework for brain plasticity: behavior, function, and structure. Neuron 76: 486–502.

Higham T, Basell L, Jacobi R et al. (2012). Testing models for the beginnings of the Aurignacian and the advent of figurative art and music: the radiocarbon chronology of Geißenklösterle. J Hum Evol 62: 664–676.

Holliman AJ, Wood C, Sheehy K (2008). Sensitivity to speech rhythm explains individual differences in reading ability independently of phonological awareness. Br J Dev Psychol 26: 357–367.

Holliman AJ, Wood C, Sheehy K (2010). Does speech rhythm sensitivity predict children's reading ability 1 year later? J Educ Psychol 102: 356.

Hornickel J, Kraus N (2013). Unstable representation of sound: a biological marker of dyslexia. J Neurosci 33: 3500–3504.

Hornickel J, Skoe E, Nicol T et al. (2009). Subcortical differentiation of stop consonants relates to reading and speech-in-noise perception. Proc Natl Acad Sci 106: 13022–13027.

Hornickel J, Anderson S, Skoe E et al. (2012). Subcortical representation of speech fine structure relates to reading ability. Neuroreport 23: 6.

Hove MJ, Risen JL (2009). It's all in the timing: interpersonal synchrony increases affiliation. Soc Cognit 27: 949–960.

Huron DB (2006). Sweet anticipation: Music and the psychology of expectation. MIT press, Cambridge, MA.

Huss M, Verney JP, Fosker T et al. (2011). Music, rhythm, rise time perception and developmental dyslexia: perception of musical meter predicts reading and phonology. Cortex 47: 674–689.

Jones MR, Moynihan H, MacKenzie N et al. (2002). Temporal aspects of stimulus-driven attending in dynamic arrays. Psychol Sci 13: 313–319.

Kim KH, Relkin NR, Lee K-M et al. (1997). Distinct cortical areas associated with native and second languages. Nature 388: 171–174.

Kirschner S, Tomasello M (2010). Joint music making promotes prosocial behavior in 4-year-old children. Evol Hum Behav 31: 354–364.

Knutson B, Adams CM, Fong GW et al. (2001). Anticipation of increasing monetary reward selectively recruits nucleus accumbens. J Neurosci 21: RC159.

Kolinsky R, Cuvelier H, Goetry V et al. (2009). Music training facilitates lexical stress processing. Music Percept 26: 235–246.

Kotz SA, Schwartze M, Schmidt-Kassow M (2009). Non-motor basal ganglia functions: a review and proposal for a model of sensory predictability in auditory language perception. Cortex 45: 982–990.

Kraus N (2012). Atypical brain oscillations: a biological basis for dyslexia? Trends Cogn Sci 16: 12–13.

Kraus N, Chandrasekaran B (2010). Music training for the development of auditory skills. Nat Rev Neurosci 11: 599–605.

Kraus N, Hornickel J (2012). Meaningful engagement with sound for strengthening communication skills. In: D Geffner, D Ross-Swain (Eds.), Auditory Processing Disorders: Assessment, Management and Treatment. Plural Publishing, San Diego.

Kraus N, Nicol T (2014). The Cognitive Auditory System: The Role of Learning in Shaping the Biology of the Auditory System. Perspectives on Auditory Research. Springer, New York.

Kraus N, Strait DL, Parbery-Clark A (2012). Cognitive factors shape brain networks for auditory skills: spotlight on auditory working memory. Ann N Y Acad Sci 1252: 100–107.

Kraus N, Slater J, Thompson EC et al. (2014a). Music enrichment programs improve the neural encoding of speech in at-risk children. J Neurosci 34: 11913–11918.

Kraus N, Slater J, Thompson EC et al. (2014b). Auditory learning through active engagement with sound: biological impact of community music lessons in at-risk children. Auditory Cognit Neurosci 8: 351.

Krishnan A, Xu Y, Gandour J et al. (2005). Encoding of pitch in the human brainstem is sensitive to language experience. Brain Res Cogn Brain Res 25: 161–168.

Krizman J, Marian V, Shook A et al. (2012). Subcortical encoding of sound is enhanced in bilinguals and relates to executive function advantages. Proc Natl Acad Sci 109: 7877–7881.

Krizman J, Skoe E, Marian V et al. (2014). Bilingualism increases neural response consistency and attentional control: evidence for sensory and cognitive coupling. Brain Lang 128: 34–40.

Krizman J, Skoe E, Marian V et al. (2014). Bilingualism increases neural stability: Evidence for the convergence of cognitive and sensory processing. Brain Lang (in press).

Kuhl PK (2004). Early language acquisition: cracking the speech code. Nat Rev Neurosci 5: 831–843.

Large EW, Jones MR (1999). The dynamics of attending: How people track time-varying events. Psychol Rev 106: 119–159.

Large EW, Snyder JS (2009). Pulse and meter as neural resonance. Ann N Y Acad Sci 1169: 46–57.

Large EW, Fink P, Kelso SJ (2002). Tracking simple and complex sequences. Psychol Res 66: 3–17.

Launay J, Dean RT, Bailes F (2013). Synchronization can influence trust following virtual interaction. Exp Psychol 60: 53.

Lee KM, Skoe E, Kraus N et al. (2009). Selective subcortical enhancement of musical intervals in musicians. J Neurosci 29: 5832–5840.

Levitin DJ (2009). The neural correlates of temporal structure in music. Music and Medicine 1: 9–13.

Levitin DJ, Chordia P, Menon V (2012). Musical rhythm spectra from Bach to Joplin obey a 1/f power law. Proc Natl Acad Sci 109: 3716–3720.

Linkenkaer-Hansen K, Nikouline VV, Palva JM et al. (2001). Long-range temporal correlations and scaling behavior in human brain oscillations. J Neurosci 21: 1370–1377.

Luo H, Poeppel D (2007). Phase patterns of neuronal responses reliably discriminate speech in human auditory cortex. Neuron 54: 1001.

Marmel F, Parbery-Clark A, Skoe E et al. (2011). Harmonic relationships influence auditory brainstem encoding of chords. Neuroreport 22: 504–508.

Mauk MD, Buonomano DV (2004). The neural basis of temporal processing. Annu Rev Neurosci 27: 307–340.

McGurk H, MacDonald J (1976). Hearing lips and seeing voices. Nature 264: 764–768.

McNealy K, Mazziotta JC, Dapretto M (2006). Cracking the language code: neural mechanisms underlying speech parsing. J Neurosci 26: 7629–7639.

Mithen SJ (2005). The singing Neanderthals: The origins of music, language, mind, and body. Harvard University Press, Boston.

Morris G, Arkadir D, Nevet A et al. (2004). Coincident but distinct messages of midbrain dopamine and striatal tonically active neurons. Neuron 43: 133–143.

Musacchia G, Sams M, Skoe E et al. (2007). Musicians have enhanced subcortical auditory and audiovisual processing of speech and music. Proc Natl Acad Sci 104: 15894–15898.

Musacchia G, Strait D, Kraus N (2008). Relationships between behavior, brainstem and cortical encoding of seen and heard speech in musicians and non-musicians. Hear Res 241: 34.

Nagarajan S, Mahncke H, Salz T et al. (1999). Cortical auditory signal processing in poor readers. Proc Natl Acad Sci 96: 6483–6488.

Nakatani LH, Schaffer JA (1978). Hearing "words" without words: Prosodic cues for word perception. J Acoust Soc Am 63: 234–245.

Nasir SM, Ostry DJ (2009). Auditory plasticity and speech motor learning. Proc Natl Acad Sci 106: 20470–20475.

Nelken I (2008). Processing of complex sounds in the auditory system. Curr Opin Neurobiol 18: 413–417.

Nelken I, Ulanovsky N (2007). Mismatch negativity and stimulus-specific adaptation in animal models. J Psychophysiol 21: 214–223.

Nozaradan S, Peretz I, Mouraux A (2012). Selective neuronal entrainment to the beat and meter embedded in a musical rhythm. J Neurosci 32: 17572–17581.

Oertel D (1999). The role of timing in the brain stem auditory nuclei of vertebrates. Annu Rev Physiol 61: 497–519.

Overy K, Nicolson RI, Fawcett AJ et al. (2003). Dyslexia and music: measuring musical timing skills. Dyslexia 9: 18–36.

Palmer C (1997). Music performance. Annu Rev Psychol 48: 115–138.

Pantev C, Roberts LE, Schulz M et al. (2001). Timbre-specific enhancement of auditory cortical representations in musicians. Neuroreport 12: 169–174.

Parbery-Clark A, Skoe E, Kraus N (2009a). Musical experience limits the degradative effects of background noise on the neural processing of sound. J Neurosci 29: 14100–14107.

Parbery-Clark A, Skoe E, Lam C et al. (2009b). Musician enhancement for speech-in-noise. Ear Hear 30: 653–661.

Parbery-Clark A, Strait D, Kraus N (2011). Context-dependent encoding in the auditory brainstem subserves enhanced speech-in-noise perception in musicians. Neuropsychologia 49: 3338–3345.

Parbery-Clark A, Anderson S, Hittner E et al. (2012a). Musical experience offsets age-related delays in neural timing. Neurobiol Aging 33: 1483.e1-4.

Parbery-Clark A, Anderson S, Hittner E et al. (2012b). Musical experience strengthens the neural representation of sounds important for communication in middle-aged adults. Front Aging Neurosci 4: 30.

Parbery-Clark A, Tierney A, Strait DL et al. (2012c). Musicians have fine-tuned neural distinction of speech syllables. Neuroscience 219: 111–119.

Patel AD (2003). Language, music, syntax and the brain. Nat Neurosci 6: 674–681.

Patel AD (2006). Musical rhythm, linguistic rhythm, and human evolution. Music Percept 24: 99–104.

Patel AD (2010). Music, language, and the brain. Oxford University Press, Oxford.

Patel AD (2011). Why would musical training benefit the neural encoding of speech? The OPERA hypothesis. Front Psychol 2: 142.

Patel AD, Iversen JR (2014). The evolutionary neuroscience of musical beat perception: the Action Simulation for Auditory Prediction (ASAP) hypothesis. Front Syst Neurosci 8: 57.

Patel AD, Peretz I (1997). Is music autonomous from language? A neuropsychological appraisal. In: I Deliège, JA Sloboda (Eds.), Perception and cognition of music. Psychology Press, Hove, pp. 191–215.

Patel AD, Iversen JR, Bregman MR et al. (2008). Investigating the human-specificity of synchronization to music. In: Proceedings of the 10th International Conference on Music and Cognition. Sapporo, Japan, 2008, 100–104.

Patel AD, Iversen JR, Bregman MR et al. (2009). Studying synchronization to a musical beat in nonhuman animals. Ann N Y Acad Sci 1169: 459–469.

Peretz I, Hyde KL (2003). What is specific to music processing? Insights from congenital amusia. Trends Cogn Sci 7: 362–367.

Phillips-Silver J, Aktipis CA, Bryant GA (2010). The ecology of entrainment: foundations of coordinated rhythmic movement. Music Percept 28: 3.

Pinker S (1999). How the mind works. Ann N Y Acad Sci 882: 119–127.

Pitt MA, Samuel AG (1990). The use of rhythm in attending to speech. J Exp Psychol Hum Percept Perform 16: 564–573.

Poeppel D (2003). The analysis of speech in different temporal integration windows: cerebral lateralization as 'asymmetric sampling in time'. Speech Comm 41: 245–255.

Quene H, Port RF (2005). Effects of timing regularity and metrical expectancy on spoken-word perception. Phonetica 62: 1–13.

Rankin SK, Large EW, Fink PW (2009). Fractal tempo fluctuation and pulse prediction. University of California Press, pp. 401–413, 26.

Reber R, Winkielman P, Schwarz N (1998). Effects of perceptual fluency on affective judgments. Psychol Sci 9: 45–48.

Reber R, Schwarz N, Winkielman P (2004). Processing fluency and aesthetic pleasure: is beauty in the perceiver's processing experience? Pers Soc Psychol Rev 8: 364–382.

Repp BH (1992). Diversity and commonality in music performance: an analysis of timing microstructure in Schumann's "Traumerei". J Acoust Soc Am 92: 2546–2568.

Repp BH, Penel A (2002). Auditory dominance in temporal processing: new evidence from synchronization with simultaneous visual and auditory sequences. J Exp Psychol Hum Percept Perform 28: 1085.

Rogalsky C, Rong F, Saberi K et al. (2011). Functional anatomy of language and music perception: temporal and structural factors investigated using functional magnetic resonance imaging. J Neurosci 31: 3843–3852.

Romberg AR, Saffran JR (2010). Statistical learning and language acquisition. Wiley Interdiscip Rev Cogn Sci 1: 906–914.

Roncaglia-Denissen MP, Schmidt-Kassow M, Kotz SA (2013). Speech rhythm facilitates syntactic ambiguity resolution: ERP evidence. PLoS One 8: e56000.

Russo NM, Nicol TG, Zecker SG et al. (2005). Auditory training improves neural timing in the human brainstem. Behav Brain Res 156: 95–103.

Saffran JR, Aslin RN, Newport EL (1996). Statistical learning by 8-month-old infants. Science 274: 1926–1928.

Saffran JR, Senghas A, Trueswell JC (2001). The acquisition of language by children. Proc Natl Acad Sci 98: 12874–12875.

Saffran J, Hauser M, Seibel R et al. (2008). Grammatical pattern learning by human infants and cotton-top tamarin monkeys. Cognition 107: 479.

Salimpoor VN, Zatorre RJ (2013). Neural interactions that give rise to musical pleasure. Psychol Aesthet Creat Arts 7: 62.

Sauseng P, Klimesch W (2008). What does phase information of oscillatory brain activity tell us about cognitive processes? Neurosci Biobehav Rev 32: 1001–1013.

Schlaug G, Marchina S, Norton A (2008). From singing to speaking: why singing may lead to recovery of expressive language function in patients with Broca's aphasia. Music Percept 25: 315.

Schmidt-Kassow M, Kotz SA (2008). Entrainment of syntactic processing? ERP-responses to predictable time intervals during syntactic reanalysis. Brain Res 1226: 144–155.

Schmidt-Kassow M, Kotz SA (2009). Attention and perceptual regularity in speech. Neuroreport 20: 1643–1647.

Schutz M, Lipscomb S (2007). Hearing gestures, seeing music: vision influences perceived tone duration. Perception 36: 888.

Scott SK, Johnsrude IS (2003). The neuroanatomical and functional organization of speech perception. Trends Neurosci 26: 100–107.

Shahin AJ, Roberts LE, Chau W et al. (2008). Music training leads to the development of timbre-specific gamma band activity. Neuroimage 41: 113–122.

Skoe E, Kraus N (2010). Auditory brain stem response to complex sounds: a tutorial. Ear Hear 31: 302–324.

Skoe E, Kraus N (2012). A little goes a long way: how the adult brain is shaped by musical training in childhood. J Neurosci 32: 11507–11510.

Skoe E, Kraus N (2013). Musical training heightens auditory brainstem function during sensitive periods in development. Front Psychol 4: 622.

Skoe E, Krizman J, Spitzer E et al. (2013). The auditory brainstem is a barometer of rapid auditory learning. Neuroscience 243: 104–114.

Slater J, Strait DL, Skoe E et al. (2014). Longitudinal effects of group music instruction on literacy skills in low-income children. PLoS One. (in press).

Slater J, Swedenborg B, Kraus N (2014). How musical expertise influences speech perception in noise: A comparison of drummers, vocalists and non-musicians. Association for Research in Otolaryngolgy Symposium, San Diego.

Stahl B, Kotz SA, Henseler I et al. (2011). Rhythm in disguise: why singing may not hold the key to recovery from aphasia. Brain: awr240.

Strait DL, Kraus N (2014). Biological impact of auditory expertise across the life span: musicians as a model of auditory learning. Hear Res 308: 109–121.

Strait DL, Kraus N, Skoe E et al. (2009). Musical experience and neural efficiency: effects of training on subcortical processing of vocal expressions of emotion. Eur J Neurosci 29: 661–668.

Strait DL, Hornickel J, Kraus N (2011). Subcortical processing of speech regularities underlies reading and music aptitude in children. Behav Brain Funct 7: 44.

Strait DL, Chan K, Ashley R et al. (2012a). Specialization among the specialized: auditory brainstem function is tuned in to timbre. Cortex 48: 360–362.

Strait DL, Parbery-Clark A, Hittner E et al. (2012b). Musical training during early childhood enhances the neural encoding of speech in noise. Brain Lang 123: 191–201.

Strait DL, Parbery-Clark A, O'Connell S et al. (2013a). Biological impact of preschool music classes on processing speech in noise. Dev Cogn Neurosci 6: 51–60.

Strait DL, O'Connell S, Parbery-Clark A et al. (2013b). Musicians' enhanced neural differentiation of speech sounds arises early in life: developmental evidence from ages 3 to 30. Cerebral Cortex. http://dx.doi.org/10.1093/cercor/bht103.

Suga N, Ma X (2003). Multiparametric corticofugal modulation and plasticity in the auditory system. Nat Rev Neurosci 4: 783–794.

Thomson JM, Goswami U (2008). Rhythmic processing in children with developmental dyslexia: auditory and motor rhythms link to reading and spelling. J Physiol Paris 102: 120–129.

Tierney A, Kraus N (2013a). The ability to move to a beat is linked to the consistency of neural responses to sound. J Neurosci 33: 14981–14988.

Tierney A, Kraus N (2013b). Music training for the development of reading skills. In: M Merzenich, M Nahum, T van Vleet (Eds.), Applying brain plasticity to advance and recover human ability progress in brain research. Elsevier, Amsterdam.

Tierney A, Krizman J, Skoe E et al. (2013c). High school music classes enhance the neural processing of speech. Front Psychol 4: 855.

Torre K, Delignières D (2008). Unraveling the finding of 1/f β noise in self-paced and synchronized tapping: a unifying mechanistic model. Biol Cybern 99: 159–170.

Trainor L (2008). Science & Music: the neural roots of music. Nature 453: 598–599.

Trainor LJ, Heinmiller BM (1998). The development of evaluative responses to music: infants prefer to listen to consonance over dissonance. Infant Behav Dev 21: 77–88.

Trainor LJ, Shahin AJ, Roberts LE (2009). Understanding the benefits of musical training: effects on oscillatory brain activity. Ann N Y Acad Sci 1169: 133–142.

Tramo MJ, Cariani PA, Delgutte B et al. (2001). Neurobiological foundations for the theory of harmony in western tonal music. Ann N Y Acad Sci 930: 92–116.

Trehub SE (2003). The developmental origins of musicality. Nat Neurosci 6: 669–673.

Tsay C-J (2013). Sight over sound in the judgment of music performance. Proc Natl Acad Sci 110: 14580–14585.

Tzounopoulos T, Kraus N (2009). Learning to encode timing: mechanisms of plasticity in the auditory brainstem. Neuron 62: 463–469.

Uhlhaas PJ, Pipa G, Lima B et al. (2009). Neural synchrony in cortical networks: history, concept and current status. Front Integr Neurosci 3: 17.

Ullman MT (2001). A neurocognitive perspective on language: the declarative/procedural model. Nat Rev Neurosci 2: 717–726.

Ullman MT (2005). A cognitive neuroscience perspective on second language acquisition: The declarative/procedural model. In: Mind and Context in Adult Second Language Acquisition, 141–178.

Von Helmholtz H (1912). On the Sensations of Tone as a Physiological Basis for the Theory of Music. Longmans, Green, and Company, London.

Vuust P, Brattico E, Seppänen M et al. (2012). Practiced musical style shapes auditory skills. Ann N Y Acad Sci 1252: 139–146.

Wan CY, Schlaug G (2010). Music making as a tool for promoting brain plasticity across the life span. Neuroscientist 16: 566–577.

Werker JF, Tees RC (1984). Cross-language speech perception: evidence for perceptual reorganization during the first year of life. Infant Behav Dev 7: 49–63.

West BJ, Shlesinger MF (1989). On the ubiquity of 1/f noise. Int J Mod Physs B 3: 795–819.

Whalley K, Hansen J (2006). The role of prosodic sensitivity in children's reading development. J Res Read 29: 288–303.

White-Schwoch T, Carr KW, Anderson S et al. (2013). Older adults benefit from music training early in life: biological evidence for long-term training-driven plasticity. J Neurosci 33: 17667–17674.

Wise RA (2004). Dopamine, learning and motivation. Nat Rev Neurosci 5: 483–494.

Wong PC, Warrier CM, Penhune VB et al. (2008). Volume of left Heschl's gyrus and linguistic pitch learning. Cereb Cortex 18: 828–836.

Wood C, Terrell C (1998). Poor readers' ability to detect speech rhythm and perceive rapid speech. Br J Dev Psychol 16: 397–413.

Woodruff Carr K, White-Schwoch T, Tierney AT et al. (2014). Beat synchronization predicts neural speech encoding and reading readiness in preschoolers. Proc Nat Acad Sci: 201406219.

Yu Y, Romero R, Lee TS (2005). Preference of sensory neural coding for 1/f signals. Phys Rev Lett 94: 108103.

Zatorre RJ, Chen JL, Penhune VB (2007). When the brain plays music: auditory-motor interactions in music perception and production. Nat Rev Neurosci 8: 547–558.

Zentner MR, Kagan J (1998). Infants' perception of consonance and dissonance in music. Infant Behav Dev 21: 483–492.

Section 2

Methodology and techniques

Handbook of Clinical Neurology, Vol. 129 (3rd series)
The Human Auditory System
G.G. Celesia and G. Hickok, Editors

Chapter 13

Invasive recordings in the human auditory cortex

KIRILL V. NOURSKI* AND MATTHEW A. HOWARD III
Department of Neurosurgery, University of Iowa, Iowa City, IA, USA

INTRODUCTION

The auditory cortex in humans resides in the temporal lobe, on superior and lateral aspects of superior temporal gyrus (STG), and is comprised of multiple anatomically and physiologically distinct fields that receive direct synaptic input from the medial geniculate body (MGB) of the thalamus (see Chapter 2, this volume). Beyond these classic fields are auditory-related areas, which include anterior pole of the STG, middle temporal gyrus (MTG), inferior frontal gyrus (IFG), and multiple other areas in the temporal, parietal, and frontal lobes. These areas do not receive direct inputs from the MGB, but are reciprocally connected with classic auditory cortical fields and are involved in higher-level auditory processing, including speech perception and production. Auditory-related cortex is also involved in multisensory convergence, integrating auditory inputs with other sensory modalities (visual, somatosensory) (Romanski et al., 1999; Kaas and Hackett, 2000).

Much of the research on auditory cortex has been performed in experimental animals, including non-human primates. This research has utilized invasive anatomic and physiologic techniques to address fundamental questions about the functional organization of auditory and auditory-related cortex. The results of studies carried out in non-human primates have given rise to a model of serial-parallel auditory cortical processing (Hackett et al., 1998; Kaas and Hackett, 2000, 2005). According to this model, the auditory cortex is divided into three hierarchically organized tiers: core, belt, and parabelt. The core region includes primary and primary-like areas that receive direct synaptic input mainly from the ventral division of the MGB. The belt region surrounds the core and includes multiple areas that receive their major input from the dorsal division of the MGB as well as from adjacent core areas. Finally, the auditory parabelt region includes areas that have reciprocal connections with adjacent auditory belt areas but not directly with the core region.

The core–belt–parabelt hierarchic model has become a useful framework for studying functional organization of the human auditory cortex (Hackett, 2003, 2007, 2008; Rauschecker and Scott, 2009). In humans, the auditory core region has been localized to the posteromedial two-thirds of the first transverse gyrus of Heschl (HG) on the superior temporal plane. The homology of auditory fields between humans and non-human primates, however, is not entirely clear and remains a matter of continued investigation (e.g., Barton et al., 2012; Baumann et al., 2013). The superior temporal plane in humans is anatomically complex, with as much as half of its gray matter hidden within its sulci (Zilles et al., 1997; Destrieux et al., 2010). Furthermore, there is a considerable variability in anatomy across individuals and between hemispheres in the same individual (Rademacher et al., 1993; Leonard et al., 1998; Hackett et al., 2001). These are serious technical challenges to studying and understanding the structure and the function of the auditory cortex in humans.

A wide range of experimental methods is available to advance the understanding of the human auditory cortex. Non-invasive approaches include electroencephalography (EEG) and magnetoencephalography (MEG), which record changes in electric and magnetic fields, respectively, that result from neuronal activity (see Chapter 14, this volume). Functional neuromaging techniques (positron emission tomography (PET); functional magnetic resonance imaging (fMRI)) measure cortical activity based on changes in cerebral metabolism and blood flow (see Chapter 15, this volume). A variant of magnetic resonance imaging – diffusion tensor imaging (DTI) – characterizes connectivity between brain areas by measuring water diffusion (see Chapter 16, this volume).

*Correspondence to: Kirill V. Nourski, Department of Neurosurgery, University of Iowa, 200 Hawkins Dr. 1815 JCP, Iowa City, IA 52242, USA. E-mail: kirill-nourski@uiowa.edu

Each of these methods has its advantages and limitations, but can be used in complementary ways to yield new information not obtainable by any single approach.

Invasive experimental methods include electrophysiologic recording, electric stimulation, and local cooling of the cortex. This research can be performed in neurosurgical patients undergoing diagnosis and treatment of medically intractable epilepsy or, more rarely, tumors. Acute experiments are carried out during surgery in the operating room. Chronic studies are performed in a specialized hospital ward equipped for long-term clinical electrocorticography (ECoG) and video monitoring. ECoG refers to recording of electric activity using electrodes placed over the cortical surface or implanted into the gray matter. The electrophysiologic signal recorded by an ECoG electrode is referred to as the local field potential (LFP). It represents the summation of voltage fluctuations produced by synaptic activity of all the neurons within the vicinity of the electrode. Additionally, electrodes implanted in the cortical gray matter can allow for recording of action potentials from single neurons or neuronal clusters (Engel et al., 2005).

Electric stimulation can be performed using the same ECoG electrodes through which electrophysiologic recordings are obtained. Electric stimulation is used to determine location of cortical areas critical for hearing, speech, and language by temporarily disrupting cortical processing around the site of stimulation, thus creating a reversible "functional lesion" (Boatman, 2004; Sinai et al., 2005). Another method of creating a functional lesion is by focal cooling, wherein synaptic transmission in a small cortical area is reversibly blocked using a specialized cooling probe (Bakken et al., 2003). Electrophysiologic recording and electric stimulation can be used together to define functional connections between auditory and auditory-related cortical areas (Liégeois-Chauvel et al., 1991; Brugge et al., 2003, 2005; Greenlee et al., 2004, 2007; Matsumoto et al., 2004, 2007; Garell et al., 2013). Invasive research using ECoG offers several important advantages over conventional non-invasive approaches:

1. ECoG is a direct measure of activity of neuronal populations. This is in contrast to functional neuroimaging methods such as fMRI or PET that measure surrogate signals – blood flow or metabolic rate – that are indirectly linked to neural activity.
2. High (on the order of millseconds) temporal resolution, better than fMRI or PET.
3. High (on the order of millimeters) spatial resolution, surpassing that of non-invasive electromagnetic recording methods (EEG and MEG).
4. High signal-to-noise ratio, allowing examination of high-frequency (>70 Hz) components of cortical activity.
5. Low susceptibility to artifacts such as those generated by eye movements and other muscle activity.

This chapter presents the evolution of invasive recordings from the human auditory cortex, followed by a description of research methodology principles and some of the specific experimental paradigms used to study the functions of multiple auditory and auditory-related fields in humans.

HISTORIC OVERVIEW

The foundations of modern invasive human brain research methods were laid by neurosurgeon Wilder Penfield at the Montreal Neurological Institute (MNI). Founded by Penfield in 1934, MNI was the first-of-a-kind collaborative, multidisciplinary center for human brain research. Clinicians and neuroscientists were integrated into a comprehensive institute, with a shared clinical care and human brain research mission. Over the years, MNI became recognized as the world's top center for the surgical treatment of patients with medically intractable partial epilepsy. Research studies performed at MNI in epilepsy patients resulted in many seminal scientific contributions, including those regarding the functional properties of human auditory cortex (Penfield and Rasmussen, 1950; Penfield and Perot, 1963).

Penfield used electric stimulation functional mapping during craniotomy procedures. The mapping procedure was performed in awake patients under local anesthesia. With the patient's brain exposed, Penfield used the electric stimulator to test functional properties of specific brain sites. For example, electric stimuli delivered to the surface of the motor cortex elicited movements of contralateral limbs. Effects of stimulation of individual sites were documented, along with intraoperative photographs of the anatomic locations of the stimulated sites. The result was a functional map that was used to guide the surgical strategy in the individual patient. In addition to its clinical utility, the electric stimulation mapping method provided direct systematic evidence of how human cerebral cortex is functionally organized.

Penfield and colleagues studied the effects of electric stimulation of the auditory cortex in a large number of patients. Temporal-lobe epilepsy is the most common form of adult focal epilepsy. Accordingly, the temporal lobe is the most commonly targeted site for resection surgery in patients with epilepsy. The lateral surface of the temporal lobe – including the auditory cortex on the lateral STG – was therefore routinely exposed during these procedures. Moreover, the superior temporal plane, including HG, was occasionally exposed during the temporal-lobe resection, allowing access to what now is defined as core auditory cortex. Auditory perceptions were only evoked by stimulation of sites on the exposed superior temporal plane or on the lateral STG. Crude auditory sensations (e.g., buzzing) were typically evoked at sites on the exposed superior temporal plane, while "psychical responses" (complex experiential hallucinations or illusions) typically occurred with stimulation of the lateral surface of the STG.

When the results of these early electric stimulation studies are interpreted, a number of physiologic considerations must be borne in mind. Most importantly, electric shocks delivered to one cortical site are likely to activate pathways connecting that site to other brain areas. The effects of electric stimulation may thus be mediated by complex interactions of excitatory and inhibitory processes over a widely distributed functional network rather than a specific cortical site underneath the stimulating electrode. Furthermore, electric stimuli of sufficient intensity can generate epileptiform activity (afterdischarges). This abnormal pattern of activation may affect the function of brain regions in a prolonged manner, well beyond the stimulated site. Indeed, more recent studies have concluded that some of the effects reported by Penfield and colleagues, particularly those of complex auditory "psychical responses," likely resulted from a complex, widely distributed pattern of activation involving brain regions beyond the auditory cortex on the STG, including the limbic system (Gloor et al., 1982). Still, even with these caveats in mind, the methodologic and scientific impact of Penfield's seminal research on electric stimulation mapping has been profound and long-lasting.

Development of electronic recording instrumentation and introduction of bioelectric signal averaging techniques made it possible to obtain high-quality ECoG recordings from the auditory cortex. Gastone Celesia and Flavio Puletti (University of Wisconsin – Madison) were among the first to systematically study the functional organization of human auditory cortex using direct electrophysiologic recordings. In those studies, simple and complex acoustic stimuli were presented to awake neurosurgical epilepsy patients, while recordings were made from the auditory cortex on the superior temporal plane and lateral STG. Recorded responses were then characterized in terms of their amplitude, latency, and spatial distribution (Celesia and Puletti, 1969; Celesia, 1976).

Other major research contributions to the field were made at the University of Washington, where a team led by Arthur Ward and, later, George Ojemann, developed an approach for obtaining single-unit recordings from the auditory cortex intraoperatively. Rigid tungsten microelectrodes were used to record from regions of the temporal lobe that were to be resected for clinical reasons. These studies generated unique data on the speech-evoked responses of individual neurons within the auditory-related cortex of the temporal pole and the MTG (Creutzfeldt and Ojemann, 1989; Creutzfeldt et al., 1989a, b).

More recent technical advances in electrode fabrication have contributed to the development of the multicontact hybrid depth electrodes (HDEs), which combine clinical and research microwire contacts, allowing for simultaneous recordings of LFPs and unit activity, respectively, in tissue that might not be resected (Howard et al., 1996a, b; Fried et al., 1999; Bitterman et al., 2008). This was achieved by modifying standard clinical depth electrodes so that microwires (25–50 µm diameter) extended from the tip or at different points along the shaft of the depth electrode. In the typical clinical scenario, depth electrodes are placed in multiple locations, some of which are in brain regions that are found not to be epileptogenic and are not resected. The fine caliber of the microwires does not alter the clinical risk profile associated with placement of the clinical depth electrodes, which are approximately 2 mm in diameter.

Advances in electrode design have been paralleled by advances in data acquisition and processing systems, which now allow for simultaneous high-resolution recording from hundreds of sites. With all this methodologic progress, the overarching scientific goal has remained unchanged since the studies of Penfield and colleagues, namely, to use recordings from multiple auditory cortical regions to gain understanding of where and how, specifically, acoustic information is processed in the human brain.

RESEARCH SUBJECTS

Invasive electrophysiologic research of the human auditory cortex is opportunistic in nature, as it takes advantage of neurosurgical patients' willingness to participate during medically necessary diagnostic and treatment procedures. Recording brain activity directly from the auditory cortex is possible in patients who undergo placement of electrodes on or in the brain for accurate localization and resection of an epilepsy seizure focus or, less frequently, resection of a brain tumor.

Epilepsy is a common chronic neurologic disorder characterized by recurrent seizures. It affects around 50 million people worldwide (World Health Organization, 2012). Individuals with epilepsy have a risk of sudden loss of consciousness and thus are often unable to drive a motor vehicle, operate machinery, or hold a job that requires them to maintain a high level of alertness. These factors, along with a social stigma often associated with the disorder, impact the quality of life for the patients and their families. A long history of seizure activity may lead to structural brain damage in these individuals (Bonilha et al., 2006; Bernhardt et al., 2009). The risk of premature death in people with epilepsy is 2.6 times higher compared to the general population (Neligan et al., 2011).

While up to 70% of newly diagnosed epilepsy patients can be successfully treated with antiepileptic drugs (AEDs) (World Health Organization, 2012), the rest are considered medically refractory, i.e., continue to have seizures despite medical treatment. A fraction of medically refractory epilepsy cases can be treated effectively by surgically resecting brain tissue that is the source of epileptic activity (the seizure focus). Candidacy for resection surgery in epilepsy patients is defined by three criteria:

1. Failure of AED treatment, typically with at least two different classes of AED over a period of years. Patients continuing to have seizures after an appropriate trial of two AEDs are highly unlikely to become seizure-free with use of additional medications (Kwan and Sperling, 2009).
2. Projected considerable improvement in the quality of life by achieving a seizure-free surgical outcome. Multiple disabling comorbidities and short life expectancy are contraindications for surgical treatment.
3. Focal nature of the disorder, confirmed by localization of the seizure focus to a circumscribed brain region.

Presurgical evaluation of patients includes structural MRI, video-EEG monitoring, and neuropsychologic testing. Eligible candidates for resection surgery are informed that they may volunteer to participate in research. The patients are given a detailed description of the planned experimental protocols and provide informed consent before becoming research subjects. Participation in the research activities does not interfere with the subjects' clinical evaluation, and subjects can rescind consent at any time without interrupting diagnostic and treatment procedures.

Research subjects then undergo additional research-related preoperative testing, which typically includes neuroimaging, audiologic, and neuropsychologic studies. Preoperative structural MRI provides a template of the subject's brain anatomy. Functional MRI studies may be performed, with their results interpreted in conjunction with subsequent direct recordings (Mukamel et al., 2005). Audiologic evaluation includes pure-tone and speech audiometry to determine possible hearing deficits such as hearing loss or tinnitus. Custom-fitted ear molds are made to be used with insert earphones for acoustic stimulus presentation. Additional psychophysical tests may also be performed at this stage to complement subsequent electrophysiologic studies.

ACUTE EXPERIMENTS

Approximately half of the epilepsy patients who fulfill the selection criteria for resection surgery undergo the procedure without additional invasive recordings. In these cases, clinical ECoG recording is performed intraoperatively to capture interictal spiking (pathologic discharges occurring in the absence of a clinically detectable seizure) and thus to obtain additional information about seizure focus localization. In patients with tumors, intraoperative electric stimulation mapping helps determine tumor resection strategy by testing for and avoiding resection of critically important (eloquent) cortex. For these clinical purposes, electrode arrays are typically placed over the exposed surface of the temporal lobe. In epilepsy cases, recordings are typically performed over two 30-minute time windows under the direction of the epileptologist. Intraoperative clinical ECoG recording sessions provide an opportunity to simultaneously conduct auditory cortex research studies without increasing the overall duration of the surgical procedure or risk to the patient.

Acute experiments are performed in the operating room, with the subject awake, alert and able to follow instructions of the research team. Wide exposure of the brain surface during surgery provides several methodologic advantages. It allows for the use of various configurations of recording arrays, specialized electric stimulation probes, and local cooling devices. These devices can be safely placed directly on the cortical surface and, if necessary for research purposes, repositioned without damaging the underlying tissue.

Limitations of acute studies include time available for research, range of research tasks, and presence of ambient noise. The 30-minute time windows impose a limit on the amount of data that can be collected in a single experiment and make preoperative research planning, equipment calibration, and testing critically important. During the surgery, the patients are covered with surgical drapes. This restricts their movement and limits interaction with the research team and equipment. Consequently, the range of tasks subjects can perform is limited to relatively simple ones, such as responding to presented auditory stimuli verbally or by pressing a button. Finally, relatively high levels of ambient noise typically present in the operating room can interfere with acoustic stimulation. Insert earphones coupled with custom-fitted ear molds can provide some degree of passive noise reduction and thus are particularly appropriate for the use in intraoperative acoustic stimulation studies.

CHRONIC EXPERIMENTS

In some epilepsy patients, candidacy for resection surgery cannot be established solely based on non-invasive studies. Invasive diagnostic ECoG monitoring may be indicated when a single, surgically resectable, seizure focus is suspected, but the results of non-invasive studies are inconclusive or conflicting. The clinical reasons for invasive monitoring are twofold: to localize the epileptic focus and to identify critical brain areas that should be spared during focus resection. Electrode coverage often includes the temporal-lobe structures, including the superior temporal plane and the lateral STG. This provides a unique research opportunity to study the auditory cortex using implanted electrodes over the course of the monitoring period, which typically is around 1–2 weeks long (Hill et al., 2012).

Electrodes used for invasive studies of human auditory cortex are either implanted to penetrate cortical gray matter (depth electrodes) or placed over the cortical surface (surface arrays). Figure 13.1 presents an overview of the types of electrode used for chronic ECoG recording. Depth electrodes are linear, thin (1–2 mm diameter), flexible, silicon-based arrays. Clinical depth electrodes have multiple circumferential platinum contacts positioned along the cylindrical electrode shaft at regular intervals. For research purposes, HDEs are often used. These arrays are outfitted with additional microwire contacts for recording both LFPs as well as action potentials from single neurons or neuronal clusters. One type of HDE has insulated platinum–iridium microwire contacts (25–50 μm diameter) placed between the clinical ECoG contacts along the electrode shaft (Howard et al., 1996b). Another type of HDE has insulated microwires that are extruded from the tip of the electrode once the electrode is implanted (Fried et al., 1999). The latter design has the potential for more efficient isolation of single neurons or neuronal clusters because unit recordings are made further away from the electrode shaft. A disadvantage of this design, however, is that the precise location of the individual microwires may be difficult to identify. HDEs may also incorporate a microdialysis probe for *in situ* sampling of neuroactive substances in the extracellular milieu (Fried et al., 1999).

Surface electrodes typically consist of platinum–iridium disc contacts embedded in a silicon membrane and arranged in a single row (strip) or multiple rows (grid) (Fig. 13.1). These arrays are placed on to the cortical surface under the dura mater (subdurally) and can be used for both intraoperative and chronic ECoG monitoring. The narrow configuration of strip electrodes makes them convenient for placement into areas beyond those directly exposed by the craniotomy, such as the inferior surface of the temporal lobe. Clinical strip and grid electrodes typically have 1-cm spacing between adjacent contacts. This spacing is adequate for seizure focus localization and electric stimulation mapping. For research studies, arrays with higher spatial resolution (4–5 mm spacing) are available (Howard et al., 2000; Brugge et al., 2008; Flinker et al., 2011). Hybrid strip/grid surface electrodes, specifically designed for coverage of the temporal pole, offer improved congruence with the cortical surface. Microgrid electrodes offer the highest spatial resolution (1 mm spacing) among surface arrays. These research electrodes can be placed on the cortical surface underneath the larger clinical grid electrodes (Kellis et al., 2010).

Electrode coverage for chronic invasive ECoG monitoring often includes multiple depth and surface electrodes (Fig. 13.2). Depth electrodes are implanted stereotactically, using a rigid slotted cannula as a guide. Two types of stereotactic method are commonly used for implantation of depth electrodes in the superior

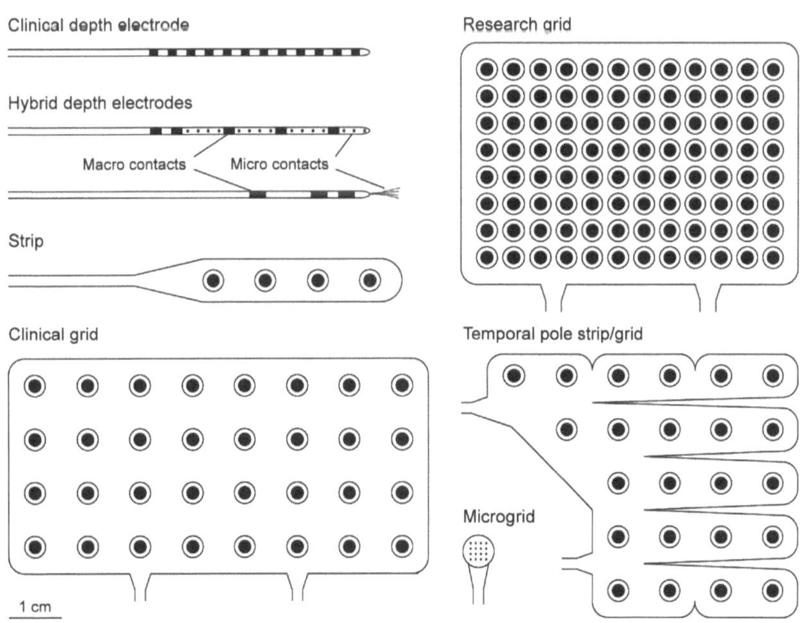

Fig. 13.1. Types of electrode used in invasive studies of human auditory cortex. See text for details.

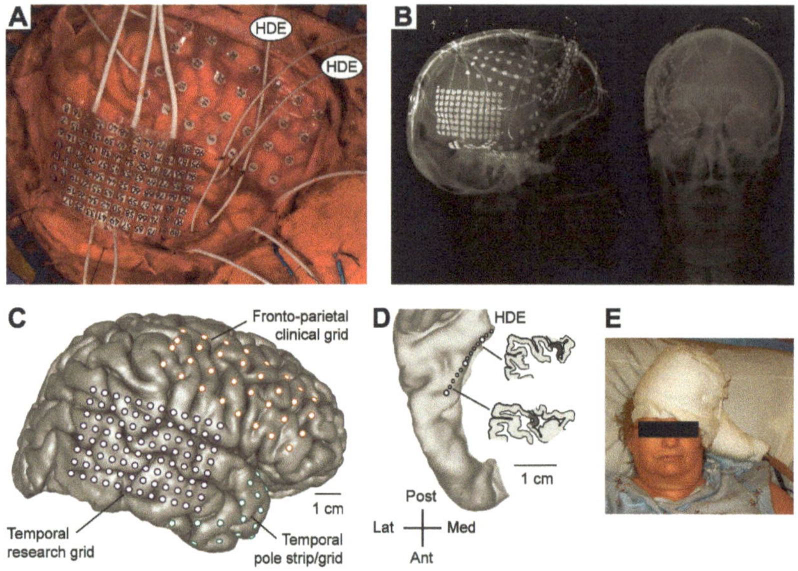

Fig. 13.2. Chronic invasive electrocorticography monitoring. (**A**) Intraoperative photograph showing the sites of entry of two hybrid depth electrodes (HDEs) and the placement of surface grid arrays. (**B**) Lateral (left) and frontal (right) postimplantation head X-ray images showing location of the implanted electrode arrays. (**C**) Projection of the surface grid contact location on to a preimplantation brain magnetic resonance imaging (MRI) scan. (**D**) Anatomic reconstruction of HDE clinical and research contacts (large and small circles, respectively) implanted in the Heschl's gyrus (HG). Left: MRI of superior temporal plane. Right: Line drawings of MRI cross-sections showing the position of three representative HDE contacts within the HG (circles). Light and dark gray shading represent cortical gray matter and the extent of HG, respectively. (**E**) Postimplantation photograph of a research subject.

temporal plane: rigid frame-based and frameless. The choice of system for implantation determines the approach to and the overall electrode coverage of the auditory cortex. In the former method, a rigid frame attached to the patient's head provides three-dimensional spatial reference points for stereotactic implantation and serves as a platform for electrode insertion devices. Preoperative evaluation for frame-based implantation typically includes high-resolution brain imaging (structural MRI or computed tomography (CT) scanning) following frame attachment. Imaging data are used to calculate electrode target coordinates within the frame reference space. Multiple depth electrodes may be inserted into the superior temporal plane along straight trajectories, parallel to one another (Fig. 13.3A). The use of multiple depth electrodes allows for simultaneous recording of auditory cortical responses from a wide area of the superior temporal plane, including HG and adjacent cortex of planum temporale and planum polare (Liégeois-Chauvel et al., 1991, 1994, 1999, 2004; Fig. 13.3B). The presence of a rigid frame, however, imposes physical restrictions on the choice of electrode trajectories, which are oblique relative to the longitudinal axis of the HG. This limits the extent to which auditory cortex on the HG, including core and adjacent belt fields, can be covered.

Frameless stereotactic methods allow for precise targeting of brain structures without the mechanic constraints of rigid frames. In these methods, three-dimensional spatial reference points are provided by MRI-compatible fiducial markers placed on the patient's head before surgery. A preoperative structural MRI scan is used to generate a three-dimensional image of the brain with the fiducial markers. During surgery, the patient's head is secured, and electrode insertion devices are co-registered with the reference space. The three-dimensional brain image is then used to guide the depth electrode to its target using specialized computer equipment and software. For depth electrodes that target superior temporal plane, the frameless system approach can provide a greater flexibility in the choice of electrode trajectory compared to rigid frame-based systems (Fig. 13.4). Specifically, the entire length of a depth electrode can be implanted within HG (Reddy et al., 2010). The electrode is inserted into the gray matter through the lateral STG and follows the HG along its longitudinal axis (Fig. 13.4A). A HDE implanted using this method can yield 18 or more recording contacts that sample the HG along its entire length. The insertion point on the lateral STG is anterior enough for simultaneous implantation of a surface grid array over the auditory cortex of lateral STG (Fig. 13.4B; see Fig. 13.2A).

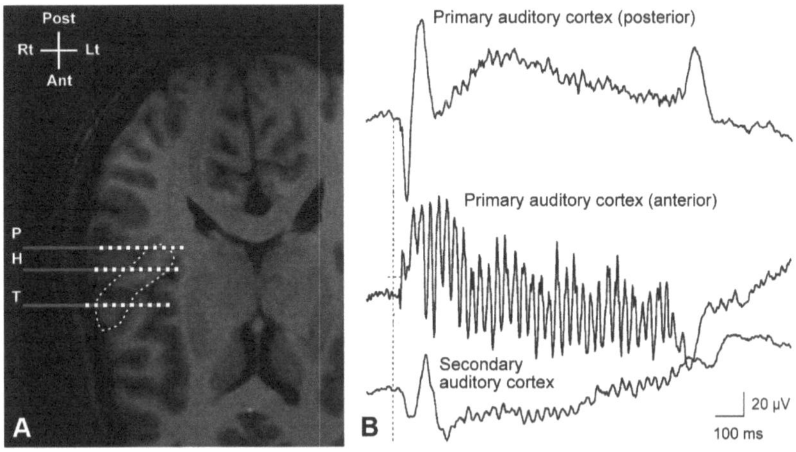

Fig. 13.3. Rigid frame-based stereotactic electrocorticography recording. (**A**) Horizontal magnetic resonance imaging through superior temporal plane showing the trajectory of three chronically implanted depth electrodes (labeled P, H, and T). Right Heschl's gyrus (HG) is outlined with a dotted line. (**B**) Averaged evoked potentials in response to 1 second 32 Hz sinusoidal amplitude-modulated noise bursts simultaneously recorded from posterior and anterior primary auditory cortex on HG, and from secondary auditory cortex. (Modified from Liégeois-Chauvel et al., 2004.)

Recording electrodes are implanted in the vicinity of the suspected seizure focus under general anesthesia, and usually no research activities are performed at this stage. Following the implantation surgery, anatomic reconstruction is performed to determine the localization of each recording contact. Pre- and postimplantation imaging data are co-registered, and stereotactic coordinates for each contact are projected on to the preimplantation MRI (see Fig. 13.2B, C). Accurate anatomic reconstruction of recording sites is challenging due to several confounding factors:

1. Implanted electrodes produce a substantial artifact, distorting the postimplantation MR image. This can be alleviated by a companion CT scan.
2. The exact position of the patient's head within the scanner is not identical during pre- and postoperative imaging. Gross anatomic skull landmarks can be used for co-registration with the standardized Talairach coordinate system to correct for this (Miller et al., 2007).
3. Placement of surface grid electrodes causes a medial displacement of the implanted cerebral hemisphere. This results in a distortion, more pronounced for superficial than deep-brain structures. To correct for this, localization of recording contacts on preimplantation MRI requires adjustments based on careful comparison of brain structures between the pre- and postimplantation MRI volumes.

Additionally, intraoperative photography during electrode implantation and removal is used to aid anatomic reconstruction of surface contacts. For depth electrodes implanted in the HG, it is important to determine not only the projection of each contact on to the superior temporal plane, but also their position relative to the cortical surface (Fig. 13.2D). This is accomplished by examining cross-sectional MR images orthogonal to the electrode shaft. The position of each depth electrode contact relative to cortical gray matter can thus be established, although the specific cortical layer(s) cannot be determined using current anatomic reconstruction methods.

Research activities typically commence 2–3 days after the implantation surgery. Patients are transferred to the epilepsy monitoring unit, and the electrodes are connected to the clinical ECoG recording system via cables that exit the bandage around the patient's head (Fig. 13.2E). Splitter connections allow for simultaneous ECoG signal recording by the research data acquisition system. During the clinical monitoring period, the patients are usually capable of participating in a wide range of experimental protocols, from simple passive-listening experiments to more complex tasks that require attention and decision making. Experimental sessions are approximately 2 hours long, including equipment set-up and data collection time. Typically, two research sessions are scheduled per day, one in the morning and one in the afternoon.

Chronic invasive recordings have several advantages over acute experiments:

1. Chronic monitoring allows for considerably longer time available for research data collection. More experiments can be scheduled, and the same experiments can be repeated multiple times over days, providing within-subject controls.

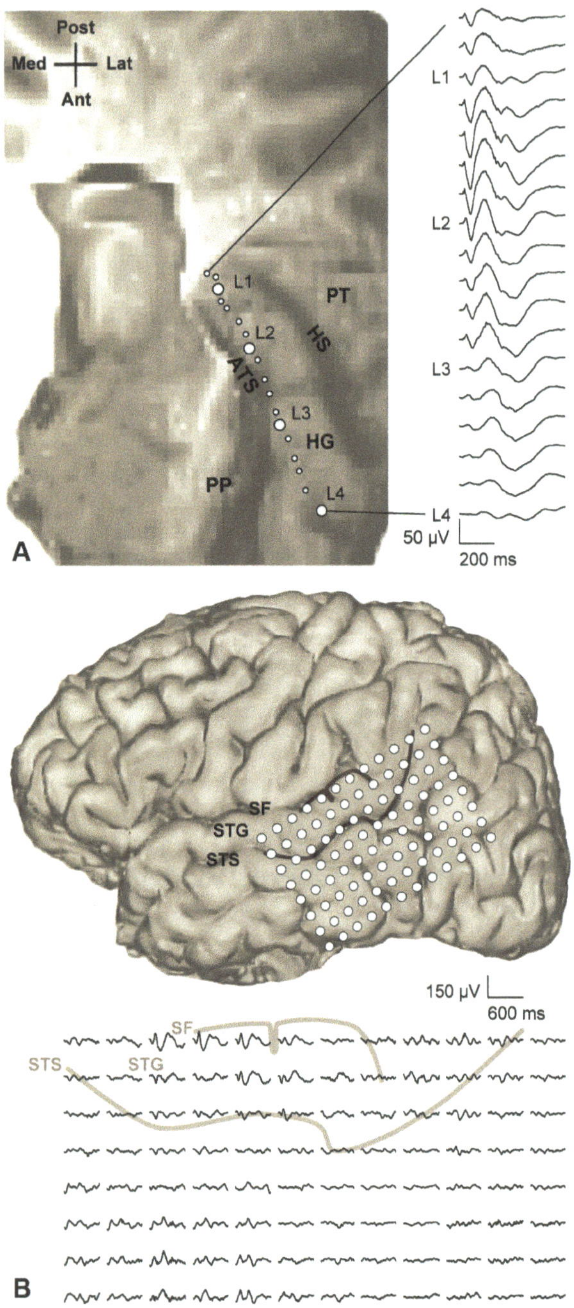

Fig. 13.4. Frameless stereotactic electrocorticography recording. (**A**) Left: Magnetic resonance imaging (MRI) of superior temporal plane in a representative subject showing the locations of clinical and research hybrid depth electrode (HDE) contacts (large and small circles, respectively), chronically implanted in Heschl's gyrus (HG). Clinical contacts are labeled as L1 through L4. Right: Averaged evoked potentials recorded from the HDE in response to an auditory stimulus (speech syllable /da/). (**B**) Top: Location of a 96-contact research grid on an MRI of the lateral brain surface. Bottom: Averaged evoked potentials recorded from these contacts. Thick gray lines indicate sulcal outlines. ATS, anterior transverse sulcus; HS, Heschl's sulcus; PP, planum polare; PT, planum temporale; SF, Sylvian fissure; STG, superior temporal gyrus; STS, superior temporal sulcus.

2. Implantation of depth electrodes as well as surface arrays provides an opportunity to study the functional organization of the auditory cortex by simultaneously recording from multiple auditory fields. Auditory fields in the superior temporal plane, particularly core auditory cortex of posteromedial HG, are typically not accessible during acute intraoperative experiments.
3. Patients are able to perform a greater variety of tasks during auditory stimulus presentation or electric stimulation studies.
4. Chronic monitoring provides for a well-controlled experimental environment, including lower ambient noise levels and better shielding from possible electromagnetic interference, compared to experimentation in the operating room.
5. Auditory activity during induction of general anesthesia can be investigated at the conclusion of the chronic monitoring period, when electrode removal and focus resection surgery are performed (Howard et al., 2000; Nourski et al., 2013b).

The most important limitation to the chronic setting compared to acute intraoperative experiments is that electrode coverage is determined solely by clinical criteria. Once implanted, the electrodes stay in place for the duration of the monitoring period and cannot be repositioned. Even in cases where extensive temporal-lobe coverage is warranted, many auditory fields remain inaccessible to the recording electrodes. This limitation is particularly important because of the considerable anatomic variability of the auditory cortex across individuals (Rademacher et al., 1993; Leonard et al., 1998). One way to address this limitation is to complement electrophysiologic experiments with fMRI studies in the same subjects under maximally similar stimulus conditions. This approach combines highly localized electrophysiologic measures provided by ECoG with a whole-brain view of cortical activity provided by fMRI (Mukamel et al., 2005).

DATA ANALYSIS

The LFP signal recorded by ECoG electrodes during auditory stimulation represents bioelectric activity elicited by the stimulus as well as ongoing cortical activity unrelated to the stimulus. To emphasize stimulus-related activity present in the LFP, identical stimuli are typically repeated many times, and responses are averaged over these multiple presentations (trials). This signal-averaging technique is illustrated in Figure 13.5. Time-domain averaging of single-trial LFP waveforms yields the averaged evoked potential (AEP), which emphasizes activity both time- and phase-locked to the stimulus that

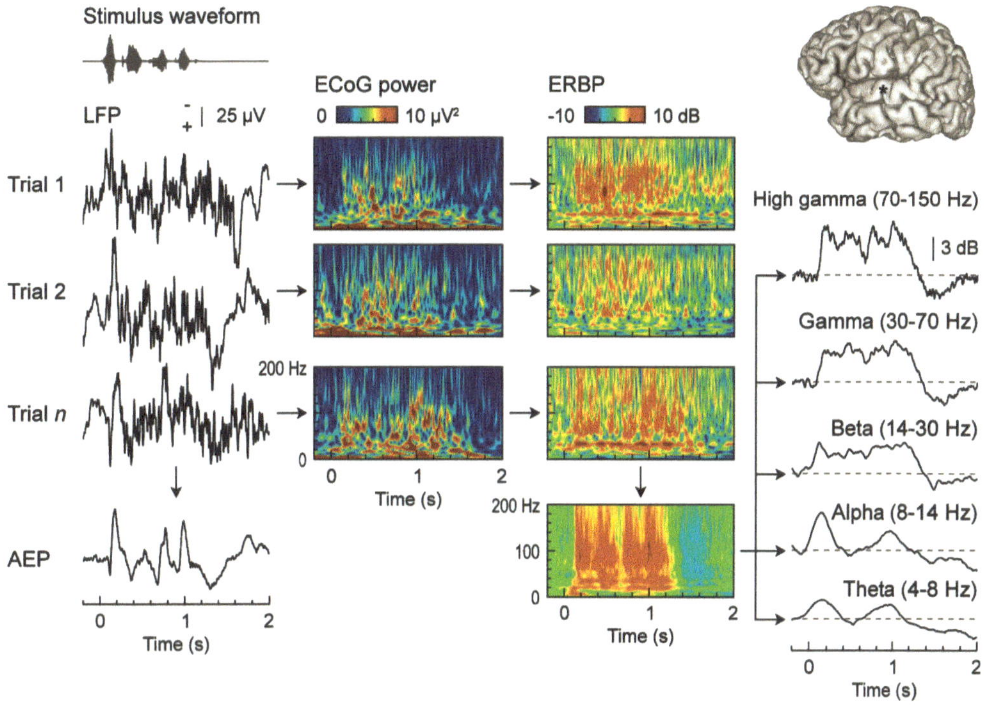

Fig. 13.5. Analysis of evoked and induced auditory cortical activity using the averaged evoked potential (AEP) and event-related band power (ERBP). Responses to a speech sentence "Black cars cannot park" (stimulus schematically shown in top left), recorded from a site on the lateral superior temporal gyrus (location of the recording electrode is shown in top right). ECoG, electrocorticography; LFP, local field potential.

is largely invariant across trials (Fig. 13.5, left column). AEP waveforms are characterized in terms of the amplitude and latency of their prominent peaks.

A limitation of simple time-domain averaging is that it does not preserve activity that may be stimulus-related yet not precisely phase-locked to the stimulus itself, and, hence, not phase-locked across multiple trials. The biologic importance of this non-phase-locked cortical activity was first established in experimental animal studies (Freeman, 1978; Gray et al., 1989; Engel et al., 1991). More recent studies have demonstrated the importance of such activity, particularly its relatively fast components (>70 Hz, the so-called high gamma band), in the auditory cortex of non-human primates as well as humans (Crone et al., 2001, 2006; Ray et al., 2008; Steinschneider et al., 2008; Brugge et al., 2009; Edwards et al., 2009; Nourski et al., 2009). Stimulus-induced power changes over the spectral range of the ECoG can be analyzed by calculating the time–frequency representation (essentially, a spectrogram) of the LFP signal on a trial-by-trial basis (second column in Fig. 13.5). The absolute power in the cortical LFP signal decreases with increase in frequency (Buszaki, 2006), whether a stimulus is presented or not. Because of this, normalization to a prestimulus reference is performed separately for different frequencies over the spectral range of the ECoG. This allows measurement of proportional changes in power related to stimulus presentation – so-called event-related band power (ERBP). ERBP can actually be greater for high-frequency components of the ECoG signal compared to low-frequency bands (Fig. 13.5, right column). When ERBP is calculated from single-trial LFP waveforms, it includes both phase-locked ("evoked") power (Pantev, 1995) as well as non-phase-locked, yet time-locked ("induced") power (Kalcher and Pfurtscheller, 1995; Pantev, 1995; Crone et al., 2001). Contribution of evoked response components to ERBP can be reduced by subtracting the AEP waveform from each single-trial LFP waveform prior to time–frequency analysis (Crone et al., 2001). Evoked and induced response components may be difficult to disambiguate, particularly when analyzing responses to periodic stimuli (e.g., click trains or amplitude-modulated noise) presented at rates that overlap with one of the conventional ECoG analysis frequency bands (e.g., high gamma range, 70–150 Hz) (Brugge et al., 2009; Nourski et al., 2013a).

Auditory cortical responses are often sufficiently robust to lend themselves to analysis on a trial-by-trial basis without averaging. Here, information-based strategies are of particular utility. Statistical classification algorithms – "classifiers" – can be trained to distinguish single-trial spatiotemporal response patterns elicited across the auditory cortex by different auditory stimuli (Chang et al., 2010; Nourski et al., 2014). Accuracy of single-trial classification based on a given stimulus feature (e.g., frequency of a pure tone or formant transitions of a consonant-vowel syllable) reflects the representation of that feature in the distributed patterns of cortical activity. Classification weights of individual recording sites reflect their relative contribution to classification accuracy and thus help localize cortical processing of specific stimulus attributes.

Experimental paradigms that rely on presentation of many repetitions of identical stimuli for subsequent across-trial averaging and/or classifier-based analysis are associated with some caveats:

1. Multiple trial-based stimulus presentations are not ecologically relevant. In real-life conditions, the auditory environment is essentially single-trial. Repeated presentation of identical stimuli in an experiment run can result in habituation phenomena, i.e., decreases in stimulus-evoked responses across trials.
2. Time efficiency of data collection can be suboptimal. Although signal-to-noise properties of ECoG are considerably better compared to non-invasive methods such as scalp EEG and MEG, typically tens of trials need to be collected in response to each stimulus for the averaging procedure to be effective.
3. Repeated stimulus presentation can cause fatigue in the research subject.

Alternative experimental paradigms include analysis of responses to sequences of unique stimuli (e.g., isolated words or sentences) (Pasley et al., 2012), or continuous presentation of a long-duration acoustic stimulus (e.g., a spoken narrative) (Kubanek et al., 2013). Stimulus sets that are sufficiently rich in their spectral content and temporal modulations can be used to define spectrotemporal receptive fields (STRFs) of individual cortical sites. STRF-based analysis (reviewed in Zhao and Zhaoping, 2011) can describe the tuning properties of cortical sites, which can then be used to predict responses to novel stimuli and, conversely, reconstruct the presented stimuli from patterns of cortical activity (Pasley et al., 2012).

As mentioned above, microwire contacts of HDEs can capture action potentials from single neurons or neuronal clusters. These can be extracted by high-pass filtering the recorded LFP signal (>500 Hz) and applying a spike-sorting algorithm (Quiroga et al., 2004). Firing patterns of auditory cortical neurons are then characterized by spike distribution histograms, dot rasters, and using reverse correlation analyses.

STUDIES OF SPECTROTEMPORAL PROCESSING: ELECTROPHYSIOLOGIC RECORDING

Invasive methods provide opportunities to better understand the functional organization of auditory and auditory-related cortex in humans using relatively simple sounds and to investigate cortical mechanisms of speech processing. The organizational framework for the study of human auditory cortex is based on the concept of multiple interconnected fields. These fields are differentiated from each other by anatomic location, connectivity patterns, and neuronal response properties, with each field contributing to auditory processing its own way. Cytoarchitectonic, connectivity, and physiologic studies have shown the existence of at least a dozen auditory cortical fields in non-human primates (Hackett, 2003, 2007, 2008), and postmortem anatomic studies have identified up to 7–8 fields on the STG in the human (Rivier and Clarke, 1997; Wallace et al., 2002; Chiry et al., 2003). Invasive recordings allow for defining the response properties of different auditory cortical fields and functional connections between them through the use of electrophysiologic recording, electric stimulation tract tracing, and functional lesioning.

One of the basic acoustic characteristics of a sound is its spectral (frequency) content. At each level in the ascending auditory pathway, many neurons exhibit frequency selectivity in their responses. Such neurons are typically most sensitive to a narrow spectral range, the center of which is referred to as the "best" or "characteristic" frequency. An orderly arrangement of neurons according to their best frequency – tonotopic organization – is a fundamental feature of auditory pathways, including individual fields of the auditory cortex (Rose et al., 1963; Aitkin and Webster, 1971; Merzenich and Brugge, 1973).

Investigation of frequency tuning on a single-neuron level is possible with direct recordings using HDEs implanted in HG (Howard et al., 1996a; Bitterman et al., 2008) (Fig. 13.6). One such study obtained frequency tuning curves similar to those recorded in auditory cortex of non-human primates and identified a high-to-low best-frequency gradient along the posteromedial portion of HG, consistent with its putative role as core auditory cortex (Fig. 13.6A). Another study (Bitterman et al., 2008) described single neurons in HG to have extraordinarily narrow tuning (Fig. 13.6B). Such sharp frequency tuning may represent a physiologic correlate for behavioral acuity of frequency discrimination in human listeners, as measured in psychophysical studies (Moore, 2012).

Non-core auditory cortex also has been found to exhibit spectrally selective response patterns (Nourski

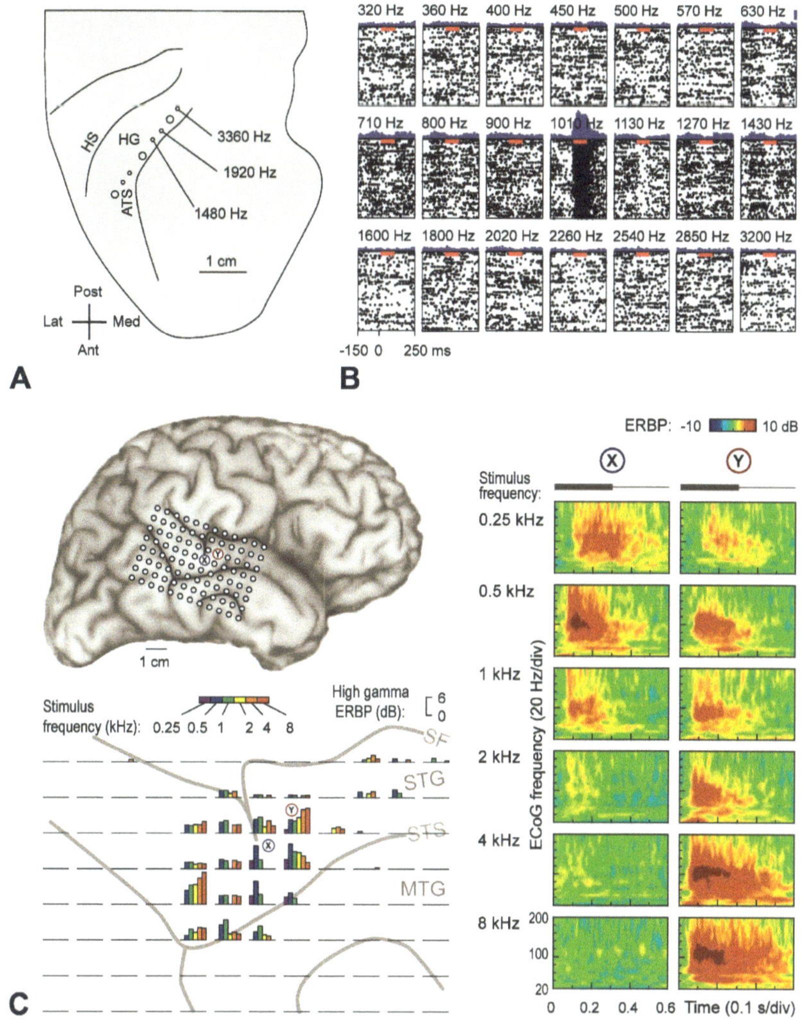

Fig. 13.6. Spectral organization of human auditory cortex. (**A**) Magnetic resonance imaging tracing of superior temporal plane showing the locations of clinical and microwire hybrid depth electrode contacts implanted in Heschl's gyrus (HG) (large and small circles, respectively). Best frequencies for units recorded at three sites within core auditory cortex are indicated. (Redrawn from Howard et al., 1996a.) (**B**) Frequency selectivity in core auditory cortex. Raster plots of responses ($n = 270$) of a unit to chords containing the frequency specified above each plot and peristimulus time histograms (blue) based on these raster plots. Vertical bar at the top right corresponds to a firing rate of 16 spikes/second. Stimulus duration (100 ms) is marked by red bars. (Modified from Bitterman et al., 2008.) (**C**) Tone-elicited responses recorded from posterolateral superior temporal gyrus (STG). Top left: location of the 96-contact subdural electrode grid. X and Y denote the locations of two sites whose responses are detailed in the right panel. Bottom left: high gamma event-related brain power (ERBP) in response to tone stimuli presented at different frequencies (color bars), simultaneously recorded from the 96-contact grid (individual plots). High gamma ERBP was averaged within the time window of 100–150 ms after stimulus onset. Right: time–frequency analysis of cortical activity elicited by pure-tone stimuli between 0.25 and 8 kHz (top to bottom), recorded from two sites on the STG (left and right columns). Stimulus is schematically shown above time–frequency plots. HS, Heschl's sulcus; ATS, anterior transverse sulcus. (Replotted from Nourski et al., 2014.)

et al., 2014). Analysis of high gamma activity recorded from surface grid arrays implanted over the temporal lobe shows that pure tones can activate a limited region within posterolateral STG (PLST). Individual sites within this area are broadly tuned, yet often exhibit frequency-selective responses. In the example shown in Figure 13.6C, site X is characterized by the largest response to the 0.5-kHz tone, whereas at an adjacent site Y, the largest responses are elicited by relatively high-frequency tones (4 and 8 kHz). Overall, frequencies are typically represented by high gamma activity in PLST in a spatially distributed manner.

Temporal modulations – variations of amplitude and frequency over time – are another basic feature of natural sounds, including speech. Indeed, sounds are events that occur over time. Repeated or periodic sounds such as clicks or bursts of modulated noise are heard differently depending on how fast they are presented. Such stimuli have been shown to elicit qualitatively different response patterns within core auditory cortex depending on their repetition rate (Brugge et al., 2009; Nourski and Brugge, 2011). Specifically, at low repetition rates (<10 Hz), clicks within a train are perceived as separate events, and each click in the train is able to evoke a separate polyphasic AEP complex. At higher click rates (>10 Hz), perceptual blending into the "flutter" and "pitch" percepts is reflected in the cortical activity by the emergence of on- and off- phasic responses at the onset and offset of the train and a sustained phase-locked frequency-following component. Yet higher click rates (above about 30–40 Hz) produce a pitch percept and elicit characteristic ECoG response patterns with increases in induced (non-phase-locked) ERBP in high gamma range. Thus, responses recorded directly from core auditory cortex in response to click trains reflect different perceptual classes of periodic sounds.

Comparison of responses to click trains simultaneously obtained from different auditory cortical fields can reveal differences in their specific physiologic response properties. An example of such a comparison is shown in Figure 13.7. Responses to click trains presented at rates between 25 and 200 Hz are shown for three cortical sites: a presumed core auditory cortex site in medial HG, a site in a non-primary field on lateral HG, and a representative PLST site, marked by X, Y, and Z, respectively. The earliest responses in the auditory cortex occur in posteromedial HG (core auditory cortex). Populations of human auditory cortex neurons here exhibit phase locking over a range of repetition rates similar to that of monkeys (Steinschneider et al., 1998; Lu et al., 2001). AEPs recorded from PLST have smaller amplitudes than those recorded from medial HG, yet are larger and have shorter latencies than AEPs recorded from lateral HG. Capacity for synchronization to the stimulus periodicity also differs across the three sites. PLST exhibits phase locking to click trains presented at 50 Hz, and rarely to higher rates, while medial HG could typically phase lock to 200-Hz click trains (Brugge et al., 2009; Nourski et al., 2013a). ERBP analysis reveals differences in magnitude and latency of ECoG power changes between the three sampled cortical areas. Putative core auditory cortex of medial HG has consistently shorter latency of high gamma response compared to PLST and lateral HG. At low rates (e.g., 25 Hz), activity recorded from medial HG features bursts of high gamma ERBP that are temporally related to each successive click in the train. Such bursts or modulation of high gamma ERBP are not evident in recordings made from either PLST or lateral HG. These data, along with spectral organization of PLST introduced above (Fig. 13.6C) provide evidence that PLST utilizes multiple strategies for representation of stimulus features and suggests that it may represent a relatively early stage in the auditory cortical hierarchy.

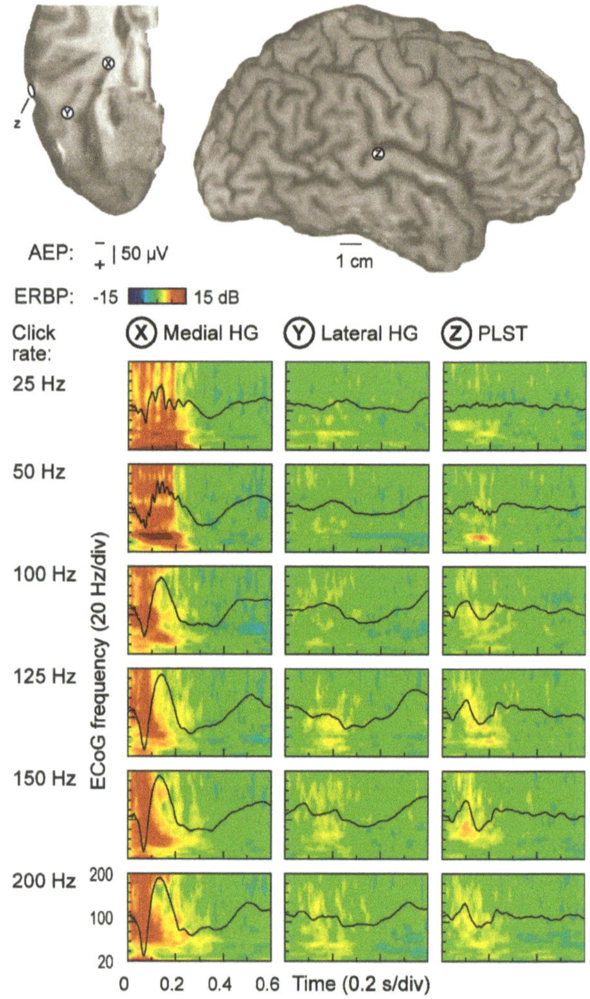

Fig. 13.7. Responses to click trains from Heschl's gyrus (HG) and posterolateral superior temporal gyrus (PLST). Top: location of three exemplary recording sites: a core cortex site in medial HG, a site in a non-core field on lateral HG and a PLST site (marked by X, Y, and Z, respectively). Bottom: averaged evoked potentials (AEPs) and event-related band power (ERBP) obtained from the three sites in response to click trains presented at rates between 25 and 200 Hz (top to bottom rows). ECoG, electrocorticography. (Modified from Nourski et al., 2013a.)

While simple sounds such as pure tones or click trains provide convenient models for the study of fundamental auditory cortical response properties, speech is the most ecologically relevant class of sounds for humans. Analysis of responses to speech sentences reveals that core auditory cortex can track temporal features of speech on multiple time scales (Fig. 13.8). Specifically, the relatively slow temporal envelope of speech sentences is

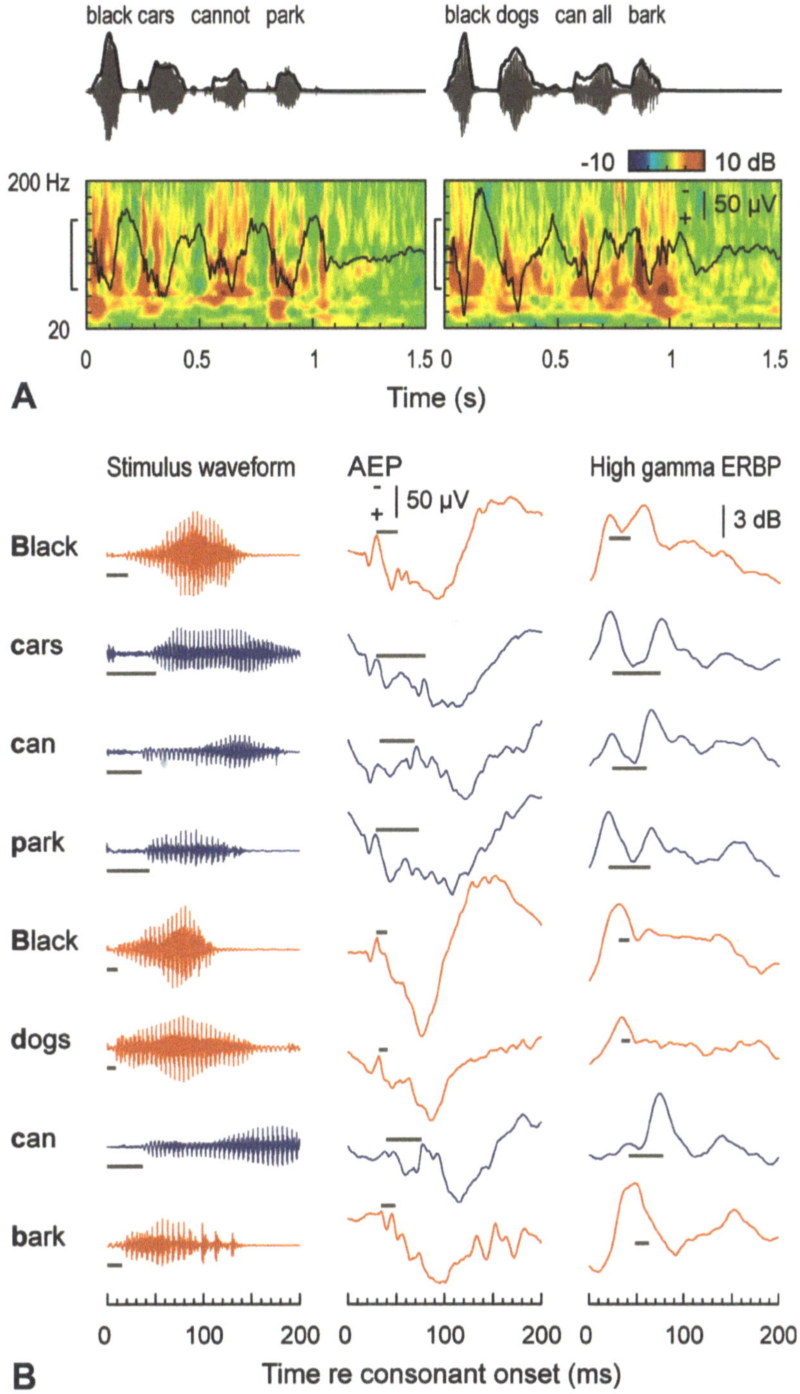

Fig. 13.8. Representation of speech temporal features in the human core auditory cortex. (**A**) Representation of temporal envelope. Top: Stimulus waveforms (gray) and temporal envelopes (black) of the two speech sentences. Bottom: averaged evoked potential (AEP) waveforms and event-related band power (ERBP) time–frequency plots. Brackets in ERBP time–frequency plots indicate high gamma frequency range. (**B**) Representation of voice onset time (VOT) of initial stop consonants. Left column: waveforms corresponding to the first 200 ms following each initial stop consonant (top to bottom) in the two sentences shown in (**A**). Middle and right column: AEP and high gamma ERBP responses, respectively, that correspond to each initial stop consonant. Red and blue plots correspond to voiced and voiceless consonants, respectively. Gray horizontal lines represent the VOT. (Modified from Steinschneider et al., 2013.)

represented by phase-locked activity evident in the AEP as well as modulation of high gamma ERBP (Nourski et al., 2009; Fig. 13.8A). Faster temporal features, such as voice onset time (VOT), are also reflected in cortical activity. Responses to voiceless consonants (long VOTs) have a different morphology from those to voiced consonants (short VOTs), featuring additional components that are time-locked to the onset of voicing. These patterns are observed in responses to isolated syllables (Steinschneider et al., 1999, 2005; Nourski and Brugge, 2011) as well as more natural running speech (Steinschneider et al., 2013; Fig. 13.8B). Temporal representation of the VOT parameter is evident when speech sentences are segmented into their component words, and both AEPs and the envelope of high gamma activity are examined. Words beginning with a voiced stop consonant ("dogs" and "bark" in Fig. 13.8) tend to elicit responses with a single peak, whereas words beginning with a voiceless stop ("cars," "can," and park" in Fig. 13.8) elicit double-peaked responses. These patterns are more prominent in high gamma activity than in the AEP. Overall, these findings indicate the utility of temporal code in cortical speech processing and highlight the similarity in auditory processing at the level of core auditory cortex between humans and non-human primates (Steinschneider et al., 1993, 1995, 2005).

FUNCTIONAL CONNECTIVITY STUDIES: MODELING AND ELECTRIC STIMULATION TRACT TRACING

Cortical auditory processing is mediated by reciprocal connections within auditory cortical fields, and between auditory cortex on STG and auditory-related areas in frontal and parietal lobes (Kaas and Hackett, 2005; Rauschecker and Scott, 2009). Coherence between patterns of sound-evoked activity recorded from different cortical sites provides information which can be used to make inferences regarding functional connectivity between these sites. Methods such as Granger causality or dynamic causal modeling are used to develop specific connectivity models (Oya et al., 2007; Gourevitch et al., 2008; Kumar et al., 2011). These models can then be tested directly using electric stimulation tract tracing.

Focal electric stimulation is an effective method to study functional connectivity of auditory cortical fields in human subjects. This method involves electric stimulation of a cortical site with brief (0.1–0.2 ms) charge-balanced current pulses with simultaneous ECoG recording from other locations. This approach has been successfully used to determine functional connectivity between core auditory cortex on HG and non-core cortex on PLST, between PLST and the IFG, between the IFG and motor cortex of the precentral gyrus, and between areas within the IFG (Liégeois-Chauvel et al., 1991; Howard et al., 2000; Brugge et al., 2003; Greenlee et al., 2004; Matsumoto et al., 2004; Garell et al., 2013). For example, electric pulses delivered directly to the auditory cortex on PLST elicit AEPs on the IFG, with largest responses localizing to its pars triangularis (Fig. 13.9). This finding suggests that a circumscribed area within human IFG may be homologous to ventral prefrontal cortex in the macaque monkey, which is known to receive direct projections from non-core auditory areas (Hackett et al., 1999; Romanski et al., 1999; Romanski and Goldman-Rakic, 2002).

Electric stimulation tract tracing can be performed in acute and chronic experiments. The acute experimental setting in the operating room is particularly well suited for this research, as custom recording and stimulating arrays can be placed on the cortical surface under visual control, repositioned as necessary, and this type of experiment does not require active participation from the subject. This method does not provide information regarding specific anatomic trajectories of the underlying neural pathways or the fine-grain detail of these connections, such as origin and destination cortical layers, or whether the connections are mono- or polysynaptic. It does, however, provide direct evidence for functional connectivity between the site of stimulation and the site(s) of recording in the living brain. Knowledge generated with this approach can be validated and further expanded using *in vivo* white-matter imaging techniques such as DTI (see Chapter 16, this volume).

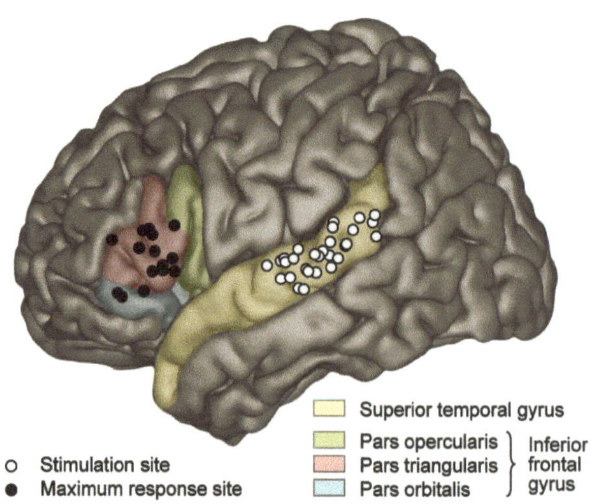

Fig. 13.9. Distribution of sites of maximal amplitude of the response on the inferior frontal gyrus to electric stimulation of sites on posterolateral superior temporal gyrus, shown on a left-hemisphere template. (Modified from Garell et al., 2013.)

FUNCTIONAL LESIONING STUDIES: ELECTRIC STIMULATION MAPPING AND CORTICAL COOLING

Functional lesioning refers to brief and reversible disruption of function of a small cortical area. Electric stimulation can, under certain stimulus conditions, produce such a temporary disruption. As described above in the historic overview, early experiments of Penfield and colleagues that investigated the functional organization of human auditory cortex involved the use of electric stimulation in the operating room (Penfield and Perot, 1963). More recently, this approach has been used by Boatman and colleagues to study the role of auditory and auditory-related cortex of the language-dominant hemisphere in speech and language processing (Boatman et al., 1995; Boatman, 2004; Sinai et al., 2005; Sinha et al., 2005).

Hemisphere language dominance can be determined preoperatively by performing an intra-arterial sodium amobarbital infusion (Wada test) and observing the effects of infusion into the left and right hemispheres on language functions. In most individuals, the left hemisphere is language-dominant, meaning that lesions affecting certain sites in the left frontal, parietal, or temporal lobe may result in a clinically significant deficit in language function. In a small number of patients, the right hemisphere is language-dominant. Some language functions, such as those linked to the prosody of speech, are thought to be mediated by the right hemisphere even in subjects with classic left-hemisphere dominance (Ross and Mesulam, 1979; Taylor and Regard, 2003). Speech sounds robustly activate the auditory cortex in both hemispheres.

Electric stimulation mapping can be performed both in acute and chronic experimental settings. In these experiments, electric pulses (0.3 ms, 10–15 mA, alternating polarity) presented as 5–10-seconds 50-Hz trains are delivered in a bipolar mode, across adjacent contacts placed on the lateral hemispheric surface or implanted in the superior temporal plane. Other contacts of the implanted arrays are used to collect ECoG recordings to capture physiologic responses to the electric stimulation and to monitor for possible afterdischarges (Sinai et al., 2005). Psychophysical testing is performed prior to and during electric stimulation mapping to identify possible deficits in auditory cortical function associated with stimulation of specific sites.

Electric stimulation of the STG can elicit auditory sensations of varying complexities (Penfield and Perot, 1963), attenuate perception of sounds (Sinha et al., 2005; Fenoy et al., 2006) and, in the language-dominant hemisphere, produce deficits at different levels of speech processing, ranging from auditory discrimination to speech sound identification to sentence comprehension (Boatman, 2004; Fig. 13.10A). Additionally, electric stimulation of the STG can elicit vestibular symptoms (such as the subjective feeling of falling) and, in patients affected by tinnitus, suppress the intensity of tinnitus (De Ridder et al., 2004; Fenoy et al., 2006; Fig. 13.10B). Taken together, these results provide a neurophysiologic framework for models of cortical speech processing and a foundation for development of novel management strategies for hearing and vestibular disorders.

Another functional lesioning method used in invasive studies of the human auditory cortex is focal cooling, which blocks synaptic transmission and provides a safe alternative to electric stimulation mapping for local and reversible disruption of cortical processing (Bakken et al., 2003). An example of a cooling probe designed specifically for human auditory cortex research is shown

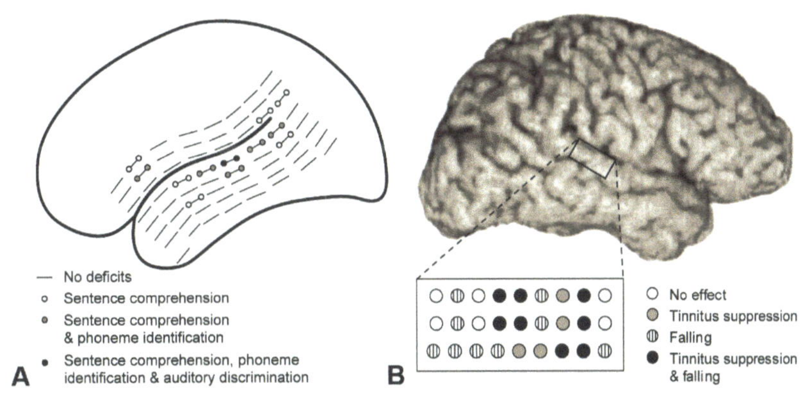

Fig. 13.10. Electric stimulation cortical mapping. (**A**) Schematic diagram of a cerebral hemisphere showing the locations of auditory discrimination deficits, induced during electric stimulation mapping, shown on a left-hemisphere template schematic. Lines represent sites where auditory discrimination was tested, but no deficits were induced. (Redrawn from Boatman, 2004.) (**B**) Tinnitus suppression induced by electric stimulation of the auditory cortex. Top: Magnetic resonance imaging showing the position of the recording array. Bottom: Effects reported by the subject in response to stimulation at each contact. (Modified from Fenoy et al., 2006.)

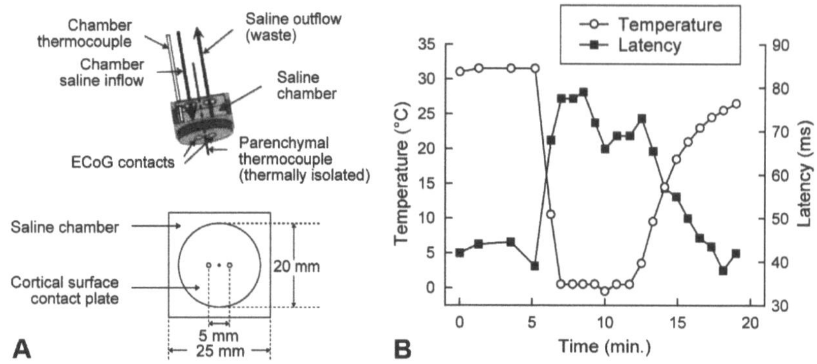

Fig. 13.11. Cortical cooling. (**A**) Top: schematic depiction of the cooling probe design. Bottom: Dimensions of the cooling probe. (**B**) Effect of cortical cooling on averaged evoked potential latency. ECoG, electrocorticography. (Replotted from Bakken et al., 2003.)

in Figure 13.11A. Here, cooling is achieved by circulation of cooled saline through a chamber that is in direct contact with the cortical surface. The cooling probe is equipped with recording contacts for electrophysiologic data collection as well as thermocouples to simultaneously measure temperature in the cooling chamber and on the cortical surface. ECoG data recorded from the probe demonstrate the reversible effect on cooling on the physiologic response properties of the underlying brain tissue (Fig. 13.11B). Advantages of cortical cooling over electric stimulation mapping include better-understood mechanisms of functional disruption, highly localized lesioning effect, and no risk of evoking seizures. Currently, the utility of focal cortical cooling is limited to acute experiments carried out in the operating room.

CAVEATS OF THE METHOD AND VALIDITY OF INVASIVE RECORDINGS

The overarching goal of invasive studies is to provide insights into the functional organization of normal human auditory cortex. The method has several limitations, which should be borne in mind when interpreting the results of invasive studies and comparing them with non-invasive human studies such as EEG, MEG, or fMRI. These caveats apply both to acute and chronic experiments and include the following:

1. Invasive recordings are only performed in patients with brain pathologies. Consequently, results may generalize to the healthy population with reservations. Surgical epilepsy patients who are eligible to become research subjects are only those diagnosed with a focal seizure disorder. The seizure focus most commonly is localized to the mesial temporal structures, away from auditory and auditory-related cortical areas. There is, however, evidence that mesial temporal epilepsy can lead to progressive atrophic changes of neocortical temporal, frontal, and parietal areas, including auditory and auditory-related cortex (Bonilha et al., 2006; Bernhardt et al., 2009, 2010). Thus, one cannot rule out the possibility that pathologic processes associated with seizure disorders influence activity recording from distant cortical sites. If seizure focus is localized to auditory or auditory-related cortex, or epileptic discharges from a distant seizure focus are detected in the data from the cortical areas under investigation, such data are excluded from research analyses.
2. The number of subjects available for research studies is much more limited compared to non-invasive studies. The subject population is relatively heterogeneous in terms of age and cognitive level. This can add variability to the research results.
3. Sampling of cortical tissue is determined by the clinical indications for monitoring and is limited by the craniotomy and implanted electrode coverage for acute and chronic experiments, respectively.
4. The use of multiple AEDs can affect the functional properties of tissue from which recordings are made. In chronic experiments, this can be controlled for by performing recordings when the patients are tapered off medication during the monitoring period. Additionally, recording from multiple subjects who receive different medication regimens can help diminish the possible effects of specific medications on the obtained electrophysiologic data.
5. Time available for experimental sessions is constrained by the clinical setup and the subjects' fatigue.

Despite these caveats, invasive recordings in human subjects provide the most direct way to measure cortical activity and, as such, essentially constitute the "gold standard" in studying neural activity within human auditory cortex. Notably, responses recorded directly from the core auditory cortex in humans have a strong resemblance to those obtained from core cortex in non-human primates (Fishman et al., 2001; Steinschneider et al., 2005, 2013; Ray et al., 2008; Brugge et al., 2009). This can be interpreted as an indication that normal functional organization and response properties at this level of auditory cortical processing are largely preserved in the studied population of epilepsy patients.

CONCLUSION

Invasive recordings from patients undergoing neurosurgical evaluation and treatment generate valuable information about the organization and function of the human auditory cortex. Knowledge gained using intracranial research methods helps bridge the gap between non-invasive imaging and neurophysiologic data obtained in experimental animal studies. Future technical advances in recording array design, miniaturization of electronic circuits, and development of wireless technologies for data transmission will enhance the patient's comfort and safety and open new research opportunities.

ACKNOWLEDGMENTS

Preparation of this chapter was supported by NIH RO1-DC004290, UL1RR024979, Hearing Health Foundation, and by the Hoover Fund. The authors would like to acknowledge Drs. Phillip Gander and Ariane Rhone for helpful comments on the manuscript.

References

Aitkin LM, Webster WR (1971). Tonotopic organization in the medial geniculate body of the cat. Brain Res 26: 402–405.

Bakken HE, Kawasaki H, Oya H et al. (2003). A device for cooling localized regions of human cerebral cortex. J Neurosurg 99: 604–608.

Barton B, Venezia JH, Saberi K et al. (2012). Orthogonal acoustic dimensions define auditory field maps in human cortex. Proc Natl Acad Sci U S A 109: 20738–20743.

Baumann S, Petkov CI, Griffiths TD (2013). A unified framework for the organization of the primate auditory cortex. Front Syst Neurosci 7: 11. http://dx.doi.org/10.3389/fnsys.2013.00011.

Bernhardt BC, Worsley KJ, Kim H et al. (2009). Longitudinal and cross-sectional analysis of atrophy in pharmacoresistant temporal lobe epilepsy. Neurology 72: 1747–1754.

Bernhardt BC, Bernasconi N, Concha L et al. (2010). Cortical thickness analysis in temporal lobe epilepsy: reproducibility and relation to outcome. Neurology 74: 1776–1784.

Bitterman Y, Mukamel R, Malach R et al. (2008). Ultra-fine frequency tuning revealed in single neurons of human auditory cortex. Nature 451: 197–201.

Boatman D (2004). Cortical bases of speech perception: evidence from functional lesion studies. Cognition 92: 47–65.

Boatman D, Lesser RP, Gordon B (1995). Auditory speech processing in the left temporal lobe: an electrical interference study. Brain Lang 51: 269–290.

Bonilha L, Rorden C, Appenzeller S et al. (2006). Gray matter atrophy associated with duration of temporal lobe epilepsy. Neuroimage 32: 1070–1079.

Brugge JF, Volkov IO, Garell PC et al. (2003). Functional connections between auditory cortex on Heschl's gyrus and on the lateral superior temporal gyrus in humans. J Neurophysiol 90: 3750–3763.

Brugge JF, Volkov IO, Reale RA et al. (2005). The posteriolateral superior temporal auditory field in humans. Functional organization and connectivity. In: R Konig, P Heil, E Budinger et al. (Eds.), The auditory cortex — toward a synthesis of human and animal research. Erlbaum, Mahwah, pp. 145–162.

Brugge JF, Volkov IO, Oya H et al. (2008). Functional localization of auditory cortical fields of human: click-train stimulation. Hear Res 238: 12–24.

Brugge JF, Nourski KV, Oya H et al. (2009). Coding of repetitive transients by auditory cortex on Heschl's gyrus. J Neurophysiol 102: 2358–2374.

Buszaki G (2006). Rhythms of the Brain. Oxford University Press, New York.

Celesia GG (1976). Organization of auditory cortical areas in man. Brain 99: 403–414.

Celesia GG, Puletti F (1969). Auditory cortical areas of man. Neurology 19: 211–220.

Chang EF, Rieger JW, Johnson K et al. (2010). Categorical speech representation in human superior temporal gyrus. Nat Neurosci 13: 1428–1432.

Chiry O, Tardif E, Magistretti PJ et al. (2003). Patterns of calcium-binding proteins support parallel and hierarchical organization of human auditory areas. Eur J Neurosci 17: 397–410.

Creutzfeldt O, Ojemann G (1989). Neuronal activity in the human lateral temporal lobe. III. Activity changes during music. Exp Brain Res 77: 490–498.

Creutzfeldt O, Ojemann G, Lettich E (1989a). Neuronal activity in the human lateral temporal lobe. I. Responses to speech. Exp Brain Res 77: 451–475.

Creutzfeldt O, Ojemann G, Lettich E (1989b). Neuronal activity in the human lateral temporal lobe. II. Responses to the subjects own voice. Exp Brain Res 77: 476–489.

Crone NE, Boatman D, Gordon B et al. (2001). Induced electrocorticographic gamma activity during auditory perception. Clin Neurophysiol 112: 565–582.

Crone NE, Sinai A, Korzeniewska A (2006). High-frequency gamma oscillations and human brain mapping with electrocorticography. Prog Brain Res 159: 275–295.

De Ridder D, De Mulder G, Walsh V et al. (2004). Magnetic and electrical stimulation of the auditory cortex for intractable tinnitus. Case report. J Neurosurg 100: 560–564.

Destrieux C, Fischl B, Dale A et al. (2010). Automatic parcellation of human cortical gyri and sulci using standard anatomical nomenclature. Neuroimage 53: 1–15.

Edwards E, Soltani M, Kim W et al. (2009). Comparison of time–frequency responses and the event-related potential to auditory speech stimuli in human cortex. J Neurophysiol 102: 377–386.

Engel AK, König P, Kreiter AK et al. (1991). Interhemispheric synchronization of oscillatory neuronal responses in cat visual cortex. Science 252: 1177–1179.

Engel AK, Moll CK, Fried I et al. (2005). Invasive recordings from the human brain: Clinical insights and beyond. Nat Rev Neurosci 6: 35–47.

Fenoy AJ, Severson MA, Volkov IO et al. (2006). Hearing suppression induced by electrical stimulation of human auditory cortex. Brain Res 1118: 75–83.

Fishman YI, Volkov IO, Noh MD et al. (2001). Consonance and dissonance of musical chords: Neural correlates in auditory cortex of monkeys and humans. J Neurophysiol 86: 2761–2788.

Flinker A, Chang EF, Barbaro NM et al. (2011). Sub-centimeter language organization in the human temporal lobe. Brain Lang 117: 103–109.

Freeman WJ (1978). Spatial properties of an EEG event in the olfactory bulb and cortex. Electroencephalogr Clin Neurophysiol 44: 586–605.

Fried I, Wilson CL, Maidment NT et al. (1999). Cerebral microdialysis combined with single-neuron and electroencephalographic recording in neurosurgical patients. Technical note. J Neurosurg 91: 697–705.

Garell PC, Bakken H, Greenlee JD et al. (2013). Functional connection between posterior superior temporal gyrus and ventrolateral prefrontal cortex in human. Cereb Cortex 10: 2309–2321.

Gloor P, Olivier A, Quesney LF et al. (1982). The role of the limbic system in experiential phenomena of temporal lobe epilepsy. Ann Neurol 12: 129–144.

Gourevitch B, Le Bouquin Jeannes R, Faucon G et al. (2008). Temporal envelope processing in the human auditory cortex: response and interconnections of auditory cortical areas. Hear Res 237: 1–18.

Gray CM, König P, Engel AK et al. (1989). Oscillatory responses in cat visual cortex exhibit inter-columnar synchronization which reflects global stimulus properties. Nature 338: 334–337.

Greenlee JD, Oya H, Kawasaki H et al. (2004). A functional connection between inferior frontal gyrus and orofacial motor cortex in human. J Neurophysiol 92: 1153–1164.

Greenlee JD, Oya H, Kawasaki H et al. (2007). Functional connections within the human inferior frontal gyrus. J Comp Neurol 503: 550–559.

Hackett TA (2003). The comparative anatomy of the primate auditory cortex. In: AA Ghazanfar (Ed.), Primate audition: Ethology and neurobiology. CRC Press, Boca Raton, pp. 199–219.

Hackett TA (2007). Organization and correspondence of the auditory cortex of humans and nonhuman primates. In: JH Kaas (Ed.), Evolution of the nervous system. Elsevier, Oxford, pp. 109–119.

Hackett TA (2008). Anatomical organization of the auditory cortex. J Am Acad Audiol 19: 774–779.

Hackett TA, Stepniewska I, Kaas JH (1998). Subdivisions of auditory cortex and ipsilateral cortical connections of the parabelt auditory cortex in macaque monkeys. J Comp Neurol 394: 475–495.

Hackett TA, Stepniewska I, Kaas JH (1999). Prefrontal connections of the parabelt auditory cortex in macaque monkeys. Brain Res 817: 45–58.

Hackett TA, Preuss TM, Kaas JH (2001). Architectonic identification of the core region in auditory cortex of macaques, chimpanzees, and humans. J Comp Neurol 441: 197–222.

Hill NJ, Gupta D, Brunner P et al. (2012). Recording human electrocorticographic (ECoG) signals for neuroscientific research and real-time functional cortical mapping. J Vis Exp 64: e3993.

Howard MA, Volkov IO, Abbas PJ et al. (1996a). A chronic microelectrode investigation of the tonotopic organization of human auditory cortex. Brain Res 724: 260–264.

Howard MA, Volkov IO, Granner MA et al. (1996b). A hybrid clinical-research depth electrode for acute and chronic in vivo microelectrode recording of human brain neurons. Technical note. J Neurosurg 84: 129–132.

Howard MA, Volkov IO, Mirsky R et al. (2000). Auditory cortex on the posterior superior temporal gyrus of human cerebral cortex. J Comp Neurol 416: 76–92.

Kaas JH, Hackett TA (2000). Subdivisions of auditory cortex and processing streams in primates. Proc Natl Acad Sci U S A 97: 11793–11799.

Kaas JH, Hackett TA (2005). Subdivisions and connections of auditory cortex in primates: A working model. In: R Konig, P Heil, E Budinger et al. (Eds.), Auditory cortex. A synthesis of human and animal research. Erlbaum, Mahwah, pp. 7–25.

Kalcher J, Pfurtscheller G (1995). Discrimination between phase-locked and non-phase-locked event-related EEG activity. Electroencephalogr Clin Neurophysiol 94: 381–384.

Kellis S, Miller K, Thomson K et al. (2010). Decoding spoken words using local field potentials recorded from the cortical surface. J Neural Eng 7: 056007.

Kubanek J, Brunner P, Gunduz A et al. (2013). The tracking of speech envelope in the human cortex. PLoS One 8: e53398.

Kumar S, Sedley W, Nourski KV et al. (2011). Predictive coding and pitch processing in the auditory cortex. J Cogn Neurosci 23: 3084–3094.

Kwan P, Sperling MR (2009). Refractory seizures: try additional antiepileptic drugs (after two have failed) or go directly to early surgery evaluation? Epilepsia 50 (Suppl 8): 57–62.

Leonard CM, Puranik C, Kuldau JM et al. (1998). Normal variation in the frequency and location of human auditory cortex landmarks. Heschl's gyrus: where is it? Cereb Cortex 8: 397–406.

Liégeois-Chauvel C, Musolino A, Chauvel P (1991). Localization of the primary auditory area in man. Brain 114: 139–151.

Liégeois-Chauvel C, Musolino A, Badier JM et al. (1994). Evoked potentials recorded from the auditory cortex in man: Evaluation and topography of the middle latency components. Electroencephalogr Clin Neurophysiol 92: 204–214.

Liégeois-Chauvel C, de Graaf JB, Laguitton V et al. (1999). Specialization of left auditory cortex for speech perception in man depends on temporal coding. Cereb Cortex 9: 484–496.

Liégeois-Chauvel C, Lorenzi C, Trebuchon A et al. (2004). Temporal envelope processing in the human left and right auditory cortices. Cereb Cortex 14: 731–740.

Lu T, Liang L, Wang X (2001). Temporal and rate representations of time-varying signals in the auditory cortex of awake primates. Nat Neurosci 4: 1131–1138.

Matsumoto R, Nair DR, LaPresto E et al. (2004). Functional connectivity in the human language system: a cortico-cortical evoked potential study. Brain 127: 2316–2330.

Matsumoto R, Nair DR, LaPresto E et al. (2007). Functional connectivity in human cortical motor system: a cortico-cortical evoked potential study. Brain 130: 181–197.

Merzenich MM, Brugge JF (1973). Representation of the cochlear partition of the superior temporal plane of the macaque monkey. Brain Res 50: 275–296.

Miller KJ, Makeig S, Hebb AO et al. (2007). Cortical electrode localization from X-rays and simple mapping for electrocorticographic research: The "Location on Cortex" (LOC) package for MATLAB. J Neurosci Methods 162: 303–308.

Moore BCJ (2012). An Introduction to the Psychology of Hearing. 6th Edn. Emerald, Bingley.

Mukamel R, Gelbard H, Arieli A et al. (2005). Coupling between neuronal firing, field potentials, and FMRI in human auditory cortex. Science 309: 951–954.

Neligan A, Bell GS, Johnson AL et al. (2011). The long-term risk of premature mortality in people with epilepsy. Brain 134: 388–395.

Nourski KV, Brugge JF (2011). Representation of temporal sound features in the human auditory cortex. Rev Neurosci 22: 187–203.

Nourski KV, Reale RA, Oya H et al. (2009). Temporal envelope of time-compressed speech represented in the human auditory cortex. J Neurosci 29: 15564–15574.

Nourski KV, Brugge JF, Reale RA et al. (2013a). Coding of repetitive transients by auditory cortex on posterolateral superior temporal gyrus in humans: an intracranial electrophysiology study. J Neurophysiol 109: 1283–1295.

Nourski KV, Etler CP, Brugge JF et al. (2013b). Direct recordings from the auditory cortex in a cochlear implant user. J Assoc Res Otolaryngol 14: 435–450.

Nourski KV, Steinschneider M, Oya H et al. (2014). Spectral organization of the human lateral superior temporal gyrus revealed by intracranial recordings. Cereb Cortex 24: 340–352.

Oya H, Poon PW, Brugge JF et al. (2007). Functional connections between auditory cortical fields in humans revealed by Granger causality analysis of intra-cranial evoked potentials to sounds: comparison of two methods. Biosystems 89: 198–207.

Pantev C (1995). Evoked and induced gamma-band activity of the human cortex. Brain Topogr 7: 321–330.

Pasley BN, David SV, Mesgarani N et al. (2012). Reconstructing speech from human auditory cortex. PLoS Biol 10: e1001251.

Penfield W, Perot P (1963). The brain's record of auditory and visual experience—a final summary and discussion. Brain 86: 595–696.

Penfield W, Rasmussen T (1950). The cerebral cortex of man—A clinical study of localization of function. Macmillan, New York.

Quiroga RQ, Nadasdy Z, Ben-Shaul Y (2004). Unsupervised spike detection and sorting with wavelets and superparamagnetic clustering. Neural Comput 16: 1661–1687.

Rademacher J, Caviness V, Steinmetz H et al. (1993). Topographical variation of the human primary cortices; implications for neuroimaging, brain mapping and neurobiology. Cereb Cortex 3: 313–329.

Rauschecker JP, Scott SK (2009). Maps and streams in the auditory cortex: Nonhuman primates illuminate human speech processing. Nat Neurosci 12: 718–724.

Ray S, Niebur E, Hsiao SS et al. (2008). High-frequency gamma activity (80–150 Hz) is increased in human cortex during selective attention. Clin Neurophysiol 119: 116–133.

Reddy CG, Dahdaleh NS, Albert G et al. (2010). A method for placing Heschl gyrus depth electrodes. J Neurosurg 112: 1301–1307.

Rivier F, Clarke S (1997). Cytochrome oxidase, acetylcholinesterase, and NADPH-diaphorase staining in human supratemporal and insular cortex: evidence for multiple auditory areas. Neuroimage 6: 288–304.

Romanski LM, Goldman-Rakic PS (2002). An auditory domain in primate prefrontal cortex. Nat Neurosci 5: 15–16.

Romanski LM, Tian B, Fritz J et al. (1999). Dual streams of auditory afferents target multiple domains in the primate prefrontal cortex. Nat Neurosci 2: 1131–1136.

Rose JE, Greenwood DD, Goldberg JM et al. (1963). Some discharge characteristics of single neurons in the inferior colliculus of the cat I: tonotopic organization, relation of spike-counts to tone intensity and firing patterns of single elements. J Neurophysiol 26: 294–320.

Ross ED, Mesulam MM (1979). Dominant language functions of the right hemisphere? Prosody and emotional gesturing. Arch Neurol 36: 144–148.

Sinai A, Bowers CW, Crainiceanu CM et al. (2005). Electrocorticographic high gamma activity versus electrical cortical stimulation mapping of naming. Brain 128: 1556–1570.

Sinha SR, Crone NE, Fotta R et al. (2005). Transient unilateral hearing loss induced by electrocortical stimulation. Neurology 64: 383–385.

Steinschneider M, Schroeder CE, Arezzo JC et al. (1993). Temporal encoding of phonetic features in auditory cortex. Ann N Y Acad Sci 682: 415–417.

Steinschneider M, Schroeder CE, Arezzo JC et al. (1995). Physiologic correlates of the voice onset time boundary in primary auditory cortex (A1) of the awake monkey: temporal response patterns. Brain Lang 48: 326–340.

Steinschneider M, Reser DH, Fishman YI et al. (1998). Click train encoding in primary auditory cortex of the awake monkey: evidence for two mechanisms subserving pitch perception. J Acoust Soc Am 104: 2935–2955.

Steinschneider M, Volkov IO, Noh MD et al. (1999). Temporal encoding of the voice onset time phonetic parameter by field potentials recorded directly from human auditory cortex. J Neurophysiol 82: 2346–2357.

Steinschneider M, Volkov IO, Fishman YI et al. (2005). Intracortical responses in human and monkey primary auditory cortex support a temporal processing mechanism for encoding of the voice onset time phonetic parameter. Cereb Cortex 15: 170–186.

Steinschneider M, Fishman YI, Arezzo JC (2008). Spectrotemporal analysis of evoked and induced electroencephalographic responses in primary auditory cortex (A1) of the awake monkey. Cereb Cortex 18: 610–625.

Steinschneider M, Nourski KV, Fishman Y (2013). Representation of speech in human auditory cortex: Is it special? Hear Res 305: 57–73.

Taylor KI, Regard M (2003). Language in the right cerebral hemisphere: contributions from reading studies. News Physiol Sci 18: 257–261.

Wallace MN, Johnston PW, Palmer AR (2002). Histochemical identification of cortical areas in the auditory region of the human brain. Exp Brain Res 143: 499–508.

World Health Organization (2012). Epilepsy. Fact sheet N°999, October 2012. Retrieved from http://www.who.int/mediacentre/factsheets/fs999

Zhao L, Zhaoping L (2011). Understanding auditory spectrotemporal receptive fields and their changes with input statistics by efficient coding principles. PLoS Comput Biol 7: e1002123.

Zilles K, Schleicher A, Langemann C et al. (1997). Quantitative analysis of sulci in the human cerebral cortex: development, regional heterogeneity, gender difference, asymmetry, intersubject variability and cortical architecture. Hum Brain Mapp 5: 218–221.

Handbook of Clinical Neurology, Vol. 129 (3rd series)
The Human Auditory System
G.G. Celesia and G. Hickok, Editors

Chapter 14

Electromagnetic recording of the auditory system

DAVID POEPPEL[1]* AND GREGORY HICKOK[2]

[1]*Department of Psychology, New York University, New York, NY, USA*

[2]*Department of Cognitive Sciences, Center for Language Science and Center for Cognitive Neuroscience, University of California, Irvine, CA, USA*

INTRODUCTION

The human auditory system must accomplish a number of seemingly unrelated computational tasks. A high-level classification that captures many of the operations executed by the auditory system includes: (1) localization of objects in the world and of oneself *vis-à-vis* the surrounding environment; (2) identification of sounds; and (3) transformation between sounds and actions. Each of these broad types of task includes internally structured and complex parts. Localization of sounds and orienting towards targets, for example, relies on subcortical structures (e.g., parts of the superior olive) as well as on cortical regions, with different functional contributions made by the anatomically and physiologically distinct regions. Similarly, the identification of sounds requires a range of operations, from calculating the acoustic attributes of a sound input to segregating such input into streams or objects to comparing the analyzed sound to stored templates (say, speech versus non-speech). In addition, being able to perform a transformation of an input sound – say a word, or a yawn, or a clap – into an action (something critical to learning and at the basis of all imitation) requires the listener to be able to translate from an auditory "format" into a motor format; in other words the manipulation of both acoustic and motoric representations in the brain. In short, what appears to us, introspectively, as automatic and seamless in the context of auditory perception is in fact constituted of a complex set of operations that are executed in different brain regions, over different timescales, and orchestrated to yield what is typically an extraordinarily sensitive, rapid, and unified perceptual experience. The speed of the underlying operations makes a high temporal resolution technique desirable in the study of audition.

Commensurate with the complexity of the tasks, the auditory system has a long "parts list." (For an extensive characterization of the structural and functional organization of human auditory cortex, see Chapters 1 and 2.) Subsequent to processing in the outer, middle, and inner ear, the different structures of the brainstem (cochlear nucleus, olive, nuclei of the lateral lemniscus, inferior colliculus (IC)) provide a sophisticated spectral and temporal analysis of input sound. One subcortical structure that has received a lot of attention in recent years in terms of its crucial contribution to hearing is the IC. In the often-made comparisons between the visual and auditory systems, there is now emerging consensus that the IC might best be seen as a structure equivalent to the primary visual cortex. That is to say, the type of representation that is constructed at this point in the auditory pathway can be interpreted as somewhat equivalent (in functional terms) to the type of visual representation generated in primary visual cortex. In the cortex, where a heavily preprocessed neural signal arrives, there exist upward of a dozen fields that process auditory input signals (Kaas and Hackett, 2000), possibly organized as orthogonal frequency (tonotopic) and time-based (periodicity) maps (Barton et al., 2012). The different cortical fields reflect segregation of function, and occasionally lateralization of function. In the context of investigating the functional architecture of the auditory system, it is essential to bear one property of the system in mind: almost without exception, the forward projections from the periphery to more central structures are balanced by feedback projections from higher-order to earlier regions (e.g., Chandrasekaran and Kraus, 2010). The surprisingly large extent to which auditory processing is conditioned by top-down mechanisms (not merely bottom-up, feedforward signal processing) is now well

*Correspondence to: David Poeppel, Department of Psychology, New York University, 6 Washington Place, New York, NY 10003, USA. E-mail: david.poeppel@nyu.edu

established (see, e.g., Bendixen et al., 2012, for a review of predictive coding in hearing; Alain and Winkler (2012) address attention, prediction, and scene analysis; Zion-Golumbic et al. (2013a, b) provide data illustrating attentional top-down control in the cocktail-party context). To begin to disentangle the feedforward and feedback operations, a recording approach with high temporal resolving power is advantageous.

Most of the cortical fields responsible for the analysis of sounds, the regions most suitable for non-invasive electrophysiologic recording, are situated in the superior temporal lobe. Heschl's gyrus, as well as the superior temporal gyrus and the planum temporale , and more ventrally the superior temporal sulcus contain the critical cortical regions. Two approaches to talking about the anatomy are dominating the literature, as discussed by Clarke and Morosan (2012). One approach builds on animal studies and distinguishes core, belt, and parabelt regions (see Chapter 2 and Fig. 2.2).

ELECTROPHYSIOLOGIC RECORDING

To adequately characterize the anatomy of the auditory system, a technique with excellent spatial resolution is required. Magnetic resonance imaging (MRI) has been the approach of choice in recent years, and we now have remarkably high-resolution anatomic images of the human auditory pathway. To investigate online neurophysiologic processing in the auditory system, recording approaches with suitable temporal resolution are necessary. Auditory perception is remarkably fast – with cortically mediated effects, such as gap detection, demonstrable in the range of a few milliseconds (sound localization is based on submillisecond processing in the brainstem) – and electrophysiologic techniques with appropriately high sampling rates are the necessary tools. In the context of studying human hearing, speech perception, and language processing, three such techniques are being used extensively: (1) electroencephalography (EEG), the oldest technology, still widely used; (2) magnetoencephalography (MEG), a newer recording technique conceptually related to EEG and very well suited to study audition (discussed in this chapter); and (3) electrocorticography (ECog), an invasive approach used only in a clinical context (presurgical evaluation of epilepsy), and recently applied increasingly in patients to the study of hearing and speech (see Chapters 13 and 17). Here we focus on the non-invasive measurements using MEG, because this recording method is especially well suited to measure a range of responses, such as stimulus evoked signals, from human cortical auditory regions with high temporal and spatial resolving power. In this chapter, we will address MEG recordings made in the context of listeners being exposed to auditory signals. The magnetoencephalogram recorded without stimuli (by analogy to the electroencephalogram) will not be discussed here. The focus is on evoked or induced responses generated by stimuli.

The principal sources of the electric and magnetic fields recorded by EEG and MEG are postsynaptic currents in cortical pyramidal cells. These primary currents cause volume currents in the surrounding conductive medium. Because pyramidal cells are laid out in a parallel geometry, when a sufficiently large number are active (10 000–100 000; for reference, approximately the number of neurons in a cortical column), the ensuing fields can be measured extracranially. MEG and EEG provide somewhat complementary information about neuronal activity because of their differing sensitivity to current sources. MEG is primarily sensitive to tangential currents closer to the recording surface (scalp); EEG is primarily sensitive to radial sources but also very sensitive to the conductive properties of brain tissue, the skull, the meninges, and the scalp. Since currents produced by neurons also generate magnetic fields, which are not distorted by the heterogeneous environment, measurements of magnetic fields can be considered to give rise to a more undistorted reflection of the underlying neural activity. In sum, MEG and EEG can be viewed as being in some sense complementary in terms of the sensitivity to underlying neural signal. Both methods have millisecond temporal resolution. Below, we focus in more detail on MEG and its contributions to the neural bases of hearing, speech, and language.

MAGNETOENCEPHALOGRAPHY

Neurophysiologic recording using MEG presents first and foremost a measurement problem. The responses from the brain are minute relative to the size of environmental electromagnetic noise, the earth's magnetic field, and other body-internal signals, for example cardiac activity. To meet this challenge, MEG uses specialized detectors, so-called superconducting quantum interference devices (SQUIDs). Each detector, or channel, acts as a high-gain low-noise amplifier of the magnetic field elicited by neuronal activity. Typical modern systems have 100–300 channels, using either magnetometer (single coils) or gradiometer (paired coils) configurations. These are embedded in "helmets" that cover as much of the skull/brain as possible as subjects are positioned either sitting or supine with their heads in the recording helmet (Fig. 14.1A). The recording systems are situated in magnetically shielded rooms. These rooms serve to maintain an electromagnetically quiet recording environment.

Each MEG channel outputs a typical electrophysiologic time-varying waveform signal (Fig. 14.1B).

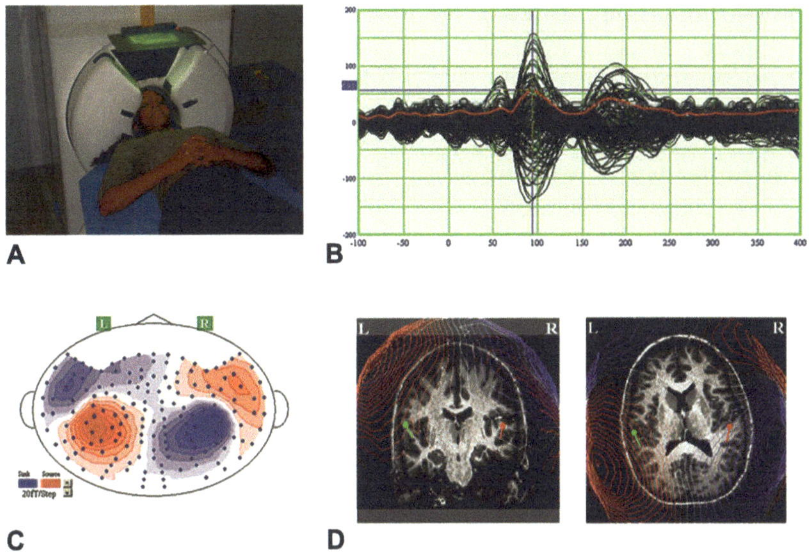

Fig. 14.1. (**A**) Magnetoencephalography (MEG) using whole-head biomagnetometers is performed inside a magnetically shielded room, with the detectors placed in a helmet-shaped device. Patients are either sitting or supine. (**B**) Each detector outputs the time-varying magnetic field at the recording surface (scalp). The presentation of a single tone, for example, generates a series of responses with a peak at ~100 ms poststimulus onset, the N100m. The data are overlayed to illustrate the peak response. Responses from the brain are in the range of 10–15 to 10–13 fT (femto-tesla), six or more orders of magnitude smaller than the surrounding magnetic fields. (**C**) The contour plot depicts the direction and amplitude of the response at the N100m peak. Each black dot represents one recording channel. The color scheme reflects whether the fields (and as a consequence, the intracranial currents) are pointing inwards or outwards, from which one can calculate the source locations. These surface contour plots can be analyzed, although typically the data are used to perform source analysis. (**D**) The coronal (left) and axial (right) structural magnetic resonance imaging slices can be coregistered with MEG data (activity visualized as contour plots outside the head, as in **C**). Simple source models of these fields, such as equivalent current dipoles (lollipop shapes in left and right auditory cortices) reflect the local activity at a time point. These sources can be calculated over time, e.g., millisecond by millisecond, and provide a dynamic view of the neural currents elicited in the task at hand.

Sampling rates of 1000 Hz and above and suitable recording bandwidths (e.g., DC to 500 Hz) permit the recording of high temporal and high-frequency resolution signals. MEG recordings thus provide data with high temporal (<1 ms), spectral (<1 Hz), and spatial (<1 cm) resolution. For a thorough and detailed book-length discussion of MEG, see Hansen et al. (2010). Discussions of auditory MEG are presented in Brignell et al. (2008), Lütkenhöner and Poeppel (2011), and Nagarajan et al. (2012).

The analysis of MEG data can be conceptualized as proceeding on two fronts. One approach deals with the sensor data directly. The neuromagnetic signals may be averaged by trial type (as is typical in EEG, when event-related potentials (ERPs) are recorded), or the data from single trials are analyzed. The latter situation is obviously more challenging from a signal-processing perspective, as the signal-to-noise ratio is highly unfavorable. Most auditory experiments analyze averaged data, generated as subjects listen to tens to hundreds of repetitions of a trial type. Such data are illustrated in Figure 14.1B, where auditory data elicited by a pure tone, from 160 recorded channels (their positions are illustrated in Figure 14.1C as black dots), are averaged and filtered (hence the smooth waveforms), and overlayed in order to show the peaks elicited by a tone (e.g., peak at 100 ms poststimulus onset). Sensor data might be analyzed by looking at peaks of responses (e.g., amplitudes, latencies), the spectral structure of the response (e.g., the relative contributions of delta, theta, alpha, beta, gamma spectral regions), or any conceivable manner in which one might analyze a time series.

A second approach to analysis deals with source space. That approach takes the sensor data recorded at the surface and aims to reconstruct the underlying neural activity that generated the recorded data. The patterns of neuromagnetic activity seen at the surface (as illustrated in Fig. 14.1C) are subjected to models that infer the neural origin, with the ultimate goal being the localizing of the neural activity in an individual subject's brain (Fig. 14.1D). The peak response shown in Figure 14.1B at 100 ms latency has spatial pattern (contour plot) depicted in Figure 14.1C. When these data are subjected to source modeling (in this case, the most simple, equivalent current dipole (ECD) modeling), the activation is localized to auditory cortex in the upper bank of the superior temporal gyrus, as in Figure 14.1D. Source modeling of neuromagnetic data is a demanding computational task because the problem is ill posed, i.e., there is no constraint-free unique solution. Dipole models, distributed current source models, beam forming, and other tomographic techniques are widely explored in solving this inverse problem. The various approaches to analyzing data in source space are ably summarized and discussed in Hansen et al. (2010) and Nagarajan et al. (2012).

AUDITORY RESEARCH USING MEG

MEG is a popular choice for auditory studies because of the fortuitous placement of many auditory areas in the lateral fissure. The geometric alignment of these parts of cortex make it very accessible to MEG-based recording. Since the early 1980s, MEG has, as a consequence, been used regularly to investigate auditory processing, speech perception, and language comprehension. In the domain of speech and language, typical studies have presented single vowels (e.g., Obleser et al., 2004; Scharinger et al., 2011) and syllables (e.g., Poeppel et al., 1996; Bonte et al., 2006; Tavabi et al., 2007), spoken words (e.g., Marinkovic et al., 2003; Uusvuori et al., 2008), and sentences (e.g., Ahissar et al., 2001; Luo and Poeppel, 2007). In the study of auditory perception and cognition, more generally, the stimuli range from tones and tone complexes (e.g., Roberts and Poeppel, 1996; Lütkenhöner et al., 2006) to tone sequences (e.g., Gutschalk et al., 2005; Xiang et al., 2010) to chords and musical phrases (e.g., Bergelson and Idsardi, 2009; Brattico et al., 2009; Boh et al., 2011).

In order to illustrate some current auditory MEG research in brief, we use the neuromagnetic response type as an organizing principle. Short examples are provided that demonstrate the utility of different response profiles in investigating auditory processing: (1) transient responses (such as the P50m, the M100, the N400m); (2) transition responses (change detection, pitch onset response (POR), mismatch negativity (MMN)) (3) auditory steady-state responses (aSSR); and (4) oscillations (evoked, induced, tracking).

Transient evoked responses: N100m (M100), P2m, M350, N400m (M400)

The energy onset associated with the beginning of a sound – whether it is a tone, chord, vowel, or sentence – elicits a cascade of transient responses. At the most elementary level, for an acoustic signal that is suprathreshold and has some minimum duration, there is an on response (e.g., Gage et al., 1998), a sustained field (Pantev et al., 1994), and an off response (e.g., Noda et al., 1998). In the context of auditory and language studies, the neural responses that have received the most attention are the large, late evoked fields: P50m, N100m/M100, P200m, M350, N400m. These transient evoked fields are the bread and butter of typical auditory MEG studies. They are quite robust, although the particular amplitude and latency of each peak are modulated very strongly by stimulus properties, the attentional requirements of the experiment, and various task demands. These responses are, of course, well known from the extensive literature on auditory ERPs, but the MEG-derived event-related fields have the added benefit that their generators can be, to a reasonable degree, localized. For example, the M100, associated with early cortical analyses of stimuli, has been argued to be generated on the planum temporale immediately adjacent to Heschl's gyrus (the medial portion of which is putative core (or primary) auditory cortex (see Lütkenhöner and Poeppel, 2011, for discussion). The N400m, in contrast, is arguably generated in the posterior aspect of the middle temporal gyrus (among, possibly, some additional sources) (Lau et al., 2008).

Two responses that are used especially often in auditory research and language research, respectively, are the M100 (or N100m) and the M400 (or N400m). Figure 14.2 shows the cascade of evoked fields elicited by simple click stimuli (Howard and Poeppel, 2009). The data illustrate the relation of the prominent M100 (Fig. 14.2A, B) to the preceding (mid-latency auditory evoked) and following peaks. It is noteworthy how precisely aligned the responses are across two independent experiments (Fig. 14.2B). Figure 14.2C demonstrates that magnetic contour pattern – the aspect of the data that forms the basis for source space analysis – is similarly consistent. Because of this high temporal and spatial reliability of this neural response (even in single trials; see Salajegheh et al., 2004), and its rather dipolar nature, it has been used to test many hypotheses in hearing and speech. For example, the precise source localization of the M100 elicited by different vowels has been employed to examine whether vowels are represented in a topographically systematic manner in auditory cortex (e.g., Obleser et al., 2004; Scharinger et al., 2011). Because vowels have a rather well-defined acoustic structure (the formant frequency structure) but also have clear linguistic (articulatory) attributes (e.g., tongue position), one can test which type of topographic map is reflected by M100 activation. Comprehensive model comparisons of the dipole locations on the superior temporal cortex show that the topographic representation in auditory cortex is not just based on acoustic bottom-up information but critically involves linguistic-featural top-down modulation (Scharinger et al., 2011). These types of data show how frequency-based excitation

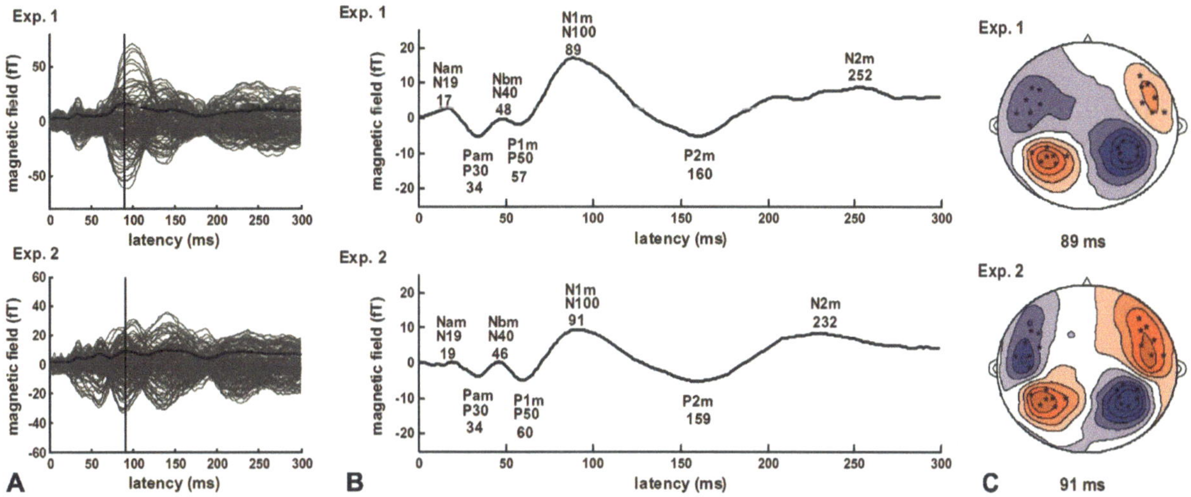

Fig. 14.2. Whole-head response to simple click stimuli to elicit auditory evoked fields. (**A**) Evoked response for experiment 1 (upper) and 2 (lower). Response shown for all 157 recorded channels (gray), the root-mean-square (RMS) waveform averaged across the 157 channels (solid black), and the N1m peak in the RMS waveform (vertical black). (**B**) Evoked response waveforms (average across all 157 channels). The major components are labeled and the latencies of component peaks (ms) are shown below labels. (**C**) Contour map of the evoked magnetic field at the N1m peak latency. Sources are shown in red, sinks in blue, and the 28 channels (seven in each quadrant) with the strongest response are indicated by black asterisks. This pattern is the canonic topography for auditory evoked fields at the M100 peak and reflects the separate (left versus right) underlying dipoles generating this field pattern. (Reproduced with permission from Howard and Poeppel, 2009.)

along the unidimensional basilar membrane of the cochlea is mapped into a higher-dimensional representation, implicating linguistic-articulatory features in auditory cortex by 100 ms.

The M400/N400m response (and the closely related M350 response implicated in many studies of morphologic processing) has played a seminal role in testing perceptual and psycholinguistic models. For example, in spoken-language processing, the N400m elicited by words has been used to test phonologic and semantic appropriateness (Helenius et al., 2002), including in dyslexic readers. Lau et al. (2009) have used the N400m response to test whether the context effects leading to an N400 are driven by lexical level processing (e.g., predictively conditioned lexical preactivation) or semantic integration of the current word into a larger context. Interestingly, Travis et al. (2011) show that the N400m effect – i.e., response amplitude modulation (AM) by degree of fit to context, or degree of priming – is even observed in 12–18-month-old children, which suggests that this late but robust response can be usefully employed to test questions of linguistic and prelinguistic processing, throughout the lifespan.

The difference between using ERP and MEG becomes very clear in the context of studies using an N400 design. The ERP literature typically sees the most robust N400 responses over Pz as well as slightly right-lateralized electrodes; on that basis, right parietal activation has been implicated. In contrast, MEG studies have used completely parallel designs and, when employing source localization approaches, typically observed left temporal-lobe neural sources. For example, Helenius et al. (2002), using dipole modeling, implicate more superior temporal cortex; Halgren et al. (2002) and Marinkovic et al. (2003), using distributed current-source modeling, identify more middle and anterior temporal sources; and a meta-analysis of both electophysiologic and imaging data suggests that middle temporal gyrus sources underlie the M400/N400m (Lau et al., 2008). The MEG-derived source estimates converge with what we know about the neural basis of language.

"Transition" responses: pitch onset response, change response, mismatch response

An interesting area of auditory research focuses on what happens as sound patterns change. Evoked fields are elicited not just by energy onsets in sounds, but also by transitions from one sound pattern to another. For example, pitch-specific responses (uncontaminated by energy-related components) have been measured by analyzing transitions from a noise to a regular interval sound (RIS) that elicits a sensation of pitch (e.g., Krumbholz et al., 2003). The noise-to-RIS transition elicited a robust POR. This is not an M100, elicited by the onset of a sound, but a response to a perceptual transition (emerging pitch) in an ongoing sound. The timing and amplitude of the POR were directly related to pitch value and pitch salience. A noise–RIS transition generates a POR that is as large as the M100. A similar result, comparing the responses to regular and irregular click trains, was reported by Gutschalk et al. (2004). MEG data testing binaural pitch converge with the POR data (Chait et al., 2007). Comparing the cortical response to Huggins pitch, a stimulus requiring binaural processing to elicit a pitch percept, with responses to tones embedded in noise – perceptually similar but physically different – support that similar pitch extraction mechanisms are involved. In these pitch studies, the stimuli are crafted to separate the responses to onset of the pitch percept (POR) from the stimulus onset (M100). The data show that, although distinct, both Huggins pitch and tones embedded in noise are mapped on to similar substrates on lateral Heschl's gyrus by 150 ms. The data provide evidence to suggest that the pitch onset response reflects central pitch mechanisms, as argued by models postulating a single, central pitch processor sensitive to abstract properties of pitch.

A different example of a transition response is revealed in studies examining how "auditory objects" might arise (Chait et al., 2007). The acoustic environment contains constant instances of the (dis)appearance of acoustic sources, often manifested as transitions in the pattern of ongoing activity. How does the auditory system detect such transitions? MEG studies suggest that temporal edge detection processes depend on the nature of the transition. Measurements of auditory cortical responses to transitions between a sequence of random frequency tone pips (disorder) and a constant tone (order), for example, show that these transitions embody key features of "auditory edges" (Fig. 14.3). Early responses (from 50 ms posttransition) show that order-to-disorder transitions, and vice versa, are mediated by related but slightly different neural mechanisms. This suggests that auditory cortex adjusts to local stimulus statistics, even when this is not required for overt behavior. Such neurophysiologic results show how remarkably sensitive the system is to local statistics and directional changes in statistics.

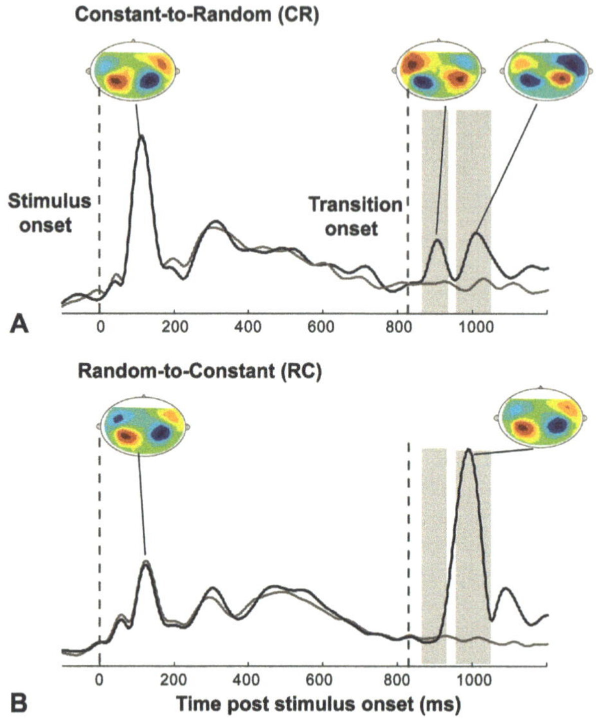

Fig. 14.3. Change detection data from Chait et al. (2007). (**A, B**) The root-mean-square (RMS) of the grand average over all subjects and all recorded channels (black) of the auditory evoked cortical responses to constant-to-random (CR) ("disappearing object") (**A**) and random-to-constant (RC) ("appearing object") (**B**) stimuli, low-passed at 10 Hz. The RMS for the control (no change) condition is plotted in gray. Contour maps at the critical time periods are provided (7.5 fT/iso contour). Red, source; blue, sink. The response to the transition was characterized by sequential increases in activity in two temporal windows at ~70 and ~150 ms after change onset. The CR condition exhibited peaks in both time windows, whereas the RC condition was characterized by a peak in the later time window only. Notably, the magnetic field pattern of the responses differed. Both peaks in the CR stimuli exhibited an M50-like dipolar pattern; the contour map corresponding to the RC response was characterized by an opposite dipolar distribution. (Reproduced with permission from Chait et al., 2007.)

One experimental approach that has been used extensively to study transition phenomena employs mismatch designs. In the mismatch field (MMF – often referred to in the EEG literature as MMN, for mismatch negativity) designs, a sequence of stimuli is presented such that one stimulus is repeated and becomes a standard while a second stimulus is presented occasionally and becomes a deviant (say, e.g., in an 8:1 ratio of standard to deviant). The difference in evoked response between deviant and standard stimuli is the so-called mismatch response. This design is easily implemented and is a reliable and highly sensitive indicator of change detection in an acoustic sequence. For example, small deviations in frequency (e.g., 1000 Hz – 1000 – 1000 – 1000 – 1004), amplitude, or timbre can be tested with the MMF design (Näätänen and Alho, 1995). Higher-order sequences can also be investigated (e.g., melodies or words or short phrases), highlighting the utility of the mismatch response to assess change detection more generally.

The MMN/MMF likely has several cortical generators, including at least one in the superior temporal cortex and one in the frontal cortex (Alho, 1995). Change detection using MMN is distinct to the responses discussed above, the prerequisite for eliciting a mismatch response being a suitable standard-to-deviant presentation ratio. In recent work on auditory processing, these experimental designs have been exploited to test hypotheses about predictive coding, since establishing standard-deviant relations permits one to construct predictive scenarios that test how strong a prediction is (e.g., Winkler and Czigler, 2012). For example, one can use such designs to test how accurately listeners predict the timing and content of a stimulus predicted by a pattern.

Auditory steady-state responses

Stimuli that repeat at a high rate, say roughly 10–100 Hz, such as click trains, or periodic modulations, such as sinusoidal AM or frequency modulation (FM), generate neuronal responses that are referred to as aSSR. These can be observed most readily by examining the frequency spectrum of the neuromagnetic data, for example via Fourier analysis. A peak in the spectrum is visible at the modulation rate, in those MEG channels that lie above auditory cortex. The aSSR persists for the duration of the sound; however, it should not be confused with the sustained response. An AM signal, for example, elicits a sustained field and an aSSR at the modulation rate. The aSSR is argued to originate in core auditory cortex, just anterior to the N100m source. The response is especially well suited to test subtle features of the temporal and frequency content of continuous stimuli; and because it is generated in very early auditory cortical fields, it reflects what sort of features are strong driving features for this part of the auditory forebrain.

The aSSR response has been used profitably in MEG research to test the processing of some elementary features of sounds, such as AM (e.g., Ross et al., 2000), FM (Millman et al., 2010), and also how simultaneous AM and FM are concurrently encoded (Luo et al., 2006; Ding and Simon, 2009). With respect to higher-order processing, the aSSR has been used to study how long acoustic sequences (typical of speech or music) are reflected in the steady-state responses (Patel and Balaban, 2004). Classic topics such as auditory scene analysis or scene segregation can also be approached systematically in aSSR designs (e.g., Elhilali et al., 2009). Ross and colleagues have done a considerable amount of research on the aSSR and established important response features. For example, Ross et al. (2000) demonstrated that, while the aSSR is evident for a wide range of AM rates, from 10 to 98 Hz, the maximum amplitude occurs around 40 Hz, a rate at which the response driven by the modulated signal overlaps with mid-latency auditory evoked responses. aSSR has also been used to establish fundamental auditory thresholds. For example, a gap detection threshold of around 3 ms is revealed in the aSSR (Ross and Pantev, 2004).

Simon and colleagues have exploited the aSSR in MEG to test a variety of issues surrounding auditory processing (e.g., Luo et al., 2006; Ding and Simon, 2009; Elhilali et al., 2009). One issue that has been of considerable interest in auditory perception and cognition is whether auditory cortex "phase locks" to critical properties of auditory stimuli. This is of particular importance for speech and music, where slow acoustic modulations below 20 Hz, of varying bandwidths, are the dominant components, carrying critical information that underpins comprehension. Wang et al. (2012) recorded the aSSR to relatively slow AM stimuli. The responses to five AM rates (1.5, 3.5, 7.5, 15.5, and 31.5 Hz) and four types of carrier (pure-tone and 1/3-, 2-, and 5-octave pink noise) were studied. A robust phase-locked aSSR was detected reliably in all conditions. The response power generally decreased with increasing modulation rate.

Response properties appear to depend only weakly on bandwidth. The neural sources of the aSSR, when approximated by a single ECD, are distinct from and medial to the ECD location of the N1m response. Such results demonstrate that the globally synchronized activity in early human auditory cortex is phase-locked to slow temporal modulations below 30 Hz. This is important because it has implications for neural coding of stimuli such as speech. If decoding relies on phase locking, it will be one way to discover which features on such signals are necessary and sufficient to drive intelligibility.

Oscillations

Neuronal oscillations are argued to play a role in various perceptual and cognitive tasks, including attention, navigation, memory, motor planning, and, in the present context, auditory perception and spoken-language comprehension (Bastiaansen and Hagoort, 2006; Schroeder et al., 2008; de Diego-Balaguer et al., 2011; Peelle and Davis, 2012). However, the specific computational functions of neuronal oscillations remain uncertain. One possible function is to coordinate the activity of distinct cortical regions and integrate activity across multiple spatial and temporal scales; an oscillatory hierarchy may serve as an organizing mechanism for such function (Fries, 2005; Schroeder et al., 2008). An oscillatory hierarchy could also serve as a central pacemaker.

As previously noted (e.g., Poeppel, 2003; Schroeder et al., 2008), there is a remarkable correspondence between average durations of speech units and the frequency ranges of cortical oscillations. Phonetic features (duration of 20–50 ms) are associated with gamma (>30 Hz) and beta (15–30 Hz) oscillations, syllables and words (mean duration of ~200 ms) with theta (4–8 Hz) oscillations, and sequences of syllables and words embedded within a prosodic phrase (300–1500 ms) with delta oscillations (<3 Hz). A systematic correlation between the acoustics of spoken language and EEG and MEG responses has been demonstrated by showing that temporal cortical responses contain enough information to discriminate single words (Suppes et al., 1997), artificial simple sentences (Suppes et al., 1998), naturalistic sentences (Luo and Poeppel, 2007), and audiovisual speech (Luo et al., 2010), or to correlate with intelligibility (Ahissar et al., 2001; Luo and Poeppel, 2007; Peelle et al., 2013). The observed cortical rhythms are principally a reflection of an underlying computational mechanism, such that the brain sets time intervals for analysis of individual speech components by intrinsic oscillations pretuned to an expected speech rate and retuned during continuous speech processing by locking to the temporal envelope (Ahissar et al., 2001; Poeppel, 2003; Giraud et al., 2007; Giraud and Poeppel, 2012). Such a computational principle is in line with the putative role of a hierarchic oscillatory array, controlling neuronal excitability and thus stimulus-related responses in neuronal ensembles (Kopell and LeMasson, 1994; Lakatos et al., 2005; Schroeder et al., 2008; Giraud and Poeppel, 2012).

Figure 14.4 illustrates data to demonstrate the tight coupling between the envelope of spoken sentences and the low-frequency oscillations measured in MEG that track the envelope, by resetting the phase to "edges" in the signal. Figure 14.4A shows the relation between the acoustic envelope of a sentence and the low-frequency MEG oscillation that tracks the envelope. Figure 14.4B provides a spectrogram-based representation, to visualize in frequency space how the acoustic and brain signals are distributed in time–frequency space. Figure 14.4C depicts spatial response patterns that show that the coupling between acoustic and MEG response (bottom) arise form a typical cortical generator (top). The interpretation of the role of these oscillations is that the low-frequency oscillations (delta, theta) serve to "parse" or chunk an input signal into temporal units; the higher-frequency beta and gamma oscillations play a (not yet well-understood) role in decoding the signals and facilitating access to stored templates.

CONCLUSION

The extremely high temporal rate at which auditory perception must operate makes it desirable to have investigative techniques that have the appropriate resolution. Beyond the ultra-high (microsecond) rate phenomena supported by brainstem structures (such as sound localization), the auditory computations executed on cortex that form the basis for much of our perception range from single milliseconds (e.g., gap detection, 2–3 ms) to tens of milliseconds (e.g., order threshold 20–30 ms, phonemes 30–80 ms) to hundreds (auditory objects, syllables 100–300 ms) and thousands of milliseconds (sentences, music, auditory context effects). Electrophysiologic techniques are well suited to study these

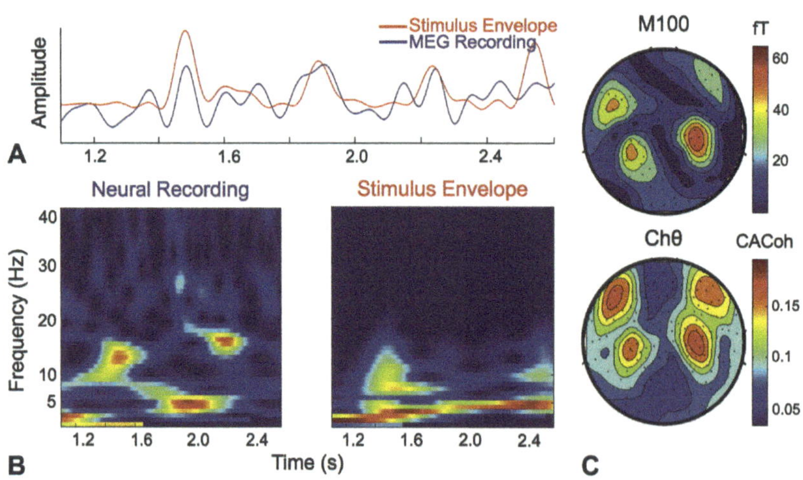

Fig. 14.4. (**A**) Representative neural data (blue) recorded during the last 1.5 seconds of a trial presenting a sentence, averaged over 20 channels. The red line shows the amplitude envelope of the stimulus presented during that same trial. There is obvious overlap between the two waveforms. (**B**) Time–frequency analysis of the example neural recording and the stimulus envelope. The time–frequency analysis is used to calculate the cerebroacoustic coherence (CACoh) at individual frequencies to assess how tightly linked the neural oscillation (here in the delta-theta band) is to a critical stimulus property. Actual analyses are performed over entire trials. (**C**) Top: Topography of an averaged M100 to single sinusoidal 1 kHz tone. Bottom: Topography of averaged CACoh values for one condition across subjects. There is a close coupling of low-frequency neural oscillations from auditory cortex to the stimulus, allowing the tracking of stimulus envelope by such neural activity. MEG, magnetoencephalography. (Reproduced with permission from Doelling et al., 2014.)

scales, but the available ones differ in their advantages and disadvantages. The highly invasive, clinically motivated ECog has an extremely high temporal, spectral, and spatial resolution, allowing investigators to probe detailed neural response properties in a well-defined (albeit clinically defined and necessarily spatially restricted) context. However, for obvious pragmatic reasons (including very restrictive experimental conditions possible for the patient participants), this is an exciting "boutique" technique that is not widely employable to study many aspects of hearing, although important phenomena like scene analysis can be investigated (e.g., Zion-Golumbic et al., 2013b). ERP recording is widely available, portable, cost-effective, and rather well understood. It therefore remains the technique of choice for auditory studies. Its disadvantage derives from some of the biophysical attributes of electric fields – in comparison to magnetic fields, they make source reconstruction very difficult – and even studies of cortical lateralization of function challenging. MEG is not widely available and somewhat costly, and, of course, not (yet) portable due to the shielding requirements. However, MEG has an especially high sensitivity to neural activity from human auditory cortex, has the appropriate temporal and suitable spatial resolution, and is rather fast to set up (for patients, children, etc.) Moreover, the rather favorable signal-to-noise ratio makes even single trial studies quite feasible (e.g., Salajegheh et al., 2004; Luo and Poeppel, 2007). As a consequence, electromagnetic recordings based on MEG have already made (and continue to make) important contributions to our understanding of auditory perception and cognition.

References

Ahissar E, Nagarajan S, Ahissar M et al. (2001). Speech comprehension is correlated with temporal response patterns recorded from auditory cortex. Proc Natl Acad Sci U S A 98: 13367–13372.

Alain C, Winkler I (2012). Recording event-related brain potentials: application to study auditory perception. In: D Poeppel, T Overath, A Popper et al. (Eds.), The Human Auditory Cortex. Springer, New York, pp. 69–98.

Alho K (1995). Cerebral generators of mismatch negativity (MMN) and its magnetic counterpart (MMNm) elicited by sound changes. Ear Hear 16: 38–51.

Barton B, Venezia JH, Saberi K et al. (2012). Orthogonal acoustic dimensions define auditory field maps in human cortex. Proc Natl Acad Sci U S A 109: 20738–20743.

Bastiaansen M, Hagoort P (2006). Oscillatory neuronal dynamics during language comprehension. Prog Brain Res 159: 179–196.

Bendixen A, San Miguel I, Schroeder E (2012). Early electrophysiological indicators for predictive processing in audition: a review. Int J Psychophysiol 83: 120–131.

Bergelson E, Idsardi WJ (2009). A neurophysiological study into the foundations of tonal harmony. Neuroreport 20: 239–244.

Boh B, Herholz SC, Lappe C et al. (2011). Processing of complex auditory patterns in musicians and nonmusicians. PLoS One 6: e21458.

Bonte M, Parviainen T, Hytönen K et al. (2006). Time course of top-down and bottom-up influences on syllable processing in the auditory cortex. Cereb Cortex 16: 115–123.

Brattico E, Pallesen KJ, Varyagina O et al. (2009). Neural discrimination of nonprototypical chords in music experts and laymen: an MEG study. J Cogn Neurosci 21: 2230–2244.

Brignell CJ, Hall DA, Witton C (2008). Mapping the auditory brain using magnetoencephalography (MEG). Brain Mapping Research Developments, 13.

Chait M, Poeppel D, de Cheveigné A et al. (2007). Processing asymmetry of transitions between order and disorder in human auditory cortex. J Neurosci 27: 5207–5214.

Chandrasekaran B, Kraus N (2010). The scalp-recorded brainstem response to speech: neural origins and plasticity. Psychophysiology 47: 236–246.

Clarke S, Morosan P (2012). Architecture, Connectivity, and Transmitter Receptors of Human Auditory Cortex. In: D Poeppel, T Overath, A Popper et al. (Eds.), The Human Auditory Cortex. Springer, New York, pp. 11–38.

de Diego-Balaguer R, Fuentemilla L, Rodriguez-Fornells A (2011). Brain dynamics sustaining rapid rule extraction from speech. J Cogn Neurosci 23: 3105–3120.

Ding N, Simon JZ (2009). Neural representations of complex temporal modulations in the human auditory cortex. J Neurophysiol 102: 2731–2743.

Doelling KB, Arnal LH, Ghitza O et al. (2014). Acoustic landmarks drive delta-theta oscillations to enable speech comprehension by facilitating perceptual parsing. Neuroimage 85: 761–768.

Elhilali M, Xiang J, Shamma SA et al. (2009). Interaction between attention and bottom-up saliency mediates the representation of foreground and background in an auditory scene. PLoS Biol 7: e1000129.

Fries P (2005). A mechanism for cognitive dynamics: neuronal communication through neuronal coherence. Trends Cogn Sci 9: 474–480.

Gage N, Poeppel D, Roberts TP et al. (1998). Auditory evoked M100 reflects onset acoustics of speech sounds. Brain Res 814: 236–239.

Giraud AL, Poeppel D (2012). Cortical oscillations and speech processing: emerging computational principles and operations. Nat Neurosci 15: 511–517.

Giraud AL, Kleinschmidt A, Poeppel D et al. (2007). Endogenous cortical rhythms determine cerebral specialization for speech perception and production. Neuron 56: 1127–1134.

Gutschalk A, Patterson RD, Scherg M et al. (2004). Temporal dynamics of pitch in human auditory cortex. Neuroimage 22: 755–766.

Gutschalk A, Micheyl C, Melcher JR et al. (2005). Neuromagnetic correlates of streaming in human auditory cortex. J Neurosci 25: 5382–5388.

Halgren E, Dhond RP, Christensen N et al. (2002). N400-like magnetoencephalography responses modulated by semantic context, word frequency, and lexical class in sentences. Neuroimage 17: 1101–1116.

Hansen PC, Kringelbach ML, Salmelin R (2010). MEG: An Introduction to Methods. Oxford University Press, Oxford.

Helenius P, Salmelin R, Service E et al. (2002). Cortical activation during spoken-word segmentation in nonreading-impaired and dyslexic adults. J Neurosci 22: 2936–2944.

Howard MF, Poeppel D (2009). Hemispheric asymmetry in mid and long latency neuromagnetic responses to single clicks. Hear Res 257: 41–52.

Kaas JH, Hackett TA (2000). Subdivisions of auditory cortex and processing streams in primates. Proc Natl Acad Sci U S A 97: 11793–11799.

Kopell N, LeMasson G (1994). Rhythmogenesis, amplitude modulation, and multiplexing in a cortical architecture. Proc Natl Acad Sci U S A 91: 10586–10590.

Krumbholz K, Patterson RD, Seither-Preisler A et al. (2003). Neuromagnetic evidence for a pitch processing center in Heschl's gyrus. Cereb Cortex 13: 765–772.

Lakatos P, Shah AS, Knuth KH et al. (2005). An oscillatory hierarchy controlling neuronal excitability and stimulus processing in the auditory cortex. J Neurophysiol 94: 1904–1911.

Lau EF, Phillips C, Poeppel D (2008). A cortical network for semantics: (de)constructing the N400. Nat Rev Neurosci 9: 920–933.

Lau E, Almeida D, Hines PC et al. (2009). A lexical basis for N400 context effects: evidence from MEG. Brain Lang 111: 161–172.

Luo H, Poeppel D (2007). Phase patterns of neuronal responses reliably discriminate speech in human auditory cortex. Neuron 54: 1001–1010.

Luo H, Wang Y, Poeppel D et al. (2006). Concurrent encoding of frequency and amplitude modulation in human auditory cortex: MEG evidence. J Neurophysiol 96: 2712–2723.

Luo H, Liu Z, Poeppel D (2010). Auditory cortex tracks both auditory and visual stimulus dynamics using low-frequency neuronal phase modulation. PLoS Biol 8: e1000445.

Lütkenhöner B, Poeppel D (2011). From Tones to Speech: Magnetoencephalographic Studies. In: C Schreiner, J Winer (Eds.), The Auditory Cortex. Springer, New York, pp. 597–615.

Lütkenhöner B, Seither-Preisler A, Seither S (2006). Piano tones evoke stronger magnetic fields than pure tones or noise, both in musicians and non-musicians. Neuroimage 30: 927–937.

Marinkovic K, Dhond RP, Dale AM et al. (2003). Spatiotemporal dynamics of modality-specific and supramodal word processing. Neuron 38: 487–497.

Millman RE, Prendergast G, Kitterick PT et al. (2010). Spatiotemporal reconstruction of the auditory steady-state response to frequency modulation using magnetoencephalography. Neuroimage 49: 745–758.

Näätänen R, Alho K (1995). Generators of electrical and magnetic mismatch responses in humans. Brain Topogr 7: 315–320.

Nagarajan S, Gabriel RA, Herman A (2012). Magnetoencephalography. In: D Poeppel, T Overath, A Popper et al. (Eds.), The Human Auditory Cortex. Springer, New York, pp. 97–128.

Noda K, Tonoike M, Doi K et al. (1998). Auditory evoked off-response: its source distribution is different from that of on-response. Neuroreport 9: 2621–2625.

Obleser J, Lahiri A, Eulitz C (2004). Magnetic brain response mirrors extraction of phonological features from spoken vowels. J Cogn Neurosci 16: 31–39.

Pantev C, Eulitz C, Elbert T et al. (1994). The auditory evoked sustained field: origin and frequency dependence. Electroencephalogr Clin Neurophysiol 90: 82–90.

Patel AD, Balaban E (2004). Human auditory cortical dynamics during perception of long acoustic sequences: phase tracking of carrier frequency by the auditory steady-state response. Cereb Cortex 14: 35–46.

Peelle JE, Davis MH (2012). Neural oscillations carry speech rhythm through to comprehension. Front Psychol 3: 320.

Peelle JE, Gross J, Davis MH (2013). Phase-locked responses to speech in human auditory cortex are enhanced during comprehension. Cereb Cortex 23: 1378–1387.

Poeppel D (2003). The analysis of speech in different temporal integration windows: cerebral lateralization as 'asymmetric sampling in time'. Speech Comm 41: 245–255.

Poeppel D, Yellin E, Phillips C et al. (1996). Task-induced asymmetry of the auditory evoked M100 neuromagnetic field elicited by speech sounds. Brain research. Cogn Brain Res 4: 231–242.

Roberts TP, Poeppel D (1996). Latency of auditory evoked M100 as a function of tone frequency. Neuroreport 7: 1138–1140.

Ross B, Pantev C (2004). Auditory steady-state responses reveal amplitude modulation gap detection thresholds. J Acoust Soc Am 115: 2193–2206.

Ross B, Borgmann C, Draganova R et al. (2000). A high-precision magnetoencephalographic study of human auditory steady-state responses to amplitude-modulated tones. J Acoust Soc Am 108: 679–691.

Salajegheh A, Link A, Elster C et al. (2004). Systematic latency variation of the auditory evoked M100: from average to single-trial data. Neuroimage 23: 288–295.

Scharinger M, Idsardi WJ, Poe S (2011). A comprehensive three-dimensional cortical map of vowel space. J Cogn Neurosci 23: 3972–3982.

Schroeder CE, Lakatos P, Kajikawa Y et al. (2008). Neuronal oscillations and visual amplification of speech. Trends Cogn Sci 12: 106–113.

Suppes P, Lu ZL, Han B (1997). Brain wave recognition of words. Proc Natl Acad Sci U S A 94: 14965–14969.

Suppes P, Han B, Lu ZL (1998). Brain-wave recognition of sentences. Proc Natl Acad Sci U S A 95: 15861–15866.

Tavabi K, Obleser J, Dobel C et al. (2007). Auditory evoked fields differentially encode speech features: an MEG investigation of the P50m and N100m time courses during syllable processing. Eur J Neurosci 25: 3155–3162.

Travis KE, Leonard MK, Brown TT et al. (2011). Spatiotemporal neural dynamics of word understanding in 12- to 18-month-old-infants. Cereb Cortex 21: 1832–1839.

Uusvuori J, Parviainen T, Inkinen M et al. (2008). Spatiotemporal interaction between sound form and meaning during spoken word perception. Cereb Cortex 18: 456–466.

Wang Y, Ding N, Ahmar N et al. (2012). Sensitivity to temporal modulation rate and spectral bandwidth in the human auditory system: MEG evidence. J Neurophysiol 107: 2033–2041.

Winkler I, Czigler I (2012). Evidence from auditory and visual event-related potential (ERP) studies of deviance detection (MMN and vMMN) linking predictive coding theories and perceptual object representations. Int J Psychophysiol 83: 132–143.

Xiang J, Simon J, Elhilali M (2010). Competing streams at the cocktail party: exploring the mechanisms of attention and temporal integration. J Neurosci 30: 12084–12093.

Zion-Golumbic E, Bickel S, Lakatos P et al. (2013a). Mechanisms underlying selective neuronal tracking of attended speech at a "cocktail party". Neuron 77: 980–991.

Zion-Golumbic E, Cogan G, Schroeder C et al. (2013b). Visual input enhances selective speech envelope tracking in auditory cortex at a 'cocktail party'. J Neurosci 33 (4): 1417–1426.

Handbook of Clinical Neurology, Vol. 129 (3rd series)
The Human Auditory System
G.G. Celesia and G. Hickok, Editors

Chapter 15

Hemodynamic imaging of the auditory cortex

DEBORAH ANN HALL[1]* AND KARIMA SUSI[2]

[1]*National Institute for Health Research, Nottingham Hearing Biomedical Research Unit, University of Nottingham, Nottingham, UK*

[2]*Division of Psychology, School of Social Sciences, Nottingham Trent University, Nottingham, UK*

INTRODUCTION

Since emerging in the 1970s, magnetic resonance imaging (MRI) has grown to become a dominant and widely used diagnostic imaging method in a variety of clinical and research domains (Fig. 15.1A). Popularity can be attributed to the success in distinguishing different types of soft biologic tissue with high resolution and high contrast (Blamire, 2008), in a non-invasive way (Fig. 15.1B). The bases of the MRI signal are protons (i.e., hydrogen nuclei) found within the water molecules that are abundant in all biologic tissue. The ability to use MRI techniques to measure functional brain activity was not discovered until the 1990s, when scientists found a way to utilize the well-known magnetic properties of blood whose state differs when the participant is engaged in a cognitively active state compared to when at rest (Ogawa et al., 1990) (Fig. 15.1C). Such differences provide an intrinsic signal contrast. In the early years of fMRI research in the mid-1990s, the auditory neuroscience community sought to expand its knowledge of the underlying physiology of hearing, while also seeking to come to grips with the inherent acoustic and magnetic disadvantages of working in the functional MRI (fMRI) environment. This chapter describes some of the major advances in auditory fMRI methodology, paying particular attention to the practical considerations involved in setting up, running, and analyzing an auditory fMRI experiment. Some of the issues relating to future directions will also be considered.

APPLICATIONS OF FMRI TO INVESTIGATE CENTRAL AUDITORY FUNCTION

Today, fMRI is often the method of choice for human neuroimaging research, enabling scientists to address questions regarding localization of function, brain connectivity, and neuroplasticity.

fMRI has good spatial sensitivity and specificity and so can be used to map auditory responses with reasonably fine topographic detail, as well as localize these responses with respect to the shape of an individual person's brain structure. The technique is suitable for multiple observations on the same individual, and so fMRI permits the investigation of longer-term dynamic processes, such as functional plasticity after disease, damage, rehabilitation retraining, or therapeutic intervention. fMRI has been applied to a range of research questions, such as localizing distinct regions of human auditory cortex that are sensitive to sound features (e.g., frequency, pitch, loudness, spatial motion). It is also being evaluated as a potential clinical tool such as reorganization of frequency representations within primary auditory cortex after deafness (e.g., Hall and Wild, 2012) and disruptions to slow-frequency oscillatory rhythms during the resting state associated with chronic tinnitus (e.g., Kim et al., 2012). Despite these exciting new developments, there are significant issues using fMRI to inform clinical decision making. One particular issue that limits the use of fMRI for

*Correspondence to: Prof. Deborah Ann Hall, NIHR Nottingham Hearing BRU, Ropewalk House, 113 The Ropewalk, Nottingham NG1 5DU, UK. Tel: +44-(0)115-823-3600, E-mail: Deborah.Hall@nottingham.ac.uk

clinical diagnosis is the large variability from person to person in terms of the extent and precise location of functional activity. One example in normal-hearing listeners is that of pitch processing. Griffiths and Hall (2012) illustrated how, in a group of 15 listeners, the pitch representations were localized in planum temporale ($n = 9$), planum polare ($n = 3$), superior temporal sulcus ($n = 2$), and inferior frontal gyrus ($n = 1$).

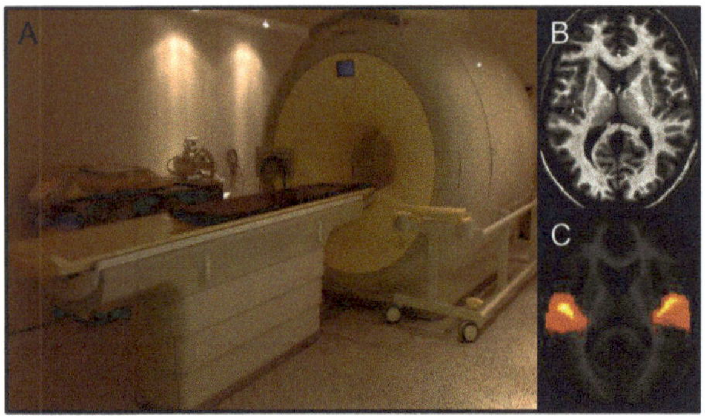

Fig. 15.1. Equipment used for magnetic resonance imaging, and some of the images produced. (**A**) Philips 3 T whole-body magnetic resonance scanner (Achieva/Intera Release) installed in the Sir Peter Mansfield Magnetic Resonance Centre, University of Nottingham, UK. (**B**) A T1-weighted anatomic image of the human brain (oblique-axial slice). (**C**) A T2*-weighted functional image depicting brain activation for incoming sound versus silence, overlaid on to a group-averaged anatomic image (oblique-axial slice).

KEY PRINCIPLES OF MAGNETIC RESONANCE IMAGING

What is MRI?

MRI exploits the weak magnetic field of spinning protons within hydrogen nuclei to create images of the body and brain. In their natural state, these protons are oriented randomly. However when subjected to an external static magnetic field, the net effect is for nuclei to align their spin axis to the direction of that external field (Fig. 15.2A). Gradient coils create small transient linear shifts in the magnetic field, the purpose of which is to encode three-dimensional (3D) space in the image (Fig. 15.2B). Whilst the hydrogen protons are in a state of alignment, the MR scanner generates a short excitation pulse of radiofrequency energy at a frequency that is absorbed by the hydrogen nuclei because the frequency exactly matches the precession frequency. This shifts them into a higher energy state and tips the proton spin axis. This creates a net magnetization component in an orientation that is different from that of the static (B0) magnetic field, within each brain slice being

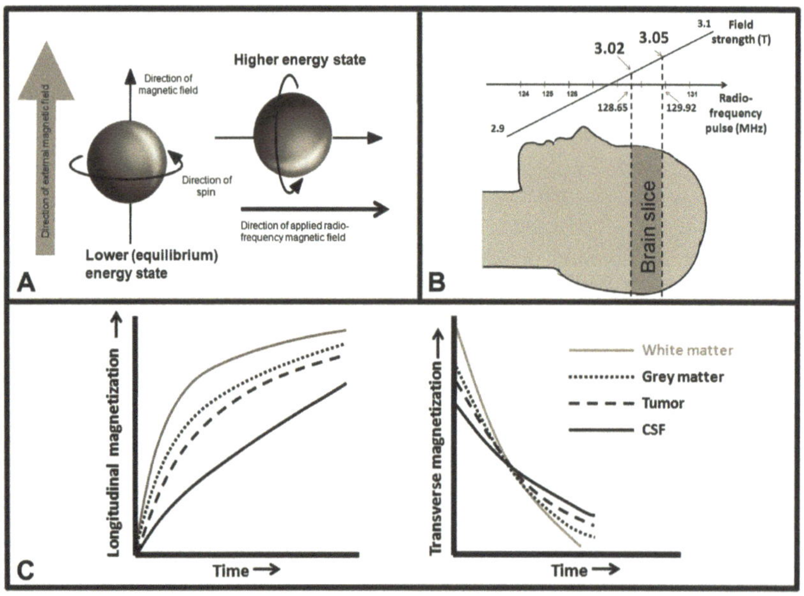

Fig. 15.2. (**A**) Net rotational direction of the spinning charge and the resulting magnetic field for hydrogen (H) nuclei in both its lower and higher energy state. The lower energy state represents the spin in the presence of the external magnetic field only. The higher energy state corresponds to the spin under the cumulative effect of both the external magnetic field and the applied radiofrequency magnetic field. (**B**) Key principles of slice excitation, namely the way in which slice orientation and thickness are determined by the strength of the magnetic field (tesla, T) and the frequency bandwidth of the applied radiofrequency pulse (MHz). (**C**) T1 recovery of longitudinal magnetization (leftmost graph) and T2 decay of transverse magnetization (rightmost graph) for white- and gray-matter tissue, cerebrospinal fluid (CSF), and a tumor. Note that T2 decay is considerably faster than T1 recovery. (From "Fundamental Principles Underlying MRI and Functional MRI" by Deborah A. Hall and Dominik C. Wild in Translational Perspectives in Auditory Neuroscience: Hearing Across the Life Span – Assessment and Disorders (p. 286) by Kelly L. Tremblay and Robert F. Burkard. Copyright © 2012 Plural Publishing, Inc. All rights reserved. Used with permission.)

acquired (Fig. 15.2B). When the excitation pulse ceases, the protons begin to relax, returning to their lower equilibrium state and emitting the electromagnetic energy that was absorbed during excitation. This emitted energy is detected by receivers in the MR scanner and the resultant signals are decoded to create 3D images of the body or brain by collating individually acquired brain slices (Fig. 15.2B).

There are two types of relaxation that determine the rate of signal decay (Fig. 15.2C). The first describes the loss of the longitudinal component of the magnetization vector in the z plane as the protons return back to their lower equilibrium state. Its decay constant is defined as T1, hence this is sometimes known as "T1 recovery." The second describes the decay of transverse magnetization in the $x \pm y$ plane due to a loss of phase coherence in proton spins, defined as "T2 decay." Phase coherence relates to the rate at which protons precess (gyrating motion) after excitation, being greatest when spins precess at the same phase and frequency. However, a loss of phase coherence inevitably occurs over time after excitation. In addition, spatial inhomogeneities in the magnetic field cause a loss in phase coherence because different spins experience different magnetic fields. T1 and T2 relaxations begin simultaneously, but contribute independently to the MR signal because their rate of recovery or decay differs (T2 has a considerably shorter time constant compared to T1, i.e., milliseconds versus seconds). The rate of recovery is also dependent on the type of tissue being scanned (Fig. 15.2C).

Required tissue contrasts can be achieved by manipulating the temporal parameters of the scanning sequence. The contrast in T1-weighted images is dependent on the T1 value of the tissue and this provides detailed anatomic information about gray and white brain matter, tumors, and vasculature abnormalities. Figure 15.1B illustrates a typical anatomic scan acquired during an experiment. Figure 15.2C illustrates how cerebrospinal fluid has the longest T1 recovery rate and hence a weaker MR signal, so it appears darker in the anatomic image than other tissue.

What is fMRI?

Unlike MRI, fMRI allows us to investigate physiologic changes in brain function over time (*in vivo*), specifically, changes in blood oxygenation caused by underlying neuronal activity (Fig. 15.1C). This is an indirect measure of neuronal activity. fMRI relies on the principle that, during perceptual or cognitive processing, the metabolic rate of active neurons increases, and so does the blood supply to those brain region(s) in order to meet the demand for energy in the form of oxygen and glucose. The endogenous concentration of oxygen-depleted deoxyhemoglobin found in the brain's vascular system is a physiologic marker of neuronal oxygen consumption and it forms the basis of the blood oxygenation level-dependent (BOLD) contrast used in fMRI. Unlike oxyhemoglobin (which is diamagnetic), deoxyhemoglobin is known to have strong paramagnetic properties that cause measurable local distortions in the surrounding magnetic field. These influence the decay characteristics of the transverse magnetization component (T2 decay) of the MR signal.

The BOLD contrast is dependent on a combination of T2 relaxation factors, a loss of phase coherence in proton spins due to spin-spin interactions (T2 decay), and inhomogeneities in the magnetic field. These relaxation factors combine to form the T2* time constant. Although constants T2 and T2* are related, their rates of decay differ (T2 decay takes longer than T2* decay) and T2*-weighted images are particularly sensitive to the relative amount of deoxyhemoglobin present. Specifically, deoxygenated brain regions result in a more rapid decay of transverse magnetization (T2*) than oxygenated regions, altering the T2* properties of nearby tissues and suppressing the MR signal. So why does neuronal activity lead to an increase in the MR signal if deoxyhemoglobin weakens the MR signal? The mechanisms underlying this BOLD response are complex (Logothetis et al., 2001), but brain activity is generally thought to trigger an "oversupply" of oxygen-rich hemoglobin to active regions, displacing deoxyhemoglobin and decreasing signal loss due to T2* effects, thus increasing the MR signal. The result is an increase in the BOLD signal compared to the resting baseline state. T2*-weighted images provide information about the relative T2* values of tissues. Voxels containing predominantly oxyhemoglobin appear slightly brighter in the functional image (greater T2* values and MR signal) than those containing more dexoyhemoglobin (Ogawa et al., 1990). This difference can be examined using analysis to determine the statistical maps which tell us about the underlying brain activity (Fig. 15.1C). We refer interested readers to several excellent textbooks describing the principles of fMRI (Jezzard et al., 2003; Huettel et al., 2009).

Although the time course of neural activity is in the millisecond range, the hemodynamic response is relatively sluggish. In the auditory cortex sound-related activity has been shown to reach its maximum about 4–5 seconds after stimulus onset and cross the baseline at about 5–8 seconds after stimulus offset (Hall et al., 2000). The imaging paradigm and statistical analysis need to take this delay into account. We return to these issues later.

Scanning parameters

Two important scanning parameters that govern the timing of key events within the pulse sequence concern the time between the transmission of the radiofrequency pulse and the data acquisition and the time between successive brain scans. These are known as time to repeat (TR) and echo time (TE), respectively. Strictly speaking, TR denotes the amount of time between two successive radiofrequency excitation pulse sequences that are applied to a given slice, while TE represents the amount of time (milliseconds) between the application of radiofrequency pulse sequence and the data acquisition of brain images.

Values of the TE and TR parameters influence the decay of the transverse magnetization (T2) components of the MR signal and the recovery of the longitudinal magnetization (T1), respectively. Optimal fMRI needs to maximize the T2* differences between tissues and minimize the T1 differences. These choices are primarily determined by the type of contrast desired and the properties of the different tissue types being imaged, as well as field homogeneity and field strength.

The greatest BOLD contrast in a given region of interest is achieved when the TE is approximately equal to the T2* values (Clare et al., 2001). There is minimal T2* contrast at both very short TEs (because little transverse magnetization is lost) and very long TEs (because nearly all the magnetization is lost). An intermediate TE is preferable because it maximizes the differences in transverse magnetization (T2* decay) between active and non-active neural tissue. T2* values typically vary across brain regions and so the choice of TE to use throughout an auditory fMRI study should be optimally selected for increasing the likelihood of detecting auditory cortical activations. For any given brain region, T2* is approximately equal to -1/(gradient of the regression line), where the regression line is plotted through data points measuring the natural logarithm of the mean signal intensity within the region of interest plotted as a function of TE (Fig. 15.3).

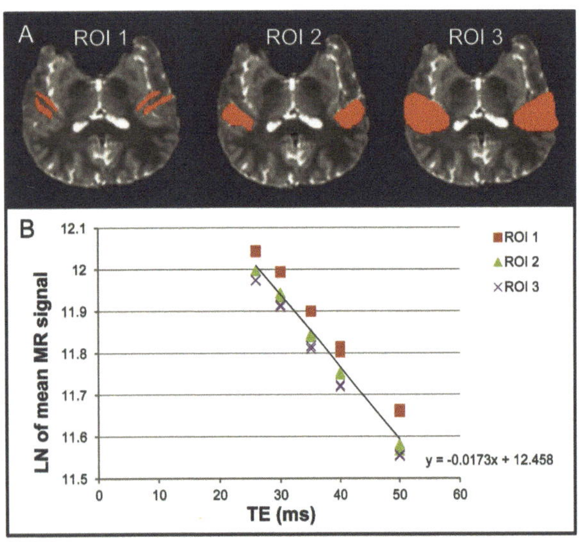

Fig. 15.3. Graph showing the calculation of optimal echo time (TE) for three different auditory cortical regions of interest (ROI). (**A**) The three ROIs (oblique-axial slice). ROI 1 delineates gray matter along Heschl's gyrus only (primary auditory cortex). ROI 2 encompasses both gray and white matter around Heschl's gyrus. ROI 3 is an expanded region which includes primary and non-primary auditory regions. (**B**) Natural logarithm (LN) of the mean signal intensity across a range of TEs. MR, magnetic resonance.

Our own data indicate that an optimal TE = 58 ms (Table 15.1). Other auditory fMRI set-ups may require different TE settings. For example, using this same procedure, Clare et al. (2001) found that in auditory cortex TE = 40 ms was optimal for their particular experimental set-up on a dedicated 3 T MR scanner at the University of Nottingham (UK), using a head gradient coil insert and standard volume coil (TR = 4 seconds). In another example, Paltoglou et al. (2009) calculated that

Table 15.1

Some of the most important functional magnetic resonance imaging (fMRI) scanning parameters and key practical considerations for optimizing fMRI

Important fMRI scanning parameters	
Time to repeat (TR)	The amount of time between two successive radiofrequency excitation pulse sequences that are applied to a given slice (measured in milliseconds to seconds)
Echo time (TE)	The amount of time between the application of radiofrequency pulse sequence and the data acquisition of brain images (measured in milliseconds)
Practical considerations for optimizing fMRI	
Signal-to-noise ratio (SNR)	The relative strength of the signal compared with noise
Field strength	The strength of the external magnetic field created within the bore of the MR scanner (measured in tesla)
Radiofrequency coils	Part of the MR scanner hardware that transmits and receives radiofrequency energy
Voxel size	A three-dimensional sampling unit that makes up the fMRI image (measured in cubic millimeters)
Number of acquisitions	The number of times a signal from a given slice has been measured during the experiment
Receiver bandwidth (rBW)	The range of frequencies that the receiver can sample accurately, which must be mapped across the field of view (measured in kiloherz)

TE = 55 ms was optimal for their auditory fMRI study using a Philips 3 T whole-body MR scanner (Achieva/Intera Release 1.2/11) equipped with an eight-channel array coil (TR = 9 seconds).

The greatest BOLD contrast in a given region of interest is achieved when the TR is long enough for complete recovery of the longitudinal magnetization (i.e., minimal T1 contrast). Here, the radiofrequency excitation pulse can apply a flip angle of 90° to tip the spinning protons fully into the x–y plane, thus maximizing the magnitude of transverse magnetization. One example can be found in the above study by Paltoglou et al. (2009), where a flip angle of 90° was selected in combination with TR = 9 seconds. A clear disadvantage of a long TR is that it either sacrifices statistical power when the overall paradigm timing is held fixed (Constable and Spencer, 2001) or extends the duration of the scanning session required to obtain a particular number of acquisitions. However, for auditory fMRI the potential benefit of long-TR imaging is to reduce the ambient scanner acoustic noise. This topic is discussed in more detail below.

Short TRs are traditionally preferable because a greater number of acquisitions can be obtained for the same duration of scanning session. Short TRs are associated with the application of smaller flip angles, to minimize T1 contrast. Although the resulting magnitude of transverse magnetization is also smaller, this is still acceptable for auditory fMRI. For example, da Costa et al. (2011) chose a flip angle of 47°, in combination with TR = 2 seconds and TE = 25 ms, to successfully map frequency-dependent activity across human primary auditory cortex.

PRACTICAL CONSIDERATIONS IN OPTIMIZING THE SIGNAL-TO-NOISE RATIO OF AUDITORY FMRI

Here we present some of the key practical considerations relating to the optimization of image quality for auditory fMRI experiments in the context of maximizing BOLD signal change by increasing the signal-to-noise ratio (SNR). SNR refers to the relative strength of the signal compared with noise. SNR can be calculated by dividing mean MR signal intensity in a small region of interest containing homogeneous tissue by the standard deviation of the MR intensity of a large region of interest (containing inhomogeneous tissue) localized elsewhere in the brain. In an fMRI experiment, changes over time in the MR signal intensity within a region of interest are particularly small when compared with the total image intensity of the MR signal (about 2–10% of the overall signal variance in a 3 T scanner: Amaro and Barker, 2006). Experimental studies are specifically interested in brain activation related to stimulus or task processing, not the unwanted fluctuations in the MR image intensity caused by confounds such as eddy currents, gradient coil vibrations, temperature fluctuations, physiologic fluctuations, and head movement (e.g., Triantafyllou et al., 2005). It is therefore important to maximize SNR as much as possible. Factors which should be considered in an auditory fMRI experiment because they are known to influence SNR include: (1) field strength; (2) coil design; (3) voxel size; (4) number of image acquisitions; and (5) receiver bandwidth.

Field strength

Field strength is measured in tesla (T). For example, an ultra-high 7 T MR scanner creates a magnetic field that is 100 000 times stronger than that of the earth's magnetic field, while the more common 3 T scanner creates a magnetic field that is 60 000 times stronger. A higher field strength creates a greater difference between high- and low-energy states of protons, meaning that more spins align with the direction of the static field and are thus "tipped" into the plane of transverse magnetization. The strength of the magnetic field therefore influences the magnitude of the net magnetization vector, increasing the amplitude of the BOLD signal (Scarabino et al., 2007). For example, measurements of the percentage signal change on activation in the motor cortex and temporal SNR at 3 T and 7 T showed that the BOLD signal contrast at 7 T increased by a factor of 1.9 (Schäfer et al., 2008).

While SNR can be optimized by scanning at ultra-high fields, experimenters often have to make choices about scanning parameters based on tradeoffs between benefits and costs. A particular problem that is relevant to fMRI studies of central auditory processing concerns image distortion due to susceptibility artifacts. Although BOLD contrast increases with field strength, so do the magnetic inhomogeneities in regions of the brain near air–tissue boundaries. The magnetic inhomogeneity results in a steeper decay constant for T2*-weighted images, leading to a greater signal loss for a given set of TR and TE parameters. This problem is particularly apparent for the temporal lobe (including the auditory cortex), which is located close to the air-filled sinuses. Magnetic field inhomogeneity can be corrected retrospectively if a map of the B0 field is acquired during the scanning session (Jezzard and Balaban, 1995). This map is based on obtaining images with slightly different echo times, and using the phase difference between them to create a map of the local field inhomogeneity to calculate the distance voxels have been erroneously shifted. In auditory cortex, field map correction has been shown to increase statistical power for group analysis (Cusack et al., 2003). Magnetic field inhomogeneities can also be reduced during the scanning session. A recent 7 T study demonstrated the feasibility of applying B(1) shimming to achieve efficient and homogeneous radiofrequency transmission in human auditory cortex (De Martino et al., 2012). This study confirmed detection of responses to a range of sound stimuli (voices, speech, animal cries, tools, and nature). As a result, this shimming method paves the way for increased spatial specificity of auditory responses, perhaps permitting in the future the *in vivo* study of functional organizations down to the columnar level of the cortex.

Radiofrequency coils

The radiofrequency coil on an imager can be likened to the lens on a camera. Radiofrequency coils transmit the excitation pulse which rotates the net magnetization in the z plane into the transverse (x–y) plane. Coils also receive the radiofrequency energy from the imaged object. Radiofrequency coils can be divided into three general categories: (1) transmit and receive coils; (2) receive only coils; and (3) transmit only coils. The sensitivity to the MRI signal as well as the amount of global brain coverage is dependent on the type of radiofrequency coil used for signal reception. Standard volume coils enable whole-head fMRI and are ideal for studies investigating distributed brain networks. For example, Degerman et al. (2006) used a quadrature head coil with a 3 T GE Signa MR scanner (General Electric, Milwaukee, USA) to investigate auditory selective attention to location and pitch dimensions of the sound stimuli, detecting activity in prefrontal, temporal, and parietal cortices.

While surface coils improve SNR, they have a reduced field of view and so are useful only when the research question has identified an *a priori* focal brain region. For example, Talavage et al. (2000) used a surface coil with a 1.5 T GE Signa MR scanner to measure frequency-dependent sound processing (tonotopy) in the superior temporal plane near Heschl's gyrus. The choice of coil was prompted by preliminary experiments using a head coil in which there was a general lack of fMRI response in auditory cortex, attributed to low SNR. In response to this challenge, the authors developed a set of bilateral imaging coils that were integrated into the circumaural ear defenders and could be used individually or as a bilateral set. The surface coil improved SNR measures from 15 to 35.

Many recent advances in auditory fMRI have been facilitated by technical improvements in coil design. Volume array coils can offer improved SNR and, as the number of array elements increases, so does the SNR advantage near those elements (Gizewski et al., 2005). Parallel MRI techniques use spatial information from a phased array of radiofrequency receiver coils to accelerate image acquisition or to improve spatial resolution. Here a volume coil transmits the radiofrequency energy while the multiple surface coils act as receiver channels. Because these coil signals have to be received, amplified, and digitized independently, the result is an increased complexity of the MRI receiver (Duyn, 2004). Currently, MRI scanners are available with 32 independent channels based on digital receiver technology, although eight-channel phased-array coils are in more widespread use. For example, da Costa et al. (2011) used an eight-channel coil (RAPID Biomedical) with a 7 T Siemens Magnetom MR scanner (Siemens Medical Solutions) to measure the two major tonotopic subfields of primary auditory cortex relative to the underlying anatomy of Heschl's gyrus.

Voxel size

The term voxel describes a 3D sampling unit that comprises fMRI images. A pixel is a 2D analogy. The typical voxel size in an auditory fMRI experiment falls within the range 1.5–3 mm^3, with the z value specifying the thickness of the image slice. Choice of voxel size can have an important impact on SNR and is a trade-off between sensitivity and specificity. For large regions of interest, where the source of activity comes from populations of neurons densely spread across several millimeters, BOLD contrast can improve as voxel size increases. However, for small regions of interest BOLD contrast improves as voxel size decreases. Here, the gains are somewhat dependent on the contribution of partial volume effects. For example, for a small volume of tissue comprising mostly sound-responsive neurons, a small voxel size would be more effective at detecting fMRI activity than a large voxel size. This is because the large voxel would include neurons unresponsive to sound, thus "diluting" the overall activation signal. The increased SNR and available BOLD signal arising from the use of 7 T systems permit the use of smaller voxel sizes in auditory fMRI. Furthermore, the BOLD signal strength of venous blood is reduced due to a shortened relaxation time, restricting activation signals to the cortical gray matter, and thus improving the spatial specificity of the BOLD signal (van der Zwaag et al., 2009). For example, da Costa et al. (2011) achieved a voxel resolution of 1.5 mm^3 in their tonotopic study, described above.

Number of acquisitions

The number of acquisitions refers to the amount of times a signal from a given slice has been measured. SNR is proportional to the square root of the total number of acquisitions, whereby if you want to increase the SNR by a given value, you need to square the SNR to get the number of acquisitions required to achieve this (e.g., for an SNR of 6, 36 image acquisitions are required). Experimenters need to find a balance between the appropriate number of acquisitions and the required scanning time of the experiment. In one example, da Costa et al. (2011) obtained a total of 240 acquisitions in each of two experimental runs using an interscan interval of 2 seconds (16 minutes overall). Such acquisition time is sufficiently short to enable use of these methods as an initial mapping step combined with an additional experiment that might be designed to probe other functional responses within specific auditory fields.

Long scanning sessions may introduce additional physiologic noise into the data by increasing the risk of subject movement. However, one of the major challenges for auditory fMRI concerns the interference of the intense ambient scanner acoustic noise on the observed patterns of sound-related activity across the auditory cortex. Many auditory fMRI researchers choose a relatively long interscan interval (8 seconds or so) as a way of circumventing this problem because the scanner does not generate a noise between image acquisitions. For example, to measure pitch representations across human auditory cortex, Hall and Plack (2009) obtained a total of 310 acquisitions using an interscan interval of 8 seconds (41 minutes overall). Such a study requires highly motivated participants to remain as still as possible for the duration of the scanning session. Considerable research effort has been devoted to reducing the impact of scanner acoustic noise in auditory fMRI experiments and some of these methods are described in more detail below.

Receiver bandwidth

Receiver bandwidth has a direct relationship to SNR, where SNR is the reciprocal of the square root of the receiver bandwidth. The receiver bandwidth depends on the steepness of the frequency-encoding gradient (Fig. 15.2B) and the amount of time required to sample (i.e., read out) the MR signal as it is received by the radiofrequency coil. More specifically, receiver bandwidth is equivalent to the reciprocal of the sampling time of the echo. In other words, the sampling time of one echo is reduced by increasing the bandwidth. Despite supporting fast imaging, high bandwidth has a trade-off because it reduces SNR due to the greater noise contributions from the outskirts of the frequency spectrum. In contrast, a smaller bandwidth improves SNR, but can cause spatial distortions.

In summary, this section has addressed a number of issues relating to quality assurance. Quality assessment should pervade throughout all stages of an fMRI study and is as essential step to identify any problems that might compromise the quality of subsequent data analysis. Quality assurance often involves the data time series to be visually examined to ensure there are no local distortions, such as those caused by magnetic field inhomogeneities in the vicinity of the temporal cortex. Some software packages provide automated ways of testing quality, providing estimates of mean signal intensity, raw SNR, and functional SNR (i.e., mean intensity divided by the standard deviation) (e.g., alternatives include Analysis of Functional NeuroImages, AFNI: http://afni.nimh.nih.gov/afni; Freesurfer: http://ftp.nmr.mgh.harvard.edu/).

PRACTICAL CONSIDERATIONS IN OPTIMIZING THE QUALITY OF AUDITORY FMRI DATA

Although SNR is an important factor for improving image quality, a number of other factors need to be taken into consideration during experimental set-up. Here we present some of the key practical considerations relating to reducing scanner noise and minimizing confounds in subcortical auditory structures.

Reducing the impact of scanner acoustic noise

Despite its successful application to human auditory neuroscience, the MR scanner is extremely noisy when in operation, creating a rather adverse environment for listening experiments. The primary source of noise comes from the switching of the gradient fields during image acquisition. Although this noise has a short burst duration (typically less than 2 seconds), it has a high-intensity tonal quality with a fundamental frequency that is typically between 0.5 and 2 kHz with higher harmonics. The sound is at its peak intensity inside the bore of the scanner (reaching about 100 dB sound pressure level; Blackman and Hall, 2011), although the precise characteristics depend on the choice of imaging sequence and the structural resonant characteristics of the machine itself.

The bursts of scanner acoustic noise are problematic for a number of reasons: (1) reduces participant comfort and tolerance of lying in the scanner bore; (2) makes the listening task more difficult (increasing the cognitive load and demands on attention; Healy et al., 2007; Blackman and Hall, 2011); and (3) reduces the dynamic range available to statistically detect the true sound-related activity. This latter effect happens because the bursts of scanner noise are present throughout, including the baseline condition of the experiment that is assumed to represent a "quiet resting state." The resulting elevated baseline level of neural activity within auditory brain regions approaches the maximum magnitude of fMRI-related activity that is possible within the auditory cortex.

The interaction between scanner acoustic and fMRI activity is a non-linear one (e.g., Olulade et al., 2011) and so it is computationally challenging to remove these confounds after data collection. Most of the effort in combating the ambient noise has therefore concentrated on reducing the noise level perceived by the participant. The dominant two methods for noise reduction are briefly outlined below.

Sparse imaging

A common preference for studies that seek to produce fine-detailed maps of auditory cortical responses is to slow down the sampling rate so that the effect of scanner acoustic noise on detecting the sound-related activation of interest is minimized. This "sparse" scanning method is described in a seminal paper (Hall et al., 1999) and it has become the standard method for auditory neuroscience. The sparse scanning method that we typically use in our fMRI studies involves acquiring a single set of brain images very rapidly and repeating this every 8 seconds or so (i.e., TR = 8 seconds). This strategy is successful because it is able to exploit the sluggish temporal characteristics of the fMRI BOLD response (notably its delay of 4–8 seconds or so; Hall et al., 2000) thus isolating activity associated with the sound stimulus of interest from that associated with the ambient noise. As shown in Figure 15.4, the BOLD response to the burst of scanner acoustic noise rises and falls within the 8-second interscan interval, approaching its true "quiet resting state" before the next set of images is obtained.

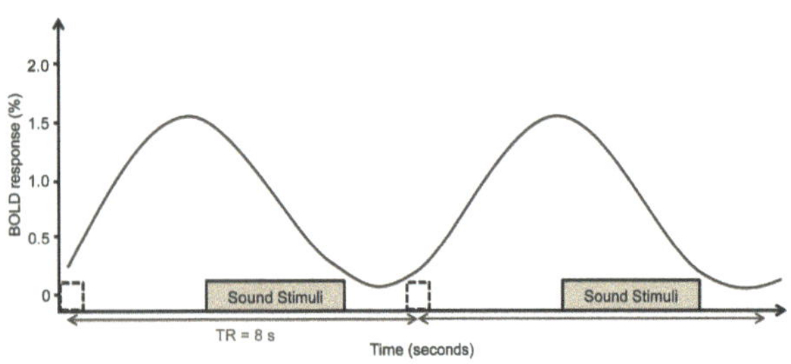

Fig. 15.4. A schematic representation of the blood oxygen level-dependent (BOLD) response to the scanner noise (gray line) and how the sound stimuli are interleaved with image acquisition during sparse imaging. The two boxes outlined in a dashed black line depict the most optimal time point in which a set of brain images should be acquired during sparse imaging because here the BOLD response to the scanner noise is lowest. At this time point, the response to the sound stimulus of interest has also reached its peak (not shown), thus minimizing any interference effects of scanner noise on the stimulus-induced auditory response. TR, time to repeat.

Typically, sequences of sound stimuli are presented in this intervening quiet period. The fMRI BOLD response to this stimulus reaches its peak at a point in time that corresponds to the lowest response to the ambient background noise.

Noise cancellation

Reduction of the ambient noise during fMRI has been enabled through innovative engineering solutions developed at the UK Medical Research Council Institute of Hearing Research. A research prototype has recently been developed that acoustically reduces the level of the background sound reaching the participant's ears using active noise cancellation (Chambers et al., 2001). The system reduces the perceived level of the background noise by producing an output ("cancellation") signal that destructively interferes with the incident sound wave produced by the scanner sound at each ear. Precise control over the frequency and phase of the cancellation signal is achieved through a digital signal-processing system that delivers the sound using a high-fidelity non-magnetic electrostatic headset. Microphone measurements at the ear demonstrate excellent sound reduction, achieving up to 35 dB of acoustic attenuation (Hall et al., 2009), improved audibility and listening quality (Blackman and Hall, 2011), and enhanced ability to detect sound-related brain activity in the hostile environment of the MR scanner (Hall et al., 2009; Blackman and Hall, 2011). Intellectual property has been licensed to Optoacoustics and the resulting OptoActive active noise-canceling headphones for MRI are the only one of their type available commercially, enabling free conversation amongst patients and clinicians. The product was launched in September 2012 as part of the interventional multichannel optical communication system.

Cardiac gating for fMRI in subcortical auditory structures

Determining how auditory information is represented within the subcortical nuclei of the ascending auditory pathway is vital for understanding the mechanisms of central auditory processing. However, until around 15 years ago, subcortical auditory activity was rarely seen in studies of auditory processing (Fig. 15.5). This is due to experimental difficulties in eliminating cardiac arterial pulsation which causes brainstem motion.

In a landmark study, Guimaraes et al. (1998) evaluated the application of cardiac gating, which is a technique that reduces the impact of pulsatile brainstem movement. This clever technique allowed Guimaraes and colleagues to synchronize image acquisitions to a specified time period within each individual listener's

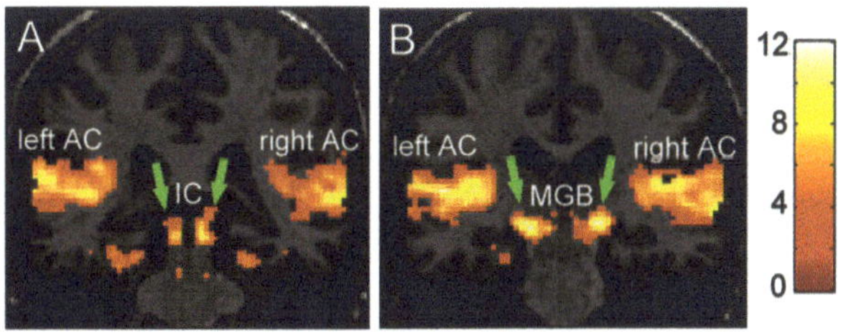

Fig. 15.5. Group functional magnetic resonance imaging (fMRI) activation using a Philips 3 T MR scanner in a standard auditory fMRI experiment without cardiac synchronization. Activation shows the response to natural environmental sounds contrasted against a silent baseline condition ($P < 0.001$, uncorrected) and superimposed on to an anatomic brain that is in standardized brain space. (**A, B**) Coronal slices: **A** corresponds to the y plane = –34 mm and **B** corresponds to the y plane = –26 mm. Auditory cortex (AC) is shown in both slices, with green arrows pointing to subcortical activation within inferior colliculus (IC) and medial geniculate body (MGB). (Unpublished data provided by Davies, Gander and Hall.)

cardiac cycle by measuring each listener's electrocardiogram and using this to trigger scanner acquisition. Postacquisition image processing then corrected for interimage variations in signal intensity caused by heart rate fluctuations. By comparing activation obtained in an equivalent standard (non-cardiac-gated) fMRI experiment, Guimaraes and colleagues found that cardiac-gated activation maps yielded much greater activation in subcortical nuclei, such as the inferior colliculus. Benefits were attributable to a reduction in signal variability, rather than an increase in percent signal change.

Subsequent fMRI studies have further corroborated this cardiac synchronization technique as a much more optimized approach, enabling improved sensitivity to detect auditory midbrain activity (Sigalovsky and Melcher, 2006). However cardiac synchronization does not appear essential for detecting subcortical activation when using ≥3 T MR scanning, as shown in Figure 15.5, which was acquired without the technique being applied. At 3 T, the strength of the magnetic field is sufficient to generate a large enough BOLD signal change to be detectable even without cardiac synchronization.

EXPERIMENTAL DESIGNS

Using fMRI, auditory neuroscientists can begin to seek answers to many questions about central auditory processing. Nevertheless, it is of upmost importance that appropriate experimental designs are employed to effectively answer hypotheses of sound representations in the auditory brain. Experimental designs can be grouped broadly into three main types: categorical, factorial, and parametric designs. A novel design based on assumptions of neural adaptation is also discussed.

Categorical designs

"Cognitive subtraction" defines the simplest experimental design in which at least two conditions (a control baseline and an experimental activation) are compared: ($B^1 - A^1$). Typically, the two conditions are matched in terms of their sensory, perceptual, and cognitive components, except for one given component of interest (Fig. 15.6A). As we shall see in the auditory fMRI example below, the assumption of perfect matching is not always met. Another assumption of the cognitive subtraction is that the component of interest can be isolated by subtracting the control baseline from the experimental activation: "pure insertion" (Friston et al., 1996a). This assumption is valid only when the additional component of interest does not influence and/or interact with the other components in the baseline condition; again this requirement for simple additivity is rarely met (Satori and Umiltà, 2000; Henson, 2007).

Since its inception, cognitive subtraction has been abundantly used in auditory fMRI investigations. For example, Hall and Plack (2009) compared a pitch-evoking stimulus to a spectrally matched random noise in a passive listening study to identify pitch-processing sites in human auditory. Attributing the differential activation to pitch coding fundamentally relies on the baseline stimulus being precisely matched to the pitch-evoking stimulus on all acoustic dimensions, except the pitch feature of interest. This is not necessarily the case. For example, one of the pitch-evoking stimuli (iterated ripple noise) contains slowly varying spectrotemporal modulations that are not found in the control stimulus and are unrelated to pitch. Hence, it is possible that the cortical response to this particular pitch stimulus is driven in part by these modulations. In a later experiment by the same authors, Barker et al. (2012) controlled for these modulations using a novel, carefully matched control noise and showed that the subtraction between pitch and noise was no longer statistically significant. These findings thus raise the possibility that a cortical response unrelated to pitch could previously have been erroneously attributed to pitch coding.

Garcia et al. (2010) found that the magnitude of the pitch-related activation is also modulated by the context of stimulus presentation. In one context, the sequence of pitch-evoking stimuli was interleaved with a spectrally matched noise burst. In the other context, the sequence of pitch-evoking stimuli was interleaved with silent bursts. The differential pitch-related activity, computed by subtracting the above pitch conditions with their matched noise control baselines, revealed much greater activity in auditory cortex when the continuous stimulation context was employed, compared to the context with silent bursts. Context effects were mostly observed in

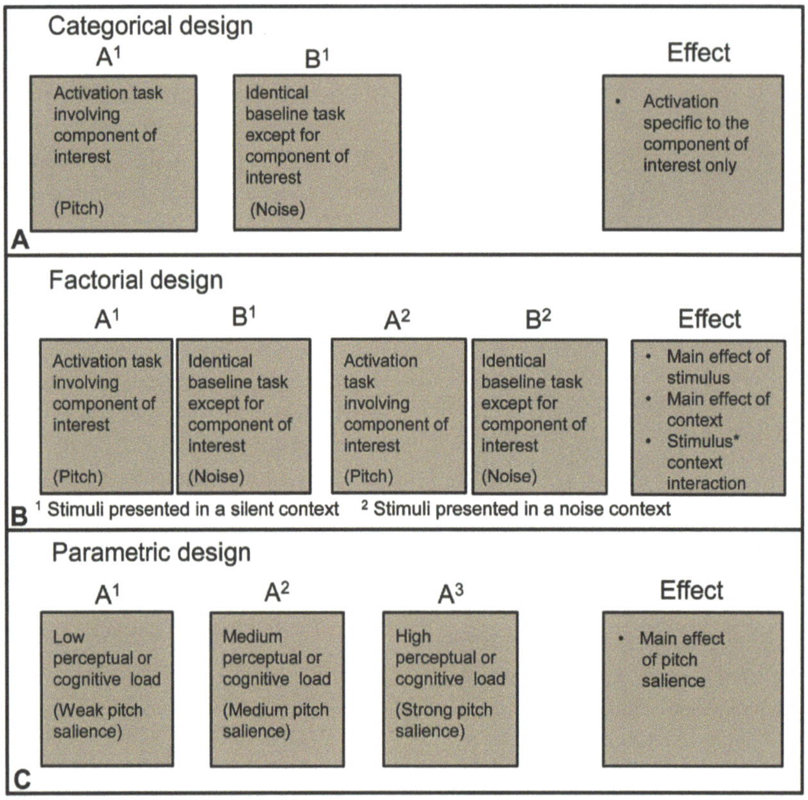

Fig. 15.6. Schematic illustrations of three major experimental designs: categorical, factorial, and parametric. (**A**) A simple cognitive subtraction design. Pitch-related activation is determined by subtracting the baseline condition (B^1) from the activation condition (A^1). (**B**) A 2-by-2 factorial design, with the two factors being signal (pitch and noise) and context (silent and noise). Activations associated with either the signal or the context can be separately determined, as can the interaction between these two factors. (**C**) Parametric design. In this example, A^1, A^2, and A^3 are all pitch conditions with three different levels of perceptual salience. Examples are given for the experiments of pitch processing described in the text.

primary auditory regions. This finding highlights the second problem of pure insertion. These data demonstrate that the pattern of pitch-related activity depends on whether or not the stimulus sequence contained repeated energy onsets. To explain this result, one infers that the same neural population is sensitive to the onsets of both energy and pitch, but the two effects are not additive (i.e., are non-linear).

Factorial designs

Factorial designs can reduce the potential problems associated with pure insertion and are argued to be a statistical, more powerful and generalizable alternative to cognitive subtraction (Friston et al., 1996a). The factorial design contains elements of the subtraction design, but uses multiple pairs of matched baseline and experimental activation conditions. Factorial designs explicitly allow for the combination of two or more factors within a single auditory fMRI experiment, enabling the "context-sensitive" interactional effects to be explicitly interrogated during statistical analysis. The simplest factorial design is a 2-by-2 design and one example is that reported in Garcia et al. (2010), described above (Fig. 15.6B). The significant interaction between signal (pitch and noise) and context allowed Garcia et al. (2010) to infer that pitch-evoked activation in auditory cortex was modulated by the context of stimulus presentation and to identify which brain regions were sensitive to such context effects. Identification of the main activation effects to signal and context alone would have not led to this conclusion (Price et al., 1997).

Cognitive conjunctions refer to an analysis approach that can be applied to data collected in a factorial design. In logic a conjunction is defined as an "AND" between truth statements. It is a more conservative way of determining a main effect than a simple cognitive subtraction because it can reduce contamination from imperfect matching of the two conditions or from context effects. Conjunction analysis computes the statistical probability of common activity differences between activation task and baseline task pairs (Nichols et al., 2005). To reach statistical significance, all the comparisons in the conjunction must be individually significant and this result holds even if the comparisons are not independent. Thus any irrelevant activation that is not common to all task pairs is excluded from the conjunction result.

The simplest factorial design enabling a conjunction analysis is a 2-by-2 design. The previous study described by Garcia et al. (2010) could lend itself to a conjunction analysis if one were interested in identifying the common pitch-related response, irrespective of context. For example, in Figure 15.6B the conjunction is defined as the following statistical relationship between the two task pairs: $(B^1 - A^1)$ AND $(B^2 - A^2)$. Another example is that reported by Bangert et al. (2006), in which the authors used a conjunction analysis to identify an auditory and sensorimotor cortical network that was more active for musicians than non-musicians during acoustic and motor tasks. In one condition, musicians and non-musicians were required to passively listen to short piano melodies. In another condition, they were required to press keys on a muted MRI-compatible piano. Piano was the component of interest common to both tasks. Despite the motor task being non-acoustic, the authors hypothesized that auditory processing would also be triggered for musicians because they would generate auditory imagery when playing the muted piano. The cognitive conjunction revealed areas of co-activation for both auditory and motor tasks distributed throughout frontal, temporal, and parietal areas and this network was greater for musicians than non-musicians. These findings suggest the presence of a core network of auditory and sensorimotor integration in musicians, which is recruited when listening to music as well as when performing musical actions.

Parametric designs

Parametric designs are based on the assumption that perceptual or cognitive tasks are processed along a continuum of varying levels of perceptual salience or difficulty (e.g., perceptual or cognitive load) that is directly associated with a change in neural activity (Büchel et al., 1996; Amaro and Barker, 2006). The stimulus parameter of interest is manipulated so that it takes a number of levels (typical three or more), and the analysis determines which brain regions show a systematic change in the BOLD response (Fig. 15.6C: A^1, A^2, A^3). For example, the change could be linear or a quadratic relationship. One of the advantages of a parametric design is that it is not necessary to include a matched baseline condition (although of course this might help with interpretation of the findings).

A number of fMRI studies of pitch processing have attempted to identify auditory cortical regions in which the pitch-related activation covaries as a function of pitch salience (e.g., Barker et al., 2011). These authors first used a psychophysical paradigm to confirm three levels of pitch salience for two types of pitch-evoking stimuli (pulse train and an unresolved harmonic complex tone). These stimuli were presented to listeners during an fMRI experiment and then a one-way analysis of variance was performed on the data for the three levels of salience. In this particular study, the F-test did not show a significant effect of salience within auditory cortex ($P > 0.05$, uncorrected).

It is possible to combine both parametric and factorial designs in order to utilize the advantages of both designs (Amaro and Barker, 2006). For example, the design used by Barker et al. (2011) could have lent itself to a conjunction analysis to determine whether any pitch salience effects were consistent for both types of pitch-evoking stimulus.

Adaptation designs

The previous experimental designs have been criticized for not being highly sensitive to detecting activation arising from sensory, perceptual, or cognitive processing in cases where neural populations coding different variables are spatially overlapping or close to one another, particularly when the overall response within a given voxel is weak (Krekelberg et al., 2006; Altmann et al., 2007). The adaptation technique is based on the principle that processing is more efficient for a repeated presentation of a stimulus feature; neural adaptation reduces activity when two successive stimuli activate the same neural subpopulation but not when they stimulate different neural subpopulations ("release from adaptation") (Grill-Spector et al., 2006). Adaptation is investigated using an initial stimulus (the adaptor), followed by a second stimulus (the probe), which varies in its relationship to the adaptor. A reduced BOLD response to the probe relative to the adaptor (i.e., adaptation) indicates that the same or overlapping neural populations are sensitive to both the adaptor and the probe stimuli (Fig. 15.7A). Equivalent BOLD responses to the probe and the adaptor (i.e., release from adaptation) indicate that a different neuronal population is recruited to process the probe change (Fig. 15.7B). Adaptation effects in the MR response have been shown to correlate with faster and more accurate performance during auditory animal discrimination tasks (Bergerbest et al., 2004).

Release from adaptation is indicative of feature-specific neural representations. Furthermore, adaptation has advantages over subtraction designs because it is not limited by the need for a precisely matched baseline; the feature of interest is present in both adaptor and probe stimuli. Adaptation can exploit the adaptor–probe relationship by varying the probe along several dimensions

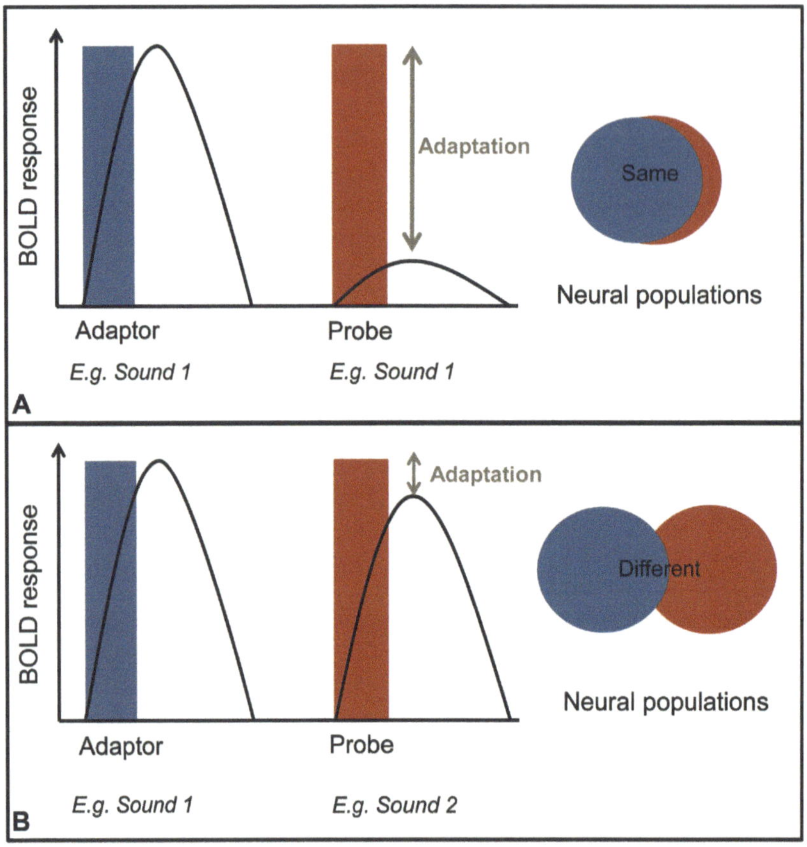

Fig. 15.7. The assumptions underlying the interpretation of adaptation designs. (**A**) Maximal adaptation after repeated stimulus presentation. (**B**) Release from adaptation.

(e.g., different animal vocalizations, periodicity, and amplitude envelope: Altmann et al., 2007).

Adaptation designs have not been widely employed in auditory fMRI experiments, but several examples can be found in the literature. For instance, Altmann et al. (2007) examined the cortical representation of complex natural sounds within auditory cortex using fMRI adaptation. They found that response magnitude in the left superior temporal gyrus was significantly weaker for repeated presentations of animal vocalizations compared to different animal vocalizations, providing evidence that this region is involved in the selective processing of animal vocalizations. Adaptation can enhance sensitivity, allowing researchers to infer the neural representations involved in coding stimulus features within and across stimulus dimensions, and detect invariant and non-invariant representations accordingly. While the neural mechanisms underlying adaptation may still be relatively unclear, this design has proven itself to be a highly robust and insightful phenomenon, whereby its continued application is likely to bring even more promise (Grill-Spector et al., 2006).

ANALYSIS

For auditory fMRI studies, the choice of analysis software, image processing, and statistical methods can have an important influence on the results obtained. A number of software packages are readily and commercially available. The most widely used has been developed by the group based at University College London: Statistical Parametric Mapping (SPM: http://www.fil.ion.ucl.ac.uk/spm). AFNI (http://afni.nimh.nih.gov/afni), originally developed at the Medical College of Wisconsin, United States, and Brain Voyager (http://brainvoyager.de), developed at Maastrict University, The Netherlands.

The two stages of data analysis are: (1) spatial preprocessing of the fMRI time series, and (2) statistical analysis. Needless to say, there is no "gold standard" for fMRI analysis and so this can make direct comparisons between study findings sometimes difficult. However there are a number of commonly used steps, which are discussed here. We refer interested readers to several excellent textbooks describing the fundamentals of fMRI analysis (Huettel et al., 2009; Poldrack et al., 2011).

Spatial preprocessing

Just like music recorded in a studio must be mixed and edited prior to its release, a number of computational procedures are required to reduce sources of within-subject variability (e.g., head movement) and between-subject variability (e.g., brain anatomy) of the fMRI time series data in preparation for statistical analysis. Broadly similar preprocessing algorithms are generally applied, irrespective of the experimental design and software used. The following is based on SPM software, which applies preprocessing in the following order to an individual subject's dataset: reorientation, realignment, slice timing, co-registration, segmentation, spatial normalization, and spatial smoothing (Friston et al., 1995; Huettel et al., 2009). Figure 15.8 shows an anatomic and functional image after these steps have been applied. All preprocessing steps are conducted separately for each listener and scanning session.

Reorientation of individual T1-weighted anatomic and T2*-weighted functional brain images ensures that they are both displayed in the same 3D orientation (i.e., in the same *x, y,* and *z* plane) as one another, as well as to the standardized brain template that is to be used in the later stages.

Within-subject realignment is intended to correct for any spatial mismatch of functional images that occurs due to head movement during the experimental session. Even the smallest movement (e.g., 2 mm) can have substantial negative effects on the fMRI data by inducing large changes in image intensity from one moment to the next which can be manifest as signal loss or spurious motion-related activation. To reduce such problems, a rigid body transformation is applied to all images in the time series so that they are realigned to match a single reference image (e.g., the middle image). In SPM, the output provides a graphical representation of the six transformations (*x, y,* and *z* plane translations, measured in millimeters, and pitch, roll, and yaw rotations, measured in degrees radians). This output enables the experimenter to visually inspect the quality of the fMRI data with respect to the estimated amount of head motion. Acceptable movements generally do not exceed values above the voxel resolution of acquired T2* images. For example, an auditory fMRI study by Paltoglou et al. (2009) acquired functional images at a voxel resolution of 3 mm^3. For quality assurance, acceptable amounts of translational and rotational head movements were limited to 3 mm and 3°, respectively.

Slice timing is an optional step in preprocessing which accounts for the acquisition timing mismatch between slices. Slice timing correction is typically implemented when scans have been acquired in equal spacing across

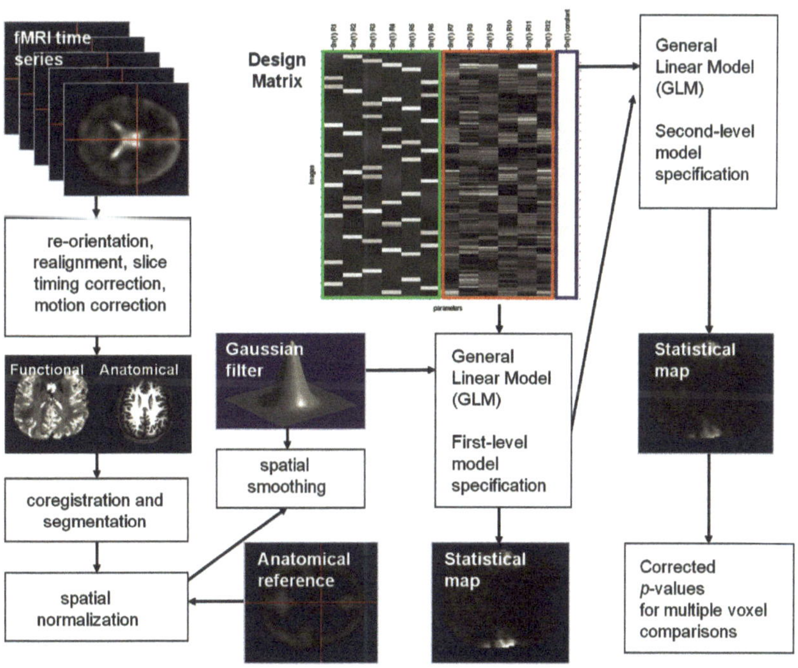

Fig. 15.8. A schematic flow diagram illustrating the different spatial preprocessing and statistical analysis steps in Statistical Parametric Mapping. Steps are described more fully in the text. In this example, the first-level design matrix depicts the experimental regressors of interest (six sound conditions: green box), the nuisance regressors (six translational and rotational head motion parameters: red box) and the β_0 constant (mean voxel intensity: blue box) for one subject's experimental run. fMRI, functional magnetic resonance imaging.

short TRs (e.g., 2 seconds; Kim et al., 2009), and when the precise timing of stimulus events is an important aspect of the analysis, as when modeling BOLD responses to single brief events. Brain slices are acquired using an ascending/descending approach (e.g., 1-2-3-4-5-6), or more commonly, an interleaved approach where all odd-number slices are collected first, followed by even-numbered slices (e.g., 1-3-5-2-4-6). Sequential slice acquisition can be problematic because adjacent slices are collected at different time points but statistical models used in later stages of analysis assume all slices in the scan were acquired at the same time point, thus causing a discrepancy between the expected and actually measured hemodynamic response (Henson et al., 1999). The order in which slice timing correction and realignment are performed can affect the nature of interactions that occur between them and subsequently affect the fMRI data. If there is little head movement between scans, slice timing correction after realignment is the most commonly used approach because, as discussed earlier, even the smallest changes in head movement can lead to extremely large intensity differences. Conversely, if there is a lot of head movement, it may be more appropriate to complete slice timing correction prior to realignment. This is because it is likely that slice timing will correct for the wrong slice, including a mixture of data acquired at different time points, if large head movements have already been corrected. However, slice time correction before realignment may cause motion-related intensity differences to be propagated through time. Alternatively, the use of multiple regressors in later statistical modeling has also been proposed to correct for temporal difference between slices (Henson et al., 1999). Temporal data interpolation is the most common correction method, in which a fixed time reference (usually the first slice) is chosen, to which all other slices are temporally realigned.

Co-registration is a fast automated procedure that matches the anatomic image to the mean functional image by maximizing mutual information across the two types of contrast images. This option is often chosen to allow for a better normalization, when computed indirectly. Segmentation is the second step in indirect normalization. It segments the T1-weighted anatomic scan into its constituent tissue types (cerebrospinal fluid, white matter, and gray matter). These calculations will be used in the next step of normalization of the functional images.

Spatial normalization enables data to be later combined across individuals for group analysis. Although there is substantial variability in the morphology of individual brains, normalization compensates for this by mathematically adjusting the individual brain. Normalization describes the non-linear transformation of anatomic and functional images from different individuals into a standardized brain space using a standard anatomic reference. The most commonly used space was created by researchers at the Montreal Neurological Institute (known as MNI brain space), where the coordinates are derived from an average of over 300 individuals (Evans et al., 1993).

Spatial smoothing applies a smoothing filter to the images. This helps with spatial differences across individual brains, especially when it comes to group analysis. Smoothing typically involves "convolving" the data with a gaussian filter (or kernel), which removes high-frequency information (Friston et al., 2000). This helps to increase SNR. A larger smoothing kernel leads to greater smoothing, but reduces spatial resolution. If there is a specific region of interest, sensitivity can be optimized by smoothing to the diameter of that region. For many auditory fMRI studies seeking to detect activity in auditory cortex, it is common practice to use a kernel within 8–12 mm full width at half maximum (e.g., Rodd et al., 2005; Garcia et al., 2010).

Statistical analyses

General linear model

The majority of fMRI data analysis tests whether the observed data fit a predicted model using assumptions of the timing and duration of the underlying stimulus- and/or task-evoked neural activity.

The analysis software packages mentioned above implement the general linear model (GLM: Worsley and Friston, 1995). A GLM is based on the principle of linear regression and assumes that the BOLD activation (y) can be modeled as a linear combination of separate factors (regressors, X_n), each with a parameter weighting (β_n), plus some uncorrelated residual gaussian noise (ϵ). Designs with more than one regressor (X_n) are easy to specify within this framework (i.e., $y = \beta_0 + \beta_1 {*} x_1 + \beta_2 {*} x_2 + \ldots + \beta_n {*} x_n + \epsilon$).

β_0 is the intercept term that reflects the total contribution of all factors that are held constant, such as the baseline signal intensity. B_n represents how much each regression factor contributes to the BOLD signal. Factors (X_n) can represent variables of interest, such as the timing of the different stimulus and/or task conditions (taking into account the relative sluggishness of the hemodynamic response), or response accuracy. Factors can also represent confounds, such as the translational and rotational values obtained during image realignment, and/or heart rate or slice timing information, to eliminate known sources of error in the MR signal. The model attempts to identify the best-fit estimates of parameter weights (β_n) that minimize the error term using a least-squares error (sum of squared residuals) cost function. The residual error is the difference between the observed data (y) and that predicted by the model ($\beta_n * x_n$), and is assumed to be independently, identically, and normally distributed, with a mean of 0 and a variance of σ^2 (i.e., $\epsilon \sim N(0,\sigma^2)$).

In SPM software, the GLM is implemented using a design matrix. A design matrix depicts the predicted time course of the hemodynamic response over an experimental session for each participant. For instance, the design matrix illustrated in Figure 15.8 was specified by the authors for a pitch-coding experiment. Each column represents a regressor and each row represents a scanning time point (TR = 8.5 seconds). There were six sound conditions (highlighted by the green box) and six head movement parameters (highlighted by the red box). For this "on/off" paradigm, each scan in which a particular sound condition was present was coded by a weighting of 1 and each scan where that condition was absent was coded by a zero. The silent baseline condition is implicitly specified in the model. Head movement parameters from the realignment step were included to minimize residual error variability and these columns contained continuous values. The last column in the design matrix (highlighted by the blue box) depicts the β_0 constant (i.e., overall mean voxel intensity). Regressors can be thought of in terms of predictors of the BOLD response, and a beta (β) value is estimated for each.

The BOLD response does not simply switch on and off like the matrix implies because it is sluggish. Regressors in the design matrix need to be reshaped or "convolved" to resemble the canonic hemodynamic response function (Friston et al., 1998). Most analysis packages, such as SPM, assume to peak at each voxel approximately 5–6 seconds after stimulus onset. Several other fMRI characteristics should also be taken into account to reduce error. For example, variable amplifier gain and scanner drift during image acquisition introduce variations of signal amplitude with every scan and between scanning sessions. These can be "corrected" at the analysis stage through proportional and grand-mean scaling of the time series and by applying high-pass filtering during implementation of the design matrix. Another well-known example is when physiologic fluctuations such as breathing and heart beat lead to serial temporal correlations with neural activity in relation to a stimulus or task, whereby activity at one time point correlates with another.

As the name implies, GLM can be used for a number of different types of analyses (e.g., t-tests, analysis of variance, correlations) and the design matrices can be flexibly applied to all of the experimental designs described in the section on experimental designs, above. The GLM is a univariate method where the analysis is computed separately at each voxel and the GLM generates a statistic from the results of this analysis, independently for each voxel. This is known as the statistical parametric map. It is important to implement design matrices that accurately model the predicted BOLD response because the statistical maps generated from these will ultimately reflect the outcome of statistical tests and hypotheses.

Individual and group-level analyses

In fMRI, the between-scan variance within an experimental session is normally much smaller than the between-subject variance across sessions. If one wishes to make an inference from a subject sample to the population, one needs a proper mixture of within- and between-subject variance. Mixed models provide this approach as the experimental factors have fixed values (fixed effects), but the "subject" factor has values that can vary (random effects). In SPM, this is achieved by a two-stage GLM procedure: (1) contrasts of parameters are estimated from a fixed-effect model for each subject; and (2) images of these contrasts become the data for a second design matrix (usually a simple t-test or analysis of variance: Holmes and Friston, 1999).

The first-level specification models the scan-to-scan variability within each participant using the GLM approach described above in the section on general linear model, which fits the model for each subject. The effect of interest for each subject is defined using a contrast vector which produces a contrast image containing the contrast of the weighted parameter estimates at each voxel. Contrast images for all participants in the experiment are fed into a second-level specification that implements a statistical test. Here the mean "activity" is compared to the distribution of activity across participants and tested for statistical significance. For example, in our pitch-coding experiment illustrated in Figure 15.8, we conducted a t-test on the first-level model specification which contrasted those pitch sound conditions against the noise control sound conditions. This contrast identified pitch-related activity (pitch > noise) across the brain. A contrast for each subject ($n = 18$) was then inputted into the second level of specification to estimate the size of the pitch-related effect across the group. Such random effects analyses have been used in a number of auditory fMRI studies such as to investigate pitch coding (Hall and Plack, 2009).

In summary, in SPM, both first- and second-level specifications use the GLM model and similar software machinery. Both produce design matrices in which the rows represent observations (first level: conditions over time; within-subject variability; and second level: subjects; between-subject variability) and the columns represent explanatory variables (first level: all conditions within the experimental design, and second level: the specific effects of interest).

Statistical inference

Statistical inference refers to the process of drawing conclusions from the model estimation. When computing the GLM, a β value is estimated for each regressor (i.e., column in the design matrix). β values can be used to compare regressors and compute activation maps by creating *t* statistics and equivalent *z* scores for each voxel in normalized brain space. The null hypothesis for fMRI images is that all βs are zero (i.e., that none of the regressors has an effect on the MR signal in the area being scanned). SPM provides a number of different levels of statistical inference for drawing conclusions about the β estimates (i.e., voxel-level and cluster-level; Friston et al., 1996b). The most commonly used is the voxel-level inference which tells us the likelihood of obtaining at least one voxel whose statistic exceeds the alpha threshold (e.g., $P < 0.05$). For example, Warren and Griffiths (2003) identified pitch- and location-sensitive regions using a voxel-level threshold of $P < 0.05$ with a small volume correction.

Plotting statistical results and multiple voxel comparisons

Statistical maps provide a graphical representation of brain activity, usually overlaid on to an anatomic image (Fig. 15.1C). The statistical map shows the output of formal significance testing by displaying those voxels passing a given probability threshold (e.g., $P < 0.05$, or 0.001). The color scale often corresponds to the statistical values and so does not represent more or less brain activity.

The problem of multiple comparisons often arises when conducting univariate analyses. A typical fMRI dataset may contain more than 100 000 voxels within the brain and so when tests are applied to each and every voxel there is a strong likelihood of false positives (type I errors). For example, using an uncorrected *P*-value of 0.05 will lead us to conclude on average that 5% of voxels are active, when they are not.

Many analysis software packages readily provide automated corrections for multiple comparison testing (Bennett et al., 2009). One way to correct for whole-brain multiple comparisons is by controlling for family-wise error rate (FWE). This refers to the probability of making one or more false discoveries, or type I errors, among all the hypotheses when performing multiple hypotheses tests and so is a very stringent correction approach. Bonferroni correction is one FWE method, but it is only exact when the tests are independent. In fMRI, although the tests are performed on separate voxels, the spatial smoothness of the data means that independence is not a valid assumption.

Random field theory is a recent body of mathematics for defining theoretic results for smoothed statistical maps and so is more relevant for fMRI (Nichols and Hayasaka, 2003). It uses the expected Euler characteristic to enable one to find a threshold in a set of data where it is not easy (or it is even impossible) to find the number of independent variables. False discovery rate (FDR) is another correction method which controls for the expected proportion of false positives amongst those voxels declared positive (Genovese et al., 2002). FDR may be more sensitive in some cases; however, it is more important to decide on a correction method *a priori*, rather than subjectively adjusting thresholds to give desirable results.

If one has a strong *a priori* hypothesis regarding the nature of activation in a particular localized brain region, then correcting the statistics for the region of interest (ROI) can be somewhat more sensitive than correcting for the whole brain (Poldrack, 2007). Here, the correction is applied to the number of voxels in the ROI (small-volume correction). For example, Warren and Griffiths (2003) used a small-volume correction to restrict analysis to the auditory cortex (superior temporal gyrus) in their study investigating pitch and spatial processing.

Comparisons between brain regions

The statistical tests described hitherto do not consider questions about how activity differs across brain regions. For auditory fMRI, this is a common question of interest, especially when seeking to understand the functional role of different non-primary auditory fields. ROI analyses seek to do just this (Fig. 15.9). For example, Garcia et al. (2010) used this approach to compare pitch-related activation in the Heschl's gyrus and planum temporale. This method was based on the β estimates from the first-level GLM specification. For each ROI, they computed the mean β value (estimates of the BOLD response magnitude) separately for each participant and sound condition. The mean β values were used as a dependent variable and subjected to formal statistical testing, using analysis of variance. Figure 15.9 illustrates another example of ROI analysis, relating to the author's pitch-coding experiment, discussed above. BOLD activation across the six sound conditions has been plotted separately for each ROI (i.e., planum temporale and Heschl's gyrus, shown in white). In this example, only Heschl's gyrus showed a sensitivity to pitch (conditions 1, 3, and 4) relative to the baseline noise (conditions 5 and 6).

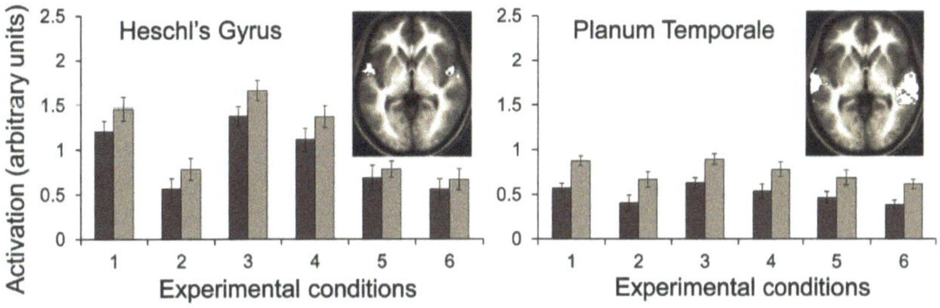

Fig. 15.9. Regions of interest (ROI) analysis of two auditory regions of interest, Heschl's gyrus and planum temporale (shown in white), which have been overlaid on to a group-averaged anatomic image for illustration. Blood oxygen level-dependent activation for sound conditions (1–6) is plotted separately for each ROI.

CONCLUDING REMARKS

As the previous sections demonstrate, fMRI using task-based or stimulus-driven experimental designs has been critical to our current understanding of central auditory processing in the human brain. Despite the strengths of fMRI, we acknowledge several limitations that are often ignored by published reports: namely, the degree of variation of responses among the normal population and the rather indirect measure of neural activity that limits fMRI to changes that are several orders of magnitude slower than the true underlying activity. Nevertheless, the application of fMRI is growing and a number of novel fMRI methods are being developed and applied to other sensory and cognitive domains, shedding light on the complexity of human brain function.

At the New Horizons in Human Brain Mapping meeting in 2013, Peter Bandettini, one of the world leaders in fMRI methods, reported on two major general trends to emerge in MRI and fMRI, both of which require algorithms for robustly characterizing individual traits from imaging neuroimaging data. The first trend is the push to use MRI and fMRI to classify individuals and assess individual variation in a systematic and rigorous manner. The second trend is to combine genetic information and fMRI data in the same subjects to define neuromechanisms linked to genetic variation. Both of these trends come together with the question of what might be most useful or instructive from MRI and fMRI information. However, these novel and innovative techniques have yet to be tested in the auditory domain.

Recent years have seen growing interest in the application of the fMRI at rest, termed resting-state or functional connectivity imaging. Resting-state fMRI investigates synchronous activations between regions that are spatially distinct, occurring in the absence of a task or stimulus. Such patterns are termed resting-state networks (RSNs). They are thought to reflect the intrinsic energy demands of neural populations that are wired together through synaptic plasticity and fire together with a common functional purpose (Cole et al., 2010). Thus RSNs reflect an integral part of information processing in the brain.

A number of distinct RSNs have been identified in normal healthy participants, including sensory networks for visual and auditory processing (e.g., Beckmann et al., 2005; Mantini et al., 2007). These studies have shown that the auditory RSN includes Heschl's gyrus, planum temporale, planum polare, and the lateral superior temporal gyrus. Investigations are now being applied to clinical populations as a potential biomarker for detecting underlying abnormalities in neural function. Most recently, resting-state imaging has been widely applied to investigate functional connectivity in people with tinnitus. Tinnitus refers to a chronic condition in which people experience a hissing, whistling, or ringing in their ears, sometimes continuously. Given that tinnitus is a constant experience, tinnitus may alter the underlying neural population activity within and between RSNs. To the best of our knowledge, no such fMRI studies had been published before 2012, but at the time of writing there are at least six publications. Kim et al. (2012) claimed to be the first group to apply this technique to tinnitus research when they reported a pilot study with 4 participants with chronic subjective tinnitus and 6 age-matched controls. Despite early promising results, subsequent studies have provided a rather mixed picture of results.

Characterization of the time course of activity within the auditory RSN has been somewhat limited by the timescale of the fMRI BOLD response. Multimodal imaging methods are now demonstrating the capability of measuring those time courses by combining the spatial resolution of fMRI with the high temporal resolution of an electroencephalograph or magnoencephalograph. Results indicate that oscillations at various frequencies form functional networks across brain regions. For example, Mantini et al. (2007) showed that the auditory RSN is characterized by power fluctuations in delta (1–4 Hz), theta (4–8 Hz), and beta (13–30 Hz) rhythms. The utility of the auditory RSN for clinical diagnosis and informing treatment choice for tinnitus is yet to be fully realized, but we remain hopeful that this will prove to be a fruitful direction for auditory fMRI research more generally.

ACKNOWLEDGMENTS

We would like to thank Dr. Sue Francis from the Sir Peter Mansfield Magnetic Resonance Centre, University of Nottingham (UK) for her assistance in the T2* mapping of our pilot fMRI data reported in Figure 15.2. We would also like to thank Mr Jeff Davies for kindly providing Figure 15.5.

References

Altmann CF, Doehrmann O, Kaiser J (2007). Selectivity for animal vocalizations in the human auditory cortex. Cereb Cortex 17: 2601–2608.

Amaro E, Barker GJ (2006). Study design in MRI: basic principles. Brain and Cogn 60: 220–232.

Bangert M, Peschel T, Schlaug G et al. (2006). Shared networks for auditory and motor processing in professional pianists: evidence from fMRI conjunction. Neuroimage 30: 917–926.

Barker D, Plack CJ, Hall DA (2011). Human auditory cortical responses to pitch and to pitch strength. Neuroreport 22: 111–115.

Barker D, Plack CJ, Hall DA (2012). Re-examining the evidence for a pitch-sensitive region: a human fMRI study using iterated ripple noise. Cereb Cortex 22: 745–753.

Beckmann CF, de Luca M, Devlin JT et al. (2005). Investigations into resting-state connectivity using independent component analysis. Philos Trans R Soc Lond B Biol Sci 360 (1457): 1001–1013.

Bennett CM, Woldord GL, Miller MB (2009). The principled control of false positives in neuroimaging. Soc Cogn Affect Neurosci 4: 417–422.

Bergerbest D, Ghahremani DG, Gabrieli JD (2004). The neural correlates of auditory repetition priming: reduced fMRI activation in the auditory cortex. J Cogn Neurosci 16 (6): 966–977.

Blackman G, Hall DA (2011). Reducing the effects of background noise during auditory functional magnetic resonance imaging of speech processing: qualitative and quantitative comparisons between two image acquisition schemes and noise cancellation. J Speech Lang Hear Res 54: 693–704.

Blamire AM (2008). The technology of MRI – the next 10 years? BJR 81: 601–617.

Büchel C, Wise RJS, Mummery CJ et al. (1996). Nonlinear regression in parametric activation studies. Neuroimage 4: 60–66.

Chambers J, Akeroyd MA, Summerfield AQ et al. (2001). Active control of the volume acquisition noise in functional magnetic resonance imaging: method and psychoacoustical evaluation. J Acoust Soc Am 110: 3041–3054.

Clare S, Francis S, Morris PG et al. (2001). Single-shot T2* measurement to establish optimum echo time for fMRI: studies of the visual, motor, and auditory cortices at 3.0 T. Magn Reson Med 45: 930–933.

Cole DM, Smith S, Beckmann CF (2010). Advances and pitfalls in the analysis and interpretation of resting-state fMRI data. Front Syst Neurosci 4 (8): 1–15.

Constable RT, Spencer DD (2001). Repetition time in echo planar functional MRI. Magn Reson Med 46 (4): 748–755.

Cusack R, Brett M, Oasswald K (2003). An evaluation of the use of magnetic field maps to undistort echo-planar images. Neuroimage 18: 127–142.

Da Costa S, van der Zwaag W, Marques JP et al. (2011). Human primary auditory cortex follows the shape of Heschl's gyrus. J Neurosci 31 (40): 14067–14075.

De Martino F, Schmitter S, Moerel M et al. (2012). Spin echo functional MRI in bilateral auditory cortices at 7 T: an application of B_1 shimming. Neuroimage 63 (3): 1313–1320.

Degerman A, Rinne T, Salmi J et al. (2006). Selective attention to sound location or pitch studied with fMRI. Brain Res 1077: 123–134.

Duyn J (2004). Receiver Technology. Second international workshop on parallel MRI. October 2004. Zurich, Switzerland. Available online at, www.mr.ethz.ch/parallelmri04/abstracts/pub/Duyn.pdf.

Evans AC, Collins DL, Mills SR et al. (1993). 3D statistical neuroanatomical models from 305 MRI volumes. In: L Klaisner (Ed.), Nuclear Science Symposium & Medical Imaging Conference. IEEE Conference Record. vol 4. 1813–1817.

Friston KJ, Ashburner J, Frith CD et al. (1995). Spatial registration and normalization of images. Hum Brain Map 3 (3): 165–189.

Friston KJ, Price CJ, Fletcher C et al. (1996a). The trouble with cognitive subtraction. Neuroimage 4: 97–104.

Friston KJ, Holmes A, Poline JB et al. (1996b). Detecting activations in PET and fMRI: levels of inference and power. Neuroimage 4 (3): 223–235.

Friston KJ, Fletcher P, Josephs O et al. (1998). Event-related fMRI: characterizing differential responses. Neuroimage 7: 30–40.

Friston KJ, Josephs O, Zarahn E et al. (2000). To smooth or not to smooth? Bias and efficiency in fMRI time series analysis. Neuroimage 12: 196–208.

Garcia D, Hall DA, Plack CJ (2010). The effect of stimulus context on pitch representations in the human auditory cortex. Neuroimage 51: 808–816.

Genovese CR, Lazar NA, Nichols TE (2002). Thresholding of statistical maps in functional neuroimaging using the false discovery rate. Neuroimage 15 (4): 870–878.

Gizewski ER, Maderwald S, Wanke I et al. (2005). Comparison of volume, four- and eight-channel head coils using standard and parallel imaging. Eur Radiol 15 (8): 1555–1562.

Griffiths TD, Hall DA (2012). Mapping pitch representation in neural ensembles with fMRI. J Neurosci 32 (39): 13343–13347.

Grill-Spector K, Henson R, Martin A (2006). Repetition and the brain: neural models of stimulus-specific effects. Trends Cogn Sci 10 (1): 14–23.

Guimaraes AR, Melcher JR, Talavage TM et al. (1998). Imaging subcortical auditory activity in humans. Hum Brain Map 6 (1): 33–41.

Hall DA, Plack CJ (2009). Pitch processing sites in human auditory cortex. Cereb Cortex 19 (3): 543–553.

Hall DA, Wild DC (2012). Fundamental principles underlying MRI and functional MRI. In: KL Tremblay, RF Burkard (Eds.), Translational Perspectives in Auditory Neuroscience: Hearing Across the Lifespan – Assessment and Disorders. CA, Plural Publishing Inc., San Diego, pp. 283–315.

Hall DA, Haggard MP, Akeroyd MA et al. (1999). Sparse temporal sampling in auditory fMRI. Hum Brain Map 7 (3): 213–223.

Hall DA, Summerfield AQ, Gonçalves MS et al. (2000). Time-course of the auditory BOLD response to scanner noise. Magn Reson Med 43 (4): 601–606.

Hall DA, Chambers J, Foster J et al. (2009). Acoustic, psychophysical, and neuroimaging measurements of the effectiveness of active cancellation during auditory functional magnetic resonance imaging. J Acoust Soc Am 125 (1): 347–359.

Healy EW, Moser DC, Morrow L et al. (2007). Speech perception in MRI scanner noise by persons with aphasia. J Speech Lang Hear Res 50: 323–334.

Henson RNA (2007). Efficient experimental design for fMRI. In: KJ Friston, JT Ashburner, SJ Kiebel et al. (Eds.), Statistical Parametric Mapping: The Analysis of Functional Brain Images. Academic Press, London, pp. 193–210.

Henson R, Büchel C, Josephs O et al. (1999). The slice-timing problem in event-related fMRI. Neuroimage 9: 125.

Holmes AP, Friston KJ (1999). Generalisability, random effects and population inference. Neuroimage 7 (4(2/3)): S754.

Huettel SA, Song AW, McCArthy G (2009). Functional magnetic resonance imaging (second edition). Sunderland, MA, USA, Sinauer Associates.

Jezzard P, Balaban RS (1995). Correction for geometric distortion in echo planar images from B0 field variations. Magn Reson Med 34 (1): 65–73.

Jezzard P, Matthews PM, Smith SM (2003). Functional MRI: an introduction to methods. Oxford University Press, Oxford, UK.

Kim DI, Mathalons DH, Ford JM (2009). Auditory oddball deficits in schizophrenia: an independent component analysis of the fMRI multisite function BIRN study. Schizophr Bull 35 (1): 67–81.

Kim JY, Kim YH, Lee S et al. (2012). Alteration of functional connectivity in tinnitus brain revealed by resting-state fMRI? A pilot study. Int J Audiol 51 (5): 413–417.

Krekelberg B, Boynton GM, van Wezel RJ (2006). Adaptation: from single cells to BOLD signals. Trends Neurosci 29 (5): 250–256.

Logothetis NK, Pauls J, Augath M et al. (2001). Neurophysiological investigation of the basis of the fMRI signal. Nature 412: 150–157.

Mantini D, Perrucci MG, Gratta DC et al. (2007). Electrophysiological signatures of resting state networks in the human brain. Proc Natl Acad Sci 104 (32): 13170–13175.

Nichols TE, Hayasaka S (2003). Controlling for familywise error rate in functional neuroimaging: a comparative review. Stat Meth Med Res 12 (5): 419–446.

Nichols T, Brett M, Andersson J et al. (2005). Valid conjunction inference with the minimum statistic. Neuroimage 25: 653–660.

Ogawa S, Lee TM, Kay AR et al. (1990). Brain magnetic resonance imaging with contrast dependent on blood oxygenation. Proc Natl Acad Sci U S A 87 (24): 9868–9872.

Olulade O, Hu S, Gonzalez-Castillo J et al. (2011). Assessment of temporal state-dependent interactions between auditory fMRI responses to desired and undesired acoustic sources. Hear Res 277 (1–2): 67–77.

Paltoglou A, Sumner CJ, Hall DA (2009). Examining the role of frequency specificity in the enhancement and suppression of human auditory cortical activity by auditory selective attention. Hear Res 257: 106–118.

Poldrack R (2007). Region of interest analysis for fMRI. Soc Cogn Affect Neurosci 2: 67–70.

Poldrack RA, Mumford JA, Nichols TE (2011). Handbook of fMRI data analysis. Cambridge University Press, Cambridge.

Price CJ, Moore CJ, Friston KJ (1997). Subtractions, conjunctions, and interactions in experimental design of activation studies. Hum Brain Map 5: 264–272.

Rodd JM, Davis MH, Johnsrude IS (2005). The neural mechanisms of speech comprehension: fMRI studies of semantic ambiguity. Cereb Cortex 15: 1261–1269.

Satori G, Umiltà C (2000). How to avoid the fallacies of cognitive subtraction in brain imaging. Brain Lang 74: 191–212.

Scarabino T, Giannatempo GM, Popolizio T et al. (2007). 3.0-T functional brain imaging: a 5-year experience. Radiol Med 112 (1): 97–112.

Schäfer A, van der Zwaag W, Francis ST et al. (2008). High resolution SE-fMRI in humans at 3 and 7 T using a motor task. Magn Reson Mater Phys 21: 113–120.

Sigalovsky IS, Melcher JR (2006). Effects of sound level on fMRI activation in human brainstem, thalamic and cortical centers. Hear Res 215 (1–2): 67–76.

Talavage TM, Ledden PJ, Benson RR et al. (2000). Frequency-dependent responses exhibited by multiple regions in human auditory cortex. Hear Res 150: 225–244.

Triantafyllou C, Hoge RD, Krueger G et al. (2005). Comparison of physiological noise at 1.5 T, 3 T and 7 T and optimization of fMRI acquisition parameters. Neuroimage 26: 243–250.

van der Zwaag W, Francis S, Head K et al. (2009). fMRI at 1.5, 3, and 7 T: characterizing BOLD signal changes. Neuroimage 47: 1425–1434.

Warren JD, Griffiths TD (2003). Distinct mechanisms for processing spatial sequences and pitch sequences in the human auditory brain. J Neurosci 23 (13): 5799–5804.

Worsley KJ, Friston KJ (1995). Analysis of fMRI data revisited—again. Neuroimage 2: 173–181.

Handbook of Clinical Neurology, Vol. 129 (3rd series)
The Human Auditory System
G.G. Celesia and G. Hickok, Editors

Chapter 16

Imaging white-matter pathways of the auditory system with diffusion imaging tractography

CHIARA MAFFEI[1], GUADALUPE SORIA[2], ALBERTO PRATS-GALINO[3], AND MARCO CATANI[1]*

[1]*NatBrainLab, Institute of Psychiatry, King's College London, London, UK*

[2]*Experimental 7T MRI Unit, Institut d'Investigacions Biomediques August Pi i Sunyer, Barcelona, Spain*

[3]*Laboratory of Surgical NeuroAnatomy, Human Anatomy and Embryology Unit, Faculty of Medicine, Universitat de Barcelona, Barcelona, Spain*

INTRODUCTION

The auditory system is a complex network of subcortical projections and corticocortical fibers whose anatomy has been widely investigated in postmortem studies of nonhuman primates using tracing methods (Mesulam and Pandya, 1973; Hackett et al., 1998; Kaas et al., 1999; Kaas and Hackett, 2000). The methods for tracing axonal connections in the animal brain are highly invasive and not applicable to the human brain. Information on the human connectional anatomy of the auditory pathways is therefore lacking.

In the last 15 years the advent of diffusion-based tractography has opened new opportunities to study the anatomic organization of white matter in the living human brain (Conturo et al., 1999; Mori et al., 1999; Basser et al., 2000). Tractography-based virtual reconstruction of white-matter tracts has the power to depict trajectories of connections in the living human brain and correlate structural interindividual differences with behavioral performances (Catani et al., 2007; Lopez-Barroso et al., 2013). This approach is a fundamental step towards understanding how the auditory system has evolved to generate uniquely human abilities (Rilling et al., 2008; Thiebaut de Schotten et al., 2012). However, diffusion imaging of the auditory system has proved to be very challenging (Behrens et al., 2007). The main obstacle is represented by the poor spatial resolution of current diffusion data and the complexity of the anatomy of the connections carrying auditory information from the brainstem to the cortex. While the visualization of the auditory projection tracts in the brainstem remains difficult, and only a few studies have been able to reconstruct *in vivo* the trajectories of the acoustic radiations, current diffusion methods have reliably identified association tracts connecting cortical regions involved in auditory processing and language. The left lateralization of the arcuate fasciculus, for example, is a distinct feature of the human brain linked to language development (Nucifora et al., 2005; Catani et al., 2007). With the progressive refinement of our diffusion methods, researchers will be able to obtain a comprehensive and quantitative map of auditory networks in the human brain to confirm simian–human equivalences and reveal unique hallmarks of our species.

This chapter is dedicated to diffusion methods and tractography applied to the auditory pathways of the human brain. We first introduce basic concepts of diffusion imaging and explain differences between the tensor model and other non-tensorial approaches. We then explain the principles behind tractography reconstruction and its application to both postmortem and *in vivo* imaging of auditory pathways. Our goal is to provide a broad overview of the methodologic advances in diffusion imaging and future developments in the field of tractography applied to the auditory networks of the human brain.

*Correspondence to: Marco Catani, NatBrainLab, PO50, Department of Forensic and Neurodevelopmental Sciences, Institute of Psychiatry, King's College London, De Crespigny Park, London SE5 8AF, UK. Tel: 0044-2078480984, E-mail: m.catani@iop.kcl.ac.uk

DIFFUSION MRI TRACTOGRAPHY

Diffusion magnetic resonance imaging

Diffusion magnetic resonance imaging (MRI) is a visualization method based on the measurement of water molecule diffusion in biologic tissue (Moseley et al., 1990; Le Bihan, 2003). In general the displacement of water molecules can be described using Einstein's equation (Einstein, 1905). This formula tells us that the mean squared displacement of molecules $<r2>$, from their starting point over time t, is directly proportional to the observation time, according to:

$$<r2> = 6Dt$$

where D is the diffusion coefficient. In diffusion imaging we do not measure the water diffusion coefficient directly, but the mean displacement of water molecules within each three-dimensional volume element, or voxel, which contains biologic tissue (Jones, 2008). In the cerebral tissue, molecules of water are not allowed to move freely because cell membranes, myelin, organelles, and so forth hinder their displacement. It follows that in a fixed period of time the overall displacement and, therefore, the diffusion coefficient (D), in brain tissue is reduced compared to the displacement in free water. This is why we generally refer to apparent diffusion coefficient (ADC), to reflect the fact that in brain tissue several biologic barriers hinder diffusion (Le Bihan, 2003).

Figure 16.1 shows a schematic representation of the diffusion of water molecules in both isotropic and anisotropic tissue. In isotropic tissue, where physical properties are identical in all directions, the diffusion of water molecules is reduced equally along all orientations. This is the case for gray matter or cerebrospinal fluid in the lateral ventricles. White matter on the contrary is anisotropic; here the well-organized axonal membranes and myelin sheet allow displacement of water molecules preferentially along the direction of the axons rather than perpendicular to them. Hence, the ADC measured along a direction perpendicular to the axonal fibers is always lower compared to the ADC measured along the direction of the fibers. Diffusion in anisotropic white matter is, therefore, characterized by having a preferential direction, which varies according to the main orientation of the underlying axons. In diffusion MRI the signal is sensitized to the displacement of water molecules along a selected direction and the ADC measurement is strictly dependent on the direction of the applied diffusion-encoding gradient (Jones, 2008).

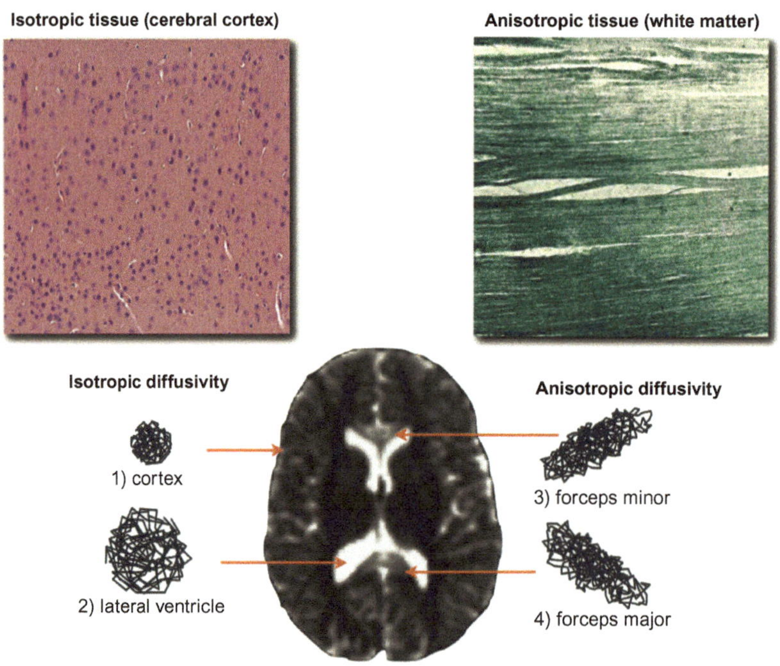

Fig. 16.1. Histologic sections of the (left) cerebral cortex and (right) white-matter fibers in the human brain. The two tissues differ as regard the microstructural organization and composition of their biologic constituents. In the middle, an axial apparent diffusion coefficient map of the human brain is shown together with the visualization of the three-dimensional displacement of a water molecule in regions with different degrees of anisotropy and diffusivity: (1) low isotropic diffusivity in the frontal cortex where the overall displacement is equally hindered in all directions; (2) high isotropic diffusivity in the cerebrospinal fluid of the lateral ventricles where the water molecules freely move in all directions; (3) oblique anisotropic diffusivity along the lateral fibers of the genu in the forceps minor; (4) oblique anisotropic diffusivity along the lateral fibers of the splenium in the forceps major.

Diffusion tensor imaging

In 1994 Peter Basser, James Mattiello, and Denis Le Bihan described diffusion tensor imaging (DTI), an MRI-based method to determine fiber tract orientation and mean particle displacements within a voxel of biologic tissue (Basser et al., 1994). With DTI it was possible, for the first time, not only to obtain useful information about molecular mobility reflective of local microstructural anatomy but also to infer orientation of axonal fibers (Dell'Acqua and Catani, 2012). The tensor model provides a mathematical way to describe the water diffusion in three-dimensional space and independently from the tissue's orientation in the laboratory frame of reference (Basser et al., 1994). The diffusion tensor can also be visualized as an ellipsoid, which describes the three-dimensional geometric profile of the distance traveled by the water molecules (Fig. 16.2A). A detailed analysis of the ellipsoids gives information about two quantitative indexes, the mean diffusivity (MD) and fractional anisotropy (FA). MD describes the average mobility of water molecules, whereas FA varies from 0 to 1 (where 0 means a totally isotropic tissue and 1 a totally anisotropic one) and represents the degree of anisotropy of the biologic tissue. The FA is a measure of how one of the three eigenvectors (the eigenvectors represent the axes of the tensor) differs from the other, providing information about the microstructural organization of the tissue. In addition to MD and FA, axial and radial diffusivity (RD) can be extracted to determine water molecule displacement along a direction parallel or perpendicular to the main orientation of fibers, respectively.

These measurements provide useful information on white-matter microstructural architecture and integrity. It is important to understand the physics behind these measurements because it impacts the interpretation of findings in the human brain. For example, during early human development, FA increases while MD and RD

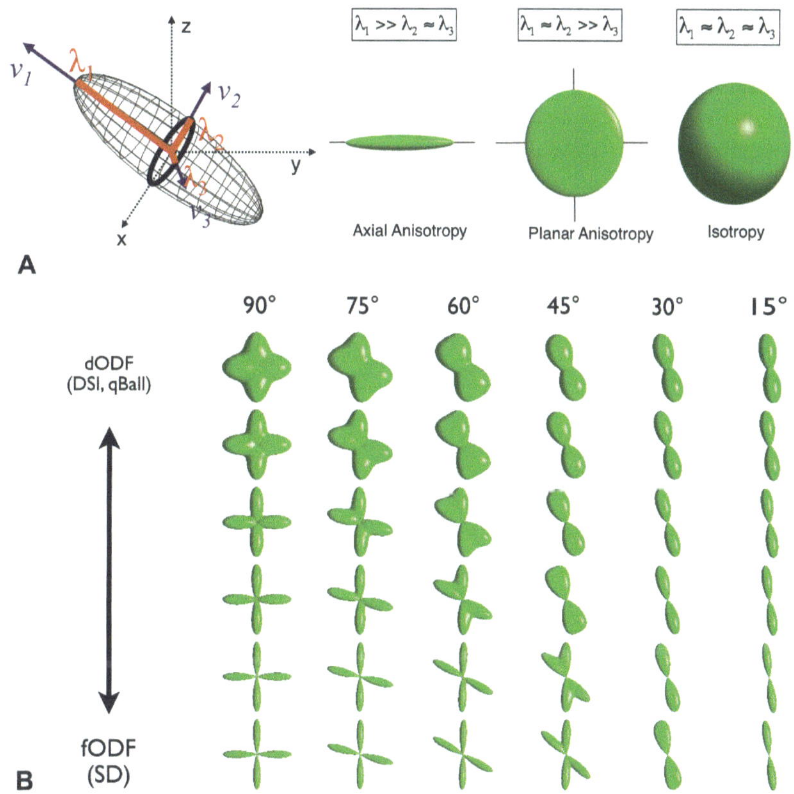

Fig. 16.2. (**A**) Visualization of the diffusion tensor as a diffusion ellipsoid. The size and the shape of the tensor are defined by the three eigenvalues (λ_1, λ_2, λ_3 in red) while the spatial orientation is described by the three eigenvectors (v_1, v_2, v_3 in blue). In biologic tissues the tensor can vary between three possible configurations: (1) axial anisotropy (where λ_1 is greater than λ_2 and λ_3), which is typical of voxels containing parallel fibers; (2) planar anisotropy (when λ_1 is similar to λ_2 and greater than λ_3), which is common in voxels containing, for example, two groups of crossing or diverging fibers; and (3) isotropy (when all three eigenvalues are similar), which is commonly observed, for example, in the gray matter. (**B**) Visualization of the possible orientation distribution function (ODF) profiles according to different advanced diffusion models methods and different crossing angles. Some diffusion ODF (dODF) methods are limited in resolving crossing below 75°, whereas fiber ODF (fODF) methods resolve crossing at 45° or lower (Catani et al., 2012).

reduce (Neil et al., 2002). These changes are usually interpreted as reflecting mainly an increase in axonal myelination during brain maturation (Neil et al., 2002; Lebel et al., 2008). In neurodegenerative disorders FA tends to reduce, whereas MD and RD increase in relation to axonal membrane damage and demyelination (Ciccarelli et al., 2008).

One complication with these measurements arises in situations when a voxel contains crossing fibers (Dell'Acqua and Catani, 2012). Unfortunately, all the diffusion measurements based on the tensor model are voxel-specific, which means that they do not take into account the anatomy of a single population of fibers within the voxel. This is not a problem if all fibers are oriented in the same direction. However, in voxels with multiple fiber population, degeneration of a single fiber component, for example, can result in diffusion changes that are opposite to what is observed in voxels with one fiber population (Pierpaoli et al., 1996). To overcome this problem alternative non-tensorial diffusion models that visualize multiple fiber components have been developed.

Non-tensorial diffusion models

In its original formulation diffusion tensor approaches proceed by first fitting a gaussian model for each voxel-wise diffusion and then extracting the principal diffusion direction (PDD) and using this as a surrogate for axon orientation within each voxel (Jbabdi and Johansen-Berg, 2011). While this may represent a valid approach for voxels with single-fiber components, in voxels containing fibers with different orientation, PDD is not a good representation of the underlying fiber orientation.

To overcome this problem, alternative methods have been developed. These include Q-ball imaging (Tuch, 2004), diffusion spectrum imaging (Wedeen et al., 2005), persistent angular structure MRI (Jansons and Alexander, 2003), hybrid (Wu and Alexander, 2007), multiple-compartment modeling (Behrens et al., 2007; Jbabdi et al., 2007), Bayesian random-effect modeling (King et al., 2009), and spherical deconvolution (Tournier et al., 2004; Dell'Acqua et al., 2010, 2012). All these alternatives to the tensor model can be broadly divided into two approaches (Fig. 16.2B): model-free approaches (or non-parametric), in which the three-dimensional displacement probability profile is obtained directly from the diffusion signal (e.g., Tuch, 2004; Wedeen et al., 2005) and model-based approaches (or multiparametric) in which the diffusion data are fitted with a chosen model that assumes *a priori* a discrete number of fiber orientations (e.g., two or more).

Non-parametric, model-independent methods characterize the water molecular displacement by using a diffusion orientation distribution function (dODF). The estimate of the dODF does not require any assumption about its form and is derived directly from the diffusion characteristics of the fibers. The distribution profile assumes the shape of a multilobe whose peaks generally correspond to a fiber orientation and provide information on the number, weight, and orientation of each fiber component. In other words, these models, instead of reconstructing the fiber orientations *per se*, attempt to recover the complete diffusion profile.

Model-based approaches, instead of modeling a single PDD (as in the tensor-based models), define a number of directions of preferred diffusion and then fit the model with the measured diffusion data. These approaches generate for each voxel an estimation of the fiber orientation distribution function (fODF) from the diffusion profile. Model-based approaches are usually described as spherical deconvolution methods (Tournier et al., 2004; Dell'acqua et al., 2010), and they generally show a higher angular resolution (i.e., the ability to resolve crossing fibers at smaller angles) compared with methods based on dODF approaches (Catani et al., 2012). Figure 16.2B shows a possible range of solutions for decreasing crossing angles for both dODF and fODF methods. The profile of the different ODFs determines the ability to separate crossing populations of fibers. For orthogonal crossing, the easiest case, all ODFs are able to distinguish two orientations, whereas for lower angles, which are more difficult to resolve, only sharper profiles are able to resolve the crossing. The ability to distinguish two or more orientations for lower angles is an important asset when advanced non-tensorial methods are used to visualize crossing pathways with tractography.

Diffusion imaging tractography

One of the major applications of DTI is the three-dimensional reconstruction of white-matter fiber trajectories (Dell'Acqua and Catani, 2012). The main assumption underpinning tractography is that the tensor aligns with the orientation of the fibers contained within each voxel. More specifically, it is assumed that the PDD, which indicates the direction of least-hindered diffusion, corresponds to the main orientation of axons (Jbabdi and Johansen-Berg, 2011).

The second assumption in tractography consists of the inference of an anatomic continuity from voxel to voxel. On the basis of these two fundamental assumptions, the tracking algorithm follows the direction of maximum diffusion from a given voxel into a neighboring voxel. This permits the description of three-dimensional trajectories of fiber bundles by piecing together with continuous lines (streamlines) discrete voxel-based estimates of the underlying fiber orientations. The reconstruction of continuous streamlines requires the determination of specific parameters for continuing or stopping the tracking propagation. For example, to avoid tracking through regions that do not contain white-matter fibers (e.g., lateral ventricles), an FA threshold is usually adopted as stopping criterion. Another common stopping rule is represented by the maximum angle between two contiguous PDD (usually $< 45°$). The aim is ultimately to prevent "unfeasible" turns of path propagation and reduce artefactual tract reconstructions (Mori et al., 1999; Conturo et al., 1999; Jones, 2008). This is particularly true in regions with crossing, kissing, splaying, or branching fibers where the tensor model does not perform so well and both false-negative and false-positive artifacts are likely to occur (Basser et al., 2000). Tractography algorithms generate false-positive reconstructions when two adjacent tensors belonging to different pathways are joined together by the tractography algorithm. False-negative artifacts occur when the tractography algorithm is unable to reconstruct pathways that truly exist. This usually occurs in regions with crossing fibers where the tensor assumes a planar or spherical shape and does not provide a clear indication of the directionality of the underlying fibers.

The availability of new diffusion models that resolve multiple fiber trajectories is crucial to describe tracts that are not visible using current DTI methods or identify new tracts. By using spherical deconvolution tractography, for example, callosal connections between temporal auditory cortex becomes visible (Fig. 16.3).

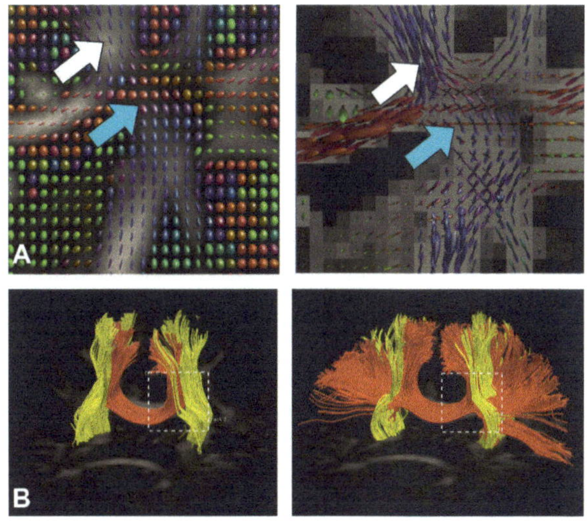

Fig. 16.3. (**A**) Visualization of the white-matter organization based on the tensor model (left) and spherical deconvolution (right) inside the corona radiata. For voxels with one fiber population, such as the corpus callosum (white arrow), both models describe orientations that are consistent with the known anatomy. However, in regions with more than one population of crossing fibers (cyan arrow), the tensor model gives an average representation of the water diffusion, whereas the spherical deconvolution model separates different fiber components and describes their individual orientations. (**B**) Diffusion tensor-based (left) and spherical deconvolution-based (right) tractography. The virtual dissections of the corpus callosum based on diffusion tensor tractography reconstruct only the most central part of the corpus callosum (red), while spherical deconvolution tractography shows several streamlines of the corpus callosum crossing the streamlines of the corticospinal tract (yellow) and reaching the lateral cortex (Catani and Thiebaut de Schotten, 2012).

Another advantage of tractography is the possibility of extracting quantitative diffusion indices, such as FA, MD, and RD, along the dissected tract. This permits one to characterize the microstructural properties of tissue in the normal and pathologic brain and provide quantitative measurements for group comparisons or individual case studies (Catani et al., 2008, 2013; Pugliese et al., 2009). However, as we have seen before, the interpretation of these indices is not always straightforward, especially in regions containing fiber crossing. An example is the increase in fractional anisotropy commonly seen in the normal-appearing white-matter regions distant to a lesioned area. It is tempting to interpret these changes as indicative of plasticity or remodeling, but alternative explanations should be taken into account (Pierpaoli et al., 1996; Dell'Acqua and Catani, 2012). In voxels containing both degenerating and normal fibers, increases in FA values are, in fact, more likely the result of the axonal degeneration of the perpendicular fibers (Fig. 16.4). The lack of specificity of current diffusion indices (i.e., diffusion changes depend on a number of biologic, biochemical, and microstructural factors) (Jones, 2010) and the intrinsic voxel-specific rather than fiber-specific information derived from current indices has stimulated scientists to work on new methods and novel diffusion models (Dell'Acqua et al., 2012; Yeh et al., 2011; Raffelt et al., 2012).

Tractometry (Jones and Pierpaoli, 2005; Catani, 2006; De Santis et al., 2014) is an interesting approach that aims to combine tractography with quantitative mapping along individual tracts of complementary neuroimaging measurements based on, for example, relaxometry, magnetization transfer ratio (Cercignani and Alexander, 2006), myelinated water fraction (Deoni et al., 2008), or multicompartmental diffusion indices (Assaf and Basser, 2005; Alexander, 2008) – methods that are beyond the scope of this review. More recently our group has also proposed novel true-tract specific indices based on spherical deconvolution that better describe the microstructural diffusion changes of individual crossing fibers within the same voxel (Dell'Acqua et al., 2012).

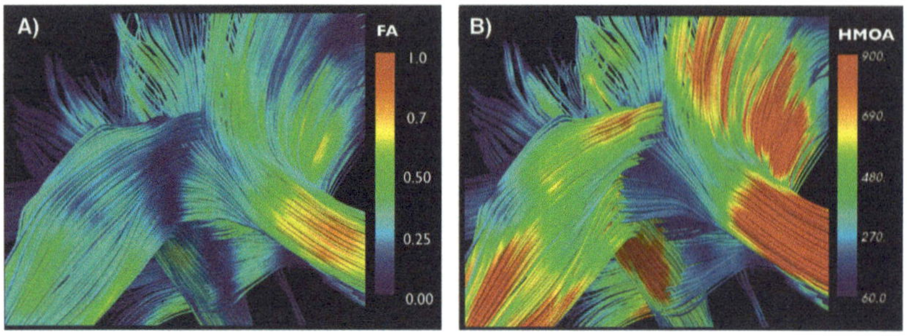

Fig. 16.4. Mapping diffusion indices along crossing white-matter tracts. The arcuate fasciculus and the lateral projections of the corpus callosum cross at the level of the corona radiata. (**A**) Fractional anisotropy (FA) values provide an average, voxel-specific description of the anisotropic properties of the selected brain region and, therefore, identical FA values are attributed to both crossing fibers. (**B**) The hindrance-modulated orientational anisotropy (HMOA) index shows distinct diffusion characteristics for the two crossing tracts, lower HMOA for the callosal projections and higher HMOA for the arcuate fibers. HMOA is therefore considered a true fiber-specific index. (Reproduced from Dell'Acqua et al., 2012.)

Hindrance-modulated orientation anisotropy (HMOA) is a newly proposed index of anisotropy along individual group of fibers within a single voxel. Changes in the HMOA have a greater sensitivity than conventional FA values to detect degeneration that occurs only in one population of fibers while the other crossing fibers remain intact. In the future, tractography will be combined also with methods able to extract even more specific tissue microstructure indices, such as axonal diameter distribution and axonal density or dispersion (Assaf et al., 2008; Zhang et al., 2012).

Two different approaches: deterministic and probabilistic tractography

The most common fiber-tracking algorithms broadly divide into deterministic and probabilistic approaches (Dell'Acqua and Catani, 2012). Deterministic tractography produces only one reconstructed trajectory per seed point (the investigator-selected region of origin for the tract) (Conturo et al., 1999; Mori et al., 1999). The estimated fiber orientation (e.g., direction of maximum diffusivity for the tensor model) is assumed to represent the best estimate to propagate streamlines. In deterministic tractography no indication of the confidence that one can assign to a reconstructed trajectory is provided.

Unlike deterministic tractography, probabilistic methods generate multiple solutions to reflect the "uncertainty" of the estimated fiber orientation (Behrens et al., 2003; Parker et al., 2005). Instead of reconstructing just a single trajectory from a given seed point they propagate multiple streamline samples per seed voxel. At every voxel a probability distribution of the orientations is estimated and connections are traced thousands of time, each time using slightly different orientations (according to their likelihood). The results are connectivity maps in which the intensity represents the number of streamlines that visit each voxel.

In other words, while deterministic models provide a single estimate of the fiber path, probabilistic methods give the probability distribution of that estimate. The latter depends on a number of parameters, which include the signal noise, partial volume effects (i.e., loss of signal due to the limited resolution of the imaging system and the inability to separate different tissues within a voxel), and the inaccuracy of the chosen diffusion model. It should be noted that the high probability of the connectivity maps does not provide confidence on the probability of the underlying anatomic connectivity. Often artifactual trajectories reconstructed with probabilistic tractography have very high probability, similar to true anatomic pathways. Paradoxically, datasets of low quality (e.g., with low signal-to-noise ratio) generate widely distributed connectivity maps in which artifacts are more likely. In general, while deterministic approaches are prone to a higher number of false-negative artifacts, probabilistic methods generate more false-positive reconstructions. Both deterministic and probabilistic diffusion tractography approaches have been applied to postmortem and *in vivo* brains to visualize white-matter trajectories of the auditory pathways.

TRACTOGRAPHY RECONSTRUCTIONS OF THE AUDITORY PATHWAYS

The auditory system is a complex multisynaptic, bilateral network composed of projection, association, and commissural connections. In the brainstem, ascending fibers from the cochlear nuclei convey auditory information to the superior olivary nucleus, the nucleus of the lateral lemniscus, the inferior colliculus and the medial geniculate nucleus. Along this chain of interconnected neurons multiple parallel pathways on both sides elaborate the auditory signal to organize tonotopic representation of the location and the identity of the auditory percepts. The lateral lemniscus, which represents the principal ascending auditory pathway in the brainstem, can be visualized with tractography. Figure 16.5 shows the tractography reconstruction of the lateral lemniscus obtained with high-resolution diffusion data (~0.4 mm) acquired from human postmortem tissue at 7 T. In its course the lateral lemniscus is lateral to the medial lemniscus and the superior cerebellar peduncle.

One of the advantages of postmortem diffusion imaging is the possibility of scanning for longer times and thus obtaining data at a higher resolution. Despite the high resolution, it is still very difficult to extract quantitative information from these images and obtain a detailed connectivity map of the different fibers that join in or branch off the lateral lemniscus. This is in part due to the paucity of the fibers that sprout off the main stem of the tract to reach the different nuclei.

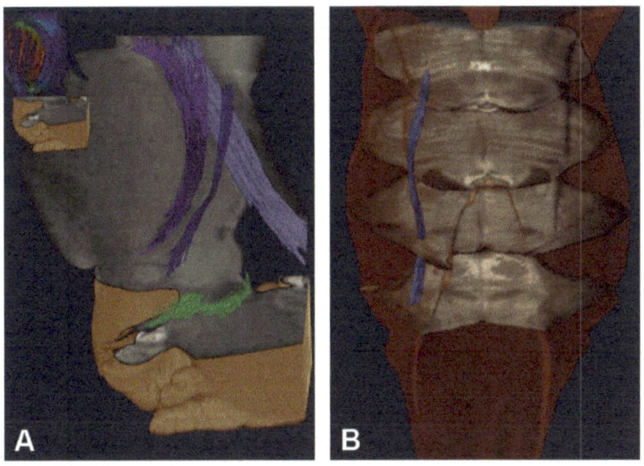

Fig. 16.5. (**A**) Tractographic reconstruction of the auditory nerve (green) and its ending at the cochlear nuclei, the lateral lemniscus (dark blue), the medial lemniscus (purple), and superior cerebellar peduncle (light blue). (**B**) Posterior view of the left lateral meniscus.

Data were obtained from a postmortem human brain after fixation by intracarotid and vertebral injection of 10% neutral buffered formalin (Soria et al., 2011; Prats-Galino et al., 2012).

More dorsally, a group of thickly myelinated fibers, the acoustic radiations, originates from the medial geniculate nucleus and projects to the ipsilateral primary auditory cortex in the Heschl gyrus. The acoustic radiations represent the only subcortical afferent connections to the cerebral cortex and are larger than the lateral lemniscus. Despite their greater diameter, the acoustic radiations are rarely visible on conventional *in vivo* diffusion tensor datasets. Here the problem is not related to the spatial resolution but rather to the anatomy of the acoustic radiations. These fibers have a tortuous S shape and cross many tracts during their course. To overcome the limitations of the tensor model, Behrens et al. (2007) used probabilistic diffusion tractography based on a non-tensorial estimation of the fiber orientation (fODF). Figure 16.6, shows the results of Behrens et al.'s study, where in all 9 subjects the single-fiber approach (tensor-based) was not able to reconstruct the acoustic radiation, whereas the multifiber model approach displayed a trajectory that is compatible with the anatomy of the acoustic radiations. Similar results are obtained with tractography based on Q-ball imaging (Berman et al., 2013) and spherical deconvolution (Fig. 16.7). While these approaches represent a valuable step towards white-matter segmentation (e.g., separation of white-matter voxels containing acoustic radiations from other tracts), they are limited when used to infer reliable maps

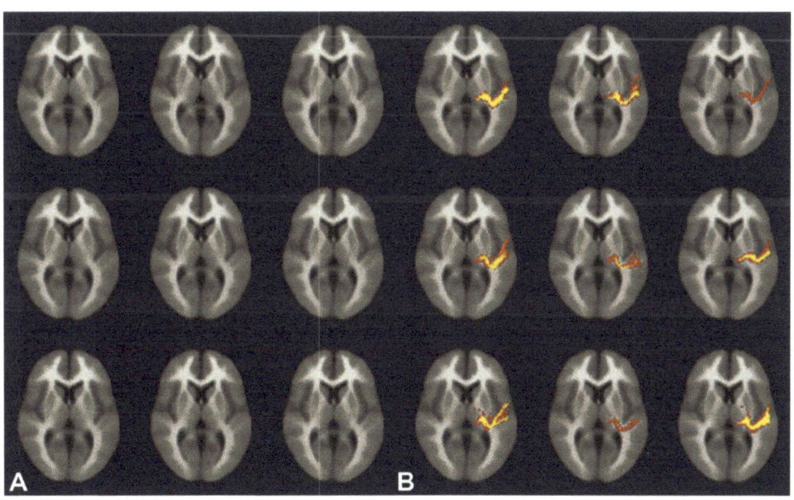

Fig. 16.6. Probabilistic tractography of the acoustic radiations between medial geniculate nucleus of thalamus and primary auditory cortex in 9 healthy subjects. (**A**) Single-fiber approach tractography shows no acoustic radiations. (**B**) On the same dataset the multifiber approach produces results that are consistent with the true anatomy of the acoustic radiations. Voxels are color-coded from 10 (red) to 50 (yellow) samples passing through the voxel. (Reproduced from Behrens et al., 2007.)

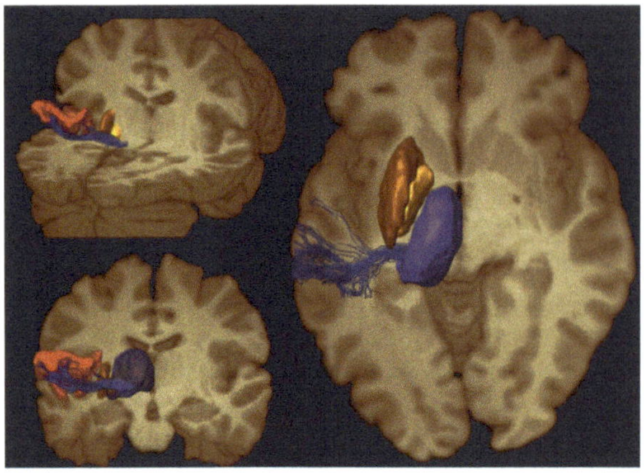

Fig. 16.7. Reconstruction of the acoustic radiations based on spherical deconvolution probabilistic tractography.

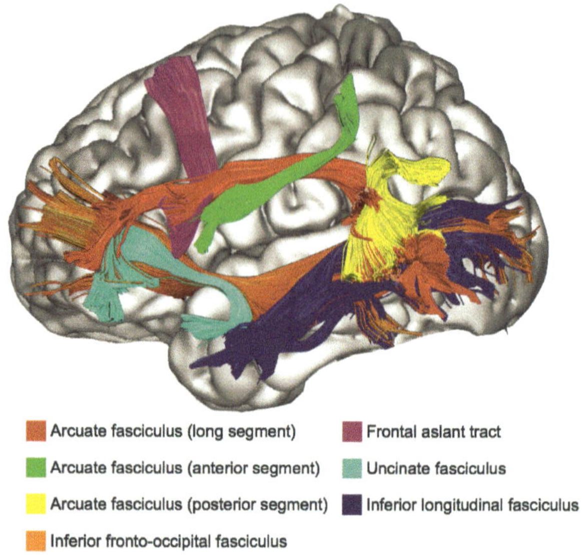

Fig. 16.8. Tractography reconstruction of the major association pathways involved in auditory processing and language.

of connectivity (e.g., studying the exact cortical projections of different components of the acoustic radiations).

The acoustic information that reaches the primary auditory cortex is then processed along corticocortical streams of short- and long-range connections within the temporal lobe and between temporal regions and parietal and frontal regions (Hickok and Poeppel, 2007; Friederici, 2009). These networks can be visualized *in vivo* using conventional and advanced diffusion approaches (Catani and Mesulam, 2008; Turken and Dronkers, 2011) (Fig. 16.8).

Short connections carry information from the Heschl gyrus to surrounding cortical areas in the superior and middle temporal gyri. From here information is projected to the inferior parietal and posterior frontal gyrus through the arcuate fasciculus. A ventral stream carries information from posterior to more anterior temporal and frontal areas through the middle/inferior longitudinal fasciculus, the uncinate fasciculus, and the inferior fronto-occipital fasciculus (Fig. 16.8).

Within the arcuate fasciculus, two parallel pathways have been distinguished using diffusion tensor tractography: the medial, direct pathway connecting Wernicke's with Broca's region (i.e., the arcuate fasciculus *sensu strictu* or long segment); and the indirect pathway, consisting of a posterior segment between Wernicke's region and Geschwind's region (inferior parietal lobule) and an anterior segment that links Geschwind's to Broca's region (Catani et al., 2005). López-Barroso et al. (2013) showed that performance in word learning correlates with microstructural properties and strength of functional connectivity of the direct connections between Wernicke's and Broca's regions (Fig. 16.9). This study demonstrates that our ability to learn new words relies on an efficient and fast communication between auditory temporal and motor frontal regions. Schulze et al. (2012) suggested that the absence of these connections in non-human primates and monkeys might explain our unique ability for word learning.

An extreme degree of leftward lateralization in the volume of the long segment of the arcuate fasciculus has been reported in approximately 60% of the normal population (Fig. 16.10) (Catani et al., 2007). The remaining 40% of the population show either a mild leftward lateralization (20%) or a bilateral, symmetric pattern (20%). Overall, females are more likely to have a bilateral pattern compared to males. The degree of lateralization of the long segment has important clinical implications. Forkel et al. (2014) showed that patients with aphasia are more likely to recover normal language at 6 months if at the time of the stroke they have a larger volume of the long segment in the right unaffected hemisphere (Fig. 16.10C).

Language areas of the temporal and frontal lobes are also connected by a set of ventral longitudinal tracts (Catani and Mesulam, 2008). These tracts are bilateral without evidence of a clear asymmetry between the left and right hemisphere (Thiebaut de Schotten et al., 2011). The inferior longitudinal fasciculus carries visual and auditory information from posterior occipital and temporal areas to the anterior temporal lobe and plays an important role in visual object recognition, reading, and linking object representations to their lexical labels (Catani et al., 2003; Turken and Dronkers, 2011). The uncinate fasciculus connects the anterior temporal lobe to the orbitofrontal region and part of the inferior frontal gyrus and may play an important role in lexical

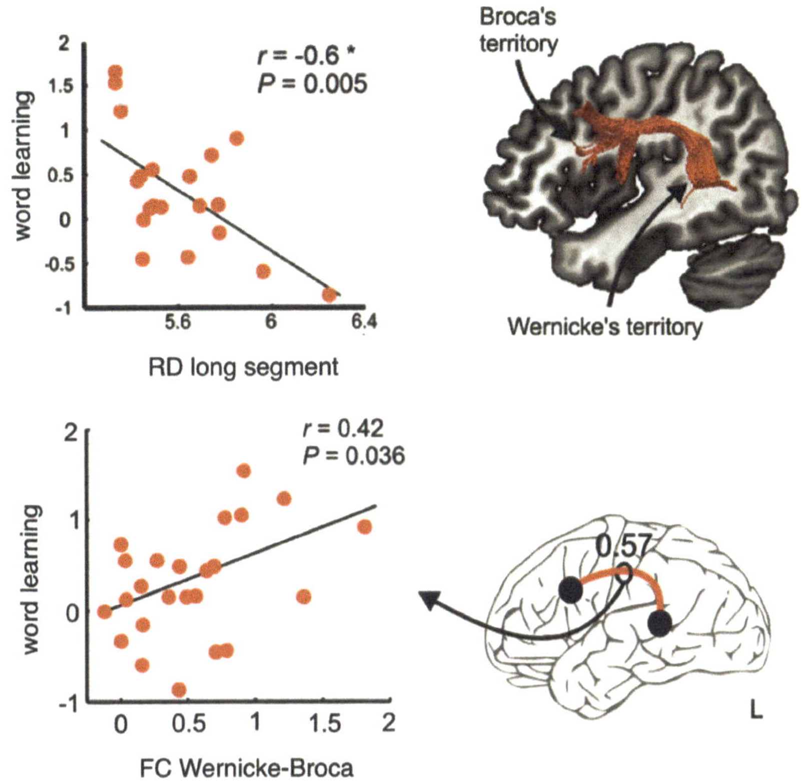

Fig. 16.9. Correlation between auditory verbal learning and both structural (diffusion tensor imaging) and functional magnetic resonance imaging (fMRI) measurements of temporal–frontal connectivity. The word-learning task consisted of a learning phase and a recognition test. The learning phase was performed during the fMRI acquisition and consisted of the auditory presentation of an artificial language stream composed of trisyllabic words separated by brief pauses. The test phase was carried out offline and consisted of the presentation of isolated words ("words" or "non-words") for which participants were required to recognize each one as correct or incorrect based on the previously heard artificial language presentation in the learning phase. Word-learning performances were measured using hit rates taking into account false alarms, calculating the *d*-prime (*d*') index, which reflects an individual's ability to discriminate words presented in the artificial language from non-presented words. RD, radial diffusivity along the long segment streamlines. RD is often related to the degree of white-matter myelination (lower RD corresponds to higher myelination). FC, functional connectivity measured as a correlation between the time series of each cortically activated region. (Reproduced from Lopez-Barroso et al., 2013.)

retrieval, semantic associations, and naming. The uncinate fasciculus is severely damaged in those patients with the semantic variant of primary progressive aphasia (Catani et al., 2013). The inferior fronto-occipital fasciculus provides direct connections between occipital and frontal cortex in the human brain. The relevance of this fasciculus to reading, writing, and other aspects of language remains to be established (Duffau et al., 2005; Forkel et al., 2012).

CONCLUSIONS AND FUTURE DIRECTIONS

The acoustic pathways represent a notable omission in diffusion-based studies of white-matter connections. The crossing fibers problem, together with the small dimension of some of the subcortical fibers and the anatomic characteristics of the acoustic radiation, limit the ability to reconstruct ascending auditory projections in the living human brain. Postmortem acquisitions at high MR fields show preliminary encouraging results, although still not comparable to classic histologic methods.

At the cortical level multifiber approaches are able to track acoustic radiations that are anatomically valid. These approaches are still biased by the high number of false-positive reconstructions, which limits their application to clinical populations. To date tractography remains a method limited to exploring surrogates of corticocortical association tracts involved in auditory processing and language. Interindividual differences in lateralization could represent important predictors of language recovery in aphasia after stroke. These findings await confirmation on larger clinical groups.

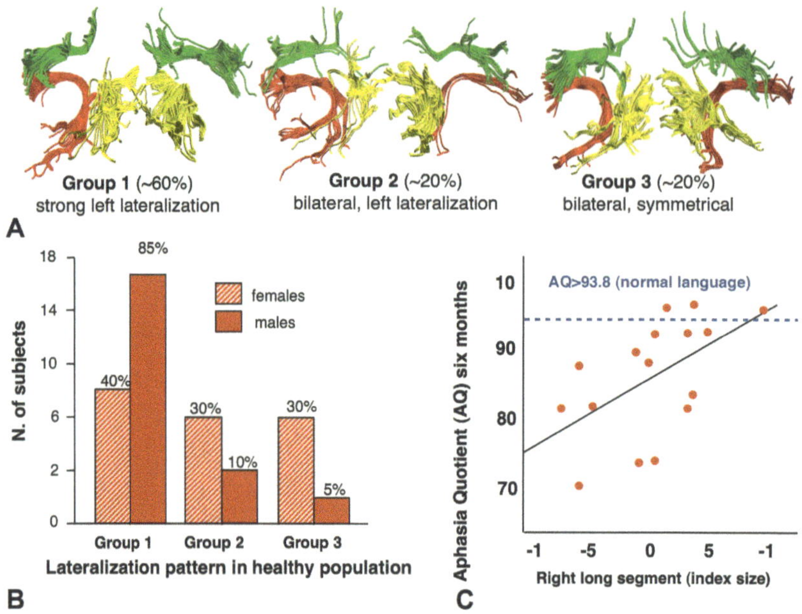

Fig. 16.10. Distribution of the pattern of lateralization of the long segment in (**A**) three groups and (**B**) differences in lateralization between genders. (Reproduced from Catani et al., 2007.) (**C**) In patients with left-hemisphere stroke, recovery of language after 6 months is correlated with the volume of the right long segment at the time of the stroke. (Reproduced from Forkel et al., 2014.)

ACKNOWLEDGMENTS

Marco Catani is the recipient of a Wellcome Trust New Investigator Award (103759/Z/14/Z). The present study has been partially supported by the Marató TV3 Grant Project (ref. 411/U/2011). We are also indebted to Eloy Martinez-Heras (Department de Neurociències, Institut d'Investigacions Biomèdiques August Pi i Sunyer (IDIBAPS) and Hospital Clinic, Barcelona, Spain) and Olga Fuentes (LSNA, Faculty of Medicine, Universitat de Barcelona, Barcelona, Spain) for their invaluable technical assistance.

REFERENCES

Alexander DC (2008). A general framework for experiment design in diffusion MRI and its application in measuring direct tissue-microstructure features. Magn Reson Med 60 (2): 439–448.

Assaf Y, Basser PJ (2005). Composite hindered and restricted model of diffusion (CHARMED) MR imaging of the human brain. Neuroimage 27 (1): 48–58.

Assaf Y, Blumenfeld-Katzir T, Yovel Y et al. (2008). AxCaliber: a method for measuring axon diameter distribution from diffusion MRI. Magn Reson Med 59 (6): 1347–1354.

Basser PJ, Mattiello J, Le Bihan D (1994). MR diffusion tensor spectroscopy and imaging. Biophys J 66 (1): 259–267.

Basser PJ, Pajevic S, Pierpaoli C et al. (2000). In vivo fiber tractography using DT-MRI data. Magn Reson Med 44 (4): 625–632.

Behrens TE, Johansen-Berg H, Woolrich MW et al. (2003). Non-invasive mapping of connections between human thalamus and cortex using diffusion imaging. Nat Neurosci 6 (7): 750–757.

Behrens TE, Berg HJ, Jbabdi S et al. (2007). Probabilistic diffusion tractography with multiple fibre orientations: what can we gain? Neuroimage 34 (1): 144–155.

Berman JI, Lanza MR, Blaskey L et al. (2013). High angular resolution diffusion imaging probabilistic tractography of the auditory radiation. AJNR Am J Neuroradiol 34 (8): 1573–1578.

Catani M (2006). Diffusion tensor magnetic resonance imaging tractography in cognitive disorders. Curr Opin Neurol 19 (6): 599–606.

Catani M, Mesulam M (2008). The arcuate fasciculus and the disconnection theme in language and aphasia: history and current state. Cortex 44 (8): 953–961.

Catani M, Jones DK, Donato R et al. (2003). Occipito-temporal connections in the human brain. Brain 126 (9): 2093–2107.

Catani M, Jones DK, ffytche DH (2005). Perisylvian language networks of the human brain. Ann Neurol 57 (1): 8–16.

Catani M, Allin MP, Husain M et al. (2007). Symmetries in human brain language pathways correlate with verbal recall. Proc Natl Acad Sci U S A 104 (43): 17163–17168.

Catani M, Jones DK, Daly E et al. (2008). Altered cerebellar feedback projections in Asperger syndrome. Neuroimage 41 (4): 1184–1191.

Catani M, Bodi I, Dell'Acqua F (2012). Comment on "The geometric structure of the brain fiber pathways". Science 337 (6102): 1605.

Catani M, Mesulam MM, Jakobsen E et al. (2013). A novel frontal pathway underlies verbal fluency in primary progressive aphasia. Brain 136 (8): 2619–2628.

Cercignani M, Alexander DC (2006). Optimal acquisition schemes for in vivo quantitative magnetization transfer MRI. Magn Reson Med 56 (4): 803–810.
Ciccarelli O, Catani M, Johansen-Berg H et al. (2008). Diffusion-based tractography in neurological disorders: concepts, applications, and future developments. Lancet Neurol 7 (8): 715–727.
Conturo TE, Lori NF, Cull TS et al. (1999). Tracking neuronal fiber pathways in the living human brain. Proc Natl Acad Sci U S A 96 (18): 10422–10427.
De Santis S, Drakesmith M, Bells S et al. (2014). Why diffusion tensor MRI does well only some of the time: variance and covariance of white matter tissue microstructure attributes in the living human brain. Neuroimage 89 (100): 35–44.
Dell'Acqua F, Catani M (2012). Structural human brain networks: hot topics in diffusion tractography. Curr Opin Neurol 25 (4): 375–383.
Dell'Acqua F, Scifo P, Rizzo G et al. (2010). A modified damped Richardson-Lucy algorithm to reduce isotropic background effects in spherical deconvolution. Neuroimage 49 (2): 1446–1458.
Dell'Acqua F, Simmons A, Williams SC et al. (2012). Can spherical deconvolution provide more information than fiber orientations? Hindrance modulated orientational anisotropy, a true-tract specific index to characterize white matter diffusion. Hum Brain Mapp.
Deoni SC, Rutt BK, Arun T et al. (2008). Gleaning multicomponent T1 and T2 information from steady-state imaging data. Magn Reson Med 60 (6): 1372–1387.
Duffau H, Gatignol P, Mandonnet E et al. (2005). New insights into the anatomo-functional connectivity of the semantic system: a study using cortico-subcortical electrostimulations. Brain 128 (4): 797–810.
Einstein A (1905). Über die von der molekularkinetischen Theorie der Wärme geforderte Bewegung .... 12.
Forkel SJ, Thiebaut de Schotten M, Dell'Acqua F et al. (2014). Anatomical predictors of aphasia recovery: a tractography study of bilateral perisylvian language networks. Brain 137 (7): 2027–2039.
Forkel S, Thiebaut de Schotten M, Kawadler J et al. (2012). Functional and clinical correlates of the occipito-frontal connections. Cortex 56 (7): 73–84.
Friederici AD (2009). Pathways to language: fiber tracts in the human brain. Trends Cogn Sci 13 (4): 175–181.
Hackett TA, Stepniewska I, Kaas JH (1998). Subdivisions of auditory cortex and ipsilateral cortical connections of the parabelt auditory cortex in macaque monkeys. J Comp Neurol 394 (4): 475–495.
Hickok G, Poeppel D (2007). The cortical organization of speech processing. Nat Rev Neurosci 8 (5): 393–402.
Jansons KM, Alexander DC (2003). Persistent angular structure: new insights from diffusion MRI data. Inf Process Med Imaging 2732 (18): 672–683.
Jbabdi S, Woolrich MW, Andersson JL et al. (2007). A Bayesian framework for global tractography. Neuroimage 37 (1): 116–129.
Jbabdi S, Johansen-Berg H (2011). Tractography: where do we go from here? Brain Connect 1 (3): 169–183.
Jones DK (2008). Studying connections in the living human brain with diffusion MRI. Cortex 44 (8): 936–952.
Jones DK (2010). Challenges and limitations of quatifying brain connectivity in vivo with diffusion MRI. Imaging Med 2 (3): 341–355.
Jones DK, Pierpaoli C (2005). Confidence mapping in diffusion tensor magnetic resonance imaging tractography using a bootstrap approach. Magn Reson Med 53 (5): 1143–1149.
Kaas JH, Hackett TA (2000). Subdivisions of auditory cortex and processing streams in primates. Proc Natl Acad Sci U S A 97 (22): 11793–11799.
Kaas JH, Hackett TA, Tramo MJ (1999). Auditory processing in primate cerebral cortex. Curr Opin Neurobiol 9 (2): 164–170.
King MD, Gadian DG, Clark CA (2009). A random effects modelling approach to the crossing-fibre problem in tractography. NeuroImage 44 (3): 753–768.
Le Bihan D (2003). Looking into the functional architecture of the brain with diffusion MRI. Nat Rev Neurosci 4 (6): 469–480.
Lebel C, Walker L, Leemans A et al. (2008). Microstructural maturation of the human brain from childhood to adulthood. Neuroimage 40 (3): 1044–1055.
Lopez-Barroso D, Catani M, Ripolles P et al. (2013). Word learning is mediated by the left arcuate fasciculus. Proc Natl Acad Sci U S A 110 (32): 13168–13173.
Mesulam MM, Pandya DN (1973). The projections of the medial geniculate complex within the sylvian fissure of the rhesus monkey. Brain Res 60 (2): 315–333.
Mori S, Crain BJ, Chacko VP et al. (1999). Three-dimensional tracking of axonal projections in the brain by magnetic resonance imaging. Annals of Neurolology 45 (2): 265–269.
Moseley ME, Cohen Y, Kucharczyk J et al. (1990). Diffusion-weighted MR imaging of anisotropic water diffusion in cat central nervous system. Radiology 176 (2): 439–445.
Neil J, Miller J, Mukherjee P et al. (2002). Diffusion tensor imaging of normal and injured developing human brain – a technical review. NMR Biomed 15 (7–8): 543–552.
Nucifora PG, Verma R, Melhem ER et al. (2005). Leftward asymmetry in relative fiber density of the arcuate fasciculus. Neuroreport 16 (8): 791–794.
Parker GJ, Luzzi S, Alexander DC et al. (2005). Lateralization of ventral and dorsal auditory-language pathways in the human brain. Neuroimage 24 (3): 656–666.
Pierpaoli C, Jezzard P, Basser PJ et al. (1996). Diffusion tensor MR imaging of the human brain. Radiology 201 (3): 637–648.
Prats-Galino A, Soria G, de Notaris M et al. (2012). Functional anatomy of subcortical circuits issuing from or integrating at the human brainstem. Clin Neurophysiol 123 (1): 4–12.
Pugliese L, Catani M, Ameis S et al. (2009). The anatomy of extended limbic pathways in Asperger syndrome: a preliminary diffusion tensor imaging tractography study. Neuroimage 47 (2): 427–434.
Raffelt D, Tournier JD, Rose S et al. (2012). Apparent fibre density: a novel measure for the analysis of diffusion-weighted magnetic resonance images. Neuroimage 59 (4): 3976–3994.
Rilling JK, Glasser MF, Preuss TM et al. (2008). The evolution of the arcuate fasciculus revealed with comparative DTI. Nat Neurosci 11 (4): 426–428.

Schulze K, Vargha-Khadem F, Mishkin M (2012). Test of a motor theory of long-term auditory memory. Proc Natl Acad Sci U S A 109 (18): 7121–7125.

Soria G, De Notaris M, Tudela R et al. (2011). Improved assessment of ex vivo brainstem neuroanatomy with high-resolution MRI and DTI at 7 Tesla. Anat Rec 294 (6): 1035–1044.

Thiebaut de Schotten M, ffytche DH, Bizzi A et al. (2011). Atlasing location, asymmetry and inter-subject variability of white matter tracts in the human brain with MR diffusion tractography. Neuroimage 54 (1): 49–59.

Thiebaut de Schotten M, Dell'Acqua F, Valabregue R et al. (2012). Monkey to human comparative anatomy of the frontal lobe association tracts. Cortex 48: 82–96.

Tournier JD, Calamante F, Gadian DG et al. (2004). Direct estimation of the fiber orientation density function from diffusion-weighted MRI data using spherical deconvolution. Neuroimage 23 (3): 1176–1185.

Tuch DS (2004). Q-Ball Imaging. Magn Reson Med 52 (6): 1358–1372.

Turken AU, Dronkers NF (2011). The neural architecture of the language comprehension network: converging evidence from lesion and connectivity analyses. Frontiers in systems neuroscience 5: 1.

Wedeen VJ, Hagmann P, Tseng WY et al. (2005). Mapping complex tissue architecture with diffusion spectrum magnetic resonance imaging. Magn Reson Med 54 (6): 1377–1386.

Wu YC, Alexander AL (2007). Hybrid diffusion imaging. Neuroimage 36 (3): 617–629.

Yeh FC, Wedeen VJ, Tseng WY (2011). Estimation of fiber orientation and spin density distribution by diffusion deconvolution. Neuroimage 55 (3): 1054–1062.

Zhang H, Schneider T, Wheeler-Kingshott CA et al. (2012). NODDI: practical in vivo neurite orientation dispersion and density imaging of the human brain. Neuroimage 61 (4): 1000–1016.

Handbook of Clinical Neurology, Vol. 129 (3rd series)
The Human Auditory System
G.G. Celesia and G. Hickok, Editors

Chapter 17

Electrophysiologic auditory tests

ALAN D. LEGATT*

Departments of Neurology and Neuroscience, Montefiore Medical Center and Albert Einstein College of Medicine, Bronx, NY, USA

INTRODUCTION

A brief acoustic stimulus can elicit a series of electric signals with latencies that range from a few milliseconds to hundreds of milliseconds (Fig. 17.1). These signals are generated within the cochlea, the eighth nerve, and the auditory pathways within the central nervous system. They can be used as neurologic and audiologic diagnostic tools, and some of them are also useful for intraoperative monitoring of the integrity of the ear and of the infratentorial auditory pathways.

The earliest electric signals comprise the electrocochleogram (ECochG), which is generated in the inner ear and is recorded close to its generator. The neurally generated signals that can be recorded from the scalp following acoustic stimulation, called auditory evoked potentials, can be divided into classes based on their latencies: short-latency auditory evoked potentials, with latencies of less than 10 ms in normal adults; long-latency auditory evoked potentials (LLAEPs), with latencies exceeding 50 ms; and middle-latency auditory evoked potentials (MLAEPs), with intermediate latencies (Fig. 17.2). These different types of auditory evoked potentials differ in their properties and in their clinical utility. The short-latency auditory evoked potentials are most often called brainstem auditory evoked potentials (BAEPs).

Auditory evoked potential components generated within the brainstem may reflect both action potentials and postsynaptic potentials. Auditory evoked neural activity becomes increasingly affected by temporal dispersion as the poststimulus latency increases and as the contribution of short-duration electrical phenomena (e.g., action potentials) is eliminated. Thus, auditory evoked potential components that are longer in latency are also wider (Figs. 17.1 and 17.2), and the MLAEP and LLAEP components are predominantly generated by postsynaptic potentials within areas of cerebral cortex that are activated by the acoustic stimulus.

Electric signals recorded from specific locations can also be classified as far-field potentials versus near-field potentials. Far-field potentials are signals that are not substantially altered by small movements of the recording electrode; the small spatial gradients reflect a recording site that is relatively far away from the generator(s) of the signal.

Many of these signals are quite small, and in the raw data they are obscured by the other electric signals that are picked up by the recording electrodes (the electroencephalogram (EEG), the electrocardiogram, electromyographic activity from nearby muscles, and various other electrical artifacts). Therefore, multiple stimuli are delivered, and the response epochs are averaged together to attenuate the noise while preserving those signals which are time-locked to the stimuli, the auditory evoked potentials. Signal averaging (Fig. 17.3) improves the signal-to-noise ratio by a factor of the square root of the number of sweeps included in the average.

Otoacoustic emissions (Janssen and Müller, 2008) are sounds generated by movements of the external hair cells of the cochlea, both spontaneously and in response to auditory stimulation. They can be used to assess hearing, hair cell function, and the function of the olivocochlear bundle. They are mechanical signals, not electric ones, however. Thus, they will not be considered further in this chapter.

ELECTROCOCHLEOGRAM

The ECochG consists of the cochlear microphonic, the cochlear summating potential, and the eighth-nerve compound action potential (CAP). Both clicks and brief tone bursts have been used to elicit the ECochG.

*Correspondence to: Alan D. Legatt, Department of Neurology, Montefiore Medical Center, 111 East 210th Street, Bronx, New York 10467. Tel: +1-718-920-6530, Fax: +1-718-920-8509, E-mail: Alegatt@montefiore.org

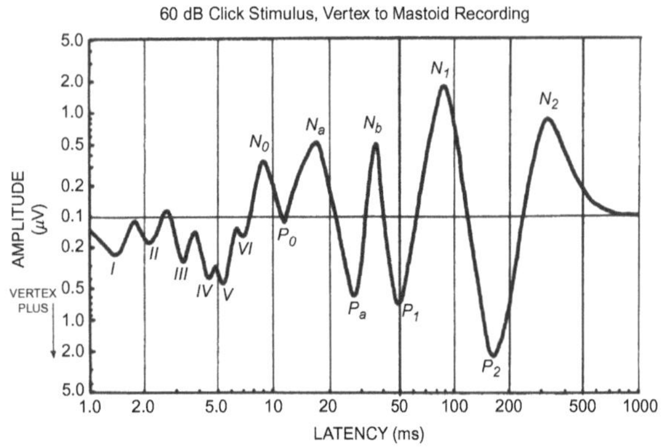

Fig. 17.1. Auditory evoked potentials to click stimulation, plotted on logarithmic time and voltage scales. Note that, as the component latency increases, the width of the component also increases, due to temporal dispersion. Mean data from 8 subjects. Vertex-to-mastoid recording, with vertex positivity plotted as a downward deflection. (Reproduced from Picton et al., 1974, with permission.)

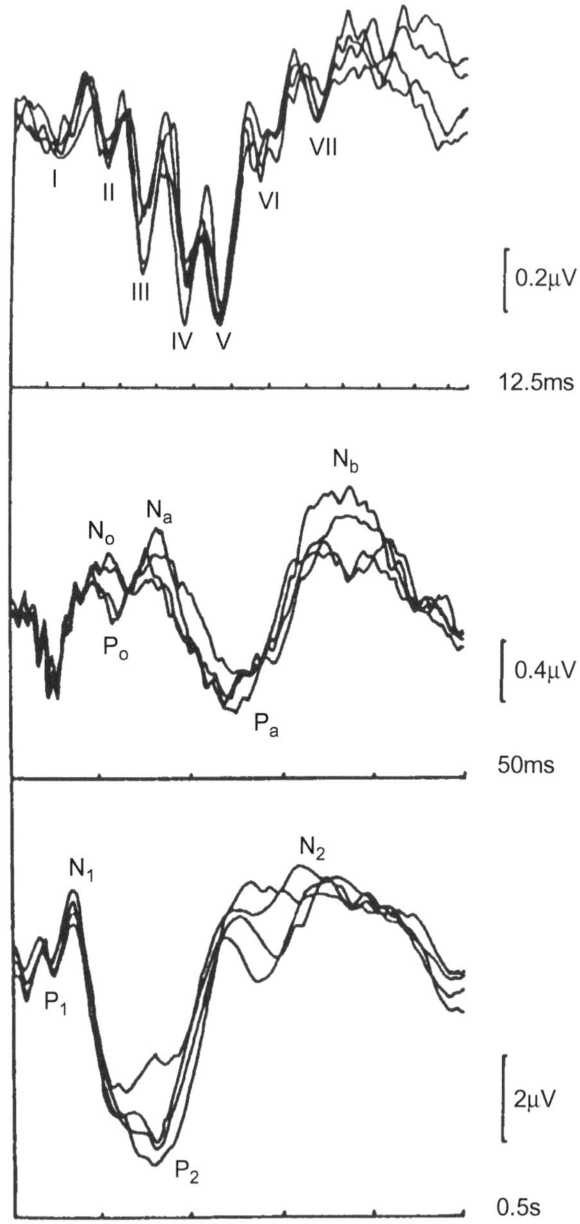

Fig. 17.2. Auditory evoked potentials to 60 dB sensation level (SL) monaural clicks presented at a rate of 1/second, shown on three different timescales to illustrate the short-latency (top), middle-latency (middle), and long-latency (bottom) components. Vertex-to-mastoid recording, with vertex positivity plotted as a downward deflection. (Reproduced from Picton et al., 1974, with permission.)

The ECochG was originally recorded from needle electrodes inserted through the tympanic membrane and placed on the promontorium of the middle ear, close to the cochlea (Ruben et al., 1961). It can also be recorded from extratympanic recording sites, such as the external ear canal (Mullatti et al., 1999; Attias et al., 2008), but at substantially smaller amplitude; signal averaging is used to record these volume-conducted signals with an adequate signal-to-noise ratio. The transtympanic approach is invasive and requires anesthesia, but may yield more accurate hearing threshold measurements and be more suitable for evaluating patients with auditory neuropathy (Santarelli and Arslan, 2013). Recordings from the tympanic membrane provide intermediate amplitudes and may be an acceptable compromise in many patients (Ferraro, 2010). Typical amplitudes for the eighth-nerve CAP for a very loud click are 15 μV from an electrode on the promontorium, 3 μV from the tympanic membrane, 1.5 μV from the wall of the canal, and 0.5 μV from a skin surface electrode on the mastoid (Picton, 2011). Sometimes the eighth-nerve CAP contains multiple peaks, reflecting the presence of multiple volleys in the eighth nerve. The first peak, N1, can be recorded at stimulus intensities down to the patient's subjective threshold. The second peak, N2, appears only at higher intensities; a third peak, N3, appears only at very high stimulus intensities (Salomon and Elberling, 1971).

Acoustic stimulation for transtympanic ECochG is usually performed with free-field stimulation, using a speaker at a distance from the subject's ear. This requires careful calibration of the stimulus intensity, using a sound level meter with its microphone placed at the same distance from the speaker. Headphones or ear insert phones are usually used when the ECochG is recorded from extratympanic locations or from the round window (Santarelli and Arslan, 2013).

Components of the electrocochleogram

Cochlear microphonic

The name of the cochlear microphonic reflects the fact that its wave shape mimics that of the acoustic stimulus.

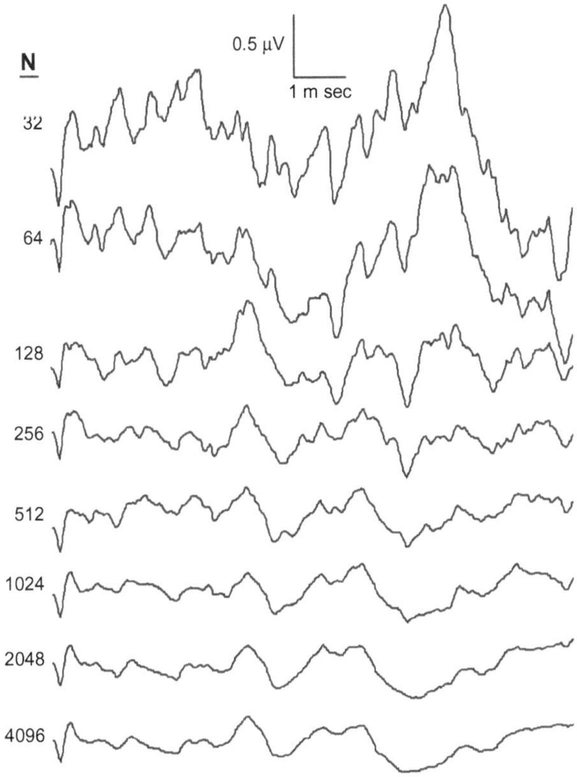

Fig. 17.3. Averaged brainstem auditory evoked potentials recorded in an awake human subject with a progressively larger number of data epochs in the averages, showing the effect of signal averaging in improving the signal-to-noise ratio. N = the number of data epochs in the average; the improvement in the signal-to-noise ratio is proportional to the square root of N. Vertex-to-ipsilateral mastoid recording, with vertex positivity plotted as an upward deflection. (Reproduced from Legatt, 2003, with permission.)

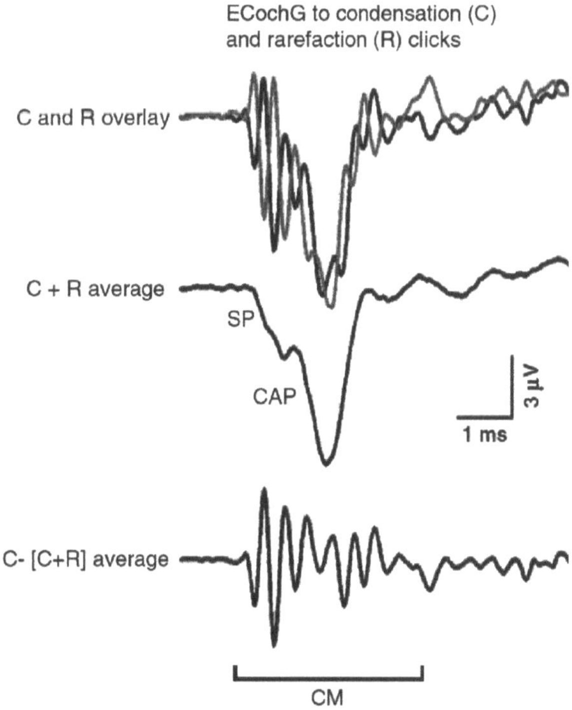

Fig. 17.4. Separation of the electrocochleogram (ECochG) components. Top: The responses to condensation and rarefaction clicks are superimposed. Middle: The average of the two responses (sum divided by 2) cancels the cochlear microphonic, leaving the cochlear summating potential (SP) and the eighth-nerve compound action potential (CAP). Bottom: The difference between the two waveforms (divided by 2) cancels the SP and the CAP, leaving the cochlear microphonic (CM). (Reprodued from Santarelli et al., 2008, with permission.)

Accordingly, it reverses in polarity when the acoustic stimulus is reversed in polarity; summation or averaging of responses to acoustic stimuli with opposite polarities is used to cancel the cochlear microphonic and record the cochlear summating potential and the eighth-nerve CAP. Conversely, the difference between the responses to the two stimulus polarities will cancel the cochlear summating potential and the eighth-nerve CAP and emphasize the cochlear microphonic (Fig. 17.4). Due to nonlinearities in the cochlea, the elimination of the cochlear summating potential and the eighth-nerve CAP may be incomplete; additional signal filtering techniques may be useful to more accurately record the cochlear microphonic (Sparacino et al., 2000). The cochlear microphonic arises from receptor potentials in the inner and outer hair cells within the cochlea, with the more numerous outer hair cells being the predominant generator (Dallos and Cheatham, 1976) and the recorded signal reflecting predominantly activity in the base of the cochlea, which is closest to the recording electrode (Withnell, 2001). The major clinical use of the cochlear microphone has been in evaluating residual hair cell function in patients with auditory neuropathy (Picton, 2011).

Cochlear summating potential

The cochlear summating potential retains the same polarity (a negativity at the promontorium) when the acoustic stimulus polarity is reversed, but its waveform varies as a function of stimulus parameters. As the stimulus intensity is progressively lowered, the summating potential disappears sooner (at a higher intensity) than does the eighth-nerve CAP (Fig. 17.5). The summating potential arises from receptor processes within the cochlea, most likely from multiple generators (Dallos et al., 1972). These receptor processes persist at high stimulus repetition rates, whereas it is difficult for auditory nerve fibers to fire consistently and synchronously at stimulus rates $\geq$100/second; thus, rapid repetitive stimulation can be used to attenuate the eighth-nerve CAP, permitting better examination of the cochlear summating potential (Eggermont, 1976; Picton, 2011)

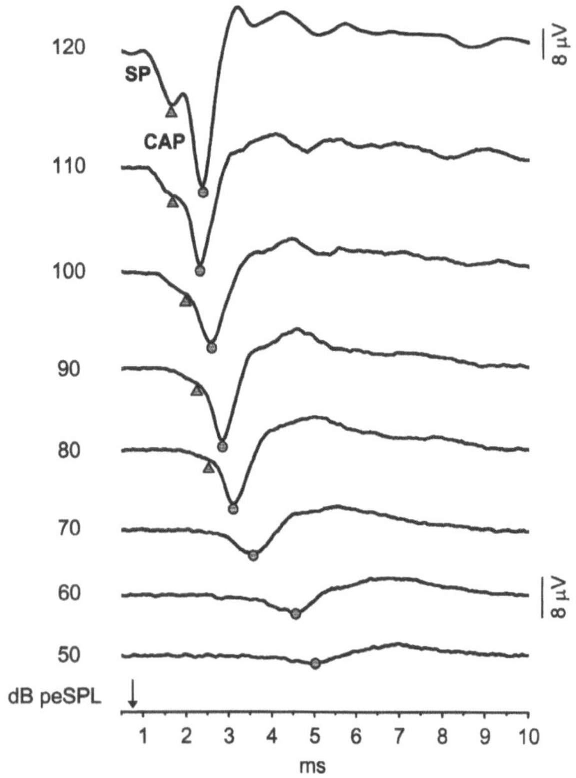

Fig. 17.5. Effect of stimulus intensity on the cochlear summating potential (SP: triangles) and the eighth-nerve compound action potential (CAP: circles) in a normal ear. The intensity of the click stimuli was progressively reduced from 120 to 50 dB peSPL (peak equivalent sound pressure level). (Reproduced from Santarelli and Arslan, 2013, with permission.)

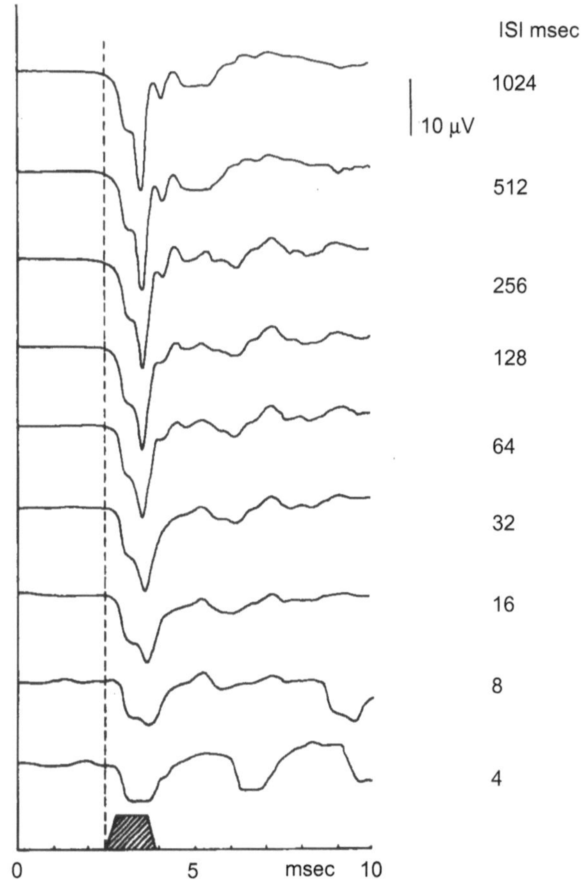

Fig. 17.6. Effect of stimulus rate on the electrocochleogram. As the interstimulus interval (ISI) was decreased from 1024 ms (approximately 1/second stimulation) to 4 ms (250/second stimulation), the eighth-nerve compound action potential became markedly attenuated but the cochlear summating potential persisted. The timing of the brief tone burst stimulus is indicated by the shaded trapezoid on the time axis. (Reproduced from Eggermont, 1976, with permission.)

(Fig. 17.6). The cochlear summating potential becomes larger than normal in Ménière's disease.

Eighth-nerve compound action potential

The eighth-nerve CAP reflects depolarization of nerve fibers in the distal (cochlear) end of the auditory nerve. With high stimulus intensities a population of fibers innervating the base of the cochlea fire in a synchronous fashion, producing a large and relatively narrow CAP waveform. At lower stimulus intensities, only lower-threshold fibers with low characteristic frequencies, innervating more apical portions of the cochlea, are activated. The time required for the sound wave to propagate to those regions of the cochlea leads to increased temporal dispersion, and the eighth-nerve CAP becomes wider as the stimulus intensity is reduced (Santarelli and Arslan, 2013) (Fig. 17.5).

When there is a cochlear hearing loss, the amplitude of the eighth-nerve CAP may be relatively normal for high stimulus intensities, but it falls off more rapidly than normal as the stimulus intensity is reduced (Santarelli and Arslan, 2013) (Fig. 17.7). This corresponds to the psychophysical phenomenon of recruitment. The threshold for recognizing the CAP is very close to the behavioral threshold during audiometry; therefore, it can provide accurate objective auditory thresholds (Picton, 2011).

Clinical applications of electrocochleography

Measurement of hearing

Electrocochleography can be used to estimate hearing thresholds in children, based on comparisons between audiometric thresholds and the thresholds for recording of the eighth-nerve CAP in cooperative adults in whom both types of hearing assessment can be performed. With click stimuli, the best correlation of the ECochG threshold is with pure-tone audiometric thresholds at 1,

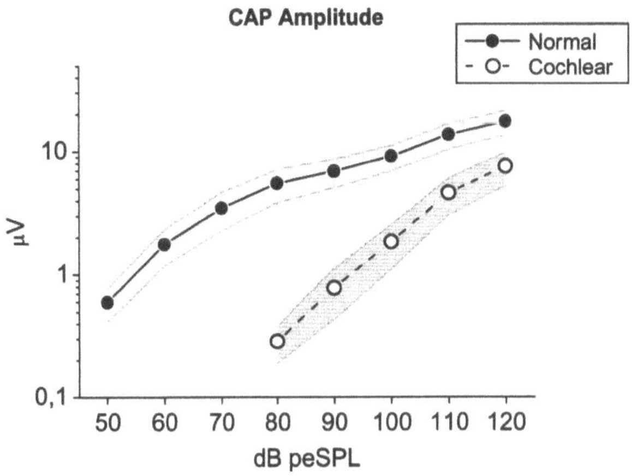

Fig. 17.7. Mean eighth-nerve compound action potential (CAP) amplitude–intensity functions from a group of ears with normal hearing (filled circles) and from a group of ears with cochlear hearing loss (open circles). Note that the slope of the curve for the cochlear hearing loss ears is steeper. (Reproduced from Santarelli and Arslan, 2013, with permission.)

2, and 4 kHz. Even closer correlation is obtained when the ECochG is performed using tone burst stimuli (Santarelli and Arslan, 2013).

Audiometry can also be performed with BAEPs (see below). In general, thresholds for wave V of the BAEP and for the eighth-nerve CAP are similar, though CAP thresholds may be lower (i.e., the CAP may be identifiable at lower stimulus intensities than the BAEPs) due to the better signal-to-noise ratio of ECochG as compared to the low-amplitude far-field BAEPs (Arslan et al., 1997). In neurologically compromised patients (including infants) in whom behavioral audiometry cannot be performed, the utility of BAEPs for objective audiometry may also be limited due to nervous system damage that affects the brainstem generators of the BAEPs (Santarelli and Arslan, 2013). Electrocochleography would be a better way of assessing hearing in such patients.

Ménière's disease

In Ménière's disease, or endolymphatic hydrops, the cochlear summating potential is typically increased in amplitude. Early investigations examined the summating potential to CAP amplitude ratio, comparing it to a single cutoff value (Coats, 1986). Because the relationship between the amplitude of the summating potential and that of the CAP is not linear, the patient's results may be plotted on a graph showing the normal amplitude relationship between these two signals (Coats, 1986). Although the amplitudes of the summating potential and the CAP are affected by the recording techniques (e.g., extratympanic vs transtympanic recordings), the relationship between them does not show substantial changes across different techniques and recording sites (Ferraro, 2010).

Other ECochG measures have been utilized in the diagnosis of Ménière's disease, including the ratio between the areas under the curves (as opposed to the peak amplitudes) of the summating potential and CAP (Devaiah et al., 2003), recording of the relationship between the summating potential and CAP using tone burst as well as click stimulation, and the modulation ("biasing") of the summating potential by changes in ear pressure produced by a low-frequency tone on which the higher-frequency stimulus tone bursts are superimposed (Iseli and Gibson, 2010).

Auditory neuropathy

In auditory neuropathy (Starr et al., 1996), there is disruption of auditory nerve function but preservation of outer hair cell function (see Chapter 28). The auditory nerve dysfunction tends to produce temporal dispersion and a loss of the timing information about the acoustic stimulus, which leads to deterioration of the BAEP waveform and impairment of speech discrimination, both of which are out of proportion to the elevation of the hearing threshold for pure tones. Because outer hair cell function is preserved, the cochlear microphonic is normal, as are otoacoustic emissions. In contrast, the eighth-nerve CAP is significantly decreased in amplitude and increased in duration, and its peak latency is significantly delayed (Santarelli et al., 2008). The cochlear summating potential is visible in some patients with auditory neuropathy and absent in others. In a small number of patients, a summating potential is present and unchanged at high stimulus rates, but there is no apparent CAP; this is thought to reflect presynaptic inner hair cell abnormalities (Santarelli and Arslan, 2013).

Intraoperative monitoring

The ECochG can be used for intraoperative neurophysiologic monitoring (IONM) of the ear and the distal eighth nerve. It may be a useful adjunct when combined with BAEP monitoring (Levine et al., 1984; Ojemann et al., 1984), because it requires less averaging (fewer epochs) than the scalp BAEPs and thus may contribute to more rapid recognition of cochlear dysfunction. However, since only components generated in the inner ear and the distal eighth nerve are assessed, monitoring of the ECochG alone may fail to detect dysfunction affecting more rostral portions of the auditory pathways, i.e., the proximal eighth nerve or the auditory pathways within the brainstem (Legatt, 2002). Thus, the ECochG is most often monitored in conjunction with BAEPs.

BRAINSTEM AUDITORY EVOKED POTENTIALS

Of the auditory evoked potentials, the short-latency components are the most useful for clinical purposes, both for extraoperative diagnostic testing and for IONM. This is because they are relatively easy to record, because their wave shapes and component peak latencies are highly consistent across normal subjects, and because they are relatively unaffected by anesthetic agents. They are most often called "brainstem auditory evoked potentials" or "auditory brainstem responses," even though they are not generated entirely within the brainstem. BAEPs are typically labeled with Roman numerals according to the convention of Jewett and Williston (1971) (Fig. 17.8). Up to seven peaks can be identified. The troughs can be labeled after the peak that they follow, e.g., IN follows wave I and VN, sometimes called the slow negativity, follows wave V (Fig. 17.8). Waves I, III, and V are the most consistent components, and the ones on which the interpretation of BAEP studies is based. Wave II, IV, and IV may be absent in some normal subjects; wave VII is absent in most BAEP studies. Waves IV and V are often fused, partially or completely, into a IV–V complex.

In extraoperative diagnostic BAEP testing, the patient's BAEPs are compared to normative data from a control population of neurologically and audiologically normal subjects. In order for this comparison to be valid, the stimuli, the recording techniques, and the state of the subjects in the patient's study and the control studies must be identical. This cannot be achieved during IONM, because the anesthetic regimen, the body temperature, the stimulus intensity reaching the inner ear, and other factors will differ from patient to patient. Therefore, during intraoperative BAEP monitoring each patient serves as his or her own control; BAEPs recorded at a time when elements of the auditory pathways are at risk are compared to those recorded earlier during the same operation (Legatt, 1991). There are also differences in the way the acoustic stimuli are delivered between the two settings.

BAEP generators

Wave I is generated in the most distal portion (cochlear end) of the auditory nerve, and arises from the first volley of action potentials in the nerve (Legatt et al., 1988a). It represents the same electrical phenomenon as the N1 component of the eighth-nerve CAP in the ECochG, as confirmed by simultaneous recordings of the ECochG and of BAEPs (Gersdorff, 1982); there may also be a small contribution from the cochlear summating potential (Chatrian et al., 1984). Its origin in the most distal portion of the nerve is demonstrated by its occasional persistence after section of the auditory nerve during resection of eighth-nerve tumors (Legatt, 2013) (Fig. 17.9).

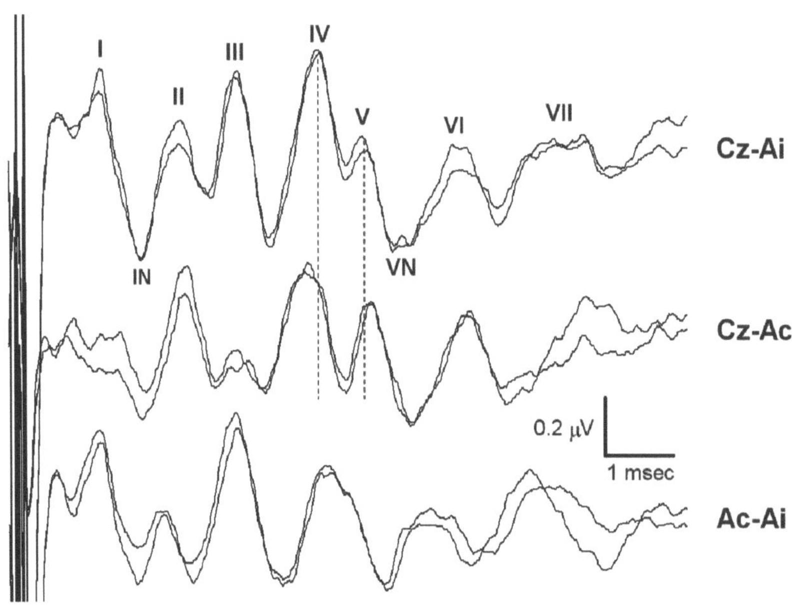

Fig. 17.8. Brainstem auditory evoked potentials recorded simultaneously from three different recording electrode linkages following monaural stimulation in a normal subject in whom all seven components are visible. The vertical dashed lines indicate the peak latencies of waves IV and V in the Cz-Ai waveforms; these peaks are typically more widely separated in the Cz-Ac waveforms. Positivity at input 1 relative to input 2 is plotted as an upward deflection. Cz, vertex; Ai, ipsilateral (stimulated) ear; Ac, contralateral ear. (Reproduced from Legatt, 2012, with permission.)

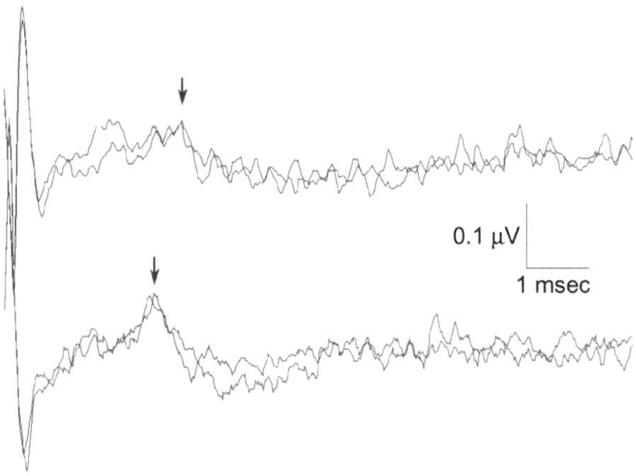

Fig. 17.9. Intraoperative brainstem auditory evoked potentials (BAEPs) to left-ear stimulation recorded during surgery for left acoustic neuromas in two different patients, showing persistence of wave I (arrows) after transection of the intracranial eighth nerve. The nerves were intentionally sacrificed to permit total resection of the tumors. Cz-Ai recording, with Cz positivity plotted as an upward deflection, in this and in subsequent BAEP figures. Cz, vertex; Ai, ipsilateral (stimulated) ear. (Reproduced from Legatt, 2002, with permission.)

Some have suggested that each BAEP peak arises from a single anatomic generator, but it has been shown that the peaks following wave I are the composites of contributions from multiple generators, as originally proposed by Jewett and Williston (1971). The complete array of generators of the BAEP components is actually quite complex (Fig. 17.10). In many cases, however, clinical–pathologic correlations suggest predominant contributions to a component from a specific anatomic area.

In the one-peak-to-one-generator models, wave II is usually associated with the cochlear nucleus. Activity at the level of the cochlear nucleus, driven by the first auditory nerve volley, does contribute to wave II (Legatt et al., 1988a). However, the second volley in the distal auditory nerve, the N2 component of the eighth-nerve CAP, occurs at the same time as this (Gersdorff, 1982), and also contributes to the scalp-recorded wave II (Legatt et al., 1988a). Thus, wave II can persist when the proximal eighth nerve has been destroyed (Legatt, 2012).

Wave III predominantly reflects activity in auditory system neurons in the caudal pontine tegmentum,

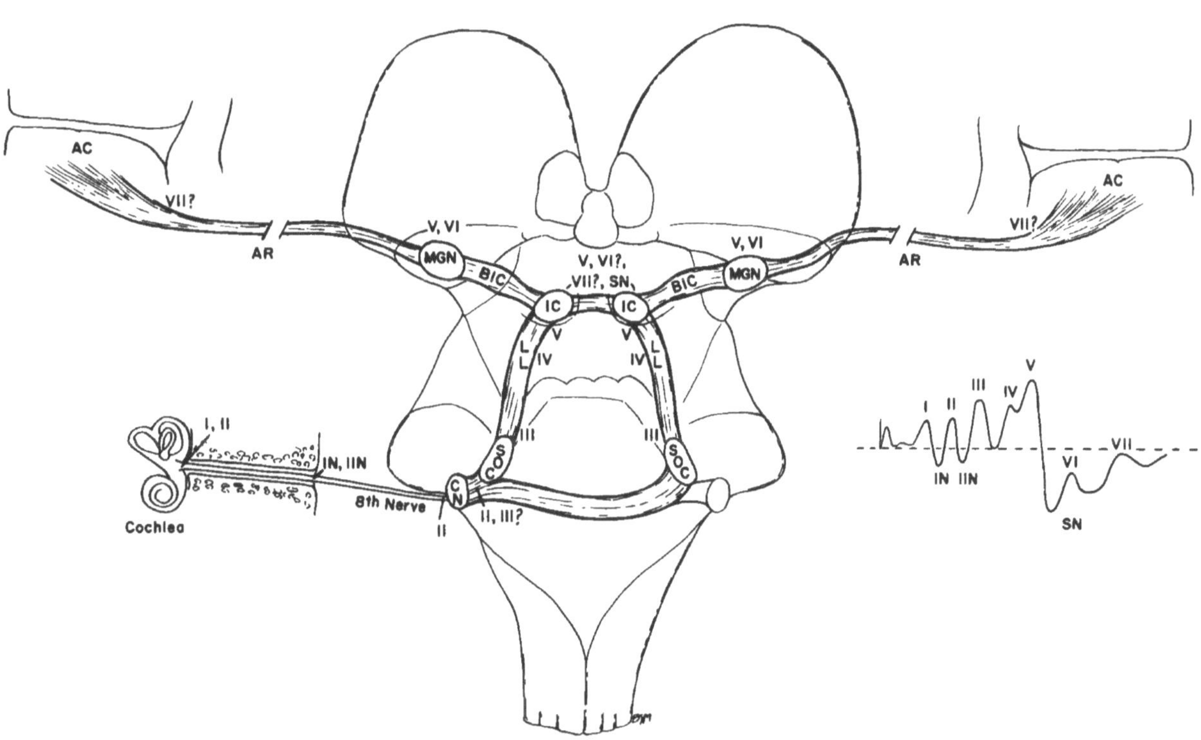

Fig. 17.10. The probable generators of the human brainstem auditory evoked potentials. AC, auditory cortex; AR, auditory radiations; BIC, brachium of the inferior colliculus; CN, cochlear nucleus; IC, inferior colliculus; LL, lateral lemniscus; MGN, medial geniculate nucleus; SN, slow negativity after wave V; SOC, superior olivary complex. (Reproduced from Legatt et al., 1988a, with permission.)

including the region of the superior olivary complex, though a contribution from continued activity at the level of the cochlear nucleus cannot be ruled out (Legatt, 2012). Since ascending projections from the cochlear nucleus are bilateral, wave III may receive contributions from brainstem auditory structures both ipsilateral and contralateral to the stimulated ear. In patients with asymmetric or unilateral brainstem lesions of the lower pons, wave III abnormalities are usually most pronounced following stimulation of the ear ipsilateral to the lesion (Brown et al., 1981; Oh et al., 1981; Faught and Oh, 1985), though occasionally they are more pronounced following contralateral stimulation (Stockard and Rossiter, 1977).

The anatomic generators of waves IV and V are most likely in close anatomic proximity or overlapping, since they are usually either both affected or both unaffected by brainstem lesions, though there are exceptions (Legatt, 2013). Wave IV appears to reflect activity predominantly in ascending auditory fibers within the dorsal and rostral pons, just caudal to the inferior colliculus, while wave V predominantly reflects activity at the level of the inferior colliculus, perhaps including activity in the rostral portion of the lateral lemniscus as it terminates in the inferior colliculus (Legatt, 2012). As is the case with wave III, wave V abnormalities due to unilateral brainstem lesions are usually most pronounced following stimulation of the ear ipsilateral to the lesion (Brown et al., 1981; Oh et al., 1981; Faught and Oh, 1985; York, 1986; Scaioli et al., 1988), though there are exceptions (Zanette et al., 1990; Fischer et al., 1995).

While waves VI and VII may in part reflect activity in more rostral structures such as the medial geniculate nucleus, they also receive contributions from activity in the inferior colliculus (Legatt, 2012). The latter generator may cause persistence of these waves in patients with auditory pathway damage rostral to the inferior colliculus. In addition, these components are absent in some normal subjects. Therefore, BAEPs cannot be used to assess or monitor the auditory pathways rostral to the mesencephalon.

BAEP recording techniques

Stimulation

BAEPs are most commonly elicited by brief acoustic click stimuli, produced by delivering monophasic square pulses of 100 μs duration to the electromechanical transducers at a rate of approximately 10 Hz. A stimulus rate of exactly 10 Hz or another submultiple of the power line frequency should be avoided; otherwise, the inevitable line frequency artifact will be time-locked to the stimuli and will not be removed by signal averaging.

A single click polarity is typically used during extraoperative diagnostic BAEP studies because the responses to rarefaction and compression clicks may differ, especially in patients with a high-frequency hearing loss (Emerson et al., 1982; Schwartz et al., 1990); the click polarity used to test the patient should be the same as that used to record the normative BAEP data. Rarefaction clicks are generally preferable because they tend to yield BAEPs with better definition of the components; this may be because the initial cochlear movements produced by rarefaction clicks are in a direction that depolarizes the hair cells (Schwartz et al., 1990).

Averaging the responses to compression and rarefaction clicks together, typically done by recording the responses to alternating-polarity clicks, may be useful in some cases to cancel a large stimulus artifact and/or the cochlear microphonic, however. Alternating-polarity clicks are commonly used during IONM, where artifacts are often problematic. During IONM of BAEPs, the same click polarity (single or alternating) should be used throughout the operation.

Audiometric headphones are typically used to deliver the stimuli during extraoperative diagnostic BAEP testing. The stimulus intensity should be loud enough to elicit a clear BAEP waveform without causing discomfort or ear damage; 60–65 dB normal hearing level (nHL) is a typical level. If hearing loss is present, the stimulus intensity may be adjusted accordingly, so that stimulation is at 60–65 dB sensation level (SL) (dB nHL is decibels relative to the threshold of the specific control population used to establish a laboratory's normative database, and dB SL is decibels relative to the threshold of the ear being tested). The subjective click threshold should be measured in all subjects in whom this is possible, to recognize hearing loss and to determine the stimulus level corresponding to 0 dB SL.

In a patient with a hearing loss, increasing the stimulus intensity may help to distinguish between different etiologies. If the stimulus intensity is increased to compensate for a conductive hearing loss and no coexisting sensorineural hearing loss is present, a normal BAEP will be recorded. In contrast, BAEPs that are delayed as a result of abnormally slowed neural conduction do not normalize when the stimulus intensity is increased (Fig. 17.11).

If the ear is stimulated at several different intensities, and the latency of wave V is graphed as a function of stimulus intensity, a latency–intensity curve is produced (Fig. 17.12). Examination of latency–intensity curves may help to classify a patient's hearing loss (Arnold, 2000). A shift of the curve to a higher-intensity level without a change in its shape suggests conductive hearing loss, whereas a change in the shape of the curve with an increased slope suggests sensorineural hearing loss.

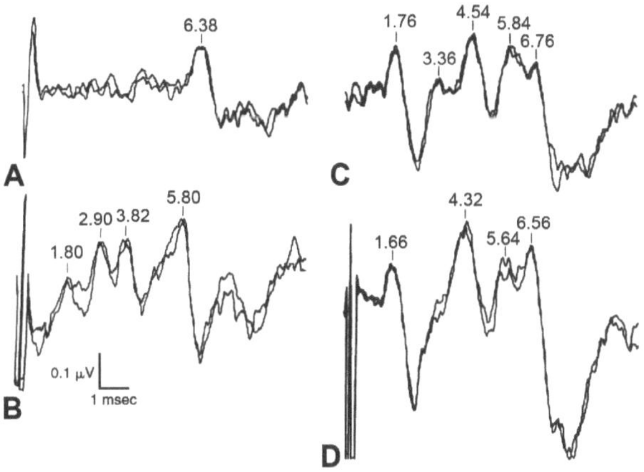

Fig. 17.11. Brainstem auditory evoked potentials (BAEPs) elicited by two different stimulus intensities in two patients. (**A**) Left-ear stimulation at 70 dB normal hearing level (nHL) elicits an abnormally delayed wave V in a 36-year-old woman with a peripheral hearing loss in the left ear. Wave I is not visible. (**B**) When the stimulus intensity is increased to 85 dB nHL, wave I becomes identifiable and the latency of wave V is markedly decreased. Interpeak intervals are normal. (**C**) Left-ear stimulation at 65 dB nHL in a 14-year-old boy with a fourth ventricular tumor elicits clear BAEPs with a normal wave I latency and an abnormally delayed wave V. The I–III and I–V interpeak intervals are both abnormally prolonged. (**D**) An increase in the stimulus intensity to 85 dB nHL causes a slight decrease in all component latencies. Wave V remains abnormally delayed, and the I–III and I–V interpeak intervals remain abnormally prolonged. (Reproduced from Legatt, 2012, with permission.)

The increased slope of the BAEP wave V latency–intensity curve may correspond to the increased slope of the eighth-nerve CAP amplitude–intensity function of the ECochG (Fig. 17.6) and to the psychophysical phenomenon of recruitment. Latency–intensity curves may reveal abnormalities that are not demonstrated by BAEP recordings at the standard, relatively high stimulus intensity typically used in BAEP studies performed for neurologic diagnosis (Legatt et al., 1988b).

Stimuli are delivered monaurally, so that a normal BAEP to stimulation of one ear does not obscure the presence of an abnormal response to stimulation of the other ear. An acoustic stimulus delivered to one ear via headphones can reach the other ear via air and bone conduction with a volume attenuation of 40–70 dB and generate an evoked potential mediated by stimulation of the contralateral ear, even if the ipsilateral ear is deaf (Chiappa, 1997; Roeser and Clark, 2000). To prevent this contralateral stimulation from occurring and possibly being misinterpreted as a BAEP arising from stimulation of the ipsilateral ear, the contralateral ear is masked with continuous white noise at an intensity 30–40 dB below that of the BAEP stimulus.

Headphones are impractical during IOMN. Instead, the acoustic stimuli are typically delivered using ear inserts, with foam cylinders that can be compressed and inserted into the ear canal; the foam then gradually expands to achieve a tight fit. The stimulus intensity that this delivers is difficult to control precisely, but this is not

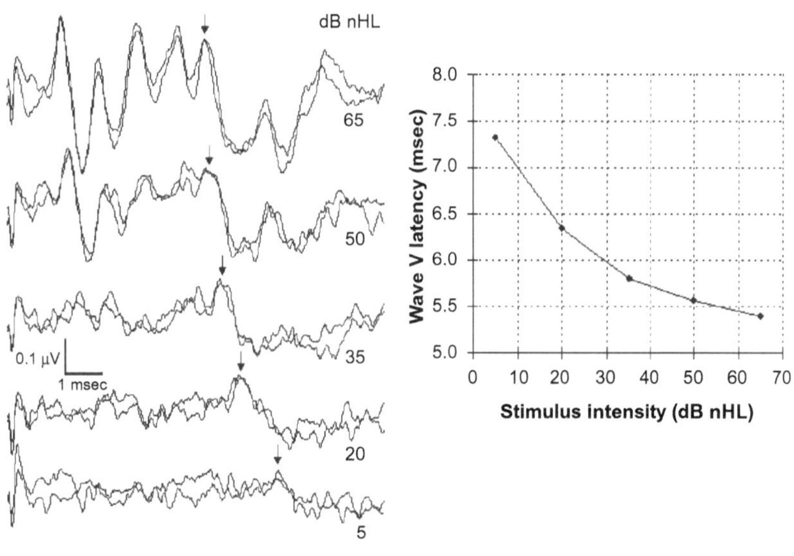

Fig. 17.12. Brainstem auditory evoked potentials (BAEPs) recorded at progressively lower stimulus intensities are shown on the left. Wave V latencies (arrows) are measured and then plotted as a function of stimulus intensity to give the latency–intensity curve on the right. This 35-year-old woman with dizziness and tinnitus had normal BAEPs and normal magnetic resonance imaging findings. (Reproduced from Legatt, 2012, with permission.)

problematic since each patient serves as his or her own control and the intensity is kept constant during the operation.

Most current IONM equipment can deliver interleaved left- and right-sided stimuli and sort the responses into separate averages, in effect acquiring averaged BAEPs to left-ear and to right-ear stimulation concurrently without actually stimulating both ears simultaneously. White-noise masking cannot be used when this is done, but this does not pose a major problem during IONM since it will not prevent recognition of the development of auditory pathway compromise when a functioning ear is stimulated. Also, the magnitude of acoustic crosstalk is less with ear-insert transducers than with headphones (Roeser and Clark, 2000).

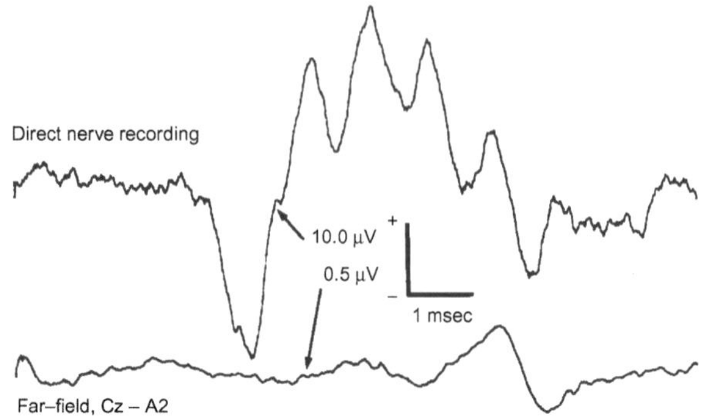

Fig. 17.13. Auditory evoked potentials to right-ear stimulation recorded directly from the right auditory nerve (top) and simultaneously from the scalp (bottom) during a selective vestibular nerve section in a 26-year-old man with Ménière's disease and intractable vertigo. Note the differing vertical scale factors; the near-field response is about 80 times larger than the far-field brainstem auditory evoked potentials.

Recording

Recording electrodes are typically placed at the vertex (position Cz of the International 10-20 System) and either at each earlobe (labeled "Ai" ipsilateral to the stimulated ear and "Ac" contralateral to it) or at each mastoid (labeled "Mi" and "Mc"). BAEPs are most often examined in a Cz-Ai/Mi waveform, in which waves I through VII are all positive in polarity (Fig. 17.8). A second recording channel, most often Cz-Ac/Mc, is useful in component identification. Wave I is present in the Cz-Ai channel but absent in the Cz-Ac channel because it is actually a near-field potential around the stimulated ear and is present in those recording channels that include the Ai electrode. Also, in the Cz-Ac recording channel, the peak latency of wave IV is typically a little shorter than, and that of wave V is a little longer than, their latencies in the Cz-Ai recording channel (see vertical dashed lines in Fig. 17.8). In patients with a fused IV–V complex, this may more clearly separate waves IV and V.

During IONM, if the cochlea or the auditory nerve is at risk (such as during resection of an eighth-nerve tumor) and the surgical exposure permits it, an electrode can be placed on the proximal eighth nerve to record a near-field CAP that is typically much larger than the far-field BAEP recorded from the scalp (Møller and Jannetta, 1983; Legatt, 1991) (Fig. 17.13). This permits signal averaging using fewer epochs and therefore more frequent assessment of the cochlea and of the auditory nerve distal to the near-field recording electrode. These near-field recordings do not monitor the auditory nerve proximal to the near-field recording electrode or the auditory pathways within the brainstem; far-field BAEPs can be recorded concurrently to monitor those structures.

A typical analog filter bandpass for BAEP recording is 100 or 150 Hz to 3000 Hz. During extraoperative diagnostic BAEP recordings in adults, an averaging epoch duration of 10 ms is typically employed initially. A longer analysis time of 15 ms may be required to record pathologically delayed BAEPs, BAEPs to lowered stimulus intensities (as when recording a latency–intensity study), BAEPs in children, and BAEPs during intraoperative monitoring.

The choice of the number of sweeps per average depends on the signal-to-noise ratio of the raw data. A value of 1000–2000 sweeps per average is typical, but more may be required if the raw data are noisy and the BAEPs are small. Near-field recording from the proximal eighth nerve (Fig. 17.13) may permit averaging using a much smaller number of sweeps.

Interpretation of BAEP studies

Component identification

The BAEP waveform typically begins with an electric stimulus artifact that is synchronous with stimulus production at the transducer. Wave I is the first major upgoing peak of the Cz-Ai waveform. The cochlear microphonic may be visible as a separate peak preceding wave I, but can be distinguished by reversing the stimulus polarity, which will reverse the polarity of the cochlear microphonic. Wave I may show a latency shift, but will not reverse polarity when the stimulus polarity is reversed.

Wave II is typically the first major upward deflection in the Cz-Ac waveform. However, it is small and difficult to identify in some normal subjects.

A substantial wave III is usually present in both the Cz-Ai and Cz-Ac channels, but is smaller in the latter. A bifid wave III, and a very small wave III in the presence of a clear wave V at normal latency, are normal variant BAEP waveform patterns.

The IV–V complex is often the most prominent component in the BAEP waveform, and is usually followed by a large negative deflection (VN or the slow negativity) that lasts several milliseconds and brings the waveform to a point below the prestimulus baseline. The morphology of the IV–V complex varies from one subject to another, and may differ between the two ears in the same person. There may be a smoothly fused waveform, two separate peaks, or one of the components may be visible as an inflection on the rising or falling phase of the other. If the latency of wave V cannot be accurately measured in the Cz-Ai waveform, then it can be measured in the Cz-Ac waveform, where waves IV and V are more separated. But this latency measurement should then be compared to normative data recorded in a Cz-Ac channel.

Modifications of the stimulus parameters can also be used to identify wave V if it is unclear when standard recording techniques are used. Wave V is the BAEP component most resistant to the effects of decreasing stimulus intensity (Fig. 17.12) or increasing stimulus rate. If either of these stimulus modifications is performed progressively until only one component remains, that peak can be identified as wave V and then traced back through the series of waveforms to identify wave V in the BAEP recorded with the standard stimulus.

Interpretation of BAEP studies, both extraoperative and intraoperative, is based predominantly on waves I, III, and V, since these are the most consistent BAEP components.

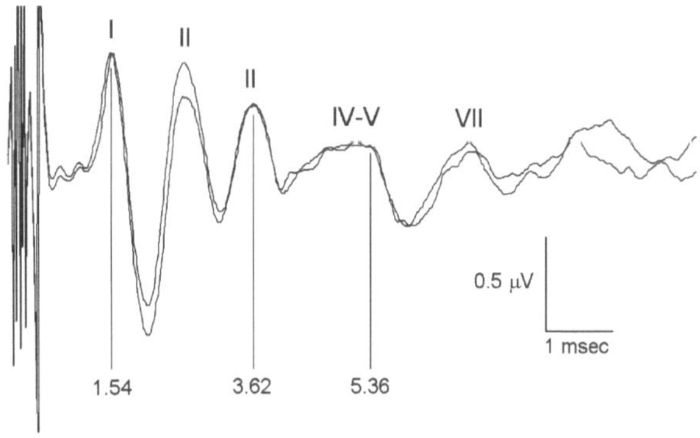

Fig. 17.14. Brainstem auditory evoked potentials recorded in a 27-year-old woman with probable multiple sclerosis. The IV–V:I amplitude ratio is 0.28 (lower limits of the normal range = 0.50); all absolute latencies and interpeak intervals are normal. The stimulus intensity was 65 dB normal hearing level (nHL). (Reproduced from Legatt, 2012, with permission.)

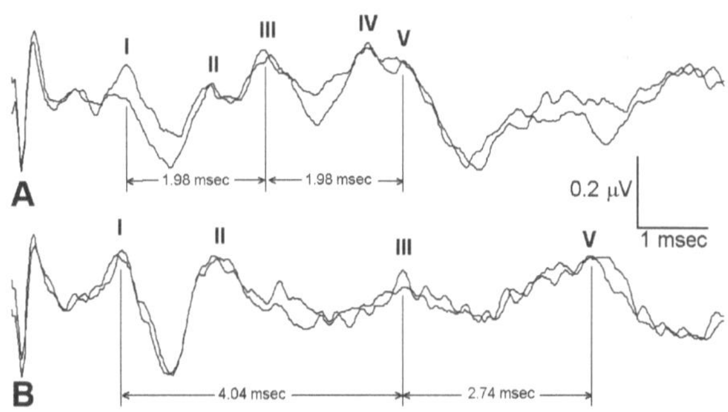

Fig. 17.15. Brainstem auditory evoked potentials (BAEPs) to stimulation of each ear in a 35-year-old woman with multiple sclerosis. (**A**) The BAEPs to left-ear stimulation are normal. (**B**) The I–III and III–V interpeak intervals are both abnormally prolonged following right-ear stimulation. (Reproduced from Legatt, 2012, with permission.)

Interpretation of extraoperative diagnostic studies

BAEP component latencies are highly consistent across subjects, whereas component amplitudes show substantial intersubject variability. Therefore, the interpretation of extraoperative diagnostic BAEP studies is based predominantly on component latencies, specifically those of the most consistent components (waves I, III, and V). The one amplitude measure that has proved to be useful is the IV–V:I amplitude ratio, the amplitude of the IV–V complex divided by the amplitude of wave I. An excessively small IV–V:I amplitude ratio (less than 0.5) can identify as abnormal some BAEP waveforms in which all component latencies and interpeak intervals are normal (Fig. 17.14).

Once waves I, III, and V have been identified and their latencies measured, the I–III, III–V, and I–V interpeak intervals are calculated. The latency of wave I has also been labeled the peripheral transmission time and the I–V interpeak interval has been called the central transmission time. Differences of component latencies and interpeak intervals between stimulation of the right and left ears are also calculated. The inter-ear latency differences typically have narrower normal ranges than do the absolute latencies; thus, examining them increases the sensitivity of the test. The amplitudes of wave I and the IV–V complex are measured, each with respect to the most negative point that follows it in the waveform, and the IV–V:I amplitude ratio is calculated.

Absence of all BAEP components or delay of wave I reflects peripheral auditory dysfunction, either a conductive or a cochlear hearing loss. Prolongation of the I–III interpeak interval (Fig. 17.15) or absence of waves III and V (Fig. 17.9) reflects dysfunction betweeen the distal eighth nerve and the lower pons, often reflecting compromise of the eighth nerve proximal to its most distal portion where wave I is generated. Prolongation of

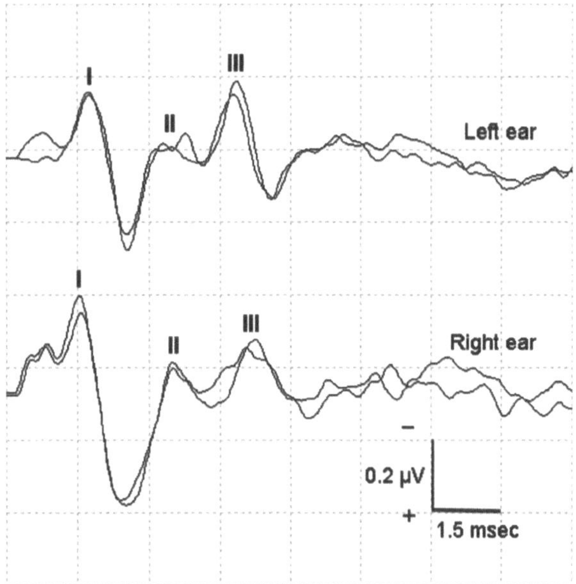

Fig. 17.16. Waves I, II, and III are present, but later components are absent, following stimulation of either ear in a 4-year-old girl with a brainstem glioma involving the rostral pons and mesencephalon.

the III–V interpeak interval (Fig. 17.15) or absence of wave V in the presence of a clear wave III (Fig. 17.16) reflects dysfunction between the lower pons and the mesencephalon.

Abnormalities in the BAEPs are not specific for etiology. They may reflect mechanical compromise of the auditory pathways, such as compression or anatomic disruption, ischemia/infarction, infection, or inflammation/demyelination. Also, BAEPs reflect activity in a subset of the brainstem auditory pathways with the shortest latencies and a high degree of synaptic security that minimizes temporal dispersion, a subset that is specialized for sound localization (Legatt et al., 1988a). Conduction slowing that leads to temporal dispersion can eliminate BAEP peaks, although the signals are still being transmitted to higher centers. Thus, pure-tone audiometry and speech discrimination may be normal in patients with grossly abnormal BAEPs; BAEP abnormalities tend to be more correlated with deficits in sound localization (Hausler and Levine, 1980; Zerlin and Mowry, 1980; Musiek and Geurkink, 1982; Kaga et al., 1986).

Interpretation of intraoperative BAEP monitoring

Whereas BAEP component amplitudes vary substantially from one subject to another, they are usually quite consistent in repeated recordings in the same subject if the recording techniques are not changed. Moreover, intraoperative compromise of neural pathways may cause amplitude changes earlier than latency changes, and in some cases components are attenuated and disappear without a significant change in their latencies. Therefore, both component amplitudes and component latencies should be followed during intraoperative BAEP monitoring. The amplitudes of the vertex-positive peaks of waves I, III, and V are measured with respect to the troughs that follow them. Typical threshold criteria for the identification of an adverse BAEP change are a 50% decrease in the amplitude of a component (most often of wave V), or a 1-ms increase in the absolute latency of wave V or in the I–V interpeak interval.

When BAEPs are abnormal, localization of the abnormality follows the same principles that were described in the preceding section. During IONM, adverse BAEP changes may reflect compression, stretching, or direct mechanical injury of tissue, thermal effects from cautery, or ischemia from compromise of the blood supply.

The cochlea receives its blood supply from the intracranial circulation via the internal auditory artery, which is usually a branch of the anterior inferior cerebellar artery. This artery, which is quite small, passes through the internal auditory canal alongside the eighth nerve (Kim et al., 1990). Damage to this artery will cause cochlear ischemia or infarction, affecting wave I and all subsequent BAEP components (Fig. 17.17).

Compromise of the eighth nerve proximal to its cochlear end will cause prolongation of the I–III interpeak interval, attenuation of waves III and V, or both. If the compromise is severe enough, waves III and V will be lost (Fig. 17.9). Cerebellar retraction to gain access to the cerebellopontine angle can stretch the eighth nerve, leading to hearing loss. During microvascular decompression surgery in the cerebellopontine angle, BAEP monitoring can detect eighth-nerve dysfunction from stretching (Fig. 17.18), letting the surgeons know when they need to adjust the retraction (Møller and Jannetta, 1983). BAEP monitoring has been shown to improve hearing outcomes following this operation (Radtke et al., 1989).

The BAEP changes that are seen with mechanical, thermal, or ischemic injury to the brainstem will depend on what parts of the brainstem are involved. BAEPs may remain unchanged in some patients who suffer brainstem damage that anatomically spares the auditory pathways (Little et al., 1987). While preservation of the BAEPs does not guarantee a good outcome, patients with significant brainstem ischemia-related BAEP changes that persist to the end of the operation almost always have new postoperative neurologic deficits (Little et al., 1983; Manninen et al., 1994).

Factors other than focal neurologic dysfunction can also cause intraoperative BAEP changes. Anesthetic agents in the usual concentrations produce only minimal changes in BAEP amplitudes and latencies (see Legatt,

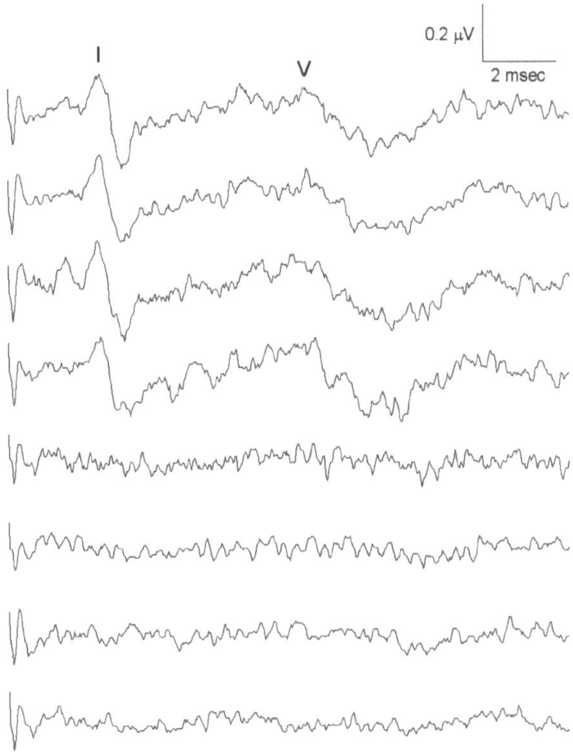

Fig. 17.17. Consecutive brainstem auditory evoked potentials (BAEPs) to left-ear stimulation (earliest waveform at the top) recorded in a patient undergoing surgery for a left vestibular schwannoma. A clear wave I and a poorly formed wave V were initially present and were stable during the initial dissection, but all BAEP components disappeared simultaneously during dissection within the internal auditory canal and remained absent through the end of the operation. This was most likely due to interruption of the blood supply to the cochlea via the internal auditory artery. (Reproduced from Legatt, 2002, with permission.)

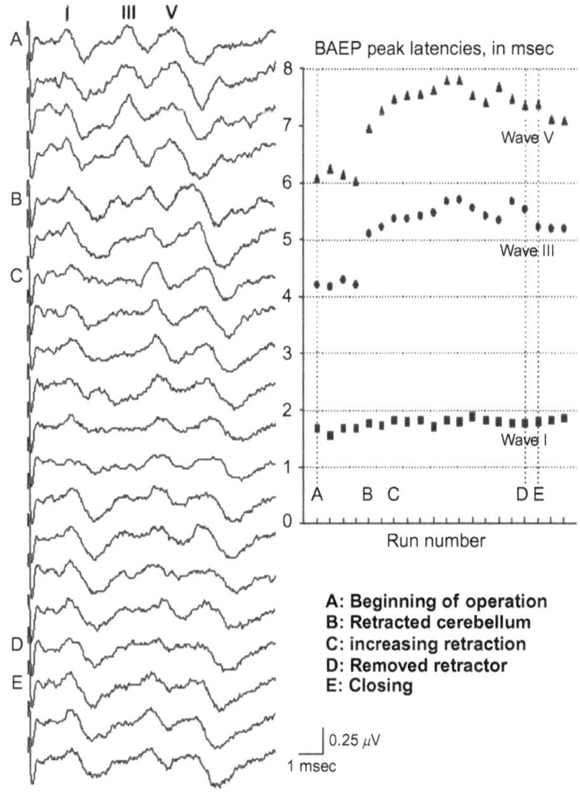

Fig. 17.18. Brainstem auditory evoked potentials (BAEPs) to right-ear stimulation during surgery for a right vestibular schwannoma in a 61-year-old woman. They were stable during the beginning of the operation. When the cerebellum was retracted (**B**), the latencies of waves III and V increased in parallel; there was no change in the latency of wave I. When the retraction was increased (**C**), there were further latency increases as well as amplitude attenuation of waves III and V. After the tumor had been resected, cerebellar retraction was released, and the latencies of waves III and V began to shorten.

2002, for more details), but BAEPs do show substantial changes due to hypothermia. Both the latency of wave I and the interpeak intervals progressively increase as the patient is cooled. Component latencies and interpeak intervals increase by about 7% for each 1 °C drop in temperature, and at 26 °C are about double their values at normal body temperature (Markand et al., 1987). The latency changes are reversible upon rewarming, but there is hysteresis – latencies at the same temperature may differ between the cooling and rewarming phases (Markand et al., 1990).

Use of a surgical drill, such as drilling off the roof of the internal auditory canal during resection of an eighth-nerve tumor, will produce high noise levels in both internal ears via bone conduction. The resultant BAEP changes, caused by acoustic masking (Levine et al., 1994), can resemble BAEP changes due to auditory system compromise. BAEP data acquisition can be paused during drilling of bone in order to prevent false alarms due to acoustic masking.

BAEPs can also be obscured by artifacts or lost due to a variety of technical factors, such as equipment malfunction, dislodged recording electrodes, and broken wires. Additional technical problems specific to IONM of BAEPs include compression or kinking of the plastic tube that carries the acoustic stimulus from the transducer to the patient's ear and the entry of liquids into the ear canal.

Frequency-following responses

Stimulation with relatively prolonged, low-frequency tone bursts (e.g., 200–500-Hz tone bursts lasting 10–15 ms each) produces an evoked potential that contains oscillatory components at the frequency of the tone (Marsh et al., 1975) (Fig. 17.19). Intracranial recordings in

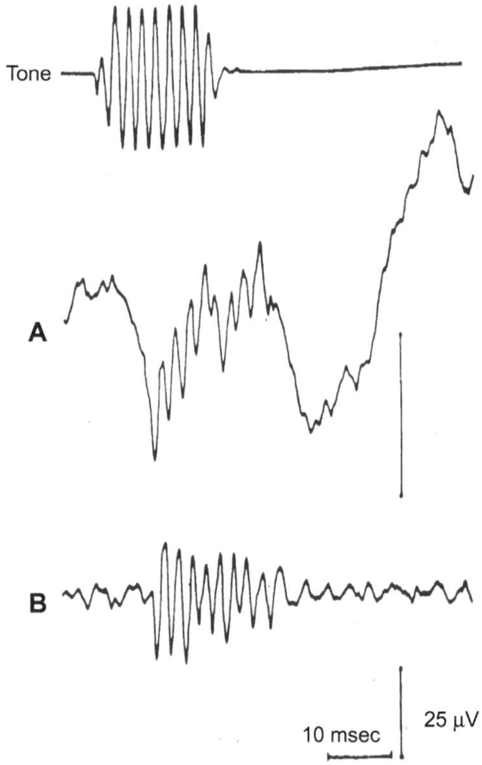

Fig. 17.19. Averaged frequency-following response to a 500-Hz tone burst, recorded between the vertex and linked ears with an amplifier filter bandpass of (A) 8 Hz to 10 000 Hz and (B) 200 Hz to 1000 Hz. The acoustic stimulus, recorded by a microphone at the subject's earpiece, is shown at the top. Note the latency shifts between the onset and the offset of the stimulus and those of the response. (Reproduced from Marsh et al., 1975, with permission.)

animals (Davis and Britt, 1984) have shown that this frequency-following response does reflect neuronal activity in the brainstem, although care must be taken to avoid contamination by the cochlear microphonic (Chimento and Schreiner, 1990). The frequency-following response and the BAEPs do not appear to have the same neural generators (Hoormann et al., 1992). Whereas the BAEPs reflect activity originating in the base of the cochlea, the frequency-following response predominantly reflects activity originating in lower-frequency regions of the cochlea. However, the clinical utility of frequency-following responses for assessing hearing in the lower frequencies has been limited by the technical difficulties associated with recording them and by their relatively high thresholds.

MIDDLE-LATENCY AUDITORY EVOKED POTENTIALS

Components and generators

The initial cortically generated MLAEPs are similar in latency to the neck and cephalic muscle contractions that

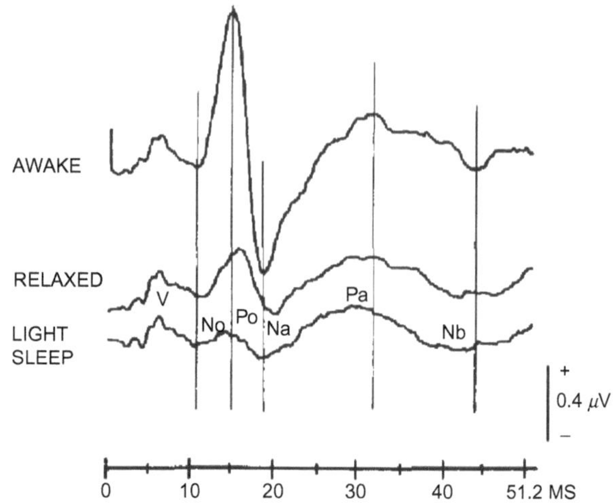

Fig. 17.20. Middle-latency auditory evoked potentials (MLAEPs) recorded in the same subject in the awake, relaxed, and asleep states. The large biphasic wave which obscures the MLAEPs in the awake state is the postauricular muscle reflex. The MLAEP components are labeled N0, P0, Na, Pa, and Nb; brainstem auditory evoked potential wave V is also labeled. Vertex-to-ear recording, vertex positivity plotted as an upward deflection. (Reproduced from Yokoyama et al., 1987, with permission.)

occur reflexively following acoustic stimulation with loud sounds, most prominently the postauricular muscle reflex. The electric correlate of these muscle contractions may be picked up by the electrodes used to record the MLAEPs and may be substantially larger than the MLAEPs themselves (Fig. 17.20). The signal that was initially reported as the neurogenic cortical MLAEP (Geisler et al., 1958) was later shown to be myogenic in origin (Bickford et al., 1964). Subsequent studies have shown that it is possible to record MLAEPs arising from cerebral cortex if care is taken to avoid myogenic contamination. The most prominent peaks of the neurogenic MLAEPs are labeled Na, Pa, Nb, and Pb (Figs 17.2 and 17.20). Small peaks preceding these are labeled N0 and P0. Pb has a peak latency of approximately 50 ms, and thus is sometimes considered as the first of the LLAEP peaks, in which case it is labeled "P1."

In the initial studies, the MLAEPs were recorded from the vertex, with a reference electrode on the inion or mastoid. However, recording from more lateral electrodes on both hemispheres may reveal lateralized MLAEP abnormalities in patients with lateralized lesions (Fig. 17.21). Recordings from arrays of scalp electrodes provide even more detailed information about the scalp topography of the MLAEPs, and suggest that there are multiple cortical generators. For example, Cacace et al. (1990) found that Pa and Pb had maxima around the vertex at latencies of about 29 ms and about 53 ms, respectively, but the detailed topographic recordings indicated an additional component with a peak latency of about

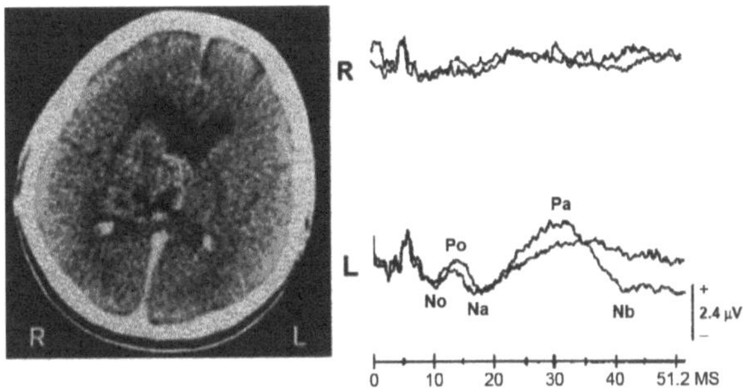

Fig. 17.21. Middle-latency auditory evoked potentials are present on the left side of the head, but absent on the right, in this patient with a glioma involving the right thalamus. The brainstem auditory evoked potentials (most prominently wave V) are visible in the first part of both recordings. (Modified from Yokoyama et al., 1987, with permission.)

41 ms which has a horizontally oriented dipole with a scalp maximum over the temporal lobes (Fig. 17.22), suggesting a lateral temporal-lobe generator. The components that are maximal around the vertex are thought to reflect the summation of bilateral vertically oriented voltage dipole sources in auditory-responsive cortex in the superior temporal plane. Clinical-pathologic correlations in patient brain lesions also localize the sources of the MLAEPs to the temporal lobes (Kileny et al., 1987; Shehata-Dieler et al., 1991).

Recordings of auditory-evoked magnetic fields (Borgmann et al., 2001; Yvert et al., 2001), intracranial recordings of MLAEPs during neurosurgical procedures (Celesia, 1976; Howard et al., 2000), and extraoperative recordings of MLAEPs from depth electrodes in candidates for epilepsy surgery who are having invasive monitoring (Liégeois-Chauvel et al., 1994; Howard et al., 2000) have localized the generators of the MLAEPs to multiple temporal-lobe regions, including primary auditory cortex within Heschl's gyri, surrounding cortex in the superior temporal plane (the dorsal surface of the temporal lobe), and some spatially restricted areas in the superior temporal gyrus on the lateral surface of the temporal lobe; the latter have been labeled the "posterior lateral superior temporal auditory area" (PLST) (Howard et al., 2000). In the magnetoencephalographic data, as the latencies of the components increased, the locations of their equivalent source dipoles tended to shift from posterior to more anterior and also from mesial (deep within the Sylvian fissure) to more lateral (Borgmann et al., 2001), suggesting sequential activation of different cortical areas.

In the intracranial studies of Howard et al. (2000), the earliest peak attributable to the PLST had a mean latency of 45 ms, longer than the earliest MLAEP components recorded from the vertex, and evoked potentials could be recorded from the PLST following electric stimulation of Heschl's gyri; these data suggest that the PLST receives corticocortical input from Heschl's gyri. This is in accordance with a hierarchic model of auditory cortical areas based on studies in primates that subdivides auditory-responsive cortex into "core," "belt," and "parabelt" regions (Hackett, 2008; Brugge, 2013), with projections from core to belt and from belt to parabelt. The core is located in Heschl's gyri (Brodmann's area 41).

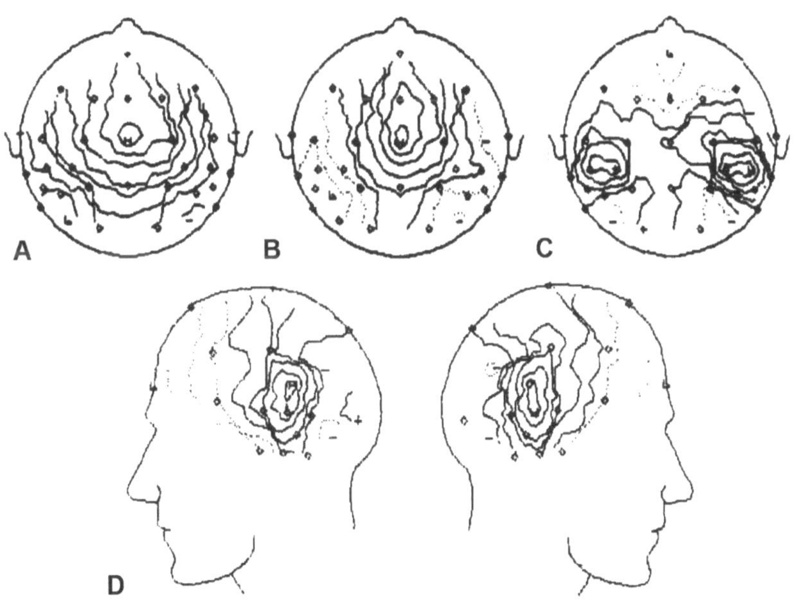

Fig. 17.22. Topographic maps of grand-average middle-latency auditory evoked potentials to binaural click stimulation at post-stimulus latencies of (**A**) 28 ms, showing the maximum of the Pa component at the vertex, (**B**) 52 ms, showing the maximum of the Pb component at the vertex, and (**C**) 42 ms, showing a separate component ("TP41") that is maximal over the temporal lobes. (**D**) Topography of the TP41 component in a different view. (Modified from Cacace et al., 1990, with permission.)

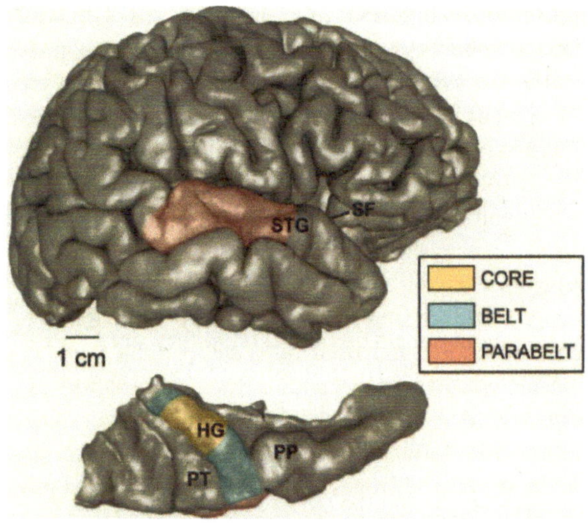

Fig. 17.23. One hypothesis regarding the location of the core, belt, and parabelt areas of auditory cortex in humans. STG, superior temporal gyrus; SF, Sylvian fissure; HG, Heschl's gyrus; PT, planum temporale; PP, planum polare. (Reproduced from Brugge, 2013, with permission.)

While the precise anatomic localization of belt and parabelt regions in humans is uncertain, most models (e.g., Fig. 17.23) suggest that the PLST is located within the parabelt region.

Further details about the anatomic organization of auditory cortical areas and about intracranial recordings of auditory evoked potentials can be found in Chapters 2 and 13 of this volume.

Clinical utility

MLAEPs can be used to assess hearing in patients in whom behavioral audiometry is impractical. The thresholds obtained are often higher than those obtained via pure-tone auditory with behavioral responses, though MLAEPs are highly sensitive in detecting cochlear hearing loss in adult patients (Kankkunen and Rosenhall, 1985). MLAEPs may have a greater utility in assessing patients with disorders of the supratentorial auditory system (Alain et al., 2013).

MLAEPs decrease in amplitude and increase in latency during sleep, and most of them disappear during slow-wave (stage 3 and 4) sleep (Picton, 2011). They can be recorded under surgical levels of anesthesia but are markedly affected by the depth of anesthesia, which limits their utility for monitoring the anatomic and functional integrity of the auditory pathways; anesthetic-related variability would cause too many false alarms. However, their sensitivity to anesthetic effects has led to trials of the MLAEPs as an indicator of the depth of anesthesia (Schneider et al., 2005; Horn et al., 2009). Some commercially available devices for estimating the depth of anesthesia assess both MLAEPs and the EEG. At present, however, the most widely used techniques for estimating the depth of anesthesia are based on analysis of the EEG alone (Bruhn et al., 2006).

LONG-LATENCY AUDITORY EVOKED POTENTIALS

The P1–N1–P2–N2 complex

LLAEPs reflect not only the auditory stimuli but also the degree to which the subject is attending to the stimuli and analyzing stimulus features (Fig. 17.24). Thus, they have also been labeled cognitive event-related potentials. These signals may also be affected by the previous auditory stimuli that had been presented to the subject, and whether the current stimulus is the same as, or different

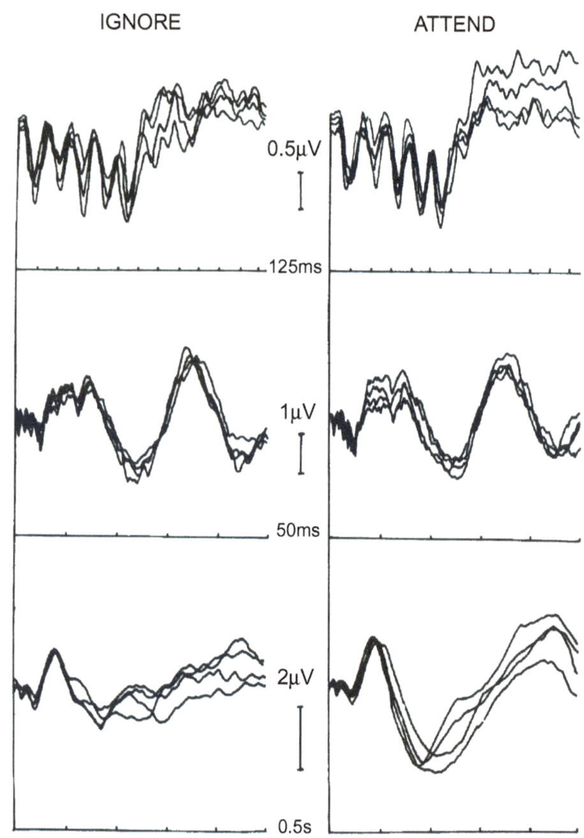

Fig. 17.24. Effect of attention on auditory evoked potentials. Trains of clicks that were occasionally lower in intensity were presented at a rate of 1/second, and the subject was asked to either ignore the clicks (left) or to attend to them and count the number of softer ones (right). Note that the brainstem auditory evoked potentials and middle-latency auditory evoked potentials are identical between the two stimulus conditions, but the long-latency auditory evoked potentials are different. Vertex-to-mastoid recording, with vertex positivity plotted as a downward deflection. (Reproduced from Picton and Hillyard, 1974, with permission.)

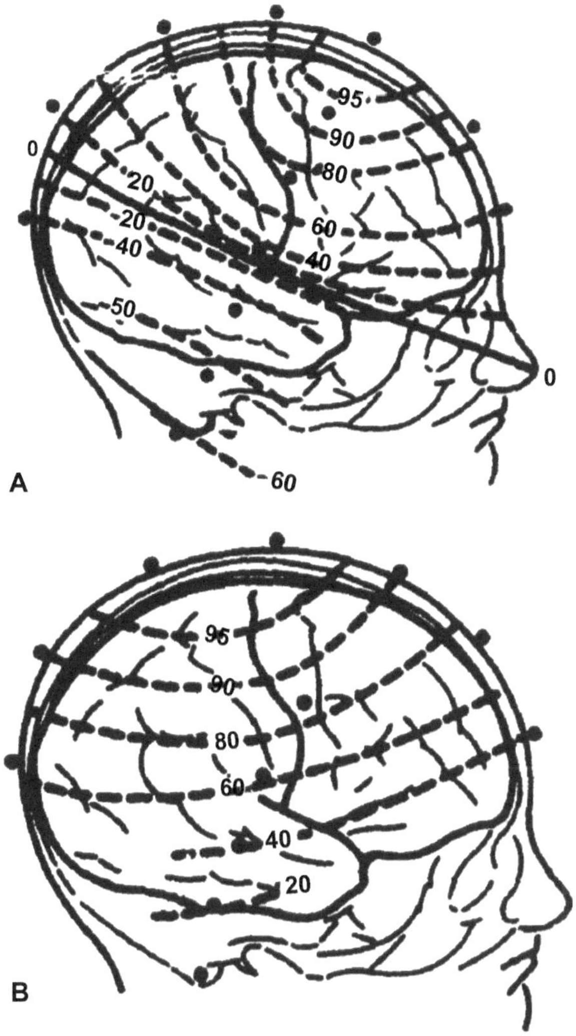

Fig. 17.25. Scalp isopotential maps for the P2 and P3 components of the long-latency auditory evoked potentials. (**A**) The P2 component is maximal at the top of the head and shows an inversion across the plane of the superior temporal plane. (**B**) The P3 component to an oddball stimulus is also maximal at the top of the head, but does not show such an inversion. (Reproduced from Vaughan and Ritter, 1970, with permission.)

from, its predecessors. The first four peaks of the LLAEPs are labeled P1, N1, P2, and N2, and have peak latencies of approximately 50, 100, 180, and 250 ms, respectively (Alain et al., 2013) (Fig. 17.2).

The P1 component, with a latency of approximately 50 ms, is the same component as the Pb component of the MLAEP. Scalp topographic mapping studies (Picton et al., 1999; Leavitt et al., 2007) and intracranial recordings (Godey et al., 2001; Korzyukov et al., 2007) suggest that P1 has multiple generators, clearly including the temporal lobe but other brain areas as well.

The N1 component is often the largest and most reliable of the LLAEPs. N1 and P2 are most likely generated in auditory cortical areas in the posterior portion of superior temporal plane, with N1 possibly having a contribution from the lateral temporal lobe and frontal motor and premotor cortices (Näätänen and Picton, 1987) as well as from nearby parietal-lobe structures (Alain et al., 2013). The voltage dipoles of these components are vertically oriented, giving the vertex maximum in the scalp topography, but the locations of the sources in the superior temporal plane cause the scalp-recorded waveform to invert across the level of the superior temporal plane (Vaughan and Ritter, 1970) (Fig. 17.25A).

Abnormalities in N1 and P2 are found in patients who have impaired speech and sound perception from a variety of etiologies (Alain et al., 2013). The P2 and N2 components are modulated by the amount of attention being given to the stimulus – they are larger when the subject is attending to the stimulus and performing a sensory discrimination task (Fig. 17.24). The LLAEPs are also affected by stimulus properties such as intensity, duration, frequency, and spatial location and may show differences related to the age of the subject (Alain et al., 2013).

The mismatch negativity

The mismatch negativity (MMN) is a signal that is generated when, during repetitive auditory stimulation, an auditory stimulus deviates in some way from recent stimuli that were similar to each other (the standard stimuli). It is recorded by subtracting the response to the standard stimuli from that to the deviant stimuli (Fig. 17.26).

The MMN is an automatic response, not depending on whether the subject is performing a sensory discrimination task to identify the oddball or deviant stimuli and not even needing the subject to attend to the stimuli (Kujala et al., 2007). An MMN can be present even if the subject does not consciously recognize that the oddball stimuli are deviant (van Zuijen et al., 2006), implying some automatic, preconscious comparator mechanism. Most MMN studies utilize standard stimuli that are identical. However, a MMN can also be generated if the standard stimuli are not identical but do follow a specific rule, such as a train of tone stimuli at various frequencies where successive stimuli are at progressively lower frequencies (Tervaniemi et al., 1994) or where tones at higher frequencies are presented at louder intensities (Paavilainen et al., 2001). The stimuli that deviate from the rule generate an MMN, showing that the auditory system has recognized more abstract features of the auditory stimuli. The system that generates the MMN may also be implicated in language processing; with phonemic stimuli, a different phoneme (e.g., /e/ as opposed to /a/) can elicit an MMN even if the standard stimulus phonemes were spoken by different people (Shestakova et al., 2002).

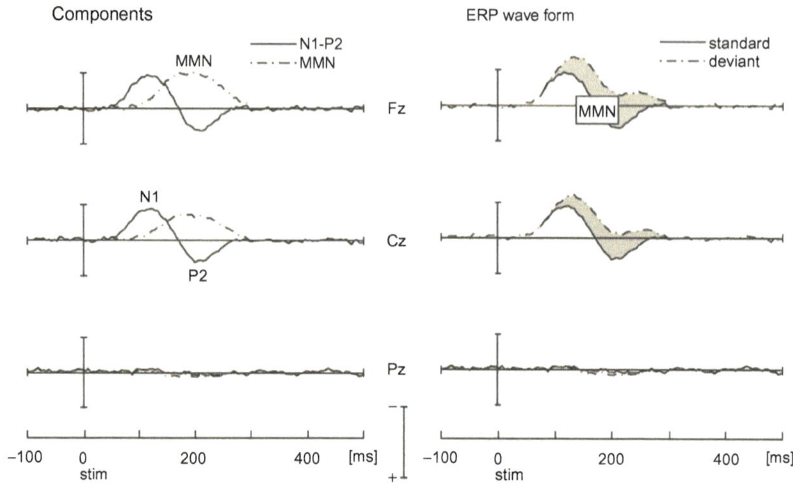

Fig. 17.26. Long-latency auditory evoked potentials to standard and deviant stimuli recorded at three midline scalp electrode positions with the subject ignoring the stimuli. The components are shown on the left, and the auditory evoked potential waveforms are shown on the right. The mismatch negativity (MMN), the difference between the responses to the deviant and standard stimuli, is shown in gray on the right. ERPs, event-related potentials. (Modified from Näätänen et al., 2011, with permission.)

The major cortical generators of the MMN are within the temporal lobes, anterior to the source of the N1 (Alho, 1995; Picton, 2011), but studies have suggested the existence of additional generators, including in the frontal lobe, in the parietal lobe, and in subcortical structures. The scalp topography of the MMN is maximal over the frontocentral region (Kujala et al., 2007). Thus, it most likely reflects predominantly bilateral vertically oriented dipole generators in the superior temporal plane.

Since the MMN is generated when an auditory stimulus differs from prior stimuli, its neural substrate must include some memory trace of the prior stimuli. The MMN has therefore been used in studies of memory. If the interstimulus interval is sufficiently prolonged, the deviant stimuli no longer generate an MMN, demonstrating decay of the memory trace. Auditory memory traces have been shown to decay faster in conditions where memory is expected to be adversely affected, such as Alzheimer's disease and alcoholism (Kujala et al., 2007).

The P3/P300 component

The MMN is generated when deviant stimuli are presented, even if the subjects are ignoring the stimuli. If, however, the subjects are asked to attend to the stimuli and identify the deviant or oddball stimuli in some way (e.g., press a reaction key or maintain a mental count of the number of oddball stimuli), then additional event-related potential components are generated (Fig. 17.27). The largest of these has been labeled the P3 component (a continuation of the P1-N1-P2-N2 sequence) or the P300 component (after its polarity and typical latency in ms). It is preceded by a negativity, N2b, which overlaps with the P2 component of the MLAEP but can be identified by subtracting the responses to the standard stimuli from those to the deviant stimuli in order to cancel out the P2.

The amplitude of the P3 is typically of the order of magnitude of 10 μV. Signal averaging is most often used to record it, but it can sometimes be identified in single-sweep raw data (Picton, 1992). This is facilitated because P3 has a more posterior scalp maximum than the other LLAEP components (Figs 17.25B and 17.27); it can be recorded over the parietal area at Pz where most of the other LLAEP components are much smaller.

The P3 has been subdivided into subcomponents. P3a, which is earlier in latency, is more anterior (frontocentral) in its scalp distribution; it is generated following an oddball stimulus embedded in a stimulus train, whether or not the subject is performing a sensory discrimination task (Polich, 2007). P3b, which is the larger of the two subcomponents, is more posterior with a parietal maximum (Fig. 17.27); it is generated following the target stimulus when the subject is performing the task of discriminating the target stimuli.

P3a most likely arises from frontal-lobe areas, and is reduced in patients with frontal-lobe lesions; it is also reduced in patients with hippocampal lesions. P3b most likely originates in multiple generators within the temporoparietal region of the brain, possibly with a frontal contribution as well. Activity at the latency of the P3b can also be recorded from depth electrodes within the hippocampus. However, studies in patients with hippocampal lesions and recordings following hippocampal resection in monkeys and in human patients suggest that the

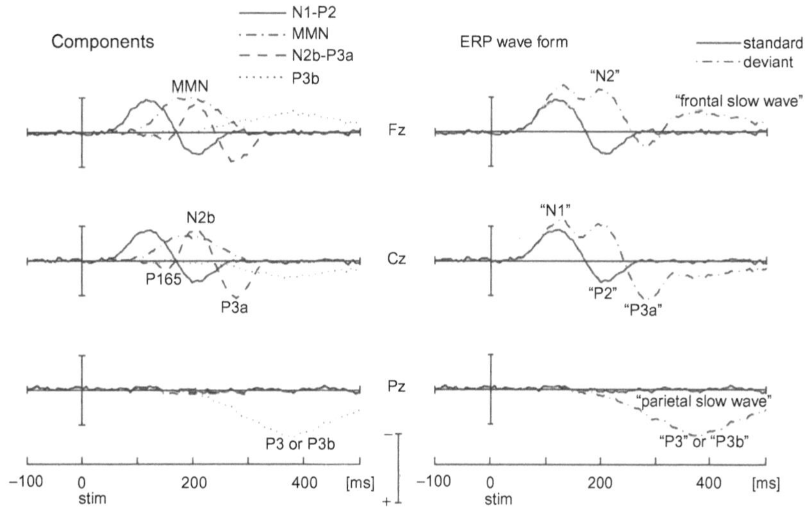

Fig. 17.27. Long-latency auditory evoked potentials to standard and deviant stimuli recorded at three midline scalp electrode positions with the subjects attending to the stimuli and discriminating the deviant or oddball stimuli. Compare to Figure 17.26; additional event-related potential (ERP) components are present in this condition as opposed to when the subjects are ignoring the stimuli. MMN, mismatch negativity. (Modified from Näätänen et al., 2011, with permission.)

activity within the hippocampus does not contribute to the scalp-recorded P3b waveform (Polich, 2007).

As the difficulty of the sensory discrimination task is increased (i.e., as the difference between the target and the standard stimuli becomes smaller), the P300 increases in latency and decreases in amplitude (Picton, 2011). Although the P3 component is elicited in a sensory discrimination task when the subject recognizes the target stimulus, it does not appear to represent the earliest recognition. In studies in which the timing of the subject's reaction to the target stimuli is measured, it is found to be correlated with the latency of P3 but clearly precedes the peak of the P3, even more so if the timing of the motor process from the brain to the finger that signals the reaction is considered (Ritter et al., 1972).

Clinical and research applications

The LLAEPs are attenuated or eliminated by surgical anesthesia; thus, they are not useful for intraoperative monitoring of the integrity of the auditory system.

In paired-stimulus experiments, the P1 component to the second stimulus is typically smaller than the first (Waldo and Freedman, 1986), a phenomenon which has been labeled "sensory gating." In group averages, the attenuation of the second stimulus is reduced in patients with schizophrenia and (to a lesser extent) in their relatives (Freedman et al., 1987), but the intersubject variability does not permit this to be used as a diagnostic test in the individual patient (Patterson et al., 2008). Decreased sensory gating is also found in other conditions, such as bipolar affective disorder (Lijffijt et al., 2009) and Alzheimer's disease (Cancelli et al., 2006).

Clinical application of the MMN and the P300 have been limited by several factors, including their large intersubject variability, the many experimental variables that can affect them, uncertainties about their anatomic generators, and the many clinical disorders that can cause attenuation of these signals (yielding a poor specificity for etiology) (Picton, 2011). The P3a and P3b components may be differentially affected in dementia, and examining them separately may increase the accuracy of identification of dementia (Juckel et al., 2008). There is evidence that LLAEPs may demonstrate abnormalities in patients with familial Alzheimer's disease many years before the subjects become demented (Golob et al., 2009).

Even if the large overlap in findings between patients and control subjects mitigates against the utility of these signals in the diagnosis of an individual subject, study of the significant group differences between normal subjects and patients with dementia, autism, various psychiatric disorders, and other conditions may provide insight into the pathophysiology of disorders of auditory, language, and cognitive functions.

Independent of clinical diagnostic use, the LLAEPs remain a research topic of great interest. They provide a window into the manner in which the brain analyzes auditory signals, automatically identifies sounds that have changed and unexpected sounds, and identifies target auditory stimuli.

References

Alain C, Rouye A, Arnott SR (2013). Middle- and long-latency auditory evoked potentials: what are they telling us on central auditory disorders? In: GG Celesia (Ed.), Disorders of Peripheral and Central Auditory Processing Handbook of Clinical Neurophysiology, Volume 10. Elsevier, Amsterdam, pp. 177–199.

Alho K (1995). Cerebral generators of mismatch negativity (MMN) and its magnetic counterpart (MMNm) elicited by sound changes. Ear Hear 16: 38–51.

Arnold SA (2000). The auditory brain stem response. In: RJ Roeser, M Valente, H Hosford-Dunn (Eds.), Audiology: Diagnosis, Thieme, New York, pp. 451–470.

Arslan E, Turrini M, Lupi G et al. (1997). Hearing threshold assessment with auditory brainstem response (ABR) and electrocochleography (ECochG) in uncooperative children. Scand Audiol Suppl 46: 32–37.

Attias J, Nageris B, Ralph J et al. (2008). Hearing preservation using combined monitoring of extra-tympanic electrocochleography and auditory brainstem responses during acoustic neuroma surgery. Int J Audiol 47: 178–184.

Bickford RG, Jacobson JL, Cody DTR (1964). Nature of average evoked potentials to sound and other stimuli in man. Ann N Y Acad Sci 112: 204–223.

Borgmann C, Ross B, Draganova R et al. (2001). Human auditory middle latency responses: influence of stimulus type and intensity. Hear Res 158: 57–64.

Brown Jr RH, Chiappa KH, Brooks E (1981). Brain stem auditory evoked responses in 22 patients with intrinsic brain stem lesions: implications for clinical interpretations. Electroencephalogr Clin Neurophysiol 52: 38P.

Brugge JF (2013). Anatomy and physiology of auditory pathways and cortex. In: GG Celesia (Ed.), Disorders of Peripheral and Central Auditory Processing Handbook of Clinical Neurophysiology, Volume 10. Elsevier, Amsterdam, pp. 25–59.

Bruhn J, Myles PS, Sneyd R et al. (2006). Depth of anaesthesia monitoring: what's available, what's validated and what's next? Br J Anaesth 97: 85–94.

Cacace AT, Satya-Murti S, Wolpaw JR (1990). Human middle-latency auditory evoked potentials: vertex and temporal components. Electroencephalogr Clin Neurophysiol 77: 6–18.

Cancelli I, Cadore IP, Merlino G et al. (2006). Sensory gating deficit assessed by P50/Pb middle latency event related potential in Alzheimer's disease. J Clin Neurophysiol 23: 421–425.

Celesia GG (1976). Organization of auditory cortical areas in man. Brain 99: 403–414.

Chatrian GE, Wirch AL, Edwards KH et al. (1984). Cochlear summating potential recorded from the external auditory meatus of normal humans. Amplitude-intensity functions and relationships to auditory nerve compound action potential. Electroencephalogr Clin Neurophysiol 59: 396–410.

Chiappa KH (1997). Brain stem auditory evoked potentials: Methodology. In: KH Chiappa (Ed.), Evoked Potentials in Clinical Medicine, 3rd edn. Raven Press, New York, pp. 157–197.

Chimento TC, Schreiner CE (1990). Selectively eliminating cochlear microphonic contamination from the frequency-following response. Electroencephalogr Clin Neurophysiol 75: 88–96.

Coats AC (1986). The normal summating potential recorded from external ear canal. Arch Otolaryngol Head Neck Surg 112: 759–768.

Dallos P, Cheatham MA (1976). Production of cochlear potentials by inner and outer hair cells. J Acoust Soc Am 60: 510–512.

Dallos P, Schoeny ZG, Cheatham MA (1972). Cochlear summating potentials. Descriptive aspects. Acta Otolaryngol Suppl 302: 1–46.

Davis RL, Britt RH (1984). Analysis of the frequency following response in the cat. Hear Res 15: 29–37.

Devaiah AK, Dawson KL, Ferraro JA et al. (2003). Utility of area curve ratio electrocochleography in early Meniere disease. Arch Otolaryngol Head Neck Surg 129: 547–551.

Eggermont JJ (1976). Summating potentials in electrocochleography: relation to hearing disorders. In: RJ Ruben, C Elberling, G Salomon (Eds.), Electrocochleography, University Park Press, Baltimore, pp. 67–87.

Emerson RG, Brooks EB, Parker SW et al. (1982). Effects of click polarity on brainstem auditory evoked potentials in normal subjects and patients: unexpected sensitivity of wave V. Ann N Y Acad Sci 388: 710–721.

Faught E, Oh SJ (1985). Brainstem auditory evoked responses in brainstem infarction. Stroke 16: 701–705.

Ferraro JA (2010). Electrocochleography: a review of recording approaches, clinical applications, and new findings in adults and children. J Am Acad Audiol 21: 145–152.

Fischer C, Bognar L, Turjman F et al. (1995). Auditory evoked potentials in a patient with a unilateral lesion of the inferior colliculus and medial geniculate body. Electroencephalogr Clin Neurophysiol 96: 261–267.

Freedman R, Adler LE, Gerhardt GA et al. (1987). Neurobiological studies of sensory gating in schizophrenia. Schizophr Bull 13: 669–678.

Geisler CD, Frishkopf LS, Rosenblith WA (1958). Extracranial responses to acoustic clicks in man. Science 128: 1210–1211.

Gersdorff MCH (1982). Simultaneous recordings of human auditory potentials: transtympanic electrocochleography (ECoG) and brainstem-evoked responses (BER). Arch Otorhinolaryngol 234: 15–20.

Godey B, Schwartz D, de Graaf JB et al. (2001). Neuromagnetic source localization of auditory evoked fields and intracerebral evoked potentials: a comparison of data in the same patients. Clin Neurophysiol 112: 1850–1859.

Golob EJ, Ringman JM, Irimajiri R et al. (2009). Cortical event-related potentials in preclinical familial Alzheimer disease. Neurology 73: 1649–1655.

Hackett TA (2008). Anatomical organization of the auditory cortex. J Am Acad Audiol 19: 774–779.

Hausler R, Levine RA (1980). Brain stem auditory evoked potentials are related to interaural time discrimination in patients with multiple sclerosis. Brain Res 191: 589–594.

Hoormann J, Falkenstein M, Hohnsbein J et al. (1992). The human frequency-following response (FFR): normal variability and relation to the click-evoked brainstem response. Hear Res 59: 179–188.

Horn B, Pilge S, Kochs EF et al. (2009). A combination of electroencephalogram and auditory evoked potentials separates different levels of anesthesia in volunteers. Anesth Analg 108: 1512–1521.

Howard MA, Volkov IO, Mirsky R et al. (2000). Auditory cortex on the human posterior superior temporal gyrus. J Comp Neurol 416: 79–92.

Iseli C, Gibson W (2010). A comparison of three methods of using transtympanic electrocochleography for the diagnosis of Meniere's disease: click summating potential measurements, tone burst summating potential amplitude measurements, and biasing of the summating potential using a low frequency tone. Acta Otolaryngol 130: 95–101.

Janssen T, Müller J (2008). Otoacoustic emissions as a diagnostic tool in a clinical context. In: G Manley, RR Fay, AN Popper (Eds.), Active Processes and Otoacoustic Emissions in Hearing. Springer Handbook of Auditory Research, Volume 30. Springer, New York, pp. 421–451.

Jewett DL, Williston JS (1971). Auditory-evoked far fields averaged from the scalp of humans. Brain 94: 681–696.

Juckel G, Clotz F, Frodl T et al. (2008). Diagnostic usefulness of cognitive auditory event-related p300 subcomponents in patients with Alzheimers disease? J Clin Neurophysiol 25: 147–152.

Kaga K, Shindo M, Tanaka Y (1986). Auditory brain stem responses and nonsense monosyllable perception test findings for patients with auditory nerve and brain stem lesions. Laryngoscope 96: 1272–1278.

Kankkunen A, Rosenhall U (1985). Comparison between thresholds obtained with pure-tone audiometry and the 40-Hz middle latency response. Scand Audiol 14: 99–104.

Kileny P, Paccioretti D, Wilson AF (1987). Effects of cortical lesions on middle-latency auditory evoked responses (MLR). Electroencephalogr Clin Neurophysiol 66: 108–120.

Kim HN, Kim YH, Park IY et al. (1990). Variability of the surgical anatomy of the neurovascular complex of the cerebellopontine angle. Ann Otol Rhinol Laryngol 99: 288–296.

Korzyukov O, Pflieger ME, Wagner M, Bowyer SM, Rosburg T, Sundaresan K et al. (2007). Generators of the intracranial P50 response in auditory sensory gating. Neuroimage 35: 814–826.

Kujala T, Tervaniemi M, Schroger E (2007). The mismatch negativity in cognitive and clinical neuroscience: theoretical and methodological considerations. Biol Psychol 74: 1–19.

Leavitt VM, Molholm S, Ritter W et al. (2007). Auditory processing in schizophrenia during the middle latency period (10–50 ms): high-density electrical mapping and source analysis reveal subcortical antecedents to early cortical deficits. J Psychiatry Neurosci 32: 339–353.

Legatt AD (1991). Intraoperative neurophysiologic monitoring. In: EAM Frost (Ed.), Clinical Anesthesia in Neurosurgery, 2nd edn. Butterworth-Heinemann, Boston, Massachusetts, pp. 63–127.

Legatt AD (2002). Mechanisms of intraoperative brainstem auditory evoked potential changes. J Clin Neurophysiol 19: 396–408.

Legatt AD (2003). Evoked potentials (EPs). In: MJ Aminoff, R Daroff (Eds.), Encyclopedia of the Neurological Sciences, Elsevier Science, San Diego, pp. 309–313.

Legatt AD (2012). Brainstem auditory evoked potentials: Methodology, interpretation, and clinical application. In: MJ Aminoff (Ed.), Aminoff's Electrodiagnosis in Clinical Neurology, 6th edn. Churchill Livingstone, New York, pp. 519–552.

Legatt AD (2013). Intraoperative monitoring of auditory evoked potentials. In: GG Celesia (Ed.), Disorders of Peripheral and Central Auditory Processing Handbook of Clinical Neurophysiology. Volume 10. Elsevier, Amsterdam, pp. 235–252.

Legatt AD, Arezzo JC, Vaughan Jr HG (1988a). The anatomic and physiologic bases of brain stem auditory evoked potentials. Neurol Clin 6: 681–704.

Legatt AD, Pedley TA, Emerson RG et al. (1988b). Normal brain-stem auditory evoked potentials with abnormal latency-intensity studies in patients with acoustic neuromas. Arch Neurol 45: 1326–1330.

Levine RA, Ojemann RG, Montgomery WW et al. (1984). Monitoring auditory evoked potentials during acoustic neuroma surgery. Insights into the mechanism of the hearing loss. Ann Otol Rhinol Laryngol 93: 116–123.

Levine RA, Ronner SF, Ojemann RG (1994). Auditory evoked potential and other neurophysiologic monitoring techniques during tumor surgery in the cerebellopontine angle. In: CM Loftus, VC Traynelis (Eds.), Intraoperative Monitoring Techniques in Neurosurgery, McGraw-Hill, New York, pp. 175–191.

Liégeois-Chauvel C, Musolino A, Badier JM et al. (1994). Evoked potentials recorded from the auditory cortex in man: evaluation and topography of the middle latency components. Electroencephalogr Clin Neurophysiol 92: 204–214.

Lijffijt M, Moeller FG, Boutros NN et al. (2009). Diminished P50, N100 and P200 auditory sensory gating in bipolar I disorder. Psychiatry Res 167: 191–201.

Little JR, Lesser RP, Lueders H et al. (1983). Brain stem auditory evoked potentials in posterior circulation surgery. Neurosurgery 12: 496–502.

Little JR, Lesser RP, Luders H (1987). Electrophysiological monitoring during basilar aneurysm operation. Neurosurgery 20: 421–427.

Manninen PH, Patterson S, Lam AM et al. (1994). Evoked potential monitoring during posterior fossa aneurysm surgery: a comparison of two modalities. Can J Anaesth 41: 92–97.

Markand ON, Lee BI, Warren C et al. (1987). Effects of hypothermia on brainstem auditory evoked potentials in humans. Ann Neurol 22: 507–513.

Markand ON, Warren C, Mallik GS et al. (1990). Temperature-dependent hysteresis in somatosensory and auditory evoked potentials. Electroencephalogr Clin Neurophysiol 77: 425–435.

Marsh JT, Brown WS, Smith JC (1975). Far-field recorded frequency-following responses: correlates of low pitch auditory perception in humans. Electroencephalogr Clin Neurophysiol 38: 113–119.

Møller AR, Jannetta PJ (1983). Monitoring auditory functions during cranial nerve microvascular decompression operations by direct recording from the eighth nerve. J Neurosurg 59: 493–499.

Mullatti N, Coakham HB, Maw AR et al. (1999). Intraoperative monitoring during surgery for acoustic neuroma: benefits of an extratympanic intrameatal electrode. J Neurol Neurosurg Psychiatry 66: 591–599.

Musiek FE, Geurkink NA (1982). Auditory brain stem response and central auditory test findings for patients with brain stem lesions: a preliminary report. Laryngoscope 92: 891–900.

Näätänen R, Picton T (1987). The N1 wave of the human electric and magnetic response to sound: a review and an analysis of the component structure. Psychophysiology 24: 375–425.

Näätänen R, Kujala T, Winkler I (2011). Auditory processing that leads to conscious perception: a unique window to central auditory processing opened by the mismatch negativity and related responses. Psychophysiology 48: 4–22.

Oh SJ, Kuba T, Soyer A, Choi IS et al. (1981). Lateralization of brainstem lesions by brainstem auditory evoked potentials. Neurology 31: 14–18.

Ojemann RG, Levine RA, Montgomery WM et al. (1984). Use of intraoperative auditory evoked potentials to preserve hearing in unilateral acoustic neuroma removal. J Neurosurg 61: 938–948.

Paavilainen P, Simola J, Jaramillo M et al. (2001). Preattentive extraction of abstract feature conjunctions from auditory stimulation as reflected by the mismatch negativity (MMN). Psychophysiology 38: 359–365.

Patterson JV, Hetrick WP, Boutros NN et al. (2008). P50 sensory gating ratios in schizophrenics and controls: a review and data analysis. Psychiatry Res 158: 226–247.

Picton TW (1992). The P300 wave of the human event-related potential. J Clin Neurophysiol 9: 456–479.

Picton TW (2011). Human Auditory Evoked Potentials, Plural Publishing, San Diego.

Picton TW, Hillyard SA (1974). Human auditory evoked potentials. II: Effects of attention. Electroencephalogr Clin Neurophysiol 36: 191–199.

Picton TW, Hillyard SA, Krausz HI et al. (1974). Human auditory evoked potentials. I. Evaluation of components. Electroencephalogr Clin Neurophysiol 36: 179–190.

Picton TW, Alain C, Woods DL et al. (1999). Intracerebral sources of human auditory-evoked potentials. Audiol Neurootol 4: 64–79.

Polich J (2007). Updating P300: an integrative theory of P3a and P3b. Clin Neurophysiol 118: 2128–2148.

Radtke RA, Erwin CW, Wilkins RH (1989). Intraoperative brainstem auditory evoked potentials: Significant decrease in postoperative morbidity. Neurology 39: 187–191.

Ritter W, Simson R, Vaughan Jr HG (1972). Association cortex potentials and reaction time in auditory discrimination. Electroencephalogr Clin Neurophysiol 33: 547–555.

Roeser RJ, Clark JL (2000). Clinical masking. In: RJ Roeser, M Valente, H Hosford-Dunn (Eds.), Audiology: Diagnosis, Thieme, New York, pp. 253–279.

Ruben RJ, Bordley JE, Lieberman AT (1961). Cochlear potentials in man. Laryngoscope 71: 1141–1164.

Salomon G, Elberling C (1971). Cochlear nerve potentials recorded from the ear canal in man. Acta Otolaryngol 71: 319–325.

Santarelli R, Arslan E (2013). Electrocochleography. In: GG Celesia (Ed.), Disorders of Peripheral and Central Auditory Processing Handbook of Clinical Neurophysiology, Volume 10. Elsevier, Amsterdam, pp. 83–113.

Santarelli R, Starr A, Michalewski HJ et al. (2008). Neural and receptor cochlear potentials obtained by transtympanic electrocochleography in auditory neuropathy. Clin Neurophysiol 119: 1028–1041.

Scaioli V, Savoiardo M, Bussone G et al. (1988). Brain-stem auditory evoked potentials (BAEPs) and magnetic resonance imaging (MRI) in a case of facial myokymia. Electroencephalogr Clin Neurophysiol 71: 153–156.

Schneider G, Hollweck R, Ningler M et al. (2005). Detection of consciousness by electroencephalogram and auditory evoked potentials. Anesthesiology 103: 934–943.

Schwartz DM, Morris MD, Spydell JD et al. (1990). Influence of click polarity on the brain-stem auditory evoked response (BAER) revisited. Electroencephalogr Clin Neurophysiol 77: 445–457.

Shehata-Dieler W, Shimizu H, Soliman SM et al. (1991). Middle latency auditory evoked potentials in temporal lobe disorders. Ear Hear 12: 377–388.

Shestakova A, Brattico E, Huotilainen M et al. (2002). Abstract phoneme representations in the left temporal cortex: magnetic mismatch negativity study. Neuroreport 13: 1813–1816.

Sparacino G, Milani S, Magnvita V et al. (2000). Electrocochleography potentials evoked by condensation and rarefaction clicks independently derived by a new numerical filtering technique. Audiol Neurootol 5: 276–291.

Starr A, Picton TW, Sininger Y et al. (1996). Auditory neuropathy. Brain 119: 741–753.

Stockard JJ, Rossiter VS (1977). Clinical and pathologic correlates of brain stem auditory response abnormalities. Neurology 27: 316–325.

Tervaniemi M, Maury S, Näätänen R (1994). Neural representations of abstract stimulus features in the human brain as reflected by the mismatch negativity. Neuroreport 5: 844–846.

van Zuijen TL, Simoens VL, Paavilainen P et al. (2006). Implicit, intuitive, and explicit knowledge of abstract regularities in a sound sequence: an event-related brain potential study. J Cogn Neurosci 18: 1292–1303.

Vaughan Jr HG, Ritter W (1970). The sources of auditory evoked responses recorded from the human scalp. Electroencephalogr Clin Neurophysiol 28: 360–367.

Waldo MC, Freedman R (1986). Gating of auditory evoked responses in normal college students. Psychiatry Res 19: 233–239.

Withnell RH (2001). Brief report: the cochlear microphonic as an indication of outer hair cell function. Ear Hear 22: 75–77.

Yokoyama T, Ryu H, Uemura K et al. (1987). Study of the constant wave form of ML-AEP in humans. Electroencephalogr Clin Neurophysiol 67: 372–378.

York DH (1986). Correlation between a unilateral midbrain-pontine lesion and abnormalities of brain-stem auditory evoked potential. Electroencephalogr Clin Neurophysiol 65: 282–288.

Yvert B, Crouzeix A, Bertrand O et al. (2001). Multiple supratemporal sources of magnetic and electric auditory evoked middle latency components in humans. Cereb Cortex 11: 411–423.

Zanette G, Carteri A, Cusumano S (1990). Reappearance of brain-stem auditory evoked potentials after surgical treatment of a brain-stem hemorrhage: contributions to the question of wave generation. Electroencephalogr Clin Neurophysiol 77: 140–144.

Zerlin S, Mowry HJ (1980). Click lateralization and the auditory brain stem response. Audiology 19: 346–354.

Handbook of Clinical Neurology, Vol. 129 (3rd series)
The Human Auditory System
G.G. Celesia and G. Hickok, Editors

Chapter 18

Psychophysical and behavioral peripheral and central auditory tests

FRANK E. MUSIEK[1] AND GAIL D. CHERMAK[2]*

[1]*Department of Communication Sciences and Department of Surgery (Otolaryngology), School of Medicine, University of Connecticut, Farmington, CT, USA*

[2]*Department of Speech and Hearing Sciences, Washington State University Spokane, Spokane, WA, USA*

INTRODUCTION

Diagnostic batteries to assess the integrity of the central auditory nervous system (CANS) include behavioral (i.e., psychophysical) tests, electrophysiologic procedures, and to some degree, electroacoustic measures. This chapter focuses on the behavioral tests which are used to identify lesions (including diffuse lesions), abnormalities, or dysfunction of the CANS, as well as identify associated functional deficits (e.g., listening in noise deficits). Accurate diagnosis of central auditory processing disorder (CAPD) requires the administration of a battery of tests with documented sensitivity to CANS lesions and/or dysfunction in a controlled acoustic environment. The peripheral auditory system is assessed using the peripheral tests discussed below prior to the administration of a central auditory test battery. The audiologist selects normed central auditory tests that provide insights regarding the presence of CAPD and the integrity of the CANS at multiple sites and levels (American Academy of Audiology, 2010). The peripheral and central auditory tests and procedures discussed in this chapter are outlined in Table 18.1. The reader is referred to the following sources for review of patient history, selection criteria, and diagnostic procedures and criteria (American Speech-Language-Hearing Association, 2005; American Academy of Audiology, 2010; Musiek and Chermak, 2014).

PERIPHERAL TESTS

Standard pure-tone thresholds and speech recognition tests

It is now well known that standard pure-tone threshold and speech recognition measures commonly used in contemporary audiology clinics are of little value in assessing the effects of CAPD. Pure-tone thresholds quantify the degree of sensorineural hearing loss, which reflects cochlear and auditory nerve function; however, these thresholds do not differentiate between the two sites of lesion. Speech recognition scores generally are poorer for auditory nerve involvement; however, these scores do not distinguish well between these two sites. As described by Bocca et al. (1954), individuals with lesions of the CANS usually present good pure-tone hearing sensitivity, despite complaints of hearing difficulties. These findings compelled Bocca and colleagues to develop new, more sensitive tests for the evaluation of individuals with CAPD. It is important to note, however, that there are instances when central lesions do affect hearing sensitivity, as revealed by reduced pure-tone thresholds. One of the exceptions was identified many years ago in studies of kernicterus (Matkin and Carhart, 1966), which is known to damage the central auditory pathway, especially the cochlear nucleus (Dublin, 1971). These early multidisciplinary studies in audiology and neuropathology yielded evidence demonstrating that damage to the cochlear nucleus could indeed affect pure-tone thresholds. To our knowledge, these data have never been disputed. Severe damage to more rostral levels of the CANS also has been shown to affect pure-tone thresholds. The extreme example of this is central deafness (see Chapter 32 in this volume), where patients may present severe hearing loss as evidenced by pure-tone thresholds or even total lack of response to pure-tone signals. In this case, severe bilateral central auditory involvement is the basis for the depressed or lack of pure-tone sensitivity. Severe unilateral central deficits also can influence the pure-tone

*Correspondence to: Gail D. Chermak, Ph.D., Department of Speech and Hearing Sciences, Washington State University Spokane, 412 E. Spokane Falls Blvd, Spokane, WA 99202, USA. Tel: +1-509-358-7598, Fax: +1-509-358-7600, E-mail: chermak@wsu.edu

Table 18.1

Peripheral and central auditory tests and procedures described in this chapter

Peripheral auditory tests	Central auditory tests
Pure-tone thresholds	Dichotic listening
Speech recognition	Binaural interaction procedures
	Masking level differences (MLD)
	Localization/lateralization tests
	Spatial listening tests (LiSN-S)
Otoacoustic emissions	Temporal processing tests
	Temporal ordering or sequencing
	Temporal resolution or discrimination
	Gaps-in-Noise (GIN)
	Random Gap Detection Test (RGDT)
	Temporal integration
	Temporal masking
Acoustic reflex	Monaural low-redundancy tests
	Low-pass filtered speech tests
	Speech-in-competition (speech-in-noise) tests
	Synthetic sentence identification test
	Pediatric speech intelligibility test
	Other speech-in-noise tests
	Time-compressed speech tests

audiogram, as supported by individual case reports (Musiek et al., 1994), as well as animal studies which reveal shifts in pure-tone thresholds in the contralateral ear subsequent to single-sided damage of the auditory cortex (Heffner and Heffner, 1986). Standard speech recognition procedures administered in quiet (i.e., in the absence of competing noise or babble) generally are not sensitive to central auditory dysfunction, although there are exceptions, as noted above for pure-tone thresholds (Musiek et al., 1994).

Otoacoustic emissions

Though primarily considered a test of cochlear function, otoacoustic emissions (OAEs), whether transient or distortion product, can assist in the diagnosis of central auditory involvement. OAEs are byproducts of preneural mechanisms of the cochlear amplifier, and in particular are linked to the normal functioning of the outer hair cells. They vary as a function of a number of variables, including age, gender, and stimulus (e.g., clicks vs pure tones). OAEs usually are absent or abnormal when cochlear hearing loss exceeds 30 dB hearing level (HL) for transient OAEs and approximately 40 dB HL for distortion product OAEs. Though it is well known that auditory nerve function is not necessary for generation of OAEs, VIIIth- nerve disorders such as acoustic tumors can disrupt OAEs. This is thought to be related to possible cochlear involvement secondary to lack of blood supply via the auditory artery that courses through the internal auditory meatus (Telischi et al., 1995). Though standard OAEs alone are not sensitive to central auditory compromise, there is at least one report of high-amplitude OAEs in patients with head injury and tinnitus (Ceranic et al., 1998). OAEs are most useful in central auditory diagnosis when a hearing loss (i.e., reduced pure-tone thresholds) is observed. That is, if the OAEs are normal in this situation, cochlear hearing loss due to outer hair cell dysfunction most likely is not the issue. It is more probable that the hearing loss arises either from the auditory (VIIIth) cranial nerve or central auditory involvement (although malingering is another possibility) (Ferber-Viart et al., 1994; Robinette and Glattke, 1997).

OAEs may offer the greatest insight into the integrity of the CANS when suppression effects are measured. Briefly, this procedure requires recording standard OAEs with and without noise in the contralateral ear. This procedure activates the medial olivocochlear bundle which results in the attenuation of the expansion and contraction of the outer hair cells. Normally, the contralateral noise results in a slight decrease in OAE level (typically approximately 1 dB) (Robinette and Glattke, 1997). In individuals with central disorder due to head injury, this suppression is lost (Ceranic et al., 1998), and is reduced as well in children diagnosed with CAPD (Muchnik et al., 2004). Because the suppression effect is small, it is likely that it would not be observed with cochlear or VIIIth-nerve compromise, making interpretation difficult; however, to our knowledge, there is no specific research which speaks to this point. Despite these reports, audiologists have not routinely incorporated OAE suppression testing clinically. This is likely related to the small size of the effect and the failure to observe it across multiple clinical populations with confirmed CAPD.

The acoustic reflex

The acoustic reflex (AR) (the reflexive contraction of the middle-ear muscles in response to sound stimulation) has a long history of clinical use in defining middle-ear, cochlear, and VIIIth nerve disorders. The AR and other audiologic tests (e.g., air and bone conduction thresholds, tympanometry, and reflex decay) can help differentiate middle-ear, cochlear, and VIIIth-nerve problems; however, these tests are not able to differentiate auditory nerve from low brainstem involvement. The AR is mediated in the lower pons, and, therefore, reflects only a small portion of the CANS. The AR, however, can provide insight as to central auditory dysfunction in the caudal brainstem.

Accurate measurement of brainstem function using the AR is dependent on normal middle-ear, cochlear, and auditory nerve function. In central (brainstem) dysfunction, the AR threshold usually is elevated or absent. Hall and Johnston's (2007) review of children with CAPD and learning problems showed mixed results across studies, revealing normal, hypersensitive, and in some cases absent or elevated AR. The AR abnormalities occurred in only a relatively small portion of the children in the studies reviewed, suggesting that the central dysfunction underlying their CAPD was rostral to the brainstem. Abnormal AR findings, however, have been reported in a number of different central auditory disorders, such as tumors of the brainstem (Jerger and Jerger, 1977), multiple sclerosis (Jerger et al., 1986), head injury (Hall et al., 1982), and recently in lead poisoning (Counter et al., 2011). In these studies, the sensitivity of the AR was good (approximately 60–80%). Therefore, even though the AR has anatomic limitations in regard to the CANS, it is useful in identifying central disorders that involve the brainstem. The AR, however, has been eclipsed by the auditory brainstem response (ABR) for audiologic diagnosis of brainstem central auditory involvement (see Chapter 29 in this volume).

CENTRAL AUDITORY TESTS

Dichotic listening

Dichotic listening refers to listening to different acoustic events presented to each ear simultaneously. Most commonly, the acoustic signals are speech, such as digits, words, consonant-vowels (CVs), or sentences. Broadbent (1954) is credited with first introducing dichotic listening; however, it was Kimura (1961a) who first provided insight regarding the clinical potential of dichotic listening by testing individuals with temporal-lobe disorders. Since these early investigations, dichotic listening has grown both in its research and clinical use, such that today it is among the most widely used behavioral tests of the CANS and central auditory processing. The most commonly used clinical tests employing the dichotic paradigm are the Staggered Spondaic Word test (Katz, 1962), Dichotic CVs (Berlin et al., 1972), Dichotic Sentences (Willeford, 1977), Dichotic Digits (Musiek, 1983), Dichotic Sentence Identification (Fifer et al., 1983), and Dichotic Words (Meyers et al., 2002).

Dichotic findings in individuals with normal peripheral and central auditory function

The right-ear advantage (REA) is perhaps the most interesting finding in normal populations. Although the REA is not obtained from all normal subjects (e.g., in some left-handed individuals), it is a consistent finding in the literature (Kimura, 1961b). Though the basis for the REA is beyond our discussion here, it is related to the more direct anatomic connection of the right ear to the left hemisphere, which in most people is specialized for language processing. The size of the REA is somewhat test-dependent, but is usually only a few percentage points. The REA can be altered by attention: for example, Hugdahl and Andersson (1986) reported increased scores in the ear to which the attention is focused. One dichotic test which is not commonly used clinically but for which directing attention has little effect is the Dichotic Rhyme Test (Shinn et al., 2005).

Methodologic factors

Early research with the dichotic paradigm demonstrated the importance of acoustic and procedural factors; however, still today these factors are not always recognized. A brief review of some of the key methodologic factors follows. For more detailed reviews, the reader is referred to Berlin and Cullen (1977) and Musiek and Pinheiro (1985).

Regarding presentation levels, it is crucial that the intensity level at each ear be the same (unless the particular test calls for a specified difference). In addition, the overall intensity level at each ear should be sufficient to allow maximum intelligibility of the dichotic stimuli. Test outcomes will vary if these factors are violated. Hearing loss and asymmetry of hearing sensitivity at the two ears are also important considerations. Hearing loss may affect intelligibility of the speech stimuli, and asymmetric hearing thresholds can cause one ear to perform differently than the other independent of central function. Temporal alignment (onsets and offsets of the stimuli) also influences test performance. If the stimulus in one ear lags in time relative to the other ear, the stimuli in the lagging ear will be more easily perceived than the stimuli in the leading ear.

The type of response required also must be considered in dichotic listening. Though there can be many kinds of report conditions when testing dichotically, two report conditions are most commonly used (Musiek and Pinheiro, 1985; Keith, 2007). The individual can be instructed to repeat stimuli directed to both ears, which requires binaural integration and is also referred to as the free recall or divided attention response mode. The second common response or report condition is binaural separation (also referred to as directed attention or directed report) in which the individual is instructed to report the stimuli presented to one ear only (directed right or directed left) and ignore stimuli presented to the other ear. Though both conditions are forms of dichotic listening, they are indeed different tasks engaging different central processes.

Stimulus familiarity and intelligibility also can influence dichotic performance. For example, digits are familiar and highly intelligible, while CVs are not; hence, if all other variables are held constant, individuals typically perform better for digits than CVs. Also, certain phonemes, because of their acoustic characteristics, are more perceptible than others, such as voiceless stops which are more easily heard than voiced stops (Berlin and Cullen, 1977). For this reason, counterbalancing list presentation is important when comparing dichotic performance across multiple conditions.

Findings in clinical populations

Various dichotic listening paradigms are sensitive to a variety of central auditory disorders. In patients with well-localized lesions, the ear contralateral to the involved hemisphere often exhibits poor performance relative to norms (intersubject) or to the other ear (intrasubject, interaural comparison) (Fig. 18.1) (Kimura, 1961b; Musiek, 1983; Musiek and Weihing, 2011; and others). If the corpus callosum is involved, a left-ear deficit is seen consistently (Musiek and Weihing, 2011). The anatomic basis for this finding stems from the more indirect routing of left-ear stimuli, which are first directed to the right hemisphere, and then are impeded in crossing (via the corpus callosum) to the left hemisphere which is required for speech processing and verbal response (Fig. 18.2). A left-ear deficit also is seen in young children (under 12 years) whose corpus callosum have not yet attained their full complement of myelin, as well as in the elderly and in individuals with diseases affecting myelin (e.g., multiple sclerosis) (Musiek et al., 1984; Musiek and Pinheiro, 1985; Musiek and Weihing, 2011) (Fig. 18.3). The left-ear dichotic listening deficit in cases of corpus callosum involvement has been studied extensively in split-brain patients (Musiek and Pinheiro, 1985). In fact, due to the anatomic basis of dichotic listening, it has become known as a test of interhemispheric transfer or corpus callosum integrity.

Binaural interaction procedures

Binaural interaction procedures require binaural processing in the CANS of stimulation presented to both ears, either sequentially or simultaneously. Despite the simultaneous, binaural stimulation, these particular procedures are not dichotic procedures. As explained above, dichotic procedures require the listener to integrate or separate different auditory stimuli presented to each

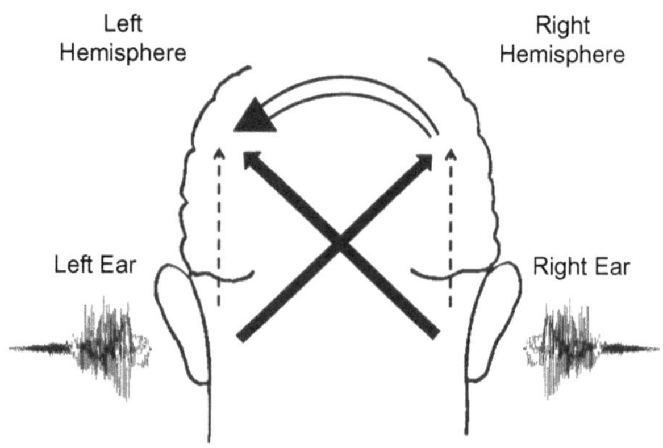

Fig. 18.2. Depiction of possible routes to the cortex during dichotic listening.

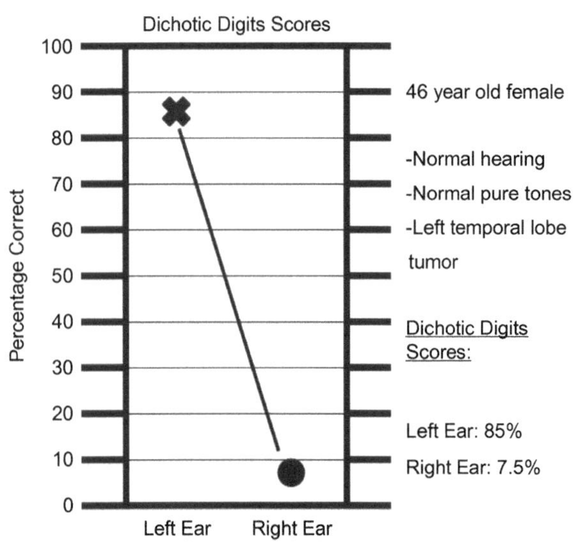

Fig. 18.1. Dichotic listening performance in a patient with left temporal-lobe tumor demonstrating the "contralateral" effect.

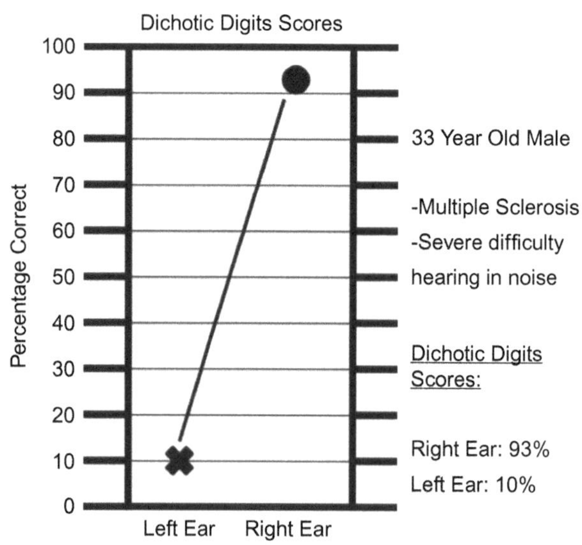

Fig. 18.3. Dichotic listening performance in a patient with multiple sclerosis.

ear simultaneously. Binaural interaction tasks require the listener to combine complementary inputs distributed between the ears, synthesizing intensity, time, or spectral differences of otherwise identical stimuli presented simultaneously or sequentially (e.g., localization and lateralization tasks) to the two ears. In this section, we focus on masking level differences (MLD) and localization tasks, with brief mention of some other procedures for reference and historic context.

Masking level differences

MLDs reflect an individual's ability to detect a target signal (usually low-frequency tones or speech) in noise when the binaural phase relationships between the signal and noise are altered. The MLD results from a "release from masking" which results in improved threshold for the target signal. For example, if the signal in one ear is out of phase by 180° relative to the other ear, there is a "release from masking" and thresholds in noise for the target signal improve. There are a variety of conditions that can be achieved by changing the phase of either the signal or noise in reference to the right or left ear (Table 18.2). Of these various conditions, two are commonly utilized: signal ($S_{\pi}N_0$) or the noise masker (S_0N_{π}) is 180° out of phase (termed the antiphasic condition). Thresholds in the antiphasic condition are compared to the homophasic condition where the noise and signal are in phase (S_0N_0) and the difference is considered the MLD or the amount of release from masking in decibels (Bamiou and McCullagh, 2014; Brown and Musiek, 2013). MLDs are greatest for low-frequency stimuli (Koehnke et al., 1986). A 500-Hz tone is most commonly used, presented at moderate intensities (e.g., 60 dB HL). In normal-hearing individuals, MLDs of 12–15 dB are commonly noted for 500-Hz tones in the $S_{\pi}N_0$ condition.

Table 18.2

Signal and noise conditions most commonly used to measure masking level differences

S_mN_m	Signal and masker are in the same ear (true monaural)
S_mN_o	Signal in one ear and noise in phase at two ears
S_mN_{π}	Signal in one ear and noise is out of phase at one ear relative to other
S_oN_o	Signal and noise in phase at two ears (homophasic)
S_oN_{π}	Signal in phase at two ears and noise 180° out of phase in one ear relative to the other (antiphasic)
$S_{\pi}N_{\pi}$	Signal and noise 180° out of phase at two ears (homophasic)
$S_{\pi}N_o$	Signal 180° out of phase at two ears and noise in phase at two ears (antiphasic)

The notation of different test conditions is as follows: "S" represents the signal, "N" represents the noise or masker, "m" refers to being presented to one ear, or monaurally, "o" refers to being in phase at the two ears, and "π" refers to being 180° out of phase between the two ears. The noise typically used as the masker is either broad-spectrum noise or a narrowband noise centered around the test frequency.

The original research on the MLD was conducted by Hirsch (1948) for pure tones and by Licklider (1948) for speech stimuli. The procedure was investigated for a number of years before its clinical potential began to be realized. Though there were likely some clinical studies earlier, it was the Noffsinger et al. (1972) study that attracted attention to the possible clinical use of the MLD. This study demonstrated a decrease in the MLD relative to norms in a large population of patients with multiple sclerosis. Since multiple sclerosis primarily affects the central auditory pathway, it became a distinct possibility that the MLD might be sensitive to central auditory disorder.

Indeed, a large number of animal studies have demonstrated the effects of central auditory lesions on localization processes which are related to MLDs (see Musiek and Baran, 2007, pp. 194–197, for a review). A study by Lynn et al. (1981) was one of the most compelling studies subsequent to Noffsinger et al. (1972) demonstrating the value of MLDs in defining central auditory dysfunction. Lynn et al. examined MLDs to speech segments (CVs) in patients with neurologic lesions in the low pons, midbrain, and auditory cortex, as well as in normal controls. The patients with low pontine lesions revealed smaller MLDs than the controls or those with lesions of the upper brainstem or cortex. This study showed the differential sensitivity of MLDs to various anatomic regions of the CANS. MLDs also have been shown to be affected by acoustic neuromas and peripheral hearing loss, especially asymmetric hearing loss; therefore, caution must be taken in applying this procedure to individuals with hearing loss or acoustic neuroma (Bamiou and McCullagh, 2014).

Localization and lateralization

Localization refers to the identification of the location of a sound source in a sound field. Lateralization is a similar process in which the individual determines the location in the head (intracranial) of a sound image presented through headphones. The effects of brain lesions on sound localization have been demonstrated extensively in animals and humans. Damage to the CANS (as well as peripheral hearing loss) reduces the ability to accurately localize a sound source. (For a review, see Bamiou, 2007.)

Clinical adoption of sound localization tests has been slow, due in part to a concern about the validity and reliability of sound localization results unless they are obtained in a well-attenuated sound room or even an anechoic chamber. Given that concern, it is interesting to note that the classic clinical studies of sound localization with central auditory disorders were performed within non-controlled acoustic environments. Sanchez-Longo and colleagues (Sanchez-Longo et al., 1957; Sanchez-Longo and Forster, 1958), using low-frequency noise bursts, showed that individuals with temporal-lobe compromise have major difficulty localizing sound in the auditory field contralateral to the damaged hemisphere. This finding, although obtained with rather rudimentary equipment in a far less than ideal acoustic environment, has been replicated in subsequent, better-controlled studies. For example, Moore et al. (1990) showed this contralateral effect for temporal-lobe lesions.

Perhaps one of the most clinically applicable localization procedures is based on the precedence effect (Cranford et al., 1990). By delaying the stimulus (often a click) to one of two speakers placed at opposite sides of a sound-treated room, one can shift the perceived position of the sound image from midline (when there is no time delay) to the left or right speaker, depending upon which speaker emits the lead sound. That is, the lead sound will result in localization to that side: the greater the time difference (or delay) between left and right speakers, the easier the localization task. This methodology or slight modifications of it was used to reliably test a variety of clinical populations. Individuals with multiple sclerosis (Cranford et al., 1990), the aging population (Cranford et al., 1993), and patients with temporal-lobe involvement (Moore et al., 1990) all demonstrated poor localization abilities on this task. Despite the simplicity and speed with which this sound localization procedure can be conducted, it has not been adopted clinically.

An interaural timing procedure has been used to examine lateralization. This task typically requires lateralization of clicks or noise bursts that are presented diotically (i.e., binaural presentation under phones) with time delays. Individuals are required to indicate the location of the intracranial acoustic image by pointing to the location on a matrix that corresponds to head position. Alternatively, the individual might be asked to simply relate in which ear the sound was heard first. This procedure has been used with normal control subjects, as well as with patients with clinical auditory disorders, including those of the CANS.

A number of studies of patients with multiple sclerosis have demonstrated interaural timing tasks to be highly sensitive to central auditory lesions, yielding abnormal results (Hausler and Levine, 1980; Levine et al., 1993; Furst et al., 1995). These studies also showed, either directly or indirectly, that this lateralization task was as sensitive as the ABR in revealing auditory brainstem dysfunction in these populations. Other studies using interaural timing tests have demonstrated their high sensitivity to vascular problems affecting the auditory brainstem pathway (Pratt et al., 1998; Furst et al., 2000). Given the sensitivity of this procedure to CANS involvement, and the accessibility of technology to program this task, it is surprising that it has not yet penetrated the clinic.

Spatial listening

The Listening in Spatialized Noise Sentences (LiSN-S) test is a new adaptive speech-in-noise test conducted under headphones that tests spatial processing. The stimuli (target) and the distracter (i.e., masker) are speech materials that have been synthesized with head-related transfer functions to generate a three-dimensional effect (Cameron and Dillon, 2007; Brown et al., 2010; Cameron et al., 2011). A listener's speech reception threshold (SRT), which is defined as the signal-to-noise ratio that yields 50% intelligibility, is established using a repetition protocol. The target stimuli (sentences) and distracter competition are presented simultaneously. The target sentences are spoken by a female voice and always appear to the listener to emanate from 0° azimuth (directly in front of the listener). The competing or distracter stimuli are looped children's stories that appear to originate from either 0° azimuth (co-located) or ±90° azimuth (a spatially separated right angle). The competing stories are spoken by either the same female speaker as the target sentences or by a different female speaker. These combinations of spatial and vocal differences result in four listening conditions: same voice at 0° (or low-cue SRT); same voice at ±90°; different voices at 0°; and different voices at ±90° (or high-cue SRT).

The LiSN-S is hypothesized to be an indicator of auditory stream segregation: the process of utilizing auditory cues (i.e., spatial location or speaker's voice) to extract meaningful acoustic signals from non-target background acoustic signals that arrive at the two ears (Cameron and Dillon, 2008; Bamiou and McCullagh, 2014). More specifically, by subtracting the thresholds obtained using the same talker versus different talkers, and for same location versus different locations, an index for the utilization of difference cues in suppressing noise or competition can be computed that is minimally affected by higher-order language, learning, and other communication processes (Cameron and Dillon, 2007; Bamiou and McCullagh, 2014).

The LiSN-S, which has been tested primarily on children suspected of CAPD and children with learning and/or attention problems (Cameron and Dillon, 2008), has revealed spatial processing deficits for the children with suspected CAPD, but not for the other groups of children. The concept of spatial listening (deficit) seems to be a promising one and more research on various well-defined populations will prove valuable. Combined with other central auditory test procedures, spatial listening may assist in better identifying those with central auditory abnormalities and help generate deficit-specific profiles.

Temporal processing tests

Auditory temporal processing refers to the perception of sound or the alteration of sound within a restricted or defined time domain (Musiek et al., 2005). Temporal processing may be the most fundamental element of most auditory processing, given that many, if not all, aspects of auditory information are in some way influenced by time (Pinheiro and Musiek, 1985a). Temporal processing can be observed at many levels, ranging from the most basic level of neural timing in the auditory nerve to cortical processing for binaural hearing and speech perception (e.g., voice onset time to prosodic detail).

Temporal processing is comprised of: (1) temporal ordering or sequencing; (2) temporal resolution or discrimination; (3) temporal integration or summation; and (4) temporal masking. Clinical measures of temporal ordering and temporal resolution are available; however, at this time, there are no clinically feasible measures of temporal integration or temporal masking. A thorough evaluation of multiple temporal processes is important not only for proper diagnosis, but for targeted intervention designed to improve specific temporal processing deficit(s).

Temporal ordering or sequencing

Temporal ordering or sequencing refers to the processing of two or more auditory stimuli in their order of occurrence in time (Pinheiro and Musiek, 1985a). The ability to properly recognize, identify, and sequence auditory patterns involves several perceptual and cognitive processes (Pinheiro and Musiek, 1985a). These processes are not restricted to one hemisphere alone; they require integration of information from both hemispheres across the corpus callosum (Musiek et al., 1980). Therefore, pattern (sequencing) tests are sensitive to hemispheric lesions, as well as interhemispheric dysfunction (Musiek and Pinheiro, 1987; Musiek et al., 1990). Information regarding contour recognition is processed in the right hemisphere and then passed via the corpus callosum to the left hemisphere where the linguistic label is applied to the signal. An individual who can hum a pattern but cannot verbally label tonal patterns most likely suffers from dysfunction in interhemispheric transfer to the left hemisphere, or dysfunction in the left hemisphere, the presumed site of the speech processor (Pinheiro and Musiek, 1985a). Given the multiple auditory areas required to verbally label these patterns, performance most often is found to be bilaterally normal or bilaterally depressed.

The most widely used clinical tests of temporal ordering are the Frequency Pattern Test (FPT) and the Duration Pattern Test (DPT) (Emanuel et al., 2011). The FPT requires individuals to verbalize (or hum if verbalization is not possible) the order of a series of three tones (Musiek, 1994). The duration of each tone is 200 ms, with a 150-ms interval between successive tones in the sequence. The tones include a low-frequency tone (880 Hz) and a high-frequency tone (1122 Hz), each with a 10-ms rise-fall time. Using three tones per pattern or trial yields six possible combinations (LLH, LHL, LLH, HLH, HLL, HHL). The DPT also is composed of three pure tones per pattern or trial. Each tone is 1000 Hz and is either 250 ms (short) or 500 ms (long) in duration, with a 300-ms interval between successive tones in the sequence. Again, there are six possible randomizations which occur (LLS, LSL, LSS, SLS, SLL, SSL). Both pattern tests utilize a 6-second interpattern interval (Musiek et al., 1990).

It is recommended that each test be administered at approximately 50 dB sensation level (SL) in reference to either the SRT or the pure-tone average (PTA). Since there is little effect of intensity on performance on these tests, they can be presented at levels as low as 20 dB SL and yield essentially the same results as that obtained at higher presentation levels (Musiek, 1994). Subjects are instructed that they will hear three tones that vary in either pitch (FPT) or duration (DPT). The subject is asked to repeat the pattern by verbally indicating either "high or low" or "long or short." It is recommended that the examiner train the subject by first using visual cues (such as finger pointing) in conjunction with the auditory stimuli. Once this has been mastered, then the visual cues should be removed and the examiner should ensure that the patient is able to complete the task in the auditory-only condition. Individuals should be encouraged to guess if they are unsure. Both pattern tests are scored in percent correct and have published normative data for both pediatric and adult populations.

Evidence dating back to the early 1960s (Milner, 1962) suggests that individuals with cerebral lesions demonstrate temporal ordering deficits (Belmont and Handler, 1971; Karaseva, 1972; Swisher and Hirsch, 1972; DeRenzi et al., 1977). Patients with compromise of the auditory areas of either hemisphere or of the interhemispheric pathways present difficulties on the FPT and DPT (Musiek et al., 1980, 1990, 1994; Pinheiro and

Musiek, 1985b; Musiek and Pinheiro, 1987). The FPT and the DPT are highly sensitive (~83% and ~86%, respectively) to cerebral lesions of the CANS; however, the sensitivity of the FPT to brainstem lesions is weaker (~45%) (Musiek and Pinheiro, 1987; Musiek et al., 1990). Most subjects with cochlear hearing loss perform within normal limits on the FPT (~88%) and the DPT (~92%) (Musiek and Pinheiro, 1987; Musiek et al., 1990), suggesting that these tests, although not completely resistant to the effects of hearing loss, can be useful in the assessment of central auditory processing in individuals with peripheral hearing loss. The FPT has been established as an excellent tool to use with young children, ages 8 years and older (Musiek, 1994).

Temporal resolution or discrimination

Temporal resolution or discrimination refers to the shortest duration of time in which an individual can discriminate between two auditory signals (Gelfand, 1998). For brief sounds, this is generally about 2–3 ms for trained subjects (Phillips, 1999). The threshold for temporal resolution is known as temporal auditory acuity or minimum integration time (Greene, 1971). Temporal resolution can be assessed utilizing a variety of methodologies. Temporal resolution can be evaluated "within channel," in which the stimuli use the same markers (i.e., same stimulus before and after the gap) or "between channel," in which the markers are different (Phillips et al., 1997). Our current clinical measures of temporal resolution have been designed to evaluate only within-channel processing. One of the most common measures of temporal resolution using behavioral methods is gap detection. Gap detection tasks require subjects to indicate whenever they hear a "silent" interval embedded in an ongoing sound or noise burst. The interstimulus interval is varied and the gap detection threshold (GDT) is defined as the shortest silent interval in a stimulus a listener is able to detect (Lister et al., 2002). A broadband stimulus may be the best stimulus for use in GDT paradigms, as it is less likely to lead to variability across different age groups or as a function of peripheral hearing status (Musiek et al., 2005).

Two measures of gap detection are commonly used clinically – the Gaps-in-Noise (GIN) test (Musiek et al., 2005) and the Random Gap Detection Test (RGDT) (Keith, 2000). Several new clinically viable gap detection procedures have been introduced, demonstrating increased interest in these kinds of temporal processing procedures and their clinical application (Griffiths et al., 2001; Lister et al., 2006). One such procedure, the Adaptive Test of Temporal Resolution (ATTR), utilizes a computerized, adaptive psychophysical methodology to evaluate temporal resolution. Thresholds obtained on the ATTR are similar (2.2 ms) to those obtained using traditional psychophysical laboratory procedures (Lister et al., 2006).

The RGDT is presented binaurally at 55 dB HL. Patients are asked to indicate whether they have heard one or two tones or clicks (1 ms white noise) either verbally or by raising one or two fingers (i.e., non-verbally). The tones are presented binaurally and are 15 ms in duration with a 1.5-ms rise–fall time. Nine gap durations ranging from 2 to 40 ms are utilized. Normative values for the RGDT tonal stimuli range from 6.0 to 7.8 ms. No normative values have been published for the click stimuli. Test administration requires approximately 10 minutes.

The GIN is composed of a series of 6-second segments of broadband noise administered monaurally. Maximum performance is achieved between 35 and 50 dB SL re PTA (Weihing et al., 2007). Each segment contains 0–3 silent intervals or "gaps". The interstimulus interval between successive noise segments is 5 seconds and the gap durations presented are 2, 3, 4, 5, 6, 8, 10, 12, 15, and 20 ms. There are six tokens of each gap duration within each list. Two measures are derived from the GIN. The approximate threshold for the GIN is defined as the shortest gap duration for which there are at least four of six correct identifications. This is an "approximate" threshold because the exact GDT is not determined in this procedure. In addition to the approximate threshold, the percentage of correct responses out of the total number of gaps (e.g., 36 out of 60 would be equal to 60% correct) can be computed for each ear. Mean thresholds for the GIN are slightly longer than traditional psychoacoustic GDTs. The mean approximate thresholds for the GIN are 4.8 and 4.9 ms for the left and right ears, respectively (Musiek et al., 2005). Differences in GDTs likely result from a number of procedural differences, including less training for subjects participating in the normative data collection for the GIN (relative to psychophysical laboratory studies), and the use of somewhat larger gap intervals in the GIN than those used in many traditional psychoacoustic experiments.

The GIN appears to be more sensitive to cortical compromise than to brainstem involvement: the sensitivity of the GIN is 67% and the specificity is 94% (Musiek et al., 2005). This sensitivity is similar to the patterning tests described above, which suggests that temporal resolution is mediated primarily in the cerebrum (Pinheiro and Musiek, 1985a). Normative data for the pediatric population indicate that that there is no developmental effect seen in children between ages 7 and 18 years; children age 7 perform comparably to adults, and similar to adults, no ear asymmetry is observed (Shinn et al., 2009), suggesting that temporal resolution is likely an auditory process that develops early and symmetrically. It also appears that children with learning disabilities demonstrate abnormal temporal resolution abilities based on gap detection procedures (Hautus et al., 2003). Though two gap detection procedures have been highlighted here, there are other procedures which also show good clinical potential (Griffiths et al., 2001; Lister et al., 2002, 2006).

Temporal integration

Temporal integration results from the summation or aggregation of neuronal activity as a function of the additional duration of sound energy (Gelfand, 1998). This summation results in threshold improvements as duration increases up to about 200 ms in normal-hearing populations (Durrant and Lovrinic, 1995). If durations are increased beyond 200 ms, little change in threshold is noted in normal-hearing subjects and those with cochlear or central auditory deficits. As duration is decreased by a factor of 10 (or one-tenth of its original duration, e.g., from 200 ms to 20 ms), an increase of approximately 10 dB in threshold is observed in normal-hearing individuals. This relationship is referred to as a time–intensity trade. While absolute thresholds for simple and static sounds such as pure tones often are undisturbed in cases of CANS involvement (Bocca and Calearo, 1963), thresholds for dynamic stimuli (e.g., patterns, speech) and stimuli of variable duration such as those used to measure temporal integration are more sensitive to cerebral lesions (Wright, 1978).

If the duration of a stimulus is decreased to less than 20 ms, threshold levels will increase in normal-hearing individuals (see Durrant and Lovrinic, 1977, p. 161). For individuals with CANS disorders, however, these short-duration stimuli result in a much larger differential (i.e., steeper integration function) between the long and short stimuli, especially when the short stimulus is less than 10 ms (Baru and Karaseva, 1972; Olsen and Matkin, 1978, p. 395; Wright, 1978). As would be expected, these increased thresholds were found to manifest mostly in the ear contralateral to the side of the brain lesion, with the ipsilateral ear performing essentially normally (Baru et al., 1964; Gershuni, 1971; Baru and Karaseva, 1972; Cranford et al., 1982). As shown in Fig. 18.4, individuals with cochlear disorders show about the same threshold differences for short and long tones (and possibly less difference – a more shallow integration function) as those seen in normally hearing individuals (Olsen and Matkin, 1978; Wright, 1978). It is important to reiterate that patients often exhibit CANS lesions despite normal pure-tone findings, which underscores the need for inclusion of central auditory tests in comprehensive auditory test batteries. Unfortunately, brief-tone audiometry, which was once used

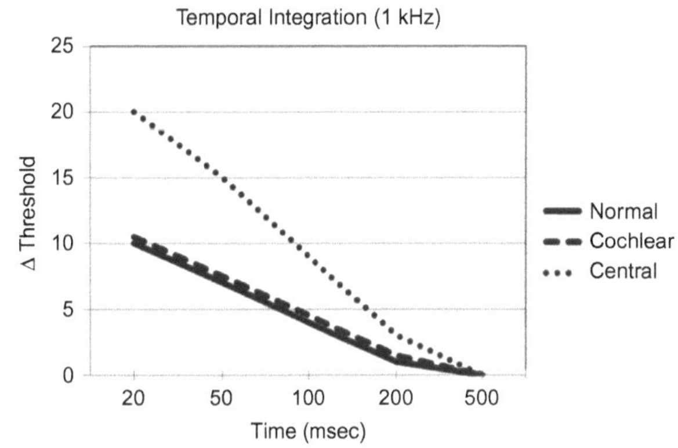

Fig. 18.4. Idealized temporal integration functions for populations with normal hearing, cochlear disorder, and central auditory deficits.

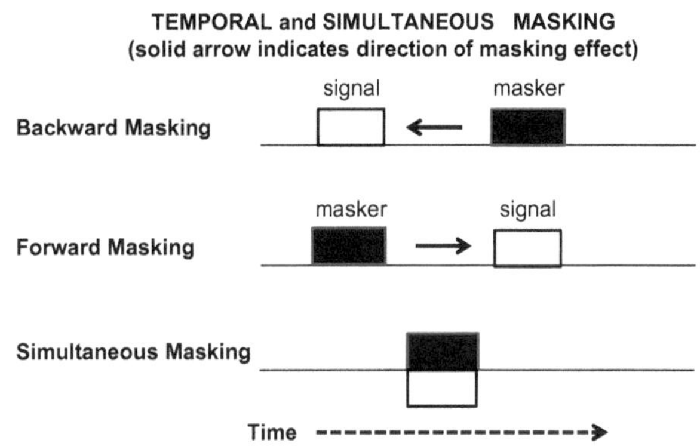

Fig. 18.5. Illustration of backward, forward, and simultaneous masking.

clinically to examine temporal integration, is not used at this time, leaving the audiologist with no feasible behavioral means to measure temporal integration in the clinic.

Temporal masking

Temporal masking refers to masking (i.e., obscuring of one sound by another) that occurs when a signal and a masker are not presented simultaneously. Backward masking occurs when the masker follows the signal; forward masking occurs when the masker precedes the signal (Fig. 18.5). Interest in temporal masking dates back to the early 1950s, and although the literature clearly demonstrates the effects of temporal masking, the exact mechanisms underlying this phenomenon are unclear (Durrant and Lovrinic, 1995).

Several parameters determine temporal masking, including the time interval between the masker and signal, masker level, masker duration, and the acoustic similarity between the masker and the signal. Assuming all other parameters are held constant, temporal masking decreases rapidly as the time interval between the masker and the signal increases. For forward masking, no masking occurs when the time interval between the masker and the signal equals or exceeds 200 ms (Durrant and Lovrinic, 1995). Interestingly, as noted above, this is the same duration at which temporal integration plateaus and minimum (best) thresholds are observed. In backward masking, minimal masking is seen when the separation between masker and signal equals or exceeds 25 ms (Durrant and Lovrinic, 1995).

Several key principles of temporal masking were demonstrated in a series of classic studies by Elliott (1962), who reported that: (1) backward masking is more effective than forward masking – given the same time interval between masker and signal, more masking occurs when the masker follows the signal; and (2) masking is more effective when the stimuli are delivered monotically as opposed to dichotically (with masker and signal presented to opposite ears). Elliott (1967) demonstrated that the duration of the masker influences forward, but not backward, masking. The spectral relationship between masker and signal also influences the degree of masking: masking is greater when the masker and signal are identical, with the amount of masking decreasing as spectral differences increase between the masker and signal (Elliott, 1967). The amount of masking increases as the level of the masker increases, although the relationship between the intensity of the masker and the amount of masking observed is nonlinear. Generally, a 10-dB increase in masking level may only result in a threshold shift of approximately 3 dB (Gelfand, 1998). The simultaneous masking condition always produces greater masking than either backward or forward masking, if all parameters other than temporal distance between signal and masker are held constant (Wilson and Carhart, 1971).

Few studies have examined the effects of CANS lesions or CAPD in the absence of frank lesions on temporal masking in children. Wright and colleagues reported that children diagnosed with specific language impairment demonstrated a severe temporal processing deficit with respect to their ability to separate one sound that is rapidly followed by another sound (backward masking) and concluded that this deficit may degrade the children's perception of brief acoustic elements of speech (Wright et al., 1997). Moore and colleagues (2011) reported large between-subject variability, particularly for backward-masking paradigms. That children with language and speech perception and processing deficits demonstrate abnormal temporal masking is not surprising when one considers the potential influence of coarticulation for speech perception. Coarticulation, an aspect of speech production resulting from the concurrent movement of multiple articulators, leads to parallel transmission of acoustic information across neighboring phonemes and, therefore, to the potential for temporal masking.

Monaural low-redundancy tests

Monaural low-redundancy speech tests are among the oldest tests used to assess the CANS (Bocca et al., 1954, 1955). These tests are administered monaurally using stimuli that have been degraded electroacoustically or digitally in the frequency/spectral, temporal, or intensity domain. Because the stimuli are degraded, the inherent redundancy of the signal is reduced and the tests are considered "sensitized" to detection of CANS pathology. In addition to some degree of sensitivity to CANS lesions, they provide insights regarding an individual's auditory closure abilities.

Even when speech is degraded (i.e., extrinsic redundancy is reduced by filtering, adding competition, or altering temporal aspects), listeners with normal intrinsic redundancy can fill in the missing information by means of their auditory closure skills and achieve normal speech recognition performance. When a listener with reduced intrinsic redundancy (CANS dysfunction) listens to unaltered speech with normal extrinsic redundancy, normal speech recognition performance is still expected; however, when speech is degraded (i.e., extrinsic redundancy is reduced), listeners with reduced intrinsic redundancy (due to CANS dysfunction) show a significant deficit in speech recognition performance.

Peripheral hearing loss, as well as CANS dysfunction, reduces the intrinsic redundancy of the auditory system. Peripheral hearing loss reduces the ability of the auditory system to resolve spectral detail. Spectral filtering and temporal resolution deficits consequent to sensorineural hearing loss can significantly influence performance on monaural low-redundancy speech tests since these tests, like other central auditory tests employing speech stimuli (e.g., CVs, words), place significant demands on cochlear processing to resolve frequency and intensity transitions of the acoustic signal. Such tests are, therefore, subject to the potential confound of peripheral hearing loss (Chermak and Musiek, 1997). Since all monaural low-redundancy tests are heavily dependent on resolution of rapid frequency–intensity interactions, the clinician should balance the value of these tests in the central auditory diagnostic battery against their potential to confound results. Results should be interpreted with caution when used as part of a central auditory test battery with individuals with known hearing loss (Baran and Musiek, 1999).

Classification of monaural low-redundancy speech tests

Monaural low-redundancy speech tests use three different methods to reduce the redundancy of the speech stimulus and are classified as: (1) low-pass filtered speech (LPFS) tests; (2) speech-in-noise (or speech-in-message competition tests); and (3) time-compressed speech tests. Monaural low-redundancy speech tests provide insight as to functional deficits (i.e., listening in noise and auditory closure problems), and therefore offer practical information for intervention (Bellis and Ferre, 1999; Bellis, 2003). However, these tests are only moderately sensitive to brainstem and cortical dysfunction (Musiek and Baran, 2002).

Low-pass filtered speech tests

The use of LPFS can be traced historically to research by Bocca et al. (1954), who reported reduced performance in the ear contralateral to the temporal-lobe lesion for most patients. Filtering using a 500-Hz cut-off frequency with an 18 dB per octave rejection rate or a 1000 Hz cut-off has been recommended for clinical use (Wilson and Mueller, 1984). LPFS tests are available on the Veterans Administration compact disk (Veterans Administration, 1992), an Auditec of St. Louis recording (Wilson and Mueller, 1984), SCAN:3A and SCAN:3C (Keith, 2009a, b), and the University of Canterbury Adaptive Speech Test (O'Beirne et al., 2012). Stimuli are presented at 50 dB above the PTA (i.e., 50 dB SL re PTA). Patients repeat the stimuli and a percent correct identification score is derived for each ear.

Limited sensitivity and specificity data for LPFS tests have been published; however, there is some evidence that these tests are more sensitive to temporal-lobe lesions than interhemispheric and brainstem lesions. Lynn and Gilroy (1972, 1977) reported that 74% of 34 patients with temporal-lobe tumors showed the expected contralateral deficit on LPFS, while 74% of 27 patients with parietal-lobe tumors presented normal performance on LPFS. Other studies reported similar outcomes: poorer scores in the ear contralateral to the temporal-lobe lesion (Bocca et al., 1955; Jerger, 1960). Karlsson and Rosenhall (1995) reported that filtered speech tests showed modest sensitivity (62–64%) to brainstem lesions and slightly higher sensitivity (65–67%) to temporal-lobe lesions. In a study examining the diagnostic accuracy of established central auditory test batteries for patients with known lesions of the central nervous system, Musiek et al. (2011) reported that the sensitivity of a one-test "battery" of LPFS alone was only 50%, and only 20% using the 2 SD cut-off recommended by the American Speech-Language Hearing Association (2005).

In contrast to the findings in temporal-lobe lesions, LPFS results have been found to be essentially normal in listeners with interhemispheric lesions (Musiek et al., 1979; Baran et al., 1986). In patients with brainstem lesions, LPFS findings vary considerably (Calearo and Antonelli, 1973; Musiek and Geurkink, 1982). Some of this variability may be due to size and location of the lesion. Large, diffuse lesions are likely to show bilateral deficits, lower brainstem lesions are likely to reveal ipsilateral deficits, and higher brainstem lesions may present with contralateral deficits (Musiek and Geurkink, 1982). Age and cognitive decline also may influence performance on LPFS tests. As reported by Krishnamurti et al. (2011), LPFS scores were significantly poorer in individuals with Alzheimer's disease than elderly listeners without cognitive issues.

In summary, LPFS tests are moderately sensitive to CANS lesions, especially those located in the temporal lobe. Since these tests simulate listening in poor (degraded) listening conditions, they are useful in the evaluation of individuals referred due to concerns about CAPD, as they can reveal functional deficits (e.g., auditory closure), even in the absence of true CANS lesion.

Speech-in-competition (speech-in-noise) tests

Common to these tests is the embedding of speech in a background of noise or speech competition (i.e., competing message) in the same ear (i.e., ipsilateral competing noise or message). Early applications of these tests can be traced back to Sinha (1959), who reported deficits in the ear contralateral to cortical lesions in patients with CANS pathology. Subsequent studies also have shown speech-in-noise deficits contralateral to the hemisphere with auditory cortex involvement (Morales-Garcia and Poole, 1972; Heilman et al., 1973). Patients with VIIIth nerve and extra-axial lesions (Dayal et al., 1966), as well as patients with intra-axial brainstem lesions (Morales-Garcia and Poole, 1972), perform poorly on these tests. Olsen et al. (1975) computed difference scores by subtracting NU-6 word recognition scores in noise from NU-6 word correct scores in quiet and found significant differences (greater than 40%) in performance in noise relative to quiet in patients with VIIIth-nerve lesions, Ménière's disease, and temporal-lobe lesions. Because peripheral hearing loss can influence results on speech-in-noise or competition tests for the reasons elaborated above, audiologists must use and interpret these tests cautiously when assessing central auditory processing in children or adults with coexisting peripheral hearing loss (Bantwal and Hall, 2011).

There are at least three standardized versions of (monaural) speech-in-competition tests currently in clinical use: (1) Auditory Figure Ground subtest of the SCAN:3A and SCAN:3C (Keith, 2009a, b); (2) Synthetic Sentence Identification (SSI) (Jerger and Jerger, 1974) with Ipsilateral Competing Message (SSI-ICM); and (3) Pediatric Speech Intelligibility Test (Jerger and Jerger, 1984) with Ipsilateral Competing Message (PSI-ICM). A key factor in using speech-in-competition tests is the signal-to-competition or message competition ratio

(MCR). Stimuli typically are presented at 40 dB SL re SRT (i.e., 40 dB above the SRT). The most commonly used MCRs range between 0 dB (which indicates that the sound pressure level of the speech is equal to the overall level of the competition) to +10 dB (i.e., the sound pressure level of the speech exceeds that of the competition by 10 dB) (Baran and Musiek, 1999). Another variable across speech-in-competition tests is the type of competition, which can render comparisons across speech-in-competition tests difficult. White noise, speech spectrum noise, multitalker babble, and even competing discourse have been used in various tests.

Synthetic sentence identification test

The SSI-ICM can be presented in two different ways: by varying the MCR (i.e., +10, 0, -10, -20) or by maintaining the MCR (usually at 0 dB), but presenting the sentences at several intensity levels from higher to lower levels. The second method allows one to identify the phenomenon of rollover, seen in patients with brainstem lesions, where speech recognition performance decreases as intensity is increased. The competition used in the SSI-ICM is connected discourse (i.e., a story about Davy Crockett). Using multiple MCRs allows the audiologist to construct performance versus intensity functions and message to competition functions. Designed to reduce a listener's reliance on linguistic skills, while preserving the syntax and temporal features of the English language, the seven-word synthetic sentences are third-order approximations of English sentences and are semantically meaningless (Jerger and Jerger, 1974). The SSI also can be presented with contralateral competition (SSI-CCM). The SSI is less affected by hearing loss as it relies on normal hearing sensitivity only through 1000 Hz, as the frequency region supporting correct identification of sentence materials centers around 750 Hz (Hall and Mueller, 1997).

Jerger and Jerger (1974) reported SSI-ICM deficits in 11 patients with intra-axial lesions. Jerger and Jerger (1975) reported that the average SSI-ICM loss approximated 40% in the ears contralateral to brainstem lesions in 10 patients with intra-axial brainstem lesions, while for patients with VIIIth-nerve lesions, the average SSI-ICM loss approximated 50% in the ear ipsilateral to the lesion. Consistent findings were seen for monosyllabic words: patients with VIIIth-nerve lesions presented a loss for the ipsilateral ear, while patients with brainstem lesions demonstrated poor scores bilaterally, with greater loss in the contralateral ear (Jerger and Jerger, 1975). The SSI-ICM appears to be sensitive to both rostral and caudal brainstem lesions, although the measure may be affected to a lesser degree by cerebral involvement (Jerger and Jerger, 1974, 1975).

Pediatric Speech Intelligibility test

The Pediatric Speech Intelligibility (PSI) test, an adaptation of the SSI, is a picture pointing (identification) task appropriate for children aged 3–6 years (Jerger and Jerger, 1984). Like the SSI, the PSI allows for the construction of performance versus intensity functions and message to competition functions. Moreover, the PSI constructs these functions based on the child's responses to both monosyllabic words (i.e., 20 simple nouns) and sentence stimuli (i.e., 10 simple sentences with animal agents). Twenty competing sentences also involve animal agents in simple contexts. Like the SSI, the PSI can be administered with competition in the ipsilateral (PSI-ICM) and the contralateral ear (PSI-CCM). In the competing conditions, the child is told to point to the picture said by "your man" and to ignore the "trick man." Like the SSI, the PSI employs a closed message set (i.e., 20 words and 10 sentences) and a closed response set (i.e., selection from among five pictures). The closed response set is particularly important when using the sentence stimuli as it reduces the potential confound of memory. The presentation level recommended for PSI-ICM is 30 dB HL, with a 0 dB MCR for sentences and +4 dB MCR for words.

Based on norms reported in the test manual, SSI-ICM scores less than 80% fall outside the 95% confidence interval for normal-hearing children (Jerger and Jerger, 1984). Sensitivity and specificity data were not reported by Jerger and Jerger (1984); however, Jerger et al. (1986) reported that children suspected of CAPD presented deficits on the PSI similar to those demonstrated by children with confirmed temporal-lobe lesions.

Other speech-in-noise tests

Speech-in-noise performance of normally hearing, hearing-impaired, and individuals with CAPD also has been examined using the Speech Perception in Noise (SPIN) test (Bilger et al., 1984) and the Hearing in Noise Test (HINT) (Nilsson et al., 1994; Elliott, 1995; Johnston et al., 2009). Tschopp and Zust (1994) reported that performance on the SPIN was not affected by peripheral hearing loss and might, therefore, assist in differentiating peripheral from central auditory deficits. Fuente and McPherson (2007) reported that workers exposed to organic solvents performed more poorly on the HINT than controls. They concluded that the HINT might be useful in revealing central auditory damage in solvent-exposed populations, in contrast to pure-tone thresholds, which are insufficient for this purpose (Fuente et al., 2006).

In summary, available data indicate that speech-in-competition or noise tests are marginally to moderately sensitive to a variety of CANS disorders (Morales-Garcia and Poole, 1972; Jerger and Jerger, 1975). Patients with low to mid-brainstem involvement, as well as cortical lesions, typically perform poorly on speech-in-noise tests. Given the variability related to stimuli, norms, and interpretation, Bantwal and Hall (2011) cautioned audiologists to avoid diagnosis of CAPD based solely on a speech-in-competition or noise test result.

Time-compressed speech tests

Time-compressed speech tests employ speech that is compressed electronically by systematically sampling and discarding segments of the signal without distorting the frequency aspects of the signal (Beasley et al., 1972; Beasley and Maki, 1976). Time compression currently is accomplished using waveform editing software that alters the rate of speech without altering the power spectrum or pitch. Artificial time compression of speech normally is performed in a linear fashion by reducing all speech segments (i.e., consonants, vowels, and pauses). By speeding up speech digitally, speech information and voice are presented without a high-pitched "Mickey Mouse-like" quality. Digitally compressed speech is preferable to naturally produced speech spoken quickly (i.e., fast speech), as the latter has been found to be much less intelligible than digitally compressed speech. When speakers speed up their rate of speech, unstressed syllables in disyllabic words are affected more in duration, relatively, than stressed syllables (Janse et al., 2003). Consonant durations are influenced differently than vowel durations, making the intelligibility of "fast speech" poorer than that seen for electronically time-compressed speech (Janse et al., 2003). Likewise, individuals perform more poorly on speech that has been compressed using the electromechanical approach introduced by Fairbanks et al. (1954) compared to their performance for digitally compressed speech. Several versions of electronically time-compressed speech tests are available (Beasley et al., 1972; Kurdziel et al., 1976; Bornstein et al., 1994; Wilson et al., 1994). Time-compressed stimuli are commonly presented at 40 dB SL re SRT. The patient typically reports the stimulus perceived.

Time-compressed speech tests assess the auditory system's capacity to process rapidly changing acoustic spectra and, therefore, probe an aspect of temporal processing, as well as the system's capacity to withstand reduced external redundancy and attain auditory closure. The amount of compression is expressed as a percentage of the original signal that is eliminated. Time compression of 45%, for example, indicates that the signal now occupies 55% of its original timeframe, with 45% of signal having been removed. Again, this is accomplished without altering the acoustical aspects (e.g., frequency, intensity) of the signal.

Time-compressed speech tests are sensitive to cortical lesions (Kurdziel et al., 1976; Karlsson and Rosenhall, 1995), with performance typically reduced in the ear contralateral to the temporal-lobe lesion (Calearo and Antonelli, 1973). Karlsson and Rosenhall (1995) reported that time-compressed speech was only moderately sensitive to brainstem lesions (62–64%); however, its sensitivity was much higher (80%) for patients with temporal-lobe lesions. Kurdziel et al. (1976) studied the effects of several degrees of time compression (0%, 40%, and 60%) on speech recognition scores in 15 patients with diffuse cortical lesions and 16 patients with anterior temporal-lobe surgical lesions. Patients with diffuse lesions showed a significant deterioration in speech recognition at 60% time compression in the ear contralateral to the lesion. In contrast, patients with discrete (i.e., anterior temporal-lobe) lesions demonstrated essentially normal performance bilaterally. Baran et al. (1985) found that compressed speech is moderately sensitive to intracranial lesions involving the temporal lobe, with the most typical performance pattern showing reduced performance in the ear contralateral to the lesion. Few studies with patients with brainstem lesions are available; however, at least one report of 9 patients with brainstem lesions found no reduction in performance for time-compressed speech (Quaranta and Cervellera, 1977).

Normative studies have explored the percentage of time compression that is most efficient in identifying those with central auditory lesion or dysfunction. Wilson et al. (1994) examined the performance of young adults with normal hearing for monosyllables compressed at 45% and 65%. These individuals exhibited no significant difficulty with 45% time compression; however, they experienced difficulty at 65% time compression. In contrast, listeners with central auditory dysfunction demonstrated reduced performance at 45% time compression (Wilson et al., 1994). Based on their work, Bellis (2003) recommended use of 45% compression for clinical evaluation. To increase the difficulty of the task, a 0.3-ms reverberation has been electronically added to the time-compressed speech on the Veterans Administration CD (1992). Norms for three standardized recordings of time-compressed speech can be found in Bellis (2003), Keith (2009a, b), and Wilson et al. (1994).

In summary, time-compressed speech tests are only moderately sensitive to brainstem lesions; however, they present much greater sensitivity to diffuse lesions of the auditory cortex (Kurdziel et al., 1976) and temporal-lobe lesions (Karlsson and Rosenhall, 1995). Sensitivity and specificity data are not available for these tests for children with CAPD diagnoses.

PSYCHOPHYSICAL TESTS IN ELECTROPHYSIOLOGIC PARADIGMS

Recording auditory evoked potentials (AEPs) in response to stimuli used in behavioral tests of central auditory function, as well as recording AEPs while subjects actually perform a central auditory test, is one of the most interesting and promising areas emerging in the diagnosis of CAPD. While the measurement of hearing thresholds using AEPs is common in clinics and labs worldwide, the recording of AEPs in response to more complex stimuli and/or while an individual concurrently performs a behavioral central auditory test has emerged recently as a potentially powerful research and clinical tool. (Although see Lasky and Rupert (1982), who coupled temporal masking behavioral paradigms with the ABR many years ago.)

Jerger et al. (2002) demonstrated the potential of this procedure using gap detection and dichotic listening tasks with twins, one of whom was diagnosed with CAPD. Palmer and Musiek (2013) reported that amplitude of the N1-P2 late potential increased as a function of increasing duration of gaps presented in broadband noise, and that the N1-P2 potential closely approximated behavioral GDTs in normal subjects. Other behavioral procedures also have been inserted into various AEP paradigms. For example, dichotic stimuli have been employed using mismatch negativity to determine hemispheric advantages in dichotic listening (Yasin, 2007). The mismatch negativity also has been used to measure auditory discrimination, yielding impressive correlations with results obtained behaviorally (Pakarinen et al., 2007).

This new paradigm provides considerable potential to advance our understanding of central auditory disorders. The CANS of individuals who successfully perform the behavioral task and produce repeatable AEPs for a given auditory task must present high auditory integrity, reflecting both the integrity of the neural substrate underlying the behavioral task as well as the neural substrate of the pathways responsible for generating the AEP. Using this procedure one could see three different outcomes: (1) both the behavioral test and the AEP might be normal; (2) both responses might be abnormal; and (3) one measure could be normal and the other abnormal. Interpretations of these three profiles are beyond the scope of this chapter, but suffice it to say that these outcomes could provide considerable insight as to CANS function.

COST-EFFECTIVENESS

In addition to efficiency (i.e., sensitivity and specificity), diagnostic tests can be compared relative to cost-effectiveness. Cost-effectiveness computations have been approached in several different ways, usually calculating the cost per patient or the cost per true positive (i.e., cost per correct identification of a patient who is positive for a disorder). Cost computations vary as a function of a number of variables, including the size of the test battery, the cost of the individual tests, test sensitivity, prevalence of the disorder, and the diagnostic criterion (e.g., lax criterion wherein a positive diagnosis is based on failure on only one test in the battery versus a strict criterion wherein a positive diagnosis requires failure on two or more tests within the battery). Costs also vary as a function of the type of test facility (e.g., university clinics generally charge lower fees than hospitals or non-residential healthcare facilities, including private clinics) and geographic area (e.g., fees may be higher in the Northeast corridor than in the Midwest). In general, the cost per true positive increases with increasing battery size. This is expected given that the cost for the battery increases with each additional test administered, and sensitivity decreases with increasing battery size when a strict criterion is applied (Musiek et al., 2011). Although not typically factored into cost computation formulas, one must also consider the additional time spent testing for larger test batteries, especially those employing a strict criterion.

Based on their examination of a selected subset of tests discussed in this chapter (dichotic digits, LPFS, frequency patterns, and competing speech), Musiek et al. (2011) calculated that for two-test batteries, which generally tended to be some of the most efficient batteries examined in their study, the cost per true positive was less than \$20 and costs were lower using a lax criterion (which tended to yield better overall sensitivities, as well as the opportunity to terminate the battery after failure of only a single test). The reader is referred to Weihing and Atcherson (2014) for discussion relative to central auditory test batteries.

SUMMARY AND CONCLUSIONS

An overview of the key behavioral tests used clinically in the assessment of CAPD was provided in this chapter. Included in this overview were tests often considered more peripheral in their application, as well as those tests sensitized by design to measure central auditory function. Pure-tone thresholds and speech recognition tests generally are not sensitive to central auditory dysfunction; however, in extreme cases, such as marked bilateral central auditory pathway involvement, patient performance on these tests can be compromised to a significant extent. Even in unilateral CANS damage, pure-tone and speech audiometry may be affected, demonstrating the contralateral effect, though this may be a fleeting effect as hearing sensitivity can recover in primates and in human subjects with central deafness. Standard OAEs are not influenced by lesions of the CANS and, therefore, serve as a good reference for cochlear function. Measuring OAE suppression effects may provide insight as to olivocochlear function, although this is not a robust finding and is therefore of marginal clinical utility at this time. The AR arc extends into the caudal pons, and therefore can reflect central disorders of the low brainstem. However, the AR typically is normal in children with CAPD whose central processing deficits likely stem from more rostral (e.g., cortical and interhemispheric) regions.

Dichotic listening tasks employing stimuli such as digits, words, and sentences have a long and useful track record both clinically and experimentally. The contralateral-ear effect for high brainstem and cortical lesions is well documented; however, the best use of dichotics may be as an index of corpus callosum function and interhemispheric transfer.

MLD, localization and lateralization tasks, and the new tests of spatial processing are classified here as tests of binaural interaction. Although the MLD, as well as localization and lateralization (i.e., interaural timing) tests, have been shown to be sensitive to compromise of the CANS, they are underutilized in the clinic. A new test of spatial processing (LiSN-S) is garnering attention and may become a key binaural interaction procedure.

Monaural low-redundancy tests (e.g., filtered speech, compressed speech, speech-in-noise, and speech-in-speech competition) have been utilized for many years. Though they generally lack high sensitivity to CANS involvement, they play a role in measuring processes such as auditory closure. Since these tests are not influenced by corpus callosum lesions, they are useful in combination with dichotic listening tests in differentiating interhemispheric versus right-hemisphere dysfunction.

Temporal processing tests, such as gap detection, sequencing (pattern perception), temporal integration, and temporal masking are perhaps the most sensitive procedures for examining the integrity of the CANS. These temporal processes, which require exquisite timing in the CANS, are pivotal to the processing of most any type of acoustic information. Table 18.3 provides a summary of the contribution of each test reviewed in this chapter to the differential diagnosis of cochlear, VIIIth-nerve, brainstem, cortical, and corpus callosum site of lesion.

Recently, gap detection has been accurately measured using evoked potentials. This coupling of behavioral and electrophysiologic procedures may usher in a new approach in the evaluation of CAPD. Electrophysiologic recordings evoked to stimuli commonly used in behavioral gap detection or other central auditory tasks would provide an objective measure of CANS function, and it would most likely require a higher degree of functional integrity of the CANS. While the authors do not specify a standard central auditory test battery as appropriate for all patient evaluations, we do offer some guidance to audiologists in selecting tests for inclusion in their test batteries. Timing is fundamental to all acoustic signals; therefore, temporal processing is integral to all central auditory processes. Moreover, the diagnostic power of temporal processing tests, especially nonspeech temporal tests, has been documented in the considerable data obtained from human and animal lesion studies. Therefore, a temporal processing test should be included in most, if not all, central auditory test batteries. Similarly, dichotic listening, one of the most well-studied measures of central auditory processing,

Table 18.3

Contributions to differential diagnosis

Test ↓	Cochlear	Eighth nerve	Brainstem	Cortex	Corpus callosum
Pure tones	●	◍	○	○	○
Standard speech recognition in quiet	◍	◍	○	○	○
Otoacoustic emissions	◍	○	○	○	○
Acoustic reflexes	◍	●	◍	○	○
Dichotics		◍	◍	●	●
Masking level difference		◍	●	○	○
Localization and lateralization			●	●	
Frequency and duration patterns			○	●	●
Gap detection			●	●	
Low-redundancy speech			◍	◍	○

Relative contribution to differential diagnosis: ○ minimal; ◍ moderate; ● considerable.

presents considerable power to elucidate the integrity of the CANS. Data from human lesion studies have confirmed the diagnostic power of dichotic measures, while revealing its neuroanatomic mechanisms, and imaging data have shown the heightened activity in auditory areas during dichotic listening. Monaural low-redundancy speech tests are less sensitive than temporal and dichotic measures; however, when used in combination with temporal and dichotic measures, low-redundancy tests assist in determining the likely involvement of the corpus callosum versus hemispheric dysfunction. Binaural interaction tests provide ecologically valid measures supported by animal lesion studies (e.g., localization), and some of the available tests (e.g., MLD) show good diagnostic power. Finally, electrophysiologic procedures may not be appropriate routinely; however, they are particularly useful as objective measures of central auditory function in cases where behavioral measures are inappropriate due to the patient's age and language and cognitive abilities.

In reviewing behavioral tests of central auditory processing, it is clear advances have been made. There remain, however, additional opportunities to enhance the clinical test battery. Given the technology, the advanced knowledge of the system, and creative approaches, we are indeed on the cusp of far greater advances than we have seen in the past.

References

American Academy of Audiology (2010). Guidelines for the Diagnosis, Treatment, and Management of Children and Adults with Central Auditory Processing Disorder. Available at, http://www.audiology.org/resources/documentlibrary/documents/capd%20guidelines%208-2010.pdf.

American Speech-Language-Hearing Association (2005). (Central) Auditory Processing Disorders. Available at, http://www.asha.org/members/deskref-journals/deskref/default.

Bamiou DE (2007). Measures of binaural interaction. In: F Musiek, G Chermak (Eds.), Handbook of (Central) Auditory Processing Disorder: Auditory Neuroscience and Diagnosis. Plural Publishing, San Diego, CA, pp. 257–286.

Bamiou DE, McCullagh J (2014). In: F Musiek, G Chermak (Eds.), Handbook of Central Auditory Processing Disorder: Auditory Neuroscience and Diagnosis. 2nd edn. Plural Publishing, San Diego, CA.

Bantwal A, Hall III JW (2011). Pediatric speech perception in noise. Curr Pediatr Rev 7: 214–226.

Baran J, Musiek F (1999). Behavioral Assessment of the Central Auditory System. In: F Musiek, W Rintelmann (Eds.), Contemporary perspectives on hearing assessment, Allyn & Bacon, Boston, MA, pp. 375–415.

Baran JA, Verkest S, Gollegly K et al. (1985). Use of time-compressed speech in the assessment of central nervous system disorders. Paper presented to the American Acoustical Society: Nashville, TN.

Baran JA, Musiek FE, Reeves AG (1986). Central auditory function following anterior sectioning of corpus callosum. Ear Hear 7: 359–362.

Baru AV, Karaseva TA (1972). The brain and hearing: Hearing disturbances associated with local brain lesions, Consultants Bureau, New York.

Baru AV, Gershuni GV, Tonkonogii IM (1964). The significance of the detection of sound signals of different durations for the diagnosis of impairments to the higher parts of the brain. Zh Nevropatol Psikhiatr 64: 481–485.

Beasley DS, Maki J (1976). Time-and frequency-altered speech. In: N Lass (Ed.), Contemporary issues in experimental phonetics, Academic Press, New York, pp. 419–458.

Beasley DS, Schwimmer S, Rintelmann WF (1972). Intelligibility of time-compressed CNC monosyllable. J Speech Hear Res 15: 340–350.

Bellis TJ (2003). Assessment and management of central auditory processing disorders in the educational setting: From science to practice. Delmar, New York.

Bellis T, Ferre J (1999). Multidimensional approach to the differential diagnosis of central auditory processing disorders in children. Am J Audiol 10: 319–328.

Belmont I, Handler A (1971). Delayed information processing and judgment of temporal order following cerebral damage. J Nerv Ment Dis 152: 353–361.

Berlin CI, Cullen JK (1977). Acoustic problems in dichotic listening tasks. In: SJ Segalowitz, FA Gruber (Eds.), Language development and neurological theory, Academic Press, New York, pp. 77–88.

Berlin CI, Lowe-Bell SS, Jannetta PJ et al. (1972). Central auditory deficits after temporal lobectomy. Arch Otolaryngol Head Neck Surg 96 (1): 4.

Bilger R, Nuetzel J, Rabinowitz W et al. (1984). Standardization of a test of speech perception in noise. J Speech Hear Res 27: 32–48.

Bocca E, Calearo C (1963). Central hearing processes. In: J Jerger (Ed.), Modern developments in audiology, Academic Press, New York, pp. 337–370.

Bocca E, Calearo C, Cassinari V (1954). A new method for testing hearing in temporal lobe tumors. Acta Otolaryngol (Stockholm) 44: 219–221.

Bocca E, Calearo C, Cassinari V et al. (1955). Testing "cortical" hearing in temporal lobe tumors. Acta Otolaryngol 42: 219–221.

Bornstein SP, Wilson RH, Cambron NK (1994). Low- and high-pass filtered Northwestern University Auditory Test No. 6 for monaural and binaural evaluation. J Am Acad Audiol 5: 259–264.

Broadbent DE (1954). The role of auditory localization in attention and memory span. J Exp Psychol 47: 191–196.

Brown M, Musiek F (2013). Pathways: The Fundamentals of Masking Level Differences for Assessing Auditory Function. Hear J 66 (1): 16–17.

Brown DK, Cameron S, Martin JS et al. (2010). The North American Listening in Spatialized NoiseSentences Test (NA LiSN-S): Normative Data and Test-Retest

Reliability Studies for Adolescents and Young Adults. J Am Acad Audiol 21 (10): 629–641.

Calearo MD, Antonelli AR (1973). Disorders of the central auditory nervous systemVol. 2. Saunders, Philadelphia.

Cameron S, Dillon H (2007). Development of the listening in spatialized noise-sentences test (LISN-S). Ear Hear 28 (2): 196–211.

Cameron S, Dillon H (2008). The listening in spatialized noise-sentences test (LISN-S): comparison to the prototype LISN and results from children with either a suspected (central) auditory processing disorder or a confirmed language disorder. J Am Acad Audiol 19 (5): 377–391.

Cameron S, Glyde H, Dillon H (2011). Listening in Spatialized Noise- Sentences Test (LiSN-S): Normative and Retest Reliability Data for Adolescents and Adults up to 60 Years of Age. J Am Acad Audiol 22 (10): 697–709.

Ceranic BJ, Prasher DK, Raglan E et al. (1998). Tinnitus after head injury: evidence from otoacoustic emissions. J Neurol Neurosurg Psychiatry 65 (4): 523–529.

Chermak GD, Musiek FE (1997). Central auditory processing disorders: New perspectives, Singular Publishing Group, San Diego.

Counter SA, Buchanan LH, Ortega F et al. (2011). Assessment of auditory brainstem function in lead-exposed children using stapedius muscle reflexes. J Neurol Sci 306 (1): 29–37.

Cranford JL, Stream RW, Rye CV et al. (1982). Detection v discrimination of brief-duration tones: findings in patients with temporal lobe damage. Arch Otolaryngol Head Neck Surg 108 (6): 350–356.

Cranford JL, Boose M, Moore CA (1990). Effects of aging on the precedence effect in sound localization. J Speech Lang Hear Res 33 (4): 654–659.

Cranford JL, Andres MA, Piatz KK et al. (1993). Influences of age and hearing loss on the precedence effect in sound localization. J Speech Lang Hear Res 36 (2): 437.

Dayal VS, Taranito L, Swisher LP (1966). Neuro-otologic studies in multiple sclerosis. Laryngoscope 76: 1798–1809.

DeRenzi E, Faglioni P, Villa P (1977). Sequential memory for figures in brain-damaged patients. Neuropsychologica 15: 43–49.

Dublin WB (1971). Fundamentals of Sensorineural Auditory Pathology, Charles C Thomas, Springfield, IL.

Durrant JD, Lovrinic JH (1977). Bases of Hearing Sciences, Williams and Williams, Baltimore.

Durrant JD, Lovrinic JH (1995). Bases of Hearing Sciences, Williams and Wilkins, Baltimore.

Elliott LL (1962). Backward masking: monotic and dichotic conditions. J Acoust Soc Am 34 (8): 1108–1115.

Elliott LL (1967). Development of auditory narrow-band frequency contours. J Acoust Soc Am 42: 143–153.

Elliott LL (1995). Verbal auditory closure and the Speech Perception in Noise (SPIN) Test. J Speech Hear Res 38: 1363–1376.

Emanuel DC, Ficca K, Korczak P (2011). Survey of the diagnosis and management of auditory processing disorders. Am J Audiol 20: 48–60.

Fairbanks G, Everitt W, Jaeger R (1954). Methods for time or frequency compression-expansion of speech. Transactions of the IRE Professional Group on Audio 1: 7–12.

Ferber-Viart C, Duclaux R, Dubreuil C et al. (1994). Otoacoustic emissions and brainstem auditory evoked potentials in children with neurological afflictions. Brain Dev 16 (3): 213–218.

Fifer RC, Jerger JF, Berlin CI et al. (1983). Development of a dichotic sentence identification test for hearing-impaired adults. Ear Hear 4 (6): 300–305.

Fuente A, McPherson B (2007). Central auditory processing effects induced by solvent exposure. Int J Occup Med Environ Health 20 (3): 271–279.

Fuente A, McPherson B, Munoz V et al. (2006). Assessment of central auditory processing in a group of workers exposed to solvents. Acta Otolaryngol 126: 1188–1194.

Furst M, Levine RA, Korczyn AD et al. (1995). Brainstem lesions and click lateralization in patients with multiple sclerosis. Hear Res 82 (1): 109–124.

Furst M, Aharonson V, Levine RA et al. (2000). Sound lateralization and interaural discrimination: effects of brainstem infarcts and multiple sclerosis lesions. Hear Res 143 (1): 29–42.

Gelfand S (1998). Hearing: An introduction to psychological and physiological acoustics. 3rd edn. Marcel Dekker, New York.

Gershuni GV (1971). Sensory processes at the neuronal and behavioral levels. Academic Press, London.

Greene D (1971). Temporal auditory acuity. Psychol Rev 78: 540–551.

Griffiths TD, Dean JL, Woods W et al. (2001). The Newcastle Auditory Battery (NAB): a temporal and spatial test battery for use on individual naïve subjects. Hear Res 154: 165–169.

Hall III JW, Johnston KN (2007). Electroacoustic and electrophysiologic auditory measures in the assessment of (central) auditory processing disorder. In: FE Musiek, GD Chermak (Eds.), Handbook of (Central) Auditory Processing Disorders: Auditory Neuroscience and Diagnosis, Plural Publishing, San Diego, pp. 287–316.

Hall III JW, Mueller III GH (1997). Audiologists desk reference. Singular, San Diego.

Hall III JW, Huang-Fu M, Gennarelli TA (1982). Auditory function in acute severe head injury. Laryngoscope 92 (8): 883–890.

Hausler R, Levine RA (1980). Brain stem auditory evoked potentials are related to interaural time discrimination in patients with multiple sclerosis. Brain Res 191 (2): 589–594.

Hautus MJ, Setchell GJ, Waldie KE et al. (2003). Age-related improvements in auditory temporal resolution in reading-impaired children. Dyslexia 9 (1): 37–45.

Heffner HE, Heffner RS (1986). Effect of unilateral and bilateral auditory cortex lesions on the discrimination of vocalizations by Japanese macaques. J Neurophysiol 56 (3): 683–701.

Heilman KM, Hammer LC, Wilder BJ (1973). An audiometric defect in temporal lobe dysfunction. Neurology 23: 384–386.

Hirsch IJ (1948). The influence of interaural phase on interaural summation and inhibition. J Acoust Soc Am 20 (4): 536–544.

Hugdahl K, Andersson L (1986). The "forced-attention paradigm" in dichotic listening to CV-syllables: a comparison between adults and children. Cortex 22 (3): 417–432.

Janse E, Noteboom S, Quene H (2003). World-level intelligibility of time-compressed speech: prosodic and segmental factors. Speech Comm 41: 287–301.

Jerger J (1960). Audiological manifestations of lesions in the auditory nervous system. Laryngoscope 70: 417–425.

Jerger J, Jerger S (1974). Auditory findings in brainstem disorders. Arch Otolaryngol 99: 342–349.

Jerger S, Jerger J (1975). Extra-and intra-axial brain stem auditory disorders. Int J Audiol 14 (2): 93–117.

Jerger S, Jerger J (1977). Diagnostic value of crossed vs uncrossed acoustic reflexes: eighth nerve and brain stem disorders. Arch Otolaryngol Head Neck Surg 103 (8): 445–453.

Jerger S, Jerger J (1984). Pediatric Speech Intelligibility Test: Manual for administration, Auditec, St. Louis.

Jerger J, Oliver TA, Rivera V et al. (1986). Abnormalities of the acoustic reflex in multiple sclerosis. Am J Otolaryngol 7 (3): 163–176.

Jerger J, Thibodeau L, Martin J et al. (2002). Behavioral and electrophysiologic evidence of auditory processing disorder: a twin study. J Am Acad Audiol 13: 438–460.

Johnston KN, John AB, Kreisman NV et al. (2009). Multiple benefits of personal FM system use by children with auditory processing disorder (APD). Int J Audiol 48: 371–383.

Karaseva T (1972). The role of temporal lobe in human auditory perception. Neuropsychologica 10: 227–231.

Karlsson A, Rosenhall U (1995). Clinical applications of distorted speech audiometry. Scand Audiol 24: 155–160.

Katz J (1962). The use of staggered spondaic words for assessing the integrity of the central auditory nervous system. J Aud Res 2: 327–337.

Keith B (2000). Random Gap Detection Test, Auditec, St. Louis.

Keith RW (2007). Diagnosing (central) auditory processing disorders in children. Audiology Diagnosis 1: 335–379.

Keith R (2009a). SCAN-3:A Tests for Auditory Processing Disorders in Adolescents and Adults, Pearson, San Antonio, TX.

Keith R (2009b). SCAN-3:C Tests for Auditory Processing Disorders for Children. Pearson, San Antonio, TX.

Kimura D (1961a). Some effects of temporal-lobe damage on auditory perception. Can J Psychol 15: 156–165.

Kimura D (1961b). Cerebral dominance and the perception of verbal stimuli. Can J Psychol 15 (3): 166–171.

Koehnke J, Colburn HS, Durlach NI (1986). Performance in several binaural-interaction experiments. J Acoust Soc Am 79: 1558–1562.

Krishnamurti S, Drake L, King J (2011). Neural network modeling of central auditory dysfunction in Alzheimer's disease. Neural Netw 24: 646–651.

Kurdziel S, Noffsinger D, Olsen W (1976). Performance by cortical lesion patients on 40% and 60% time-compressed materials. J Am Audiol Soc 2: 3–7.

Lasky RE, Rupert AL (1982). Temporal masking of auditory evoked brainstem responses in human newborns and adults. Hear Res 6 (3): 315–334.

Levine RA, Gardner JC, Fullerton BC et al. (1993). Effects of multiple sclerosis brainstem lesions on sound lateralization and brainstem auditory evoked potentials. Hear Res 68 (1): 73–88.

Licklider JCR (1948). The influence of interaural phase relations upon the masking of speech by white noise. J Acoust Soc Am 20 (2): 150–159.

Lister J, Besing J, Koehnke J (2002). Effects of age and frequency disparity on gap discrimination. J Acoust Soc Am 111: 2793–2800.

Lister JJ, Roberts RA, Shackelford J et al. (2006). An adaptive clinical test of temporal resolution. Am J Audiol 15 (2): 133–140.

Lynn GE, Gilroy J (1972). Neuro-audiological abnormalities in patients with temporal lobe tumors. J Neurol Sci 17: 167–184.

Lynn GE, Gilroy J (1977). Evaluation of central auditory dysfunction in patients with neurological disorders. In: RW Keith (Ed.), Central auditory dysfunction, Grune & Stratton, New York, pp. 177–221.

Lynn GE, Gilroy J, Taylor PC et al. (1981). Binaural masking-level differences in neurological disorders. Arch Otolaryngol Head Neck Surg 107 (6): 357–362.

Matkin ND, Carhart R (1966). Auditory profiles associated with Rh incompatibility. Arch Otolaryngol Head Neck Surg 84 (5): 502–513.

Meyers JE, Roberts RJ, Bayless JD et al. (2002). Dichotic listening: expanded norms and clinical application. Arch Clin Neuropsychol 17 (1): 79–90.

Milner B (1962). Laterality effects in audition. In: V Mountcastle (Ed.), Interhemispheric relations and cerebral dominance, Johns Hopkins Press, Baltimore, pp. 117–195.

Moore C, Cranford J, Rahn A (1990). Tracking a "moving" fused auditory image under conditions that elicit the precedence effect. J Speech Hear Res 33: 141–148.

Moore D, Cowan J, Riley A et al. (2011). Development of auditory processing in 6- to 11-yr-old children. Ear Hear 32 (3): 269–285.

Morales-Garcia C, Poole JO (1972). Masked speech audiometry in central deafness. Acta Otolaryngol 74: 307–316.

Muchnik C, Othman-Jebara R, Putter-Katz H et al. (2004). Reduced medial olivocochlear bundle system function in children with auditory processing disorders. Audiol Neurotol 9 (2): 107–114.

Musiek FE (1983). Assessment of central auditory dysfunction: the dichotic digit test revisited. Ear Hear 4: 79–83.

Musiek FE (1994). Frequency (Pitch) and Duration Pattern Tests. J Am Acad Audiol 5: 265–286.

Musiek FE, Baran JA (2002). Central auditory evaluation of patients with neurological involvement. In: J Katz (Ed.), Handbook of clinical audiology, 5th edn. Lippincott Williams and Wilkins, Baltimore, pp. 532–544.

Musiek FE, Baran JA (2007). The Auditory System, Anatomy, Physiology, and Clinical Correlates. Pearson Education, Inc., Boston, MA.

Musiek FE, Chermak GD (2014). Handbook of (Central) Auditory Processing Disorder: Auditory Neuroscience and Diagnosis. 2nd edn. Plural Publishing, San Diego, CA.

Musiek FE, Geurkink NA (1982). Auditory brain response and central auditory test findings for patients with brain stem lesions: a preliminary report. Laryngoscope 92: 891–990.

Musiek FE, Pinheiro ML (1985). Dichotic speech tests in the detection of central auditory dysfunction. In: ML Pinheiro (Ed.), Assessment of Central Auditory Dysfunction: Foundations and Clinical Correlates, Williams & Wilkins, Baltimore, pp. 201–217.

Musiek FE, Pinheiro M (1987). Frequency patterns in cochlear, brainstem and cerebral lesions. Audiology 26: 78–88.

Musiek FE, Weihing J (2011). Perspectives on dichotic listening and the corpus callosum. Brain Cogn 76: 225–232.

Musiek FE, Wilson DH, Pinheiro ML (1979). Audiological manifestations in "split-brain" patients. J Am Audiol Soc 5: 25–29.

Musiek FE, Pinheiro M, Wilson D (1980). Auditory pattern perception in 'split brain' patients. Arch Otolaryngol 106: 610–612.

Musiek FE, Kibbe K, Baran J (1984). Neuraudiological results from split-brain patients. Semin Hear 5: 219–229.

Musiek FE, Baran JA, Pinheiro M (1990). Duration pattern recognition in normal subjects and patients with cerebral and cochlear lesions. Audiology 29: 304–313.

Musiek FE, Baran JA, Pinheiro ML (1994). Neuroaudiology: Case Studies, Singular, San Diego.

Musiek FE, Shinn J, Jirsa R et al. (2005). The GIN© (Gaps in Noise) Test performance in subjects with confirmed central auditory nervous system involvement. Ear Hear 26: 608–618.

Musiek FE, Chermak GD, Weihing J et al. (2011). Diagnostic accuracy of established central auditory processing test batteries in patients with documented brain lesions. J Am Acad Audiol 22 (6): 342–358.

Nilsson M, Soli S, Sullivan J (1994). Development of the Hearing In Noise Test for the measurement of speech reception thresholds in quiet and in noise. J Acoust Soc Am 95 (2): 1085–1099.

Noffsinger D, Olsen WO, Carhart R et al. (1972). Auditory and vestibular aberrations in multiple sclerosis. Acta Otolyaryngol Suppl 303: 1–63.

O'Beirne GA, McGaffin AJ, Rickard NA (2012). Development of an adaptive low-pass filtered speech test for the identification of auditory processing disorders. Int J Pediatr Otorhinolaryngol 76: 777–782.

Olsen WO, Matkin ND (1978). Differential audiology. In: DE Rose (Ed.), Audiological Assessment, Prentice-Hall, Englewood Cliffs, pp. 368–419.

Olsen WO, Noffsinger D, Kurdziel S (1975). Speech discrimination in noise by patients with peripheral and central lesions. Acta Otolaryngol 80: 375–382.

Pakarinen S, Takegata R, Rinne T et al. (2007). Measurement of extensive auditory discrimination profiles using mismatch negativity (MMN) of the auditory event-related potential. Clin Neurophysiol 118: 177–185.

Palmer SB, Musiek FE (2013). N1-p2 recordings to gaps in broadband noise. J Am Acad Audiol 24 (1): 37–45.

Phillips D (1999). Auditory gap detection, perceptual channels and temporal resolution in speech perception. J Am Acad Audiol 10: 343–354.

Phillips D, Taylor TL, Hall SE et al. (1997). Detection of silent intervals between noises activating different perceptual channels: some properties of "central" auditory gap detection. J Acoust Soc Am 101 (6): 3694–3705.

Pinheiro ML, Musiek FE (1985a). Sequencing and temporal ordering in the auditory system. In: ML Pinheiro, FE Musiek (Eds.), Assessment of central auditory dysfunction: Foundations and clinical correlates. Williams and Wilkins, Baltimore.

Pinheiro M, Musiek F (1985b). Assessment of Central Auditory Dysfunction: Foundations and Clinical Correlates. Williams and Wilkins, Baltimore.

Pratt H, Polyakov A, Aharonson V et al. (1998). Effects of localized pontine lesions on auditory brain-stem evoked potentials and binaural processing in humans. Electroencephalogr Clin Neurophysiol 108 (5): 511–520.

Quaranta A, Cervellera G (1977). Masking level differences in central nervous system diseases. Arch Otolaryngol 103: 483–484.

Robinette MS, Glattke TJ (1997). Otoacoustic Emissions: Clinical Applications. Thieme, New York.

Sanchez-Longo LP, Forster FM (1958). Clinical significance of impairment of sound localization. Neurology 8 (2): 119.

Sanchez-Longo LP, Forster FM, Auth TL (1957). A clinical test for sound localization and its applications. Neurology 7 (9): 655–663.

Shinn JB, Baran JA, Moncrieff DW et al. (2005). Differential attention effects on dichotic listening. J Am Acad Audiol 16: 205–218.

Shinn JB, Chermak GD, Musiek FE (2009). GIN (Gaps-In-Noise) performance in the pediatric population. J Am Acad Audiol 20 (4): 229–238.

Sinha S (1959). The role of the temporal lobe in hearing, Unpublished master's thesis, McGill University, Montreal, Quebec.

Swisher L, Hirsch I (1972). Brain damage and the ordering of two temporally successive stimuli. Neuropsychologica 10: 137–152.

Telischi F, Roth J, Stagner B et al. (1995). Patterns of evoked otoacoustic emissions associated with acoustic neuromas. Laryngoscope 105: 675–682.

Tschopp K, Zust H (1994). Performance of normally hearing and hearing-impaired listeners using a German version of the SPIN Test. Scand Audiol 23 (4): 241–247.

Veterans Administration (1992). Tonal and speech materials for auditory perceptual assessment. Research and Development Service, Veterans' Administration Central Office, Long Beach, CA.

Weihing J, Atcherson S (2014). Primer on clinical decision analysis. In: F Musiek, G Chermak (Eds.), Handbook of central auditory processing disorder: Auditory neuroscience and diagnosis. 2nd edn. Plural Publishing, San Diego.

Weihing J, Musiek FE, Shinn J (2007). The effect of presentation level on the Gaps-In-Noise (GIN) test. J Am Acad Audiol 18: 141–150.

Willeford J (1977). Assessing central auditory behavior in children: A test battery approach. In: Central auditory dysfunction, 43–72.

Wilson RH, Carhart R (1971). Forward and backward masking: interactions and additivity. J Acoust Soc Am 49: 1254–1263.

Wilson L, Mueller HG (1984). Performance on normal hearing individuals on Auditec filtered speech tests. American Speech and Hearing Association 27: 189.

Wilson RH, Preece JP, Salamon DL et al. (1994). Effects of time compression and time compression plus reverberation on the intelligibility of Northwestern University Auditory Test No. 6. J Am Acad Audiol 5 (4): 269–277.

Wright HN (1978). Brief tone audiometry. In: J Katz (Ed.), Handbook of clinical audiology, 2nd edn. Williams & Wilkins, Baltimore, MD.

Wright BA, Lombardino LJ, King WM et al. (1997). Deficits in auditory temporal and spectral resolution in language-impaired children. Nature 387: 176–178.

Yasin I (2007). Hemispheric differences in processing dichotic meaningful and non-meaningful words. Neuropsychologia 45 (12): 2718–2729.

Section 3

Disorders of the auditory system

Handbook of Clinical Neurology, Vol. 129 (3rd series)
The Human Auditory System
G.G. Celesia and G. Hickok, Editors

Chapter 19

Neurocognitive development in congenitally deaf children

ELIZABETH FITZPATRICK*

Faculty of Health Sciences, University of Ottawa and Children's Hospital of Eastern Ontario Research Institute, Otttawa, ON, Canada

INTRODUCTION

Permanent hearing disorders affect neurocognitive development in children across a number of related domains, including auditory, linguistic, cognitive, social, literacy, and academic functioning. The most consistent, notable, and striking consequence of congenital hearing loss is related to typical spoken language acquisition. Although earlier identification of hearing loss and access to hearing technology (hearing aids, cochlear implants, assistive devices) have improved the opportunities for communication development, much remains to be learned about the preferred intervention methods to facilitate neurocognitive development and about the factors underlying the wide variability in children's development. There have been increased opportunities for developing spoken language in the last two decades through knowledge about the developing brain, technology advances, and the heightened interest in spoken language acquisition. Given this shift, coupled with the fact that more than 95% of children with hearing loss have parents with normal hearing, this chapter will focus primarily on spoken language rather than visual language (e.g., sign language) development.

PREVALENCE AND EPIDEMIOLOGY

Permanent childhood hearing loss is recognized as one of the most common pediatric disorders, affecting 2–3 per 1000 live births (Prieve and Stevens, 2000; Fortnum et al., 2001; Watkin and Baldwin, 2011). Variations in reported prevalence are related to discrepancies in the definition of hearing disorders in terms of severity and laterality. In particular, when mild degrees (20–40 dB) are included, these together account for approximately 40% of permanent childhood hearing disorders (Fitzpatrick et al., 2010, 2014). Hearing loss is categorized according to severity of hearing loss, ranging from mild to profound degrees. Table 19.1 presents the definitions applied in this chapter.

Hearing loss affects children from all socioeconomic levels but is more prevalent in developing countries due to consanguinity, adverse perinatal conditions, and socioeconomic challenges which contribute to risk factors associated with hearing loss, including low birth weight as well as maternal and child diseases. For example, the lack of immunizations for diseases such as mumps, rubella, and meningitis is associated with increased risks for hearing disorders (Olusanya et al., 2008).

Onset of hearing loss

Onset of hearing loss is typically described as congenital, delayed-onset, or acquired (see Table 19.1 for definitions). In the absence of population-based newborn hearing screening, it was difficult to pinpoint the onset of hearing disorders and a substantial number of children were assumed to have congenital disorders based on their case histories at the time of referral. Recent reports from universal newborn hearing screening (UNHS) programs suggest that more children than previously assumed experience delayed-onset hearing loss in early childhood (Watkin and Baldwin, 2011; Fitzpatrick et al., 2014). For example, in a recent study of a newborn screening cohort, almost one-third of children who were identified with mild bilateral or unilateral hearing loss were found to have delayed-onset hearing loss (Fitzpatrick et al., 2014). Figure 19.1 provides an example of population-based data from a cohort of 893 Canadian children identified over more than a 20-year period from 1990 to 2012. Of the 491 children whose onset was known, 64% had congenital and 36% had delayed-onset

*Correspondence to: Elizabeth Fitzpatrick, Ph.D., Associate Professor, Faculty of Health Sciences, University of Ottawa, Senior Scientist, Children's Hospital of Eastern Ontario Research Institute, Faculty of Sciences, University of Ottawa, 451 Smyth Road, Ottawa, ON K1H 8M5, Canada. Tel: +1-613-562-5428, E-mail: elizabeth.fitzpatrick@uottawa.ca

Table 19.1

Definitions of hearing loss characteristics

Congenital	Present at or very shortly after birth
Delayed-onset	Develops postnatally; cannot be attributed to an exogenous cause
Acquired	Develops postnatally; attributable to an exogenous cause
Early-onset	Develops within 3 months of birth
Prelingual	Develops before language is well established (<3 years of age)
Progressive	Audiometric thresholds deteriorate over time
Mild	PTA average (500, 1000, 2000 kHz) ≥20–40 dB HL
Moderate	PTA average 41–55 dB HL
Moderately severe	PTA average 56–70 dB HL
Severe	PTA average 71–90 dB HL
Profound	PTA >90 dB HL

PTA, pure-tone average; HL, hearing level.

loss. These data show that 92% of the known congenital group had bilateral loss. Results for the delayed-onset group were similar, with 89% bilateral loss.

Progressive hearing loss

It is important to note that a substantial number of children with congenital loss experience deterioration in hearing levels over time. For example, 22% of children first identified through audiology with mild bilateral or unilateral hearing loss showed deterioration in hearing thresholds. In the subgroup of children who were exposed to newborn hearing screening, 32% showed progression of hearing loss (Fitzpatrick et al., 2014). Similar to the situation with delayed-onset hearing loss, until the widespread implementation of UNHS, accurate documentation of progressive hearing loss from birth was challenging. Children presumed to have congenital loss were identified well beyond infancy and might have experienced deterioration in thresholds prior to the initial diagnosis.

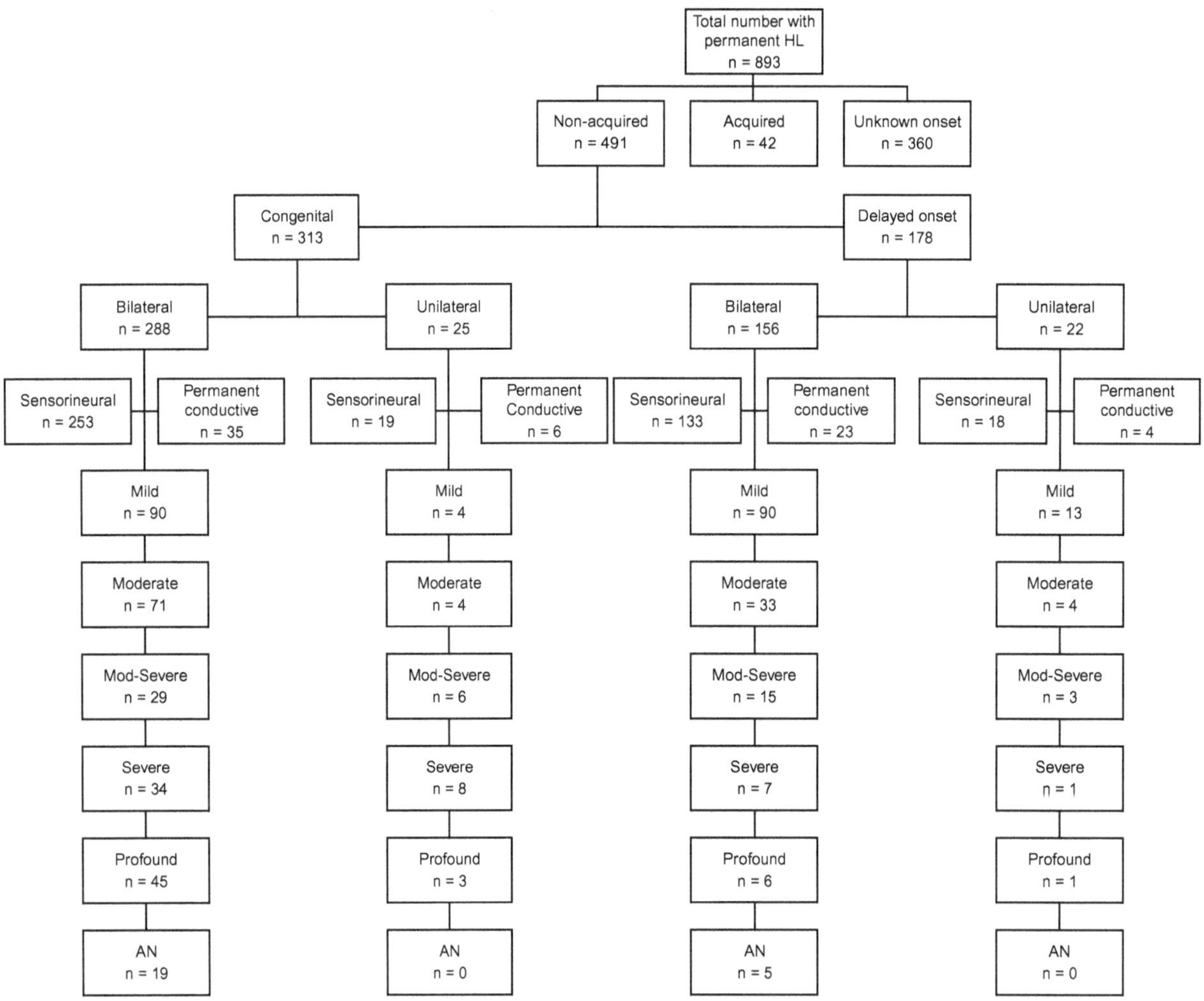

Fig. 19.1. Example of characteristics of a population of children with hearing loss. AN, auditory neuropathy.

Type of hearing loss

As shown in Figure 19.1, most congenital hearing loss is sensorineural in nature, with a small proportion of permanent conductive loss. Within the broader category of sensorineural hearing loss, since the 1990s, clinical diagnosis of a hearing disorder described as auditory neuropathy (now referred to as auditory neuropathy spectrum disorder (ANSD)) has been described (Starr et al., 1996). It is estimated that 5–10% of children with permanent hearing loss present with ANSD (Rance and Starr, 2011). The clinical diagnostic profile has been well described and includes the presence of cochlear emissions, as measured by otoacoustic emission testing and abnormal auditory brainstem responses (Starr et al., 1996; Berlin et al., 2003) (see Chapter 28, this volume, for further information). Clinical characteristic of ANSD can be variable and include severity ranging from mild to profound, fluctuating or unstable audiometric thresholds, poor speech perception results relative to typical sensorineural loss, as well as normal hearing levels and spontaneous improvement in a minority of cases (Berlin et al., 2003; Raveh et al., 2007; Rance and Starr, 2011).

Severity of hearing loss

Hearing loss is classified according to severity; classifications for severity used in this chapter (Table 19.1) are based on a three-frequency (500, 1000, 2000 Hz) pure-tone average. The population-based data from one regional center presented in Figure 19.1 provide an overview of the distribution of severity in a pediatric population. These data appear to be consistent with other reports in the developing world (Watkin and Baldwin, 2011) and with earlier population data from the program used as an illustrative example in Figure 19.1 (Durieux-Smith and Whittingham, 2000; Durieux-Smith et al., 2008). This snapshot suggests that about 66% of children with congenital bilateral sensorineural hearing loss have mild to moderately severe degrees of hearing loss (20–70 dB hearing level (HL)) at diagnosis. Approximately 27% have severe or profound loss (>70 dB HL), essentially preventing them from detecting speech sounds without specialized hearing technology (hearing aids or cochlear implants). In this cohort, 7% presented with ANSD. Consideration of the subgroup of 285 children identified since the implementation of UNHS reveals strikingly similar results.

This chapter is concerned with congenital hearing loss of a permanent nature, the overwhelming majority being sensorineural in origin, but there is one important caveat. In studies conducted prior to UNHS, children diagnosed with hearing loss before age 2–3 years were frequently assumed to have congenital disorders in the absence of any specific information regarding delayed onset. Thus, specifically isolating the effects of congenital hearing loss is difficult until sufficient studies become available from newborn screening cohorts that report data separately for children with congenital and later-onset hearing loss.

Hearing loss and additional disabilities

One of the complexities of determining the effects of hearing loss on development is that it does not necessarily occur in isolation, independently of other conditions that affect children's neurodevelopment outcomes. Several reports indicate that up to 30–40% of children with hearing disorders have at least one additional disability (Van Naarden et al., 1999; Picard, 2004; Gallaudet Research Institute, 2011). A recent review of population-based clinical data showed that 82 (23%) of 363 children identified with permanent hearing disorders in a 10-year period had at least one additional developmental disability; disabilities were associated with a syndrome for 26 (32%) children (Fitzpatrick and Doucet, 2013). Table 19.2 provides an overview of the broad range of developmental conditions documented for the children with non-syndromic deafness. In the presence of other disabilities, particularly when associated with cognitive delay, disentangling the specific consequences of hearing loss on spoken language and other neurodevelopmental areas is complicated. For example, the majority of children with Down syndrome have some level of hearing loss (permanent conductive loss due to middle-ear disorders or sensorineural loss) which persists throughout childhood and which can have an additive or even multiplicative effect for cognitive and speech language development (Rasore Quartino, 2011).

There is now an extensive literature on children with additional disabilities associated with developmental delay who have profound deafness and were fitted with cochlear implants. There is evidence to suggest that these

Table 19.2

Clinical profile of 56 children with additional documented developmental disabilities

Developmental condition	Percent
Motor disability	33.9%
Communication-related disability	10.8%
Autism spectrum disorder	7.1%
Cognitive disability	7.1%
Attention-deficit hyperactivity disorder	3.6%
Learning disability	3.6%
Other (various conditions)	33.9%

children achieve lower language skills than would be expected according to their cognitive abilities (Meinzen-Derr et al., 2011). Overall, children with severe cognitive delay have shown minimal progress in the development of auditory skills and spoken language (Olds et al., 2005; Meinzen-Derr et al., 2011).

EFFECTS OF PERMANENT CHILDHOOD HEARING LOSS ON THE CHILD AND FAMILY

The most apparent and understandable effect of congenital hearing loss is disruption in the typical acquisition of spoken language. Consequences of delayed language skills are also manifested, at least for some children, in later difficulties in literacy, academic and social functioning (Wake et al., 2004; Geers, 2006). In addition to the more obvious cascade effects of deafness resulting in language, educational, and quality-of-life concerns, early auditory access seems to contribute substantially to the beginning stages of the child's perceptual and cognitive development (Ruff et al., 1992; Perone et al., 2008; Fagan and Pisoni, 2009). Therefore, in addition to providing an important foundation for spoken language development, hearing plays an important role in the development of cognitive abilities in domains such as attention, learning, and memory (Pisoni et al., 2008). It follows, then, that children with congenital loss are especially susceptible to disruptions in language development due to a lack of auditory stimulation in early infancy, a sensitive period for language acquisition.

Auditory plasticity

Evidence has accumulated from neuroscience to show that the developing brain depends on multiple sensory inputs for typical maturation. There is evidence from animal studies that deafness during the early years of life interferes with or disrupts typical auditory cortical development by delaying or preventing the formation of neuronal connections and the maturation of cortical layers (Kral and O'Donoghue, 2010; Kral and Sharma, 2012). Research points to a sensitive period for auditory cortical development, suggesting that after this period the brain reorganizes itself, that is, cross-modal reorganization takes place to adapt to information from other sensory modalities (Kral, 2007; Kral and O'Donoghue, 2010). Consequently, auditory cortical areas become reassigned to other types of non-auditory sensory input, decreasing the brain's efficiency in processing auditory input (Sharma et al., 2009).

A series of studies using cortical auditory evoked potentials in children with profound deafness fitted with cochlear implants at various ages lend support to a "sensitive" period for maximal auditory development (Sharma et al., 2002, 2005; Sharma and Dorman, 2006). By measuring cortical auditory evoked potential (P1) responses in children who received cochlear implants at different ages, Sharma and Dorman (2006) summarized convincing evidence for the importance of early auditory stimulation. In studies of 107 individuals (104 children) with either congenital deafness or onset in the first year, age 3.5 years emerged as the approximate upper boundary for normal cortical responses while children implanted after 7.0 years demonstrated abnormal responses following several years of stimulation with their implant (Fig. 19.2). Children fitted with cochlear implants between these two ages showed a range of response latencies. Studies in individuals with long-term deafness suggest that visual processing may occur in cortical regions usually dedicated to auditory processing. For example, as shown in Figure 19.3 from Neville et al. (1998), in adults with congenital deafness who were primarily users of sign rather than spoken language, processing of sign language (visual stimuli) activates areas in the auditory cortex. Functional studies of individuals with congenital deafness who were

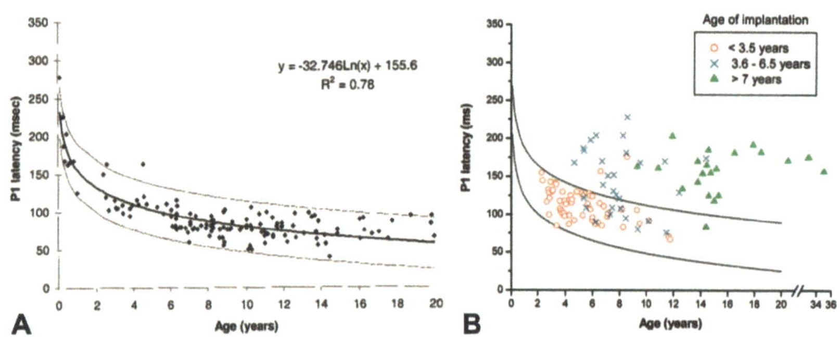

Fig. 19.2. (**A**) P1 latencies as a function of age for normal-hearing children. The line of best fit and the 95% confidence interval are superimposed on the raw data. (**B**) P1 latencies as a function of chronologic age for children with cochlear implants. The solid functions are the 95% confidence limits for normal-hearing children. P1 latencies for children implanted before age 3.5 years (early-implanted group) are shown as circles. P1 latencies for children implanted between age 3.5 years and 6.5 years (middle-implanted group) are shown as crosses. P1 latencies for children implanted after age 7 years (late-implanted group) are shown as triangles. (Reproduced from Sharma et al., 2002.)

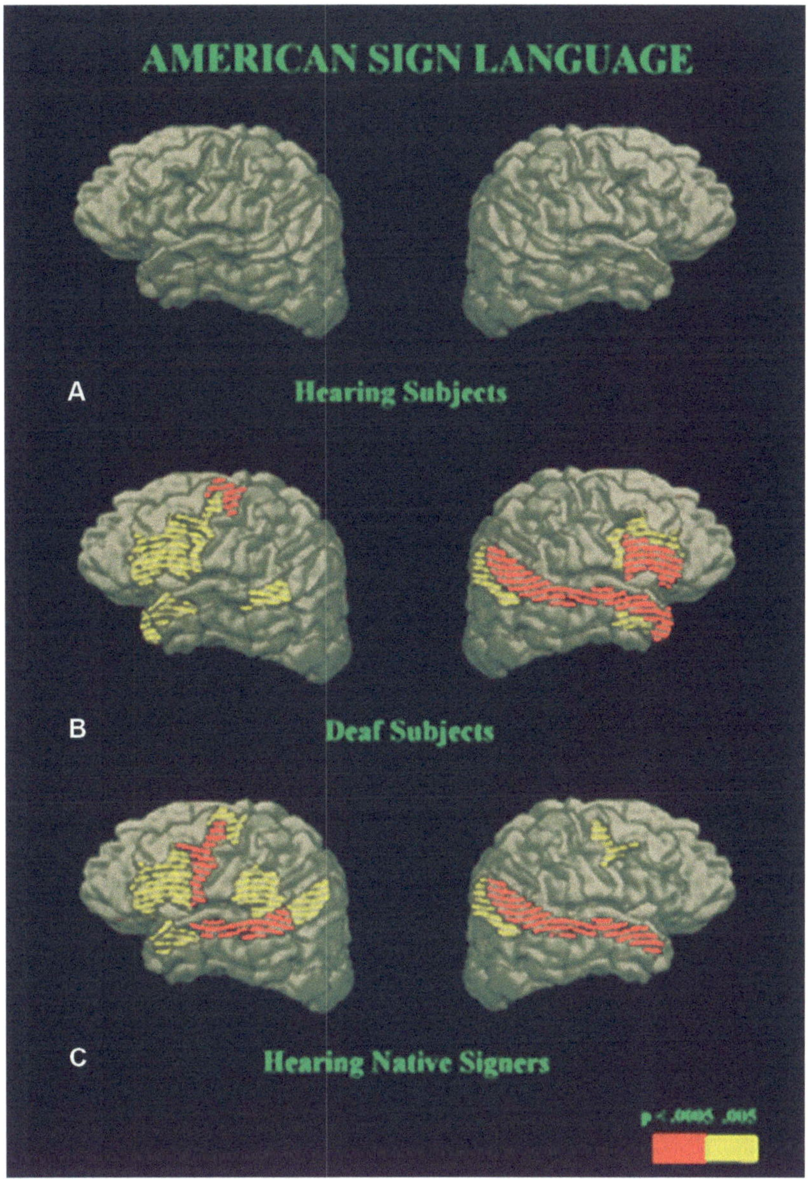

Fig. 19.3. Cortical areas displaying activation ($P < 0.005$) for American Sign Language (ASL) sentences (versus non-signs) for three subject groups. These results (from Neville et al., 1998) suggest that spoken language acquisition is not a requirement for establishing specialized language systems in the left hemisphere. When processing sentences in ASL versus non-signs, normal-hearing adults who were not sign language users showed no differences in activation (**A**); deaf adult users of ASL showed significant left-hemisphere activation as well as right-hemisphere activation (**B**); similarly, normal hearing adults who were native signers also showed significant left-hemisphere activation as well as right-hemisphere recruitment (**C**). (Reproduced with permission from Neville et al., 1998. Copyright (1998) National Academy of Sciences, USA.)

implanted as adolescents or adults show limited ability for auditory-only speech recognition for words and phrases relative to post-linguistically deafened adults or early-implanted children (Schramm et al., 2002; Waltzman et al., 2002).

Sensitive period for language learning

In addition to the concept of a "critical period" specifically for auditory stimulation, children seem to be organically ready to develop language (Chomsky, 1986), the length of the period varying depending on the particular aspect of language (Ruben, 1997). Language acquisition occurs at remarkably similar timelines across children from different social and cultural backgrounds. Arshavsky (2009) proposes that early language experience is required to "activate" genetic programs that underlie language acquisition during early brain development. The existence of a time window for language development is supported by naturally occurring experiments of children who were disconnected from linguistic environments in the early years and who, when discovered as adolescents, were unable to acquire normal language (Curtis, 1977; Singh and Zingg, 1996).

Studies in infant speech perception have revealed that, from the first few months, infants are programmed to listen to speech and prefer speech over non-speech sounds (Vouloumanos and Werker, 2007; Vouloumanos et al., 2009). Research indicates that, during the first 6 months, infants essentially function as universal language receptors; that is, they appear to differentiate between all phonemic categories (for example, pa versus ba contrast), even those not used in their native language (Werker and Tees, 2005). However, by their first birthday, children's brains undergo reorganization to discriminate phonemes specific to the languages to which they are exposed in their daily environments (Werker and Yeung, 2005). During this perceptual fine-tuning period, the infant appears to "sharpen" the phonetic contrasts required for language and to ignore differences that are not relevant in the language spoken around the infant. Most children with congenital hearing loss will not have normal access to speech and therefore cannot fully experience this foundation period for language learning unless hearing technology is provided within the first days or weeks of life.

In typically developing infants, perception of speech sounds, phonologic development, and vocabulary interact are viewed as cognitive-linguistic blocks on which later receptive and expressive language and even literacy are built (Jusczyk, 1997). It seems that, for most children with hearing loss, there is likely at least some disconnect in this early synergy among various components of language development. Capitalizing on these periods when children are developmentally "programmed" to learn language appears to offer the greatest chance for natural language acquisition.

Impact on child's experience in the family

As described above, reduced hearing that leads to delayed and degraded acoustic inputs can be expected to have repercussions for the developing child by interfering with brain maturation of the auditory cortex at a sensitive developmental period. Also of paramount importance is that hearing loss not only impacts the child, but also affects how the family interacts with the child in this early learning period. There is some evidence showing that families, primarily mothers, of children with hearing loss alter how they interact with their children (Cole, 1992; Calderon and Greenberg, 2003; Cole and Flexer, 2011). For example, in the absence of typical feedback from a child, parents may spend less time talking with the child, be less engaged, and use more directive speech, that is, use more requests and commands in communicating with their child rather than usual infant-directed talk.

There is considerable evidence from studies related to children with normal hearing in early development programs that mothers' responsiveness is a key factor in early linguistic development. In fact, some studies have shown that responsiveness has a far greater impact on early child development than specific therapeutic interventions undertaken to enhance language development, even stronger than the intervention specifically on the child's development. There is strong support from other literature for the role that parents play in enhancing cognitive development. For example, studies have shown that parent responsiveness is an important predictor of children's developmental age and intelligence quotient scores (Fewell et al., 1996; Landry et al., 1997). For example, Landry et al., in their study of 299 full-term and low-birth-weight children, reported that children had higher cognitive-language ages at 18 months of age if their mothers' interactions were characterized by high proportions of maintaining interest ($P = 0.002$) (e.g., engaging and responding to child's attempt to attract attention). In contrast, mothers' use of more restrictive interactions ($P = 0.002$) such as attempting to stop their child's attempts at communicating, was highly associated with lower cognitive-linguistic ages. Furthermore, higher levels of restrictive behaviors were related to lower social initiating ages for these children, particularly for the 73 children categorized as medically high-risk. These parent behaviors have also emerged as critical in research related to children with developmental disabilities (Mahoney et al., 2004).

Mahoney et al. (2007) suggested parent responsiveness increases the frequency that children use certain pivotal behaviors, behaviors considered as foundations for social and cognitive developmental learning. Their study of 45 mother–child dyads (pairs) showed that, in children under age 3 with developmental delays, heightened mother responsiveness enhanced the use of the child's pivotal behaviors such as attention, initiation, persistence, cooperation, and joint attention. Furthermore, more frequent use of these behaviors was associated with higher rate of social communication and cognitive development. The authors conclude that, regardless of type of developmental disability, parents' ability to respond to their child is key to nurturing children's overall learning and development. Other research with children with disabilities has shown that intervention was in fact ineffective if it did not enhance parents' responsiveness with their children (Mahoney et al., 2004). A study of 32 mothers and their preschool children with cochlear implants revealed that mothers' involvement and their sense of self-efficacy were highly related to

Table 19.3

Associations between maternal self-efficacy and involvement and facilitative language techniques

Facilitative language techniques	Maternal self-efficacy cochlear implant	Maternal involvement cochlear implant	Maternal self-efficacy speech language	Maternal involvement speech language
Parallel talk	−0.03	0.04	0.45*	0.56†
Expansion	−0.11	0.08	0.22	0.39*
Recast	−0.35	−0.23	0.08	−0.31
Open–ended question	0.27	0.14	0.10	0.30
Linguistic mapping	−0.36	−0.09	−0.63†	−0.37
Closed-ended question	0.13	−0.35	0.12	−0.05
Imitation	−0.01	0.36	−0.63*	−0.56*
Label	−0.07	0.08	−0.29	−0.51*
Directive	−0.18	−0.39	−0.52*	−0.53*
Comment	0.15	−0.03	−0.39	−0.40*

*$P < 0.05$; †$P < 0.01$.

Reproduced from Desjardin and Eisenberg (2007).

the quantity and quality of linguistic input they provided (Desjardin and Eisenberg, 2007). Table 19.3 details the associations reported in this study between mothers' use of facilitative language techniques and their sense of self-efficacy and involvement with both their child's cochlear implant device and speech-language development.

Taken together, these studies lead to the conclusion that any interference with natural experiences, including reduced linguistic and social exposure from parents, can have a negative "experiential" effect on communication, learning, and social development. It seems reasonable to infer that young children with hearing loss, whose parents learn about the disorder in the early developing years, are particularly at risk. Not only are affected children receiving reduced or poorer-quality sensory information through hearing technology but it can be a difficult emotional time for caregivers. Fewer than 10% of children with hearing loss have parents who are themselves deaf (Mitchell and Karchmer, 2004), so most families are adjusting to the shock and diagnosis of hearing loss (English, 2011). Parent interviews about their early experiences with learning about deafness suggest that it can indeed be a fragile period wrought with emotions such as guilt and grief (Young and Tattersall, 2007; Fitzpatrick et al., 2008) and the early months and years can be characterized by a period of particular vulnerability, guilt, and grief (Luterman, 2008). Of importance is that parents are living these poignant experiences and learning about deafness and new technology in a period that is understood to be a sensitive learning period for all children acquiring spoken language. As reflected in parent interviews (Fitzpatrick et al., 2007b), this can be a particularly taxing time. This information suggests that investment in informing, teaching, coaching, and emotionally supporting parents during this period is extremely important to encourage auditory and general cognitive stimulation of the brain centers (Fitzpatrick and Doucet, 2013). Interestingly, what many parents report that they need most at this period is support from other parents (Fitzpatrick et al., 2008; English, 2011).

A recent international consensus statement (Moeller et al., 2013) supports family-centered practice as the best approach to facilitating enriching developmental experiences in the early years. One of the best practice principles outlined in the statement is that service providers and families work together to establish optimal learning environments. This implies that families learn to enhance their child's communication and developmental skills through natural techniques during routine daily experiences. The risk for families, and by default for the child, is well recognized in the Joint Committee on Infant Hearing (2007) guidelines that recommend that all families of children with any degree of hearing loss should be considered eligible for early intervention support. This recognizes the prominent role the family plays in fostering the child's overall neurocognitive development and that rehabilitation for a young child cannot be considered separately from the family context.

One of the greatest challenges for families in the early months after diagnosis is achieving full-time use of hearing technology (Moeller et al., 2009), an outcome considered so necessary to maximizing auditory input, language, and early cognitive skills. Particularly vulnerable groups are children with mild and unilateral hearing loss, with studies showing 40–50% not achieving consistent use of amplification devices (MacKay et al., 2008; Fitzpatrick et al., 2010).

To conclude, it is well accepted that the trajectory for language and literacy is established in infancy. That is, quality stimulation in the early years is requisite for laying the groundwork for later cognitive, linguistic, literacy, academic, and social competence (Hart and Risley, 1995; Mustard, 2006). For example, in a landmark longitudinal study of families and children from different socioeconomic strata, Hart and Risley demonstrated the incredible impact that the quantity of language exposure has on children's verbal skills. Embellished experiences are likely even more critical for children with congenital hearing loss who do not have a "buffer." Recognition of the neurophysiologic and range of cognitive benefits from enriched input in the early years has provided a compelling argument for investments in early identification of hearing loss to facilitate early access to auditory and linguistic stimulation.

CURRENT CONTEXT FOR CHILDREN WITH HEARING LOSS

Two key developments in the last two decades are largely credited with dramatically transforming neurodevelopmental outcomes and expectations for future progress for children with congenital hearing loss: early intervention in the first few months of life and the availability of improved hearing technology, particularly cochlear implants for children with severe to profound hearing loss.

Neonatal hearing screening

Early intervention has become possible because of worldwide UNHS initiatives. This movement has been fueled by comparison studies that showed improved developmental outcomes in children who benefitted from early detection (Yoshinaga-Itano et al., 1998a; Moeller, 2000). Until the 1990s, the average age of identification of hearing loss was 2–2.5 years (Durieux-Smith et al., 2008), with an inverse relationship between severity of loss and age of identification. Developments in screening technologies, notably otoacoustic emissions testing and automation of auditory brainstem response testing, contributed to the feasibility, accuracy, and affordability of population-based screening. These events, coupled with convincing evidence from neuroscience about the effects of auditory deprivation on the developing brain, have significantly changed the landscape of deafness.

The overarching goal of screening as defined by the Joint Committee on Infant Hearing (2007) is to reduce or prevent delays in development. The recommended timelines are screening by 1 month of age, identification by 3 months, and intervention by 6 months. While there is good evidence that UNHS has significantly lowered the age of detection for children with moderate and greater hearing losses (Nelson et al., 2008), documenting evidence that early detection translates to improved developmental outcomes, as discussed in a subsequent section, is more tenuous (Nelson et al., 2008; Colgan et al., 2012).

Cochlear implantation

The second key development, pediatric cochlear implantation, has unequivocally dramatically improved the amount and quality of auditory information available to children with severe to profound hearing loss (Thoutenhoofd et al., 2005; Bond et al., 2009). Since first approved for children in the United States in 1990, pediatric cochlear implantation has become standard care for this population of children. While in the first decade unilateral cochlear implantation was common practice, there has been a trend toward bilateral implants in recent years (Papsin and Gordon, 2008). As will be discussed in subsequent sections of this chapter, cochlear implantation has provided a unique opportunity to examine neurodevelopmental outcomes in children who previously had little or no auditory stimulation.

Since the widespread use of cochlear implants, degree of hearing loss is no longer a barrier to accessing spoken language, even for congenital deafness. Clinical candidacy criteria have expanded to include less severe hearing loss (Fitzpatrick et al., 2009; Leigh et al., 2011) and younger ages at implantation (Dettman et al., 2007). Several researchers have equated hearing via cochlear implants to be roughly equivalent to hearing by children with moderate or moderately severe hearing loss using hearing aids (Sarant et al., 2009; Spencer and Marschark, 2010). However, cochlear implantation does not constitute a cure for hearing loss and it is well recognized that even children who undergo this intervention very early require specialized rehabilitation therapy and that families require ongoing guidance to optimize neurocognitive development.

Expectations

A third but and perhaps less perceptible shift shaping the current context is professionals' (e.g., otolaryngologists', audiologists' and language, and other health and education specialists') expectations that auditory speech reception and spoken language development are now a possibility for the overwhelming majority of children with congenital hearing loss, irrespective of degree of hearing loss. These attitudes directly and indirectly influence environmental factors, such as policies, funding, and type of intervention offered. There is now an "investment" in making speech accessible to all children from an early age if parents choose hearing technology options. Early intervention for children with hearing loss encompass comprehensive services that include audiologic, medical, language, cognitive, and social interventions that aim to improve the development of the child and family (Jerger et al., 2001; Joint Committee on Infant Hearing, 2007). Despite greater and easier access to "hearing" than ever before, hearing has not been restored.

Summary

To summarize, in the current context of neonatal screening and early intervention, there is a consensus that:

- Intervention should begin as early as possible to take advantage of neuroplasticity of the auditory system and to provide optimal stimulation synchrony with the typical language acquisition period in the first 3 years of life.
- Hearing technology alone is insufficient for most children and children need specialized intervention.
- Families need specific guidance to provide optimal stimulation, and therefore intervention should be family-centered to provide children with embellished experiences for language learning.
- Increasingly, children with normal cognitive abilities who receive early technology and neurocognitive rehabilitation have the potential to develop language skills comparable to those of their hearing peers.

FACTORS AFFECTING NEUROCOGNITIVE DEVELOPMENT

Figure 19.4 provides a framework (Fitzpatrick, 2010), adapted from the World Health Organization (WHO) *International Classification of Functioning and Disability* (World Health Organization, 2001), of the multitude of factors potentially affecting neurocognitive development in children with hearing loss. In the WHO model, body functions and structure (e.g., hearing loss) affect an individual's outcomes that in turn impact participation in society. Outcomes in this expanded model include not only neurodevelopmental outcomes such as communication and cognition, but also process and family outcomes, all of which have emerged in research with families as important aspects to consider (Tattersall and Young, 2006; Fitzpatrick et al., 2007b; Nelson et al., 2008). Consistent with the WHO model, multiple

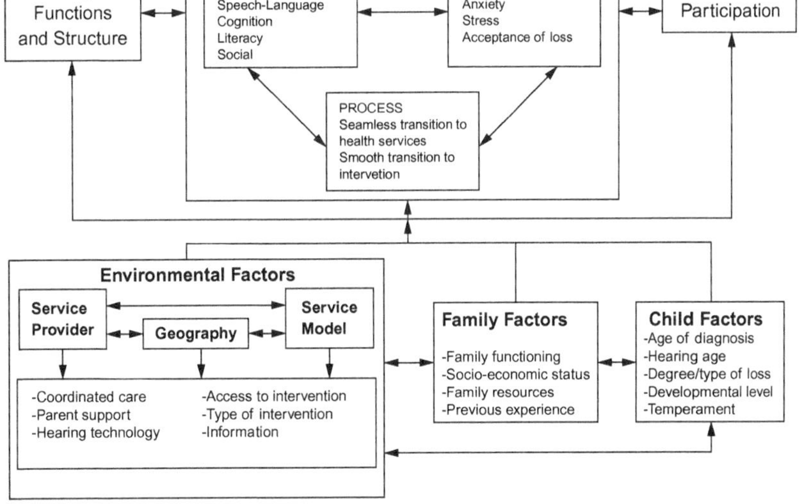

Fig. 19.4. Conceptual framework of outcomes and factors related to development in children with hearing loss. (Adapted with permission from Fitzpatrick, 2010.)

environmental and personal (both family and child) factors interact to shape the child's experience and affect developmental outcomes.

Approaches to neurodevelopmental rehabilitation

Despite the important gains in research on language acquisition in children with hearing loss, particularly since the availability of cochlear implants, many outstanding questions remain about how spoken communication develops, the most effective intervention methods for facilitating language development, and the contribution of various factors to language outcomes. Overall, there is considerable optimism about neurocognitive development opportunities for the current generation of children with early-identified hearing disorders. Since most children now have the potential to develop spoken language, the subsequent sections will focus on neurocognitive development where spoken language is the goal of intervention. Historically, there has been no consensus on how best to achieve desired outcomes in children with hearing loss whose parents choose spoken language. Intervention methods range from auditory-based approaches such as auditory-verbal practice, focused on listening and inclusion of the child into regular education programs, to programs that include sign language to support spoken language development (Durieux-Smith and Fitzpatrick, 2011). Despite a long history of approaches and techniques, the evidence supporting the effectiveness of any particular intervention approach is relatively weak due to the lack of good-quality comparative studies (Schachter et al., 2002; Spencer and Marschark, 2010).

AUDITION AND SPOKEN LANGUAGE

The overwhelming focus of research on neurocognitive development is related to speech recognition (e.g., identification of words, sentences) and language outcomes, not a surprising finding given the important and marked impact of hearing disorders on early auditory development and subsequent spoken-language skills. Since the 1990s, much of pediatric hearing outcomes research has been divided into two somewhat distinct categories determined by the type of hearing technology used, hearing aids or cochlear implants, which roughly corresponds to mild to severe and severe to profound hearing loss. There has been a proliferation of research documenting the effectiveness of cochlear implants. In the early years, considerable research was focused on speech recognition measures such as the percentage of phonemes, words, or sentences correctly identified, as the best proxy of hearing restoration with a cochlear implant. This specific focus on auditory benefit was driven by a need to demonstrate the efficacy and effectiveness of a new biomedical device. With convincing evidence that even children with congenital and early-onset profound deafness could process information through audition alone, researchers also became interested in examining the benefits of this new intervention for language development. More recently, as described in subsequent sections, there has been increasing interest in examining other aspects of development beyond hearing and language and in investigating other neurocognitive functions to better understand variability in outcomes and to provide children who use cochlear implants with more customized interventions (Peterson et al., 2010).

Children with mild to severe degrees of hearing loss

Children with mild to severe degrees of hearing loss account for 70% or more of clinical populations with bilateral hearing loss. Clinical practices for children with moderate to severe loss are well defined (Bagatto et al., 2010; King, 2010) and they are typically fitted with acoustic hearing aids to improve their access to speech and environmental sounds. Although hearing aids and specialized systems that facilitate listening in difficult listening situations such as the classroom have less dramatically impacted auditory and speech-language outcomes than cochlear implants, technologic advances in hearing devices have occurred and pediatric-focused techniques have been developed (Bagatto et al., 2011).

Moeller et al. (2007) called attention to the lack of research on this new generation of children who do not use cochlear implants but who have benefitted from earlier intervention and modern hearing technologies. In a series of papers in a special issue, these authors highlighted that hearing loss across the spectrum of severity affects children's development in multiple domains. A review of language outcomes at that time for this population of children found that most of the literature pertained to children who were late identified, making it difficult to define realistic expectations. Similarly, although there is an extensive literature on literacy development in children with severe to profound hearing impairment, again Moeller et al. (2007) found very limited information, but the few studies suggested that children lag behind their hearing peers.

A number of studies that have investigated outcomes in this population in recent years combine results for less severe children with those for children with cochlear implants, making it difficult to draw conclusions specifically about the group with less severe loss. Data from 51 children with congenital or assumed congenital losses recruited from three Canadian auditory-verbal clinics

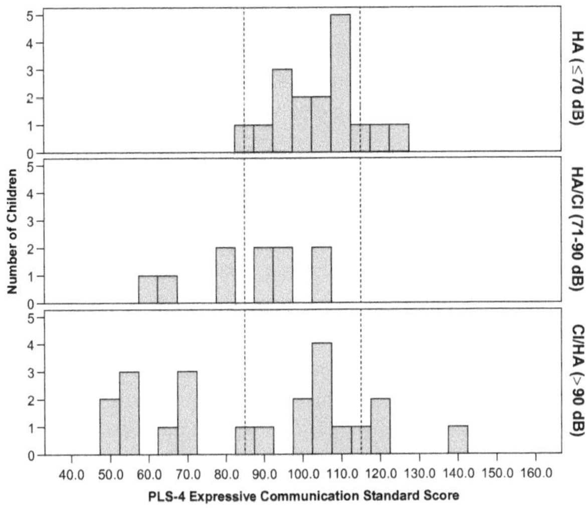

Fig. 19.5. Distribution of expressive language scores (Preschool Language Scale (PLS), Expressive Communication) for children according to severity of hearing loss. Dotted lines indicate scores within 1 SD of test normative data. HA, hearing aids; CI, cochlear implant. (Reproduced from Fitzpatrick et al., 2011a.)

(focus on developing spoken language through audition) showed that 94–100% of children with hearing loss <70 dB obtained scores on various standardized speech-language tests in the average range for children with normal hearing. An example of results recorded for expressive communication (test mean 100, SD 15) is shown in Figure 19.5. However, children in the 70–90-dB range with hearing aids and those >90 dB with cochlear implants showed more variable results than their hearing peers (Fitzpatrick et al., 2011a). Overall, 65–85% (depending on the test measure) of the 51 children obtained scores within 1 SD of their normal-hearing peers. Figure 19.6 provides an example of clinical language comprehension results for 22 children identified at a mean age of 13.4 months followed in one family-centered rehabilitation program with a strong emphasis on auditory development. As shown, average group scores at 3, 4, and 5 years of age were within 1 SD of the normative mean (mean 100, SD 15 for this measure). An examination of individual results for the subgroup of 10 children with known congenital loss shows considerable individual variability in outcomes over the three test intervals (Fig. 19.7). This example serves to illustrate the patterns of language development that can be observed in children with hearing loss. Children who are following a typical developmental trajectory would be expected to obtain average test scores (mean 100, SD 15 for the measure shown) at each age. In contrast, these children show a wide range of trajectories, with most lagging behind normal-hearing peers at age 3 years, but eventually catching up.

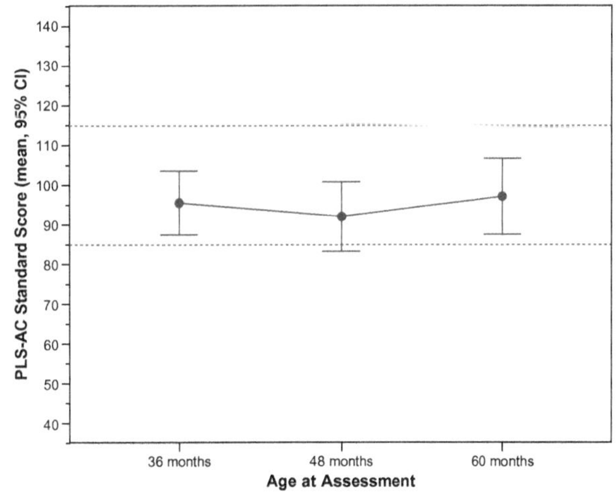

Fig. 19.6. Mean scores for language comprehension (Preschool Language Scale, Auditory Comprehension (PLS-AC)) at age 3, 4, and 5 years. Dotted lines indicate scores within 1 SD of test normative data.

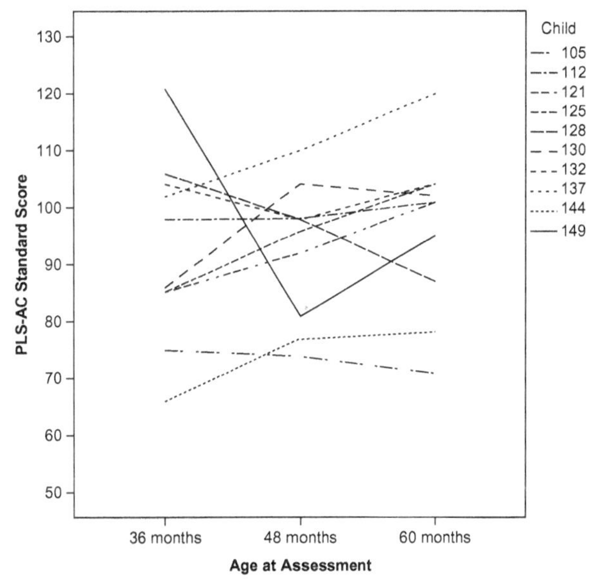

Fig. 19.7. Individual profiles in language comprehension (Preschool Language Scale, Auditory Comprehension (PLS-AC)) for 10 children with congenital hearing loss at age 3, 4, and 5 years.

Recent reports on early-identified children are beginning to contribute further understanding of the difficulties related to hearing loss of lesser severity. Koehlinger et al. (2013) examined grammatic aspects of language for 100 children with mild to severe hearing loss, 60 assessed at age 3, and 40 at age 6 recruited through Early Hearing Detection and Intervention programs in the United States. Results were compared to 23 (age 3) and 17 (age 6) normal-hearing children. The authors reported that, on average, children lagged behind normal-hearing peers based on analysis of mean length of utterance (MLU) and percent correct of morphologic markers for verbs (e.g., use of past tense -ed, plurals -s) from spontaneous language samples. MLU is well accepted as a robust marker of early language acquisition (Brown, 1973). More than 50% of the children achieved below the 25th percentile for typically developing children. Similar gaps in grammatic development were documented for 3- and 6-year-olds compared to normal-hearing children, leading to concerns that these children are at risk for later difficulties.

In an earlier study examining vocalizations of young children with hearing loss, 24-month-olds were found to produce significantly fewer words than normal-hearing children. Language samples of the children with hearing loss were described as comprised of primarily unintelligible utterances compared to normal-hearing 24-month olds who used words and phrases (Moeller et al., 2007). Taken together, these studies suggest that age-appropriate spoken-language skills in children with mild to severe hearing loss, while attainable, cannot be taken for granted and, despite early intervention, children require significant focused rehabilitation.

Additional information can be gleaned from studies in the last decade that have examined outcomes in early- versus late-identified children. These studies are shedding more light on developmental outcomes across the spectrum of hearing loss and confirm that congenital and early-onset deafness negatively impact children's development in multiple domains. Interestingly, while age at identification was a dominant factor in early newborn screening outcome studies (Yoshinaga-Itano et al., 1998a, 2001), the specific contribution of age at identification or intervention has become less clear. In the last decade, age of identification has been associated with some areas of development measured, but only in some studies (Kennedy et al., 2006; Sininger et al., 2010), but has not emerged as a significant predictor in other studies (Fitzpatrick et al., 2007a; Korver et al., 2010; Ching et al., 2013). A second notable point emerging from an examination of several of these studies is that severity of hearing loss has a significant effect on several language development outcomes, such as vocabulary, and receptive and expressive language (Wake et al., 2004; Fitzpatrick et al., 2007a; Sininger et al., 2010; Ching et al., 2013).

A third important finding from these early-identification studies is that there are a substantial number of children who do not achieve language levels aligned with their hearing peers (Ching et al., 2013). It is common practice in most studies to exclude children with additional disabilities that would be expected to interfere with typical language development. Population-level research in childhood hearing loss has a relatively recent history. In the largest population-based study to date of 451 children, Ching et al. (2013) found that the presence of additional disabilities, severity of hearing loss, and maternal education, but not age at hearing-aid fitting, predicted spoken-language development at 3 years of age. Age at cochlear implant activation was however associated with better developmental outcomes in 134 children with cochlear implants; by age 3 years, cochlear implants were used by most children with hearing loss greater than 75 dB HL. On average, this population of children had global outcome scores (comprised of auditory function, receptive and expressive language, and speech production measures) 1 SD below the normative mean even after adjusting for additional disabilities. On individual test measures, including a social functional measure, children's abilities were at or below 1 SD of the mean. As noted previously, about 30–40% of children with permanent hearing loss are affected with additional disabilities that can reduce the auditory and language gains from hearing technology (Edwards, 2007; Meinzen-Derr et al., 2011; Eze et al., 2013). Lower non-verbal cognitive abilities are associated with poorer language skills (Sarant et al., 2009).

Based more on expert opinion rather than empiric evidence, many language specialists working with children stress the need for continued intervention even for children whose standardized test scores suggest their skills are aligned with those of hearing peers (Fitzpatrick and Doucet, 2012). In other words, individuals working with these children on a day-to-day basis in real-world environments judge them to be at risk for difficulties in terms of auditory and linguistic skills and observe that they require ongoing practice to maintain and solidify these skills, to access advanced academic concepts, and to foster social development (Fitzpatrick and Olds, 2014).

Children with mild bilateral and unilateral hearing loss

Children with mild bilateral and unilateral hearing loss represent a subgroup of children with hearing loss that merit a separate discussion point because: (1) they are frequently not a targeted group for early identification; (2) they receive amplification later, even when early identified; and (3) the effect of early-identified hearing loss of this degree is largely unknown. Children with mild bilateral or unilateral hearing loss have historically been identified late, often at school age, when the effects of reduced hearing access may become more evident in noisy environments (Fitzpatrick et al., 2014). Studies of late-identified children with milder losses showed that they were not immune to the effects of hearing disorders on development and that they were at risk for language,

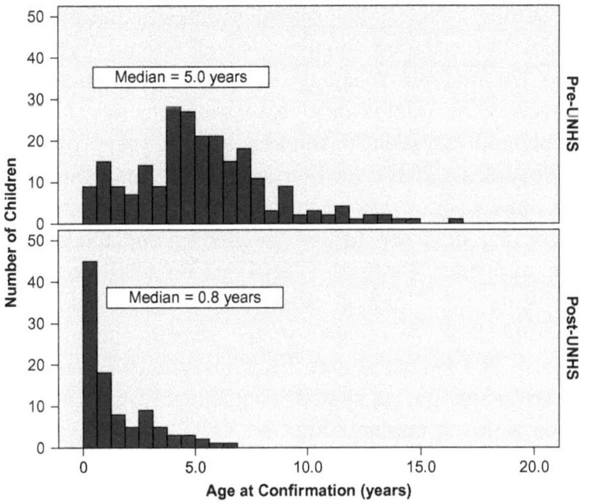

Fig. 19.8. Age at identification of mild bilateral and unilateral hearing loss the year before ($n = 244$) and after universal newborn hearing screening (UNHS) ($n = 102$).

cognitive, and behavioral difficulties (Tharpe, 2008; Porter and Bess, 2011). Figure 19.8 shows the dramatic difference in age at identification for this particular population when milder degrees of hearing loss <40 dB are targeted through newborn screening initiatives. However, the benefits of early identification related to neurocognitive development in these children are still unclear and best practices, including decisions about fitting hearing technology, are not well established. This lack of information, coupled with the technical difficulties and costs related to lowering the degree of the targeted disorder, have likely prompted many programs throughout the world to establish >40 dB hearing loss as the cut-off for newborn hearing screening programs.

There is some evidence that even early-identified children in this category may experience difficulties compared to their hearing peers. For example, up to one-third of children have been reported to experience some degree of language delay in the early years (Yoshinaga-Itano et al., 2008). Children with unilateral hearing loss have been reported to have poorer language scores than their siblings and to be receiving speech-language services (Lieu et al., 2010), and are considered to be at risk for later difficulties. Furthermore, recent examination of newborn screening data showed that 32% of children with mild bilateral or unilateral loss experienced deterioration in hearing from diagnosis to later years and that 11% with unilateral loss progressed to bilateral loss (Fitzpatrick et al., 2014). There is considerable uncertainty about best practices in managing these children's losses through technology. Data show that approximately 90% eventually receive amplification but that there is considerable uncertainty around when to fit technology such that time from identification to amplification is much longer than for children with moderate or greater losses. These results suggest that questions remain about the effects of this degree of hearing loss on neurocognitive development and whether amplification can prevent difficulties observed in the past.

Children with cochlear implants

As noted, there is now a vast body of literature on children with severe to profound hearing loss who have received cochlear implants over the last two decades. Overall, research indicates that children with cochlear implants develop better spoken language and with substantially greater ease than previous generations of children with profound loss who used hearing aids (Geers et al., 2008; Geers and Nicholas, 2013; Tobey et al., 2013). Convincing evidence has accumulated to suggest that, while cochlear implants provide only partial access to auditory cues important for speech processing, children develop high-level spoken language skills, some achieving scores on standardized language measures comparable to peers with normal hearing (Nicholas and Geers, 2007; Fitzpatrick et al., 2011a; Leigh et al., 2013). However, overall, findings from large studies show that on average their language skills are not comparable to their hearing peers (Niparko et al., 2010; Ching et al., 2013; Tobey et al., 2013) and delays persist throughout the school years (Geers and Nicholas, 2013). Common to most pediatric cochlear implantation outcomes are two key points: (1) average development in multiple language domains tends to show gaps compared to normal-hearing children; and (2) there is great variability in performance on language tests.

The important and unexplained variability in results has led researchers to examine factors predictive of neurodevelopmental outcomes. Key factors associated with differences reported in the vast cochlear implant literature include:

- age of identification of hearing loss
- age at implantation
- amount of "hearing" experience before implant
- type of rehabilitation/communication
- language experience and family involvement.

The influence receiving the greatest attention has been age at implantation. Providing auditory input during the sensitive language development period for children with normal hearing is associated with better language outcomes (Ching et al., 2013; Tobey et al., 2013). However, age at implantation did not emerge as a predictor in one of the largest cochlear implants studies of 181 8- and 9-year-olds who received implants prior to age 5. Reports on whether early implantation effects are lasting are mixed. For example, Geers and Nicholas

(2013) found that the effects of early implantation on language scores recorded at 4.5 years were maintained at 10.5 years. In contrast, Dunn et al. (2014) reported that the positive effects of age of implantation in preschool were no longer significant for receptive and expressive language at ages 7–10 years, depending on the outcome measured.

The concept of a sensitive period and promising outcomes from early implantation has led some investigators to advocate for early cochlear implantation before the Food and Drug Administration approved 12-month lower age-limit criteria. There is accumulating evidence that children implanted before their first birthday can attain typical developmental milestones in speech-language. Studies show advantages in language and speech production for children implanted before 1 year of age compared to those implanted between 1 and 2 years (Holman et al., 2013; Leigh et al., 2013). These findings have resulted in expectations that very early-implanted children can develop language skills equivalent to children with normal hearing.

However, Hof et al. (2013) have raised a note of caution regarding overly aggressive cochlear implant management of preterm infants during the first year of life. In a study of 14 neonates who were diagnosed with hearing loss ranging from 40 to 105 dB after referral from newborn screening, 9 of 14 infants' hearing improved (to normal hearing for 4 and to moderate loss for 5). These findings led the authors to raise concerns related to delayed maturation of the auditory system in this population and to recommend ongoing audiologic assessment until at least 80 weeks' gestational age before proceeding with surgical decisions.

Reasons for late implantation

While age at implantation is widely advocated to take advantage of the neurodevelopmental window, a review of studies indicates that overall average age of implantation in developed countries is well beyond 1 year of age. For example, mean age at cochlear implant activation was 29 months in a large study of 160 children implanted between 6 months and 5 years of age (Tobey et al., 2013). An examination of implantation in a Canadian program since the availability of newborn screening indicated that, of 163 children implanted after the implementation of UNHS, 61.3% (100) received implants more than 12 months after identification of hearing loss. A review of the reasons detailed in Figure 19.9 indicated that the primary reason was progressive hearing, which was documented for 58% of children with late implantation.

Figure 19.9 provides an overview of the multiple reasons accounting for late implantation in children. The findings that children with additional and complex disabilities also receive implants later are consistent with other reports in the literature as the decision-making and counseling process for these children is frequently more complicated, given complex medical issues and the need to adjust expectations for language-related outcomes (Fitzpatrick et al., 2011b; McCracken and Turner, 2012). Of importance in this discussion is that later implantation does not necessarily equate to lack of auditory stimulation if children were receiving adequate benefit from hearing aids prior to deterioration of hearing. This information may help explain why, in some studies, age at implantation has not necessarily been associated with greater linguistic benefits. These findings further confirm the difficulty of disentangling the true effects of age of implantation and to what extent some amount of residual hearing may impact early speech-language development. They also point to the importance of carefully monitoring all children with hearing loss to decide at what point cochlear implant intervention offers the best option to maximize hearing potential.

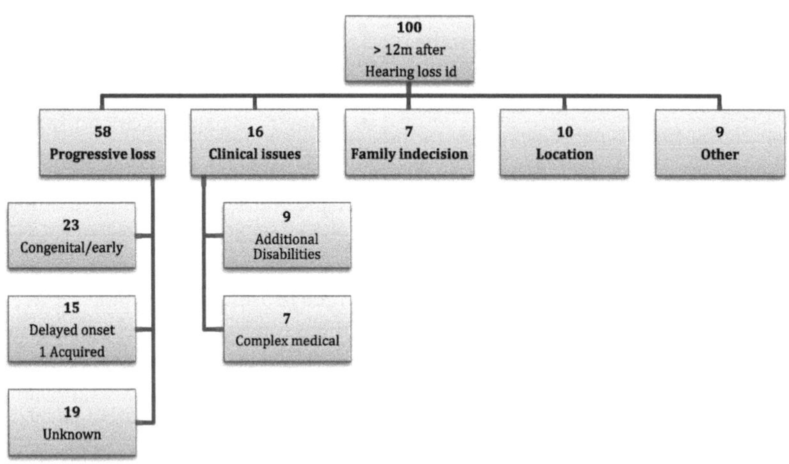

Fig. 19.9. Reasons for late implantation (>12 months after identification of hearing loss). id, identification.

Literacy development

Language acquisition is viewed as an important pathway to literacy, academics, social development, and employment. A comprehensive description of the impact of hearing loss on these areas of functioning is beyond the scope of this chapter. However, it is important to highlight that historically achievement levels in reading have shown important gaps, with reading skills of hearing peers with average grade 4 reading levels documented at high-school graduation (Babbidge, 1965; Traxler, 2000). It is well established in normal-hearing children that solid language abilities, phonologic awareness, and print awareness are associated with good reading skills (National Reading Panel, 2000; Robertson, 2014). Children with hearing loss appear to follow the same sequence of literacy development as normal-hearing children (Williams, 2004; Mayer, 2007). They are at risk for delayed reading skills because late or poor-quality access to hearing in the early years weakens these underlying language abilities. In particular, hearing difficulties can interfere with decoding skills, a fundamental skill for reading. Furthermore, difficulties with distance hearing and incidental learning interfere with the development of conceptual knowledge, known to be associated with reading comprehension (Robertson, 2014). In various studies, difficulties in phonologic processing, vocabulary, sentence structure, and conversation have been associated with reading delays in children with hearing disorders (Perfetti and Sandak, 2000; Moeller et al., 2007; Geers and Hayes, 2011). For an indepth discussion of literacy and hearing loss, the reader is referred to Robertson (2014).

COGNITION AND LEARNING

Starting in infancy, children develop cognitive abilities through a range of early sensory experiences, including visual, tactile, and auditory interactions with their surrounding environment. As described by Fagan and Pisoni (2009), typically developing infants receive multimodal sensory input through their everyday learning environments. As early as the first months of life, infants learn to assimilate various inputs to process information and to initiate the beginnings of expressive communication, for example, smiling, cooing, and later babbling. In contrast, in infants with hearing loss, the auditory component of multimodal processing may be seriously compromised or even absent in the case of severe and profound hearing loss prior to intervention. This can result in absent or reduced sound awareness (knowing sounds are present), sound association (for example, associating a sound pattern with mother's versus father's voice, associating sounds with environmental occurrences), sound localization, and delayed vocal behavior (for example, babbling at 6–9 months of age).

As described above, even children with congenital deafness, when provided with early access to sound through cochlear implantation, show auditory cortical responses similar to those of hearing children. However, little is known about the broader and long-term effects of early or extended periods of auditory deprivation on language and overall neurocognitive development. It is important to remember that not only are children receiving delayed access to sound but that the quality, and in some cases the quantity, of auditory input is not comparable to that experienced by normal-hearing infants. Language inexperience (relative to adults) means that, even in normal-hearing children, phonemic representation is not robust. Furthermore, they are particularly vulnerable when there is disruption of the auditory signal, for example in background noise (Neuman et al., 2010). At this point, it is not clear whether the gaps in development between normal-hearing children and those with hearing loss are related to environmental deficits and simply time lost that needs to be "made up," or whether there are neural changes caused by weaker early sensory experiences that have a lasting effect on other cognitive areas of development. In other words, are cognitive processes altered, due to later and poor-quality auditory stimulation and if so, are these changes amenable to treatment? New information on early neurocognitive development may help to better "customize" intervention and improve spoken language and overall development (Pisoni et al., 2008).

Researchers have attended to the connection between cognitive functioning and language for many years, deafness being a prime example of how these two concepts are interwoven (Marschark and Hauser, 2008; Spencer and Marschark, 2010). Differences in play behaviors, for example, have been documented in preschool children compared to children with hearing (Yoshinaga-Itano et al., 1998b; Spencer and Marschark, 2010). Lower levels of language functioning were felt to interfere with mothers' abilities to structure symbolic play and other cognitive abilities. It has been suggested that the additional time and visual input required by children with hearing loss to follow play interactions and sequences could interfere with typical cognitive development unless measures are taken to overcome them (Spencer and Marschark, 2010). This points to the importance of family-centered intervention and the impact of parents on the early development of the child with hearing loss.

Poor average academic achievements such as low reading levels reported historically for children who are deaf (Traxler, 2000) have undoubtedly motivated ongoing examination of cognitive functioning. Marked differences between students who are deaf and those who are hearing have been revealed in areas such as visual perception, memory, and problem solving

(Marschark and Hauser, 2008). However, Marschark and Hauser (2008) underscore that research has not had the expected contribution as it has been largely undertaken in a piecemeal approach in independent domains, including cognitive, social, and academic outcomes, limiting understanding of how these various aspects of learning are intertwined.

It is common knowledge that verbal cognitive measures frequently show lower scores, a result that is generally attributed to differences in language abilities rather than cognitive functioning (Spencer and Marschark, 2010). While it has been well recognized that deafness does not directly interfere with intelligence, based on non-verbal scores of cognitive abilities, Marschark and Hauser (2008) suggest that, regardless of whether children's primary communication mode has been visual or oral language, they may have different knowledge and learning experiences. As summarized by Marschark and Hauser (2008), there is some evidence from studies of children who are deaf and use sign language as their primary communication that they have conceptual knowledge that is less richly anchored than their hearing peers. In turn, these different experiences potentially result in cognitive differences such as visual perception, memory, and problem solving. For example, children with significant hearing loss have fewer opportunities for overhearing, incidental learning, following group conversations, and accessing information in a fast-paced classroom environment. They frequently require additional processing time to understand speech. In addition, as children mature, more of their learning necessarily takes place in degraded acoustic environments such as classrooms (Bradley and Sato, 2008; Klatte et al., 2010). Studies showing that normal-hearing children's listening comprehension is poorer in background speech or in babble than in other background noise (Klatte et al., 2010; Prodi et al., 2013) may be indicative that higher-level cognitive processes are affected and that cognitive and auditory processes are integrated. Children with hearing loss are more vulnerable in adverse listening conditions. Ignoring irrelevant speech likely involves even more cognitive resources and greater listening effort than in their normal-hearing peers. They may therefore have different learning experiences that in turn may lead to different cognitive organization and strategies in learning.

Cognitive functioning in children with cochlear implants

In the past decade, considerable attention has been accorded to the relationship between neurocognitive functions and language acquisition in children with cochlear implants. In particular, a team of researchers at the Indiana University School of Medicine have contributed an extensive body of research to this field (Pisoni et al., 2008). It seems reasonable to assume that children with various degrees of hearing loss severity are affected by many of the same factors as those who use cochlear implants. However, the tremendous difference between deaf and "hearing" that seemed to result from cochlear implants and the attention captured in the medical community have likely triggered a strong interest in this population. Researchers have used the opportunities presented through cochlear implantation to further study neurocognitive processes such as working memory and attention. New investigations and perspectives appear to be largely fueled by the incredible variability observed in both auditory and language-based outcomes documented in children with cochlear implants (Peterson et al., 2010). Pisoni et al. (2008, p. 54) describe the typical speech recognition and language outcomes collected in cochlear implant studies as the "endpoint of a long series of neural and neurocognitive processes." Achieving a more integrated understanding of these differences might in turn assist in shaping more focused treatments for children.

Variability in language development in children with normal hearing and understanding its causes and consequences are central issues in the field of language disorders more generally (Marchman and Fernald, 2013). For example, early processing speed of familiar words by 2-year-old children has been found to be significantly related to early receptive and expressive vocabulary skills at age 2 (Fernald et al., 2006) and to be predictive of long-term cognitive (working memory) and language skills at 8 years of age (Marchman and Fernald, 2008). These and other findings suggest that differences in early speech processing capacities may have implications for later language development (Marchman and Fernald, 2013). Although the source of this variability is unknown, the investigators propose there is some evidence to suggest that early language experience may be at least partly responsible. Given that children with hearing loss have later and potentially lower-quality early language experiences, they are particularly at risk.

As cochlear implantation has become standard care, there has been increasing interest in learning more about cognitive correlates that affect information processing that in turn is relevant to language learning. Research in children with normal hearing has established the link between cognitive processes such as working memory and linguistic-related tasks, including word recognition, vocabulary and language development (Gathercole and Baddeley, 1990; Gathercole et al., 1997). There is accumulating evidence in cochlear implant studies to suggest that deafness, weaker auditory experiences, and possibly disrupted early learning experiences affect neurocognitive functions in addition to spoken language skills (Pisoni et al., 2008).

For example, using digit span forward and digit span backward tests to measure aspects of working memory in school-age children with cochlear implants, researchers have shown that children with cochlear implants have shorter memory and working memory skills compared to children with normal hearing (Pisoni et al., 2008). One of the early studies more than a decade ago investigated memory capacity in 176 children with cochlear implants who were part of a large study investigating multiple outcomes of speech, language, and reading in 8- and 9-year-old children (Pisoni and Geers, 2001) implanted before age 5. Significantly shorter digit spans were measured for the children with cochlear implants compared to normal-hearing age-matched peers. In the same study, verbal rehearsal speed (speaking rate) was found to account for much of the variability in speech and language scores in the cochlear implant group. Several subsequent studies described in various reports (Peterson et al., 2010; Houston et al., 2012) provided further evidence for the interference of deafness in neurocognitive processes such as working memory, verbal rehearsal speech, and perceptual encoding, suggesting that a period of auditory deprivation and delayed language interfered with the development of foundation cognitive processes which in turn affected children's speech perception, language development, and learning after cochlear implantation.

Researchers in this area of study propose that these foundation neurocognitive functions are a prime area for targeted intervention as part of rehabilitation programs. For further details concerning the large body of research, readers are referred to research summaries from the University of Indiana group (Pisoni et al., 2008; Peterson et al., 2010; Houston et al., 2012).

Overall, this promising area of research suggests that auditory and spoken-language development relies heavily on neurocognitive skills and that these underlying skills may account for at least some of the variability measured in traditional cochlear implant outcomes. What is unknown and difficult to disentangle from heterogeneous groups of children with cochlear implants with different onsets of profound loss and different ages at implantation is the extent to which these differences are related to sensory deprivation prior to cochlear implantation. Listening and learning with degraded input through electric or acoustic or combined stimulation modes and exposure to even more degraded signals in classroom learning situations with auditory stressors such as reverberation, noise, and distance may also contribute to these differences. As evidence accumulates related to children who have benefitted from early identification and recent technologies, a more integrated understanding of the neurodevelopmental effects of congenital deafness on these developmental areas should emerge. Sensory deprivation and less than optimal quality of sensory input when it becomes available seem to have long-lasting consequences for cognitive, language, and social development. To date, substantial evidence has been collected on children who use cochlear implants. Given that children with severe hearing loss now receive implants (Fitzpatrick et al., 2009; Ching et al., 2013), it seems reasonable to assume that children with less severe hearing loss and hearing aids experience some of the same difficulties.

Theory of mind and social development

Another area of early cognitive development that has received some attention is theory of mind (ToM). Considered an important indicator of cognitive development in the preschool years, ToM refers to the ability to move beyond one's own beliefs and realize that others have different desires and perspectives (Nelson, 2007). It has been described as a social-cognitive achievement that is observed in typically developing children by about 4–5 years of age. In other words, young children through the course of normal interactions learn to attach meaning to what other people around them feel and think. Also referred to as "mentalizing" in typically developing children, ToM is considered an important metric that requires a rich language context. The ability to recognize other perspectives broadens opportunities for social interactions and is one of the core constructs that allow children to participate in their cultural world (Nelson et al., 1998; Nelson, 2007). Children's mastery of mental state tasks is related to the quantity of experience with mental state conversations; in other words, how much parents and siblings use advanced levels of cognitive talk with them (Perner et al., 1994).

Studies have shown that children with hearing loss of hearing parents are delayed relative to typically developing children in developing ToM abilities (Moeller and Schick, 2006; Schick et al., 2007). In contrast, the development of this mental state in children who have deaf parents appears to be more closely aligned with that of normal-hearing peers (Courtin, 2000; Schick et al., 2007). These differences between children with and without hearing and between children with deaf parents have been attributed to limited language delays and reasoning skills and to less exposure to talk about mental states (Moeller and Schick, 2006; Schick et al., 2007). Children who sign and have deaf parents are richly immersed in conversations in the course of their natural environments.

Moeller and Schick (2006) examined the extent to which 28 hearing mothers with children with hearing loss (using sign and spoken language) and 26 with normal-hearing children converse with their children about mental states. This study supported the notion that mothers of children with normal hearing used a greater quantity of and more diverse mental terms than hearing mothers of children with hearing loss. Better use of mental state verbs by mothers was positively associated with children's ToM development scores. Therefore, children of hearing parents who are developing either spoken or sign language or some combination of both are at risk of having fewer opportunities to naturally learn these skills (Moeller and Schick, 2006). Children who are learning sign are often exposed to less fluent and less confident sign language users. Children with hearing loss learning spoken language appear to be disadvantaged because of missed opportunities to hear at a distance, overhear, or "drop in" on conversations, all of which can limit social interactions. Given that incidental learning and social interactions are more limited, they likely have fewer opportunities to naturally engage in or learn about mental state conversations when developing spoken language.

SUMMARY

Large national and population-based studies are beginning to provide a clearer snapshot of broader developmental outcomes in children with hearing loss. Smaller clinical studies add useful context-specific information that may help sort out the myriad of factors impacting neurocognitive development. Despite the growing evidence that early interventions and improved technology have improved neurocognitive developmental outcomes in children with hearing loss, there remain considerable language gaps. Although there is a large body of literature substantiating the effectiveness of cochlear implantation in developing auditory and language functions, many children still do not achieve language proficiency aligned with their hearing peers. Much remains to be learned in relation to underlying cognitive skills that intersect with language development for children across the entire spectrum of hearing disorders.

Early intervention has two principal goals: to reduce and prevent, if possible, irreversible damage or disruption to the typically developing auditory system. To date, the bulk of the evidence on development suggests that early hearing access through timely detection and advanced hearing technologies has made important progress in reducing the impact of congenital hearing loss. With ongoing technologic advances coupled with increased understanding from neuroscience, it seems reasonable to expect that preventing difficulties in language acquisition is a reasonable goal.

ACKNOWLEDGMENTS

Preparation of this chapter and research reported were partly supported by a Canadian Institutes of Health Research New Investigator award and a Canadian Child Health Clinician Scientist Career Enhancement award. The author is grateful to Julia Ham and JoAnne Whittingham for assistance with compiling data and figures.

References

Arshavsky YI (2009). Two functions of early language experience. Brain Res Rev 2009: 327–340.

Babbidge H (1965). Education of the deaf. A report to the Secretary of Health, Education, and Welfare by his Advisory Committee on the Education of the Deaf. Ref. No. 0-765-119, Government Printing Office, Washington, DC.

Bagatto M, Scollie SD, Hyde M et al. (2010). Protocol for the provision of amplification within the Ontario Infant hearing program. Int J Audiol 49: S70–S79.

Bagatto MP, Moodie ST, Malandrino AC et al. (2011). The University of Western Ontario Pediatric Audiological Monitoring Protocol (UWO PedAMP). Trends Amplif 15: 57–76.

Berlin CI, Morlet T, Hood LJ (2003). Auditory neuropathy/dys-synchrony: its diagnosis and management. Pediatr Clin North Am 50: 331–340.

Bond M, Elston J, Mealing S et al. (2009). Effectiveness of multi-channel unilateral cochlear implants for profoundly deaf children: a systematic review. Clin Otolaryngol 34: 199–211.

Bradley JS, Sato H (2008). The intelligibility of speech in elementary school classrooms. J Acoust Soc Am 123: 2078–2086.

Brown R (1973). A First Language: the early stages, Harvard University Press, Cambridge, MA.

Calderon R, Greenberg M (2003). Social and emotional development of deaf children. In: M Marschak, P Spencer (Eds.), The Oxford Handbook of Deaf Studies, Language and Education, Oxford University Press, New York, pp. 177–189.

Ching TY, Dillon H, Marnane V et al. (2013). Outcomes of early- and late-identified children at 3 years of age: findings from a prospective population-based study. Ear and Hear 34: 535–552.

Chomsky N (1986). Knowledge and Language: its nature, origin, and use, praeger, New York.

Cole EB (1992). Listening and Talking: a guide to promoting spoken language in young hearing-impaired children. Alexander Graham Bell Association for the Deaf, Washington, DC.

Cole EB, Flexer C (2011). Children with Hearing Loss: developing listening and talking, plural publishing, San Diego, CA, Birth to Six.

Colgan S, Gold L, Wirth K et al. (2012). The cost-effectiveness of universal newborn screening for bilateral permanent congenital hearing impairment: systematic review. Acad Pediatr 171–180.

Courtin C (2000). The impact of sign language on the cognitive development of deaf children: the case of theories of mind. J Deaf Stud Deaf Educ 5: 266–276.

Curtis S (1977). Genie: a psycholinguistic study of a modern-day 'wild child'. Academic Press, New York.

Desjardin JL, Eisenberg LS (2007). Maternal contributions: supporting language development in young children with cochlear implants. Ear Hear 28: 456–469.

Dettman SJ, Pinder D, Briggs RJ et al. (2007). Communication development in children who receive the cochlear implant younger than 12 months: risks versus benefits. Ear Hear 28: 11S–18S.

Dunn CC, Walker EA, Oleson J et al. (2014). Longitudinal speech perception and language performance in pediatric cochlear implant users: the effect of age at implantation. Ear Hear 35: 148–160.

Durieux-Smith A, Fitzpatrick EM (2011). History of the management of hearing loss in children. In: R Seewald, A-M Tharpe AM (Eds.), Comprehensive Handbook of Pediatric Audiology, Plural Publishing, San Diego, CA, pp. 617–629.

Durieux-Smith A, Whittingham J (2000). The rationale for neonatal hearing screening. J Speech-Lang Pathol Audiol 24: 59–67.

Durieux-Smith A, Fitzpatrick E, Whittingham J (2008). Universal newborn hearing screening: a question of evidence. Int J Audiol 47: 1–10.

Edwards LC (2007). Children with cochlear implants and complex needs: a review of outcome research and psychological practice. J Deaf Studies Deaf Educ 12: 258–268.

English K (2011). Family informational and support counseling. In: R Seewald, A-M Tharpe (Eds.), Comprehensive Handbook of Pediatric Audiology, Plural Publishing, San Diego, CA, pp. 767–776.

Eze N, Ofo E, Jiang D et al. (2013). Systematic review of cochlear implantation in children with developmental disability. Otol Neurotol 34: 1385–1393.

Fagan MK, Pisoni DB (2009). Perspectives on multisensory experience and cognitive development in infants with cochlear implants. Scand J Psychol 50: 457–462.

Fernald A, Perfors A, Marchman VA (2006). Picking up speed in understanding: speech processing efficiency and vocabularly growth across the second year. Dev Psychol 42: 98–116.

Fewell RR, Casal SG, Glick MP et al. (1996). Maternal education and maternal responsiveness as predictors of play competence in low birth weight, premature infants: a preliminary report. Dev Behav Pediatr 17: 100–104.

Fitzpatrick E (2010). A framework for research and practice in infant hearing. Can J Speech-Langu Pathol Audiol 34: 25–32.

Fitzpatrick EM, Doucet SP (2012). When should children be discharged from an auditory-verbal program. In: W Estabrooks (Ed.), 101 Frequenty Asked Questions About Auditory-verbal Practice, AG Bell Association, Washington, DC.

Fitzpatrick EM, Doucet SD (2013). Pediatric Audiologic Rehabilitation: from infancy to adolescence, Thieme Medical Publishers, New York.

Fitzpatrick EM, Olds J (2014). Practitioners' perspectives on the functioning of school age children with cochlear implants. Cochlear Implants Int 35 (1): 10–18.

Fitzpatrick E, Durieux-Smith A, Eriks-Brophy A et al. (2007a). The impact of newborn hearing screening on communication development. J Med Screen 14: 123–131.

Fitzpatrick E, Graham ID, Durieux-Smith A et al. (2007b). Parents' perspectives on the impact of the early diagnosis of childhood hearing loss. Int J Audiol 46: 97–106.

Fitzpatrick E, Angus D, Durieux-Smith A et al. (2008). Parents needs following identification of childhood hearing loss. Am J Audiol 17: 1–12.

Fitzpatrick E, Olds J, McCrae R et al. (2009). Pediatric cochlear implantation: how much hearing is too much? Int J Audiol 48: 101–107.

Fitzpatrick EM, Durieux-Smith A, Whittingham J (2010). Clinical practice for children with mild bilateral and unilateral hearing loss. Ear Hear 31: 392–400.

Fitzpatrick EM, Crawford L, Ni A et al. (2011a). A descriptive analysis of language and speech skills in 4- to 5-yr-old children with hearing loss. Ear Hear 32: 605–616.

Fitzpatrick EM, Johnson E, Durieux-Smith A (2011b). Exploring factors that affect the age of cochlear implantation in children. Int J Pediatr Otorhinolaryngol 75: 1082–1087.

Fitzpatrick EM, Whittingham J, Durieux-Smith A (2014). Mild bilateral and unilateral hearing loss in children: a 20 year view of characteristics and practices. Ear Hear 35: 10–18.

Fortnum HM, Summerfield AQ, Marshall DH et al. (2001). Prevalence of permanent childhood hearing impairment in the United Kingdom and implications for universal neonatal hearing screening: questionnaire based ascertainment study. BMJ 323: 536–540.

Gallaudet Research Institute (2011). Regional and national summary report of data from the 2009–2010 Annual Survey of Deaf and Hard of Hearing Children and Youth, GRI, Gallaudet University, Washington, DC.

Gathercole SE, Hitch GJ, Service E et al. (1997). Phonological short-term memory and new word learning in children. Dev Psychol 33: 966–979.

Gathercole SE, Baddeley AD (1990). Phonological memory deficits in language disordered children: is there a causal connection? J Mem Lang 29: 336–360.

Geers AE (2006). Spoken language in children with cochlear implants. In: M Marschark (Ed.), Spencer PE. Advances in the Spoken Language Development of Deaf and Hard-of-hearing Children. Oxford University Press, New York, pp. 50–65.

Geers AE, Hayes H (2011). Reading, writing, and phonological processing skills of adolescents with 10 or more years of cochlear implant experience. Ear Hear 32: 48S–59S.

Geers AE, Nicholas JG (2013). Enduring advantages of early cochlear implantation for spoken language development. J Speech Lang Hear Res 56: 643–653.

Geers A, Tobey E, Moog J et al. (2008). Long-term outcomes of cochlear implantation in the preschool years: from elementary grades to high school. Int J Audiol 47 (Suppl 2): S21–S30.

Hart B, Risley TR (1995). Meaningful Differences in the Everyday Experience of Young American Children, Paul H, Brookes, Baltimore, MA.

Hof JR, Stokroos RJ, Wox E et al. (2013). Auditory maturation in premature infants: a potential pitfall for early cochlear implantation. Otol Neurotol 34: 251–258.

Holman MA, Carlson ML, Driscoll CLW et al. (2013). Cochlear implantation in children 12 months of age and younger. Otol Neurotol 34: 251–258.

Houston DM, Beer J, Bergeson TR et al. (2012). The ear is connected to the brain: some new directions in the study of children with cochlear implants at Indiana University. J Am Acad Audiol 23: 446–463.

Jerger J, Roeser RJ, Tobey EA (2001). Management of hearing loss in infants: the UTD/Callier Center position statement. J Am Acad Audiol 12: 329–336.

Joint Committee on Infant Hearing (2007). Year 2007 position statement: principles and guidelines for early hearing detection and intervention. Retrieved Nov 10, 2013 from http://www.asha.org/policy.

Jusczyk P (1997). The Discovery of Spoken Language, MIT Press, Cambridge, MA.

Kennedy CR, McCann DC, Campbell MJ et al. (2006). Language ability after early detection of permanent childhood hearing impairment. N Engl J Med 354: 2131–2141.

King AM (2010). The national protocol for paediatric amplification in Australia. Int J Audiol 49 (S1): S64–S69.

Klatte M, Lachmann T, Meis M (2010). Effects of noise and reverberation on speech perception and listening comprehension of children and adults in a classroom-like setting. Noise Health 12: 270–282.

Koehlinger KM, Owen Van Horne A, Moeller MP (2013). Grammatical outcomes of 3- and 6-year-old children who are hard of hearing. J Speech Lang Hear Res 56: 1701–1714.

Korver AM, Konings S, Dekker FW et al. (2010). Newborn hearing screening vs later hearing screening and developmental outcomes in children with permanent childhood hearing impairment. JAMA 304: 1701–1708.

Kral A (2007). Unimodal and cross-modal plasticity in the 'deaf' auditory cortex. Int J Audiol 46: 479–493.

Kral A, O'Donoghue GM (2010). Profound deafness in childhood. N Engl J Med 363: 1438–1450.

Kral A, Sharma A (2012). Developmental neuroplasticity after cochlear implantation. Trends Neurosci 35: 111–122.

Landry SH, Smith KE, Miller Loncar CL et al. (1997). Predicting cognitive-lanauge and social growth curves from early maternal behaviors in children at varying degrees of biological risk. Dev Psychol 33: 1040–1053.

Leigh J, Dettman S, Dowell R et al. (2011). Evidence-based approach for making cochlear implant recommendations for infants with residual hearing. Ear Hear 32: 313–322.

Leigh J, Dettman S, Dowell R et al. (2013). Communication development in children who receive a cochlear implant in children who receive a cochlear implant by 12 months of age. Otol Neurotol 34: 443–450.

Lieu JEC, Tye-Murray N, Karzon RK et al. (2010). Unilateral hearing loss is associated with worse speech-language scores in children. Pediatrics 125: e1348–e1355.

Luterman DL (2008). Counseling families with communication disorders and their families, Pro-Ed, Austin, TX.

MacKay S, Gravel JS, Tharpe AM (2008). Amplification considerations for children with minimal or mild bilateral hearing loss and unilateral hearing loss. Trends in Amplif 12: 43–54.

Mahoney G, Wheeden CA, Perales F (2004). Relationship of preschool special education outcomes to instructional practices and parent-child interaction. Res Dev Disabil 25: 493–595.

Mahoney GJ, Kim JM, Lin CS (2007). The pivotal behavior model of developmental learning. Infants Young Child 20: 311–325.

Marchman VA, Fernald A (2008). Speed of word recognition and vocabulary knowledge in infancy predict cognitive and language outcomes in later childhood. Dev Sci 11: F9–F16.

Marchman VA, Fernald A (2013). Variability in real-time spoken language processing in typically developing and late-talking toddlers. In: LA Rescorla, PS Dale (Eds.), Late Talkers: language development, interventions, and outcomes, Paul H. Brookes Publishing Co., Baltimore, MA, pp. 145–151.

Marschark M, Hauser PC (2008). Cognitive underpinnings of learning by deaf and hard-of-hearing students. In: M Marschark, PC Hauser (Eds.), Deaf Cognition: foundations and outcomes, Oxford University Press, New York, pp. 3–23.

Mayer C (2007). What really matters in the early literacy development of deaf child. J Deaf Studies Deaf Educ 12: 411–431.

McCracken W, Turner O (2012). Deaf children with complex needs: parental experience of access to cochlear implants and ongoing support. Deaf Deaf Educ Int 14: 22–35.

Meinzen-Derr J, Wiley S, Grether S et al. (2011). Children with cochlear implants and developmental disabilities: a language skills study with developmentally matched hearing peers. Res Dev Disabil 32: 757–767.

Mitchell R, Karchmer M (2004). Chasing the mythical ten percent: parental hearing status of deaf and hard of hearing students in the United States. Sign Lang Studies 4: 138–163.

Moeller MP (2000). Early intervention and language development in children who are deaf and hard of hearing. Pediatrics 106: E43.

Moeller MP, Schick B (2006). Relations between maternal input and theory of mind understanding in deaf children. Child Dev 77: 751–766.

Moeller MP, Tomblin JB, Yoshinaga-Itano C et al. (2007). Current state of knowledge: language and literacy of children with hearing impairment. Ear Hear 28: 740–753.

Moeller MP, Hoover B, Peterson B et al. (2009). Consistency of hearing aid use in infants with early-identified hearing loss. Am J Audiol 18: 14–23.

Moeller MP, Carr G, Seaver L et al. (2013). Best practices in family-centered early intervention for children who are deaf of hard of hearing: an international consensus statement. J Deaf Studies Deaf Educ 18: 429–445.

Mustard JF (2006). Early Child Development and Experience-based Brain Development: the scientific underpinnings of the importance of early child development in a globalized world, The Brookings Institution, Washington, DC.

National Reading Panel (2000). Report on the national reading panel: teaching children to read: an evidence-based assessment of the scientific research literature on reading and its implications for reading instruction. National Institute for Literacy at ED Pubs, Jessup, MD.

Nelson K (2007). Young Minds in Social Worlds: Experience, Meaning, and Memory, Harvard University Press, Cambridge, MA.

Nelson K, Plesa D, Henseler S (1998). Children's theory of mind. An experiential interpretation. Hum Dev 41: 7–29.

Nelson HD, Bougatsos C, Nygren P (2008). Universal newborn hearing screening: systematic review to update the 2001 US Preventive Services Task Force recommendation. Pediatrics 122: e266–e276.

Neuman AC, Wroblewski M, Hajicek J et al. (2010). Combined effects of noise and reverberation on speech recognition performance of norm-hearing children and adults. Ear Hear 31: 336–344.

Neville HJ, Bavelier D, Corinas D et al. (1998). Cerebral organization for language in deaf and hearing subjects: biological constraints and effects of experience. Proc Natl Acad Sci 92: 922–929.

Nicholas J, Geers AE (2007). Will they catch up? The role of age at cochlear implantation in the spoken language development of children with severe to profound hearing loss. J Speech Lang Hear Res 50: 1048–1062.

Niparko JK, Tobey EA, Thal DJ et al. (2010). CDaCI Investigative Team. Spoken language development in children following cochlear implantation. JAMA 303: 1498–1506.

Olds J, Fitzpatrick E, Schramm D et al. (2005). Outcome of cochlear implantation in children with complex disabilities. Ninth Symposium on Cochlear Implants in Children. Dallas, TX.

Olusanya BO, Somefun AO, Swanepoel de W (2008). The need for standardization of methods for worldwide infant hearing screening: a systematic review. Laryngoscope 118: 1830–1836.

Papsin BC, Gordon KA (2008). Bilateral cochlear implants should be the standard for children with bilateral sensorineural deafness. Curr Opin Otolaryngol Head Neck Surg 16: 69–74.

Perfetti CA, Sandak R (2000). Reading optimally builds on spoken language: implication for deaf readers. J Deaf Studies Deaf Educ 5: 32–50.

Perner J, Ruffman T, Leekam S (1994). Theory of mind is contagious: you catch it from your sibs. Child Dev 65: 1228–1238.

Perone S, Modole KL, Ross-Sheehy S et al. (2008). The relation between infant's activity with objects and attention to object appearance. Dev Psychol 44: 1242–1248.

Peterson NR, Pisoni DB, Miyamoto RT (2010). Cochlear implants and spoken language processing abilities: review and assessment of the literature. Restorative Neurol Neurosci 28: 237–250.

Picard M (2004). Children with permanent hearing loss and associated disabilities: revisiting current epidemiological data and causes of deafness. Volta Rev 104: 221–236.

Pisoni DB, Geers A (2001). Working memory in deaf children with cochlear implants: correlations between digit span and measures of spoken language processing. Annals Otol Rhinol Laryngol 109: 92–93.

Pisoni DB, Conway CM, Kronenberger DL et al. (2008). Efficacy and effectiveness of cochlear implants in deaf children. In: M Marchark, PC Hauser (Eds.), Deaf Cognition: foundations and outcomes, Oxford University Press, New York, pp. 237–250.

Porter H, Bess FH (2011). Children with unilateral hearing loss. In: A-M Tharpe (Ed.), R Seewald R. Comprehensive Handbook of Pediatric Audiology. Plural Publishing, San Diego, CA, pp. 175–191.

Prieve BA, Stevens F (2000). The New York State universal newborn hearing screening demonstration project: introduction and overview. Ear Hear 21: 85–91.

Prodi N, Visentin C, Feletti A (2013). On the perception of speech in primary school classrooms: ranking of noise interference and of age influence. J Acoust Soc Am 133: 255–268.

Rance G, Starr A (2011). Auditory neuropathy/dys-synchrony type hearing loss. In: A-M Tharpe (Ed.), R Seewald, Comprehensive Handbook of Pediatric Audiology. Plural Publishing, San Diego, CA, pp. 225–242.

Rasore Quartino A (2011). Early medical caretaking and follow-up. In: J-A Rondal, J Perera, D Spiker (Eds.), Neurocognitive Rehabilitation of Down Syndrome, The Early Years. Cambridge University Press, New York, pp. 118–127.

Raveh E, Buller N, Badrana O et al. (2007). Auditory neuropathy: clinical characteristics and therapeutic approach. Am J Otolaryngol 28: 302–308.

Robertson L (2014). Literacy and Deafness: listening and spoken language, plural publishing, San Diego, CA.

Ruben RJ (1997). A time frame of critical/sensitive periods of language development. Acta Oto-laryngol 117: 202–205.

Ruff HA, Saltarelli LM, Capozzoli M et al. (1992). The differentiation of activity in infants' exploration of objects. Dev Psychol 28: 851–861.

Sarant JZ, Holt CM, Dowell RC et al. (2009). Spoken language development in oral preschool children with permanent childhood deafness. J Deaf Studies Deaf Educ 14: 205–217.

Schachter HM, Clifford TJ, Fitzpatrick E et al. (2002). Systematic review of interventions for hearing loss in children, Health Canada, Ottawa.

Schick B, de Villiers P, de Villiers J et al. (2007). Language and theory of mind: a study of deaf children. Child Dev 78: 376–396.

Schramm D, Fitzpatrick E, Séguin C (2002). Cochlear implantation in adolescents and adults with pre-linguistic deafness. Otol Neurotol 23: 698–703.

Sharma A, Dorman M (2006). Central auditory development in children with cochlear implants: clinical implications. Advances Otorhinolaryngol 64: 66–88.

Sharma A, Dorman MF, Spahr AJ (2002). A sensitive period for the development of the central auditory system in children with cochlear implants: implications for age of implantation. Ear Hear 23: 532–539.

Sharma A, Dorman MF, Kral A (2005). The influence of a sensitive period on central auditory development in children with unilateral and bilateral cochlear implants. Hear Res 203: 134–143.

Sharma A, Nash AA, Dorman M (2009). Coritcal development, plasticity and re-organization in children with cochlear implants. J Com Dis 42: 272–279.

Singh JA, Zingg RM (1996). Wolf-Children and Feral Man, Archon Books, Hamden.

Sininger YS, Grimes A, Christensen E (2010). Auditory development in early amplified children: factors influencing auditory-based communication outcomes in children with hearing loss. Ear Hear 31: 1–20.

Spencer PE, Marschark M (2010). Evidence-based practice in educating deaf and hard-of-hearing students, Oxford University Press, New York.

Starr A, Picton TW, Sininger YS et al. (1996). Auditory neuropathy. Brain 119: 741–753.

Tattersall H, Young A (2006). Deaf children identified through newborn hearing: parents' experiences of the diagnostic process. Child Care Health Dev 32: 33–45.

Tharpe AM (2008). Unilateral and mild bilateral hearing loss in children: past and current perspectives. Trends Amplif 12: 7–15.

Thoutenhoofd ED, Archbold SM, Gregory S et al. (2005). Paediatric Cochlear Implantation: evaluating outcomes, Whurr Publishers, London.

Tobey EA, Thal D, Niparko JK et al. (2013). Influence of implantation age on school-age language performance in pediatric cochlear implant users. Int Journal Audiol 52: 219–229.

Traxler CB (2000). Measuring up to performance standards in reading and mathematics: achievement of selected deaf and hard-of-hearing students in the national norming of the 9th Edition Stanford Achievement Test. J Deaf Studies Deaf Educ 5: 337–348.

Van Naarden K, Decouflé P, Caldwell K (1999). Prevalence and characteristics of children with serious hearing impairment in metropolitain Atlanta, 1991–1993. Pediatrics 103: 570–575.

Vouloumanos A, Werker JF (2007). Listening to language at birth: evidence for a bias for speech in neonates. Dev Sci 10: 159–171.

Vouloumanos A, Hauser MD, Werker JF et al. (2009). The tuning of human neonate's preference for speech. Child Dev 61: 1584–1595.

Wake M, Hughes EK, Poulakis Z et al. (2004). Outcomes of children with mild-profound congenital hearing loss at 7 to 8 years: a population study. Ear Hear 25: 1–8.

Waltzman SB, Roland JT, Cohen N (2002). Delayed implantation in congenitally deaf children and adults. Otol Neurotol 23: 333–340.

Watkin PM, Baldwin M (2011). Identifying deafness in ealy childhood: requirements after the newborn hearing screen. Arch Dis Child 96: 62–66.

Werker JF, Tees RC (2005). Speech perception as a window for understanding plasticity and commitment in language systems of the brain. Dev Psychobiol 46: 233–251.

Werker JF, Yeung HH (2005). Speech perception bootstraps word learning in infancy. Trends Cog Sci 9: 519–527.

Williams C (2004). Emergent literacy of deaf children. J Deaf Studies Deaf Educ 9: 352–365.

World Health Organization (2001). International classification of functioning, disability and health: ICF, WHO, Geneva.

Yoshinaga-Itano C, Sedey AL, Coulter DK et al. (1998a). Language of early- and later-identified children with hearing loss. Pediatrics 102: 1161–1171.

Yoshinaga-Itano C, Snyder L, Day D (1998b). The relationship of language and symbolic play in children with hearing loss. Volta Rev 100: 135–164.

Yoshinaga-Itano C, Coulter D, Thomson V (2001). Developmental outcomes of children with hearing loss born in Colorado hospitals with and without universal newborn hearing screening programs. Semin Neonatol 6: 521–529.

Yoshinaga-Itano C, Johnson CD, Carpenter K et al. (2008). Outcomes of children with mild bilateral hearing loss and unilateral hearing loss. Semin Hear 29: 196–211.

Young A, Tattersall H (2007). Universal newborn hearing screening and early identification of deafness: parents' responses to knowing early and their expectations of child communication development. J Deaf Studies Deaf Educ 12: 209–220.

Handbook of Clinical Neurology, Vol. 129 (3rd series)
The Human Auditory System
G.G. Celesia and G. Hickok, Editors

Chapter 20

Aging of the auditory system

THOMAS NICOLAS ROTH*
Department of Otorhinolaryngology, Head and Neck Surgery, University Hospital of Zurich, Zurich, Switzerland

GENERAL ASPECTS OF AGING

From a biologic point of view, an individual life cycle begins with the fusion of two chromosome sets, continues with proliferation and differentiation, and ends with the loss of cell and organ function, leading to death. Hence, life can be considered a genetic program depending on innumerable factors and interacting with just as many promoting, as well as potentially destroying, influences. This process of development is controlled meticulously by mechanisms responding to the need of the cell and the organ.

These mechanisms oscillate between proliferation and degeneration in order to maintain a dynamic equilibrium in the continuous cycle of metabolism.

The ability of proliferation requires the capability to deplete the ancient state of function and to create a new condition that permits mutation in another state. Hence, in this mutation process, various actions, such as inhibition and stimulation, agonists and antagonists, degeneration and proliferation, are all present at the same time throughout the body and interact in a multitude of feedback mechanisms.

In a more general aspect the growth of any species is a controlled development of different organs and structures that on the one hand depend on one another and on the other hand permit the species to interact with other species.

In large, complex, multicellular organisms, each part of the body has its specific function, leading to a balanced and interconnected system able to survive in the prevailing circumstances and to respond to both internal and external factors.

As proliferation is a genetically controlled process that depends in a certain way on the senescence of the antecedent cellular state, aging is in a comparable way a genetic process that affects cells, organs, or the whole organism itself. The rare syndrome of progeria that describes the premature senescence emphasizes the genetic impact in the process of aging (Coppedè, 2013).

Illnesses and pathologies develop due to disturbed regulation or overwhelmed adaptation mechanisms. Pathologic decline of specific organs may entail deleterious knock-on effects on other organs. For example, diabetes mellitus, whilst being a primarily exocrine pancreatic problem, leads to a multitude of distant effects, including accelerated arteriosclerosis and polyneuropathy.

To what extent aging itself is "pathologic" is debatable. Certain age-related changes are normally regarded as physiologic, for example the growth spurt of puberty, the involution of the thymus, or even the menopause. Such changes may be viewed as positive for the body and have protecting reasons, e.g., for the hosting female organism and for the potentially new embryo as well.

The specific aging mechanisms of the inner ear include genetic and environmental/systemic factors. leading to a progressive hearing loss of significant individually variation (Walters and Zuo, 2013). Beyond the cochlea itself, changes in the auditory pathway and brain also significantly affect our sense of hearing.

In this chapter, general and epidemiologic aspects as well as histologic changes based on Schuknecht's classification are discussed. Furthermore, some molecular biologic considerations, e.g., oxidative stress, endocochlear potential, the role of mitochondria and supporting elements such as fibrocytes and their effect on presbycusis, are discussed with the aim of giving further insights to the age-related cellular metabolism that may explain histologic changes and the clinical characteristics of age-related hearing loss (ARHL). Furthermore, some aspects of age-related changes in the central nervous system are also considered. Finally, management of the consequences of ARHL is discussed.

*Correspondence to: Thomas Nicolas Roth, Department of Otorhinolaryngology, Head and Neck Surgery, University Hospital of Zurich, Frauenklinikstr. 24, 8091 Zurich, Switzerland. E-mail: t.n.roth@bluewin.ch

DEFINITION AND TERMINOLOGY

Presbyacusis is a term that literally translates to "hearing in the older age." It includes all factors and structural changes leading to hearing loss in elderly people. Nowadays the term "age-related hearing loss," abbreviated ARHL, is used preferably. The disorder is characterized by high-frequency-dominated hearing loss, reduced speech understanding (particularly in noisy environments), slowed central processing of acoustic information, and impaired localization of sound sources. These difficulties have a direct impact on lifestyle and social activities by impairing communication (Gates and Mills, 2005).

Additional terms may complement the description of the clinical presentation of ARHL. The term "sensory hearing loss" refers to an elevated hearing threshold level as measured in pure-tone audiometry demanding a higher acoustic level to excite the organ of Corti, while "sensorineural hearing loss" includes the signal transduction of sensory cells, their synapses, and the cochlear nerve. ARHL seems to belong mainly to this latter category. The term "perceptual" derives from the Latin verb *percipere*, signifying to understand or to hold something, and includes perceiving or catching on to the meaning of spoken words or other acoustic messages; this term may include a disturbed central auditory process in conjunction with age-related changes of the sensory and neural parts of the auditory system. Finally, the term "central" restricts the origin of the hearing disorder to the central nervous system, including the auditory cortex.

Patients with presbycusis typically describe two types of difficulty: on the one hand they complain, "I do not hear well" or "My hearing has become worse," whereas on the other hand, "I can hear, but I can't understand in a noisy environment." The observation of these two different aspects implies that it is not only the peripheral cochlear part with decreased hearing threshold that plays an important role in the aging process of the auditory pathway, but that other elements in neurophysiologic signal transduction – including the central neural framework – are involved, leading to hearing difficulties, especially in noisy situations (Otto and McCandless, 1982). Consequently, auditory perception presupposes both a well-functioning peripheral part and a well-interconnected central framework of modulating and sensory areas.

The age-related decline in function of the peripheral hearing organ is related to an individual's genetics background and the cumulative damage of noise, middle-ear inflammation, medication, exposure to toxins including nicotine, and cardiovascular and rheumatologic factors. Indeed, innumerable factors influence the severity and onset of ARHL, and there is no generally accepted single etiologic factor. Even a multifactorial analysis is extremely complex and likely an oversimplification, though some of the factors thought to be most significant are discussed below in more detail.

Aging also affects central nervous function in general by slowing down processing speed and making task-related inhibition less effective. Poorer speech understanding in ARHL often has both peripheral and central components and it is often difficult to assess which is the most significant contributor (Otto and McCandless, 1982; Gates and Popelka, 1992). Moreover, patients with disease-related impairment of the central nervous system typically exhibit deficits in other cognitive functions affecting daily life more severely than patients with an isolated hearing loss.

EPIDEMIOLOGY

It is difficult to assess the prevalence and epidemiology of ARHL because of manifold influences and different factors contributing to hearing loss and due to a variety of sociocultural, behavioral, and psychologic aspects. Furthermore, difficulties arise because of a lack of standardization and different selection criteria that have their impact on hearing function. The lack of consistent definitions prevents the compiling of accurate data on hearing loss.

Epidemiologic data are easy to measure for certain types of information, for example mortality, or well-defined and measurable states, such as whether a pathohistologic diagnosis of cancer exists. In contrast, ARHL is on a continuous scale without a clear border between normal and impaired function. Furthermore, such objective measures do not necessarily correlate with subjective complaints and perceived handicaps. For example, a professional musician will notice hearing impairment more strongly than someone who does not rely on hearing for a profession.

Any definition of ARHL is unavoidably subjected to arbitrary limits due to the vast variety of factors, and this gives rise to methodologic problems. An evident and primary difficulty in comparing reported prevalence data is the different measures and cut-offs for hearing impairment. There is also a difficulty in defining chronologic "age" itself related to the prevalence of ARHL. While several international systems of hearing loss classification exist, they do not include age in their definitions. Even if a chronologic age may have limited biologically relevance, definitions are necessary to obtain clear epidemiologic data. International classification systems such as those of the European Union or the World Health Organization (WHO) differ considerably. Standard age limits may be even more difficult to define than limits of

Table 20.1

Standardized hearing loss categories

Categorization	WHO classification
Normal	dBHL < 26
Mild	26 ≤ dBHL < 40
Moderate	41 ≤ dBHL < 60*
Severe	61 ≤ dBHL < 80
Profound	80 < dBHL

*According to the World Health Organization (WHO), hearing impairment of the better ear >41 dB has been defined as disabling.

hearing loss. For example, ISO-7029 is a statistical distribution of hearing threshold as a function of age from the International Organization for Standardization (ISO); this standard is based on a linear model for ages 18–70 years (ISO, 2000), but does not include older ages with their non-linear increase in hearing loss. A single and widely accepted universal classification of ARHL is needed to collect epidemiologic data and to compare prevalence in different countries. Because no system has clear advantages over the other, the WHO classification system is recommended to be used for future data collection (Table 20.1).

In addition to problems associated with a lack of standardized methods for defining ARHL, a recent systematic review of epidemiologic data on the prevalence of ARHL in Europe revealed significant information gaps, for example, heterogeneity of selection criteria and differences in cut-offs for grades of hearing impairment (Roth et al., 2011). Neither geographic distributions nor developments over time could be extracted to a reasonable degree. Nevertheless, the studies reflect the well-known patterns of a non-linear increase of hearing loss, with age affecting men more than women. If the data are crudely averaged and interpolated, roughly 30% of men and 20% of women in Europe have a hearing loss of 30 dB HL or more at age 70 years, and 55% of men and 45% of women at age 80 years. The geographic distribution of European studies showed a relative paucity of studies in the middle, southern, and eastern parts of Europe; this may reflect differences in public health systems which attach more or less importance to ARHL in different regions (Table 20.2).

Differences in the language spoken in different regions may also impact epidemiologic data for ARHL. For example, vowels have a distinct perceptual advantage over consonants in determining speech intelligibility (Kewley-Port et al., 2007; Fogerty and Kewley-Port, 2009), which could affect the prevalence of self-reported hearing impairment across different languages (Roth et al., 2011).

In the United States, the overall prevalence of ARHL, defined as hearing loss of more than 25 dB HL in speech frequency, was reported to be 39% for men and 17% for women at the age range of 60–69 years, and 63% for men and 48% for women at the age range of 70–79 years (Agrawal et al., 2008). Another estimate reported a prevalence of 23% of American people aged between 65 and 75 years and 40% of people over 75 years of age, illustrating the vastly different results obtained in different studies (Seidman et al., 2002). Considerations about the difference between men and women are discussed in the section below. Another interesting aspect of the American data is that Afro-Americans seem to be considerably less affected by ARHL than white people (Agrawal et al., 2008; Lin et al., 2011).

Table 20.2

Prevalence of age-related hearing loss in Europe*

Reference	Country	Prevalence with nearest cut-offs for HL and age
Bergmann and Rosenhall, 2001	Sweden	> 19% (70 years at 30–39 dB)
Borchgrevink et al., 2005	Norway	> 14.2% (60–64 years at 35 dB)
Davis, 1989	UK	> 7.4% (61–70 years at 45 dB)
Davis, 1995	UK	> 24.5% (61–70 years at 30 dB)
Hietanen et al., 2005	Denmark, Sweden, Finland	>16.5% (75 years at 40–69 dB)
Hietanen et al., 2004	Finland	>28.3% (80 years at 40–69 dB)
Johansson and Arlinger, 2003	Sweden	> 8.8% (60–70 years at 35 dB)
Moller, 1981	Sweden	9% (70 years at 35 dB)
Quaranta et al., 1996	Italy	6.7% (61–70 years at 45 dB)
Rahko et al., 1985	Finland	10.3% (65 years at 30 dB)
Rosenhall and Karlsson Espmark, 2003	Sweden	24% (70 years at 30–39 dB)
Wilson et al., 1993	UK	54.3% (≥65 years at 35 dB)

*This table lists the minimum prevalence in men and women for minimum better-ear hearing loss (HL) of 30 dB and lower age interval border of 60 years.

These epidemiologic data confirm that ARHL is a major health concern in the aging population. They also demonstrate the need for standardized collection of epidemiologic data on hearing loss.

Epidemiology and etiology are intimately related in that population-based endogenous factors will be exposed to geographically varying exogenous factors. Given the variable epidemiologic data and the complex interaction of etiologic factors and their influence on ARHL, it is not surprising that meaningful connections between epidemiology and etiologic factors are difficult to be established.

The most relevant factors in the etiology of ARHL seem to be heredity (that may contribute up to 50%), noise exposure, history of chronic middle-ear inflammation, and cardiovascular factors, including diabetes, smoking, and hypertension. Additional relevant factors are hormones (including gender-related differences), exposure to ototoxic medication or chemicals and comorbidities, such as rheumatologic disease (Murdin et al., 2008; George and Pradhan, 2009). All these factors may cause oxidative stress as a common end molecular pathway. Such stress leads to alteration of cochlear homeostasis and to changes of the different anatomic structures accelerating the aging process.

HISTOPATHOLOGIC CHANGES AND SCHUKNECHT'S CLASSIFICATION

Different components of the cochlea have specific functions in inner-ear homeostasis and are prone to age-related changes; for example, the spiral ligament with its cellular and fibrous part as well as the important role of fibroblasts on the basilar membrane or the stria vascularis, considered as the "cochlear battery," with its impact on the electrolyte homeostasis in the endolymphatic liquid. The inner and outer hair cells with their stereocilia covered by the tectorial membrane are also vulnerable to aging.

Beyond the organ of Corti, the poorly understood neural framework of afferent and efferent fibers, the spiral ganglion, and the cochlear nucleus of the brainstem exhibit age-related changes (Frisina and Walton, 2006; Makary et al., 2011). Furthermore, there is evidence that age-related pathologies of the central nervous system have a negative impact on the peripheral cochlear organ (Otto and McCandless, 1982; Eckert et al., 2012).

The classification of Schuknecht is based on the age-related changes of the cochlea. In correlation with rather crude audiometric data archived in his laboratory, Schuknecht described in various temporal bone studies degeneration or age-related changes of the organ of Corti, ganglion cell, stria vascularis and basilar membrane. This led to a classification of ARHL in sensory, neural, strial or metabolic and cochlear-conductive categories. Two more categories, indeterminate and mixed, were also added. The latter was reported to be responsible for up to 25% of cases. This classification is called the Schuknecht's typology of ARHL and has been revised recently by Merchant and Nadol (Nadol, 2010) as follows.

Sensory presbycusis

Histopathologically, this diagnosis rests on the finding of hair cell loss at the beginning of the basal end of the cochlea. Schuknecht connected this kind of histopathologic finding to audiograms showing typically an abrupt high-tone hearing loss (Fig. 20.1) (Schuknecht and Gacek, 1993). The earliest notable degenerative changes in hair cells affected the sterocilia, followed by slightly distorted and flattened organ of Corti, and subsequently loss of hair cells and supporting cells. Both cell types accumulate lipofuscin in the apical cytoplasm. This accumulation correlates with the presence of lysosomes, indicating exhausted enzyme activity.

Neural presbycusis

When the loss of cochlear neurons was more than 50%, the term neural presbycusis was used. The progressive loss of neurons was observed throughout the cochlea (Fig. 20.2). There is a debate about the causes of spiral ganglion cell loss and whether it is primary or secondary. Loss of inner hair cells, supporting elements, and injury to dendritic fibers are proposed factors for secondary degeneration, but primary degeneration has been identified in different entities, e.g., sudden deafness, Friedreich's ataxia, Ménière's disease, or Usher's syndrome, and age-related primary degeneration has been reported in both humans and animals.

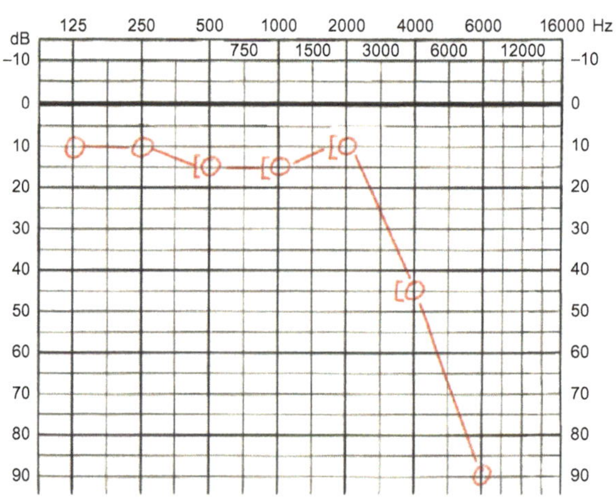

Fig. 20.1. Sensory presbycusis.

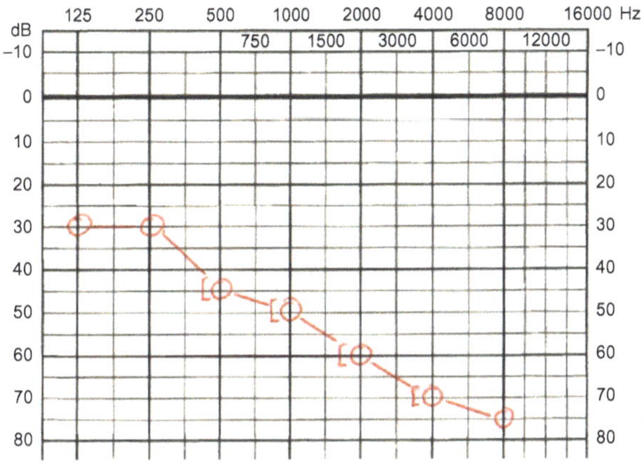

Fig. 20.2. Neural presbycusis.

The characteristic clinical finding is a progressive loss of word discrimination, termed phonemic regression. Elderly people with rapidly progressive neural presbycusis exhibit signs of general central neurodegeneration, including motor weakness, coordination problems, irritability, loss of memory, and intellectual deterioration.

Strial presbycusis

Strial presbycusis is defined as a loss of 30% or more of the strial tissue. When strial loss exceeded this percentage, threshold levels deteriorated in the tone audiogram. The stria vascularis is considered to have an important functional role in inner-ear homeostasis, particularly in the generation and maintenance of the endocochlear potential. Strial atrophy was noted to have a familial correlation, and hence may have a genetic predisposition. The main audiometric characteristic is a flat or slightly descending threshold level (Fig. 20.3). Usually patients do not complain of discomfort with loud sounds. The histologic changes are predominantly in the apical and basal turn of the cochlea (Figs 20.4 and 20.5).

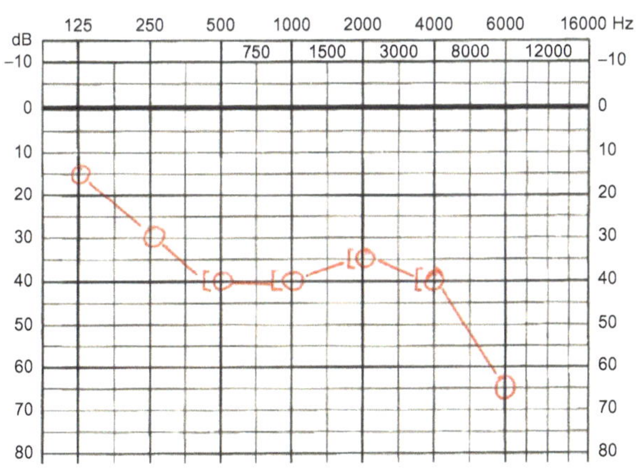

Fig. 20.3. Strial presbycusis.

Atrophy of the spiral ligament

Schuknecht observed a degeneration of the spiral ligament as a function of age, beginning in childhood. The atrophy was most severe in the apical region of the cochlea. Degeneration of the enzyme-producing spiral ligament seems to play an important role in the development of presbycusis in animal studies. Degeneration of fibrocytes preceded the loss of hair cells and neurons. In humans as well, the earliest alteration is the loss of fibrocytes; these changes were more frequent in ears exhibiting descending audiometric patterns. Figures 20.6–20.9 show differences between a young and an aged gerbil in the spiral ligament.

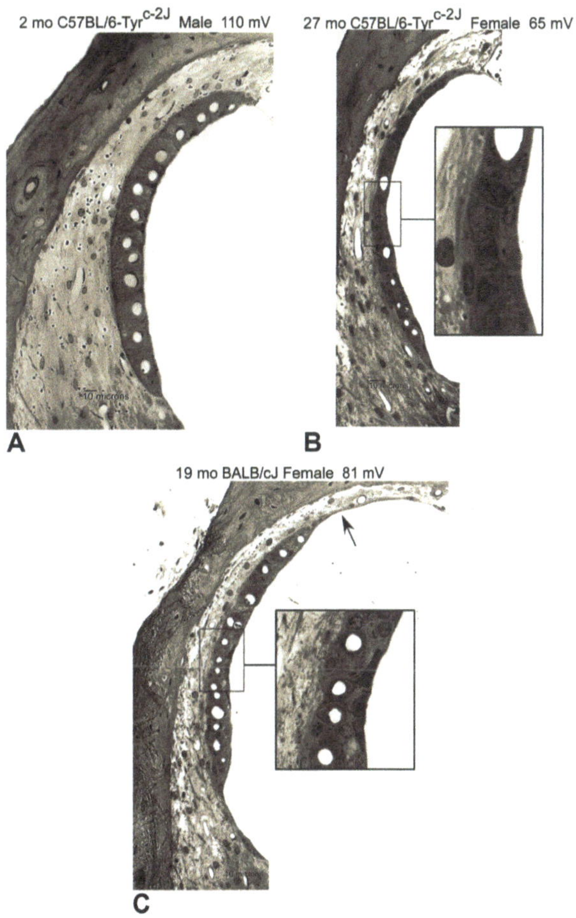

Fig. 20.4. Strial atrophy and spiral ligament pathologies. Examples of lateral wall in the lower cochlear base of young B6 albino (**A**), old B6 albino (**B**), and old BALB/c (**C**). Age, gender, and basal turn endocochlear potential are indicated). Aging was associated with strial thinning and marginal cell loss in both strains. Severity of capillary loss and ligament thinning in (**B**) are atypical for this strain. BALB (**C**) features greater loss of marginal cells along the luminal surface. Marginal cells in (**C**) inset show dense staining and retraction of processes. Arrow in (**C**) denotes somewhat unusual strial atrophy. (Reproduced from Ohlemiller, 2009.)

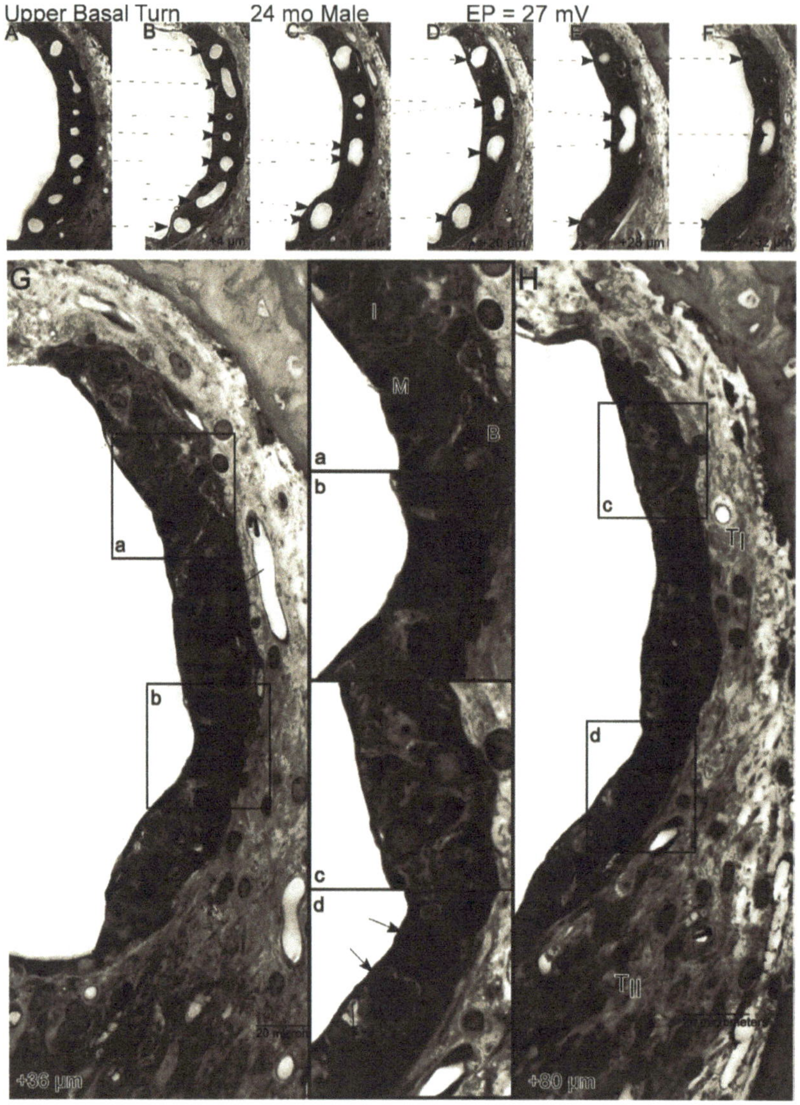

Fig. 20.5. Strial microvascular pathologies in 24-month-old male gerbil. Sequential sections (**A–H**) proceeding apically from the upper base of a 24-month-old male gerbil, showing progressive merging of strial capillaries, leaving the stria completely avascular over a span of 80 μm. Dashed arrows in **A–F** show how capillaries progressively merge, ultimately forming a blind loop. The first completely avascular segment (**G**) features well-organized regions (inset a), as well as regions with increased numbers of poorly differentiated cells (inset b). The arrow in **G** shows the wall of the last capillary loop. By 80 μm (**H**), the stria is still present but thin, and shows hyperplasia of undefined cell types over most of its length (inset c). Traces of degenerated capillaries can still be seen (inset d, arrows). The ligament at this location appears normal. TI, type I fibrocytes; TII, type II fibrocytes; B, basal cells; I, intermediate cells; M, marginal cells; EP, endocochlear potential. (Reproduced from Ohlemiller et al., 2008.)

Indeterminant presbycusis

When no significant age-related change of any structure was present, the term "indeterminant presbycusis" was used. Twenty-five percent of all clinical cases of human presbycusis do not exhibit cochlear changes by light microscopy. However, degeneration of the dendritic arborization in the spiral ganglion cells and alteration of the tip links on hair cells have been described in electron microscopy. Changes of altered enzymatic pathways at molecular levels on the lateral cochlear wall have been put forward as a possible pathophysiologic explanation. These pathways seem to exhibit an important role in inner-ear homeostasis and maintenance of cellular function. Other hypotheses are age-related changes affecting more central parts of the hearing pathway, e.g., the cochlear nuclei.

Mixed presbycusis

When significant age-related changes are observed in more than one structure, they are classified as mixed

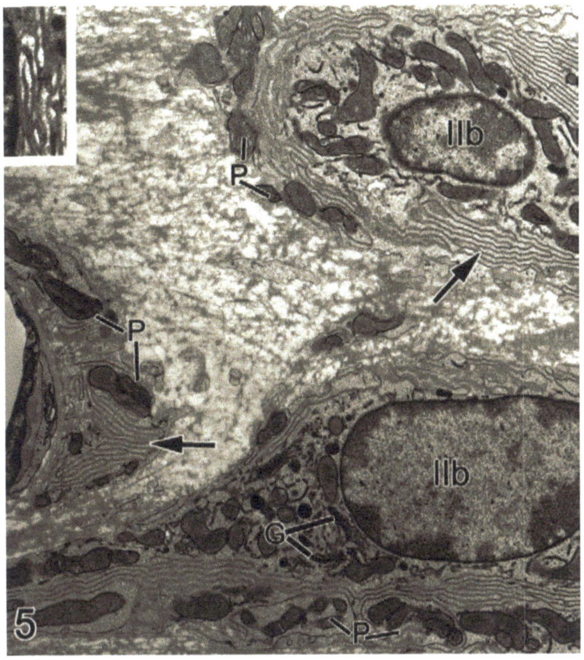

Fig. 20.6. Spiral ligament of a young gerbil. In the spiral ligament of a young control, elongated type IIb fibrocytes display elaborately pleated foldings that greatly amplify the cell surface (arrows and inset). The fibrocytes enclose numerous large mitochondria bordering the nucleus and infiltrating primary processes (P), together with prominent Golgi zones (G). Magnification × 6000. Basal region. (Reproduced from Spicer and Schulte, 2002.)

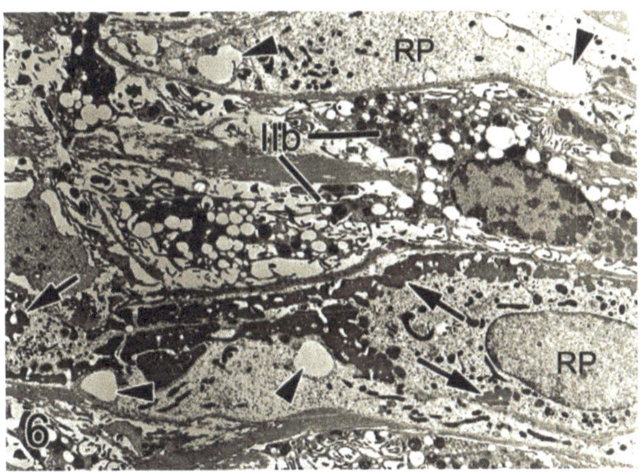

Fig. 20.7. Spiral ligament pathologies of an aged gerbil. Type IIb fibrocytes from aged gerbil 1 exhibit shrunken cell bodies heavily infiltrated with variable-sized clear vacuoles. Empty-looking spaces separate the cells and their plasmalemmal projections. A thick layer of dense substance lines the periphery (arrows) of a root process (RP) paralleling a vacuolated fibrocyte. Neighboring root processes enclosing a few vacuole-like spaces (arrowheads) lack the marginal dense deposits. Magnification × 2800. Basal region. (Reproduced from Spicer and Schulte, 2002.)

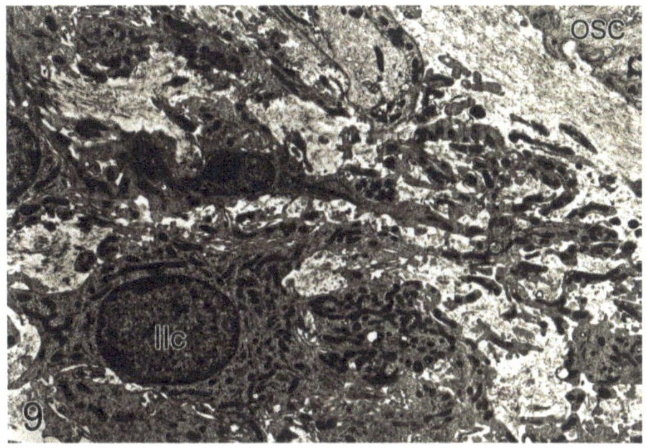

Fig. 20.8. Spiral ligament of a young adult gerbil. The spiral ligament under outer sulcus cells (OSC) of a young adult gerbil exhibits congregated type IIc fibrocytes containing many mitochondria. Abundant processes of these rather stellate cells intermingle in loose stroma devoid of the dense bands that border IIb fibrocytes. Magnification × 4800. Apical region. (Reproduced from Spicer and Schulte, 2002.)

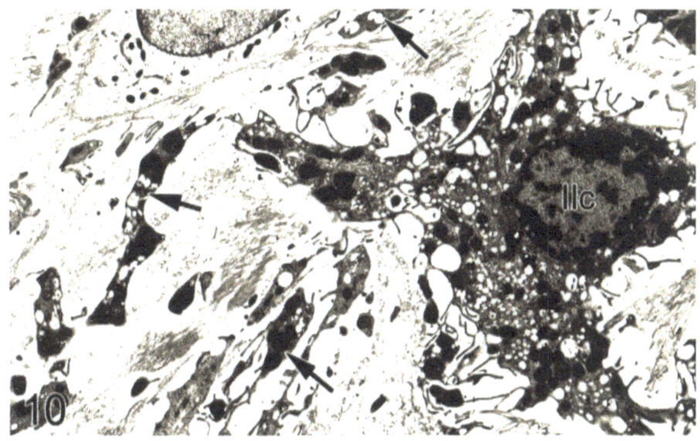

Fig. 20.9. Spiral ligament pathologies of an aged gerbil. An empty-looking zone encircles the stellate profile of a type IIc fibrocyte from a gerbil. Numerous small to large vacuoles populate the cytosol in the cell body and in nearby cell processes (arrows). Magnification × 4800. Basal region. (Reproduced from Spicer and Schulte, 2002.)

presbycusis. Unfortunately, this type exhibits no consistent audiometric patterns.

Cochlear conductive presbycusis

Cochlear conductive presbycusis is based on both clinical and pathologic criteria. It can be considered an interesting hypothesis of a mechanism leading to sensorineural hearing loss, but there is, however, no convincing evidence of a cochlear conductive defect. Hence this remains an unproven theory. Schuknecht described age-related thickening and hyaline degeneration of the basilar membrane and deposition of calcium salts in the basal portion of the membrane. A possible explanation concerning these findings will be considered in the section below, focusing on the role of fibrocytes and inner-ear homeostasis.

Even though the classic types of presbycusis, i.e., sensory, neural, and strial persbycusis could be correlated with defined shapes of audiograms (Figs 20.1–20.3), a clear pathophysiologic explanation or pathohistologic pattern for the other types is lacking. Hence, Schuknecht's classification is considered vague and imprecise because most aging human and animal cochleas exhibit a mix of pathologies; furthermore, it was somewhat criticized because the classification concentrates on morphology of the peripheral hearing organ, excluding many functional and central aspects of hearing processing (Ohlemiller, 2004).

Therefore, and due to the fact that several causes overlap and interfere with each other, further explanations may be given by molecular biology and genetic analyses, opening a vast field in the modern area of research on ARHL.

A brief overview of some of these mechanisms is presented in the next section, with the aim of contributing some insights to the underlying cellular processes, which may lead to the above-mentioned histopathologic alterations of the cochlea.

AGING OF THE PERIPHERAL AUDITORY SYSTEM

Aging is not only determined by genetics, but also influenced by oxidative stress (Fig. 20.10). Several factors,

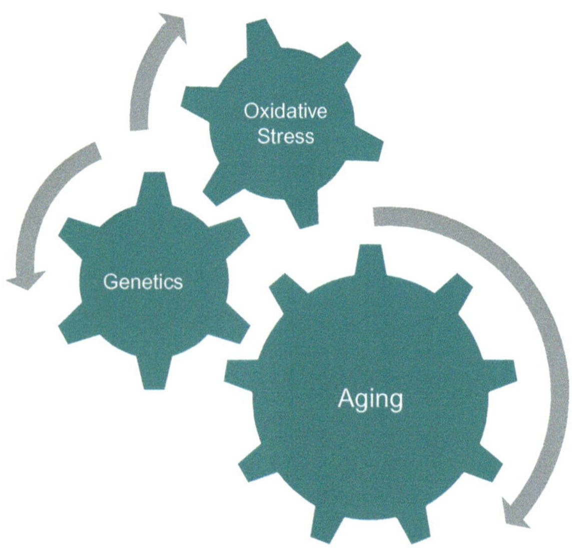

Fig. 20.10. Aging is a lifelong process which can be visualized as a wheel that is promoted by genetics. Various endogenous and exogenous factors lead to oxidative stress, which may overwhelm compensatory mechanisms and accelerate aging.

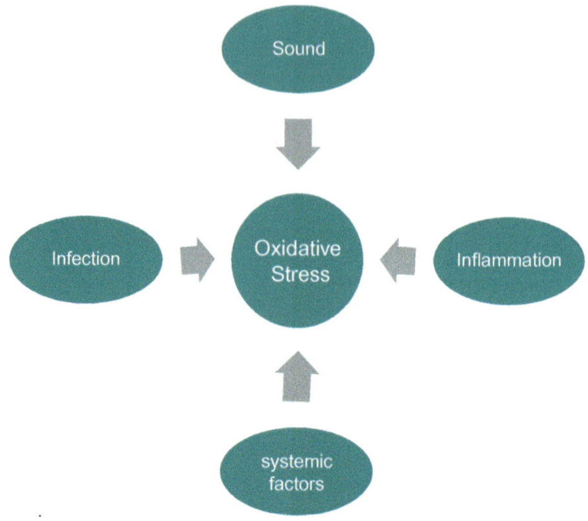

Fig. 20.11. Causes of oxidative stress in the inner ear. Many factors lead to disturbed intracellular homeostasis and therefore to oxidative stress. Overexposure to sound causes oxidative stress in the inner ear as well as systemic factors, e.g., diabetes, vascular and rheumatic factors, and some medication. Furthermore, infection, chronic inflammation, and irradiation therapy have a negative effect on inner-ear homeostasis.

e.g., infection of the middle ear, noise, and systemic diseases, cause oxidative stress that damages inner-ear structures and leads to premature aging (Fig. 20.11). Based on the above-mentioned histologic changes, some of the cellular parts of the inner ear are pointed out with respect to age-related changes. First, the role of oxidative stress and the damaging free radicals is highlighted and then some genetic aspects are summarized with respect to the damage caused by the oxidative stress. Furthermore some mechanisms of inner-ear homeostasis and metabolic factors that influence cell aging are described.

Oxidative stress

Oxidative stress is determined by the imbalance between the production of oxygen metabolites and the cellular ability to detoxify reactive metabolites. These reactive oxygen metabolites and free radicals are highly toxic to many cellular components, even though some of them act as cellular messengers. Thus, if the cell is unable to repair the damage or to restore normal homeostasis, oxidative stress disturbs normal cellular signaling and leads to cell injury or even to cell death (Fig. 20.12).

Mitochondria exhibit an important function in maintaining the production and consumption of these toxic metabolites. Reduced activity of mitochondria increases the reactive metabolites and causes DNA damage that in turn reduces the mitochondrial activity on its own (Seidman, 2000).

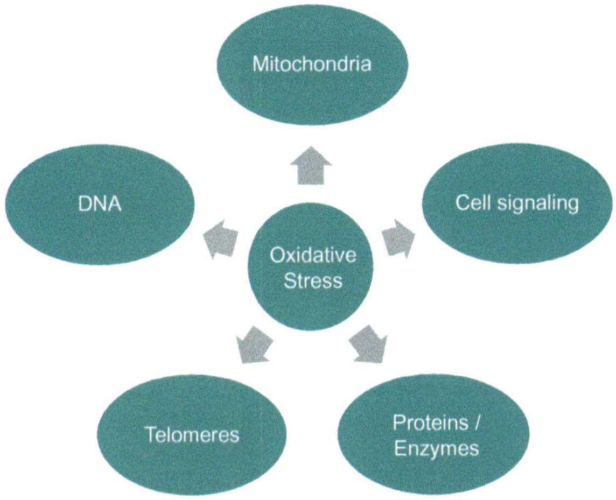

Fig. 20.12. Oxidative stress is due to the accumulation of free radicals during cell respiration. When the cell is unable to detoxify, cell metabolism is injured by these free radicals, leading to cell aging or even cell death. Mitochondria have an important role in detoxifying the cell, but mitochondria on their own are vulnerable to overwhelming oxidative stress. The disturbed calcium regulation due to mitochondrial injury leads to altered cell signaling with adverse effects on the surrounding cells.

Many factors cause oxidative stress (Fig. 20.11). In terms of cochlear pathology, bacterial toxins and noise overexposure cause acute oxidative stress in the cochlea and labyrinth and disturb inner-ear homeostasis with subsequent cell damage, resulting in sensorineural hearing loss and vertigo. Chronic inflammation or infection leads to oxidative stress and negatively affects the cochlea. The accumulation of free radicals in stressed cells on one hand has a toxic effect to bacteria that is advantageous for survival of the organ, but on the other hand damages cellular structures and in some way accelerates the genetic process of aging.

Genetics

As mentioned above, progeria as well as other pathologies, e.g., muscular dystrophies, scleroderma, and cardiomyopathies, belong to the group called laminopathies. They are determined by a genetic defect of the LMNA gene, producing aberrant nuclear proteins (lamin). These proteins lead to abnormal shape of the cellular nucleus and, therefore, have a wide range of effects on cellular function, leading to a broad presentation of clinical symptoms and the manifestation of premature aging (Zhang et al., 2013). To date, no studies are available in the literature investigating the effect of laminopathies on the inner ear.

Another protein, called klotho, was discovered in 1997 and acts as an aging suppressor gene (Kuro-o et al., 1997). It was named after the Greek goddess who spins the thread of life. The gene mutation in the klotho mouse exhibited multiple disorders resembling human premature aging (Kuro-o, 2008), but the protein acts in a different manner than those causing the syndromes of laminopathies. The klotho gene encodes for a secreted protein and has therefore an extracellular function which seems to be involved in the regulation of oxidative stress and senescence.

Klotho is expressed predominantly in tissues involved in the regulation of calcium homeostasis and was found in the stria vascularis and the vestibular dark cells of the inner ear (Kamemori et al., 2002; Takumida et al., 2009). Due to the distribution of klotho close to the Na-K-Cl co-transporter and the involvement of klotho with other proteins that regulate calcium homeostasis, e.g., transient receptor potential (TRPV), it was suggested that klotho participates in the regulation of the ionic composition of the endolymph; furthermore, the expression of klotho in the inner-ear sensory cell of the mouse led to the hypothesis that klotho is involved in signal transduction in the inner ear (Takumida et al., 2009). The expression of klotho was found to be closely related to the lifespan of the organism and klotho seems to have a protective effect against oxidative stress and to increase cell resistance to the subsequent damage (Yamamoto et al., 2005; Ikushima et al., 2006).

Telomeres consist of a series of non-coding, hexanucleotide repeats found at the end of chromosomes, and serve to ensure genomic stability during cell division. Telomeres and telomerase have been investigated not only in cancer development but also in their role of cellular senescence in mitotic cells (Smith et al., 2013). The corresponding enzyme, telomerase, is responsible for maintaining telomere length. It has been reported that telomeres and telomerase are a preferential target of DNA damage under oxidative and genotoxic conditions (Hewitt et al., 2012). Hence, defective DNA leads to a defective cell cycle, with the well-known adverse effects that gradually affect cell and organ structure and function, including cellular aging as well.

Mitochondria

As a consequence of the mechanisms mentioned above, an important aspect of aging can be attributed to mitochondria. Mitochondria exhibit a key role in cell cycle: on the one hand they provide energy and on the other hand they clean reactive oxygen species.

Mitochondria are involved in the aging process. Their interaction with calcium homeostasis and their important role in reducing oxidative stress lead to the mitochondrial clock theory of aging, also referred to as the membrane hypothesis of aging (Seidman et al., 2004).

Mitochondria have a high transmembrane gradient that is vulnerable to oxidative stress. Thus, oxidative stress not only damages the mitochondrial membrane but also leads to a disturbed calcium regulation and hence to altered intracellular signal processes that entail disturbed cell metabolism and finally lead to cell death (Smaili et al., 2013).

Mitochondrial pathologies play an important role in both inherited and acquired hearing loss. Genetic mutation of mitochondria, e.g., the MELAS syndrome (syndrome of mitochondrial encephalopathy, lactic acidosis, and stroke-like episodes) presents clinically in a high percentage with sensorineural hearing loss caused mainly by dysfunction of the stria vascularis, which accounts for the cochlear battery, as mentioned above (Takahashi et al., 2003).

ARHL seems to be caused by mitochondrial dysfunction. Several factors cause oxidative stress that leads to the accumulation of acquired mitochondrial DNA mutations. With preponderance of DNA damage, mitochondria are damaged and the cell becomes bioenergetically deficient (Fischel-Ghodsian et al., 2004). Prior to hair cell injury, mitochondria seem to be affected by oxidative stress first. Altered mitochondria were found in the afferent nerve endings (Omata and Schätzle, 1984), and elevation of reactive oxygen species was observed after acute noise exposure (Ohlemiller et al., 1999).

Furthermore, mitochondria are providing the energy for the transmembrane ion pumps that maintain the high endocochlear potential.

Endocochlear potential

Histological changes in ARHL are predominantly in the basal end of the cochlea, progressing to the upper base. They were investigated with particular interest in the stria vascularis (Ohlemiller et al., 2008). The stria vascularis shows a high density of mitochondria and contains marginal cells, which are characterized by a unique complement of potassium channels and pumps responsible for maintaining the high endocochlear potential. The potassium exchange is supported by fibrocytes, which exhibit an important function in inner-ear homeostasis (García Berrocal et al., 2008). Typically, ARHL is combined with degeneration of the stria vascularis and reduced endocochlear potential (Fig. 20.13) (Schuknecht et al., 1974; Ohlemiller, 2009).

The changes in the stria vascularis are not only marked by loss of marginal cells and fibrocytes, but also in reduced density of mitochondria. Both these changes lead to reduced endocochlear potential. Given the observation that all these parts of the stria vascularis, i.e., mitochondria, fibrocytes and endocholear potential, are reduced in ARHL, this suggests that a closely related and interdependent connection meticulously tunes the electrolyte exchange, particularly sodium, potassium, and calcium (Fig. 20.14) (Goto et al., 1999; Ichimiya et al., 2000; Wangemann, 2002; Spicer and Schulte, 2002; Ohlemiller et al., 2006; Mann et al., 2009; Smaili et al., 2013).

Reduced endocochlear potential itself influences inner-ear homeostasis by increasing oxidative stress, causing further cell damage and thereby accelerating the aging process. This age-related alteration is most apparent in the basal turn of the cochlea, the region with the highest density of mitochondria. Whether the basal turn has high energy demands or is the most important and powerful region in inner-ear homeostasis is a difficult question to answer.

In addition to the mitochondrial aging theory or other local cellular factors, strial insufficiency may also be caused by microvascular pathologies due to systemic changes, such as hypertension, diabetes mellitus, hyperlipoproteinemia, hyperlipidemia, or autoimmune diseases, which all have an impact on ARHL. Correlated with age and duration of diabetes, significant changes in the cochlea were described: thickening of the vessels and basilar membrane and atrophy of stria vascularis were seen in all turns of the cochlea, while loss of outer hair cell was found only on the lower basal turn (Fukushima et al., 2005).

In a similar manner, acute otitis media with labyrinthitis shows a decreasing slope in higher frequencies in the tone audiogram, similar to Figure 20.1. This is explained by bacterial toxins and inflammatory mediators penetrating through the oval window and leading to oxidative stress and consequently to a disturbed homeostasis damaging the adjacent structures, i.e., in the basal turn of the cochlea (Cureoglu et al., 2004). The destruction depends on bacterial virulence and local immune conditions.

Independently of the pathophysiologic origin of energy loss, it seems conceivable that age-related reduction of the endocochlear potential will be particularly distinctive in the basal turn due to the high density of mitochondria. This corresponds with the clinical observation that many entities affecting the inner ear, e.g., inflammation, noise, or infection, lead to a sensory hearing loss (Fig. 20.1). The significant histopathologic changes of the cochlea, i.e., decrease in outer and inner hair cells and decrease in stria vascularis and the spiral ligament in the basal turn, were shown to be induced by chronic otitis media (Cureoglu et al., 2004). Cochlear inflammation is supposed to disrupt fibrocyte function of the spiral ligament; the stimulated fibrocytes produce chemoattractant mediators that induce prolonged inflammation and consequently lead to structural

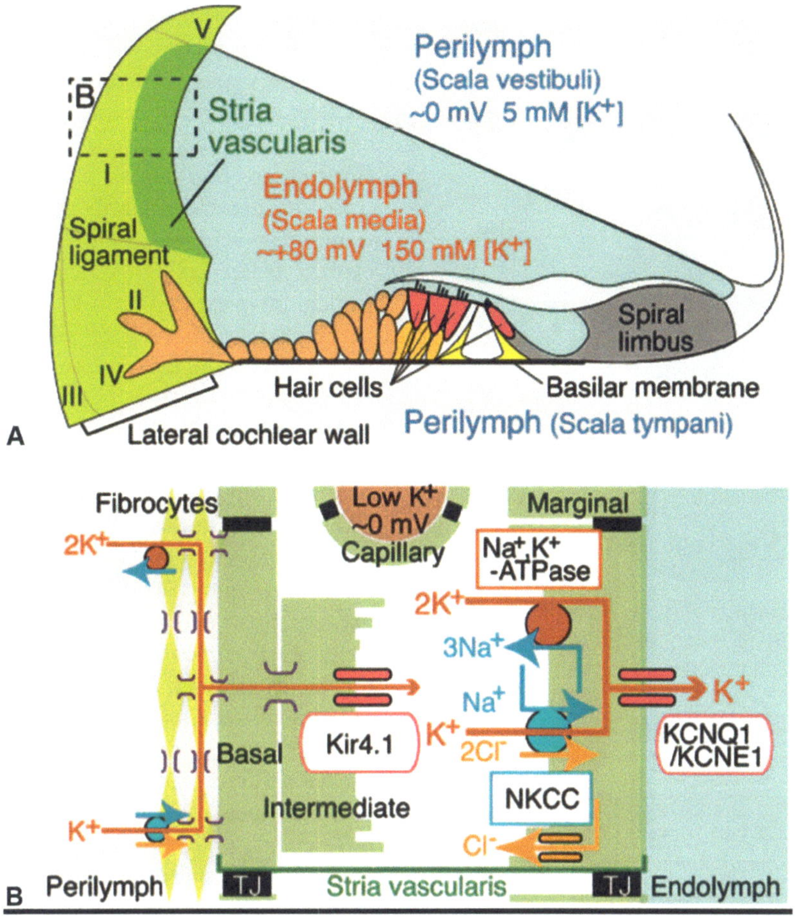

Fig. 20.13. Potassium exchange in the lateral wall of the cochlea. Reprinted from Ohlemiller (2009) and Nin et al. (2008), with permission. Copyright (2008) National Academy of Sciences, U.S.A. Schematic structure of the cochlear duct with the lateral wall. (**A**) Appropriate ionic composition of the endolymph and the endocochlear potential (EP) require that an ion barrier line scala media, the intrastrial space, and strial capillaries. A nearly continuous cellular network guides K^+ from the organ of Corti through finger-like root cells into (primarily) type II and I fibrocytes. Fibrocytes must take up K^+ from the extracellular space around root cells. Locations of five major types of fibrocytes are indicated by roman numerals. (**B**) Schematic enlargement of the boxed area in (**A**) depicts the cells and components required to generate the EP. Type I fibrocytes, strial basal cells, intermediate cells, as well as capillary pericytes (not shown) are joined by gap junctions composed mostly of connexins 26 and 30. K^+ enters the intrastrial space through Kir4.1, then through marginal cells via Na^+/K^+-ATPase, the NKCC ion exchanger, and KCNQ1/KCNE1 channel complexes. TJ, tight junction. (Reproduced from Ohlemiller, 2009 and Nin et al., 2008.)

changes, as mentioned above (Ichimiya et al., 2000; Moon et al., 2006).

The reduced endocochlear potential not only affects the sensory part of the inner ear but also leads to alteration of the supporting structures, e.g., fibrocytes and connective tissue.

Fibrocytes and fibroblast growth factor

Fibrocytes are not only important in the embryologic development of the cochlea; they continue to exhibit a crucial function in "backstage work," providing inner-ear homeostasis; as mentioned above, they are involved in potassium ion exchange. Five different types of fibrocytes are present in the spiral ligament, enabling the Corti organ to conduct signal transduction (Suko et al., 2000) (Fig. 20.15). Moreover, fibrocytes are involved in repairing function during acute or chronic inflammation (Ichimiya et al., 2000); their mediating role in inflammation seems to be decreased and inhibited by corticosteroids (Moriyama et al., 2007).

Fibrocytes are important contributors to forming and supporting cochlear structures, such as delivering components of microtubules to the hair cells (Szarama et al., 2012). Microtubules of the Corti organ also show age-related changes and may contribute to the alteration of micromechanical properties in signal transduction (Saha and Slepecky, 2000). Normal ciliar function

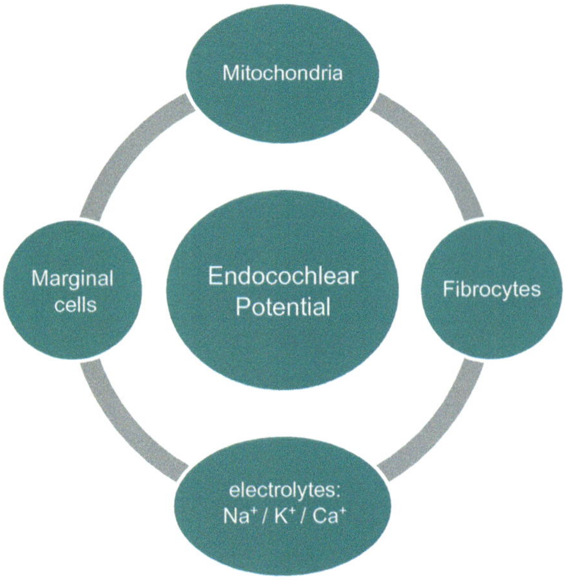

Fig. 20.14. The maintenance of the endocochlear potential demands a high quantity of energy that is supplied by mitochondria. Mitochondria are particularly present in the marginal cells. Marginal cells and fibrocytes are also supporting the high potassium gradient of the inner ear.

depends also on calcium homeostasis (Lumpkin and Hudspeth, 1998). Disturbed function of the cilia in hair cells contributes to altered signal transduction and consequently to hearing difficulties.

Age-related altered fibrocytes not only have decreased efficiency in maintaining inner-ear homeostasis with respect to the demanding endocochlear potential, but probably secrete altered microfilaments and other microstructures, which may influence cochlear dynamics by reducing the mechanical elasticity of the basilar membrane (Spicer and Schulte, 2002). Given these observations, it may be hypothesized that these changes are primarily due to altered function of supporting inner-ear cells, e.g., fibrocytes. It may explain Schuknecht's difficulties in correlating the audiogram with the variable histopathologic changes leading to the somewhat unclear or unproven classification of indeterminant, mixed, or cochlear-conductive hearing loss.

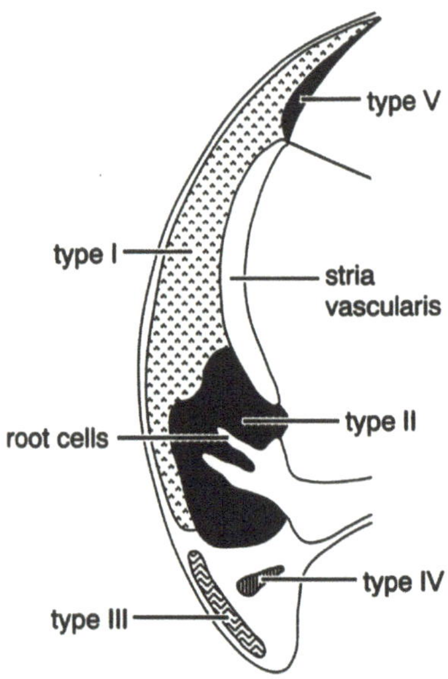

Fig. 20.15. Schematic showing localization of the spiral ligament fibrocyte types, as classified by Spicer and Schulte. The type II and type V fibrocytes shown here are both considered type II fibrocytes according to the Takahashi and Kimura classification system. (Reproduced from Ichimiya et al., 2000.)

Fibroblast growth factor (FGF) is expressed by neurons in the spiral ganglion and in the cochlear nucleus. FGF is not only responsible for normal fibrocyte differentiation and function, but also exhibits an important role in myelination of the cochlear nerve. Two cell types are involved in myelination: Schwann cells myelinate the peripheral part and oligodendrocytes the central part of the cochlear nerve (Wang et al., 2009). Disturbed function of myelination, independently of central or peripheral origin, leads to reduced signal transmission in the cochlear nerve visible in longer latencies of auditory evoked potentials (Makary et al., 2011).

Furthermore, FGF has an important role in hair cell differentiation and repairing function of the inner ear after injury (Jacques et al., 2012). Due to the production of FGF in spiral ganglion cells, it may be hypothesized that spiral ganglion loss results in hypo- or demyelination of the cochlear nerve and has adverse effects on sensory and supporting cells of the cochlea. Loss of spiral ganglion cells was also shown to be a direct result of acoustic overexposure, not just a consequence of hair cell loss (Makary et al., 2011).

Gender and hormones

Epidemiologically, men are more affected by ARHL than women. Consequently, hormones, especially estrogen, have to be considered not only in presbyacusis but also in other pathologies. Gender differences have been reported in auditory brainstem response, showing that women have shorter latencies than men. Women with Turner's syndrome, being biologically estrogen-deficient, have longer latencies in auditory brainstem response and show early presbyacusis (Hultcrantz et al., 2006). Hormone replacement therapy was supposed to have a slightly positive effect on hearing thresholds in menopausal women, but contradictory observations exist, reporting that contraceptive medication and hormone replacement therapy are associated with sudden hearing loss (Hanna, 1986; Strachan, 1996). Moreover, the known prothrombotic effect of conceptive pills might play a role. Nevertheless, human studies investigating the effect of sex hormones on hearing show an influence of estrogen (Horner, 2009).

The estrogen receptors, alpha and beta, are localized where electric impulses are transmitted (inner and outer hair cells, spiral ganglion) and where inner-ear homeostasis is maintained (stria vascularis, spiral ligament) (Stenberg et al., 2002); therefore, estrogens seem to have a role in signal transmission and cochlear homeostasis.

AGING OF THE CENTRAL AUDITORY SYSTEM

Auditory cortex

Speech recognition is a frequent complaint of older adults, particularly in complex and demanding listening conditions. When differentiated hearing is impaired, the age-related changes not only involve the above-mentioned cellular structures of the cochlea, but also affect the central nervous system, in particular the auditory cortex and the frontal lobe. The latter seems to be progressively involved with increased age and to exhibit an important role in speech recognition when central auditory regions show a declining structural integrity (Eckert et al., 2008).

ARHL has been described as correlating with age-related changes in the auditory cortex (Eckert et al., 2012). A lower gray-matter volume of the auditory cortex has been observed in older adults with mild to moderate hearing loss (Husain et al., 2011), but it remains a question for further research whether these findings are causatively connected. A study investigated the gender-related morphologic symmetries of brain structures and described differences in the auditory cortex volume, which was less in men than in women (Rademacher et al., 2001). Given the epidemiologic observation that women are less affected by ARHL than men, the question about causality between epidemiology and the structural difference in the auditory cortex arises and the role and influence of sex hormones have to be considered.

Theory of cortical disconnection

Before the era of functional imaging the classic model of hearing was focused on the Broca's and Wernicke's area on the left hemisphere. The hemispheric specialization of language was supported for many years by differences in dichotic listening paradigms. But recent studies using functional neuroimaging revealed that speech recognition is not restricted to the left hemisphere, but seems to be bilaterally organized (Rogalsky et al., 2008). Furthermore, speech perception is not only confined to the temporal lobe, but also involves the frontal lobe. These findings open the discussion about a new and more sophisticated concept than the classic model of understanding human language and central auditory system (Poeppel et al., 2012).

Despite the controversies concerning the hemispheric asymmetry or connectivity and the modulating role of the frontal lobe, it has been shown that, in elderly patients, particularly in patients with Alzheimer's disease, the asymmetry in dichotic listening is accentuated (Bouma and Gootjes, 2011). These observations of age-related and pathologic changes of the brain with progressive decline of structural areas and tracts may lead to partial or gradual intrahemispheric and interhemispheric disconnectivity, due to disturbed signal transmission. Thus, the theory of cortical "disconnection," i.e., the disruption of subcortical white-matter tracts, has been put forward as a mechanism of age-related cognitive decline. The cortical disconnection seems to involve the central auditory system and impair speech perception (Gootjes et al., 2007).

Calcium homeostasis

Calcium was shown on the one hand to have an important function in intracellular processes and cell signaling and on the other hand to be closely involved in genetics of aging (Takumida et al., 2009; Smaili et al., 2013). Furthermore, calcium not only provides the synaptic potential that permits a finely tuned signal transmission, but also has an important impact on neural viability and synaptic plasticity (Foster, 2007).

As mentioned above, calcium is intimately linked to the activity of mitochondria that are implicated in the clearance of reactive oxygen metabolites. The high membrane potential of mitochondria is maintained by a calcium gradient that is tightly regulated by specific calcium transport mechanisms. Oxidative stress leads to defective mitochondrial membrane proteins and in turn to a disturbed calcium homeostasis and changed intracellular activity with its pathophysiologic consequences.

Contrariwise, age-related reduced calcium regulation alters mitochondrial activity that increases oxidative stress, leading to cell damage (Ouda et al., 2012; Smaili et al., 2013). Calcium dysregulation decreases neural transmission properties and leads to impaired or disrupted information exchange between the connected brain areas

The connection between mitochondrial activity and intracellular calcium regulation points to the importance of both components in aging.

Progressive demyelination of the central nervous system due to age-related changes in calcium homeostasis or mitochondrial activity impairs precise and rapid signal transmission. Impaired signal transduction reduces the viability of synapses and neurons. As explained in the section above, loss of spiral ganglion cells disturbs the activity of fibrocytes, affecting the endocochlear potential and the highly differentiated microstructures of the Corti organ.

Comprehension of these interacting and interconnected mechanisms is important to understand the age-related changes in the clinical examination, discussed in the section below.

DIAGNOSIS AND CLINICAL MANIFESTATION

Patients often complain about hearing loss in a nonspecific way, but a detailed history can point out the circumstances or the social environment where hearing poses a problem. Given the above-mentioned age-related changes of the entire auditory pathway from the peripheral organ to the auditory cortex, all parts of the auditory system have to be considered and analyzed separately, but none the less, in relation to the entire auditory system. According to the site of the lesion, patients' complaints may vary in degree and characteristics. The clinical evaluation of ARHL should not focus only on the cochlea but also pay attention to the central nervous system. Additionally, psychologic aspects have to be considered, together with social personal activities when aiming to provide comprehensive holistic medical care.

One of the main characteristics in ARHL is the progressive difficulty in speech recognition, particularly in noisy environment. Some patients note that intelligibility depends on the frequency and phonetic resonance of the speaker's voice. This complaint may correspond to the above-mentioned age-related changes in the cochlea with its altered structures, mainly in the basal turn of the Corti organ. But diminished speech recognition may also be a sign of affected auditory cortex. It is quite a demanding clinical task to differentiate the involved site of ARHL and to determine whether the peripheral, the central, or both parts are predominantly affected.

Several clinical tests are used in the topodiagnostic evaluation of hearing loss. These are complemented by other tests, such as auditory evoked potentials, functional magnetic resonance imaging, and motor evoked potential, which can document slowing of the central nervous system.

There are two main components of the central nervous system that are affected by age: cognitive ability and temporal processing. Temporal processing becomes slower and cognitive ability decreases progressively. The latter is mainly influenced by altered central selection mechanisms, consisting of impairment of information transduction. In the aging central nervous system these modulating mechanisms, apart from being slower and less efficient, are shifted to the frontal cortex and hence, associative or memory function becomes more important in hearing processing and speech recognition.

The tone audiogram is the basic test to determine hearing loss (Figs 20.1–20.3). ARHL is typically characterized by symmetric high-frequency loss. This test depends on the patient's attention and cooperation. The tone audiogram provides quite a good evaluation of peripheral audition, including its mechanical components.

A central component is included in speech audiometry, since speech recognition depends on central auditory pathways and requires cognitive abilities. Central auditory testing by dichotic sentence or digit identification is recommended in the evaluation of elderly with hearing impairment (Gates et al., 2008).

Otoacoustic emissions play a less important role in the clinical evaluation of ARHL, because they exhibit an individual variability and therefore it is impossible to get a standard reference (Probst et al., 1991). It represents a dynamic exam that only with longitudinal testing can reveal age-related changes of the individually measured cochlear mechanisms. These changes are less prominent compared with threshold level and seem to depend on the decreasing endocochlear potential that is not measurable in clinical routine exam.

Auditory evoked potentials give important information about the central component of the auditory system. According to the age-related changes mentioned above, latencies are prolonged and the potential diminished. In the clinical exam they are mainly used to answer the question of whether there is a retrocochlear hearing disturbance. As with otoacoustic emissions, there is no standard reference, for the same reason of individual variability.

REHABILITATION

Current hearing rehabilitation is focused on the peripheral auditory part by applying hearing aids. With the continuous development of technical possibilities, the industry tries to improve the fine-tuning of the devices to compensate for age-related deficits. Unfortunately, they show limited success due to the above-mentioned variety of structural changes in the hearing pathway. The more the central auditory pathway and central nervous system are affected by structural changes, the less the success of hearing aids. In contrast, peripheral deficits respond well to the application of hearing aids.

In case of profound hearing loss unamenable to standard hearing aids, a cochlear implant (CI) can be evaluated. CIs in the elderly showed satisfying results, even if the difference in hearing improvement was less marked than in children. Hence, if the peripheral auditory deficit can be compensated for sufficiently by CI, the central part of auditory deficit seems to be of less importance (Lenarz et al., 2012; Blamey et al., 2013).

When age-related changes mainly affect the central components, peripheral efforts will fail and auditory training should be attempted together with psychomotor or general daily activity.

CONSEQUENCES OF ARHL

Based on the above-mentioned epidemiologic considerations, ARHL is the most common chronic disabling condition in older adults. Hearing loss affects one of the most important human capabilities: verbal communication.

The disabling severity of ARHL is shown by a wide range of negative consequences caused by difficulties in understanding. These age-related changes in hearing lead to a reduced quality of life that affects all parts of human activity. Social isolation, emotional frustration, cognitive dysfunction, and behavioral changes may be further consequences of hearing impairment.

Moreover, the central nervous system, social activity, or lifestyle and psychologic factors have to be considered when dealing with hearing impairment. Patients with hearing impairment risk being isolated and isolating themselves, and therefore are prone to developing depression.

Premature aging of the central nervous system seems to affect the hearing process, on the one hand due to declined psychomotor activity and social interest, and on the other hand because reduced central activity may have a negative impact on auditory processing and possibly also on the peripheral auditory organ.

Hearing loss also affects negatively work productivity and has an impact on the economy. In this context, hearing loss seems to entail negative consequences for the social state, health, and survival (Barnett and Franks, 1999). Furthermore, ARHL is estimated to have a considerable economic weight in the public healthcare system.

For all these reasons, prevention (for example, through public health campaigns and work legislation on noise exposure), early identification, and appropriate management of AHRL are all important.

References

Agrawal Y, Platz EA, Niparko JK (2008). Prevalence of hearing loss and differences by demographic characteristics among US adults: data from the National Health and Nutrition Examination Survey, 1999-2004. Arch Intern Med 168 (14): 1522–1530.

Barnett S, Franks P (1999). Deafness and mortality: analyses of linked data from the National Health Interview Survey and National Death Index. Public Health Rep 114 (4): 330–336.

Bergmann B, Rosenhall U (2001). Vision and hearing in old age. Scand Audiol 30 (4): 255–263.

Blamey P, Artieres F, Başkent D et al. (2013). Factors affecting auditory performance of postlinguistically deaf adults using cochlear implants: an update with 2251 patients. Audiol Neurootol 18 (1): 36–47.

Borchgrevink HM, Tambs K, Hoffman HJ (2005). The Nord-Trondelag Norway Audiometric Survey 1996–98: unscreened thresholds and prevalence of hearing impairment for adults > 20 years. Noise Health 7 (28): 1–15.

Bouma A, Gootjes L (2011). Effects of attention on dichotic listening in elderly and patients with dementia of the Alzheimer type. Brain Cogn 76 (2): 286–293.

Coppedè F (2013). The epidemiology of premature aging and associated comorbidities. Clin Interv Aging 8: 1023–1032.

Cureoglu S, Schachern PA, Paparella MM et al. (2004). Cochlear changes in chronic otitis media. Laryngoscope 114 (4): 622–626.

Davis AC (1989). The prevalence of hearing impairment and reported hearing disability among adults in Great Britain. Int J Epidemiol 18 (4): 911–917.

Davis A (1995). Hearing in adults. Whurr, London.

Eckert MA, Walczak A, Ahlstrom J et al. (2008). Age-related effects on word recognition: reliance on cognitive control systems with structural declines in speech-responsive cortex. J Assoc Res Otolaryngol 9 (2): 252–259.

Eckert MA, Cute SL, Vaden KI et al. (2012). Auditory cortex signs of age-related hearing loss. J Assoc Res Otolaryngol 13 (5): 703–713.

Fischel-Ghodsian N, Kopke RD, Ge X (2004). Mitochondrial dysfunction in hearing loss. Mitochondrion 4 (5–6): 675–694.

Fogerty D, Kewley-Port D (2009). Perceptual contributions of the consonant-vowel boundary to sentence intelligibility. J Acoust Soc Am 126 (2): 847–857.

Foster TC (2007). Calcium homeostasis and modulation of synaptic plasticity in the aged brain. Aging Cell 6 (3): 319–325.

Frisina RD, Walton JP (2006). Age-related structural and functional changes in the cochlear nucleus. Hear Res 216–217: 216–223.

Fukushima H, Cureoglu S, Schachern PA et al. (2005). Cochlear changes in patients with type 1 diabetes mellitus. Otolaryngol Head Neck Surg 133 (1): 100–106.

García Berrocal JR, Méndez-Benegassi I, Martí C et al. (2008). Intervention of spiral ligament fibrocytes in the metabolic regulation of the inner ear. Acta Otorrinolaringol Esp 59 (10): 494–499.

Gates GA, Mills JH (2005). Presbycusis. Lancet 366 (9491): 1111–1120.

Gates GA, Popelka GR (1992). Neural presbycusis: a diagnostic dilemma. Am J Otol 13 (4): 313–317.

Gates GA, Anderson ML, Feeney MP et al. (2008). Central auditory dysfunction in older persons with memory impairment or Alzheimer dementia. Arch Otolaryngol Head Neck Surg 134 (7): 771–777.

George DL, Pradhan S (2009). Idiopathic sensorineural hearing disorders in adults – a pragmatic approach. Nat Rev Rheumatol 5 (9): 505–512.

Gootjes L, Scheltens P, Van Strien JW et al. (2007). Subcortical white matter pathology as a mediating factor for age-related decreased performance in dichotic listening. Neuropsychologia 45 (10): 2322–2332.

Goto S, Oshima T, Ikeda K et al. (1999). Expression and localization of the Na^+-H^+ exchanger in the guinea pig cochlea. Hear Res 128 (1–2): 89–96.

Hanna GS (1986). Sudden deafness and the contraceptive pill. J Laryngol Otol 100 (6): 701–706.

Hewitt G, Jurk D, Marques FD et al. (2012). Telomeres are favoured targets of a persistent DNA damage response in ageing and stress-induced senescence. Nat Commun 3: 708.

Hietanen A, Era P, Sorri M et al. (2004). Changes in hearing in 80-year-old people: a 10-year follow-up study. Int J Audiol 43 (3): 126–135.

Hietanen A, Era P, Henrichsen J et al. (2005). Hearing among 75-year-old people in three Nordic localities: a comparative study. Int J Audiol 44 (9): 500–508.

Horner KC (2009). The effect of sex hormones on bone metabolism of the otic capsule – an overview. Hear Res 252 (1–2): 56–60.

Hultcrantz M, Simonoska R, Stenberg AE (2006). Estrogen and hearing: a summary of recent investigations. Acta Otolaryngol 126 (1): 10–14.

Husain FT, Medina RE, Davis CW et al. (2011). Neuroanatomical changes due to hearing loss and chronic tinnitus: a combined VBM and DTI study. Brain Res 1369: 74–88.

Ichimiya I, Yoshida K, Hirano T et al. (2000). Significance of spiral ligament fibrocytes with cochlear inflammation. Int J Pediatr Otorhinolaryngol 56 (1): 45–51.

Ikushima M, Rakugi H, Ishikawa K et al. (2006). Anti-apoptotic and anti-senescence effects of Klotho on vascular endothelial cells. Biochem Biophys Res Commun 339 (3): 827–832.

International Organization for Standardization (ISO) (2000). Acoustics – Statistical distribution of hearing thresolds as a function of age. ISO, Geneva.

Jacques BE, Dabdoub A, Kelley MW (2012). Fgf signaling regulates development and transdifferentiation of hair cells and supporting cells in the basilar papilla. Hear Res 289 (1–2): 27–39.

Johansson MS, Arlinger SD (2003). Prevalence of hearing impairment in a population in Sweden. Int J Audiol 42 (1): 18–28.

Kamemori M, Ohyama Y, Kurabayashi M et al. (2002). Expression of Klotho protein in the inner ear. Hear Res 171 (1–2): 103–110.

Kewley-Port D, Burkle TZ, Lee JH (2007). Contribution of consonant versus vowel information to sentence intelligibility for young normal-hearing and elderly hearing-impaired listeners. J Acoust Soc Am 122 (4): 2365–2375.

Kuro-o M (2008). Klotho as a regulator of oxidative stress and senescence. Biol Chem 389 (3): 233–241.

Kuro-o M, Matsumura Y, Aizawa H et al. (1997). Mutation of the mouse klotho gene leads to a syndrome resembling ageing. Nature 390 (6655): 45–51.

Lenarz M, Sönmez H, Joseph G et al. (2012). Cochlear implant performance in geriatric patients. Laryngoscope 122 (6): 1361–1365.

Lin FR, Niparko JK, Ferrucci L (2011). Hearing loss prevalence in the United States. Arch Intern Med 171 (20): 1851–1852.

Lumpkin EA, Hudspeth AJ (1998). Regulation of free Ca^{2+} concentration in hair-cell stereocilia. J Neurosci 18 (16): 6300–6318.

Makary CA, Shin J, Kujawa SG et al. (2011). Age-related primary cochlear neuronal degeneration in human temporal bones. J Assoc Res Otolaryngol 12 (6): 711–717.

Mann ZF, Duchen MR, Gale JE (2009). Mitochondria modulate the spatio-temporal properties of intra- and intercellular Ca^{2+} signals in cochlear supporting cells. Cell Calcium 46 (2): 136–146.

Moller MB (1981). Hearing in 70 and 75 year old people: results from a cross sectional and longitudinal population study. Am J Otolaryngol 2 (1): 22–29.

Moon SK, Park R, Lee HY et al. (2006). Spiral ligament fibrocytes release chemokines in response to otitis media pathogens. Acta Otolaryngol 126 (6): 564–569.

Moriyama M, Yoshida K, Ichimiya I et al. (2007). Nitric oxide production from cultured spiral ligament fibrocytes: effects of corticosteroids. Acta Otolaryngol 127 (7): 676–681.

Murdin L, Patel S, Walmsley J et al. (2008). Hearing difficulties are common in patients with rheumatoid arthritis. Clin Rheumatol 27 (5): 637–640.

Nadol Jr JB (2010). Disorders of Aging. In: SN Merchant, NJ Nadol (Eds.), Schuknecht's Pathology of the Ear. People's Medical Publishing House-USA, Shelton, CT.

Nin F, Hibino H, Doi K et al. (2008). The endocochlear potential depends on two K^+ diffusion potentials and an electrical barrier in the stria vascularis of the inner ear. Proc Natl Acad Sci U S A 105 (5): 1751–1756.

Ohlemiller KK (2004). Age-related hearing loss: the status of Schuknecht's typology. Curr Opin Otolaryngol Head Neck Surg 12 (5): 439–443.

Ohlemiller KK (2009). Mechanisms and genes in human strial presbycusis from animal models. Brain Res 1277: 70–83.

Ohlemiller KK, Wright JS, Dugan LL (1999). Early elevation of cochlear reactive oxygen species following noise exposure. Audiol Neurootol 4 (5): 229–236.

Ohlemiller KK, Lett JM, Gagnon PM (2006). Cellular correlates of age-related endocochlear potential reduction in a mouse model. Hear Res 220 (1–2): 10–26.

Ohlemiller KK, Rice ME, Gagnon PM (2008). Strial microvascular pathology and age-associated endocochlear potential decline in NOD congenic mice. Hear Res 244 (1–2): 85–97.

Omata T, Schätzle W (1984). Electron microscopical studies on the effect of lapsed time on the nerve endings of the outer hair cells in acoustically exposed rabbits. Arch Otorhinolaryngol 240 (2): 175–183.

Otto WC, McCandless GA (1982). Aging and auditory site of lesion. Ear Hear 3 (3): 110–117.

Ouda L, Burianova J, Syka J (2012). Age-related changes in calbindin and calretinin immunoreactivity in the central auditory system of the rat. Exp Gerontol 47 (7): 497–506.

Poeppel D, Emmorey K, Hickok G et al. (2012). Towards a new neurobiology of language. J Neurosci 32 (41): 14125–14131.

Probst R, Lonsbury-Martin BL, Martin GK (1991). A review of otoacoustic emissions. J Acoust Soc Am 89 (5): 2027–2067.

Quaranta A, Assennato G, Sallustio V (1996). Epidemiology of hearing problems among adults in Italy. Scand Audiol Suppl 42: 9–13.

Rademacher J, Morosan P, Schleicher A et al. (2001). Human primary auditory cortex in women and men. Neuroreport 12 (8): 1561–1565.

Rahko T, Kallio V, Kataja M et al. (1985). Prevalence of handicapping hearing loss in an aging population. Ann Otol Rhinol Laryngol 94 (2 Pt 1): 140–144.

Rogalsky C, Pitz E, Hillis AE et al. (2008). Auditory word comprehension impairment in acute stroke: relative contribution of phonemic versus semantic factors. Brain Lang 107 (2): 167–169.

Rosenhall U, Karlsson Espmark AK (2003). Hearing aid rehabilitation: what do older people want, and what does the audiogram tell? Int J Audiol 42 (Suppl. 2): 2S53–2S57.

Roth TN, Hanebuth D, Probst R (2011). Prevalence of age-related hearing loss in Europe: a review. Eur Arch Otorhinolaryngol 268 (8): 1101–1107.

Saha S, Slepecky NB (2000). Age-related changes in microtubules in the guinea pig organ of Corti. Tubulin isoform shifts with increasing age suggest changes in micromechanical properties of the sensory epithelium. Cell Tissue Res 300 (1): 29–46.

Schuknecht HF, Gacek MR (1993). Cochlear pathology in presbycusis. Ann Otol Rhinol Laryngol 102 (1 Pt 2): 1–16.

Schuknecht HF, Watanuki K, Takahashi T et al. (1974). Atrophy of the stria vascularis, a common cause for hearing loss. Laryngoscope 84 (10): 1777–1821.

Seidman MD (2000). Effects of dietary restriction and antioxidants on presbyacusis. Laryngoscope 110 (5 Pt 1): 727–738.

Seidman MD, Ahmad N, Bai U (2002). Molecular mechanisms of age-related hearing loss. Ageing Res Rev 1 (3): 331–343.

Seidman MD, Ahmad N, Joshi D et al. (2004). Age-related hearing loss and its association with reactive oxygen species and mitochondrial DNA damage. Acta Otolaryngol Suppl 552: 16–24.

Smaili SS, Pereira GJ, Costa MM et al. (2013). The role of calcium stores in apoptosis and autophagy. Curr Mol Med 13 (2): 252–265.

Smith JA, Park S, Krause JS et al. (2013). Oxidative stress, DNA damage, and the telomeric complex as therapeutic targets in acute neurodegeneration. Neurochem Int 62 (5): 764–775.

Spicer SS, Schulte BA (2002). Spiral ligament pathology in quiet-aged gerbils. Hear Res 172 (1–2): 172–185.

Stenberg AE, Wang H, Sahlin L et al. (2002). Estrogen receptors alpha and beta in the inner ear of the 'Turner mouse' and an estrogen receptor beta knockout mouse. Hear Res 166 (1–2): 1–8.

Strachan D (1996). Sudden sensorineural deafness and hormone replacement therapy. J Laryngol Otol 110 (12): 1148–1150.

Suko T, Ichimiya I, Yoshida K et al. (2000). Classification and culture of spiral ligament fibrocytes from mice. Hear Res 140 (1–2): 137–144.

Szarama KB, Stepanyan R, Petralia RS et al. (2012). Fibroblast growth factor receptor 3 regulates microtubule formation and cell surface mechanical properties in the developing organ of Corti. Bioarchitecture 2 (6): 214–219.

Takahashi K, Merchant SN, Miyazawa T et al. (2003). Temporal bone histopathological and quantitative analysis of mitochondrial DNA in MELAS. Laryngoscope 113 (8): 1362–1368.

Takumida M, Ishibashi T, Hamamoto T et al. (2009). Age-dependent changes in the expression of klotho protein, TRPV5 and TRPV6 in mouse inner ear. Acta Otolaryngol 129 (12): 1340–1350.

Walters BJ, Zuo J (2013). Postnatal development, maturation and aging in the mouse cochlea and their effects on hair cell regeneration. Hear Res 297: 68–83.

Wang SJ, Furusho M, D'Sa C et al. (2009). Inactivation of fibroblast growth factor receptor signaling in myelinating glial cells results in significant loss of adult spiral ganglion neurons accompanied by age-related hearing impairment. J Neurosci Res 87 (15): 3428–3437.

Wangemann P (2002). K(+) cycling and its regulation in the cochlea and the vestibular labyrinth. Audiol Neurootol 7 (4): 199–205.

Wilson PS, Fleming DM, Donaldson I (1993). Prevalence of hearing loss among people aged 65 years and over: screening and hearing aid provision. Br J Gen Pract 43 (375): 406–409.

Yamamoto M, Clark JD, Pastor JV et al. (2005). Regulation of oxidative stress by the anti-aging hormone klotho. J Biol Chem 280 (45): 38029–38034.

Zhang H, Kieckhaefer JE, Cao K (2013). Mouse models of laminopathies. Aging Cell 12 (1): 2–10.

Handbook of Clinical Neurology, Vol. 129 (3rd series)
The Human Auditory System
G.G. Celesia and G. Hickok, Editors

Chapter 21

Decreased sound tolerance: hyperacusis, misophonia, diplacousis, and polyacousis

PAWEL J. JASTREBOFF[1]* AND MARGARET M. JASTREBOFF[2]
[1]*Department of Otolaryngology, Emory University School of Medicine, Atlanta, GA, USA*
[2]*JHDF, Inc., Ellicott City, MD, USA*

INTRODUCTION

Decreased sound tolerance, diplacousis, and polyacousis are dysfunctions of auditory perception. Decreased sound tolerance can be subdivided into hyperacusis and misophonia (Jastreboff and Jastreboff, 2002, 2013). While both phenomena are frequently observed in otolaryngologic and audiologic clinics, our knowledge of decreased sound tolerance is very limited and its proposed mechanisms are unproven and speculative. Nevertheless, available information suggests that decreased sound tolerance results from dysfunction of the central auditory pathways and their connections with the central nervous system rather than from dysfunction of the inner ear.

There is no consensus on the testing of decreased sound tolerance. Limited information is available regarding normative data for loudness discomfort levels (LDLs), which are typically used to assess the presence of hyperacusis (Sherlock and Formby, 2005). Questionnaires have been proposed to evaluate hyperacusis severity (Khalfa et al., 2002; Dauman and Bouscau-Faure, 2005); however, their validity needs to be confirmed. Moreover, the questionnaires do not address the presence and extent of misophonia. Reported data on the prevalence and epidemiology of hyperacusis are highly diverse. Hyperacusis can occur alone or as an adjunct to complex medical conditions (Jastreboff and Jastreboff, 2009). In spite of all the limitations listed above, the treatment of decreased sound tolerance can be highly successful (Jastreboff and Jastreboff, 2000, 2013; Jastreboff and Hazell, 2004; Formby, 2007; Formby et al., 2007).

Diplacousis was first reported in 1907 by Shambaugh, Sr. (Albers and Wilson, 1968a) and is much less frequently diagnosed than decreased sound tolerance. Diplacousis occurs when a subject who is exposed to a single-frequency tone perceives multiple tones, roughness, and beats (Ward, 1952, 1955; Flottorp, 1953; Zurek, 1981; Bacon and Viemeister, 1985). A characteristic feature of diplacousis is that the distortions are perceived only for low levels of sound. When the intensity of the external tone increases, the distortions disappear and the subject perceives a single pure tone. The mechanism of diplacousis is linked to dysfunction of the inner ear (Knight, 2004), and only single reports link it to dysfunction of the central auditory pathways (Ghosh, 1990). Polyacousis occurs when more than two tones are perceived. This term is seldom used and only one paper on the subject can be found in MedLine (Corliss et al., 1968). An internet search, using various search engines, yielded no hits for polyacousis.

DEFINITIONS

A variety of terms (e.g., hyperacousia, auditory hyperesthesia, recruitment, dysacousis, auditory dysesthesia, odynacusis, and auditory allodynia) (Baguley and McFerran, 2010) have been proposed to describe decreased sound tolerance, with the word "hyperacusis" used most frequently (Jastreboff and Jastreboff, 2004). One approach proposes further dividing hyperacusis into loudness hyperacusis, annoyance hyperacusis, and fear hyperacusis (Tyler et al., 2009).

Stedman's Medical Dictionary defines hyperacusis as "Abnormal acuteness of hearing due to increased

*Correspondence to: Pawel J. Jastreboff, Ph.D., Sc.D., Department of Otolaryngology, Emory University School of Medicine, 550 Peachtree St. NE, Medical Office Tower 11th Floor, Suite 1130, Atlanta, GA 30308, USA. Tel: +1-404-778-3388, Fax: +1-404-778-3382, E-mail: pjastre@emory.edu

irritability of the sensory neural mechanism. Syn: auditory hyperesthesia," with hyperesthesia defined as "Abnormal acuteness of sensitivity to touch, pain, or other sensory stimuli" (Anon., 1997). *American Heritage Dictionary* defines hyperacusis as "An abnormal or pathological increase in sensitivity to sensory stimuli, [as of the skin to touch or] the ear to sound" (Anon., 1994).

The definitions promoted in this chapter are based on reports from patients and on potential physiologic mechanisms, in which the auditory system (both peripheral and central parts) and the limbic and autonomic nervous systems are the main systems involved. An analysis of the negative reactions of subjects who suffer from decreased sound tolerance points to the involvement of the limbic and autonomic nervous systems in cases of clinically significant decreased sound tolerance (i.e., where a subject's difficulty with this condition is substantial enough to seek professional help). A wide spectrum of emotional reactions is observed (e.g., discomfort, dislike, distress, annoyance, anxiety, and fear), as well as a variety of negative sensations (e.g., pain, physical discomfort, fullness in the ear) and reactions associated with the overstimulation of the sympathetic part of the autonomic nervous system (e.g., anxiety, panic, decreased ability to enjoy life activities, sleep problems, problem with digestion) (Jastreboff and Hazell, 2004; Jastreboff and Jastreboff, 2004, 2009, 2013; Schroder et al., 2013).

Decreased sound tolerance is defined as present when a subject exhibits negative reactions when exposed to sound that would not evoke the same response in an average listener (Jastreboff and Jastreboff, 2002, 2013; Jastreboff and Hazell, 2004). These negative reactions can be observed when the subject is exposed to low-, moderate-, or high-level sound. The notion that sound must be loud, or louder than a certain level, is not a prerequisite for, and is not used as a criterion to, diagnose decreased sound tolerance. A patient's negative reactions may occur in response to any (including low) level of sound. It is postulated that decreased sound tolerance consists of two components: hyperacusis and misophonia, which frequently occur together.

Hyperacusis is defined as present when negative reactions to a sound depend only on its physical characteristics (i.e., its spectrum and intensity). The sound's meaning and the context in which it occurs are irrelevant (Jastreboff and Jastreboff, 2002, 2013; Jastreboff and Hazell, 2004).

Misophonia is defined as an abnormally strong reaction to a sound with a specific pattern and meaning to a given subject. The physical characteristics of the sound are secondary. Reactions to the sound depend on the subject's past history and on non-auditory factors such as the subject's previous evaluation of the sound, his/her belief that the sound is a potential threat or that exposure to it will be harmful, the subject's psychologic profile, and the context in which the sound is presented (Jastreboff and Jastreboff, 2002, 2013; Jastreboff and Hazell, 2004). The strength of the reaction is only partially determined by the sound's physical characteristics. A specific category of misophonia occurs when fear is the dominant emotion and patients are afraid of the sound (i.e., phonophobia: *phobia* = fear) (Jastreboff and Jastreboff, 2002, 2013; Jastreboff and Hazell, 2004).

Neither hyperacusis nor misophonia has any relation to the threshold of hearing, which can be normal or can involve hearing loss. Therefore, the term "recruitment" is not related to decreased sound tolerance. Recruitment refers to an unusually rapid growth of loudness as the level of a tone is increased. It occurs in association with hearing loss and is a purely cochlear-based phenomenon (Moore, 1995). Recruitment might coexist with decreased sound tolerance, but there is no functional link between the two conditions (Jastreboff and Jastreboff, 2004). Out of 740 patients treated at the Emory Tinnitus and Hyperacusis Center, during the first visit 497 patients reported a problem with decreased sound tolerance and 73% of patients in this group reported a problem with hearing as well. A total of 126 patients (17%) had hyperacusis that required specific treatment and out of that group of patients, 84 (67%) had concurrent problems with hearing. On the other hand, 476 out of 740 patients (64%) reported a problem with hearing. In all, 362 patients (76%) from this group also reported a problem with decreased sound tolerance and 70 patients (15%) required specific treatment for hyperacusis.

There is more than one definition of diplacousis (the spelling "diplacusis" is also used). One of the older definitions is that diplacousis is a perception of rough or noisy sound, or the perception of two or more sounds when a subject is exposed to a single pure tone (Ward, 1952). Currently, diplacousis is defined as an "abnormal perception of sound either in time or in pitch, such that one sound is heard as two" (Anon., 1997). The name originates from a combination of two Greek words: *diplous*, meaning double and *akousis*, meaning hearing. While specific epidemiologic studies are lacking, it is believed that diplacousis is present in many people; however, if the pitch difference is less than a half-tone and the subject does not have musical training it is not noticed. Diplacousis predominantly affects musicians as they are often trained to distinguish between small pitch differences.

A broader definition is that diplacousis is present when exposure to a pure tone results in the perception of several tones in addition to the presented tone (Ward, 1955). A tone may be perceived as harmonics of a fundamental sound or the subject's voice (Brookler, 2009). Diplacousis depends on the frequency of the evoking sound (Brookler, 2009) and on the sound's intensity. When the intensity increases, diplacousis decreases (Burns, 1982).

Several subtypes of diplacousis can be distinguished (Bauman, 2013). Diplacousis binauralis occurs when one sound is perceived dissimilarly in each ear and may differ in pitch or in time. Diplacousis dysharmonica occurs when the sound is perceived normally in one ear and at a different pitch in the other ear (i.e., an interaural pitch difference (IPD)). Diplacousis echoica occurs when exposure to a sound results in the perception of the sound followed by the repetition or echo of the sound. Diplacousis monauralis occurs when exposure to a sound results in the perception of two different sounds in one ear. Diplacousis dysharmonica/IPD (d_L) is the most common type of diplacousis. IPD is defined as $d_L = (f_R - f_L)/f_L = f_R/f_L - 1$, where f_R/f_L denotes the frequencies at the left and right ear, obtained by the adjustment of either frequency to achieve the pitch match (Terhardt, 2000). Diplacousis is typically associated with hearing loss. A variety of presumed mechanisms for diplacousis are associated with its different subtypes and are discussed below.

Polyacousis is a specific case of diplacousis that occurs when more than two tones are perceived. This term is rarely used.

DECREASED SOUND TOLERANCE (HYPERACUSIS AND MISOPHONIA), DIPLACOUSIS, AND POLYACOUSIS AS A PROBLEM

Decreased sound tolerance can have an extremely strong impact on patients' lives. Any sound can become offensive and evoke a negative reaction. With hyperacusis, any sound which is louder than acceptable to a patient evokes negative reactions. For misophonia, a broad variety of sounds have been reported to create problems. These sounds can be loud or soft, such as street sounds, car brakes, lawnmower, vacuum cleaner, garbage disposal, flushing toilet, cutlery and plates, keys rattling, hair dryer, school bell, announcements in a metro station or on a train or airplane, typing on a computer keyboard, the sound of a computer, refrigerator, supermarket freezer, the hum of electricity, a swimming-pool pump, shoveling cement, a husband/wife breathing in bed, snoring, sounds of eating, one's own voice, a crying baby, a musical instrument, singing, other people's headphones, fetal heart sounds, or the sound of drawing with a felt-tipped pen (Hazell et al., 2002; Jastreboff and Jastreboff, 2004, 2013). On the other hand, sounds such as bird song, running water, wind, and rain are rarely reported as negative (Hazell et al., 2002).

Decreased sound tolerance can prevent people from working and interacting socially. In extreme cases, patients do not leave their homes, and their lives and their family's lives revolve around (and are controlled by) the avoidance of offensive sounds. Misophonia evokes the same reactions as hyperacusis and further enhances the effects of hyperacusis.

Diplacousis and polyacousis are rare and are infrequently reported as causing a problem for subjects who experience them. The exception is some musicians, as diplacousis may have an impact on their ability to work by decreasing their perception of music (Kaharit et al., 2003; Jansen et al., 2009). Jansen et al. (2009) performed a study on 241 professional musicians in a symphony orchestra, aged from 23 to 64 years, and assessed their hearing loss and other hearing-related complaints. They found that musicians often complained about hyperacusis and tinnitus (79% and 51%, respectively), while diplacousis diagnosed by the presence of IPD > 1% was measured in 44% of musicians, but reported as a problem for only 7%. Interestingly, 24% of subjects complained about the distortion of tones.

Considering that differences in pitch sensation between 1.6% and 2.3% are common in non-musicians (Brink van den, 1970; Burns, 1982), it is notable in this study of diplacousis that IPD larger than 2% was observed in 18% of subjects and IPD larger than 3% was present in 3% of subjects.

Jansen et al. (2009) reported only a few very sensitive people who experienced diplacousis. This postulate is further corroborated by their finding that, when the subjective results on diplacousis were compared to the results of diplacousis matching, no significant correlation was found for any of the tested frequencies. Consequently, even when diplacousis is documented by IPD, it may neither be noticed nor create a problem, even for musicians.

Kaharit et al. (2003) evaluated 139 rock/jazz musicians for hearing disorders (Table 21.1). The presence of tinnitus, hyperacusis, distortion, or diplacousis was established by the subject's answers on a questionnaire. The authors defined tinnitus, hyperacusis, distortion, and diplacousis in the following non-standard manner:

> *Tinnitus was defined as a spontaneous or evoked sensation of sounds, e.g. ringing or buzzing, often combined with pure tones that occur in the absence of an external sound source. The different sounds could be uni- or bilaterally located in the ears, or experienced and located somewhere in the head. Hyperacusis was defined as hypersensitivity to the loudness of sounds, including a*

Table 21.1

Type of hearing disorders reported in 103 affected musicians

Groups of single or >1 hearing disorder	Hearing loss *n* (%)	Tinnitus *n* (%)	Hyperacusis *n* (%)	Distortion *n* (%)	Diplacusis *n* (%)
Single hearing disorder	11 (10.7)	7 (6.8)	9 (8.7)	1 (1)	0 (0)
>1 Hearing disorder	57 (55.3)	60 (58.3)	54 (52.4)	24 (23.3)	4 (3.9)

Data modified from Kaharit et al. (2003).

decreased pure tone and uncomfortable loudness level of specific sounds normally not experienced as loud, uncomfortable or annoying. Distortion was defined as frequencies, overtones and/or harmonics that were not experienced in their true original form but as distorted, unclear, fuzzy and out of tune. Diplacusis was defined as a pathological matching of frequency and pitch that may involve dissonance or a sudden change of pitch when a change in loudness occurs.

On the basis of audiometry and the questionnaire, to determine the presence of hearing loss the subjects were first divided into two subgroups: those who were not affected by tinnitus, hyperacusis, distortion, or diplacousis ($n = 36$ or 26%) and those who were affected by at least one of these disorders ($n = 103$ or 74%). The latter group was further subdivided into a group that had one of these disorders and a group that had more than one disorder, as shown in Table 21.1.

Hearing disorders have been observed in 74% of subjects, with tinnitus, hyperacusis, distortion, and diplacousis reported either as a single or combined disorder (Table 21.1).

DIAGNOSIS

Diagnosis of decreased sound tolerance is complex and there is no agreement on how it should be done. For hyperacusis, there is a general agreement that pure-tone LDLs must have decreased values. Low values by themselves, however, do not prove the presence of hyperacusis as these values may be due to misophonia.

There is no consensus on how LDLs should be evaluated. Varied results depend on the specific method used to administer LDLs (e.g., stimuli: pure tone, warble tone, noise; presentation: free field, insert earphones, headphones) and the instructions given to patients (Hawkins et al., 1987; Byrne and Dirks, 1996; Ricketts and Bentler, 1996; Cox et al., 1997; Sherlock and Formby, 2005). Several studies indicate LDLs are in the range of 90–110 dB sound pressure level (SPL) in the normal population. The results tend to cluster within 95–110 dB SPL for frequencies from 500 to 8000 Hz, which corresponds to approximately 90–100 dB hearing level (HL) (Hood and Poole, 1966; Stephens and Anderson, 1971; Sherlock and Formby, 2005). A study aimed specifically at this issue showed that the average LDL value for subjects who do not have a sound tolerance problem was 100 dB HL (Sherlock and Formby, 2005). The study has been performed on 59 adults with normal hearing and no problem with sound tolerance. In addition to evaluating LDLs, loudness growth function was measured by categoric scaling judgments on a subgroup of 18 subjects. There were no differences between absolute (LDL) and relative (categoric scaling) judgment of loudness discomfort, intersubject variability, or intrasubject test–retest reliability. The authors concluded that administering LDLs to estimate loudness discomfort is an efficient and valid clinical measure to characterize the "threshold of discomfort."

When hyperacusis is present, LDLs are lower, typically in the 60–85 dB HL range (Jastreboff and Jastreboff, 2002). In cases of pure misophonia, LDL values from 30 to 120 dB HL can be observed. Therefore, LDLs alone are insufficient to diagnose hyperacusis or misophonia, and a specific detailed interview is crucial to diagnose and assess the relative contribution of hyperacusis and misophonia to decreased sound tolerance. Notably, hyperacusis and misophonia frequently occur together. Indeed, misophonia is inevitable in cases of severe hyperacusis. However, misophonia does not induce hyperacusis.

When interviewing prospective patients, it is important for the clinician to identify sounds that evoke negative reactions as well as sounds that are well tolerated in order to detect any discrepancies between reactions and the intensity of the sound. While a patient with normal LDLs does not have hyperacusis, decreased sound tolerance can still be present due to misophonia. Consequently, diagnosis of decreased sound tolerance is subjective and based predominantly on the patient's report.

To decrease the impact of misophonia on LDLs, we promote a modification of the standard procedure for administering LDLs (Hood and Poole, 1966) so that the results are dominated by hyperacusis and the effects of the misophonic component of decreased sound tolerance are kept to a minimum. To achieve this, patients are provided full control during testing and given the power to stop it at any time (Jastreboff et al., 1996; Jastreboff and Jastreboff, 2004).

There are no established and validated questionnaires to determine the severity of decreased sound tolerance. Two questionnaires have been proposed to assess the extent of hyperacusis – the Hyperacusis Questionnaire (Table 21.2) (Khalfa et al., 2002) and the Multiple-Activity Scale for Hyperacusis (MASH) (Table 21.3) (Dauman and Bouscau-Faure, 2005) – but their specificity and selectivity need to be further evaluated. The Hyperacusis Questionnaire was created by testing that was performed only on the general population of subjects who did not complain of hyperacusis. The questionnaire focuses on the psychologic and social aspects of hearing rather than on an indication of hyperacusis. Only four of 14 questions (rated on a scale from 1 to 4) are related to hyperacusis while the remaining questions appear to be related to other aspects, including hearing loss ("5. Do you have difficulty listening to conversations in noisy places?") (Khalfa et al., 2002). MASH consists of a list of 14 activities, and patients are asked to indicate their level of annoyance related to a given activity on a scale from 0 to 10 (Dauman and Bouscau-Faure, 2005). Both questionnaires have been reported to be effective in the evaluation of hyperacusis, but interestingly, there was no correlation of their scores with audiologic measurements of discomfort levels (i.e., LDL and speech discomfort level) (Dauman and Bouscau-Faure, 2005). Notably, neither of these questionnaires differentiates between hyperacusis and misophonia. There is no validated questionnaire for misophonia.

Diagnosis of diplacousis and polyacousis is based on an interview with the patient and on pitch matching of

Table 21.2

Khalfa's Hyperacusis Questionnaire

Surname, first name:
Sex:
Age:
Profession or studies:
Place (town or area) of residence:
Telephone:
Are you or have you been exposed to noise?
Do you tolerate noise less well as compared to a few years ago?
Have you ever had hearing problems? If so, of what kind?
In the following questionnaire, put a cross in the box corresponding to the answer which best applies to you: No; Yes, a little; Yes, quite a lot; Yes, a lot

1. Do you ever use earplugs or earmuffs to reduce your noise perception (do not consider the use of hearing protection during abnormally high noise exposure situations)?
2. Do you find it harder to ignore sounds around you in everyday situations?
3. Do you have trouble reading in a noisy or loud environment?
4. Do you have trouble concentrating in noisy surroundings?
5. Do you have difficulty listening to conversations in noisy places?
6. Has anyone you know ever told you that you tolerate noise or certain kinds of sound badly?
7. Are you particularly sensitive to or bothered by street noise?
8. Do you find the noise unpleasant in certain social situations (e.g., night clubs, pubs or bars, concerts, firework displays, cocktail receptions)?
9. When someone suggests doing something (going out, to the cinema, to a concert, etc.), do you immediately think about the noise you are going to have to put up with?
10. Do you ever turn down an invitation or not go out because of the noise you would have to face?
11. Do noises or particular sounds bother you more in a quiet place than in a slightly noisy room?
12. Do stress and tiredness reduce your ability to concentrate in noise?
13. Are you less able to concentrate in noise towards the end of the day?
14. Do noise and certain sounds cause you stress and irritation?

Table 21.3

The Multiple-Activity Scale for Hyperacusis (MASH)

Concert
Shopping center
Cinema/TV
Work
Restaurant
Driving a car
Sport
Church
Housework
Children
Social activities
Pottering about
Gardening
Others

For each individual activity the patient was asked to provide a score from 0 to 10 to indicate the level of annoyance caused. If the subject did not attend concerts, because of a dislike of or lack of interest in live music, his/her reaction to loud music was taken into consideration. When the scores for cinema and TV differed, the highest was selected. When the patient felt unable to indicate a score for a given item, the item was deleted, even though the activity may have been relevant to him/her. The most frequently mentioned "other" noises (open-set) were small motorcycles and emergency vehicle sirens. The mean MASH score was calculated by dividing the total score by the number of relevant activities.

the perceived tone in both ears (Kaharit et al., 2003; Ogura et al., 2003; Jansen et al., 2009). For example, Jansen et al. (2009) used an adaptive procedure to compare the pitch of pure tones presented alternatively to the right and left ear by headphones using three different frequencies: 1, 2, and 4 kHz. Specifically, in the first step, subjects were asked to match the loudness of a tone in the left ear to a 60-dB HL tone presented in the right ear using increments of 1 dB. Subjects were then asked to match the pitch of the tone in the left ear to that in the right ear using increments of 1 Hz. The procedure was repeated by alternating ears and changing the frequency of the test tone. The result of diplacousis matching was determined by the deviation between the ears and was expressed as a percentage of the measured frequency (Jansen et al., 2009). There are no reported methods to assess the severity and impact of diplacousis and polyacousis on a patient's life; severity is judged only on the basis of the patient's report.

PREVALENCE AND EPIDEMIOLOGY

The lack of an objective method to determine the presence of decreased sound tolerance and the variety of epidemiologically oriented questionnaires used to do so yield limited data on the prevalence of decreased sound tolerance in the general population. Notably, the data obtained from 10 349 randomly selected subjects showed that 15.3% reported hyperacusis (Fabijanska et al., 1999). Diagnosis of hyperacusis is made even more complex because LDL measurements are not part of a routine audiologic evaluation. Moreover, a careful inspection of the reports revealed that, in some cases, misophonia rather than hyperacusis was present (e.g., reports on Williams syndrome) (Klein et al., 1990; Blomberg et al., 2006).

An estimation of the prevalence of decreased sound tolerance can be obtained by analyzing its co-appearance with tinnitus, and more detailed and extensive data are available in that regard. Approximately 60% of tinnitus patients have significant decreased sound tolerance. Of these, about 30% of patients suffer from hyperacusis and require treatment for it (Jastreboff et al., 1999; Jastreboff and Jastreboff, 2002; Jastreboff and Hazell, 2004). Since, on average, the prevalence of bothersome tinnitus reported in the literature is about 4% (Hoffman and Reed, 2004), it is possible to estimate that 1.2% of the general population has tinnitus and hyperacusis. On the other hand, it has been reported that 86% of patients with hyperacusis suffer from tinnitus (Anari et al., 1999). It is therefore possible to calculate that 14% of subjects with hyperacusis (i.e., 0.56% of the general population) do not have tinnitus. Taking the above information into account, it is possible to estimate that clinically significant hyperacusis exists in approximately 1.75% of the general population. As the data indicate that about half the patients with decreased sound tolerance have hyperacusis, the prevalence of decreased sound tolerance in the general population can be estimated at 3.5%.

Limited data exist on the prevalence of diplacousis and polyacousis. A wide variety of ototoxic factors have been linked to diplacousis, with noise exposure being the most common (Albers and Wilson, 1968b). Nevertheless, clinical observations indicate that diplacousis is reported infrequently, and mainly by musicians or people to whom the accurate perception of music is crucial for their work. To date, only a couple of studies in the literature report on the epidemiology of diplacousis and both studies analyze the population of musicians. In the first study, performed on 139 rock/jazz musicians and described earlier in this chapter, 4 subjects (2.9%) reported diplacousis (Kaharit et al., 2003). A more recent study of diplacousis prevalence amongst 241 symphony orchestra musicians revealed that diplacousis was generally not reported as a problem and affected only 7% of subjects (Jansen et al., 2009). Subjects in this study mainly complained about tinnitus and hyperacusis. A total of 79% of the musicians complained about hyperacusis, 51% complained about tinnitus, and 24% complained about tone distortion (Jansen et al., 2009). Interestingly, while only 7% of the musicians perceived an interaural difference in pitch perception as a problem, the results of diplacousis matching revealed that 18% experienced an IPD of more than 2%. Notably, the subjective perception of diplacousis did not correlate with the extent of differences in pitch matching and there was no difference between males and females with regard to the subjective rating of diplacousis (Jansen et al., 2009).

MECHANISMS

The mechanisms of hyperacusis are speculative. The lack of an animal model of hyperacusis makes it difficult to prove the validity of any postulated mechanisms responsible for this condition. Existing theories suggest a potential involvement of both peripheral and/or central mechanisms (Wrinch, 1909; Jastreboff and Jastreboff, 2004; Baguley and Andersson, 2007; Niu et al., 2013).

At the peripheral level, the abnormal enhancement of cochlear basilar membrane vibration by the outer hair cells (OHCs) might result in the overstimulation of the inner hair cells, and therefore result in hyperacusis (Jastreboff, 1990; Jastreboff and Hazell, 2004). Indeed, in rare cases, it is possible to observe high-amplitude distortion product otoacoustic emissions and distortion products evoked by low-level primaries (Jastreboff and Mattox, 1998). The presence of asymmetric hyperacusis would indicate a peripheral mechanism because the involvement of central mechanisms would more likely act similarly on both sides. However, in nearly all cases decreased sound tolerance is symmetric, which argues against the dominant role of the peripheral mechanisms (Jastreboff et al., 1999).

Substantial data support the presence of central mechanisms in hyperacusis. Animal research has shown that damage to the cochlea or a decrease in auditory input results in a decrease of the response threshold in a significant proportion of neurons in the ventral cochlear nucleus and inferior colliculus (Boettcher and Salvi, 1993). Studies on evoked potentials indicate an abnormal increase of gain in the auditory pathways after such manipulations are applied (Gerken, 1993). The notion of increased gain within the central part of the auditory pathways has been discussed and promoted in recent approaches to the mechanisms of tinnitus and hyperacusis (Norena and Farley, 2013).

A number of medical conditions have been linked to decreased sound tolerance (e.g., tinnitus, Williams syndrome, Bell's palsy, Lyme disease, Ramsay Hunt syndrome, poststapedectomy, perilymphatic fistula, head injury, migraine, depression, withdrawal from benzodiazepines, cerebrospinal fluid high pressure, Addison's disease, translabyrinthine excision of a vestibular schwannoma) (Adour and Wingerd, 1974; Klein et al., 1990; Wayman et al., 1990; Lader, 1994; Nields et al., 1999; Gopal et al., 2000; Jastreboff and Hazell, 2004; Blomberg et al., 2006). These conditions can be linked to the central processing of signals and to the modification of the level of neuromodulators as possible factors that induce or enhance hyperacusis. Moreover, serotonin has been implicated in hyperacusis on the basis of indirect reasoning that some conditions occur with hyperacusis as a symptom (e.g., migraine, depression, pyridoxine deficiency, benzodiazepine dependence, and postviral fatigue syndrome) and involve a disturbance in serotonin activity (Marriage and Barnes, 1995). These authors speculated that, as serotonin is considered to have an inhibitory role in sensory modulation at a central level, a reduction in forebrain serotonin activity is therefore the most likely underlying pathology that causes central hyperacusis. The authors have not proposed any more specific mechanisms of serotonin involvement, and have stated that the increase or decrease of serotonin may be linked to hyperacusis. Additionally, they labeled decreased sound tolerance as hyperacusis, with stress on phonophobia, which may have different mechanisms than hyperacusis, and is inconsistent with current classifications (Jastreboff and Jastreboff, 2002, 2013; Tyler et al., 2009).

Serotonin involvement in hyperacusis has not been confirmed. There is only one case presentation which supports the proposed hypothesis by showing an improvement in hyperacusis, difficulty understanding speech, withdrawn depression, lethargy, and hypersensitivity to touch, pressure, and light after following treatment with selective serotonin reuptake inhibitors (Gopal et al., 2000).

A high prevalence of decreased sound tolerance in people with Williams syndrome (over 80%) suggests a genetic basis of decreased sound tolerance in those subjects (Nigam and Samuel, 1994; Gothelf et al., 2006).

The mechanisms of misophonia could involve the enhancement of the functional links between the auditory and limbic systems at both the cognitive and subconscious levels (Jastreboff, 1990; Jastreboff and Hazell, 2004). Alternatively, a tonic high level of activation of the limbic and autonomic nervous systems may result in strong behavioral reactions to moderate sounds (Jastreboff and Hazell, 2004). A recent study supported the proposed mechanisms of misophonia by showing the enhanced autonomic reactivity to a sound, but not to other sensory stimuli in misophonic patients (Edelstein et al., 2013). In January 2013, Schroder et al. redefined misophonia based on their work in a psychiatric center and proposed to classify the condition as a new psychiatric disorder. In our opinion, Schroder et al. studied a population of psychiatric patients who happened to have misophonia as well. In our clinical work, we have seen 318 misophonic patients (compared to 42 cases reported by Schroder et al.), all evaluated by physicians, and in only 7 cases (2.2%) did patients exhibit psychiatric problems. Moreover, our misophonic patients showed significant improvement when treated with a combination of counseling and a specific version of sound therapy (described below), without any need for psychiatric intervention.

The mechanisms of diplacousis are hypothetic. Nearly all the proposed mechanisms of diplacousis involve the cochlea, and only one paper describes diplacousis of (presumably) central origin linked to a lesion in the posterior thalamus (Ghosh, 1990). The first class of proposed mechanisms links diplacousis to hearing loss and OHC damage in the cochlea. OHCs work as a mechanical amplifier within the cochlea and are responsible for sharp tuning of the traveling wave in the cochlea. Notably, the OHCs amplify sounds of lower intensities only – below 60 dB SPL – which corresponds roughly to half the dynamic range of hearing. OHCs provide gradually less amplification when the level of a sound increases and become inactive for sound intensities higher than 60 dB SPL.

The mechanism for diplacousis linked to hearing loss is discussed below. The hypothesis is based on basic science regarding the functioning of the cochlea and the mechanisms of pitch perception. There is a strict relationship between a given place on the cochlea's basilar membrane and the frequency of tone that evokes it. Neuronal activity in the auditory nerve is correlated with the phase of the incoming pure tone only for frequencies below 1000 Hz. For higher frequencies, the place on the basilar membrane where the maximal amplitude of the traveling wave develops determines the perceived pitch (Moore, 1995). When a group of OHCs is damaged, the tuning curve becomes broader and its peak shifts in frequency. When compared to an ear with undamaged OHCs for this specific frequency region, the shift in place occurs where maximal stimulation of the basilar membrane occurs. Consequently, the subject perceives a different pitch in one ear than in the other ear. Note that, as OHCs do not function for sound intensities higher than 60 dB SPL, diplacousis based on this mechanism decreases with the increase of a sound's intensity and disappears when the intensity exceeds 60 dB SPL. Diplacousis may or may not appear depending on the difference in the damage of OHC systems in one ear versus the other ear for a given frequency range.

This theory explains the clinical observation that diplacousis is particularly linked to unilateral/asymmetric hearing loss, the intensity of a sound to which a subject is exposed, and the relation of the affected pitch to the regions of hearing loss. The theory also explains instances of temporal or permanent diplacousis evoked by a loud noise, as reported in the literature (Knight, 2004; Jansen et al., 2009). The temporal dysfunction of OHCs occurs following exposure to a loud noise (i.e., the cilia of the OHC becomes disorganized, making a cell temporarily dysfunctional). Depending on the asymmetry of damage and pre-existing damage of the OHC system, diplacousis may temporarily appear. If the exposure to a sound results in permanent damage to a group of OHCs, permanent diplacousis emerges. Local damage of OHCs causes a loss of sharp tuning of stimulation of IHCs and a shift of frequency where maximal stimulation occurs. This results in the perception of a different frequency than the one to which the subject is exposed (Fig. 21.1). Broadening of the curve may result in the perception of a fuzzy sound.

A question arises regarding the mechanisms of diplacousis in subjects without hearing loss. First, a normal audiogram can be seen in subjects who lost up to 30% of their OHCs (Harding and Bohne, 2007, 2009; Chen et al., 2008), and yet for these subjects the proposed mechanisms described above are still applicable. An evaluation of the functional status of OHCs by high-frequency resolution distortion product otoacoustic emissions (DPOAE) measurement is needed to determine whether groups of OHCs are damaged. A study of a sudden hearing loss case supports the proposed mechanisms: while there was no clear relation of diplacousis with hearing threshold and transient otoacoustic emissions, the observed frequency shifts in the DPAOE fine structure were in close agreement with the changes in diplacousis (Knight, 2004).

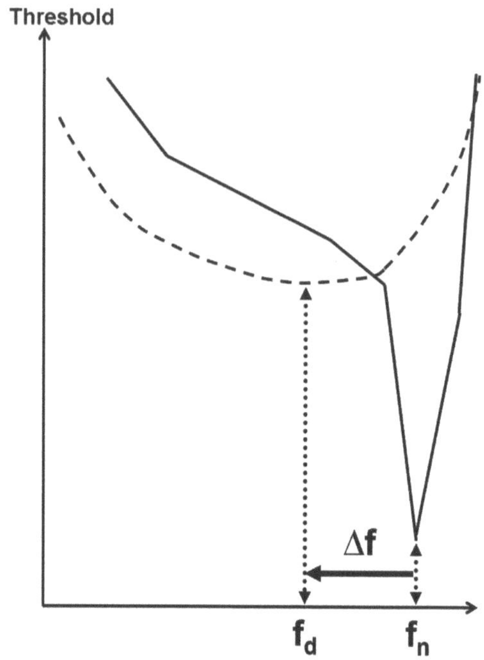

Fig. 21.1. Potential mechanisms of diplacousis as a result of outer hair cell (OHC) dysfunction or loss. Damage of OHC results in a broadening response curve and shifting frequency where maximal stimulation occurs. Horizontal axis, frequency; vertical axis, threshold of stimulation of inner hair cell (IHC); solid line, intact OHC; dashed line, OHC are dysfunctional or damaged in an area on the basilar membrane; f_n perceived frequency in normal cochlea; f_d perceived diplacousis frequency in cochlear with dysfunctional OHC; Δf diplacousis frequency shift.

Another mechanism may be involved in subjects with diplacousis who have normal hearing. This mechanism may also explain the presence of monaural diplacousis. An indication for this mechanism arises from two observations: (1) in diplacousis related to hearing loss, the pitch shift is pronounced and exists in a broad frequency range, while in subjects with normal hearing the shift is typically only ±2% and is distributed in a random manner with a mean shift close to zero; (2) spontaneous otoacoustic emissions (SPOAE) are commonly observed in people with normal hearing; these emissions consist of a number of pure-tone/very narrow noise bands and may have relatively high intensity.

If a person with SPOAE is exposed to a low-level tone, then due to the non-linear properties of the cochlea, the tone will interact with SPOAE idiotones and create a number of distortions with frequencies that follow the equation $f_d = \pm m f_{et} \pm n f_{SPi}$, where f denotes the frequency, subscript d denotes distortion, et refers to an external tone, SPi is an "i" component of SPOAE, and n and m are natural numbers (i.e., 1, 2, 3, etc.). The condition $f_d > 0$ has to be fulfilled. As SPOAE may contain many frequencies, the resulting perception of a sound can be complex. Reported properties of diplacousis support this hypothesis, while pointing to the complex nature of the proposed interactions (Formby and Gjerdingen, 1981; Long, 1998).

THE NEUROPHYSIOLOGICAL MODEL OF DECREASED SOUND TOLERANCE

The neurophysiologic model of tinnitus, decreased sound tolerance, and the treatment approach based on the model, known as tinnitus retraining therapy (TRT), were introduced in 1990 (Jastreboff, 2000, 2007b, 2010; Jastreboff and Jastreboff, 2001, 2006). An analysis of negative reactions experienced by patients led to the proposition that the auditory system plays a secondary role in clinically significant decreased sound tolerance, and that other systems in the brain are dominant. While a number of systems in the brain are important (e.g., prefrontal cortex, attention networks, systems involved in memory), the limbic and autonomic nervous systems seem to be crucial. The block diagram of the neurophysiological model of tinnitus as applied to decreased sound tolerance has been described in detail elsewhere (Jastreboff, 2004; Jastreboff and Hazell, 2004; Jastreboff and Jastreboff, 2004).

The model can be used to discuss the potential mechanisms of hyperacusis and misophonia. First, the reactions experienced by patients are the same in both hyperacusis and misophonia. This supports the postulate that the same systems in the brain are responsible for these reactions. In turn, the type of negative reactions supports the notion that the limbic and autonomic nervous systems are dominant in the emergence of these reactions. The activation of these systems in the brain occurs in a different manner for hyperacusis than it does for misophonia. With hyperacusis, the presumed mechanism is an abnormally high amplification occurring within the auditory pathways (predominantly at the subconscious level) that yields a strong neuronal activation evoked by moderate or weak sound. Similar activation is evoked in a normal subject following exposure to a high level of sound. The limbic and autonomic nervous systems are activated as a consequence of the high level of auditory system activation. Thus, patients with hyperacusis experience the same problems while exposed to moderate or weak sound that a person without hyperacusis does when subjected to a very high level of sound (Fig. 21.2A) (Jastreboff and Jastreboff, 2004).

Proposed mechanisms of misophonia involve a high level of activation of the limbic and autonomic nervous systems due to enhanced functional connections between the auditory system and other systems in the brain (mainly the limbic and autonomic nervous systems for specific patterns of a sound only) (Jastreboff and Jastreboff, 2009, 2013). At the same time, the auditory system functions normally (Fig. 21.2B). Note that the presumed neurophysiological mechanisms are distinctively dissimilar for hyperacusis and misophonia, and consequently the treatments are different as well. Therefore, it is important to assess the presence and extent of both phenomena in a patient as each condition needs to be treated using different approaches. Audiologic evaluation provides only partial help in this regard and therefore a detailed interview is crucial.

TREATMENTS

Treatment for hyperacusis has taken two opposite directions. The most common approach is to advise patients to avoid sounds and use ear protection. This is based on the reasoning that, because patients became sensitive to a sound, they are supposedly more susceptible to sound exposure and consequently need extra protection. Patients easily embrace this reasoning and begin to

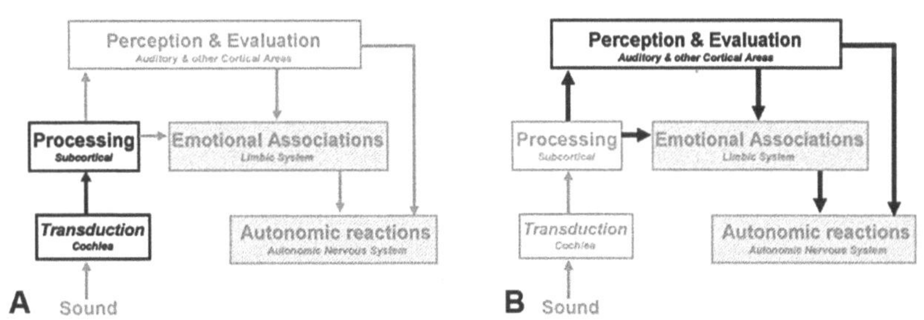

Fig. 21.2. Block diagram of potential mechanisms responsible for decreased sound tolerance. (**A**) Block diagram of potential mechanisms responsible for hyperacusis. Hyperacusis results from an abnormal amplification of sound-evoked activity occurring within the auditory pathways (marked by black lines). (**B**) Block diagram of potential mechanisms responsible for misophonia. Misophonia results from enhanced functional connections between the auditory and the limbic and autonomic nervous systems for a specific pattern of sound-evoked activity (marked by black lines). Note the overamplification of the limbic and autonomic system occurring in both conditions and marked by cross pattern. (Modified from Jastreboff and Jastreboff, 2004.)

protect their ears, even to the extent of using earplugs in quiet environments. Unfortunately, this well-intended approach makes the auditory system even more sensitive to sound and further exacerbates hyperacusis (Vernon and Press, 1998; Formby et al., 2003, 2007, 2013; Hawley et al., 2007).

The opposite approach to treating hyperacusis involves desensitization, wherein patients are exposed to a variety of sounds. The desensitization approach has been promoted for some time with a number of protocols and types of sounds used. For example, sounds with certain frequencies removed, short exposure to moderately loud sound, or prolonged exposure to relatively low-level sounds have been used (Vernon and Press, 1998; Jastreboff and Jastreboff, 2000). "Pink noise therapy," as proposed by Vernon and Press (1998), has gained some recognition. In this approach, patients were advised to stop ear overprotection and to listen to pink noise through headphones set to the highest comfortable level for 2 hours a day. A group of patients were provided with pink-noise cassette tapes and questionnaires were mailed to 30 participants. Out of 20 patients who responded, 13 reported using these tapes in a systematic manner and 7 (54%) reported improvement in their hyperacusis (Vernon and Press, 1998). A version of pink-noise therapy has also recently been proposed for tinnitus and hyperacusis by Johnson (2014). The protocol combines the use of pink noise with a 16-week cognitive exercise program divided into sections devoted to different topics, such as recognition and relief. To date, no results have been published in the peer-reviewed literature regarding the effectiveness of this approach.

There are no published treatments for diplacousis and polyacousis. In one patient's case, the use of aspirin resulted in a reduction of SPOAE into the noise floor and eliminated monaural diplacousis (Long, 1998).

Tinnitus retraining therapy (TRT) for decreased sound tolerance

TRT can help patients with tinnitus and hyperacusis. The presence of hyperacusis is one of the key factors in the categorization of TRT patients and in determining the protocol for treatment (Jastreboff, 1999, 2010; Jastreboff and Hazell, 2004). It is recommended that if hyperacusis is present it must be treated first. For some patients with decreased sound tolerance, it is possible to completely remove hyperacusis and misophonia and to effectively provide a cure for these conditions (Jastreboff and Jastreboff, 2001, 2013).

According to the principles of the neurophysiological model of tinnitus, prolonged exposure to relatively low-level sound is recommended and is used as a part of TRT. Independent results support the use of TRT for the treatment of hyperacusis and have shown that the desensitization approach used in TRT has a statistically and clinically significant impact on hyperacusis (Formby, 2007; Formby et al., 2007, 2013). Therefore, for hyperacusis patients, avoiding silence and being continually exposed to sound are crucially important. The sound level should be well controlled during treatment and should never induce discomfort or annoyance. Ear-level, wearable sound generators, which facilitate keeping the sound at a constant, well-controlled level, are often employed.

Unfortunately, patients with decreased sound tolerance (particularly misophonia) tend to set the sound level of their devices close to or at the threshold of hearing, and this is considerably less efficient for tinnitus and hyperacusis treatment. The use of real-ear measurements as a guide in setting and checking the sound level for all patients who use instrumentation to treat decreased sound tolerance is very helpful to patients and should be performed during their initial and follow-up visits with the TRT clinician. The recommended range of sound is from 6 to 16 dB sensation level (SL).

Desensitization works on the auditory system. Therefore, this approach will not affect misophonia, which needs to be addressed via the active extinction of conditioned reflexes between the auditory system and the limbic and autonomic nervous systems. This can be achieved by specific counseling and protocols for working with sound (Jastreboff, 2007a; Jastreboff and Jastreboff, 2002, 2004, 2013). In particular, patients are advised to systematically engage in pleasant activities they enjoy where sounds play an indispensable role, such as listening actively to one's favorite music or to audiobooks following a specific protocol (Jastreboff and Jastreboff, 2002). Other activities include shopping in a mall, going to parties, dining in restaurants, and attending movies. The main idea is to create an association of a given sound with a pleasant situation (implementing the active extinction of conditioned reflexes).

There are four classes of protocols for misophonia. The specific protocol used is tailored to the individual patient. Frequently, more than one protocol is used to treat misophonia. Each protocol is geared to create a positive association with a sound, but the protocols differ with respect to the extent of control the patient has over the sound environment and, in the case of protocol category 4, employ the use of sounds with a positive association together with bothersome sounds which evoke negative reactions (Jastreboff and Jastreboff, 2013).

Protocol category 1 provides the patient with full control over the selection of sound, its level, and duration. As such, it can be used even in cases of coexisting significant hyperacusis and can be implemented from the start of treatment.

Protocol category 2 affords the patient full control over the type of sound, but only partial, indirect control over the sound level, by yielding control to someone close to the patient who is instructed to set the sound volume to a level he or she thinks the patient will accept. After a listening session, the patient should provide feedback as to whether the sound level was too high, too low, or just fine.

Protocol category 3 enables the patient to select the type of sound, but the sound level is fully out of the patient's control. This protocol can be used only when significant hyperacusis is absent or has been eliminated by treatment. Therefore, the introduction of this protocol is frequently delayed.

Protocol category 4 uses the concept of complex conditioned stimuli and combines the exposure to sounds which evoke negative reactions with the simultaneous exposure to sound the patient regards as highly positive and enjoyable. The ratio of sound levels of positive-to-negative sound is gradually decreased. The environment where this protocol is used is taken into account, as many patients react differently depending on where they are exposed to a bothersome sound (i.e., home, school, a public place, a restaurant, a friend's home). The multisensory aspect of stimuli is also taken into account, as some patients react even to seeing someone produce offensive sounds (e.g., eating).

An analysis was made of 201 consecutive patients diagnosed with decreased sound tolerance and treated with TRT at Emory Tinnitus and Hyperacusis Center. A total of 184 patients (92%) had misophonia; 17 patients (8%) had hyperacusis alone, and 56 patients (28%) had hyperacusis and misophonia concurrently. The proportion of patients with significant hyperacusis (with or without misophonia) who required specific treatment is similar to the 25–30% reported in the literature (Jastreboff et al., 1999; Hazell et al., 2002; Herraiz et al., 2003, 2006; Jastreboff and Jastreboff, 2004) and to the 26% we have reported previously (Jastreboff and Jastreboff, 2002).

Improvement in hyperacusis was judged on the basis of changes in LDLs combined with responses obtained during the structured interviews, while improvement in misophonia was solely based on the interviews. This decision was based on the observation that, in cases of misophonia, any LDL values could be seen and the values were not correlated with patients' judgment of the problems due to decreased sound tolerance.

Of 201 patients with decreased sound tolerance, 165 patients (82%) showed significant improvement. For 56 patients with hyperacusis (with or without misophonia), 45 patients (80%) showed significant improvement. The effectiveness of treatment for misophonia with or without hyperacusis was identical (152 of 184 patients with misophonia accompanied by hyperacusis, or 83%, and 139 of 167 patients with misophonia alone, or 83%). As noted earlier, in some cases it is possible to achieve a cure for misophonia as well as for hyperacusis (Jastreboff and Jastreboff, 2001, 2013). Furthermore, our clinical observations demonstrated that treatment of misophonia is crucial to achieving a successful outcome for tinnitus treatment (Jastreboff and Jastreboff, 2012).

CONCLUSIONS

Decreased sound tolerance (hyperacusis and misophonia) remains a challenging topic to study and to treat. Many questions are unanswered. The mechanisms are speculative and unproven. Even less is known about diplacousis and polyacousis.

The neurophysiological model of tinnitus and TRT provide an approach that helps patients and may ultimately result in a better understanding of decreased sound tolerance. There is a need to investigate the potential mechanisms of diplacousis and polyacousis as our knowledge of these phenomena is severely limited and effective treatments for these problems have not been established.

References

Adour KK, Wingerd J (1974). Idiopathic facial paralysis (Bell's palsy): factors affecting severity and outcome in 446 patients. Neurology 24 (12): 1112–1116.

Albers GD, Wilson WH (1968a). Diplacusis. I. Historical review. Arch Otolaryngol 87 (6): 601–603.

Albers GD, Wilson WH (1968b). Diplacusis. II. Etiology. Arch Otolaryngol 87 (6): 604–606.

Anari M, Axelsson A, Elies W et al. (1999). Hypersensitivity to sound – questionnaire data, audiometry and classification. Scand Audiol 28 (4): 219–230.

Anon (1994). The American Heritage Dictionary, International, SoftKey.

Anon (1997). Stedman's Concise Medical Dictionary, 3rd ed. Lippencott, Williams & Wilkins, Baltimore, pp. 1–1091.

Bacon SP, Viemeister NF (1985). A case study of monaural diplacusis. Hear Res 19 (1): 49–56.

Baguley D, Andersson G (2007). Hyperacusis mechanisms, diagnosis, and therapies, Plural Publishing, San Diego.

Baguley D, McFerran D (2010). Hyperacusis and disorders of loudness perception. In: A Moller, T Kleinjung, B Langguth, D De Ridder (Eds.), Textbook of Tinnitus, Springer, New York, Dordrecht, Heidelberg, London, pp. 13–23.

Bauman N (2013). Diplacusis – The strange world of people with double hearing. http://www.hearinglosshelp.com/articles/diplacusis.htm.

Blomberg S, Rosander M, Andersson G (2006). Fears, hyperacusis and musicality in Williams syndrome. Res Dev Disabil 27 (6): 668–680.

Boettcher FA, Salvi RJ (1993). Functional changes in the ventral cochlear nucleus following acute acoustic overstimulation. J Acoust Soc Am 94: 2123–2134.

Brink van den G (1970). Experiments on binaural diplacusis and toned perception. In: R Plomp, GF Smoorenburg (Eds.), Frequency analysis and periodicity detection in hearing, A.W. Sijthoff, Leiden, pp. 362–374.

Brookler KH (2009). Monaural diplacusis with tinnitus, aural fullness, hyperacusis, and sensorineural hearing loss. Ear Nose Throat J 88 (2): 772–774.

Burns EM (1982). Pure-tone anomalies. I. Pitch-intensity effects and diplacusis in normal ears. J Acoust Soc Am 72 (5): 1394–1402.

Byrne D, Dirks D (1996). Effects of acclimatization and deprivation on non-speech auditory abilities. Ear Hear 17 (3): 29S–37S.

Chen GD, Tanaka C, Henderson D (2008). Relation between outer hair cell loss and hearing loss in rats exposed to styrene. Hear Res 243 (1–2): 28–34.

Corliss EL, Burnett ED, Stimson HF (1968). "Polyacusis", a hearing anomaly. J Acoust Soc Am 43 (6): 1231–1236.

Cox RM, Alexander GC, Taylor IM et al. (1997). The Countour test of loudness perception. Ear Hear 18 (5): 388–400.

Dauman R, Bouscau-Faure F (2005). Assessment and amelioration of hyperacusis in tinnitus patients. Acta Otolaryngol 125 (5): 503–509.

Edelstein M, Brang D, Rouw R et al. (2013). Misophonia: physiological investigations and case descriptions. Front Hum Neurosci 7: 296.

Fabijanska A, Rogowski M, Bartnik G et al. (1999). Epidemiology of tinnitus and hyperacusis in Poland. 1999, Cambridge, UK, In: JWP Hazell (Ed.), Proceedings of the Sixth International Tinnitus Seminar, THC, London, UK, pp. 569–571.

Flottorp G (1953). Pure-tone tinnitus evoked by acoustic stimulation: the idiophonic effect. Acta Otolaryngol 43 (4–5): 396–415.

Formby C (2007). Hyperacusis and related sound tolerance complains: differential diagnosis, treatment effects, and models. Seminars in Hearing 28: 227–259.

Formby C, Gjerdingen DB (1981). Some systematic observations on monaural diplacusis. Audiology 20 (3): 219–233.

Formby C, Sherlock LP, Gold SL (2003). Adaptive plasticity of loudness induced by chronic attenuation and enhancement of the acoustic background. J Acoust Soc Am 114 (1): 55–58.

Formby C, Gold SL, Keaser ML et al. (2007). Secondary benefits from Tinnitus Retraining Therapy: clinically signficant increase in Loudness Discomfort level and expansion of the Auditory Dynamic Range. Seminars in Hearing 28: 227–260.

Formby C, Hawley M, Sherlock LP et al. (2013). Intervention for restricted dynamic range and reduced sound tolerance: clinical trial using a Tinnitus Retraining Therapy protocol for hyperacusis. J Acoust Soc Am 133 (5): 3382–3383.

Gerken GM (1993). Alteration of central auditory processing of brief stimuli: a review and a neural model. J Acoust Soc Am 93: 2038–2049.

Ghosh P (1990). Central diplacusis. Eur Arch Otorhinolaryngol 247 (1): 48–50.

Gopal KV, Daly DM, Daniloff RG et al. (2000). Effects of selective serotonin reuptake inhibitors on auditory processing: case study. J Am Acad Audiol 11 (8): 454–463.

Gothelf D, Farber N, Raveh E et al. (2006). Hyperacusis in Williams syndrome: characteristics and associated neuroaudiologic abnormalities. Neurology 66 (3): 390–395.

Harding GW, Bohne BA (2007). Distribution of focal lesions in the chinchilla organ of Corti following exposure to a 4-kHz or a 0.5-kHz octave band of noise. Hear Res 225 (1–2): 50–59.

Harding GW, Bohne BA (2009). Relation of focal hair-cell lesions to noise-exposure parameters from a 4- or a 0.5-kHz octave band of noise. Hear Res 254 (1–2): 54–63.

Hawkins DB, Walden BE, Montgomery A et al. (1987). Description and validation of an LDL procedure designed to select SSPL90. Ear Hear 8 (3): 162–169.

Hawley ML, Keaser ML, Formby C (2007). Predicting hyperacusis in tinnitus patients. Seminars in Hearing 28 (4): 261–275.

Hazell JWP, Sheldrake JB, Graham RL (2002). Decreased sound tolerance: Predisposing/triggering factors and treatment outcome following tinnitus retraining therapy (TRT). In: 7th International Tinnitus Seminar, pp. 255–261, Fremantle, Western Australia.

Herraiz C, Hernandez CJ, Plaza G et al. (2003). [Study of hyperacusis at a tinnitus unit]. Acta Otorrinolaringol Esp 54 (9): 617–622.

Herraiz C, Plaza G, Aparicio JM (2006). [Mechanisms and management of hyperacusis (decreased sound tolerance)]. Acta Otorrinolaringol Esp 57 (8): 373–377.

Hoffman HJ, Reed GW (2004). Epidemiology of Tinnitus. In: JB Snow (Ed.), Tinnitus: Theory and Management, BC Decker, Hamilton, pp. 16–41.

Hood JD, Poole JP (1966). Tolerable limit to loudness: it's clinical and physiological significance. J Acoust Soc Am 40: 47–53.

Jansen EJ, Helleman HW, Dreschler WA et al. (2009). Noise induced hearing loss and other hearing complaints among musicians of symphony orchestras. Int Arch Occup Environ Health 82 (2): 153–164.

Jastreboff PJ (1990). Phantom auditory perception (tinnitus): mechanisms of generation and perception. Neurosci Res 8: 221–254.

Jastreboff PJ (1999). Categories of the patients and the treatment outcome. In: JWP Hazell (Ed.), Proceedings of the Sixth International Tinnitus Seminar, 1999, Cambridge, UK, THC, London, UK, pp. 394–398.

Jastreboff PJ (2000). Tinnitus Habituation Therapy (THT) and Tinnitus Retraining Therapy (TRT). In: R Tyler (Ed.), Tinnitus Handbook, Singular, Thomson Learning, San Diego, pp. 357–376.

Jastreboff PJ (2004). The neurophysiological model of tinnitus. In: JB Snow (Ed.), Tinnitus: Theory and Management, BC Decker, Hamilton, London, pp. 96–106.

Jastreboff PJ (2007a). Tinnitus retraining therapy. Prog Brain Res 166: 415–423.

Jastreboff MM (2007b). Sound therapies for tinnitus management. In: B Langguth, G Hajak, T Kleinjung et al. (Eds.), Tinnitus: Pathophysiology and Treatment, Elsevier, 449–454. New York, Dordrecht, Heidelberg, London.

Jastreboff PJ (2010). Tinnitus Retraining Therapy. In: A Moller, T Kleinjung, B Langguth et al. (Eds.), Textbook of Tinnitus, Springer, New York, Dordrecht, Heidelberg, London, pp. 575–596.

Jastreboff PJ, Hazell JWP (2004). Tinnitus Retraining Therapy: Implementing the Neurophysiological Model, Cambridge University Press, Cambridge.

Jastreboff PJ, Jastreboff MM (2000). Tinnitus Retraining Therapy (TRT) as a method for treatment of tinnitus and hyperacusis patients. J Am Acad Audiol 11 (3): 156–161.

Jastreboff PJ, Jastreboff MM (2001). Tinnitus Retraining Therapy. In: D Baguley (Ed.), Perspectives in Tinnitus Management, Thieme, New York, pp. 51–63.

Jastreboff MM, Jastreboff PJ (2002). Decreased sound tolerance and Tinnitus Retraining Therapy (TRT). Australian and New Zealand Journal of Audiology 21 (2): 74–81.

Jastreboff PJ, Jastreboff MM (2004). Decreased sound tolerance. In: JB Snow (Ed.), Tinnitus: Theory and Management, BC Decker, Hamilton, London, pp. 8–15.

Jastreboff PJ, Jastreboff MM (2006). Tinnitus retraining therapy: a different view on tinnitus. ORL J Otorhinolaryngol Relat Spec 68 (1): 23–29.

Jastreboff PJ, Jastreboff MM (2009). Tinnitus and Decreased Sound Tolerance. In: JJ Ballenger, JB Snow, WP Ashley (Eds.), Ballenger's Otorhinolaryngology Head and Neck Surgery, Singular Publishing, San Diego, pp. 351–362.

Jastreboff PJ, Jastreboff MM (2012). Decreased Sound Tolerance (Hyperacusis and Misophonia): Theory and Clinical Practice. Meeting of the American Academy of Audiology, Boston.

Jastreboff PJ, Jastreboff MM (2013). Using TRT to treat hyperacusis, misophonia and phonophobia. ENT & audiology news 21 (6): 88–90.

Jastreboff PJ, Mattox DE (1998). Treatment of hyperacusis by aspirin, Assoc. Res, Otolaryngol.

Jastreboff PJ, Gray WC, Gold SL (1996). Neurophysiological approach to tinnitus patients. Am J Otol 17: 236–240.

Jastreboff PJ, Jastreboff MM, Sheldrake JB (1999). Audiometrical characterization of hyperacusis patients before and during TRT. In: JWP Hazell (Ed.), Proceedings of the Sixth International Tinnitus Seminar, 1999, Cambridge, UK, THC, London, UK, pp. 495–498.

Johnson M (2014). Pink Sound Protocol For Tinnitus & Hyperacusis, http://www.pinksound.info/.

Kaharit K, Zachau G, Eklof M et al. (2003). Assessment of hearing and hearing disorders in rock/jazz musicians. Int J Audiol 42 (5): 279–288.

Khalfa S, Dubal S, Veuillet E et al. (2002). Psychometric normalization of a hyperacusis questionnaire. ORL J Otorhinolaryngol Relat Spec 64 (6): 436–442.

Klein AJ, Armstrong BL, Greer MK et al. (1990). Hyperacusis and otitis media in individuals with Williams syndrome. J Speech Hear Disord 55: 339–344.

Knight RD (2004). Diplacusis, hearing threshold and otoacoustic emissions in an episode of sudden, unilateral cochlear hearing loss. Int J Audiol 43 (1): 45–53.

Lader M (1994). Anxiolytic drugs: dependence, addiction and abuse. Eur Neuropsychopharmacol 4 (2): 85–91.

Long G (1998). Perceptual consequences of the interactions between spontaneous otoacoustic emissions and external tones. I. Monaural diplacusis and aftertones. Hear Res 119 (1–2): 49–60.

Marriage J, Barnes NM (1995). Is central hyperacusis a symptom of 5-hydroxytryptamine (5-HT) dysfunction? J Laryngol Otol 109 (10): 915–921.

Moore BCJ (1995). An Introduction to the Psychology of Hearing, Academic Press, San Diego.

Nields JA, Fallon BA, Jastreboff PJ (1999). Carbamazepine in the treatment of Lyme disease-induced hyperacusi. J Neuropsychiatry Clin Neurosci 11 (1): 97–99.

Nigam A, Samuel PR (1994). Hyperacusis and Williams syndrome. J Laryngol Otol 108 (6): 494–496.

Niu Y, Kumaraguru A, Wang R et al. (2013). Hyperexcitability of inferior colliculus neurons caused by acute noise exposure. J Neurosci Res 91 (2): 292–299.

Norena AJ, Farley BJ (2013). Tinnitus-related neural activity: theories of generation, propagation, and centralization. Hear Res 295: 161–171.

Ogura M, Kawase T, Kobayashi T et al. (2003). Modified binaural pitch-matching test for the assessment of diplacusis. Int J Audiol 42 (6): 297–302.

Ricketts TA, Bentler RA (1996). The effect of test signal type and bandwidth on the categorical scaling of loudness. J Acoust Soc Am 99: 2281–2287.

Schroder A, Vulink N, Denys D (2013). Misophonia: diagnostic criteria for a new psychiatric disorder. PLoS One 8 (1): e54706.

Sherlock LP, Formby C (2005). Estimates of loudness, loudness discomfort, and the auditory dynamic range: normative estimates, comparison of procedures, and test-retest reliability. J Am Acad Audiol 16 (2): 85–100.

Stephens SD, Anderson C (1971). Experimental studies on the uncomfortable loudness level. J Speech Hear Res 14: 262–270.

Terhardt E (2000). Diplacusis binauralis (IPD), Available online at: http://www.mmk.ei.tum.de/persons/ter/top/diplacusis.html.

Tyler RS, Noble W, Coelho C et al. (2009). Tinnitus and Hyperacusis. In: J Katz, L Medwetsky, R Burkard et al. (Eds.), Handbook of Clinical Audiology, Lippincott Williams & Williams, Baltimore, pp. 726–742.

Vernon J, Press L (1998). Treatment for hyperacusis. In: JA Vernon (Ed.), Tinnitus Treatment and Relief, Allyn and Bacon, Boston, pp. 223–227.

Ward WD (1952). A Case of Tonal Uniaural Diplacusis. J Acoust Soc Am 24 (4): 449.

Ward WD (1955). Tonal Monaural Diplacusis. J Acoust Soc Am 27 (2): 365–372.

Wayman DM, Pham HN, Byl FM et al. (1990). Audiological manifestations of Ramsay Hunt syndrome. J Laryngol Otol 104 (2): 104–108.

Wrinch FS (1909). A case of diplacusis depending upon the tympanic mechanism. Science 30 (762): 186–187.

Zurek PM (1981). Spontaneous narrowband acoustic signals emitted by human ears. J Acoust Soc Am 69: 514–523.

Handbook of Clinical Neurology, Vol. 129 (3rd series)
The Human Auditory System
G.G. Celesia and G. Hickok, Editors

Chapter 22

Auditory synesthesias

PEGAH AFRA*
Department of Neurology, School of Medicine, University of Utah, Salt Lake City, UT, USA

INTRODUCTION

The term synesthesia is composed of two parts, *syn* = together and *aesthesia* = sensation, literally meaning "experiencing together" (Baron-Cohen et al., 1987). It denotes that sensory stimulation of one sensory modality (the inducer) will elicit an involuntary or automatic sensation (the concurrent). The sensory experience of concurrent can be in a different aspect of the same sensory modality as inducer or in another sensory modality. The inducer does not normally elicit the concurrent in normal subjects and the perception of concurrent does not represent the events in the external world. For example, in auditory-visual synesthesia, hearing music or environmental sounds can trigger color experience, or in auditory-gustatory synesthesia, hearing spoken words may trigger taste. As mentioned above, the concurrent can be in the same modality as inducer (for example, in grapheme-color synesthesia, the letter "A" can be seen as "red" regardless of its ink color), or the inducer and concurrent can be in two different modalities (for example, in auditory-visual synesthesia).

The terminology of synesthesia was rather ambiguous until 2001, when Grossenbacher and Lovelace introduced the terms inducer and concurrent. They suggested the inducer–concurrent pairing to be used as standard terminology for synesthesia (i.e., instead of earlier terms of "colored-hearing" synesthesia, the term "sound-color" or "auditory-visual" synesthesia should be used).

Regarding the inducer phenomenology, synesthetic perception occurs when the concurrent is induced by perceiving particular sensory stimuli, while synesthetic conception is when the concurrent is induced by thinking about a particular concept (Grossenbacher and Lovelace, 2001). In regard to concurrent phenomenology, associator (internal) synesthetes perceive the concurrent by the "mind's eye" while projector (external) synesthetes perceive the concurrent in the external space (Dixon et al., 2004).

TYPES OF SYNESTHESIA

Three types of synesthesia – developmental, acquired, and induced – have been recognized (Grossenbacher and Lovelace, 2001; Sinke et al., 2012). Developmental synesthesias are lifelong experiences and subjects remember them back to their childhood. Acquired synesthesias usually happen subsequent to sensory deafferentation or pathologic insult to the brain (although a single idiopathic case has been described). They persist for a variable duration of time, from days to months or even years. Induced synesthesias are transient phenomena and are experienced temporarily in the setting of sensory deprivation or acute intoxication.

Sinke et al. (2012) have reviewed the phenomenologic differences between the different types of synesthesia. Some of these are discussed below.

Consistency

The developmental synesthesias are highly consistent experiences throughout one's life, i.e., the same inducer will always induce the same concurrent in the same individual (low intraindividual variability). They are also highly idiosyncratic experiences, i.e., the same inducer can elicit different concurrents in different individuals (high interindividual variability). Consistency has not been studied in acquired forms. Drug-induced synesthesias are highly variable experiences both inter- and intraindividually.

Automaticity

Developmental synesthesias are highly automatic processes. As soon as the subject recognizes the inducer,

*Correspondence to: Pegah Afra, M.D., Department of Neurology, University of Utah, 175 N. Medical Drive East, 5th Floor, Salt Lake City, UT 84132, USA. E-mail: pegah.afra@hsc.utah.edu

the concurrent is involuntarily experienced. Acquired synesthesias are less automatic in that synesthetic experience can depend on the state of the subject, i.e., synesthesia is usually experienced during drowsiness, calm, and still states (Lessell and Cohen, 1979; Jacobs et al., 1981). Drug-induced synesthesias are not automatic and are highly dependent on the emotional state of the subject.

Dynamicity

In developmental synesthesias the dynamicity of the concurrent mimics the inducer. Specifically, when the inducer is music, then both the inducer and concurrent are quite dynamic. Acquired synesthesias are not dynamic. Drug-induced synesthesias are highly dynamic and complex, with different inducers eliciting the same concurrent and constantly changing concurrents.

Affectivity

In developmental and acquired synesthesias the subject's emotional state does not influence the synesthetic experience. Developmental synesthetes find their experience enjoyable and aesthetically appealing and synesthetic artists use their synesthesia as inspiration for their artwork. The synesthesias in acquired form can be disturbing to the subject. In drug-induced synesthesia affectivity plays a central role.

The three types of auditory synesthesia are summarized in Table 22.1 and discussed below.

Developmental auditory synesthesias

The developmental synesthesias are not a neurologic disorder but rather a different way of experiencing one's environment. These are involuntary and highly consistent experiences throughout one's life and the subjects remember them back to their childhood (Rich and Mattingley, 2002). Several types of developmental auditory synesthesia have been studied scientifically and are discussed in this chapter (Rich and Mattingley, 2002; Rich et al., 2005; Novich et al., 2011). These are summarized in Table 22.1. Additionally, many more types have been self-reported to The Synesthesia List, an international e-mail forum for synesthetes and those researching or interested in synesthesia (Day, 2005). These are summarized in Tables 22.2–22.4. The wide variation in presentation of synesthesia is due to the presence of multiple types of auditory stimuli triggering a variety of concurrents (Tables 22.2 and 22.3) and multiple types of inducer that can trigger auditory synesthetic perception (Table 22.4).

Table 22.1

Auditory synesthesias that have been scientifically reported

Types of auditory synesthesia
Developmental auditory synesthesia
Unidirectional auditory synesthesia
Auditory synesthesia with auditory stimuli as inducer
Auditory-visual synesthesia
Auditory-olfactory synesthesia
Auditory-gustatory synesthesia
Auditory synesthesia with auditory concurrent
"Hearing-motion" synesthesia
Bidirectional auditory synesthesia
Auditory-visual and visual-auditory synesthesia
Acquired auditory synesthesia
Lesional
Pathological affection of anterior optic pathways (i.e., visual deafferentation)
Pathological affection of the central nervous system with intact visual pathways
Non-lesional
Associated with epilepsy
Associated with migraines
Idiopathic
Induced auditory synesthesia
Induced by sensory deprivation
Drug-induced

Auditory synesthesia with auditory stimuli as inducer (Tables 22.2 and 22.3)

Auditory-visual synesthesia

Auditory-visual synesthesia is the most common type of developmental auditory synesthesia and is reported to make up 18% of a large self-referral sample of synesthetes (Day, 2005). It occurs when auditory stimuli elicit a definite and reliable visual experience. A variety of auditory stimuli have been reported as inducers. These could be classified as lexical-phoneme (i.e., spoken words) and non-lexical auditory stimuli (i.e., noise and sound stimuli, including clicking sounds, environmental sounds, vowels, voices, musical sounds, notes, chords, intervals, and timbre). Different characteristics of the auditory stimuli have been associated with different characteristics of visual concurrent. The pitch of the auditory stimuli has been associated with luminance of visual concurrent (higher-pitch sounds are lighter, while lower-pitch sounds are darker), vertical position or spatial coordinate (higher-pitch sounds are higher in space, while lower-pitch sounds are lower in space), shape (higher-pitch sounds are angular, while lower-pitch sounds are more rounded), and size (higher-pitch sounds are smaller, while lower-pitch sounds are larger)

Table 22.2

Reported inducer–concurrent couplings in developmental auditory synesthesias with auditory inducer and visual concurrent

Auditory inducer	Visual concurrent							Authors
	Color	Color and texture	Size or shape	Luminescence	Spatial coordinate	Movement	Number	
Noise			x				x	Jackson and Sandramouli, 2012
Sound (pitch or loudness)	x		x	x			x	Marks, 1975; Day, 2005; Jackson and Sandramouli, 2012
Musical sounds, notes, or chords	x	x	x		x	x		Marks, 1975; Mills et al., 2003; Beeli et al., 2005; Day, 2005; Rich et al., 2005; Novich et al., 2011
Music interval	x	x						Mills et al., 2003
Timbre	x	x	x					Marks, 1975; Mills et al., 2003; Novich et al., 2011
Genre of music	x	x	x			x		Marks, 1975; Mills et al., 2003
Vowels	x							Marks, 1975
Lexicon-phoneme	x							Day, 2005
Voice	x		x					Rich et al., 2005; Jackson and Sandramouli, 2012

Table 22.3

Reported inducer–concurrent couplings in developmental auditory synesthesias with auditory inducer and non-visual concurrent

Auditory inducer	Non-visual concurrent								Authors
	Olfactory	Gustatory		Somatosensory				Emotional	
	Smell	Taste	Food quality	Tactile/ touch	Temperature	Pain	Kinetic	Emotion	
Noise	x		x			x			Pierce, 1907; Cytowic, 2002; Jackson and Sandramouli, 2012
Sound (pitch or loudness)	x	x	x	x	x		x		Day, 2005; Day website; Jackson and Sandramouli, 2012; Pierce, 1907; Novich et al., 2011; Richer et al., 2011
Musical sounds, notes, or chords	x	x	x	x				x	Pierce, 1907; Day website; Rich et al., 2005; Richer et al., 2011
Music interval		x							Beeli et al., 2005
Timbre									
Genre of music									
Vowels		x	x						Pierce, 1907
Lexicon-phoneme		x	x						Pierce, 1907; Richer et al., 2011
Voice			x						Rich et al., 2005; Richer et al., 2011

Day website: types of synesthesia.

Table 22.4

Reported inducer–concurrent couplings in developmental auditory synesthesias with auditory concurrent

Inducers	Auditory concurrent					Author(s)
	Noise	Sound	Music	Phoneme	Voice	
Olfactory						
Smell		x				Day, 2005; Day website; Rich et al., 2005
Gustatory						
Taste/flavor		x				Day, 2005; Day website; Rich et al., 2005
Visual						
Vision		x				Day, 2005; Day website; Rich et al., 2005; Novich et al., 2011
Flash/motion		x				Saenz and Koch, 2008
Complex/painting			x			Rich et al., 2005
Somatosensory						
Tactile, touch		x				Day, 2005; Day website
Temperature		x				Day, 2005; Day website
Pain		x				Rich et al., 2005
Kinetic		x				Day website
Time unit		x				Day website
Emotion		x				Day website

Day website: types of synesthesia.

(Karowski et al., 1942; Ward et al., 2006; Chiou et al., 2012; Fernay et al., 2012). Timbre has been associated with color saturation or richness of colors (Ward et al., 2006).

The visual concurrents are usually colors or geometric shapes, although movement and texture have also been reported (Mills et al., 2003). Color is the most common visual concurrent, and its properties can depend on the acoustic properties of the inducing stimuli, as described above. However, when spoken words are the inducing stimuli, then color can depend on the linguistic properties of the auditory stimuli (Paulesu et al., 1995). In such cases the term lexical-color synesthesia is suggested (Rich et al., 2005), pointing to the fact that color concurrent can be induced by the graphemic (i.e., visual) or phonemic (i.e., auditory) component of the words. Colors can be either within the visible spectrum or non-real "Martian colors" (Ramachandran and Hubbard, 2001; Sinke et al., 2012). When within the visible spectrum, synesthetes can have difficulty explaining their concurrent synesthetic color. In case of non-real "Martian colors" they claim that they have never seen their concurrent synesthetic color in reality (Sinke et al., 2012). When music is the inducer, the colors can be accompanied by form and structure that can move and change with the music (i.e., the concurrent mimics the inducer in its dynamics). Auditory-visual synesthesia can occur in combination with other types of synesthesia. These are discussed below.

Auditory-olfactory and auditory-gustatory synesthesia

A case of auditory-visual synesthesia coexisting with auditory-olfactory synesthesia has been reported (Jackson and Sandramouli, 2012). In this case, the auditory-visual component consisted of voices causing shapes with different locations in the visual field, and other sounds (like clicking sounds, treading on grass or gravel, ripping paper, or untuned television) causing numbers, shapes, or colors. For the auditory-olfactory component noises and certain sounds elicited different smells (for example, the sound of a drill smelled like bleach, vacuum cleaner like vomit, and music like food). The synesthetic smell could be overpowering (i.e., the smell of bleach could cause vomiting). A unique case of a musician with auditory-visual and auditory-gustatory (music interval-taste) synesthesia who made use of the synesthetic sensation in the complex task of tone-interval identification has also been reported (Beeli et al., 2005). In the auditory-visual component, the inducers were tones and concurrent synesthetic perceptions were colors (for example, C is red and F-sharp is violet). In regard to the auditory-gustatory component, the inducers were musical-tone intervals and the concurrent synesthetic perceptions were tastes (for example, minor second is sour and major third is sweet).

Historically the earliest cases of auditory-gustatory synesthesia were reported in the early 20th century. Pierce (1907) reported a case of auditory-gustatory synesthesia in an anosmic subject with partial deafness. The inducers were reported to be a wide range of auditory stimuli, including spoken words, phonemes, vocal sounds (in the form of nonsense syllables) and non-vocal sounds (in the form of musical notes, tuning-fork vibrations, as well as noises). The concurrents were the four basic tastes, and food characteristics like temperature, texture, consistency, and pressure. Auditory-gustatory synesthesia had also been reported in two Italian speakers with intact olfaction with and without associated auditory-olfactory synesthesia (Ferrari, 1907, 1910).

In recent years, the term lexical-gustatory synesthesia has been suggested to replace auditory-gustatory synesthesia. This term denotes that the linguistic inducer of the taste can be auditory (i.e., spoken words and phonemes) or visual (i.e., written words and graphemes), although in nearly all studied cases of lexical-gustatory synesthesia the inducers are auditory i.e., phonemes (Ward et al., 2005; Simner et al., 2006). The concurrent gustatory experiences are complex and detailed. As with the cases discussed above, they include not only taste but also food characteristics such as texture and temperature.

Multiple other developmental auditory synesthesias are summarized in Table 22.3. These were self-reported to The Synesthesia List (Day, 2005).

Auditory synesthesia with auditory concurrent (Table 22.4)

A unique case series of 4 subjects in whom visual motion (i.e., inducer) elicited sounds (i.e., concurrent) were reported and called by authors as "hearing-motion synesthesia."

Almost all other cases of auditory synesthetic perception have been self-reported to The Synesthesia List (Day, 2005), with a variety of inducer–concurrent pairings, as summarized in Table 22.4.

Bidirectional auditory synesthesia

Bidirectional synesthesias are rare and bidirectional auditory synesthesias have been occasionally reported. These include reports of 2 subjects who have concomitant auditory-visual and visual-auditory synesthesia (Goller et al., 2009), and a very unusual case of bidirectional lexical-gustatory synesthesia (Richer et al., 2011).

Prevalence and genetics of developmental synesthesias

The reported prevalence of developmental synesthesias varies from 1 in 10 to 1 in 100 000 depending on the definitional criteria of the particular study (Simner et al., 2006). On one self-referral-based study the prevalence of synesthesia is reported to be 1 in 2000 (Baron-Cohen et al., 1996). In another study the prevalence of synesthesia was found to be 1.1–4.4% (Simner et al., 2006). In this latter study the definition of synesthesia was more liberal, large samples of the population were screened, and synesthesia was verified with objective measures of consistency. The reported prevalence of synesthesia within synesthetic families is 16% (Ward and Simner, 2005). Additionally, about 36–44% of synesthetes report at least one other family member with synesthesia (Rich et al., 2005; Ward and Simner, 2005).

The female-to-male ratio is also variable across different studies, ranging from 6:1 to 1:1 (Baron-Cohen et al., 1996; Rich et al., 2005; Simner et al., 2006; Barnett et al., 2008; Ward, 2013). Earlier studies had shown a higher female-to-male ratio and familial segregation, therefore at one time synesthesia was thought to be an X-linked dominant trait with lethality in males (Ward and Simner, 2005). Subsequent studies did not show the female preponderance and the earlier-reported female bias is thought to be due to sex differences in self-disclosure (Simner et al., 2006). There is continued evidence for strong genetic predisposition with complex modes of inheritance. Whole-genome linkage studies in auditory-visual synesthesia suggested an oligogenic disorder subject to multiple modes of inheritance and locus heterogeneity with linkage to chromosomes 5q33, 6p12, and 12p12 (Asher et al., 2009).

Neurobiologic investigation of developmental auditory synesthesias

The neurobiologic investigations of developmental auditory synesthesias can be divided into three groups: structural studies, functional studies, and neurophysiologic studies. These are discussed below.

Structural studies

There is a single study (Hanggi et al., 2008) investigating the neuroanatomic basis of interval-taste and tone-color synesthesia in a synesthetic musician in comparison to 20 controls and 17 non-synesthetic musicians (10 with absolute pitch and 7 with relative pitch). In the synesthetic subject with interval-taste and tone-color synesthesia, anatomic differences were found in the inducer auditory area for both tone and musical interval (i.e., Heschl's gyrus/planum temporale) and the concurrent areas for taste (insular cortex) and color (occipital cortex). In comparison with 20 controls and 17 non-synesthetic musicians, there was an increase in white-matter volumes in the right Heschl's gyrus/planum temporale and a decrease in gray-matter volumes in left superior temporal gyrus (i.e., the inducer area for both tone and musical interval). Additionally there was a decrease in gray-matter volume in left insula (i.e., concurrent areas for taste) and an increase in gray-matter volume in occipital areas (i.e., the concurrent area in tone-color synesthesia). Fractional anisotropy revealed evidence of hyperconnectivity in bilateral perisylvian-insular regions (i.e., between inducer and concurrent areas involved in the interval-taste synesthesia) in the synesthetic subject compared to 10 non-synesthetic musicians with absolute pitch.

Functional studies

Using the spatial resolution of functional magnetic resonance imaging (fMRI) and positron emission tomography (PET), these studies aid in functional localization of auditory synesthesias, by identifying the anatomic areas that are functionally different in synesthetes compared to controls.

There is a single fMRI study of auditory-visual synesthesia with 14 synesthetic subjects and 14 controls (Neufeld et al., 2012). This study used six different classes of auditory tone (i.e., not spoken words) and found stronger connectivity of the left inferior parietal cortex with left primary auditory cortex and right primary visual cortex in synesthetic subjects compared to controls (Fig. 22.1).

Several functional studies investigate lexical-color synesthesia (i.e., experience of color triggered by spoken words). Overall there are reports of activation of primary visual areas (V1), color areas (V4/V8), visual association cortices (posterior inferior temporal (PIT) cortex, and parieto-occipital junction). Due to some inconsistencies between the different studies, they are described individually below.

In a PET study, six synesthetes were compared with six controls in the task of listening to spoken words (Paulesu et al., 1995). These synesthetes had color sensation elicited only during listening to spoken words and not to tones, suggesting lexical-color synesthesia. The inducer brain areas (i.e., perisylvian regions) were activated in both synesthetes and controls, while the visual association cortices (i.e., PIT and parieto-occipital junction) were activated only in synesthetes. In this study there was no activation of the primary visual areas (V1 and V2) or the color area (V4).

The fMRI studies have higher spatial resolution and sensitivity compared to PET studies. In two fMRI studies the task of passively listening to spoken words was used, pointing to the fact that their synesthetic subjects had lexical-color synesthesia in response to spoken words. One of the studies (Aleman et al., 2001) reported activation of V1 in addition to PIT in a single synesthetic subject. The other study (Nunn et al., 2002) reported activation of left-hemispheric color areas (V4/V8) in 13 synesthetes compared to controls in the condition of listening to spoken words versus pure tones. However, no activation was found in V1 (Fig. 22.2). The reason for the inconsistencies is unknown, but may be related to the baseline condition of listening to pure tones in the latter study.

Neurophysiologic studies

With their higher temporal resolution, these studies evaluate the neurophysiologic basis of auditory synesthesias and the timing of synesthetic perception. Table 22.5 summarizes the results of these studies.

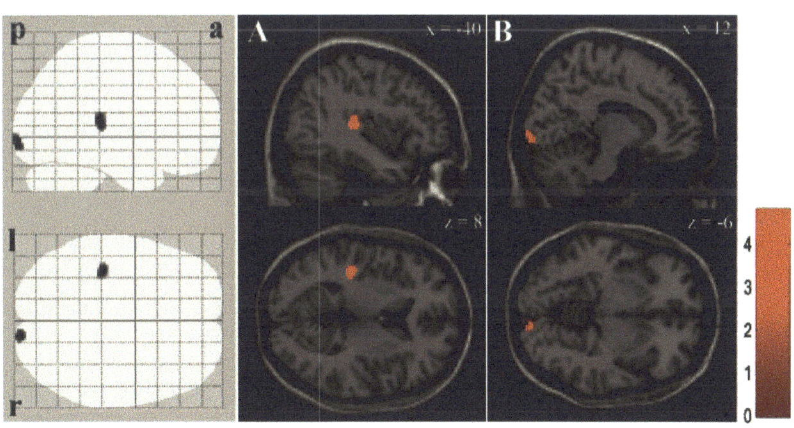

Fig. 22.1. Significant group differences in functional connectivity of the inferior parietal cortex (IPC). The connectivity analysis revealed two brain areas exhibiting significantly more connectivity to the IPC in synesthetes compared to controls: (**A**) a cluster in the left temporal cortex (center of mass at MNI coordinates $xyz = -40, -26, 8$), identified as the primary auditory cortex (Brodmann area (BA) 41) and (**B**) a cluster in the left occipital cortex (center of mass at MNI coordinates $xyz = 12, -94, -6$), identified as the primary visual cortex (Brodmann area 17). p, posterior; a, anterior; r, right; l, left; color bars indicate the strength of activation. (Reproduced from Neufeld et al., 2012.)

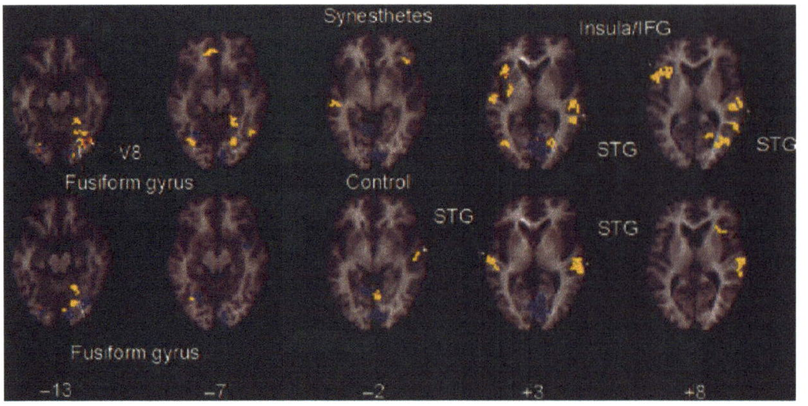

Fig. 22.2. Yellow: Activation maps in the condition of listening to spoken words versus baseline test of listening to pure tones (words minus tones) in synesthetes (top row) versus controls (bottom row). Blue: Color activation maps in normal non-synesthetic subjects (top and bottom row). Red: cluster common to both conditions. Note activation of language areas (i.e., bilateral superior temporal gyrus (STG) and left inferior frontal gyrus (IFG)) in both synesthetes and controls. Left-hemispheric colored areas were activated only in synesthetes. (Reproduced from Nunn et al., 2002.)

Table 22.5

Summary of neurophysiologic studies in developmental auditory synesthesia

Authors	Number of subjects	Modality/stimulus	Results
Rizzo and Eslinger, 1989	1 auditory-visual synesthesia	BAEP/clicks LLAEP/clicks	Normal BAEP (wave I–V) Normal LLAEPs No evidence of auditory evoked visual potential (in O1, O2, and Oz)
Goller et al., 2009	8 auditory-visual synesthesia 2 auditory-visual synesthesia + visual-auditory synesthesia 10 controls	LLAEP/tones	No report on BAEPs Lower amplitude of N1 in long-latency AEPs No evidence of auditory evoked visual potential No evidence of late negative deflections (i.e., slow wave around 400–800 ms)
Beeli et al., 2008	13 auditory-visual synesthesia 13 controls	LLAEP/words, pseudowords and letters	No report on BAEPs Lower amplitude and longer latency of N1 and P2 in LLAEPs
		LORETA	N1 (at 122 ms): enhanced activation in left PIT P2 (at 218 ms): (a) Enhanced activation in ventromedial OFC and PIT (for words) (b) Enhanced activation left SFG, left PC, right IPS (for letters)
		Averaged activation over 2000 ms (simulating time resolution of fMRI)	Left V4, mesial and left lateral orbitofrontal areas
Jancke et al., 2012	11 auditory-visual synesthesia 11 controls	Preattentive MMN /tones	Increased MMN amplitude in response to deviant tones Higher P3a amplitudes
		LORETA	Increased intracerebral current densities from auditory cortex, parietal cortex, and ventral visual areas

BAEP, brainstem auditory evoked potentials; LLAEP, long-latency auditory evoked potentials; LORETA, low-resolution brain electromagnetic tomography; MMN, mismatch negativity; OFC, orbitofrontal cortex; PIT, posterior inferior temporal; SFG, superior frontal gyrus; PC, precuneus; IPS, intraparietal sulcus; fMRI, functional magnetic resonance imaging.

To sum up their results, brainstem auditory evoked potentials (BAEPs) were normal (Rizzo and Eslinger, 1989). The results of long-latency auditory evoked potentials (LLAEPs) have been variable. When auditory clicks were used, the LLAEPs were normal (Rizzo and Eslinger, 1989), when tones were used as stimulus, the N1 component of the LLAEPs had longer latency (Goller et al., 2009), and when words and letters were listened to, there was lower amplitude and longer latency of the N1 and P2 component of LLAEPs (Beeli et al., 2008).

Regardless of the type of stimulus used, there was no evidence of auditory induced visual evoked potentials (i.e., no occipital activation was seen). The intracerebral source of the N1 and P2 components was estimated with low-resolution brain electromagnetic tomography. This revealed that in synesthetes there was stronger activation in the left PIT at 122 ms (i.e., N1 component) for both words and letters, and in the orbitofrontal cortex and left PIT for words, and in left superior frontal gyrus, left precuneus, and right intraparietal sulcus for letters. This indicates that the timecourse of processing of the synesthetic colors in the brain is early and similar to visualized colors. When the activation was averaged over 2000 ms to simulate resolution time of fMRI, local maxima of averaged activation were seen in left V4 and left mesial and lateral orbitofrontal areas for both words and letters.

One study used mismatch negativity (MMN), a preattentive component of event-related potentials, in the task of passive listening to the standard tones that were rarely interspersed with deviant tones (Jancke et al., 2012). In synesthetes, the deviant tones induced a change in color perception of the standard tone. This caused an increase in MMN amplitude between 100 and 150 ms after stimulus onset in synesthetes compared to controls. This indicates that the tones and concurrent color perception are processed early and automatically in the processing stream (i.e., preattentively, at 100–150 ms after tone presentation).

Additionally, a resting-state electroencephalograph (EEG) study in auditory-visual synesthesia has shown strong hub regions (i.e., brain regions with strong interconnections) in the right-sided auditory cortex and the left-sided parietal area (Jancke and Langer, 2011).

Neural models of developmental synesthesias

Synesthesia is an umbrella term that encompasses several distinct groups with different underlying mechanisms (Novich et al., 2011). Therefore different neuropsychologic models have been proposed to explain the neural basis of synesthesia. These models have been developed based on behavioral observation and/or neurophysiologic and imaging investigations of synesthesia. Some of the models have been used to explain auditory synesthesias (Ward et al., 2006) and are discussed below.

The cross-activation model (Fig. 22.3A and B) provides the best explanation for the neural basis of synesthesias in which inducer and concurrent brain areas are adjacent to each other, although the adjacency is not necessary (Hubbard et al., 2011). This model proposes that: (1) the neural representations of inducer and concurrent should lie in densely interconnected regions; (2) activation should pass directly from neurons that code for the inducer to the neurons that code for the concurrent; and (3) genetic factors lead to a decrease in pruning and the resulting anatomic differences are responsible for synesthetic experience (Ramachandran and Hubbard, 2001; Rich and Mattingley, 2002; Hubbard and Ramachandran, 2005; Bargary and Mitchell, 2008). This theory was originally developed to explain the neural basis of grapheme-color synesthesia (a non-auditory synesthesia that at the time was thought to be the most common form of synesthesia). Grapheme-color synesthesia could be considered a form of lexical-color synesthesia that occurs in response to graphemes. Music interval-taste synesthesia is a rare example of auditory synesthesia in which there is adjacency between inducer and concurrent areas, and using fractional anisotropy, hyperconnectivity in bilateral perisylvian-insular regions (i.e., between inducer and concurrent areas) has been reported (Hanggi et al., 2008).

The re-entrant feedback model (Fig. 22.3C and D) suggests cross-talk between the inducer and concurrent brain areas mediated via feedback from a multisensory nexus area (Grossenbacher and Lovelace, 2001; Smilek et al., 2001; Bargary and Mitchell, 2008).

The disinhibited feedback model (Fig. 22.3E and F) suggests that synesthesia results from failure of mechanisms that normally inhibit cross-talk between the inducer and concurrent brain areas mediated via feedback from a multisensory nexus area. This theory differs from other theories in that it suggests synesthesia is due to already existing connections in the brain (Grossenbacher and Lovelace, 2001; Ward et al., 2006), i.e., not due to atypical anatomic connections not found in non-synesthetes. There are several observations in support of this model (Sagiv and Ward, 2006), particularly for the developmental auditory-visual synesthesias.

There is evidence of auditory-visual connections in human infants. In the first month of life auditory evoked potentials (in response to spoken language) are recorded not only over the temporal cortex but also over the occipital areas (Maurer, 1993). These auditory evoked visual potentials persist until 6 months of age. After 6 months

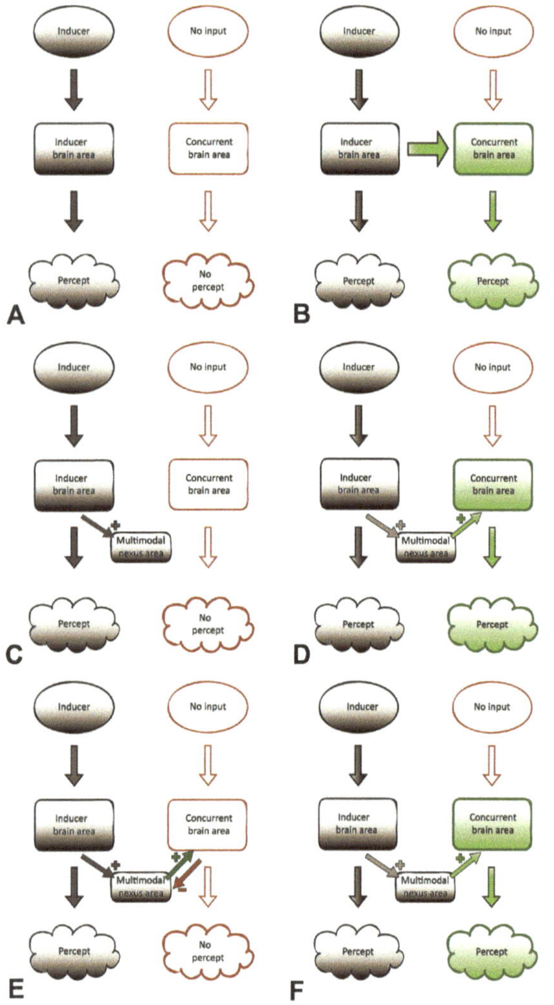

Fig. 22.3. (**A and B**) Oversimplified schematic drawing of the cross-activation model. (**A**) Neural representation of inducer and concurrent pathways in a normal non-synesthetic individual. (**B**) Neural representation of inducer and concurrent pathways in a synesthetic individual, demonstrating the presence of direct cross-activation from inducer to concurrent area, resulting in concurrent perception. (**C** and **D**) Oversimplified schematic drawing of the re-entrant feedback model. (**C**) Neural representation of inducer and concurrent pathways in normal non-synesthetic individual. (**D**) Neural representation of inducer and concurrent pathways in synesthetic individual, demonstrating cross-talk between the inducer and concurrent brain areas mediated via feedback from a multimodal nexus area. (**E** and **F**) Oversimplified schematic drawing of the disinhibited feedback model. (**E**) Neural representation of inducer and concurrent pathways in normal non-synesthetic individual. (**F**) Neural representation of inducer and concurrent pathways in synesthetic individual, demonstrating cross-talk between the inducer and concurrent brain areas mediated via disinhibited feedback from a multimodal nexus area. (Modified from Bargary and Mitchell, 2008.)

of age the auditory evoked responses over the visual areas gradually decrease until they disappear at 36 months (Neville, 1995). These observations have led to development of a neonatal synesthesia hypothesis which suggests that human infants are born with intercortical connections between sensory areas and synesthesia resulting from failure of normal pruning processes that eliminate these connections (Baron-Cohen, 1996).

There is observational evidence of the persistence of auditory-visual connections in adult humans with intact visual pathways that could be unmasked chemically with hallucinogenic drugs (Hollister, 1968) or with acute visual deprivation due to experimental or surgical blindfolding (Lessell and Cohen, 1979; Merabet et al., 2004), resulting in emergent auditory-visual synesthesia.

There is neurophysiologic and functional neuroimaging evidence of the persistence of these auditory-visual connections in the late blinds, when auditory evoked potentials are recorded over the occipital areas (Rao et al., 2007), or when there is functional evidence of auditory processing by fMRI in the occipital cortex (Gougoux et al., 2005; Fig. 22.4).

Neurophysiologic studies in normal human adults have shown the sound-induced illusory flash effect, i.e., when a single visual flash is accompanied by multiple auditory beeps, the single flash is perceived as multiple flashes. The brain evoked potentials recorded from occipital electrodes (O1, O2, and Oz) were similar for a physical flash and illusory flash (Shams et al., 2001).

Comparison between synesthetes and non-synesthetes has shown shared cross-modal audiovisual associations (between pitch and lightness, or timbre and saturation). However, the synesthetic experience is automatic, precise, conscious, and explicit, while the non-synesthetic associations are not (Ward et al., 2006).

Despite all the above, neurophysiologic studies to date have not shown auditory-induced visual evoked potentials in developmental auditory-visual synesthesias, although auditory-induced visual evoked potentials have been reported in neonates, in individuals who become blind late in life, and in acquired synesthesia (Maurer, 1993; Rao et al., 2007; Afra et al., 2012). Therefore it is quite possible that the neurophysiologic basis of developmental auditory-visual synesthesias is different from the synesthesia in neonates, individuals who become blind late in life, or any other form of acquired synesthesia. This topic is subject to future research.

The hyperbinding model suggests that synesthesia arises due to anomalous binding of information from inducer and concurrent brain regions and results from overactivation of parietal binding mechanisms (Robertson, 2003; Hubbard, 2007). Because the binding mechanisms require a feature upon which to act, a

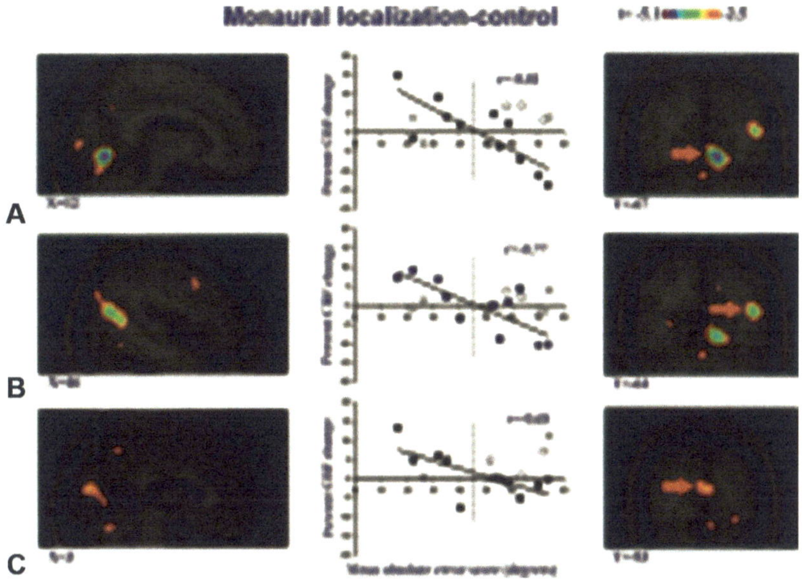

Fig. 22.4. Functional evidence of auditory processing by functional magnetic resonance imaging in occipital cortex during monaural sound localization in 12 early blind subjects versus 7 healthy sighted subjects. These data show the correlational analysis between performance (mean absolute error) in a pointing task to monaurally presented sounds and cerebral blood flow (CBF) in a group of blind subjects. The two columns of brain images (left image series, sagittal sections; right image series, coronal sections) illustrate the statistical parametric map of the correlation, which is maximal in the ventral extrastriate cortex (**A**) but also significant in dorsal extrastriate (**B**) and striate (**C**) cortices. The red arrows in the coronal slices indicate the focus selected for the respective sagittal slices. The scattergram shows the individual values extracted from each of these regions; closed circles indicate blind subjects; open circles indicate the sighted control group. (Reproduced from Gougoux et al., 2005.)

two-stage cross-activation/hyperbinding model is suggested (Hubbard, 2007; Jancke et al., 2012). A number of functional and neurophysiologic studies are supportive of this theory, including PET and fMRI studies in which activation of PIT is seen (Paulesu et al., 1995; Aleman et al., 2001), and resting-state EEG study showing left-sided parietal area as a strong hub region (Jancke and Langer, 2011).

The limbic mediation hypothesis (Cytowic, 2002) suggests that synesthesia is mediated by the hippocampus and the limbic system.

Acquired auditory synesthesias

The acquired auditory synesthesias are symptoms of neurologic disease; they usually emerge subsequent to neuropathologic insult to the brain. Cases of acquired auditory-visual and auditory-tactile synesthesias have been described in the neurologic literature and auditory-gustatory synesthesia has been described in the psychiatric literature.

Acquired auditory-visual synesthesia

The acquired auditory-visual synesthesias have been reported in the presence as well as absence of a recognizable neuropathologic lesion, with the most commonly reported cases being in the setting of visual deafferentation. Therefore the reported cases of acquired auditory-visual synesthesia are discussed below in two categories of lesional and non-lesional (Table 22.1). In the lesional category the reported cases could be further divided depending on the presence or absence of visual deafferentation (i.e., pathological affection of anterior optic pathways versus pathological affection of the central nervous system with intact visual pathways (Table 22.1).

In the non-lesional category the auditory-visual synesthesias have been reported in association with epilepsy and migraine, as well as in a single idiopathic case (Table 22.1).

Lesional acquired auditory-visual synesthesia

Lesional acquired auditory-visual synesthesia with pathological affection of anterior optic pathways (i.e., visual deafferentation). There have been about 19 reported cases across different case series and reports (Bender, 1977; Lessell and Cohen, 1979; Jacobs et al., 1981; Page et al., 1982; Kim et al., 2006; Rao et al., 2007), making this the most common type of acquired auditory synesthesia. In these cases the auditory stimuli are experienced ectopically in the deafferented visual system. Concomitant positive visual phenomena that were not sound-induced and were not perceived concomitant with synesthesia have been reported in some cases (Jacobs et al., 1981; Page et al., 1982).

Table 22.6 summarizes the clinical characteristics of the 19 reported cases of acquired auditory-visual synesthesia (Afra et al., 2009). The neuropathologic abnormalities involve the anterior optic pathways (i.e., optic nerve, chiasm, or both) and consist of demyelination, ischemia, or tumors as well as posttraumatic total ocular blindness, with the most commonly reported neuropathology being optic neuritis followed by ischemic optic neuropathy. The degree of deafferentation has been variable across different reported cases, ranging from mild to severe partial deafferentation as well as near-complete and complete blindness (Afra et al., 2009). There seems to be no relationship between the onset of synesthesia and the underlying neuropathology. The onset of synesthetic visual sensations in relation to the clinical onset of the visual symptoms (due to deafferentation) has been variable, ranging from acute (1–3 days), and subacute (1–4 weeks) to chronic (months). There has been a single reported case of synesthetic visual sensations preceding the clinical onset of the visual symptoms (Page et al., 1982).

The reported synesthetic visual sensations are not well formed and rather simple, like light flashes, kaleidoscope, and color (Afra et al., 2009). Eight of the 19 reported cases have commented on the ambient light during experience of synesthesia: five were reported to experience their synesthesia in dark, two in light, and one in both light and dark. Eighteen of the 19 reported cases commented on the state of the patient: nine experienced their synesthesia in a drowsy state, seven in a relaxed state, one in an awake state and two when they were still (Lessell and Cohen, 1979; Jacobs et al., 1981; Page et al., 1982; Kim et al., 2006; Rao et al., 2007; Afra et al., 2009). The induced visual sensations are usually ipsilateral to the side of the perceived auditory stimuli and, regardless of laterality, they are perceived in the deafferented eye, and at times in the area of scotoma itself (Jacobs et al., 1981). In one case the induced visual phenomena were reported to occur in front of the eye (Page et al., 1982). The duration of synesthesia was reported in 16/19 of cases. It ranged from 2 to 49 months in 7/19 of the reported cases, to being persistent in 9/19 of the cases (Jacobs et al., 1981; Page et al., 1982; Kim et al., 2006; Rao et al., 2007; Afra et al., 2009).

Neurophysiologic investigations. A variety of neurophysiologic studies have been used in the assessment of acquired auditory visual synesthesia.

1. Visual evoked potentials show abnormalities of anterior optic pathways as expected (Jacobs et al., 1981; Page et al., 1982; Rao et al., 2007).
2. Electroretinogram (ERG) has not shown any ERG activation after sound stimuli (Page et al., 1982).
3. BAEPs have shown intact auditory pathways (Jacobs et al., 1981; Page et al., 1982). Rao et al. (2007) reported occipital activation (i.e., auditory evoked visual potentials) 40–80 ms (P1 component) and 100–120 ms (N1 component) following auditory stimulation.

Table 22.6

Summary of reported cases of acquired auditory-visual synesthesia due to deafferentation

Pathology	No	Deafferentation	Onset	Laterality	Author(s)
Vascular occlusive disease	1	Partial	Acute	Ipsilateral	Jacobs et al., 1981
Temporal arteritis	2	Near complete	Acute	Ipsilateral	Jacobs et al., 1981
Chiasmal tumor	2	Partial	Acute	Ipsi/bilateral	Jacobs et al., 1981
Optic neuritis	6	5 Partial, 1 near complete	Acute/ subacute	4 Ipsilateral; 2 not reported	Bender, 1977; Lessell and Cohen, 1979; Jacobs et al., 1981; Page et al., 1982
ION	3	Partial	Preceding/ subacute	Not reported	Page et al., 1982
Arteritic ION	1	Partial	Subacute	Not reported	Page et al., 1982
Neurilemoma	1	Partial	Chronic	Ipsilateral	Lessell and Cohen, 1979
Melanocytoma	1	Partial	Chronic	Not reported	Kim et al., 2006
PSA	1	Complete	Chronic	Ipsilateral	Jacobs et al., 1981
PT TOB/B	1	Complete	Chronic	Ipsi/bilateral	Rao et al., 2007

No, number of reported patients; ION, ischemic optic neuropathy; PSA, postsurgical amaurosis; PT TOB, posttraumatic total ocular blindness; Ipsi, ipsilateral; B, bilateral.

Lesional auditory-visual synesthesia associated with pathological affection of the central nervous system with intact visual pathways. Two cases of lesional acquired auditory-visual synesthesia with pathologic effect on the central nervous system with intact visual pathways have been reported. In one case (Vike et al., 1984) a gliotic mass involved the left temporal area and adjacent midbrain. The synesthesia was reported to be experienced bilaterally but more on the left (ipsilateral side). BAEPs and visual evoked potentials were normal with no commentary about auditory evoked visual potentials (i.e., occipital activation following auditory stimulation). Synesthesia was aborted after the removal of tumor.

In the second case (Fornazzari et al., 2012) a hemorrhagic stroke involving the left lateral posterior nucleus of the thalamus was reported. The stroke initially caused right hemiparesis, hemisensory deficit, hemianopsia, slurred speech with normal comprehension, and a transient alien-limb phenomenon. The synesthesia developed 9 months after the stroke, at which time the patient had residual right hemiparetic gait, hemisensory deficits, and full visual fields. The synestheisa was multimodal with auditory-visual, auditory-tactile, and grapheme-gustatory components as well as conceptual synesthesia. The auditory-visual component consisted of the appearance of light blue colors in the left halves of both visual fields when listening to high-tone sounds such as Hindi, Inuit, or Chinese music. The auditory-tactile component is discussed further below (under acquired auditory-tactile synesthesia).

Non-lesional auditory-visual synesthesia

Non-lesional auditory-visual synesthesia associated with epilepsy. Synesthesia-like symptoms, called "internal synesthesia," were reported in a single case of left temporal-lobe epilepsy (Jacome and Gumnit, 1979). During his aura the patient experienced sudden onset of pain on the right side of his face, and simultaneously heard the word "five" in both ears and saw the number "5" or "five" on a gray background before his eyes.

Non-lesional auditory-visual synesthesia associated with migraines. There have been two reported cases of auditory-visual synesthesia occurring in the aural phase of migraine and one reported case in the headache phase. In one case (Sacks, 1992), the patient had experienced "synesthetic equivalence between auditory stimuli and visual images," i.e., in the midst of migraine aura (with migratory paresthesias, sense of timelessness, and *déjà vu*), the subject experienced the hum of crickets translated into the hum of color when eyes were closed. In the second case (Alstadhaug and Benjaminsen, 2010), multimodal synesthesia was reported in the aural phase of migraine headache. Interestingly, the patient distinguished the aural component from the synesthesia component: "the flickering is present all along and seems to come from within," but "the other is caused by external stimuli and may be avoided." In regard to multimodal synesthesia, the auditory-visual component (audio-induced central scotoma) was intermixed with a visual-gustatory component (staring at bright light induced intense taste of lemon with flow from the salivary glands). What is unusual and unique about this case of multimodal synesthesia is the simultaneous presence of negative visual phenomena in the auditory-visual component with positive phenomena in the visual-gustatory component.

In a third case, the patient had dual diagnosis of basilar migraines and migraine aura without headache (Podoll and Robinson, 2002). The auditory-visual synesthesia occurred in the headache phase on three separate occasions: when the patient had been woken up by the alarm clock experiencing migraine headache and a "colored optical pattern" pulsating in correspondence to the alarm sound in the center of her visual field.

Idiopathic non-lesional auditory-visual synesthesia. A single case of adult-onset acquired auditory-visual synesthesia has been reported for which no cause could be identified despite extensive workup (Afra et al., 2012). Neurophysiologic investigation included magnetoencephalography. Auditory stimuli (700 Hz, 50 ms duration, 0.5 second interstimulus interval) were presented binaurally at 60 dB above the hearing threshold in a dark room. The patient had bilateral symmetric auditory evoked neuromagnetic responses in the auditory cortices followed by an occipital evoked field 16.3 ms later (Figs 22.5 and 22.6). It was suggested that activation of occipital cortex following auditory stimuli may represent recruitment of existing cross-modal sensory pathways.

Mechanisms of acquired auditory-visual synesthesia

The pathophysiologic basis of acquired auditory-visual synesthesia is not known. The anatomic, imaging, and neurophysiologic evidence for the presence of early cross-modal interactions may explain potential mechanisms and is discussed briefly.

There is anatomic evidence of the presence of connections between auditory and primary visual areas in primates by tracer injection studies (Falchier et al., 2002; Rockland and Ojima, 2003). It is not known if the same connections exist in humans. As discussed above, there is neurophysiologic evidence of the existence of these connections in human neonates, when auditory evoked responses extend over the occipital areas in the first

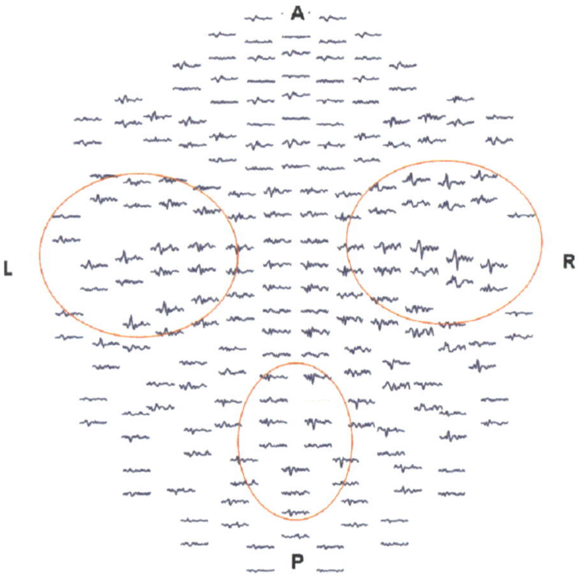

Fig. 22.5. Topographic magnetoencephalography channel plot of averaged auditory evoked fields (–50 ms to 400 ms). The encircled selections (red) represent the different activated areas: left and right auditory cortices as well as midline occipital cortex. Data of 204 planar gradiometer are displayed. A, anterior; P, posterior; L, left; R, right. (Reproduced from Afra et al., 2012.)

6 months of life (Maurer, 1993), followed by a gradual decrease until disappearance at 36 months (Neville, 1995). In normal human adults there is functional evidence of the persistence of these auditory-visual connections with the presence of bilateral blood oxygen level-dependent responses in the primary visual cortices subsequent to auditory stimuli (Martuzzi et al., 2007). In individuals who were blind early in life, there is functional evidence of auditory processing by fMRI in the occipital cortex (Gougoux et al., 2005). Also in an individual who became blind late in life with posttraumatic total ocular blindness, auditory evoked potentials have been recorded over the occipital areas, providing neurophysiologic evidence of the persistence of these connections (Rao et al., 2007). An additional piece of neurophysiologic evidence is the presence of a sound-induced illusory flash effect in normal humans (Shams et al., 2001).

When synesthesia occurs as a result of neuropathologic insult to the anterior optic pathways, the region of the primary visual cortex corresponding to the affected axons will be deafferented. The imbalance between excitation and inhibition will result in increased excitability in and surrounding the deafferented area (Eysel et al., 1999). There will be reorganization and

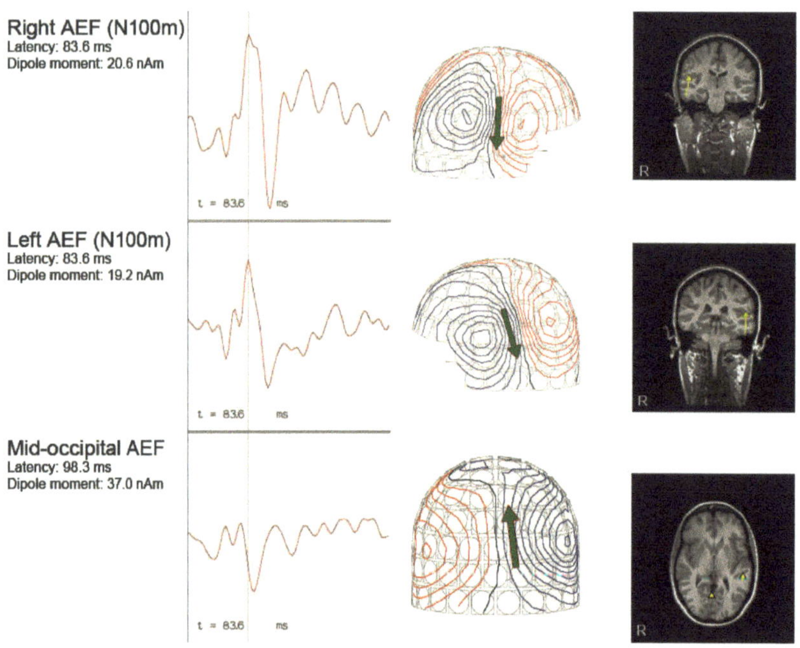

Fig. 22.6. Auditory evoked magnetic responses, magnetic field maps, and source localizations. Left: Right, left, and midline occipital auditory evoked magnetic fields (AEFs) in response to binaural stimulation at 60 dB above hearing threshold (auditory tone of 700 Hz, 50 ms in duration at a rate of 2 Hz). Center: AEF maps over the expected right and left temporal areas as well as the occipital region. Right: Magnetic source localizations of the AEFs spatially aligned with magnetic resonance imaging. (Reproduced from Afra et al., 2012.)

change in weight of synaptic neurons before recovery. It has been shown that recovery from optic neuritis to normal or near-normal vision is associated with a change in the distribution of cerebral response to visual stimulation in fMRI (Werring et al., 2000). There is extensive extrastriate activation that may be evidence of long-term changes in the networks involved in neuronal processing of visual stimuli.

The onset of synesthesia is variable across different patients, with some patients developing it preceding or acutely after visual symptoms (in the subclinical or clinical denervation supersensitivity period), while others develop it subacutely to chronically during or after recovery (i.e., during or after any cortical reorganization or plasticity). Also the degree of deafferentation is variable, with some patients experiencing synesthesia with mild degrees of deafferentation and some with severe degrees of deafferentation, and some never experiencing it at all. The reason for this variability in presentation is unknown but may be due to interindividual differences between patients and their central sensory synapses.

The temporal lobe harbors the primary auditory cortex and auditory association areas in the Heschl's and superior temporal gyri as well as higher-order association areas in the lateral temporal lobe. In the case of temporal-lobe epilepsy, these areas could be easily recruited by seizures of temporal-lobe origin. In the case of temporal-lobe space-occupying lesions, synesthesia could have been due to a mass effect on these areas, as it disappeared after removal of the lesion.

It is interesting that synesthesias associated with migraine headaches have been reported in both aural and headache phases. The mechanisms of these synesthesias are not understood but may be due to alterations of cross-modal networks by cortical spreading depression (Alstadhaug and Benjaminsen, 2010) or the neurovascular changes of the headaches phase.

Acquired auditory-tactile synesthesia

The two reported cases of auditory-tactile synesthesia are in association with thalamic stroke. In one case, there was hemorrhagic stroke affecting the left lateral posterior nucleus of the thalamus (Fornazzari et al., 2012). In this case the auditory-tactile synesthesia emerged as a component of multimodal synesthesia 9 months after the stroke (as described above, under lesional auditory-visual synesthesia). The auditory-tactile component was reported as extracorporeal sensations of "riding the music, surfing the music or flying on the top of the music."

In another case a lacunar infarct affected the right ventrolateral nucleus of the thalamus, resulting in residual multisensory neglect for vision and touch and tactile and visual antiextinction at 1 year poststroke (Ro et al., 2007). Auditory-tactile synesthesia, contralateral to the thalamic lesion, was experienced at 3 years poststroke. The neural basis of synesthesia was suggested to be abnormal connectivity from right ventrolateral to the cortex in the form of disorganized and fewer fiber bundles originating from the right thalamus.

Acquired auditory-gustatory synesthesia

A case of auditory-gustatory synesthesia in combination with olfactory-gustatory synesthesia has been reported in a patient with postpartum depression. The synesthesia was considered part of a perceptual disorder associated with depression and was resistant to tricyclic antidepressants and electroconvulsive therapy, but was treated successfully with phenelzine (McKane and Hughes, 1988).

Induced synesthesias

Induced auditory synesthesias are emergent phenomenon that have been reported in the setting of sensory deprivation (i.e., physical deafferentaion) as well as recreational drug use (i.e., chemical deafferentation).

Auditory synesthesia induced by sensory deprivation

Auditory-visual synesthesias have been reported in the context of visual deprivation in a single case of postsurgical blindfolding for keratoconus surgery (Lessell and Cohen, 1979) and two of 13 subjects in visual deprivation experiments (Merabet et al., 2004). In all three cases the reported synesthesia was of acute onset and emergent. The triggering auditory stimuli in the postsurgical case were unexpected auditory sounds and noises and in the visual deprivation experiments were TV and Mozart's Requiem. The duration of synesthesia in the postsurgical case was 1 week, despite blindfolding for several weeks.

Drug-induced auditory synesthesia (use of hallucinogens and psychedelics)

Drug-induced synethesias are highly variable experiences with no consistent inducer–concurrent couplings. The experienced synesthesias are not automatic, i.e., the same stimulus may induce synesthetic perception in one occasion but not another. In the majority of drug-induced cases the inducers are auditory (music and sound), although any sensory input (haptic, visual, gustatory, olfactory, pain, or emotional stimuli) can act as inducer (Shanon, 2002; Sinke et al., 2012). The concurrents are almost always visual, although a single case of drug-induced synesthesia with auditory concurrent has been reported. In this reported case of visual-auditory synesthesia due to methamphetamine abuse, the subject was reported to hear voices of colors that he saw in the carpet (Ahmadi et al., 2011).

Auditory-visual synesthesias have been reported with a variety of drugs, including LSD (Hollister, 1968), ayahuasca (Shanon, 2002), mescaline (Simpson and McKellar, 1955), hashish (Marks, 1975), and a psychotropic mint called *Salvia divinorum* (Babu et al., 2008). Additionally auditory-tactile synesthesia as well as auditory-algesic synesthesia in association with auditory-visual synesthesia has been reported with mescaline (Simpson and McKellar, 1955). The synesthetic experience with hallucinogenic drugs happens as an accompaniment to other symptoms of intensified sensory perception, including illusions, pseudohallucinations, and true hallucinations (Leuner, 1962), all of which can be well formed and complex (Hollister, 1968). *Salvia divinorum* also produces visual hallucinations in addition to synesthesia (Babu et al., 2008).

The experienced drug-induced synesthesias are highly dependent on the state of the subject and can be experienced with both closed and open eyes (Sinke et al., 2012). With closed eyes it is experienced in any kind of inner screen while with open eyes it is in the form of synesthetic illusion (i.e., superimposed over real things in the outer world). The synesthetic experience can be well formed and complex (Hollister, 1968), as well as very intense and dynamic with emotional involvement (Sinke et al., 2012). They occur in a dream-like state of consciousness, although the subject is aware of the synesthetic experience (Sinke et al., 2012).

Mechanism of induced auditory synesthesias

Acute visual deprivation (by blindfolding) causes cortical excitability, evidenced by decreased phosphene threshold to transcranial magnetic stimulation in the occipital regions and enhanced activation of visual cortex to incoming input, as measured by fMRI (Boroojerdi et al., 2000). In synesthesia secondary to surgical or experimental blindfolding, there is complete and sudden deafferentation with no neuronal injury. In these cases the onset of synesthesia is acute (Lessell and Cohen, 1979; Merabet et al., 2004), pointing to the presence of some existing auditory-visual pathways that could be unmasked due to hyperexcitability of the cortex deprived of visual stimuli.

A variety of mechanisms have been proposed as the basis of drug-induced synesthesias. The cause of hallucinogenic drug-induced auditory-visual synesthesia is thought to be chemical deafferentation of the lateral geniculate body, which deprives the occipital cortex of visual stimuli (Hollister, 1968).

Brang and Ramachandran (2008) hypothesized that serotonin 2a receptors are the "synesthesia receptor." LSD is known to selectively activate serotonin 2a receptor. Additionally mescaline and ayahuasca are serotonin 2a agonists. Ayahuasca contains the serotonergic psychedelic *N,N*-dimethyltryptamine (DMT) and beta-carboline, a monoamine oxidase inhibitor (Riba et al., 2003). DMT is a partial agonist of serotonin 2a receptors and beta-carboline inhibits the visceral metabolism of DMT, increasing its systemic levels.

Salvinorin A, the active ingredient of *Salvia divinorum*, is a selective kappa-opioid receptor agonist (Babu et al., 2008), and hence points to a different mechanism of action compared to serotonin 2a receptor agonists. Delta-9-tetrahydrocannabinol, the active component of hashish, exerts its central effects through the CB1 cannabinoid receptor (Iversen, 2003).

EFFECT OF DRUGS ON DEVELOPMENTAL SYNESTHESIA

LSD has been reported to cause a new type of synesthesia. A subject with developmental grapheme-color synesthesia was reported to have experienced a new type of synesthesia in the form of auditory-visual synesthesia under LSD (Sinke et al., 2012).

LSD and cannabis have also been reported to alter or overpower inducer–concurrent pairings, causing "false synesthesia." Following their use a subject with auditory-visual synesthesia was reported to experience musical tones in the wrong color (Sinke et al., 2012).

Cannabis has been reported to intensify pre-existing auditory-visual synesthesia (Mayer-Gross, 1931) and mescaline has been reported to augment the vividness of synesthetic experience in two patients, one with auditory-visual and the other with auditory-tactile synesthesia (Simpson and McKellar, 1955).

AUDITORY SYNESTHESIA: HISTORY AND ART

Synesthesia has been an old accompaniment to the human race and reports of synesthesia date back to ancient China and Persia. An ancient Persian scheme provides correspondence between musical tones and colors: D violet, C blue-black, G bright blue, A green, E yellow, B rose, F black (Day's synesthesia website: History). Reports of auditory-visual synesthesia date back to Pythagoras, sixth century BC and Aristotle,fourth century BC, and a steady and continuous stream of reported cases from the 16th century up to today (Marks, 1975). The very first known scientifically reported case of synesthesia, published by Dr. G.T.L. Sachs in 1812 in Latin (Ione and Tyler, 2004), was a medical dissertation on two albinos with synesthesia (the author himself and his sister). The author refers to his colored sensations evoked by vowels, consonants, musical notes, sounds of instruments, numbers, dates, days of the week, periods of history, and stages of human life (Ione and Tyler, 2004). Another well-known early scientific report of synesthesia is an essay by Galton in 1883. In this essay he referred to published earlier cases, including ones by Bleuler and Lehman in 1881.

The famous writer, Vladimir Nabakov, and several famous musicians/composers have been reported to have had auditory-visual synesthesia. These include Jean Sibelius, Franz Liszt, and Tony DeCaprio (Pearce, 2007). A full account of famous artists and writers can be found in Day's synesthesia website (please see under web references).

CONCLUSION

Auditory synesthesias are anomalous perceptual experiences induced by auditory stimulation or due to auditory synesthetic perception. The history of these synesthetic perceptual experiences dates back to ancient times. Auditory synesthesias can be lifelong and normal variations of human perceptual and cognitive function. Alternatively, they can be emergent and a symptom of neurologic or psychiatric disease, or the result of recreational drug intake. Neural mechanisms of developmental synesthesias provide a window into human cognitive processes and are the subject of research in the field of human neuropsychology, while the acquired forms of synesthesias represent an unusual symptom of neurologic and psychiatric disease and warrant neurophysiologic and neuroimaging investigation.

References

Afra P, Funke M, Matsuo F (2009). Acquired auditory-visual synesthesia: a window to early cross-modal sensory interactions. Psychol Res Behav Manag 2: 31–37.

Afra P, Anderson J, Funke M et al. (2012). Neurophysiological investigation of idiopathic acquired auditory-visual synesthesia. Neurocase 18 (4): 323–329.

Ahmadi J, Keshtkar M, Pridmore S (2011). Methamphetamine induced synesthesia: a case report. Am J Addict 20 (3): 306.

Aleman A, Rutten GJ, Sitskoorn MM et al. (2001). Activation of striate cortex in the absence of visual stimulation: an fMRI study of synesthesia. Neuroreport 12 (13): 2827–2830.

Alstadhaug KB, Benjaminsen E (2010). Synesthesia and migraine: case report. BMC Neurol 10: 121.

Asher JE, Lamb JA, Brocklebank D et al. (2009). A whole-genome scan and fine-mapping linkage study of auditory-visual synesthesia eveals evidence of linkage to chromosome 2q24, 5q33, 6p12 and 12p12. Am J Hum Genet 84 (2): 279–285.

Babu KM, McCurdy CR, Boyer EW (2008). Opioid receptors and legal highs: *Salvia divinorum* and Kratom. Clinical Toxicol (Phila) 46 (2): 146–152.

Bargary G, Mitchell KJ (2008). Synaesthesia and cortical connectivity. Trends Neurosci 31 (7): 335–342.

Barnett KJ, Finucane C, Asher JE et al. (2008). Familial patterns and the origins of individual differences in synaesthesia. Cognition 106 (2): 871–893.

Baron-Cohen S (1996). Is there a normal phase of synesthesia in development? Psyche 2. Available online at, http://psyche.cs.monash.edu.au/.

Baron-Cohen S, Wyke MA, Binnie C (1987). Hearing words and seeing colours: an experimental investigation of a case of synesthesia. Perception 16 (6): 761–767.

Baron-Cohen S, Burt L, Smith-Laittan F et al. (1996). Synaesthesia: prevalence and familiarity. Perception 25 (9): 1073–1079.

Beeli G, Esslen M, Jancke L (2005). Synaesthesia: when colored sounds taste sweet. Nature 434 (7029): 38.

Beeli G, Esslen M, Jancke L (2008). Time course of neural activity correlated with colored-hearing synesthesia. Cereb Cortex 18 (2): 379–385.

Bender MB (1977). Neuro-ophtalmology. In: AB Baker, LH Baker (Eds.), In: Clinical Neurology. Vol. 1. Harper and Row, Hagerstown, MD, p. 37.

Boroojerdi B, Bushara KO, Corwell B et al. (2000). Enhanced excitability of the human visual cortex induced by short-term light deprivation. Cereb Cortex 10 (5): 529–534.

Brang D, Ramachandran VS (2008). Psychopharmacology of synesthesia; the role of serotonin S2a receptor activation. Med Hypotheses 70 (4): 903–904.

Chiou R, Stelter M, Rich AN (2012). Beyond color perception: auditory-visual synaesthesia induces experiences of geometric objects in specific locations. Cortex 49 (6): 1750–1763.

Cytowic RE (2002). Synesthesia: A union of senses, 2nd edn. MIT Press, Cambridge, MA.

Day S (2005). Some demographic and socio-cultural aspects of synesthesia. In: LC Robertson, N Sagiv (Eds.), Synesthesia: Perspectives From Cognitive Neuroscience, Oxford University Press, New York, pp. 11–33.

Dixon MJ, Smilek D, Merikle PM (2004). Not all synaesthetes are created equal: projector versus associator synaesthetes. Cogn Affect Behav Neurosci 4 (3): 335–343.

Eysel UT, Schweigart G, Mittmann T et al. (1999). Reorganization in the visual cortex after retinal and cortical damage. Restor Neurol Neurosci 15 (2–3): 153–164.

Falchier A, Clavagnier S, Barone P et al. (2002). Anatomical evidence of multimodal integration in primate striate cortex. J Neurosci 22 (13): 5749–5759.

Fernay L, Reby D, Ward J (2012). Visualized voices: a case study of audio-visual synesthesia. Neurocase 18 (1): 50–56.

Ferrari GC (1907). Una varieta nuova di sinestesia. Rivista Psicologia 3: 297–317.

Ferrari GC (1910). Un nuovo caso di sinestesia uditivo-gustativa. Rivista Psicologia 6: 101–104.

Fornazzari L, Fischer CE, Ringer L et al. (2012). Blue is music to my ears: multimodal synesthesia after a thalamic stroke. Neurocase 18 (4): 318–322.

Galton F (1883). Inquiries into human faculty and its development, Macmillan, London.

Goller AI, Otten LJ, Ward J (2009). Seeing sounds and hearing colors: an event-related potential study of auditory-visual synesthesia. J Cogn Neurosci 21 (10): 1869–1881.

Gougoux F, Zatorre RJ, Lassonde M et al. (2005). A functional neuroimaging study of sound localization: visual cortex activity predicts performance in early-blind individuals. PLoS Biol 3 (2): e27.

Grossenbacher PG, Lovelace CT (2001). Mechanisms of synesthesia: cognitive and physiological constraints. Trends Cogn Sci 5 (1): 36–41.

Hanggi J, Beeli G, Oechslin MS et al. (2008). The multiple synaesthete E.S.: neuroanatomical basis of interval-taste and tone colour synaesthesia. Neuroimage 43 (2): 192–203.

Hollister LE (1968). Chemical psychoses: LSD and related drugs, Charles C Thomas, Springfield.

Hubbard EM (2007). Neurophysiology of synesthesia. Curr Psychiatry Rep 9 (3): 193–199.

Hubbard EM, Ramachandran VS (2005). Neurocognitive mechanisms of synesthesia. Neuron 48 (3): 509–520.

Hubbard EM, Brand D, Ramachandran VS (2011). The cross activation theory at 10. J Neuropsychol 5 (2): 152–177.

Ione A, Tyler C (2004). Neuroscience, history and the arts. Synesthesia: is F-sharp colored violet? J His Neurosci 13 (1): 58–65.

Iversen L (2003). Cannabis and the brain. Brain 126 (Pt 6): 1252–1270.

Jackson TE, Sandramouli S (2012). Auditory synesthesia coexisting with auditory-visual synesthesia. J Neuroophthalmol 32 (3): 221–223.

Jacobs L, Karpik A, Bozian D et al. (1981). Auditory-visual synesthesia: sound-induced photisms. Arch Neurol 38 (4): 211–216.

Jacome DE, Gumnit RJ (1979). Audioalgesic and audiovisuoalgesic synesthesias: epileptic manifestation. Neurology 29 (7): 1050–1053.

Jancke L, Langer N (2011). A strong parietal hub in the small-world network of coloured-hearing synesthetes during resting state EEG. J Neuropsychol 5 (2): 178–202.

Jancke L, Rogenmoser L, Meyer M et al. (2012). Pre-attentive modulation of brain responses to tones in coloured-hearing synesthetes. BMC Neurosci 13: 151.

Karowski TF, Odbert HS, Osgood CE (1942). Studies in synesthetic thinking. II. The role of form in visual responses to music. J Gen Psychol 26 (2): 199–222.

Kim IK, Dryja TP, Lessell S et al. (2006). Melanocytoma of the optic nerve associated with sound-induced phosphenes. Arch Ophthalmol 124 (2): 273–277.

Lessell S, Cohen MM (1979). Phosphenes induced by sound. Neurology 29 (11): 1524–1526.

Leuner H (1962). Die experimentelle Psychose, Springer, Berlin.

Marks LE (1975). On colored-hearing synesthesia: cross-modal translations of sensory dimensions. Psychol Bull 82 (3): 303–331.

Martuzzi R, Murray MM, Michel CM et al. (2007). Multisensory interactions within human primary cortices revealed by BOLD dynamics. Cereb Cortex 17 (7): 1672–1679.

Maurer D (1993). Neonatal synesthesia: implications for the processing of speech and faces. In: B de Boysson-Bardies, S de Schonen, P Jusczyk et al. (Eds.), Developmental neurocognition: speech and face processing in the first year of life, Kluwer, Dordrecht, pp. 109–124.

Mayer-Gross W (1931). Uber Synasthesien im Meskalinrausch. In: G Anschutz (Ed.), Ferbe-ton-forschungen, vol. III. Psychologischasthetische Forchungsgesellschaft, Hamburg, pp. 266–277.

McKane JP, Hughes AM (1988). Synaesthesia and major affective disorder. Acta Psychiatr Scand 77 (4): 493–494.

Merabet LB, Maquire D, Warde A et al. (2004). Visual hallucinations during prolonged blindfolding in sighted subjects. J Neuroophthalmol 24 (2): 109–113.

Mills CB, Boteler EH, Larcombe GK (2003). "Seeing things in my head": a synesthete's images for music and notes. Perception 32 (11): 1359–1376.

Neufeld J, Sinke C, Zedler M et al. (2012). Disinhibited feedback as a cause of synesthesia: evidence from a functional connectivity study on auditory-visual synesthetes. Neuropsychologia 50 (7): 1471–1477.

Neville HJ (1995). Developmental specificity in neurocognitive development on humans. In: M Gazzaniga (Ed.), The Cognitive Neurosciences, MIT Press, Cambridge, pp. 219–231.

Novich S, Cheng S, Eagleman DM (2011). Is synaesthesia one condition or many? A large scale analysis reveals subgroups. J Neuropsychol 5 (2): 353–371.

Nunn JA, Gregory LJ, Brammer M et al. (2002). Functional magnetic resonance imaging of synesthesia: activation of V4/V8 by spoken words. Nature Neurosci 5 (4): 371–375.

Page NG, Bolger JP, Sanders MD (1982). Auditory evoked phosphenes in optic nerve disease. J Neurol Neurosurg Psychiatry 45 (1): 7–12.

Paulesu E, Harrison J, Baron-Cohen S et al. (1995). The physiology of coloured hearing. A PET activation study of colour–word synaesthesia. Brain 118: 661–676.

Pearce JMS (2007). Synaesthesia. Eur Neurol 57 (2): 120–124.

Pierce AH (1907). Gustatory audition: a hitherto undescribed variety of synaesthesia. Am J Psychol 18 (3): 341–352.

Podoll K, Robinson D (2002). Auditory-visual synesthesia in a patient with basilar migraine. J Neurol 249 (4): 476–477.

Ramachandran VS, Hubbard EM (2001). Psychophysical investigation into the neural basis of synesthesia. Proc Biol Sci 268 (1470): 979–998.

Rao A, Nobre AC, Alexander I et al. (2007). Auditory evoked visual awareness following sudden ocular blindness: an EEG and TMS investigation. Exp Brain Res 176 (2): 288–298.

Riba J, Valle M, Urbano G et al. (2003). Human pharmacology of ayahuasca: subjective and cardiovascular effects, monoamine metabolite excretion, and pharmacokinetics. J Pharmacol Exp Ther 306 (1): 73–83.

Rich AN, Mattingley JB (2002). Anomalous perception in synaesthesia: a cognitive neuroscience perspective. Nat Rev Neurosci 3 (1): 43–52.

Rich AN, Bradshaw JL, Mattingley JB (2005). A systematic, large-scale study of synaesthesia: implications for the role of early experience in lexical-colour associations. Cognition 98 (1): 53–84.

Richer F, Beaufils GA, Poirier S (2011). Bidirectional lexical-gustatory synesthesia. Conscious Cogn 20 (4): 1738–1743.

Rizzo M, Eslinger PJ (1989). Colored hearing synesthesia: an investigation of neural factors. Neurology 39 (6): 781–784.

Ro T, Farne A, Johnson RM et al. (2007). Feeling sounds after a thalamic lesion. Ann Neurol 62 (5): 433–441.

Robertson LC (2003). Binding, spatial attention and perceptual awareness. Nat Rev Neurosci 4 (2): 93–102.

Rockland KS, Ojima H (2003). Multisensory convergence in calcarine visual cortex. Int J Psychophysiol 50 (1–2): 19–26.

Sacks OW (1992). Migraine: Revised and expanded, University of California Press, Berkeley, pp. 86–87.

Saenz M, Koch C (2008). The sound of change: visually-induced auditory synesthesia. Curr Biol 18 (15): R650–R651.

Sagiv N, Ward J (2006). Crossmodal interactions: lessons from synesthesia. Prog Brain Res 155: 259–271.

Shams L, Kamitani Y, Thompson S et al. (2001). Sound alters visual evoked potentials in humans. Neuroreport 12 (17): 3849–3852.

Shanon B (2002). The antipodes of the mind: Charting the phenomenology of Ayahuasca Experience, Oxford University Press, Oxford.

Simner J, Mulvenna C, Sagive N et al. (2006). Synesthesia: the prevalence of atypical cross –modal experience. Perception 35 (8): 1024–1033.

Simpson L, McKellar P (1955). Types of synaesthesia. J Ment Sci 101: 141–147.

Sinke C, Halpern JH, Zedler M et al. (2012). Genuine and drug induced synesthesia: a comparison. Conscious Cogn 21 (3): 1419–1434.

Smilek D, Dixon MJ, Cudahy C et al. (2001). Synaesthetic photism influence visual perception. J Cogn Neursci 13 (7): 930–936.

Vike J, Jabbari B, Maitland CG (1984). Auditory-visual synesthesia. Report of a case with intact visual pathways. Arch Neurol 41 (6): 680–681.

Ward J (2013). Synesthesia. Annu Rev Psychol 64: 49–75.

Ward J, Simner J (2005). Is synaesthesia an X-linked dominant trait with lethality in males? Perception 34 (5): 611–623.

Ward J, Simner J, Auyeung V (2005). A comparison of lexical-gustatory and grapheme-colour synesthesia. Cogn Neuropsychol 22 (1): 28–41.

Ward J, Huckstep B, Tsakanikos E (2006). Sound-color synesthesia: to what extent does it use cross-modal mechanisms common to us all? Cortex 42 (2): 264–280.

Werring DJ, Bullmore ET, Toosy AT et al. (2000). Recovery from optic neuritis is associated with a change in the distribution of cerebral response to visual stimulation: a functional magnetic resonance imaging study. J Neurol Neurosurg Psychiatry 68 (4): 441–449.

Web references

http://www.daysyn.com/Types-of-Syn.html (last accessed 06/30/2013).

http://www.daysyn.com/History.html (last accessed 06/30/2013).

http://www.daysyn.com/Famous-synesthetes.html (last accessed 06/30/2013).

Handbook of Clinical Neurology, Vol. 129 (3rd series)
The Human Auditory System
G.G. Celesia and G. Hickok, Editors

Chapter 23

Tinnitus

ROBERT A. LEVINE[1]* AND YAHAV ORON[2]

[1]*Department of Ear, Nose and Throat and Head and Neck Surgery, Tel Aviv Sourasky Medical Center, Tel Aviv, Israel*

[2]*Department of Otolaryngology, Head and Neck Surgery, E. Wolfson Medical Centre, Holon, Israel*

INTRODUCTION

Tinnitus, the perception of sound in the absence of an external sound, usually results from a disorder of: (1) the auditory system (peripheral/central); (2) the somatosensory system (head and neck); or (3) a combination of the two. The purpose of this chapter is twofold: (1) to describe an approach for determining its cause using the presenting complaint; and (2) to present the major features of its pathophysiology.

DETERMINING TINNITUS ETIOLOGY

The history must pay close attention to not only hearing issues, but also cervical, dental, and head issues, including their relationship to the tinnitus onset, such as new neck stress from altering an exercise program. The most important features of the percept that must be ascertained are: (1) its quality, particularly whether or not it is ever pulsatile or has a clicking component, even if very minor; (2) its location, whether it is heard on one side or not; (3) its variability, whether it is constant, fluctuating, or intermittent; (4) whether it is predominantly low- or high-pitched; and (5) whether the patient can do something to modulate it. The physical exam should include inspection of the teeth for evidence of wear, listening around the ear and neck for sounds similar to their tinnitus, palpation of the craniocervical musculature for trigger points (discrete regions with increased muscle tension and tenderness) with special attention to asymmetries, and probing whether the tinnitus percept can be modulated with "somatic testing": maximal isometric contraction of the head and neck muscles or strong pressure upon these same muscles and the auricle. All subjects should have a recent audiogram.

The tinnitus characteristics will help in determining its etiology (Table 23.1), as described below.

Specific quality

Some types of tinnitus have such characteristic features that the description alone is the major determinant of the direction of the diagnostic approach. These types of tinnitus can be strictly in one ear (lateralized) or non-lateralized. Patients use a variety of expressions to describe non-lateralized tinnitus, such as: "both ears about equally," "both ears but worse in one ear," "in the head but not strongly toward either side," or "in the head toward one side." Tinnitus with a characteristic quality includes sudden, brief, unilateral, tapering tinnitus (SBUTTs), hallucinations (musical or auditory), exploding-head syndrome (auditory sleep starts), staccato irregular unilateral tinnitus ("typewriter tinnitus"), and "somatosounds" – physical sounds generated by the body and heard by one or both ears.

SPECIFIC AND ALWAYS LATERALIZED

Sudden, brief, unilateral, tapering tinnitus (SBUTT)

About 75% of adults have experienced the sudden onset of a tone or noise in one ear that begins abruptly and fades away within seconds with no definite trigger. It is sometimes associated with a fullness, pressure, blocking, or hearing loss of the same ear. Once it begins, it remains at a constant loudness for several seconds then wanes until it is no longer perceived. All the while, the quality of the tinnitus otherwise remains unchanged. Typically it lasts less than a minute with no permanent hearing change. Their frequency of occurrence ranges

*Correspondence to: Robert Aaron Levine, M.D., Department of Ear, Nose and Throat and Head and Neck Surgery, Tel Aviv Sourasky Medical Center, 6 Weizmann Street, Tel Aviv 64239, Israel. Tel: 617-383-4955, Fax: 617-963-7130, E-mail: ObbyLevine@Yahoo.com

Table 23.1

Tinnitus characteristics

Tinnitus with specific quality

Lateralized
- Sudden, brief, unilateral, tapering tinnitus (SBUTT)
- Coarse intermittent sounds coincident with jaw or head movements
- Fluttering

Non-lateralized
- Exploding-head syndrome

Sometimes, but not always, unilateral
- Autophony (echoing of the voice), or blowing tinnitus
- Hallucinations (non-verbal, stereotyped, repetitive)
- Clicking
  - Non-lateralized
  - Lateralized
    - Unilateral staccato irregular intermittent (typewriter tinnitus)
- Pulsatile (cardiac synchronous)
  - Non-lateralized
  - Lateralized
    - Auditory nerve vascular compression
  - Lateralized or Non-lateralized: Somatosensory pulsatile tinnitus syndrome

Tinnitus with non-specific quality

Non-lateralized
- Chronic progressive symmetric hearing loss (chronic acoustic trauma, presbycusis, hereditary hearing loss)
- Autoimmune inner-ear disease
- Central nervous system disorder – rostral to trapezoid body
- Medication-related (including withdrawal syndromes)

Lateralized
- Never with vestibular symptoms
  - Conductive hearing loss
  - Otoacoustic emissions
  - Auditory nerve compression (usually vascular)
- May be with vestibular symptoms
  - Ménière's syndrome
  - Perilymphatic fistula
  - Semicircular canal dehiscence (SSCD)
  - Herpes zoster oticus (Ramsay-Hunt syndrome)
  - Cerebellopontine angle tumors
  - Central nervous system disorder – caudal to trapezoid body
  - Sudden idiopathic hearing loss

May be lateralized or non-lateralized
- Acute acoustic trauma
- Somatic (head or upper cervical)
- Head trauma
- Postinfectious
- Idiopathic

from less than one a year to more than twice a week. They never occur simultaneously in both ears; the right is more common. In the majority they are always in the same ear, but for some the ear can vary. SBUTTs are benign. They are not associated with the development of hearing loss or chronic tinnitus (Oron et al., 2011).

Coarse intermittent sounds coincident with jaw or head movements

This is typical of a foreign body such as cerumen, hair, or a liquid (usually water) resting against the tympanic membrane.

Fluttering

Stapedius muscle contractions tend to be described as a fluttering. If the fluttering is associated with facial movements, then stapedial contraction is highly likely to be causing the fluttering sound. This is most commonly seen after recovery from Bell's palsy, unilateral facial paralysis. When the affected side of the face contracts, the ipsilateral stapedius muscle also contracts (synkinesis) due to aberrant facial nerve regeneration. Abnormalities in the pattern of the stapedial reflex or acoustic impedance measurements corresponding to the characteristics of the patient's tinnitus can occur (Marchiando et al., 1983). The normal stapedius reflex occurs only for high-intensity sounds (>85 dB hearing loss (HL)). Fluttering that is evoked by high-intensity sounds is usually due to stapedius reflex dysfunction and can be treated by releasing the stapedius tendon.

Specific and always non-lateralized

Exploding-head syndrome

Typically subjects report that as they are dropping off to sleep they hear a brief loud sound (e.g., "bang," "explosion," "shotgun," "cymbals") throughout their head. By the time the subject is wide awake it stops. Most commonly it occurs with initiation of sleep, but it can also happen during the night or even with napping. One sleep study found it occurring during the transition from non-rapid-eye movement stage 1 sleep to wake (Palikh and Vaughn, 2010). It can occur in any age group, including childhood. Often the condition remits for prolonged intervals, even several months (Pearce, 1989). Reassurance is usually sufficient. If not, benefit from clonazepam, clomipramine, nifedipine, and topiramate has been suggested from isolated case reports.

Specific and sometimes, but not always, unilateral

Autophony (echoing of the voice), or blowing tinnitus

The characteristics of this type of tinnitus are so unique that the history alone virtually makes the diagnosis, namely a patulous (freely open) eustachian tube. Patients describe blowing sound with respiration and an echoing quality to their own voice. The condition can be unilateral or bilateral. Confirmatory features include (1) disappearance of their complaints when their head is dependent and (2) abnormally large changes in the tympanic membrane acoustic impedance with respirations (Poe, 2007).

Hallucinations (non-verbal, stereotyped repetitive)

Unlike the hallucinations associated with psychoses, these patients have no associated thought disorder, and the hallucinations do not have personally relevant content. Rather the hallucinations are either "musical," in which patients report hearing one or a series of familiar tunes incessantly, or "auditory," in which a variety of different sounds are described (Baurier and Tuca, 1996).

Typically the strictly musical hallucinations occur in elderly patients (more commonly in women) with a longstanding progressive moderate to severe bilateral hearing loss (see Chapter 24). The tunes can be vocal and/or instrumental. While they are usually bilateral, they can be unilateral even with a bilateral hearing loss. Occasionally they can be precipitated by a new medication, and the hallucinations resolve when the medication is stopped. If the presentation is typical, no brain imaging is necessary.

Auditory hallucinations differ from musical hallucinations in several respects. They are usually abrupt in onset and associated with focal neurologic findings due to a brainstem stroke or space-occupying lesion. There usually is no major pre-existing chronic hearing loss and the hallucinations are usually not only musical, but may have a variety of other sounds such as bells or a waterfall. They often are transient (Baurier and Tuca, 1996). Brain imaging is required.

Clicking

Non-lateralized. Very few conditions cause synchronous clicking, i.e., both ears at the same time. Palatal myoclonus is the only consideration. Often the clicking can be heard by the examiner listening at the patient's ear or with use of stethoscope or recording with a microphone in the external auditory canal. Inspection of the palate will usually reveal the problem. Occasionally nasal endoscopy will be necessary. Medications that have been reported to improve palatal myoclonus include clonazepam, lamotrigine, piracetam, and valproate. Botox injections provide temporary relief (4–12 months) (Krause et al., 2006).

Lateralized. Unilateral clicking tinnitus can have two etiologies. It may be a unilateral form of palatal myoclonus due to contractions of tensor tympani or the eustachian tube muscles, which can be observed or heard by the examiner. Sometimes acoustic impedance measurements have abnormalities that correspond to this type of tinnitus or visualization of the tympanic membrane can reveal movements coincident with the clicking.

Unilateral staccato irregular intermittent (typewriter tinnitus). Probably the most common type of unilateral clicking is "typewriter" tinnitus, characterized by unilateral staccato irregular intermittent clicking. Besides "typewriter," other terms that patients have used are "popcorn," "Morse code," "nail clipper," "tapping," "snapping," "crackling," and "machine gun." In some the clicking can be triggered by a specific class of external sounds (e.g., running water). Most have no associated unilateral hearing loss. The irregular clicking can occur with no other symptoms, including no other tinnitus and no hearing loss. It always responds to carbamazepine usually at low dose; it can also respond to other trigeminal neuralgia medications but less predictably (Levine, 2006; Brantberg, 2010).

Like the other unilateral staccato irregular intermittent cranial nerve syndromes (trigeminal and glossopharyngeal neuralgia, hemifacial spasm, and vestibular paroxysmia), typewriter tinnitus typically occurs in people above the age of 50, probably because it is most commonly due to vascular compression of the VIIIth nerve (Fig. 23.1); however, any condition that distorts the VIIIth nerve can cause typewriter tinnitus, probably because the VIIIth nerve becomes hyperexcitable (Nam et al., 2010).

Variants of pure typewriter tinnitus occur. There may be another tinnitus: ipsilateral constant hissing usually

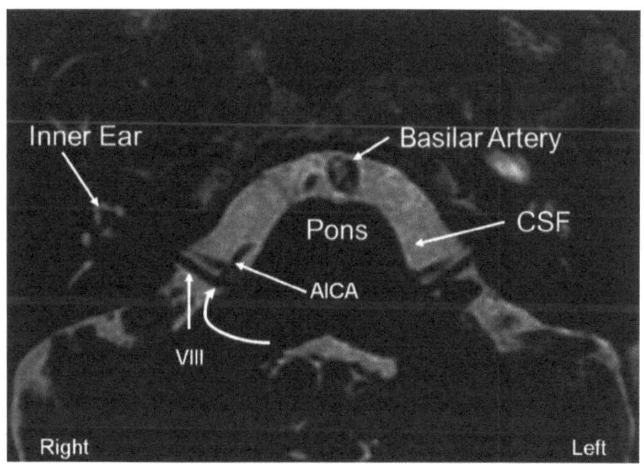

Fig. 23.1. Vascular compression of right VIIIth nerve. Magnetic resonance imaging scan from a 60-year-old whose right cardiac synchronous tinnitus alternated between loud pulsatile "whooshing" and quieter pulsatile "clicking." The point of contact between right anterior inferior cerebellar artery (AICA) and right VIIIth nerve (curved arrow) results in deviation of the VIIIth nerve from the normal straight path, as can be seen for the contralateral VIIIth nerve. CSF, cerebrospinal fluid.

associated with a sensorineural hearing loss; it can even be pulsatile. An example is shown in Figure 23.1; this patient had definite vascular compression ipsilateral to her pulsatile tinnitus that alternated between typewriter ("clicking") and non-typewriter ("whooshing"). In general, the typewriter tinnitus can begin at the same time, precede, or follow the constant tinnitus by months or years. Only the typewriter tinnitus component responds to carbamazepine (Levine, 2006). Another variant is bursts of typewriter tinnitus occurring with vertigo (vestibular paroxysmia) (Russell and Baloh, 2009).

Pulsatile (cardiac synchronous)

The first step is to determine whether the pulsations are cardiac synchronous. This can be evaluated by comparing the examiner's silent count of the cardiac pulse with the patient's silent count of the tinnitus pulsations. If the two counts are discordant then the tinnitus is not cardiac-related. Once it has been established that the pulsatile tinnitus is cardiac synchronous, then the next property that needs to be established is where the percept is perceived.

Non-lateralized. If it is "heard" throughout the head, then very few conditions can cause it. Two different unilateral sources, one from each side, can be easily excluded because their onset will have been at different times. A high cardiac output state such as anemia or hyperthyroidism can be checked with a complete blood count and thyroid profile. A central somatosound is another possibility, such as from a carotid-cavernous fistula or aortic stenosis. Once these uncommon possibilities are excluded, then somatosensory pulsatile tinnitus syndrome becomes most likely.

Lateralized or Non-lateralized: Somatosensory pulsatile tinnitus syndrome. Its percept is cardiac synchronous, usually high-pitched, and can be either lateralized or non-lateralized. Its defining feature is that the pulsations and/or the tinnitus can be momentarily abolished by somatic testing: a strong muscle contraction of the head or neck or a strong pressure applied to these same muscles (Levine et al., 2008; Levine, 2013). Usually the tinnitus is constant but in some it is intermittent. Continuous auricular stimulation with electric pulses can quiet the tinnitus of many who have this syndrome (Cardarelli et al., 2010).

Lateralized. The patient's history can give clues to the source of the pulsatile tinnitus (Sismanis, 2011). An association with headaches, blurring of vision, and menstrual irregularities in an obese young woman is suspicious for benign intracranial hypertension (pseudotumor cerebri). Abrupt onset with unilateral neck or head pains suggests a carotid dissection. A change in tinnitus intensity with head turning suggests a venous source for the tinnitus, from a source ipsilateral to the direction that decreases the tinnitus. This can be confirmed with jugular compression. If the patient can obliterate the tinnitus with mild localized pressure in the periauricular region then an emissary vein is probably accounting for the tinnitus.

The physical exam can be revealing. Abolishing the tinnitus with ipsilateral jugular compression points to a venous source. A crescentic purple coloration to the tympanic membrane is diagnostic of a glomus jugulare tumor. A red mass behind the tympanic membrane is evidence for an aberrant carotid artery, dehiscent jugular bulb, or a vascular tumor. A unilateral conductive hearing loss in association with ipsilateral pulsatile tinnitus and an otherwise normal exam suggests otosclerosis, as does Schwartze's sign (a red hue behind the tympanic membrane on otoscopy). Detection of an ipsilateral bruit suggests the tinnitus is from the source of the bruit. A bruit in the region of the carotid artery bifurcation suggests fibromuscular dysplasia, carotid stenosis, or carotid dissection; ipsilateral Horner's syndrome suggests a carotid dissection.

If the bruit is more widely distributed, such as throughout the periauricular region or even more widespread, a dural arteriovenous fistula becomes likely. If heard over the globe, a carotid-cavernous sinus fistula is suspected, particularly if there is proptosis. If venous pulsations are seen within at least one of the optic fundi, then cerebrospinal fluid pressure is normal and raised intracranial pressure can be ruled out. If "somatic testing" eliminates the pulsations or the tinnitus then the somatosensory pulsatile tinnitus syndrome becomes the working diagnosis, particularly if all diagnostic testing including imaging is negative (see above under non-lateralized pulsatile tinnitus). If somatic testing does not stop the pulsations, then VIIIth-nerve vascular compression is likely (see below and Fig. 23.1).

The diagnostic studies following the initial visit will be guided by the findings of the clinical evaluation and laboratory studies (Fig. 23.2). Because high-cardiac-output states such as anemia or hyperthyroidism can cause pulsatile tinnitus (usually bilateral), all patients should have a thyroid profile and a hematocrit. If a carotid lesion is suspected then either a duplex ultrasound study of the carotid, computed tomography (CT) angiography, or magnetic resonance angiography (MRA) should be performed. If a retrotympanic mass is suspected, then a high-resolution contrast-enhanced CT scan of the temporal bones should be obtained. Otherwise a contrast-enhanced magnetic resonance imaging (MRI) scan of the temporal bone and cranium

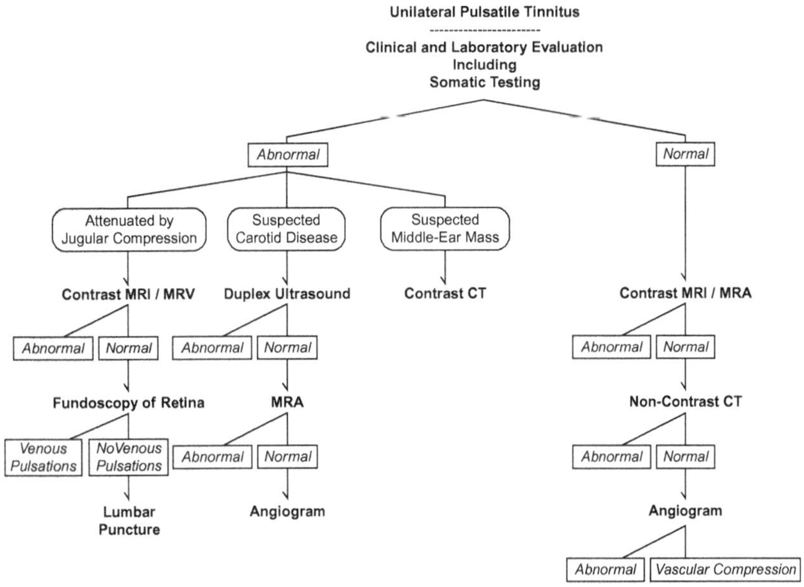

Fig. 23.2. Pulsatile tinnitus: diagnostic algorithm. Somatic testing and laboratory studies (hematocrit and thyroid profile) should be done before imaging. If somatic testing does not suppress pulsations and diagnostic testing reveals no abnormality, then vascular compression of the auditory nerve is very likely. MRI, magnetic resonance imaging; MRA/MRV, magnetic resonance arteriography/venography of cervical and intracranial vasculature; CT, thin-section computed tomography of temporal bone.

should be obtained. The MRI scan may not detect anomalous arterial patterns such as a persistent stapedial artery, so a non-contrast high-resolution CT scan of the temporal bone is performed if the MRI scan is normal. If still no etiology is apparent and neither papilledema nor retinal venous pulsations were observed (by the examiner and/or a neuro-ophthalmologic consultant) then cerebrospinal fluid pressure should be measured via a lumbar puncture. If all the above non-invasive imaging studies have been unremarkable and raised intracranial pressure has been ruled out, then cerebral angiography, which can have rare but significant morbidity, should be considered, because a dural arteriovenous malformation can sometimes go undetected by any other diagnostic study, even though there may or may not be a thrill or bruit on physical examination (Madani and Connor, 2009).

Eighth-nerve vascular compression. That VIIIth-nerve vascular compression can cause unilateral tinnitus which in some cases is pulsatile has been well established by Ryu and colleagues. They made their observations in subjects whose primary complaint was hemifacial spasm, not tinnitus. From surgically decompressing the facial nerve in the cerebellopontine angle, it was observed that 100% of those with preoperative ipsilateral tinnitus (half of whom were pulsatile) had VIIIth-nerve compression, and in those without tinnitus, only 6% had VIIIth-nerve compression. Furthermore, with decompression 70% of those with tinnitus had their tinnitus abolished and another 10% were markedly improved. Eighty percent of those with pulsatile tinnitus had their tinnitus resolved (Ryu et al., 1998). Hence, vascular compression of the auditory nerve can cause unilateral tinnitus, including pulsatile tinnitus.

With rare exceptions (Fig. 23.1), determining with certainty that an individual's unilateral tinnitus, whether pulsatile or non-pulsatile, is from VIIIth-nerve vascular compression has not yet been possible. A comprehensive study of vascular decompression surgery using very conservative auditory brainstem response (ABR) and MRI criteria ("VIII neurovascular conflict defined by an orthogonal contact between the nerve and the blood vessel which provoked a nerve's pathway distortion," such as in Fig. 23.1) found that, after more than 5 years of postoperative follow-up, 3 of 5 patients with vertebral artery vascular compression had no tinnitus. In contrast, none with anterior or posterior inferior cerebellar artery compression were abolished; but 40% were improved (Guevara et al., 2008).

From analogy with trigeminal neuralgia and with some supporting imaging data we have suggested previously that typewriter tinnitus is also due to VIIIth-nerve vascular compression. The symptoms of the patient, whose MRI (Fig. 23.1) revealed clear-cut right auditory nerve vascular compression, provide more supporting evidence for: (1) pulsatile tinnitus and (2) clicking tinnitus being related to VIIIth-nerve vascular compression, since (1) her tinnitus was always pulsatile and (2) when quieter it was clicking pulsatile. When louder it was whooshing pulsatile. She never heard clicking and whooshing together.

Non-specific quality

Included in this category are a variety of descriptors of tinnitus, such as buzzing, tonal, hissing, humming, ringing, roaring, rushing, whistling and whooshing, cicadas, and crickets. None of these descriptors points to a specific diagnosis. Roaring while non-specific, when unilateral, is often associated with Ménière's syndrome. Since the quality of the tinnitus is non-specific, aids in making a diagnosis must come from sources other than the tinnitus quality. Associated symptoms, ameliorating and exacerbating factors, as well as circumstances surrounding the tinnitus onset, are some of the pointers to the tinnitus etiology.

In attempting to arrive at a diagnosis for tinnitus in which the characteristics of tinnitus are non-specific, there are several important considerations that must be kept in mind (Table 23.2). The first is that tinnitus is very common in the general population. Many people have tinnitus but have never complained of it or sought medical attention for it. The landmark 1953 study found such tinnitus in 94% of people (Heller and Bergman, 1953). More recent studies of normal-hearing subjects have also found large percentages (52–92%) reporting such tinnitus (Levine et al., 2003; Tucker et al., 2005; Del Bo et al., 2008; Knobel and Sanchez, 2008).

In our study in which normal subjects were placed in a low-noise room, 20% had previously been aware that they had tinnitus in the quiet. Another 35% for the first time noticed that they had tinnitus in the quiet (Levine et al., 2003). Seventy-five percent of our subjects reported having experienced SBUTTs, another type of "normal" tinnitus; namely, unilateral tonal/noise tinnitus typically lasting less than a minute (Oron et al., 2011). Another type of transient tinnitus, which is common, follows exposure to loud sound. About 50% of our study subjects recalled such tinnitus, lasting from a few minutes to several hours or even days (Abel and Levine, 2004). In fact, patients have occasionally presented to our clinic complaining of one of these types of "normal" tinnitus.

Table 23.2

Non-specific tinnitus: why problematic to establish a definitive etiology?

1. Tinnitus is common in the general population
2. For any disease, not all subjects will develop tinnitus
3. The coexistence of tinnitus and a disease does not imply that the two are related
4. Tinnitus can be multifactorial

A second important consideration is that with any pathologic process associated with tinnitus, not all subjects will develop tinnitus. In surveys of profoundly deaf subjects, about 80% will have tinnitus, but 20% will not (Levine, 1999); similarly for idiopathic sudden sensorineural hearing loss (Huy and Sauvaget, 2005). Hence, the presence of tinnitus and a pathologic process by itself does not imply that the two are related. Because there is no obligatory association between the pathologic process and tinnitus, it is possible that, even though a tinnitus patient has a chronic condition associated with tinnitus, the tinnitus may not be related to the chronic condition; rather the tinnitus and the pathologic process could coexist, but be unrelated (Levine et al., 2007).

Because tinnitus is common in the general population (35% of whom are unaware of it), a third consideration is that the pathologic process only draws the patient's attention to the pre-existing tinnitus. Furthermore, if the disease in and of itself did not cause the tinnitus but hearing loss is part of the disease, the hearing loss could be "unmasking" the pre-existing tinnitus. Just as bringing a normal subject into a low-noise environment can make subjects aware of tinnitus they had not appreciated previously, so could a hearing loss unmask tinnitus (Levine et al., 2003). Therefore, the question always remains in any patient with non-specific tinnitus: was the tinnitus pre-existing and unmasked because of loss of hearing, or because the patient's attention was drawn to his or her hearing?

A striking example was a young adult with no hearing complaints (including tinnitus) who awoke with new onset of distressing tinnitus the day after learning that she had a bilateral major high-frequency sensorineural hearing loss (60 dB loss at 2 kHz and above) (Levine, 2013).

Casual attention to tinnitus of which people were previously unaware does not appear to be associated with the development of disturbing tinnitus, since following the Heller and Bergman type of studies no subjects developed troublesome tinnitus. Clinical anecdotes suggest that, when attention to tinnitus, of which they were previously unaware, is emotionally charged, then the tinnitus can become troublesome (Levine, 2013).

Since on the one hand non-specific tinnitus can be physiologic, and on the other hand non-specific tinnitus is not obligatorily associated with any pathologic process, when a condition is very slowly progressive (e.g., presbycusis), establishing the etiology of non-specific tinnitus is more problematic than for most medical symptoms.

In general, there is a level of confidence associated with any diagnosis that might account for a symptom. For tinnitus some relationships increase this confidence. The first is a temporal association between the tinnitus and the diagnostic consideration. For example, in Ménière's syndrome, if the tinnitus fluctuates with the hearing loss and vertigo, then this strengthens the confidence of the association between the tinnitus and Ménière's syndrome. Another way the confidence of the diagnosis is strengthened is if the pitch of the tinnitus corresponds to the audiometric hearing loss pattern. Considering again Ménière's syndrome, where early in the illness low-frequency hearing loss often predominates, if the tinnitus is described as roaring and/or the pitch match is a low-frequency one, then the confidence of the relationship between the Ménière's syndrome and the tinnitus is strengthened. Likewise, changes in the tinnitus percept that are closely coupled to changes in physical findings strengthen the association between the two.

Another important consideration in establishing the tinnitus etiology is that its cause may be multifactorial. Tinnitus can be considered a threshold phenomenon (see section "Tinnitus is a threshold phenomenon", below), such that, while any one factor such as chronic progressive hearing loss may not be sufficient to elicit a tinnitus complaint, two or more factors may synergistically lead to the tinnitus becoming symptomatic (Levine et al., 2007).

Closely related to the threshold idea is the concept of "triggering factors" that can lead to symptomatic chronic tinnitus (Fowler, 1943; Coles, 1996). Such factors include psychosocial stress (see young adult case described above in this section), viral infection or the postviral state, medications or withdrawal from medications, and head and neck somatic factors (e.g., whiplash, temporomandibular joint syndrome, ear syringing, trauma not involving the head). Clinical experience suggests that the clinical problem of tinnitus can be precipitated by one (or more) of these triggering factors. While a triggering factor may appear to be responsible for initiating the tinnitus, sometimes the tinnitus will persist despite resolution of the triggering factor.

With these considerations in mind, we next consider diagnostic entities that appear to be associated with non-specific tinnitus.

Always non-lateralized

Chronic progressive symmetric hearing loss (presbycusis, chronic acoustic trauma, hereditary hearing loss)

These three conditions can be considered together, since they affect only hearing, generally have a symmetric hearing loss, and are slowly progressive, albeit at different rates. They are the most common type of tinnitus encountered in a tinnitus clinic. When their tinnitus begins, it is non-lateralized. Presently it cannot be predicted when, during the course of their progressive hearing loss, their non-lateralized tinnitus will begin, if ever, since 20% of such people will never develop tinnitus despite a profound hearing loss. If: (1) the audiogram is asymmetric; (2) the tinnitus begins first in one ear and only later in the other ear; or (3) the tinnitus is lateralized, then one or more other factors, such as acoustic neuroma, must be sought to account for the asymmetry.

The establishment of a causal relationship between chronic progressive hearing loss and tinnitus is problematic, because there is no perceptible change in the hearing or audiogram associated with the onset of the tinnitus. What has been well established is that the prevalence and reported loudness of tinnitus increase with increasing hearing loss (Chung et al., 1984). However, for any patient with chronic progressive symmetric hearing loss and recent onset of non-lateralized tinnitus, a triggering factor or other cause for the tinnitus must be sought, even though the yield has been low in the authors' experience. The association between tinnitus and chronic progressive hearing loss must be considered tenuous.

Autoimmune inner-ear disease

This condition is like chronic progressive hearing loss, except the progression of the bilateral hearing loss is measured in weeks or months rather than years. It can fluctuate and be asymmetric. Due to its more rapid time course, the association of tinnitus with the disease is more compelling. Positive blood tests for general autoimmune disease or inner-ear antibodies support the diagnosis.

Central nervous system disorder – rostral to trapezoid body

The evidence available indicates that, for central nervous system lesions rostral to the trapezoid body (such as involving the inferior colliculus), tinnitus is usually transient and bilateral (see section The tinnitus pathway rostral to DCN, below) (Hausler and Levine, 2000).

Medication-related (including withdrawal syndromes)

The temporal association of the onset of the tinnitus with exposure to a toxin establishes the diagnosis particularly if the tinnitus resolves when the toxin is withdrawn. An association is less clear when tinnitus begins just after a new medication was begun but does not remit when discontinued. Such instances raise the possibility that the new medication acted as a trigger factor for the tinnitus. Cisplatin, aminoglycoside antibotics, and loop diuretics can cause permanent hearing loss and tinnitus. High-dose aspirin and quinine can cause a reversible hearing impairment and tinnitus. Transient or, rarely, long-lasting tinnitus can be a part of a sedative withdrawal syndrome (Ashton, 1991) which, in the case of benzodiazepines, can be precipitated by fluoroquinolones, because these antibiotics cause a pharmacologic withdrawal. Despite no change in the daily benzodiazepine consumption, they compete directly with benzodiazepines for the benzodiazepine receptor site, displacing benzodiazepines and thereby precipitating a pharmacologic benzodiazepine withdrawal (McConnell, 2008).

Always lateralized

Never with vestibular symptoms

Conductive hearing loss. Any type of unilateral conductive hearing loss, such as cerumen impaction, ossicular discontinuity, or otosclerosis, can be associated with tinnitus of that ear. The tinnitus may be related at least in part to an unmasking of a "normal" underlying tinnitus, as discussed above. Otosclerosis may sometimes be associated with inner-ear involvement, which could be contributing to the tinnitus as well.

Otoacoustic emissions. While spontaneous otoacoustic emissions are common (75% of females and 45% of males with normal or near-normal ears), tinnitus due to spontaneous otoacoustic emissions is uncommon. It is said to be accounting for the tinnitus of 1–2% of the patients of one British tinnitus clinic, but in our experience it has been much less frequent (Penner, 1992). The diagnosis is made by measuring an emission and showing that its suppression abolishes the tinnitus. The emission can be suppressed in one of two ways: presentation of a low-level tone near the emission frequency or the use of aspirin (Penner and Coles, 1992).

Eighth-nerve compression (usually vascular). As discussed earlier, the tinnitus of VIIIth-nerve compression is unilateral, rarely associated with vestibular symptoms, and can take three forms: (1) non-specific (hissing, buzzing, etc.); (2) pulsatile; and (3) staccato, irregular intermittent clicking (typewriter). All those with unilateral non-specific tinnitus should be questioned closely about whether there ever is a clicking or pulsating component to their non-specific tinnitus, even if very rarely. If so, then the possibility of VIIIth-nerve compression becomes more likely and should be investigated using a heavily T2-weighted MRI (constructive interference in steady state (CISS) or fast imaging employing steady-state acquisition (FIESTA)) of the cerebellopontine angle.

May be with vestibular symptoms

Ménière's syndrome. As a syndrome this condition has tinnitus as one of its defining features. The full-blown picture consists of episodic attacks of intense vertigo persisting hours, fluctuating unilateral hearing loss typically involving the lower frequencies (in the early stages), ear fullness, and a roaring low-frequency tinnitus. Early in its course there may be no persistent symptomatology; however, with recurrent episodes any or the entire symptoms can persist and cumulatively progress with each recurrence. While the tinnitus is often described as roaring early in the illness, with more advanced stages of the syndrome, the tinnitus tends to become more variable in its description. There are no definitive tests to establish the diagnosis; however, electrocochleography can be supportive of the diagnosis, if there is a large ratio of summating potential to action potential amplitudes. Documentation of the fluctuation in hearing with serial audiograms is supportive of the diagnosis. The fluorescent treponemal antibody absorption (FTA-Abs) test can sometimes be positive in this syndrome.

Formes frustes of this syndrome may occur. In particular, episodic low-frequency fluctuating hearing loss with a contemporaneous roaring tinnitus and aural fullness may occur without vertigo.

Perilymphatic fistula

Like Ménière's syndrome, perilymphatic fistula can cause hearing loss, vertigo, and tinnitus. However, the tinnitus and hearing loss tend to be high-frequency (hissing, crickets, etc.) with no recovery. The fistula consists of a communication between the perilymph of the inner-ear fluids and the middle ear through a round or oval window defect or sometimes a defect of the bony labyrinth. The defect can be caused by barotrauma (e.g., airplane descent or scuba ascent), head trauma, Valsalva maneuver, erosion of the bony labyrinth due to an inflammatory or neoplastic process, or following middle-ear surgery such as stapedectomy. The diagnosis can be suggested by the "fistula test" – the induction of nystagmus by positive or negative pressure applied to the external auditory canal. If symptoms persist and the findings are suggestive then the middle ear can be explored for a fistula with patching of the round and oval windows. Generally the hearing loss and tinnitus are not improved by patching the oval and round windows whether or not a leak is found at surgery.

Superior semicircular canal dehiscence (SSCD)

SSCD is closely related to perilymphatic fistula. Whereas the perilymphatic fistula is a defect of the otic capsule into the middle ear, SSCD is a defect of the otic capsule into the floor of the middle cranial fossa. While vestibular symptoms from loud sounds or straining are more commonly the presenting SSCD symptom, in one series

35% of patients reported unilateral high-pitched tinnitus as one feature of their presentation (Yuen et al., 2009). Other auditory symptoms include monaural heightened bodily sounds (eye movements, heartbeat, footsteps, or voice). The diagnosis is suspected from an air–bone gap with bone thresholds exceptionally good, intact acoustic reflexes, and ipsilateral low thresholds of their vestibular evoked myogenic potential. CT scan can demonstrate the dehiscence.

Much rarer are dehiscences involving the posterior or lateral semicircular canals. A variety of surgically induced dehiscences can also occur.

Herpes zoster oticus (Ramsay Hunt syndrome)

Intense ear pain followed by ipsilateral tinnitus, hearing loss, vertigo, and facial paralysis will be recognized as due to herpes zoster once vesicles appear on the pinna, external auditory canal, or tympanic membrane.

Cerebellopontine angle tumors

The most common presentation of such tumors is a gradual unilateral sensorineural hearing loss with minimal, if any, vestibular complaints. Dizziness and facial weakness, in general, are either non-existent or very minor accompaniments of acoustic neuromas at any time. The audiometric pattern is variable. They are more likely to have poor speech discrimination, acoustic reflex decay, and pure-tone decay. Other presentations do occur, including unilateral tinnitus only or sudden hearing loss with or without subsequent recovery.

Any unilateral tinnitus with or without unilateral sensorineural hearing loss must be considered suspect for a cerebellopontine angle tumor, even though only about 3% of all patients with acoustic neuromas first present complaining of unilateral tinnitus only (Dornhoffer et al., 1994; Morrison and Sterkers, 1996). Therefore, should every such patient have a contrast-enhanced or high-resolution T2-weighted MRI scan to exclude the diagnosis, since such imaging has a sensitivity "approaching, if not reaching 100%" (Curtin, 1997; Fortnum et al., 2009)? In 25 years of evaluating all patients with unilateral tinnitus for a cerebellopontine angle lesion, we have detected just two lesions of the cerebellopontine angle. Dawes and Basiouny (1999) have reported a similar experience: one acoustic neuroma was detected in 174 patients with unilateral tinnitus Alternatives to obtaining the "gold standard" contrast-enhanced MRI scan immediately are using the short-latency brainstem auditory evoked potentials (especially stacked ABR: Don et al., 2012), a good but imperfect test, to decide whether or not to proceed to the MRI scan, or following the patient with revisits and repeat audiograms at 6-month intervals, looking for development of an ipsilateral hearing loss before going to the MRI scan. The issue ultimately becomes one of cost-effectiveness and availability of MRI scanning, since there is virtually no morbidity associated with the MRI scan.

Central nervous system disorder – caudal to trapezoid body

The hallmark of tinnitus due to a disorder of the central nervous system is other neurologic system involvement. If the central auditory nervous system involvement occurs between the ear and the trapezoid body (where the auditory inputs from the two sides intermix and cross) then there may be an associated ipsilateral hearing loss with unilateral tinnitus. The type of associated neurologic involvement will depend crucially upon the precise location of the lesion. With unilateral tinnitus it can include dizziness, diplopia, limb ataxia, ipsilateral facial weakness, ipsilateral facial paresthesias, but contralateral-limb paresthesias. The tinnitus is often transient. Stroke, intrinsic or extrinsic neoplasms, demyelinating disease, inflammatory diseases, and meningitides can all lead to unilateral tinnitus and hearing loss. The diagnosis will be established by the pattern of neurologic system involvement, the temporal profile of the illness, and the results of ancillary diagnostic studies such as MRI scanning, cerebrospinal fluid examination, or arteriography.

Sudden idiopathic hearing loss

Patients can present with an abrupt unilateral tinnitus as their only complaint and upon evaluation are found to have a corresponding unilateral hearing loss of which they may have been unaware. More typically they complain of both the hearing loss and tinnitus or hearing loss alone. The hearing loss and tinnitus are abrupt in onset and always unilateral. The quality of the tinnitus usually is closely related to the pattern of the pure-tone audiogram. Vestibular symptoms, should they occur, are usually not prominent. Once other rare but identifiable causes have been considered, such as Ménière's syndrome, cerebellopontine angle tumor, cochlear ischemia, syphilis, or herpes zoster, then the diagnosis of sudden idiopathic hearing loss is secure.

May or may not be lateralized

Acute acoustic trauma

There is no difficulty establishing the diagnosis of tinnitus due to acoustic trauma when the history is one of immediate development of hearing loss and tinnitus following an intense sound exposure with partial or complete recovery of hearing (temporary threshold shift) over a few days. On the other hand, if the audiogram is normal or the tinnitus does not immediately follow a unilateral intense sound exposure, then other causes for the tinnitus must be sought.

Somatic (head or upper cervical)

Observations abound supporting the notion that head and neck somatic events can be associated with tinnitus (Biesinger, 1997). The patient is not aware of any hearing loss with its onset, but there may or may not be a preexisting hearing loss. Prior to their first visit, about 20% of patients in our clinic have noticed that they can modulate their tinnitus somatically, such as by clenching the teeth or pushing on various places on the head (Levine, 2004). For 2 years we systematically examined our patients with a battery of isometric head and neck contractions. More than 75% of patients could modulate their tinnitus. A variety of changes can occur. Most frequently the tinnitus gets louder (72%), but less often the tinnitus becomes quieter (25%) (Levine et al., 2003; Abel and Levine, 2004). In 18% the tinnitus became quieter for some maneuvers and louder for others (Levine et al., 2008). Less frequently, patients describe pitch or location changes.

Tinnitus is generally included amongst the features associated with pain in the temporal or preauricular region that goes by various names, such as Costen's syndrome, craniomandibular disorder, and temporomandibular joint syndrome. Well-designed studies have shown a higher incidence of tinnitus in normal-hearing subjects with temporomandibular joint syndrome than in controls (Chole and Parker, 1992). The same is true for whiplash (Tjell et al., 1999; Tranter and Graham, 2009). From multiple other observations and case reports, the concept of tinnitus associated with whiplash and temporomandibular joint syndrome can be generalized to include tinnitus associated with any disorder of the upper cervical region or head, including dental pain.

The tinnitus temporally associated with unilateral somatic disorders is localized to the ipsilateral ear (Levine, 1999; Reisshauer et al., 2006). Hence unilateral tinnitus with no associated auditory or vestibular symptoms such as hearing loss must be suspect for an ipsilateral head or neck somatic disorder. The physical examination should include: (1) inspection of the teeth for wear as evidence of bruxism; (2) palpation of the head and neck musculature for tender muscles with increased tension (trigger points) (Teachey et al., 2012); and (3) assessment of the effects upon the patient's tinnitus from "somatic testing": forceful systematic isometric contraction of muscle groups of the head and neck and deep palpation of these same muscles and the auricle.

At least four factors have been associated with changes in tinnitus attributes: somatic modulation (SM), stress, exposure to loud sound, and use of benzodiazepines. As described above, it is clear that to formal systematic testing most subjects can somatically modulate their tinnitus. It may be that in day-to-day activities everyone can somatically modulate their tinnitus. Patients consistently describe that they are more bothered by their tinnitus when stressed. Whether this is due to the patient focusing attention upon the tinnitus or actual changes in the tinnitus loudness usually cannot be distinguished by the patient. In fact, it is very possible that stress acts through SM to increase tinnitus loudness, since there is a very well-recognized association between stress and oral parafunctional behavior such as tooth grinding, jaw clenching, fingernail biting, and pencil chewing, as well as furrowing the brow, or grimacing. Hence one way by which stress may lead to increased tinnitus loudness is through increasing head and neck muscle contractions which in turn lead to louder tinnitus by the somatic mechanism. Some subjects clearly associate an increase in their tinnitus loudness with exposure to loud sound, and in some the louder tinnitus can persist for hours after the exposure has discontinued. Finally use of benzodiazepines can cause fluctuations in tinnitus since they are commonly used for insomnia and anxiety, both of which are often associated with tinnitus, and benzodiazepines quieten tinnitus to some degree in the majority of people (Johnson et al., 1993; Vernon and Meikle, 2003; Jalali et al., 2009; Han et al., 2012). Thus, if a patient reports that tinnitus is intermittent or has wide fluctuations in loudness or other qualities, and there is no exposure to intense sound, evidence for stress, or use of benzodiazepines, then the SM mechanism must be suspected. A history of variations in tinnitus loudness then raises the suspicion for a somatic factor modulating the percept's loudness (Table 23.3). At an extreme are patients who describe that they have periods when their tinnitus cannot be heard, even in the quiet while focusing on their hearing. Others report wide variations in the loudness of their tinnitus. For still others, their tinnitus is unilateral when it is relatively quiet but becomes non-lateralized when the tinnitus is louder. Such

Table 23.3

Tinnitus properties suggesting a somatic component

1. Large fluctuations in loudness, including intermittent tinnitus, and diurnal tinnitus
2. Variability of location
3. No hearing loss but head or neck disorder

phenomena suggest that there are ongoing somatically mediated factors modulating the tinnitus percept.

Diurnal fluctuations in the tinnitus percept also suggest that SM is operative. Patients who describe their tinnitus as louder upon awakening raise the possibility that somatic factors (such as bruxism, or grinding of the teeth) are active during sleep and are causing an increase in tinnitus loudness. Others describe that their tinnitus has usually vanished by the time they awaken and then returns a few hours into the day; this scenario suggests that during the day they are reactivating their tinnitus through somatic mechanisms, such as the tonic muscle contractions required to support the head in an upright position or oral parafunctional activities related to the stress of daily activities. Finally, others describe that their tinnitus is louder after awakening from a nap in a chair; this may relate to somatic factors such as stretching of the neck muscles when their head passively falls forward while dozing in a sitting position.

In our experience, while a somatic factor on its own can cause tinnitus, much more frequently somatic factors combine with other factors (such as chronic hearing loss) to act as trigger factors or modulators (Levine et al., 2007).

Head trauma

The contemporaneous association between head trauma, hearing loss, and tinnitus makes the diagnosis straightforward. The association with trauma becomes less certain when some of these elements are missing, such as a delay between the trauma and the onset of the tinnitus. The longer the delay, the less is the confidence of any association between the two. Tinnitus but no hearing loss following trauma, likewise, makes the association more tenuous and raises the possibility that trauma is causing tinnitus through a brainstem mechanism or indirectly such as through a somatic mechanism. Abnormalities of otoacoustic emissions, including those involving the brainstem (the contralateral suppression of otoacoustic emissions), have been found in subjects with tinnitus and head injury, including some with normal audiograms, raising the likelihood that a brainstem mechanism is accounting for the tinnitus associated with head trauma (Ceranic et al., 1998; Attias et al., 2005). Hyperbaric oxygen has recently been reported to provide major benefit to some patients with tinnitus following mild traumatic brain injury (Shlamkovitch, 2012; Boussi-Gross et al., 2013).

Postinfectious

Occasionally patients report the onset of tinnitus following an upper respiratory infection. Whether this is cause and effect has never been established. Since upper respiratory infections are common, and idiopathic tinnitus not uncommon, chance association of the two is to be expected.

Idiopathic

Sometimes, despite an exhaustive evaluation of nonspecific tinnitus, no specific diagnosis can be made with any high degree of confidence. Hence, the diagnosis of "idiopathic" is tentatively made.

TREATMENT OF TINNITUS

Once the etiology of the tinnitus is tentatively identified then treating the underlying cause of the tinnitus can lead to its resolution. A preliminary study using a somatic approach to tinnitus treatment by needling head and neck myofascial trigger points resulted in complete cessation of tinnitus for 25% of patients and decided improvement in another 25% (Teachey et al., 2012).

If the pathologic process for the tinnitus cannot be reversed, then suppression of the tinnitus acoustically, pharmacologically, or electrically can be attempted. For people with hearing loss, hearing aids can often act as maskers of their tinnitus. Additionally, when hearing is improved, patients no longer perceive the tinnitus as interfering with their hearing. Consequently they pay less attention to their tinnitus once their ability to hear external sounds is improved.

With programmable digital hearing aids, it has been reported that there was a more than 50% tinnitus improvement in 65% of subjects who received one hearing aid and in 80% of subjects who received bilateral hearing aids (Trotter and Donaldson, 2008). With hearing aids that incorporate linear octave frequency transposition, a recent study reported a large number of subjects who completely suppressed their tinnitus; these same subjects were said to have had no benefit from hearing aids using classic amplification or non-linear frequency compression (Peltier et al., 2012).

Anecdotal tinnitus suppression has been reported for a large number of medications (fluoxetine, sertraline, paroxetine, risperidone, ginkgo, scopolamine patch, famotidine, and gabapentin). Only benzodiazepines have been shown to significantly quieten tinnitus in multiple well-designed studies. All three benzodiazepine studies report that in about 75% of subjects the benzodiazepine quietens tinnitus to some degree (Johnson et al., 1993; Vernon and Meikle, 2003; Han et al., 2012). A fourth study found a "significant improvement in VAS [visual analog scale] score in the benzodiazepine group compared with the placebo group ($P < 0.001$)" (Jalali et al., 2009). In our clonazepam study using a strict protocol, 7% had their tinnitus completely suppressed. It is likely that in clinical practice the percentages who suppress their tinnitus completely will be much higher, because the dose can be individualized and other benzodiazepines can be tried.

Acamprosate was shown to quieten tinnitus in two well-designed studies (Azevedo and Figueiredo, 2005; Sharma et al., 2012) but was said not to be effective in a third as yet unpublished study (Martin, 2011). One of the studies found that at 90 days the tinnitus was not heard in more than 10% of subjects (Sharma et al., 2012).

The medications for trigeminal neuralgia, especially carbamazepine, suppress completely staccato irregular unilateral tinnitus ("typewriter tinnitus").

Auditory nerve electric stimulation, whether by a cochlear implant or from the middle-ear promontory, has been repeatedly shown to suppress tinnitus in many subjects, but no promontory stimulation device is currently commercially available (Matsushima et al., 1997a; Van de Heyning et al., 2008).

If the tinnitus cannot be quietened satisfactorily, then behavioral methods or antidepressants can improve the subject's tinnitus tolerance by shifting the subject's attention away from the tinnitus and/or improving the associated depression (Sullivan et al., 1993). While a review of randomized controlled trials concluded that "only studies examining cognitive behavioral therapy were numerous and similar enough to perform meta-analysis, from which the efficacy of cognitive behavioral therapy (moderate effect size) appears to be reasonably established," in the hands of a proficient practitioner the behavioral methods may all be about equally effective (Hoare et al., 2011). Some of the other behavioral techniques include tinnitus retraining therapy, neuromonics, mindfulness, binaural beats, and self-hypnosis (Attias et al., 1993; Caffier et al., 2006; Davis et al., 2008; David et al., 2010; Hesser et al., 2011; Kreuzer et al., 2012).

HYPERACUSIS

Hyperacusis refers to reduced sound tolerance (see Chapter 21). People with hyperacusis report that they find sound intensities considered comfortable by most people to be unbearably loud (Baguley, 2003). There is a close relationship between hyperacusis and tinnitus; about 40% of patients with tinnitus report some degree of hyperacusis. Furthermore, while less than 10% of people in the general population have hyperacusis, more than 80% of those with hyperacusis report coexisting tinnitus. In a small number of individuals hyperacusis is their prime concern; tinnitus is of secondary importance.

There is evidence supporting the notion that sound avoidance as a reaction to new-onset tinnitus creates a negative-feedback situation leading to hyperacusis. It has been shown in short-term experiments with people wearing ear plugs that this form of sound avoidance leads to decreased sound tolerance (Florentine, 1976; Formby et al., 2003; Blaesing and Kroener-Herwig, 2012). The negative feedback then is as follows: sound avoidance because of tinnitus leads to more sound intolerance, which in turn leads to more sound avoidance. Further support for this concept has come from the success in treating hyperacusis with desensitization programs (Vernon, 1987; Jastreboff and Jastreboff, 2003; Andersson et al., 2005; Norena and Chery-Croze, 2007). These desensitization programs all have in common the gradual introduction of sounds within a supportive environment that had been previously poorly tolerated, thereby breaking the vicious cycle of negative feedback.

THE NEUROLOGY OF TINNITUS

Over the last 45 years, based upon interactions between clinical observations and animal models, a comprehensive understanding of the neurology of tinnitus has emerged. Since tinnitus may have several different etiologies, so too there are different ways that the tinnitus percept may develop in the central nervous system. Presently the most compelling evidence for the most common types of tinnitus is the DCN hypothesis, which will be presented in detail.

Hearing loss and somatic tinnitus: the unilateral dorsal cochlear nucleus hypothesis

HEARING LOSS

The first insights into the neurology of hearing-loss tinnitus came from auditory nerve fiber recordings in experimental sensorineural hearing loss. In these cases auditory nerve fibers lost all spontaneous activity (normally spontaneous activity ranges between 1 and 100 per second) and became unresponsive to acoustic stimulation (Kiang et al., 1970). Over 25 years later it was shown that, with a cochlear hearing loss, the DCN develops elevated multiunit spontaneous activity limited to the region of the DCN tonotopically corresponding to the damaged region of the cochlea (Kaltenbach and McCaslin, 1996). Subsequent studies demonstrated this finding in multiple species, including animals with behavioral evidence of "tinnitus" (Brozoski et al., 2002; Kaltenbach et al., 2005; Dehmel et al., 2012) and found that it is the fusiform cell which has elevated spontaneous activity (Brozoski et al., 2002; Finlayson and Kaltenbach, 2009). More recently it has been shown that (1) elevated spontaneous activity involves only one type of fusiform cell, those that respond to tone bursts with a "buildup" response pattern (Dehmel et al., 2012) and (2) only the subpopulation of DCN neurons that can be excited by the somatosensory system show increased spontaneous rates following hearing loss.

This DCN subpopulation also developed increased sensitivity to trigeminal electrical stimulation and a change in their response pattern to sounds interacting with trigeminal electric stimulation (Koehler and Shore, 2013). This increased sensitivity finding is consistent with functional imaging findings in tinnitus patients. It was observed that the only significant difference between tinnitus and non-tinnitus subjects to somatosensory activation (jaw motion) was increased responsiveness in the brainstem nuclei (cochlear nucleus and inferior colliculus) of tinnitus patients (Lanting et al., 2010).

This change in the response pattern to sounds when interacting with trigeminal electric stimulation follows upon *in vitro* studies of the DCN (Tzounopoulos, 2008). The changes occurred only in the animals with "tinnitus." Despite being exposed to the same tinnitus-inducing stimulus, the animals that did not develop "tinnitus" from the tinnitus-inducing stimulus also did not show a change in their auditory–somatosensory interaction pattern. This latter finding is highly significant because it opens the way to providing insights into why some people with the same disorder, whether somatosensory or auditory, develop tinnitus and others do not. These findings are consistent with clinical observations and support the notion that alteration in the auditory–somatosensory interaction at the level of the DCN fusiform cell is a fundamental mechanism of tinnitus.

In support of the DCN hypothesis, lesions interrupting rostral DCN output prevented the development of "tinnitus" if they occurred prior to the "tinnitus" inducer (high-level noise) and were bilateral (Brozoski et al., 2012). On the other hand the full story appears to be more complex. If such lesions were unilateral, they did not prevent "tinnitus," whether ipsilateral or contralateral to the noise exposure. When contralateral the experimental animal's tinnitus behavior was exacerbated. Also if bilateral lesions were made after "tinnitus" onset there was no effect upon the "tinnitus" (Brozoski and Bauer, 2005).

The somatosensory system

After hearing loss, tinnitus from a somatic disorder of the head and neck is the other major cause of tinnitus (Levine, 2004). That it can cause tinnitus is now beyond doubt, as previously discussed. In 1997 Biesinger (in the German literature) raised the possibility that "dizziness, tinnitus, sudden hearing loss, otalgia, globus pharyngeus and cephalgia" may originate from the cervical spine and that manual treatment of the upper cervical spine (C2 and C3) could treat these conditions. He postulated a neurologic basis for the tinnitus that was related to projections from the C2 and C3 dorsal nerve roots to the cochlear nuclei (Biesinger, 1997).

Unaware of Biesinger's work, Levine (1999) came to the same conclusion, but went further. Using existing anatomic and physiologic evidence combined with the properties of either the somatic tinnitus syndrome alone or SM of tinnitus alone, he elaborated and detailed how both somatic aspects of tinnitus lead to the conclusion that the ipsilateral DCN is the likely site of somatic–auditory interaction and the initiating site of somatic tinnitus (Fig. 23.3). The details of the arguments that led to these conclusions can be found in the 1999 paper (Levine, 1999).

In summary, consideration of the clinical features of the somatic tinnitus syndrome as well as the properties of SM of tinnitus, along with experimental neuroanatomy and electrophysiology, has led to the hypothesis that SM of tinnitus and the somatic tinnitus syndrome occur through modulation of the pathway from somatosensory nucleus of the medulla to the ipsilateral DCN (Fig. 23.3). If increased activity in the fusiform cells, the primary output cells of the DCN, is associated with tinnitus, as has been suggested for hearing loss tinnitus, then the speculation can be taken a step further to suggest that inhibition of this pathway (somatosensory nucleus of the medulla to ipsilateral DCN) could lead to tinnitus through disinhibition of the DCN fusiform cells (Brozoski et al., 2002; Finlayson and Kaltenbach, 2009).

Based upon clinical observations Levine et al. concluded that the somatosensory inputs appear to originate from motor afferents (Golgi tendon organs and muscle spindles) of the proprioceptive system and almost exclusively from the ipsilateral upper cervical and head regions (Levine et al., 2003). Two more recent cases have provided unequivocal support for SM being mediated by the posterior column-medial lemniscus exclusively (Table 23.4). Both subjects had unilateral tinnitus and symmetric audiograms. Both could modulate their tinnitus with standard somatic testing (strong muscle contractions of the head and neck), but in addition both had roughly the same area of their scalp in which their tinnitus could be modulated. The very circumscribed region was located in their temporal scalp above and just rostral to the pinna, as shown in Table 23.4. Within this region their tinnitus loudness could be raised by stimuli from the posterior column-medial lemniscus class of stimuli (deep pressure and vibration), could not be modified by stimuli from other somatosensory classes (light touch, temperature, pain), and could not be modified by any of these stimuli when applied outside the outlined temporal regions.

Soon after Levine's 1999 paper more support for this hypothesis of disinhibition of the ipsilateral DCN through the muscle and joint receptors of the posterior

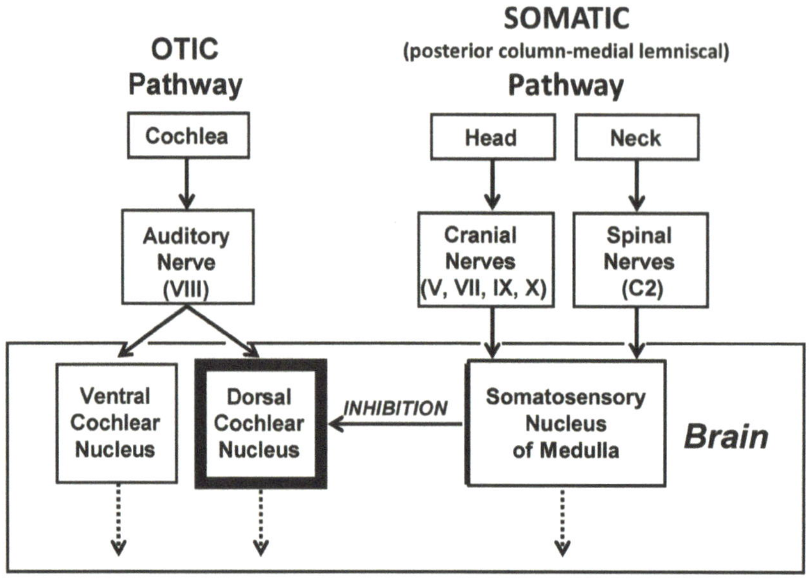

Fig. 23.3. Schematic depiction of the anatomic basis for the dorsal cochlear nucleus (DCN) hypothesis: both somatic and otic (ear) tinnitus occur owing to net disinhibition of the DCN. In both cases, tinnitus is due to increased activity in the output of the DCN, which projects to the higher centers and eventually leads to activation of the auditory perceptual machinery responsible for tinnitus. For somatic tinnitus ipsilateral sensory inputs from (1) the face via the trigeminal (V) nerve in the spinal trigeminal tract; (2) the external and middle ears via the common spinal tract of the facial (VII), glossopharyngeal (IX), and vagus (X) cranial nerves; and (3) the neck via the C2 dorsal spinal root converge to a common region of the lower part of the medulla, the somatosensory nucleus of the medulla, from which fibers project to the ipsilateral DCN. Modulation of activity in the somatosensory nucleus of the medulla to DCN pathway results in net disinhibition of the DCN. For otic tinnitus, loss of input (spontaneous activity) from the auditory (VIII) nerve leads to net disinhibition of the dorsal cochlear nucleus.

Table 23.4

Changes in tinnitus loudness (visual analog scale) for 2 subjects with similar temporal regions of sensitivity (outlined by triangles) ipsilateral to their unilateral tinnitus

		Subject AR	Subject AM
		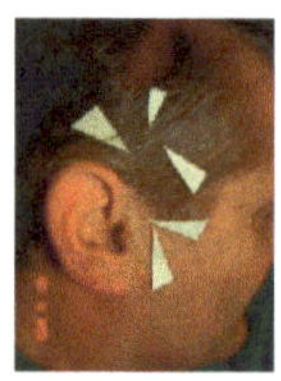	
Baseline visual analog scale loudness (0–10)		6 R / 0 L	0 R / 6 L
System	Modality		
Cutaneous	Light touch	NC	NC
	Pulling hair	NC	NC
	Stretch skin	NC	NC
Spinothalamic	Cooling	NC	NC
	Pinprick	NC	NC
Posterior column	Vibration	10 R / 0 L	0 R / 8 L
	Deep pressure	10 R / 0 L	0 R / 8 L

R, right; L, left; NC, no change.

column-medial lemniscus system came from an *in vivo* animal physiologic study (Kanold and Young, 2001). Kanold and Young (Fig. 23.4) showed that stimulation of the pinna with modalities subserved by the posterior column-medial lemniscus system and not stimuli from other sensory systems modulates ipsilateral DCN single cell activity. Just as in clinical observations regarding SM of tinnitus and somatic tinnitus they reported that "only

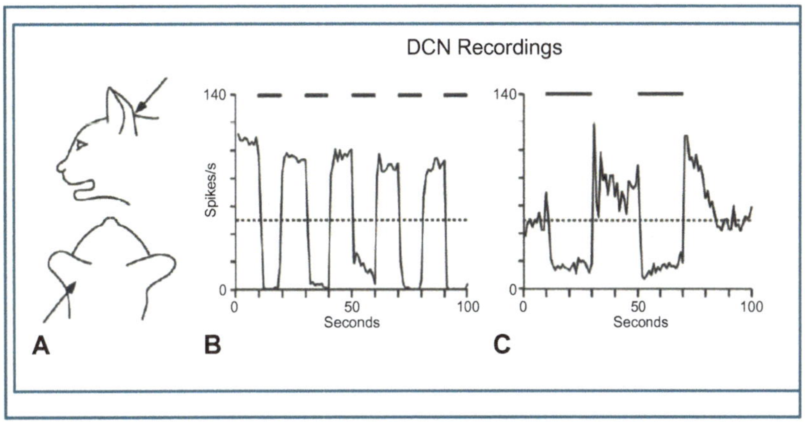

Fig. 23.4. Responses of dorsal cochlear nucleus (DCN) fusiform cells to stretch of cat pinna muscles. (**B** and **C**) The responses of two different neurons to manual pressure (indicated by the bars above the plots) applied to the ipsilateral pinna, as shown in (**A**). The effect of the pinna pressure was continuous inhibition, maintained as long as the pressure was applied. The effect of pinna pressure is mediated through the following neural pathway: C2 spinal nerve to fasciculus cuneatus to the somatosensory nucleus of the medulla to ipsilateral dorsal cochlear nucleus (Fig. 23.3). Muscle stretch or vibration was the most potent modulator of DCN activity. Cutaneous stimulation (light touch, brushing of hairs, or stretching of skin) had no effect. (Reproduced from Kanold and Young, 2001.)

stimuli that activate . . . muscle receptors, such as stretch or vibration of the muscles connected to the pinna, were effective in driving DCN units, whereas cutaneous stimuli such as light touch, brushing of hairs, and stretching of skin were ineffective" (Kanold and Young, 2001). A more recent brain slice study provides evidence that it is a decrease in gamma-aminobutyric acid (GABA)-mediated inhibition that is accounting for the tinnitus-related DCN disinhibition (Middleton et al., 2011).

Levine's group went on to show that SM of auditory perception: (1) was not a phenomenon restricted to people with clinical tinnitus but was just as prevalent in people whose tinnitus was not a clinical problem; (2) was present in about 60% of people who did not have tinnitus; and (3) was present in people with non-functioning ears whether or not they had tinnitus (Table 23.5) (Levine et al., 2003). This result in the profoundly deaf made the neurologic model iron-clad. Later confirmatory support for these findings came from a functional imaging study, which found that somatosensory input enhances activity throughout the auditory pathway in all subjects whether or not they had tinnitus (Lanting et al., 2010).

Consistent with the unilateral DCN hypothesis regarding hearing loss and somatic tinnitus is a report of (a) a significant correlation between ipsilateral tinnitus and unilaterally diminished cervical rotation and (b) abolishment of unilateral tinnitus with blocking the ipsilateral C2–3 facet joint (Reisshauer et al., 2006; Gritsenko et al., 2014).

Besides changes in auditory perception as related to strong muscle contractions, evidence for auditory–posterior column-medial lemniscus interaction in humans has been coming from other very different approaches. As previously mentioned, a functional imaging study has reported somatosensory enhancement of the auditory pathway, including the cochlear nucleus, but also Lewald and colleagues have shown how auditory spatial perception in the horizontal plane could be altered by neck vibration, a potent posterior column-medial lemniscus stimulus (Lewald et al., 1999; Lanting et al., 2010). Consistent with these studies is a study reporting poorer sound lateralization performance in people with tinnitus, especially on the side with unilateral tinnitus (An et al., 2012).

Why does the somatosensory system communicate with the DCN?

The answer to this question relates to the general principle that the DCN, as a cerebellar-like structure, is a pattern recognizer (Bell et al., 1997; Steuber et al., 2007).

Neck proprioception

The DCN is highly involved in auditory spatial perception in the vertical plane (i.e., front–back and up–down). By using modifications in the acoustic spectra produced by the interactions of sound with the pinna, the DCN can determine the sound location with respect to the ear (Oertel and Young, 2004). In order to know the position of the sound in space, the DCN also needs to know where the ears are located – in other words, the head's position.

The position of the head (ears) is provided by the cervical posterior column-medial lemniscus system (proprioceptive input coming from the joints, muscles, and tendons of the upper neck). By integrating the auditory and proprioceptive information the central nervous system can infer from where in space a sound originates.

Jaw proprioception

This sound location argument easily accounts for why the neck somatosensory system projects to DCN, but does not account for the jaw proprioceptive inputs. (This involves a more involved argument that again involves the unitary principle that the DCN is a pattern recognizer.) The DCN cytoarchitectonics (the cellular structure and the arrangement of its cells) is similar to that of another sensory nucleus, the electrosensory nucleus of the mormyrid electric fish. The argument then is as follows: similar structure implies similar function. If the function of the electrosensory nucleus for electric stimuli is known, then by analogy the function of the DCN will be something similar but for acoustic stimuli.

There is strong evidence that the function of this fish's electrosensory nucleus is to perform a "sensory subtraction." The self-generated electric signals (such as from its muscles, including its heart) are subtracted from the total signal in the electrosensory nucleus to obtain the environmental electric signals (Bell et al., 1997). This leads to the hypothesis that the DCN functions as a "sensory subtractor" for acoustic signals. The self-generated auditory signals are subtracted from the total signal in the DCN to obtain the environmental auditory signals. And what are the self-generated auditory signals? They are sounds generated from our breathing, chewing, vocalizations, and heart beats (Haenggeli et al., 2005; Shore, 2005). The generation of many of these sounds involves the neck and jaw muscles and therefore to optimize their subtraction the DCN uses the proprioceptive input from the neck and jaw muscles.

Recall that the hallmark of the somatosensory pulsatile tinnitus syndrome is that somatic testing suppresses pulsations (heart beats). This feature suggests that this type of pulsatile tinnitus may be due to a failure of the DCN to suppress the normal self-generated sound of our heart beats. Suppression of pulsatile tinnitus with somatosensory activation from somatic testing may represent a temporary correction of this malfunction (Levine et al., 2008).

The dorsal cochlear nucleus hypothesis and non-lateralized tinnitus

Support for DCN involvement in non-lateralized tinnitus is less compelling; however, the giant cells in the deep layers of the DCN do project to the contralateral dorsal, anteroventral, and posteroventral cochlear nucleus (Cant and Gaston, 1982; Doucet et al., 2009; Brown et al., 2012). Hence these direct synaptic connections from one cochlear nucleus to the other could play a significant role in non-lateralized tinnitus. These commissural connections between the cochlear nuclei of the two sides, as well as the connections between the different divisions of each cochlear nucleus, raise numerous possibilities besides the simple DCN model described above. Bringing home this point is a recent demonstration that in human subjects the auditory nerve compound action potential is reduced in tinnitus subjects (Gu et al., 2012).

This commissural pathway could also account for the unusual property of the tinnitus of some people deafened from acoustic neuroma surgery. While they are deaf and have tinnitus on the side from which the tumor was removed, sound presented to their contralateral normal functioning ear transiently intensifies their deaf ear's tinnitus and the intensification can persist for hours.

Implications of the DCN hypothesis

The hypothesis can account for why, with any degree of hearing loss, some people develop tinnitus and others do not. A schematic diagram of the fusiform cell of the DCN along with its inputs is shown in Figure 23.5A. While it is a complex picture, it can be simplified by segregating the inputs to the fusiform cell by whether they are via the apical or basal dendrite (Fig. 23.5B) and whether they are excitatory or inhibitory (Fig. 23.5C). The output of the DCN is principally via the fusiform cell axon, which projects to the contralateral inferior colliculus (Fig. 23.5B and C).

The apical dendrites receive no direct inputs from outside the DCN. Their inputs are from the parallel fibers of the "granule cell domain," whose inputs are multiple. They include type II auditory nerve fibers from cochlear outer hair cells, the head and neck posterior column-medial lemniscus system, auditory cortex, vestibular system, and pontine nuclei. In contrast, the basal dendrite receives direct inputs from the type I auditory nerve fibers from cochlear inner hair cells and its inhibitory interneuron, the vertical cell. By rearranging the diagram into excitatory and inhibitory inputs to the fusiform cell, it can be appreciated that tinnitus results when excitation outweighs inhibition at the level of the DCN fusiform cell. By the same token, there will be no tinnitus when inhibition outweighs excitation.

This framework can account then for a fundamental unanswered question regarding tinnitus and hearing loss, namely, with any degree of hearing loss, why do some people develop tinnitus and others do not (Chung et al., 1984), and likewise with animal models (Brozoski et al., 2010; Dehmel et al., 2012; Bauer et al.,

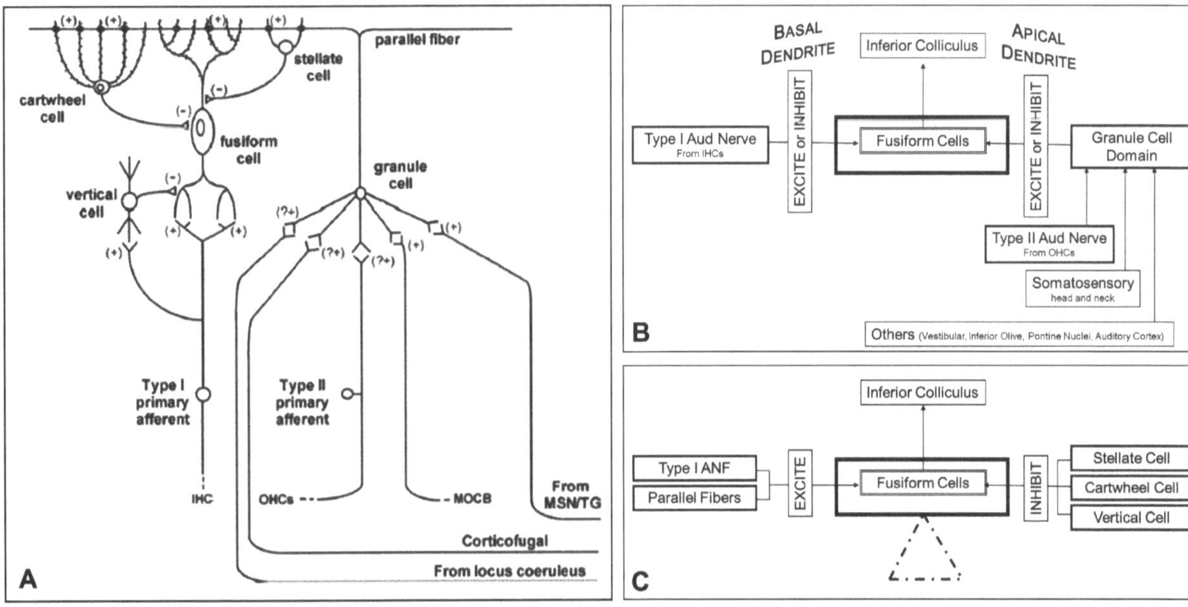

Fig. 23.5. (**A**) A schematic representation of a dorsal cochlear nucleus fusiform cell with its inputs. (Reproduced from Kaltenbach et al., 2005.) (**B**) Block diagram of (**A**) based upon whether the input is to the apical or basal dendrite. (**C**) Rearrangement of the inputs to the fusiform cell according to whether they are excitatory or inhibitory. ANF, auditory nerve fiber; IHC, inner hair cell; OHC, outer hair cell; MOCB, medial olivocochlear bundle; MSN, somatosensory nucleus of medulla; TG, trigeminal ganglion; + indicates an excitatory synapse; – indicates an inhibitatory synapse.

2013; Koehler and Shore, 2013)? While the prevalence of tinnitus increases with increasing hearing loss, it never exceeds 80% (e.g., sensorineural hearing loss (Table 23.5), including sudden idiopathic hearing loss (Levine, 1999; Huy and Sauvaget, 2005)). From transection of the auditory nerve some (50–75%) but not all subjects get relief of their unilateral tinnitus (Perry and Gantz, 2000). Whether or not any individual will develop tinnitus from a tinnitus-inducing insult, then, will depend upon the insult's net effect upon the fusiform cell (Fig. 23.5C) as well as a second related factor, the individual's "tinnitus threshold" (see section Tinnitus is a threshold phenomenon, below).

Table 23.5

Effect of somatic testing on the auditory perception of profoundly deaf subjects

Subject type	Totals	Modulate?	
		Yes	No
With tinnitus	11	6 (54%)	5 (45%)
Without tinnitus	3	2 (67%)	1 (33%)
Totals	14	8 (57%)	6 (43%)

The hypothesis can account for why potentiators of GABA quieten tinnitus. As can be appreciated from Figure 23.5C, anything that promotes inhibition of the fusiform cell or reduces its excitation will diminish the output of the fusiform cell and thereby tinnitus.

This can explain why GABAergic drugs quieten tinnitus, as has been shown in both human and animal studies. The synapse between the stellate cell and the fusiform cell (Fig. 23.5A) is mediated by the inhibitory neurotransmitter, GABA. Hence, medications that enhance this synapse will reduce fusiform cell activity and thereby quieten tinnitus. In fact, four well-designed studies have shown that the GABA potentiators, the benzodiazepines, quieten tinnitus in about 75% of people (Johnson et al., 1993; Vernon and Meikle, 2003; Jalali et al., 2009; Han et al., 2012) and suppress tinnitus completely in 7% of people (Han et al., 2012). Two other GABA-potentiating drugs, primidone (Mysoline) and ethanol, have also been shown to quieten tinnitus in humans, albeit in less rigorous studies (Shea and Emmett, 1981). In animals vigabatrin, a GABA potentiator, reversibly abolishes "tinnitus" (Brozoski et al., 2007).

The hypothesis can account for why cochlear electric stimulation can quieten tinnitus. Multiple human studies have demonstrated that electric stimulation of the cochlea quietens tinnitus (Matsushima et al., 1997b; Rubinstein et al., 2003; Van de Heyning et al.,

2008). According to the DCN hypothesis, this can occur in multiple ways, such as: (1) activation of inhibition (vertical, cartwheel, or stellate cells); (2) reduction in excitation of the fusiform cell (type I auditory nerve fiber or parallel fibers); or (3) some combination of the two.

The tinnitus pathway rostral to DCN

The tinnitus pathway rostral to DCN is likely very complex. For example, a recent report implicates involvement of the paraflocculus of the cerebellum in a rat model of tinnitus. Established "tinnitus" could be eliminated by paraflocculus inactivation; however, its ablation prior to "tinnitus" induction did not eliminate the rats' "tinnitus" (Bauer et al., 2013).

The primary output of the DCN is from the fusiform cell, whose axon projects via the dorsal acoustic stria to the contralateral inferior colliculus (Fig. 23.5) (Osen, 1972; Romand and Avan, 1997; Alibardi, 2000). Inferior colliculus recordings contralateral to the traumatized ear in animals with "tinnitus" have found an increased number of neurons with bursting (epochs of very regular 1/ms neuronal discharges) (Bauer et al., 2008). Elevated spontaneous activity and increased cross-fiber synchrony were found in the inferior colliculus for some, but not all, types of experimental hearing loss.

In human observations a 2-cm infarction involving the left caudate, putamen, and neighboring deep white matter totally eliminated the tinnitus of a 63-year-old otolaryngologist with 40 years of bilateral high-pitched tinnitus. Similarly, a discrete infarction involving the junction of the head and body of the left caudate nucleus of a 56-year-old woman with refractory Parkinson's disease resulted in bilateral tinnitus suppression (total for the ipsilateral ear, partial for the contralateral ear) (Larson and Cheung, 2013). Electric stimulation in this same locus can modulate (usually decrease) tinnitus loudness bilaterally (Cheung and Larson, 2010). With electric stimulation of the ventralis intermedius nucleus of the thalamus, 2 of 4 subjects reported a transient decrease in their tinnitus loudness for 15–20 minutes (Shi et al., 2009).

Chronic electric stimulation of primary or secondary auditory cortex can suppress tinnitus to some degree in about two-thirds of subjects, selected by their response to transcranial magnetic stimulation (TMS). However not only does TMS not predict who will respond to such chronic electric stimulation, but no other predictors have been identified (Langguth et al., 2012). TMS has transiently suppressed tinnitus in some subjects from a variety of skull locations, including the ipsilateral and contralateral temporoparietal region and frontal region (dorsolateral prefrontal lobe).

Tinnitus is a threshold phenomenon

Evidence that tinnitus is a threshold phenomenon has emerged from observations of differences between subjects with and without chronic tinnitus regarding the three ways transient tinnitus commonly occurs: (1) from SM, strong pressure on or contraction of the muscles of the head and neck; (2) from loud sounds (LS); and (3) SBUTTs. For all three phenomena there is evidence that a change in auditory perception is more frequent for people with chronic tinnitus as opposed to those without. This contrasts with Lanting's functional imaging data which revealed "a remarkable result . . . for the majority of auditory brain areas, the response to jaw protrusion is similar between" tinnitus and non-tinnitus subjects (Lanting et al., 2010). The Lanting result is very different than the result for SM and LS, where a significant difference between tinnitus and non-tinnitus subjects was found for both. In tinnitus subjects about 80% can modulate their auditory perception with SM as opposed to 60% in non-tinnitus subjects. For LS the comparable percentages were 63% vs 32% (Abel and Levine, 2004). A similar trend was found for SBUTTs (Oron et al., 2011).

The reason these findings are evidence for a tinnitus threshold will be explained from the SM data, since it was not dependent upon the subject's memory. With SM tinnitus can become louder (SM(+)) or quieter (SM(–)) and/or change its quality. If there were not a threshold, then it would be predicted that, for SM, the occurrence of a change in auditory perception with SM(+) would be the same whether or not people had ongoing tinnitus, as was found with Lanting's functional imaging. Since this is not the case, a straightforward interpretation is that neural activity must exceed a neural threshold in order to be perceived as a sound (Fig. 23.6). Following the Lanting result, the hypothesis is that for SM(+) the strong muscle pressure or contraction produces an elevation in neural activity within the auditory pathway (Fig. 23.6A). For any degree of elevated neural activity, the tinnitus subject would be more likely than the non-tinnitus subject to detect the change in neural activity, since it will be manifested as an increase in tinnitus loudness. On the other hand, the non-tinnitus subject will sometimes not detect the change in neural activity, because the overall rise in neural activity with strong muscle contraction will not be sufficient to cross the non-tinnitus subject's neural threshold for auditory perception. All other things being equal, then, the prevalence of a change in auditory perception with SM(+) will be higher in the tinnitus subject than in the non-tinnitus subject. For SM there is another factor at play as well, namely SM(–), that is, quieting of tinnitus with strong muscle contraction (Fig. 23.6B). Presumably

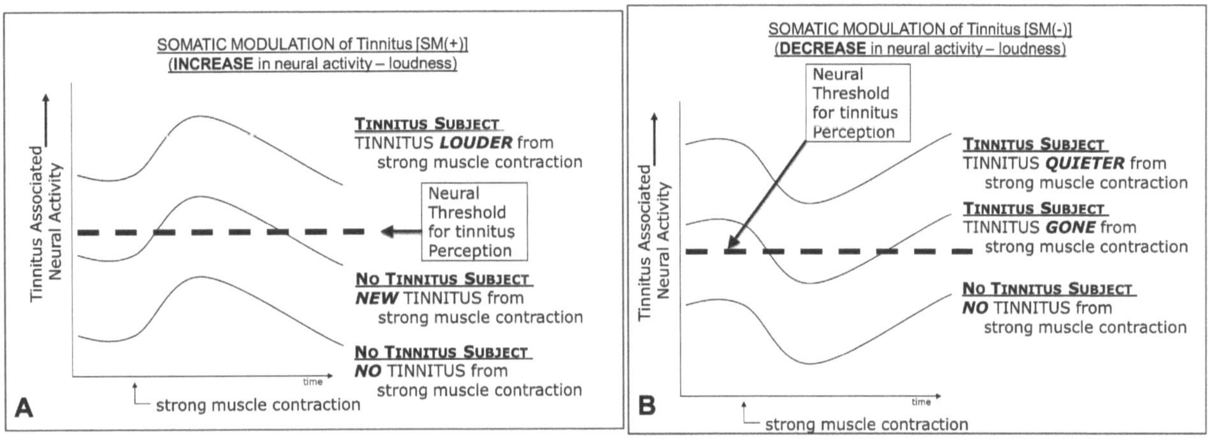

Fig. 23.6. How the threshold concept for tinnitus accounts for higher incidence of somatic modulation of tinnitus in people with than without ongoing tinnitus. The dashed line represents the neural threshold for tinnitus perception, i.e., tinnitus is perceived when neural activity exceeds this threshold. People with tinnitus are represented by the curve whose neural activity prior to a strong muscle contraction is above the dashed line and those without tinnitus are below the dashed line. Part (**A**) considers the case where somatic testing results in a net increase in neural activity. For the same degree of increase in neural activity the change will always be detected by the ongoing tinnitus subjects, but not all the non-tinnitus subjects, if the degree of change is insufficient to exceed the neural threshold. Likewise, in (**B**) for somatic testing that results in a net decrease in neural activity, no change in auditory perception will occur for any non-tinnitus subject since, unlike the tinnitus subjects with a strong muscle contraction, their neural activity remains below the neural threshold for tinnitus.

SM(–) corresponds to a decrease in neural activity within the auditory pathway. Since a non-tinnitus subject has no ongoing auditory perception, a decrease in neural activity within the auditory pathway will not be perceived as a change in auditory perception. Hence, because of the threshold effect and no detection for SM(–), the subject without ongoing tinnitus has a lower incidence of transient tinnitus from SM than does the tinnitus subject.

Typewriter tinnitus: cross-talk between auditory nerve fibers

To understand how typewriter tinnitus likely occurs requires an understanding of how we hear faint sounds. In a healthy ear, all 30 000 auditory nerve fibers have "spontaneous activity." In the absence of sound they are continuously transmitting nerve impulses from the inner ear to the cochlear nuclei (at a rate of 1–100/second) (Liberman, 1978). Sound is not perceived because the spontaneous activity is random and the nerve fibers discharge their spikes independently of each other (Johnson and Kiang, 1976). The spike-firing pattern of one auditory nerve fiber is different from the spike-firing pattern of all other auditory nerve fibers. There is no synchrony between the discharge patterns of any of the auditory nerve fibers.

When a faint sound is heard there is no change in the firing rate of the individual nerve fibers, but they become synchronized to the faint sound and thereby to each other. Faint sounds are heard because of neural synchrony.

Typewriter tinnitus appears to be caused by distortion of the auditory nerve, most commonly from vascular compression of the VIIIth nerve (Fig. 23.1), probably because the insulating myelin sheath of the auditory nerve fibers becomes thinned. This leads to ephaptic transmission or "cross-talk" between adjacent single auditory nerve fibers. At the points where the myelin has been critically thinned, an original spike in one nerve will cause a new synchronous spike in its neighbor. Since auditory nerve fiber synchrony is responsible for auditory perception, this synchrony due to cross-talk will be perceived as a click. This cross-talk hypothesis explains not only why the clicks of typewriter tinnitus can occur spontaneously (ergo "tinnitus") but also why sometimes clicking can be elicited by an external sound. External sounds generate auditory nerve activity and this auditory nerve activity leads to cross-talk which causes nerve fiber synchrony that then causes the perception of a click.

Medications that suppress typewriter tinnitus, such as carbamazepine, do so by decreasing nerve excitability, which reduces the likelihood of cross-talk, and thereby reduces the clicking.

CONCLUSIONS

In some surveys about 10% of adults report having chronic tinnitus and one out of 20 of them feels the tinnitus interferes with the ability to lead a normal life. On the other hand, in dead silence somewhere between 50 and 90% of adults report experiencing an auditory percept. About 75% of adults have experienced sudden brief unilateral tapering tinnitus, typically lasting about a minute or less. About 50% of adults have experienced transient tinnitus following loud noise exposure.

A wide range of disorders can cause tinnitus. Many involve a disturbance of the peripheral auditory system, including various types of hearing loss. However one of the major, but underappreciated, causes of tinnitus has no involvement with the ear but causes tinnitus through its connections to the auditory system within the central nervous system. Myofascial disorders of the head and neck cause tinnitus by modulating the activity of the somatosensory system which projects to the central auditory system at multiple levels, including the DCN and inferior colliculus.

While myofascial disorders or hearing loss alone can cause tinnitus, they also can combine to cause tinnitus. For this reason all tinnitus patients should have an audiogram and be questioned and examined for a myofascial disorder of the head and neck. Specifically the history must cover the topics of headache, bruxism, dental disorders, neck and facial pain, trauma, and changes in physical activities such as a new or altered exercise program. The physical exam must include an examination of the head and neck for myofascial disorders and for evidence of tinnitus modulation from activation of the head and neck muscles.

If there is evidence that the tinnitus is related to a myofascial disorder of the head and neck, then a well-designed treatment program directed toward the myofascial disorder can result in a major improvement of the tinnitus in about half of such individuals.

Two features of tinnitus must be specifically questioned, since they are often very minor and so not spontaneously described by the patient, yet they can have important diagnostic significance. If unilateral clicking is present even if very infrequent and not troubling, the possibility of auditory nerve compression must be considered. Likewise, it must be ascertained whether the tinnitus sometimes has a cardiac-synchronous (pulsatile) character. If bilateral and sometimes but not always pulsatile, then a myofascial etiology is very likely; if unilateral and intermittent then, besides myofascial, auditory nerve vascular compression must also be considered.

An interaction between findings in the experimental animal and clinical observations has led to the DCN hypothesis as the most compelling model that can account for the major features of the common tinnitus syndromes. The DCN hypothesis proposes that increased activity from the major output cell of the DCN can cause tinnitus when a threshold is exceeded due to changes in the auditory inputs to the DCN, the somatosensory inputs to the DCN, or a combination of the two. Unilateral staccato irregular intermittent clicking (typewriter tinnitus), like other vascular compression syndromes, appears to be related to ephaptic transmission between adjacent single auditory nerve fibers and, like the other vascular compression syndromes, is usually suppressed by membrane-stabilizing drugs such as carbamazepine.

References

Abel MD, Levine RA (2004). Muscle contractions and auditory perception in tinnitus patients and nonclinical subjects. Cranio 22 (3): 181–191.

Alibardi L (2000). Cytology of large neurons in the guinea pig dorsal cochlear nucleus contacting the inferior colliculus. Eur J Histochem 44 (4): 365–375.

An YH, Lee LH, Yoon SW et al. (2012). Does tinnitus affect the sound localization ability? Otol Neurotol 33 (5): 692–698.

Andersson G, Juris L, Kaldo V et al. (2005). Hyperacusis – an unexplored field. Cognitive behavior therapy can relieve problems in auditory intolerance, a condition with many questions. Lakartidningen 102 (44): 3210–3212.

Ashton H (1991). Protracted withdrawal syndromes from benzodiazepines. J Subst Abuse Treat 8 (1–2): 19–28.

Attias J, Shemesh Z, Sohmer H et al. (1993). Comparison between self-hypnosis, masking and attentiveness for alleviation of chronic tinnitus. Audiology 32 (3): 205–212.

Attias J, Zwecker-Lazar I, Nageris B et al. (2005). Dysfunction of the auditory efferent system in patients with traumatic brain injuries with tinnitus and hyperacusis. J Basic Clin Physiol Pharmacol 16 (2–3): 117–126.

Azevedo AA, Figueiredo RR (2005). Tinnitus treatment with acamprosate: double-blind study. Braz J Otorhinolaryngol 71 (5): 618–623.

Baguley DM (2003). Hyperacusis. J R Soc Med 96 (12): 582–585.

Bauer CA, Turner JG, Caspary DM et al. (2008). Tinnitus and inferior colliculus activity in chinchillas related to three distinct patterns of cochlear trauma. J Neurosci Res 86 (11): 2564–2578.

Bauer CA, Kurt W, Sybert LT et al. (2013). The cerebellum as a novel tinnitus generator. Hear Res 295: 130–139.

Baurier VO, Tuca JO (1996). Auditory hallucinosis in brainstem lesions and musical hallucination in deafness. Analogies and differences. Eur J Neurol 3 (3): 203–211.

Bell C, Bodznick D, Montgomery J et al. (1997). The generation and subtraction of sensory expectations within cerebellum-like structures. Brain Behav Evol 50 (Suppl 1): 17–31.

Biesinger E (1997). C2 and C3 cervical nerve root syndrome: the influence of cervical spine dysfunction on ENT symptoms. Manuel Med 35: 12–19.

Blaesing L, Kroener-Herwig B (2012). Self-reported and behavioral sound avoidance in tinnitus and hyperacusis subjects, and association with anxiety ratings. Int J Audiol 51 (8): 611–617.

Boussi-Gross R, Golan H, Fishlev G et al. (2013). Hyperbaric oxygen therapy can improve post concussion syndrome years after mild traumatic brain injury – randomized prospective trial. PLoS One 8 (11): e79995.

Brantberg K (2010). Paroxysmal staccato tinnitus: a carbamazepine responsive hyperactivity dysfunction symptom of the eighth cranial nerve. J Neurol Neurosurg Psychiatry 81 (4): 451–455.

Brown MC, Drottar M, Benson TE et al. (2012). Commissural axons of the mouse cochlear nucleus. J Comp Neurol 521 (7): 1683–1696.

Brozoski TJ, Bauer CA (2005). The effect of dorsal cochlear nucleus ablation on tinnitus in rats. Hear Res 206 (1–2): 227–236.

Brozoski TJ, Bauer CA, Caspary DM (2002). Elevated fusiform cell activity in the dorsal cochlear nucleus of chinchillas with psychophysical evidence of tinnitus. J Neurosci 22 (6): 2383–2390.

Brozoski TJ, Spires TJ, Bauer CA (2007). Vigabatrin, a GABA transaminase inhibitor, reversibly eliminates tinnitus in an animal model. J Assoc Res Otolaryngol 8 (1): 105–118.

Brozoski TJ, Caspary DM, Bauer CA et al. (2010). The effect of supplemental dietary taurine on tinnitus and auditory discrimination in an animal model. Hear Res 270 (1–2): 71–80.

Brozoski TJ, Wisner KW, Sybert LT et al. (2012). Bilateral dorsal cochlear nucleus lesions prevent acoustic-trauma induced tinnitus in an animal model. J Assoc Res Otolaryngol 13 (1): 55–66.

Caffier PP, Haupt H, Scherer H et al. (2006). Outcomes of long-term outpatient tinnitus-coping therapy: psychometric changes and value of tinnitus-control instruments. Ear Hear 27 (6): 619–627.

Cant NB, Gaston KC (1982). Pathways connecting the right and left cochlear nuclei. J Comp Neurol 212 (3): 313–326.

Cardarelli F, Melcher JR, Szeles JC et al. (2010). Continuous auricular electrical stimulation quiets the tinnitus of the somatosensory pulsatile tinnitus syndrome. In: 4th International TRI Tinnitus Conference, Frontiers in Tinnitus Research, Dallas.

Ceranic BJ, Prasher DK, Raglan E et al. (1998). Tinnitus after head injury: evidence from otoacoustic emissions. J Neurol Neurosurg Psychiatry 65 (4): 523–529.

Cheung SW, Larson PS (2010). Tinnitus modulation by deep brain stimulation in locus of caudate neurons (area LC). Neuroscience 169 (4): 1768–1778.

Chole RA, Parker WS (1992). Tinnitus and vertigo in patients with temporomandibular disorder. Arch Otolaryngol Head Neck Surg 118 (8): 817–821.

Chung DY, Gannon RP, Mason K (1984). Factors affecting the prevalence of tinnitus. Audiology 23 (5): 441–452.

Coles RR (1996). Tinnitus. In: D Stephens (Ed.), Adult Audiology, Butterworth Heinemann, Guilford (UK).

Curtin HD (1997). Rule out eighth nerve tumor: contrast-enhanced T1-weighted or high-resolution T2-weighted MR? AJNR Am J Neuroradiol 18 (10): 1834–1838.

David JB, Naftali A, Katz A (2010). Tinntrain: a multifactorial treatment for tinnitus using binaural beats. Hear J 63: 25–28.

Davis PB, Wilde RA, Steed LG et al. (2008). Treatment of tinnitus with a customized acoustic neural stimulus: a controlled clinical study. Ear Nose Throat J 87 (6): 330–339.

Dawes PJ, Basiouny HE (1999). Outcome of using magnetic resonance imaging as an initial screen to exclude vestibular schwannoma in patients presenting with unilateral tinnitus. J Laryngol Otol 113 (9): 818–822.

Dehmel S, Eisinger D, Shore SE (2012). Gap prepulse inhibition and auditory brainstem-evoked potentials as objective measures for tinnitus in guinea pigs. Front Syst Neurosci 6: 42.

Del Bo L, Forti S, Ambrosetti U et al. (2008). Tinnitus aurium in persons with normal hearing: 55 years later. Otolaryngol Head Neck Surg 139 (3): 391–394.

Don M, Kwong B, Tanaka C (2012). Interaural stacked auditory brainstem response measures for detecting small unilateral acoustic tumors. Audiol Neurootol 17 (1): 54–68.

Dornhoffer JL, Helms J, Hoehmann DH (1994). Presentation and diagnosis of small acoustic tumors. Otolaryngol Head Neck Surg 111 (3 Pt 1): 232–235.

Doucet JR, Lenihan NM, May BJ (2009). Commissural neurons in the rat ventral cochlear nucleus. J Assoc Res Otolaryngol 10 (2): 269–280.

Finlayson PG, Kaltenbach JA (2009). Alterations in the spontaneous discharge patterns of single units in the dorsal cochlear nucleus following intense sound exposure. Hear Res 256 (1–2): 104–117.

Florentine M (1976). Relation between lateralization and loudness in asymmetrical hearing losses. J Am Audiol Soc 1 (6): 243–251.

Formby C, Sherlock LP, Gold SL (2003). Adaptive plasticity of loudness induced by chronic attenuation and enhancement of the acoustic background. J Acoust Soc Am 114 (1): 55–58.

Fortnum H, O'Neill C, Taylor R et al. (2009). The role of magnetic resonance imaging in the identification of suspected acoustic neuroma: a systematic review of clinical and cost effectiveness and natural history. Health Technol Assess 13 (18): iii–iv, ix–xi, 1–154.

Fowler EP (1943). Control of head noises: their illusions of loudness and timbre. Arch Otolaryngol 37: 391–398.

Gritsenko K, Caldwell W, Shaparin N et al. (2014). Resolution of long standing tinnitus following radiofrequency ablation of C2-C3 medial branches – a case report. Pain Physician 17 (1): E95–E98.

Gu JW, Herrmann BS, Levine RA et al. (2012). Brainstem auditory evoked potentials suggest a role for the ventral cochlear nucleus in tinnitus. J Assoc Res Otolaryngol 13 (6): 819–833.

Guevara N, Deveze A, Buza V et al. (2008). Microvascular decompression of cochlear nerve for tinnitus incapacity: pre-surgical data, surgical analyses and long-term follow-up of 15 patients. Eur Arch Otorhinolaryngol 265 (4): 397–401.

Haenggeli CA, Pongstaporn T, Doucet JR et al. (2005). Projections from the spinal trigeminal nucleus to the cochlear nucleus in the rat. J Comp Neurol 484 (2): 191–205.

Han SS, Nam EC, Won JY et al. (2012). Clonazepam quiets tinnitus: a randomised crossover study with Ginkgo biloba. J Neurol Neurosurg Psychiatry 83 (8): 821–827.

Hausler R, Levine RA (2000). Auditory dysfunction in stroke. Acta Otolaryngol 120 (6): 689–703.

Heller MF, Bergman M (1953). Tinnitus aurium in normally hearing persons. Ann Otol Rhinol Laryngol 62 (1): 73–83.

Hesser H, Weise C, Westin VZ et al. (2011). A systematic review and meta-analysis of randomized controlled trials of cognitive-behavioral therapy for tinnitus distress. Clin Psychol Rev 31 (4): 545–553.

Hoare DJ, Kowalkowski VL, Kang S et al. (2011). Systematic review and meta-analyses of randomized controlled trials examining tinnitus management. Laryngoscope 121 (7): 1555–1564.

Huy PT, Sauvaget E (2005). Idiopathic sudden sensorineural hearing loss is not an otologic emergency. Otol Neurotol 26 (5): 896–902.

Jalali MM, Kousha A, Naghavi SE et al. (2009). The effects of alprazolam on tinnitus: a cross-over randomized clinical trial. Med Sci Monit 15 (11): PI55–PI60.

Jastreboff PJ, Jastreboff MM (2003). Tinnitus retraining therapy for patients with tinnitus and decreased sound tolerance. Otolaryngol Clin North Am 36 (2): 321–336.

Johnson DH, Kiang NY (1976). Analysis of discharges recorded simultaneously from pairs of auditory nerve fibers. Biophys J 16 (7): 719–734.

Johnson RM, Brummett R, Schleuning A (1993). Use of alprazolam for relief of tinnitus. A double-blind study. Arch Otolaryngol Head Neck Surg 119 (8): 842–845.

Kaltenbach JA, McCaslin DL (1996). Increases in spontaneous activity of dorsal cochlear nucleus neurons following intense sound exposure. Auditory Neurosci 3: 57–78.

Kaltenbach JA, Zhang J, Finlayson P (2005). Tinnitus as a plastic phenomenon and its possible neural underpinnings in the dorsal cochlear nucleus. Hear Res 206 (1–2): 200–226.

Kanold PO, Young ED (2001). Proprioceptive information from the pinna provides somatosensory input to cat dorsal cochlear nucleus. J Neurosci 21 (19): 7848–7858.

Kiang NY, Moxon EC, Levine RA (1970). Auditory-nerve activity in cats with normal and abnormal cochleas. In: Sensorineural hearing loss. Ciba Found Symp 241–273.

Knobel KA, Sanchez TG (2008). Influence of silence and attention on tinnitus perception. Otolaryngol Head Neck Surg 138 (1): 18–22.

Koehler SD, Shore SE (2013). Stimulus timing-dependent plasticity in dorsal cochlear nucleus is altered in tinnitus. J Neurosci 33 (50): 19647–19656.

Krause E, Leunig A, Klopstock T et al. (2006). Treatment of essential palatal myoclonus in a 10-year-old girl with botulinum neurotoxin. Otol Neurotol 27 (5): 672–675.

Kreuzer PM, Goetz M, Holl M et al. (2012). Mindfulness- and body-psychotherapy-based group treatment of chronic tinnitus: a randomized controlled pilot study. BMC Complement Altern Med 12: 235.

Langguth B, Schecklmann M, Lehner A et al. (2012). Neuroimaging and neuromodulation: complementary approaches for identifying the neuronal correlates of tinnitus. Front Syst Neurosci 6: 15.

Lanting CP, de Kleine E, Eppinga RN et al. (2010). Neural correlates of human somatosensory integration in tinnitus. Hear Res 267 (1–2): 78–88.

Larson PS, Cheung SW (2013). A stroke of silence: tinnitus suppression following placement of a deep brain stimulation electrode with infarction in area LC. J Neurosurg 118 (1): 192–194.

Levine RA (1999). Somatic (craniocervical) tinnitus and the dorsal cochlear nucleus hypothesis. Am J Otolaryngol 20 (6): 351–362.

Levine RA (2004). Somatic tinnitus. In: JB Snow (Ed.), Tinnitus: Theory and Management, BC Decker, Hamilton, Ontario, pp. 108–124.

Levine RA (2006). Typewriter tinnitus: a carbamazepine-responsive syndrome related to auditory nerve vascular compression. ORL J Otorhinolaryngol Relat Spec 68 (1): 43–47.

Levine RA (2013). Tinnitus: diagnostic approach leading to treatment. Semin Neurol 33 (3): 256–269.

Levine RA, Abel M, Cheng H (2003). CNS somatosensory-auditory interactions elicit or modulate tinnitus. Exp Brain Res 153 (4): 643–648.

Levine RA, Nam EC, Oron Y et al. (2007). Evidence for a tinnitus subgroup responsive to somatosensory based treatment modalities. Prog Brain Res 166: 195–207.

Levine RA, Nam EC, Melcher J (2008). Somatosensory pulsatile tinnitus syndrome: somatic testing identifies a pulsatile tinnitus subtype that implicates the somatosensory system. Trends Amplif 12 (3): 242–253.

Lewald J, Karnath HO, Ehrenstein WH (1999). Neck-proprioceptive influence on auditory lateralization. Exp Brain Res 125 (4): 389–396.

Liberman MC (1978). Auditory-nerve response from cats raised in a low-noise chamber. J Acoust Soc Am 63 (2): 442–455.

Madani G, Connor SE (2009). Imaging in pulsatile tinnitus. Clin Radiol 64 (3): 319–328.

Marchiando A, Per-Lee JH, Jackson RT (1983). Tinnitus due to idiopathic stapedial muscle spasm. Ear Nose Throat J 62 (1): 8–13.

Martin WH (2011). Acamprosate trial for tinnitus. In: 10th International Tinnitus Seminar, Florianopolis, Brazil, March 16-19, 2011.

Matsushima JI, Sakai N, Uemi N et al. (1997a). Evaluation of implanted tinnitus suppressor based on tinnitus stress test. Int Tinnitus J 3 (2): 123–131.

Matsushima J, Kumagai M, Kamada T et al. (1997b). Preliminary study of improved perception of words with the same sound but different intonation in tinnitus patients following electrical stimulation of the ear. Acta Otolaryngol Suppl 532: 112–114.

McConnell JG (2008). Benzodiazepine tolerance, dependency, and withdrawal syndromes and interactions with fluoroquinolone antimicrobials. Br J Gen Pract 58 (550): 365–366.

Middleton JW, Kiritani T, Pedersen C et al. (2011). Mice with behavioral evidence of tinnitus exhibit dorsal cochlear nucleus hyperactivity because of decreased GABAergic inhibition. Proc Natl Acad Sci U S A 108 (18): 7601–7606.

Morrison GA, Sterkers JM (1996). Unusual presentations of acoustic tumours. Clin Otolaryngol Allied Sci 21 (1): 80–83.

Nam EC, Handzel O, Levine RA (2010). Carbamazepine responsive typewriter tinnitus from basilar invagination. J Neurol Neurosurg Psychiatry 81 (4): 456–458.

Norena AJ, Chery-Croze S (2007). Enriched acoustic environment rescales auditory sensitivity. Neuroreport 18 (12): 1251–1255.

Oertel D, Young ED (2004). What's a cerebellar circuit doing in the auditory system? Trends Neurosci 27 (2): 104–110.

Oron Y, Roth Y, Levine RA (2011). Sudden brief unilateral tapering tinnitus: prevalence and properties. Otol Neurotol 32 (9): 1409–1414.

Osen KK (1972). Projection of the cochlear nuclei on the inferior colliculus in the cat. J Comp Neurol 144 (3): 355–372.

Palikh GM, Vaughn BV (2010). Topiramate responsive exploding head syndrome. J Clin Sleep Med 6 (4): 382–383.

Pearce JM (1989). Clinical features of the exploding head syndrome. J Neurol Neurosurg Psychiatry 52 (7): 907–910.

Peltier E, Peltier C, Tahar S et al. (2012). Long-term tinnitus suppression with linear octave frequency transposition hearing AIDS. PLoS One 7 (12): e51915.

Penner MJ (1992). Linking spontaneous otoacoustic emissions and tinnitus. Br J Audiol 26 (2): 115–123.

Penner MJ, Coles RR (1992). Indications for aspirin as a palliative for tinnitus caused by SOAEs: a case study. Br J Audiol 26 (2): 91–96.

Perry BP, Gantz BJ (2000). Medical and surgical evaluation and management of tinnitus. In: R Tyler (Ed.), Tinnitus Handbook, Singular, San Diego, pp. 221–242.

Poe DS (2007). Diagnosis and management of the patulous eustachian tube. Otol Neurotol 28 (5): 668–677.

Reisshauer A, Mathiske-Schmidt K, Kuchler I et al. (2006). Functional disturbances of the cervical spine in tinnitus. HNO 54 (2): 125–131.

Romand R, Avan P (1997). Anatomical and functional aspect of the cochlear nucleus. In: G Ehret, R Romand (Eds.), The central auditory system, Oxford, New York, pp. 97–191.

Rubinstein JT, Tyler RS, Johnson A et al. (2003). Electrical suppression of tinnitus with high-rate pulse trains. Otol Neurotol 24 (3): 478–485.

Russell D, Baloh RW (2009). Gabapentin responsive audiovestibular paroxysmia. J Neurol Sci 281 (1–2): 99–100.

Ryu H, Yamamoto S, Sugiyama K et al. (1998). Neurovascular decompression of the eighth cranial nerve in patients with hemifacial spasm and incidental tinnitus: an alternative way to study tinnitus. J Neurosurg 88 (2): 232–236.

Sharma DK, Kaur S, Singh J et al. (2012). Role of acamprosate in sensorineural tinnitus. Indian J Pharmacol 44 (1): 93–96.

Shea JJ, Emmett JR (1981). The medical treatment of tinnitus. J Laryngol Otol Suppl 4: 130–138.

Shi Y, Burchiel KJ, Anderson VC et al. (2009). Deep brain stimulation effects in patients with tinnitus. Otolaryngol Head Neck Surg 141 (2): 285–287.

Shlamkovitch N (2012). The effect of hyperbaric oxygen treatment on post traumatic central-type chronic disabling tinnitus. In: 6th International TRI Tinnitus Conference, Tinnitus: The Art and Science of Innovation, Bruges.

Shore SE (2005). Multisensory integration in the dorsal cochlear nucleus: unit responses to acoustic and trigeminal ganglion stimulation. Eur J Neurosci 21 (12): 3334–3348.

Sismanis A (2011). Pulsatile tinnitus: contemporary assessment and management. Curr Opin Otolaryngol Head Neck Surg 19 (5): 348–357.

Steuber V, Mittmann W, Hoebeek FE et al. (2007). Cerebellar LTD and pattern recognition by Purkinje cells. Neuron 54 (1): 121–136.

Sullivan M, Katon W, Russo J et al. (1993). A randomized trial of nortriptyline for severe chronic tinnitus. Effects on depression, disability, and tinnitus symptoms. Arch Intern Med 153 (19): 2251–2259.

Teachey WS, Wijtmans EH, Cardarelli F et al. (2012). Tinnitus of myofascial origin. Int Tinnitus J 17 (1): 70–73.

Tjell C, Tenenbaum A, Rosenhall U (1999). Auditory function in whiplash-associated disorders. Scand Audiol 28 (4): 203–209.

Tranter RM, Graham JR (2009). A review of the otological aspects of whiplash injury. J Forensic Leg Med 16 (2): 53–55.

Trotter MI, Donaldson I (2008). Hearing aids and tinnitus therapy: a 25-year experience. J Laryngol Otol 122 (10): 1052–1056.

Tucker DA, Phillips SL, Ruth RA et al. (2005). The effect of silence on tinnitus perception. Otolaryngol Head Neck Surg 132 (1): 20–24.

Tzounopoulos T (2008). Mechanisms of synaptic plasticity in the dorsal cochlear nucleus: plasticity-induced changes that could underlie tinnitus. Am J Audiol 17 (2): S170–S175.

Van de Heyning P, Vermeire K, Diebl M et al. (2008). Incapacitating unilateral tinnitus in single-sided deafness treated by cochlear implantation. Ann Otol Rhinol Laryngol 117 (9): 645–652.

Vernon JA (1987). Pathophysiology of tinnitus: a special case–hyperacusis and a proposed treatment. Am J Otol 8 (3): 201–202.

Vernon JA, Meikle MB (2003). Masking devices and alprazolam treatment for tinnitus. Otolaryngol Clin North Am 36 (2): 307–320, vii.

Yuen HW, Eikelboom RH, Atlas MD (2009). Auditory manifestations of superior semicircular canal dehiscence. Otol Neurotol 30 (3): 280–285.

Handbook of Clinical Neurology, Vol. 129 (3rd series)
The Human Auditory System
G.G. Celesia and G. Hickok, Editors

Chapter 24

Auditory hallucinations

JAN DIRK BLOM*

Parnassia Psychiatric Institute, The Hague and University of Groningen, Groningen, The Netherlands

INTRODUCTION

Our knowledge of the variegated group of phenomena that fall under the term auditory hallucinations has rapidly expanded over the past two decades. This is mainly due to the possibility of measuring and visualizing hallucinatory activity in the brain with the aid of functional imaging techniques, and to insights derived from the philosophy of science which radically altered the way we conceptualize the auditory hallucinatory experience as well as the hearing of actual sounds. The two phenomena are now considered products of gamma-band oscillations of approximately 40 Hz resonating within the thalamocortical network, and hence as closed, intrinsic functional brain states which only differ as regards the degree of restraint exerted by information from the inner ear.

The brain areas responsible for the mediation of those percepts, as traditionally conjectured on the basis of lesion studies, neuropsychologic experiments, and clinical reasoning, have by and large been empirically confirmed with the aid of functional and structural neuroimaging. While recently many other brain areas have been added to the traditional list, we ceased to conceptualize those areas in terms of – relatively isolated – modules, and instead started granting them, in conformity with the principles of network science, the role of hubs in the structural and functional networks responsible for mediating the auditory hallucinatory experience. It is thus that functional neuroimaging, the philosophy of science, and network science have revolutionized hallucinations research, and in their wake fueled studies on more traditional aspects, such as the phenomenology, epidemiology, neuroanatomy, neurophysiology, and neuropsychiatry of auditory hallucinations.

Due to the intensification of empiric research throughout the past two decades, a rather detailed body of knowledge on auditory hallucinations is now available. In fact our understanding of those phenomena now surpasses that of the other types of hallucinations, perhaps even including that of the visual hallucinations. The sobering and rather pragmatic reason for this head start is that auditory hallucinations are often experienced in the form of discrete epochs of activity with "silent" episodes in between, both with a duration in the order of minutes, which makes them particularly well suited for research with the aid of functional neuroimaging techniques. As the success of such hallucinatory state studies stands and falls with the presence of a sufficiently contrasting change of signal during the time of scanning, other types of hallucination – which are mostly experienced either sporadically or for prolonged periods of time – tend to defy a similar approach.

On the theoretic side, various explanatory models have been devised for specific types of auditory hallucinations, and network science has been heralded as the prime candidate for a truly unifying model. The present chapter will summarize the current state of affairs in this rapidly expanding area, and propose various directions for future research. As we shall see, the source material is rich, voluminous, and revealing. However, throughout the present chapter it should be borne in mind that the area is still very much in flux. Despite all the new insights, auditory hallucinations go on to constitute an elusive group of phenomena with many challenging aspects yet to be fathomed, and novel treatment paradigms yet to come to fruition.

DEFINITION, CONCEPTUALIZATION, AND CLASSIFICATION

Definition

Auditory hallucinations are also known as acoustic hallucinations, aural hallucinations, hallucinations of

*Correspondence to: Jan Dirk Blom, M.D., Ph.D., Parnassia Academy, Kiwistraat 43, 2552 DH The Hague, The Netherlands. Tel: 0031-88-3570322, E-mail: jd.blom@parnassia.nl

hearing, and voices. They are traditionally defined as sounds experienced by a waking individual in the absence of a corresponding stimulus from the external world. Sometimes the alternative term auditory hallucinosis is used. Since its introduction by Wernicke (1900, p. 270), hallucinosis has variously been defined as: (1) a mental state characterized by continual hallucinations; (2) an abnormal condition characterized by hallucinations; (3) a psychiatric disorder involving hallucinations; (4) a syndrome, usually of organic origin, characterized by more or less persistent hallucinations; (5) a transient hallucinatory state accompanied by a clear sensorium and/or proper insight into its nature; and (6) a delusional state triggered by hallucinations (Blom, 2010, p. 229). Especially in clinical neurology, the term is sometimes used to denote hallucinations with an organic origin, but because of its multiple connotations, psychiatrists and hallucinations researchers do not employ it or, for that matter, the related term auditory hallucinosic syndrome. For the sake of continuity with the mainstream literature on the subject, the present chapter will also refrain from using those terms and stick to the more common term, auditory hallucination.

Conceptualization and demarcation

Auditory hallucinations are traditionally conceptualized as percepts – i.e., auditory percepts – experienced during the waking state. As such they are distinguished from auditory percepts experienced during the dream state, auditory percepts experienced while falling asleep or waking up (i.e., hypnagogia), auditory imagery (i.e., sounds imagined or remembered), and auditory pareidolia (also known as auditory illusion, i.e., the hearing of actual sounds which are either misperceived or misinterpreted). Although proper differentiation may not always be feasible in clinical practice, they are also distinguished from non-perceptual phenomena such as thought insertions and obsessive thoughts.

Alternatively, auditory hallucinations have occasionally been conceptualized as "reported auditory sensations" (Myers and Murphy, 1960). Or, as Berrios (2005) puts it, a class of utterances reporting subjective experiences "putatively perceptual in nature" which occur in the absence of an adequate external stimulus. Although our access to other people's hallucinations does indeed depend on third-hand verbal self-reports, in clinical practice it is customary to take for granted that those reports refer to actual perceptual phenomena. Therefore the present chapter will also designate auditory hallucinations as percepts rather than as reports of percepts.

Classification

The classification of auditory hallucinations has yielded numerous types and subtypes (for an overview, see Blom and Sommer, 2010). From a taxonomic point of view it is debatable whether also various closely related phenomena should be included, such as echo of reading, *Gedankenlautwerden*, polyacousis (Chapter 21), certain forms of auditory synesthesia (Chapter 22), tinnitus (Chapter 23), and palinacousis (Chapter 25). As all those auditory phenomena involve the hearing of sounds which are not – or no longer – objectively present, they would seem to fit in well with the class of auditory hallucinations. Alternatively, they are sometimes conceptualized as lying on a continuum of endogenously mediated auditory percepts, i.e., a continuum which also encompasses the auditory hallucination.

Irrespective of whether we adhere to a broad or a narrow definition, the class of auditory hallucinations constitutes a patchwork of interrelated phenomena which for various reasons have been singled out as clinically and/or scientifically meaningful. Proceeding from a broad definition, Table 24.1 offers an overview of those types of hallucination. In addition, it indicates their occurrence in the context of various clinical and nonclinical conditions.

PHENOMENOLOGIC CHARACTERISTICS

Perceived location

Auditory hallucinations are perceived as originating in the extracorporeal world (i.e., external auditory hallucinations), inside the head, or from an indefinite location. In the latter two cases, they are known as internal auditory hallucinations (Copolov et al., 2004). External auditory hallucinations are usually perceived on both sides, in which case they are called bilateral hallucinations, and occasionally from one side alone, when they are called unilateral hallucinations. But they may also be perceived as stemming from aberrant locations such as the belly, the knee, the nose, or the shoe, in the sense that they are described as being heard by the belly, the knee, the nose, or the shoe respectively, as if the voice hearer's ear were located in those places. Such bizarre percepts are appropriately called extracampine hallucinations (i.e., hallucinations occurring outside the regular field of perception; Bleuler, 1903). In a study by McCarthy-Jones et al. (2014) among 199 psychotic individuals experiencing auditory hallucinations, 47% reported internal hallucinations, 38% external ones, and 15% internal as well as external ones. A study by Kingdon et al. (2010) and one by Slotema et al. (2012) among a total number of 148 patients diagnosed with

Table 24.1

Types of auditory hallucination in relation to various conditions

	Psychotic disorder	Mood disorder	Hearing loss	Substance-induced	Alcoholic hallucinosis	Hypnagogic state	Normal wakefulness
Auditory sleep start							
Auditory synesthesia	+	+		+	+		+
Bilateral hallucination	+	+		+	+		+
Command hallucination	+	+	+	+	+		+
Echo of reading	+			+			+
External hallucination	+	+	+	+	+		+
Extracampine hallucination	+			+			
Functional hallucination	+	+	+	+	+		+
Gedankenlautwerden	+			+	+		+
Hypnagogic hallucination						+	
Hypnopompic hallucination						+	
Indirect *Gedankenlautwerden*	+	+		+	+		+
Internal hallucination	+	+	+	+	+		
Musical hallucination	+	+	+	+	+		+
Musical tinnitus			+				+
Non-verbal auditory hallucination	+	+	+	+	+		+
Palinacousis	+						
Psychomotor verbal hallucination (i.e., subvocalization)	+						+
Pulsatile tinnitus			+				+
Tinnitus			+				+
Unilateral hallucination	+		+				+
Verbal auditory hallucination	+	+	+	+	+		+

+ indicates that this type of hallucination is known to occur in the context of the condition listed.
Adapted from Blom and Sommer (2010).

psychotic disorder and/or borderline personality disorder found roughly comparable rates for internal and external hallucinations, with internal auditory hallucinations being reported by slightly more than 50% of the patients in both groups (Slotema and Kingdon, 2012). Although clinicians have long believed that voices experienced by individuals diagnosed with borderline personality disorder tend to be of the internal type – or even that they constitute some sort of imagery, designated sometimes as "pseudohallucination" or "quasi-hallucination" – those two studies indicate that their phenomenologic characteristics are similar to those experienced by psychotic individuals, and that the percepts at hand therefore fully deserve the predicate of hallucinations proper. Extracampine hallucinations have been described almost exclusively in the context of psychosis. Their prevalence is unknown, but they are believed to be extremely rare (Sato and Berrios, 2003). The same holds true for unilateral auditory hallucinations, which, due to their clinical association with ipsilateral ear lesions and contralateral brain lesions, always require careful neurologic as well as auxiliary examination (Almeida et al., 1993; Takebayashi et al., 2002).

Verbal auditory hallucinations

As regards their content, auditory hallucinations are traditionally divided into verbal and non-verbal ones. Verbal auditory hallucinations are also known as auditory verbal hallucinations, voice hallucinations, phonemes, hallucinated speech, and voices. They consist of hallucinated sounds which feature a linguistic content.

That content may vary from hearing one's name called out (a phenomenon usually designated as a simple misperception, due to its relatively low complexity) to complete sentences or even choruses of multiple voices which may or may not be continually present. Some special types of verbal auditory hallucination are command hallucinations (which convey an incentive or command; Erkwoh and Willmes, 2002), functional hallucinations (which are triggered by a shift of attention towards an actual stimulus such as running water or the hissing of steam pipes; Jaspers, 1997, p. 66), *Gedankenlautwerden* (hallucinated sounds which are either recognized as one's own thoughts or literally echo one's conscious thoughts; Sommer et al., 2009), echo of reading (a condition in which the reading of words or sentences is accompanied by a hallucinated echo of the same linguistic content; Morel, 1933), and psychomotor verbal hallucinations (also known as motor hallucinations, motor verbal hallucinations, muscular verbal hallucinations, and subvocalization), in which verbal auditory hallucinations may co-occur with subtle instances of motor activity within the larynx and/or vocal cords (Green and Preston, 1981). Among psychotic individuals who experience auditory hallucinations, command hallucinations have been reported by 67% of the patients (McCarthy-Jones et al., 2014). Functional hallucinations, *Gedankenlautwerden*, echo of reading, and indirect *Gedankenlautwerden* (a type of illusion in which actual, but often unintelligible, sounds are experienced as conveying one's own thoughts) would seem to be much rarer, but their prevalence has never been assessed systematically. When psychomotor verbal hallucinations were first described it was hoped that the laryngeal activity typical of those phenomena might hold the key to understanding all types of auditory hallucination, but later studies indicated that they constitute a relatively rare subgroup of the class as a whole.

Non-verbal auditory hallucinations

Non-verbal auditory hallucinations are also known as acoasms, acousmata, non-verbal hallucinations, and non-vocal auditory hallucinations. They lack the linguistic content characteristic of verbal auditory hallucinations, consisting instead of sounds ranging from ticking, clicking, or humming to the barking of dogs, the drone of airplane engines or all-out classic symphonies. In their study among psychotic individuals experiencing auditory hallucinations, McCarthy-Jones et al. (2014) found that non-verbal auditory hallucinations were reported in 32% of cases. Forty-six percent of those cases involved musical hallucinations, 43% ringing, 29% animal sounds, 27% clicks, 24% humming, 10% water, and 56% other sounds; some individuals experienced more than one type of non-verbal auditory hallucination. The substantial prevalence of musical hallucinations in that study is remarkable, especially if we consider that those phenomena have traditionally been regarded as relatively rare. However, it also confirms the clinical impression that they may well be more prevalent than the literature has until recently suggested (Berrios, 1990; Sacks and Blom, 2012).

Non-verbal auditory hallucinations occupy the end of the spectrum closest to the vicinity of tinnitus. As tinnitus involves the hearing of amorphous sounds not actually present, one might argue that it too belongs to the class of auditory hallucinations (Blom and Sommer, 2010). But even when conceptualized as a separate phenomenon, there are many instances in which proper demarcation may prove a challenging task. For instance, when a noise-type tinnitus or a tonal tinnitus gradually takes on a musical quality, it is referred to as musical tinnitus, even though phenomenologically it is then indistinguishable from a musical hallucination. Likewise, there are many instances of non-verbal sounds which can be classified at once as non-verbal auditory hallucinations or as tinnitus. This may perhaps strike one as a somewhat academic classificatory issue, but in clinical practice it may have consequences for the treatment of choice.

A notable exception to the rule of thumb that auditory hallucinations are experienced during wakefulness is the auditory sleep start or exploding-head syndrome, a rare phenomenon typically experienced within an hour or two of falling asleep (Pearce, 1988). It consists of a loud bang, explosion, roar, or ringing noise which appears to originate from a place deep inside the head. It is typically accompanied by a flash of light, called the visual sleep start. The cause of either phenomenon is unknown. Pearce (1988) was the first to describe this phenomenon, and in his paper he speculated that "a momentary (almost ictal) disinhibition of the cochlea or its central connections in the temporal lobes, or a sudden involuntary movement of the tympanum or tensor tympani, might be the explanation." But with empiric studies in this area yet to be initiated, other possible causes cannot be ruled out.

Musical hallucinations

Musical hallucinations consist of songs, tunes, melodies, harmonics, rhythms, and/or timbres which are perceived in the absence of any actual music. Individuals experiencing those phenomena initially tend to believe that there is a band or a radio playing nearby, and they often go around asking whether someone could turn down the volume. After that initial phase of surprise and disbelief, most people come to accept within a couple of days that the music originates in their head. But out of fear for being perceived as "crazy" or demented, most of them keep the experience to themselves. This may partly explain why the reported prevalence of musical hallucinations has historically been so low (another explanation being that only few clinicians have the habit of systematically assessing their presence). Musical hallucinations differ from "earworms" (Sacks, 2006), or tunes that go round in the head – familiar to us all, by their unmistakable perceptual quality. Moreover, they tend to resist termination or alteration by force of will. They are often continuously present, and although their content does not tend to be threatening or frightening in nature, they may interfere significantly with one's daily functioning. They can be heard inside the head, but more often they are described as if coming from outside the head. When they are repetitive and stereotyped in nature they are called stable musical hallucinations, as opposed to the group of complex musical hallucinations which are more elaborate and more prone to change. Thus various types of musical hallucination can be discerned. Table 24.2 provides an overview of those types, as well as an overview of related phenomena which involve the hearing of music.

ETIOLOGY

Auditory hallucinations are experienced in the context of various psychiatric and neurologic conditions, but also in the context of various intoxications, withdrawal syndromes, metabolic conditions, and even in the absence of any demonstrable pathology. Table 24.3 provides a schematic overview of those various conditions. It might be tempting to designate those conditions as etiologic factors, but the etiology of auditory hallucinations is a rather complex issue that is all but fully understood. As auditory hallucinations can be experienced in the context of numerous clinical and non-clinical conditions, their occurrence in the context of any given condition does not necessarily imply that that condition can *de facto* be considered the agent responsible for initiating them. For example, a verbal auditory hallucination experienced by a patient diagnosed with schizophrenia may be mediated by the hyperactive dopaminergic signal transduction considered characteristic of that condition, but it may also be triggered by the smoking of marijuana, prolonged social isolation, or the recent death of a parent. Moreover, as auditory hallucinations are also experienced by otherwise healthy individuals, it may be that there are etiologic mechanisms at play which have nothing to do whatsoever with the clinical condition (or conditions) diagnosed in this particular patient. In addition, it should be noted that auditory hallucinations tend to show little respect for the boundaries we draw between various clinical conditions. As a consequence, it is not always clear whether known associations with clinical conditions are fully deserving of the predicate "etiologies." With that important restriction in mind, the following can be said about the etiology of auditory hallucinations.

Auditory hallucinations, notably those of the verbal type, tend to be associated primarily with psychosis. And yet, although it is true that some 70% of all individuals diagnosed with schizophrenia report verbal auditory hallucinations (Mueser et al., 1990), they may also be

Table 24.2

Glossary of musical hallucinations and related phenomena

Phenomenon	Characterization
Musical hallucination (musical hallucinosis, musical-ear syndrome)	An auditory hallucination characterized by songs, tunes, melodies, harmonics, rhythms, and/or timbres
Auditory Charles Bonnet syndrome	A musical hallucination in the presence of severe deafness and in the absence of other pathology
Idiopathic musical hallucination	A musical hallucination occurring in the absence of any type of associated pathology other than hearing loss
Symptomatic musical hallucination	A musical hallucination occurring in the presence of any type of associated pathology other than hearing loss
Auditory illusion (auditory pareidolia)	An auditory percept resulting from a misrepresentation or misinterpretation of random auditory stimuli, which may or may not have a musical quality
Musical illusion	A paradoxic auditory illusion created with the aid of actual musical sounds
Musical tinnitus	A musical hallucination with evolution from "ringing in the ears"
Musical palinacousis	The continued perception or reperception of an actual piece of music
Musical imagery	An imagined or remembered piece of music, having no perceptual characteristics
Earworm	A tune going round in the head, having no perceptual characteristics

After Sacks and Blom (2012).

Table 24.3

Schematic overview of various clinical and non-clinical conditions in the context of which auditory hallucinations are experienced

Psychiatric disorders	Neurologic disorders	Otologic disorders	Systemic disease	Substance intoxication and/or withdrawal	Physiologic
Psychotic disorders	Delirium	Presbycusis	Neurosyphilis	Alcohol	Idiopathic
Bipolar disorder	Delirium tremens	Tinnitus	AIDS	Cannabis	Hypnagogia
Depressive disorder	Dementia	Severe deafness	Thyrotoxicity	Cocaine	Bereavement
Dissociative disorder	Parkinson's disease		Cushing's disease	Amphetamines	Social and/or sensory isolation
Borderline personality disorder	Brain tumors		Hyperhomocysteinuria	Hallucinogens	Intensive meditation
	Paraneoplastic neurologic syndromes		22q11.1 deletion syndrome	Antidepressants	Posthypnotic
	Limbic encephalitis		Sarcoidosis	Anticholinergics	Sleep deprivation
	Prion disease			Antibiotics	Aging
	Intracerebral arteriovenous malformations				
	Epilepsy				
	Exploding-head syndrome				

AIDS, acquired immunodeficiency syndrome.

experienced in the context of unipolar depressive disorder, bipolar disorder, dementia, dissociative disorder, borderline personality disorder, and occasionally other psychiatric disorders (Larøi et al., 2012). In addition, they may be experienced in the context of bereavement, sensory or social isolation, sleep deprivation, intensive meditation, psychotic disorder due to alcohol abuse, alcohol withdrawal, delirium, delirium tremens, the use of cannabis or other illicit substances, and as a side-effect of antidepressants, anticholinergics, and numerous other pharmaceuticals (Larøi et al., 2012; Sommer et al., 2012a). Auditory hallucinations may also be experienced in the context of systemic somatic disease (i.e., organic psychosis, a condition which constitutes 3% of all the psychoses; Falkai, 1996). Some examples of those systemic somatic diseases are neurosyphilis, human immunodeficiency virus (HIV) infection, thyrotoxicity, and Cushing's disease. Other associations with somatic conditions include brain tumors, paraneoplastic neurologic syndromes, limbic encephalitis, prion disease, hyperhomocysteinuria, 22q11.1 deletion syndrome, sarcoidosis, and intracerebral arteriovenous malformations (Sommer et al., 2012a). Hallucinations occurring in the context of Parkinson's disease tend to be visual in nature, but auditory ones are reported in no less than 20% of cases (Fénelon and Alves, 2010). In the context of epilepsy they tend to occur more often, either in the form of an aura preceding an epileptic seizure or in the context of post- or interictal psychosis (see Chapter 26 for additional details). In most of those cases they take the form of non-verbal auditory hallucinations, but verbal auditory hallucinations have also been reported (Kasper et al., 2010). Most of them are simple and repetitive in nature, and yet more extensive linguistic utterances have also been described, as well as musical hallucinations (Sacks and Blom, 2012).

Occurrence in the absence of pathology

Auditory hallucinations primarily evoke associations with psychiatric and, to a lesser extent, neurologic and otologic disease, but they are also experienced by

Table 24.4

Prevalence rates of hallucinations in the general population, as found in 10 cross-sectional studies

Study	Number of participants	Prevalence of hallucinations
United Kingdom: SPR (Sidgwick et al., 1894)*	17 000	9.9%
France (Marillier, 1894)*	3393	20.01%
United States: ASPR (James, 1894)*	6311	13.5%
Germany (Von Schrenck-Notzing, 1894)*	625	8.16%
United States: ECA Program baseline (Tien, 1991)	18 572	13.0%
United States: ECA Program follow-up (Tien, 1991)	15 258	11.1%
United Kingdom/Germany/Italy (Ohayon, 2000)	13 057	21.2%
United States (Olfson et al., 2002)	1005	13.0%
United Kingdom (Johns et al., 2004)	8580	4.2%
The Netherlands: NEMESIS (Van Os et al., 2005)	7076	6.1%

*Summarized in Parish (1897).
For full details of the references listed in this table, see Blom (2012).
Reproduced from Blom (2012).

10–15% of all individuals in the general population, i.e., in the absence of any demonstrable pathology (Beavan et al., 2011). That lifetime prevalence rate has been quite consistent throughout the past 120 years, during which time 10 large epidemiologic surveys of hallucinations were carried out in the general population of various western countries (Table 24.4). The frequency of auditory hallucinations in healthy voice hearers tends to be low, but they may also be experienced throughout the day and for years on end (Daalman et al., 2011). In either case the majority of those phenomena are appraised as benevolent in nature. By comparing 111 healthy voice hearers with 118 voice hearers diagnosed with a schizophrenia spectrum disorder, Daalman et al. (2011) found that healthy voice hearers have a higher degree of control over their voices, and that they tend to experience them from a significantly younger age onwards. The authors also found that a younger age of debut, a higher degree of control over the voices, a relatively low frequency, and a benevolent appraisal of their content can in 92% of the cases accurately predict the absence of psychotic disorder (i.e., in terms of the criteria listed in the *Diagnostic and Statistical Manual of Mental Disorders*: American Psychiatric Association, 2013), while benevolent appraisal alone can do so in 88% of cases (Daalman et al., 2011). Other phenomenologic characteristics of auditory hallucinations such as loudness, the number of voices, and perceived location were quite similar among healthy and psychotic voice hearers, and proved to be of little aid in predicting the presence or absence of psychotic disorder.

A group of phenomena experienced by an even larger number of individuals in the healthy population are hypnagogic and hypnopompic hallucinations, collectively known as hypnagogia. Hypnagogic hallucinations, i.e., perceptual phenomena experienced during the transitional phase from wakefulness to sleep, have reported incidence rates of 75% and higher in the general population (Mavromatis, 1987, p. 5). They tend to consist of elevator feelings, geometric visual hallucinations, and facial hallucinations (hence the synonym "faces in the dark") but they may also feature non-verbal and even verbal auditory phenomena, such as replays of recent conversations, hearing one's name called out, the sound of one's child calling or crying, sentences as if read from a book, and indistinct talking. Occasionally musical hallucinations have also been described. Hypnopompic hallucinations, on the other hand, which occur during the transitional phase from sleep to wakefulness, would seem to be less prevalent, and to consist mainly of dream images that linger on during the first seconds or minutes of wakefulness. It appears to be quite rare for them to consist of auditory percepts, although hypnopompic musical hallucinations have also been described. Because of their association with sleep, neither hypnagogic nor hypnopompic hallucinations tend to be counted as hallucinations proper, and in the absence of any additional indicators of parasomnia their clinical relevance is considered negligible.

PATHOPHYSIOLOGY

As one might expect with such a variegated group of phenomena, the pathophysiology of auditory hallucinations is far from straightforward. And yet, regardless of all the variation, a central role in their mediation is attributed to the auditory nervous system. It has long been suspected that many parts – if not any part – of that system may be involved in the mediation of auditory hallucinations, and hence historically various explanatory models

Table 24.5

Historic hypotheses regarding the mediation of auditory hallucinations (AHs)

Hypothesis	Characterization	Reference
Centripetal model	Suggests that AHs may arise from any part of the auditory system, and emphasizes the possible involvement of the peripheral auditory system	Braun et al. (2003)
Reperception model	Conceptualizes AHs as reperceptions of previously experienced auditory percepts; emphasizes the involvement of memory function and limbic structures	Copolov et al. (2003)
Brainstem auditory hallucinosis model	Conceptualizes AHs as a variant of peduncular hallucinosis; emphasizes the involvement of the pedunculus cerebri, midbrain, pons, or diencephalon	Cascino and Adams (1986)
Deafferentiation model	Attributes AHs to spontaneous activity within deafferentiated cortical areas, notably in conjunction with hearing loss; also known as the auditory Charles Bonnet syndrome	Cole et al. (2002)
Aura model	Attributes AHs occurring in the context of migraine or epilepsy to paroxysmal neuronal activity affecting the superior temporal gyrus	Kechid et al. (2008)
Inner-speech model	Conceptualizes AHs in terms of a misattribution of one's own thoughts to an external source; emphasizes the involvement of Wernicke's and Broca's areas	Mechelli et al. (2007)
Corollary discharge model	Basically a variant of the inner-speech model; attributes the mediation of AHs to a default in the corollary discharge signal from frontal speech production areas towards speech reception areas	Ford et al. (2001)
Lateralization model	Attributes AHs to a disinhibition of non-dominant speech areas, notably within the right hemisphere, due to aberrant cerebral lateralization	Sommer et al. (2008)

Adapted from Blom and Sommer (2010).

(Table 24.5) have been competing for first place, notably the perceptual release theory, the reperception model, the deafferentiation hypothesis, the cortical excitation hypothesis, the inner-speech hypothesis, and the lateralization hypothesis (Blom and Sommer, 2010; McCarthy-Jones, 2012). Throughout the next section it should be held in mind that the majority of the findings to be presented stem from studies among voice hearers diagnosed with psychotic disorder, and that it is as yet unclear whether those findings allow for proper generalization to other patient groups and/or types of auditory hallucination. Moreover, it should be noted that the pathophysiology of musical hallucinations has not been studied very extensively, and that the literature on other non-verbal auditory hallucinations largely overlaps with that on tinnitus (Chapter 23).

Evidence for a functional auditory network

The brain network responsible for the mediation of verbal auditory hallucinations comprises the classic and non-classic ascending pathways, the descending systems, as well as a number of anatomic structures connected with the non-classic pathways. As summarized by Jardri et al. (2011) in their meta-analysis of functional neuroimaging findings, the areas involved include Broca's area and its contralateral homolog, the anterior insula, the frontal operculum, the precentral gyrus, the middle temporal gyrus, the superior temporal gyrus (i.e., Wernicke's area) and its contralateral homolog, the bilateral inferior parietal lobule, and the hippocampal and parahippocampal regions. Figure 24.1 provides a graphic overview of those regions, based on a model-based analysis of functional magnetic resonance imaging (fMRI) blood oxygen level-dependent (BOLD) signals obtained from individuals who were diagnosed with psychotic disorder, and who experienced intermittent verbal auditory hallucinations (Looijestijn et al., 2013). Figure 24.2 provides a somewhat similar overview, but one that is based on a model-free analysis of those same fMRI data (Looijestijn et al., submitted). The two activation maps show a significant overlap, but due to its increased number of degrees of freedom, the model-free analysis yields a larger number of structures involved. The significance of this will be discussed below.

In terms of the network science paradigm, the networks outlined in Figures 24.1 and 24.2 are designated as functional networks. The fMRI BOLD signal reflects local increases in cerebral oxygen consumption, and the neurovascular-coupling hypothesis stipulates that such increases are indicative of changes in brain function (Villringer, 1997). When those changes co-occur with reported verbal auditory hallucinations (in this case, signaled with the aid of balloon presses by the subjects inside the scanner, indicating the onset as well as the

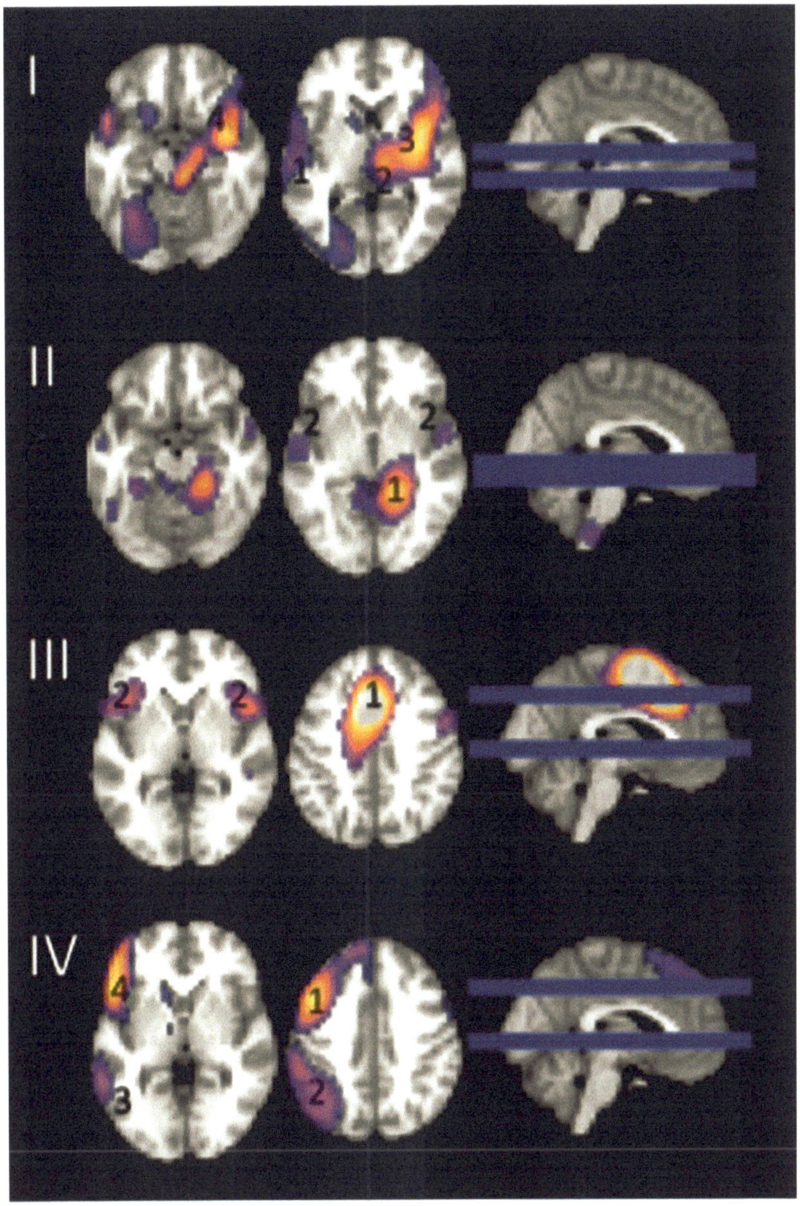

Fig. 24.1. Brain regions active during the conscious experience of verbal auditory hallucinations, as derived from a model-based analysis of functional magnetic resonance imaging (fMRI) blood oxygen level-dependent (BOLD) signals during within-scanner reporting of hallucinations with the aid of balloon presses. The purple-yellow color coding indicates hallucinatory activity, with the lighter color referring towards Z_{max}. Multiple brain regions were found to be involved, including bilateral inferior and middle frontal areas (1), bilateral insula (2), anterior cingulate gyrus (3), and predominantly left-sided superior temporal gyrus (4), as well as a motor network which included left motor cortex (5) and the right cerebellum (6), the latter most likely corresponding with the balloon presses. (Courtesy of Dr. Jasper Looijestijn, Parnassia Psychiatric Institute, The Hague, the Netherlands, based on fMRI scans obtained at the University Medical Center Utrecht, Utrecht, the Netherlands.)

cessation of hallucinatory epochs), they are taken as reliable indicators of hallucinatory activity.

Given the prominent role of the auditory cerebral cortex in the conscious perception of sounds, it is perhaps unsurprising to find that that bilateral structure also plays a significant role in the perception of hallucinated sounds. With the aid of fMRI recordings, Dierks et al. (1999) demonstrated activity in the primary auditory cortex concurring with verbal auditory hallucinations, and Shergill et al. (2000) in the bilateral temporal cortices. However, Copolov et al. (2003) were unable to confirm those results in their positron emission tomography (PET) study, and instead found activity in the parahippocampal gyri. Diederen et al. (2010) also found significant

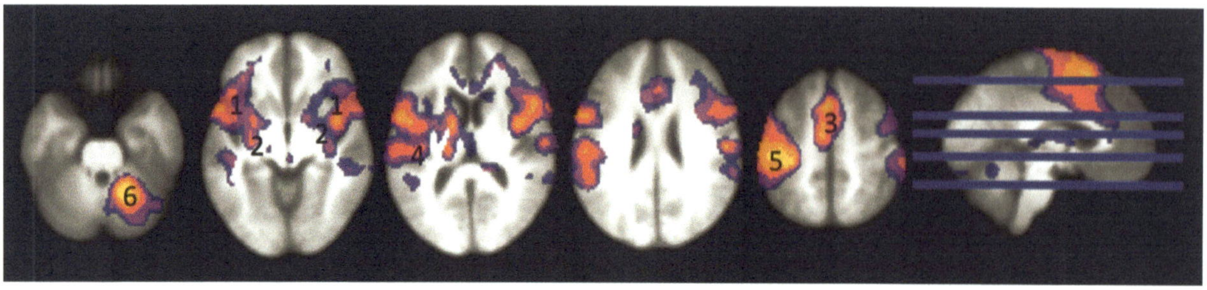

Fig. 24.2. Brain regions active during the conscious experience of verbal auditory hallucinations, as derived from a model-free analysis of functional magnetic resonance imaging (fMRI) blood oxygen level-dependent (BOLD) signals during within-scanner reporting of hallucinations with the aid of balloon presses. The method employed was a data-driven independent-component analysis (ICA). It indicates the presence of four functional networks associated with the mediation of verbal auditory hallucinations, derived from *post hoc* linkage of group ICA, functional networks with hallucination timings, and psychometrics. The purple-yellow color coding indicates hallucinatory activity, with the lighter color referring towards Z_{max}. Component I consists of predominantly left-sided superior temporal gyrus (1), and a strongly right-oriented medial temporal (2), insula, putamen, thalamus (3), and inferior frontal network (4). Component II consists of left-sided (para)hippocampal regions (1), and bilateral superior temporal regions (2). Component III consists of anterior cingulate cortex (1) and bilateral Broca's area (2). Component IV consists of left-oriented dorsolateral prefrontal cortex (1), widespread parietal cortex (2), middle temporal regions (3), and inferior frontal gyrus (4). Components I and III show large overlap with the activation patterns in Figure 24.1. (Courtesy of Dr. Jasper Looijestijn, Parnassia Psychiatric Institute, The Hague, the Netherlands, based on fMRI scans obtained at the University Medical Center Utrecht, Utrecht, the Netherlands.)

activity in the parahippocampal gyrus, but that activity typically preceded the conscious experience of verbal auditory hallucinations. Careful reconstructions of the time series of fMRI findings in their patients indicated that deactivation of the parahippocampal gyri takes place some 6 seconds prior to the conscious experience of verbal auditory hallucinations, and that the signal then spreads into the direction of the language areas and, notably, their contralateral homologs (Diederen et al., 2010). Part of those findings confirmed the prior results of a single-photon emission computed tomography (SPECT) study by McGuire et al. (1993), which found increased activity in Broca's area, the anterior cingulate, and left temporal cortex, an fMRI study by Hoffman et al. (2007), which indicated that the conscious perception of verbal auditory hallucinations coincides with bilateral activity in frontal and temporal areas, and a study by Sommer et al. (2008), which found marked activity in the right inferior frontal area and the right superior temporal gyrus. By comparing the fMRI findings of 21 healthy voice hearers with those of 21 voice hearers diagnosed with psychotic disorder (who were matched for the frequency and duration of hallucinatory epochs), Diederen et al. (2011) found similar patterns of activity in the language areas and their contralateral homologs, thus indicating that the verbal auditory hallucinations experienced by healthy voice hearers are physiologically similar to those of patients diagnosed with psychotic disorder, and, more specifically, that the mediation of verbal auditory hallucinations would seem to depend on the very resources involved in the normal processing of linguistic information.

The speech perception and speech production areas, usually located on the left in right-handed individuals, are held responsible for providing the linguistic content of verbal auditory hallucinations. The same holds true for their contralateral homologs, even though the latter presumably have relatively limited capacities for linguistic processing, and in spite of the fact that their activity normally tends to be inhibited by the language areas via reciprocal callosal connections (Bloom and Hynd, 2005) during the execution of language functions such as reading, speaking, and listening to speech. As proposed by Sommer et al. (2008), the involvement of the non-dominant language areas may well hint at an insufficient inhibition by the dominant language areas, and perhaps also at a structural problem of language lateralization. The ensuing lateralization hypothesis serves to explain why many voice hearers diagnosed with psychotic disorder experience relatively simple and stereotyped propositions (being mediated by the right hemisphere with its limited linguistic capacities), especially during the advanced stages of their illness, and also why engaging in conversation or listening to music – activities that promote a competitive use of auditory areas – may aid to suppress those hallucinated voices. The involvement of the parahippocampal gyrus, on the other hand, would seem to indicate that memory, and presumably also emotion, are important factors during the preparatory phase of verbal auditory hallucinations in individuals diagnosed with psychotic disorder.

The role of memory in the mediation of hallucinatory activity complies with the reperception model as proposed by classic authorities such as Kahlbaum (1886) and Penfield (1975). And yet the majority of functional imaging findings would seem to be in line with Hughlings-Jackson's Darwinian thesis that normal activity of the brain's "lower" (i.e., evolutionary "older") centers may present in the form of positive symptoms when "higher" (i.e., evolutionary "younger") structures cease to exert their inhibitory function (Hughlings-Jackson, 1888). Devoid of its Darwinian connotations, that model was reintroduced by West (1962), and given the name perceptual release theory (also known as dream intrusion, dual-input model, and seepage theory) to designate hallucinations as release phenomena originating from aberrant activity in the limbic system, the diencephalon, and other subcortical centers. As indicated above, especially the work by Diederen et al. (2010) on the preparatory deactivation of the parahippocampal gyri can be interpreted as an empiric corroboration of that hypothesis. In a similar vein, Hoffman et al. (2008) found prehallucination activity in the left anterior insula and the right middle temporal gyrus, as well as deactivation of the anterior cingulate and parahippocampal gyri, whereas Shergill et al. (2004) failed to find any such deactivation, but instead found increased activity 6–9 seconds prior to the onset of verbal auditory hallucinations in the left frontal and right middle temporal gyri, which the authors interpreted as an indication of inner speech being generated prior to the conscious experience of verbal auditory hallucinations. Other subcortical structures, including the thalamus, were found to be involved by Silbersweig et al. (1995) and others.

Although the results from studies such as these are not unequivocal, they do indicate that the mediation of verbal auditory hallucinations may well be equally dependent on the temporal course of prehallucinatory activity as on the activity registered in cortical areas during hallucination epochs. More research is needed to establish which time courses play a role under which circumstances, but the very notion of time courses would seem to be paramount to our understanding of hallucinatory activity.

The components of the auditory hallucination network identified with the aid of neuroimaging studies largely overlap with the modular components targeted as candidate regions in classic models. And yet neuroimaging studies have added various novel components to the classic canon, including the insula (Sommer et al., 2008), the inner-speech network (Vercammen et al., 2011), and the auditory "where" pathway, which comprises the left planum temporale and prefrontal regions. As indicated by Looijestijn et al. (2013), in an fMRI study among 52 right-handed, psychotic voice hearers, the exteriorization of verbal auditory hallucinations would seem to depend on increased activity of the left-sided middle planum temporale and the right-sided middle frontal gyrus. Although the left planum temporale is also known for its role in phonologic processing, its role in projecting hallucinated voices outwards into auditory space – and thus in lending them the perceptual quality of externally mediated sounds – is in conformity with the structure's previously established function in the categoric analysis of auditory signals.

Evidence for a structural auditory network

The functional auditory network presented above currently constitutes our most reliable indicator of the brain network involved in the mediation of verbal auditory hallucinations. The involvement of that network has been largely corroborated by structural imaging studies, in the sense that they also show up on scans obtained with the aid of techniques such as MRI and diffusion tensor imaging. It might be tempting to infer from those findings that the functional aberrations involved in the mediation of verbal auditory hallucinations are therefore caused by structural aberrations in the neural networks themselves, but to draw that conclusion would be premature at this stage. A review of studies on postlesion hallucinations by Braun et al. (2003) indicates that the lesions causing hallucinations tend to be located in the pathways of the sensory modality involved (i.e., lesions in the auditory pathways in the case of auditory hallucinations, and so on), and the authors suggest that the damaged tissue must therefore originally have had an inhibitory function that prevented the pathway from engaging in the mediation of hallucinatory activity. But structural imaging studies of individual voice hearers do not tend to show any signs of such structural pathology. As a consequence, the mapping of structural abnormalities associated with auditory hallucinations has mainly been a matter of comparing brain scans of groups of individuals with and without auditory hallucinations (Allen and Modinos, 2012). The findings thus obtained in groups of psychotic individuals experiencing verbal auditory hallucinations include volumetric decreases in the superior temporal sulcus and other sensory areas, in non-sensory areas such as prefrontal cortex, insula, amygdala, anterior cingulate cortex, and the parahippocampal gyri, as well as an enhanced structural integrity of white-matter tracts (as deduced from enhanced fractional anisotropy found in diffusion tensor imaging studies) such as the arcuate fasciculus and inferior longitudinal fasciculus, which both play a role in language processing (Allen and Modinos, 2012). By and large, those findings are consistent with the functional networks outlined in the previous section. They would seem to justify hypotheses involving a reduced connectivity between frontal and temporal regions, notably the auditory cortical areas, language-related areas, and areas that have a function in memory retrieval (Clos et al., 2014). But as most of the structural studies carried out so far involve region-of-interest studies, structural aberrations located outside those regions may well have been systematically neglected. As a consequence, the structural network responsible for the mediation of auditory hallucinations may well be larger than we currently believe it to be, although at present it would seem to comply roughly with the functional network identified so far.

Pathophysiologic processes

It would seem that pathophysiologic models of auditory hallucinatory activity as formulated throughout the history of the neurosciences all deserve a role in the current auditory hallucination paradigm. As a corollary, perceptual release, reperception, deafferentiation, cortical excitation, inner speech, and default lateralization are no longer envisaged as competing doctrines or mutually exclusive mechanisms, but rather as mechanisms applicable to certain patients and to certain types of auditory hallucination. As regards the pathophysiologic processes that govern those mechanisms, relatively little is certain as yet. Cortical excitation is by definition a matter of aberrant electrophysiologic activity, but the other mechanisms may well be indebted to a variety of processes. As suggested by the dopamine hypothesis of schizophrenia, hallucinations occurring in the context of psychosis may well be mediated by hyperactive dopaminergic signal transduction, notably in the mesolimbic regions (Carlsson and Carlsson, 1990). Although there is a voluminous body of work on the role of dopamine and its various receptor subtypes in hallucinations and other psychotic symptoms (Madras, 2013), and although the efficacy of phenothiazines and other substances marketed as antipsychotics would seem to suggest that this neuromodulator does indeed play a role in their mediation, the evidence for this is indirect and in need of further research. Likewise, the role of other neurotransmitters and neuromodulators is in need of further research.

The same holds true for the involvement of default corollary discharge signals. As originally suggested by Feinberg (1978), a default functioning of the corollary discharge mechanism may well play a role in the mediation of auditory hallucinations. Ford et al. (2001, 2010), among others, studied this basic feedforward system with the aid of electroencephalographs and magnetoencephalography in patients diagnosed with psychotic disorder, and noted that those experiencing verbal auditory hallucinations show disruptions in the signal sent from frontal areas involved in thought generation to temporal areas involved in speech reception, thus failing to tag those signals as self-generated (van Lutterveld and Ford, 2012). That failure of the corollary discharge signal has been interpreted as supportive of the inner-speech model, which suggests that verbal auditory hallucinations may result from an inability to recognize one's own thoughts as inner speech. Unfortunately, however, it is as yet unclear in how far those findings can be generalized to other types of auditory hallucination and to different population groups.

Pathophysiology of musical hallucinations

The pathophysiology of musical hallucinations has not been studied on a scale comparable to that of verbal auditory hallucinations. This is mainly due to the fact that musical hallucinations have a significantly lower prevalence, and that it is therefore more difficult to find large-enough groups of patients for functional neuroimaging studies. That said, the few functional imaging studies carried out so far with the aid of PET and fMRI indicate that musical hallucinations, not unlike actual musical perception, co-occur with activity in an extensive network of brain areas which includes auditory areas, motor cortex, visual areas, basal ganglia, brainstem, pons, tegmentum, cerebellum, hippocampi, amygdala, and perhaps even the peripheral auditory system (Gordon, 1999; Sacks and Blom, 2012).

Apparently the network as a whole – which may well include many additional brain areas not yet identified as relevant – is capable of promoting and/or mediating musical hallucinations (Stewart et al., 2006). The factors that precede and promote musical hallucinations may not be as diverse as those in verbal auditory hallucinations, but nevertheless they show considerable variation. In temporal-lobe epilepsy, for example, the musical network is stimulated directly by an irritable focus. Especially instances of repetitive, stereotyped musical hallucination tend to be attributed to that mechanism. But what seems to occur in most cases of musical hallucination is a release of activity in the musical network when normal mechanisms of inhibition or constraint are weakened (Keshavan et al., 1992). In this sense the pathophysiologic mechanism underlying the majority of musical hallucinations is considered more or less similar to that underlying many types of verbal auditory hallucination. In either case, however, we should take heed not to draw any premature conclusions, and acknowledge that future research may well shed a different light on the alleged role of those mechanisms.

Contrary to the situation in verbal auditory hallucinations, the most prevalent clinical condition associated with musical hallucinations is auditory deprivation or deafness. Thus the musical hallucinations of – notably the elderly – deaf are considered analogous to the visual hallucinations of the visually handicapped, also known as Charles Bonnet syndrome. In both cases, it is as if the networks normally employed in musical or visual perception escape into a spontaneous and autonomous activity of their own if there is no longer adequate perception to constrain them. In that sense, most (although certainly not all) instances of musical hallucination would seem to fulfill the criteria of deafferentiation phenomena.

TREATMENT

Auditory hallucinations do not always require treatment. They are experienced by a substantial number of individuals who lack any demonstrable pathology, and many of those individuals value their hallucinations as positive or at least find that the burden caused by them does not outweigh the risks and side-effects of treatment (Sommer et al., 2010). As a consequence, a first step should always be to carefully assess the need for treatment (Sommer and Blom, 2012). When there is no such need, it may suffice to offer reassurance and some basic psychoeducation aimed at normalization, acceptance, and awareness of any circumstances that should prompt the voice hearer to apply for re-evaluation. In addition, it may be helpful to provide some information on neurobiologic models of hearing voices, on their occurrence in the general population, and on the numerous ways in which others have succeeded in coping with them (see, for example, Romme et al., 2009). As advocated by the Hearing Voices Movement (see section on non-pharmacologic treatment methods, below), there are even ways to help people find peace with their auditory hallucinations without necessarily attempting to diminish their frequency or intensity.

However, if there is a need – and/or a medical necessity or legal obligation – to treat auditory hallucinations, the method of choice is determined primarily by the clinical disorder or syndrome in the context of which they are experienced. When they occur in the context of intoxications or neurologic or other somatic conditions, treatment will primarily be targeted at alleviating those conditions. Thus hallucinations co-occurring with epilepsy are treated differently from those co-occurring with cannabis use, AIDS, Parkinson's disease, or Lewy body disease. But even in such cases symptomatic treatment may be required as an augmentation strategy. Currently various methods are being explored for the symptomatic treatment of auditory hallucinations in their own right, notably repetitive transcranial magnetic stimulation. In what follows, however, the various conventional treatment options for auditory hallucinations will be summarized. As we shall see, most of those treatment methods are known for their effects on psychotic symptoms in general, and only few of them have been assessed in double-blind, randomized controlled trials aimed at establishing their efficacy on auditory hallucinations *per se*.

Pharmacotherapy

The method of choice for the symptomatic treatment of auditory hallucinations is antipsychotic medication, especially when the hallucinations at hand are experienced in the context of psychotic disorder, mood disorder, or delirium. Although clinical trials that compare the efficacy of various antipsychotic drugs for the sole and specific indication of auditory hallucinations are lacking, results from the European First Episode Schizophrenia Trial (EUFEST) allow for comparison of the effects of five antipsychotic agents on the severity of hallucinations experienced by patients diagnosed with psychotic disorder (Kahn et al., 2008). As suggested by the EUFEST data, which were obtained from 498 individuals with a first psychotic episode, there are no significant differences in efficacy between haloperidol, olanzapine, ziprasidone, quetiapine, and amisulpride in their potential to reduce the frequency or severity of hallucinations. They all yield a strong reduction in symptom severity during the first month, with further reductions taking place during the first 12 months of treatment. It should be noted, however, that haloperidol was found to yield a slightly less pronounced effect than the other antipsychotic agents (Sommer et al., 2012b). Regarding their side-effects, antipsychotics roughly fall into two groups, i.e., those predominantly inducing weight gain, metabolic syndrome, and sedation (i.e., quetiapine, olanzapine, and clozapine), and those primarily associated with extrapyramidal symptoms, such as acathisia, dystonia, and parkinsonism (i.e., haloperidol and ziprasidone). As a consequence, the choice for a specific type of antipsychotic is determined more often on the basis of potential side-effects than on any alleged differences in therapeutic efficacy.

If the drug of first choice fails to yield a significant decrease in symptom severity during the first 2–3 weeks, a switch is recommended. Traditionally such a switch is advocated 4–6 weeks into treatment, but relatively recent studies indicate that antipsychotics start to exert their efficacy during the first hours to days rather than weeks, and that prolonged periods without switching do not tend to yield any additional results (Agid et al., 2007). If necessary, a different antipsychotic agent tends to be chosen from a group with a different receptor profile, even though there is no direct evidence that that strategy will lead to superior results (Buchanan et al., 2010). When the second antipsychotic also fails to yield any significant results, a switch to clozapine is warranted. Apart from the risk of sedation and metabolic syndrome as side-effects, significant drawbacks of the use of clozapine are an increased risk of leukocytopenia and agranulocytosis (which in case of infectious diseases may lead to unnecessary and even fatal complications) and heart disease (notably pericarditis, myocarditis, and cardiomyopathy). Therefore treatment with the aid of clozapine is only considered safe when blood tests are run during the full course of treatment, and when patients are being monitored for any signs of clozapine-induced heart disease, especially during the first 15 days of treatment (Layland et al., 2009). In those cases where hallucinations prove to be resistant even to clozapine at adequate therapeutic blood levels for a prolonged period of time, clinicians tend to advocate various pharmacologic augmentation strategies with the aid of therapeutics ranging from selective serotonin reuptake inhibitors to lithium, sodium valproate, lamotrigine, benzodiazepines, risperidone, haloperidol, quetiapine, and aripiprazole. Case reports and randomized controlled trials among modest numbers of patients suggest a superior result of such combinations. A recent meta-analysis which quantitatively summarized 29 randomized controlled trials involving the pharmacologic augmentation of clozapine among a total number of 1066 patients indicates a modest improvement of total symptom severity for augmentation with lamotrigine, sulpiride, citalopram, and the glutamatergic agonist CX516 (Sommer et al., 2011). On the basis of the results of that study we must conclude, however, that pharmacologic augmentation in clozapine therapy is not yet supported by much convincing evidence from the literature (Sommer and Blom, 2012).

Cases of tinnitus (mostly experienced in the absence of any signs of psychosis) are treated differently (see Chapter 23), and the same holds true for musical hallucinations. In cases of musical hallucination, any underlying pathology, if present, should be treated first. In the absence of any identifiable pathology, the symptomatic treatment of musical hallucinations tends to involve the prescription of anticonvulsant, antidepressant, or antipsychotic drugs. In the literature a few case reports can be found on successful treatments with the aid of the anticonvulsants gabapentin, carbamazepine, and valproic acid, respectively, as well as with the antidepressant clomipramine. However, none of those drugs has yet shown a broad usefulness, and the majority of the cases of musical hallucination appear to be quite refractory to current treatment methods (Sacks and Blom, 2012).

Non-pharmacologic treatment methods

Non-pharmacologic treatment methods for auditory hallucinations include psychotherapy, various protocols aimed at teaching patients to cope with their voices, self-help groups, and electroconvulsive therapy (ECT). Some academic institutions also offer treatments with the aid of repetitive transcranial magnetic stimulation, but transcranial magnetic stimulation treatment paradigms are still under development and currently the method has only been registered in a limited number of countries for the treatment of depressive disorder (Slotema et al., 2011). Usually non-pharmacologic treatment methods are offered in combination with pharmacotherapy, but in milder cases it is sometimes possible to start with psychotherapy alone, or with a coping-with-voices course or participation in the meetings of a self-help group.

As regards psychotherapy, the method of choice tends to be cognitive-behavioral therapy. Starting from the premise that the impact of auditory hallucinations depends primarily on the way they are appraised and dealt with psychologically, cognitive-behavioral therapy targets five specific aspects of the hallucinatory experience: (1) the voices' often compellingly realistic nature; (2) the personal meaning of their content; (3) the more or less fixed identity of the voices; (4) the voice hearer's relationship with them; and (5) the impact of the voices on the voice hearer's life (Beavan, 2011). The cognitive strand of cognitive-behavioral therapy involves putting those appraisals in perspective and offering alternative explanations for their origin and meaning. The behavioral strand involves the testing of novel ways to deal with situations where the voices play a negative role, as well as relabeling their emotional appraisal (Sommer et al., 2012b). As soon as the first signs of doubt begin to show regarding the alleged reality and power of the voices, patients are encouraged to limit the time they give them their attention, to pick up the routines of their daily lives, and to reclaim meaningful social roles. In addition, they are encouraged to diminish any safety routines they may have developed, such as the avoidance of social contacts, sports, education, or labor out of a fearful anticipation of possible disastrous consequences.

Cognitive-behavioral therapy has been developed for voice hearers in general, but also for specific subgroups such as those suffering from command hallucinations, critical voices, humiliating voices, or reperceptive hallucinations. Effect sizes for those treatment programs range from 0.64 to 1.16 (Sommer et al., 2012b).

Coping-with-voices protocols can either be offered in the context of cognitive-behavioral therapy programs or as separate treatment methods (van Gastel and Daalman, 2012). The same holds true for the meetings offered by any of the self-help groups which have been founded in the majority of western countries throughout the past 25 years, the most renowned one being the international Hearing Voices Movement (Romme and Escher, 2012). Those groups tend to proceed from a set of principles aimed at toning down the medicalization of auditory hallucinations, at destigmatizing and normalizing the experience *per se*, and at exploring and validating possible links with prior traumatic experiences. By focusing on acceptance, emancipation, and coping skills rather than medical diagnosis and treatment, they tend to encourage empowerment rather than symptom reduction *per se*.

A final non-pharmacologic treatment method available for – especially medication-resistant – auditory hallucinations is ECT. Although the efficacy of ECT has never been studied for the exclusive purpose of treating auditory hallucinations, and consensus on its usefulness in such cases is lacking, in clinical practice ECT tends to be used mainly when hallucinations occur in association with psychotic depression or catatonia, i.e., the two standard indications for ECT (Sommer and Blom, 2012). In a systematic meta-analysis of double-blind, randomized studies comparing ECT and antipsychotic medication with sham and medication, Tharyan and Adams (2005) included 10 randomized controlled trials carried out among a total number of 392 patients, and concluded that the relative risk for clinical improvement was 0.78 in favor of ECT. Although a significant finding, the studies on which the meta-analysis was based failed to reveal anything specific about the reaction of auditory hallucinations to ECT.

METATHEORETIC CONSIDERATIONS

Structural and especially functional neuroimaging findings have revolutionized our understanding of auditory hallucinations by linking them to an extensive network of brain areas involved in their mediation. In that sense, our insights into their nature and origin have increased significantly over the past 20 years. Meanwhile the conceptualization of auditory hallucinations – as well as hallucinations in general – has silently undergone a revolution of its own, partly due to those neuroimaging findings, and partly due to ideas that stem from the philosophy of science and from network science. Those conceptual developments and their theoretical implications will now be summarized.

Insights from the philosophy of science

As regards hallucinations and so-called veridical perceptions – and any differences between the two – much is to be learned from philosophers. A classic epistemologic conundrum is whether hallucinations differ qualitatively from so-called veridical perceptions. From what was said so far, the differences between the two may appear self-evident, but there are many critical reflections on this issue to be found in the philosophical literature, ranging from Sextus Empiricus (*c.* 160–210 AD) to Nelson Goodman (1906–1988) and beyond. Especially those concerned with the nature of existence (ontology) and the nature of knowledge (epistemology) consider the hallucinatory experience a fertile test case for paradigms ranging from naïve realism to idealism, skepticism, and constructionism (de Haan, 2009). In addition, there are various empiric findings that invite us to examine the conceptualization of hallucinations as percepts-without-an-external-source in some more detail, and it is to those that we shall now turn.

In the first place, the phenomenologic characteristics of sensory and hallucinatory percepts appear to be quite similar. Voice hearers tell us that their verbal auditory hallucinations tend to sound like actual voices, and individuals experiencing musical hallucinations that they sound like actual music. The hallucinations at hand may not invariably possess all the phenomenologic characteristics of so-called veridical perceptions (as in the case of internal auditory hallucinations and unilateral types), but they certainly can, and frequently they do. What is more, especially external auditory hallucinations tend to blend in seamlessly with environmental sounds, and thus constitute an integral part of the voice hearer's auditory landscape.

Secondly, we may ask ourselves what we mean exactly when we use the term veridical perception. As we all know, regular sense perception involves something more than a straightforward representation of stimuli we assume present in the external world. Whatever we may think of the ontologic status of those stimuli, as soon as we accept that they constitute the basis of our sense perceptions, we also accept that sense perception is not a purely linear process, and acknowledge instead that it leaves us some room for manufacturing our own idiosyncratic version of that which we call reality. In the case of audition, we tend to rely on contextualization and the "filling in" of unheard and ambiguous sounds to arrive at meaningful phrases which may or may not correspond with what was spoken. Meanwhile preoccupations, attentional biases, wishes, fears, and many other psychologic factors exert a considerable influence on all that we (think that we) are hearing. So we may ask ourselves how "veridical" our veridical percepts actually are.

In the third place, it is as yet uncertain whether there are any neurophysiologic differences between sensory and hallucinatory percepts. As suggested by the neuroimaging findings discussed above, auditory hallucinations and actual sounds appear to depend on common brain networks for their mediation. Apparently the networks at hand are capable of mediating auditory percepts on their own, irrespective of the presence or absence of external auditory stimuli. As a consequence, it may not always be clear to us whether that which we perceive has an ontologic substrate in the extracorporeal world or not.

Taken together, those findings beg the ancient philosophic question of whether a reliable distinction can be made between hallucinations and that which we call veridical perceptions. According to Descartes (1637), we are fundamentally unable to make that distinction (unless we acknowledge that it is God who tells us which is which). Kant (1781) likewise stated that we are unable to gain access to the world itself, the *Ding an sich*, and said that all we ever experience are our own internally mediated percepts. Descartes and Kant lived centuries ago, but the epistemologic issues they raised never lost any of their significance. Ironically, perhaps, similar issues are now being raised by empiric scientists such as hallucinations researchers and those working in the field of perceptual neuroscience.

Fish (2009), in a scholarly defense of realism, offers us the following solution to the question of how hallucinations differ from veridical perceptions. According to him, the perception of objects or events in the outside world involves our becoming acquainted with them, thus rendering the experience a phenomenal character, whereas hallucinations are mistakenly supposed to possess such a phenomenal character whilst in fact they lack them. Among philosophers that solution is known as disjunctivism, a term coined by Hinton (1973). As proposed by Hinton, the qualities of perceived objects are instantiated in cases of genuine sense perception, whereas in cases of hallucinatory perception they are merely represented.

With their disjunctivist approach, authors such as Hinton and Fish would seem to voice the traditional point of view that hallucinations can be distinguished from sense perceptions because of our knowledge of what is actually going on in the external world. Their opponents, however, notably those who adhere to the qualia theory, stress the Kantian point of view that our attempts to gain access to the external world through the senses are ultimately in vain, and that the qualitative properties of the ensuing percepts are indistinguishable from those of hallucinations (see, for example, Crane, 2006). They offer a somewhat different solution, but one that has significant implications for the conceptualization of both phenomena.

During the 1980s it was suggested that all perception depends on gamma-band oscillations of approximately 40 Hz resonating in the thalamocortical system (Llinás and Ribary, 1994). Although the exact mechanism is still in need of further elucidation, those oscillations are held responsible for engaging widely disseminated parts of the thalamocortical system in synchronous electrophysiologic activity, thus ensuring their collective contribution to the mediation of percepts. Judging by the continuous presence of those gamma-band oscillations, it would seem that the mediation of percepts is an ongoing process. Apparently many – if not most – of the ensuing percepts never breach the threshold of our consciousness, but when we dream, daydream, or experience microsleeps we are able to catch glimpses of them. As soon as we are fully awake, however, and start optimizing the use of our sense organs, those oscillations are modulated in such a way that they provide us with significantly more information about what is going on in the outside world than during sleep or other states of lowered consciousness. The information originating from the sense organs is therefore believed to temporarily restrict the thalamocortical system's potential, in the sense that it limits its ability to create percepts the way it does in its natural, more or less freewheeling, state (Behrendt and Young, 2004; Collerton et al., 2005). As regards the auditory modality, this means that the number of degrees of freedom of the thalamocortical system is being curtailed by information provided by the inner ear, thus forcing it to refrain from mediating its auditory percepts more or less at random, and instead forcing it to modulate them in such a way that they show certain parallels with the external auditory stimuli to which the system is exposed.

On the basis of this line of thought it has been suggested that auditory perception – like all perception – comes down to a closed, intrinsic functional brain state (Llinás and Ribary, 1994), irrespective of whether it is experienced during a dream, in a hallucinatory state, or in an unclouded state of wakeful consciousness. Following through this line of thought, actual sounds can then be characterized as products of the thalamocortical system which are restrained by information from the inner ear, whereas auditory hallucinations can be characterized as similar products which are unrestrained by information from the inner ear, and auditory pareidolias (i.e., illusions) as similar products which are underrestrained by information from the inner ear. If anything, this account does justice to the role of neurophysiologic and psychologic factors in audition, and hence downplays the notion that voice hearers might be the only ones exposed to endogenous modulation of their percepts. In addition, it would seem to make sense in terms of the brain's economic use of resources. However, what this account sacrifices is the clear – albeit somewhat naïve – distinction between sensory and hallucinatory percepts we are acquainted with. Whether that is a sacrifice worth making is up to the reader to decide.

Insights from network science

A further development that has fundamentally altered the way we think about hallucinatory activity stems from the relatively young field of network science. Having emerged from earlier mathematic disciplines such as graph theory, cybernetics, and probabilistic theory, network science is a means of analyzing complex relational data which is applicable, due to its scale-free nature, to any aspect of the universe, and hence also to brain structures and the way they are interrelated at a structural as well as a functional level. Because of the huge number of factors involved in the mediation of hallucinations, until recently it has appeared nigh impossible to develop a global view of their mediation, let alone indicate which factors stand in causal relations to others, and to determine the weight and importance of each factor under varying circumstances. The sections above have indicated that we are still far removed from realizing such a global view of hallucinatory activity, but network science counts as a serious candidate for bringing that goal within our reach during the years to come.

As explained by Goekoop and Looijestijn (2012) in their scholarly introduction to the subject (upon which most of the present section is based), network science starts from the deceptively simple premise that interactions between agents existing separately in space can be studied with the aid of graphic representations called network graphs (for an example of such a network graph, see Fig. 24.3). Those network graphs consist of nodes, and of links interconnecting those nodes. Together, the nodes and links constitute the network's structure. Information traveling from one node to another along the links represents the network's function. Like all biologic systems, cerebral networks are conceptualized as possessing a small-world structure, which means that they are believed to consist of many nodes which possess relatively few connections, and relatively few nodes which possess a large number of connections (Meunier et al., 2010). The richly connected nodes are called hubs, and those hubs are connected to other hubs in such a manner that information traveling from one side of the brain to the other can take place in a maximum of six steps, irrespective of location and function. It is believed that the brain's small-world topology thus serves to maximize the efficiency of information transfer, and that it does so in the most economic way, i.e., by using the smallest possible number of links in its network.

As regards the study of hallucinations, network science offers the possibility to examine physiologic as well as pathologic changes in neural network structure and function with the aid of computer models of neural networks, in which have been incorporated data deriving from anatomic, physiologic, electrophysiologic, biochemical, molecular genetic, or neuroimaging studies, or in fact any other type of empiric data relevant to the subject. Because of the method's scale-free character, the ensuing network graphs can even be linked to networks graphs representing phenomenologic, psychologic, and social as well as other environmental factors, thus allowing for weighted analyses at levels of conceptualization that extend beyond the brain itself.

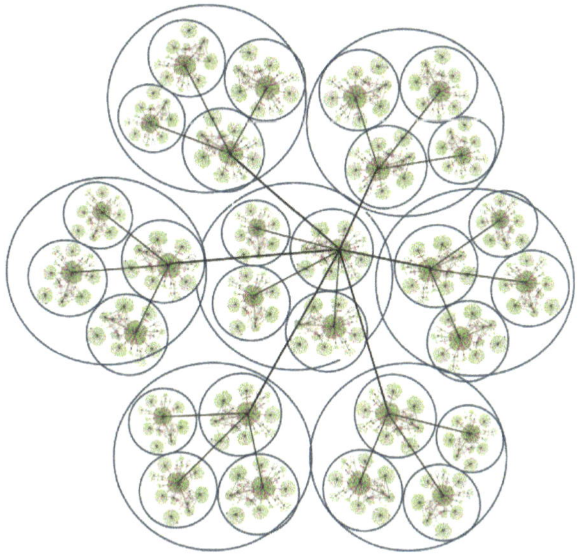

Fig. 24.3. Network graph showing an example of a "small-world" network structure. Hubs (i.e., high-degree nodes) connect clusters of lesser-degree nodes into superclusters (marked by circles), and so on. This hierarchic network structure is called scale-free or fractal-like, since a similar structure can be found in nature at all spatial-scale levels of organization, including networks of genes, proteins, metabolites, organelles, neurons, brain areas, social networks, and markets. (Reproduced from Goekoop and Looijestijn, 2012, with permission.)

Some of the models currently used for that purpose are attractor network models (Brunel and Wang, 2001). An attractor network comprises a network of neurons (such as pyramidal cells in the primary auditory cortex) which receives exogenous dendritic input (originating from the inner ear, for example) which it modulates in such a way that it can serve as an output signal for axons that connect the network to another attractor network (such as the auditory association cortex). Neurons located within each attractor network have excitatory collaterals which feed back to their dendritic connections (serving to produce collateral excitation) and to inhibitory interneurons which, on average, suppress the activity in networks lying outside the attractor network (serving to produce collateral inhibition) (Brunel and Wang, 2001). In this way the excitatory and inhibitory collaterals constitute positive- and negative-feedback loops, which are believed to operate via *N*-methyl-D-aspartic-acid (NMDA) and gamma-aminobutyric-acid (GABA) receptors, respectively (Goekoop and Looijestijn, 2012). That basic layout can be used to explain how the auditory network is able to mediate all types of auditory perception, irrespective of whether the ensuing percepts originate from signals in the environment or from within the brain itself.

In terms of the network science paradigm, auditory hallucinations are conceptualized as attractor states (i.e., active or passive states) which are produced – and often sustained – in the absence of an external source. As proposed by Goekoop and Looijestijn (2012), such false-positive attractor states can be produced in a normal brain under certain extreme circumstances (such as sensory or social deprivation, deafness, extreme anxiety, or substance abuse), as well as under normal circumstances in a sensitive brain. As the authors explain, in sensory deprivation the lack of external stimuli may force attractor networks to enter their resting states, and to prevent collateral inhibition from suppressing any "noise" deriving from within the brain itself. Meanwhile the low stimulus intensities may lead to neural adaptation, thus further reducing the number of GABAergic inhibitory currents, and decreasing the firing thresholds of neurons within the attractor network (Goekoop and Looijestijn, 2012). As a result, the attractor network at hand will become extremely sensitive, and require only little energy to assume a false-positive state. In terms of energy landscapes, the attractor state thus constitutes a local minimum or "hole" from which trajectories of network activity cannot easily escape. As a consequence, the aberrant neurophysiologic activity has a tendency to persist, and, when altered, to return time and again to its aberrant state.

In the case of a sensitive brain, the network attractor model aids to explain the mediation of auditory hallucinations under normal circumstances. It does so by describing pathologic differences in the structure and/or function of the attractor network itself. Such differences may involve any level of excitation, inhibition, and/or neuromodulation governing the activity of the attractor network, notably those regarding the NMDA receptor, deficits in GABAergic inhibition, and elevated ratios of D2 versus D1 dopamine receptors (Goekoop and Looijestijn, 2012).

As noted by Jardri and Denève (2013), the attractor network model is a means of investigating the functional stability of networks, and, more specifically, to study how at the synaptic level alterations in ion channels and neurotransmitters may affect spiking activity within such networks. In addition, it offers an unprecedented opportunity to study the mediation of auditory hallucinations by examining the interrelationships between networks and biochemical agents, structural and functional connectivity, phenomenologic characteristics of hallucinations, psychologic factors, and the social – as well as wider environmental – context. Figure 24.4 offers a graphic representation of the ensuing coupled network graphs. With network attractor models becoming increasingly sophisticated, it is expected that computer simulations will soon be able to aid in identifying brain areas to be targeted for the purpose of symptom reduction.

Functional neuroimaging studies, meanwhile, may aid in identifying the networks and hub regions to be studied in network attractor models. Traditionally the results of such neuroimaging studies are based on model-based analyses (see, for example, van Lutterveld et al., 2014), which means that the brain areas are selected with a preconceived set of notions in mind regarding their possible involvement. An alternative approach is model-free analysis, a type of analysis which refrains from using any such preconceived ideas. Model-free analyses tend to yield a substantially larger number of brain areas potentially involved in the mediation of hallucinations. Some of those areas may turn out to be false-positive findings, whereas others may constitute valuable additions to the hallucination network previously overlooked in model-based analyses. Because of the overwhelmingly large number of data to be analyzed in model-free analyses of whole-brain scans, it used to be hard – if not technically impossible – to attempt such analyses. But with the aid of network science such analyses have now become feasible, and various studies using this technique are currently underway.

CONCLUSION

With the above outline of network science as a method potentially useful for the study of auditory hallucinations, we have gradually moved from discussing rather robust empiric findings to what, strictly speaking, deserves little more than the status of a "future development."

Summing up, the present chapter demonstrated that our knowledge of auditory hallucinations has transcended the rich phenomenology and taxonomy inherited from classic neuroscientific discourses to encompass the even richer harvest of neuroimaging findings obtained during the past two decades, thus making it possible to gain an impression of a structural as well as a functional auditory network in the brain, and to relate various aspects of the auditory hallucinatory experience to

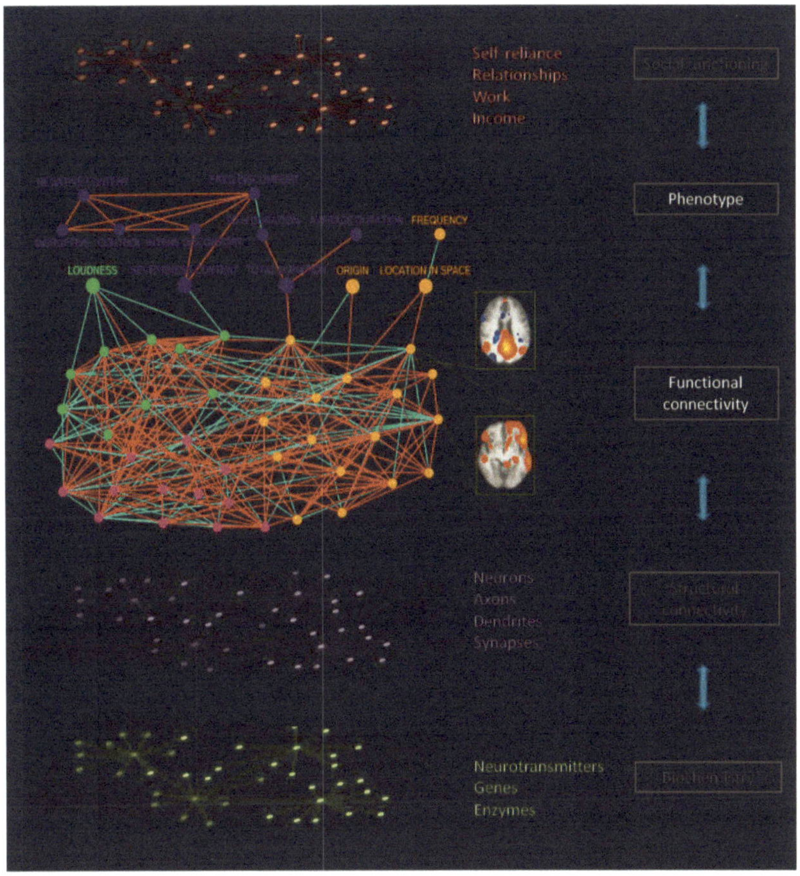

Fig. 24.4. Coupled network graphs of functional connectivity in relation to a network structure of symptoms of patients who experience verbal auditory hallucinations (Looijestijn et al., data in preparation). Measured data are shown in color. Functional connectivity: network of neurophysiologic data (independent components of functional magnetic resonance signal). Phenotype: subjective symptoms as rated with the aid of a questionnaire (PSYRATS) and registered hallucination timings during magnetic resonance imaging. Hypothetic additional levels of organization are shown in gray, and can be added to the model at later stages. Based on knowledge of the relationships between the various factors that relate to the origin and phenomenologic expression of hallucinations, therapeutic intervention foci can be predicted at each level of network organization (i.e., medication, transcranial magnetic stimulation, psychotherapy, work). (Reproduced from Goekoop and Looijestijn, 2012, with permission.)

important nodes and hubs located in those networks. In the case of verbal auditory hallucinations experienced in the context of psychosis, subcortical structures such as the parahippocampal gyri and the insula were thus found to play a dominant role in the initiation of preparatory hallucinatory activity, the language centers, and their contralateral homologs in furnishing them with a linguistic content, the auditory cortices in promoting awareness of the percepts involved, and the auditory "where" pathway in lending them the phenomenologic quality of either an internal or external percept. Although admittedly a crude simplification of what must actually be going on in the brain before and during the conscious experience of a verbal auditory hallucination, that characterization allowed us to catch a glimpse of the temporal sequences leading up to that hallucinatory experience, and of the widely disseminated brain areas involved in their mediation.

In addition, the present chapter indicated that recent developments such as network science and model-free analysis hold the promise of a further – and possibly exponential – growth of knowledge in this already rapidly expanding area of research. Meanwhile insights derived from the philosophy of science compelled us to adjust our conceptualization of auditory hallucinations – as well as that of the hearing of actual sounds – in such a way that they do justice to our empiric findings. As a consequence, we now conceive of auditory hallucinations as closed, intrinsic functional brain states produced by the thalamocortical system in such a way that they are unrestrained by information from the inner ear; of auditory illusions as similar brain states underrestrained by information from the inner ear; and of actual sounds as similar brain states restrained by information from the inner ear. Thus the three types of auditory percept are conceptualized in conformity with the qualia theory as being qualitatively similar, which also makes sense from the point of view that they are all indebted to the same auditory network for their mediation.

In a word, the previous sections indicated that our knowledge of the auditory hallucinatory experience has increased considerably since the advent of functional neuroimaging research. On the downside, we saw that our knowledge of the neurophysiologic basis of auditory hallucinations is almost exclusively restricted to the subgroup of verbal auditory hallucinations. The various other subgroups – largely neglected from the 1950s onwards by major psychiatric classifications – are now back in the center of scientific attention, but it will probably take some time before they will be investigated as extensively as the subgroup of hallucinated voices. Another significant limitation of our current knowledge is that it has so far failed to yield any new, successful therapeutic strategies. There are various novel treatment methods on the horizon, including asenapine and other drugs marketed as antipsychotics, as well as cannabidiol – a non-psychoactive compound of cannabis – but on the basis of our current insights into the role of specific hubs in the auditory network it should become possible in the near future to target those areas more precisely, for example with the aid of direct-current stimulation, deep-brain stimulation or transcranial magnetic stimulation. Transcranial magnetic stimulation has been researched extensively for its potential to target relatively circumscript cortical – and sometimes even subcortical – areas involved in the mediation of auditory hallucinations without yielding any significant side-effects, but treatment paradigms in this area are still in need of further development for the method to become applicable in clinical practice (Aleman and Larøi, 2011; Sommer et al., 2012c).

References

Agid O, Remington G, Kapur S et al. (2007). (2007). Early use of clozapine for poorly responding first-episode psychosis. J Clin Psychopharmacol 27: 369–373.

Aleman A, Larøi F (2011). Insights into hallucinations in schizophrenia: novel treatment approaches. Expert Rev Neurother 11: 1007–1015.

Allen P, Modinos G (2012). Structural neuroimaging in psychotic patients with auditory verbal hallucinations. In: JD Blom, IEC Sommer (Eds.), Hallucinations, Research and Practice. Springer, New York, NY, pp. 251–265.

Almeida OP, Forstl H, Howard R et al. (1993). Unilateral auditory hallucinations. Br J Psychiatry 162: 262–264.

American Psychiatric Association (2013). Diagnostic and Statistical Manual of Mental Disorders. Fifth Edition. American Psychiatric Association, Washington, DC.

Beavan V (2011). Towards a definition of 'hearing voices': a phenomenological approach. Psychosis 3: 63–73.

Beavan V, Read J, Cartwright C (2011). The prevalence of voice-hearers in the general population: a literature review. J Ment Health 20: 281–292.

Behrendt RP, Young C (2004). Hallucinations in schizophrenia, sensory impairment and brain disease: a unifying model. Behav Brain Sci 27: 771–787.

Berrios GE (1990). Musical hallucinations. A historical and clinical study. Br J Psychiatry 156: 188–194.

Berrios GE (2005). On the fantastic apparitions of vision by Johannes Müller. Hist Psychiatry 16: 229–246.

Bleuler E (1903). Extracampine Hallucinationen. Psychiatrisch-neurologische Wochenschrift 25: 261–264.

Blom JD (2010). A Dictionary of Hallucinations, Springer, New York, NY.

Blom JD (2012). Hallucinations and other sensory deceptions in psychiatric disorders. In: R Jardri, A Cachia, P Thomas et al. (Eds.), The Neuroscience of Hallucinations, Springer, New York, NY, pp. 43–58.

Blom JD, Sommer IEC (2010). Auditory hallucinations: nomenclature and classification. Cogn Behav Neurol 23: 55–62.

Bloom JS, Hynd GW (2005). The role of the corpus callosum in interhemispheric transfer of information: excitation or inhibition? Neuropsychol Rev 15: 59–71.

Braun CMJ, Dumont M, Duval J et al. (2003). Brain modules of hallucination: an analysis of multiple patients with brain lesions. J Psychiatry Neurosci 28: 432–449.

Brunel N, Wang XJ (2001). Effects of neuromodulation in a cortical network model of object working memory dominated by recurrent inhibition. J Comput Neurosci 11: 63–85.

Buchanan RW, Kreyenbuhl J, Kelly DL et al. (2010). The 2009 schizophrenia PORT psychopharmacological treatment recommendations and summary statements. Schizophr Bull 36: 71–93.

Carlsson M, Carlsson A (1990). Schizophrenia: a subcortical neurotransmitter imbalance syndrome? Schizophr Bull 16: 425–430.

Cascino GD, Adams RD (1986). Brainstem auditory hallucinosis. Neurology 36: 1042–1047.

Clos M, Diederen KM, Meijering AL et al. (2014). Aberrant connectivity of areas for decoding degraded speech in patients with auditory verbal hallucinations. Brain Struct Funct 219: 581–594.

Cole MG, Dowson L, Dendukuri N et al. (2002). The prevalence and phenomenology of auditory hallucinations among elderly subjects attending an audiology clinic. Int J Geriatr Psychiatry 17: 444–452.

Collerton D, Perry E, McKeith I (2005). Why people see things that are not there: a novel perception and attention deficit model for recurrent complex visual hallucinations. Behav Brain Sci 28: 737–794.

Copolov DL, Seal ML, Maruff P et al. (2003). Cortical activation associated with the experience of auditory hallucinations and perception of human speech in schizophrenia: A PET correlation study. Psychiatry Res 122: 139–152.

Copolov D, Trauer T, Mackinnon A (2004). On the non-significance of internal versus external auditory hallucinations. Schizophr Res 69: 1–6.

Crane T (2006). Is there a perceptual relation? In: TS Gendler, J Hawthorne (Eds.), Perceptual Experience, Clarendon Press, Oxford, pp. 126–146.

Daalman K, Boks MP, Diederen KM et al. (2011). The same or different? A phenomenological comparison of auditory verbal hallucinations in healthy and psychotic individuals. J Clin Psychiatry 72: 320–325.

De Haan S (2009). Philosophical interpretations and existential effects of hallucinations. Eur Psychiatry 24: S153.

Descartes R (1637/1984). Discourse on the method. In The Philosophical Writings of Descartes, Volume I. In: J Cottingham, R Stoothoff, D Murdoch (Eds.), Cambridge University Press, Cambridge, pp. 111–151.

Diederen KMJ, Neggers SFW, Daalman K et al. (2010). Deactivation of the parahippocampal gyrus preceding auditory hallucinations in schizophrenia. Am J Psychiatry 167: 427–435.

Diederen KM, Daalman K, de Weijer AD et al. (2011). Auditory hallucinations elicit similar brain activation in psychotic and nonpsychotic individuals. Schizophr Bull 38: 1074–1082.

Dierks T, Linden DE, Jandl M et al. (1999). Activation of Heschl's gyrus during auditory hallucinations. Neuron 22: 615–621.

Erkwoh R, Willmes K (2002). Command hallucinations: Who obeys and who resists when? Psychopathology 35: 272–279.

Falkai P (1996). Differential diagnosis in acute psychotic episode. Int Clin Psychopharmacol 11 (Suppl 2): 13–17.

Feinberg J (1978). Efference copy and corollary discharge: implications for thinking and its disorders. Schizophr Bull 4: 636–640.

Fénelon G, Alves G (2010). Epidemiology of psychosis in Parkinson's disease. J Neurol Sci 15: 12–17.

Fish W (2009). Perception, Hallucination, and Illusion, Oxford University Press, Oxford.

Ford JM, Mathalon DH, Heinks T et al. (2001). Neurophysiological evidence of corollary discharge dysfunction in schizophrenia. Am J Psychiatry 158: 2069–2071.

Ford JM, Roach BJ, Mathalon DH (2010). Assessing corollary discharge in humans using noninvasive neurophysiological methods. Nat Protoc 5: 1160–1168.

Goekoop R, Looijestijn J (2012). A network model of hallucinations. In: JD Blom, IEC Sommer (Eds.), Hallucinations, Research and Practice. Springer, New York, NY, pp. 33–54.

Gordon AG (1999). Do musical hallucinations always arise from the inner ear? Med Hypotheses 53: 89–90.

Green P, Preston M (1981). Reinforcement of vocal correlates of auditory hallucinations by auditory feedback: a case study. Br J Psychiatry 139: 204–208.

Hinton JM (1973). Experiences: An Inquiry into Some Ambiguities, Clarendon Press, Oxford.

Hoffman RE, Hampson M, Wu K et al. (2007). Probing the pathophysiology of auditory/verbal hallucinations by combining functional magnetic resonance imaging and transcranial magnetic stimulation. Cereb Cortex 7: 2733–2743.

Hoffman RE, Anderson AW, Varanko M et al. (2008). Time course of regional brain activation associated with onset of auditory verbal hallucinations. Br J Psychiatry 193: 424–425.

Hughlings-Jackson J (1888). Remarks on evolution and dissolution of the nervous system. Am J Psychol 1: 336–338.

Jardri R, Denève S (2013). Computational models of hallucinations. In: R Jardri, A Cachia, P Thomas et al. (Eds.), The Neuroscience of Hallucinations, Springer, New York, NY, pp. 289–313.

Jardri R, Pouchet A, Pins D et al. (2011). Cortical activations during auditory verbal hallucinations in schizophrenia: A coordinate-based meta-analysis. Am J Psychiatry 168: 73–81.

Jaspers K (1997). General Psychopathology, Vol. 1. Johns Hopkins University Press, Baltimore, MA, Translated by Hoenig J.

Kahlbaum K (1886). Die Sinnesdelirien. Allgemeine Zeitschrift für Psychiatrie und psychisch-gerichtliche Medizin 23: 56–78.

Kahn RS, Fleischhacker WW, Boter H et al. (2008). Effectiveness of antipsychotic drugs in first-episode schizophrenia and schizophreniform disorder: An open randomised clinical trial. Lancet 2008 (29): 1085–1097.

Kant I (1781). Kritik der Reinen Vernunft, Johann Friedrich Hartknoch, Riga.

Kasper BS, Kasper EM, Pauli E et al. (2010). Phenomenology of hallucinations, illusions, and delusions as part of seizure semiology. Epilepsy Behav 18: 13–23.

Kechid G, Auvin S, Jardri R et al. (2008). Hearing hallucinations in a 12-year-old child: psychotic disorders or temporal epilepsy? Prim Care Companion J Clin Psychiatry 10: 328–329.

Keshavan MS, Davis AS, Steingard S et al. (1992). Musical hallucinations: a review and synthesis. Neuropsychiatry Neuropsychol Behav Neurol 5: 211–223.

Kingdon DG, Ashcroft K, Bhandari B et al. (2010). Schizophrenia and borderline personality disorder. Similarities and differences in the experience of auditory hallucinations, paranoia and childhood trauma. J Nerv Ment Dis 198: 399–403.

Larøi F, Sommer IE, Blom JD et al. (2012). The characteristic features of auditory verbal hallucinations in clinical and nonclinical groups: state-of-the-art overview and future directions. Schizophr Bull 38: 724–733.

Layland JJ, Liew D, Prior DL (2009). Clozapine-induced cardiotoxicity: a clinical update. Med J Aust 16: 190–192.

Llinás RR, Ribary U (1994). Perception as an oneiric-like state modulated by the senses. In: C Koch, JL Davis (Eds.), Large-Scale Neuronal Theories of the Brain, MIT Press, Cambridge, MA, pp. 111–124.

Looijestijn J, Diederen KMJ, Goekoop R et al. (2013). The auditory dorsal stream plays a crucial role in projecting voices into external space. Schizophr Res 146: 314–319.

Looijestijn J, Goekoop R, Blom JD, et al. (submitted). Verbal auditory hallucinations and their relationship to the multimodular, hierarchic organization of brain function: A data-driven approach.

Madras BK (2013). History of the discovery of the antipsychotic dopamine D2 receptor: a basis for the dopamine hypothesis of schizophrenia. J Hist Neurosci 22: 62–78.

Mavromatis A (1987). Hypnagogia. The Unique State of Consciousness Between Wakefulness and Sleep, Routledge, London.

McCarthy-Jones S (2012). Hearing Voices. The Histories, Causes and Meanings of Auditory Verbal Hallucinations, Cambridge University Press, Cambridge.

McCarthy-Jones S, Trauer T, Mackinnon A et al. (2014). A new phenomenological survey of auditory hallucinations: evidence for subtypes and implications for theory and practice. Schizophr Bull 40: 231–235.

McGuire PK, Shah GM, Murray RM (1993). Increased blood flow in Broca's area during auditory hallucinations in schizophrenia. Lancet 342: 703–706.

Mechelli A, Allen P, Amaro E et al. (2007). Misattribution of speech and impaired connectivity in patients with auditory verbal hallucinations. Hum Brain Mapp 28: 1213–1222.

Meunier D, Lambiotte R, Bullmore ET (2010). Modular and hierarchically modular organization of brain networks. Front Neurosci 4: 200.

Morel F (1933). Écho de la lecture. Contribution à l'étude des hallucinations auditives verbales l'Encéphale 28: 169–183.

Mueser KT, Bellack AS, Brady EU (1990). Hallucinations in schizophrenia. Acta Psychiatr Scand 82: 26–29.

Myers TI, Murphy DB (1960). Reported visual sensation during brief exposure to reduced sensory input. Research memo from the Human Resources Research Office, George Washington University, Alexandria, VA.

Parish E (1897). Hallucinations and Illusions, A Study of the Fallacies of Perception. London, Walter Scott.

Pearce JMS (1988). Exploding head syndrome. Lancet 332: 270–271.

Penfield W (1975). The Mystery of the Mind, Princeton University Press, Princeton, NJ, A Critical Study of Consciousness and the Human Brain.

Romme M, Escher E (2012). The Hearing Voices Movement. In: JD Blom, IEC Sommer (Eds.), Hallucinations, Research and Practice. Springer, New York, NY, pp. 385–400.

Romme M, Escher E, Dillon J et al. (2009). 50 Stories of Recovery. Living with Voices. Ross-on-Wye, PCCS Books.

Sacks O (2006). The power of music. Brain 129: 2528–2532.

Sacks O, Blom JD (2012). Musical hallucinations. In: JD Blom, IEC Sommer (Eds.), Hallucinations, Research and Practice. Springer, New York, NY, pp. 133–142.

Sato Y, Berrios G (2003). Extracampine hallucinations. Lancet 361: 1479–1480.

Shergill SS, Brammer MJ, Williams SC et al. (2000). Mapping auditory hallucinations in schizophrenia using functional magnetic resonance imaging. Arch Gen Psychiatry 57: 1033–1038.

Shergill SS, Brammer MJ, Amaro E et al. (2004). Temporal course of auditory hallucinations. Br J Psychiatry 185: 516–517.

Silbersweig DA, Stern E, Frith C et al. (1995). A functional neuroanatomy of hallucinations in schizophrenia. Nature 378: 176–179.

Slotema CW, Kingdon DG (2012). Auditory verbal hallucinations in patients with borderline personality disorder. In: JD Blom, IEC Sommer (Eds.), Hallucinations, Research and Practice. Springer, New York, NY, pp. 125–132.

Slotema CW, Blom JD, de Weijer AD et al. (2011). Can low-frequency repetitive transcranial magnetic stimulation really relieve medication-resistant auditory verbal hallucinations? Negative results from a large randomized controlled trial. Biol Psychiatry 69: 450–456.

Slotema CW, Daalman K, Blom JD et al. (2012). Auditory verbal hallucinations in patients with borderline personality disorder are similar to those in schizophrenia. Psychol Med 42: 1873–1878.

Sommer IEC, Blom JD (2012). Classical somatic treatments: Pharmacotherapy and ECT. In: JD Blom, IEC Sommer (Eds.), Hallucinations, Research and Practice. Springer, New York, NY, pp. 331–348.

Sommer IEC, Diederen KMJ, Blom J-D et al. (2008). Auditory verbal hallucinations predominantly activate the *right* inferior frontal area. Brain 131: 3169–3177.

Sommer IEC, Diederen KMJ, Selten J-P et al. (2009). Dissecting auditory verbal hallucinations into two components: audibility (Gedankenlautwerden) and alienation (thought insertion). Psychopathology 43: 137–140.

Sommer IEC, Daalman K, Rietkerk T et al. (2010). Healthy individuals with auditory verbal hallucinations; who are they? Psychiatric assessments of a selected sample of 103 subjects. Schizophr Bull 36: 633–641.

Sommer IEC, Begemann MJ, Temmerman A et al. (2011). Pharmacological augmentation strategies for schizophrenia patients with insufficient response to clozapine: A quantitative literature review. Schizophr Bull 38: 1003–1011.

Sommer IEC, Koops S, Blom JD (2012a). Comparison of auditory hallucinations across different disorders and syndromes. Neuropsychiatry 2: 57–68.

Sommer IEC, Slotema CW, Daskalakis ZJ et al. (2012b). The treatment of hallucinations in schizophrenia spectrum disorders. Schizophr Bull 38: 704–714.

Sommer IEC, Aleman A, Slotema CW et al. (2012c). Transcranial stimulation for psychosis: the relationship between effect size and published findings. Am J Psychiatry 169: 1211.

Stewart L, von Kriegstein K, Warren JD et al. (2006). Music and the brain: disorders of musical listening. Brain 129: 2533–2553.

Takebayashi H, Takei N, Mori N et al. (2002). Unilateral auditory hallucinations in schizophrenia after damage to the right hippocampus. Schizophr Res 58: 329–331.

Tharyan P, Adams CE (2005). Electroconvulsive therapy for schizophrenia. Cochrane Database Syst Rev 18, CD000076.

Van Gastel WA, Daalman K (2012). Groundwork for the treatment of voices: Introducing the *Coping-With-Voices Protocol*. In: JD Blom, IEC Sommer (Eds.), Hallucinations. Research and Practice. Springer, New York, NY, pp. 375–384.

Van Lutterveld R, Ford JM (2012). Neurophysiological research: EEG and MEG. In: JD Blom, IEC Sommer

(Eds.), Hallucinations, Research and Practice. Springer, New York, NY, pp. 283–295.

Van Lutterveld R, Diederen KM, Otte WM et al. (2014). Network analysis of auditory hallucinations in nonpsychotic individuals. Hum Brain Mapp 35: 1436–1445.

Vercammen A, Knegtering H, Bruggeman R et al. (2011). Subjective loudness and reality of verbal auditory hallucinations and activation of the inner speech processing network. Schizophr Bull 37: 1009–1016.

Villringer A (1997). Understanding functional neuroimaging methods based on neurovascular coupling. Adv Exp Med Biol 413: 177–193.

Wernicke C (1900). Grundriss der Psychiatrie, Verlag von Georg Thieme, Leipzig.

West LJ (1962). A general theory of hallucinations and dreams. In: LJ West (Ed.), Hallucinations, Grune & Stratton, New York, NY, pp. 275–291.

Handbook of Clinical Neurology, Vol. 129 (3rd series)
The Human Auditory System
G.G. Celesia and G. Hickok, Editors

Chapter 25

Palinacousis

MADELINE C. FIELDS* AND LARA V. MARCUSE
Department of Neurology, Icahn School of Medicine at Mount Sinai, Mount Sinai Hospital, New York, NY, USA

INTRODUCTION

Palinacousis is derived from the Greek words *palin*, which means again or anew, and *acousis*, meaning hearing. It was first described by Jacobs et al. (1971), who defined the phenomenon as "an auditory illusion of perseveration or persistence of sound impressions for seconds, minutes, or hours after the cessation of auditory stimulation." One of the salient features of palinacousis is that it does not occur spontaneously but is triggered by something in the environment.

The first published case of auditory perseveration can be found as early as the 1930s in a boy after a right temporal lobectomy (Rylander, 1939). In the 1950s, Mullen and Penfield (1959) were able to reproducibly elicit auditory perseveration by direct electric stimulation of the temporal lobe. In the 1970s, Jacobs et al. described the first 7 cases of "palinacousis," most of whom had recent seizures and lesions of the temporal lobe. However, over the last 75 years there have been only a limited number of case reports in the literature.

From the time of its original description, palinacousis has been associated with seizure phenomena (Jacobs et al., 1971). Electroencephalographic (EEG) recordings from patients with palinacousis have often shown the brain region to be hyperexcitable but frank electrographic seizures during palinacousis have not been captured. Despite even more modern technologies like magnetic resonance imaging (MRI), positron emission tomography (PET), and single-photon emission computed tomography (SPECT) scans, deciphering the etiology of palinacousis remains somewhat of a diagnostic conundrum.

DEFINING CHARACTERISTICS

Perseveration

The definition of perseveration is "the continuation or recurrence of an experience or activity without the appropriate stimulus" (MacNalty, 1961; Dorland, 1965). The term perseveration was first coined by Neisser in 1895. Goldstein (1948) described perseveration as an inability to inhibit a previous thought. When Jasper (1931) helped to define perseveration he wrote that the cause of this phenomenon is "the tendency of a set of neurons, once excited, to persist in the state of excitement autonomously, showing resistance to any change in this state." Allison (1966) went on to describe perseveration as "the continuance or recurrence of a purposeful response which is more appropriate to a preceding stimulus than to the succeeding one which has just been given, and which is essential to provoke it." Perseveration is involuntary. By definition it does not occur without previous stimulation and is not a spontaneous phenomenon (Allison, 1966).

Perseveration can occur in a variety of conditions, including epilepsy, psychoses, dementia, Parkinson's disease, traumatic brain injury, and stroke. Perseveration of movements, thoughts, and speech have all been well described. In Alzheimer's disease perseveration worsens as the disease progresses (Pekkala et al., 2007). Perseveration of thought is seen in patients with psychosis or obsessive-compulsive disorder and perseveration of speech can be seen in aphasic patients after a stroke. Palinacousis is a sound or phrase from the environment that replays in the person's head – auditory perseveration. Palinacousis can result in perseveration of speech. The perseveration of speech occurs because the individual is responding to an auditory perseveration as if it were real.

Illusion

Palinacousis is a form of illusion. Our senses of hearing, touch, sight, smell, and taste provide the brain with primary sensory data that the brain integrates, creating a perception that is experienced. An illusion is a

*Correspondence to: Madeline C. Fields, MD, Department of Neurology, Icahn School of Medicine at Mount Sinai, Mount Sinai Hospital, New York, NY 10029, USA. E-mail: madeline.fields@mssm.edu

misperception of real external stimuli. Illusions are not rare in human experience. A common example is our perception of three-dimensional space while looking at a two-dimensional picture of a road that travels off into the distance. Warren and Warren (1970) investigated the presence of auditory illusions in normal subjects in the 1970s. In their experiments, a phoneme or speech sound was deleted from an audiotape and replaced by another sound or cough of the same duration. Subjects heard the missing speech sound as clearly as any of the phonemes that were actually present. Additionally, after being told that a sound was missing, subjects could not distinguish the illusionary sound from the real one. If the phoneme was replaced by a period or silence rather than a cough the deletion became obvious. Misperceptions of sound can occur in normal healthy subjects, as just described. In palinacousis these misperceptions are recurring auditory phenomena when in fact none exists.

Illusions are different than either hallucinations or delusions. A hallucination is a false sensory perception experienced without real external stimuli. For example, a schizophrenic person may perceive voices talking or hear a voice speaking his/her thoughts when in fact none exists. However, these voices occur spontaneously, not in response to an environmental stimulus. Alternatively, delusions are fixed, false beliefs (not perceptions) that cannot be corrected by logic. Palinacousis is an auditory illusion. It is the persistence or repetition of a sound or phrase (auditory perseveration) when in fact the original sound or phrase no longer exists.

HISTORY

In the late 1800s Hughlings Jackson recognized that auditory hallucinations occur during epileptic attacks and could be localized to the "temporosphenodial" lobe.

In 1939 Rylander described a boy's status post partial right temporal lobectomy. "Sometimes he [the patient] repeated the same sentence. . .in a way typical of an epileptic person." When given psychologic tests "the patient sometimes repeated the answer to the question directly preceding the one put to him." It was thought that the boy repeated the answer to the question because he heard the question again. This is in fact the first recorded description of palinacousis.

In one of the cases of electric stimulation performed by Mahl et al. in 1964, the experimenter said "findings" as part of a sentence seconds prior to electric stimulation. During temporal-lobe stimulation the patient heard the word "findings" over and over.

In the 1950s and 1960s electric stimulation of the brain in epileptic patients helped to elucidate the anatomy and physiology of the brain as well as characterize epileptic seizures. Penfield and Perot (1963) found that auditory illusions were triggered by stimulating either cerebral hemisphere, primarily in the first temporal convolution, which is now more commonly referred to as the superior temporal gyrus (Fig. 25.1).

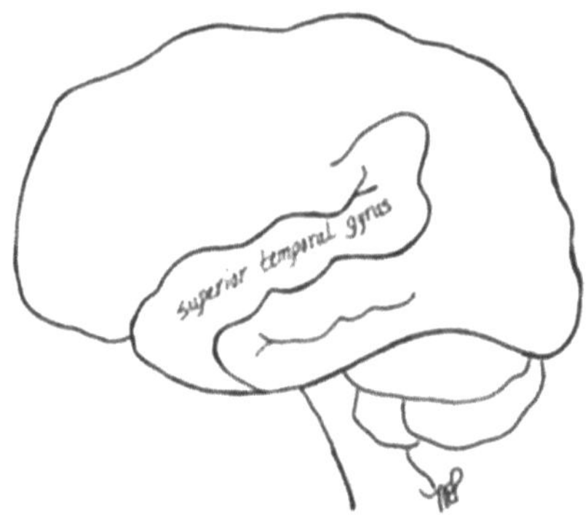

Fig. 25.1. Superior temporal gyrus. (Courtesy of Nora Patrone.)

Penfield describes "flashbacks" or "experiential responses" as those induced by electric stimulation of the cortex via implanted electrodes. These experiences were elicited in 40 of 520 patients whose temporal lobes were electrically stimulated. Over 60% of these patients experienced a sound and the accompanying emotions of a time period identified by the patient as coming from his/her past (Penfield and Perot, 1963). In case 66, the "operator's voice echoed." In other cases the quality of the operator's voice was altered. During electric stimulation, auditory sensations were most often referable to the ear contralateral to the temporal lobe being stimulated (Penfield and Rasmussen, 1958).

ANATOMY

For a detailed explanation of the auditory pathway, please see Chapters 1 and 2.

In the literature there is a single illustrative case of palinacousis caused by a lesion in the medial geniculate body (MGB). The MGB is the primary sensory input to the auditory cortex. The MBG has subsequently been considered the auditory relay nucleus and receives both auditory and somatosensory inputs. The MGB may be involved with sound pattern recognition, auditory memory and localization of sound in space (Pandya, 1995). In 1998 Fukutake and Hattori described a 49-year-old male with untreated hypertension who was also an 80-pack-year smoker. The patient was watching a gunshot scene on television when suddenly the sounds appeared louder and began to echo, predominantly on the right side. He shut the television off but then yelled at his wife to again shut the television off because he continued to hear sounds coming from the television. These illusionary sounds lasted for 10 minutes. Additionally, he was noted to have transient slurred speech, left mouth numbness, and hypesthesia in the left thigh. A computed tomography (CT) scan of the head showed a low-density lesion in the right posterior inferior thalamus and an MRI showed a hemorrhagic infarction in the right posterior inferior thalamus, including the MGB. This case describes auditory illusions caused by a relatively discrete lesion involving the MGB.

The primary auditory cortex (PAC) (Brodmann area 41) can be identified along the mediolateral axis of Heschl gyrus (HG), which is located in the superior temporal gyrus (Fig. 25.2A). Penfield and Perot (1963) found that stimulation of HG produced alterations in interpretation of sounds being heard. The PAC occupies most of HG (Fig. 25.2B). PAC is tonotopically organized with low-frequency stimuli more lateral and high-frequency stimuli more medial. There is still an ongoing argument as to the lateralization or asymmetries following acoustic stimulation (Meyer, 2011). Brodmann area 42, which is inferior to area 41, is also involved in the detection and recognition of speech. Brodmann areas 21 and 22 are the auditory association areas.

Dysfunction of the temporal-lobe association areas have been implicated in some cases as the cause of palinacousis (Auzou et al., 1995, 1997). These areas are referred to as Wernicke's area and deficits in the region can result in Wernicke's aphasia or fluent aphasia. Individuals suffering from Wernicke's aphasia speak fluently but they do not make sense. They are not able to repeat simple phrases and they are most often unaware of their deficit. They have great difficulty naming objects. Furthermore, they cannot make sense of auditory commands and will not follow commands well. The deficit in Wernicke's aphasia centers around the comprehension of speech and language and is more closely linked to audition than a Broca's aphasia. Broca's aphasia is the result of lesions in the frontal lobe. These individuals can understand language but they cannot produce it well and they are highly frustrated.

In 11 out of the 24 cases of palinacousis described in the literature, aphasia or a language disturbance is a part of the presenting complaint. This is likely due to the fact that the areas involved in auditory processing, comprehension, and memory are all centered within a limited area of the temporal lobe. Typically in patients experiencing palinacousis, once the aphasia improves or resolves they are able to describe auditory perseveration. For example, in 2007 Kim et al. described a 67-year-old woman with hypertension who suddenly developed a Wernicke-type aphasia. Three days after admission, when her symptoms were improving she "occasionally heard voices repeatedly when they actually were no longer produced." For example, when the physician-in-charge said, "Your symptoms are much improved," she heard "much improved" repeatedly afterward. This patient typically heard fragments of the latter part of the sentence. Sometimes, the repetitious sounds occurred a few hours after the actual conversation. When her symptoms first started she replied to the voice because she thought that somebody nearby actually talked to her. She soon realized that she was experiencing an illusion. MRI revealed an intracerebral hemorrhage in the left middle temporal gyrus. These authors concluded that, because the lesion itself did not directly involve the superior temporal gyrus

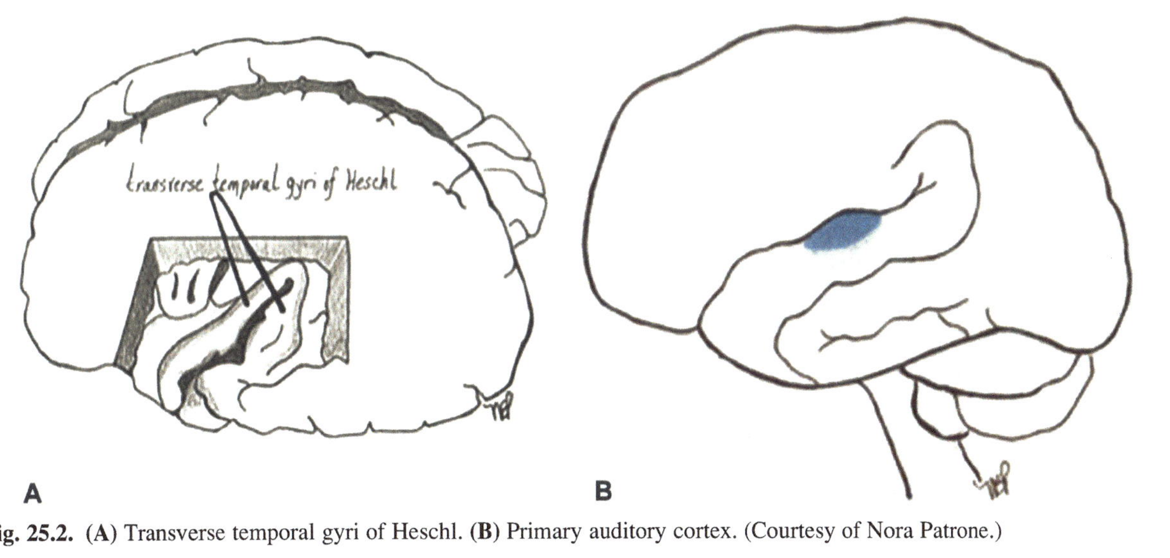

Fig. 25.2. **(A)** Transverse temporal gyri of Heschl. **(B)** Primary auditory cortex. (Courtesy of Nora Patrone.)

(Wernicke area or HG), there was a rapid improvement in the Wernicke aphasia and palinacousis. Additionally, hyperemia in the area surrounding the intracerebral hemorrhage may have resulted in either hyperexcitation of surrounding auditory pathways or transient loss of inhibitory fibers to the auditory system.

AUDITORY MEMORY

Comprehension of noise or discourse depends upon auditory short-term memory. Auditory short-term memory involves the posterior superior temporal gyrus and sulcus (Penhune et al., 1996; Leff et al., 2009) and the inferior parietal lobe (Becker et al., 1999; Buchsbaum et al., 2001). Based on a recent study looking at patients with brain lesions it has been proposed that verbal short-term memory is mediated by a left-hemisphere inferior frontal/posterior temporal circuit and that manipulation or rearrangement of information within short-term memory recruits posterior parietal cortices. Additionally, there is evidence that verbal short-term memory and language processing share overlapping neural substrates (Koenigs et al., 2011).

The temporal lobe is responsible for recalling past experiences as well as interpreting present experiences. A seizure in the temporal lobe can therefore trigger previously stored auditory experiences (Jacobs et al., 1973). During seizures or electric stimulation in the operating room, as described above, a patient may hear a familiar sound/voice. This auditory precept is thought to be stored and retrieved from one's memory. In palinacousis the brain registers and stores an auditory memory, which then misfires. Jacobs et al. (1973) suggest that the retrieval of stored auditory sounds by abnormal electric stimulation of the brain is similar to electric excitation of the temporal region which may revive stored thoughts/emotions which then appear as reality.

CLINICAL CHARACTERISTICS OF PALINACOUSIS

Content

Human voices tend to be the most frequent source of perseverated auditory stimuli. Typically the voice is one of someone nearby. For example, a 16-year-old girl described by Jacobs et al. (1971) heard the word "Eisenhower" in a discussion of government policies. The girl continued to hear "Eisenhower" repetitively over the next 10 minutes. The perceived sound almost always originates from the external environment. There is only one case where a woman heard her own thoughts perseverated and was thought to suffer from palinacousis (Terao and Matsunaga, 1999). In this case the patient thought "I am going to bed" and heard "going to bed" repeated about 100 times.

Single words or phrases seem to perseverate. Most often a fragment of the original sentence is heard repeatedly (Jacobs et al., 1973; Auzou et al., 1995, 1997; Mohamed et al., 2012). Non-verbal stimuli can also recur, including ringing of a doorbell, a telephone, sirens, whispering, grinding of subway wheels, toilet flushing, gunshots from a television, and pages of a book turning (Jacobs et al, 1971; Fukutake and Hattori, 1998; Wustmann and Gutmann, 2007). Another feature of palinacousis is that the content of the voice/sound is neutral and not disturbing. This fact is important when differentiating palinacousis from the persecutory voices of psychotic illness.

Quality

The perseverated sound/sentence may seem louder or softer. It may have a muffled quality. A few of the patients described by Jacobs et al. in 1973 mentioned feeling a "clogged ear" prior to experiencing palinacousis as well as during palinacousis. Often times the utterance sounds distorted to the patient. By far the most common quality described by patients with palinacousis is that of an "echo." For example, in one of the cases described by DiDio et al. (2007), a 72-year-old woman with a history of a left temporal meningioma suffered from recurrent aphasic seizures. On one occasion she presented with global aphasia on exam and was found to be perseverative. This patient "complained of hearing people's voices echoing in the room emanating from over her right shoulder." During this time the patient could be found turning and looking to the right after a presented auditory stimulus. She was noted to perseverate on a previous action or command. The patient was ultimately able to say that she heard the examiners' voices echo phrases over and over again, from beyond her right side.

The repetitious sound typically blocks out other sounds. This results in the repetitious sound having the same impact as the original sounds. Very often the perseverated sound seems so real that the patient responds by behaviorally searching for the source in the environment (Jacobs et al., 1973; Auzou et al., 1995; DiDio et al., 2007).

In a single case, covering the ear suppressed the perseverated sound (Jacobs et al., 1973). In this case the patient was listening to the television and could not understand what was occurring because he heard "the sound repetitively bouncing off the wall on his right." When he put his hand over his right ear he could decipher the voices on the television with the left ear. Other attempts at covering the ear to eliminate the palinacousis have not been reproducible (Patterson et al., 1988;

Prueter et al., 2002; Bega et al., 2014). In the example by Patterson a previously healthy 50 year-old woman presented with complaints of a headache 2 days prior to admission as well as nausea, vomiting, and pain in her right thigh. In the emergency room she was confused with global aphasia. Shortly thereafter she had two tonic-clonic seizures. A day later she was more alert but with a fluent aphasia (Wernicke's aphasia). On the third day she complained of "echoing voices" in her right ear. "These did not arise spontaneously, but represented recurring words or fragments of sentences that the patient had heard during the preceding minutes." Sounds other than speech also "echoed" in her right ear. The sounds were not altered by occlusion of either or both ears. The patient was found to have a large contrast-enhancing lesion in the left temporoparietal region with mass effect by CT scan.

Latency

The perseverated sound typically occurs immediately following the original stimulus. For example, if the physician asks "How are you?" the patient will hear "How are you? How are you? How are you?" immediately after the question is asked. In other patients the perseverated sound is triggered by another sound.

Rarely, palinacousis appears with a latency of minutes, hours, or days after the original stimulus (Jacobs et al., 1971; Kim et al., 2007; Mohamed et al., 2012). In one case described by Jacobs et al. (1973), the patient heard the perseverated sound 24 hours after the original stimulus. This sound would then recur periodically thereafter for a period of 4 days. This certainly makes the diagnosis much more difficult.

Lateralization of sound

In 11 cases (46%) in the literature, the auditory illusion is heard in the opposite auditory space or in the opposite ear to the cerebral lesion (Jacobs et al., 1973; Patterson et al., 1988; Auzou et al., 1995, 1997; Prueter et al., 2002; DiDio et al., 2007; Mohamed et al., 2012) (Table 25.1). In 2 (8%) documented cases the perseverated stimulus was heard ipsilateral to the lesion (Jacobs et al, 1973; Fukutake and Hattori, 1998). In the remaining cases no lesion was found, the patients could not lateralize/localize the sound, or it seemed to perseverate from the initial stimulus location (Jacobs et al., 1971; Bekeny and Peter, 1972; Malone and Leiman, 1983; Masson et al., 1993; Freudenreich and McEvoy, 1996; Terao and Matsunaga, 1999; DiDio et al., 2007; Kim et al., 2007; Wustmann et al., 2010; Bega et al., 2014). DiDio et al. (2007) described a 49-year-old woman with a history of depression, anxiety, and known left parietal cavernous angioma. She presented a number of times to the hospital with recurrent episodes of aphasia. While the patient was aphasic she retrospectively was able to describe experiencing an echo in her head immediately after a stimulus was presented. This original stimulus would echo numerous times, sometimes lasting for more than 10 minutes. This echoing remained at a constant intensity. As her aphasia resolved she perceived a more complex echoing phenomenon: she perceived a short-lived echo localized in space to a subsequently presented auditory stimulus. The echo was heard only when a second auditory stimulus was presented. For example, she heard the first sentence presented by the doctor echo only after the doctor uttered a second sentence. The second auditory stimulus was extinguished by the resurgence of the first stimulus, in the form of an echo. At times this echoing phenomenon was even more complicated. If three auditory stimuli were presented one after another, the third stimulus would be extinguished by the first two stimuli echoing simultaneously. Additionally, this patient heard her own voice echoing from the mouth of the doctor as he answered her questions. This echoing would last for seconds.

Lesion location

In 12 cases (50%) the lesion was on the left side. In 5 cases (21%) the lesion was on the right side. The remainder of the cases either had bilateral lesions or no lesions (Table 25.1).

The most common location for a cerebral lesion is the temporal lobe (Jacobs et al., 1973; Patterson et al., 1988; Auzou et al., 1995, 1997; DiDio et al., 2007; Kim et al., 2007; Bega et al., 2014) (Table 25.1). Other areas of the brain that can produce palinacousis include the MGB (Fukutake and Hattori, 1998), and parietal lobe (Masson et al., 1993; DiDio et al., 2007; Mohamed et al., 2012; Bega et al., 2014). All brain areas documented to produce palinacousis are known to process auditory stimuli.

DIFFERENTIAL DIAGNOSIS

Auditory hallucinations of psychotic illness

One of the important distinctions to make is between palinacousis and auditory hallucinations caused by an underlying psychiatric condition (see Chapter 24). Auditory hallucinations are the most common form of hallucination, often occurring in schizophrenia and in bipolar disorders (Andreasen, 1997). In fact, up to 75% of schizophrenic patients experience auditory hallucinations (Nayani and David, 1996). Thought echoing, also known as Schneider's symptoms, is characteristic of schizophrenia. Both palinacousis and auditory hallucinations are false perceptions of sound. However, there are

Table 25.1

List of patients with palinacousis

Case	Author	Sex and age	Lesion lobe	Lesion laterality	Auditory space	Lesion etiology	Recent seizure and/or abnormal EEG
1	Jacobs et al. (1971)	F40	Temporal	R	L	Schwannoma resected	
2		F16	Temporal-parietal-occipital	L	R		Yes
3		UK	Temporal-occipital	R	L		Yes
4		UK	Temporal	L	R		Yes
5		M60	Temporal	L	R		Yes
6		UK	Indefinite		L/R		Yes
7		M64	Temporal-occipital	R	R	Meningioma	No
8	Bekeny and Peter (1972)	F38	Frontal-temporal	L	None	Cystercercosis	Yes
9	Malone and Leiman (1983)	F75	None	None	None	None	Yes
10	Patterson et al. (1988)	F50	Temporal-parietal	L	R	Grade IV astrocytoma	Yes
11	Masson et al. (1993)	M60	Temporal-parietal	R	None	ICH	Yes
12	Auzou et al. (1995)	M78	Temporal-parietal	L	R	GBM	Yes
13	Freudenreich and McEvoy (1996)	M32	None	None	None	None	No
14	Auzou et al. (1997)	F51	Temporal-parietal	L	R	GBM	Yes
15	Fukutake and Hattori (1998)	M49	MGB	R	R	Hemorrhagic stroke	N/a
16	Terao and Matsunaga (1999)	F75	None	None	None	None	No
17	Prueter et al. (2002)	F20	Temporal	L	R	None	Yes
18	Wustmann and Gutmann (2007)	M49	None	None	L	None	No
19	Kim et al. (2007)	F67	Temporal	L	None	ICH	Yes
20	DiDio et al. (2007)	F49	Parietal	L	*	Cavernous angioma	Yes
21		F72	Temporal	L	R	Meningioma	Yes
22	Wustmann et al. (2010)	M18	None	None	None	None	No
23	Mohamed et al. (2012)	W68	Parietal	L	R	Metastases	Yes
24	Bega et al. (2014)	M61	Temporal-parietal/temporal	R/L	None	Metastases	Yes

EEG, electroencephalogram; M, male; F, female; L, left; R, right; GBM, glioblastoma multiforme; ICH, intracerebral hemorrhage; MGB, Medial geniculate body; N/a, not applicable; UK, unknown.

*Localized to subsequently presented auditory stimuli.

Table 25.2

Palinacousis vs auditory hallucinations of psychotic illness

Palinacousis	Auditory hallucinations of psychotic Illness
Neutral content	Persecutory
Replicas or fragments of original environmental sounds	Highly personalized
No other psychotic symptoms	One in a constellation of psychotic symptoms
Triggered by external stimuli	No external stimuli

a number of differences between palinacousis and auditory hallucinations of psychotic illness (Table 25.2). Auditory hallucinations tend to be persecutory and personalized, and are typically found in patients with other psychotic symptoms. Auditory hallucinations occur without an external stimulus, whereas palinacousis is neutral in content, environmentally provoked, and usually seen in the absence of psychotic symptoms.

There have been 4 cases described in the literature where patients with schizophrenia concomitantly suffered from palinacousis (Malone and Leiman, 1983; Freudenreich and McEvoy, 1996; Prueter et al., 2002; Wustmann et al., 2010).

The first case of palinacousis in a patient with schizophrenia was described in 1983 by Malone and Leiman. A 75-year-old woman who 15 years earlier had been diagnosed with schizophrenia began to complain of repeatedly hearing her own voice, voices of people around her, and noises from the television or radio. She described these voices/noises repeating to her exactly as she heard them the first time. The voice/noise would repeat until another external sound occurred, at which point the new sound would be repeated. Seven months after these symptoms started she presented to the hospital. A head CT showed degenerative changes without any focal lesion. An EEG showed generalized slowing with sharp waves in the left temporal regions, which is not an expected finding in schizophrenia. Seven days after starting phenytoin her palinacousis ceased. A repeat EEG 5 days later showed only disorganization, no left temporal irritability. The palinacousis was thought to be related to ongoing perhaps subclinical seizures.

In 1996 Freudenreich and McEvoy described a 32-year-old male with a 10-year history of schizophrenia with new-onset palinacousis. In this patient there was a history of traumatic brain injury and suspected seizure disorder previously treated with phenytoin. At the time of his palinacousis he described "an echo; you hear something, and it repeats the same thing several times until it goes away." These symptoms were seen in conjunction with auras of bad smells as well as sensory and visual hallucinations. This patient had two normal EEGs, one of which was sleep-deprived (not clearly performed at the time of an actual event). He also had a normal MRI. Despite the normal EEG, the clinicians suspected seizures and carbamazepine was initiated. The palinacousis then resolved. (Note that a normal EEG does not rule out the presence of seizures.)

In 2002 Prueter et al. reported a 20-year-old woman with a 1-year history of schizophrenia who described "echoing voices" in her right ear. Her echoes consisted of words other people said or even words she herself had said. Head CT, MRI, and fluorodeoxyglucose (FDG)-PET were normal. EEGs showed a normal background with slowing over the left temporal region. Dipole modeling of the brain waves demonstrated a left mesiotemporal focus. Four weeks after starting carbamazepine the palinacousis resolved.

Lastly, in a patient with newly diagnosed schizophrenia the withdrawal of risperdone triggered palinacousis as well as music hallucinations (Wustmann et al., 2010). This patient described hearing contents of conversations after they had ended. Both EEG and MRI were normal. After 6 weeks (the last 2 of which he was treated with the maximum dose of an antipsychotic (sertindole)), the patient's palinacousis and musical hallucinations resolved.

Postictal psychosis

Auditory hallucinations have been known to occur in epileptics either as part of their ictal phenomenon or as part of a postictal psychosis (PIP). PIP consists of a lucid interval following a seizure prior to the appearance of psychotic symptoms (Kanemoto et al., 2010). The lucid interval after a seizure may last from 12 hours to 6 days. Psychosis typically occurs within a week of normal functioning after a seizure. The psychosis may last from a day to more than 3 months (Logsdail and Toone, 1988).

Auditory hallucinations have been described in people with PIP (Slater and Bread, 1963; Logsdail and Toone, 1988; Kanemoto et al., 1996). However these auditory hallucinations have a more menacing quality (Slater and Bread, 1963).

Echolalia

Echolalia is derived from Greek *echo*, "to repeat," and *laliá*, meaning "speech" or "talk." Echolalia is the meaningless repetition of words or phrases immediately after their occurrence. This phenomenon is seen normally in children and pathologically in conditions such as Tourette's and autism. This behavior is automatic and unintentional. Echolalia and palinacousis are similar in that there is a repetition of heard environmental stimulation. They are different in that echolalia is a motor event with speech repetition and palinacousis involves auditory perseveration. Furthermore, in echolalia, the patient will most often repeat the most recent phrase. In palinacousis the patient typically hears the original presenting sound/phrase over and over again. Palinacousis occurs after environmental auditory stimulation; however, it tends to be less changeable, and people often perceive the same sound over and over again despite new sounds in the environment.

Palilalia

Palilalia is derived from the Greek word *pálin,* meaning "again," and *laliá,* meaning "speech" or "to talk." Palilalia was originally described in 1908 by Alexandre-Achille Souques. He found this condition in a patient with a right brain stroke leading to left-sided hemiplegia. In palilalia the patient repeats the last one or two words of a sentence, often with increasing rapidity and decreasing volume. Palilalia is often seen in stroke patients, especially with pseudobulbar palsy and in postencephalitic parkinsonism. Palilalia occurs as readily in response to the first as to the second stimulus.

Tinnitus

Tinnitus is a peripheral auditory phenomenon. There is typically an intense or prolonged antecedent acoustic exposure (acoustic trauma). For example, many people have experienced a temporary ringing in the ears after a loud rock concert. Unlike palinacousis, the sound is usually stereotyped, with no resemblance in quality to the actual stimulus. Additionally, the sound is usually experienced in the head or bilateral ears, unlike palinacousis, which is experienced as coming from the external environment, often on one particular side (Jacobs et al., 1973). There is a single case of tinnitus occurring with palinacousis (Bega et al., 2014). In this patient the palinacousis was distinguishable from the tinnitus.

PATHOPHYSIOLOGY OF PALINACOUSIS

The pathophysiology of palinacousis remains unclear, but it does seem to typically occur in the setting of seizures. The debate continues as to whether palinacousis is a direct symptom of an ongoing seizure (ictal) or a kind of release phenomenon of postictal brain tissue.

Ictal

The original reports by Penfield and Perot in 1963 found that auditory inputs could be pathologically retrieved from memory by electric stimulation. A number of authors have suggested that palinacousis is an ictal (abnormal electric) phenomenon (Jacobs et al., 1973; Malone and Leiman, 1983; Patterson et al., 1988; Auzou et al., 1997; Prueter et al., 2002; DiDio et al., 2007). This conclusion is based on a number of characteristics, including, first, its close temporal relationship to seizures. In 6 out of 7 patients seen by Jacobs et al. (1973), palinacousis appeared just prior to a convulsion or minutes to hours following a seizure. Auzou et al. (1995) described a patient with recurrent partial seizures over 4 days who experienced palinacousis during the same time period but at different moments. Auzou et al. (1997) described another patient with a glioblastoma multiforme who developed partial status epilepticus after a brain biopsy and shortly thereafter developed palinacousis. Patterson et al. (1988) had a patient with palinacousis after a generalized tonic-clonic seizure. Additionally, DiDio et al. (2007) observed palinacousis in 2 patients after aphasic seizures.

Second, in a number of the cases reported in the literature there was a positive response to antiseizure medications (Malone and Leiman, 1983; Patterson et al., 1988; Freudenreich and McEvoy, 1996; Prueter et al., 2002).

Lastly, although not always performed exactly at the time patients were experiencing palinacousis, EEG data often revealed an abnormal focus (Jacobs et al., 1971; Bekeny and Peter, 1972; Malone and Leiman, 1983; Patterson et al., 1988; Masson et al., 1993; Auzou et al., 1995, 1997; Prueter et al., 2002; DiDio et al., 2007; Kim et al., 2007; Mohamed et al., 2012; Bega et al., 2014) (Table 25.1). This abnormal focus consisted of focal slowing, spikes, or sharp waves. Focal slowing is indicative of underlying cerebral dysfunction and sometimes can be seen in patients with structural abnormalities. Spikes and sharp waves are seen in patients with seizure disorders or in those who are predisposed to seizures. In the article by DiDio et al. (2007), periodic spikes and/or sharp waves (lateralized periodic discharges (LPD), previously known as periodic lateralized epileptiform discharges) were seen in both patients described (Figs 25.3 and 25.4). LPD can be seen in a variety of conditions, including stroke, tumors, and meningioencephalitis. Although considered an interictal pattern, they are associated with epileptic seizures and even status epilepticus (Baykan et al., 2000).

Although no electrographic seizures up to this point have been recorded during an episode of palinacousis, there are a number of possibilities to explain this lack of sensitivity. As previously mentioned, the EEG was not always performed while the patients were experiencing palinacousis. Most of the EEGs were routine recordings (20–40-minute studies) and may not have been sufficient to capture a seizure. The epileptic focus may have been too small to be detected by scalp electrodes. An epileptic focus has to involve at least 6 cm^2 to be seen with scalp electrodes.

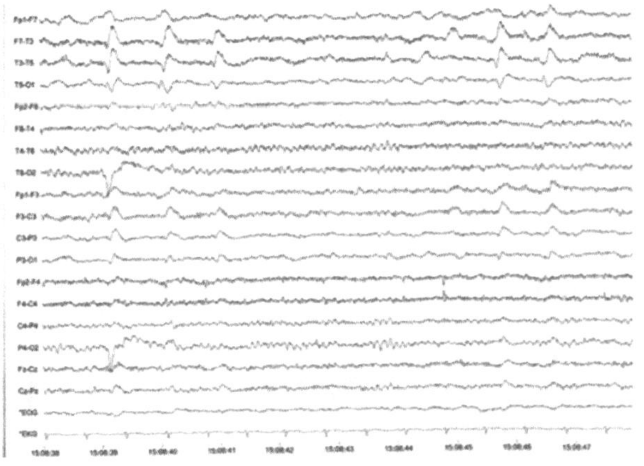

Fig. 25.3. Electroencephalograph showing short runs of spikes and sharp waves in the left temporal region. (Reproduced from DiDio et al., 2007.)

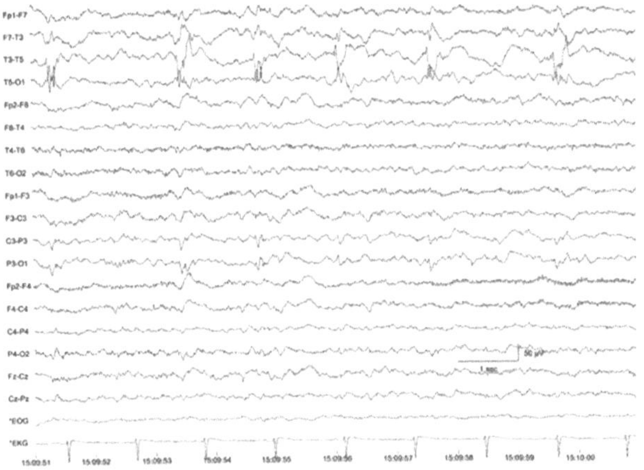

Fig. 25.4. Electroencephalograph showing lateralized periodic discharges seen maximally in the left posterior temporal lobe. (Reproduced from DiDio et al., 2007.)

Postictal

Alternatively, some authors have suggested that palinacousis is a postictal or interictal phenomenon. Below is a case with an extremely detailed workup supporting this hypothesis.

Mohamed et al. (2012) described a 68-year-old woman who presented to medical attention due to a change in mental status and unintelligible speech. On exam she was found to have components of expressive and receptive aphasia as well as fluent speech with paraphasic errors. The remainder of her neurologic examination was normal. A brain MRI demonstrated a 10-mm enhancing mass in the left inferior parietal lobe as well as increased T2 signal and diffusion restriction in the left hippocampus, pulvinar, and insular cortex without enhancement. A chest CT revealed a mass in the left upper lung, which turned out to be an invasive non-small-cell carcinoma on biopsy. Her brain lesion was felt most likely to be a metastatic lesion. A routine EEG showed LPD in the left temporocentral region. She was given intravenous levetiracetam.

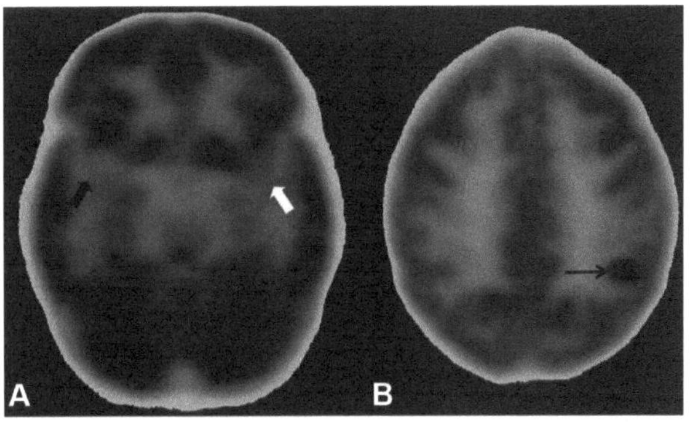

Fig. 25.5. Selective images of fluorodeoxyglucose positron emission tomography (FDG-PET) scan obtained during events of palinacousis. (**A**) Hypometabolism in bilateral medial temporal lobes, left (white arrow) worse than right (black arrow). (**B**) Hypermetabolism in the left parietal mass (black arrow). (Reproduced from Mohamed et al., 2012.)

The following day her aphasia resolved but she complained of palinacousis. She heard words or phrases from the end of a sentence or command lasting for minutes to hours arising from her right ear. She would also repeatedly hear music from the television played the previous night or conversations between people. After she mentioned this phenomenon to her physicians she was connected to a continuous EEG monitor. She experienced palinacousis while on the EEG, which showed left temporal slowing but no seizures or epileptiform waves. Her dose of levetiracetam was increased without any benefit. Additionally, the patient underwent FDG-PET during an episode of palinacousis, which revealed hypometabolism of the bilateral medial temporal cortices which was worse on the left. The metastatic lesion showed increased tracer uptake (Fig. 25.5). A repeat MRI showed decrease in the T2 hyperintensities in the left temporal cortex and near-complete resolution of the diffusion restriction. The authors concluded that palinacousis is a postictal phenomenon because it started after levetiracetam was administered and did not improve when this medication was increased. Her reversible MRI findings were thought to be secondary to status epilepticus. The EEG performed during the patient's episode of palinacousis only showed left temporal slowing. Additionally, there was decreased FDG uptake on PET scan consistent with postictal hypometabolism.

Below is an e-mail recently received by one of the author's patients. This note was written by a woman who is status post resection of a right temporal arteriovenous malformation and had suffered from a recent seizure with left-arm numbness and a left-sided dense hemianopsia.

> *I'm a little concerned I am not only hearing sounds in my left ear but more worrying I hear an echo of people talking around me for 20 minutes to longer even after when it is quiet around. It is hard to ignore. I know I am not crazy but it started in the airplane the day after the seizure. I am in Vegas now and in the plane on the way in I started hearing sounds and echos. Is this a phenomenon that shall pass?*

ETIOLOGY

Various types of lesions have been noted to cause palinacousis. These include vascular malformations, intracerebral hemorrhages, primary brain tumors, including meningiomas, astrocytomas, and glioblastomas, metastatic tumors, stroke, traumatic brain injury, and meningioencephalitis caused by cystercercosis (Bekeny and Peter, 1972; Jacobs et al., 1973; Patterson et al., 1988; Masson et al., 1993; Freudenreich and McEvoy, 1996; Fukutake and Hattori, 1998; DiDio et al., 2007; Kim et al., 2007; Mohamed et al., 2012; Bega et al., 2014). As noted by Mohamed et al. (2012), various types of lesions result in a similar phenomenology and therefore it is more likely the location rather than the etiology that results in palinacousis. Additionally, there are a number of cases where no lesion was identified yet palinacousis was attributed to alcohol withdrawal (Wustmann and Gutmann, 2007), seizures themselves, or antipsychotic medication withdrawal (Wustmann et al., 2010).

ASSOCIATED PHENOMENA

Palinopsia

Palinopsia is derived from the Greek word *palin*, for "again," and *opsia*, for "seeing." Palinopsia is a perceptual illusion in which images persist or recur after the viewed object has been removed (Critchley, 1951). In both palinacousis and palinopsia typically only a segment of the original perceptual event is perseverated. Both of these phenomena occur around the time of seizures. There have been a number of cases where palinacousis was seen in conjunction with palinopsia (Jacobs et al., 1973; Auzou et al., 1997). Palinopsia is thought to be the visual equivalent of palinacousis.

Music hallucinations

There are 2 cases of palinacousis occurring in conjunction with music hallucinations (Terao and Matsunaga, 1999; Wustmann et al., 2010). Music hallucinations are characterized by perceiving music without an external source. In Terao's case a 75-year-old woman heard old Japanese songs repeatedly as well as the voice of conversations exactly as if they were being repeated over and over. She had no clear seizure history. Her EEG was unremarkable. Her MRI showed possible age-related changes. She did have a SPECT scan of the brain, which showed hypoperfusion in the bilateral thalamus and basal ganglia. This woman was found to have syphilis (not neurosyphilis, as lumbar puncture was negative). Carbamazepine caused itching and anorexia and was subsequently discontinued. She was treated with penicillin and her symptoms gradually improved.

CONCLUSIONS

Palinacousis is a rare phenomenon but likely more common than is currently recognized and reported. Verbal perseveration may be a clinical sign that a patient is experiencing palinacousis as a patient may respond to the auditory stimulation as if it were real. Palinacousis typically is perceived from the side of space opposite the brain lesion and is less commonly seen ipsilateral or from the source of the environmental trigger. The brain lesion or dysfunction most often localizes to the temporal lobe, and it appears to be more common with left-sided lesions. This may be because language function is most often on the left, and perhaps the left brain is more vulnerable to auditory perseveration. While the underlying mechanism of palinacousis is still not clear, it nearly always occurs in the clinical setting of recent or ongoing seizures.

References

Allison RS (1966). Perseveration as a sign of diffuse and focal brain damage. I. Br Med J 2: 1027–1032.

Andreasen NC (1997). Linking mind and brain in the study of mental illness: a project for a scientific psychopathology. Science 275: 1586–1593.

Auzou P, Hannequin D, Cochin JP et al. (1995). Palinacousie avec hémianacouse par lésion temporale gauche. Rev Neurol 151: 129–131.

Auzou P, Parain D, Ozsancak C et al. (1997). Enregistrements EEG contemporains d'épisodes de palinacousie et de palinopsie. Rev Neurol 153: 687–689.

Baykan B, Kinay D, Gokyigit A et al. (2000). Periodic lateralized epileptiform discharges: association with seizures. Seizure 9: 402–406.

Becker JT, MacAndrew DK, Fiez JA (1999). A comment on the functional localization of the phonological storage subsystem of working memory. Brain Cogn 41: 27–38.

Bega D, Wang N, Klein J (2014). Reversible palinacousis from intracranial metastases. Neurohospitalist 4: 22–26.

Bekeny G, Peter A (1972). Cysticercus meningoencephalitis hallasi perseveratioval. Orv Hetil 113: 3083–3086.

Buchsbaum BR, Hickok G, Humphries C (2001). Role of left superior temporal gyrus in phonological processing for speech perception and production. Cogn Sci 26: 663–378.

Critchley M (1951). Types of visual perseveration; "palinopsia" and illusory visual spread. Brain 74: 267–299.

DiDio AS, Fields MC, Rowan AJ (2007). Palinacousis – auditory perseveration: two cases and a review of the literature. Epilepsia 48: 1801–1806.

Dorland J (1965). Dorland's Illustrated Medical Dictionary, 24th edn. Saunders, London.

Freudenreich O, McEvoy JP (1996). Palincousis after close-head injury in a patient with schizophrenia. J Clin Psychopharmacol 16: 94.

Fukutake T, Hattori T (1998). Auditory illusions caused by a small lesion in the right medical geniculate body. Neurology 51: 1469–1471.

Goldstein K (1948). Language and Language Disturbances; aphasic symptom complexes and their significance for medicine and theory of language, Grune & Stratton, New York.

Jacobs L, Feldman M, Bender M (1971). Palinacousis or persistent auditory sensations. Trans Amer Neurol Assoc 96: 123–126.

Jacobs L, Feldman M, Diamond SP et al. (1973). Palincousis: persistent or recurring auditory sensations. Cortex 9: 275–287.

Jasper HH (1931). Is perseveration a functional unit participating in all behavior processes? J Soc Psychol 2: 28–50.

Kanemoto K, Kawasaki J, Kawai I (1996). Postictal psychosis: a comparison with acute interictal and chronic psychoses. Epilepsia 37: 551–556.

Kanemoto K, Tadokoro Y, Oshima T (2010). Violence and postictal psychosis: a comparison of postictal psychosis, interictal psychosis, and postictal confusion. Epilepsy Behav 19: 162–166.

Kim J, Kwon M, Jung J (2007). Palinacousis in temporal lobe intracerebral hemorrhage. Neurology 68: 1321–1322.

Koenigs M, Acheson DJ, Barbey AK et al. (2011). Areas of left perisylvian cortex mediate auditory-verbal short-term memory. Neuropsychologia 49: 3621–3619.

Leff AP, Schofield TM, Crinion JT et al. (2009). The left superior temporal gyrus is a shared substrate for auditory short-term memory and speech comprehension: evidence from 210 patients with stroke. Brain 132: 3401–3410.

Logsdail S, Toone BK (1988). Post-ictal psychosis. A clinical and phenomenological description. Br J Psychiatry 152: 246–262.

MacNalty AS (1961). The British Medical Dictionary, Caxton, London.

Mahl GF, Rothenberg A, Delgado JM et al. (1964). Psychological responses in the human to intracerebral electrical stimulation. Psychosom Med 26: 337–368.

Malone GL, Leiman HI (1983). Differential diagnosis of palinacousis in a psychiatric patient. Am J Psychiatry 140: 1067–1068.

Masson C, Sztern A, Cambier J et al. (1993). Palinacousie en relation avec une hémorrhagie temporo-pariétale droite. Presse Med 22: 12.

Meyer K (2011). Primary sensory cortices, top-down projections and conscious experience. Prog Neurobiol 94: 408–417.

Mohamed W, Ahuja N, Shah A (2012). Palinacousis – evidence to suggest a post-ictal phenomenon. J Neurol Sci 31: 6–12.

Mullen S, Penfield W (1959). Illusions of comparative interpretation and emotion. Production by epileptic discharge and by electrical stimulation in the temporal cortex. Arch Neurol Psychiatry 81: 269–284.

Nayani TH, David AS (1996). The auditory hallucination: a phenomenologic survey. Psychol Med 26: 177–189.

Neisser A (1895). Krankenvorstellung (Fall von 'Asymbolie'). Allgemeine Zeitschrift Für Psychiatrie 51: 1061–1021.

Pandya DN (1995). Anatomy of the auditory cortex. Rev Neurol (Paris) 151: 486–494.

Patterson MC, Tomlinson FH, Stuart GC (1988). Palinacousis: a Case Report. Neurosurgery 22: 1088–1090.

Pekkala S, Albert ML, Spiro 3rd A, et al. (2007). Perseveration in Alzheimer's disease. Dement Geriatr Cogn Disord 26: 109–114.

Penfield W, Perot P (1963). The brain's record of auditory and visual experience. A final summary and discussion. Brain 86: 595–696.

Penfield W, Rasmussen T (1958). Cortical stimulation, psychical responses. The excitable cortex in conscious man, MacMillan, New York.

Penhune VP, Zatorre RJ, MacDonald JD et al. (1996). Interhemispheric anatomical differences in human primary auditory cortex: probabilistic mapping and volume measurement form magnetic resonance scans. Cereb Cortex 6: 661–672.

Prueter C, Waberski TD, Norra C et al. (2002). Palinacousis leading to the diagnosis of temporal lobe seizures in a patient with schizophrenia. Seizure 11: 198–200.

Rylander G (1939). Personality Changes after Operations on Frontal Lobes, Oxford University Press, London.

Slater E, Bread W (1963). Postictal psychosis typically occurs with a lucid interval, after seizures or a cluster of seizures and is characterized by mild confusion and subsequent amnesia. Br J Psychiatry 109: 95–112.

Souques A (1908). Palilalie. Rev Neurol 16: 340–342.

Terao T, Matsunaga K (1999). Musical hallucinations and palinacousis. Psychopathology 32: 57–59.

Warren RM, Warren RP (1970). Auditory illusions and confusions. Sci Am 223: 30–36.

Wustmann T, Gutmann P (2007). Palinakusis bei Alkoholhalluzinose. Psychiat Prax 34: 302–304.

Wustmann T, Roettig S, Marneros A (2010). Palinacousis as discontinuation symptom after antipsychotic treatment in a patient with first-episode schizophrenia. Psychopathology 43: 248–249.

Handbook of Clinical Neurology, Vol. 129 (3rd series)
The Human Auditory System
G.G. Celesia and G. Hickok, Editors

Chapter 26

Musicogenic epilepsy

JOHN STERN*
Department of Neurology, Geffen School of Medicine, University of California, Los Angeles, CA, USA

SEIZURES AND REFLEX EPILEPSIES

Musicogenic epilepsy is a form of reflex epilepsy, that is, an epilepsy with seizures that are consistently triggered by a specific stimulus. For musicogenic epilepsy, the stimulus is music, but the seizures are epileptic seizures with the other features typical of epileptic seizures. As a whole, epileptic seizures are episodes of cerebral dysfunction due to transiently abnormal neuronal synchronization and excitation, and this dysfunction produces a behavioral change that may be subjective or objective. Subjective experiences during a seizure can include hallucinations (see Chapter 24), and paramnestic phenomena such as *déjá vu*, but a large number of specific experiences occur. Objective experiences may be a loss of responsiveness with surroundings or a generalized convulsion with bilateral stiffening, falling, and uncontrolled movements. How the seizure manifests depends on where the seizure starts and where the epileptic abnormality then propagates within the brain. Seizures that originate focally within one cerebral hemisphere may produce a subjective experience and such seizures also can propagate to produce a generalized convulsion. These are termed focal-onset seizures and contrast with generalized-onset seizures, which originate with a bihemispheric distribution on scalp electroencephalogram (EEG).

Epilepsies are divided into syndromes that each characterize a clinical condition comprising the seizures, other neurologic features, natural history, and treatment options. Reflex epilepsies are one group of epilepsy syndromes, and this group includes both epilepsies with focal-onset seizures and epilepsies with generalized-onset seizures. Moreover, reflex seizures can occur in epilepsies that are due to diffuse brain dysfunction with accompanying intellectual disability and also epilepsies without significant abnormality other than seizures. As such, reflex epilepsies include a broad range of epilepsy syndromes and are distinct only because of the association of seizure occurrences with particular triggers. These triggers are almost always either a sensory stimulus or a cognitive state.

Epilepsies that are not considered reflex epilepsies also may demonstrate sensitivity to triggers, but the triggers are ones that impact more global aspects of brain function. Examples of non-specific triggers are sleep deprivation, fever, and sedative withdrawal (Engel, 2001, 2006). The fact that these non-specific factors, when sufficiently extreme, can produce seizures in people without epilepsy is one justification for considering them non-specific triggers and therefore not a reflex epilepsy trigger. Another reason is that a majority of people with epilepsy will identify a non-specific factor as a seizure trigger and these individuals encompass the full range of epilepsy syndromes. As such, the non-specific triggers may be affecting the seizure threshold and are not a component of a particular seizure onset or epilepsy type. However, some overlap exists between reflex epilepsies and non-reflex epilepsies with provoked seizures (Illingworth and Ring, 2013).

Approximately 6.5% of people with epilepsy have exclusively reflex seizures, as was found in a series of 1000 consecutive patients (Symonds, 1959). The recognized precipitating stimuli for reflex seizures include visual stimuli, somatosensory stimuli, hot-water immersion, arithmetic, reading, other specific cognitive tasks or states, and music. Among the reflex epilepsies in this series, approximately 80% are syndromes with sensitivity to a visual trigger (Symonds, 1959; Kasteleijn-Nolst Trenite, 2012). The visual triggers may be simple or complex stimuli, and this is also true for non-visual triggers. Flickering light is an example of a simple visual stimulus, and a specific scene or image with multiple features is an example of a complex visual stimulus. For an individual

*Correspondence to: John Stern, MD, Department of Neurology, Geffen School of Medicine, University of California, Los Angeles, CA 90095, USA. Tel: +1-310-825-5745, E-mail: jstern@ucla.edu

with reflex epilepsy, the response to a trigger is affected by idiosyncratic sensitivity, duration of exposure, and whether a seizure recently occurred. A refractory period of insensitivity commonly follows a reflex seizure, and this period may be as short as seconds but it more often lasts minutes to hours (Forster et al., 1963).

One important aspect of the clinical burden of epilepsy is the unpredictable timing of seizures. For the vast majority of people with epilepsy, seizures occur at apparently random times and much of the risk of injury or death from seizures relates to the lack of adequate opportunity to maximize safety. The risk of an unexpected seizure is far less for individuals with reflex epilepsy if avoidance of the seizure trigger is possible. Avoidance of the trigger is routinely sought, and avoidance may have a great impact on daily life, depending on what is being avoided. This possibility of avoidance produces a different clinical burden for people with reflex epilepsies compared to other epilepsies. People with epilepsy often have to restrict their lives to minimize risks if a seizure occurs, but those with reflex epilepsies also typically restrict their lives to avoid exposures to what are innocuous situations to others, including those with non-reflex epilepsies. This can produce considerable compromise to the breadth of daily activities. For example, an individual with medication-resistant non-reflex epilepsy will not drive a car or engage in recreational activities that would pose a risk if a seizure occurred. However, an individual with reflex epilepsy will not allow exposure to the particular trigger, which may mean avoiding situations with flashing lights, such as at a dance club, or situations with particular forms of music, such as the ambient music in stores or the sound track to some television commercials.

CLINICAL ASPECTS OF MUSICOGENIC SEIZURES

Musicogenic epilepsy is especially rare and has a prevalence in the general population as low as 1 in 10 million (Critchley, 1977). Nevertheless, it has been well described and clearly defined. Musicogenic seizures are distinguished from simple stimulus audiogenic seizures by clinical features of the epilepsy and by the complexity of the auditory trigger. Musicogenic epilepsy typically first develops later in life than epilepsy with audiogenic seizures, and the average onset age for musicogenic epilepsy is 28 years. Instances of childhood- and even infant-onset musicogenic epilepsy are known to occur, but they are not the norm. Onset in youth is more common for audiogenic epilepsy (Wieser et al., 1997; Lin et al., 2003; Tayah et al., 2006). Audiogenic seizures are less specific to a type of sound and most commonly are triggered by sudden sounds and occur with a short latency between the sound and the seizure onset. In contrast, musicogenic seizures are associated with sounds that are in melodic or harmonic combination with complex features that make specific sound triggers for each individual. Musicogenic seizures also typically occur with a longer latency, that can be several minutes of stimulation (Shaw and Hill, 1947; Zifkin and Zatorre, 1998). The musical complexity of the sound and the longer latency indicate a cognitive or emotional aspect to the musicogenic trigger, which differs from the more primary response when random sounds or noise are a trigger (Kaplan, 2003). For most individuals with musicogenic epilepsy, the trigger is a specific piece or type of music and 14 of a case series of 83 patients (17%) had seizures only when triggered by music (Wieser et al., 1997). That is, this 17% are consistently seizure-free when not exposed to music.

Although the musicogenic trigger is specific for the affected individual, triggers vary broadly across individuals with musicogenic epilepsy. Triggers can be a sequence of simple tones, voices of particular singers, the complex non-musical sound of a vacuum cleaner, background music that is minimally heard, only music that is actively heard, musical pieces when heard or imagined, and only when performing a specific piece of music, that is, hearing the piece or silently pretending to perform the piece is not a trigger (Critchley, 1937; Poskanzer et al., 1962; Sutherling et al., 1980; Brust, 2001; Carlson and St. Louis, 2004).

DIAGNOSTIC EVALUATION OF MUSICOGENIC SEIZURES

Musicogenic epilepsy is a syndrome with focal-onset seizures and a temporal-lobe epileptogenic region is present in about 56 of 71 (79%) (Wieser et al., 1997) (Table 26.1). Lateralization to the right occurs in about 60%. The precise localization of musicogenic seizures has varied across the reported patients, which may be due to both the differences in localization techniques and the heterogeneity of the patients. Localizing the epileptogenic region is impacted by the complexity of the normal neural response to music. This response, which has been theorized to include the limbic circuits beyond the regions of acoustic perception in Heschl's gyrus, may be the reason for varied localizations and the delay from stimulus to seizure onset.

Musicogenic seizure localization usually depends upon functional tests because most patients do not have a structural lesion (Kaplan, 2003). Interictal EEG has been reported to demonstrate abnormal temporal slowing or sharp waves that evolve with an increase in frequency during seizure occurrence (Daly and Barry, 1957; Kaplan, 2003). Source localization of interictal

Table 26.1

Reported localization of musicogenic seizures for 71 patients

Localization	Total	Temporal	Extratemporal
Unilateral focal	**50**		
Right focal	34	29	5
Left focal	12	9	3
Uncertain	3	3	
Independent Right and left	1	1	
Bilateral	**19**		
R>L	4	4	
L>R	8	5	3
Uncertain	7	5	2
Generalized	**2**		

Data from Wieser et al. (1997), Nakano et al. (1998), Genc et al. (2001), Gelisse et al. (2003), Lin et al. (2003), Morocz et al. (2003), Shibata et al. (2006), Tayah et al. (2006), Stern et al. (2006), Cho et al. (2007), Pittau et al. (2008), Marrosu et al. (2009), Mehta et al. (2009), Duanyu et al. (2009).

anterior and midtemporal spikes identified equivalent current dipoles in the posterior transverse temporal gyrus (Shibata et al., 2006).

Three patients have been reported to have undergone intracranial EEG monitoring of musicogenic seizures, and the findings are similar to the described patient (Tayah et al., 2006). All three had subdural grids or strips placed across the mesial and lateral temporal lobes with varying additional coverage of the frontal or posterior temporal regions. The electrographic ictal onset for each seizure preceded the behavioral onset and was found to be right mesial temporal for 1 patient, right lateral temporal for 1 patient, and independent right and left mesial temporal for the third patient.

Reports include description of the surgical treatment of 3 patients with musicogenic epilepsy in addition to the described patient. All 3 reported patients underwent a temporal-lobe resection with 2 who had an anterior-mesial resection (similar to the one described above) and 1 who had a posterior neocortical resection (Tayah et al., 2006; Duanyu et al., 2009). Although all 4 patients eventually became seizure-free, the patient who had a neocortical resection had early postoperative seizures.

ILLUSTRATIVE CASE

A 46-year-old right-handed woman had the onset of seizures when she was 21 years old. She had no epilepsy risk factors, including a negative family history for epilepsy or seizures, and no history of head injury. For the first few years after the onset, the seizures were triggered by hearing singing. Over the subsequent few years, the triggers became a broader range of other specific triggers that included, at different times, electronic music, heavy-metal rock music, flute music, or what the patient described as "squeaky" high-pitched music. During this time, which spanned approximately 10 years, classical music could be used to abort a seizure and headphones playing classical music were routinely worn while in public as a means of protection from seizures that would be triggered by music that happened to be playing in the immediate environment. Classical music in headphones ceased to be protective about 10 years ago prior to her presentation, when classical music became another trigger. At the time of presentation, all forms of music could be a trigger; however, not all forms were equally likely to provoke a seizure and the music of the performer Madonna was the trigger most likely to produce a seizure.

Overall, seizure frequency depended entirely on whether there was exposure to music because most exposures triggered a seizure. A refractory period with minimal seizure risk was present for about an hour after a seizure. The seizures always followed an aura of palpitations, followed by an indescribable feeling in the head, and then an auditory illusion of the triggering music repeating in a loop. The seizure behavior was a loss of awareness with right-hand purposeless movements (automatisms) and sometimes falling due to loss of postural tone. Generalization, with progression into a tonic-clonic convulsion, never occurred.

The patient's EEG between seizures (interictal) had intermittent abnormal slowing across the right temporal region and occasional right anterior and midtemporal epileptiform discharges, which indicates dysfunction in the temporal region with an increased predisposition for focal seizures within the temporal lobe. Four seizures were recorded with simultaneous video and EEG and these were stereotyped in behavior and EEG pattern. The behavior matched what was previously described and the ictal EEG demonstrated increased right anterior temporal spike discharge occurrences followed by broad right-sided slowing that evolved into a focal rhythm of up to 5 Hz over the right anterior temporal lobe. Interictal magnetoencephalography (MEG) identified right inferior frontal and midtemporal epileptiform dipoles that corresponded to the scalp EEG's anterior and midtemporal epileptiform discharges. Ictal MEG identified a right temporal onset with spread to the right frontal region.

Structural magnetic resonance imaging (MRI) and fluorodeoxyglucose (FDG) positron emission tomography (PET) were normal. Neuropsychologic testing found strength in verbal memory and severe impairment of the non-verbal domains of memory, problem solving, conceptualization, and organization, which indicated right-sided dysfunction based on her right-hand dominance. Overall, this non-invasive evaluation indicated a high likelihood for right temporal-lobe epilepsy, but the extent of the epileptogenic zone and its overlap with auditory function regions were not known.

To continue with the evaluation for treatment with resective epilepsy surgery, subdural grid electrodes were placed across the right lateral temporal region with extension of the electrodes into the right mesial temporal and inferior parietal and frontal regions (Fig. 26.1). Recording with these intracranial electrodes identified an increase in interictal epileptiform discharge frequency across the mesial and superior temporal regions when listening to music. Seizures occurred when the music exposure was greater, and the seizure onset zone co-localized to this region. After a resection of the mesial temporal structures and the anterior 6 cm of the lateral temporal lobe, music did not induce epileptiform discharges. The histopathologic evaluation did not identify abnormality; however, the patient has been seizure-free for the approximately 21 months since surgery.

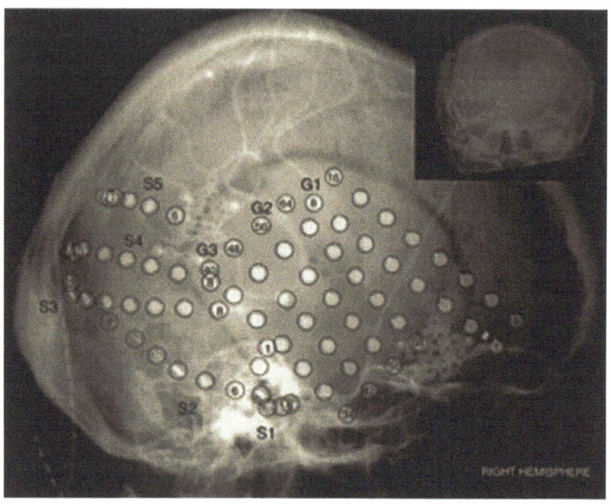

Fig. 26.1. X-ray of subdural electrode placement.

FUNCTIONAL IMAGING OF MUSICOGENIC SEIZURES

Functional imaging research techniques have been used to identify the region of seizure onset in musicogenic epilepsy, and the patient described in the illustrative case had a seizure during functional MRI (fMRI) prior to resective epilepsy surgery. Scalp EEG was recorded during the fMRI using 64 electrodes that covered the full scalp along with ocular and cardiac regions. Subtraction of fMRI artifact on the EEG was performed using techniques that have been described (Goldman et al., 2000; Stern et al., 2006). Music identified by the patient to likely trigger a seizure was played through headphones at the beginning of functional imaging. The first EEG evidence of a seizure occurred approximately 3 minutes into the recording and was the development of rhythmic 2–3-Hz activity across the right hemisphere (Fig. 26.2). This activity evolved with increasing amplitude as the co-localized background activity diminished. Approximately 2 minutes later, the patient reported an aura and then manifested unresponsiveness with her typical ictal automatisms. fMRI was terminated at this time. Because the patient was motionless through the initial 2 minutes of the seizure, the EEG and fMRI onsets were recorded without movement artifact.

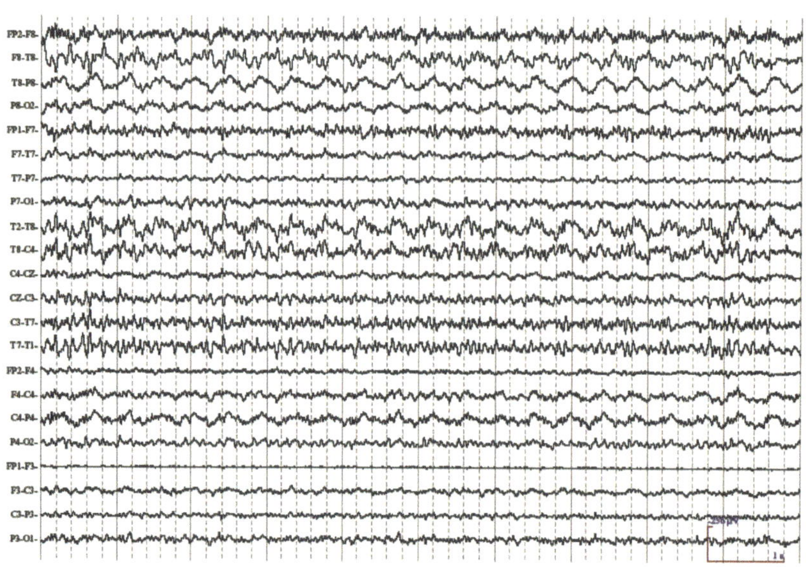

Fig. 26.2. Electroencephalogram during functional magnetic resonance imaging demonstrating ictal rhythm across the right temporal region.

Image analysis was performed using FEAT (fMRI Expert Analysis Tool) version 5.1 as part of FSL (fMRIB's Software Library, Oxford, UK) version 3.2 (Forman et al., 1995; Woolrich et al., 2001). Statistical analysis included a canonic hemodynamic response function and a cluster threshold determined by a minimum z score of 1.6 and cluster significance of $P = 0.05$. Analyses were performed using the entire image file, which was 307.5 seconds.

The seizure recording with simultaneous EEG and fMRI lasted 132.5 seconds and ended about 15 seconds after the aura occurred and around the time of loss of awareness. Ictal fMRI analysis that used the EEG onset time as the seizure's beginning and included the entire recorded seizure in the analysis localized the seizure to bilateral, anterior frontal lobes and cerebellum. The ictal fMRI results differed when the seizure onset time was moved to 10 seconds prior to EEG onset based on the recognition that the true seizure onset precedes the scalp EEG onset when scalp EEG is compared to intracranial EEG. Scalp EEG is not as sensitive to the earliest changes as intracranial EEG. The addition of 10 seconds prior to EEG onset resulted in a seizure localization within the right frontal and temporal lobes (Fig. 26.3).

Subsequent analyses divided the seizure into 10-second periods with the intention to observe the seizure propagation. This produced a sequence of localizations. The first period began 20 seconds before EEG onset (–20 to – 10 seconds) and did not have a significant fMRI finding. The next period (–10 to 0 seconds) identified a localization across a large region in the mid and posterior right temporal lobe, the left cerebellar hemisphere, and bilateral inferior parietal/perisylvian regions. During the first 10 seconds after EEG onset (0–10 seconds), localization was within the right anterior and midtemporal lobe and the right perisylvian region immediately superior to the temporal-lobe activity (Fig. 26.4). At the time of the aura, which was a 30-second period at the end of the imaging, fMRI localization was bilateral frontal lobes, bilateral cerebellum, and the right temporal lobe.

The onset localization identified with fMRI for the illustrative case patient is consistent with most of the published reports using ictal single-photon emission computed tomography (SPECT). Of the six ictal SPECTs reported, four were right temporal, one was left temporal, and one did not identify a region of hyperperfusion (Wieser et al., 1997; Nakano et al., 1998; Genc et al., 2001; Lin et al., 2003; Gelisse et al., 2003; Cho et al., 2007). Two reports of the right temporal hyperperfusion also included description of mesial temporal hyperperfusion. The one description of left temporal hyperperfusion was of an infant with musicogenic seizures, so the atypical age of the patient raises a question of the differences due to brain development. In one instance, the trigger was found to be unilateral, with a seizure occurring only when monoauricular stimulation was given to the left ear, which was contralateral to the ictal onset on SPECT

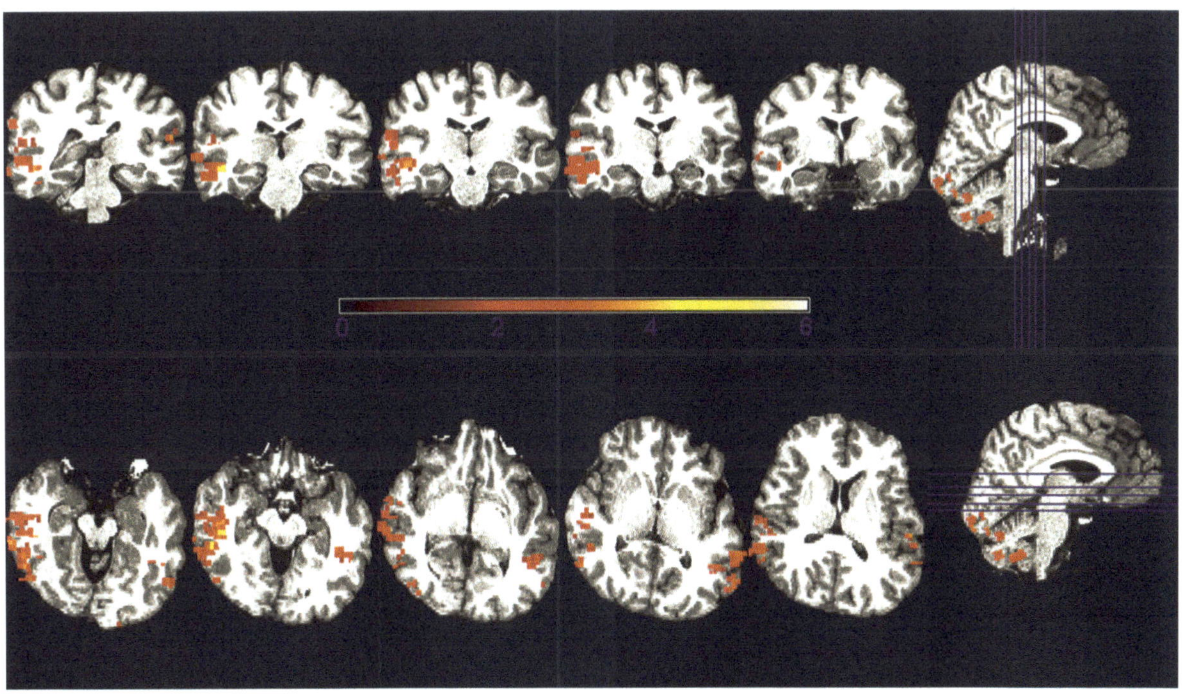

Fig. 26.3. Ictal functional magnetic resonance imaging with seizure defined as starting 10 seconds before onset on electroencephalogram and lasting until imaging ended due to seizure-related movements. Threshold set to $z \geq 1.6$.

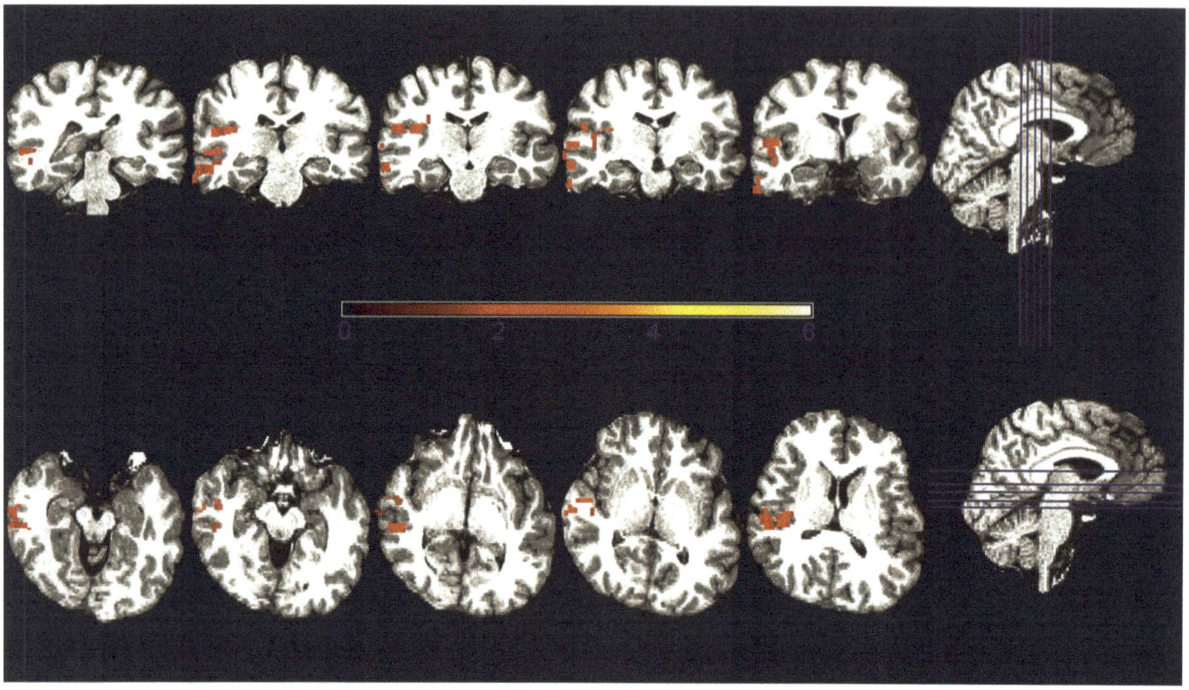

Fig. 26.4. Ictal functional magnetic resonance imaging of first 10 seconds after seizure onset on electroencephalogram. Threshold set to $z \geq 1.6$.

(Gelisse et al., 2003). Interictal FDG-PET has identified right temporal hypometabolism in musicogenic epilepsy (Wieser et al., 1997; Cho et al., 2007).

Reports of simultaneous EEG and fMRI during a musical trigger for musicogenic seizures have identified signal change within the bilateral frontal and right temporal lobes (Morocz et al., 2003; Pittau et al., 2008; Marrosu et al., 2009). The frontal regions included parts of the insula, middle cingulate cortex, orbitofrontal cortex, accumbens, and frontal pole with the right frontal lobe containing larger regions of signal change than the left. The imaging methodology included alternating music stimulus and no stimulus until a seizure occurred, with image results that combined seizure onset and spread. Overall, musicogenic epilepsy has a mostly consistent lateralization and localization with early functional imaging changes in the posterior temporal lobe and subsequent spread through the limbic system and anterior frontal regions.

RELEVANCE OF MUSICOGENIC EPILEPSY TO EPILEPSY AND MUSIC

Reflex epilepsies provide opportunities to elucidate the mechanisms of seizure generation and spread because of the predictable seizure occurrence through an association with a particular trigger (Avanzini, 2003). Although the seizures of epilepsies without triggers may be recorded as spontaneous episodes with video-EEG and with ictal SPECT scanning, predictable occurrence is needed to record a seizure with ictal fMRI, which has considerable advantages by providing high-resolution, tomographic localization of metabolic changes associated with the seizure. Reflex epilepsies also lend opportunities for insight into the integration of the epileptogenic zone into normal brain function by comparing patients' cerebral responses to the trigger when a seizure occurs to when no seizure occurs, and by comparing the response to the trigger in patients to the response in healthy control individuals.

Despite the opportunity and value of ictal fMRI in reflex epilepsies, recordings of either auras or behavioral seizures have not been widely obtained because of the difficulties in identifying the time of seizure onset and obtaining reliable images during seizure-related movement. Original reports used behavioral observation to determine the seizure onset time and identified regions of fMRI signal increase that were concordant with the behavioral indicators of the seizure focus (Jackson et al., 1994; Detre et al., 1996; Krings et al., 2000). These reports relied upon the seizures not producing sufficient movement to cause problematic imaging artifact.

The issue of seizure-produced movement artifact was avoided in a more recent report of 3 patients who had seizures with significant movement during fMRI when the image analysis compared the 1-minute period prior to the seizure to a 1-minute period several minutes earlier (Federico et al., 2005). Significant fMRI signal increase was observed with localizations that were concordant with the results of some or all of each patient's standard diagnostic tests. With or without significant movement, the ictal fMRI for all of these investigations depended on the seizure's behavioral onset, which may follow the electrophysiologic onset by seconds to minutes. Therefore, the results may not indicate the earliest epileptic changes, but they do support fMRI's sensitivity for early ictal changes. In contrast, reflex epilepsies have an advantage over spontaneous seizures for investigating early seizure changes by limiting the earliest seizure onset time to the time of the trigger, whereas an earliest possible seizure onset time cannot be defined for spontaneous seizures.

EEG that is recorded during fMRI provides a marker of seizure propagation after the trigger is delivered, but it does not provide the time of seizure onset. Comparisons of scalp and intracranial EEG have shown that extensive propagation can be present before the first epileptiform abnormality develops on scalp EEG. Therefore, the occurrence of fMRI signal changes prior to the seizure's EEG onset presents a question of whether the changes are due to normal function or the seizure. This is due to fMRI's lack of specificity for epilepsy, which contrasts with EEG's high specificity for epilepsy.

The earliest fMRI changes in the illustrative case included bilateral inferior parietal and perisylvian regions, which are outside the identified epileptogenic zone, but these regions are adjacent to auditory cortex that may be important in music perception, engaged in musicogenic epilepsy without being within the epileptogenic region, or both. If engaged in musicogenic epilepsy, the regions may be part of an antiseizure homeostatic response to the seizure onset. Regardless of these considerations, the seizure control that has been achieved from resections sparing the parietal and perisylvian regions indicates that these regions are not within the epileptogenic region.

The progression of ictal fMRI seizure localization through the seizure supports fMRI's potential to monitor a seizure's evolution from preictal changes through the EEG onset to the behavioral onset. This surpasses the capacity of ictal SPECT, which provides one image of the ictus. Nevertheless, the illustrative case, 5 of the reported cases with ictal SPECT, and the 3 reported cases of ictal fMRI are concordant in indicating a temporal-lobe localization for musicogenic epilepsy. fMRI's superior sensitivity may explain its inclusion of the frontal lobes, and the serial images from the presented case's analyses may explain the frontal-lobe involvement as propagation from a temporal-lobe origin. Observation of this propagation through intracranial EEG would not be possible for the 4 patients who had intracranial EEG because the electrode coverage did not include the anterior frontal regions demonstrating the fMRI findings. Overall, the fMRI localizations for the musicogenic seizures are concordant with the intracranial EEG's localizations and the fMRI provides supplementary information about a larger extent of involvement.

The fMRI results also provide better understanding of the phenomenon of musicogenic seizures. Morocz and colleagues (2003) report the gyrus rectus as associated with the response to ictogenic music and both the orbitofrontal and anterior temporal lobes as associated with the occurrence of musicogenic auras.. This supports the concept of musicogenic epilepsy as a limbic epilepsy that includes both temporal and frontal limbic regions. In the illustrative case, fMRI signal change was present in the gyrus rectus and not the temporal lobe at the time just prior to the aura, which agrees with Morocz and colleagues' findings. However, the presented patient's aura occurred after more than 2 minutes of epileptiform abnormality on the EEG, so the gyrus rectus finding may be relevant to the epilepsy and not necessarily a component of music perception.

Another sign of seizure propagation is evident in the frontal-lobe signal changes that occur with the evolution of the seizure into its behavioral manifestation. At the seizure's onset, the frontal-lobe signal was limited and right-sided. In the final 30 seconds of imaging, the fMRI signal change expands frontally to become bilateral and more widespread. As expected, this is accompanied by cerebellar signal change that also becomes bilateral and more widespread. This greater and bilateral frontal-lobe involvement with behavioral onset agrees with the observation of widespread intracranial EEG ictal activity at the time of behavioral onset. It also demonstrates fMRI's superior sensitivity to scalp EEG because the simultaneous EEG demonstrated only minor evolution. The EEG progressed during the seizure with an increase in the ictal rhythm's amplitude and a decrease in the amplitude of the background activity, but it did not indicate greater frontal-lobe involvement. Although this seizure had electrographic abnormality for a longer time than most temporal-lobe seizures before behavioral onset, the significant frontal-lobe changes occurred only at the end of the recorded period, so this propagation pattern also may be observable for shorter temporal-lobe seizures. The progression to include bilateral regions is the expected anatomic correlate to the loss of awareness or responsiveness, which is conventionally understood to occur once the epileptic abnormality becomes bilateral on intracranial EEG.

Mesial temporal-lobe epilepsy is the most common form of focal epilepsy. The epileptogenic zone typically includes the hippocampus, but the syndrome may be considered more broadly as a limbic epilepsy. As such, olfactory, emotional, and mnestic experiences are common.

Auditory hallucinations, which arise from neocortical regions of the temporal lobes, are not as common (see Chapter 24), and this also supports the consideration of mesial temporal-lobe epilepsy as a limbic epilepsy. Nevertheless, seizures with neocortical onset may propagate into the limbic system and produce behavioral features of limbic seizures. Musicogenic seizures demonstrate this by most commonly manifesting as focal dyscognitive seizures with automatisms. However, musicogenic seizures also demonstrate the limbic system's epileptic sensitivity to sensory input through neocortex. Apparently, the limbic system's subsequent processing of the sound experience produces the seizure. Memory and emotion may both be involved, but the occurrence in an infant supports the possibility of an early-life brain's intrinsic sensitivity to music even without the experiences and learning that music produces through development and over years of adulthood.

Overall, the limbic system's common involvement in musicogenic seizures and the successful treatment of musicogenic seizures with anteromesial temporal-lobe resections support the understanding that musicogenic epilepsy is a limbic epilepsy. As such, primary auditory cortex is not the typical generator of musicogenic seizures. With an anatomic basis in the limbic system, the affective experience of music is integral to the seizure development. This affective experience may be for a specific musical piece or musical theme, regardless of the instrument or voice performing it, or for a specific musical quality that an instrument or voice produces, regardless of the music piece (Brien and Murray, 1984). Cerebral processing of music includes both thematic and mechanic aspects, and clinical experience has shown that each may be integrated into epileptic dysfunction.

ACKNOWLEDGMENTS

This study was supported by NINDS K23 grant NS044936 and the Leff Family Foundation.

References

Avanzini G (2003). Musicogenic seizures. Ann N Y Acad Sci 999: 95–102.

Brien SE, Murray TJ (1984). Musicogenic epilepsy. Can Med Assoc J 131: 1255–1258.

Brust J (2001). Music and the neurologist. Ann N Y Acad Sci 930: 143–152.

Carlson C, St. Louis EK (2004). Vacuum cleaner epilepsy. Neurology 63: 190–191.

Cho J-W, Seo DW, Joo EY et al. (2007). Neural correlates of musicogenic epilepsy: SISCOM and FDG-PET. Epilepsy Res 77: 169–173.

Critchley M (1937). Musicogenic epilepsy. Brain 60: 13–27.

Critchley M (1977). Musicogenic epilepsy. In: M Critchley, RA Hensen (Eds.), Music and the Brain, Heinemann, London.

Daly DD, Barry MJ (1957). Musicogenic epilepsy: report of three cases. Psychosom Med 19: 399–408.

Detre JA, Alsop DC, Aguirre GK et al. (1996). Coupling of cortical and thalamic ictal activity in human partial epilepsy: demonstration by functional magnetic resonance imaging. Epilepsia 37: 657–661.

Duanyu N, Yongjie L, Guojun Z et al. (2009). Surgical treatment for musicogenic epilepsy. J Clin Neurosci 17: 127–129.

Engel Jr J (2001). A proposed diagnostic scheme for people with epileptic seizures and with epilepsy: report of the ILAE task force on classification and terminology. Epilepsia 42: 796–803.

Engel Jr J (2006). ILAE classification of epilepsy syndromes. Epilepsy Res 70S: S5–S10.

Federico P, Abbott DF, Briellmann RS et al. (2005). Functional MRI of the pre-ictal state. Brain 128: 1811–1817.

Forman SD, Cohen JD, Fitzgerald M et al. (1995). Improved assessment of significant activiation in functional magnetic resonance imaging (fMRI): use of cluster-size threshold. Magn Reson Med 33: 636–647.

Forster FM, Chun RWM, Forster MB (1963). Conditioned changes in focal epilepsy: I. In animals with intact central nervous system. Arch Neurol 9: 188–193.

Gelisse P, Thomas P, Padovani R et al. (2003). Ictal SPECT in a case of pure musicogenic epilepsy. Epileptic Disord 5: 133–137.

Genc BO, Genc E, Tastekin G et al. (2001). Musicogenic epilepsy with ictal single photon emission computed tomography (SPECT): could these cases contribute to our knowledge of music processing? Eur J Neurol 8: 191–194.

Goldman RI, Stern JM, Engel Jr J et al. (2000). Acquiring simultaneous EEG and functional MRI. Clin Neurophysiol 111: 1974–1980.

Illingworth JL, Ring H (2013). Conceptual distinctions between reflex and nonreflex precipitated seizures in the epilepsies: a systematic review of definitions employed in the research literature. Epilepsia 54: 2036–2047.

Jackson GC, Connelly A, Cross JH et al. (1994). Functional magnetic resonance imaging of focal seizures. Neurology 44: 850–856.

Kaplan PW (2003). Musicogenic epilepsy and epileptic music: a seizure's song. Epilepsy Behav 4: 464–473.

Kasteleijn-Nolst Trenite DGA (2012). Provoked and reflex seizures: Surprising or common? Epilepsia 53 (Suppl. 4): 105–113.

Krings T, Topper R, Reinges MH et al. (2000). Hemodynamic changes in simple partial epilepsy: a functional MRI study. Neurology 54: 524–527.

Lin K-L, Wang H-S, Kao P-F (2003). A young infant with musicogenic epilepsy. Pediatr Neurol 28: 379–381.

Marrosu F, Barberini L, Puligheddu M et al. (2009). Combined EEG/fMRI recording in musicogenic epilepsy. Epilepsy Res 84: 77–81.

Mehta AD, Ettinger AB, Perrine K et al. (2009). Seizure propagation in a patient with musicogenic epilepsy. Epilepsy Behav 14: 421–424.

Morocz IA, Karni A, Haut S et al. (2003). fMRI of triggerable aurae in musicogenic epilepsy. Neurology 60: 705–709.

Nakano M, Takase Y, Tatsumi C (1998). A case of musicogenic epilepsy induced by listening to American pop music. Rinsho Shinkeigaku 31: 1067–1069.

Pittau F, Tinuper P, Bisulli F et al. (2008). Videopolygraphic and functionalMRI study of musicogenic epilepsy. A case report and literature review. Epilepsy Behav 13: 685–692.

Poskanzer DC, Brown AE, Miller H (1962). Musicogenic epilepsy caused only by a discrete frequency band of church bells. Brain 85: 77–92.

Shaw D, Hill D (1947). A case of musicogenic epilepsy. J Neurol Neurosurg Psychiatry 10: 107–117.

Shibata N, Kubota F, Kikuchi S (2006). The origin of the focal spike in musicogenic epilepsy. Epileptic Disord 8: 131–135.

Stern JM, Tripathi M, Akhtari M et al. (2006). Musicogenic seizure localization with simultaneous EEG and functional MRI. Neurology 66 (Suppl. 2): 90.

Sutherling WW, Hershman LM, Miller JQ et al. (1980). Seizures induced by playing music. Neurology 30: 1001–1004.

Symonds C (1959). Excitation and inhibition in epilepsy. Proc R Soc Med 52: 305–402.

Tayah TF, Abou-Khalil B, Gilliam FG et al. (2006). Musicogenic seizures can arise from multiple temporal lobe foci: intracranial EEG analyses of three patients. Epilepsia 47: 1402–1406.

Wieser HG, Hungerbuhler H, Siegel AM et al. (1997). Musicogenic epilepsy: review of the literature and case report with ictal single photon emission computed tomography. Epilepsia 38: 200–207.

Woolrich MW, Ripley BD, Brady M et al. (2001). Temporal autocorrelation in univariate linear modeling of FMRI data. Neuroimage 14: 1370–1386.

Zifkin BG, Zatorre RJ (1998). Musicogenic epilepsy. In: BG Zifkin, F Andermann, A Beaumanoir et al. (Eds.), Reflex epilepsies and reflex seizures, Advances in Neurology, Vol. 75. Lippincott-Raven, Philadelphia.

Handbook of Clinical Neurology, Vol. 129 (3rd series)
The Human Auditory System
G.G. Celesia and G. Hickok, Editors

Chapter 27

Deafness in cochlear and auditory nerve disorders

KATHRYN HOPKINS*
School of Psychological Sciences, University of Manchester, Manchester, UK

INTRODUCTION

Sensorineural hearing loss is the collective term for hearing damage to the cochlea and auditory nerve, and is by far the most common type of hearing loss in adults, accounting for over 90% of all cases (Davis, 1995; Cruickshanks et al., 1998; Wilson et al., 1998). Sensorineural hearing loss is a heterogeneous disorder, which can arise due to damage to several structures in the peripheral auditory system and has many causes. This chapter describes the pathologies that affect the cochlea and the auditory nerve and their perceptual consequences.

STRUCTURES AFFECTED BY SENSORINEURAL HEARING LOSS

Sensorineural hearing loss can occur due to damage to one or several structures in the cochlea or auditory nerve. Here the role of each structure is briefly discussed, together with the results of animal studies investigating the effects of damage to each of these structures on auditory coding.

Outer hair cells

Outer hair cells amplify basilar membrane motion (Ashmore, 1987). The amount of amplification is greatest at low input levels and at frequencies close to the characteristic frequency of the place on the basilar membrane where the hair cell is located (Rhode, 1971; Sellick et al., 1982). These properties mean that outer hair cells improve hearing sensitivity and lead to a compressive basilar membrane growth function, which is reflected in neural activity in the auditory nerve (Yates et al., 1990). If outer hair cells are damaged, this compression is lost and detection thresholds are elevated (Ryan and Dallos, 1975). The basilar membrane response becomes more linear, and a reduced range of sound levels can be encoded (Patuzzi et al., 1989). Outer hair cells also substantially improve the frequency selectivity of the basilar membrane (the ability to separate complex sounds into their constituent frequencies). Outer hair cell dysfunction therefore also results in a reduction in the sharpness of basilar membrane tuning (Ruggero and Rich, 1991).

Inner hair cells

Inner hair cells transduce basilar membrane vibration into electrical activity. Basilar membrane vibration causes stereocilia on the surface of inner hair cells to bend (see Fig. 1.1, Chapter 1). Tip links that connect adjacent stereocilia stretch, opening cation channels and allowing potassium ions to enter the cells, causing depolarization. Inner hair cells synapse with type I auditory nerve fibers, and depolarization of the inner hair cells increases the probability of action potential generation in these fibers. This provides the main route for transmission of information along the auditory nerve to the central auditory system. Several animal studies have investigated the effects of inner hair cell loss on auditory coding using carboplatin, which causes selective inner hair cell loss in some species. Takeno et al. (1994) found that thresholds for the compound action potential (which is a measure of neural activity in the auditory nerve) was reduced in proportion with the amount of inner hair cell loss caused by carboplatin administration. Wang et al. (1997) examined responses from individual auditory nerve fibers following selective inner hair cell loss as well as compound action potential thresholds. They found, like Takeno et al., that severe inner hair cell loss caused elevation of compound action potential thresholds, but that thresholds for many single auditory nerve fibers were normal, and had normal frequency tuning. Less

*Correspondence to: Kathryn Hopkins, University of Manchester, School of Psychological Sciences, Manchester, M13 9PL, UK. E-mail: kathryn.hopkins-2@manchester.ac.uk

severe inner hair cell loss (15–28% along the length of the cochlea) did not result in significant elevation of compound action potential thresholds, although it did lead to reduced compound action potential amplitude at higher levels. These data are shown in Figure 27.1. The top panel shows compound action potential amplitudes for auditory nerve fibers at 8 kHz for control animals and animals subject to a single dose of carboplatin. The bottom panel shows compound action potential thresholds as a function of frequency for the same groups. The effect of inner hair cell loss on auditory coding at higher levels of the auditory system is as yet unknown.

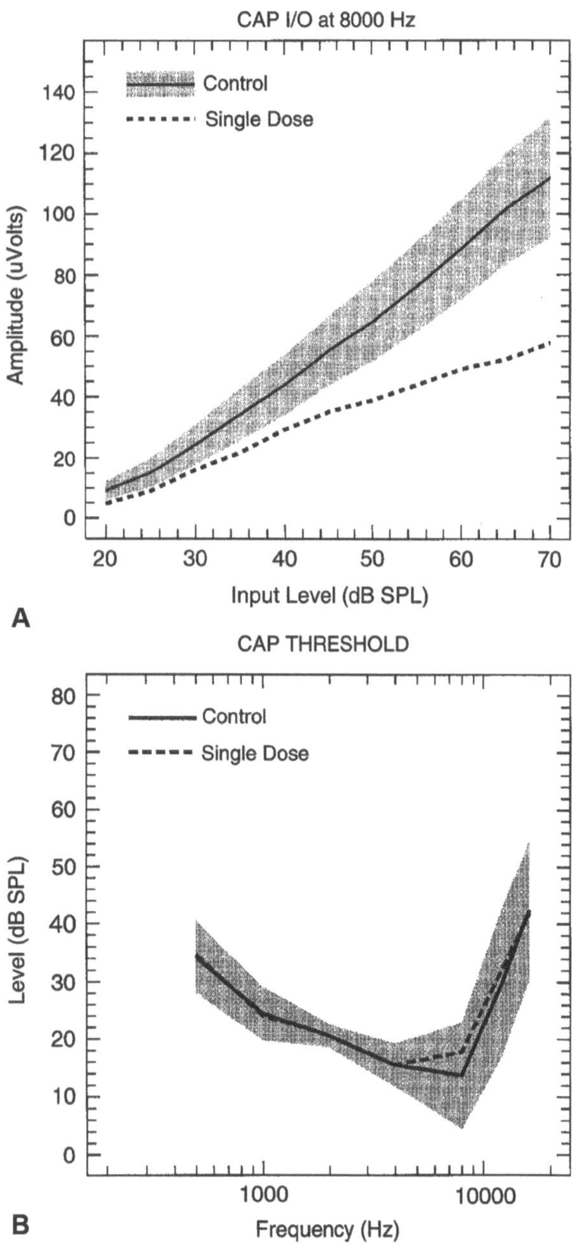

Fig. 27.1. (**A**) Mean compound action potential (CAP) input/output (I/O) functions at 8 kHz of the single-dose group ($n=8$ ears) and the normal control group ($n=29$; shaded area is 95% confidence interval). (**B**) Mean ($n=8$) CAP thresholds of single-dose carboplatin group and control group ($n=29$; shaded area is 95% confidence interval). SPL, sound pressure level. (Reproduced from Wang et al., 1997.)

Stria vascularis

The stria vascularis maintains the ionic composition of endolymph, which is found in the scala media and surrounds the inner and outer hair cells. Unlike most extracellular fluid, endolymph has a high concentration of potassium ions, and a low concentration of sodium ions, leading to a positive endocochlear potential. The correct ionic composition of endolymph is essential for normal cochlear function, because both the outer and inner hair cells rely on entry of potassium ions to cause depolarization; this leads to mechanic shortening in the case of outer hair cells, and auditory transduction in the case of inner hair cells if the stria vascularis is not functioning correctly, and the concentration of potassium ions in the scala media is insufficient.

Sewell (1984) studied the effects of furosemide administration on auditory nerve responses. Furosemide blocks the Na^+/K^+ transporter in the stria vascularis that maintains the endocochlear potential, so this provides a way of examining the effects of stria vascularis dysfunction independently. Sewell found that furosemide administration led to elevated thresholds, both in single auditory nerve fibers and in the compound action potential. Tuning of individual auditory nerve fibers became broader, in a similar way as occurs in the case of outer hair cell loss. However, the tuning curves were qualitatively different to those reported following selective outer hair cell loss, in that the sensitivity to tones remote from the characteristic frequency of the nerve fiber was reduced (in the "tail" of the tuning curve), whereas this was not the case for tuning curves in the case of selective outer hair cell loss. A possible explanation for this observation is that dysfunction of the stria vascularis and endocochlear potential affects both outer and inner hair cells. Examples of tuning curves before and after furosemide treatment are shown in Figure 27.2.

Auditory nerve

Damage to the auditory nerve can take two forms: auditory nerve fibers can degenerate completely, or alternatively, demyelination can occur. Degeneration can occur secondary to inner hair cell loss – auditory nerve fibers degenerate if input from the cochlea is reduced or lost (Webster and Webster, 1981; Simpson, 2009). More recent evidence has shown that auditory nerve degeneration can also occur in the absence of cochlear damage (so-called primary auditory nerve degeneration). Kujawa and Liberman (2009) found that up to 50% of

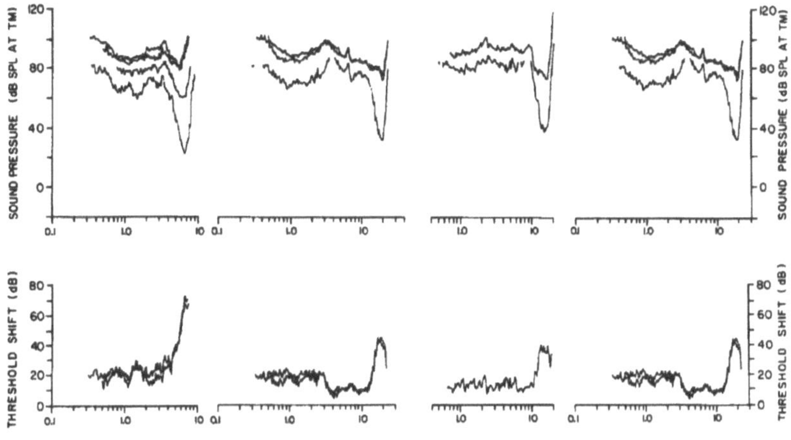

Fig. 27.2. Representative examples of the changes in single-fiber tuning curves following intravenous furosemide administration. In all cases, the preinjection (control) tuning curve is that with the lowest threshold. The lower panels contain plots of the difference between the preinjection tuning curve and those obtained following furosemide administration. SPL, sound pressure level; TM, tympanic membrane. (Reproduced from Sewell, 1984.)

auditory nerve fibers degenerated following a single period of noise exposure that did not cause permanent outer or inner hair cell damage in mice. The auditory brainstem response thresholds recovered to normal (following a period of temporary threshold shift) even though the loss of auditory nerve fibers was permanent, suggesting that sensitivity to weak sounds can be unaffected by substantial auditory nerve fiber loss. The amplitude of the auditory brainstem response to high-level sounds was, however, reduced when compared with control animals; the authors hypothesized that coding of above-threshold sounds would be affected, even though detection thresholds appeared to be normal.

Auditory nerve fibers are normally surrounded by a myelin sheath, which increases the speed of transmission of action potentials along the auditory nerve. El-Badry et al. (2007) examined the effects of myelin damage on the auditory nerve compound action potential and auditory brainstem response amplitudes. They found that both mild and severe myelin damage increased the transmission latency to the central auditory system, and that severe (but not mild) myelin damage causes a reduction in both the compound action potential and auditory brainstem response. The authors hypothesized that this reduction in amplitude occurred due to desynchronization of activity in different auditory nerve fibers, which might be expected if myelin damage is not uniform.

ETIOLOGY

Many causes of sensorineural hearing loss have been documented; some of the most common are summarized in Table 27.1 and are discussed here.

Hereditary causes

Mutations in many genes can cause sensorineural hearing loss (for a review, see Makary et al., 2011). Several genetic syndromes, such as Usher syndrome (Bentler, 2005), Jervell and Lange-Nielsen syndrome (Wilson et al., 1999), and Pendred syndrome (He et al., 2007), have sensorineural hearing loss as a symptom. Additionally hundreds of mutations have been identified that cause sensorineural hearing loss with no other associated symptoms (so-called non-syndromic hereditary hearing loss). The structures affected by hereditary hearing loss depend on the nature of the mutation; mutations have been identified that affect all of the structures described above. It should also be noted that there is emerging evidence for a genetic contribution to a susceptibility to many forms of acquired hearing loss. For example, susceptibility to noise-induced hearing loss (Herman et al., 1977), ototoxic drugs (Marmel et al., 2013), and age-related hearing loss (Weiss and Rose, 1988) have all been shown to have a large genetic component.

Noise

Exposure to high noise levels causes damage to the cochlear and auditory nerve, leading to sensorineural hearing loss. The degree of noise-induced hearing loss is related to both the level of the noise and the duration of exposure.

Several mechanisms for noise-induced hearing damage have been identified. Outer hair cells can be damaged or lost as a result of mechanic damage, or due to the generation of reactive oxygen species, which can lead to cell death (Lebo and Reddell, 1972). Reactive oxygen species are produced as a normal product of

Table 27.1

Common etiologies of sensorineural hearing loss

Etiology	Examples	Structures affected	Mechanism
Hereditary	Usher syndrome, Lange-Nielsen syndrome, Pendred syndrome, others	Heterogeneous	Heterogeneous
Ototoxic drugs	Aminoglycoside antibiotics, furosemide, salicylic acid, carboplatin, others	Hair cells, stria vascularis	See Table 27.2
Hypoxia	Birth complications, cardiovascular problems, cigarette smoking	Hair cells	Reactive oxygen species lead to cell death
		Stria vascularis	Reduced endocochlear potential (reversible)
Noise	Occupational, recreational	Cochlear partition	Direct mechanic
		Hair cells	Reactive oxygen species lead to cell death
		Auditory nerve	Excitotoxicity
Aging	–	Hair cells, stria vascularis, auditory nerve	Strial degeneration, accumulated insults (e.g., noise, hypoxia, otoxic drugs)

metabolism, but during noise exposure, reactive oxygen species production increases, and causes damage to several outer hair cell structures, including DNA. Likewise, reactive oxygen species can also damage inner hair cells, although inner hair cells appear less susceptible to damage. Glutamate excitoxicity is another mechanism of noise-induced hearing damage. Glutamate is an excitatory neurotransmitter, which is released at the synapse between inner hair cells and auditory nerve fibers. Excess glutamate is toxic and causes disruption of the synapse (Moore et al., 2012). This excitotoxicity was originally thought to cause only temporary damage, but more recent studies have found evidence of permanent degeneration of auditory nerve fibers following excessive noise exposure (Kujawa and Liberman, 2009; Lin et al., 2011b).

Ototoxic drugs

Many drugs have been found to be associated with sensorineural hearing loss, and different substances are associated with damage to different structures. Table 27.2 lists some of the commonly used drugs that are known to be ototoxic, together with the hypothesized mechanisms of damage.

Hypoxia

Hypoxia (either local or systemic) can occur for a variety of reasons, for example, due to birth complications, cardiovascular problems such as atherosclerosis, or cigarette smoking. Damage to the auditory system depends on hypoxia severity and duration, but has been reported to affect both the inner and outer hair cells of the cochlea (Mazurek et al., 2006).

Aging

There is no doubt that the prevalence of sensorineural hearing loss increases dramatically with age (see Chapter 20). Figure 27.3 shows the prevalence of hearing loss estimated as a function of age for the British (Davis, 1989, 1995) and Australian (Wilson et al., 1998, 1999) populations. A similar increase in prevalence of hearing loss with age has been reported in the United States (Lin et al., 2011a). Both Davis (1995) and Wilson et al. (1999) reported that sensorineural hearing loss accounted for over 90% of all cases of hearing loss, and so most of the increase in prevalence with age can be attributed to sensorineural hearing loss.

Age-related hearing loss is characterized by a progressive increase in thresholds at high frequencies. Schuknecht and Gacek (1993) described a series of histologic postmortem studies on human subjects with age-related hearing loss. Schuknecht and Gacek reported that age-related hearing loss was associated with damage to the hair cells, stria vascularis, and auditory nerve, but that the relative damage to these structures varied among individuals. More recent work has found that primary auditory nerve degeneration occurs with age, even among individuals with no damage to the cochlea (Makary et al., 2011).

The extent to which age-related hearing loss occurs due to purely age-related effects is, however, unclear. It is possible that hearing damage associated with aging may occur at least partly due to cumulative damage caused by exposure to environmental insults such as noise and ototoxic drugs. Animal studies can shed some light on this question by assessing the effects of aging on the auditory system while carefully controlling exposure

Table 27.2

Hypothesized mechanisms of action for some of the most commonly used ototoxic drugs

Otoxic drugs	Structures affected	Mechanism
Salicylic acid	Hair cells (in particular outer hair cells)	Oxidative stress causing hair cell loss Reduced endocochlear potential
Cisplatin	Outer hair cells	Oxidative stress causing outer hair cell loss
Furosemide	Stria vascularis	Reversible reduction in the endocochlear potential
Carboplatin	Selective inner hair cell damage in chinchillas, but affects both inner and outer hair cells in humans	Inner hair cell loss in chinchillas Uncertain mechanism in humans

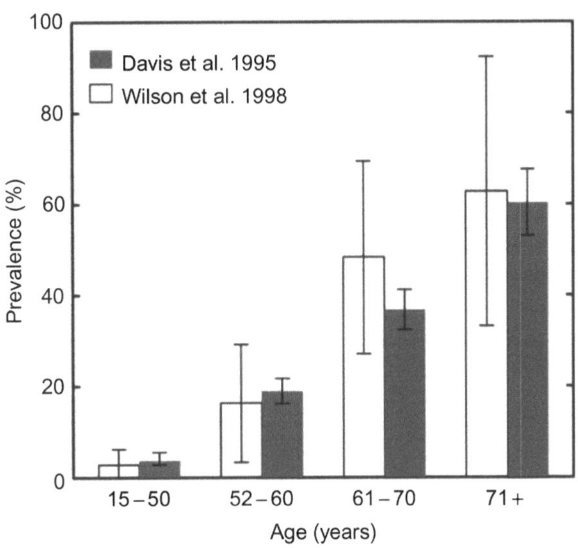

Fig. 27.3. Estimated hearing loss (HL) prevalence as a function of age for the Australian and British populations. HL was defined as mean audiometric thresholds at octave frequencies between 0.5 and 4 kHz greater than 25 dB HL. Sensorineural HL accounted for over 90% of cases across age groups. (Plotted from data reported by Davis, 1989, 1995, and Wilson et al., 1998, 1999.)

to noise and other insults. Mills et al. (1990) found that auditory evoked potential thresholds measured for quiet-reared gerbils (analogous to audiometric thresholds for human listeners) were elevated compared with young gerbils, providing evidence to support the idea that aging causes hearing loss independently of external otologic insults. Schulte and Schmiedt (1992) reported that the endocochlear potential and the activity of ion pumps in the stria vascularis were reduced in quiet-aged gerbils, providing a possible mechanism for age-related hearing loss.

DIAGNOSIS

The following section discusses the tests used to diagnose sensorineural hearing loss. No single test can definitively diagnose damage to a single structure, although by combining the results of several tests, the structures affected can be more accurately determined. Table 27.3 shows the most commonly used diagnostic tests, together with possible findings and interpretations.

Threshold measurement

Sensorineural hearing loss is usually diagnosed using pure-tone audiometry, for which the quietest detectible sounds are measured as a function of frequency (Hughson and Westlake, 1944). Patients are asked to indicate when they hear a sound, and the sound level is varied adaptively to determine a threshold value. Measured thresholds are compared with normative threshold values measured using young listeners with no known otologic abnormalities, to give a measure of threshold elevation in decibel hearing level (dB HL) (ANSI, 2004). Sensorineural hearing loss is differentiated from conductive hearing loss (where hearing loss occurs due to impaired transmission of mechanical vibration to the cochlea) by measuring audiometric thresholds via both air and bone conduction (Hood, 1957). Air conduction audiometry delivers sounds to the eardrum using either headphones or a loudspeaker; bone conduction audiometry uses a bone conductor on the skull (usually on the mastoid prominence or forehead) to stimulate the cochlea directly, bypassing the outer and middle ear. Sensorineural hearing loss is indicated if air and bone conduction thresholds are elevated by a similar amount; a conductive hearing loss is indicated if air conduction thresholds are elevated more than bone conduction thresholds.

Table 27.3

Diagnostic tests commonly used to diagnose cochlear and auditory nerve disorders

Test	Finding	Indication
Air conduction audiometry	Elevated thresholds	Conductive or sensorineural loss
Bone conduction audiometry	Elevated thresholds	Sensorineural loss
Otoacoustic emissions	Reduced or absent emissions	Outer hair cell dysfunction or conductive loss
	Reduced or absent emissions with evidence of normal middle-ear function	Outer hair cell dysfunction
Auditory brainstem response	Elevated thresholds	Conductive or sensorineural hearing loss
	Absent or severely disordered response at all levels with evidence of normal outer hair cell function	Inner hair cell or auditory nerve dysfunction

Analogous electrophysiologic techniques can also be used to estimate detection thresholds; the most commonly used procedures are measurement of the auditory brainstem response or the auditory steady-state response. Both of these techniques are used to determine the level at which a response that is synchronized with presentation of a sound stimulus can just be detected in electric activity recorded using electrodes placed on the scalp. The advantage of these procedures for estimating thresholds is that they do not require the cooperation of the patient, and so are often used when testing patients who are unable to cooperate with behavioral testing, such as young children.

Differentiating damage to structures in the cochlear and auditory nerve

Measurement of detection thresholds using behavioral audiometry or analogous electrophysiologic techniques cannot alone be used to differentiate damage to different structures in the cochlear and auditory nerve. Further tests can give some information about the particular structures affected, although diagnostic techniques capable of definitively identifying damage to different structures in the human peripheral auditory system are not yet available. Furthermore, there is evidence that mild inner hair loss or even substantial auditory nerve fiber loss is not manifest by a change in detection threshold in animals (see above). Such damage is therefore not expected to be reflected in threshold measurements in humans, and so would go undetected with current diagnostic techniques.

Otoacoustic emissions

Otoacoustic emissions (Kemp, 2002) can be used to assess the integrity of outer hair cells. Otoacoustic emissions are non-linear distortion products generated in the cochlea and transmitted back through the middle ear to the outer ear, where they can be detected as sound using a sensitive microphone. Otoacoustic emissions are generated by outer hair cells, and so can be used to determine if outer hair cells are functioning normally. This allows hearing loss due to damage to the outer hair cells and other structures such as the inner hair cells and auditory nerve to be differentiated to some degree. However, the amplitude of otoacoustic emissions is only weakly correlated with degree of outer hair cell dysfunction (Bernstein et al., 2013), so they are only really useful in determining whether outer-cell damage is present or absent.

Diagnosing auditory nerve dysfunction

Auditory nerve damage is suspected when there is evidence of normal cochlear function (for example, through the presence of otoacoustic emissions), but the auditory brainstem response is absent or highly disordered. Patients who meet these criteria are diagnosed with auditory neuropathy spectrum disorder (for a detailed review, see Chapter 28). However, this pattern of results could also occur due to selective inner hair cell loss (for example, due to an otoferlin mutation), or due to damage to the auditory brainstem. Likewise, these diagnostic criteria are unlikely to detect mild or moderate auditory nerve fiber loss, or auditory nerve damage that is coupled with cochlear dysfunction. Consequently, it is currently not possible to definitively diagnose auditory nerve damage in humans.

THE EFFECTS OF SENSORINEURAL HEARING LOSS ON AUDITORY CODING

Sounds can be characterized by their changing frequency spectrum. Coding of this information in the peripheral auditory system can occur by two mechanisms: rate place coding and temporal coding. Rate place coding occurs when information is transmitted according to the profile of neural activity across an array of auditory neurons, where each neuron is "tuned" to a characteristic frequency. Temporal coding occurs when information is transmitted by the timing of neural spikes in relation to each other, either in a single neuron or a population of neurons. Auditory nerve fibers have a greater probability of firing at peaks in a sound waveform, a phenomenon that is known as "phase locking" (Rose et al., 1967). Phase locking occurs most robustly at low frequencies, and becomes less accurate as frequency increases. The exact frequency limit of phase locking varies among species (and is unknown in humans), but in most mammals this limit is thought to be around 4000–5000 Hz (Palmer and Russell, 1986). The relative usefulness of place and temporal information in normal auditory coding is still debated. For example, information about the frequency of a pure tone could be determined from auditory nerve firing patterns either by determining the characteristic frequency of the nerve fibers with the greatest firing rate, or by determining the reciprocal of the most common interspike interval of the neural responses (assuming the pure tone is below the limit of phase locking). Both types of information are available in auditory nerve-firing patterns and it is not yet clear how the information is used by the central auditory system. There is evidence that both rate place and temporal coding are affected by sensorineural hearing loss in humans; the effects of cochlear and auditory nerve damage on auditory coding (so far as they can be differentiated in human listeners) will be considered separately.

The effects of cochlear damage on auditory coding

Rate place coding

Rate place coding in the auditory nerve relies on separation of a sound into its component frequencies. The ability of the auditory system to separate a complex sound into its constituent frequency components is known as frequency selectivity.

Several studies have assessed the effects of sensorineural hearing loss on frequency selectivity, and all have shown a significant correlation between degree of hearing loss (as measured by the detection threshold) and the broadness of auditory filters, with more substantial hearing loss associated with broader auditory filters (Martin, 1974; Florentine et al., 1980; Peters and Moore, 1992; Hopkins and Moore, 2011). Figure 27.4 shows auditory filter bandwidths measured for 40 listeners with a range of audiometric thresholds at the center frequency of the filter. Filter bandwidths were estimated using a "notched noise" technique (Patterson, 1976). Thresholds were measured for a tone in the presence of a noise with a spectral notch centered on the tone frequency. Thresholds for several notch widths were measured, and these data used to derive an estimate for the auditory filter bandwidth at the tone frequency. Note that, although there is a strong correlation between audiometric threshold and filter bandwidth, there is considerable variability between participants with similar audiometric thresholds, and the amount of variability increases with increasing audiometric threshold.

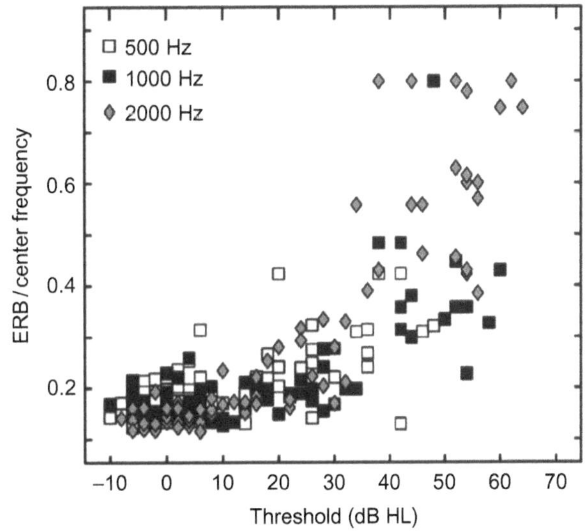

Fig. 27.4. Auditory filter bandwidths (plotted as equivalent rectangular bandwidths (ERB)/filter center frequency) as a function of audiometric threshold for 40 listeners with a range of audiometric thresholds at the filter center frequency. HL, hearing loss. (Adapted from Hopkins and Moore, 2011.)

Rate place coding would also be expected to be affected by substantial inner hair cell loss. If all inner hair cells at a particular place in the cochlea are absent, the pattern of firing across the array of auditory nerve fibers will no longer provide an accurate representation of the frequency spectrum.

Temporal coding

It has also been suggested that temporal coding may be impaired in listeners with sensorineural hearing loss. Listeners with cochlear hearing loss are poorer than normal-hearing listeners at tasks that are thought to rely on temporal coding, such as interaural phase discrimination (Lacher-Fougère and Demany, 2005; Strelcyk and Dau, 2009), low-frequency modulation detection (Lacher-Fougère and Demany, 1998; Buss et al., 2004), and the discrimination of harmonic and frequency-shifted tones (Hopkins and Moore, 2007, 2011).

The mechanism of the deficit is uncertain. There are several possibilities:

- Cochlear hearing loss may be associated with a phase-locking deficit. Woolf et al. (1981) found that phase locking to pure-tone stimuli was impaired in guinea pigs with cochlear hearing loss, although this effect has not been replicated in other mammalian species (Harrison and Evans, 1979; Kale and Heinz, 2010).
- The temporal processing difficulties may arise because of the poorer frequency selectivity associated with cochlear hearing loss (see above). The pattern of temporal information at the output of broadened filters is more complex, and may be uninterpretable by the central auditory system. In a situation where a narrowband signal is to be detected or discriminated in the presence of background noise, broadened auditory filters would pass more noise than narrower filters, so reducing the signal-to-noise ratio at the output of the filter, and reducing the fidelity of the temporal representation of the signal (Henry and Heinz, 2012).
- Some temporal models suggest that temporal information is extracted by comparing the phase at different points along the basilar membrane (Shamma, 1985; de Cheveigne and Pressnitzer, 2006). The phase response is altered in cochlear hearing loss (Ruggero et al., 1996) and so this could explain why listeners with sensorineural hearing loss are poor at tasks thought to rely on temporal coding.

The effects of auditory nerve damage on auditory coding

Due to the problems identifying auditory nerve function in human listeners described in the diagnosis section, above, few studies have assessed the effect of auditory nerve damage on auditory coding directly. However, from a theoretic perspective, auditory nerve damage is likely to cause deficits in auditory coding. Similarly to inner hair cell loss, if auditory nerve fibers innervating a particular place on the basilar membrane are absent, the pattern of neural activity across the array of auditory nerve fibers will not accurately represent the frequency spectrum, so presumably affecting rate place coding. Temporal coding is also likely to be affected, even by less widespread auditory nerve fiber loss. Auditory neurons do not fire on every cycle of a stimulus; instead an accurate representation of a waveform is coded by a combination of neural activity across many nerve fibers. Hence, loss of nerve fibers is expected to reduce the fidelity of temporal coding. Demyelination of auditory nerve fibers is also expected to have an adverse effect on temporal coding; demyelination is unlikely to be uniform across fibers, so differing degrees of myelin damage would be expected to have a desynchronizing effect on auditory nerve responses.

PERCEPTUAL CONSEQUENCES OF SENSORINEURAL HEARING LOSS

As discussed in the sections on structures affected by sensorineural hearing loss and etiology, above, sensorineural hearing loss is a heterogeneous condition, which can occur due to damage to several structures in the peripheral auditory system. As discussed in the diagnosis section, above, it is currently not possible to conclusively diagnose damage to specific structures in the peripheral auditory in humans, and peripheral hearing damage is tentatively classified as of cochlear or neural origin based on the pattern of diagnostic tests and suspected etiology. Here, the perceptual consequences of hearing loss of suspected cochlear origin are discussed. However, the limitations of diagnostic tests in categorizing damage to different structures should be remembered. The perceptual consequences of auditory neuropathy spectrum disorder, which is often thought to be caused by damage to the auditory nerve, are discussed in Chapter 28. The perceptual consequences of cortical and brainstem lesions are discussed in Chapters 6, 9, and 29–33.

Perceptual consequences of cochlear damage

Loudness perception

Loudness is the perceptual attribute of sound related to intensity. Cochlear hearing loss is associated with abnormal loudness perception; detection thresholds are elevated, but the level of sound that is found uncomfortably loud is elevated by a smaller amount (Kamm et al., 1978). This means that the dynamic range of hearing is reduced, an effect known as loudness recruitment. Figure 27.5 shows mean loudness ratings for pure tones as a function of intensity for a group of normal-hearing listeners and three groups of listeners with cochlear hearing loss of 50, 55, and 60 dB HL at the tone frequency. The slope of the loudness function is steeper for the hearing-impaired listeners, and the size of the slope increases with increasing hearing loss. Loudness recruitment is likely to arise directly as a result of outer hair cell damage. As discussed in the section on structures affected by sensorineural hearing loss, above, outer hair cells apply gain at low levels and result in a shallow, non-linear basilar membrane response. The slopes of the loudness functions plotted in Figure 27.5 correspond well with basilar membrane growth functions for animals with normal hearing and various degrees of outer hair cell loss.

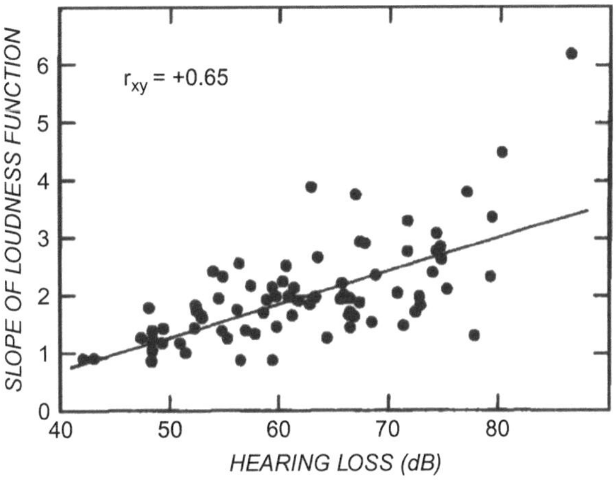

Fig. 27.5. Relation between the slope of the loudness function and the degree of hearing loss for 78 listeners with noise-induced losses. The linear function was obtained by the least-squares method. (Reproduced from Hellman and Meiselman, 1990.)

Frequency and pitch perception

Pitch is the attribute of auditory sensation that allows sounds to be ordered on a music scale. It is important for the perception of speech and music. For example, pitch is used to convey non-semantic information, such as distinguishing a question from a statement, and to provide emphasis. In tonal languages, pitch conveys semantic information, with different word meaning being signified by different pitch contours.

Pitch-evoking stimuli usually consist of a series of harmonics that are integer multiples of a fundamental frequency (F0). A pitch is heard that corresponds to F0. The mechanisms underlying the perception of pitch are still debated, but fall into two categories: pattern recognition and temporal models. Pattern recognition models are based on the idea that the auditory system extracts information about the frequencies of individual harmonics, and derives a pitch based on the best-fitting harmonic series. Temporal models are based on the idea that information about pitch is extracted from the inter-spike intervals in the auditory nerve. The perception of pitch has usually been assessed by either assessing the ability to recognize musical melodies, or by measuring the discrimination thresholds for pitch-evoking stimuli, such as pure or complex tones.

Most studies investigating pitch perception among listeners with cochlear hearing loss have found that, on average, pitch perception is poorer among those with a hearing loss. Frequency discrimination thresholds for pure tones (Turner et al., 1983; Nelson and Freyman, 1986; Simon and Yund, 1993) and complex tones (Moore and Peters, 1992; Arehart, 1994; Moore et al., 2006) are elevated and melody recognition is poorer for listeners with cochlear hearing loss. However,

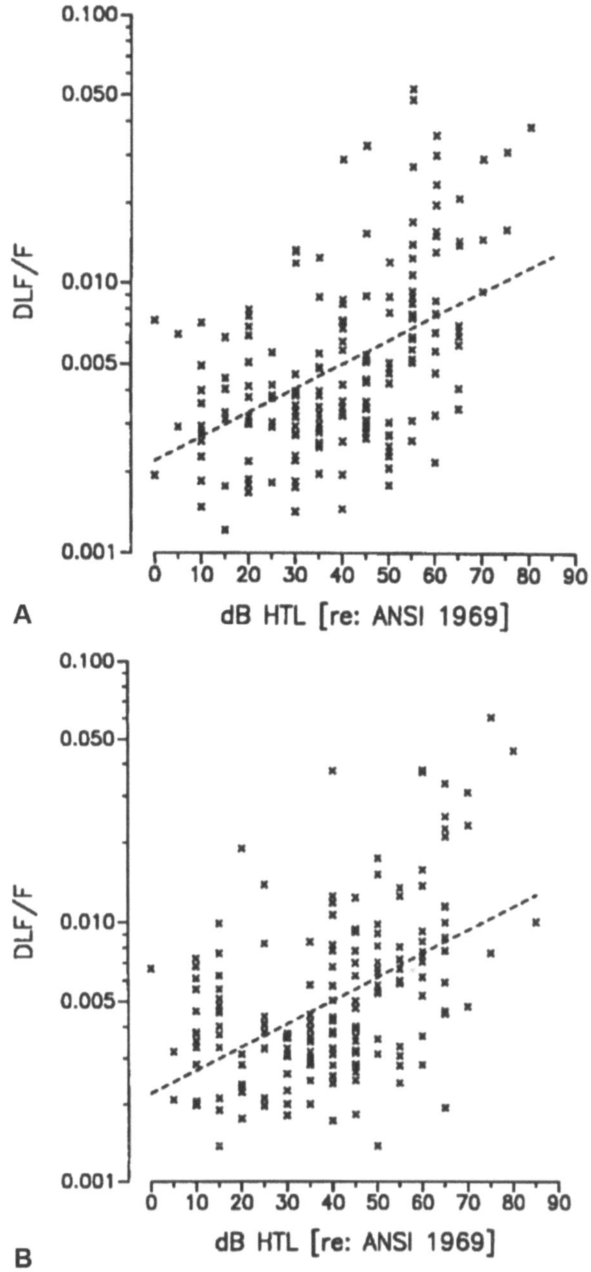

Fig. 27.6. The relation between frequency discrimination and the absolute hearing threshold level (HTL) in right (**A**) and left (**B**) ears. Frequency discrimination threshold (DLF/F) is plotted versus HTL for each subject at each frequency where frequency discrimination was measurable in both ears. A regression line was fitted to the data. (Reproduced from Simon and Yund, 1993.)

considerable individual differences have been reported, with performance as good as normal-hearing listeners in some cases. For example, Figure 27.6 shows frequency discrimination thresholds plotted as a function of audiometric threshold for 34 listeners. There was a significant correlation between audiometric thresholds and frequency discrimination, but the correlation was only moderate, and some listeners with a substantial hearing loss performed as well as those with audiometric thresholds within the normal range.

The mechanisms that underlie the poor pitch perception of listeners with cochlear hearing loss are uncertain. Two main ideas have been proposed: poor frequency selectivity and poor temporal coding.

Poor frequency selectivity

Frequency selectivity is associated with audiometric thresholds, as described in the section on the effects of cochlear damage on auditory coding, above. Most models of pitch perception predict that pitch perception will be adversely affected by reduced frequency selectivity. "Pattern recognition" models of pitch perception rely on the identification of resolved harmonics, which are fed into a central pattern recognizer that identifies the fundamental frequency that best fits the harmonic series. Reduced frequency selectivity is expected to reduce the number of resolved harmonics present in a complex tone, and so would be expected to impair pitch perception. Some temporal models of pitch perception also predict that reduced frequency selectivity would impair the perception of pitch. For example, Moore (1982) suggested that pitch is extracted by combining information from resolved and unresolved harmonics and determining the most common periodicity in the pattern of firing. Resolved harmonics are argued to convey more precise temporal information than unresolved harmonics, and so if frequency selectivity is reduced (so reducing the number of resolved harmonics), this is expected to impair perception of pitch.

If the poor pitch perception of hearing-impaired listeners arises due to poor frequency selectivity, a strong correlation between tasks measuring pitch perception and frequency selectivity would be expected. Tyler et al. (1983) measured psychophysical tuning curves and frequency discrimination limens for normal-hearing subjects and listeners with a sensorineural hearing loss ranging from mild to moderate. Frequency selectivity (as measured by the high-frequency slope of psychophysical tuning curves) and frequency discrimination limens were poorer for the hearing-impaired than the normal-hearing subjects and there were moderate correlations between frequency selectivity and frequency discrimination limens for pure tones. However, these correlations became small and non-significant when audiometric threshold was partialed out, suggesting that impaired frequency selectivity could not entirely explain the poor frequency discrimination of the hearing-impaired listeners.

Moore and Peters (1992) conducted a similar experiment investigating the relation between frequency selectivity and frequency discrimination for pure and complex tones. Similarly to Tyler et al. (1983), they found that correlations between frequency selectivity and frequency discrimination thresholds for pure tones were very small and non-significant after audiometric thresholds were partialed out. However, correlations between frequency selectivity and fundamental frequency difference limens for complex tones were stronger, at least at some frequencies, and in some cases remained significant when audiometric thresholds were partialed out.

Bernstein and Oxenham (2006) measured fundamental frequency difference limens for complex tones for normal-hearing listeners and listeners with sensorineural hearing loss. The lowest harmonic present in the complexes was varied. Fundamental frequency difference limens increased as the lowest harmonic number increased (consistent with previous studies), and they found that frequency selectivity, measured in the same listeners, was significantly correlated with the harmonic number at which F0 discrimination transitioned from good to poor. Together, these results suggest that poor frequency selectivity may at least partly explain why listeners with cochlear hearing loss have poorer pitch perception than normal-hearing listeners.

Poor temporal coding

A temporal coding deficit provides another possible explanation for the poor pitch perception experienced by hearing-impaired listeners. Again, poor temporal coding might be predicted to lead to poor pitch perception based on either pattern recognition or temporal models. Pattern recognition models rely on the coding of information about the frequencies of individual resolved harmonics, but do not make any assumption about whether the individual frequencies are coded using a rate place or temporal mechanism. If the latter of these two mechanisms is used, then pitch perception would be expected to be worse if temporal coding is impaired. Similarly, temporal models would predict impaired pitch perception if temporal coding was impaired. However, as noted in the section on the effects of cochlear damage on auditory coding, above, the apparent temporal coding deficit experienced by listeners with cochlear hearing loss may arise as a result of poor frequency selectivity, and so the effects are difficult to disentangle.

Spatial hearing

Spatial hearing refers to the ability to perceive and exploit information about the spatial position of a sound or sounds. Localization is the ability to determine the position of a sound – an essential skill for successful negotiation of our auditory environment. Localization in the horizontal plane relies mainly on binaural cues.

Interaural level differences (ILDs) arise because of the attenuation of sound with distance, and the "shadowing" effect of the head, which is particularly prominent at high frequencies. Interaural time differences (ITDs) arise because sounds outside the median plane arrive at one ear before the other. Localization in the vertical plane relies on detection of peaks and dips in the frequency spectrum of broadband sounds imposed by the external and torso, mainly at high frequencies, that vary according to elevation.

Listeners with cochlear hearing loss report localization difficulties in everyday life (Noble et al., 1995), and listeners with an asymmetric hearing loss report more problems than those with a symmetric loss (Noble and Gatehouse, 2004). Consistent with this, Noble et al. (1994) found that localization of noise bursts in both the vertical and horizontal planes was moderately correlated with degree of hearing loss in a group of 66 listeners with varying degrees of cochlear hearing loss. The effect of sensorineural hearing loss on localization is greater when more than one sound source is present simultaneously. Figure 27.7 shows localization errors made by normal-hearing listeners and hearing-impaired listeners for two conditions: one in which a single target word was presented and the second in which five target words were presented simultaneously and listeners were required to localize a single target word. The difference between normal-hearing and hearing-impaired groups was significantly larger when many sounds were presented simultaneously.

In order to understand the localization deficits experienced by hearing-impaired listeners it is helpful to consider the effect of cochlear hearing loss on discrimination of ITD and ILD cues. Hawkins and Wightman (1980) found that listeners with sensorineural hearing loss were poorer at ITD discrimination for narrow-band sounds at both low and high frequencies, and at low and high levels. Several more recent studies have found that interaural phase discrimination (closely related to ITD discrimination) is poorer for listeners with a cochlear hearing loss than listeners with normal hearing (Lacher-Fougère and Demany, 2005; Strelcyk and Dau, 2009; Hopkins and Moore, 2011), an effect that has been interpreted as evidence for a temporal processing deficit among these listeners. Few studies have examined the effect of hearing loss on interaural level discrimination, although there is some evidence that interaural level discrimination is also affected by cochlear hearing loss (Koehnke et al., 1995). However, the extent to which these elevations in interaural discrimination thresholds might explain the effect of hearing loss on localization is unknown.

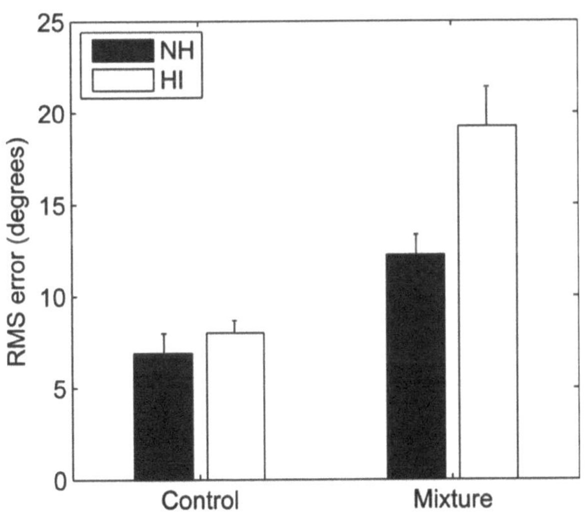

Fig. 27.7. Mean root mean squared (RMS) errors in the control and mixture conditions for the two listener groups (normal hearing (NH) and sensorineural hearing loss (HI)). Error bars show standard errors of the mean. (Reproduced from Best et al., 2011.)

Binaural unmasking

Binaural unmasking refers to the increase in detectability or discriminability of a signal in noise that occurs due to differences in ITD or ILDs between the signal and masker. For normal-hearing listeners, this effect can be substantial. For example, Jeffress et al. (1952) reported that thresholds for detecting a tone in background noise were up to 14 dB better when the ITDs for the tone and noise were different compared with when they were the same (a phenomenon known as the binaural masking level difference). A similar effect has been reported for speech understanding in a spatially separated background; speech reception thresholds in noise are substantially better when the speech and noise are spatially separated (giving rise to differences in binaural cues) than when they are co-located (Hawley et al., 1999; Freyman et al., 2001). Many studies have found that the binaural masking level difference for pure tones is reduced among listeners with a cochlear hearing loss (Hall et al., 1984; Jerger et al., 1984; Strelcyk and Dau, 2009). Similarly, the amount of speech intelligibility benefit derived from spatial separation of speech and masking noise has been reported to be smaller for listeners with a cochlear hearing loss than listeners with normal hearing (Arbogast et al., 2005; Marrone et al., 2008).

Speech perception

The perception of speech demands special attention because of its importance for communication, and the amount of research devoted to its perception. Speech perception with and without the presence of background noise is poorer for listeners with cochlear hearing loss

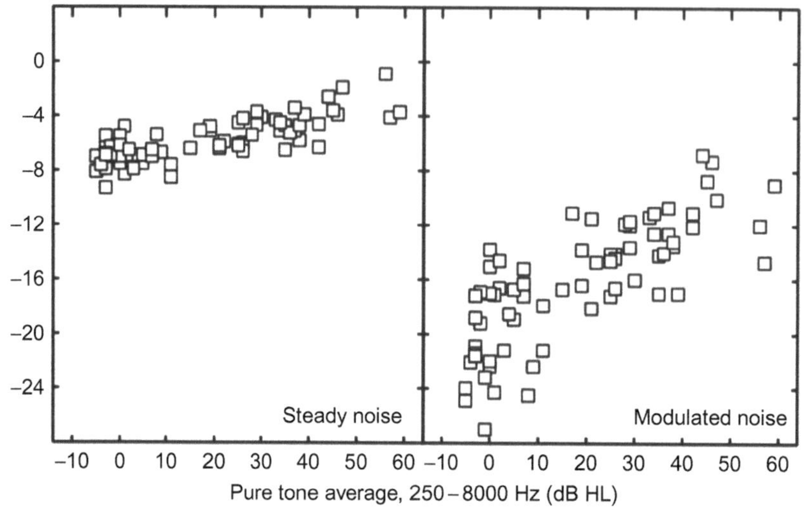

Fig. 27.8. Speech reception thresholds for speech in steady (left panel) and modulated (right panel) background noise for 40 listeners with a range of audiometric thresholds. Speech reception thresholds are plotted as a function of mean audiometric threshold at octave frequencies between 0.25 and 8 kHz. HL, hearing loss.

than listeners with normal hearing (Carhart and Tillman, 1970; Bonding, 1979; Bronkhorst and Plomp, 1989; Barrenäs and Wikström, 2000). This can partly be explained by the reduced audibility caused by elevated audiometric thresholds; the greatest difference between normal-hearing and hearing-impaired listeners is observed when speech is presented at a low level (Humes and Roberts, 1990; Ching et al., 1998), and using masking noise to match audibility between normal-hearing and hearing-impaired listeners reduces the difference in performance between the two groups (Humes and Roberts, 1990). However, audibility cannot entirely explain the differences in speech intelligibility between normal-hearing and hearing-impaired listeners. Hearing-impaired listeners perform more poorly that would be predicted from the speech intelligibility index, which predicts intelligibility based on the audibility of the signal, and is very accurate for normal-hearing listeners (Festen and Plomp, 1990; Smoorenburg, 1992; Ching et al., 1998). Speech perception deficits persist even when speech is fully audible (Turner and Cummings, 1999; Summers and Molis, 2004), and the magnitude of the remaining deficit is related to the degree of audiometric threshold elevation (Turner and Cummings, 1999; Hopkins and Moore, 2011). Speech perception in quiet appears to be less adversely affected than speech perception in background noise, especially for people with a mild hearing loss, although it is not clear to what extent this relates to ceiling effects for task performance when speech intelligibility is measured in quiet.

The effects of hearing loss on speech perception in noise appear to be greater for noise that has temporal modulations than noise that is steady (Festen and Plomp, 1990; Lorenzi et al., 2006; Hopkins and Moore, 2011). Figure 27.8 shows speech reception thresholds for 50% correct word identification in background noise for listeners with a range of audiometric thresholds. The correlation between audiometric thresholds and speech reception thresholds for modulated noise were similar, but the effect of hearing loss (in dB signal-to-noise ratio) was greater for the modulated noise than the steady noise. Normal-hearing listeners benefited from dips in the modulated masker, but the amount of benefit reduced as hearing loss increased.

Many studies have investigated possible reasons for the speech perception deficits experienced by listeners with a cochlear hearing loss. Deficits in both rate place (Patterson et al., 1982; Glasberg and Moore, 1989; Dubno and Schaefer, 1995) and temporal (Lorenzi et al., 2006; Hopkins et al., 2008; Hopkins and Moore, 2011) coding have been proposed as possible explanations.

TREATMENT OF SENSORINEURAL HEARING LOSS

In most cases, sensorineural hearing loss is irreversible, given the limits of current knowledge. Consequently, treatment focuses on management of the hearing loss rather than cure.

Hearing aids

Hearing aids are the most common treatment for hearing impairment. They consist of a microphone, amplifier, and speaker, and the goal is to amplify sounds made inaudible by the hearing loss and deliver these to the patient. Hearing aids are relatively successful in restoring audibility for most types of hearing loss. However, hearing aids do not restore auditory perception to normal in cases of sensorineural hearing loss, because of the suprathreshold perceptual deficits described in this chapter. Some attempts have been made to compensate for the perceptual problems encountered by hearing-impaired listeners with hearing aids. The most successful and widely implemented of these is the use of amplitude compression to compensate for the effects of loudness recruitment (Gatehouse et al., 2006). As described in the section on perceptual consequences of cochlear damage, above, listeners with cochlear hearing loss usually experience loudness recruitment, whereby weak sounds are inaudible, but intense sounds are just as loud as for people with normal hearing, meaning that the dynamic range of hearing is reduced. Amplitude compression attempts to compensate for this problem by amplifying low-level sounds more than high-level sounds.

Another strategy for improving the perception of speech for hearing-aid users with sensorineural hearing loss is to attempt to improve the speech-to-noise ratio. As mentioned above, listeners with sensorineural hearing loss encounter particular problems when listening to speech in background noise, and these problems persist even if sound is amplified to a level that is audible. Speech intelligibility is strongly related to signal-to-noise ratio for both normal-hearing and hearing-impaired listeners, and so if the signal-to-noise ratio can be increased, an increase in speech intelligibility is expected. Several strategies have been proposed. The most successful so far is to combine the outputs of two microphones with differing directionality (the most typical configuration is a directional microphone oriented directly ahead, and an omnidirectional microphone). Assuming that the speech and background are spatially separated, this can result in an improvement in signal-to-noise ratio, and a corresponding improvement in speech intelligibility (Bentler, 2005).

Cochlear implants

Cochlear implants are used to treat severe to profound cochlear hearing loss. Cochlear implants consist of a microphone and processor (usually located behind the ear), a transmitter (located on the skull behind the ear, usually held in place with a magnet), a receiver (implanted under the skin), and an array of electrodes in the cochlea. Each electrode stimulates a different place in the cochlea, which corresponds to a different characteristic frequency. Sounds are processed by filtering into frequency channels; signals from each channel are used to stimulate different electrodes along the length of the electrode array. Similar noise reduction techniques can be employed as for hearing aids; the noise reduction occurs before the main cochlear implant processing.

SUMMARY

- Sensorineural hearing loss arises because of damage to the inner ear. Often several structures are affected, including the inner and outer hair cells of the cochlea, the stria vascularis, and the auditory nerve.
- Sensorineural hearing loss has many causes, including genetic mutations affecting the structures of the inner ear and environmental insults such as noise, ototoxic substances, and hypoxia. The prevalence of sensorineural hearing loss increases dramatically with age.
- Clinical diagnosis is most commonly accomplished by measuring detection thresholds (audiometric thresholds), and comparing these to normative values to determine the degree of hearing loss.
- Often it is not clinically possible to distinguish between hearing loss caused by damage to the cochlear and auditory nerve.
- The most obvious perceptual consequence of sensorineural hearing loss is insensitivity to weak sounds. Other adverse perceptual consequences include loudness recruitment, poor perception of pitch and auditory space, and difficulty understanding speech, particularly in the presence of background noise.
- Sensorineural hearing loss is usually incurable. Treatment focuses on restoring the audibility of sounds made inaudible by hearing loss using either hearing aids or cochlear implants.

References

ANSI (2004). ANSI S3.6-2004 Specification for audiometers, American National Standards Institute, New York.

Arbogast TL, Mason CR, Kidd Jr G (2005). The effect of spatial separation on informational masking of speech in normal-hearing and hearing-impaired listeners. J Acoust Soc Am 117: 2169–2180.

Arehart KH (1994). Effects of harmonic content on complex-tone fundamental-frequency discrimination in hearing-impaired listeners. J Acoust Soc Am 95: 3574–3585.

Ashmore JF (1987). A fast motile response in guinea pig outer hair cells: the cellular basis of the cochlear amplifier. J Physiol 388: 323–347.

Barrenäs ML, Wikström I (2000). The influence of hearing and age on speech recognition scores in noise in audiological patients and in the general population. Ear Hear 21: 569–577.

Bentler RA (2005). Effectiveness of directional microphones and noise reduction schemes in hearing aids: a systematic review of the evidence. J Am Acad Audiol 16: 473–484.

Bernstein JG, Oxenham AJ (2006). The relationship between frequency selectivity and pitch discrimination: sensorineural hearing loss. J Acoust Soc Am 120: 3929–3945.

Bernstein JGW, Summers V, Grassi E et al. (2013). Auditory models of suprathreshold distortion and speech intelligibility in persons with impaired hearing. J Am Acad Audiol 24: 307–328.

Best V, Carlile S, Kopco N et al. (2011). Localization in speech mixtures by listeners with hearing loss. J Acoust Soc Am 129: E210–E215.

Bonding P (1979). Frequency selectivity and speech discrimination in sensorineural hearing loss. Scand Audiol 8: 205–216.

Bronkhorst AW, Plomp R (1989). Binaural speech intelligibility in noise for hearing-impaired listeners. J Acoust Soc Am 86: 1374–1383.

Buss E, Hall 3rd JW, Grose JH (2004). Temporal fine-structure cues to speech and pure tone modulation in observers with sensorineural hearing loss. Ear Hear 25: 242–250.

Carhart RC, Tillman TW (1970). Interaction of competing speech signals with hearing losses. Arch Otolaryngol 91: 273–279.

Ching T, Dillon H, Byrne D (1998). Speech recognition of hearing-impaired listeners: predictions from audibility and the limited role of high-frequency amplification. J Acoust Soc Am 103: 1128–1140.

Cruickshanks KJ, Wiley TL, Tweed TS et al. (1998). Prevalence of hearing loss in older adults in Beaver Dam. Wisconsin The Epidemiology of Hearing Loss Study. Am J Epidemiol 148: 879–886.

Davis AC (1989). The prevalence of hearing impairment and reported hearing disability among adults in Great Britain. Int J Epidemiol 18: 911–917.

Davis A (1995). Hearing in Adults, Whurr, London.

de Cheveigne A, Pressnitzer D (2006). The case of the missing delay lines: synthetic delays obtained by cross-channel phase interaction. J Acoust Soc Am 119: 3908–3918.

Dubno JR, Schaefer AB (1995). Frequency selectivity and consonant recognition for hearing-impaired and normal-hearing listeners with equivalent masked thresholds. J Acoust Soc Am 97: 1165–1174.

El-Badry MM, Ding DL, McFadden SL et al. (2007). Physiological effects of auditory nerve myelinopathy in chinchillas. Eur J Neurosci 25: 1437–1446.

Festen JM, Plomp R (1990). Effects of fluctuating noise and interfering speech on the speech-reception threshold for impaired and normal hearing. J Acoust Soc Am 88: 1725–1736.

Florentine M, Buus S, Scharf B et al. (1980). Frequency selectivity in normally-hearing and hearing-impaired observers. J Speech Hear Res 23: 643–669.

Freyman RL, Balakrishnan U, Helfer KS (2001). Spatial release from informational masking in speech recognition. J Acoust Soc Am 109: 2112–2122.

Gatehouse S, Naylor G, Elberling C (2006). Linear and non-linear hearing aid fittings – 1. Patterns of benefit. Int J Audiol 45: 130–152.

Glasberg BR, Moore BCJ (1989). Psychoacoustic abilities of subjects with unilateral and bilateral cochlear impairments and their relationship to the ability to understand speech. Scand Audiol Suppl 32: 1–25.

Hall JW, Tyler RS, Fernandes MA (1984). Factors influencing the masking level difference in cochlear hearing-impaired and normal-hearing listeners. J Speech Hear Res 27: 145–154.

Harrison RV, Evans EF (1979). Some aspects of temporal coding by single cochlear fibres from regions of cochlear hair cell degeneration in the guinea pig. Arch Otolaryngol 224: 71–78.

Hawkins DB, Wightman FL (1980). Interaural time discrimination ability of listeners with sensori-neural hearing loss. Audiol 19: 495–507.

Hawley ML, Litovsky RY, Colburn HS (1999). Speech intelligibility and localization in a multi-source environment. J Acoust Soc Am 105: 3436–3448.

He NJ, Mills JH, Dubno JR (2007). Frequency modulation detection: effects of age, psychophysical method, and modulation waveform. J Acoust Soc Am 122: 467–477.

Hellman RP, Meiselman CH (1990). Loudness relations for individuals and groups in normal and impaired hearing. J Acoust Soc Am 88: 2596–2606.

Henry KS, Heinz MG (2012). Diminished temporal coding with sensorineural hearing loss emerges in background noise. Nat Neurosci 15: 1362–1364.

Herman GE, Warren LR, Wagener JW (1977). Auditory lateralization – age-differences in sensitivity to dichotic time and amplitude cues. J Gerontol 32: 187–191.

Hood JD (1957). The principles and practice of bone conduction audiometry: a review of the present position. Proc R Soc Med 50: 689–697.

Hopkins K, Moore BCJ (2007). Moderate cochlear hearing loss leads to a reduced ability to use temporal fine structure information. J Acoust Soc Am 122: 1055–1068.

Hopkins K, Moore BCJ (2011). The effects of age and cochlear hearing loss on temporal fine structure sensitivity, frequency selectivity, and speech reception in noise. J Acoust Soc Am 130: 334–349.

Hopkins K, Moore BCJ, Stone MA (2008). The effects of moderate cochlear hearing loss on the ability to benefit from temporal fine structure information in speech. J Acoust Soc Am 123: 1140–1153.

Hughson W, Westlake A (1944). Manual for program outline for rehabilitation of aural casualties both military and civilian. Trans Am Acad Ophthalmol Otolaryngol Suppl 48: 1–15.

Humes LE, Roberts L (1990). Speech-recognition difficulties of the hearing-impaired elderly: the contributions of audibility. J Speech Hear Res 33: 726–735.

Jeffress LA, Blodgett HC, Deatherage BH (1952). The masking of tones by white noise as a function of the interaural phases of both components: I. 500 cycles. J Acoust Soc Am 24: 523–527.

Jerger J, Brown D, Smith S (1984). Effect of peripheral hearing loss on the MLD. Arch Otolaryngol 110: 290–296.

Kale S, Heinz MG (2010). Envelope coding in auditory nerve fibers following noise-induced hearing loss. J Assoc Res Otolaryngol 11: 657–673.

Kamm C, Dirks DD, Mickey MR (1978). Effect of sensorineural hearing loss on loudness discomfort level and most comfortable loudness judgments. J Speech Lang Hear Res 21: 668–681.

Kemp DT (2002). Oto-acoustic emissions, their origin in cochlear function, and use. Br Med Bull 63: 223–241.

Koehnke J, Culotta CP, Hawley ML et al. (1995). Effects of reference interaural time and intensity differences on binaural performance in listeners with normal and impaired hearing. Ear Hear 16: 331–353.

Kujawa SG, Liberman MC (2009). Adding insult to injury: cochlear nerve degeneration after "temporary" noise-induced hearing loss. J Neurosci 29: 14077–14085.

Lacher-Fougère S, Demany L (1998). Modulation detection by normal and hearing-impaired listeners. Audiology 37: 109–121.

Lacher-Fougère S, Demany L (2005). Consequences of cochlear damage for the detection of interaural phase differences. J Acoust Soc Am 118: 2519–2526.

Lebo CP, Reddell RC (1972). Presbycusis component in occupational hearing-loss. Laryngoscope 82: 1399.

Lin FR, Niparko JK, Ferrucci L (2011a). Hearing loss prevalence in the United States. Arch Intern Med 171: 1851–1852.

Lin HW, Furman AC, Kujawa SG et al. (2011b). Primary neural degeneration in the guinea pig cochlea after reversible noise-induced threshold shift. J Assoc Res Otolaryngol 12: 605–616.

Lorenzi C, Gilbert G, Carn H et al. (2006). Speech perception problems of the hearing impaired reflect inability to use temporal fine structure. Proc Natl Acad Sci U S A 103: 18866–18869.

Makary CA, Shin J, Kujawa SG et al. (2011). Age-related primary cochlear neuronal degeneration in human temporal bones. J Assoc Res Otolaryngol 12: 711–717.

Marmel F, Linley D, Carlyon RP et al. (2013). Subcortical neural synchrony and absolute thresholds predict frequency discrimination independently. J Assoc Res Otolaryngol 14: 757–766.

Marrone N, Mason CR, Kidd Jr G (2008). The effects of hearing loss and age on the benefit of spatial separation between multiple talkers in reverberant rooms. J Acoust Soc Am 124: 3064–3075.

Martin MC (1974). Critical bands in sensorineural hearing loss. Scand Audiol 3: 133–140.

Mazurek B, Haupt H, Georgiewa P et al. (2006). A model of peripherally developing hearing loss and tinnitus based on the role of hypoxia and ischemia. Med Hypotheses 67: 892–899.

Mills JH, Schmiedt RA, Kulish LF (1990). Age-related changes in auditory potentials of Mongolian gerbil. Hear Res 46: 201–210.

Moore BCJ (1982). An Introduction to the Psychology of Hearing, 2nd Edn. Academic Press, London.

Moore BCJ, Peters RW (1992). Pitch discrimination and phase sensitivity in young and elderly subjects and its relationship to frequency selectivity. J Acoust Soc Am 91: 2881–2893.

Moore BCJ, Glasberg BR, Hopkins K (2006). Frequency discrimination of complex tones by hearing-impaired subjects: evidence for loss of ability to use temporal fine structure. Hear Res 222: 16–27.

Moore BCJ, Glasberg BR, Stoev M et al. (2012). The influence of age and high-frequency hearing loss on sensitivity to temporal fine structure at low frequencies (L). J Acoust Soc Am 131: 1003–1006.

Nelson DA, Freyman RL (1986). Psychometric functions for frequency discrimination from listeners with sensorineural hearing loss. J Acoust Soc Am 79: 799–805.

Noble W, Gatehouse S (2004). Interaural asymmetry of hearing loss, Speech, Spatial and Qualities of Hearing Scale (SSQ) disabilities, and handicap. Int J Audiol 43: 100–114.

Noble W, Byrne D, Lepage B (1994). Effects on sound localization of configuration and type of hearing impairment. J Acoust Soc Am 95: 992–1005.

Noble W, Ter-Horst K, Byrne D (1995). Disabilities and handicaps associated with impaired auditory localization. J Am Acad Audiol 6: 129–140.

Palmer AR, Russell IJ (1986). Phase-locking in the cochlear nerve of the guinea-pig and its relation to the receptor potential of inner hair-cells. Hear Res 24: 1–15.

Patterson RD (1976). Auditory filter shapes derived with noise stimuli. J Acoust Soc Am 59: 640–654.

Patterson RD, Nimmo-Smith I, Weber DL et al. (1982). The deterioration of hearing with age: frequency selectivity, the critical ratio, the audiogram, and speech threshold. J Acoust Soc Am 72: 1788–1803.

Patuzzi RB, Yates GK, Johnstone B (1989). Outer hair cell receptor current and sensorineural hearing loss. Hear Res 42: 47–72.

Peters RW, Moore BCJ (1992). Auditory filter shapes at low center frequencies in young and elderly hearing-impaired subjects. J Acoust Soc Am 91: 256–266.

Rhode WS (1971). Observations of the vibration of the basilar membrane in squirrel monkeys using the Mössbauer technique. J Acoust Soc Am 49: 1218–1231.

Rose JE, Brugge JF, Anderson DJ et al. (1967). Phase-locked response to low-frequency tones in single auditory nerve fibers of the squirrel monkey. J Neurophysiol 30: 769–793.

Ruggero MA, Rich NC (1991). Furosemide alters organ of Corti mechanics: evidence for feedback of outer hair cells upon the basilar membrane. J Neurosci 11: 1057–1067.

Ruggero MA, Rich NC, Robles L et al. (1996). The effects of acoustic trauma, other cochlea injury and death on basilar membrane responses to sound. In: A Axelsson, H Borchgrevink, RP Hamernik et al. (Eds.), Scientific Basis of Noise-Induced Hearing Loss, Thieme, Stockholm, pp. 23–35.

Ryan A, Dallos P (1975). Absence of cochlear outer hair cells: effect on behavioral auditory threshold. Nature 253: 44–46.

Schuknecht HF, Gacek MR (1993). Cochlear pathology in presbycusis. Ann Otol Rhinol Laryngol 102 (Suppl. 158): 1–16.

Schulte BA, Schmiedt RA (1992). Lateral wall Na, K-ATPase and endocochlear potentials decline with age in quiet-reared gerbils. Hear Res 61: 35–46.

Sellick PM, Patuzzi R, Johnstone BM (1982). Measurement of basilar membrane motion in the guinea pig using the Mössbauer technique. J Acoust Soc Am 72: 131–141.

Sewell WF (1984). The effects of furosemide on the endocochlear potential and auditory-nerve fiber tuning curves in cats. Hear Res 14: 305–314.

Shamma SA (1985). Speech processing in the auditory system II: Lateral inhibition and the central processing of speech evoked activity in the auditory nerve. J Acoust Soc Am 78: 1622–1632.

Simon HJ, Yund EW (1993). Frequency discrimination in listeners with sensorineural hearing loss. Ear Hear 14: 190–199.

Simpson A (2009). Frequency-lowering devices for managing high-frequency hearing loss: a review. Trends Amplif 13: 87–106.

Smoorenburg GF (1992). Speech reception in quiet and in noisy conditions by individuals with noise-induced hearing loss in relation to their tone audiogram. J Acoust Soc Am 91: 421–437.

Strelcyk O, Dau T (2009). Relations between frequency selectivity, temporal fine-structure processing, and speech reception in impaired hearing. J Acoust Soc Am 125: 3328–3345.

Summers V, Molis MR (2004). Speech recognition in fluctuating and continuous maskers: effects of hearing loss and presentation level. J Speech Lang Hear Res 47: 245–256.

Takeno S, Harrison R, Ibrahim W et al. (1994). Cochlear function after selective inner hair cell degeneration by carboplatin. Hear Res 75: 93–102.

Turner CW, Cummings KJ (1999). Speech audibility for listeners with high-frequency hearing loss. Am J Audiol 8: 47–56.

Turner CW, Burns EM, Nelson DA (1983). Pure tone pitch perception and low-frequency hearing loss. J Acoust Soc Am 73: 966–975.

Tyler RS, Wood EJ, Fernandes MA (1983). Frequency resolution and discrimination of constant and dynamic tones in normal and hearing-impaired listeners. J Acoust Soc Am 74: 1190–1199.

Wang J, Powers NL, Hofstetter P et al. (1997). Effects of selective inner hair cell loss on auditory nerve fiber threshold, tuning and spontaneous and driven discharge rate. Hear Res 107: 67–82.

Webster M, Webster DB (1981). Spiral ganglion neuron loss following organ of Corti loss: a quantitative study. Brain Res 212: 17–30.

Weiss TF, Rose C (1988). A comparison of synchronization filters in different auditory receptor organs. Hear Res 33: 175–179.

Wilson D, Walsh PG, Sanchez L et al. (1998). Hearing impairment in an Australian population, Centre for Population Studies in Epidemiology, Department of Human Services, Adelaide.

Wilson DH, Walsh PG, Sanchez L et al. (1999). The epidemiology of hearing impairment in an Australian adult population. Int J Epidemiol 28: 247–252.

Woolf NK, Ryan AF, Bone RC (1981). Neural phase-locking properties in the absence of outer hair cells. Hear Res 4: 335–346.

Yates GK, Winter IM, Robertson D (1990). Basilar membrane nonlinearity determines auditory nerve rate-intensity functions and cochlear dynamic range. Hear Res 45: 203–219.

Handbook of Clinical Neurology, Vol. 129 (3rd series)
The Human Auditory System
G.G. Celesia and G. Hickok, Editors

Chapter 28

Auditory neuropathy

ARNOLD STARR[1]* AND GARY RANCE[2]
[1]*Departments of Neurology and Neurobiology, University of California, Irvine, CA, USA*
[2]*School of Audiology, University of Melbourne, Melbourne, Australia*

INTRODUCTION

Objective measures of hearing thresholds using auditory brainstem responses (ABRs) were championed by Galambos and his associates four decades ago (Hecox and Galambos, 1974; Galambos and Despland, 1980) and are now used routinely to provide objective measures of auditory nerve and brainstem responses in neonates and adults (Norton et al., 2000).

While the tests were of great benefit for objectively defining "deafness," there were exceptions that tested this assumption. Some subjects with "hearing problems" had relatively normal audiometric thresholds but absent or severely abnormal ABRs (Kaga and Tanaka, 1980; Worthington and Peters, 1980; Hildesheimer et al., 1985; Satya-Murti et al., 1983; Kraus et al., 1984). We identified that their deafness was due to abnormal auditory nerve function in the presence of normal-functioning cochlear sensory hair cells. Figure 28.1 displays measures (audiogram, distortion product otoacoustic emissions (OAEs), ABRs to clicks and to tones, and auditory cortical potentials) from an individual with auditory neuropathy (AN) and a normal-hearing control (normal).

The AN subject in Figure 28.1 had a mild loss of audibility, impaired speech perception out of proportion to the audibility changes, ABRs with absent waves I–III, and barely detectable wave V that was delayed in latency. In contrast, the cochlear hair cell measures (DPOAEs) and cochlear microphonics (CMs) were normal. Cortical N100 potentials were present and delayed in latency (Starr et al, 1991, 1996). The pattern of these results was consistent with a hearing disorder (Starr et al., 1991) due to abnormal auditory nerve functions. Some of the subjects were found to also have cranial nerve (vestibular, optic nerve) and/or peripheral neuropathies, supporting the idea that the auditory nerve was also affected by neuropathic disorders. The term "auditory neuropathy" was used to describe the group of patients we studied (Starr et al., 1996). We also recognized that inner hair cell disorders affecting ribbon synapse function would display similar clinical features.

The hearing disorder of AN affected processing of acoustic temporal cues that are essential for: (1) speech comprehension; (2) localization of sounds; and (3) separating signals from background noise (Starr et al., 1991; Zeng et al., 2005).

Examination of temporal bones from subjects dying with AN showed inner and outer hair cells to be normal in number and appearance, whereas auditory ganglion cells and nerve fibers were both reduced in number and demyelinated (Starr et al., 2003). Loss of auditory nerve fibers would attenuate neural input while demyelination would affect the synchrony of neural conduction. We consider that both the loss of auditory nerve fibers and altered neural transmission contribute to the abnormalities of both ABRs and hearing.

Abnorml ABRs and normal hair cell measures were identified in neonates with otoferlin (OTOF) mutations that affected glutamate neurotransmitter release from inner hair cell ribbon synapses. The locus of the auditory nerve disorder in this mutation was presynaptic and has been shown to affect ribbon synaptic function (Varga et al., 2006; Rodríguez-Ballesteros et al., 2008).

We anticipate that patients with abnormal ABRs and normal cochlear hair measures will also be identified in disorders of neurotransmitter reuptake and auditory dendritic receptor.

We have been asked to comment on other terminologies used for individuals with abnormal ABRs and preserved cochlear hair cell activities. The most common ones are "auditory neuropathy spectrum disorder" and

*Correspondence to: Arnold Starr, Department of Neurology, University of California, 154 Med Surge I, Mail Code: 4290, Irvine, CA 92697, USA. Tel: +1-949-230-7269, E-mail: astarr@uci.edu

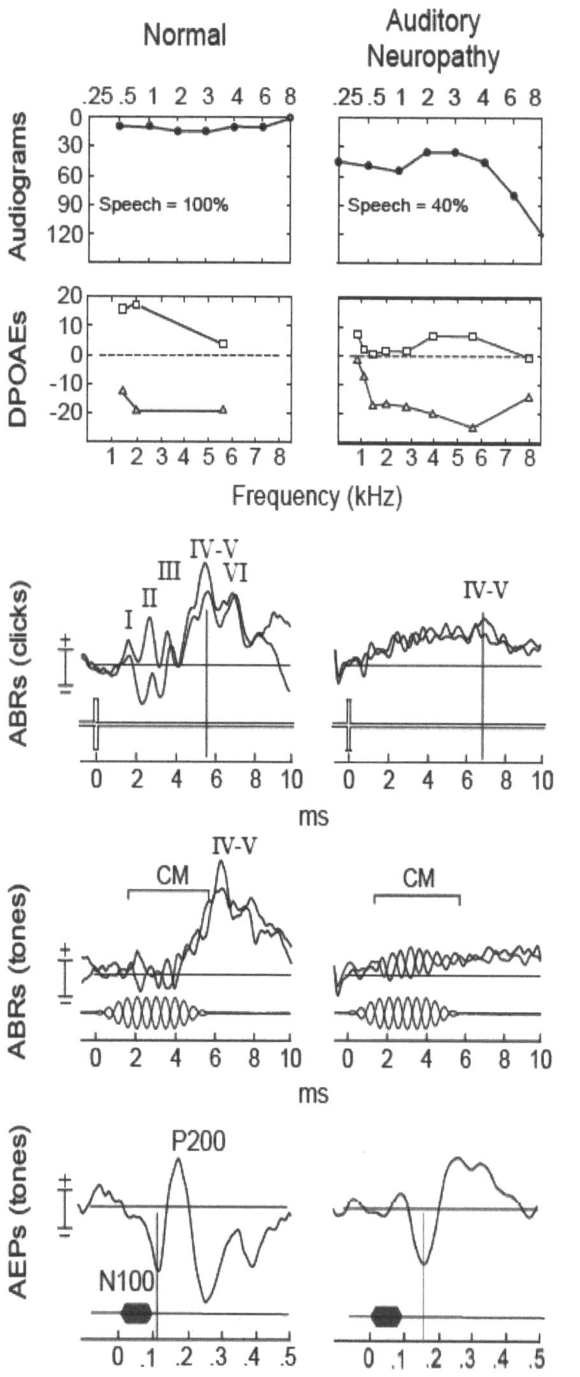

Fig. 28.1. Measures of auditory function in a normal control and a subject with auditory neuropathy. DPOAEs, cochlear hair cell measures; ABRs, auditory brainstem responses; AEPs, auditory evoked potentials; CM, cochlear microphonics.

"auditory neuropathy/auditory dyssynchrony." The term "spectrum disorder" is used when there is both a paucity of objective measures and knowledge of their etiology. Spectrum disorder is inappropriate for AN since we do use objective measures (e.g., ABRs) that can localize the sites of auditory nerve dysfunction as affecting auditory nerve and inner hair cell ribbon synapses (Moser et al., 2013). Most importantly, there are many specific etiologies that have been identified as causing AN (Starr et al., 2001; Santarelli, 2010). We suggest that the term postsynaptic AN be used when there is loss and/or demyelination of auditory nerves. When inner hair cells are affected the term presynaptic AN is appropriate.

The term "auditory neuropathy/auditory dyssynchrony" is used to indicate that there is reduction of neural synchrony of auditory nerve fibers in patients with AN. The concept is attractive but as yet there is no quantitative measure of the degrees of dyssynchrony. We suggest that ABRs may be able to provide such measures of changes of neural synchrony. For instance, in patients with postsynaptic AN the ABR to the initial click could be normal but was then delayed and lost to subsequent clicks in the train (e.g., conduction slowing: Wynne et al., 2013). Dyssynchrony in this instance had a time course of expression. Moreover, both a reduction in signal intensity and a decrease in signal-to-noise ratio in normal-hearing subjects can result in both ABR and psychoacoustic measures that are similar to AN. We now depend on psychoacoustic methods in trained observers to quantify the effects of dyssynchrony of auditory perceptions. Objective measures of both brainstem and cortical potentials may prove a way of quantifying the magnitude change in central auditory processing in normal hearing and in auditory neuropathies (Kraus et al., 2000; Michalewski et al., 2005).

We will review below diagnostic features of AN, including audiologic and psychoacoustics, etiologies (e.g., developmental, genetic, metabolic, degenerative, iatrogenic), their associated pathologies, and the effects of specific therapies.

DIAGNOSIS

The diagnosis of AN relies on electrophysiologic tests with ancillary support from neuroimaging and audiologic assessment.

Electrophysiologic procedures

ABRs in AN are absent or attenuated in amplitude as well as delayed in latency (Table 28.1) (Starr et al., 1996). The conduction velocity of the auditory nerve can be measured only when both waves I and II are present (Butinar et al., 2008). Some of the details of performing these measures are detailed below.

Cochlear hair cell microphonics are typically preserved and can be defined by separately averaging ABRs to condensation and to rarefaction stimuli. Subtracting

Table 28.1

Objective electrophysiologic findings in auditory neuropathy

Test	Results
ABRs	Absent or abnormal
Acoustic evoked middle-ear muscle reflexes	Typically absent
Cochlear microphonics	Normal
Otoacoustic emissions	Normal
Electrocochleography	CAP broad and low amplitude
Neurophonics	Absent

ABRs, auditory brainstem responses; CAP, compound action potential.

the ABR condensation from rarefaction clicks attenuates the neural components, revealing CM that are of opposite polarity to condensation and rarefaction stimuli. Addition of the ABRs to condensation and rarefaction stimuli will cancel CMs and enhance neural components (Starr et al., 1991; Berlin et al., 1998).

Both inner hair cell summating potentials and the compound action potential (CAP) of auditory nerve (wave I of the ABR) can be most clearly identified using electrocochleography (Santarelli and Arslan, 2002). The method helps to localize the sites of auditory nerve dysfunction as involving inner hair cells and/or auditory nerve (Santarelli, 2010). This information can assist the clinician in assessing the likely benefits of cochlear implants. Typically, if the dysfunction is limited to the inner hair cells, cochlear implants will be very beneficial. The technique requires general anesthesia for infants and children but can be done under local anesthesia in adults.

Auditory cortical potentials

A cortical potential can be recorded to the onset of acoustic signals at a peak latency of approximately 100 ms. The potential is of positive polarity in children and changes to a negative polarity by around 8 years of age. The N100/P100 is present in AN even when ABRs are absent. Rance and colleagues (2002) showed that 50% of school-aged children with AN were without an N100 or P100 and its absence was highly correlated with impaired speech perceptual abilities. The presence of a preserved N100 of normal latency provided an objective measure of neural synchrony at the cortical level (Fig. 28.2).

In adults, the latency of N100 to brief tones is also sensitive to the onset of temporal features of the stimulus. N100 latency is prolonged as stimulus onset is

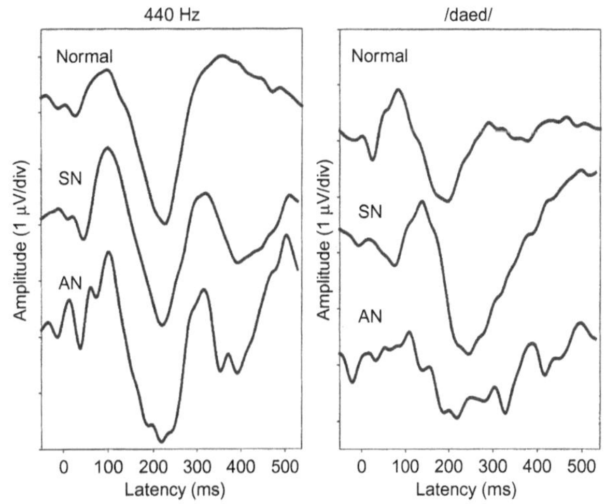

Fig. 28.2. Grand mean cortical event-related potential waveforms in response to tones (left panel) and to speech (right panel) for children with normal hearing (top traces), sensorineural (SN) hearing loss (440 Hz: $n=17$; /dæd/: $n=15$, middle traces), and auditory neuropathy (AN) ($n=11$, bottom traces). Daed is the phonetic representation of the word "dad." (Reproduced from Rance, 2005, with permission.)

slowed (Onishi and Davis, 1968) but is relatively independent of stimulus intensity. In AN subjects, N100 latencies to brief tone bursts, while present, are abnormally delayed in latency (approximately 40 ms) and the magnitude of the delay is independent of audibility changes (Michalewski et al., 2009). Cortical measures can be used to quantify temporal processing deficits of auditory nerve and/or inner hair cell ribbon synapses (Wynne et al., 2013).

Currently it is difficult to distinguish between disorders of auditory dendrites (noise trauma), auditory axons accompanying neurologic disorders (both a form of deafferentation), altered neural conduction, and ribbon synaptic abnormalities. All these conditions affect auditory nerve and brainstem activity, reflected by ABR. We anticipate that further studies of changes of ABR latency and amplitude of ABR components will help to distinguish among etiologies. For instance, AN accompanying degenerative etiology such as Charcot–Marie–Tooth (CMT) will show progressive latency delays and reduction in amplitudes of ABR components (Starr et al., 2003). These measures also may be used to quantify the course of the disorder and provide objective measures of therapies in clinical trials (Rance et al., 2010).

ABR abnormalities (Fig. 28.3) can be helpful in localizing the site of auditory nerve abnormality. Wave I, generated by distal portions of auditory nerve, is typically preserved in proximal disorders of the auditory nerve (e.g., acoustic neuromas) but absent in both presynaptic (OTOF-related ribbon synaptic disorders) and neuropathic disorders of auditory nerve. Conduction times between waves I and II (the latter generated by proximal auditory nerve; Martin et al., 1995) can reveal abnormal slowing of auditory nerve conduction (Butinar et al., 2008). Prolongation of latency difference between waves I and V (>4.5 ms) reflects slowed conduction between auditory nerve and lateral lemnisci.

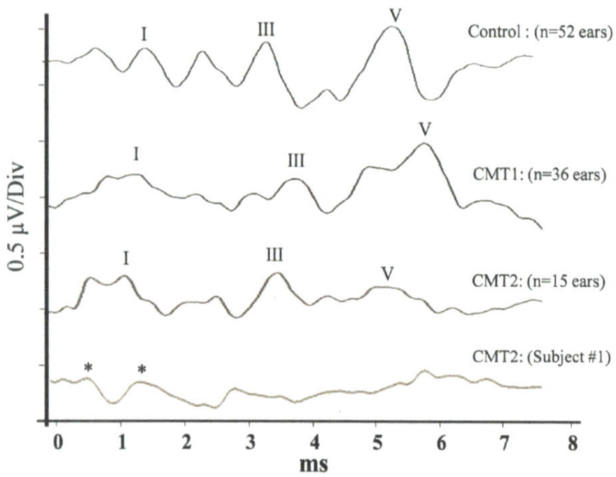

Fig. 28.3. Auditory brainstem response waveforms obtained for children with Charcot–Marie–Tooth (CMT) disease. The top tracing shows the combined waveform for a normal control group. The second is the averaged waveform for children showing slow conduction between waves I (auditory nerve) and III (cochlear nucleus) consistent with demyelination of auditory nerve (CMT1). The third is the averaged waveform for children with CMT2, showing axonal loss and reflected by normal conduction times between waves I and III and V, but reduced amplitude of wave V. The bottom tracing is for a single child who showed no repeatable components I, II, and V to stimuli at maximum presentation levels (90 dB nHL) but whose audibility was impaired. The asterisks in this case are cochlear microphonics. (Reproduced from Rance et al., 2012a.)

The relative amplitude of wave V to wave I is normally > 1. When this ratio is less than 0.5, the site of auditory nerve disorder is proximal rather than distal (Starr and Achor, 1975; Rance et al., 2010). Cochlear hair cell activities are normal. These include microphonics (CMs), reflecting outer and inner hair cell intracellular potential changes; OAEs, reflecting contractions of outer hair cells, are present in newborns with AN (Table 28.1). They become absent in approximately 30% of these infants by 2 years of age (Rance et al., 1999). CMs are preserved in those subjects that have lost OAEs (Starr et al., 1998). Thus, for diagnosing AN, the CMs are a reliable measure of preserved hair cell activities (Starr et al., 1991).

Neurophonics or frequency-following responses are field potentials of auditory brainstem structures to low-frequency tones (<500 Hz) that can be recorded using the same ABR recording methods (Fig. 28.3). Their onset occurs approximately 3–5 ms after the onset of CM (Fig. 28.1) and is consistent with brainstem generation. Neurophonics have not been recorded routinely in AN but, in the limited number of subjects tested, neurophonics are absent. An example of one such recording to consonant-vowels by Nina Kraus in a 10-year-old with AN is shown in Figure 28.4. In this subject, neurophonics were absent. The measurement of neurophonics provides an objective measure of low-frequency temporal processing.

Electrocochleography is a method allowing near-field recording of cochlear and nerve potentials using a needle electrode that is passed through the tympanic membrane to rest on the bony cochlea (Eggermont and Odenthal, 1974). The recordings reveal high-amplitude inner hair cell summating receptor potentials, CAPs of auditory nerve, and CMs. The procedure requires local anesthesia in adults and general anesthesia in children and provides

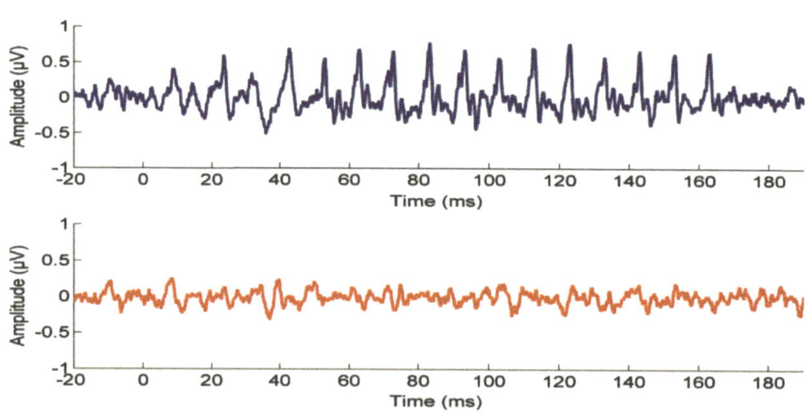

Fig. 28.4. This figure contains auditory brainstem responses (ABRs) to repeated presentations of /da/. The upper trace (blue) is from a normal hearing subject showing averaged potentials over 180 ms. containing repetitive positive potentials occurring at approximately 100 cycles/s. These repetitive potentials are also called "frequency following responses" or (FFRs) and reflect a temporal neural encoding of speech components. The brainstem potentials from an 8 year old subject with auditory neuropathy are in red and do not appear to contain frequency following responses and the child had marked difficulty identifying differences between speech sounds but could hear the speech clearly. We thank Nina Kraus PhD for providing this figure.

objective measures of both inner hair cell summating potentials and auditory neural responses. ABRs do not usually include a summating potential. Figure 28.5 shows electrocochleography from 8 normal and 8 AN subjects. All controls had two distinct negative deflections: the first, an increasingly negative summating potential with onset latency approximately 0.3 ms after CM. The summating potential is interrupted by a second negative/positive potential of high amplitude and short duration, returning to baseline approximately 3 ms after onset of the summating potential.

In contrast to the controls, AN subjects had either absent CAP or a low-amplitude CAP (identified in only 2 AN subjects, #3, 4, circled in Fig. 28.5), whereas a summating potential was present in all. All AN subjects showed a prolonged negative potential of large amplitude, even when the CAP was absent. The generators for the abnormally prolonged negative potentials in AN may reflect dyssynchronous generation of summating potential and/or dyssynchronous discharges of nerve fibers.

Usually the onset of the CAP at high intensity is similar in both normal and AN subjects, whereas the CAP is broad and of low amplitude in AN subjects.

Neuroimaging

Magnetic resonance imaging (MRI) of the brainstem and auditory nerve can assist in localizing the site(s) of auditory nerve involvement. Congenital atrophy of auditory nerve occurs in neonates and children with normal cochlear hair cell measures (Buchman et al., 2006). The degree of atrophy may be a factor affecting the outcome of a cochlear implant. MRI and computed tomography scans are now the methods of choice for detecting structural abnormalities of auditory nerve causing AN, such as acoustic neuromas and brainstem tumors (see Chapter 29).

Audiology and psychoacoustics

Audiologic measures

Diagnostic features of AN include: (1) impaired speech perception out of proportion to audibility changes; (2) impaired ability to detect rapid changes in intensity; (3) inability to utilize interaural differences of temporal cues; and (4) abnormal ability to process acoustic signals in the presence of noise masking (Starr et al., 1991, 1996;

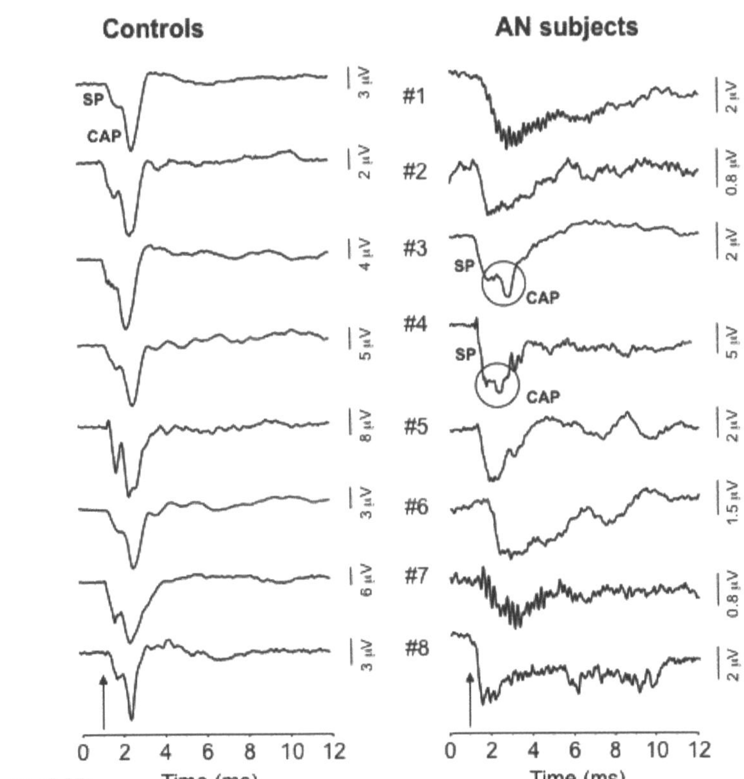

Fig. 28.5. Electrocochleography in normal-hearing (left column) and auditory neuropathy (AN) subjects (right column). Summating potential (SP) and compound action potential (CAP) in response to high-intensity clicks (110 dB) recorded from 8 controls (left) and AN subjects (right). The separation of the CAP from the SP is clear in controls, while a CAP can only be identified in 2 AN subjects (#3, #4; circled). In the remaining AN ears the SP continues and a boundary between CAP and SP cannot be identified. Arrows at the bottom indicate CM onset. (Reproduced from Santarelli et al., 2008.)

Rance et al., 1999; Zeng et al., 2005). Details of hearing abnormalities are described in the section on new directions, below.

ETIOLOGY

The etiologies of AN phenotype are multiple (see Starr et al., 1998, for review). These include developmental disorders (e.g., hypoplasia of auditory nerve: Buchman et al., 2006), toxic-metabolic disorders (e.g., hyperbilirubinemia: Shapiro, 2003); infections (e.g., meningitis), inflammation (e.g., siderosis), neoplasms (e.g., acoustic neuroma), genetic mutations affecting neural functions (e.g., CMT disease: Starr et al., 2003) and ribbon synapse function (Moser et al., 2013), optic neuropathy (e.g., Huang et al., 2009), mitochondrial disorders such as Friedreich's ataxia (Cacace and Pinheiro, 2011), autoimmune disorders (e.g., Guillain–Barré syndrome), nutritional disorders (Attias et al., 2012), and degenerative changes accompanying aging (Masuda and Kaga, 2011).

Neurologic examinations can identify neurologic and medical conditions that are associated with AN. These disorders include genetic mutations affecting the optic nerve (Opa1 gene) and/or vestibular nerve (Fujikawa and Starr, 2000; Santarelli et al., 2008; Huang et al., 2009), disorders of peripheral nerves (e.g., Starr et al., 2003; Butinar et al., 2008), metabolic disorders causing neuropathies (e.g., Acar et al., 2012), immune disorders e.g., Guillain–Barré syndrome (Ueda and Kuroiwa, 2008), and inflammatory disorders such as rupture of aneurysm producing siderosis (Nadol et al., 2011) or as a consequence of meningitis (Celis-Aguilar et al., 2012).

PATHOLOGIES OF AUDITORY NEUROPATHY

Auditory nerve

Examination of human temporal bones ($n=5$) from subjects with hereditary disorders of auditory neural function (Spoendlin, 1974; Hallpike et al., 1980; Merchant et al., 2001; Starr et al., 2003; Bahmad et al., 2007) showed marked loss of auditory nerve ganglion cells, nerve fibers and demyelination of many of the remaining fibers (Fig. 28.6). Both the number and appearance of inner and outer hair cells were normal (Fig. 28.7). The neuropathology is consistent with deafferentation and impaired neural conduction to affect both the magnitude of auditory nerve input and the synchrony between fibers (Wynne et al., 2013). Patients with AN due to compression from acoustic neuromas also have loss of nerve fibers and demyelination in proximity to the region of compression. The pathology would lead to neural

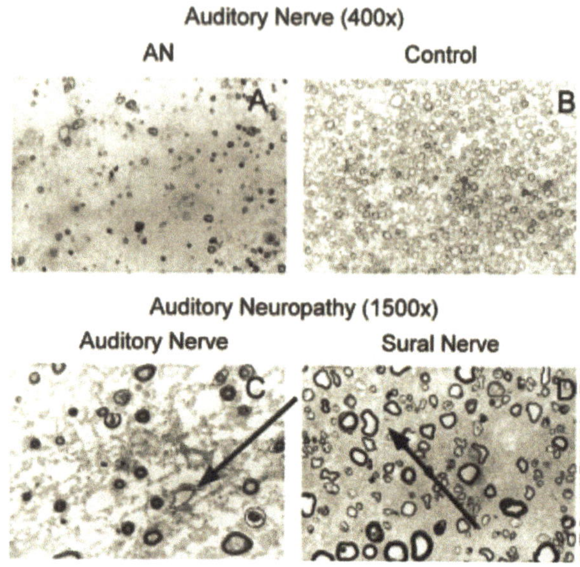

Fig. 28.6. Auditory nerve adjacent to brainstem in the top panel from auditory neuropathy (AN) subject (**A**) and age-matched control (**B**) (osmium tetroxide, ×400). Lower panel from the AN subject shows high-power view (×1500) of auditory nerve (**C**) and sural nerve (**D**). The arrows are directed to thinly myelinated fibers in both nerves, reflecting incomplete remyelination. (Reproduced from Starr et al., 2003, with permission.)

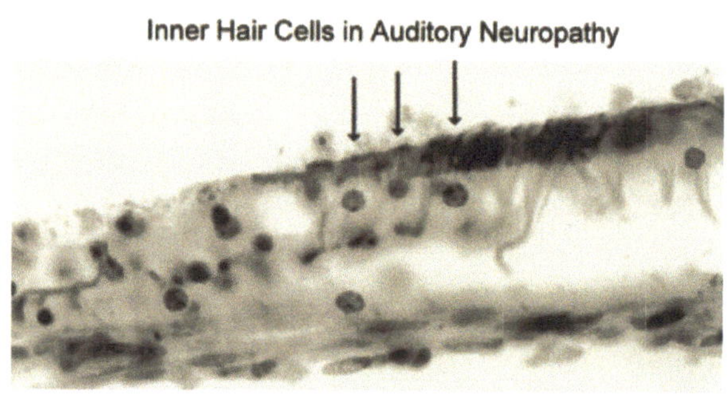

Fig. 28.7. Intact inner hair cells (arrows) as seen between the inner sulcus and inner pillar cells in a section cut on the bias (osmium tetraoxide, hematoxylin and eosin, ×650). (Reproduced from Starr et al., 2003, with permission.)

conduction blocks and could possibly compromise cochlear blood supply to affect hair cell function (Matsunaga and Kanzaki, 2000).

Inner hair cells, but not ganglion cells, have been shown to be selectively depleted in temporal bones of some premature infants (Amatuzzi et al., 2011). This selective inner hair cell loss, if expressed in term infants, would be expected to show absent or marked attenuation of inner hair cell summating receptor potentials but intact neural potentials to electric stimulation of auditory nerve.

Auditory nerve damage accompanying acoustic trauma

Attention to details of cochlear damage accompanying excessive noise exposure has revealed immediate damage to auditory nerve dendrites relative to the expression of noise damage to hair cells and auditory nerve dendrites (Furman et al., 2013; Moser et al., 2013). The dendritic effects result from excessive neurotransmitter release causing prolonged excitation of the dendrites and their subsequent retraction from inner hair cells by neurotransmitter release (glutamate) from inner hair cell ribbon synapses. These effects are of greater magnitude with high- than low-threshold auditory nerve fibers. Noise exposure can cause both delayed damage to auditory ganglion cells and inner hair cell ribbon synapses. These data suggest that noise-induced hearing disorders such as tinnitus are due, in part, to AN affecting both auditory nerve dendrites, a postsynaptic AN, and inner hair cells, a presynaptic disorder AN.

Inner hair cell ribbon synapses

Physiologic studies of experimental animals with homozygous OTOF mutations have marked impairment of glutamate neurotransmitter release (Moser et al., 2013). A compound mutation of OTOF mutations in humans is expressed as a temperature-sensitive deafness (Starr et al., 1998; Varga et al., 2006; Marlin et al., 2010). Experimental animal models accompanying gene mutations also have impaired inner hair cell neurotransmitter release (Jing et al., 2013). In humans with temperature-sensitive AN due to OTOF mutation, auditory nerve activity is reduced during continuous sound stimulation accompanied by abnormally rapid adaptation of loudness to continuous auditory signals (Wynne et al., 2013).

In summary, the clinical picture of dysfunction of auditory nerve (absent/abnormal ABRs in the presence of preserved cochlear hair cell measures (OAEs and/or CMs) has been shown to accompany: (1) disordered presynaptic release of neurotransmitter by ribbon synapses (Starr et al., 1998; Marlin et al., 2010); (2) deafferentation accompanying loss of auditory nerve fibers (Starr et al., 2003); and (3) conduction blocks (Wynne et al., 2013) accompanying demyelination of nerve fibers.

The distinction between both etiology and site of the disorder affecting auditory nerve function requires expertise in several disciplines, including audiology, genetics, neurology, psychoacoustics, and speech processing.

CLINICAL EXPRESSION OF AUDITORY NEUROPATHY

The prevalence of AN in adult populations is difficult to determine as the physiologic assessments used to identify the condition (ABR/CM/OAE) are not routinely undertaken unless there are specific clinical indicators for retrocochlear abnormality (asymmetric hearing/unusually poor speech understanding) or a prior diagnosis of genetic disorder involving peripheral neuropathy (e.g., Freidreich's ataxia, CMT disorders). In pediatric populations, on the other hand, electrophysiologic assessment is the cornerstone of infant hearing testing and universal screening has revealed that AN is a relatively high-incidence condition, particularly amongst babies in the neonatal intensive care unit. Data from our laboratory over a 5-year period (1991–1996), for example, found that 1 in every 423 (0.23%) special care unit graduates presented with the AN result pattern (Rance et al., 1999). We estimate that AN and ribbon synapse disorders accounts for approximately 10% of all permanent pediatric hearing loss.

In newborns in neonatal intensive care units (NICU) there is increased incidence of AN in the presence of metabolic disorders (e.g., acidosis, hypoxia) and infections. The latter may reflect the common use in the ICU of antibiotics with cochlear toxicities. MRIs of the brain commonly show diffuse periventricular brain lesions. ABR abnormalities in many of these neonates do resolve.

In healthy neonates abnormal ABRs occur at increased incidence in families with a history of hearing loss. If measures of hair cell function are preserved, AN is a likely diagnosis. Several genetic mutations have been identified as causal in these children. Atrophy of the auditory nerve, either unilaterally or bilaterally, has been identified by MRI and is a cause of AN (Buchman et al., 2006).

In school-age children audiograms may show normal or mild hearing loss in the presence of impaired speech comprehension. Many of these children utilize lip reading without their realization and definition of the disorder requires awareness of this compensatory adjustment.

In adults the hearing disorder affects speech comprehension out of proportion to the audiogram loss. These patients are aware of their deficits and that they become exaggerated by background noise. Neurologic exam can identify the site of the disorder as being neural by the finding of neuropathies affecting other peripheral or cranial nerves (most commonly, vestibular and/or optic) that may be asymptomatic.

In an aging population AN is common and may be related to the frequent occurrence of other cranial (particularly vestibular, optic) and peripheral neuropathies. These patients typically have abnormal cochlear outer hair cell measures (OAEs, CMs) consistent with a sensory hearing loss. When abnormalities of speech comprehension are beyond that seen with cochlear hearing loss, AN may also be likely. The clinical and objective assessment of peripheral and other cranial nerve functions may help identify the hearing disorder as being affected by neural disorder. Cochlear implantation has been shown to be of benefit for speech comprehension in these elderly subjects (Orabi et al., 2006; Poissant et al., 2008; Carlson et al., 2010).

Noise-induced hearing loss is a form of AN initially affecting auditory nerve dendritic connections with inner hair cells.

HEARING DISORDERS ACCOMPANYING AUDITORY NEUROPATHY

Among the various hearing difficulties experienced by AN patients, two are particularly important: sound detection thresholds and speech perception.

Sound detection thresholds

Hearing threshold levels in individuals with AN vary across the audiometric range independently of whether the ABR is present or absent. Sound detection thresholds in both adult and pediatric populations are evenly distributed and range from normal to profound levels (Rance et al., 1999; Sininger and Oba, 2001; Berlin et al., 2010).

Importantly, discrimination of complex acoustic signals (e.g., speech) is not significantly correlated with audibility in individuals with AN (Zeng et al., 2001; Rance et al., 2002). In contrast, our experience has been that perception in cases with sensory hearing loss is closely related to audibility, with those individuals presenting with profoundly impaired sound detection thresholds showing the poorest perception (Yellin et al., 1989; Rance et al., 2002). AN listeners with hearing thresholds in the profound range also suffer extreme discrimination deficits, but so do individuals with normal or near-normal sound detection (Starr et al., 1996; Zeng et al., 2005; Rance et al., 2008). As such, it is often the degree to which the neural representation of sound is distorted rather than audibility in this group that determines perceptual ability.

Speech perception

Speech comprehension difficulty is a consistently reported feature of AN. Most affected adults have shown perceptual deficits greater than predicted from their audiogram (Starr et al., 1996, 2000; Zeng et al., 2001; Rance et al., 2008). Results in children have been more variable. At best, young listeners with AN show speech perception test results comparable to their peers with sensory (cochlear) hearing loss. At worst, they show no functional hearing ability at all despite (in many cases) enjoying complete access to the normal speech spectrum. Figure 28.8 shows this broad spread of perceptual performance, reflecting the lack of a significant relationship between speech understanding and audibility in both adults and children with AN. As can be seen from these data, approximately 50% of AN cases show perceptual ability poorer than the expected minimum for sensory hearing loss of equivalent degree.

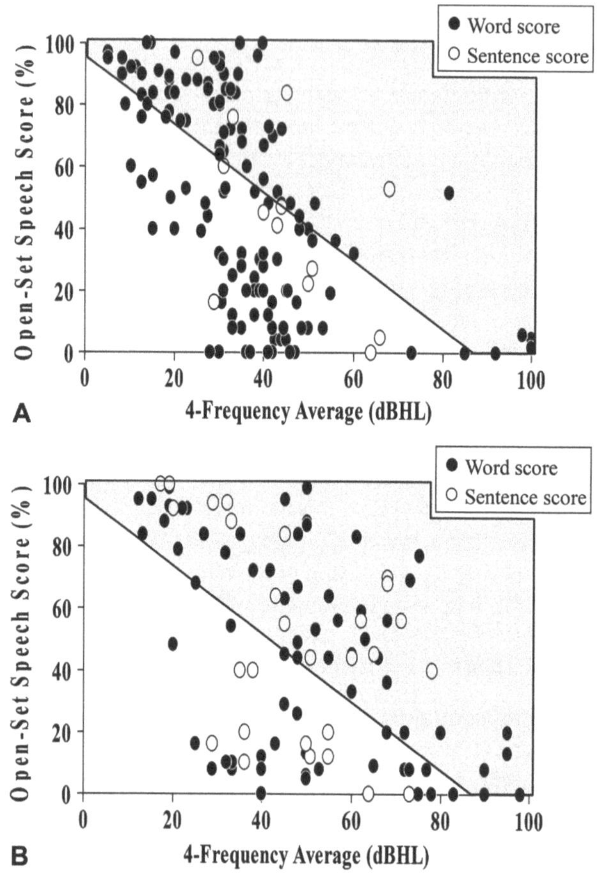

Fig. 28.8. Open-set speech perception scores (in quiet) plotted against average hearing level (HL) for adults (**A**) and children (**B**) with auditory neuropathy who have been presented in the literature. The filled data points represent findings from open-set word tests and the open points show open-set sentence test results. The gray area represents the expected performance range for ears with sensory hearing loss. (Reproduced from Yellin et al., 1989, with permission.)

In addition to these perceptual limitations in quiet, speech understanding in noise is a particular problem for listeners with AN-type hearing loss (Zeng and Liu, 2006; Rance et al., 2008). As demonstrated in Figure 28.9 (which shows speech perception findings for a group of patients with Friedreich ataxia), even those individuals with AN who enjoy relatively normal perception in favorable (quiet) conditions struggle to

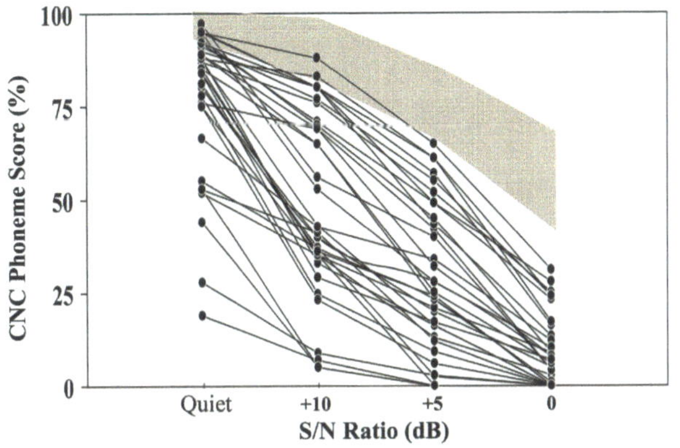

Fig. 28.9. Speech perception in Friedreich ataxia. Results were obtained in quiet listening conditions and in background noise at three signal-to-noise (S/N) ratios. The shaded area represents the 95% confidence range for a matched control group. CNC, Consonant-Nucleus-Consonant. (Reproduced from Rance et al., 2008, with permission.)

discriminate speech in background noise. For instance, signal-to-noise ratios of around 0–3 dB are typical and accepted in "real-life" listening environments such as school classrooms and open-plan offices. In contrast, the majority of AN subjects report comprehension difficulties when exposed to the typical signal-to-noise ratios of <3 dB that affect most of their day-to-day interactions. Psychophysical studies have shown the same degree of impaired comprehension in both simultaneous (where the signal is presented within the noise) and non-simultaneous (where the noise occurs immediately before or after the signal) masking experiments (Kraus et al., 2000; Zeng et al., 2005; Vinay and Moore, 2007). As such, it appears that listeners with AN are less able to separate sounds occurring successively, a temporal processing disorder. In an everyday listening context, where the level of background noise fluctuates, this temporal processing deficit might impair the listener's ability to use brief gaps in the noise to optimize speech understanding (Fig. 28.9).

AUDITORY PROCESSING IN AUDITORY NEUROPATHY

The mismatch between audibility and speech understanding in listeners with AN suggests that distortions of suprathreshold cues is the limiting factor in perceptual performance (Starr et al., 1996, 2000; Rance et al., 1999, 2002, 2012a). Speech perception ability in fact appears closely related to the degree of neural disruption in VIIIth-nerve and central auditory pathways (Kraus et al., 2000; Rance et al., 2002; Zeng et al., 2005). Psychophysical studies investigating the ways in which this disruption affects basic auditory processing have revealed that AN produces a pattern of effects quite distinct from that caused by sensory hearing loss. For example, as the cochlea is responsible for the initial processing of spectral cues (through the precise tonotopic arrangement of signals along the basilar membrane), sensory hearing loss is usually associated with impaired "frequency resolution" (the ability to perceive (resolve) the frequency components of a complex sound) (Rance et al., 2008). Listeners with AN on the other hand (who typically show evidence of normal cochlear (outer hair cell) function) generally enjoy normal cochlear frequency processing for high but not low frequencies (Cacace et al., 1983; Starr et al., 1991; Rance et al., 2004; Zeng et al., 2005; Vinay and Moore, 2007). Similarly, intensity discrimination (the capacity to detect level differences between sounds) which is also largely determined by cochlear-level processing, is relatively normal in AN listeners (Zeng et al., 2005).

Temperature-sensitive forms of AN due to mutations of OTOF (Starr et al., 1998; Marlin et al., 2010; Wynne et al., 2013) show clinical features that vary in degree as body temperature changes. When afebrile, their temporal processing ability (e.g., gap detection threshold) and speech comprehension may be normal. However, for sustained stimuli they experience marked adaptation of loudness similar to that described for acoustic neuromas. When body temperature rises their ability to comprehend speech is markedly impaired. It appears that temperature-sensitive AN degrades ribbon synaptic neurotransmitter release to be both reduced and dyssynchronous, even at stimulus onset, and then both impairments increase during continuous stimulation. In contrast, neural forms of AN show normal adaptation to sustained stimulation (Wynne et al., 2013). Abnormal adaptation is a hallmark of presynaptic AN due to ribbon synaptic disorder. (Santarelli et al., 2008) The impairment of neurotransmitter release likely participates in affecting speech perception in noise.

In contrast, by disrupting the integrity of the temporal neural code, AN affects perception based upon timing cues. In particular, temporal resolution (the ability to perceive rapid changes in auditory signals over time) and the temporal aspects of pitch discrimination can be severely compromised. In AN, temporal resolution deficits have been demonstrated both in "gap detection" tasks where listeners with AN typically require a silent period 2–5 times longer than normal controls before they become aware of a change in a continuous signal (Starr et al., 1991; Zeng et al., 2005) and in "amplitude modulation detection," where they show an impaired ability to track rapid signal envelope changes (Rance et al., 2004, 2010, 2012a; Zeng et al., 2005).

As well as demonstrating monaural temporal processing limitations, individuals with AN show impaired ability to integrate binaural difference cues. Abnormal masking-level difference (MLD) results, for example, are a consistently reported finding (Starr et al., 1991, 1996; Hood, 1999). MLD assessment measures the release from masking obtained when a signal or noise is presented out of phase with a competing signal in the contralateral ear. Normal subjects typically show an MLD (with dichotic phase inversion) of $\approx$ 10 dB, indicating that phase information from each ear has been accurately represented at the level of the lower brainstem (Licklider, 1948). AN subjects, in contrast, typically show little or no masking release, suggesting an inability to accurately combine the neural code from each ear. This impaired ability to generate normal neural responses to monaural signals also accounts for their marked impairment to judge sound direction. When a sound emanates from any point other than directly in front of, or behind, the listener, the signal travels an unequal distance to each ear, giving rise to subtle (<1 ms) interaural time differences. Furthermore, the head acts as a barrier when the source is not directly in front or behind, which results in interaural level differences (of $\approx$ 10 dB). These difference cues are initially processed in the superior olivary complex of the brainstem where neural impulses originating in left and right auditory nerves converge on the same neural elements. These neurons are tuned to subtle binaural time and intensity differences essential for location (and movement) (Riedel and Kollmeier, 2002). While localization based on intensity cues is unimpaired in AN listeners, a number of studies have found that even gross interaural timing differences (>0.5 ms) are not interpreted as changes in sound direction (Starr et al., 1991; Zeng et al., 2005).

Sound localization has obvious benefits, alerting the normal listener to possible information sources and environmental dangers. In addition, sound localization cues may be used to improve perception in background noise when a target signal (such as speech) and competing noise arise from different directions (Micheyl et al., 2007) and or different spectral sources. AN affects the ability to selectively attend to a particular voice based on its location and, as a result, affected listeners typically require much higher signal-to-noise ratios (or lower noise levels) than their normal peers to comprehend speech and communicate effectively (Rance et al., 2012b).

TREATMENT OF AUDITORY NEUROPATHY

There are two main approaches to reducing the functional hearing deficit in individuals with hearing impairment. The first is to improve the listening environment. The second is to amplify/modify the signal reaching the listener's ear to maximize access to its salient cues.

Signal clarity

Listening-in-noise difficulty is a cardinal feature of AN, and as such, any approach that improves the signal-to-noise ratio (i.e., the relative loudness of speech to background noise) is likely to be important. Simple "listening tactics," whereby the individual physically structures the communication environment to reduce ambient noise and maximize the clarity of the speaker's voice, may be beneficial for general communication. Sound-field amplification systems, which amplify the speaker's voice (via loudspeakers), may also be useful for structured listening situations such as the school classroom or auditorium (Johnstone et al., 2009).

Another approach involves the use of "FM-listening" systems. These devices, which have been used extensively in children and adults with sensory hearing loss, transmit speech signals (detected by a lapel-worn microphone) via radio waves to ear-level receivers worn by the listener. As a result, the listener obtains a signal-to-noise ratio advantage from the proximity of the speaker's mouth to the transmitter microphone. Recent data from our laboratory have demonstrated significant speech perception and general communication benefits in a group of children and adults with AN due to Friedreich ataxia (Rance et al., 2010).

Amplification (hearing aid)

The basic function of conventional hearing aids is to amplify sound. As such, they can make environmental sounds and speech louder (allowing access to AN listeners with impaired sound detection) but they cannot improve the clarity of the sounds. Amplification outcomes in patients have been mixed. Some children (perhaps those with lesser degrees of temporal disruption) have responded well to hearing aids and have shown aided speech perception abilities consistent with their sensorineural counterparts (Rance et al., 2002; Roush et al., 2011; Ching et al., 2013). In many youngsters and almost all affected adults, however, conventional amplification has been of little or no benefit (Starr et al., 1996; Rance, 2005; Berlin et al., 2010). As a result, digital signal processing hearing aids designed to accentuate temporal and/or amplitude differences in the acoustic signal have been considered (Zeng et al., 2001; Narne and Vanaja, 2008) and may, in the future, improve outcomes in some cases.

Digital signal processing hearing aids can be programmed to process complex sounds (such as speech) in real time. They are not limited to making sounds louder, but can additionally (or alternatively) enhance the acoustic information that contributes most to intelligibility for people with a hearing impairment. Furthermore, they can reduce the deleterious effects of unwanted background noise. For people who have specific difficulties with the perception of temporal information, sound-processing schemes have been devised that make timing cues more salient. For example, exaggerating the changes in level that occur in natural acoustic signals over a relatively short time scale can enhance temporal information. This process is known as amplitude expansion.

Cochlear electric implants

Cochlear implantation is currently the intervention option of choice for AN patients with severe auditory processing difficulties but not for children less than a year. Most reported cases have shown significant perceptual benefits and speech perception performance equivalent to that obtained by implantees with cochlear hearing loss (Trautwein et al., 2001; Madden et al., 2002; Shallop, 2002; Mason et al., 2003; Zeng and Liu, 2006; Teagle et al., 2010). These findings may seem counterintuitive, as the signal provided by the implant must still pass through a pathologic inner hair cell and/or a pathologic auditory nerve. Of the >200 AN cases reported, only a handful have had poor outcomes (Miyamoto et al., 1999; Rance et al., 1999; Teagle et al., 2010; Roush et al., 2011). Significantly, most subjects with cochlear implants show normal ABRs to electric stimulation (where previously to acoustic stimulation they had not). This suggests either an increase in the number of neural elements contributing to the evoked response (perhaps as a result of bypassing peripheral abnormality and stimulating the spiral ganglion and/or first node of Ranvier in auditory dendrites directly) or an improvement in the synchrony of neural firing. Whatever the explanation, it would seem that cochlear implants offer a viable means of improving functional hearing in most individuals with AN.

In the selection of candidates for cochlear implantation, the first criterion is the preservation of a normal-sized auditory nerve, as shown by MRI.

The audiologic measures in cochlear implantation candidates usually fall into two categories:

1. Those with hearing thresholds in the severe/profound range. In such cases the benefit of a hearing aid is limited and implantation should be undertaken as soon as reliable sound detection thresholds can be obtained. In children this is typically possible from $\approx$8 to 12 months of age.
2. Those with hearing thresholds in the normal to moderate range, but in whom poor discrimination may pose a problem. In children with AN, speech perception ability cannot be reliably evaluated until $\approx$3–4 years of age. This is obviously later than desirable. Research is needed to develop objective measures that can predict long-term auditory capacity and allow cochlear implantation candidature decisions to be made in infancy. In contrast, adults with AN who have impaired speech perception but relatively normal audiograms should be considered for implantation.

NEW DIRECTIONS

The quantification of auditory function in AN may be useful biomarkers of changes in function of certain hereditary or metabolic disorders affecting cranial or peripheral nerves (e.g., Friedreich's ataxia (Rance et al., 2008, 2012b) and CMT (Rance et al., 2012a)). The ABR measures of latency are stable and the effects of variables such as core temperature, gender, or cochlear damage are well known. The ability to define conduction times between the distal and proximal portions of the auditory nerve (Butinar et al., 2008) and between auditory brainstem structures (Starr and Achor, 1975) enhances the detection of slight alterations of function before symptoms or neurologic signs of progression occur.

The use of cortical electrophysiologic measures (Wynne et al., 2013) may provide information as to the capacity of the central auditory system to reorganize and deal with the disordered ribbon synapses and auditory nerve fibers.

References

Acar M, Aycan Z, Acar B et al. (2012). Audiologic evaluation in pediatric patients with type 1 diabetes mellitus. J Pediatr Endocrinol Metab 25 (5–6): 503–508.

Amatuzzi M, Liberman MC, Northrop C (2011). Selective inner hair cell loss in prematurity: a temporal bone study of infants from a neonatal intensive care unit. J Assoc Res Otolaryngol 12 (5): 595–604.

Attias J, Raveh E, Aizer-Dannon A et al. (2012). Auditory system dysfunction due to infantile thiamine deficiency: long-term auditory sequelae. Audiol Neurootol 17 (5): 309–320.

Bahmad Jr F, Merchant SN, Nadol Jr JB et al. (2007). Otopathology in Mohr-Tranebjaerg syndrome. Laryngoscope 117 (7): 1202–1208.

Berlin CI, Bordelon J, St John P et al. (1998). Reversing click polarity may uncover auditory neuropathy in infants. Ear Hear 19 (1): 37–47.

Berlin CI, Hood LJ, Morlet T et al. (2010). Multi-site diagnosis and management of 260 patients with auditory neuropathy/dys-synchrony (auditory neuropathy spectrum disorder). Int J Audiol 49 (1): 30–43.

Buchman CA, Roush PA, Teagle HFB et al. (2006). Auditory neuropathy characteristics in children with cochlear nerve deficiency. Ear Hear 27: 399–408.

Butinar D, Starr A, Zidar J et al. (2008). Auditory nerve is affected in one of two different point mutations of the neurofilament light gene. Clin Neurophysiol 119 (2): 367–375.

Cacace AT, Pinheiro JM (2011). The mitochondrial connection in auditory neuropathy. Audiol Neurootol 16 (6): 398–413.

Cacace AT, Satya Murti S, Grimes CT (1983). Frequency selectivity and temporal processing in Friedreich's ataxia. Ann Otol Rhinol Laryngol 92: 276–280.

Carlson ML, Breen JT, Gifford RH et al. (2010). Cochlear implantation in the octogenarian and nonagenarian. Otol Neurotol 31 (8): 1343–1349.

Celis-Aguilar E, Macias-Valle L, Coutinho-De Toledo H (2012). Auditory neuropathy secondary to cryptococcal central nervous system infection in 2 immunocompromised patients. Otolaryngol Head Neck Surg 147 (3): 597–598.

Ching T, Day J, Dillon H et al. (2013). Impact of the presence of auditory neuropathy spectrum disorder (ANSD) on outcomes of children at 3 years of age. Int J Audiol 52 (Suppl 2): S55–S64.

Eggermont JJ, Odenthal D (1974). Methods in eletrtocochleograohy. Acta Otolaryngol Suppl 316: 17–24.

Fujikawa S, Starr A (2000). Vestibular neuropathy accompanying auditory and peripheral neuropathies. Arch Otolaryngol Head Neck Surg 126 (12): 1453–1456.

Furman AC, Kujawa SG, Liberman MC (2013). Noise-induced cochlear neuropathy is selective for fibers with low spontaneous rates. J Neurophysiol 110 (3): 577–586.

Galambos R, Despland PA (1980). The auditory brainstem response (ABR) evaluates risk factors for hearing loss in the newborn. Pediatr Res 14 (2): 159–163.

Hallpike CS, Harriman DG, Wells CE (1980). A case of afferent neuropathy and deafness. J Laryngol Otol 94 (8): 945–964.

Hecox K, Galambos R (1974). Brain stem auditory evoked responses in human infants and adults. Arch Otolaryngol 99 (1): 30–33.

Hildesheimer M, Muchnik C, Rubinstein M (1985). Problems in interpretation of brainstem-evoked response audiometry results. Audiology 24: 374–379.

Hood LJ (1999). A review of objective methods of evaluating neural pathways. Laryngoscope 109: 1745–1748.

Huang T, Santarelli R, Starr A (2009). Mutation of OPA1 gene causes deafness by affecting function of auditory nerve terminals. Brain Res 1300: 97–104.

Jing Z, Rutherford MA, Takago H et al. (2013). Disruption of the presynaptic cytomatrix protein bassoon degrades ribbon anchorage, multiquantal release, and sound encoding at the hair cell afferent synapse. J Neurosci 33 (10): 4456–4467.

Johnstone KN, John AB, Hall JW et al. (2009). Multiple benefits of personal FM system use by children with auditory processing disorder (APD). Int J Audiol 48: 371–383.

Kaga K, Tanaka Y (1980). Auditory brainstem response and behavioral audiometry. Developmental correlates. Arch Otolaryngol 106 (9): 564–566.

Kraus N, Ozdamar O, Stein L et al. (1984). Absent auditory brain stem response: peripheral hearing loss or brain stem dysfunction. Laryngoscope 94 (3): 400–406.

Kraus N, Bradlow AR, Cheatham J et al. (2000). Consequences of neural asynchrony: a case of auditory neuropathy. J Assoc Res Otolaryngol 1 (1): 33–45.

Licklider JCR (1948). The influence of interaural phase relation upon the masking of speech by white noise. J Acoust Soc Am 20: 150–159.

Madden C, Rutter M, Hilbert L et al. (2002). Clinical and audiological features in auditory neuropathy. Arch Otolaryngol Head Neck Surg 128: 1026–1030.

Marlin S, Feldmann D, Nguyen Y et al. (2010). Temperature-sensitive auditory neuropathy associated with an otoferlin mutation: Deafening fever! Biochem Biophys Res Commun 394 (3): 737–742.

Martin WH, Pratt H, Schwegler JW (1995). The origin of the human auditory brain-stem response wave II. Electroencephalogr Clin Neurophysiol 96 (4): 357–370.

Mason JC, De Michele A, Stevens C et al. (2003). Cochlear implantation in patients with auditory neuropathy of varied etiologies. Laryngoscope 113 (1): 45–49.

Masuda T, Kaga K (2011). Influence of aging over 10 years on auditory and vestibular functions in three patients with auditory neuropathy. Acta Otolaryngol 131 (5): 562–568.

Matsunaga T, Kanzaki J (2000). Morphological evidence that impaired intraneural microcirculation is a possible mechanism of eighth nerve conduction block in acoustic neuromas. Eur Arch Otorhinolaryngol 257 (8): 412–417.

Merchant SN, McKenna MJ, Nadol Jr JB et al. (2001). Temporal bone histopathologic and genetic studies in Mohr-Tranebjaerg syndrome (DFN-1). Otol Neurotol 22 (4): 506–511.

Michalewski HW, Starr A, Nguyen TT et al. (2005). Auditory temporal processes in normal hearing individuals and in patients with auditory neuropathy. Clin Neurophysiol 116 (3): 669–680.

Michalewski HJ, Starr A, Zeng FG et al. (2009). N100 cortical potentials accompanying disrupted auditory nerve activity in auditory neuropathy (AN): effects of signal intensity and continuous noise. Clin Neurophysiol 120 (7): 1352–1363.

Micheyl C, Carlyon RP, Gutshalk A et al. (2007). The role of auditory cortex in the formation of auditory streams. Hear Res 229: 116–131.

Miyamoto RT, Iler Kirk K, Renshaw J et al. (1999). Cochlear implantation in auditory neuropathy. Laryngoscope 109: 181–185.

Moser T, Predoehl F, Starr A (2013). Review of hair cell synapse defects in sensorineural hearing impairment. Otol Neurotol 34 (6): 995–1004.

Nadol Jr JB, Adams JC, O'Malley JT (2011). Temporal bone histopathology in a case of sensorineural hearing loss caused by superficial siderosis of the central nervous system and treated by cochlear implantation. Otol Neurotol 32 (5): 748–755.

Narne VK, Vanaja CS (2008). Speech identification and cortical potentials in individuals with auditory neuropathy. Behav Brain Funct 4: 15.

Norton SJ, Gorga MP, Widen JE et al. (2000). Identification of neonatal hearing impairment: summary and recommendations. Ear Hear 21 (5): 529–535.

Onishi S, Davis H (1968). Effects of duration and rise time of tone bursts on evoked V potentials. J Acoust Soc Am 44 (2): 582–591.

Orabi AA, Mawman D, Al-Zoubi F et al. (2006). Cochlear implant outcomes and quality of life in the elderly: Manchester experience over 13 years. Clin Otolaryngol 31 (2): 116–122.

Poissant SF, Beaudoin F, Huang J et al. (2008). Impact of cochlear implantation on speech understanding, depression, and loneliness in the elderly. J Otolaryngol Head Neck Surg 37 (4): 488–494.

Rance G (2005). Auditory neuropathy/dys-synchrony and its perceptual consequences. Trends Amplif 9 (1): 1–433.

Rance G, Beer DE, Cone-Wesson B et al. (1999). Clinical findings for a group of infants and young children with auditory neuropathy. Ear Hear 20 (3): 238–252.

Rance G, Cone-Wesson B, Wunderlich J et al. (2002). Speech perception and cortical event related potentials in children with auditory neuropathy. Ear Hear 23: 239–253.

Rance G, McKay C, Grayden D (2004). Perceptual characterisation of children with auditory neuropathy. Ear Hear 25: 34–46.

Rance G, Fava R, Baldock H et al. (2008). Speech perception ability in individuals with Friedreich ataxia. Brain 131 (Pt 8): 2002–2012.

Rance G, Corben LA, DuBourg E et al. (2010). Successful treatment of auditory perceptual disorder in individuals with Friedreich ataxia. Neuroscience 171: 552–555.

Rance G, Ryan MM, Bayliss K et al. (2012a). Auditory function in children with Charcot-Marie-Tooth disease. Brain 135: 1412–1422.

Rance G, O'Hare F, O'Leary S et al. (2012b). Auditory processing deficits in individuals with primary open-angle glaucoma. Int J Audiol 51 (1): 10–15.

Riedel H, Kollmeier B (2002). Comparison of binaural auditory brainstem responses and the binaural difference potential evoked by chirps and clicks. Hear Res 169 (1–2): 85–96.

Rodríguez-Ballesteros M, Reynoso R, Olarte M et al. (2008). A multicenter study on the prevalence and spectrum of mutations in the otoferlin gene (OTOF) in subjects with nonsyndromic hearing impairment and auditory neuropathy. Hum Mutat 29 (6): 823–831.

Roush P, Frymark T, Venedtiktov R et al. (2011). Audiologic management of auditory neuropathy spectrum disorder in children: a systematic review of the literature. Am J Audiol 20: 159–170.

Santarelli R (2010). Information from cochlear potentials and genetic mutations helps localize the lesion site in auditory neuropathy. Genome Med 2 (12): 91.

Santarelli R, Arslan E (2002). Electrocochleography in auditory neuropathy. Hear Res 170 (1–2): 32–47.

Santarelli R, Starr A, Michalewski HJ et al. (2008). Neural and receptor cochlear potentials obtained by transtympanic electrocochleography in auditory neuropathy. Clin Neurophysiol 119 (5): 1028–1041.

Satya-Murti S, Wolpaw JR, Cacace AT et al. (1983). Late auditory evoked potentials can occur without brain stem potentials. Electroencephalogr Clin Neurophysiol 56 (4): 304–308.

Shallop JK (2002). Auditory neuropathy spectrum disorder in adults and children. Seminars in Hearing 23 (3): 215–223.

Shapiro SM (2003). Bilirubin toxicity in the developing nervous system. Pediatr Neurol 29 (5): 410–421.

Sininger YS, Oba S (2001). Patients with auditory neuropathy: Who are they and what can they hear? In: YS Sininger, A Starr (Eds.), Auditory Neuropathy, Singular Publishing, San Diego, pp. 15–36.

Spoendlin H (1974). Optic cochleovestibular degenerations in hereditary ataxias. II. Temporal bone pathology in two cases of Friedreich's ataxia with vestibulo-cochlear disorders. Brain 97 (1): 41–48.

Starr A, Achor J (1975). Auditory brain stem responses in neurological disease. Arch Neurol 32 (11): 761–768.

Starr A, McPherson D, Patterson J et al. (1991). Absence of timing cues. Brain 114 (Pt 3): 1157–1180.

Starr A, Picton TW, Sininger YS et al. (1996). Auditory neuropathy. Brain 119 (3): 741–753.

Starr A, Sininger Y, Winter M et al. (1998). Transient deafness due to temperature-sensitive auditory neuropathy. Ear Hear 19 (3): 169–179.

Starr A, Sininger YS, Pratt H (2000). The varieties of auditory neuropathy. J Basic Clin Physiol Pharmacol 11 (3): 215–230.

Starr A, Sininger Y, Nguyen T et al. (2001). Cochlear receptor (microphonic and summating potentials, otoacoustic emissions) and auditory pathway (auditory brain stem potentials) activity in auditory neuropathy. Ear Hear 22 (2): 91–99.

Starr A, Michalewski HJ, Zeng FG et al. (2003). Pathology and physiology of auditory neuropathy with a novel mutation in the MPZ gene (Tyr145->Ser). Brain 126 (Pt 7): 1604–1619, Erratum in: Brain 2003; 126(Pt 7):1718.

Teagle H, Roush P, Woodward JS et al. (2010). Cochlear implantation in children with auditory neuropathy spectrum disorder. Am J Audiol 20: 159–170.

Trautwein P, Shallop J, Fabry L et al. (2001). Cochlear implantation of patients with auditory neuropathy. In: YS Sininger, A Starr (Eds.), Auditory Neuropathy, Singular Publishing, San Diego, CA, pp. 203–232.

Ueda N, Kuroiwa Y (2008). Sensorineural deafness in Guillain-Barré syndrome. Brain Nerve 60 (10): 1181–1186.

Varga R, Avenarius MR, Kelley PM et al. (2006). OTOF mutations revealed by genetic analysis of hearing loss families including a potential temperature sensitive auditory neuropathy allele. J Med Genet 43 (7): 576–581.

Vinay, Moore BCJ (2007). Ten(HL)-test results and psychophysical tuning curves for subjects with auditory neuropathy. Int J Audiol 46 (1): 39–46.

Worthington DW, Peters JF (1980). Quantifiable hearing and no ABR: paradox or error? Ear Hear 1 (5): 281–285.

Wynne DP, Zeng FG, Bhatt S et al. (2013). Loudness adaptation accompanying ribbon synapse and auditory nerve disorders. Brain 136 (Pt 5): 1626–1638.

Yellin MW, Jerger J, Fifer RC (1989). Norms for disproportionate loss in speech intelligibility. Ear Hear 10 (4): 231–234.

Zeng F-G, Liu S (2006). Speech perception in individuals with auditory neuropathy. J Speech Lang Hear Res 49: 36.

Zeng F-G, Oba S, Starr A (2001). Supra threshold processing deficits due to desynchronous neural activities in auditory neuropathy. In: DJ Breebaart, AJM Houstma, A Kohlrausch et al. (Eds.), Physiological and Psychophysical bases of Auditory Function, Shaker Publishing, Maastricht, Netherlands, pp. 365–372.

Zeng FG, Kong YY, Michalewski HJ et al. (2005). Perceptual consequences of disrupted auditory nerve activity. J Neurophysiol 93 (6): 3050–3063.

Handbook of Clinical Neurology, Vol. 129 (3rd series)
The Human Auditory System
G.G. Celesia and G. Hickok, Editors

Chapter 29

Hearing disorders in brainstem lesions

GASTONE G. CELESIA*
Department of Neurology, Loyola University of Chicago; Chicago Council for Science and Technology, Chicago, IL, USA

INTRODUCTION

Brainstem lesions involving auditory pathways and/or nuclei should result in hearing abnormalities, yet often the involvement of the auditory system is either overlooked or the lesions are silent (Lidén, 1969; Korsen-Bengtsen, 1970; Levine et al., 1994). Hausler and Levine (2000) noted that, in strokes involving the brainstem "vestibular signs and symptoms are common" while "hearing disturbances are much less frequent"; a similar observation has been reported in multiple sclerosis (MS).

The purpose of this chapter is to provide a review of the brainstem auditory system and the effects of brainstem lesions on auditory processing. We will address the issue of whether brainstem lesions are truly silent and whether in these instances classic audiometry is adequate in assessing hearing.

BRAINSTEM AUDITORY SYSTEM

The details of the anatomy and physiology of the auditory pathways are described in Chapter 1. Here we briefly summarize the points pertinent to the development of clinical symptomatology.

The acoustic fibers of the VIIIth nerve enter the brainstem at the pontomedullary junction and terminate in the cochlear nuclei. Their outputs (Fig. 1.10, Chapter 1) project to the ipsilateral and contralateral superior olivary complexes, trapezoid nuclei, lateral lemniscus nuclei, inferior colliculi (IC), and middle geniculate bodies (Moore, 2000; Brugge, 2013). Tonotopic representation from the cochlea is preserved in the cochlear nuclei (Fig. 29.1) and at each successive auditory relay station, including the auditory cortex (Moore and Osen, 1979; Lauter et al., 1985; Brugge, 1994; Talavage et al., 2004). The cochlear nuclei are the first station of arrival of information transmitted by primary cochlear afferents via the auditory nerve. In a postmortem examination of an infant who died from erythroblastosis due to an Rh-immune disorder, Dublin (1974) reported a tonotopic representation in the cochlear nucleus. In this case the organ of Corti and spiral ganglion were within normal limits. The dorsal cochlear nucleus was essentially unaltered. The ventral nucleus, especially the superior division, showed degeneration of spheroid cells in particular. The cell degeneration was seen in the dorsal portion of the ventral nucleus and corresponded with the audiographic dip encountered characteristically in posterythroblastotic hearing.

The tonotopic representation of the human cochlear nucleus has since been confirmed by studies to develop auditory prostheses for the profoundly deaf (McCreery et al., 1998; McCreery, 2008). The cochlear nuclei output is transmitted over parallel pathways both ipsilateral and contralateral to a wide number of nuclear groups of the pons, midbrain, and thalamus (Møller, 2003; Brugge, 2013); consequently neurons in the brainstem auditory pathways receive input from both ears, with the contralateral input being greater than the ipsilateral one (Puletti and Celesia, 1970; Møller, 2003). Binaural interaction occurs at three levels of the brain: the superior olivary complex, the nuclei of the lateral meniscus, and the IC (Moore, 1991, 2000). Interaural time and interaural level differences have a critical role in the ability to localize the source of sound in space (Moore, 2000; Brugge, 2013). Among the various auditory brainstem nuclei there are extensive anatomic connections via the commissure of Probst and the IC commissure.

Two important features of the auditory pathways are as follows:

1. Structures rostral to the cochlear nuclei process binaural information; the superior olivary complex is the first station in the central auditory pathways

*Correspondence to: Gastone G. Celesia, MD, FAAN, 3016 Heritage Oak Lane, Oakbrook, IL 60523, USA. E-mail: g.celesia@comcast.net

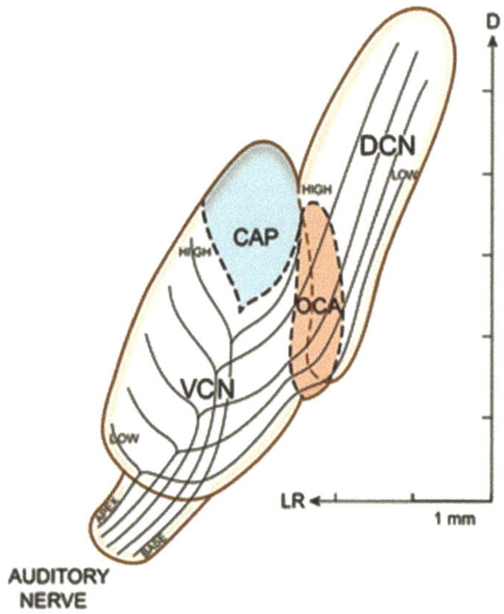

Fig. 29.1. Structure of the human cochlear nuclei showing the orderly cochlear base-to-apex bifurcation of incoming auditory nerve axons. The cochleotopic projection to the ventral cochlear nucleus (VCN) and dorsal cochlear nucleus (DCN) imposes a high-to-low frequency tonotopic organization to both nuclei. CPA represents small cap cells and OCA region of octopus cells. (From Brugge 2013 as adapted from Moore and Osen, 1979 with permission.)

where the information from the two cochlear nuclei converges. Dysfunction below or within the cochlear nuclei will produce ipsilateral abnormalities, while disorders rostral to the cochlear nuclei may result in bilateral abnormalities.

2. The multiplicity of parallel and overlapping auditory pathways and the extensive connections of the auditory brainstem nuclei not only are responsible for binaural interaction but also assure redundancy in the system. This redundancy may explain why small brainstem lesions are often clinically silent.

The brainstem receives its blood supply via two vertebral arteries, the basilar artery and the two posterior cerebral arteries. The medulla, pons, and midbrain (Fig. 29.2) have multiple blood supply (Haymaker, 1969; Duvernoy, 1999; Takahashi, 2010) with anastomoses among the various arteries that assure a redundancy of vascularization for the brainstem. The cochlear and vestibular nuclei in the upper medulla and lower pons are vascularized by the anterior inferior cerebellar artery (AICA), a branch of the vertebral or basilar artery. The cochlea receives its blood supply from the internal auditory artery (Fig. 29.3) (also called the labyrinthine artery) that most often arises from the AICA (Kim et al., 1999), although in some individuals is a branch of the basilar artery (Mazzoni, 1972; Adams and Liberman, 2010). The auditory nuclei and pathways of the upper pons and lower midbrain are vascularized by the superior cerebellar artery (SCA) and the lateral pontile arteries, all branches of the basilar artery. The medial geniculate body is vascularized by the laterosuperior midbrain arteries branches of the posterior cerebral artery (Haymaker, 1969).

PREVALENCE OF HEARING DISORDERS IN BRAINSTEM DISEASES

Any disorder of the brainstem (neoplasms, vascular disorders, infections, trauma, demyelinating disorders, neurodegenerative diseases, malformations) that involves the auditory pathways and/or centers may result in hearing abnormalities. It is not known how often a brainstem lesion causes hearing dysfunction; we are not aware of any prospective studies on the prevalence of such abnormalities. The only report on the prevalence of hearing loss in brainstem disorders corrected for age and sex is a retrospective study by Luxon (1980) from the National Hospital Queen Square, London. Luxon studied 309 patients with brainstem disorders examined from 1941 to 1976, and noted that 181 or 58.6% had objective evidence of hearing loss shown by pure-tone audiograms (Table 29.1), while 128 or 41.4% of patients had no hearing loss. To exclude aging related hearing disorders, the authors removed age-corrected audiograms with hearing levels under 10 dB at all frequencies. The hearing deficit was unilateral in 49 (15.9%) and bilateral in 132 (42.7%) patients. Only 26 subjects (8.4%) complained of hearing difficulties. The etiology varied, including brainstem malignant tumors (84), vertebrobasilar insufficiency (57), MS (52), and other disorders (116), such as syringobulbia, angiomas, abscesses, collagen diseases, encephalomyelitis, degenerative disorders, tuberculosis, and benign cysts. Smaller studies limited to specific disorders of the brainstem have been reported in the literature and are summarized in Table 29.1.

Retrospective studies (Luxon, 1980; Nakashima et al., 1999) yielded fewer hearing abnormalities than prospective ones (Jerger and Jerger, 1974; Musiek and Pinheiro, 1987; van der Poel et al., 1988; Levine et al., 1994; Aharonson et al., 1998; Lee et al., 2002; Rance et al., 2010), an occurrence of 52.8% versus 65.8% respectively. However, these prospective studies were small and contained a selection bias, being limited to some diseases seen in specific referral centers. The occurrence of auditory abnormalities in the studies of Luxon (1980) and Nakashima et al. (1999) may have been underestimated because psychoacoustic methods to evaluate central auditory function were not used. The studies of Musiek and Pinheiro (1987), Levine et al. (1994), and Aharonson et al. (1998) employed only one specific auditory test, therefore not evaluating the full spectrum of hearing. The importance of the type of test used is noted

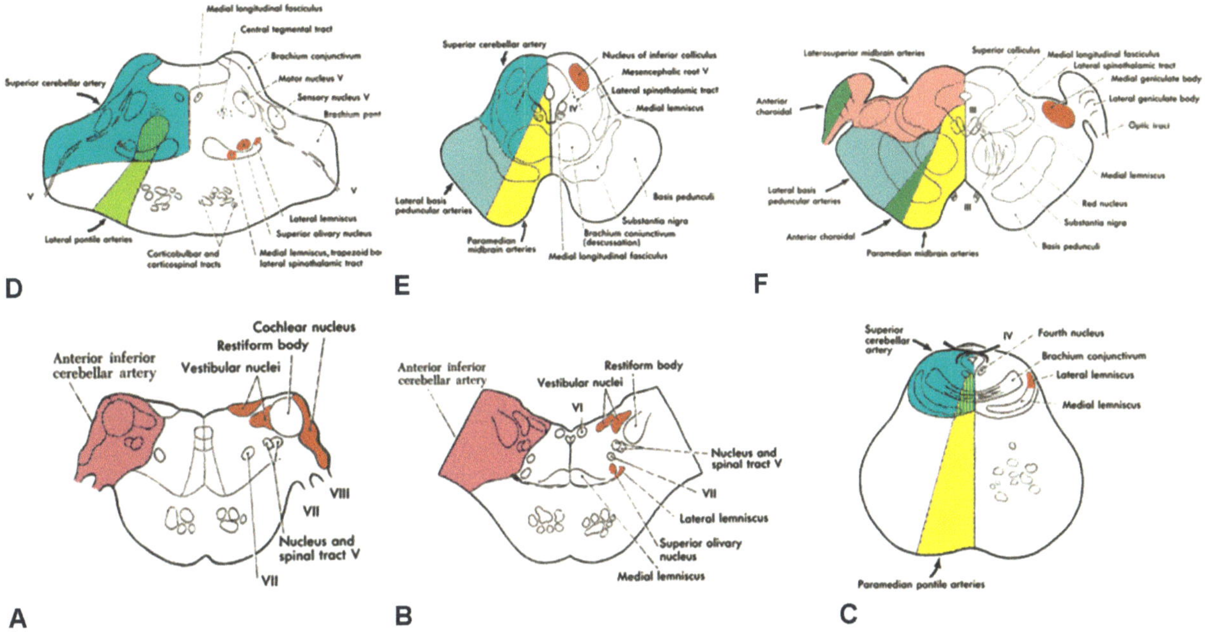

Fig. 29.2. Arterial supply of the brainstem. (**A**) Upper medulla; (**B**) lower pons; (**C**) middle pons at the level of the fifth nerve; and (**D**) the upper pons at the level of the decussation of the fourth nerve. (**E**) Midbrain at the level of the inferior colliculus; (**F**) midbrain at the level of superior colliculus. The anterior inferior cerebellar artery is a branch of the vertebral or basilar artery; the superior cerebellar artery, the paramedian and the lateral pontile arteries are branches of the basilar artery. The laterosuperior midbrain arteries and the lateral basis peduncular arteries are branches of the posterior cerebral artery. The paramedian midbrain arteries are branches of the posterior cerebral and basilar arteries. The areas supplied by various arteries are drawn with different colors (purple for the anterior inferior cerebellar artery, blue for the superior cerebellar artery, light blue for the lateral basis peduncular arteries, yellow for the paramedian pontile arteries, light green for the lateral pontile arteries, and dark green for the anterior choroidal artery). The auditory nuclei and pathways are drawn in red, while hatched regions represent areas of overlapping vascularization. (Modified from Haymaker, 1969.)

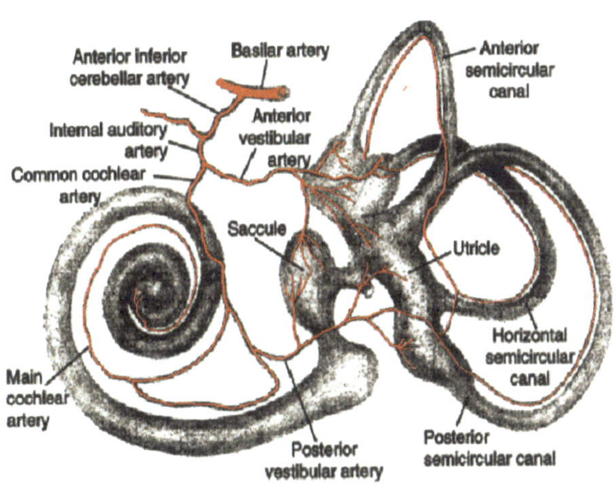

Fig. 29.3. Arterial circulation to the inner ear. (Modified from Kim et al., 1999.)

in the study of van der Poel et al. (1988); these authors evaluated two tests of binaural hearing: (1) intensity discrimination of alternating monaural clicks and (2) interaural time difference discrimination (IATD) of binaural clicks in 28 patients with definite MS and 12 patients with an isolated brainstem lesion compatible with demyelination. Intensity discrimination defects were found in 53.5% of MS and 25% of isolated brainstem demyelination patients; however, when IATD discrimination test was employed, the deficits increased to 82% and 33% respectively. It is clear that the assessment of central auditory processing cannot be limited to classic audiograms but requires the use of a battery of behavioral and physiologic tests (Bellis, 2003; Musiek and Chermak, 2006; Schow and Seikel, 2006).

EVALUATION OF BRAINSTEM-RELATED HEARING DISORDERS

Deficits of auditory function from disorders affecting the auditory pathways past the cochlear nuclei may not be detected by classic audiometry but require a battery of behavioral tests that assess specific central auditory processes (Jerger and Hayes, 1976; Musiek and Pinheiro, 1987; Levine et al., 1994; Griffiths, 2002; Bellis, 2003; Musiek and Chermak, 2006; Schow and Seikel, 2006; Shemesh, 2008; Musiek et al., 2011). The American Academy of Audiology (2010) suggests test batteries that include: "sound localization and lateralization, auditory discrimination, auditory temporal processing, auditory pattern processing, dichotic listening, auditory performance in competing acoustic signals,

Table 29.1

Prevalence of hearing disorders in brainstem diseases

Author(s), year	Disorder	Total number	Hearing loss	
			Number	Percent
Jerger and Jerger, 1974	Intra-axial brainstem	16	10	62.5
Luxon, 1980	Mix (MS, stroke, tumors, etc.)	309	181	58.6
Musiek and Pinheiro, 1987	Mix (MS, stroke, tumors, etc.)*	22	10	45
Levine et al., 1994	MS	38	27	71
van der Poel et al., 1988	MS	40	27	67.5
Aharonson et al., 1998	MS and pontine strokes†	22	16	72.7
Nakashima et al., 1999	MS	43	5	11.5
Lee et al., 2002	Unilateral AICA strokes‡	12	11	92
Rance et al., 2002	Friedreich's ataxia	14	9	64.3
Total		516	296	57.4

*Test limited to performance on frequency patterns.
†Test limited to lateralizing abnormalities.
‡The deficit was cochlear in 6 and mixed cochlear and retrocochlear in 4.
MS, multiple sclerosis; AICA, anterior inferior cerebellar artery.

Table 29.2

Testing battery for brainstem disorders

Audiometry
- Pure-tone audiogram

Psychoacoustic tests
- *Tests of monaural low-redundancy speech perception*
  - Filtering selected frequencies (e.g., high- or low-pass filtering)
  - Time compression of the speech signals
  - Embedding speech in background noise or verbal competition
- *Tests of temporal processes*
  - The Gaps-in-Noise (GIN)
  - Frequency (or Pitch) Pattern Sequence Test
  - Duration Patterns Test
- *Dichotic listening (speech) tests*
- *Tests of localization/lateralization and other binaural interactions*
  - Masking level difference (MLD)
  - Listening in Spatialized Noise-Sentences Test

Electrophysiologic tests
- *Otoacoustic emission*
- *Auditory brainstem evoked responses*

and auditory performance with degraded acoustic signals." A comprehensive battery of tests is proposed in Table 29.2.

Pure-tone audiogram is "essential" (Griffiths, 2002) to exclude an inner-ear hearing disorder. The American Academy of Audiology (2010) suggests the additional utilization of distortion product otoacoustic emissions to similarly exclude cochlear (outer hair cell) dysfunction. The presence of an inner-ear deficit does not exclude involvement of the brainstem; Griffiths (2002) pointed out that it is also "possible for a degree of cochlear" deafness to "co-exist with other disorders." Furthermore, ipsilateral hearing loss is common with lesions of the auditory nerve and/or cochlear nuclei. Brainstem auditory evoked potentials (BAEPs), although quite sensitive in detecting brainstem lesions and pinpointing their locations, correlate poorly with auditory function, as abnormal BAEPs have been reported in subjects with normal hearing (Legatt et al., 1988; Hu et al., 1997; Vitte et al., 2002; Celesia, 2011).

Psychoacoustic testing

As binaural interaction and sound localization are first processed in the medulla and pons, psychophysical testing emphasizing these properties should be employed (Table 29.2) in suspected brainstem disorders. Psychoacoustic tests allow "the demonstration of deficits that may be related to the symptoms from which the patients suffer" (Griffiths, 2002). Because these tests rely on a subjective response, they require repeated responses to achieve reliable parameters. To improve the diagnostic yield, a test battery approach has been advocated (Jerger and Hayes, 1976; Chermak and Musiek, 1997; Musiek and Chermak, 2006; American Academy of Audiology, 2010). This approach is very time consuming and further raises concern about cost-effectiveness. Can we reduce the number of tests without reducing their sensitivity and specificity? Bellis (2003) advocates the use of masking level difference (MLD) as the: "most sensitive behavior technique for assessing brainstem integrity." Abnormalities of MLD have been reported in lesions of the pontomedullary junction by several authors (Olsen et al., 1976; Lynn et al., 1981; Musiek et al., 1989; Pillion et al., 2008). Others have claimed that dichotic listening testing is an effective alternative to analyze brainstem auditory function (Hariri et al., 1994; Fischer et al., 1995; Cho et al., 2005).

Musiek et al. (2011) studied the sensitivity, specificity, and efficiency of commonly used behavioral central auditory processing test batteries in 20 individuals with known lesions of the central auditory nervous system and related auditory symptoms. The four tests used were dichotic digits (DD), competing sentences, frequency patterns (FP), and low-pass filtered speech. They found that the test battery providing the best balance between sensitivity and specificity was a two-test battery comprising dichotic digits and frequency patterns. They estimated the cost of DD-FP testing to be less than $20. These results need independent confirmation; nevertheless they suggest that a slimmed-down battery of tests may be satisfactory for the diagnosis of central auditory processing disorders and specifically brainstem auditory disorders.

Clinically we are faced with two possible scenarios: patients with brainstem disorders without auditory complaints and patients with brainstem disorders and hearing loss.

Patients with brainstem disorders without auditory complaints

This group consists of patients with a diagnosis of diseases affecting the brainstem, such as brainstem infarcts, hemorrhages, neoplasms, infections, and demyelinating disorders. The issue in these cases is whether there is an additional auditory disorder increasing the burden of known neurologic deficits. Classic examples are MS patients with signs of ataxia and/or other brainstem deficits (internuclear ophthalmoplegia, dysarthria, cranial nerve deficits) but without auditory complaints, that when tested showed disorders of sound localization and of IATD (Levine et al., 1993, 1994; Furst et al., 1995, 2000).

As shown in Table 29.3, abnormal lateralization of sound was also present in patients with stroke. Of 29 patients studied (16 with MS and 13 with strokes), Aharonson et al. (1998) found that 72.4% had abnormalities of sound lateralization. When the lesion is in the superior olivary complex, the trapezoid body, and/or the lateral lemniscus the hearing deficit is often bilateral, whereas in lesions of the VIIIth nerve and cochlear nuclei the deficit is ipsilateral. Sometimes these patients may complain of difficulties listening in background noise, or "understanding rapid or degraded speech" (Bamiou et al., 2001), although they have normal peripheral hearing; in short, they show evidence of central auditory processing disorders.

Patients with brainstem disorders and hearing loss

This group of patients developed either gradual hearing loss or sudden hearing loss. In either case they show a sensorineural deficit on pure-tone audiometry (Fig. 29.4). The deficit is usually unilateral and can occur from either extra- or intra-axial lesion of the brainstem. Jerger and Jerger (1975) studied 45 patients with expanding lesions of the brainstem, including 28 extra-axial and 17 intra-axial lesions. Each patient was tested with pure-tone sensitivity, Békésy audiometry, acoustic reflex, rollover of the performance-intensity function for phonemically balanced words and synthetic sentence identification test. Although there was considerable overlap between the two groups, patients with intra-axial brainstem disorders usually had normal acoustic reflex, pure-tone, and Békésy audiograms and the deficits were either bilateral or to the contralateral ear only (Fig. 29.5). Patients with extra-axial lesions had abnormal acoustic reflex, pure-tone, and Békésy audiograms and the deficit was to the ipsilateral ear only (Jerger and Jerger, 1975).

While auditory tests are indispensable in determining the nature of the hearing deficits they are unsatisfactory in establishing the location and nature of the disorders. Neuroimaging techniques are necessary to establish the etiologic diagnosis in most cases.

Table 29.3

Sound lateralization in multiple sclerosis and strokes

	Total	Normal	Abnormal	Center oriented	Side-oriented
Multiple sclerosis	16	5	11	7	4
Stroke	13	3	10	7	3
Total	29	8	21	14	7

Data from Aharonson et al. (1998).

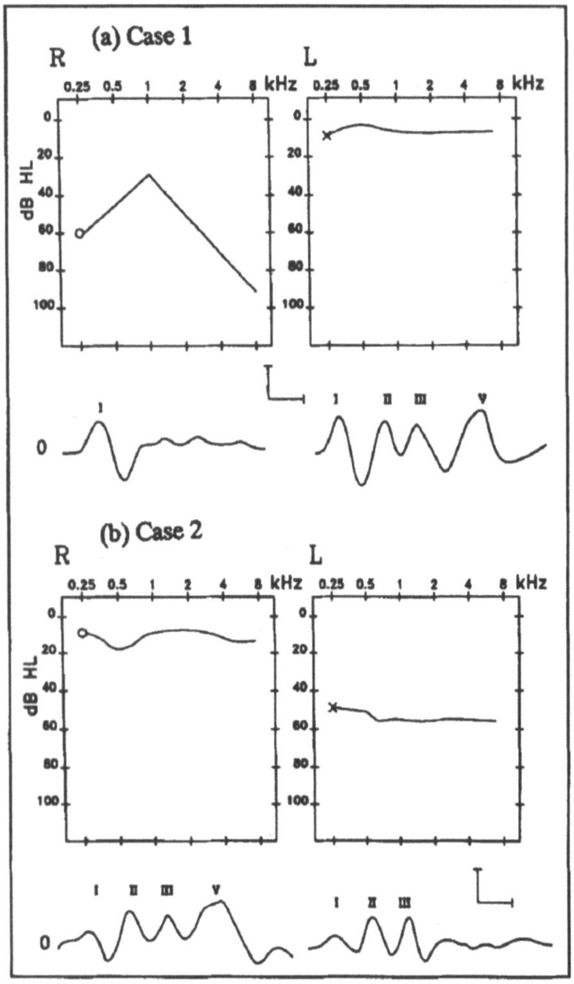

Fig. 29.4. Pure-tone audiograms and brainstem auditory evoked potentials in 2 patients with unilateral sudden hearing loss as the first monosymptomatic manifestation of multiple sclerosis. R, right; L, left. The horizontal scale represents 1 ms, while the vertical scale represents 0.3 μV. (Reproduced from Drulovic et al., 1993, with permission.)

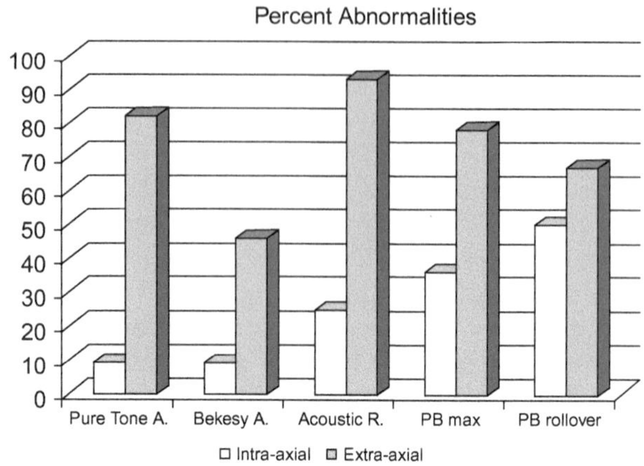

Fig. 29.5. Summary of auditory test results for the intra-axial and extra-axial brainstem groups. The percentage of patients with abnormal results on the involved ear is presented for each test. (Modified from Jerger and Jerger, 1975.)

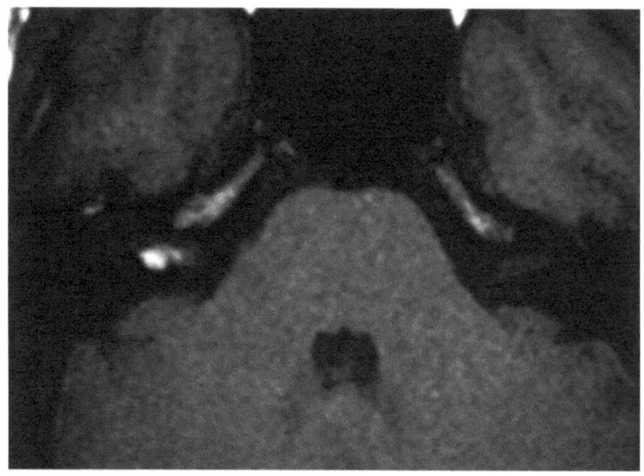

Fig. 29.6. Axial magnetic resonance imaging, T1 W image revealing a hyperintense lesion within the right intra-auricular meatus (IAM), consistent with an IAM lipoma. This signal was suppressed on fat-saturated sequences and was static at 5-year follow-up. (Reproduced from Sriskandan and Connor, 2011, with permission.)

Neuroimaging

Magnetic resonance imaging (MRI) of the brain, with special emphasis on the brainstem, is the preferred method to diagnose brainstem disorders. Bognar et al. (1994) compared computed tomography (CT) and MRI scans in 12 patients with histologically proved tectal plate gliomas. MRI showed the lesion in all cases, while CT scans failed to show the tumor in 4 patients. Cueva (2004) studied 312 patients (between the ages of 18 and 87) with asymmetric sensorineural hearing loss (SNHL) and prospectively compared MRI and BAEP results. Thirty-one (9.94%) patients of the study population of 312 were found on MRI to have lesions causing their SNHL. Twenty-two of the 31 patients had abnormal BAEPs while 9 patients (7 with small vestibular schwannomas) had normal BAEPs. The author concluded that MRI is "the screening test of choice" for the evaluation of asymmetric SNHL (Cueva, 2004). MRI is now a recognized "gold standard" test for symptoms pertaining to the inner ear and for the detection of retrocochlear pathology in both adults (Fig. 29.6) and children (Schick et al., 2001; Powell and Choa, 2010; Pennings et al., 2011; Poussaint et al., 2011). MRI can visualize a small acoustic neuroma within the inner auditory canal and is now the standard to screen for small tumors (Newton et al., 2010; Yoshida et al., 2011).

Vascular lesions of the brainstem are also detected by MRI (Kim et al., 2006; Hendrikse et al., 2009; Lee et al., 2009; Gállego Culleré and Erro Aguirre, 2011). Similarly,

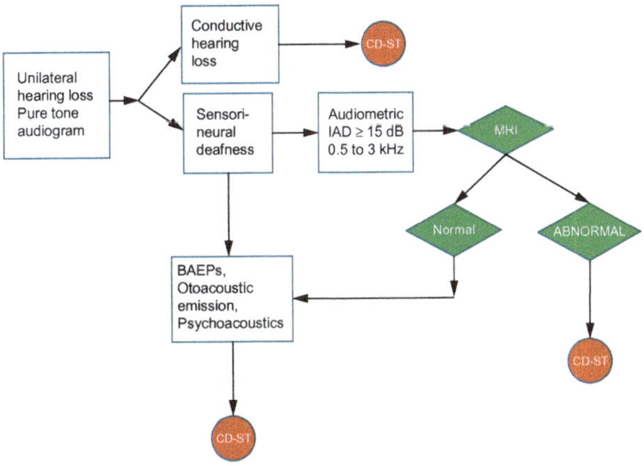

Fig. 29.7. Algorithm for the diagnosis of unilateral sensorineural hearing loss. IAD, interaural difference; CD-ST, clinical diagnosis-start therapy; MRI, magnetic resonance imaging; BAEPs, brainstem auditory evoked potentials.

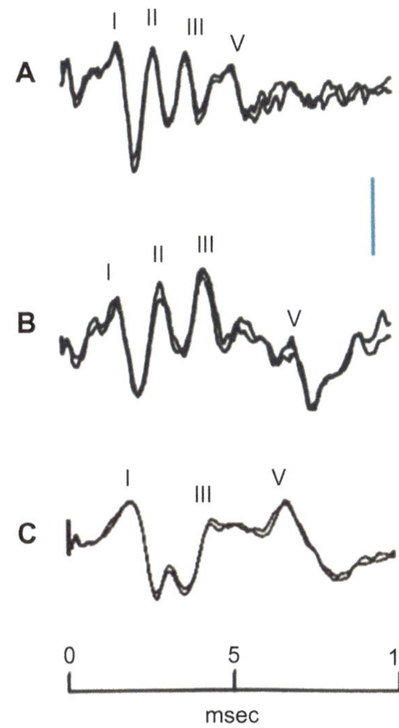

Fig. 29.8. Brainstem auditory evoked potentials (BAEPs) in brainstem disorders. (**A**) Normal BAEP. (**B**) BAEP of a 31-year-old male with a pontine glioma. Wave III and V are delayed and interval I–V is prolonged at 5.4 ms. Wave V is also small in amplitude. (**C**) BAEP in a 24-year-old male with a left acoustic neuroma. Wave I was within the upper limits of normal (latency of 1.88 ms); wave V was delayed (6.88 ms) and the interval I–V prolonged at 5 ms. The vertical calibration bar represents 0.25 μV.

MRI is the preferred diagnostic test to confirm the diagnosis of MS (McDonald et al., 2001; Polman et al., 2005).

Although MRI has a clear role in the diagnosis of retrocochlear hearing disorders, its high cost precludes its use as a screening test. Cheng and Wareing (2012) reported the MRI findings in 7494 cases of sensorineural hearing disorders "across ENT clinics" and noted that MRIs were abnormal in 2088 (27.8%) patients, including 284 (3.8%) vestibular schwannomas and 209 (2.8%) ischemic infarcts. Similarly, in sudden cases of hearing loss the diagnostic yield of MRI is low, varying from 4% to 34.5%. with vestibular schwannomas representing only 2.5–3.8% of all cases (Fitzgerald and Mark, 1998; Stokroos et al., 1998; Schick et al., 2001; Cadoni et al., 2006; Chau et al., 2010). Because of this low diagnostic yield and high cost, the need has arisen to formulate guidelines as part of "decision making for MRI screening of acoustic tumors" (Cheng and Wareing, 2012) and other unilateral SNHL. In Figure 29.7 we propose an algorithm for patients presenting with unilateral hearing loss based on an audiometric protocol. In cases of sensorineural deafness with interaural difference greater than 15 dB at 0.5–3 kHz, the subjects should undergo an MRI. If the MRI is negative, additional audiologic tests are warranted, including otacoustic emission (abnormal in cochlear disorders affecting the cochlear outer hair cell), BAEPs (sensitive to auditory nerve and brainstem disorders), and other psychoacoustic tests. The latter may better assess the central nature of the hearing disorders and suggest remedial interventions.

Brainstem auditory evoked potentials

BAEPs are a tool to study the function of the brainstem auditory nuclei and tracts and they provide "a look at physiological anatomy" (Chiappa, 1997). However, abnormalities of BAEPs do not imply the presence of a hearing deficit (Hu et al., 1997; Vitte et al., 2002; Celesia, 2011) and are etiologically non-specific (Chiappa, 1997). In view of these limitations, what is the role of BAEPs in brainstem disorders? As is made clear in the paragraphs to follow, opinions vary and there is currently no consensus regarding when to use BAEPs.

Hendrix et al. (1990) studied 225 consecutive cases of asymmetric SNHL with a variety of diagnostic tests. Eighty-six percent of these patients had a cochlear dysfunction, while the remaining 14% had a retrocochlear lesion. All 31 patients with retrocochlear abnormalities had an "abnormal auditory brainstem response." These authors suggested that MRI should be performed in patients with BAEP abnormalities indicating a retrocochlear lesion, rather than in all patients with asymmetric SNHL.

BAEPs are very sensitive to brainstem abnormalities (Fig. 29.8), thus they are useful in evaluating subclinical or yet undetected damage to the auditory system (Hosford-Dunn, 1985; Lisowska et al., 2001; Church et al., 2007; Pessin et al., 2008; Pillion et al., 2008;

Jiang et al., 2010). Martinez-Arizala et al. (1983) suggest that BAEPs could be used as "a screening tool for patients whose clinical picture is suggestive of structural abnormalities of the posterior fossa." Hosford-Dunn (1985) suggests that BAEPs are a screening test for the detection of brainstem disorders.

BAEPs are useful in assessing the auditory system in young children because they do not require their cooperation. Church et al. (2007) reported that, in cases of craniostenosis, BAEPs showed prolongation of I–III interpeak interval in 65% of children with normal hearing. They recommend that "auditory brainstem response diagnostics become standard clinical care for this patient group as the best way to detect auditory nerve compression." Wellman et al. (2003) tested 68 children survivors of bacterial meningitis and reported that 22 cases (32.3%) had abnormal BAEPs; 11 of these patients had a verified SNHL. Because early identification and remedial amplification lead to better academic and language outcomes, these authors advocate routine inpatient audiologic screening of postmeningitic children.

HEARING ABNORMALITIES IN ISOLATED LESIONS OF AUDITORY BRAINSTEM CENTERS

Diseases of the brainstem usually involve many structures, producing a plethora of neurologic deficits (Haymaker, 1969; Voordecker et al., 1988; Urban and Caplan, 2011), often relegating "auditory symptoms in the background" (Cho et al., 2005). Rarely, a small lesion is limited to the auditory system. Voordecker et al. (1988) observed that: "intrinsic brain-stem lesions that affect only unilaterally a portion of the auditory pathways without distortion or pressure on adjacent structures are exceptional." Tracts and nuclei in the medulla-pons are in close proximity to each other, thus rarely a disease state affects only one structure. These limiting factors must be considered in our understanding of the published reports of isolated lesions of the brainstem.

Disorders limited to the auditory nerve

Disorders limited to the auditory nerve include neuropathies and intracanalicular vestibular schwannoma (also called acoustic neuroma). When the disease is limited to the auditory nerve without involvement of the cochlea, the cochlear microphonic, the summating potentials of the electrocochleogram and the otoacoustic emission are normal (Starr et al., 1996; Santarelli and Arslan, 2002; Berlin et al., 2003; Rance, 2005). However, acoustic reflexes, pure-tone audiogram, speech discrimination, speech detection threshold, tone decay test, dichotic listening, and BAEPs are abnormal (Fig. 29.6 and Table 29.4) ipsilateral to the lesion (Johnson, 1977; Starr et al., 1996; Ottaviani et al., 2002; Berlin et al., 2003; Rance, 2005; Pennings et al., 2011). In auditory neuropathy BAEPs are absent but the middle- and long-latency auditory evoked responses have been reported to be within the normal limit (Kraus et al., 2000; Rance et al., 2002).

Unilateral lesions of the cochlear nuclei

Unilateral lesions of the cochlear nuclei are reported in occlusion of the AICA and in cerebellopontine angle tumors compressing the upper medulla. In both cases the vestibular nuclei and other structures are also implicated. The auditory-processing abnormalities are ipsilateral to the lesion (Table 29.4) with abnormal pure-tone audiogram, speech detection threshold, tone decay test, dichotic listening, and BAEPs (Hausler and Levine, 2000; Lee et al., 2002). The deficit profile of lesions of the cochlear nuclei is similar to the VIIIth-nerve involvement.

Unilateral lesions of the superior olivary complex

The superior olivary complex is the initial station of the binaural signal integration (Illing et al., 2000; Moore, 2000; Yin, 2002), thus a unilateral lesion may result in impairment of interaural time and level differences and sound localization (Table 29.4). Levine et al. (1993) reported isolated demyelinating lesions of the superior olivary complex in MS. The acoustic abnormalities (hearing loss, elevated threshold on audiograms, impaired speech discrimination) were predominantly ipsilateral to the lesion, whereas the otoacoustic emission tests were abnormal contralaterally, and the acoustic reflexes were either normal or abnormal contralateral to the lesion. Ipsilateral sound localization deficit and impaired binaural fusion were also noted (Levine et al., 1993; Levine and Hausler, 2001). BAEPs showed delayed waves III and V with prolonged I–V interval ipsilateral to the lesion (Levine et al., 1993).

Unilateral lesions of the lateral lemniscus

Unilateral lesions of the lateral lemniscus have been reported to be clinically silent on standard testing, but their effects can be revealed using careful psychoacoustic testing (Table 29.4), which shows abnormal sound localization and interaural time perception (Levine et al., 1994; Hausler and Levine, 2000; Cho et al., 2005). Lee et al. (2008) described normal speech and pure-tone audiometry, normal tympanometry, and acoustic reflexes but hyperacusis in a case of right pontine hemorrhage involving the lateral lemniscus;

Table 29.4

Auditory-processing abnormalities in unilateral brainstem lesions

Auditory process	Eighth nerve	Medulla	Pons	Pons	Pons	Midbrain
		Cochlear nuclei	Superior olivary complex	Trapezoid body	Lateral lemniscus	Inferior colliculus
Clinical findings						
Reported symptoms	Loss of hearing ipsilateral Ipsilateral tinnitus, vertigo No tinnitus/ vertigo in auditory neuropathy	Loss of hearing ipsilateral Sometimes tinnitus ipsilateral	Loss of hearing ipsilateral Sometimes auditory hallucinations	None, sometimes auditory hallucinations	None or contralateral tinnitus "Not hearing as well" Sometimes hyperacusis	Sometimes tinnitus (contralateral) and/or hyperacusis
Neuro-otologic findings	Ipsilateral deafness	Ipsilateral deafness	Ipsilateral deafness	Normal	Normal	Normal
Electrophysiologic testing						
Audiogram	Elevated threshold (ipsilateral) Retrocochlear hearing loss (ipsilateral)	Elevated threshold (ipsilateral)	Normal or elevated threshold (ipsilateral)	Normal	Normal	Normal
Acoustic reflexes	Absent ipsilateral Normal contralateral	Absent ipsilateral Normal contralateral	Normal or mildly impaired (contralateral)	Absent contralaterally Normal ipsilateral	Normal	Normal
Cochlear microphonic	Normal	Normal	Normal	Normal	Normal	Normal
Otoacoustic emissions	Normal	Normal	Abnormal contralateral	Abnormal contralateral Normal ipsilateral	Normal	Normal
BAEPs	Ipsilateral wave I and V prolonged or wave I normal wave V and I–V interval prolonged Absent in auditory neuropathy	Ipsilateral wave I normal, II and V prolonged I–V intervals prolonged	Delayed wave III and V (ipsilateral) Prolonged I–V interval (ipsilateral)	Wave V bilaterally Delayed	Delayed and attenuated wave V; prolonged I–V interval bilateral worse contralateral	Delayed wave V (contralateral or bilateral) I–V and III–V intervals prolonged (contralateral or bilateral); sometimes wave V depressed or absent (contralateral)
Middle-latency ERP	Normal				Normal	Abnormal contralateral

Continued

Table 29.4—Continued

Auditory process	Eighth nerve	Medulla	Pons	Pons	Pons	Midbrain
		Cochlear nuclei	Superior olivary complex	Trapezoid body	Lateral lemniscus	Inferior colliculus
Psychoacoustic testing						
Speech detection threshold	Impaired ipsilateral	Impaired ipsilateral	Normal or impaired ipsilateral		Normal	Normal
Speech discrimination	Impaired ipsilateral	Impaired ipsilateral	Sometimes impaired (ipsilateral)	Normal	Normal	Impaired with competing ipsilateral signals
Dichotic listening	Impaired ipsilateral	Impaired ipsilateral	Impaired	Impaired contralateral	Impaired contralateral	Abnormal contralateral
Binaural fusion	Impaired ipsilateral	Impaired ipsilateral	Impaired	Impaired contralateral	Impaired contralateral	Impaired contralateral
Binaural interaction (masking level difference: MLD)	Impaired ipsilateral	Impaired ipsilateral	Impaired		Impaired	Normal
Frequency pattern recognition			Impaired			Normal
Frequency pattern sequence						Normal
Duration pattern sequence test						Abnormal to stimulation contralateral ear
Sound-source localization in space			Impaired	Impaired (center-biased)	Impaired contralateral	Impaired for source localized in contralateral hemifield

BAEPs, brainstem auditory evoked potentials; ERP, event-related potential.

however they did not test for sound localization. The patient also had uncomfortable paresthesia in her left body. Cho et al. (2005) in a pontine hemorrhage noted normal speech and pure-tone audiometry while dichotic listening testing revealed an extinction of contralateral ear input. Levine et al. (1993) noted that a unilateral lesion in the region of the lateral lemniscus affected IATD for low-frequency stimuli. However the deficit was less severe than bilateral lesions of the lateral lemniscus.

Unilateral lesions of the trapezoid body

Furst et al. (1995, 2000) demonstrated in patients with MS that lesions of the trapezoid body affect sound localization: patients tended to hear all the sounds toward the middle of the head (center-oriented or biased). Cambier et al. (1987) reported that one patient with a lesion of the trapezoid body had "central type hypoacusis and a severe disorder of localization of sounds." Levine et al. (1993) noted that a unilateral lesion of the trapezoid body produced an interaural level discrimination abnormality.

Unilateral lesions to the inferior colliculus

Unilateral lesions to the IC are rare (Durrant et al., 1994; Litovsky et al., 2002; Champoux et al., 2007a, b; Stimmer et al., 2009). Litovsky et al. (2002) studied a 48-year-old man with a small traumatic hemorrhage of the right dorsal midbrain, including the IC. They reported normal audiograms bilaterally, abnormal BAEPs with a decrease

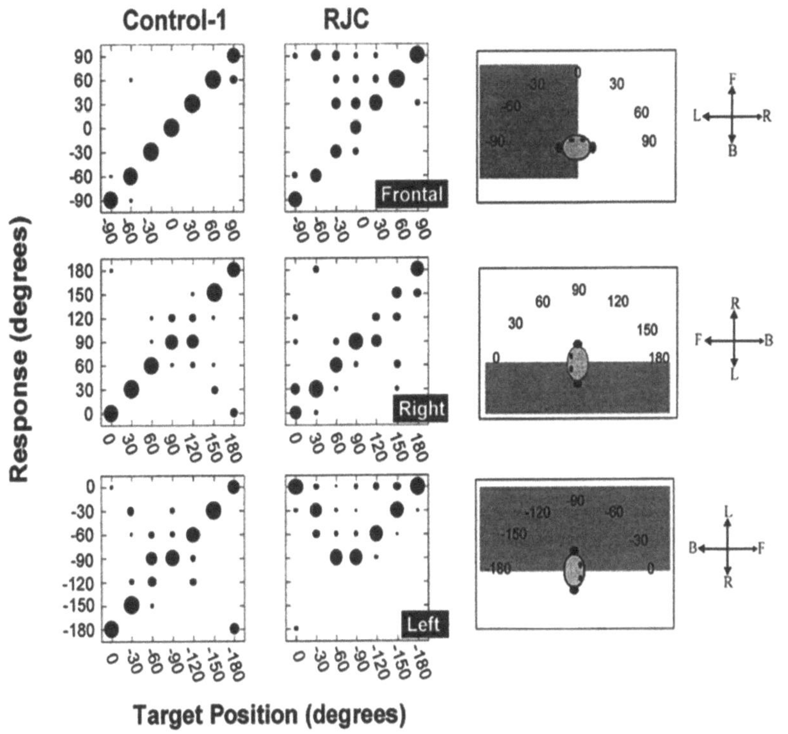

Fig. 29.9. Single-source localization results. Data for one normal control subject and for patient RJC are shown in the left and middle columns, respectively, for three stimulus configurations depicted in the right column. RJC is a 36-year-old right-handed man, who suffered a small traumatic hemorrhage of the right brainstem involving the inferior colliculus. From top to bottom, these are the frontal, right, and left conditions. The numbers in each configuration represent the angles of the speakers in the room relative to the center of the subject's head. Gray areas denote the hemifield contralateral to the lesion site. Each plot shows the proportion of trials on which responses occurred at each position, as a function of the true (target) source positions, demarcated by dot size. (Reproduced from Litovsky et al., 2002, with permission.)

in wave V amplitude following left (contralateral)-ear stimulation. The patient had a deficit in sound localization contralateral to his IC lesion (Fig. 29.9). Echo suppression was abnormally weak compared with that seen in control subjects, but only for sources contralateral to the lesion. Although speech discrimination and detection threshold tests were normal, the patient complained of having trouble understanding speech in noisy environments. Stimmer et al. (2009) reported on a 48-year-old man with an acute circumscribed hemorrhagic lesion of the left IC. The patient complained of right-ear tinnitus but no other otoneurologic signs or symptoms. Auditory brainstem response revealed prolonged III–V interpeak latencies from stimulating either ear.

Champoux et al. (2007a, b) reported on a 12-year-old boy with a "circumscribed lesion of the right IC without additional neurological damage"; the subject suffered a traumatic hemorrhagic lesion at age 9 years. An extensive battery of psychophysical hearing tests revealed normal peripheral auditory functioning bilaterally (Table 29.4); dichotic listening, duration pattern sequence test, and sound source localization were abnormal contralaterally (Fig. 29.10); speech recognition was impaired in the presence of a competing ipsilateral signal (Champoux et al., 2007a, b). The authors concluded that "recognition of low-redundancy speech presented monaurally, recognition of tone duration patterns, binaural separation and integration, as well as sound-source localization in space depend on the integrity of the bilateral auditory pathways at the IC level" (Champoux et al., 2006, 2007b; Paiement et al., 2009).

A follow-up in the same patient showed that cortical activation in response to binaural stimulus presentations as monitored by functional MRI (fMRI) produced a stronger activation in the contralateral hemisphere rather than the well-balanced bilateral activation patterns seen in normal subjects. Auditory pathways failed to re-establish normal cortical activation patterns in response to binaural stimulation following the unilateral lesion of the IC. The authors concluded that the IC is an essential auditory relay and that its loss cannot be significantly compensated (Paiement et al., 2009). Durrant et al. (1994) limited their study to three-channel Lissajous trajectories of the auditory brainstem responses in a human with a focal lesion of the right IC. They found an intact wave IV but a negligible, if not totally absent, wave V with stimulation of the left (contralateral) ear.

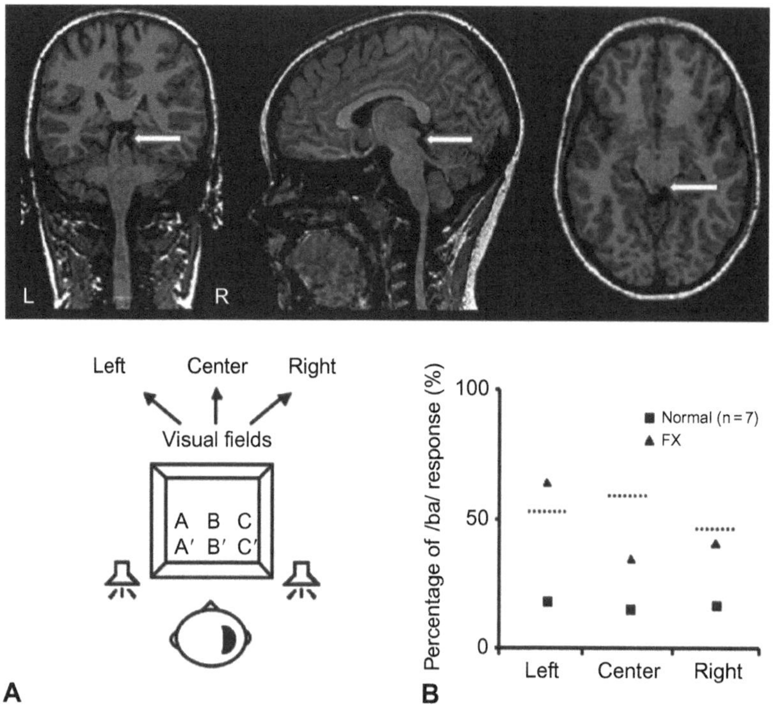

Fig. 29.10. Illustration of subject FX. The upper illustration shows the coronal, sagittal and axial magnetic resonance imaging scans showing a small lesion in the right inferior colliculus (arrow). R, right; L, left. The lower illustration part (**A**) is the schematic representation of the experimental setup. In all trials the auditory stimuli were presented in the free field via loudspeakers positioned on each side of the participant. The visual stimuli were presented at various locations: directly in front of the subject (B-B'), five degrees to the left (A-A') or to the right (CC') of the point of fixation. The area in black drawn on the participant's stylized head represents the site of FX's lesion at the level of the right inferior colliculus. Part (**B**) shows the performance for the incongruent stimuli. Data are shown for three experimental conditions in which the visual stimuli were presented either in the center of the monitor or in the participant's left or right hemifield. The percentage of /ba/ responses given are plotted for FX (triangle) and for the group of matched controls (square). Dashed lines represent values 2 SD above the mean for the control group, indicating abnormal audiovisual speech integration. (Modified from Champoux et al., 2006.)

Isolated bilateral lesion of the inferior colliculi

Isolated bilateral lesion of the IC have also been described (Johkura et al., 1988; Jani et al., 1991; Hu et al., 1997; Vitte et al., 2002; Hoistad and Hain, 2003; Kimiskidis et al., 2004; Musiek et al., 2004), with some inconsistent results. Johkura et al. (1988) reported a case of traumatic small hemorrhage involving the IC bilaterally and without evidence of a temporal-lobe lesion. The patient was unable to comprehend spoken words, although he had intact speech production, reading and writing abilities. Comprehension of environmental sounds was also affected. Among the receptive musical abilities, discrimination of intensity, tone, and rhythm was preserved, while recognition of melody was impaired. Pure-tone audiometry was normal. BAEP waves I–IV were normal, but wave V had reduced amplitude and prolonged latency bilaterally. Musiek et al. (2004) reported on a 21-year-old male with a subarachnoid bleed affecting both ICs who experienced central deafness. Dichotic listening was abnormal bilaterally, BAEPs were normal, whereas middle and late auditory evoked potentials were significantly abnormal.

Hu et al. (1997) described a 48-year-old woman with bilateral contusions of the IC (Fig. 29.11). The patient was able to speak, read, and write; her cognitive functions were normal, but she was unable to hear. She did not respond to either pure-tone or speech stimuli. Impedance audiometry revealed normal middle-ear pressure and mobility bilaterally. The crossed and uncrossed stapedial reflexes and the BAEPs were normal bilaterally. Hoistad and Hain (2003) presented a 43-year-old man with bilateral lesions of the IC due to central nervous system lymphoma. He had normal pure-tone audiogram but severely reduced bilateral word recognition scores. Kimiskidis et al. (2004) studied a 48-year-old woman who suffered a hemorrhage localized at the quadrigeminal plate (including the IC). Her pure-tone audiogram revealed bilateral mild SNHL and speech audiometry showed persistent word deafness. Acoustic reflexes

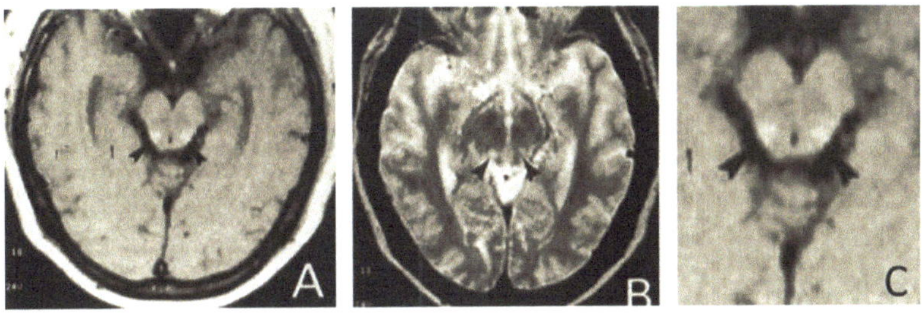

Fig. 29.11. Magnetic resonance imaging (MRI) of a 48-year-old woman who became totally deaf after a head injury that caused contusions of both inferior colliculi. (**A**) Axial T1-weighted MRI with Gd-DTPA enhancement (slice thickness = 7 mm; interval = 8 mm) showed enhanced lesions in both inferior colliculi, consistent with subacute contusional hemorrhage (arrows). (**B**) Axial T2-weighted MRI revealed abnormal signals in the inferior colliculi (arrows). (**C**) Enlarged view of midbrain from (**A**). (Modified from Hu et al., 1997.)

were normal bilaterally. Transient evoked otoacoustic emissions were normal from both ears, with however a failure for their amplitude to be suppressed with contralateral sound stimulation. BAEPs were of normal amplitude and latencies bilaterally. Vitte et al. (2002) described two cases with bilateral symmetric lesions of the IC. Both cases had MRI showing either an infarct or a hemorrhage. Both were able to speak, read, and write, but were unable to hear. They had a mild bilateral SNHL, normal impedance audiometry, stapedial reflexes, and BAEPs.

Isolated bilateral lesion of pontomedullary region

There are very few reports of bilateral lesions of the pontomedullary region predominantly affecting the auditory system. Egan et al. (1996) described a 64-year-old woman with a midline pontine hematoma who had bilateral deafness, right abducens paresis, loss of proprioception in both arms and in both legs, with normal touch and pinprick sensation. She had limb ataxia and bilateral extensor plantar responses. Pure-tone audiometry showed bilateral moderately severe SNHL. Speech discrimination was absent bilaterally, ipsilateral acoustic reflexes were preserved, whereas contralateral reflexes were absent. BAEPs showed bilaterally preserved wave IV and V. The authors interpreted the findings as related to a lesion of the "ventral acoustic stria decussating in the trapezoid body." Significant hearing loss was also noted by Cohen et al. (1996) in cases of central pontine hemorrhage.

AUDITORY HALLUCINATIONS IN BRAINSTEM DISORDERS

Hallucinations caused by brainstem lesions were first noted by Lhermitte in 1922 as purely visual phenomena. Cascino and Adams (1986) were the first to report auditory hallucinations in 3 patients with brainstem disorders (Table 29.5). These authors described the phenomena varying from simple buzzing and "clanging like motors of machines" or "like sheets of metal being banged together" to music like "bells chiming" or "organ notes." Auditory hallucinations differ from tinnitus because they are more complex than a simple buzzing or ringing and often are perceived as real. However, the boundary between tinnitus and auditory hallucinations is blurred and often authors include cases of tinnitus among their cases of hallucinations. Additional reports have been published since, under variable terminology: peduncular hallucinations, brainstem hallucinosis, and auditory illusions (Table 29.5). Sometimes the hallucinations are musical, as in the case reported by Silva and Brucki (2010), consisting of a continuous repetition of a *Carmen Habanera* aria. In a few cases the patient experienced both auditory and visual hallucinations (Cambier et al., 1987; Taylor et al., 2005; Silva and Brucki, 2010), although the auditory component is the predominant one. Of the reported 20 cases, 14 lesions (70%) were localized in the pontine tegmentum (Fig. 29.12), four (20%) in the midbrain (IC), one in the middle geniculate body, and one in the thalamus, probably involving the middle geniculate body (Table 29.5). All these lesions involved auditory structures (cochlear nuclei, acoustic stria, superior olive, trapezoid body, lateral lemniscus, and middle geniculate body), in agreement with Shergill et al. (2001) that: "hallucinations in a given modality seem to involve areas that normally process sensory information in that modality."

The etiology varied from hemorrhage to trauma, infections, neoplasm, and demyelinating disorders. Hearing loss was frequent, but not necessary to the development of the hallucinations. The pathophysiology of these auditory hallucinations is uncertain. Cascino and Adams (1986) favored "a release mechanism" in their cases. Cambier et al. (1987) suggested that "Auditory deafferentiation could predispose to auditory hallucinosis"; in these cases presumably the structural lesion producing deafferentiation resulted in release of inhibition, leading to activation of the auditory cortical system. Cambier et al. (1987) suggested that auditory

Table 29.5

Brainstem auditory hallucinations

Authors	Age	Sex	Type of hallucinations	Side of hallucinations	Duration of hallucinations	Hearing loss	Associated visual hallucinations	Other neurologic phenomena	Site of lesion	Etiology
Cascino and Adams, 1986	33	F	Sounds, musical (piano playing)	Bilateral	>18 months	Unilateral right	No	Left VI, left face hypoesthesia, right hemihypesthesia	Left wall fourth ventricle and pontine tegmentum	AVM, subarachnoid hemorrhage
	42	F	Sounds, musical (bells chiming)	Right	>8 years but intermittent	Unilateral right	No	Right Horner, VI and VII, right ataxia, left hemihypoesthesia	Right pontine tegmentum	Hemorrhage
	53	F	Sounds (humming, metal banging)	Bilateral	6 months, till death	Bilateral	No	Confusion, bilateral ptosis, diplopia, left hemiparesis	Tectum of midbrain (inferior colliculi) right cerebellum	Metastatic tumor
Cambier et al., 1987	56	F	Musical (melodies, songs)	Bilateral	Few weeks	Bilateral and loss of sound localization	No	Left hemisensory loss, right VII	Right pontine tegmentum	Infarct
	55	F	Sounds, musical (melodies, songs)	Bilateral	24 days	Bilateral and loss of sound localization	No	Right VI, left hemianesthesia	Right paramedian pontine tegmentum	Infarct
	74	F	Musical, verbal, sounds	Bilateral	20 days	Bilateral and loss of sound localization	No	Right VI,VII, left hemianesthesia, left-limb ataxia	Right paramedian pontine tegmentum	Infarct
	61	M	Musical (melodies) sounds	Bilateral, worse on the left	15–20 days	Bilateral and loss of sound localization	Yes	Right hemisensory loss and right ataxia Internuclear ophthalmoplegia, left VII	Left paramedian pontine tegmentum	Hemorrhage
	54	M	Musical (bells) sounds	Bilateral	31 days	none	No	Internuclear ophthalmoplegia, ataxia Left hemisensory loss	Inferior colliculi and pontine tegmentum	Infarct

Lanska et al., 1987	55	M	Musical and verbal	Bilateral	Several months	Bilateral	No	left internal opthalmoplegia, left VI, left Horner impaired sound localization	Left dorsal pons	Hemorrhage
Forcadas and Zarranz, 1994	72	F	Musical (childhood songs)	Bilateral	Few months	hyperacusia	Yes	Right VI, VII, ataxia Left hemisensory loss, left hemiparesis	Right paramedian pontomedulla	Infarction
	62	F	Not stated	Not stated	Few months	Not stated	Yes	Left hemisensory loss, left hemiparesis Internuclear ophthalmoplegia, ataxia	Right paramedian pontine tegmentum	Hemorrhage
Inzelberg et al., 1993	75	F	Musical (melody)	Bilateral	Few weeks	Bilateral	No	Right hemiparesis, dysphasia	Left thalamus	Ischemic infarct
Murata et al., 1994	55	M	Musical (chirping of cicadas, songs)	Right	18 days	Bilateral	No	Left hemiparesis and hemisensory loss, right VII	Right dorsal pontine tegmentum extended to left pontine tegmentum	Hemorrhage
Baurier and Tuca, 1996	47	M	Sound (child crying, water falling)	Bilateral, more to the right	5 days	Bilateral more to the right	No	Left upper-extremity weakness	Right paramedian pons	Hemorrhage
Douen and Bourque, 1997	43	M	Musical (orchestral melodies, and songs)	Left	1 month	Left	No	Left V, VII, VIII, IX, X and left-limb ataxia	Left middle cerebellar peduncule	Listeria rhombencephalitis
Fukutake and Hattori, 1998	49	M	Sounds (roaring waves and bells)	Right	10 minutes	Bilateral hyperacusis	No	Transient dysarthria and hypoesthesia left thigh	Right middle geniculate body	Hemorrhagic infart
Schielke et al., 2000	57	M	Musical (boy singing songs)	Right	5 weeks	Bilateral hyperacusis	No	Right hemiparesis and hemihypoesthesia Left 1 and 1/2 syndrome, left VII	Left pontine tegmentum	Brainstem abscess

Continued

Table 29.5—**Continued**

Authors	Age	Sex	Type of hallucinations	Side of hallucinations	Duration of hallucinations	Hearing loss	Associated visual hallucinations	Other neurologic phenomena	Site of lesion	Etiology
Taylor et al., 2005	24	F	Verbal	Bilateral	2 weeks	None	Yes	Internuclear ophthalmoplegia, ataxia	Periacqueductal lower midbrain	Multiple sclerosis
Silva and Brucki, 2010	43	F	Musical (aria of Carmen Habanera)	Bilateral	10 days	None	Yes	Diplopia, right dysmetria, left VII, right V gait ataxia	Right cerebellar peduncle and pontine tegmentum	Brainstem encephalitis
Lo et al., 2011	22	M	Verbal	Not stated	Transient during migraine aura	None	No	Right frontal headaches	Left posterolateral mesencephalic peduncle	Migraine

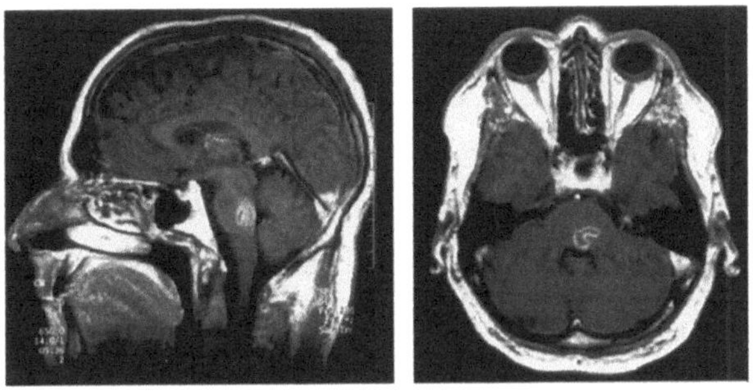

Fig. 29.12. Sagittal (left) and transverse (right) magnetic resonance imaging on hospital day 17 (during the time the patient experienced musical hallucinosis) shows an abscess in the left paramedian pontine tegmentum. (Modified from: Schielke et al., 2000.)

hallucinations are related to hypnagogic hallucinations as the dream mechanisms are released by the normal inhibitory control. The alternative hypothesis is that these phenomena are due to excitation of neurologic structures as caused by seizure discharges; however, there has been no example of proven brainstem paroxysmal excitation.

BRAINSTEM CLINICAL DISORDERS

Many diseases can involve the brainstem, as shown in Table 29.6, and whenever they include the auditory pathways and/or centers they will produce hearing deficits. Clinically the hearing loss is sensorineural and, as we discussed previously, if the lesion is extra-axial the deafness is ipsilateral to the lesion, whereas intra-axial lesions result most often in bilateral or contralateral hearing loss (Jerger and Jerger, 1975; Egan et al., 1996). In the section on hearing abnormalities in isolated lesions of auditory brainstem centers, above, we presented the auditory deficits induced by the rare occurrence of small lesions of the brainstem; however, most frequently brainstem disorders involve large areas and result in more severe auditory deficits as well as deficits in other modalities (Fig. 29.13). Because therapy is related to the etiology of the disorders a correct diagnosis is mandatory. A careful clinical history and a full neuro-otologic examination are prerequisites of any diagnostic consideration. Among the many disorders that may involve the brainstem, the most frequent pathologies are cerebellopontine angle tumors, brainstem strokes, and MS. We will briefly discuss their clinical features.

Vestibular schwannomas and cerebellopontine angle tumors

The most frequent extra-axial brainstem neoplasms are localized in the cerebellopontine angle. In all, 96–98% of these cerebellopontine angle tumors are benign (Valvassori, 1988; Nadol and Martuza, 2005; Régis and Roche, 2008), and 75–90% of these benign tumors are represented by vestibular schwannomas (these tumors are also called acoustic neuromas or neurinomas). Brackmann and Bartels (1980) reported that in a series of 1354 cerebellopontine angle tumors 90% were vestibular schwannomas. Similarly, Moffat etal. (1993) found that, out of 305 cerebellopontine angle tumors, 246 (80.7%) were vestibular schwannomas. Stangerup et al. (2010a) evaluated the incidence of sporadic unilateral vestibular schwannomas in Denmark from 1976 to 2008. A total of 2283 new cases of vestibular schwannomas were diagnosed and registered in a national database covering 5.0–5.5 million inhabitants. The incidence of vestibular schwannomas appears to be about 19 tumors per million per year. Of the other benign cerebellopontine angle tumors, 5–13% are meningiomas, 3–6% cholesteatomas, the reminder representing schwannomas of the VIIth and other cranial nerves, arachnoid cyst, lipomas (Fig. 29.6), and granulomas (Lalwani, 1992; Long, 1992; Tekkok et al., 1992; Nadol and Martuza, 2005; Springborg et al., 2008). Only approximately 2% of cerebellopontine angle tumors are malignant, either primary or metastatic (Brackmann and Bartels, 1980; Nakada et al., 1983; Link et al., 2002; Nadol and Martuza, 2005; Akamatsu et al., 2012), and they are characterized by rapid development of symptoms and by invasion of other structures in the area (Brackmann and Bartels, 1980; Hanabusa et al., 2001; Nadol and Martuza, 2005).

Symptomatology

The most frequent presenting symptoms of vestibular schwannomas (and, by extrapolation, of other cerebellopontine tumors) are unilateral hearing loss and tinnitus. As shown in Table 29.7 there is considerable variation in the reported symptoms of vestibular schwannomas. Johnson (1977) stated that 95% of the 500 patients with surgical verified vestibular schwannomas had "initial symptoms of varying degrees of hearing loss." More recent studies show a lower percentage of hearing impairment. Caye-Thomasen et al. (2007) studied 156 patients with purely intracanalicular vestibular schwannoma and found that the mean pure-tone average (PTA) was 51 dB hearing level (HL) and the mean speech discrimination score was 60% at diagnosis. Similarly, Stangerup et al. (2010b) reported that, of 1144 patients with vestibular schwannoma, 491 patients (53%) presented with good hearing. In their patients, speech discrimination was better than 70%. Pennings et al. (2011) noted unilateral hearing loss in 87% of 47 patients with intracanalicular vestibular schwannomas; their mean pure-tone audiogram was 37.5 dB HL and the mean word recognition score was 66.2%. Surprisingly, there was no reliable correlation between tumor size or location and hearing loss (Nadol et al., 1996;

Table 29.6

Brainstem disorders

Disorder	Syndrome	Clinical symptoms
Eighth-nerve disorders		
	Acoustic/vestibular schwannoma (neuroma)	Unilateral ipsilateral hearing loss and tinnitus, vertigo; if compression of brainstem ataxia and other cranial nerve (VII, V) symptoms and signs
	Auditory neuropathy	Bilateral hearing loss
	Diabetic neuropathy	Bilateral hearing abnormalities
	Cochlear nerve aplasia	Unilateral ipsilateral sensorineural hearing loss
	Arachnoid cyst of the internal auditory canal	Unilateral ipsilateral hearing loss
	Meningeal carcinomatosis	Unilateral or bilateral hearing loss, vertigo, and imbalance
Brainstem strokes		
	Occlusion internal auditory artery (also called labyrinthine artery) (branch of AICA)	Sudden ipsilateral deafness; tinnitus and vertigo
	Lateral inferior pontine syndrome (occlusion of AICA)	Ipsilateral ataxia, nystagmus, vertigo, tinnitus ipsilateral deafness, ipsilateral seventh-nerve palsy, ipsilateral hypoesthesia of face
	Lateral superior pontine syndrome (occlusion of superior cerebellar artery)	Severe ipsilateral ataxia, Horner's syndrome, contralateral impairment, pain temperature of face and body, impairment of dichotic listening and sound localization
	Intralabyrinthine haemorrhage	Sudden ipsilateral deafness and intense vertigo
	Brainstem haemorrhage	Sudden deafness and vertigo, other symptoms related to affected areas
	Midbrain haemorrhage	Bilateral tinnitus, hyperacusis and abnormal dichotic listening and neuro-ophthalmologic signs if superior colliculi involved
Brainstem neoplasms		
	Cerebellopontine angle tumors	Unilateral ipsilateral deafness, ipsilateral ataxia, tinnitus Other symptoms related to extent of brain compression
	Brainstem gliomas (diffuse pontine glioma, focal brainstem glioma)	Partial deafness, impaired sound localization, limb ataxia, other symptoms may be present depending on extent of the tumor
	Metastatic neoplasms	Ipsilateral, contralateral, or bilateral deafness; other symptoms depend on extent of tumor
	Cavernous angioma of brainstem	Sudden deafness, fluctuating hearing loss, Ménière-like vertigo
Brainstem demyelinating disorders		
	Multiple sclerosis affecting the pons or mesencephalon	Partial ipsilateral deafness, impaired sound localization, other symptoms related to extent of the lesion or lesions
Brainstem infections		
	Brainstem abscess	Frequent sixth- and seventh-nerve involvement, partial hearing loss, symptoms vary depending on location of lesions
	Brainstem encephalitis	Brainstem and cerebellar signs, mild hearing involvement Impaired sensorium
Brainstem trauma		
	Quadrigeminal plate trauma (contusion)	Bilateral deafness
Other brainstem disorders		
	Cholesterol cyst of petrous apex	Ipsilateral deafness, ipsilateral ataxia, tinnitus, other symptoms related to extent of brain compression and other nerve involvement
	Capillary telangiectasia of the pons	Sudden-onset tinnitus and sensorineural hearing loss
	Pontocerebellar hypoplasia type III (CLAM)	Developmental delay, short stature, microcephaly, hearing impairment, optic atrophy
	Mucopolysaccharidosis VI	Progressive bilateral hearing impairment
	Mitochondrial encephalomyopathies	Progressive bilateral hearing impairment

AICA, anterior inferior cerebellar artery; CLAM, cerebellar atrophy with progressive microcephaly.

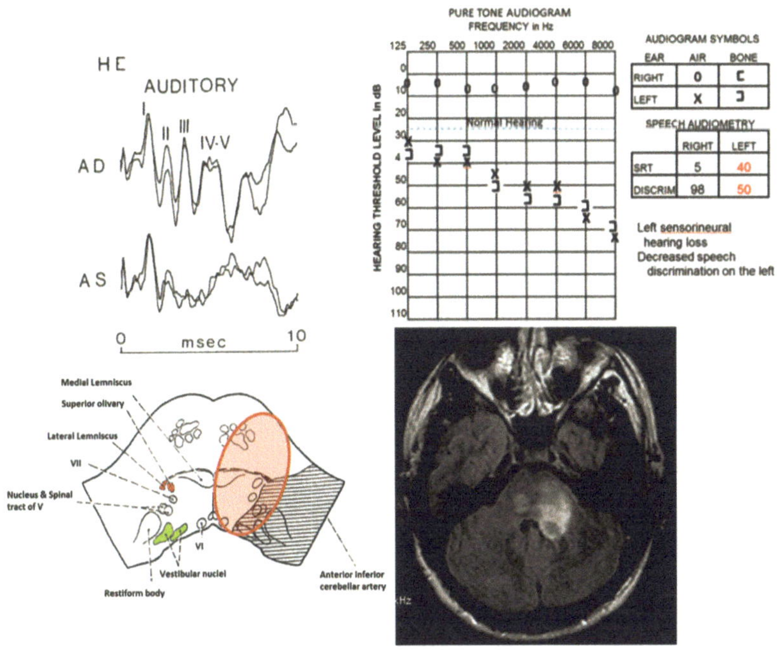

Fig. 29.13. Case HE is a 41 year-old female who presented with sudden onset of deafness in the left ear. On examination she also had a mild ataxia with tandem gait. The upper part of the figure shows the pure-tone audiogram indicating a mild sensorineural hearing loss in the left ear with impaired speech audiometry in the same side. Brainstem auditory evoked potentials showed a normal wave I, a wave II of low amplitude but normal latency, and a very prolonged I–V interval, suggesting impaired auditory processing past the cochlear nuclei in the left brainstem. Magnetic resonance imagingI showed a large demyelinating lesion in the pons. She also had two smaller demyelinating lesions in the white matter of the left hemisphere. She was diagnosed with multiple sclerosis. AD, right ear; AS, left ear.

Table 29.7

Symptoms of vestibular schwannomas

Symptoms at diagnosis	Percent
Hearing loss	50–95
Sudden hearing loss	5–22
Tinnitus	15–80
Disturbance of balance (vertigo, dizziness, ataxia)	15–75
Headaches	25–33
Facial numbness	35–50
Facial weakness	16–50
Lower cranial nerve involvement	2
Hydrocephalus	10

Caye-Thomasen et al., 2007). The hearing loss is usually gradual, with occasionally the patient being unaware of it. Johnson (1977) noted that only 68% of 425 patients with surgically verified vestibular schwannoma were aware of their hearing impairment. Sudden unilateral hearing loss has also being described in 5–22% of cases (Chow and Garcia, 1985; Berg et al., 1986; Ogawa et al., 1991; Yanagihara and Asai, 1993; Saunders et al., 1995; Nadol and Martuza, 2005). Vertigo or some other form of unsteadiness has been reported in about 50% of cases. Caloric testing showed a higher occurrence of vestibular hypofunction (96% of 46 cases) (Ojemann et al., 1972). Other symptoms are related to the size of the tumor and the extent of involvement of other structures. Involvement of the VIIth nerve is particularly frequent, with various degree of facial weakness (Table 29.7).

Audiometry and other diagnostic testing

While the hearing deficit in vestibular schwannoma is sensorineural, pure-tone audiograms show a great variety of abnormalities (Johnson and House, 1964; Ottaviani et al., 1999). Some authors (Eggston and Wolff, 1947; Cambon and Guilford, 1958) have suggested that audiograms showing low-frequency loss and near normal high-frequency hearing may indicate the possibility of vestibular schwannomas. Ottaviani et al. (1999) reported U-shaped SNHL on pure-tone audiometry and mildly impaired speech discrimination with small tumors and severe low- and middle-frequency hearing loss with impaired speech discrimination with larger tumors. Several authors have pointed out that, when the hearing threshold level for pure tones is minimally affected, while speech discrimination is disproportionately reduced, a retrocochlear pathology should be suspected, and acoustic neurinoma would be the most likely etiology (Canty, 1978; Van Dijk et al., 2000).

Increased detection-threshold or no detectable acoustic emission was noted in VIIIth-nerve tumors by Bonfils and Uziel (1988), while Lidén and Korsan-Bengtsen (1973) stated that an elevated stapedius reflex threshold was a "consistent finding in acoustic neuromas."

Prior to the routine availability of MRI, brainstem auditory testing was used to screen for suspicion of cerebellopontine tumors. As we described in the section on brainstem auditory evoked potentials, above, BAEPs can only indicate the presence of a sensorineural deficit and additionally suggest involvement of intramedullary structures (Legatt et al., 1988; Chiappa, 1997; Montaguti et al., 2007; Celesia, 2011).

Physiopathology

The sudden onset of hearing loss, the cochlear involvement, and the lack of correlation between tumor size and hearing loss in some vestibular schwannomas cast doubts that the only cause of symptoms is by compression of the VIIIth nerve. Cochlear involvement has been demonstrated in approximately 50% of vestibular schwannomas (Igarashi et al., 1974; Bonfils and Uziel, 1988; Prasher et al., 1995; Norman et al., 1996; Telischi, 2000). It is now acknowledged that the symptoms of intracanalicular vestibular schwannomas are related to two possible mechanisms (Igarashi et al., 1974; Prasher et al., 1995; Takhur et al., 2012):

1. mechanical compression and/or infiltration of:
   (a) the cochlea (Fig. 29.14)
   (b) the auditory (and vestibular) nerve
   (c) the internal auditory artery
   (d) the efferent olivocochlear bundle
   (e) hair cell degeneration secondary to neuronal loss in the VIIIth nerve.
2. neurotoxicity:
   (a) toxic tumor metabolites (proteinaceous deposit in inner-ear fluid caused by tumor metabolites)
   (b) alterations of biochemic properties of inner-ear fluids.

If the tumor expands into the cerebellopontine angle, compression of the brainstem may occur, resulting in additional involvement of nervous structures and if there is interference with cerebrospinal fluid flow, development of hydrocephalus may occur.

Neuroimaging

The role of neuroimaging in the diagnosis of brainstem disorders and cerebellopontine angle tumors has been discussed already in the section on neuroimaging, above. Here we briefly summarize the importance of neuroimaging in the diagnosis of vestibular schwannomas (Fig. 29.15). Two neuroimaging modalities have been used: high-resolution CT and MRI.

A consensus is emerging that, for most of the cases, magnetic resonance is the imaging modality of choice for lesions of the cerebellopontine angle and internal auditory canal (Zamani, 2000; Adunka et al., 2007; Saliba et al., 2010; Sriskandan and Connor, 2011; Yoshida et al., 2011). However, when visualization of the internal auditory canal and/or of the bony cochlear nerve canal is needed, then high-resolution CT is best as it allows better evaluation of bony structures (Adunka et al., 2007).

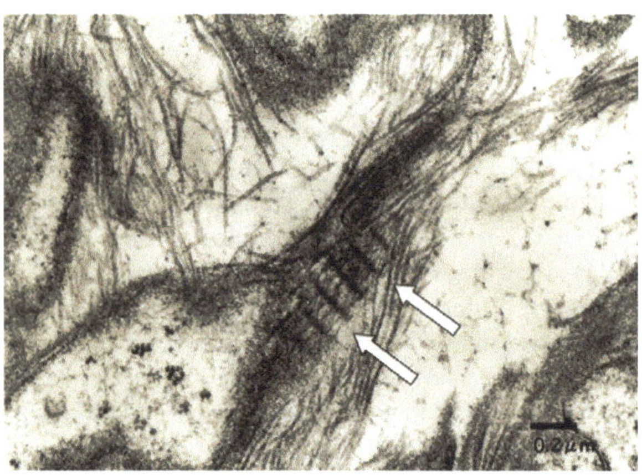

Fig. 29.14. Transmission electron micrograph of intracochlear schwannoma cells demonstrating the presence of bundles of extracellular fibrous longspacing collagen ("Luse bodies," arrows). (Reprinted from Yoshida et al., 2011, with permission.)

Fig. 29.15. Coronal magnetic resonance imaging, T1 W, postcontrast image of the right intra-auricular meatus (IAM), which contains an enhancing schwannoma with further abnormal enhancement of the cochlea and vestibule, consistent with transmodiolar (small arrow) and transmacular (large arrow) extension, respectively. (Reproduced from Sriskandan and Connor, 2011, with permission.)

Neurofibromatosis type 2 (NF2)

Approximately 7% of patients with vestibular schwannomas have NF2 (Evans et al., 2005). NF2 is a genetic disorder due to a defective tumor suppressor gene located on chromosome 22q12.2. The defective protein produced by the gene is "called merlin or schwannomin" (Asthagiri et al., 2009; Houshmandi et al., 2009; Schulz et al., 2013). NF2 is rare, with a prevalence of about 1 in 60 000 (Gareth and Evans, 2009). Although NF2 is a genetic disorder, approximately 40% of the identified cases are sporadic, related to new mutations (Evans et al., 2005; Gareth and Evans, 2009). Some of the milder cases and up to 20% of cases without a family history of the disease are mosaic (Ruttledge et al., 1966; Gareth and Evans, 2009). Approximately 25% of all constitutional mutations in the NF2 tumor suppressor gene are splicing mutations (Ellis et al., 2011).

The clinical presentation may be variable, but the presence of bilateral vestibular schwannoma is always indicative of NF2. In 1988 the National Institutes of Health gathered a conference to establish the criteria for the diagnosis of neurofibromatosis. Additional criteria have been added since then and are described in Table 29.8 (National Institutes of Health Consensus Development Conference, 1988, Gutmann et al., 1997; Gareth and Evans, 2009). Clearly the presence of more than one central nervous system tumor should raise the diagnosis of NF2.

Treatment modalities

The aim of the treatment is to remove the tumor while preserving hearing and other neurologic functions. In any therapeutic consideration the cost/benefit ratio requires careful consideration. The size of the tumor and its growth rate are important variables in the management of vestibular schwannoma (Agrawal et al., 2010; Stangerup et al., 2010b; Arthurs et al., 2011; Pennings et al., 2011). Surgical intervention is recommended in tumors showing rapid growth and in tumors growing outside the internal auditory canal impinging in the surrounding neurologic structures (Nadol and Martuza, 2005; Régis and Roche, 2008; Noudel et al., 2009; Whitmore et al., 2010).

The treatment for NF2 is more complex, related to the multiplicity of tumors and/or lesions (Asthagiri et al., 2009; Odat et al., 2011). Asthagiri et al. (2009) suggest that the optimum treatment should be multidisciplinary. Recently novel therapies have been proposed aimed at inducing inhibition of angiogenesis and regulation of the epidermal growth factor receptor family of receptors (Fong et al., 2011; Blakeley, 2012; Plotkin et al., 2012; Yener et al., 2012). The effectiveness of this chemotherapy is still under evaluation.

Brainstem strokes affecting the auditory system

Strokes affecting the brainstem auditory system are represented predominantly by the following syndromes: occlusion of the internal auditory artery syndrome, lateral inferior pontine syndrome, and lateral superior pontine syndrome.

Occlusion of the internal auditory artery

Occlusion of the internal auditory artery is very rare and it occurs secondary to arteriosclerosis in the vertebrobasilar system or embolism. The artery supplies both the cochlea and the labyrinth (Fig. 29.3) and the occlusion is characterized by sudden unilateral deafness, tinnitus, and vertigo (Millikan et al., 1959; Kim et al., 1999; Levine and Hausler, 2001) (Table 29.9).

Table 29.8

Diagnostic criteria for neurofibromatosis type 2 (NF2)

Definite NF2	Probable NF2
Bilateral vestibular schwannoma (VS) or	Unilateral VS plus any two of: meningioma, glioma, neurofibroma, schwannoma, posterior subcapsular lenticular opacities
Family history of NF2 or Unilateral VS diagnosed before age 30 or Any two of the following: Meningioma, glioma, neurofibroma, schwannoma, posterior subcapsular lenticular opacities	Multiple meningiomas (two or more) plus any two of: meningioma, glioma, neurofibroma, schwannoma, and cataract

Modified from National Institutes of Health Consensus Development Conference (1988), Gutmann et al. (1997), and Garreth and Evans (2009).

Table 29.9

Classic stroke syndromes affecting the brainstem auditory system

Syndrome	Artery	Clinical signs	Ipsilateral/ contralateral
Infarction of internal auditory artery	Internal auditory artery	Sudden profound sensorineural deafness	Ipsilateral
		Vertigo and tinnitus	
Lateral inferior pontine syndrome	Anterior inferior cerebellar artery (AICA)	Hearing impairment	Ipsilateral
		Horner's	Ipsilateral
		Facial weakness	Ipsilateral
		Ataxia of limbs	Ipsilateral
		Impaired pain and temperature of face	Ipsilateral
		Impaired pain and temperature of body	Contralateral
Lateral superior pontine syndrome	Superior cerebellar artery	Horner's	Ipsilateral
		Ataxia of limbs	Ipsilateral
		Impaired gait (falling to side of lesion)	
		Impaired pain and temperature of face and body	Contralateral
		Sometimes central facial weakness	Contralateral
		Impaired dichotic listening	Contralateral
		Impaired sound localization	Contralateral
		Vertigo and tinnitus	

Table 29.10

Neuro-otologic findings in 82 patients with infarction of anterior inferior cerebellar artery

Symptoms	Number abnormal	Percent abnormal
Presented with vertigo	80	97.6
Associated with other neurologic symptoms or signs	81	98.8
Associated with ocular motor dysfunction	79	96.3
Combined audiovestibular loss	49	59.8
Normal audiovestibular function	24	29.3
Isolated auditory loss	3	3.7
Isolated vestibular loss	4	4.9
Prodromal audiovestibular disturbance	13	15.9

Data from Lee et al. (2009).

The lateral inferior pontine syndrome

The lateral inferior pontine syndrome is caused by occlusion of the AICA and related ischemic infarct of the lateral inferior pons (Fig. 29.2). Occlusion of AICA is the most frequent of the brainstem strokes involving the auditory system.

The syndrome is characterized by sudden onset of vertigo, nystagmus, ipsilateral facial weakness, deafness, and tinnitus, ipsilateral ataxia and hypoesthesia of the face (Adams, 1943; Amarenco et al., 1993; Lee et al., 2002; Choi et al., 2006) (Table 29.10). Lee et al. (2009) prospectively identified 82 consecutive patients with AICA-territory infarction diagnosed by MRI. Each patient underwent a neuro-otologic evaluation, including bithermal caloric tests, pure-tone audiogram, stapedial reflex threshold evaluation, speech discrimination, and BAEPs. The most common "pattern of audiovestibular dysfunction was the combined loss of auditory and vestibular function ($n = 49$ [60%]). A selective loss of vestibular ($n = 4$ or 5%) or cochlear ($n = 3$ or 4%) function (Table 29.10) was rarely observed" (Lee et al., 2009). In 6 patients the site of the auditory deficit was localized in the cochlea, in 1 patient in the retrocochlear region (presumably the cochlear nuclei), while in 4 patients the severity of the hearing loss precluded any localization. One patient had no hearing deficit.

The lateral superior pontine syndrome

The lateral superior pontine syndrome is caused by occlusion of the SCA and is characterized by sudden onset of ataxia of limbs and gait with falling to the side of the lesion, ipsilateral Horner's syndrome, horizontal nystagmus, contralateral face and body impairment of pain and temperature, and vertigo (Davison et al., 1935; Luhan and Pollack, 1953) (Table 29.9). Luhan and Pollack (1953) reported on 6 cases of occlusion of the SCA, 3 with postmortem verification. Hearing impairment was noted in 3 cases, was not reported in 2, and was "grossly normal" in 1. The hearing abnormalities were ipsilateral in 2 patients and contralateral in 1. In none of the cases was detailed audiologic testing performed. They commented:

> *The lateral lemniscus lies in close proximity to the pain pathways in the rostral pons and may be damaged. The lateral lemniscus is mainly a crossed tract but many of its fibers arise from the cochlear nuclei of the same side. Individual variations undoubtedly exist. This bilateral projection from the cochlear nuclei would tend to mask any hearing impairment resulting from slight damage to the lateral lemniscus (Luhan and Pollack, 1953).*

Amarenco and Hauw (1990) commented that, in spite of infarction of the lateral lemniscus and the IC in more than half of the reported cases, hearing complaints were not described. In a study of 23 patients with infarcts of the SCA without involvement of the territory of other cerebellar arteries from the Raymond Escourolle Laboratory of Neuropathology at the Hôpital de la Salpetrière, Amarenco and Hauw (1990) found that the "classic" syndrome of occlusion of SCA is "very rare." Their autoptic findings showed considerable variability in brainstem tissue involvement: for instance, the lateral lemniscus was infarcted in only 12 cases (52%) and the IC in 3 (13%), while cerebellar involvement occurred in all cases. These authors further commented that occlusion of the SCA was often associated with occlusion of the distal basal or vertebral artery. They stated: "This pathologic involvement resulted in either a clinical 'rostral basilar artery syndrome,' close to the 'top of the basilar syndrome' or a deep coma often associated with tetraplegia; these syndromes overshadowed signs of cerebellar involvement" and we add, overshadowed signs of acoustic deficits.

More recent studies of the effect of unilateral lesions of the lateral lemniscus indicate that, while pure-tone and speech audiometry may be normal, there is contralateral impaired dichotic listening (Cho et al., 2005).

Multiple sclerosis hearing abnormalities

In MS hearing abnormalities are often overlooked because they can only be detected by psychoacoustic hearing tests dependent upon brainstem processing (Levine et al., 1993, 1994; Furst et al., 1995). As we described in the section on prevalence of hearing disorders in brainstem diseases, hearing abnormalities in MS may be as frequent as 70%. Levine et al. (1994), in their studies of MS, concluded that "multiple sclerosis lesions of the auditory pons are not silent." The reader is referred to Chapter 36 of this volume for more details.

MANAGEMENT OF BRAINSTEM HEARING DISORDERS

The neurologic deficits in brainstem lesions are often overwhelming, pushing the hearing deficits "in the background" (Cho et al., 2005). The clinician needs to address both the therapy of the disease causing the lesion and the rehabilitation of all neurologic deficits, including the auditory problems. Once the acute intervention has improved and/or stabilized the etiologic disorders (e.g., infarct, infection, neoplasm, demyelinating disorder), we need to address the auditory concerns. As we discussed in the section on evaluation of brainstem-related hearing disorders, above, a detailed evaluation is essential to determine the nature and extent of the auditory deficit. A multidisciplinary team, including audiologists, speech pathologists, and neuropsychologists, may be the best approach to assure a comprehensive management (American Academy of Audiology, 2010). The team will determine whether hearing aids are necessary and/or sufficient and/or whether additional intervention is indicated. Environmental modifications (such as improved room acoustics, assistive listening systems), auditory training (such as dichotic listening training, clear speech), and language strategies may be recommended.

The aim of the intervention is to improve auditory processing and related cognitive and behavioral changes.

References

Adams R (1943). Occlusion of the anterior inferior cerebellar artery. Arch Neurol Psychiatr 49: 765–770.

Adams J, Liberman MC (2010). Anatomy. In: SN Merchant, JB Nadol (Eds.), Schucknect's Pathology of the Ear. People's Medical Publishing House-USA, Shelton, pp. 53–96.

Adunka OF, Jewells V, Buchman CA (2007). Value of computed tomography in the evaluation of children with cochlear nerve deficiency. Otol Neurotol 28: 597–604.

Agrawal Y, Clark JH, Limb CJ et al. (2010). Predictors of vestibular schwannoma growth and clinical implications. Otol Neurotol 31: 807–812.

Aharonson V, Furst M, Levine RA et al. (1998). Lateralization and binaural discrimination of patients with pontine lesions. J Acoust Soc Am 103: 2624–2683.

Akamatsu Y, Murakami K, Watanabe M et al. (2012). Malignant peripheral nerve sheath tumor arising from benign vestibular schwannoma treated by gamma knife radiosurgery after two previous surgeries: a case report with surgical and pathological observations. World Neurosurg 73: 751–754.

Amarenco P, Hauw JJ (1990). Cerebellar infarction in the territory of the superior cerebellar artery: a clinicopathological study of 33 cases. Neurology 40: 1383–1390.

Amarenco P, Rosengart A, DeWitt LD et al. (1993). Anterior inferior cerebellar artery territory infarcts. Mechanisms and clinical features. Arch Neurol 50: 154–161.

American Academy of Audiology (2010). Clinical Practice Guidelines: Diagnosis, Treatment and Management of Children and Adults with Central Auditory Processing Disorder. Retrieved from, http://www.audiology.org/resources/documentlibrary/Documents/central-auditory-processing-disorder.

Arthurs BJ, Fairbanks RK, Demakas JJ et al. (2011). A review of treatment modalities for vestibular schwannoma. Neurosurg Rev 34: 265–277.

Asthagiri AR, Parry DM, Butman JA et al. (2009). Neurofibromatosis type 2. Lancet 373: 1974–1986.

Bamiou DE, Musiek FE, Luxon LM (2001). Aetiology and clinical presentations of auditory processing disorders—a review. Arch Dis Child 85: 361–365.

Baurier VO, Tuca JO (1996). Auditory hallucinosis in brainstem lesions and musical hallucination in deafness. Analogies and differences. Eur J Neurol 3: 203–211.

Bellis T (2003). Assessment and management of central auditory processing disorders in the educational setting: from science to practice. Delmar Learning, Clifton Park, NY, p. 533.

Berg HM, Cohen NL, Hammerschlag PE et al. (1986). Acoustic neuroma presenting as sudden hearing loss with recovery. Otolaryngol Head Neck Surg 94: 15–22.

Berlin CI, Morlet T, Hood LJ (2003). Auditory neuropathy/dyssynchrony: its diagnosis and management. Pediatr Clin North Am 50: 331–340.

Blakeley J (2012). Development of drug treatments for neurofibromatosis type 2-associated vestibular schwannoma. Curr Opin Otolaryngol Head Neck Surg 20: 372–379.

Bognar L, Turjman F, Villanyi E et al. (1994). Tectal plate gliomas. Part II: CT scans and MR imaging of tectal gliomas. Acta Neurochir (Wien) 127: 48–54.

Bonfils P, Uziel A (1988). Evoked otoacoustic emissions in patients with acoustic neuromas. Am J Otol 9: 412–417.

Brackmann DE, Bartels LJ (1980). Rare tumors of the cerebellopontine angle. Otolaryngol Head Neck Surg 88: 555–559.

Brugge JF (1994). An overview of central auditory processing. In: A Popper, RR Fay (Eds.), The mammalian auditory pathways: Neurophysiology. Springer-Verlag, New York, pp. 1–33.

Brugge JF (2013). Anatomy and physiology of auditory pathways and cortex. In: GG Celesia (Ed.), Disorders of Peripheral and Central Auditory Processing. Elsevier, Amsterdam, pp. 25–59.

Cadoni G, Cianfoni A, Agostino S et al. (2006). Magnetic resonance imaging findings in sudden sensorineural hearing loss. J Otolaryngol 35: 310–316.

Cambier J, Decroix JP, Masson C (1987). Auditory hallucinations in lesions of the brain stem. Rev Neurol (Paris) 143: 255–262.

Cambon K, Guilford FR (1958). Acoustic neurilemoma. Arch Otolaryngol 67: 302–312.

Canty P (1978). Speech audiometry in acoustic neuroma. J Laryngol Otol 92: 843–851.

Cascino GD, Adams RD (1986). Brainstem auditory hallucinosis. Neurology 36: 1042–1047.

Caye-Thomasen P, Dethloff T, Hansen S et al. (2007). Hearing in patients with intracanalicular vestibular schwannomas. Audiol Neurootol 12: 1–12.

Celesia G (2011). Brain stem auditory evoked potentials (BAEPs) and other auditory evoked potentials. In: DI Schomer, FH Lopes da Silva (Eds.), Niedemeyer's Electroencephalography. Wolters Kluver, Philadelphia, pp. 975–1002.

Champoux F, Tremblay C, Mercier C et al. (2006). A role for the inferior colliculus in multisensory speech integration. Neuroreport 17: 1607–1610.

Champoux F, Paiement P, Mercier C et al. (2007a). Auditory processing in a patient with a unilateral lesion of the inferior colliculus. Eur J Neurosci 25: 291–297.

Champoux F, Paiement P, Vannasing P et al. (2007b). Auditory scene analysis following unilateral inferior colliculus infarct. Neuroreport 18: 1793–1796.

Chau JK, Lin JR, Atashband S et al. (2010). Systematic review of the evidence for the etiology of adult sudden sensorineural hearing loss. Laryngoscope 120: 1011–1021.

Cheng TC, Wareing MJ (2012). Three-year ear, nose, and throat cross-sectional analysis of audiometric protocols for magnetic resonance imaging screening of acoustic tumors. Otolaryngol Head Neck Surg 146: 438–447.

Chermak G, Musiek F (1997). Central Auditory Processing Disorders: New Perspectives, Singular Publishing Group, San Diego, p. 374.

Chiappa K (1997). Evoked Potentials in Clinical Medicine. Lippincott/Raven, Philadelphia, p.713.

Cho TH, Fischer C, Nighoghossian N et al. (2005). Auditory and electrophysiological patterns of a unilateral lesion of the lateral lemniscus. Audiol Neurootol 10: 153–158.

Choi KD, Chun JU, Han MG et al. (2006). Embolic internal auditory artery infarction from vertebral artery dissection. J Neurol Sci 246: 169–172.

Chow JM, Garcia J (1985). Acoustic neuroma presenting as sudden hearing loss. Am J Otolaryngol 6: 115–119.

Church MW, Parent-Jenkins L, Rozzelle AA et al. (2007). Auditory brainstem response abnormalities and hearing loss in children with craniosynostosis. Pediatrics 119: e1351–e1360.

Cohen M, Luxon L, Rudge P (1996). Auditory deficits and hearing loss associated with focal brainstem haemorrhage. Scand Audiol 25: 133–141.

Cueva RA (2004). Auditory brainstem response versus magnetic resonance imaging for the evaluation of asymmetric sensorineural hearing loss. Laryngoscope 114: 1686–1692.

Davison C, Goodhart SP, Savitsky N (1935). The syndrome of the superior cerebellar artery and its branches. Arch Neurol Psychiat 33: 1143–1172.

Douen A, Bourque PR (1997). Musical auditory hallucinosis from *Listeria* rhombencephalitis. Can J Neurol Sci 24: 70–72.

Drulovic B, Ribaric-Jankes K, Kostic VS et al. (1993). Sudden hearing loss as the initial monosymptom of multiple sclerosis. Neurology 43: 2703–2705.

Dublin WB (1974). Cytoarchitecture of the cochlear nuclei: Report of an illustrative case of erythroblastosis. Arch Otolaryngol Head Neck Surg 100: 355–359.

Durrant JD, Martin WH, Hirsch B et al. (1994). 3CLT ABR analyses in a human subject with unilateral extirpation of the inferior colliculus. Hear Res 72: 99–107.

Duvernoy H (1999). Human brain stem vessels: including the pineal gland and information on brain stem infarction. Springer, Berlin, p. 261.

Egan CA, Davies L, Halmagyi GM (1996). Bilateral total deafness due to pontine haematoma. J Neurol Neurosurg Psychiatr 61: 628–631.

Eggston A, Wolff D (1947). Histopathology of the ear, nose and throat. Williams and Wilkins, Baltimore, p. 1080.

Ellis J, Heinrich B, Mautner VF et al. (2011). Effects of splicing mutations on NF2-transcripts: transcript analysis and information theoretic predictions. Gene Chromosome Canc 50: 571–584.

Evans DGR, Moran A, King A et al. (2005). Incidence of vestibular schwannoma and neurofibromatosis 2 in the north west of England over a 10-year period: higher incidence than previously thought. Otol Neurotol 26: 93–97.

Fischer C, Bognar L, Turjman F et al. (1995). Auditory evoked potentials in a patient with a unilateral lesion of the inferior colliculus and medial geniculate body. Electroencephalogr Clin Neurophysiol 96: 261–267.

Fitzgerald D, Mark AS (1998). Sudden hearing loss: frequency of abnormal findings on contrast-enhanced MR studies. Am J Neuroradiol 19: 1433–1436.

Fong B, Barkhoudarian G, Pezeshkian P et al. (2011). The molecular biology and novel treatments of vestibular schwannomas. J Neurosurg 115: 906–914.

Forcadas MI, Zarranz JJ (1994). Insomnia and hallucinations from vascular lesions of the pontine tegmentum in men. Neurologia 9: 211–223.

Fukutake T, Hattori T (1998). Auditory illusions caused by a small lesion in the right medial geniculate body. Neurology 51: 1469–1471.

Furst M, Levine RA, Korczyn AD et al. (1995). Brainstem lesions and click lateralization in patients with multiple sclerosis. Hear Res 82: 109–124.

Furst M, Aharonson V, Levine RA et al. (2000). Sound lateralization and interaural discrimination. Effects of brainstem infarcts and multiple sclerosis lesions. Hear Res 143: 29–42.

Gállego Culleré J, Erro Aguirre ME (2011). Basilar branch occlusion. Curr Treat Options Cardiovasc Med 13: 247–260.

Gareth D, Evans R (2009). Neurofibromatosis type 2 (NF2): A clinical and molecular review. Orphanet J Rare Dis 4. http://www.ojrd.com/content/4/1/16.

Griffiths TD (2002). Central auditory pathologies. Brit Med Bull 63: 107–120.

Gutmann D, Aylsworth A, Carey JC et al. (1997). The diagnostic evaluation and multidisciplinary management of neurofibromatosis 1 and neurofibromatosis 2. JAMA 278: 51–57.

Hanabusa K, Morikawa A, Murata T et al. (2001). Acoustic neuroma with malignant transformation. J Neurosurg 95: 518–521.

Hariri MA, Lakshmi MV, Larner S et al. (1994). Auditory problems in elderly patients with stroke. Age Ageing 23: 312–316.

Hausler R, Levine RA (2000). Auditory dysfunction in stroke. Acta Otolaryngol 120: 689–703.

Haymaker W (1969). Bing's local diagnosis in neurological diseases. C.V. Mosby, St Louis, p. 600.

Hendrikse J, Petersen ET, Cheze A et al. (2009). Relation between cerebral perfusion territories and location of cerebral infarcts. Stroke 40: 1617–1622.

Hendrix RA, DeDio RM, Sclafani AP (1990). The use of diagnostic testing in asymmetric sensorineural hearing loss. Otolaryngol Head Neck Surg 103: 593–598.

Hoistad DL, Hain TC (2003). Central hearing loss with a bilateral inferior colliculus lesion. Audiol Neurootol 8: 111–113.

Hosford-Dunn H (1985). Auditory brainstem response audiometry. Applications in central disorders. Otolaryngol Clin North Am 18: 257–284.

Houshmandi SS, Emnett RJ, Giovannini M et al. (2009). The neurofibromatosis 2 protein, merlin, regulates glial cell growth in an ErbB2- and Src-dependent manner. Mol Cell Biol 29: 1472–1486.

Hu CJ, Chan KY, Lin TJ et al. (1997). Traumatic brainstem deafness with normal brainstem auditory evoked potentials. Neurology 48: 1448–1451.

Igarashi M, Jergerm J, Alford BR et al. (1974). Functional and histological findings in acoustic tumor. Archives of Otolaryngology and Head and Neck Surgery 99: 379–384.

Illing RB, Kraus KS, Michler SA (2000). Plasticity of the superior olivary complex. Microsc Res Tech 51: 364–381.

Inzelberg R, Vishnievskaya S, Korczyn AD (1993). Transient musical hallucinosis. J Neurol Neurosurg Psychiatry 56: 833.

Jani NN, Laureno R, Mark AS et al. (1991). Deafness after bilateral midbrain contusion: a correlation of magnetic resonance imaging with auditory brain stem evoked responses. Neurosurg 29: 106–108.

Nadol Jr JB, Martuza RL (2005). Cerebellopontine angle tumors. In: IB Nadol, MJ McKenna (Eds.), Surgery of the ear and temporal bone. Lippincott Williams and Wilkins, Philadelphia, pp. 523–540.

Nadol Jr JB, Diamond PF, Thornton AR (1996). Correlation of hearing loss and radiologic dimensions of vestibular schwannomas (acoustic neuromas). Am J Otol 17: 312–316.

Jerger J, Hayes D (1976). The cross-check principle in pediatric audiometry. Arch Otolaryngol Head Neck Surg 102: 614–620.

Jerger J, Jerger S (1974). Auditory findings in brain stem disorders. Arch Otolaryngol Head Neck Surg 99: 342–350.

Jerger S, Jerger J (1975). Extra- and intra-axial brain stem auditory disorders. Int J Audiol 14: 93–117.

Jiang ZD, Brosi DM, Chen C et al. (2010). Brainstem response amplitudes in neonatal chronic lung disease and differences from perinatal asphyxia. Clin Neurophysiol 120: 967–973.

Johkura K, Matsumoto S, Hasegawa O et al. (1988). Defective auditory recognition after small hemorrhage in the inferior colliculi. J Neurol Sci 161: 91–96.

Johnson E (1977). Auditory test results in 500 cases of acoustic neuroma. Arch Otolaryngol 103: 152–158.

Johnson EW, House WF (1964). Auditory findings in 53 cases of acoustic neuromas. Arch Otolaryngol 80: 667–677.

Kim JS, Lopez I, DiPatre PL et al. (1999). Internal auditory artery infarction: clinicopathologic correlation. Neurology 52: 40–44.

Kim HA, Lee H, Sohn SI et al. (2006). Bilateral infarcts in the territory of the superior cerebellar artery: clinical presentation, presumed cause, and outcome. J Neurol Sci 246: 103–109.

Kimiskidis VK, Lalaki P, Papagiannopoulos S et al. (2004). Sensorineural hearing loss and word deafness caused by a mesencephalic lesion: clinicoelectrophysiologic correlations. Otol Neurotol 25: 178–182.

Korsen-Bengtsen M (1970). Auditory patterns in diseases of the acoustic nerve, brain stem, and cortex. In: C Rojskaer (Ed.), Speech Audiometry. Second Danavox Symposium, Odense, Denmark, pp. 184–190.

Kraus N, Bradlow AR, Cheatham J et al. (2000). Consequences of neural asynchrony: a case of auditory neuropathy. J Assoc Res Otolaryngol 1: 33–45.

Lalwani AK (1992). Meningiomas, epidermoids, and other nonacoustic tumors of the cerebellopontine angle. Otolaryngol Clin North Am 25: 707–728.

Lanska DJ, Lanska MJ, Mendez MF (1987). Brainstem auditory hallucinosis. Neurology 37: 1685.

Lauter J, Herscovitch P, Formby C et al. (1985). Tonotopic organization in human auditory cortex revealed by positron emission tomography. Hear Res 20: 199–205.

Lee H, Sohn SI, Jung DK et al. (2002). Sudden deafness and anterior inferior cerebellar artery infarction. Stroke 33: 2807–2812.

Lee E, Sohn HY, Kwon M et al. (2008). Contralateral hyperacusis in unilateral pontine hemorrhage. Neurology 70: 2413–2415.

Lee H, Kim JS, Chung E-J et al. (2009). Infarction in the territory of anterior inferior cerebellar artery: spectrum of audiovestibular loss. Stroke 40: 3745–3751.

Legatt AD, Pedley TA, Emerson RG et al. (1988). Normal brain-stem auditory evoked potentials with abnormal latency-intensity studies in patients with acoustic neuromas. Arch Neurol 45: 1326–1330.

Levine R, Hausler R (2001). Auditory disorders in stroke. In: J Bogousslavsky, LR Caplan (Eds.), Stroke syndromes. Cambridge University Press, Cambridge, UK, pp. 144–161.

Levine RA, Gardner JC, Fullerton BC et al. (1993). Effects of multiple sclerosis brainstem lesions on sound lateralization and brainstem auditory evoked potentials. Hear Res 68: 73–88.

Levine RA, Gardner JC, Fullerton BC et al. (1994). Multiple sclerosis lesions of the auditory pons are not silent. Brain 117: 1127–1141.

Lhermitte J (1922). Syndrome de la calotte du pédoncule cerebral. Les troubles psycho-sensoriels dans les lesions du mésencéphale. Rev Neurol (Paris) 38: 1359–1365.

Lidén G (1969). The scope and application of current audiometric tests. J Laryngol Otol 83: 507–520.

Lidén G, Korsan-Bengtsen M (1973). Audiometric manifestations of retrocochlear lesions. Scand Audiol 2: 29–40.

Link MJ, Driscoll CL, Giannini C (2002). Isolated, giant cerebellopontine angle craniopharyngioma in a patient with Gardner syndrome: case report. Neurosurg 51: 221–225.

Lisowska G, Namyslowski G, Morawski K et al. (2001). Early identification of hearing impairment in patients with type 1 diabetes mellitus. Otol Neurotol 22: 316–320.

Litovsky RY, Fligor BJ, Tramo MJ (2002). Functional role of the human inferior colliculus in binaural hearing. Hear Res 165: 177–188.

Lo YL, Hameed S, Rumpel H et al. (2011). Auditory hallucinations and migraine of possible brainstem origin. J Headache Pain 12: 573–575.

Long DM (1992). Non-glial tumors of the brain: tumors of the cerebellopontine angle and meningiomas. Curr Opin Neurol Neurosurg 5: 813–817.

Luhan JA, Pollack SL (1953). Occlusion of the superior cerebellar artery. Neurology 3: 77–89.

Luxon LM (1980). Hearing loss in brainstem disorders. J Neurol Neurosurg Psychiat 43: 510–515.

Lynn GE, Gilroy J, Taylor PC et al. (1981). Binaural masking-level differences in neurological disorders. Arch Otolaryngol 107: 357–362.

Martinez-Arizala A, Brown CD, McCarty GE (1983). Auditory evoked potentials in gliomas of the brain stem. Arch Neurol 40: 769–771.

Mazzoni A (1972). Internal auditory artery supply to the petrous bone. Ann Otol Rhinol Laryngol 81: 13–21.

McCreery DB (2008). Cochlear nucleus auditory prostheses. Hear Res 242: 64–73.

McCreery DB, Shannon RV, Moore JK et al. (1998). Accessing the tonotopic organization of the ventral cochlear nucleus by intranuclear microstimulation. Rehabilitation Engineering, IEEE Transact 6: 391–399.

McDonald WI, Compston A, Edan G et al. (2001). Recommended diagnostic criteria for multiple sclerosis: guidelines from the international panel on the diagnosis of multiple sclerosis. Ann Neurol 50: 121–127.

Millikan CH, Siekert RG, Whisnant JP (1959). The syndrome of occlusion of the labyrinthine division of the internal auditory artery. Trans Am Neurol Assoc 84: 11.

Moffat D, Saunders JE, McElveen Jr JT et al. (1993). Unusual cerebello-pontine angle tumours. J Laryngol Otol 107: 1087–1098.

Møller A (2003). Sensory systems: anatomy and physiology, Elsevier, Amsterdam, p. 469.

Montaguti M, Bergonzoni C, Zanetti MA et al. (2007). Comparative evaluation of ABR abnormalities in patients

with and without neurinoma of VIII cranial nerve. Acta Otorhinolaryngol Ital 27: 68–72.
Moore DR (1991). Anatomy and physiology of binaural hearing. Audiol 30: 125–134.
Moore JK (2000). Organization of the human superior olivary complex. Microsc Res Tech 51: 403–412.
Moore JK, Osen KK (1979). The cochlear nuclei in man. Am J Anat 154: 393–418.
Murata S, Naritomi H, Sawada T (1994). Musical auditory hallucinations caused by a brainstem lesion. Neurology 44: 156–158.
Musiek FE, Chermak GD (2006). Handbook of (Central) Auditory Processing Disorders Vol. 1: Auditory Neuroscience and Diagnosis. Plural Publishing, San Diego p, 502.
Musiek FE, Pinheiro ML (1987). Frequency patterns in cochlear, brainstem, and cerebral lesions. Audiol 26: 79–88.
Musiek FE, Gollegly KM, Kibbe KS et al. (1989). Electrophysiologic and behavioral auditory findings in multiple sclerosis. Am J Otol 10: 343–350.
Musiek FE, Charette L, Morse D et al. (2004). Central deafness associated with a midbrain lesion. J Am Acad Audiol 15: 133–151.
Musiek FE, Chermak GD, Weihing J et al. (2011). Diagnostic accuracy of established central auditory processing test batteries in patients with documented brain lesions. J Am Acad Audiol 22: 342–358.
Nakada T, St. John JN, Knight RT (1983). Solitary metastasis of systemic malignant lymphoma to the cerebellopontine angle. Neurorad 24: 225–228.
Nakashima I, Fujihara K, Okita N et al. (1999). Clinical and MRI study of brain stem and cerebellar involvement in Japanese patients with multiple sclerosis. J Neurol Neurosurg Psychiat 67: 153–157.
National Institutes of Health Consensus Development Conference: Neurofibromatosis Conference Statement (1988). Arch Neurol 45: 575–578.
Newton JR, Shakel M, Flatman S et al. (2010). Magnetic resonance imaging screening in acoustic neuroma. Am J Otolaryngol 31: 173–180.
Norman M, Thornton AR, Phillips AJ et al. (1996). Otoacoustic emissions recorded at high rates in patients with confirmed acoustic neuromas. Am J Otol 17: 763–772.
Noudel R, Gomis P, Duntze J et al. (2009). Hearing preservation and facial nerve function after microsurgery for intracanalicular vestibular schwannomas: comparison of middle fossa and retrosigmoid approaches. Acta Neurochir (Wien) 151: 935–944.
Odat HA, Piccirillo E, Sequino G et al. (2011). Management strategy of vestibular schwannoma in neurofibromatosis type 2. Otol Neurotol 32: 1163–1170.
Ogawa K, Kanzaki J, Ogawa S et al. (1991). Acoustic neuromas presenting as sudden hearing loss. Acta Otolaryngol Suppl 487: 138–143.
Ojemann RG, Montgomery WW, Weiss AD (1972). Evaluation and surgical treatment of acoustic neuroma. N Engl J Med 287: 895–899.
Olsen W, Noffsinger D, Carhart R (1976). Masking level difference encountered in clinical populations. Audiol 15: 287–301.
Ottaviani F, Almadori G, Maurizi M (1999). Eight nerve neuroma: audiological correlations. In: M Barbara (Ed.), Acoustic neuroma: trends and controversies. Kugler, The Hague, pp. 29–36.
Ottaviani F, Neglia C, Ventrella L et al. (2002). Hearing loss and changes in transient evoked otoacoustic emissions after gamma knife radiosurgery for acoustic neuromas. Arch Otolaryng Head Neck Surg 128: 1308–1312.
Paiement P, Champoux F, Bacon BA et al. (2009). Auditory pathways fail to re-establish normal cortical activation patterns in response to binaural stimulation following a unilateral lesion of the inferior colliculus. Neurocase 15: 89–96.
Pennings RJE, Morris DP, Clarke L et al. (2011). Natural history of hearing deterioration in intracanalicular vestibular schwannoma. Neurosurg 68: 68–77.
Pessin AB, Martins RH, Pimenta Wde P et al. (2008). Auditory evaluation in patients with type 1 diabetes. Ann Otol Rhinol Laryngol 117: 366–370.
Pillion JP, Moser HW, Raymond GV (2008). Auditory function in adrenomyeloneuropathy. J Neurol Sci 269: 24–29.
Plotkin SR, Merker VL, Halpin C et al. (2012). Bevacizumab for progressive vestibular schwannoma in neurofibromatosis type 2: a retrospective review of 31 patients. Otol Neurotol 33: 1046–1052.
Polman CH, Reingold SC, Edan G (2005). Diagnostic criteria for multiple sclerosis: 2005 revisions to the "McDonald criteria." Ann Neurol 58: 840–846.
Poussaint TY, Kocak M, Vajapeyam S et al. (2011). MRI as a central component of clinical trials analysis in brainstem glioma: a report from the Pediatric Brain Tumor Consortium (PBTC). Neuro Oncol 13: 417–427.
Powell HR, Choa DI (2010). Should all patients referred for magnetic resonance imaging scans of their internal auditory meatus be followed up in ENT clinics? Eur Arch Otorhinolaryngol 267: 1361–1366.
Prasher DK, Tun T, Brookes GB et al. (1995). Mechanisms of hearing loss in acoustic neuroma: an otoacoustic emission study. Acta Otolaryngol 115: 375–381.
Puletti F, Celesia GG (1970). Functional properties of the primary cortical auditory area in man. J Neurosurg 32: 244–247.
Rance G (2005). Auditory neuropathy/dys-synchrony and its perceptual consequences. Trends Amplif 9: 1–43.
Rance G, Cone-Wesson B, Wunderlich J et al. (2002). Speech perception and cortical event related potentials in children with auditory neuropathy. Ear Hear 21: 239–253.
Rance G, Corben L, Barker E et al. (2010). Auditory perception in individuals with Friedreich's ataxia. Audiol Neurootol 15: 229–240.
Régis J, Roche P-H (2008). Modern Management of Acoustic Neuroma. Karger Publishers, Basel, p. 261.
Ruttledge M, Andermann AA, Phelan CM et al. (1966). Type of mutation in the neurofibromatosis type 2 gene (NF2) frequently determines severity of disease. Am J Hum Genet 59: 331–342.

Saliba I, Bergeron M, Martineau G et al. (2010). Rule 3,000: a more reliable precursor to perceive vestibular schwannoma on MRI in screened asymmetric sensorineural hearing loss. Eur Arch Otorhinolaryngol 268: 207–212.

Santarelli R, Arslan E (2002). Electrocochleography in auditory neuropathy. Hear Res 170: 32–47.

Saunders JE, Luxford WM, Devgan KK et al. (1995). Sudden hearing loss in acoustic neuroma patients. Otolaryngol Head Neck Surg 113: 23–31.

Schick B, Brors D, Koch O et al. (2001). Magnetic resonance imaging in patients with sudden hearing loss, tinnitus and vertigo. Otol Neurotol 22: 808–812.

Schielke E, Reuter U, Hoffmann O et al. (2000). Musical hallucinations with dorsal pontine lesions. Neurology 55: 454–455.

Schow RL, Seikel AJ (2006). Screening for central auditory processing disorder. In: G Chermak, FE Musiek (Eds.), Handbook of (central) auditory processing disorder. In: Auditory neuroscience and diagnosis. 1. Plural Publishing, San Diego, pp. 137–161.

Schulz A, Kyselyova A, Baader SL et al. (2013). Neuronal merlin influences ERBB2 receptor expression on Schwann cells through neuregulin 1 type III signalling. Brain 137: 420–432.

Shemesh R (2008). Psychoacoustic tests for central auditory processing: normative data. J Basic Clin Physiol Pharmacol 19: 249–259.

Shergill SS, Cameron LA, Brammer MJ et al. (2001). Modality specific neural correlates of auditory and somatic hallucinations. J Neurol Neurosurg Psychiat 71: 688–690.

Silva M, Brucki SMD (2010). Multiple hallucinations due to brainstem injury: a case report. Dementia Neuropsychol 4: 353–355.

Springborg JB, Poulsgaard L, Thomsen J (2008). Nonvestibular schwannoma tumors in the cerebellopontine angle: a structured approach and management guidelines. Skull Base 18: 217–227.

Sriskandan N, Connor SE (2011). The role of radiology in the diagnosis and management of vestibular schwannoma. Clin Radiol 66: 357–365.

Stangerup SE, Tos M, Thomsen J et al. (2010a). True incidence of vestibular schwannoma? Neurosurg 67: 1335–1340.

Stangerup SE, Tos M, Thomsen J et al. (2010b). Hearing outcomes of vestibular schwannoma patients managed with 'wait and scan': predictive value of hearing level at diagnosis. J Laryngol Otol 124: 490–494.

Starr A, Picton TW, Sininger Y et al. (1996). Auditory neuropathy. Brain 119: 741–753.

Stimmer H, Borrmann A, Löer C et al. (2009). Monaural tinnitus from a contralateral inferior colliculus hemorrhage. Audiol Neurootol 14: 35–38.

Stokroos RJ, Albers FW, Krikke AP et al. (1998). Magnetic resonance imaging of the inner ear in patients with idiopathic sudden sensorineural hearing loss. Eur Arch Otorhinolaryngol 255: 433–436.

Takahashi S (2010). Neurovascular Imaging: MRI and Microangiography. Springer, Berlin, p. 515.

Talavage T, Sereno MI, Melcher JR et al. (2004). Tonotopic organization in human auditory cortex revealed by progressions of frequency sensitivity. J Neurophysiol 91: 1282–1296.

Taylor K, Brugger P, Schwarz U (2005). Audiovisual peduncular hallucinations: a release of cross-modal integration sites? Cogn Behav Neurol 18: 135–136.

Tekkok IH, Suzer T, Erbengi A (1992). Non-acoustic tumors of the cerebellopontine angle. Neurosurg Rev 15: 117–123.

Telischi F (2000). An objective method of analyzing cochlear versus noncochlear patterns of distortion-product otoacoustic emissions in patients with acoustic neuromas. Laryngoscope 110: 553–562.

Urban P, Caplan L (2011). Brainstem Disorders. Springer, Heidelberg, p. 360.

Valvassori GE (1988). Cerebellopontine angle tumors. Otolaryngol Clin North Am 21: 337–348.

van der Poel JC, Jones SJ, Miller DH (1988). Sound lateralization, brainstem auditory evoked potentials and magnetic resonance imaging in multiple sclerosis. Brain 111: 1453–1474.

Van Dijk JE, Duijndam J, Graamans K (2000). Acoustic neuroma: deterioration of speech discrimination related to thresholds in pure-tone audiometry. Acta Otolaryngol 120: 627–632.

Vitte E, Tankere F, Bernat I et al. (2002). Midbrain deafness with normal brainstem auditory evoked potentials. Neurology 58: 970–973.

Voordecker P, Brunko E, de Beyl Z (1988). Selective unilateral absence or attenuation of wave V or brain stem auditory evoked potentials with intrinsic brain stem lesions. Arch Neurol 45: 1272–1276.

Wellman MB, Sommer DD, McKenna J (2003). Sensorineural hearing loss in postmeningitic children. Otol Neurotol 24: 907–912.

Whitmore RG, Urban C, Church E et al. (2010). Decision analysis of treatment options for vestibular schwannoma. J Neurosurg 114: 400–413.

Yanagihara N, Asai M (1993). Sudden hearing loss induced by acoustic neuroma: significance of small tumors. Laryngoscope 103: 308–311.

Yener U, Avsar T, Akgün E et al. (2012). Assessment of antiangiogenic effect of imatinib mesylate on vestibular schwannoma tumors using in vivo corneal angiogenesis assay. J Neurosurg 117: 697–704.

Yin T (2002). Neural mechanisms of encoding binaural localization cues in the auditory brainstem. In: D Oertel, RR Fay, AN Popper (Eds.), Integrative Functions in the Mammalian Auditory Pathway. Springer-Verlag, New York, pp. 99–159.

Yoshida T, Sone M, Naganawa S et al. (2011). Accuracy of 3.0 Tesla magnetic resonance imaging in the diagnosis of intracochlear schwannoma. Auris Nasus Larynx 38: 551–554.

Zamani AA (2000). Cerebellopontine angle tumors: role of magnetic resonance imaging. Top Magn Reson Imag 11: 98–107.

Handbook of Clinical Neurology, Vol. 129 (3rd series)
The Human Auditory System
G.G. Celesia and G. Hickok, Editors

Chapter 30

Central auditory processing disorders in children and adults

TERI JAMES BELLIS[1]* AND JENNIFER D. BELLIS[2]

[1]*Department of Communication Sciences and Disorders, USD Speech-Language-Hearing Clinics and Division of Basic Biomedical Sciences, University of South Dakota, Vermillion, SD, USA*

[2]*Auditory Neuroscience Laboratory, Department of Communication Sciences and Disorders, University of South Dakota, Vermillion, SD, USA*

INTRODUCTION

Central auditory processing disorders (CAPD) can affect children and adults of all ages. Etiologies ranging from trauma to neurotoxic substances and neurologic disease or insult may cause dysfunction in the central auditory nervous system (CANS), resulting in a CAPD (American Academy of Audiology, 2010). In addition, CAPD may coexist with and/or mimic other disorders that affect listening, learning, and communication, particularly in children. Musiek et al. (1985; Musiek and Gollegly, 1988) postulated three primary underlying etiologies of CAPD in children: (1) neuromorphologic disorders, including atypical left/right anatomic asymmetries in planum temporale, ectopic areas, and polymicrogyri (65–70%); (2) neuromaturational delay (25–30%); and (3) neurologic disorder, disease, or insult (<5%). Although frank neuropathology may be relatively rare in children, the effects of acquired neurologic disorders on central auditory function in this population have nonetheless garnered recent attention (Chermak and Musiek, 2011). Similarly, in recent years, evidence of atypical neuromorphology in children with CAPD has been presented (e.g., Boscariol et al., 2009, 2010, 2011; Fig. 30.1). It has been estimated that CAPD affects 2–5% of the school-aged population, or approximately half of all children with learning disorders, and up to 76% of elderly adults, depending on the criteria used for diagnosis (Cooper and Gates, 1991; Chermak and Musiek, 1997; Bamiou et al., 2001; Golding et al., 2004).

More recently, increasing attention has been focused on the prevalence of CAPD in adults with acquired traumatic injuries, including returning war veterans and individuals involved in contact sports (e.g., Okie, 2005; Martin et al., 2008; Malloy and Bellis, 2013). In a recent study by Gallun et al. (2012), behavioral and electrophysiologic tests were administered to 36 blast-exposed veterans and 29 controls. Results indicated that the blast-exposed veterans exhibited a threefold increase in abnormalities on tests of central auditory function and speech in noise, despite normal peripheral hearing sensitivity. Similarly, Malloy and Bellis (2013) studied eight Division 1 football players with no history of diagnosed concussion. When compared to age- and education-matched controls with no history of participation in contact sports at the high-school level or above, the football players exhibited significantly smaller amplitudes and longer response latencies on several auditory evoked potential measures. These results indicated that the types of recurrent impulses and impacts experienced by those involved in contact sports such as football may result in central auditory deficits.

In short, CAPD is a distinct clinical disorder that has gained increasing attention in the literature over the past two decades. This chapter will provide an overview of current definitions and conceptualizations, methods of diagnosis of, and intervention for, CAPD. The chapter will culminate with a case study illustrating pre- and posttreatment behavioral and electrophysiologic diagnostic findings.

CURRENT DEFINITIONS AND CONCEPTUALIZATIONS OF CAPD

Since Myklebust (1954) emphasized the need to address auditory perceptual disorders not attributable to

---

*Correspondence to: Prof. Teri James Bellis, Ph.D., CCC-A, FAAA, F-ASHA, Department of Communication Sciences and Disorders, University of South Dakota, 414 East Clark Street, Vermillion, SD 57069, USA. Tel: +1-605-677-6201, Fax: +1-605-677-5767, E-mail: Teri.Bellis@usd.edu

Test	Results
Auditory processing	
Dichotic Digits Testing	RE=abnormal
	LE=normal
Random Gap Detection Test	Abnormal
Non Verbal Dichotic Testing	
Free Report	RE=normal
	LE=normal
Forced-Right Report	RE=abnormal
	LE=abnormal
Forced-Left Report	RE=abnormal
	LE=abnormal
Language	
Phonology	Impaired
Syntax	Impaired
Semantics	Normal
Pragmatics	Normal
Reading and Written Language (including Phonological Awareness)	Low Performance/Abnormal

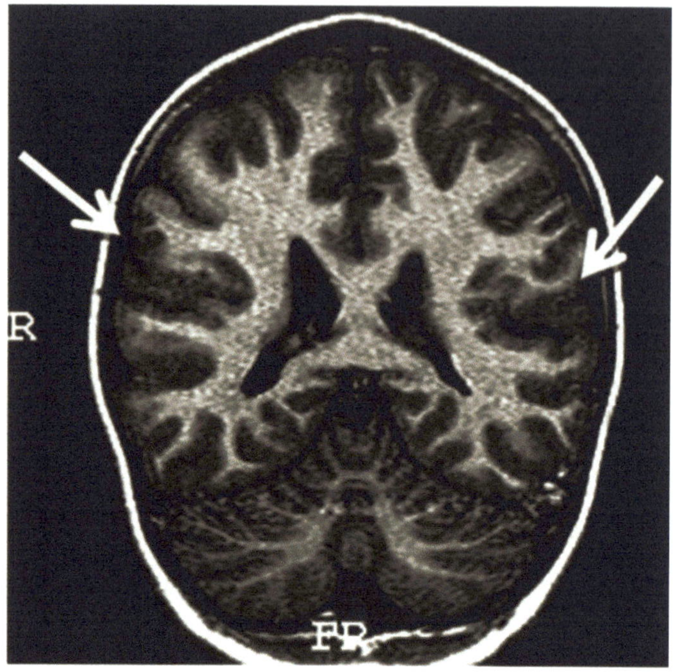

Fig. 30.1. Left: the performance on central auditory, language, and reading/written language assessments of a 14-year-old boy. Right: T1 magnetic resonance imaging (MRI) image of the same boy. The MRI demonstrates bilateral posterior parietal polymicrogyria (white arrows). (Modified from Boscariol et al., 2009.)

peripheral hearing loss, various definitions of CAPD have been offered which range from the very simple (e.g., "What we do with what we hear," Katz et al., 1992, p. 5) to the very complex (e.g., American Speech-Language-Hearing Association, 2005a, b; American Academy of Audiology, 2010).

The American Speech-Language-Hearing Association Working Group on Auditory Processing Disorders (American Speech-Language-Hearing Association, 2005a, b) defines central auditory processing as the "efficiency and effectiveness by which the central nervous system (CNS) utilizes auditory information [as well as] the perceptual processing of auditory information in the CNS and the neurobiologic activity that underlies that processing and gives rise to the electrophysiologic auditory potentials" (American Speech-Language-Hearing Association, 2005a, p. 2). Therefore, it follows that CAPD is a neurobiologic deficit in the CANS that affects those mechanisms that underlie fundamental auditory perception, including localization and lateralization; discrimination of speech and non-speech sounds; auditory pattern recognition; temporal aspects of audition, including integration, ordering, and masking; and auditory performance with competing and/or degraded acoustic signals (American Speech-Language-Hearing Association, 2005a, b).

Alternate definitions of CAPD have been proposed in recent years. For example, the British Society of Audiology (British Society of Audiology, 2011; Moore et al., 2013) asserts that developmental CAPD, the type that seems to be most common amongst children and for which no frank neuropathology can be identified, is a component of a generalized cognitive or neurodevelopmental delay rather than an auditory disorder. This is in stark contrast to the American Speech-Language-Hearing Association (2005a, b) and American Academy of Audiology (2010) definitions of CAPD, which emphasize that CAPD is not attributable to higher-order, global cognitive or related disorders. The British Society of Audiology (2011) viewpoint, however, fails to explain the presence of patterns of performance across central auditory and other measures in children with CAPD that precisely mirror those of known central auditory dysfunction. Further, while the known influence of top-down factors such as cognition, attention, language, and memory in auditory processing is important to consider, the British Society of Audiology (2011) contention that CAPD in children is most likely part of a general cognitive delay is not in keeping with the evidence demonstrating central auditory deficits in the presence of normal cognitive function (Weihing et al., 2012b).

The Canadian Inter-organizational Steering Group for Speech-Language Pathology and Audiology (CISGSLPA) recently published the *Canadian Guidelines on Auditory Processing Disorder in Children and Adults* (CISGSLPA, 2013). These guidelines present a comprehensive overview of diagnosis and management of CAPD from a World Health Organization (WHO) *International Classification of Functioning, Disability and Health* perspective (WHO, 2002). The Canadian perspective embraces the conceptualization of CAPD as a neurobiologic disorder requiring the use of sensitized tests of central auditory function for diagnosis while, at the same time, focusing on the complex interaction between cognition and auditory function.

In 2010, the American Academy of Audiology Task Force on Central Auditory Processing Disorders (American Academy of Audiology, 2010) re-endorsed the American Speech-Language-Hearing Association (2005a, b) definition of CAPD and provided the first set of levels-of-evidence-based guidelines for diagnosis of, and intervention for, CAPD to undergo both select and widespread peer review.

Several key points are inherent in the American Speech-Language-Hearing Association (2005a, b) and American Academy of Audiology (2010) definitions of CAPD. As CAPD is fundamentally a disorder of the CANS, diagnosis of the disorder requires the administration of tests that have been shown to be valid for the identification of CANS dysfunction. There is a plethora of behavioral, electrophysiologic, and neuroimaging evidence to support a neurobiologic basis for CAPD. These include findings of abnormal neurophysiologic representation of auditory (including speech) stimuli, atypical hemispheric asymmetries of auditory/language functions, inefficient interhemispheric transfer of auditory signals, and other findings that can be attributed neither to peripheral hearing loss nor to higher-order cognitive or related confounds (Musiek et al., 1985; Kraus et al., 1996; Bellis et al., 2000, 2008, 2011; Bellis and Wilber, 2001; Purdy et al., 2002; Tremblay and Kraus, 2002; Johnson et al., 2005; Filippini and Schochat, 2009; Bellis and Ross, 2011; Chermak and Musiek, 2011; Weihing et al., 2012b). When taken together, these findings provide unequivocal support for the presence of underlying CANS dysfunction in CAPD.

Although it is recognized that central auditory dysfunction may coexist with other disorders, particularly those that rely on shared neuroanatomic substrates or on functions related to auditory processing, CAPD is conceptualized as a sensory-based auditory disorder. Therefore, it should manifest itself primarily in the auditory modality (American Speech-Language-Hearing Association, 2005a). However, the expectation that CAPD is a completely auditory-modality-specific disorder is inconsistent with information processing in the CNS (Musiek et al., 2005). All information, including auditory, processing relies on the interaction of parallel, distributed networks throughout cortical and subcortical brain regions. Moreover, although some areas of the brain have traditionally been considered to be sensory-specific (i.e., auditory brainstem, primary auditory cortex), it is now generally recognized that non-auditory (i.e., visual) input can alter neurophysiologic responses and neural activity in these regions in humans. Similarly, additional areas not previously thought to be involved in "pure" sensory tasks have been found to be recruited for even basic sensory encoding (e.g., Sams et al., 1991; Calvert et al., 1997; Musacchia et al., 2006). Auditory processing is complex and this interaction amongst brain regions and functions contributes significantly to the oft-cited comorbidity of CAPD with disorders such as attention-deficit hyperactivity disorder (ADHD), learning disabilities, and others (Bellis, 2002, 2003; Musiek et al., 2005; Sharma et al., 2006, 2009; Banai et al., 2007; Billiet and Bellis, 2011).

Top-down processing and resource allocation are two additional concepts that must be taken into account when considering processing of auditory input. Cognitive and language function, attention abilities, memory, and other higher-order, pansensory factors will affect an individual's ability to process, comprehend, remember, and act upon what has been heard. This is true for any behavioral task, including the seemingly simple task of responding to pure-tone signals during audiologic threshold testing, and these top-down influences must be taken into account when considering auditory-based disorders. For example, higher-level linguistic and related representations support auditory processing by providing context for the incoming auditory input. It is easier for a listener to achieve auditory closure when he or she is familiar with the vocabulary, language, and topic. Therefore, individuals with higher-order disorders involving language, attention, and cognition may present clinically with "listening" difficulties, even in the absence of an auditory deficit.

Conversely, degradation of an incoming auditory signal, resulting either from environmental factors such as noise or poor acoustic signal quality or from factors internal to the listener such as peripheral or central auditory dysfunction, will result in a disproportionate amount of resources being allocated to basic signal encoding. This, in turn, will leave fewer resources available for higher-order functions such as comprehension and memory (Pichora-Fuller et al., 1995; Peelle et al., 2010). For example, Peelle et al. (2010), in a functional magnetic resonance imaging study of 20 elderly participants, found decreased activation in specialized auditory processing regions, and a limited ability to coordinate activity between regions, resulting in older adults' difficulty with sentence comprehension under difficult listening conditions (Fig. 30.2). This complex interaction of bottom-up and top-down processing holds significant implications for both diagnosis of, and intervention for, CAPD.

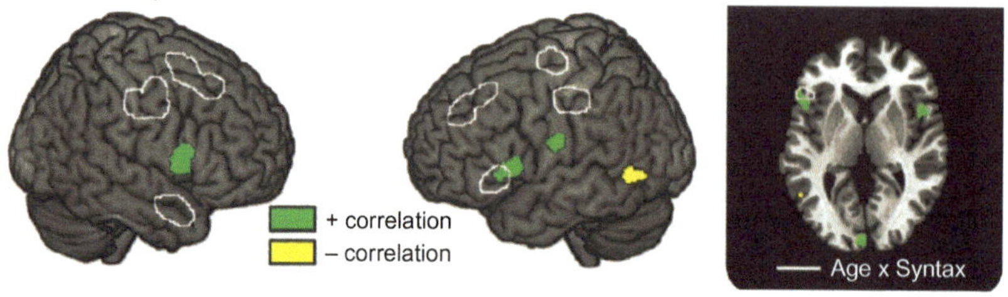

Fig. 30.2. Functional magnetic resonance imaging (fMRI) of brain areas showing responses related to processing syntactically complex sentences. The fMRI results show the correlation between syntax-related activity at the fast speech rate and accuracy at this rate in the older adults. Green regions indicate a positive correlation; yellow regions indicate a negative correlation. The regions outlined in white are those that demonstrate a significant age–syntax interaction. Older adults showed recruitment of inferior frontal regions relative to young adults. (Modified from Peelle et al., 2010.)

Finally, this definition of CAPD enumerates several fundamental auditory mechanisms that underlie more complex auditory-related tasks. Thus, for example, localization of sound is a critical aspect of auditory stream segregation and, ultimately, the ability to extract an auditory signal from a noisy background. It is, however, only one of many functions important to understanding speech in noise and dysfunction in various subcortical and cortical regions can affect localization ability. Clinicians must take the inherent interaction of multiple functions into account when attempting to disentangle the potential contribution of central auditory dysfunction to an individual's listening, learning, and communication complaints. Further, the unique confluence of an individual's top-down and bottom-up abilities will determine the nature of the functional, real-world difficulties he or she exhibits, leading to substantial heterogeneity in clinical presentations of CAPD.

Relationship of CAPD to language, learning, and communication

According to the American Speech-Language-Hearing Association (2005a), functions such as phonologic awareness; attention to, memory for, and synthesis of auditory information; and comprehension and interpretation of auditory-verbal input are presumably reliant, at least in part, on auditory processing. Nonetheless, these skills are considered to be higher-order language and related functions and are not in and of themselves considered to be part of the CAPD definition. Instead, CAPD is defined as a deficit in the neural processing of auditory stimuli that is not due to higher-order language, cognitive, attention, or related dysfunction or disorder. From a clinical perspective, this means that the diagnostic "label" of CAPD should not be applied to the generalized listening difficulties exhibited by children and adults with disorders such as primary language disorder, ADHD, decreased intellectual capacity, and autism until and unless specific dysfunction in the CANS can be demonstrated through the use of appropriately sensitized psychophysical and/or electrophysiologic diagnostic measures.

Listening complaints, such as difficulty understanding speech in a background of noise and "mishearing" auditory-verbal input, are well-accepted functional symptoms associated with CAPD. However, less well accepted is the degree to which CAPD may cause or be associated with difficulties in higher-order functions such as reading and spelling, speech-sound production, language, and other learning and communication abilities. Some researchers have demonstrated a lack of significant correlation between fundamental auditory functions such as gap detection and higher-order outcomes such as reading and spelling ability. This has been taken to indicate that CAPD, therefore, cannot at present be considered a "meaningful" disorder affecting real-world activities (Watson and Kidd, 2002, 2009; Cacace and McFarland, 2005). Other researchers, however, have demonstrated that the relationship of CAPD to language, learning, and related outcomes is highly individualized, and depends on the nature of the central auditory dysfunction that is present in combination with the individual's unique top-down and bottom-up characteristics (Bellis and Ferre, 1999; Cestnick and Jerger, 2000; Bellis, 2002, 2003).

Further, auditory processing is but one factor that contributes to complex learning and language abilities. As a result, there is substantial heterogeneity in both CAPD and language/learning/communication disorders that precludes simplistic predictions regarding functional consequences of a particular central auditory deficit. This means that valid diagnostic measures must be employed to determine a given individual's specific areas of functional deficit so that appropriate and customized intervention plans can be programmed.

Drawing upon a substantial body of literature in the auditory and cognitive neurosciences, neuropsychology, cognitive science, and related fields, Bellis and Ferre (1999; Bellis, 2002, 2003) have proposed a functional deficit profiling model of CAPD that synthesizes documented brain–behavior relationships across modalities. In this model, well-established patterns of performance on central auditory diagnostic tests resulting from lesions and/or dysfunction in various brain regions are combined with additional performance patterns across multidisciplinary listening, learning, and communicative domains. These relationships are summarized in Table 30.1 and are described below.

Dysfunction in non-language-dominant (usually right) cortical regions (pattern 1, Table 30.1) is more likely to affect functions such as music perception, mathematics calculation, receptive and expressive prosody, abstract reasoning, and non-verbal cognitive skills, among others. These difficulties have been associated with patterns of electrophysiologic and behavioral central auditory diagnostic test results that are consistent with dysfunction in right-hemisphere cortical sites affecting auditory processes such as temporal sequencing and perception of acoustic contours involving frequency and duration changes that may underlie comprehension of prosodic elements of speech, such as rhythm, stress, intonation, and other tone-of-voice cues.

Sequelae associated with language-dominant (usually left) cortex insult injury, or dysfunction (pattern 2, Table 30.1) have been well established in the literature, as well, and include language-based deficits, speech-sound production difficulties, phonologic (speech-sound)-based spelling disorders, and similar complaints. Central auditory test patterns in these cases may include temporal resolution deficits, difficulty achieving auditory closure with degraded speech, speech-sound representation, and discrimination deficits (especially for those speech-sounds involving rapid spectrotemporal acoustic changes), and significant difficulty processing speech in conditions of dichotic competition. These behavioral findings may be accompanied by abnormal electrophysiologic responses over left-hemisphere electrode sites, particularly when speech stimuli are used.

Table 30.1

Functional deficit profiling patterns

Central auditory test patterns	Possible auditory and related deficits	Presumed region of dysfunction
Pattern 1		
Dichotic speech tests – left-ear deficit Temporal patterning tests – poor labeling and humming responses Reduced electrophysiologic responses (MLR, corticals) over affected hemisphere	Difficulty interpreting prosody Flat affect/monotonic speech Difficulty extracting key words/main idea Mathematics calculation concerns Poor music perception Visual/spatial concerns Poor sight-word reading/spelling	Non-language-dominant (usually right) hemisphere
Pattern 2		
Dichotic speech tests – bilateral or right-ear deficit Monaural low-redundancy speech tests – bilateral or right-ear deficit Gap detection thresholds – elevated Reduced electrophysiologic responses (MLR, corticals) over affected hemisphere	Auditory closure/hearing-in-noise difficulties Poor speech-sound discrimination Poor spelling/reading decoding Possible language/vocabulary concerns Phonologic-based articulation concerns	Language-dominant (usually left) hemisphere, primary auditory cortex
Pattern 3		
Dichotic speech tests – left-ear deficit Temporal patterning tests – poor labeling response; humming within normal limits Electrophysiology (MLR, corticals) often normal	Hearing-in-noise difficulties Difficulty with tasks requiring bimanual/bipedal coordination Increased visual-motor interhemispheric transfer time Poor reading speed/fluency	Interhemispheric pathways (corpus callosum)

MLR, middle-latency response.

Adapted from Bellis and Ferre (1999) and Bellis (2003).

The impact of interhemispheric pathway (corpus callosum) dysfunction (pattern 3, Table 30.1) on multimodal and central auditory function has been demonstrated convincingly, as well. Subtle bimanual and bipedal difficulties, increased visual-motor interhemispheric transfer time, significant speech-in-noise difficulties, and inefficient processing of auditory stimuli presented to the left ear in the presence of right-ear competition all have been shown to be hallmarks of interhemispheric pathway compromise. Further, delayed neuromaturation of corpus callosal tracts has been shown to be a common cause of auditory and related complaints in children (Musiek et al., 1984). Similarly, aging effects on corpus callosum function also result in auditory complaints in adults, even when peripheral hearing and cognitive function are normal (Bellis and Wilber, 2001).

Other findings, such as the presence of abnormal brainstem neurophysiologic representation of speech stimuli, have been associated with reading, spelling, and speech-in-noise deficits. Banai et al. (2007) pooled the data from several preliminary studies of auditory brainstem responses (ABR) to speech stimuli in children with language-based learning (including reading) disorder and concomitant speech-in-noise difficulties. Their results indicated that approximately 30–40% of these children exhibited abnormal speech-evoked ABR activity. Different brain regions subserve different listening, learning, and related functions, and patterns of dysfunction in these regions may be demonstrated across multidisciplinary measures and correlated with distinct patterns of performance on central auditory diagnostic tests.

However, two caveats must be made at this juncture. First, despite the documentation of these patterns, one should not assume that all CAPDs lead to or are associated with deficits in language, learning, or other functions, nor should one assume that all language, learning, and related difficulties are associated with a CAPD. Given the highly individualized nature of CNS functioning and the heterogeneity of such disorders, it is critical to emphasize that functional deficit profiling models cannot be used as cookie-cutter guides to diagnosis of and intervention for CAPD, a concept that is true for language, learning, and other complex disorders, as well.

Second, one must not assume causality when a documented central auditory deficit coexists with a language, learning, or related disorder. It is true that, at least in some cases, a key contributing factor to the higher-order disorder may, indeed, be inefficient neural processing of auditory input. For example, poor temporal resolution resulting in inaccurate speech-sound representation may be a causal factor in the phonologic decoding difficulties seen in some children with reading and spelling disorders. In these cases, auditory-based intervention may result in improvement in both listening and reading/spelling skills. Conversely, it would not be logical to assume that the bimanual/bipedal or visual-motor difficulties exhibited by individuals with inefficient interhemispheric transfer are related causally to the hearing-in-noise and other auditory symptoms that also are associated with corpus callosum dysfunction. Instead, these are comorbid symptoms that arise due to a deficit in shared neuroanatomic substrates (Bellis and Wilber, 2001; Bellis, 2002, 2003; Bellis et al., 2008, 2011). This means that, in this case, treatment of the auditory disorder could be expected to improve the auditory skills but would not be expected to improve the non-auditory-based, associated sequelae, which would require separate intervention.

The observed relationships among central auditory processing, language, learning, and communication underscore the need for a multidisciplinary approach to differential diagnosis of and intervention for CAPD. In those cases in which an underlying central auditory deficit can be confirmed as a contributing factor to a child's or adult's functional complaints, it is critical that the deficit be identified and appropriate auditory-based intervention begun. The case study presented in the final section of this chapter will illustrate these concepts.

DIAGNOSIS OF CAPD

The purpose of a central auditory diagnostic test battery is to evaluate the integrity of the CANS, determine if a CAPD exists, and describe its parameters so that deficit-specific intervention can be programmed. This involves the administration of behavioral and/or electrophysiologic audiologic tests that have been shown to be sensitive and specific to dysfunction of the CANS.

Behavioral tests for CAPD diagnosis

Comprehensive guidelines for test selection have been published elsewhere (Bellis, 2003; Musiek and Chermak, 2007; American Academy of Audiology, 2010). In addition, Chapter 18 of this volume provides a detailed discussion of the tests introduced in this section. The behavioral test battery used to diagnose CAPD should be complaint- rather than test-driven. That is, consistent with the recommendations of Dillon et al. (2012), the tests chosen to evaluate an individual's central auditory function should be selected based on the clinical presentation of the child or adult as well as his or her age, linguistic proficiency, peripheral hearing status, and other factors. In addition, the behavioral CAP test battery should include both speech and non-speech tests, should target multiple levels and loci within the CANS, and should provide an assessment of a variety of auditory processes or mechanisms. The American Speech-Language-Hearing Association (2005a) and American Academy of Audiology (2010) recommend that behavioral tests of central auditory function be selected from the following general categories.

Dichotic speech tests

These tests involve the presentation of disparate verbal stimuli to each ear simultaneously. The listener's task may require binaural separation (i.e., directed or forced report), in which the listener is instructed to attend to one ear and ignore the other. Alternatively, it may require binaural integration (i.e., free report), in which the listener is instructed to attend to and report the stimuli delivered to both ears. A third type of dichotic process is binaural fusion, in which the listener perceives and reports only one stimulus despite being provided with two stimuli, one to each ear.

Kompus et al. (2012) studied 113 subjects using a dichotic listening paradigm consisting of pairwise presentations of consonant-vowel syllables. The participants were required to respond in three different conditions: (1) forced-right response, in which they were instructed to repeat the stimuli presented to the right ear and ignore the stimuli presented to the left ear; (2) forced-left response, in which left-ear stimuli were to be repeated; and non-forced, or free report, in which stimuli presented to both ears were to be repeated. The forced-left condition resulted in distinct activations in the left inferior prefrontal gyrus and caudate nucleus, while the right inferior frontal gyrus and caudate were activated in all three conditions. These findings suggest that the directed report condition (particularly under forced-left conditions) recruits additional brain regions responsible for attention and executive function, whereas the free-report condition appears to be more reflective of true auditory perceptual asymmetries (Fig. 30.3).

Shinn et al. (2005) evaluated 20 normally hearing participants on various dichotic report conditions and demonstrated that the binaural fusion condition is least impacted by higher-order functions such as attention and memory as compared to both the directed report and the free-report conditions. Therefore, it could be argued that a central auditory test battery should include more than one dichotic test, each of which assesses different processes, in order to tap different auditory and related processes and mechanisms. In this manner, deficit-specific intervention may be designed to address the individual's complaints in a more targeted fashion.

Temporal processing tests

These tests assess one or more elements of timing in the auditory system. Tests may include measures of perception of acoustic contours, or temporal patterns over time, as well as tests of temporal resolution, such as gap detection measures. In general, temporal processing tests utilize non-speech stimuli.

Monaural low-redundancy speech tests

This category of tests assesses the process of auditory closure, or the ability to fill in missing components of the signal when the input is less than clear. Methods of degrading the signal may include filtering (e.g., low-pass filtering), time compression, addition of reverberation, or presentation of speech in the presence of competing ipsilateral noise.

Binaural interaction tests

These tests focus on the ability of the ears to work together at the level of the auditory brainstem. They include measures such as the masking level difference and binaural fusion. Also included in this category are tests of localization and/or lateralization. Unfortunately, there is a marked lack of commercially available tests of binaural interaction at this time (American Academy of Audiology, 2010). However, a recent test of spatial hearing which appears to assess elements of localization has recently been developed – the Listening in Spatialized Noise – Sentence (LISN-S) test (Cameron and Dillon, 2008). Although this test shows promise for the evaluation of binaural interaction abilities, the degree to which the LISN-S is sensitive and specific to CANS dysfunction has yet to be determined.

Auditory discrimination tests

Both speech and non-speech discrimination abilities are fundamental to auditory processing. However, although tests of the ability to detect small differences between signals are an important addition to the auditory processing test battery, there is a dearth of commercially available, norm-referenced, audiologic tests for this purpose at the present time. Nonetheless, some measures of auditory discrimination are available within the speech-language armamentarium.

These test categories, along with the processes assessed by each and examples of specific tests within each category, are summarized in Table 30.2.

Selection of behavioral central auditory tests

In order to assess various regions/loci within the CANS and to probe a variety of auditory processes, the central

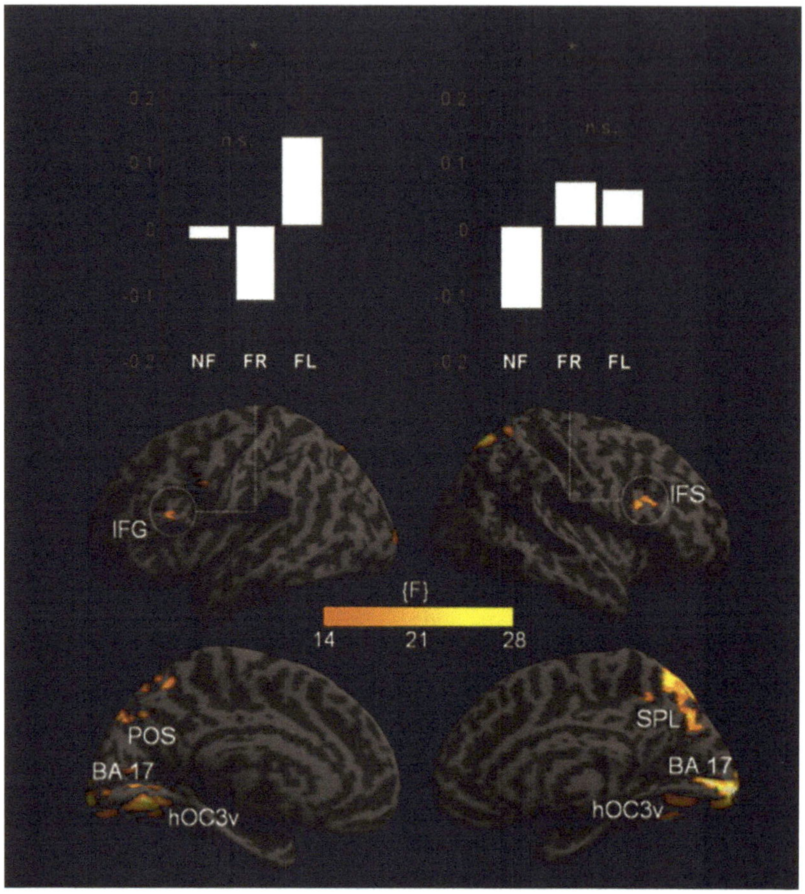

Fig. 30.3. Whole-brain functional magnetic resonance imaging analysis. The lower panel shows the statistical parameter maps (F-values) representing the main effect of condition overlaid on inflated mesh brain templates as provided by SPM8 (family-wise error correction to $P < 0.05$), left hemisphere presented on the left side. A significant main effect was found bilaterally in the inferior frontal cortex, situated in the inferior frontal gyrus (IFG) in the left and inferior frontal sulcus (IFS) in the right hemisphere; the right medial superior parietal lobule (SPL); bilaterally in the calcarine gyrus (BA17), extending ventrally into the lingual gyrus (area hOC3v); as well as in the left medial parieto-occipital sulcus (POS). The upper panel shows graphs representing the mean activation (in arbitary units, *y*-axis) for the three attention conditions (non-forced, NF; forced-right, FR; and forced-left conditions, FL) in the left and right hemispheric inferior frontal cortex. (Reproduced from Kompus et al., 2012.)

auditory test battery must include more than one test. However, it must be noted that the inclusion of additional tests often compromises specificity while enhancing sensitivity. It is important to attempt to limit the measures to the battery of tests that will provide the best sensitivity and specificity (efficiency) for the identification of CANS dysfunction while, at the same time, providing sufficient process-based information for intervention-planning purposes.

Musiek and Weihing (2011) studied 20 participants with known lesions of the CNS and compared their performance on tests of central auditory function to 29 normal controls. They evaluated the sensitivity, specificity, and efficiency of commonly used behavioral central auditory tests. The authors demonstrated that a two-test battery consisting of Dichotic Digits (Musiek, 1983) and Frequency Patterns (Pinheiro and Ptacek, 1971) provided the best balance of sensitivity and specificity for diagnosis of CANS dysfunction, at least when a strict criterion (i.e., cut-off of >2 SD below the mean on each test) was used. While this may provide clinicians with a minimum test battery to diagnose the presence of CAPD in the least amount of time, thus resulting in the most beneficial cost/benefit ratio, it should be noted that this approach may not be sufficient for intervention-planning purposes.

Similarly, in a series of studies examining performance on central auditory tests and corresponding visual analogs in normally hearing adults ($n = 10$), normally hearing children ($n = 10$), children with ADHD ($n = 10$), and children who meet current consensus definitions for diagnosis of CAPD (American Speech-Language-Hearing Association, 2005a; American Academy of Audiology, 2010; $n = 7$), Bellis and colleagues

Table 30.2

Categories of behavioral central auditory tests, processes assessed, and selected examples of each

Test category	Processes assessed	Selected examples
Dichotic speech tests	Binaural separation Binaural integration Binaural fusion	Dichotic Digits Dichotic Consonant-Vowels Staggered Spondaic Word test (SSW) Competing Sentences Dichotic Rhyme
Temporal patterning/processing tests	Temporal ordering/sequencing Frequency discrimination Duration discrimination Temporal resolution	Frequency Patterns Duration Patterns Gaps-In-Noise (GIN)
Monaural low-redundancy speech tests	Auditory closure	Low-Pass Filtered Speech Time Compressed Speech with and without Reverberation
Binaural interaction tests	Binaural interaction Localization/lateralization (informally)	Binaural Fusion tests Masking Level Difference (MLD) Listening in Spatialized Noise Sentence Test (LISN-S)
Auditory discrimination tests	Auditory discrimination	Frequency and Duration just-noticeable differences (JNDs) Speech-language measures of speech-sound discrimination

demonstrated that each of the central auditory tests alone was sufficient to differentiate among groups when intratest comparison measures were employed (Bellis et al., 2008, 2011; Bellis and Ross, 2011). Specifically, when overall test performance was considered, children in the CAPD and ADHD groups performed similarly, exhibiting poorer performance than normal controls. However, when intratest ear comparisons of ear advantage (dichotic sentences and digits) and response condition (temporal patterning) were considered, performance for each individual central auditory test differentiated the CAPD group from the non-CAPD groups, with the CAPD group demonstrating a significantly larger right-ear advantage on the dichotic speech tests and a significantly larger humming–labeling difference on the temporal patterning tests (Fig. 30.4). The addition of visual analogs added no diagnostic advantage, despite previous suggestions that the inclusion of such analogs is critical for the differential diagnosis of CAPD from higher-order, supramodal disorders such as ADHD (Cacace and McFarland, 2005).

Although a single test – when interpreted appropriately – may be sufficient to differentiate children with CAPD from those with attention-related deficits or from typically developing children, it is unlikely that one test would provide sufficient process-based information to inform intervention efforts. It is for this reason that the American Academy of Audiology (2010) guidelines suggest a test battery approach consisting of measures that assess a variety of processes. Our clinical experience demonstrates that, even when five or six tests are included in the battery, behavioral central auditory testing rarely takes longer than 45 minutes to 1 hour to complete. These observations are confirmed by Musiek (personal communication).

The American Speech-Language-Hearing Association (2005a) and American Academy of Audiology (2010) stress that diagnostic tests of central auditory function should be demonstrated to be both sensitive and specific for the identification of CANS lesions. Although few would argue that such a lesion-based approach would be appropriate for the testing of adults (and children) with known or suspected neurologic disease or disorder, some have questioned the utility of using confirmed lesions of the CANS as the "gold standard" for diagnostic tests of central auditory processing in children with "developmental CAPD." It may be that future research using electrophysiologic and/or imaging techniques may provide a more appropriate gold standard against which to compare this population. However, it should also be emphasized that, at the

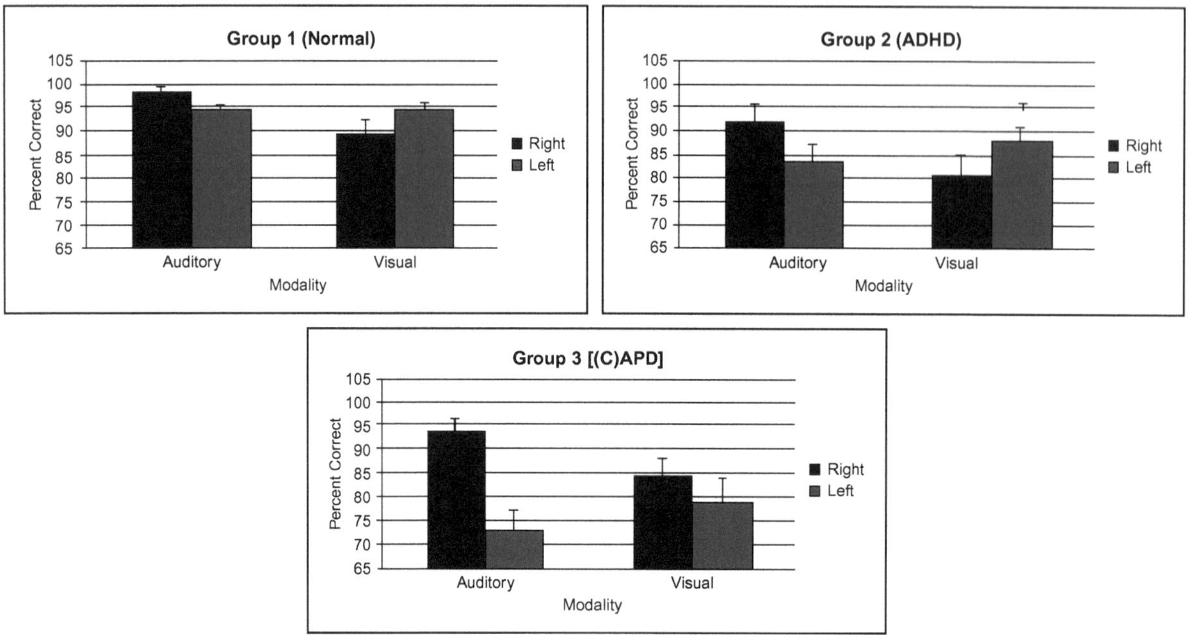

Fig. 30.4. Responses from three groups of children: normal ($n = 10$), attention-deficit hyperactivity disorder (ADHD) ($n = 10$), and central auditory processing disorders (CAPD) ($n = 7$) on three tests of central auditory function and their corresponding visual analogs. Children with ADHD and CAPD performed more poorly overall than typically developing children on all tasks when overall performance is considered. However, when intratest analyses (i.e., ear/visual-field differences, humming–labeling response differences) are considered, children with CAPD can be differentiated from those with ADHD and from normal controls. (Reproduced from Bellis et al., 2011.)

present time, the evaluation of central auditory function using tests shown to be valid for the identification of documented CANS dysfunction provides the best guide for test selection. Children with "developmental CAPD" demonstrate patterns of performance across these tests that are identical to those of children and adults with known CANS lesions (Bellis and Ferre, 1999; Bellis, 2002, 2003). Similarly, Boscariol et al. (2009, 2010, 2011) demonstrated that children with previously undetected neuromorphologic abnormalities also present with these same patterns.

A final point to be considered is that, due to the variability in neuromaturation (particularly of the corpus callosum) in younger children, most behavioral central auditory diagnostic tests cannot be reliably administered and interpreted until the age of approximately 7 or 8 years. After this age, age-specific normative values are available and must be used for interpretation purposes. In younger children, the American Academy of Audiology (2010) guidelines suggest that the presenting complaints, perhaps combined with electrophysiologic measures and/or screening tests, may provide a guide for the identification of functional auditory strengths and weaknesses. This guide may help to determine the need for auditory-based intervention, but should not be taken as evidence of a diagnosis of CAPD until a valid diagnostic test battery can be administered.

Interpretation of behavioral central auditory tests

Interpretation of the behavioral central auditory test battery initially involves comparing the individual's performance against age-matched normative values. However, because higher-order, pansensory or global disorders, or motivational factors may affect overall performance, inter- and intratest pattern analysis is critical for the accurate diagnosis of CAPD (American Speech-Language-Hearing Association, 2005a; American Academy of Audiology, 2010; Bellis and Ross, 2011; Bellis et al., 2011). The American Speech-Language-Hearing Association (2005a) and American Academy of Audiology (2010) suggest that a CAPD be diagnosed only when performance is abnormal relative to age-specific normative values on at least two tests of central auditory function and a consistent inter- and/or intratest pattern of performance indicative of CANS dysfunction can be demonstrated. Poor performance across all test measures, or an inconsistent pattern of performance (e.g., left-ear deficit on one test combined with a right-ear deficit on another test, improvement of performance with reinforcement or reinstruction, significantly poorer performance near the end of a test as compared to the beginning) should be taken as evidence of a higher-order cognitive, attention, motivation, or related confound rather than a CAPD.

A multidisciplinary approach to assessment is critical for both differential diagnosis and intervention purposes, because of possible comorbidity of CAPD with other disorders involving attention, learning, language, and communication. Only when the behavioral central auditory test results of a given individual are placed within the context of his or her functioning across other domains can the relative contribution of a CAPD to real-world complaints be estimated and a comprehensive, ecologically valid, individualized intervention program be developed (Bellis, 2002, 2003; American Speech-Language-Hearing Association, 2005a; Chermak and Musiek, 2007, in press; American Academy of Audiology, 2010).

Electrophysiologic tests for CAPD diagnosis

Although the standard, click-evoked ABR has not been shown to be particularly sensitive to the identification of CAPD in most instances, it is very useful in cases of trauma, pathology, or other dysfunction of the brainstem auditory pathways. In contrast, recent research has demonstrated that the measurement of ABR in response to complex signals, such as speech, instead of to simple click or tone-burst stimuli may have immense utility in illuminating central auditory pathway abnormalities in children and adults with auditory-based learning concerns, including CAPD, and determining the need for auditory intervention (e.g., Kraus et al., 1996; Banai et al., 2005, 2007). Billiet and Bellis (2011) studied 10 children diagnosed with dyslexia who exhibited normal speech-evoked ABRs, 10 children diagnosed with dyslexia who exhibited abnormal speech-evoked ABRs, and 10 age-matched controls. Their results indicated that, of the children diagnosed with dyslexia who exhibited normal speech-evoked ABRs, 100% exhibited behavioral central auditory test findings indicative of CAPD. In contrast, the majority of children who exhibited abnormal speech-evoked ABR performed within the normal range on the behavioral CAPD test battery. These findings suggest that the speech-evoked ABR may identify CANS dysfunction and drive intervention for some children who may be missed by the standard behavioral central auditory test battery (Billiet and Bellis, 2011; Fig. 30.5).

In recent years, the middle-latency response (MLR) has been shown to be of substantial utility in the identification of involvement of the thalamocortical central auditory pathways. However, due to substantial intersubject variability in amplitude and latency of MLR response components, the most useful indices for abnormalities appear to be intrasubject comparisons of response amplitude across multiple electrode sites (e.g., left hemisphere, right hemisphere, midline) and between stimulus ears (Musiek et al., 1989; Schochat et al., 2010; Weihing et al., 2012a). Cortical event-related auditory potentials, including the obligatory cortical potentials and the P300, also have been shown to be useful in the identification of dysfunction in auditory regions of the brain, including temporal lobe (Chermak and Musiek, 2011).

There is little general agreement regarding when auditory electrophysiologic measures should be included in a central auditory test battery. Some have suggested that they always be included (e.g., Jerger and Musiek, 2000); however, this approach is likely not feasible in most clinical settings. The American Academy of Audiology (2010) suggests that electrophysiologic evaluation be conducted when traditional behavioral testing cannot be completed (e.g., due to age or native language)

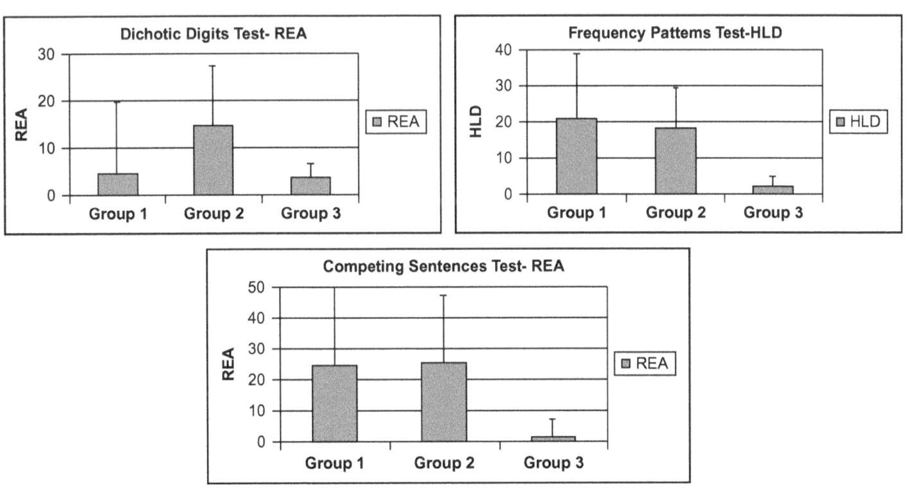

Fig. 30.5. Intratest analysis measures on three central auditory tests from three groups of children: those with dyslexia and abnormal brainstem timing as measured by speech-evoked auditory brainstem responses (group 1, $n = 12$), those with dyslexia and normal brainstem timing (group 2, $n = 20$), and normal controls (group 3, $n = 10$). Results indicate that children with dyslexia exhibited significantly larger right-ear advantages (REA) on dichotic speech tasks and significantly larger humming-labeling differentials (HLD) on frequency patterns testing when compared to normal controls. (Modified from Billiet and Bellis, 2011.)

or results of such testing are inconclusive, when a neurologic disorder or disease requiring medical follow-up is suspected, or when objective documentation of central auditory involvement is needed. In addition, these measures may provide data suggestive of treatment prognosis as well as documenting treatment efficacy in some individuals (Tremblay et al., 2001; Tremblay and Kraus, 2002; Tremblay, 2007; Sharma et al., 2009; Schochat et al., 2010; Kraus, 2012).

INTERVENTION FOR CAPD

Intervention recommendations for CAPD are based on the demonstrated auditory processing deficits and related listening and related complaints (Bellis, 2003; American Academy of Audiology, 2010). Further, just as auditory processing involves both bottom-up and top-down factors, intervention for CAPD should include both bottom-up and top-down activities. The individual's diagnostic test findings, case history, and information from other multidisciplinary professionals such as speech-language and psychoeducational specialists should be taken into consideration when developing a comprehensive, ecologically valid intervention plan for CAPD. Most importantly, the intervention plan should be deficit-specific and individualized to the child or adult in question. The ultimate goals of intervention for CAPD should be the remediation of the demonstrated auditory deficits and the minimization of the impact of the disorder on the individual's daily life functions. This is typically done through focusing on three areas: environmental modifications, compensatory strategies/central resources training, and direct skills remediation (Bellis, 2002, 2003; American Speech-Language-Hearing Association, 2005a, b; Chermak and Musiek, 2007, in press; American Academy of Audiology, 2010). Although intervention recommendations should be individualized and deficit-specific, some common interventions that may be useful for different profiles of CAPD are enumerated in Table 30.3.

Environmental modifications

Environmental modifications include both bottom-up and top-down approaches in that they address methods of enhancing access to the auditory signal as well as strategies for imparting information in order to render the signal more easily comprehensible. For example, bottom-up (i.e., signal-driven) environmental accommodations may include, but are not limited to, preferential seating, clear speech, and use of hearing-assistive technology. Top-down (i.e., strategy-driven) modifications may include activities such as pre-teaching new information/vocabulary to assist auditory closure, repetition or rephrasing of information, and use of visual aids (American Speech-Language-Hearing Association, 2005a; American Academy of Audiology, 2010). It is interesting to note that, although environmental modifications are generally designed to improve access to information rather than to treat the disorder *per se*, recent research has suggested that the use of hearing-assistive technology may, in fact, lead to improvements in listening and related (e.g., reading, spelling) skills (Hornickel et al., 2012). In this study, the authors examined reading performance in 38 normally hearing children diagnosed with dyslexia, half of whom had been fitted with an assistive listening device in the classroom for a period of 1 year. Results indicated that those children who had utilized hearing-assistive technology improved on measures of phonologic awareness and reading whereas those who were not fitted with such technology did not demonstrate these improvements. These results indicate that the use of hearing-assistive technology may drive auditory neuroplasticity in children (Fig. 30.6).

Most importantly, as with all aspects of intervention, decisions regarding which environmental modifications to implement in a given situation should be individualized and specific to the child's or adult's presenting difficulties and diagnostic findings. There is no "one-size-fits-all" list of environmental accommodations appropriate for all individuals with CAPD, although the basic principles of improving the signal-to-noise ratio and using clear speech benefit all listeners, particularly those with disorders such as CAPD.

Central resources training

Compensatory strategies/central resources training utilize a top-down treatment approach. The goals of central resources training are to minimize the impact of the CAPD through recruiting and strengthening higher-order, top-down cognitive, attention, language, and related processes and to address possible secondary motivational and related deficit areas. These strategies may include metalinguistic and metacognitive techniques such as self-instruction, cognitive problem solving, assertiveness/attribution training, auditory memory enhancement, and other activities. Strategies such as these provide benefits allowing for enhanced listening, communication, social, and learning outcomes not directly addressed through auditory training (Katz, 1983; Musiek and Chermak, 1995; Sloan, 1995; Chermak and Musiek, 1997, in press; Chermak, 1998; Musiek, 1999; Bellis, 2002, 2003; American Speech-Language-Hearing Association, 2005a). It is important to note that, although specific studies regarding efficacy of central resources training in cases of CAPD *per se* are

Table 30.3

Common intervention techniques for different subtypes of CAPD

CAPD deficit profile	Environmental modifications	Compensatory strategies/ central resource training activities	Direct remediation activities
Right hemisphere (non-language-dominant cortex)	Placement with "animated" teacher; avoid hints/subtleties in conversation; avoid abstract language	Metamemory and metalinguistic techniques; schema induction using social scripts; key word extraction	Non-speech (e.g., frequency, duration) discrimination training Temporal patterning training Prosody training
Left hemisphere (primary auditory cortex)	Preferential seating with visual access to teacher; classroom acoustic modifications; hearing-assistive technology; repetition rather than rephrasing of key information; pre-teaching of new vocabulary	Auditory closure; vocabulary building; problem solving; attribution training; active listening techniques	Temporal resolution training Speech-sound discrimination Phonologic awareness activities Speech-to-print skills Dichotic listening training (to address speech-in-noise skills)
Corpus callosum/ interhemispheric	Avoid use of multimodal cues; classroom acoustic modifications; hearing-assistive technology	Attribution training; active listening techniques; problem solving	Dichotic listening training Interhemispheric activities Localization training

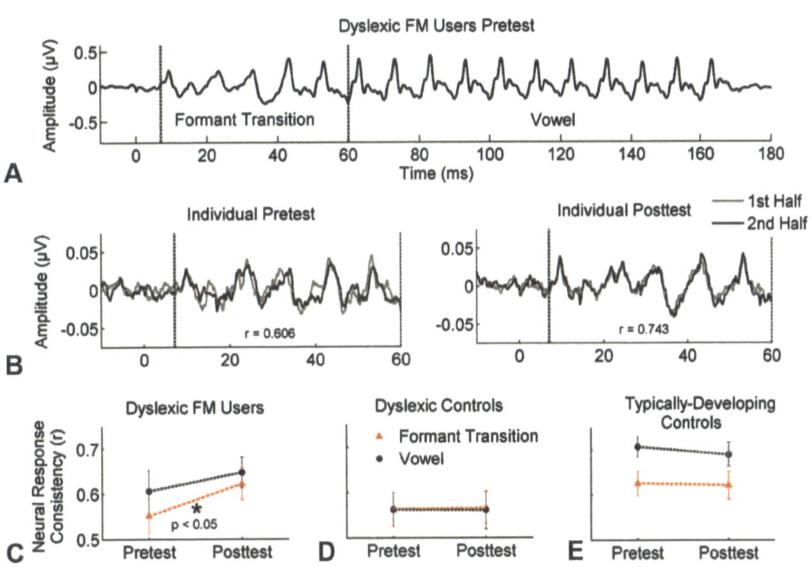

Fig. 30.6. Frequency-modulated (FM) system use increases the consistency of auditory brainstem representation of speech. (**A**) Responses from dyslexic children before FM system use with the formant transition (7–60 ms) and vowel (60–180 ms) portions of the speech syllables marked. (**B**) Response consistency, quantified as the correlation between responses collected during the first half of the recording session (light gray) and those collected in the second half of the recording session (dark gray), improved with FM system use during the formant transition for this representative individual. The more consistent response at posttest is reflected by a higher r-value. (**C**) After FM system use for one school year, children with dyslexia had more consistent speech-evoked brainstem responses, particularly for the response to the formant transition (red triangles; trending at $P = 0.036$). Significant changes were not seen in response to the vowel portion (black circles). (**D**) Control children with dyslexia who did not use the FM systems show no change in response consistency in either the formant transition (red triangles) or the vowel (black circles) portions of the response. (**E**) Response consistency also does not change for typically developing controls in either the formant transition (red triangles) or the vowel (black circles) portions. (Reprinted from Hornickel et al., 2012, with permission.)

lacking, research into resource allocation, discussed previously, clearly indicates that top-down processes provide important scaffolding for bottom-up auditory skills. Thus, improving "upstream" cognitive and related skills and abilities allows for greater resources to be available "downstream" for acoustic signal processing (Pichora-Fuller et al., 1995).

Direct remediation

The goal of direct skills remediation is to improve the specific auditory deficits that have been identified by a battery of validated tests of central auditory function (American Speech-Language-Hearing Association, 2005a, b; Chermak and Musiek, 2007, in press; American Academy of Audiology, 2010; Bellis et al., 2012). This is done through targeted, behavioral auditory training activities focused on the identified areas of deficit. Auditory training may be either formal or informal. Formal auditory training activities typically take place within a clinical setting and include well-controlled acoustic stimuli presented through audiologic equipment. Informal auditory training may be done outside the clinic, including at school and in the patient's home. Because of the need for frequent treatment sessions, informal auditory training is particularly important in rural areas where patients may live some distance from the nearest CAPD clinic. In addition, there are several computer software programs currently available that address various aspects of auditory training.

Auditory training activities may include. but are not limited to, procedures targeting speech- and non-speech-sound discrimination; temporal resolution; auditory pattern recognition; localization/lateralization of sound; and speech-in-competition (e.g., speech-in-noise and/or dichotic listening training). Additionally, exercises that address multimodal interhemispheric transfer may be indicated in some cases. Auditory discrimination and sequencing, which are essential components of language and reading, are often areas of difficulty for a child diagnosed with CAPD and thus should be considered during a course of auditory training (Wright et al., 2000; Foxton et al., 2004).

Auditory training has been shown to result in both functional and structural improvements in the CANS in both animals and humans (Galbraith et al., 1995; Jancke et al., 2001; Tremblay and Kraus, 2002; Hayes et al., 2003; Bao et al., 2004; Foxton et al., 2004; Johnson et al., 2005, 2008; Russo et al., 2005; Chermak and Musiek, 2007; Chermak et al., 2013 in press; Alonso and Schochat, 2009; de Villers-Sidani et al., 2010; Schochat et al., 2010; Strait and Kraus, 2011; Bellis et al., 2012; Kraus, 2012). After just a week of direct auditory training, some studies have shown behavioral and electrophysiologic benefits in normal subjects (Kraus et al., 1996; Tremblay et al., 2001).

Tremblay and colleagues (2001) investigated the ability of 10 normally hearing young adults to learn to discriminate variants of the consonant-vowel syllable /ba/ that differed in voice-onset time (10 ms versus 20 ms). Initially, the participants could not discriminate between the two signals; however, using an adaptive training paradigm, they were able to learn to differentiate the signals. Concurrently, the N1-P1 complex of the cortical event-related potential, evoked by the same stimuli, demonstrated training-induced increases in amplitude that corresponded to the participants' performance improvements. Alonso and Schochat (2009) also documented changes in auditory event-related potentials (P300) and central auditory performance following formal, clinic-based auditory training in a group of 29 children diagnosed with CAPD (Fig. 30.7).

Studies demonstrating the efficacy of other forms of auditory training also have been emerging in the literature. For example, Moncrieff and Wertz (2008) administered a quasi-dichotic listening therapy to 8 (experiment 1) and 13 (experiment 2) children with diagnosed CAPD. Different stimuli were routed through each of two speakers in the sound field and interspeaker intensity was gradually adjusted over time. Treatment sessions occurred several times per week for a total of 12–24 sessions, depending on the participant. Posttreatment results demonstrated significant improvements in dichotic listening abilities and concomitant listening comprehension measures.

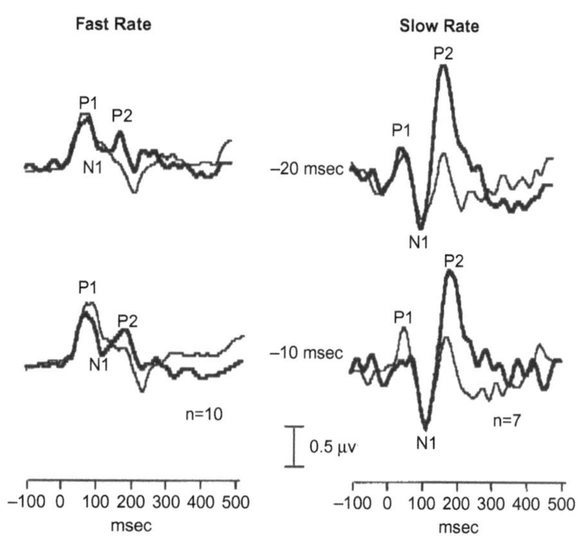

Fig. 30.7. Pre- and posttraining grand averaged waveforms measured from electrode Cz. Pretraining waveforms are thin. Posttraining waveforms are thick. As subjects learned to identify the difference between the 220 and 210 ms voice-onset time (VOT) stimuli, N1-P2 peak-to-peak amplitude increased. (Reprinted from Tremblay et al., 2001, with permission.)

Musiek et al. (2008) also demonstrated the effectiveness of a dichotic listening therapy in a group of 14 children diagnosed with CAPD. In their study, the dichotic stimuli were varied in interaural temporal lag time, with shorter temporal lags being more challenging than longer lags. After 10 weeks of training, a significant improvement in dichotic listening abilities was noted and was accompanied by parent and teacher reports of hearing-in-noise improvements.

Similar posttreatment improvements have been noted following frequency and duration discrimination training (McArthur et al., 2008), computer-based auditory skills training (Sharma et al., 2012), and acoustic contour (pitch and duration patterns) training (Foxton et al., 2004). In addition, recent research has begun to demonstrate the impact of music training on auditory and learning abilities (Schon et al., 2004; Marques et al., 2007; Moreno et al., 2009; Parbery-Clark et al., 2009; Pantev and Herholz, 2011; Kraus, 2012). Chapter 12 of this volume provides additional information on the effect of musical experience and training on auditory and related processing.

Notwithstanding these and similar studies indicating effectiveness of auditory training in addressing auditory complaints of those with CAPD, there is a clear need for additional, well-controlled treatment efficacy research using both children and adults with precise, consensus-based CAPD diagnoses. Following is a case study illustrating the diagnostic and intervention concepts introduced in this chapter.

CASE STUDY

The patient, KA, was a 17-year-old healthy female who presented at our clinic with primary complaints of significant difficulty hearing in noise, "mishearing" auditory input leading to misunderstanding of auditory-verbal communications, and reading and spelling difficulties. She had been an accomplished musician (piano, violin) from a very early age.

At the age of 10, KA underwent a stereotactic volumetric resection of a right posterior temporal-lobe glioma. Prior to the surgery, KA complained of headaches and visual disturbances; however, she did not endorse any auditory, learning, or related complaints and was performing at age level in all benchmarks. Postoperatively, KA complained of hyperacusis (particularly in the right ear), headaches in the right side of the head, and episodes of syncope. No postsurgical visual concerns were reported. A postoperative magnetic resonance imaging scan was unremarkable.

When she arrived for CAPD evaluation, KA and her parents reported that her musical abilities had not been affected by the surgical removal of the lesion. Instead, her reading and spelling skills (which had been intact and progressing at an age-appropriate rate preoperatively) had "stalled" at the fifth-grade level, showing little improvement thereafter. These findings were accompanied by complaints of significant difficulties hearing in noise and substituting similar-sounding phonemes in perception of verbal input, all of which started following the surgery at age 10 and had become progressively more noticeable. Receptive and expressive language abilities, as well as cognitive abilities, were demonstrated to be age-appropriate through standardized tests.

A comprehensive central auditory test battery was administered, consisting of both psychophysical and electrophysiologic measures. KA was cooperative throughout testing and no indicators of attention, cognitive, or linguistic confounds were apparent. Five behavioral central auditory tests were administered: (1) Dichotic Rhyme (Musiek et al., 1989); (2) Dichotic Digits (Musiek, 1983); (3) Competing Sentences (Auditec version); (4) Time-Compressed Speech with and without Reverberation (Tonal and Speech Materials for Auditory Perceptual Assessment, v. 2.0); and (5) Frequency Patterns (Pinheiro and Ptacek, 1971; Ptacek and Pinheiro, 1971).

Results indicated that KA exhibited abnormal performance on four of the five tests administered, including a left-ear deficit on Dichotic Rhyme, a bilateral deficit on Dichotic Digits and Competing Sentences, and a bilateral deficit on Time-Compressed Speech, which demonstrated significant worsening with the addition of reverberation. Conversely, her performance in non-sensitized conditions (i.e., repetition of number sequences and sentences in non-competing conditions, repetition of monosyllabic words in non-degraded conditions) was well within normal limits, indicating that the abnormalities observed could be attributed to the acoustic modifications rather than to generalized language, memory, or related confounds.

When taken together, results of behavioral testing met American Speech-Language-Hearing Association (2005a) and American Academy of Audiology (2010) diagnostic criteria for CAPD, including performance outside of the normal range for age on at least two tests of central auditory function and an inter- and intratest pattern of performance consistent with CANS dysfunction. The observed central auditory deficits appeared to be affecting the processes of binaural separation, binaural integration, and auditory closure, all of which are critical for successful auditory perception in competition and with degraded auditory stimuli.

Following the behavioral central auditory evaluation, a comprehensive auditory electrophysiologic evaluation was also completed. Click-evoked ABR testing revealed essentially normal neural transmission of auditory stimuli through brainstem auditory pathways. In contrast, MLR recordings were normal with right-ear stimulation, but were significantly reduced in amplitude or absent over left- and right-hemisphere electrode sites with left-ear stimulation. This finding indicated abnormal transmission of auditory signals through the thalamocortical auditory pathways with left-ear stimulation, resulting in an "ear effect" for the ear contralateral to the site of dysfunction (Musiek et al., 1989; Weihing et al., 2012a). Cortical auditory event-related potentials, including P300, also were obtained and were within normal limits across all electrode sites with both left- and right-ear stimulation.

Recommendations for intervention were made, including environmental modifications, central resources training, and direct auditory remediation. To target the deficient processes of binaural separation/integration, direct remediation in the form of dichotic listening training was recommended (Bellis, 2002, 2003).

Twelve weeks following the initial evaluation, KA returned for a central auditory re-evaluation. She had undergone the intensive dichotic listening training for the recommended 6 weeks. At this time, both KA and her mother reported a marked improvement in hearing-in-noise and other listening skills. School records supplied by her mother confirmed a significant improvement in KA's reading and spelling skills, and she was currently achieving grades of A and B in her courses.

Posttreatment behavioral central auditory test results revealed performance within the normal range for age for all measures (Fig. 30.8). These results indicated significant improvement in binaural separation and integration abilities. Interestingly, KA's auditory closure skills, as measured by her performance on time-compressed speech testing with and without reverberation, also demonstrated significant improvement, despite not having been explicitly trained.

Posttreatment MLR testing mirrored the behavioral results, indicating normal thalamocortical transmission of auditory signals with both right-ear and left-ear stimulation (Fig. 30.9). These findings provided objective evidence of improved neural integrity underlying central auditory function. Given the short training period (6–12 weeks), the age of the patient (17 years), and the absence of other intervention activities during the time since initial evaluation, these improvements could not be attributed to maturation, environmental factors, or other contributors. Instead, when taken together, the results of both behavioral and electrophysiologic testing suggested significant improvement in both auditory behaviors and underlying neural substrates resulting from targeted, intensive auditory training.

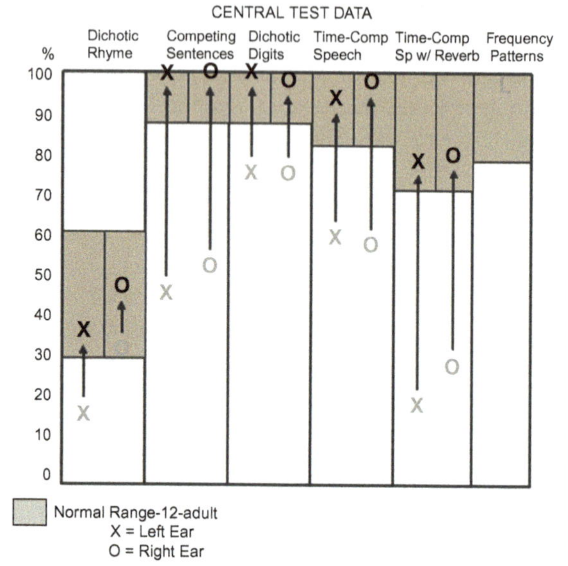

Fig. 30.8. Pretherapy (gray) and posttherapy (black) behavioral central auditory test results from right ear (O) and left ear (X). Shaded areas represent normative ranges for age. Note the improved performance after therapy.

This case study illustrates the use of diagnostic central auditory measures to guide deficit-specific intervention and demonstrates both behavioral and neurophysiologic improvements following targeted, intensive auditory training. In our clinic, KA is not an atypical case. We have seen nearly identical results with a host of other children and adults both with and without frank neuropathology but with diagnosed CAPD. Further, in those cases in which reading, spelling, and other learning difficulties coexisted with the CAPD, improvements in those skills also were apparent, despite not having been explicitly addressed. We hope that future research can assist in illuminating the underlying neural mechanisms responsible for these improvements, particularly in non-auditory, learning-related abilities.

CONCLUSIONS

In conclusion, CAPD can affect anyone of any age. Various etiologies for CAPD may include neurologic disorders, genetic diseases, neurotoxins, and trauma; however, in many cases, no identifiable neuropathology may be evident, particularly in children. Nonetheless, the most widely accepted definitions of CAPD emphasize that the disorder is neurobiologic in origin and affects perceptual processing (and neural transmission) of auditory information in the CANS.

Diagnosis of CAPD requires the administration of behavioral and electrophysiologic central auditory tests that have been shown to be sensitive, specific, and efficient for the identification of dysfunction in various

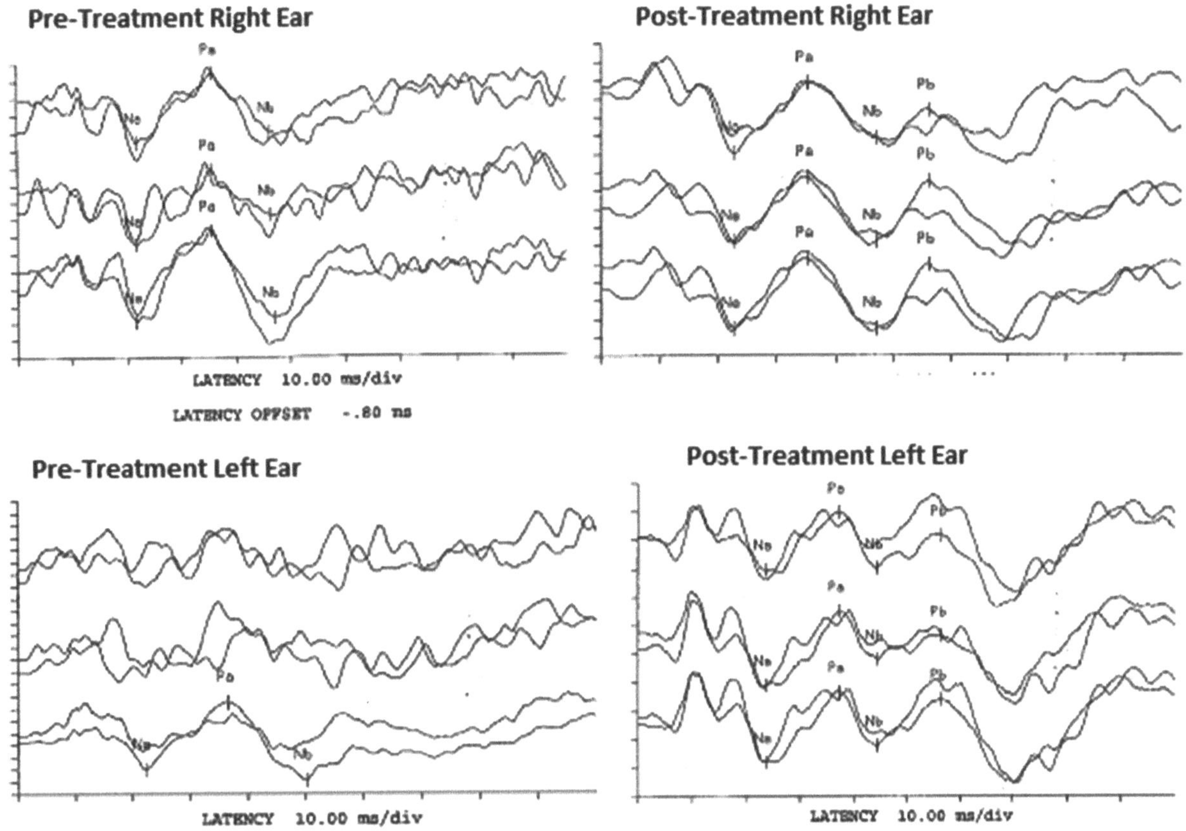

Fig. 30.9. Middle-latency auditory evoked responses pre- and posttherapy. Note the improved responses following therapy. In each set of responses, the top tracing indicates responses obtained at C3, the middle tracing indicates responses obtained at C4, and the bottom tracing indicates responses at Cz.

regions of the CANS. Intervention for CAPD should be individualized, deficit-specific, and multidisciplinary in nature, and should include bottom-up and top-down approaches to improve access to information in the listening/learning environment; recruit and strengthen higher-order cognitive, language, and related skills through central resources training; and directly remediate the auditory deficits identified through diagnostic testing.

References

Alonso R, Schochat E (2009). The efficacy of formal auditory training in children with (central) auditory processing disorder: behavioral and electrophysiological evaluation. Braz J Otorhinolaryngol 75 (5): 726–732.

American Academy of Audiology (2010). Guidelines for the diagnosis, treatment, and management of children and adults with central auditory processing disorder. Available online at, http:/www.audiology.org/resources/documentlibrary/Documents/ait.pdf, Retrieved: May 15, 2013.

American Speech-Language-Hearing Association (2005a). (Central) auditory processing disorders [technical report]. Available online at, http:/www.asha.org/docs/html/TR2005-00043.html, Retrieved: May 15, 2013.

American Speech-Language-Hearing Association (2005b). (Central) auditory processing disorders – The role of the audiologist [position statement]. Available online at http:/www.asha.org/docs/html/PS2005-00114.html Retrieved: May 15, 2013.

Bamiou DE, Musiek FE, Luxon LM (2001). Aetiology and clinical presentation of auditory processing disorders – a review. Arch Dis Child 85: 361–365.

Banai K, Nicol T, Zecker S et al. (2005). Brainstem timing: implications for cortical processing and literacy. J Neurosci 25 (43): 9850–9857.

Banai K, Abrams D, Kraus N (2007). Sensory-based learning disability: insights from brainstem processing of speech sounds. Int J Audiol 46: 524–532.

Bao S, Chang EF, Woods J et al. (2004). Temporal plasticity in the primary auditory cortex induced by operant perceptual learning. Nat Neurosci 7: 974–981.

Bellis T (2002). When the brain can't hear, Atria Books, New York.

Bellis T (2003). Assessment and management of central auditory processing disorders in the educational setting, 2nd edn. Delmar Learning, Clifton Park, NY.

Bellis T, Ferre J (1999). Multidimensional approach to the differential diagnosis of central auditory processing disorders in children. J Am Acad Audiol 10: 319–328.

Bellis T, Ross J (2011). Performance of normal adults and children on central auditory diagnostic tests and their corresponding visual analogs. J Am Acad Audiol 22 (8): 491–500.

Bellis T, Wilber L (2001). Effects of aging and gender on interhemispheric function. J Speech Lang Hear Res 44: 24–263.

Bellis T, Nicol T, Kraus N (2000). Aging affects hemispheric asymmetry in the neural representation of speech sounds. J Neurosci 20 (2): 791–797.

Bellis T, Billiet C, Ross J (2008). Hemispheric lateralization of bilaterally presented homologous visual and auditory stimuli in normal adults, normal children, and children with central auditory dysfunction. Brain Cogn 66: 280–289.

Bellis T, Billiet C, Ross J (2011). The utility of visual analogs of central auditory tests in the differential diagnosis of (central) auditory processing disorder and attention deficit hyperactivity disorder. J Am Acad Audiol 22 (8): 501–514.

Bellis T, Chermak G, Weihing J et al. (2012). Efficacy of auditory interventions for central auditory processing disorder: a response to Fey et al. (2011). Lang Speech Hear Serv Sch 43: 381–385.

Billiet C, Bellis T (2011). The relationship between brainstem temporal processing and performance on tests of central auditory function in children with reading disorders. J Speech Lang Hear Res 54: 228–242.

Boscariol M, Garcia VL, Guimaraes CA et al. (2009). Auditory processing disorders in twins with perisylvian polymicrogyria. Arq Neuropsiquiatr 67 (2B): 499–501.

Boscariol M, Garcia VL, Guimaraes CA et al. (2010). Auditory processing disorder in perisylvian syndrome. Brain Dev 32 (4): 299–304.

Boscariol M, Guimaraes CA, Hage SRV et al. (2011). Auditory processing disorder in patients with language-learning impairment and correlation with malformation of cortical development. Brain Dev 33 (10): 82–831.

British Society of Audiology (2011). Auditory processing disorder [position statement]. Available online at, http:/www.thebsa.org.uk/docs/docsfromold/BSA_APD_PositionPaper_31March11_FINAL.pdf, Retrieved: May 15, 2013.

Cacace A, McFarland D (2005). The importance of modality specificity in diagnosing central auditory processing disorder. Am J Audiol 14: 112–123.

Calvert GA, Bullmore ET, Brammer MJ et al. (1997). Activation of auditory cortex during silent lipreading. Science 276: 593–596.

Cameron S, Dillon H (2008). The listening in spatialized noise-sentences tests (LISN-S): comparison to the prototype LISN and results from children with either a suspected (central) auditory processing disorder or a confirmed language disorder. J Am Acad Audiol 19: 377–391.

Cestnick L, Jerger J (2000). Auditory temporal processing and lexical/nonlexical reading in developmental dyslexia. J Am Acad Audiol 11: 501–513.

Chermak GD (1998). Managing central auditory processing disorders: metalinguistic and metacognitive approaches. Semin Hear 19: 379–392.

Chermak GD, Musiek FE (1997). Central auditory processing disorders. New perspectives, Singular, San Diego.

Chermak GD, Musiek FE (Eds.), (2007). Handbook of (central) auditory processing disorder: Vol. 2. Comprehensive intervention, Plural Publishing, San Diego.

Chermak GD, Musiek FE (2011). Neurological substrate of central auditory processing deficits in children. Curr Pediatr Rev 7 (3): 241–251.

Chermak GD, Musiek FE (Eds.) (in press). Handbook of central auditory processing disorder: Vol. 2. Comprehensive intervention. 2nd edn. Plural Publishing, San Diego, CA.

Chermak GD, Bellis TJ, Musiek FE (2013 in press). Neurobiology, cognitive science, and intervention (2nd edn). In: GD Chermak, FE Musiek (Eds.), Handbook of central auditory processing disorder: Vol. 2: Comprehensive intervention, Plural Publishing, San Diego, CA.

CISGSLPA (2013). Canadian guidelines on auditory processing disorder in children and adults: Assessment and intervention. Available online at, http://www.cshhpbc.org/docs/canadian_guidelines_on_auditory_processing_disorder_in_children_and_adults_english_final_2012.pdf, Retrieved: May 13, 2013.

Cooper JC, Gates GA (1991). Hearing in the elderly – the Framingham cohort, 1983–1985. Part II. Prevalence of central auditory processing disorders. Ear Hear 12: 304–311.

deVillers-Sidani E, Alzghoul L, Zhou X et al. (2010). Recovery of functional and structural age-related changes in the rat primary auditory cortex with operant training. Proc Natl Acad Sci 107: 13900–13905.

Dillon H, Cameron S, Glyde H et al. (2012). An opinion on the assessment of people who may have an auditory processing disorder. J Am Acad Audiol 23: 97–105.

Filippini R, Schochat E (2009). Brainstem evoked auditory potentials with speech stimulus in the auditory processing disorder. Braz J Otorhinolaryngol 75 (3): 449–455.

Foxton J, Brown A, Chambers S et al. (2004). Training improves acoustic pattern perception. Curr Biol 14 (4): 322–325.

Galbraith G, Arbagey P, Branski R et al. (1995). Intelligible speech encoded in the human brainstem frequency following response. NeuroReport 6: 2363–2367.

Gallun FJ, Diedesch AC, Kubli LR et al. (2012). Performance on tests of central auditory processing by individuals exposed to high-intensity blasts. J Rehabil Res Dev 49 (7): 1005–1024.

Golding M, Carter N, Mitchell P et al. (2004). Prevalence of central auditory processing (CAP) abnormality in older Australian population: The Blue Mountains hearing study. J Am Acad Audiol 15: 633–642.

Hayes E, Warrier C, Nicol T et al. (2003). Neural plasticity following auditory training in children with learning problems. Clin Neurophysiol 114: 673–684.

Hornickel J, Stecker SG, Bradlow AR et al. (2012). Assistive listening devices drive neuroplasticity in children with dyslexia. Proc Natl Acad Sci 109: 16731–16736.

Jancke L, Gaab N, Wustenberg H et al. (2001). Short-term functional plasticity in the human auditory cortex: an fMRI study. Cogn Brain Res 12: 479–485.

Jerger J, Musiek F (2000). Report of the consensus conference on the diagnosis of auditory processing disorders in school-aged children. J Am Acad Audiol 11: 467–474.

Johnson K, Nicol T, Kraus N (2005). Brainstem response to speech: a biological marker of auditory processing. Ear Hear 26 (5): 424–434.

Johnson KL, Nicol T, Zecker SG et al. (2008). Developmental plasticity in the human auditory brainstem. J Neurosci 28: 4000–4007.

Katz J (1983). Phonemic synthesis. In: E Lasky, J Katz (Eds.), Central auditory processing disorders: Problems of speech, language, and learning, University Park Press, Baltimore, pp. 269–272.

Katz J, Stecker N, Henderson D (1992). Central auditory processing: A transdisciplinary view, Mosby Year Book, St. Louis.

Kompus K, Specht K, Ersland L et al. (2012). A forced-attention dichotic listening fMRI study on 113 subjects. Brain Lang 121 (3): 240–247.

Kraus N (2012). Biological impact of music and software-based auditory training. J Comm Disord 45: 403–410.

Kraus N, McGee T, Carrell T et al. (1996). Auditory neurophysiologic responses and discrimination deficits in children with learning problems. Science 273 (5277): 971–973.

Malloy L, Bellis T (2013) The effect of recurrent impacts and impulses on electrophysiologic measures in collegiate football players. Poster presented at the American Academy of Audiology annual conference, Anaheim, CA.

Marques C, Moreno S, Castro S et al. (2007). Musicians detect pitch violation in a foreign language better than nonmusicians: Behavioral and electrophysiological evidence. J Cogn Neurosci 19: 1453–1463.

Martin E, Lu W, Helmick K et al. (2008). Traumatic brain injuries sustained in the Afghanistan and Iraq wars. J Trauma Nurs 15 (3): 94–99.

McArthur G, Ellis D, Atkinson C et al. (2008). Auditory processing deficits in children with reading and language impairments: can they (and should they) be treated? Cognition 107 (3): 946–977.

Moncrieff D, Wertz D (2008). Auditory rehabilitation for interaural asymmetry: preliminary evidence of improved dichotic listening performance following intensive training. Int J Audiol 47 (2): 84–97.

Moore D, Rosen S, Bamiou D et al. (2013). Evolving concepts of developmental auditory processing disorder (APD): a British society of audiology APD special interest group 'white paper.' Int J Audiol 52: 3–13.

Moreno S, Marques C, Santos A et al. (2009). Musical training influences linguistic abilities in 8-year old children: more evidence for brain plasticity. Cereb Cortex 19: 712–723.

Musacchia G, Sams M, Nicol T et al. (2006). Seeing speech affects acoustic information processing in the human brainstem. Exp Brain Res 168: 1–10.

Musiek FE (1983). Assessment of central auditory dysfunction: The Dichotic Digits Test revisited. Ear Hear 4: 79–83.

Musiek FE (1999). Habilitation and management of auditory processing disorders: overview of selected procedures. J Am Acad Audiol 10: 329–342.

Musiek F, Chermak G (1995). Three commonly asked questions about central auditory processing disorders: management. Am J Audiol 4: 15–18.

Musiek FE, Chermak G (Eds.), (2007). Handbook of central auditory processing disorder: Volume 1. Diagnosis, 2nd edn. Plural Publishing Inc., San Diego.

Musiek F, Gollegly K (1988). Maturational considerations in the neuroauditory evaluation of children. In: F Bess (Ed.), Hearing impairment in children, York Press, Maryland, pp. 231–252.

Musiek F, Weihing J (2011). Perspective on dichotic listening and the corpus callosum. Brain Cogn 76: 225–232.

Musiek F, Gollegly K, Baran J (1984). Myelination of the corpus callosum and auditory processing problems in children: theoretical and clinical correlates. Semin Hear 5: 231–241.

Musiek F, Gollegly K, Ross M (1985). Profiles of types of auditory processing disorders in children with learning disabilities. J Child Commun Disord 9: 43.

Musiek F, Kurdzeil-Schwan S, Kibbe K et al. (1989). The dichotic rhyme task: results in split-brain patients. Ear Hear 10: 33–39.

Musiek F, Bellis T, Chermak G (2005). Nonmodularity of the central auditory nervous system: implications for (central) auditory processing disorder. Am J Audiol 14: 128–138.

Musiek F, Weihing J, Lau C (2008). Dichotic interaural intensity difference (DIID) training: a review of existing research and future directions. J Acad Rehab Audiol 41: 51–65.

Myklebust H (1954). Auditory disorders in children, Grune and Stratton, New York.

Okie S (2005). Traumatic brain injury in the war zone. N Engl J Med 352: 2043–2047.

Pantev C, Herholz S (2011). Plasticity of the human auditory cortex related to muscical training. Neurosci Biobehav Rev 35: 2140–2154.

Parbery-Clark A, Skoe E, Lam C et al. (2009). Musician enhancement for speech-in-noise. Ear Hear 30: 653–661.

Peelle J, Trojani V, Wingfield A et al. (2010). Neural processing during older adults' comprehension of spoken sentences: age differences in resource allocation and connectivity. Cereb Cortex 20: 773–782.

Pichora-Fuller M, Schneider B, Daneman M (1995). How young and old adults listen to and remember speech in noise. J Acoust Soc Am 97: 593–608.

Pinheiro M, Ptacek P (1971). Reversals in the perception of noise and tone patterns. J Acoust Soc Am 49 (6): 1778–1883.

Ptacek P, Pinheiro M (1971). Pattern reversal in auditory perception. J Acoust Soc Am 49: 493–498.

Purdy SC, Kelly AS, Davies MG (2002). Auditory brainstem response, middle latency response, and late cortical evoked potentials in children with learning disabilities. J Am Acad Audiol 13 (7): 367–382.

Russo N, Nicol T, Zecker S et al. (2005). Auditory training improves neural timing in the human brainstem. Behav Brain Res 156: 95–103.

Sams M, Aulanko R, Hamalainen M et al. (1991). Seeing speech: visual information from lip movements modifies activity in the human auditory cortex. Neurosci Lett 127: 141–145.

Schochat E, Musiek F, Alonso R et al. (2010). Effect of auditory training on the middle latency response in children with (central) auditory processing disorder. Braz J Med Biol Res 43 (8): 777–785.

Schon D, Magne C, Besson M (2004). The music of speech: music training facilitates pitch processing in both music and language. Psychophysiology 41: 341–349.

Sharma M, Purdy SD, Newall P et al. (2006). Electrophysiological and behavioral evidence of auditory processing deficits in children with reading disorder. Clin Neurophysiol 117: 1130–1144.

Sharma M, Purdy S, Kelly A (2009). Comorbidity of auditory processing, language, and reading disorders. J Speech Lang Hear Res 52: 706–722.

Sharma M, Purdy S, Kelly A (2012). A randomized control trial of interventions in school-aged children with auditory processing disorders. Int J Audiol 51: 506–518.

Shinn J, Baran J, Moncrieff D et al. (2005). Differential attention effects on dichotic listening. J Am Acad Audiol 16 (4): 205–218.

Sloan C (1995). Treating auditory processing difficulties in children, Singular, San Diego.

Strait D, Kraus N (2011). Can you hear me now? Musical training shapes functional brain networks for selective auditory attention and hearing speech in noise. Front Psychol 2: 113.

Tremblay K (2007). Training-related changes in the brain: evidence from human auditory evoked potentials. Semin Hear 28: 120–132.

Tremblay K, Kraus N (2002). Beyond the ear: central auditory plasticity. Otorinolaringologia 52: 93–100.

Tremblay K, Kraus N, McGee T et al. (2001). Central auditory plasticity: changes in the N1-P2 complex after speech-sound training. Ear Hear 22 (2): 79–90.

Watson C, Kidd G (2002). On the lack of association between basic auditory abilities, speech processing, and other cognitive skills. Semin Hear 23: 83–93.

Watson C, Kidd G (2009). Associations between auditory abilities, reading, and other language skills in children and adults. In: T Cacace, D McFarland (Eds.), Current controversies in central auditory processing disorder (CAPD), Plural Publishing, San Diego, CA, pp. 217–242.

Weihing J, Schochat E, Musiek F (2012a). Ear and electrode effects reduce within-group variability in middle latency response amplitude measures. Int J Audiol 51 (5): 405–412.

Weihing J, Bellis T, Chermak G et al. (2012b). Current issues in the diagnosis and treatment of CAPD in children. In: D Geffner, D Swain (Eds.), Auditory processing disorders: Assessment, management, and treatment, Plural Publishing, San Diego, CA, pp. 3–32.

World Health Organization (2002). Towards a common language for functioning, disability, and health: ICF. World Health Organization, Geneva.

Wright B, Bowen R, Zecker S (2000). Nonlinguistic deficits associated with reading and language disorders. Curr Opin Neurobiol 10 (4): 482–486.

Handbook of Clinical Neurology, Vol. 129 (3rd series)
The Human Auditory System
G.G. Celesia and G. Hickok, Editors

Chapter 31

Auditory neglect and related disorders

ALEXANDER GUTSCHALK* AND ANDREW DYKSTRA
Department of Neurology, Ruprecht-Karls-Universität Heidelberg, Heidelberg, Germany

INTRODUCTION

Clinical presentation of neglect

Neglect is a neurologic syndrome that is frequently observed after right medial cerebral artery (MCA) stroke and other right-hemisphere lesions. Patients with acute, full-blown neglect are easily recognized by their gaze deviation towards the ipsilesional – typically right – side (Becker and Karnath, 2010), and their lack of awareness of contralesional – typically left-sided – events in their environment. When addressed from the left, patients with neglect often orient wrongly towards their right side. Lack of contralesional awareness may also include the patient's own body such that, even in the absence of hemiplegia, they may cease to move their left arm and leg. Finally, patients are often unaware of their condition and its severity, a symptom that is known as anosognosia. A comprehensive review of the general clinical aspects of neglect is outside the scope of the current chapter, and we refer the reader to one of the many review papers and textbook chapters on the topic (Heilman et al., 2003; Husain, 2008; Corbetta and Shulman, 2011).

While early reports of neglect symptoms date back to the 19th century (Oppenheim, 1885), the right-hemisphere dominance of the syndrome was not generally acknowledged until later (Brain, 1941). Since then, many studies have evaluated the perceptual manifestations of neglect, especially in vision. In bedside tests, the most severely affected patients may be completely unaware of stimuli presented to the neglected visual hemifield, making it difficult to decide whether hemianopia is present in addition to hemispatial neglect (Husain, 2008). Detecting subtler cases of visual neglect may require more formal examination. So-called cancellation tests, in which patients are required to denote certain targets (e.g., the letter "A" (Weintraub and Mesulam, 1985) or a schematic bell (Gauthier et al., 1989)) amongst a cluttered array, have proven particularly useful: the more severe the neglect, the more target items on the left, contralesional side will be missed (Rorden and Karnath, 2010). Another option to detect less severe deficits is the test introduced by Posner: here, patients are cued to the left or right before the presentation of a visual target, and patients suffering from mild neglect typically show prolonged reaction time for targets presented on the left, most prominently when invalidly cued to the right side (Posner et al., 1984). The Posner task is more time consuming, but also more sensitive than cancelation tasks in chronic and less severely affected patients (Rengachary et al., 2009).

While visual deficits have clearly been the focus of neglect research to date, neglect is a multimodal disorder and similarly involves the auditory and somatosensory domains. We use the term "auditory neglect" in this chapter to refer to the auditory manifestations of a supramodal neurologic disorder (i.e., neglect), and do not mean to imply the existence of a unique disorder that manifests itself only in the auditory domain. Most auditory signs that are observed in neglect are lateralized to the left hemifield. The first of these signs is auditory extinction (Heilman and Valenstein, 1972), though it should be noted that unimodal auditory extinction (Fig. 31.1) can also be caused by lesions to auditory cortex. The second classic sign of auditory neglect is disruption of spatial perception with reduced accuracy of spatial perception on the left and a bias of sound lateralization towards the right. In our view, however, the most distinctive sign of auditory neglect is alloacusis (Brain, 1941), where patients attribute the source of all sounds to their right. More recently, it has been shown that neglect is also associated with non-spatial auditory deficits, in particular in tasks that require sustained auditory attention (Robertson et al., 1997), or the comparison

*Correspondence to: Alexander Gutschalk, Department of Neurology, University of Heidelberg, Im Neuenheimer Feld 400, 69120 Heidelberg, Germany. Tel: +49-6221-56 36811, Fax: +49-6221-56 5258, E-mail: Alexander.Gutschalk@med.uni-heidelberg.de

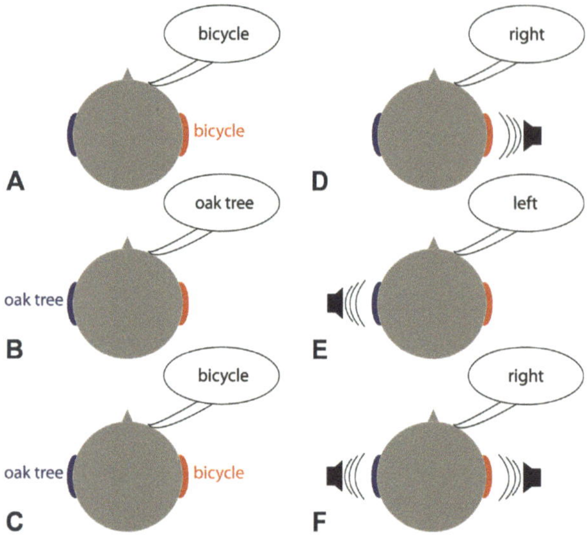

Fig. 31.1. Schematic of auditory extinction. Patients with left-sided auditory extinction can generally report single words presented to the right (**A**) or left (**B**), but report only the right-ear word with dichotic stimulation (**C**). Similarly, patients are able to locate single sounds to the left or right ear, but locate synchronous left- and right-ear stimuli to the right.

of subsequently presented sounds (Cusack et al., 2000). Maybe with the exception of alloacusis, none of these auditory deficits is sufficiently specific for neglect, but alloacusis is more typical in severely affected, acute cases and subsides in most patients after a while. As we suggest below, the combination of auditory extinction or auditory spatial bias and impaired sustained auditory attention might provide a more specific clinical definition of neglect in the auditory domain, which can also be applied to patients with chronic or less severe forms of neglect where alloacusis has subsided or is not present.

Extinction

Extinction is a potential clinical indication of neglect, although the exact relationship between the two is disputed: patients who show sensory extinction are still able to detect and identify single stimuli presented either to the left or right side, but often fail to detect a left-sided (i.e., contralesional) stimulus when it is presented simultaneously with a right-sided, ipsilesional stimulus. Two different sets of procedures and instructions are typically used to test extinction: one is to use similar stimuli on both sides and ask the patient from which direction the stimulus was perceived. Alternatively, two different stimuli may be presented and the patient is then asked to identify the perceived stimuli (Fig. 31.1).

Extinction can be observed with visual, auditory, and somatosensory stimulation (Oppenheim, 1885), and can also be observed between modalities, such that a right-sided visual stimulus may extinguish a left-sided auditory stimulus (Diamond and Bender, 1965). Some authors consider extinction a milder clinical presentation of neglect that occurs in patients with smaller lesions or where other symptoms of neglect may have subsided over time (Hier et al., 1983; Posner et al., 1984; Rengachary et al., 2011). Others have suggested that extinction is a distinct syndrome that may often co-occur with neglect, but that visual neglect and extinction can be doubly dissociated (Vallar et al., 1994; Vossel et al., 2011). A study that investigated the incidence of visual neglect and extinction in patients with acute MCA stroke found that both are clearly right-hemisphere syndromes and that visual extinction is not observed more frequently with left-sided lesions than neglect (Becker and Karnath, 2007). In their study, 26% of acute right-hemisphere stroke patients ($n=42$) showed signs of visual neglect and 24% showed visual extinction; in left-hemisphere patients ($n=51$) the percentage was 2.4% and 4.9%, respectively. As we will see in the next section, the situation is somewhat more obscure in the auditory domain, where extinction can also be observed after lesions of auditory cortex in the absence of accompanying clinical signs of neglect (Gutschalk et al., 2012).

AUDITORY BEHAVIORAL DEFICITS IN NEGLECT AND SIMILAR DISORDERS

Auditory extinction

An early report of auditory extinction by Oppenheim described a patient with left MCA stroke and good hearing in both ears who did not detect the ticking of a watch held next to her right ear when another watch was simultaneously held next to her left ear (Oppenheim, 1885). Auditory extinction was also observed in a series of patients in which small bells, whistles, jangling keys, finger snapping, and other small objects were used to create different sounds at the left and right ears while listeners were asked to identify the sounds (Diamond and Bender, 1965). The authors also reported that extinction of left-sided sound by right-sided visual or somatosensory stimulation was especially prevalent in patients with neglect, which they termed "left hemispatial disorientation syndrome." The first study to provide detailed localization information regarding the lesion sites typically associated with neglect used technetium-99 scans in 17 patients with auditory extinction and other signs of clinical neglect; 9 patients had lesions in the right inferior parietal lobe, one in the left frontal lobe, and 7 had lesion sites that could not be determined (Heilman and

Valenstein, 1972). The potential involvement of damage to the inferior parietal lobe in neglect was indirectly confirmed in four rhesus monkeys showing auditory extinction after removal of the lower parietal lobe and posterior superior temporal gyrus (STG) (Heilman et al., 1971). Note, however, that the possible impact of the STG lesion on auditory cortex was not discussed by the authors. A later study reported that the frequency of auditory extinction was not significantly different between acute left- and right-hemisphere lesions, but persisted more frequently in right-hemisphere lesions when the same patients were evaluated again after 30 days or more (De Renzi et al., 1984). The authors studied 144 patients (69 right- and 75 left-hemisphere strokes) and found contralateral extinction in 45% of right- and 36% of left-hemisphere patients. A number of patients also showed ipsilesional extinction, the source of which is unclear at this point.

Especially in left-hemisphere patients, but also in some right-hemisphere patients, auditory extinction was not generally coupled to neglect or visual extinction. CT scans showed that in most cases of auditory extinction associated with cortical lesions, auditory cortex and parietal cortex were involved (compare illustration in Fig. 31.2). Subcortical lesions associated with persistent auditory extinction consistently extended into the sublenticular portion of the internal capsule, likely destroying major parts of the auditory thalamocortical radiation. The authors therefore concluded that extinction is, at least in a subset of patients, caused by an imbalance within the afferent auditory pathway, and may therefore be a sensory phenomenon (De Renzi et al., 1984).

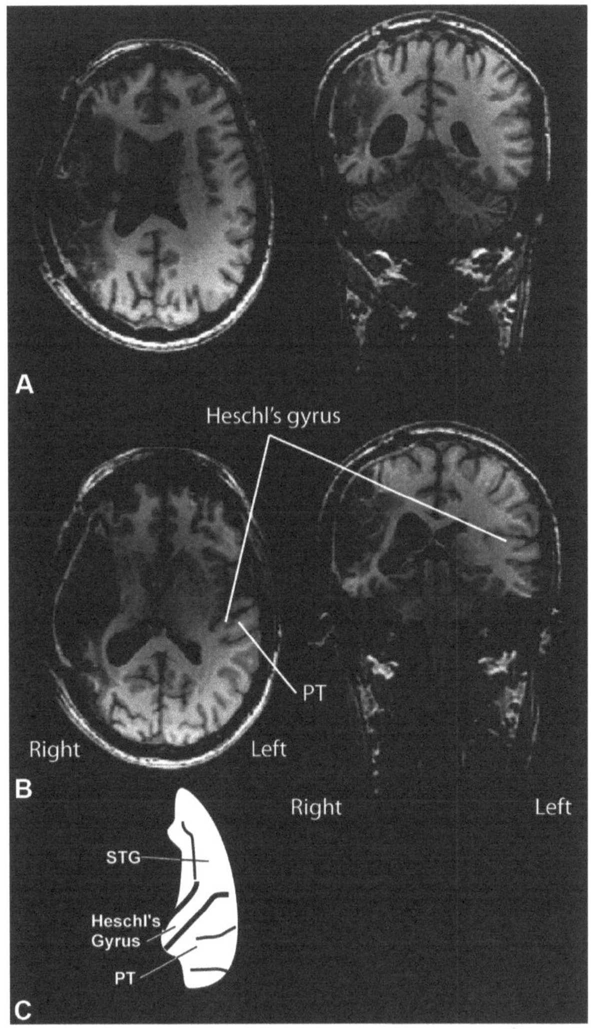

Fig. 31.2. Magnetic resonance imaging of a patient with auditory neglect 1 year after a subtotal right medial cerebral artery (MCA) stroke (the patient had received hemicraniectomy for malignant MCA infarction). The lesion comprises substantial parts of the lower parietal lobe (**A**), parts of prefrontal cortex, and the superior temporal plane. As in many patients with neglect, the auditory cortex (**B**) is also affected, whereas the auditory cortex in the left hemisphere is still intact. Major parts of primary auditory cortex reside in Heschl's gyrus (**B, C**), secondary areas are in planum temporale (PT) and the posterior end of the superior temporal gyrus (STG). Auditory test results of this patient have been previously reported (Gutschalk et al., 2012).

Auditory extinction in dichotic speech tests

Dichotic speech tests (see Chapter 18), in which listeners are asked to identify different words, spoken digits, or phrases presented independently to the left and right ears via headphones, were introduced with the availability of sound recordings (Broadbent, 1952; Cherry, 1953). In the context of functional neuroanatomy, these tests were first applied in a population of patients with temporal-lobe lesions who showed contralesional extinction that was attributed to involvement of the auditory pathway (Kimura, 1961; Schulhoff and Goodglas, 1969). Dichotic speech tests have more recently been used as a diagnostic tool for auditory neglect (Hugdahl et al., 1991; Jacquin-Courtois et al., 2010), sparking a discussion as to whether auditory extinction in dichotic speech tests can be used to diagnose auditory neglect (Beaton and McCarthy, 1993; Hugdahl and Wester, 1994). In an attempt to make the dichotic speech test more specific for neglect, as opposed to auditory cortex lesions, it has been suggested to lateralize the two speech elements by interaural time differences (ITD) instead of by ear (Bellmann et al., 2001; Spierer et al., 2007), based on the contralateral processing bias for monaural stimuli. These authors showed that extinction in ITD-lateralized tests was not observed in all patients who demonstrated extinction in classic dichotic tests, and suggested that extinction in the ITD-lateralized test indicated a deficit more specific to spatial representation, supposedly explained at a processing level beyond auditory cortex and more specifically related to neglect. This partial dissociation was first demonstrated in 4 patients (Bellmann et al., 2001) and then in a series of 15 more patients. However, when a similar ITD-based test was applied to a patient sample that included patients with auditory cortex lesions with and without neglect, dichotic and ITD-lateralized tests were found to be highly correlated (Gutschalk et al., 2012). The latter study demonstrated that auditory extinction can be observed in patients with auditory cortex lesions of either hemisphere (Fig. 31.3), both with and without neglect, as well as in patients with neglect in whom auditory cortex is intact. Thus, while the dissociation of monaural and ITD-lateralized dichotic tests remains of interest for further research (Fig. 31.4), its diagnostic value is currently unclear.

A third group of patients who show extinction in dichotic speech tests have lesions affecting the transcallosal tract connecting the auditory cortices on the left and right. Patients with left-hemisphere speech dominance with such lesions show extinction of left-ear speech, as was first shown in a patient who had undergone callosotomy for epilepsy (Sparks and Geschwind, 1968). Note that lesions of the transcallosal pathway may occur in the left, right, or middle, and that the position of the lesion has no effect on the side of extinction in dichotic speech test. This is thought to be due to a stronger contralateral processing bias in the presence of competing dichotic stimuli such that information presented to the

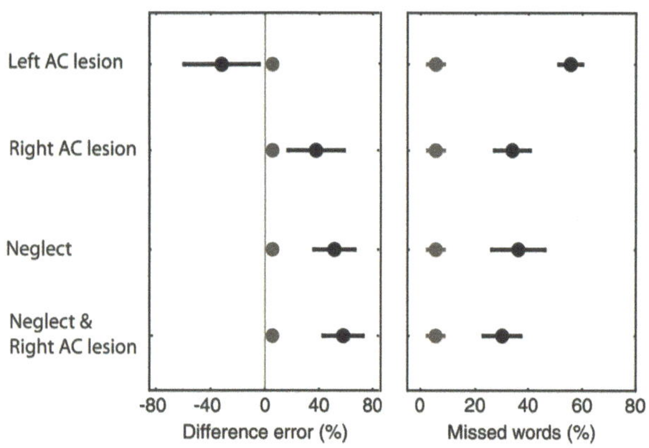

Fig. 31.3. Auditory extinction in dichotic speech test. Listeners reported the words they heard in the left and right ear (mean ± standard error). Control data ($n=25$) are plotted in gray; the patient data are plotted in black. Patients either had a lesion of the left or right auditory cortex (AC), or they were included because visual neglect was present. The neglect group was again separated into patients with and without a lesion of the right auditory cortex ($n=7$ in each of the four groups). Listeners generally extinguished more words contralateral to the lesion (difference error, left panel). However, 2 patients in the left AC group showed ipsilesional extinction (see Fig. 31.4), reducing the mean effect. The overall error rate (right panel) was higher in patients with left-sided lesions, probably because some aphasia was present. There was otherwise no statistical difference in the amount of extinction between patient groups. (Modified from Gutschalk et al., 2012.)

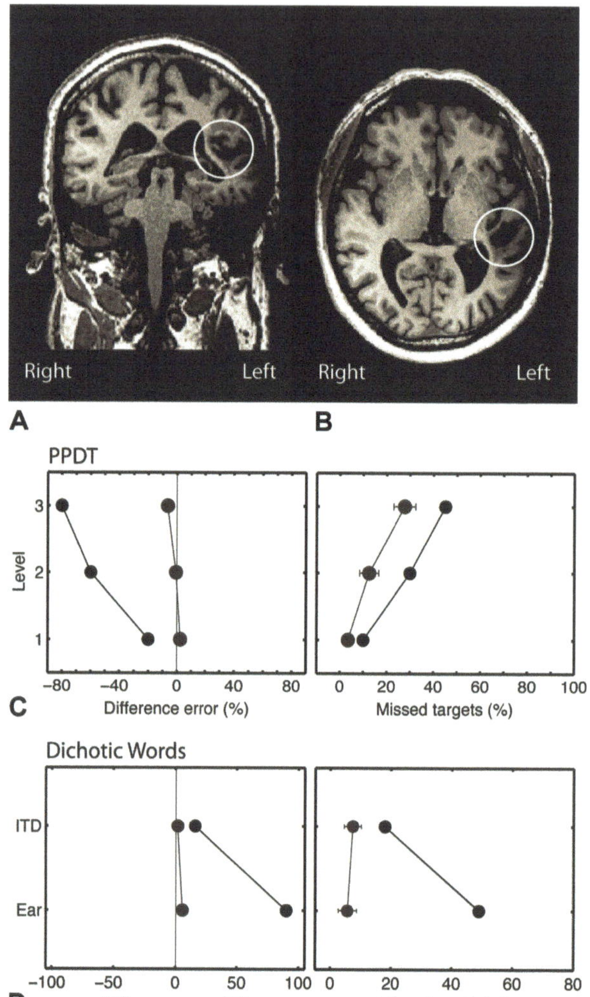

Fig. 31.4. Patient with an infarction in the left medial cerebral artery territory that destroyed (**A**) part of medial Heschl's gyrus (including the primary auditory cortex) and adjacent planum temporale (Gutschalk et al., 2012). The subcortical part of the lesion extends to the lateral ventricle and up to the parietal lobe, thereby dissecting the transcallosal auditory pathway. (**B**) Lateral parts of the left auditory cortex remain unaffected. (**C**) The patient missed many contralateral, right-ear targets in the psychoacoustic pattern discrimination test (PPDT) (see Fig. 31.5). (**D**) In contrast, extinction for words was ipsilateral to the lesion in a dichotic setup and only mildly affected in an interaural time difference-based setup. Thus, there is a double dissociation of a contralateral deficit in the PPDT, supposedly related to the auditory cortex lesion, and an ispilateral deficit in the dichotic word test, supposedly related to the lesion of the transcallosal auditory pathway.

left ear is processed initially in the right hemisphere and, due to the lesion to the transcallosal tract, is not transferred to the left hemisphere's speech regions. Ipsilesional extinction, most frequently after left-sided lesions of the transcallosal auditory pathway, has been termed "paradoxical extinction" (Sparks and Geschwind, 1968; Damasio and Damasio, 1979). The latter authors studied 20 patients, of which 13 showed extinction. Four of 8 patients with lesions in the left hemisphere, and 1 of 5 patients with right-hemisphere lesions showed ipsilesional extinction. Notably, the one right-hemisphere patient with ipsilesional extinction was shown to have right-hemisphere dominance for language in the Wada test.

When there is an additional lesion of the auditory cortex, patients with lesions of the transcallosal pathway show left-ear extinction in speech and right-ear deficits in non-speech dichotic tests (Gutschalk et al., 2012), a pattern that was found in 2 of 7 patients with left auditory-cortex lesions (Fig. 31.4). This double dissociation of speech and non-speech auditory extinction clearly demonstrates that auditory extinction is a symptom that can be caused by at least two independent underlying pathologies and that auditory extinction is more heterogeneous than visual extinction.

In summary, dichotic speech tests are sensitive and easy to apply, but they are usually not sufficient to differentiate neglect from other conditions. In particular, patients with right-hemisphere lesions may show contralesion extinction due to an auditory cortex lesion, neglect, a transcallosal lesion, or any combination thereof. In left-hemisphere lesions, the test may be limited by aphasia, and even patients with mild aphasia (as those shown in Fig. 31.3) have overall higher error rates on both sides.

Extinction and temporal order

In patients with auditory extinction, a perceived time delay has been shown for contralesional stimuli: when listeners are required to adjust the time delay between left and right ear tones such that they hear the two synchronously, they typically adjust the contralesional stimulus to be presented earlier by 50–300 ms (Karnath et al., 2002; Sinnett et al., 2007; Barrett et al., 2010), as has been demonstrated in overall 13 patients with right-hemisphere lesions.

This comparison requires that listeners actually perceive both stimuli. The relationship between onset asynchrony and extinction has been evaluated in more detail in a study that showed that temporal asynchrony can mitigate extinction (Witte et al., 2012). For right-hemisphere lesions ($n = 17$), contralesional stimuli were typically extinguished irrespective of which sound was presented first. A different pattern was observed for left-hemisphere lesions ($n = 17$), where extinction was more frequently observed for the trailing sounds, such that ipsilesional extinction was observed more frequently in this case. Since the patient selection did not control for lesions of the auditory cortex or underlying white matter, it is conceivable that the pattern for left-hemisphere cases could primarily reflect lesions of auditory cortex, whereas the pattern in right-hemisphere lesions could reflect auditory extinction in a predominantly multimodal context, although most patients did not show clinical signs of visual neglect at the time of the investigation.

Target detection tasks

Dichotic auditory target detection tasks have been applied in order to diagnose either auditory neglect (De Renzi et al., 1989) or lesions of the telencephalic auditory pathway (Blaettner et al., 1989): A target detection task proposed to be specific for neglect is the detection of 8–20-ms-long gaps in an ongoing, 4-minute-long pure tone presented via headphones (De Renzi et al., 1989). The test was first presented in a monaural variant, where all gaps occurred in the same ear. Thereafter, the tone was presented in both ears, while the interruption occurred randomly in either the left or right ear. Overall, 30 patients with right- and 15 patients with left-hemisphere lesions of various etiologies were tested. Predominantly contralesional misses were observed for nine right-hemisphere lesions, particularly when the stimuli were dichotic, but also in some cases for monaural stimuli. Most patients also showed auditory extinction, but 7 more patients with right-hemisphere lesions showed auditory extinction and normal target detection. Moreover, 2 patients without auditory extinction who showed auditory neglect as defined by the gap detection task showed no signs of visual neglect (although one showed visual and tactile extinction). An unpublished replication of their test using 200-ms-long gaps in ongoing noise confirmed a high rate of missed gaps in patients with neglect; however, this deficit was not strongly lateralized (Leifert-Fiebach, 2008).

Another dichotic target detection task was conceived to study auditory telencephalic hearing disorders. In this psychoacoustic pattern discrimination test (PPDT), participants must detect monaural, randomly presented intensity (Fig. 31.5A) or duration deviants amongst a stream of uniform dichotic noise bursts or click trains (Blaettner et al., 1989). The authors studied 62 patients who had suffered a stroke at least 3 months ago (42 in the left and 15 in the right hemisphere) and 48 controls. Contralesional misses in the PPDT were associated with anatomic lesions of the auditory cortex

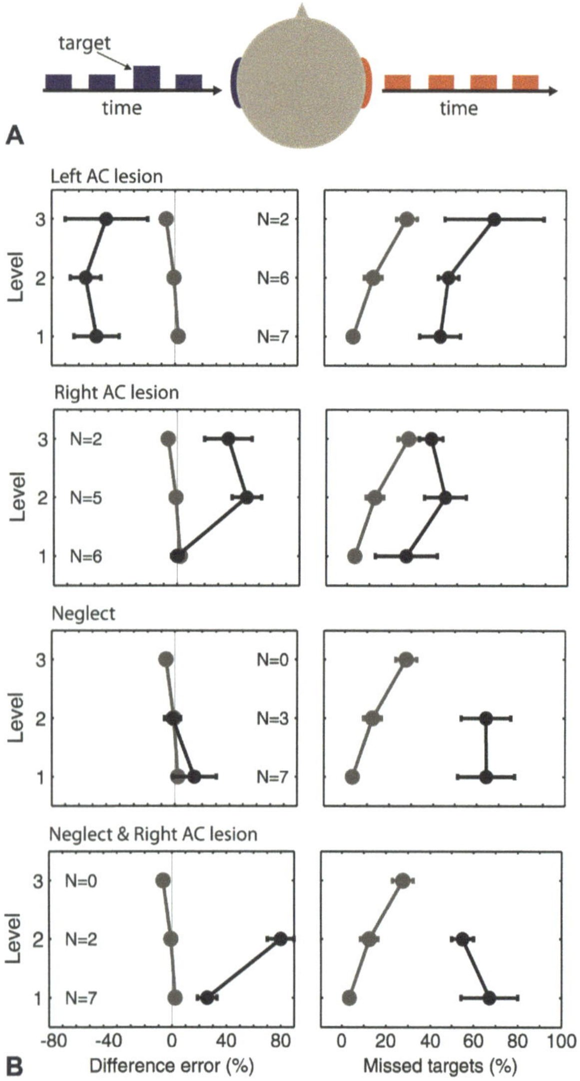

Fig. 31.5. (A) Schematic of the psychoacoustic pattern discrimination test (Blaettner et al., 1989). Every 600 ms, 200-ms noise bursts are presented to both ears. Targets, where the level is 10, 6, or 4 dB higher in one ear, occur randomly after 6–10 repetitions. Listeners are instructed to press a response button as soon as they detect a target among the stream of standards. (**B**) Data of the control group ($n=26$) are plotted in gray; patient data are plotted in black (mean ±standard error). The number of subjects is indicated separately for each level because the procedure adapts to patient performance. Patients with auditory cortex (AC) lesions show predominantly contralateral deficits (difference error, left column). Patients with neglect showed a higher error rate (missed targets, right column), but no significant lateralization when their AC was intact. (Modified from Gutschalk et al., 2012.)

and thalamocortical radiation in cranial computed tomography (CCT) and auditory evoked potentials. Of the 46 patients who had such a lesion in CCT, 74% were correctly classified based on the PPDT test score.

The association of lateralized deficits in the PPDT with auditory cortex lesions was recently confirmed in a study that also applied the test to a group of patients with clinical signs of neglect (Gutschalk et al., 2012). While 13 of 14 patients with neglect had basic difficulty performing the task and missed many targets on both sides, their deficit was consistently lateralized to the contralesional ear only when a lesion involving the auditory cortex was additionally present (Fig. 31.5B). Patients with auditory cortex lesions who did not show clinical signs of neglect only showed such general problems performing the task in 2 of 14 cases (one left- and one right-hemisphere lesion).

Non-spatial auditory deficits in neglect

A general deficit of sustained auditory attention has also been found with a non-lateralized auditory target detection test (Robertson et al., 1997). In these so-called "elevator counting tests" (Wilkins et al., 1987), listeners are required to count the numbers of tones in subsequently presented sound sequences, a task that is very simple for normal listeners, but which produced considerable problems for the patients showing signs of visual neglect. A correlation between visual neglect tests and auditory target detection was evaluated in a group of 44 patients who had a right-hemisphere stroke the previous year. Taken together with the results discussed earlier, it seems likely that problems with auditory target detection constitute a non-spatial auditory deficit of neglect, whereas strongly lateralized deficits, at least in the PPDT (Blaettner et al., 1989; Gutschalk et al., 2012), could indicate additional lesions to auditory cortex.

However, there is another auditory deficit that has been observed in patients with visual neglect that is not readily explained by the above observations (Cusack et al., 2000). This study showed that, while patients with signs of visual neglect ($n=9$) performed well for tasks in which a sound feature can be identified within a single interval, they are severely impaired for tasks that require the comparison of two or more sounds. The authors interpreted their result to indicate a deficit of allocating attention between auditory objects. By using different time intervals between sounds, they could show that this deficit is not explained by the prominently prolonged duration of the attentional blink observed for visual target detection in patients with neglect (Husain et al., 1997). In spite of the fact that these tests showed such large discriminability between patients and controls, these non-spatial auditory tests have, to the best of our knowledge, not yet been utilized in clinical settings.

Alloacusis

Alloacusis is a unique clinical sign of auditory neglect that can be easily assessed in bedside tests alongside auditory extinction: patients showing alloacusis attribute the source of all sounds to the ipsilesional hemifield. Thus, the detection of alloacusis requires asking the patient "where" a stimulus is perceived. The answer in patients with right-hemisphere lesions is uniformly "on the right," irrespective of whether the sound is presented to the left ear, right ear, or both ears simultaneously.

Alloacusis was first reported in the 1940s (Brain, 1941), and has since been replicated in more than 15 patients who also showed other signs of neglect (Denny-Brown et al., 1952; Diamond and Bender, 1965; Soroker et al., 1997; Clarke et al., 2000; Bellmann et al., 2001; Zimmer et al., 2003; Gutschalk et al., 2012).

In our own experience, alloacusis is often observed in patients with acute, severe right-MCA stroke, who typically also show full-blown visual neglect. When these severely affected patients are addressed from the left, they often orient to the right in search of the speaker. This observation might be explained by the finding of alloacusis, i.e., the patients detect the sound but misattribute the sound source to the right. In our experience, severe alloacusis subsides within a few days in a majority of cases but can persist for months or years in some patients.

It is currently unclear whether alloacusis can be dissociated from extinction or if auditory extinction is generally present in patients with alloacusis. Patients with alloacusis who were also tested with dichotic speech tests additionally showed extinction (Bellmann et al., 2001; Gutschalk et al., 2012), but it cannot be excluded that the latter could have been caused by auditory cortex or transcallosal lesions. It has been reported that occasional alloacusis in a free-field target identification task was associated with lower contralesional identification rates with single as well as with double stimulation (Soroker et al., 1997). This combination of symptoms was typically found in right-hemisphere patients only, but was not correlated with the presence of visual neglect.

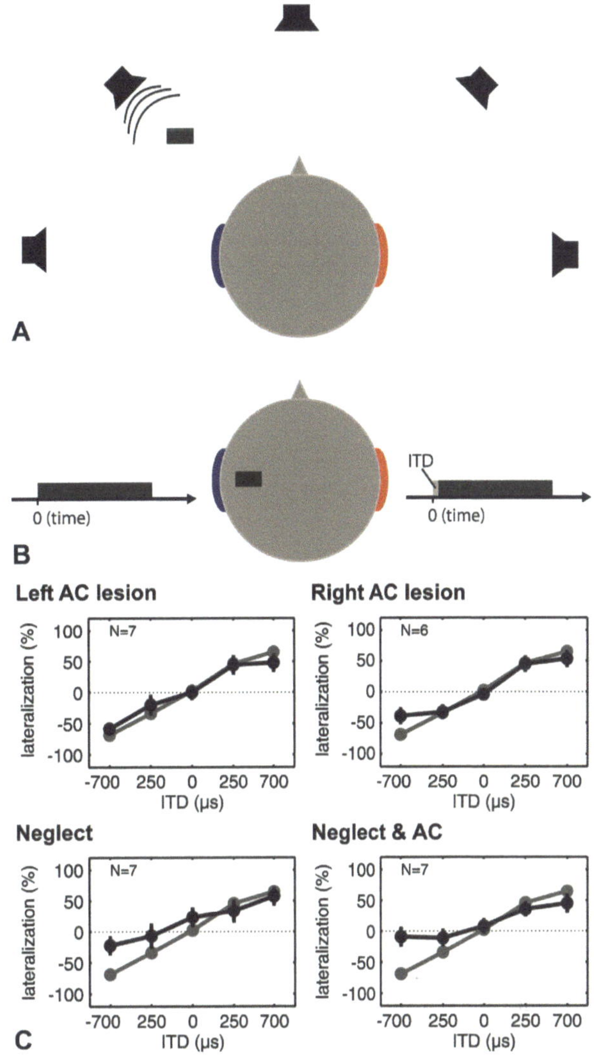

Fig. 31.6. Lateralization perception can be tested in the free field (**A**) with a set of loudspeakers arranged in a circle, usually made invisible to the listener. The sound is then played from one speaker and the listener is asked to indicate the position where he or she perceives the sound. In this setting, sounds are perceived outside the head. Simpler setups use headphones (**B**), and lateralize the sound for example with interaural time differences (ITD). Here, the sound is perceived inside the head, and the listener is typically asked to indicate the lateralization of the sound. (**C**) Data obtained with a setup similar to panel **B** for the same patients and controls shown in Figures 31.3 and 31.5 (Gutschalk et al., 2012). Perceived lateralization of the neglect patients was typically more to the right (black) compared to the rating of the control group (gray). Note that the variability in this test is quite high; for example, patients with alloacusis rate all sounds at the far right. AC, auditory cortex.

Sound lateralization

Disturbance of auditory spatial perception is not always as prominent as in alloacusis. A number of tests (Fig. 31.6) have been used to reveal more subtle deficits of auditory lateralization using free-field stimulation (Ruff et al., 1981; Vallar et al., 1995; Zatorre and Penhune, 2001) or headphones (Bisiach et al., 1984; Kerkhoff et al., 1999; Tanaka et al., 1999; Clarke et al., 2000; Pavani et al., 2001; Zimmer et al., 2003). In one such study, listeners were required to indicate the position from where they perceived lateralized sounds (Ruff et al., 1981; Bisiach et al., 1984; Clarke et al., 2000; Bellmann et al., 2001; Zatorre and Penhune, 2001; Zimmer et al., 2003; Spierer et al., 2009;

Gutschalk et al., 2012). Based on these studies, there is converging evidence that disturbance of auditory lateralization is more frequent and more severe after right-hemisphere lesions. Across studies, the most consistent finding was that patients displaced lateralized sounds towards the side of the lesion, i.e., most frequently towards the right. These findings are based on a sample of 267 patients comprising 119 patients with left- (L) and 148 patients with right-hemisphere (R) lesions (Ruff et al., 1981: L = 14, R = 15; Bisiach et al., 1984: L = 51, R = 56; Tunaka et al., 1999: L = 15, R = 29; Zimmer et al., 2003: L = 9, R = 15; Zatorre and Penhune, 2001: L = 5, R = 8; Spierer et al., 2009: L = 25, R = 25).

A number of authors have tried to derive the subjective straight ahead in these patients, i.e., the physical stimulus lateralization in binaural stimulation (free field or headphone) that patients perceive as arising from the midline. Based on the observation that sounds emanating from the left or midline are perceptually displaced towards the right, one may assume that auditory midline perception has been shifted towards the left (Tanaka et al., 1999; Zimmer et al., 2003), i.e., a physically left-lateralized sound is perceptually rated as midline. In contrast, two studies found the auditory midline shifted towards the right in neglect patients (Vallar et al., 1995; Kerkhoff et al., 1999). It has been suggested that the reason for this discrepancy might lie in the instruction, and that the displacement of sound to the left is only with reference to the more strongly right-lateralized visual midline perception (Zimmer et al., 2003). Another suggestion (Pavani et al., 2004) is that the difference depends on whether the lateralized percept is perceived within the head (as with simple interaural time or frequency differences) or outside the head (with free-field listening or head-related transfer functions). However, there is at least one study that used free-field listening and found ipsilesional lateralization bias in patients with right-hemisphere lesions and neglect (Pinek et al., 1989).

Another type of test to reveal deficits in lateralization required listeners to decide whether two subsequent sounds emanated from the same or different sources in space (Pavani et al., 2001; Zatorre and Penhune, 2001). The results confirmed that deficits of lateralization perception are typically observed in patients with right-hemisphere lesions. (Note that performance in these two-interval tasks may be additionally reduced because of non-spatial deficits (Cusack et al., 2000). The lateralization of the behavioral deficit to contralesional space suggests, however, that there is also a specific spatial deficit.)

The most frequently reported lesion site to produce the above-summarized deficits is in the right inferior parietal lobe and right temporoparietal junction (Ruff et al., 1981; Bisiach et al., 1984; Bellmann et al., 2001; Pavani et al., 2001; Zimmer et al., 2003; Spierer et al., 2009). In many cases, there appears to be a close relationship between visual neglect and deficits in auditory lateralization (Zimmer et al., 2003; Pavani et al., 2004; Gutschalk et al., 2012). However, deficits in auditory lateralization have also been observed, albeit rarely, in patients with left-hemisphere lesions (Clarke et al., 2000; Zimmer et al., 2003) and lesions to the right anterior temporal lobe (Zatorre and Penhune, 2001). No signs of visual neglect were documented in these patients, such that a general equivalence between deficits in auditory lateralization and visual neglect cannot be derived from the available data. There is good evidence, however, that lesions of the primary auditory cortex on both sides do not cause the severe deficits that are observed with right parietal lesions (Zatorre and Penhune, 2001; Gutschalk et al., 2012).

A general issue in comparing the incidence of behavioral deficits between right- and left-hemisphere lesions is aphasia, which may be so severe that informed consent cannot be obtained and task instruction is impossible. Since many of these patients may have large left-sided lesions, there could be a bias for larger lesions on the right or a lack of patients with lesions in the left temporoparietal junction. However, given the large number of patients studied with lateralization tests in the past, the finding of a right-hemisphere dominance for pathologic findings in these tests is now well established. One study additionally showed that the lesions in their left- and right-hemisphere patients were highly similar (Spierer et al., 2009). Finally, it has been demonstrated that other auditory tasks, discussed in previous sections, revealed similar deficits after lesions in either hemisphere (de Renzi et al., 1984; Blaettner et al., 1989; Clarke et al., 2000; Gutschalk et al., 2012), despite excluding some patients because of aphasia.

Other deficits of spatial hearing

The acuity of auditory spatial perception in patients with visual neglect is reduced not only with respect to lateralization, but also to localization of sound elevation (Pavani et al., 2002), with the deficit being more prominent for sounds in the contralesional (left) hemifield ($n = 4$). The performance in this task has been shown to depend on gaze direction, such that the contralesional deficit was reduced when the patients' gaze was directed towards the left (Pavani et al., 2005).

It is also the case that the perception of sound movement may be reduced after right parietal lesions (Griffiths et al., 1996); the patient in this case report could not detect sound movement, but was able to dissociate whether a tone was presented to the left or right ear. Nevertheless, it is noted that the patient also had biased lateralization perception for static sound sources, and subsequent studies that used static as well as moving sounds to test for deficits of spatial hearing usually found deficits in both tests in patients with right parietal lesions (Bellmann et al., 2001; Spierer et al., 2009).

NEURAL BASIS OF NEGLECT

Lesion sites

Auditory neglect has been most typically reported with lesions of inferior right parietal cortex and the temporoparietal junction (Fig. 31.7) in studies focusing on auditory extinction (Heilman and Valenstein, 1972) as well as those that focused on disturbances to auditory spatial perception (Ruff et al., 1981; Spierer et al., 2007). Some authors have also stressed the importance of the right posterior STG (Zimmer et al., 2003), whereas disturbed auditory lateralization without other signs of neglect has also been observed with lesions of the anterior STG (Zatorre and Penhune, 2001). Lesions of the STG may frequently extend into the more medial auditory core and belt regions (see above), which can cause auditory extinction (Gutschalk et al., 2012) that is not necessarily associated with neglect.

Among the reported cases with signs of auditory neglect, there are also patients with lesions in the frontal lobe (Hugdahl et al., 1991) or basal ganglia (Bellmann et al., 2001), highlighting the fact that the types of lesions that can cause auditory neglect are as diverse as those known to cause visual neglect (Damasio et al., 1980; Vallar and Perani, 1986; Karnath et al., 2001; Mort et al., 2003; Verdon et al., 2010).

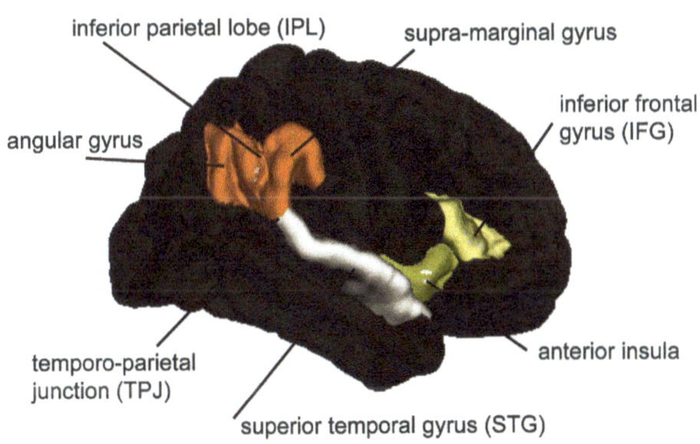

Fig. 31.7. Topography of perisylvian areas in the right hemisphere thought to cause neglect when destroyed. The inferior parietal lobe (IPL) is the area most traditionally associated with neglect. Its dorsal limitation is the intraparietal sulcus. The IPL comprises the supramarginal gyrus and the angular gyrus. Another area thought to be important in the context of neglect and auditory lateralization perception is the superior temporal gyrus (STG). The transition from STG to the supramarginal gyrus is the temporoparietal junction (TPJ), another lesion focus in neglect patients. Less frequently, neglect has been reported with isolated lesions of the lower frontal lobe (marked is the inferior temporal gyrus) or the anterior insula.

Models of neglect

Most models to explain neglect are based on results observed in vision and suggest that neglect is a disorder of spatial attention. An early model suggested a balance between attentional vectors in the left and right hemisphere, and that their imbalance is stronger after right-sided lesions (Kinsbourne, 1970). Others have suggested a general specialization of the right hemisphere for attention and arousal (Heilman et al., 1978; Mesulam, 1981); this type of model can also explain some of the non-spatial deficits observed with auditory neglect (Robertson et al., 1997). To explain core spatial deficits in neglect, it has been assumed that the right hemisphere could orient attention to both sides in space, whereas the left hemisphere would predominantly orient to contralateral space (Mesulam, 1981), but this assumption has been challenged more recently (Corbetta and Shulman, 2011).

Finally, a problem of disengaging attention from the ipsilesional side towards the contralesional side has been suggested based on cuing tasks (Posner et al., 1984). In this model, attention cannot be directed towards the left hemifield because it cannot be disengaged from the right. Disengaging attention has been attributed to the right temporoparietal junction. Thus, directing attention is not the disturbed mechanism in this model. In the context of auditory neglect, attention-based models are particularly well suited to explain the extinction phenomena reviewed above.

Another line of models has considered neglect a primary disorder of spatial representation (Bisiach et al., 1981) rather than attention in space. One suggestion is that mapping of spatial information between different coordinate systems – such as retinal, egocentric, and allocentric (Karnath, 1997) – is defective. As a result, the coordinate system would be rotated towards the right with shrinkage of the left hemifield, which causes the deficit of contralesional exploration typically observed in neglect.

Functional Imaging

A recent model of neglect (Corbetta and Shulman, 2011) is based on functional imaging studies in healthy volunteers and patients with neglect: in this model a dorsal attention system, required to voluntarily direct attention in space, is dissociated from a ventral, stimulus-driven attention system. While the dorsal attention system is thought to possess a topographic representation of space, this representation appears to be mostly contralateral, i.e., each hemisphere predominantly associated with its respective contralateral side of space. In contrast, the ventral attention system is strongly right-lateralized and is comprised of the same brain areas often associated with neglect (the temporoparietal junction, ventrolateral prefrontal cortex, and the anterior insular) (Corbetta and Shulman, 2002). Moreover, functional imaging studies in patients with neglect show that patients with lesions to the structures associated with the ventral attention system show abnormally low task-evoked activation (Corbetta et al., 2005), as well as modulated functional connectivity (He et al., 2007), in the dorsal attention system, more prominently in the right hemisphere. Follow-up studies with the same patients demonstrated that when activity in the dorsal attention system had normalized, the spatial neglect symptoms subsided (Fig. 31.8). This suggests that the degraded spatial processing characteristic of neglect may reflect the ventral lesion's functional impact on the dorsal attention system, rather than being a direct consequence of structural damage to the ventral system. This model (Corbetta and Shulman, 2011) can elegantly explain the heterogeneity of lesion sites in patients with neglect as well as their tendency to recover behaviorally. In contrast, the non-spatial deficits often experienced by neglect patients (Robertson et al., 1997) might be a more direct result of damage to the ventral attention system.

Only a handful of electroencephalograph (EEG) and magnetoencephalograph (MEG) studies have investigated patients with auditory neglect. One such EEG study used free-field stimulation and an oddball paradigm in which rare deviant stimuli are interspersed amongst more frequent standard stimuli. The additional activity evoked by the deviants compared to the standards in the auditory cortex is known classically as the mismatch negativity, or MMN (Näätänen et al., 2007). In patients with neglect, the MMN was observed to be smaller when the stimulus stream was presented to the left side compared to when it was presented to the right side, a result not observed in control participants (Deouell et al., 2000). In the same study, the N1 evoked

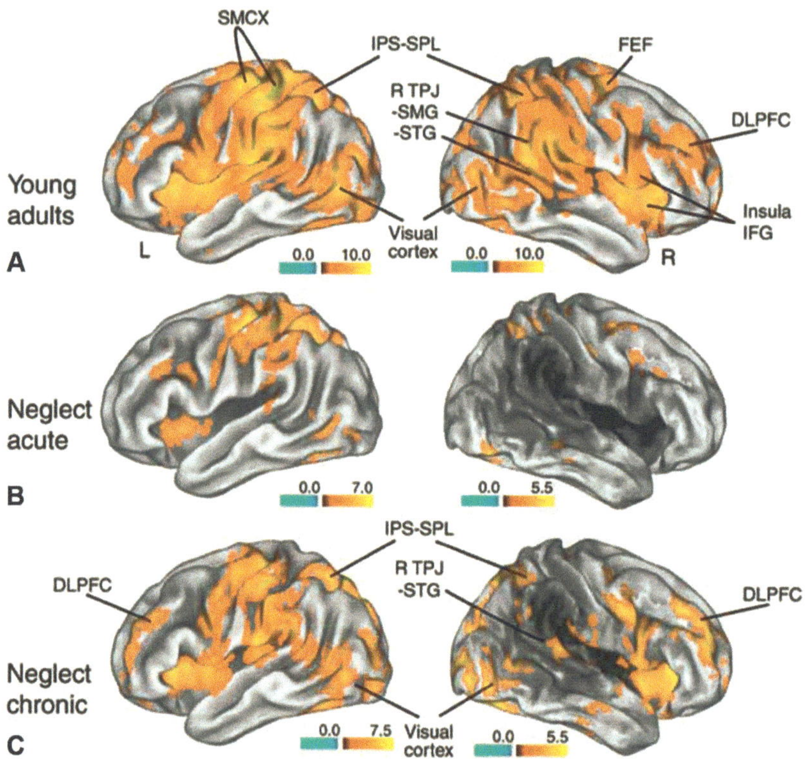

Fig. 31.8. Functional magnetic resonance imaging (fMRI) activation maps in a Posner task (visual stimulus presentation) from the study by Corbetta et al. (2005). (**A**) The task produces widespread activation in a normal control group. In patients with neglect, the activity is prominently reduced. (**B**) While the lesions in the patient group are mostly confined to more ventral areas, activity in more dorsal parietal and frontal areas is also reduced. (**C**) When the same patients were studied again after 39 weeks, their left-sided attention deficits were substantially improved. This goes along with recovered activation levels in fMRI in areas that are not part of the lesion in the right, but also in the left hemisphere. SMCX, somatosensory and motor cortex; IPS, intraparietal sulcus; STL, superior temporal lobe; FEF, frontal eye field; R TPJ, right temporoparietal junction; SMG, supramarginal gyrus; STG, superior temporal gyrus; DLPC, dorsolateral prefrontal cortex; IFG, inferior frontal gyrus. (Reproduced from Corbetta et al., 2005, with permission from Nature publishing group.)

by the standard stimuli was smaller over the right compared to the left hemisphere, but did not differ based on side of presentation. A subsequent evaluation of the auditory-evoked N1 in patients with neglect showed that stimuli presented to the left ear evoked a smaller N1 than the same stimuli presented to the right ear (Tarkka et al., 2011). Note, however, that neither of these studies controlled whether auditory cortex was intact in their patients, and part of their findings might therefore be related to auditory cortex lesions. Nevertheless, the typical rightwards lateralization of auditory cortex activity (Ross et al., 2005; Shaw et al., 2013) was found to be reduced in patients with neglect and structurally intact auditory cortex (Gutschalk et al., 2012), independently of stimulus lateralization. In summary, there is some evidence to suggest that auditory neglect might involve modulation of early processing at the level of the auditory cortex, a finding that finds a parallel in reduced activity in the right visual cortex in patients with visual neglect (Corbetta et al., 2005; Vuilleumier et al., 2008).

CONCLUSION

What is auditory neglect?

The heterogeneity of lesion sites and broad spectrum of behavioral deficits in patients with neglect have prompted some authors to suggest that neglect is a heterogeneous disorder, with individual behavioral profiles directly related to the anatomic loci of the lesion(s) found in each patient (Verdon et al., 2010). Others have emphasized a core of neglect symptoms (Corbetta and Shulman, 2011; Karnath and Rorden, 2012), including an attentional bias towards the egocentric ipsilesional (right) side. In vision, this deficit is reliably measured by the amount of contralesional (left) omissions in cancellation tasks (Karnath and Rorden, 2012).

Little has been said about whether auditory manifestations of neglect should be considered core features of neglect. In our opinion, alloacusis is the most obvious auditory manifestation of clinical neglect. While alloacusis is not present in all neglect patients, to the best of our knowledge there is no report of a patient with severe alloacusis who did not also show other core symptoms of neglect. (Note, however, that occasional alloacusis has been observed in patients who did not show signs of visual neglect at the time of testing (Soroker et al., 1997)). This is informally confirmed by our own clinical experience. Disturbance of lateralization is probably a less severe variant of similar origin and is also strongly correlated with neglect. However, the diagnosis of biased lateralization requires standardized tests that are not readily available at the patient's bedside. Moreover, such deficits have also been observed in the absence of core neglect symptoms, making them a less reliable sign of auditory neglect.

Auditory extinction is present in many patients with core neglect in our experience, but suffers from a lack of specificity since there are other conditions, such as lesions of the auditory cortex or the auditory transcallosal pathway, which can also cause auditory extinction. It is therefore difficult based solely on behavioral testing to differentiate whether a patient with neglect additionally suffers from a lesion of the auditory cortex. This difficulty parallels the problem of testing for hemianopia in patients with full-blown visual neglect, because single contralesional stimuli can be missed due solely to neglect without the additional requirement of a lesion to visual cortex. In contrast, a lesion of the primary visual cortex can be easily excluded in patients exhibiting extinction instead of full-blown neglect. Because of this heterogeneity of auditory extinction, we would expect that, while auditory extinction should still be more frequently found with right- than left-hemisphere lesions, auditory extinction will be more frequently observed after left-sided lesions than visual extinction (Becker and Karnath, 2007).

Thus, while auditory extinction and disturbed auditory localization are frequent clinical findings in patients with neglect, the diagnosis of neglect cannot be confirmed based on these symptoms alone: imagine a patient with a lesion of the right anterior STG extending more posteriorly on the superior temporal plane and including the primary auditory cortex. Such a lesion could cause both disturbed auditory lateralization and extinction without other signs of neglect, whereas a similar lesion in the left would only cause auditory extinction. We would not classify this patient as suffering from auditory neglect.

One then wonders whether diagnosis of neglect can be made purely by examination conducted in the auditory modality, or whether neglect, as it is canonically known from the visual literature, could ever manifest itself exclusively in the auditory domain? One potential tool for differentially diagnosing neglect outside of vision could be the additional presence of a non-spatial deficit in auditory target detection (Robertson et al., 1997; Gutschalk et al., 2012). If diminished target detection is taken as a core feature of neglect, then this, in the presence of the two other auditory characteristics of neglect, could suffice, although this requires further evaluation.

Which workup should be used?

Bedside testing for auditory neglect can be accomplished by simple finger snapping next to the patient's ear and asking whether the snapping was heard at the left, right, or both ears. This test is most sensitive for alloacusis, where the patient will label all three stimuli as emanating from the right. If alloacusis is not present, this test can be used to test for auditory extinction, where snapping next to the left and right ear is correctly classified, but snapping synchronously next to both ears will typically be attributed to the ipsilesional ear.

While this test is easy to apply, other more rigorous tasks might be more sensitive. Dichotic speech tests are not specific for auditory neglect, but they are quite sensitive in our experience and take comparatively little time. An additional test for spatial perception should be applied to pick up disturbance of spatial perception. We use a simple task in which noise bursts are presented randomly with five different ITDs, and listeners are required to indicate the perceived lateralization to the left or right on a sliding scale (Gutschalk et al., 2012).

To test for extinction-like deficits that are unrelated to neglect, in particular after lesions to auditory cortex, the psychoacoustic pattern discrimination test introduced above can be used (Blaettner et al., 1989). It appears that neglect patients without auditory cortex lesions do not show lateralized deficits in this test, but a more general deficit in auditory target detection might be picked up instead (Robertson et al., 1997; Gutschalk et al., 2012). The deficit of sustained attention and target detection might alternatively be picked up more simply and specifically with the elevator counting test (Wilkins et al., 1987; Robertson et al., 1997).

Can auditory neglect be treated?

A number of approaches have been explored for rehabilitation of neglect patients (Kerkhoff and Schenk, 2012), most commonly utilizing visual exploration with feedback. More recent techniques have sought to reorient patients towards their left side using optokinetic stimulation, neck muscle vibration, vestibular stimulation, or prism adaptation. Another line of recent experimental therapies applies non-invasive neurostimulation with repetitive transcranial magnetic stimulation and transcranial direct current stimulation (Hesse et al., 2011). Most of these approaches are under exploration and their effect has not yet been proven in larger controlled studies. While auditory neglect has not been tested in most therapy studies, it is often assumed that therapy effects transfer to other modalities because of the multimodal nature of the disorder, but this requires explicit confirmation in our opinion.

Only a few studies have specifically reported rehabilitation effects on auditory aspects of neglect. Two studies investigated the effect of prism adaptation, in which prism glasses are used that displace the optic image by 10–15° to the right, on auditory neglect. Patients undergoing prism adaptation are trained to make reaching movements until they are adapted to the prism displacement. When the glasses are subsequently removed, there is an after-effect towards the left, which has been shown to transiently improve performance in patients with neglect (Rossetti et al., 1998). One study found that prism adaptation significantly reduces auditory extinction as measured in a dichotic speech test (Jacquin-Courtois et al., 2010). The other study used a target detection task and found improvement in both visual and auditory target detection that was similar for ipsi- and contralesional stimuli (Eramudugolla et al., 2010). The effect of prism adaptation on auditory extinction lasted for 2 hours (Jacquin-Courtois et al., 2010). Longer-lasting effects on visual measures have been reported for repetitive application of prism adaptation (Frassinetti et al., 2002).

Another technique that produces an after-effect is optokinetic stimulation. This technique has been shown to normalize the auditory midline perception in patients with neglect for 30 minutes after a single application, and longer effects after repetitive application (Kerkhoff et al., 2012). Effects on sound lateralization in normal listeners have also been found for neck muscle vibration (Lewald et al., 1999), providing indirect evidence that this technique might also show an effect on auditory neglect.

Finally, auditory phasic alerting has been used to ameliorate visual neglect on the left, irrespective of whether the auditory stimulus was applied to the left or right side (Robertson et al., 1998). Similar effects on visual neglect are achieved with auditory sustained attention training (Van Vleet and DeGutis, 2013), further stressing the multimodal nature of neglect. Studies with larger patient samples will be necessary to evaluate these techniques towards an evidence-based rehabilitation of neglect that includes the auditory modality.

Open questions

While a profile of auditory neglect is emerging from the literature reviewed in this chapter, many questions must first be answered in order to gain a more complete understanding of auditory neglect. One is the nature of auditory extinction, which apparently is a more heterogeneous phenomenon than visual extinction. Further studies are required to dissociate the subtypes of auditory extinction and explore, for example, whether perceived temporal asynchrony (Karnath et al., 2002; Witte et al., 2012) or extinction across modalities (Diamond and Bender, 1965) can be used to define more exactly auditory extinction in the context of neglect (or, more generally, supramodal extinction), and dissociate it more clearly from other causes of auditory extinction. Moreover, it will be interesting to see if auditory extinction requires a bilateral and synchronous stimulus presentation, or if other situations of perceptual competition, for example informational masking (Gutschalk et al., 2008; Kidd et al., 2008), may more strongly reduce auditory awareness in neglect.

Finally, non-spatial deficits in auditory neglect (Robertson et al., 1997; Cusack et al., 2000) deserve more attention by future studies. Not only may these fundamental deficits be used to develop new tests for clinical evaluation of auditory neglect, it is also conceivable that these deficits are more important for the patients' rehabilitation than are the spatial signs of auditory neglect, which are rarely a cause of severe disability in chronic neglect patients, first, because sustained auditory attention is clearly important in everyday life, and second, because non-spatial deficits may be more directly related to the primary lesion site and might therefore require more time to subside.

ACKNOWLEDGMENTS

Supported by Bundesministerium für Bildung und Forschung (BMBF) program "junior research group bioimaging," grant 01EV0712. We thank Dr. Maurizio Corbetta for granting permission to reproduce Figure 31.8.

References

Barrett DJ, Edmondson-Jones AM, Hall DA (2010). Attention in neglect and extinction: assessing the degree of correspondence between visual and auditory impairments using matched tasks. J Clin Exp Neuropsychol 32: 71–80.

Beaton A, McCarthy M (1993). "Auditory neglect after right frontal lobe and right pulvinar thalamic lesions": comments on Hugdahl, Wester, and Asbjornsen (1991) and some preliminary findings. Brain Lang 44: 121–126.

Becker E, Karnath HO (2007). Incidence of visual extinction after left versus right hemisphere stroke. Stroke 38: 3172–3174.

Becker E, Karnath HO (2010). Neuroimaging of eye position reveals spatial neglect. Brain 133: 909–914.

Bellmann A, Meuli R, Clarke S (2001). Two types of auditory neglect. Brain 124: 676–687.

Bisiach E, Capitani E, Luzzatti C et al. (1981). Brain and conscious representation of outside reality. Neuropsychologia 19: 543–551.

Bisiach E, Cornacchia L, Sterzi R et al. (1984). Disorders of perceived auditory lateralization after lesions of the right hemisphere. Brain 107 (Pt 1): 37–52.

Blaettner U, Scherg M, von Cramon D (1989). Diagnosis of unilateral telencephalic hearing disorders. Evaluation of a simple psychoacoustic pattern discrimination test. Brain 112 (Pt 1): 177–195.

Brain WR (1941). Visual disorientation with special reference to lesions of the right cerebral hemisphere. Brain 64: 43–62.

Broadbent DE (1952). Listening to one of two synchronous messages. J Exp Psychol 44: 51–55.

Cherry C (1953). Some experiments on the recognition of speech, with one and two ears. J Acoust Soc Am 25: 975–981.

Clarke S, Bellmann A, Meuli RA et al. (2000). Auditory agnosia and auditory spatial deficits following left hemispheric lesions: evidence for distinct processing pathways. Neuropsychologia 38: 797–807.

Corbetta M, Shulman GL (2002). Control of goal-directed and stimulus-driven attention in the brain. Nat Rev Neurosci 3: 201–215.

Corbetta M, Shulman GL (2011). Spatial neglect and attention networks. Annu Rev Neurosci 34: 569–599.

Corbetta M, Kincade MJ, Lewis C et al. (2005). Neural basis and recovery of spatial attention deficits in spatial neglect. Nat Neurosci 8: 1603–1610.

Cusack R, Carlyon RP, Robertson IH (2000). Neglect between but not within auditory objects. J Cogn Neurosci 12: 1056–1065.

Damasio H, Damasio A (1979). "Paradoxic" ear extinction in dichotic listening: possible anatomic significance. Neurology 29: 644–653.

Damasio AR, Damasio H, Chui HC (1980). Neglect following damage to frontal lobe or basal ganglia. Neuropsychologia 18: 123–132.

De Renzi E, Gentilini M, Pattacini F (1984). Auditory extinction following hemisphere damage. Neuropsychologia 22: 733–744.

De Renzi E, Gentilini M, Barbieri C (1989). Auditory neglect. J Neurol Neurosurg Psychiatry 52: 613–617.

Denny-Brown D, Meyer JS, Horenstein S (1952). The significance of perceptual rivalry resulting from parietal lesion. Brain 75: 433–471.

Deouell LY, Bentin S, Soroker N (2000). Electrophysiological evidence for an early (pre-attentive) information processing deficit in patients with right hemisphere damage and unilateral neglect. Brain 123 (Pt 2): 353–365.

Diamond SP, Bender MB (1965). On auditory extinction and alloacusis. Trans Am Neurol Assoc 90: 154–157.

Eramudugolla R, Boyce A, Irvine DR et al. (2010). Effects of prismatic adaptation on spatial gradients in unilateral neglect: a comparison of visual and auditory target detection with central attentional load. Neuropsychologia 48: 2681–2692.

Frassinetti F, Angeli V, Meneghello F et al. (2002). Long-lasting amelioration of visuospatial neglect by prism adaptation. Brain 125: 608–623.

Gauthier L, Dehaut F, Joanette Y (1989). The bells test: a quantitative and qualitative test for visual neglect. Int J Clin Neuropsychol 11: 49–54.

Griffiths TD, Rees A, Witton C et al. (1996). Evidence for a sound movement area in the human cerebral cortex. Nature 383: 425–427.

Gutschalk A, Micheyl C, Oxenham AJ (2008). Neural correlates of auditory perceptual awareness under informational masking. PLoS Biol 6: e138.

Gutschalk A, Brandt T, Bartsch A et al. (2012). Comparison of auditory deficits associated with neglect and auditory cortex lesions. Neuropsychologia 50: 926–938.

He BJ, Snyder AZ, Vincent JL et al. (2007). Breakdown of functional connectivity in frontoparietal networks underlies behavioral deficits in spatial neglect. Neuron 53: 905–918.

Heilman KM, Valenstein E (1972). Auditory neglect in man. Arch Neurol 26: 32–35.

Heilman KM, Pandya DN, Karol EA et al. (1971). Auditory inattention. Arch Neurol 24: 323–325.

Heilman KM, Schwartz HD, Watson RT (1978). Hypoarousal in patients with the neglect syndrome and emotional indifference. Neurology 28: 229–232.

Heilman KM, Watson RT, Valenstein E (2003). Neglect: Clinical and anatomical issues. In: TE Feinberg, MJ Farah (Eds.), Behavioral Neurology & Neuropsychology, 2nd edn. McGraw-Hill, New York, pp. 303–312.

Hesse MD, Sparing R, Fink GR (2011). Ameliorating spatial neglect with non-invasive brain stimulation: from pathophysiological concepts to novel treatment strategies. Neuropsychol Rehabil 21: 676–702.

Hier DB, Mondlock J, Caplan LR (1983). Recovery of behavioral abnormalities after right hemisphere stroke. Neurology 33: 345–350.

Hugdahl K, Wester K (1994). Auditory neglect and the ear extinction effect in dichotic listening: a reply to Beaton and McCarthy (1993). Brain Lang 46: 166–173.

Hugdahl K, Wester K, Asbjornsen A (1991). Auditory neglect after right frontal lobe and right pulvinar thalamic lesions. Brain Lang 41: 465–473.

Husain M (2008). Hemispatial neglect. Handb Clin Neurol 88: 359–372.

Husain M, Shapiro K, Martin J et al. (1997). Abnormal temporal dynamics of visual attention in spatial neglect patients. Nature 385: 154–156.

Jacquin-Courtois S, Rode G, Pavani F et al. (2010). Effect of prism adaptation on left dichotic listening deficit in neglect patients: glasses to hear better? Brain 133: 895–908.

Karnath HO (1997). Spatial orientation and the representation of space with parietal lobe lesions. Philos Trans R Soc Lond B Biol Sci 352: 1411–1419.

Karnath HO, Rorden C (2012). The anatomy of spatial neglect. Neuropsychologia 50: 1010–1017.

Karnath HO, Ferber S, Himmelbach M (2001). Spatial awareness is a function of the temporal not the posterior parietal lobe. Nature 411: 950–953.

Karnath HO, Zimmer U, Lewald J (2002). Impaired perception of temporal order in auditory extinction. Neuropsychologia 40: 1977–1982.

Kerkhoff G, Schenk T (2012). Rehabilitation of neglect: an update. Neuropsychologia 50: 1072–1079.

Kerkhoff G, Artinger F, Ziegler W (1999). Contrasting spatial hearing deficits in hemianopia and spatial neglect. Neuroreport 10: 3555–3560.

Kerkhoff G, Keller I, Artinger F et al. (2012). Recovery from auditory and visual neglect after optokinetic stimulation with pursuit eye movements – transient modulation and enduring treatment effects. Neuropsychologia 50: 1164–1177.

Kidd G, Mason CR, Richards VM et al. (2008). Informational Masking. In: WA Yost, AN Popper, RR Fay (Eds.), Auditory Perception of Sound Sources, Springer, New York.

Kimura D (1961). Some effects of temporal-lobe damage on auditory perception. Can J Psychol 15: 156–165.

Kinsbourne M (1970). A model for the mechanism of unilateral neglect of space. Trans Am Neurol Assoc 95: 143–147.

Leifert-Fiebach G (2008). Auditorischer Neglekt und auditorische Extinktion bei Patienten mit visuell-räumlichem Neglekt, Universität Koblenz, Landau.

Lewald J, Karnath HO, Ehrenstein WH (1999). Neck-proprioceptive influence on auditory lateralization. Exp Brain Res 125: 389–396.

Mesulam MM (1981). A cortical network for directed attention and unilateral neglect. Ann Neurol 10: 309–325.

Mort DJ, Malhotra P, Mannan SK et al. (2003). The anatomy of visual neglect. Brain 126: 1986–1997.

Näätänen R, Paavilainen P, Rinne T et al. (2007). The mismatch negativity (MMN) in basic research of central auditory processing: a review. Clin Neurophysiol 118: 2544–2590.

Oppenheim H (1885). Ueber eine durch eine klinisch bisher nicht verwerthete Untersuchungsmethode ermittelte Form der Sensibilitätsstörung bei einseitigen Erkrankungen des Grosshirns. Neurologisches Centralblatt 4: 529–533.

Pavani F, Meneghello F, Ladavas E (2001). Deficit of auditory space perception in patients with visuospatial neglect. Neuropsychologia 39: 1401–1409.

Pavani F, Ladavas E, Driver J (2002). Selective deficit of auditory localisation in patients with visuospatial neglect. Neuropsychologia 40: 291–301.

Pavani F, Husain M, Ladavas E et al. (2004). Auditory deficits in visuospatial neglect patients. Cortex 40: 347–365.

Pavani F, Ladavas E, Driver J (2005). Gaze direction modulates auditory spatial deficits in stroke patients with neglect. Cortex 41: 181–188.

Pinek B, Duhamel JR, Cave C et al. (1989). Audio-spatial deficits in humans: differential effects associated with left versus right hemisphere parietal damage. Cortex 25: 175–186.

Posner MI, Walker JA, Friedrich FJ et al. (1984). Effects of parietal injury on covert orienting of attention. J Neurosci 4: 1863–1874.

Rengachary J, d'Avossa G, Sapir A et al. (2009). Is the Posner reaction time test more accurate than clinical tests in detecting left neglect in acute and chronic stroke? Arch Phys Med Rehabil 90: 2081–2088.

Rengachary J, He BJ, Shulman GL et al. (2011). A behavioral analysis of spatial neglect and its recovery after stroke. Front Hum Neurosci 5: 29.

Robertson IH, Manly T, Beschin N et al. (1997). Auditory sustained attention is a marker of unilateral spatial neglect. Neuropsychologia 35: 1527–1532.

Robertson IH, Mattingley JB, Rorden C et al. (1998). Phasic alerting of neglect patients overcomes their spatial deficit in visual awareness. Nature 395: 169–172.

Rorden C, Karnath HO (2010). A simple measure of neglect severity. Neuropsychologia 48: 2758–2763.

Ross B, Herdman AT, Pantev C (2005). Right hemispheric laterality of human 40 Hz auditory steady-state responses. Cereb Cortex 15: 2029–2039.

Rossetti Y, Rode G, Pisella L et al. (1998). Prism adaptation to a rightward optical deviation rehabilitates left hemispatial neglect. Nature 395: 166–169.

Ruff RM, Hersh NA, Pribram KH (1981). Auditory spatial deficits in the personal and extrapersonal frames of reference due to cortical lesions. Neuropsychologia 19: 435–443.

Schulhoff C, Goodglas H (1969). Dichotic listening, side of brain injury and cerebral dominance. Neuropsychologia 7: 149–160.

Shaw ME, Hämäläinen MS, Gutschalk A (2013). How anatomical asymmetry of human auditory cortex can lead to a rightward bias in auditory evoked fields. Neuroimage 74: 22–29.

Sinnett S, Juncadella M, Rafal R et al. (2007). A dissociation between visual and auditory hemi-inattention: evidence from temporal order judgements. Neuropsychologia 45: 552–560.

Soroker N, Calamaro N, Glicksohn J et al. (1997). Auditory inattention in right-hemisphere-damaged patients with and without visual neglect. Neuropsychologia 35: 249–256.

Sparks R, Geschwind N (1968). Dichotic listening in man after section of neocortical comissures. Cortex 4: 3–16.

Spierer L, Meuli R, Clarke S (2007). Extinction of auditory stimuli in hemineglect: Space versus ear. Neuropsychologia 45: 540–551.

Spierer L, Bellmann-Thiran A, Maeder P et al. (2009). Hemispheric competence for auditory spatial representation. Brain 132: 1953–1966.

Tanaka H, Hachisuka K, Ogata H (1999). Sound lateralisation in patients with left or right cerebral hemispheric lesions: relation with unilateral visuospatial neglect. J Neurol Neurosurg Psychiatry 67: 481–486.

Tarkka IM, Luukkainen-Markkula R, Pitkanen K et al. (2011). Alterations in visual and auditory processing in hemispatial neglect: an evoked potential follow-up study. Int J Psychophysiol 79: 272–279.

Vallar G, Perani D (1986). The anatomy of unilateral neglect after right-hemisphere stroke lesions. A clinical/CT-scan correlation study in man. Neuropsychologia 24: 609–622.

Vallar G, Rusconi ML, Bignamini L et al. (1994). Anatomical correlates of visual and tactile extinction in humans: a clinical CT scan study. J Neurol Neurosurg Psychiatry 57: 464–470.

Vallar G, Guariglia C, Nico D et al. (1995). Spatial hemine glect in back space. Brain 118 (Pt 2): 467–472.

Van Vleet TM, DeGutis JM (2013). Cross-training in hemispatial neglect: auditory sustained attention training ameliorates visual attention deficits. Cortex 49: 679–690.

Verdon V, Schwartz S, Lovblad KO et al. (2010). Neuroanatomy of hemispatial neglect and its functional components: a study using voxel-based lesion-symptom mapping. Brain 133: 880–894.

Vossel S, Eschenbeck P, Weiss PH et al. (2011). Visual extinction in relation to visuospatial neglect after right-hemispheric stroke: quantitative assessment and statistical lesion-symptom mapping. J Neurol Neurosurg Psychiatry 82: 862–868.

Vuilleumier P, Schwartz S, Verdon V et al. (2008). Abnormal attentional modulation of retinotopic cortex in parietal patients with spatial neglect. Curr Biol 18: 1525–1529.

Weintraub S, Mesulam MM (1985). Mental state assessment of young and elderly adults in behavioral neurology. In: MM Mesulam (Ed.), Principles of behavioral neurology, David Company, Philadelphia, pp. 71–123.

Wilkins AJ, Shallice T, McCarthy R (1987). Frontal lesions and sustained attention. Neuropsychologia 25: 359–365.

Witte C, Grube M, Cramon DY et al. (2012). Auditory extinction and spatio-temporal order judgment in patients with left- and right-hemisphere lesions. Neuropsychologia 50: 892–903.

Zatorre RJ, Penhune VB (2001). Spatial localization after excision of human auditory cortex. J Neurosci 21: 6321–6328.

Zimmer U, Lewald J, Karnath HO (2003). Disturbed sound lateralization in patients with spatial neglect. J Cogn Neurosci 15: 694–703.

Handbook of Clinical Neurology, Vol. 129 (3rd series)
The Human Auditory System
G.G. Celesia and G. Hickok, Editors

Chapter 32

Auditory agnosia

L. ROBERT SLEVC* AND ALISON R. SHELL
Department of Psychology, University of Maryland, College Park, MD, USA

INTRODUCTION

Imagine chatting with a friend at a crowded outdoor café, background music playing, a couple arguing at the next table, and cars driving by. The auditory information reaching your ears in this situation is a mixture from all these sources (and more), yet ultimately yields a representation of the auditory scene in which you are able to successfully follow your conversation. Although the ability to identify and process these sounds may seem nearly effortless in normal circumstances, the complexity of our auditory world becomes clear in light of neurologic conditions that can lead to strikingly distinct deficits in auditory perception.

These impairments of auditory perception and/or recognition that are not attributable to hearing or cognitive deficits are termed auditory agnosia. Auditory agnosia is distinct from peripheral hearing loss and also distinct from impaired hearing acuity as a result of bilateral damage to primary auditory cortex – a condition termed cortical deafness (although perhaps more appropriately called central deafness or cerebral deafness given that it may result from damage to the projections to auditory cortex rather than cortical damage *per se*; Tanaka et al., 1991; Griffiths, 2002). Whereas cortically deaf patients typically behave as if they were deaf (despite normal peripheral hearing and brainstem evoked potentials), patients with auditory agnosia are aware of sounds but have difficulty with sound identification. Auditory agnosia is also distinct from language perception deficits associated with the posterior (sensory) aphasias, although both cortical deafness and sensory aphasia can evolve into auditory agnosia (e.g., Mendez and Geehan, 1988; Slevc et al., 2011).

DEFINITIONS

Auditory agnosia has not received as much attention as the visual agnosias; however it is similar in that it can be a general deficit affecting all domains of auditory perception, or can be relatively specific, largely affecting only one type of sound perception/recognition. Patients can have agnosia specific to speech perception, called verbal auditory agnosia or pure word deafness; agnosia specific to music perception, called amusia; or agnosia specific to non-speech environmental sounds, called non-verbal auditory agnosia or environmental sound agnosia.

There is, unfortunately, considerable terminological variability in the literature; here we use auditory agnosia to refer broadly to all subtypes of agnosia for auditory stimuli; however note that "auditory agnosia" is sometimes used more narrowly to refer specifically to environmental sound agnosia (i.e., agnosia for environmental sounds in the presence of preserved perception of speech and (sometimes) music; e.g., Vignolo, 1982, 2003). These specific deficits of auditory perception have garnered much interest despite their rarity because they may reveal the distinct cognitive and anatomic units of auditory perception and thus inform the extent to which auditory perception relies on a modular, and possibly domain-specific, architecture. However the extent to which auditory agnosia reflects separable domains of auditory perception is still debated: claims that the agnosias reflect damage to a modular system, where separate domains of auditory stimuli are processed independently (e.g., Peretz et al., 1994), contrast with claims that the agnosias result from deficits of varying severity to a domain of a general auditory processing system (e.g., Griffiths et al., 1999).

The generalist approach gains some support from the fact that dissociations between auditory processing domains in these agnosia subtypes are rarely "pure"; instead most cases show at least some degree of overlapping deficits. Although this is widely acknowledged, the prevalence of overlapping (and non-overlapping)

*Correspondence to: L. Robert Slevc, Department of Psychology, University of Maryland, College Park, MD 20742, USA. E-mail: slevc@umd.edu deficits is not well documented, perhaps due simply to the practical constraints of carrying out extensive testing across multiple types of stimuli. In addition, it is not yet clear exactly what constitutes an affected domain in these subtypes. For example, is verbal auditory agnosia specific to speech sounds or does it instead reflect a deficit in some more basic (non-speech) aspect of auditory analysis that is disproportionately involved in speech perception? Overlapping deficits are to be expected if they reflect difficulty with particular aspects of auditory analysis that differentially, but not exclusively, affect particular domains of sound processing (Griffiths et al., 1999). As detailed below, work that has carefully examined different aspects of auditory analysis suggests that cases of auditory agnosia do not tend to support domain specificity at the level of language, music, or environmental sounds (although there are cases with remarkable dissociations of these types, e.g., Metz-Lutz and Dahl, 1984; Peretz et al., 1994). Instead, these cases can illuminate the types of distinct mechanisms involved in sound processing, thereby informing our understanding about the basic cognitive and neural processes involved in auditory perception more generally (e.g., Zatorre et al., 2002; Scott and Wise, 2004; Hickok and Poeppel, 2007; Giraud and Poeppel, 2012).

Auditory agnosia typically results from cerebrovascular accidents, although other causes are possible, including herpes simplex encephalitis (Buchman et al., 1986), Landau–Kleffner syndrome (Metz-Lutz, 2009; Stefanatos, 2011), and dementia (Otsuki et al., 1998; Goll et al., 2010a; Gibbons et al., 2012). There are also congenital forms of agnosia, which have been documented mostly for musical perception (congenital amusia; see Peretz and Hyde, 2003; Stewart, 2011; and Chapter 33, this volume). To arrive at a diagnosis of auditory agnosia, it is necessary to first establish that a patient's sound-processing deficit is not due to peripheral hearing problems via audiometric testing (pure-tone, speech, or immittance audiometry), otoacoustic emission testing, brainstem auditory evoked potentials, or other audiometric methods (Hall and Mueller, 1997; see Chapters 17 and 18 of this volume). Of course, as auditory agnosia typically occurs in older adults, some degree of age-related hearing loss is to be expected (see Chapter 20, this volume); however a diagnosis of auditory agnosia requires any hearing loss to be sufficiently mild to not explain the patient's deficit.

It is also necessary to rule out possible "top-down" deficits that could affect auditory perception, for example, attentional, linguistic, or memory deficits, as assessed by standard neuropsychological tests. If both peripheral hearing and general cognitive and linguistic abilities are intact, then additional testing can be done to determine the extent and domain of agnosia. This likely requires an *ad hoc* approach designed for a particular patient to determine which (if any) specific subtype of auditory agnosia (as detailed below) is the appropriate diagnosis.

Although it is difficult to make a precise estimate, auditory agnosia appears to be quite rare and, accordingly, the specific subtypes are even less common. Polster and Rose (1998) point out several reasons why agnosia might be less commonly observed in the auditory than in the visual domain. One such reason is that there is greater neuroanatomic redundancy in the auditory than in the visual system; in contrast to vision, where each hemisphere receives information from only half of the visual field, auditory information arrives to each hemisphere from the entire auditory scene. Because of this, unilateral lesions to auditory areas often leave sound processing relatively functional overall and thus auditory agnosia typically requires bilateral lesions to the lateral temporal lobes (which usually only occurs with two separate cerebrovascular accidents). A second reason for the rarity of auditory agnosia diagnoses is that auditory agnosia may often be obscured by other deficits; for example, verbal auditory agnosia may not be obviously distinguishable from other forms of posterior aphasia (especially as auditory agnosia often evolves from initial presentations of sensory aphasia). In any case, there is a relative scarcity of data on auditory agnosia. This, combined with the considerable variability in clinical presentation and in exactly what is tested across cases, means that there is not yet a consensus on the specific etiologies, or even the specific symptoms, associated with these subtypes of auditory agnosia.

An influential distinction has been made between apperceptive and associative forms of auditory agnosia (e.g., Vignolo, 1982; Buchtel and Stewart, 1989). This apperceptive/associative distinction comes from work on visual agnosia, where Lissauer (1890; cited in Farah, 2004) suggested that object recognition could be impaired via an apperceptive deficit, due to problems with visual object processing, or via an associative deficit, due to problems associating intact perceptual information with object representations. The division of apperceptive and associative deficits is less straightforward in audition than in vision, as it is not entirely obvious what constitutes an auditory object (Griffiths and Warren, 2004). Nevertheless, early conceptions of auditory agnosia were primarily associative (e.g., Lichtheim, 1885; Bastian, 1897), assuming that basic perceptual processing (apperception) was intact and the deficit resulted from an inability to link a perceived pattern to an appropriate auditory object, or sound meaning. Other theories assume that auditory agnosia is best characterized as an apperceptive disorder, pointing to the fact that agnosia subtypes typically overlap, which likely reflects underlying problems with sound analysis (Goldstein, 1974; Buchtel and Stewart, 1989; Griffiths et al., 1999; Griffiths, 2002). Still, while considerable work argues that auditory agnosia (and its subtypes) may indeed reflect relatively basic problems perceiving and/or analyzing sound patterns (e.g., Albert and Bear, 1974; Buchtel and Stewart, 1989; Phillips and Farmer, 1990; Poeppel, 2001; Vignolo, 2003; Stefanatos, 2008; Slevc et al., 2011), there are cases that seem better characterized as associative deficits (e.g., Franklin et al., 1996; Saygin et al., 2010). This apperceptive/associative distinction might thus be expected to link different forms of auditory agnosia to different anatomic etiologies (e.g., Buchtel and Stewart, 1989; Goll et al., 2010b).

This chapter does not take apperceptive/associative as its primary organizational principle (but see Goll et al., 2010b, for such an approach). Instead, we divide auditory agnosia into four major categories – (1) general auditory agnosia; (2) agnosia for speech (verbal auditory agnosia or pure word deafness); (3) agnosia for environmental sounds; and (4) agnosia for music (amusia) – and discuss apperceptive and associative variants within each subtype (for other reviews, see Polster and Rose, 1998; Griffiths et al., 1999, 2012; Simons and Lambon Ralph, 1999; Griffiths, 2002; Bauer and McDonald, 2003; Lechevalier et al., 2003; Goll et al., 2010b; Stefanatos and DeMarco, 2012). This taxonomy of auditory agnosia subtypes is, of course, somewhat problematic as even the rare cases of domain specificity do not typically show perfect dissociations (if multiple domains are even assessed). In addition, these categories do not make up an exhaustive set of auditory perceptual deficits; for example, one auditory perceptual deficit not addressed here is selective to sound localization rather than identification (e.g., Clarke et al., 2000, 2002; Adriani et al., 2003).

Although this could be considered as a type of auditory agnosia, our focus is on deficits in sound identification and/or discrimination so we do not discuss deficits in auditory spatial perception here (instead, see Chapter 31, this volume). Finally, it is important to acknowledge that cases with similar types of impairment may or may not involve disruption of the same underlying process(es), thus cases should be evaluated individually instead of attempting to make conclusions from the average performance of a group from a particular clinical category (Caramazza, 1984). That said, a division into a tractable number of subtypes may be a necessary simplification in order to make any generalizations from the highly variable and unique series of individual case studies that make up our knowledge of auditory agnosia. Table 32.1 thus indicates the common pattern of spared

Table 32.1

Subtypes of auditory agnosia

Typical clinical features	General	Verbal		Phonagnosia		Non-verbal	Amusia	
		Apperceptive	Associative	Apperceptive	Associative		Apperceptive	Associative
Hearing ability (audiometric sensitivity)	+	+	+	+	+	+	+	+
Speech comprehension	–	–	–	+	+	+	+	+
Speech repetition	–	–	+	+	+	+	+	+
Non-speech sound recognition	–	+	+	+	+	–	+/–	+
Non-speech sound discrimination	–	+	+	+	+	+/–	+/–	+
Pitch perception	–	+/–	+	+/–	+	+/–	–	+
Rhythm perception	–	+/–	+	+/–	+	+/–	–	+
Timbre perception	–	+/–*	+	+/–	+	+/–	–	+
Recognition of familiar melodies	–	+/–	+/–*	+	+	+/–	–	–
Voice discrimination	–	+/–	+	–	+	+	+	+
Recognition of familiar voices	–	+/–	+	–	–	+	+	+

Typical perceptual abilities in auditory agnosia subtypes. + intact abilities; – impaired abilities; +/– mixed findings; *likely patterns, but based on insufficient evidence.

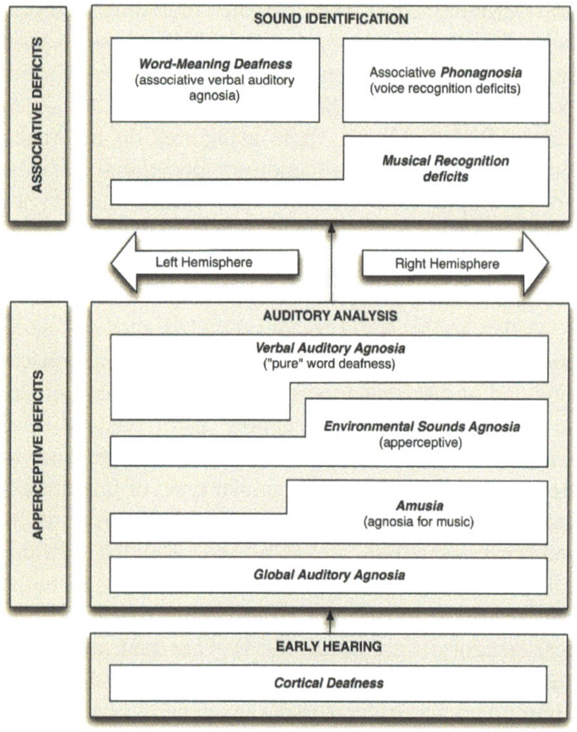

Fig. 32.1. Typical lesion laterality for auditory agnosia subtypes, divided into deficits of auditory analysis (apperceptive agnosias; bottom) and deficits of sound identification (associative agnosias; top) deficits. Most cases involve lesions to left, right, or both temporal cortices, although many cases involve subcortical and/or more extensive cortical lesions.

and impaired abilities associated with different subtypes of auditory agnosia, and Figure 32.1 indicates the typical laterality of lesions underlying different agnosia subtypes.

GENERAL AUDITORY AGNOSIA

Although much of the interest in auditory agnosia comes from specific subtypes, the most common presentation is of general (or global) agnosia across all auditory domains (e.g., Oppenheimer and Newcombe, 1978; Miceli et al., 1980; Auerbach et al., 1982; Miceli, 1982; Kazui et al., 1990). General auditory agnosia can develop from what initially appears to be cortical deafness (Ziegler, 1952; Mendez and Geehan, 1988; Godefroy et al., 1995), and can sometimes resolve into more specific subtypes of auditory agnosia (e.g., Motomura et al., 1986; Engelien et al., 1995). General auditory agnosia is characterized by impaired recognition of all types of auditory stimuli, including speech, environmental sounds, and music. This broad deficit suggests an underlying problem with sound recognition that affects all domains of auditory perception, and indeed general auditory agnosia is associated with deficits in both temporal and spectral auditory analysis (Tanaka et al., 1987; Mendez and Geehan, 1988). General auditory agnosia is most often associated with bilateral temporal damage: of 27 cases cited here that report both a clear deficit in at least two domains of auditory perception and reasonably clear lesion localization, 17 involved bilateral temporal damage (including insular cortex) and 5 involved other types of bilateral damage (mostly subcortical). Although general auditory agnosia is likely the most common type of auditory agnosia, most investigations have focused on the specific subtypes (as will be evident below). In some cases, this truly reflects investigations of patients with relatively specific deficits, but in others it is rather a matter of terminology, where a patient who might be best diagnosed with general auditory agnosia is instead labeled as a case of some specific subtype due to insufficient testing to reveal the extent of cross-domain deficits. For example, in a careful analysis of a patient previously diagnosed with pure word deafness (verbal auditory agnosia), Pinard et al. (2002) found considerable deficits also in environmental sound and musical perception.

VERBAL AUDITORY AGNOSIA (WORD DEAFNESS)

Agnosia specific to speech sounds is often called pure word deafness (and sometimes also called word sound deafness or just word deafness). This is, critically, distinct from posterior (sensory) aphasia in that other domains of linguistic processing, including reading, writing, and speaking, are relatively intact, although agnosia is often still accompanied by some degree of aphasic symptoms (for reviews, see Goldstein, 1974; Buchman et al., 1986; Poeppel, 2001; Badecker, 2005; Stefanatos et al., 2005b). Verbal auditory agnosia often evolves from Wernicke's aphasia, in that paraphasias and reading and writing deficits resolve without corresponding improvement of speech perception and repetition (e.g., Albert and Bear, 1974; Slevc et al., 2011). The term pure word deafness is somewhat misleading as the syndrome does not typically seem to be an agnosia for words *per se*, but rather agnosia for speech sounds. That is, verbal auditory agnosia is most often apperceptive in nature. In addition, word deafness is rarely (if ever) pure; in a review of 37 reports of pure word deafness, Buchman et al. (1986) argued that word deafness never occurred without other auditory processing impairments (but see Coslett et al., 1984; Metz-Lutz and Dahl, 1984; Yaqub et al., 1988; Takahashi et al., 1992). We thus use the term verbal auditory agnosia here to emphasize its relationship to other auditory agnosias (Wang et al., 2000). Although there is still relatively little research given the rarity of this syndrome, verbal auditory agnosia is arguably the most well-studied subtype of auditory agnosia. This greater concentration of work on verbal auditory agnosia does not necessarily reflect a greater prevalence compared to other agnosia subtypes; rather, it likely reflects greater rates of diagnosis and referrals because speech perception deficits can be particularly detrimental to overall functioning.

Verbal auditory agnosia is sometimes grouped with the aphasias as a language disorder, and taken as evidence for speech specificity at an early stage of auditory perception. This fits with claims that even the very early stages of speech sound perception occur in a language-specific way (e.g., Liberman and Mattingly, 1989; Liberman and Whalen, 2000) and with evidence for very early left lateralization of speech perception (e.g., Näätänen et al., 1997; Hornickel et al., 2008). However there is also considerable evidence that the early stages of speech processing occur bilaterally (for discussion, see Hickok and Poeppel, 2004, 2007, and Chapter 9, this volume), and that lateralization differences may instead reflect hemispheric specializations for more basic auditory processes. Indeed, verbal auditory agnosia typically results from bilateral temporal lesions in the superior temporal cortex (or subcortically, affecting the relevant projecting auditory radiations; see Chapter 1, Fig. 1.9): of 63 reports with sufficiently detailed lesion data (Wang et al., 2000; Jörgens et al., 2008; Slevc et al., 2011; and earlier cases reviewed by Poeppel, 2001), 43, or almost 70%, involve bilateral lesions. Patients with verbal auditory agnosia are aware of speech but describe it as sounding like a foreign language (Albert and Bear, 1974), as distorted and cartoonlike (Wee and Menard, 1999), as rapidly fading (Klein and Harper, 1956), or with descriptions like "voice comes but no words" (Hemphill and Stengel, 1940). Speech production in verbal auditory agnosia is relatively normal (Table 32.1), although in some cases production is overly loud and has abnormal prosody (e.g., Otsuki et al., 1998; Jörgens et al., 2008). Speech perception is typically improved by relying on cross-modal information such as lip reading and from "top-down" contextual information (e.g., Saffran et al., 1976; Coslett et al., 1984; Buchtel and Stewart, 1989; Slevc et al., 2011; Robson et al., 2012). Perception is also sometimes improved by dramatically slowing the speech signal (Albert and Bear, 1974; Stefanatos et al., 2005a), perhaps by ameliorating some patients' particular difficulty with rapid temporal aspects of the speech signal (see below).

Although verbal auditory agnosia is often claimed to only rarely result from unilateral damage (e.g., Poeppel, 2001; Bauer and McDonald, 2003), unilateral cases are by no means vanishingly rare. In fact, the first described cases of verbal auditory agnosia were in patients with unilateral left temporal lesions (Kussmaul, 1877; Lichtheim, 1885) and, as mentioned above, approximately 30% of cases with clearly identifiable lesions have only unilateral damage. These unilateral lesions are overwhelmingly left-lateralized; there is only a single report of a right-hemisphere lesion associated with verbal auditory agnosia (Roberts et al., 1987). Interestingly, this was not obviously a case where language was right-lateralized: the patient showed a strong right-hand bias and had no family history of left-handedness. (Note also that this patient's deficit was probably not specific to speech – he reported an acquired inability to appreciate music or recognize familiar melodies as well as some difficulty recognizing environmental sounds – although it is unclear how severe these other deficits were as non-speech perception was not formally tested.)

A common way to reconcile these two anatomic etiologies is to think of verbal auditory agnosia as a disconnection syndrome (Lichtheim, 1885; Geschwind, 1965; Gazzaniga et al., 1973; Takahashi et al., 1992), as illustrated in Figure 32.2. By this account, bilateral temporal damage prevents auditory input from either hemisphere from reaching speech-processing mechanisms (i.e., Wernicke's area; for example, see Fig. 32.2A). Unilateral damage must not only interrupt ipsilateral transmission of left-hemisphere auditory input to speech-processing areas, but also destroy contralateral projections from right-hemisphere auditory analysis (Fig. 32.2B and C). In fact, this is essentially the early conception of verbal auditory agnosia as a primarily associative deficit (Lichtheim, 1885; Bastian, 1897; see Goldstein, 1974, for a review), in which the "left auditory word center" (i.e., Wernicke's area) is dissociated from auditory input (by this account, the "word center" itself was assumed to be spared given preserved speech production). If verbal auditory agnosia is a disconnection syndrome, then unilateral cases of verbal auditory agnosia should be accompanied by damage to cross-hemispheric white-matter pathways from primary auditory cortex in the

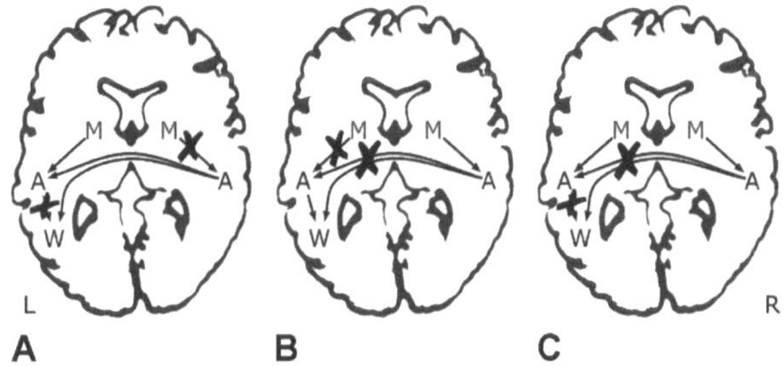

Fig. 32.2. Three hypothesized lesion profiles that could result in verbal auditory agnosia / word deafness, according to a disconnection approach. A, auditory area; M, medial geniculate body; W, Wernicke's area; X, lesion. (Adapted from Takahashi et al., 1992; brain outlines courtesy of Akira O'Connor.)

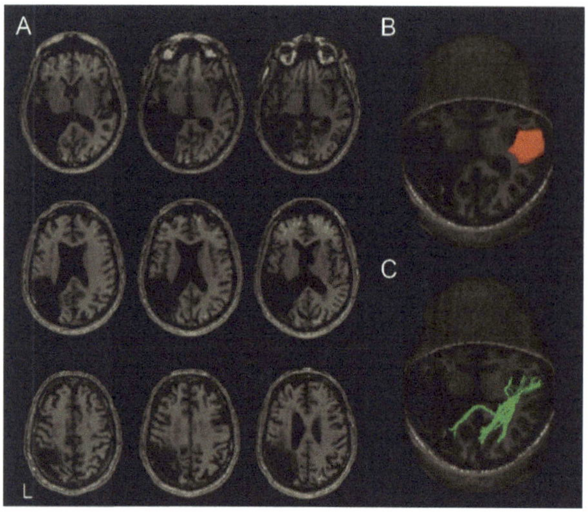

Fig. 32.3. (**A**) Anatomic magnetic resonance imaging images from a patient with verbal auditory agnosia following a left-hemisphere lesion. (**B**) Region of interest (ROI) used to define Heschl's convolutions in the right hemisphere, and (**C**) white-matter pathways passing through the ROI shown in (**B**) as assessed with diffusion tensor imaging (DTI). (Reproduced from Slevc et al., 2011.)

undamaged hemisphere. So far, however, the only study that has investigated white-matter connectivity patterns in a patient with unilateral verbal auditory agnosia (using diffusion tensor imaging; see Chapter 16, this volume) found preserved cross-hemispheric white-matter pathways (Slevc et al., 2011; Fig. 32.3). These preserved pathways suggest that such a disconnection is not the only cause of verbal auditory agnosia with unilateral damage (although note that the preserved existence of white-matter tracts does not necessarily imply their preserved functionality). Instead, it may be that verbal auditory agnosia (at least in cases with unilateral damage) is not an associative disorder, but rather an apperceptive deficit in some aspect of auditory processing that is both left-lateralized and disproportionately involved in speech perception.

Apperceptive verbal auditory agnosia

As noted above, considerable work suggests that verbal auditory agnosia may be best characterized as an apperceptive disorder, reflecting an auditory processing deficit that disproportionately affects speech (Albert and Bear, 1974; Buchtel and Stewart, 1989; Phillips and Farmer, 1990; Griffiths et al., 1999; Poeppel, 2001; Griffiths, 2002). A particularly influential proposal is that verbal auditory agnosia might result from an underlying deficit in rapid temporal processing that is particularly important for the perception of some speech sounds, especially for some consonants that are characterized by rapid temporal transitions on the order of tens of milliseconds (e.g., Shannon et al., 1995). A rapid temporal processing deficit could disproportionately affect speech perception as the processing of environmental sounds and of music likely relies more on pitch and spectral cues (de Cheveigné, 2005) and on temporal cues occurring over relatively longer time periods.

Evidence for a rapid temporal processing deficit in verbal auditory agnosia comes from the observation that individuals with verbal auditory agnosia often have particular difficulty with temporally dynamic aspects of speech stimuli, such as place of articulation and voicing contrasts in consonants, compared to temporally protracted aspects of speech such as vowels (Saffran et al., 1976; Auerbach et al., 1982; Miceli, 1982; Yaqub et al., 1988; Praamstra et al., 1991; Wang et al., 2000; Stefanatos et al., 2005a; Slevc et al., 2011). Verbal auditory agnosia also is associated with pronounced deficits in perception of rapid temporal changes in non-speech stimuli (Fig. 32.4), including most (5 out of 6) cases that have evaluated click fusion and gap detection thresholds (Fig. 32.4C; Albert and Bear, 1974; Tanaka et al., 1987; Otsuki et al., 1998; Yaqub et al., 1988; Jörgens et al., 2008; but see Stefanatos et al., 2005b) as well as 3 cases with impaired discrimination of non-speech sine-wave stimuli differing in rapid temporal transitions (Fig. 32.4B; Wang et al., 2000; Stefanatos et al., 2005b; Slevc et al., 2011). Temporal processing deficits are not, however, uniquely associated with verbal auditory agnosia; impairments on click fusion tasks have also been reported in other forms of auditory agnosia (in 9 of 10 tested cases: Albert et al., 1972; Auerbach et al., 1982; Motomura et al., 1986; Buchtel and Stewart, 1989; Kazui et al., 1990; Best and Howard, 1994; Godefroy et al., 1995; Mendez, 2001; Stefanatos, 2008; but see Lambert et al., 1989). In 3 of these cases, a gradual improvement in temporal processing (as assessed with click fusion thresholds) accompanied recovery from auditory agnosia (Motomura et al., 1986; Best and Howard, 1994; Godefroy et al., 1995), lending further (although correlational) support to this relationship.

It is not yet clear how such an apperceptive deficit in rapid temporal processing relates to the different lesion profiles of verbal auditory agnosia. One proposal is that an apperceptive temporal processing deficit, or a prephonemic variant of verbal auditory agnosia, reflects an underlying deficit in temporal acuity due to bilateral lesions, whereas an associative, phonemic variant results from left-hemisphere damage (Auerbach et al., 1982; Poeppel, 2001). However, other theories point to a left-hemisphere specialization for rapid temporal analysis, fitting with general theories of auditory processing that propose language lateralization reflects an underlying left-hemispheric specialization for rapid temporal

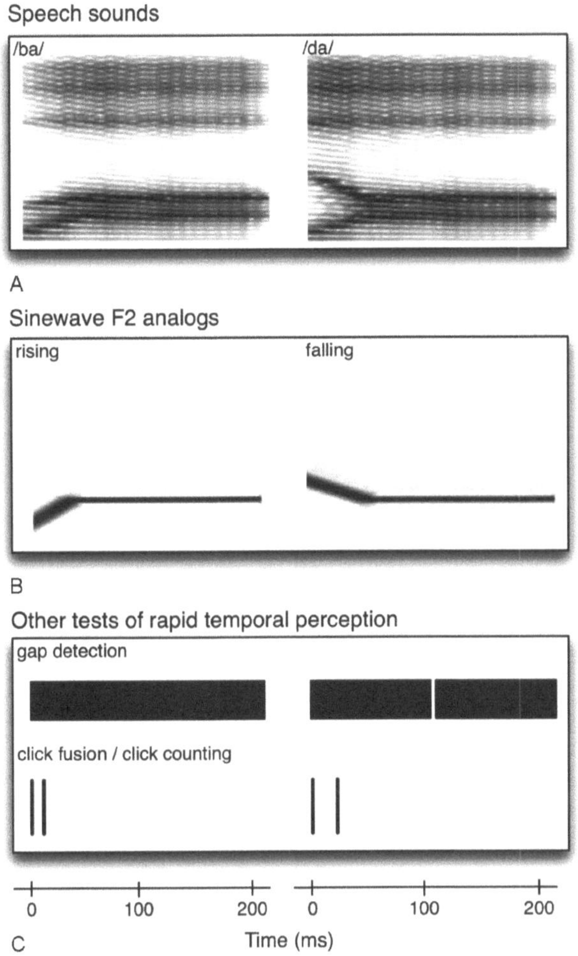

Fig. 32.4. Example stimuli for tests used to assess rapid temporal processing in (verbal) auditory agnosia. (**A**) Synthesized speech stimuli from a /ba/-/da/ continuum, differing in the rapid (~40 ms) onset transition in the second formant (F2) and (**B**) non-speech analogs of F2 in (**A**). (Adapted from Slevc et al., 2011). (**C**) Schematic of gap detection and click fusion tasks.

analysis and a right-hemispheric specialization for either spectral analysis or for temporal analysis over a longer time window (e.g., Poeppel, 2001; Zatorre and Belin, 2001; Boemio et al., 2005; Hickok and Poeppel, 2007; Poeppel and Monahan, 2008; but see McGettigan and Scott, 2012). This distinction implies that cases of verbal auditory agnosia due to unilateral (left-hemisphere) damage might reflect an underlying deficit in rapid temporal processing, and, indeed, at least three case reports of unilateral verbal auditory agnosia have found particular difficulties with rapidly changing stimuli (Wang et al., 2000; Stefanatos et al., 2005b; Slevc et al., 2011). Stefanatos et al.'s (2005b) patient NH showed impaired perception of rapid temporal transitions but did not show a complementary deficit in basic click fusion threshold (click fusion data were not available in these other cases), raising the possibility that unilateral cases of verbal auditory agnosia might reflect a somewhat "higher-level" deficit of temporal processing despite preserved "lower-level" temporal perception.

While these findings have been taken as evidence that speech perception deficits in verbal auditory agnosia often do result from an underlying deficit in rapid temporal processing (e.g., Phillips and Farmer, 1990; cf. work on developmental language deficits, e.g., Merzenich et al., 1993; Tallal, 2004), it is important to note that not all cases of verbal auditory agnosia are associated with impairments in basic temporal acuity (e.g., Stefanatos et al., 2005b), and some show deficits even with relatively slow temporal changes on the order of hundreds of milliseconds (e.g., Wang et al., 2000; Jörgens et al., 2008). Furthermore, an attempt to improve rapid temporal processing via a training paradigm in one verbal auditory agnosia patient led to improvements only on discrimination of non-speech stimuli (Slevc et al., 2011), supporting claims that the demands speech perception places on rapid temporal processing are somewhat overstated (McGettigan and Scott, 2012). Nevertheless, the bulk of evidence suggests that rapid temporal processing deficits may underlie at least some cases of auditory agnosia. A better understanding of the precise role of temporal processing deficits in auditory agnosia will depend on more consistent testing of complex sound perception in agnosia patients using well-validated psychoacoustic techniques (e.g., Griffiths et al., 1999).

Associative verbal auditory agnosia

Although early conceptions of verbal auditory agnosia were associative in nature (see above discussion on the disconnection approach), there are very few reports of clearly associative forms of verbal auditory agnosia. The cases that do exist are often termed word-meaning deafness, first reported by Bramwell (1897; reprinted in Ellis, 1984), or occasionally described as a variant of transcortical sensory aphasia (e.g., Heilman et al., 1981). Word-meaning deafness is characterized by intact ability to repeat speech, discriminate phonemes, and perform auditory lexical decisions, but impaired ability to comprehend spoken words (Kohn and Friedman, 1986; Franklin et al., 1994, 1996; Hall and Riddoch, 1997; Plasencia et al., 2006; Wirkowski et al., 2006). These patients thus may have the somewhat surprising ability to write down a word or phrase that they do not understand and then successfully understand what they have written (Hall and Riddoch, 1997; Francis et al., 2001). Auditory comprehension of abstract words may be worse than of concrete words (Franklin et al., 1994, 1996), although this seemingly conceptually based dissociation has also been explained in terms of an apperceptive deficit that essentially causes "noisy" input to the semantic system (Tyler and Moss, 1997). "Pure" cases of word-meaning deafness are very rare, with only two prototypical patients in the literature – KW (Hall and Riddoch, 1997; Francis et al., 2001) and DrO (Franklin et al., 1996; Tyler and Moss, 1997) – and only a handful more when including less clear-cut cases (Heilman et al., 1981; Ellis, 1984; Kohn and Friedman, 1986; Franklin et al., 1994; Plasencia et al., 2006; Wirkowski et al., 2006). KW, DrO, and 5 of the 7 other cases appear to have suffered unilateral left cortical lesions, although note that similar lesions can also result in an apperceptive form of verbal auditory agnosia (e.g., Wang et al., 2000; Slevc et al., 2011). Unfortunately, word-meaning deafness is not yet well understood even by the standards of auditory agnosia, and the few reports available suggest substantial heterogeneity in the associated perceptual/cognitive deficits.

Table 32.2

Summary of experimental neuropsychological profiles in phonagnosia patients QR and KL

	Domain	Case QR	Case KL
Voices	Identification	↓	↓
	Familiarity	↓	↓
	Emotion recognition	N	N
	Perception	N	N
Other sounds	Musical instrument matching	↓	N
	Environmental sound recognition	N	N
Faces	Recognition	N*	↓↓
	Perception	N	N

↓impaired performance relative to controls.
↓↓impaired performance relative to both controls and other case.
*When matched to voices for difficulty.
N, normal performance.
Reproduced from Hailstone et al. (2010).

PHONAGNOSIA

Agnosias for speech are not always lexical in nature; phonagnosia is an apparent auditory parallel to prosopagnosia (impaired visual recognition of faces; e.g., Gainotti and Marra, 2011), characterized by difficulty recognizing familiar voices (Van Lancker and Kreiman, 1987; Van Lancker et al., 1989; Hailstone et al., 2010). Phonagnosia can occur in an apperceptive form, characterized by difficulty discriminating between voices after either left or right temporal damage, and an associative form, characterized by an inability to recognize voices after right parietal damage (Van Lancker et al., 1989; Lang et al., 2009). Phonagnosia can also accompany frontotemporal lobar degeneration (Hailstone et al., 2010) or can arise as a developmental deficit (Garrido et al., 2009). The extent to which phonagnosia can occur in the absence of other auditory perceptual or person identification deficits remains unclear; some patients with phonagnosia do appear to be able to successfully recognize speech and environmental sounds; however the deficit is more often associated with other forms of auditory agnosia and/or with other types of person recognition deficits such as prosopagnosia (see, e.g., Table 32.2; Neuner and Schweinberger, 2000).

AGNOSIA FOR ENVIRONMENTAL SOUNDS

Auditory agnosia can also occur with (relatively) intact speech perception, but impairment in perceiving and/or comprehending non-speech environmental sounds (Spreen et al., 1965; Albert et al., 1972; Motomura et al., 1986; Fujii et al., 1990; Taniwaki et al., 2000). For example, Spreen et al.'s (1965) patient identified the sound of scissors as "a clock ticking" and rattled keys as "a chain jingling or a bell ringing." This selective non-speech perceptual deficit is sometimes called auditory agnosia or non-verbal auditory agnosia (e.g., Vignolo, 1982, 2003; Saygin et al., 2010); however, here we use the more specific term environmental sound agnosia, referring specifically to difficulty in identifying familiar non-speech sounds. Many cases of environmental sound agnosia also involve deficits of musical perception (e.g., Spreen et al., 1965; Lambert et al., 1989; Eustache et al., 1990; Vignolo, 2003), although this is difficult to evaluate precisely as musical perception is often not tested at all (e.g., Clarke et al., 2000) or only tested in a very general way (e.g., successful recognition of familiar songs (Eustache et al., 1990; Saygin et al., 2010) may be possible even in the presence of abnormal musical processing). Nevertheless, at least two reports suggest that environmental sound agnosia can exist independently of both musical and speech perception deficits (Motomura et al., 1986; Saygin et al., 2010). There are relatively few investigations of environmental sound agnosia in the absence of verbal auditory agnosia, perhaps because such deficits are less detrimental to everyday functioning than verbal auditory agnosia, and so tend to go unreported. Vignolo (1982) points out the additional reason why careful tests of environmental sound perception have rarely been conducted: only recently have batteries of recorded environmental sounds been easily available for testing (e.g., Saygin et al., 2003), and many cases simply involve the observation that a patient identifies or even just reacts appropriately to some small set of easily produced sounds.

Environmental sound agnosia has been associated with a variety of lesion locations. Typically, it occurs as a result of cortical damage, which can be unilateral (to the left or right hemisphere) or bilateral (Schnider et al., 1994). However there are also cases arising from subcortical lesions to the auditory radiations bilaterally (Godefroy et al., 1995; Taniwaki et al., 2000). These different lesion etiologies may be linked to different forms of environmental sound agnosia: an apperceptive form, where identification errors are acoustically similar to the target (e.g., mistaking cat meows for someone singing), has been associated with right-hemisphere lesions (Fujii et al., 1990), and an associative form, where targets are additionally confused with semantically related sources (e.g., choosing a dog as the source of a cat meow), has been associated with left-hemisphere lesions (Vignolo, 1982; Schnider et al., 1994; but see Clarke et al., 1996).

Other work suggests more complicated dissociations in associative environmental sound processing. For example, Trumpp et al. (2013) report a patient with damage to left posterior superior and middle temporal gyrus who showed difficulty identifying everyday object sounds with high acoustic relevance (e.g., telephone), but normal identification of sounds from animals and musical instruments. This patient showed a similar dissociation in lexical decision and in verbal fluency (abnormally slow lexical decisions and generating abnormally few exemplars only for sound-related objects), suggesting that an associative environmental sound deficit may also impair aspects of conceptual representations, and fitting with the idea that modality-specific perceptual information is an important aspect of conceptual knowledge (e.g., Barsalou, 2008).

AMUSIA

The testing of musical sound analysis after cerebrovascular accident (or after any other cause of acquired auditory agnosia) is more challenging than testing speech or environmental sound perception because, unlike in these other domains, experience and training with music vary widely across the population and so it is not possible to assume a common baseline level of musical ability. A majority of work that has attempted to directly test for amusia has relied on a standardized diagnostic battery: the Montreal Battery for Evaluation of Amusia (Peretz et al., 2003). Here, we only briefly discuss acquired agnosia for musical perception; for more detailed treatment of acquired amusia, see Chapter 34 and for discussion of congenital amusia, see Chapter 33.

In brief, amusia can also occur in apperceptive and associative forms. Apperceptive cases of acquired amusia are characterized by perceptual deficits in musical perception in the face of preserved musical production (Griffiths et al., 1997; Ayotte et al., 2000). Such sensory amusia generally results from right-hemisphere lesions (Stewart et al., 2006), although cases have also been reported with bilateral damage (e.g., Peretz et al., 1994). Music is a complex auditory stimulus, and while relatively little work has contrasted different aspects of musical perception, it does appear that agnosia can be relatively specific to different musical features. For example, there are reports of amusia confined to perception of pitch (Peretz et al., 1994), timbre (Kohlmetz et al., 2003), or rhythm (Mavlov, 1980). In addition, some aspects of musical perception can be selectively spared: Mendez (2001) reports a patient with global agnosia who developed an increased appreciation for music, perhaps relying on spared rhythm perception. An associative form of amusia was reported by Peretz (1996), whose patient CN showed preserved processing of pitch, timbre, and rhythm, but an impaired ability to recognize melodies and sing from memory (despite preserved ability to recognize familiar song lyrics) following bilateral temporal lesions. Such associative cases appear to be rare; however, it is difficult to assess the prevalence of musical agnosia (of any type) as such deficits do not often greatly disrupt everyday life, and so are likely to go unreported.

AUDITORY AFFECTIVE AGNOSIA

Auditory affective agnosia or sensory aprosodia refers to selective impairments in recognizing affective information in speech despite relatively normal production of emotional prosody (Heilman et al., 1975; Darby, 1993). This is sometimes classified as a paralinguistic variant of verbal auditory agnosia (e.g., Bauer and McDonald, 2003). However the deficit seems likely to reflect problems understanding how variations in pitch, timbre, and volume reflect aspects of emotional communication and/or the speaker's emotional state, and so may be better thought of as a form of associative amusia. Auditory affective agnosia is often associated with right posterior damage, especially in the right temporal operculum (Gorelick and Ross, 1987; Ross and Monnot, 2008), and is likely associated with other kinds of pragmatic impairments associated with right-hemisphere damage (e.g., Brownell et al., 1983). Alternatively, auditory affective agnosia might reflect an apperceptive deficit in perceiving the musical aspects of speech, which would then be closely related to apperceptive amusia (Patel, 2008). These possibilities have not yet been directly contrasted, as patients with impairments in auditory affective perception have not been carefully tested on other domains of affective perception or on perception of the musical features corresponding to affective speech.

Thus it remains unclear whether auditory affective agnosia is a separate form of associative auditory agnosia, a more general consequence of affective deficits, or a subtype of apperceptive amusia.

TREATMENT FOR AUDITORY AGNOSIA

There have been few attempts to treat the perceptual deficits associated with auditory agnosia; those that exist have investigated a diverse set of patients and have met with only mixed success. Essentially all treatment studies so far focus on the remediation of speech perception deficits, either in patients with verbal auditory agnosia specifically or, more commonly, in patients with a variety of perceptual and/or aphasic symptoms. The most common approach along these lines has simply been to provide training on minimal-pair discrimination or phoneme identification tasks, sometimes supported by visual cues (Morris et al., 1996; Wee and Menard, 1999; Maneta et al., 2001; Tessier et al., 2007). Morris et al.'s (1996) patient had severe difficulties with speech perception (but intact environmental sound perception) in the context of broader aphasic symptoms. They administered an adaptive syllable discrimination training, where the patient first practiced discriminating between syllables differing in three distinctive features (voice, place, and manner of articulation) then, as performance improved, syllables differing on two of these features, then syllables differing in only one distinctive feature. Although their patient did improve on minimal-pair discrimination and repetition following 6 weeks (12 sessions) of this training, he did not show commensurate improvements in comprehension.

Maneta et al. (2001) reported a treatment study for another aphasic patient with a particularly severe speech perception deficit (perception of non-speech sounds was not tested). This treatment program used a similar program of minimal-pair discrimination training, plus additional training on lip reading (via diagrams of lip shapes plus exaggerated lip movements during the discrimination tasks) and on cued speech (hand signals corresponding to voicing, manner, and place of articulation); however they found no improvements in speech sound discrimination after 12 sessions. Tessier et al. (2007) found more encouraging results for a patient with general auditory agnosia: phoneme recognition, discrimination, and speech comprehension were improved after a treatment program that involved intensive phoneme discrimination and phoneme recognition tasks accompanied by visual cues (written representations of the phonemes) that were gradually delayed and removed. Francis et al. (2001) report a treatment program for a patient with word-meaning deafness that involved a series of reading and auditory comprehension exercises. This met with some success, although improvements may have partially reflected an adopted strategy of covertly spelling and visualizing the target words.

Another treatment approach, based on theories of a rapid temporal processing deficit underlying verbal auditory agnosia, is to employ treatment specifically designed to improve temporal processing. Stefanatos et al. (2005b) found that extended formant transitions in digitally synthesized speech stimuli did not lead to notable improvements in phonemic discrimination in their verbal auditory agnosia patient, but did show that temporally extending the duration of naturally produced words and sentences led to some improvement in comprehension (Albert and Bear, 1974). Slevc et al. (2011) employed a similar sound manipulation strategy in a training task, using a treatment designed to improve rapid temporal processing in developmental language deficits (Merzenich et al., 1993; see Tallal, 2004, for a review) over a relatively long treatment period (43 sessions over 2.5 months). Although their verbal auditory agnosia patient improved in the ability to discriminate rapid temporal transitions in non-speech stimuli (sine-wave sweeps; Fig. 32.4B), this was not accompanied by any improvements in discrimination of synthesized speech stimuli or in speech perception more generally. Thus there is little evidence overall for successful treatment programs for speech perception in auditory agnosia, although this may simply reflect the limited number of treatment studies to date.

Even less work has investigated treatment of other types of sound processing in auditory agnosia; in fact, there do not appear to be any documented treatment studies targeting deficits in environmental sound or music perception. In terms of spontaneous recovery, some patients recover from auditory agnosia altogether (e.g., Fujii et al. 1990), while other cases show either no recovery or persistence of specific perceptual deficits despite recovery of other forms of sound identification. Saygin et al. (2010) present an interesting case of a patient with a unilateral left-hemisphere lesion to Wernicke's area (Fig. 32.5) who recovered his speech perception abilities but was left with severe environmental sound agnosia. They suggest that speech perception may have recruited intact auditory processing areas typically associated with non-verbal processing (i.e., speech perception may have "taken over" environmental sound perception) – a hypothesis supported by functional magnetic resonance imaging evidence showing his selective responses to speech in areas associated with general (non-speech-specific) auditory processing in controls (specifically, right anterior temporal cortex and left perilesional temporal cortex; see also Engelien et al., 1995). While speculative, this suggests the possibility of functional reorganization of the auditory system

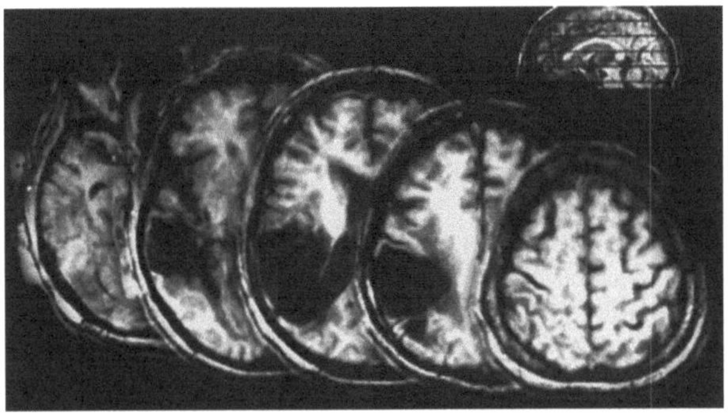

Fig. 32.5. Magnetic resonance imaging of patient M's lesion. Five selected axial images are shown to depict the extent of lesion. Neurologic convention. (Reproduced from Saygin et al., 2010.)

and highlights the need to consider how functional reorganization after brain damage may alter correlations between brain and behavior (Musso, 1999; Leff et al., 2002).

CONCLUSION

Much of the interest in auditory agnosia comes from the apparent specificity of the agnosia subtypes, including verbal auditory agnosia (or word deafness), environmental sound agnosia, and amusia. Auditory agnosia thus plays an important role in a larger debate about the underlying nature of auditory perception. It is certainly the case, however, that clear dissociations between disordered perception of speech, of environmental sounds, and of music are quite rare. This may indicate that the auditory agnosias are not, in fact, distinct deficits at the level of language, music, and environmental sounds, but instead reflect relative degrees of impairment of underlying aspects of auditory analysis that disproportionately affect different domains of processing (e.g., Griffiths et al., 1999; Poeppel, 2001). However, it is also possible that this taxonomy does reflect distinct deficits that are simply difficult to observe in the "natural experiments" that the neuropsychologic literature provides. That is, patients rarely, if ever, have circumscribed lesions in specific regions of interest; instead, lesions often affect large areas of neural tissue, and lesion locations are a function of vascular anatomy and other physiologic constraints. In addition, lesions do not simply disrupt the lesioned region, but also affect a much broader set of functional neural networks. Thus, while lesion evidence can demonstrate that a brain region is necessary for some function, it cannot show that the region is sufficient, as that function may depend as much or more on other aspects of a larger network.

These limitations are not specific to the study of auditory agnosia, but are part of a general set of caveats common to most neuropsychologic studies. A few other such caveats deserve mention. For one, behavioral deficits also change over time, both as a function of physiologic recovery and as patients develop compensatory cognitive processes. Because of this, it is not always trivial to separate normal functioning from abnormal (but perhaps effective) processing based on alternative adopted strategies (Francis et al., 2001; Saygin et al., 2010). A second caveat is that the interpretation of any particular pattern of performance may also be confounded due to comorbidity with other deficits. Auditory agnosia refers to a deficit in auditory perception in the face of relatively preserved hearing and cognitive/linguistic functioning; however most cases show at least some degree of impairment in other areas. In addition, most studies lack premorbid measures of the relevant abilities, making it difficult to objectively evaluate the extent of impairment. Finally, few studies comprehensively evaluate different aspects of auditory processing, resulting in cases of apparent domain specificity without appropriate evidence (for example, a doctor's jangling of keys is not a comprehensive test of environmental sound perception; cf. Pinard et al., 2002).

Fortunately, progress is being made on many of these fronts. Significant advances in the understanding of auditory agnosia are likely to accompany the use of more sophisticated behavioral assessments; for example, the use of careful psychoacoustic methods to investigate complex sound perception at levels intermediate between pure-tone audiometry and complex speech sounds (Griffiths et al., 1999). Advances will also likely accompany more sophisticated methods for brain imaging; for example, measures of white-matter tractography (see Chapter 16, this volume) provide a straightforward way to investigate disconnection accounts (Slevc et al., 2011). Additional advances will likely come with more comprehensive theories of auditory perception and cognition, informed both by the auditory agnosias and by neuroimaging work on non-disordered populations.

The different types of auditory perceptual deficits grouped under the term auditory agnosia provide an important window into our understanding of auditory perception and cognition. While there may yet be more questions than answers, a growing body of work investigating auditory agnosia is increasing our understanding of our remarkable ability to process complex sounds.

References

Adriani M, Maeder P, Meuli R et al. (2003). Sound recognition and localization in man: specialized cortical networks and effects of acute circumscribed lesions. Exp Brain Res 153: 591–604.

Albert ML, Bear D (1974). Time to understand: a case study of word deafness with reference to the role of time in auditory comprehension. Brain 97: 373–384.

Albert ML, Sparks R, von Stockert T et al. (1972). A case study of auditory agnosia: linguistic and non-linguistic processing. Cortex 8: 427–443.

Auerbach S, Allard T, Naeser M et al. (1982). Pure word deafness. Analysis of a case with bilateral lesions and a defect at the prephonemic level. Brain 105: 271–300.

Ayotte J, Peretz I, Rousseau I et al. (2000). Patterns of music agnosia associated with middle cerebral artery infarcts. Brain 123 (Pt 9): 1926–1938.

Badecker W (2005). Speech Perception Following Focal Brain Injury. In: DB Pisoni, RE Remez (Eds.), The Handbook of Speech Perception, Blackwell Publishing, Malden, MA, pp. 524–545.

Barsalou LW (2008). Grounded cognition. Annu Rev Psychol 59: 617–645.

Bastian H (1897). The Lumleian Lectures on some problems in connection with Aphasia and other speech defects: delivered before the Royal College of Physicians, April 1st, 1897. Br Med J 1 (1892): 833.

Bauer R, McDonald C (2003). Auditory agnosia and amusia. In: MJ Farah, TE Feinberg (Eds.), Patient-Based Approaches to Cognitive Neuroscience, MIT Press, Cambridge, MA, pp. 133–146.

Best W, Howard D (1994). Word sound deafness resolved? Aphasiology 8: 223–256.

Boemio A, Fromm S, Braun A (2005). Hierarchical and asymmetric temporal sensitivity in human auditory cortices. Nat Neurosci 8 (3): 389–395.

Brownell H, Michel D, Powelson A et al. (1983). Surprise but not coherence: sensitivity to verbal humor in right hemisphere patients. Brain Lang 18: 20–27.

Buchman A, Garron D, Trost-Cardamone J et al. (1986). Word deafness: one hundred years later. J Neurol Neurosurg Psychiatry 49 (5): 489–499.

Buchtel H, Stewart JD (1989). Auditory agnosia: apperceptive or associative disorder? Brain Lang 37 (1): 12–25.

Caramazza A (1984). The logic of neuropsychological research and the problem of patient classification in Aphasia. Brain Lang 21: 9–20.

Clarke S, Thiran A, De Ribaupierre F et al. (1996). Nonverbal auditory recognition in normal subjects and brain-damaged patients: evidence for parallel processing. Neuropsychologia 34 (6): 587–603.

Clarke S, Bellman A, Meuli R et al. (2000). Auditory agnosia and auditory spatial deficits following left hemispheric lesions: evidence for distinct processing pathways. Neuropsychologia 38 (6): 797–807.

Clarke S, Thiran A, Maeder P et al. (2002). What and where in human audition: selective deficits following focal hemispheric lesions. Exp Brain Res 147 (1): 8–15.

Coslett H, Brashear H, Heilman K (1984). Pure word deafness after bilateral primary auditory cortex infarcts. Neurology 34 (3): 347–352.

Darby D (1993). Sensory aprosodia: a clinical clue to lesions of the inferior division of the right middle cerebral artery? Neurology 43 (3): 567–572.

De Cheveigné A et al. (2005). Pitch perception models. In: C Plack, A Oxenham, RR Fay (Eds.), Pitch, Springer, New York, NY, pp. 169–233.

Ellis A (1984). Introduction to Byron Bramwell's (1897) Case of word meaning deafness. Cogn Neuropsychol 1 (3): 245–258.

Engelien A, Silbersweig D, Stern E et al. (1995). The functional anatomy of recovery from auditory agnosia. A PET study of sound categorization in a neurological patient and normal controls. Brain 118 (Pt 6): 1395–1409.

Eustache F, Lechvalier B, Viader F et al. (1990). Identification and discrimination disorders in auditory perception: a report on two cases. Neuropsychologia 28 (3): 257–270.

Farah MJ (2004). Visual Agnosia, MIT Press, Cambridge, MA.

Francis D, Riddoch M, Humphreys G (2001). Cognitive rehabilitation of word meaning deafness. Aphasiology 15 (8): 749–766.

Franklin S, Howard D, Patterson K (1994). Abstract word meaning deafness. Cogn Neuropsychol 11 (1): 1–34.

Franklin S, Turner J, Ralph M et al. (1996). A distinctive case of word meaning deafness? Cogn Neuropsychol 13 (8): 1139–1162.

Fujii T, Fukatsu R, Shin-ichi W et al. (1990). Auditory sound agnosia without aphasia following a right temporal lobe lesion. Cortex 26 (263–268).

Gainotti G, Marra C (2011). Differential contribution of right and left temporo-occipital and anterior temporal lesions to face recognition disorders. Front Hum Neurosci 5: 55.

Garrido L, Eisner F, McGettigan C et al. (2009). Developmental phonagnosia: a selective deficit of vocal identity recognition. Neuropsychologia 47 (1): 123–131.

Gazzaniga M, Velletri Glass A, Sarno M et al. (1973). Pure word deafness and hemispheric dynamics: a case history. Cortex 9 (1): 136–143.

Geschwind N (1965). Disconnection syndromes in man and animals. Brain 88: 237–294.

Gibbons C, Oken B, Fried-Oken M (2012). Augmented input reveals word deafness in a man with frontotemporal dementia. Behavioral Neurobiology 25: 151–154.

Giraud A-L, Poeppel D (2012). Cortical oscillations and speech processing: emerging computational principles and operations. Nat Neurosci 15 (4): 511–517.

Godefroy O, Leys D, Furby A et al. (1995). Psychoacoustical deficits related to bilateral subcortical hemorrhages. A case with apperceptive auditory agnosia. Cortex 31 (1): 149–159.

Goldstein M (1974). Auditory agnosia for speech ('pure word deafness'); A historical review with current implications. Brain Lang 1: 195–204.

Goll J, Crutch S, Loo J et al. (2010a). Non-verbal sound processing in the primary progressive aphasias. Brain 133 (Pt 1): 272–285.

Goll J, Crutch S, Warren JD (2010b). Central auditory disorders: toward a neuropsychology of auditory objects. Curr Opin Neurol 23 (6): 617–627.

Gorelick P, Ross E (1987). The aprosodias: further functional-anatomical evidence for the organisation of affective language in the right hemisphere. J Neurol Neurosurg Psychiatry 50 (5): 553–560.

Griffiths T (2002). Central auditory pathologies. Br Med Bull 63: 107–120.

Griffiths T, Warren J (2004). What is an auditory object? Nat Rev Neurosci 5: 887–892.

Griffiths T, Rees A, Witton C et al. (1997). Spatial and temporal auditory processing deficits following right hemisphere infarction: a psychophysical study. Brain 120: 785–794.

Griffiths T, Rees A, Green G (1999). Disorders of human complex sound processing. Neurocase 5 (5): 365–378.

Griffiths T, Bamiou D-E, Warren J (2012). Disorders of the auditory brain. In: AR Palmer, A Rees (Eds.), The Oxford Handbook of Auditory Science: The Auditory Brain, Oxford University Press, New York.

Hailstone J, Crutch S, Vestergaard M et al. (2010). Progressive associative phonagnosia: a neuropsychological analysis. Neuropsychologia 48 (4): 1104–1114.

Hall III JW, Mueller HG (1997). The Audiologists' Desk Reference, Vol. I Singular Publishing Group, San Diego.

Hall D, Riddoch MJ (1997). Word meaning deafness: spelling words that are not understood. Cogn Neuropsychol 14 (8): 1131–1164.

Heilman KM, Scholes R, Watson RT (1975). Auditory affective agnosia. Disturbed comprehension of affective speech. J Neurol Neurosurg Psychiatry 38 (1): 69–72.

Heilman KM, Rothi L, McFarling D et al. (1981). Transcortical sensory aphasia with relatively spared spontaneous speech and naming. Arch Neurol 38 (4): 236–239.

Hemphill RE, Stengel E (1940). A study on pure word-deafness. Journal of Neurology and Psychiatry 3 (3): 251–262.

Hickok G, Poeppel D (2004). Dorsal and ventral streams: a framework for understanding aspects of the functional anatomy of language. Cognition 92 (1–2): 67–99.

Hickok G, Poeppel D (2007). The cortical organization of speech processing. Nat Rev Neurosci 8: 303–402.

Hornickel J, Skoe E, Kraus N (2008). Subcortical laterality of speech encoding. Audiology and Neurotology 14 (3): 198–207.

Jörgens S, Biermann-Ruben K, Kurz M et al. (2008). Word deafness as a cortical auditory processing deficit: a case report with MEG. Neurocase 14 (4): 307–316.

Kazui S, Naritomi H, Sawada T et al. (1990). Subcortical auditory agnosia. Brain Lang 38 (4): 476–487.

Klein R, Harper J (1956). The problem of agnosia in the light of a case of pure word deafness. Br J Psychiatry 102 (426): 112–120.

Kohlmetz C, Muller S, Nage W (2003). Selective loss of timbre perception for keyboard and percussion instruments following a right temporal lesion. Neurocase 9 (1): 86–93.

Kohn S, Friedman R (1986). Word-meaning deafness: a phonological-semantic dissociation. Cogn Neuropsychol 3 (3): 291–308.

Kussmaul A (1877). Disturbances of Speech. In: H von Ziemssen (Ed.), Cyclopedia of the Practice of Medicine, vol. 14. William Wood, New York, pp. 581–875.

Lambert J, Eustache F, Lechevalier B et al. (1989). Auditory agnosia with relative sparing of speech perception. Cortex 25 (1): 71–82.

Lang C, Kneidl O, Hielscher-Fastabend M et al. (2009). Voice recognition in aphasic and non-aphasic stroke patients. J Neurol 256 (8): 1303–1306.

Lechevalier B, Lambert J, Moreau S et al. (2003). Auditory disorders related to strokes. In: O Godefroy, J Bogousslavsky (Eds.), Behavioral and Cognitive Neurology of Stroke, Cambridge University Press, West Nyack, NY, pp. 348–365.

Leff A, Crinion J, Scott S et al. (2002). A physiological change in the homotopic cortex following left posterior temporal lobe infarction. Ann Neurol 51 (5): 553–558.

Liberman AM, Mattingly I (1989). A specialization for speech perception. Science 243 (4890): 489–494.

Liberman AM, Whalen DH (2000). On the relation of speech to language. Trends Cogn Sci 4 (5): 187–196.

Lichtheim L (1885). On aphasia. Brain 7 (4): 433–484.

Maneta A, Marshall J, Lindsay J (2001). Direct and indirect therapy for word sound deafness. Int J Lang Commun Disord 36 (1): 91–106.

Mavlov L (1980). Amusia due to rhythm agnosia in a musician with left hemisphere damage: a non-auditory supramodal defect. Cortex 16: 331–338.

McGettigan C, Scott SK (2012). Cortical asymmetries in speech perception: what's wrong, what's right and what's left? Trends Cogn Sci 16 (5): 269–276.

Mendez MF (2001). Generalized auditory agnosia with spared music recognition in a left-hander. Analysis of a case with a right temporal stroke. Cortex 37 (1): 139–150.

Mendez MF, Geehan G (1988). Cortical auditory disorders: clinical and psychoacoustic features. J Neurol Neurosurg Psychiatry 51 (June): 1–9.

Merzenich M, Jenkins W, Johnston P (1993). Temporal processing deficits of language-learning impaired children ameliorated by training. Science 271–281.

Metz-Lutz M-N (2009). The assessment of auditory function in CSWS: lessons from long-term outcome. Epilepsia 50 (Suppl 7): 73–76.

Metz-Lutz M-N, Dahl E (1984). Analysis of word comprehension in a case of pure word deafness. Brain Lang 23: 13–25.

Miceli G (1982). The processing of speech sounds in a patient with cortical auditory disorder. Neuropsychologia 20 (1): 5–20.

Miceli G, Gainotti G, Caltagirone C et al. (1980). Some aspects of phonological impairment in aphasia. Brain Lang 11 (1): 159–169.

Morris J, Franklin S, Ellis A et al. (1996). Remediating a speech perception deficit in an aphasic patient. Aphasiology 19 (2): 137–158.

Motomura N, Yamdori A, Mori E et al. (1986). Auditory agnosia: analysis of a case with bilateral subcortical lesions. Brain 109 (3): 379–391.

Musso M (1999). Training-induced brain plasticity in aphasia. Brain 122 (9): 1781–1790.

Näätänen R, Lehtokoski A, Lennes M et al. (1997). Language-specific phoneme representations revealed by electric and magnetic brain responses. Nature 385: 432–434.

Neuner F, Schweinberger S (2000). Neuropsychological impairments in the recognition of faces, voices, and personal names. Brain Cogn 44 (3): 342–366.

Oppenheimer DR, Newcombe F (1978). Clinical and anatomic findings in a case of auditory agnosia. Arch Neurol 35 (11): 712–719.

Otsuki M, Soma Y, Sato M et al. (1998). Slowly progressive pure word deafness. Eur Neurol 951: 135–140.

Patel AD (2008). A neurobiological strategy for exploring links between emotion and speech. Behav Brain Sci 31: 589–590.

Peretz I (1996). Can we lose memory for music? A case of music agnosia in a nonmusician. J Cogn Neurosci 8 (6): 481–496.

Peretz I, Hyde KL (2003). What is specific to music processing? Insights from congenital amusia. Trends Cogn Sci 7 (8): 362–367.

Peretz I, Regine K, Tramo M et al. (1994). Functional dissociations following bilateral lesions of auditory cortex. Brain 117 (Pt 6): 1283–1301.

Peretz I, Champod AS, Hyde K (2003). Varieties of musical disorders. Ann N Y Acad Sci 999 (1): 58–75.

Phillips DP, Farmer ME (1990). Acquired word deafness, and the temporal grain of sound representation in the primary auditory cortex. Behav Brain Res 40 (2): 85–94.

Pinard M, Chertkow H, Black S et al. (2002). A case study of pure word deafness: modularity in auditory processing? Neurocase 8 (1–2): 40–55.

Plasencia P, Dorado J, Serrano Rodiríguez J et al. (2006). Neuropsychological evidence for "word-meaning deafness" in a Spanish-speaking patient. Brain Lang 97 (2): 214–218.

Poeppel D (2001). Pure word deafness and the bilateral processing of the speech code. Cognit Sci 25 (5): 679–693.

Poeppel D, Monahan P (2008). Speech perception cognitive foundations and cortical implementation. Current Directions in Psychological Science 17 (2): 80–85.

Polster M, Rose S (1998). Disorders of auditory processing: evidence for modularity in audition. Cortex 34 (1): 47–65.

Praamstra P, Hagoort P, Maassen B et al. (1991). Word deafness and auditory cortical function: a case history and hypothesis. Brain 114 (3): 1197–1225.

Roberts M, Sandercock P, Ghadiali E (1987). Pure word deafness and unilateral right temporo-parietal lesion: a case report. J Neurol Neurosurg Psychiatry 50 (12): 1708–1709.

Robson H, Davies S, Ralph M et al. (2012). Facilitating and disrupting speech perception in word deafness. Aphasiology 26 (2): 37–41.

Ross E, Monnot M (2008). Neurology of affective prosody and its functional-anatomic organization in right hemisphere. Brain Lang 104 (1): 51–74.

Saffran E, Marin O, Yeni-Komshian G (1976). An analysis of speech perception in word deafness. Brain Lang 3 (2): 209–228.

Saygin AP, Dick F, Wilson S et al. (2003). Neural resources for processing language and environmental sounds Evidence from aphasia. Brain 126: 928–945.

Saygin AP, Leech R, Dick F (2010). Nonverbal auditory agnosia with lesion to Wernicke's area. Neuropsychologia 48 (1): 107–113.

Schnider A, Benson D, Alexander N et al. (1994). Non-verbal environmental sound recognition after unilateral hemispheric stroke. Brain 117 (2): 281–287.

Scott SK, Wise RJS (2004). The functional neuroanatomy of prelexical processing in speech perception. Cognition 92 (1): 13–45.

Shannon R, Zeng F, Kamath V et al. (1995). Speech recognition with primarily temporal cues. Science 270 (5234): 303–304.

Simons J, Lambon Ralph M (1999). The auditory agnosias. Neurocase 5 (5): 379–406.

Slevc LR, Martin RC, Hamilton AC et al. (2011). Speech perception, rapid temporal processing, and the left hemisphere: a case study of unilateral pure word deafness. Neuropsychologia 49 (2): 216–230.

Spreen O, Benton AL, Fincham RW (1965). Auditory agnosia without aphasia. Arch Neurol 13: 84–92.

Stefanatos GA (2008). Speech perceived through a damaged temporal window: lessons from word deafness and aphasia. Semin Speech Lang 3 (3): 239–252.

Stefanatos GA (2011). Changing perspectives on Landau-Kleffner syndrome. Clin Neuropsychol 25 (6): 963–988.

Stefanatos GA, Demarco AT (2012). Central auditory processing disorders. Encyclopedia of Human Behavior, pp. 441–452.

Stefanatos GA, Gershkoff A, Madigan S (2005a). Computer-mediated tools for the investigation and rehabilitation of auditory and phonological processing in aphasia. Aphasiology 19 (10–11): 955–964.

Stefanatos GA, Gershkoff A, Madigan S (2005b). On pure word deafness, temporal processing, and the left hemisphere. J Int Neuropsychol Soc 11: 456–470.

Stewart L (2011). Characterizing congenital amusia. Q J Exp Psychol (Colchester) 64 (4): 625–638.

Stewart L, von Kriegstein K, Warren J et al. (2006). Music and the brain: disorders of musical listening. Brain 129 (Pt 10): 2533–2553.

Takahashi N, Kawamura M, Shinotou H et al. (1992). Pure word deafness due to left hemisphere damage. Cortex 28 (2): 295–303.

Tallal P (2004). Improving language and literacy is a matter of time. Nat Rev Neurosci 5 (9): 721–728.

Tanaka Y, Yamadori A, Mori E (1987). Pure word deafness following bilateral lesions a psychophysical analysis. Brain 110 (2): 381–403.

Tanaka Y, Tsutomu K, Mitusuo Y et al. (1991). "So-called" cortical deafness. Brain 114: 2385–2401.

Taniwaki T, Tagawa K, Sato F et al. (2000). Auditory agnosia restricted to environmental sounds following cortical deafness and generalized auditory agnosia. Clin Neurol Neurosurg 102 (3): 156–162.

Tessier C, Weill-Chounlamountry A, Michelot N et al. (2007). Rehabilitation of word deafness due to auditory analysis disorder. Brain Inj 21 (11): 1165–1174.

Trumpp N, Kliese D, Hoenig K et al. (2013). Losing the sound of concepts: damage to auditory association cortex impairs the processing of sound-related concepts. Cortex 49 (2): 474–486.

Tyler L, Moss H (1997). Imageability and category-specificity. Cogn Neuropsychol 14 (2): 293–318.

Van Lancker D, Kreiman J (1987). Voice discrimination and recognition are separate abilities. Neuropsychologia 25 (5): 829–834.

Van Lancker D, Kreiman J, Cummings J (1989). Voice perception deficits: neuroanatomical correlates of phonagnosia. J Clin Exp Neuropsychol 11 (5): 665–674.

Vignolo L (1982). Auditory agnosia. Philos Trans R Soc Lond B Biol Sci 298: 49–57.

Vignolo L (2003). Music agnosia and auditory agnosia. Ann N Y Acad Sci 999 (1): 50–57.

Wang E, Peach R, Xu Y et al. (2000). Perception of dynamic acoustic patterns by an individual with unilateral verbal auditory agnosia. Brain Lang 73 (3): 442–455.

Wee J, Menard M (1999). "Pure word deafness": implications for assessment and management in communication disorder – a report of two cases. Arch Phys Med Rehabil 80 (9): 1106–1109.

Wirkowski E, Echausse N, Overby C et al. (2006). I can hear you yet cannot comprehend: a case of pure word deafness. J Emerg Med 30 (1): 53–55.

Yaqub B, Gascon G, Alnosha M et al. (1988). Pure word deafness (acquired verbal auditory agnosia) in an Arabic speaking patient. Brain 111 (2): 357–466.

Zatorre RJ, Belin P (2001). Spectral and temporal processing in human auditory cortex. Cereb Cortex 11 (10): 946–953.

Zatorre RJ, Belin P, Penhune V (2002). Structure and function of auditory cortex: music and speech. Trends Cogn Sci 6 (1): 37–46.

Ziegler D (1952). Word deafness and Wenicke's aphasia: report of cases and discussion of the syndrome. AMA Arch Neurol Psychiatry 67 (3): 323–331.

Handbook of Clinical Neurology, Vol. 129 (3rd series)
The Human Auditory System
G.G. Celesia and G. Hickok, Editors

Chapter 33

Congenital amusias

B. TILLMANN[1,3]*, P. ALBOUY[1,2,3], AND A. CACLIN[2,3]
[1]*Auditory Cognition and Psychoacoustics Team, Lyon Neuroscience Research Center, Lyon, France*
[2]*Brain Dynamics and Cognition Team, Lyon Neuroscience Research Center, Lyon, France*
[3]*University Lyon 1, Lyon, France*

INTRODUCTION

Numerous research studies in music cognition have provided evidence for sophisticated music processing in the general population, even without any explicit musical training. Via mere exposure to music in everyday life, non-musicians acquire knowledge about the musical system of their culture, which then shapes music perception and memory (e.g., Francès, 1958; Tillmann et al., 2000; Bigand and Poulin-Charronnat, 2006). Recent research has also shown this musical capacity for production, notably for non-musicians' singing, which is considerably more precise than generally assumed (Dalla Bella et al., 2007). Together with research on music perception capacities in infants (e.g., Hannon and Trehub, 2005), these findings have also been discussed in terms of humans' innate predisposition for music.

However, for some individuals, a deficit of music processing has been reported and suggested as a lifelong deficit. The affected individuals have difficulty detecting when someone sings out of tune (including themselves), recognizing familiar tunes without lyrics, detecting a wrong or out-of-tune note, and memorizing even short melodies. This deficit has been referred to as tone deafness, dysmusia, dysmelodia, and more recently, as congenital amusia, thus reflecting the hypothesis that individuals may be born with this music-processing deficit (e.g., Kalmus and Fry, 1980; Peretz and Hyde, 2003). Note that currently, most research refers to congenital amusia, while some to tone deafness. In the present chapter, we will cover research using both labels, but refer more generally to congenital amusia.

In contrast to cases of acquired amusia following brain damage (see Chapter 34; Peretz et al., 1997; Tillmann et al., 2007), congenital amusia occurs without brain damage or sensory deficits. This musical disorder occurs despite normal performance on tests of intelligence, auditory processing, cognitive functioning, and language processing, and it is not due to a lack of environmental stimulation to music (see Ayotte et al., 2002; Peretz et al., 2002; Foxton et al., 2004, for extensive testing).

Congenital amusia has been described as a deficit in music production and perception. It thus needs to be distinguished from the phenomenon that has been termed "poor singing" (e.g., Dalla Bella et al., 2007; Pfordresher and Brown, 2007), referring to individuals who only have production deficits, but who do not have perceptual difficulties. It needs also to be distinguished from the general non-musician population who might be lacking confidence in their musical capacities and refer to themselves as tone-deaf, but who are actually not (Cuddy et al., 2005; Sloboda et al., 2005).

Even though the disorder has been recognized for a long time (Allen, 1878), it has been systematically studied only relatively recently (Ayotte et al., 2002), mainly thanks to the development of the Montreal Battery for the Evaluation of Amusia (MBEA) (Peretz et al., 2003; available at http:/www.brams.umontreal.ca/plab/publications/article/57). In the MBEA, six subtests address various components of music perception and memory, notably the pitch dimension (detection of an out-of-key note, a contour violation, or interval changes), the time dimension (rhythm and meter perception), and incidental memory (i.e., melodies used in

*Correspondence to: Barbara Tillmann, CNRS UMR5292, INSERM U1028, Centre de Recherche en Neurosciences de Lyon, Equipe Cognition Auditive et Psychoacoustique, 50 Avenue Tony Garnier, F-69366 Lyon cedex 01, France. E-mail: btillmann@olfac.univ-lyon1.fr

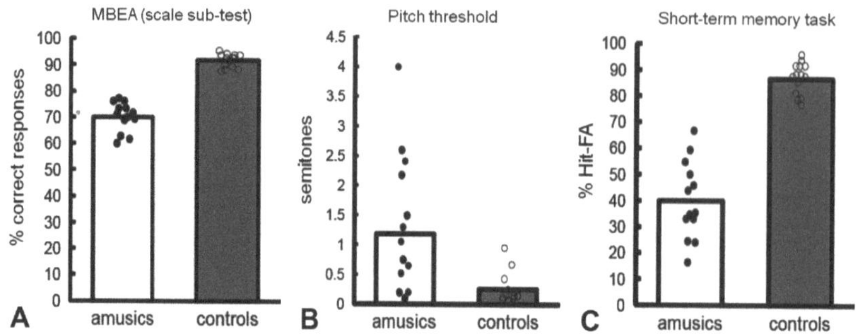

Fig. 33.1. Group data for amusic and control participants together with individual data points (13 amusic participants and 14 control participants) for three tasks requiring the processing of the pitch dimension. (**A**) Performance (percentage of correct responses) for the scale subtest of the Montreal Battery for the Evaluation of Amusia (MBEA) (Peretz et al., 2003). (**B**) Pitch detection threshold (task measuring the smallest pitch difference, in semitones, that participants can detect; see procedure in Tillmann et al., 2009). (**C**) Performance (hits minus false alarms (FAs)) in a short-term memory task, indicating whether two six-tone melodies (S1, S2) were the same or different, with different trials being changed on the melodic contour (Albouy et al., 2013a). Note that results of the amusic group (average performance) are significantly worse than the results of the control group for all three tasks, while individual data show considerable overlap between the two groups in the pitch threshold task, but not in the MBEA and the short-term memory task. Note that the same participants from our pool in Lyon have performed the three tasks (**A, B, C**).

preceding subtests) (see Fig. 33.1A for example performances on the MBEA). An adaptation of this battery has been proposed recently for the testing of children (Peretz et al., 2013). Another test used to define a population with a musical deficit is the Distorted Tune Test. This test investigates participants' capacity to discriminate intervals between tones in a melody (Kalmus and Fry, 1980; Drayna et al., 2001; Jones et al., 2009a). However, it uses well-known tunes (in North America) and is thus dependent on participants' long-term memory knowledge of a cultural musical repertoire. The MBEA is based on newly composed melodies, and it has been used successfully across various countries and cultures (North America, Europe (e.g., France, United Kingdom, Greece), China, and New Zealand). As the MBEA is time consuming (taking approximately 1 hour 30 to 2 hours), a shorter online test has been proposed more recently (Peretz et al., 2007). Similarly, the test battery for children (the MBEMA) is proposed in a long and a short version (Peretz et al., 2013).

The musical deficit has been estimated to affect about 4% of the general population (Kalmus and Fry, 1980). This estimate has been criticized as it depends on the test used, how the cut-off scores are defined, and the skew in distributions (Henry and McAuley, 2010). This criticism is not specific to congenital amusia, but it also affects prevalence reports for other disorders, such as dyslexia, dyscalculia, or developmental prosopagnosia (Henry and McAuley, 2010). Furthermore, there exists considerable uncertainty when assessments are only based on verbal reports without direct testing, notably because of individuals' underestimation of their capacities as well as response bias due to difficulty in admitting that they do not like or understand music.

The hypothesis of a genetic origin of this disorder has received support from a family aggregation study (Peretz et al., 2007) and a twin study (Drayna et al., 2001). Peretz et al. (2007) tested nine amusic families and 10 control families: 39% of first-degree relatives in the amusic families showed the same disorder (in contrast to only 3% in the control families). The incidence of amusia as a heritable disorder is of similar magnitude compared to that reported for other cognitive functions (e.g., specific language impairment, absolute pitch). Note that the calculation of the risk factor is based on the information from the siblings and the hypothesis of a population prevalence of 4% (thus also subjected to the criticism of Henry and McAuley, 2010). Nevertheless, these findings provide the basis for future research aiming to map the genetic loci for hereditary amusia. Recently, congenital amusia has been also reported in childhood, thus providing some further evidence for the definition as a lifelong disorder (Lebrun et al., 2012).

The research domain interested in congenital amusia is currently expanding. In addition to behavioral methods, neurophysiologic methods are now increasingly used (voxel-based morphometry (VBM), diffusion tensor imaging, electro- or magnetoencephalography (EEG/MEG), and functional magnetic resonance imaging (fMRI)), aiming to define anatomic and functional brain anomalies of this deficit, thus going beyond a phenomenologic description of the deficit and contributing also to our understanding of unimpaired neural circuitry underlying music processing (perception, memory, production). Furthermore, investigating congenital amusia has been promoted as a possibility to study the interaction between influences due to genes, environment, and behavior, in particular for music cognition and its relation to other cognitive capacities (Peretz, 2008).

Our chapter aims to provide a state-of-the-art overview of the current understanding of this disorder and also to present the open questions and challenges of this research domain. The present chapter is entitled "congenital amusias." The term congenital amusia was coined by Peretz (2001), and its use in the plural aims to acknowledge the possibility of various forms of congenital amusia, similar to the various forms that exist for acquired amusia (due to accidental brain damage). The potential variety of this disorder is further supported by the recent progress of the research domain, now suggesting different subtypes of congenital amusia.

A DEFICIT ON THE PITCH DIMENSION: PERCEPTION AND MEMORY

Currently, the most extensively investigated hypothesis about the main deficit in congenital amusia concerns the pitch dimension, notably as pitch is a major component of music structures. As we shall see below, this predominance of pitch-processing studies in the congenital amusia literature also originates from seminal studies showing that pitch was more affected than rhythm (e.g., Allen, 1878; Peretz et al., 2003). While the first studies have focused on pitch perception deficits (i.e., pitch discrimination), more recent studies have pointed to deficits regarding pitch memory – even in cases with unimpaired pitch discrimination capacities.

Pitch perception

Seminal studies of congenital amusia (Ayotte et al., 2002; Peretz et al., 2002; Foxton et al., 2004; Hyde and Peretz, 2004) led to the assumption that this condition is based on a disorder of fine-grained pitch processing, notably pointing to fundamental deficits in pitch discrimination.

Documenting the first reported case of congenital amusia, Peretz et al. (2002) showed difficulties in detecting pitch changes as large as two semitones in tone sequences and tone pairs. The first group study of individuals afflicted with this developmental disorder (Ayotte et al., 2002) further defined this deficit by studying the discrimination, perception, and memory of musical material as compared to other materials (i.e., language, environmental sounds). While amusic participants performed as well as controls for non-pitch-related material (i.e., language, environmental sounds; see below), all amusic participants were impaired in detecting pitch changes in both familiar and unfamiliar melodies. For example, amusics' pitch-processing abilities were impaired in an anomalous pitch detection task: participants were asked to detect a pitch anomaly within a melody (a note shifted by a semitone, thus creating an out-of-key violation). To investigate the pitch-processing deficit outside a music context, Hyde and Peretz (2004) required amusic individuals to detect a pitch change (of 0.25–3 semitones) within the context of a five-tone sequence (monotonic and isochronous). Amusics' performance was impaired for pitch changes smaller than two semitones, while controls showed ceiling performance for changes as small as 0.25 semitones (Fig. 33.2A and B).

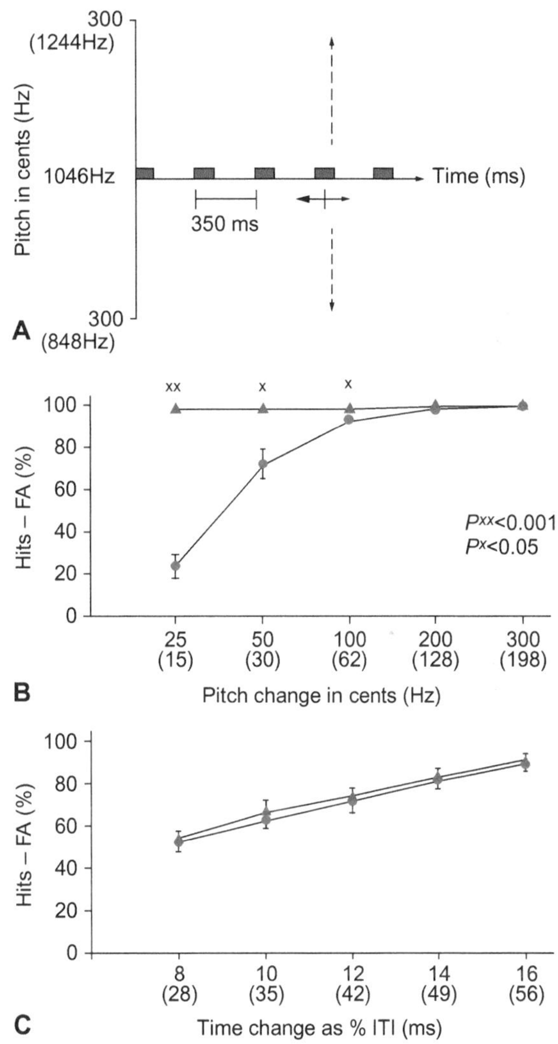

Fig. 33.2. (**A**) Presentation of the experimental task for pitch and time dimensions used in Hyde and Peretz (2004). For the pitch task, the fourth tone was altered by one of five pitch distances, ranging from 25 to 300 cents (100 cents corresponding to one semitone). For the time task, the fourth tone was changed by temporal increments from 8 to 16%. (**B**) Hits minus false alarm (FA) rates for amusic and control participants in the pitch condition. (**C**) Hits minus FA rates for amusic and control participants in the time condition. ITI, intertone interval. (Reproduced from Peretz and Hyde, 2003.)

These seminal studies led to the hypothesis that abnormalities in pitch perception underlie the phenomenon of congenital amusia (Peretz and Hyde, 2003). The developmental disorder would arise from the failure to encode pitch with sufficient precision to allow for the acquisition of knowledge about the pitch structure of the musical system (Tillmann et al., 2000, 2007), thus impairing tonal coding and also leading to general difficulties in pitch-related processing (i.e., amusics' inability to perceive dissonance: Ayotte et al., 2002; Cousineau et al., 2012).

To further investigate this hypothesis, subsequent studies have used auditory threshold-based testing with other cohorts of amusics (Foxton et al., 2004; Jones et al., 2009a; Tillmann et al., 2009; Liu et al., 2010; Williamson and Stewart, 2010; Albouy et al., 2013a, b). Foxton et al. (2004) have used a range of pitch tasks with forced-choice procedures to evaluate amusics' deficit in pitch perception. Deficits were identified in "low-level" pitch perception tasks (detection of segmented and gliding frequency changes) and in pitch pattern perception tasks, which ranged from determining the direction of pitch changes to detecting differences between pitch sequences. These later findings have also led to the hypothesis of a major deficit concerning the processing of pitch direction (Stewart, 2011). In contrast, the perceptual organization of sounds into different streams on the basis of pitch was found to be unimpaired in amusic individuals (Foxton et al., 2004).

These data have provided evidence for pitch perception deficits in congenital amusia, notably by showing that amusic individuals exhibit elevated pitch discrimination thresholds (in comparison to controls). However, several studies reported overlap in pitch thresholds between amusic and control groups (for pitch discrimination: Foxton et al., 2004; Jones et al., 2009b; Tillmann et al., 2009; Liu et al., 2010; Albouy et al., 2013a, b; and for pitch direction: Foxton et al., 2004; Liu et al., 2010; Williamson and Stewart, 2010), showing that some amusics have thresholds within the range of control participants (Fig. 33.1B), and thus suggesting the need to consider additional impaired processes in this condition.

Memory for pitch

A first hypothesis has suggested that amusics' elevated pitch discrimination thresholds would lead to impaired detection of small pitch changes in tone sequences, such as in the melodies of the MBEA (Peretz et al., 2003), the battery used as a diagnostic tool for amusia. However, it is relevant to point out that the pitch tests of the MBEA (scale, contour, and interval tests, respectively) also require short-term memory, as two melodies separated by a silent retention period need to be compared. Several studies have provided evidence for amusics' impairment of short-term memory for pitch, even for amusics without elevated pitch discrimination thresholds, suggesting that the core deficit in congenital amusia might be related to impaired pitch memory (Fig. 33.1B and C).

Memory deficits for pitch patterns in congenital amusia were explored by Foxton et al. (2004), who asked amusic and control groups to compare two pitch sequences (of four or five pure tones) that were separated by a 1100-ms gap (same/different paradigm). Using three different tasks for which the comparison sequences were contour-violated, contour-preserved, or contour-violated within a transposed sequence (shifted up or down by half an octave, participants being requested to ignore the transposition), they demonstrated that amusics were strongly impaired in short-term memory performance in comparison to controls, without overlap in performance between the two groups (thus contrasting with the observed overlap in the pitch discrimination thresholds).

Subsequent studies used comparison tasks for single tones (Gosselin et al., 2009; Williamson et al., 2010) and for tone sequences (Gosselin et al., 2009; Tillmann et al., 2009; Albouy et al., 2013a, b) to further investigate the extent of the deficit in short-term memory for pitch.

For single tones, these studies asked amusic participants to compare two tones separated by a silent retention period, the delay between the to-be-compared tones being either unfilled or filled with irrelevant tones (Gosselin et al., 2009; Williamson et al., 2010). While control participants showed good performance for all pitch intervals (one to three tones in Gosselin et al., 2009; and one tone in Williamson et al., 2010) and in both conditions (filled, unfilled delays), congenital amusics performed worse than controls when irrelevant tones were presented during the delay. A short-term memory deficit for single tones was also shown when testing the effect of the duration of the silent retention period between the two to-be-compared tones (0–15 seconds): amusic participants showed faster decline in performance with increasing retention delay than did controls (Williamson et al., 2010). Further evidence for amusics' short-term memory deficit for pitch, suggesting a weaker and more unstable memory trace, was revealed with amusics' sensitivity to memory load: when participants had to compare two tone sequences (separated by a retention interval of 2 seconds), amusics' performance was lower for five-tone sequences than for three-tone sequences, while controls' performance was only slightly affected by load (Gosselin et al., 2009).

The short-term memory deficit was observed for music material (tones, timbres), but did not affect spoken words: when requested to compare sequences of tones, timbres, or words (separated by a retention delay of 3 seconds), amusic participants performed worse than controls for tone and timbre sequences, but not for words (Tillmann et al., 2009). The observed deficit was stronger for the tone material than for the timbre material. Impaired memory for timbre was confirmed by Marin et al. (2012), who replicated results of Tillmann et al. (2009) using timbre sequences (composed of four to eight events), and further showed that this impairment is not related to a deficit in timbre discrimination (Fig. 33.3). The specificity of the deficit for musical material has been further supported by Williamson and Stewart (2010), comparing a digit span test and a tone span test: congenital amusics had significantly smaller tone spans in comparison to controls, whereas their digits spans were of similar size to those of controls.

Importantly, in all these studies, using either isolated tones (Gosselin et al., 2009; Williamson et al., 2010) or tone sequences (Gosselin et al., 2009; Tillmann et al., 2009), the pitch changes exceeded amusic participants' psychophysically measured pitch difference detection thresholds, pointing to a memory-based deficit for pitch that extends beyond a perceptual discrimination impairment. In addition, it is worth noting that, whereas some amusics exhibited normal pitch discrimination thresholds, all amusics displayed impairments in pitch memory (Fig. 33.1C) (Foxton et al., 2004; Tillmann et al., 2009; Albouy et al., 2013a).

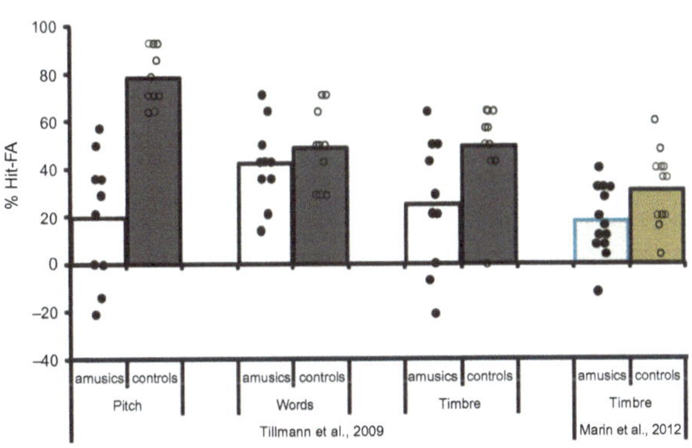

Fig. 33.3. Performance of amusic and control individuals in terms of hits minus false alarms (FA), presented as a function of the material used in a short-term memory task (pitch, timbre, words). Left: Data from Tillmann et al. (2009) (10 participants per group); right: data from Marin et al., 2012 (13 participants per group) (collapsed across sequence lengths from four to eight items). Note that for the same sequence length (five items), as used in Tillmann et al. (2009), average performances of amusics were similar in Marin et al. (2012). (Hits minus FA rates were 25% in Tillmann et al., 2009, and 18% in Marin et al., 2012.) Bars indicate group average, and circles individual data points. Note that the amusic group was significantly impaired in comparison to the control group for pitch and timbre tasks, but not for verbal material. While individual data show overlap between the two groups in the timbre short-term memory tasks, no overlap was observed for the pitch task (as in Fig. 33.1).

By considering the deficits in fine-grained pitch processing and in short-term memory for pitch, one question now is to determine whether deficits in pitch discrimination and pitch memory constitute separated or combined elements in congenital amusia. One hypothesis would suggest that the pitch discrimination deficit might be related to a pitch memory deficit, but with amusics differing in the extent of this memory impairment. As pitch discrimination measures require the comparison of two or more events (or testing intervals), a strong pitch memory deficit could also affect more perceptual measures, such as discrimination thresholds.

NEURAL CORRELATES OF CONGENITAL AMUSIA: ANATOMIC AND FUNCTIONAL DATA

In parallel to the investigation of behavioral expressions of the deficit, brain-imaging studies have started to decipher the cerebral correlates of this neurodevelopmental disorder using anatomic and functional investigations. This research not only contributes to our understanding of the deficits underlying congenital amusia, but also furthers our understanding of neural correlates underlying normal functioning of music perception and memory.

Anatomic correlates

The first studies have primarily investigated anatomic correlates of congenital amusia. The study of Hyde et al. (2006) used VBM in two independent cohorts of amusics to explore whether structural abnormalities for white- and gray-matter concentration could underlie the disorder. In comparison to controls, the amusic brain showed decreased white-matter concentration and increased gray-matter concentration in the right inferior frontal gyrus (rIFG: Brodmann area (BA) 47; see Fig. 33.4, upper panel). This result was confirmed with VBM data for a third independent group of amusics (Albouy et al., 2013a). To extend this observation with another method, Hyde et al. (2007) used cortical thickness measures and revealed abnormalities in the amusic brain in terms of gray-matter integrity in the rIFG (BA 47) and in the right superior temporal gyrus (rSTG): amusics showed thicker cortex than controls in these regions. Note, however, that VBM revealed decreased gray-matter concentration in the rSTG for amusics (Albouy et al., 2013a); this might be due to methodologic differences between cortical thickness and VBM. Nevertheless, both studies are in agreement in revealing

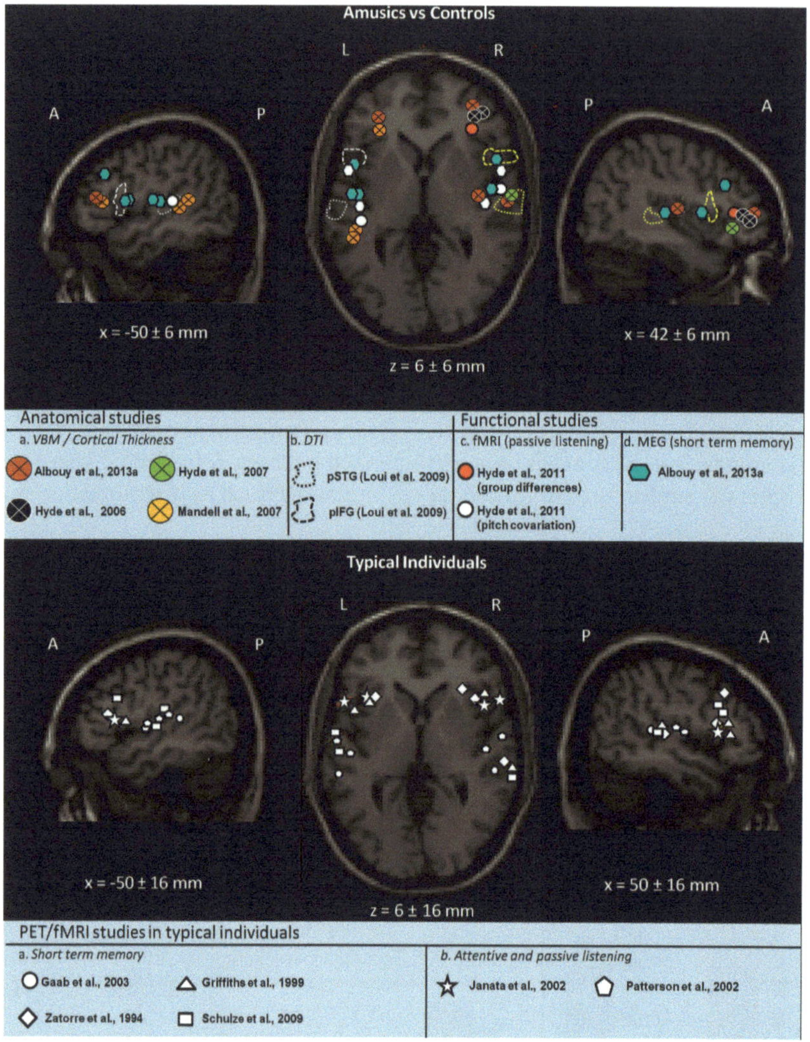

Fig. 33.4. Upper panel: Brain regions observed in anatomic and functional studies comparing control participants and amusics (white: no group difference; color: group difference). (**a**) Coordinates of the anatomic group differences observed with voxel-based morphometry (Hyde et al., 2006 (black); Mandell et al., 2007 (orange); Albouy et al., 2013a (red)) and cortical thickness (Hyde et al., 2007 (green)) approaches. (**b**) Approximate location of the posterior inferior frontal gyrus (pIFG) and posterior superior temporal gyrus (pSTG) seed regions used in the tractography study of the arcuate fasciculus by Loui et al. (2009) (bold dotted lines and plain dotted line, respectively). Note that these regions were manually selected by a neurologist who was blind concerning the protocol and previously reported activations. Group differences were observed only in the right hemisphere (yellow color). (**c**) Coordinates of the functional group differences observed with functional magnetic resonance imaging (fMRI) in the right inferior frontal gyrus (Hyde et al., 2011, purple circle), and of areas in the auditory cortices where both amusics and control exhibit similar blood oxygen level-dependent activity to pitch variations (Hyde et al., 2011, white circles). (**d**) Coordinates of the magnetoencephalography group differences during a short-term memory task (Albouy et al., 2013a, hexagons).

Lower panel: Activation peaks observed in studies using positron emission tomography or fMRI to investigate music processing in typical individuals. Coordinates of activations obtained with pitch material for (**a**) auditory short-term memory tasks (Zatorre et al., 1994, diamonds; Griffths et al., 1999, triangles; Gaab et al., 2003, circles; Schulze et al., 2009, squares); (**b**) attentive listening (Janata et al., 2002, stars), and passive listening (Patterson et al., 2002, pentagons).

All activations are displayed on the single-subject T1 template provided by SPM8. The figure displays activations with coordinates that were 6 mm up or down the central slice and 6 mm left or right of the lateral slices for the upper panel, and ±16 mm for the lower panel. R, right; L, left; A, anterior; P, posterior.

structural abnormalities in the rSTG of the amusic brain for gray matter (Hyde et al., 2007; Albouy et al., 2013a; see Fig. 33.4, upper panel).

These group differences were also supported by correlation analyses between anatomic measures and behavioral data: the performance in music tests (scale and memory test of the MBEA) correlated positively with white-matter concentration in the IFG (Hyde et al., 2006); and the global score of the MBEA correlated negatively with cortical thickness in the rIFG (and rSTG) (Hyde et al., 2007). In another sample of amusics (Mandell et al., 2007), positive correlation between MBEA scores and gray-matter concentration were observed in the left IFG (BA 47, for the melodic subtests) and left STG (BA 22, for the rhythmic subtest).

The findings suggesting abnormalities in white- and/or gray-matter concentrations are in line with structural anomalies in adult brains having experienced long-lasting deficiencies (such as language and speech disorders; see Watkins et al., 2002). It has thus been argued that the observed abnormalities in the amusic brain have a genetic basis (Peretz et al., 2007), leading to cortical malformations that may compromise the normal development of a right frontotemporal pathway. In agreement with this hypothesis, abnormalities of white matter have been observed with diffusion tensor imaging, showing reduced right arcuate fasciculus connectivity in amusics (Loui et al., 2009). In addition, the volume of the superior arcuate fasciculus branches correlated positively with pitch discrimination abilities in controls and amusics.

The observed structural differences suggest cortical malformations of a right frontotemporal pathway in the amusic brain, resulting in a pitch-processing impairment. This hypothesis is consistent with patient data on acquired amusia (see Chapter 34) and with neuroimaging data of the normal, non-amusic brain showing that the right superior temporal cortex is involved in perceptual analyses of melodies and that pitch processing recruits neural networks, including the right frontal cortex (Zatorre et al., 1994; Griffiths et al., 1999; Maess et al., 2001; Tillmann et al., 2003, 2006; Peretz and Zatorre, 2005; see Fig. 33.4, lower panel).

Functional investigations

To further link the anatomic abnormalities to the behavioral manifestations of the deficit, recent studies investigated the cerebral correlates of pitch perception and memory in congenital amusia on a functional level (using EEG, fMRI, and MEG). In Peretz et al. (2005), amusic and control participants performed a pitch change detection task during EEG recording, and auditory evoked responses were analyzed for the deviant and standard tones. Amusics' N1 component was normal and its topography was consistent with a localization of the generators in the secondary auditory cortex (Näätänen and Picton, 1987). Investigating the neural processing of musical pitch incongruities in congenital amusia, Peretz et al. (2009) showed that the amusic brain can track small pitch changes (i.e., a quarter tone), exhibiting an early right-lateralized negative brain response, even though the amusic participants mostly failed to report explicitly these pitch incongruities. Moreover, a recent mismatch negativity (MMN) study (Moreau et al., 2013) provided congruent evidence of normal preattentive pitch change detection in amusics.

These findings suggest near-normal cerebral processing of musical material in the amusic brain, notably showing a near-normal functioning of the auditory cortex, despite explicit deficits on pitch tasks at a behavioral level. This was further supported by fMRI data showing that amusics' auditory cortices respond normally to pitch during passive listening to pure-tone melody-like patterns: activity increased in the auditory cortex as a function of increasing pitch distance (from 0.25 to 2 semitones) (Hyde et al., 2011). In addition, abnormal blood oxygen level-dependent deactivation in the rIFG was associated with decreased right frontotemporal functional connectivity between the rIFG and the rSTG in amusics in comparison to controls. These findings thus led to the hypothesis that the functional correlates in congenital amusia are related to impaired anatomic and functional frontotemporal connectivity, with intact functioning of the auditory cortex.

However, these studies used only passive listening and pitch perception tasks to investigate the functional correlates of congenital amusia. To extend our knowledge of the functional underpinnings of amusics' deficits, Albouy et al. (2013a) investigated the neural bases of short-term memory for tones in congenital amusia with MEG recordings. Using a classic short-term memory task with melodies, functional abnormalities were revealed during encoding, retention, and retrieval of the melodic information in the amusic brain. During encoding of melodies, the amusic brain elicited abnormal (decreased and delayed) N100m components in bilateral IFG (pars opercularis) and STG. These altered responses were interpreted as reflecting impaired high-level processing and stimulus representations in the amusic brain. Furthermore, these abnormal responses were associated with abnormal intrinsic connectivity (as revealed by Dynamic Causal Modeling: David et al., 2006) in both auditory cortices, associated with increased lateral connectivity between left and right auditory cortices as well as a decreased right frontotemporal backward connectivity in amusics relative to controls (see also Hyde et al., 2011, for converging evidence). The abnormal functioning of this frontotemporal network was also shown during the retention interval as well as the retrieval of melodic information. These findings were consistent with the involvement of right temporofrontal regions in pitch and melody processing and memory, as observed in normal participants with functional cerebral imaging (Zatorre et al., 1994, 2002; Griffiths, 1999; Griffiths et al., 1999; Janata et al., 2002; Gaab et al., 2003; Tillmann et al., 2003, 2006; Koelsch et al., 2005, 2009; Peretz and Zatorre, 2005; Schulze et al., 2009, 2011a, b, see Fig. 33.4, lower panel) and are congruent with anatomic anomalies observed along the auditory-frontal pathway in the amusic brain (Hyde et al., 2006, 2007, 2011; Mandell et al., 2007; Loui et al., 2009; Albouy et al., 2013a) (see Fig. 33.4, upper panel).

Taken together, these studies suggest that the amusic brain presents abnormalities in the processing of musical material both in the auditory cortices and the inferior frontal cortices, associated with decreased functional connectivity between the structures.

ARE THE PROCESSING DEFICITS IN CONGENITAL AMUSIA PITCH-SPECIFIC?

Testing amusics' processing of the time dimension in music, of language, and of space

The dominant hypothesis of amusics' deficits refers to the pitch dimension, notably with impaired pitch perception and memory. This focus can be explained by the fact that pitch information is the main form-bearing dimension in western tonal music. Some research has addressed amusics' processing capacities for other information, either related to music (temporal structures) or pitch processing related to other materials (speech, space).

Temporal processing

In addition to the pitch dimension, the second form-bearing dimension of the western tonal musical system is the time dimension, that is, the rhythmic and metric structures (McAdams, 1989). Temporal processing is also tested in the MBEA, revealing that, in contrast to the subtests of the pitch dimension (scale, contour, interval), performance in the temporal subtests (i.e., rhythm and meter processing) is more variable, with about half of the amusics showing a deficit (Ayotte et al., 2002; Peretz et al., 2003).

Some amusics are impaired in discriminating melodies by their rhythms, and have difficulties in keeping time to a musical beat (Ayotte et al., 2002). Ayotte et al. (2002) proposed the hypothesis that this deficit in temporal processing might be a cascade effect of impaired pitch processing and not a temporal processing deficit *per se*. This hypothesis has received partial support: when tested for temporal processing in a non-musical context (i.e., detecting a time deviation from isochrony in a monotonous sequence), amusics were as accurate as control participants – even though these same amusics were severely impaired when testing pitch processing in a non-musical context (detecting a pitch deviation from monotony in an isochronous sequence) (Hyde and Peretz, 2004; see Fig. 33.2). Converging evidence has been reported by Foxton et al. (2006): When amusics had to compare two sound sequences, which were either monotonic or randomly varying in pitch, for the presence of a changed temporal interval, the pitch changes interfered with the processing of the rhythmic patterns. Amusics' performance was decreased for the sequences that varied in pitch compared to the sequences that did not, while controls' performance did not differ between the two conditions. Pfeuty and Peretz (2010) further investigated the pitch and time processing in a non-musical context; their data suggest that amusics also have a deficit in the integration of pitch and time information in a combined mental representation.

While cases of acquired amusia have shown selective deficits of either the pitch or the time dimension (e.g., Peretz and Kolinsky, 1993; see Chapter 34), only recently, selective deficits on the time dimensions have been reported for congenital amusia (Phillips-Silver et al., 2011; see also Thompson, 2007). Phillips-Silver et al. (2011) reported a case study with a deficit in beat finding in the context of music, while showing near-normal synchronization with a metronome (i.e., no pitch changes). The participant (who did not show any other cognitive or perceptual deficits, without any neurologic or psychiatric anomalies) scored in the normal range of the MBEA, except for the meter subtest. His data pattern led the authors to propose a new form of congenital amusia based on time deficits, which calls for future investigations and extensions to group studies.

Speech processing

Both acquired and congenital amusias represent an opportunity to investigate the hypothesis of the modularity of music and language processing (e.g., Peretz et al., 1997; Piccirilli et al., 2000; see Chapter 34): Are neural correlates independent or shared for the processing of these two structured systems? Individuals with congenital amusia have been reported to be unimpaired in language tasks, such as learning and recognizing lyrics (Ayotte et al., 2002; Peretz et al., 2002), as well as short-term memory tests for words or digits (Tillmann et al., 2009; Williamson and Stewart, 2010). In earlier studies, amusic individuals have been also reported as unimpaired for prosody tasks, such as classifying a spoken sentence as statement or question based on final (falling or rising) pitch information as well as identifying or discriminating stressed words in sentences (Ayotte et al., 2002; Peretz et al., 2002; Patel et al., 2005). More recent studies, however, revealed deficits also for prosody processing in amusia (see below).

Because of the initially reported spared intonation processing, Peretz and Hyde (2003) suggested that the difference between speech and music perception might be linked to the size of relevant pitch variations (Fig. 33.5). Indeed, in speech intonation of non-tonal languages, variations in fundamental frequency (F0), in particular those indicating statements and questions, are typically coarse (up to more than 12 semitones for the pitch rise of the final word in a question; see Patel et al., 2008). In music, however, the pitch variations are typically more fine-grained (one or two semitones; see Vos and Troost, 1989). Accordingly, amusics' pitch deficit would affect music more than speech because music is more demanding in pitch resolution, and congenital amusia would represent a music-relevant deficit, not necessarily a music-specific deficit.

However, amusics' deficit persisted for large pitch differences when these were derived from speech intonation patterns and presented as musical analogs (imitating gliding pitch changes or transformed into discrete steps). Amusics were impaired when asked to discriminate imitations of spoken intonation (without words). The deficit was limited to the non-speech context because amusics were normal at discriminating the same changes in the context of speech (Ayotte et al., 2002; Patel et al., 2005). Ayotte et al. (2002) suggested a beneficial effect of the linguistic cues that could serve as anchoring points for poor pitch-processing abilities. Another possible account for this speech advantage is that labeling the word with the salient pitch change might provide a strategy that decreases the memory load of the task – a strategy that is not possible for the tone analogs. In the latter case, the pitch memory problem experienced by amusics (Gosselin et al., 2009; Tillmann et al., 2009; Williamson et al., 2010) may explain why they fail to discriminate intonation patterns in musical analogs even though they contain relatively large pitch changes.

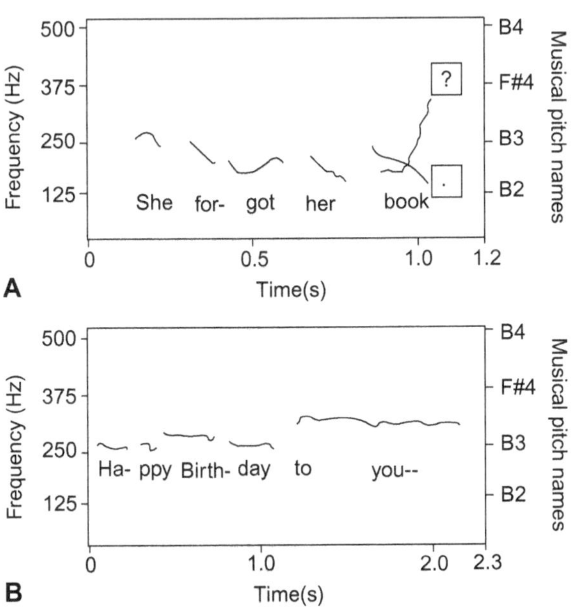

Fig. 33.5. (**A**) Spectrograms illustrating pitch variations in two sentences (a question, a statement) spoken by a female English speaker. (**B**) Spectrogram illustrating the pitch variations in the song "Happy birthday" sung by the same voice as spoken material in (**A**). (Reproduced from Peretz and Hyde, 2003.)

The hypothesis that amusics cannot fully compensate for their pitch deficit by using speech-based strategies is supported by other data sets, revealing subtle deficits in speech processing. Amusics showed mild deficits in processing speech intonation (questions vs statement) in their native language (English or French; Patel et al., 2008; Liu et al., 2010), or processing pitch contrasts in tone language words (Mandarin or Thai), whether for native Mandarin speakers (Nan et al., 2010; Jiang et al., 2010, 2012) or native French speakers (Nguyen et al., 2009; Tillmann et al., 2011a). When tested with natural speech, which involves multiple acoustic cues, Mandarin amusics were not impaired (Liu et al., 2012a, b). Note also that it has been recently suggested that amusics who are native speakers of Hong Kong Cantonese show better pitch-processing abilities than amusics who are native speakers of non-tonal languages (English, French: Wong et al., 2012). For non-tonal language speakers, it has also been shown that amusics performed better with speech than with musical analogs, especially among those amusics with high pitch discrimination thresholds (over one semitone: Tillmann et al., 2011a, b) (Fig. 33.6). Thus, speech may enhance pitch processing in amusia, even though it does not necessarily restore normal processing.

As the main deficit of congenital amusics seems to concern pitch processing, amusic individuals have been tested not only for the processing of speech intonation, but also for emotional processing, which was shown to be impaired too (Thompson et al., 2012). Pitch-processing capacities and their deficits have been recently linked to phonologic and phonemic awareness abilities (Jones et al., 2009b; Loui et al., 2011), leading to the hypothesis that there might be a shared or common neural basis for pitch-related impairments in amusia/tone deafness and phonemic awareness (i.e., the ability to manipulate phonemes and syllables in spoken words) in dyslexia (Loui et al., 2011).

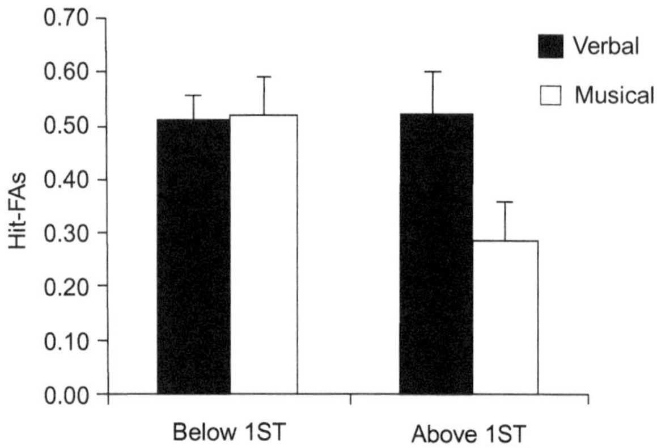

Fig. 33.6. Performance (expressed as hits – false alarms (FA)) for amusic participants in experiment 2 of Tillmann et al. (2011a), tested with parts of the Thai speech material and its musical analogs. Amusic participants were separated into two groups: eight amusics with thresholds below one semitone (1ST) (mean of 0.68, ranging from 0.13 to 0.97) and seven amusics with thresholds above 1ST (mean of 2.41, ranging from 1.3 to 4). For amusics with thresholds above 1ST, performance was significantly better for the verbal material than the musical material; six of the seven amusic participants showed higher mean performance for the verbal material. This advantage was not observed for amusics with thresholds below one semitone.

Space processing

As exposed above, congenital amusia is thought to result from a pitch-processing deficit. This prevailing view has been challenged by Douglas and Bilkey (2007), who reported that amusics were impaired in a classic mental rotation task and were less influenced by the spatial layout of response keys in a pitch judgment task, compared to non-amusic participants. The hypothesis of a link between pitch and space processing has been previously supported by interactions between pitch processing and visual space representations (e.g., stimulus–response compatibility effects: Rusconi et al., 2006; Lidji et al., 2007) as well as enhanced visuospatial processing in musicians (e.g., Brochard et al., 2004). According to Douglas and Bilkey (2007), amusics' deficit of spatial processing suggests that pitch processing "depends on the cognitive mechanisms that are used to process spatial representations in other modalities" (p. 915) and that amusics' disorder might arise as a consequence of a general spatial processing deficit.

However, this study has been criticized, leading to a call for "a clear need for replication and extension before such a view is accepted" (Stewart and Walsh, 2007, p. R893), and two further studies investigating spatial processing in congenital amusia failed to show any spatial processing deficit in congenital amusics with a variety of spatial tasks (e.g., Shepard-Metzler mental rotation task, bisection task, Corsi block task, visual patterns test; Tillmann et al., 2010; Williamson et al., 2011). These findings thus suggest that amusics' deficit in pitch processing does not co-occur with a deficit in spatial processing, and further suggest largely independent pitch representations and visuospatial representations.

DISORDERS OF PITCH PRODUCTION AND MUSIC PRODUCTION IN CONGENITAL AMUSIA: A DISSOCIATION FROM PERCEPTION?

Congenital amusia is characterized by pitch perception and memory deficits, more or less strongly accompanied by deficits along the time dimension (i.e., deficits in the processing of rhythm or meter). These perception and memory deficits impact most strongly on musical material, with some consequences also in the speech realm. However, the colloquial understanding of amusia is that people suffering from this condition notoriously sing out of tune. Some studies have thus attempted to characterize musical production deficits in congenital amusia: Initial studies of musical production have shown that amusic participants are impaired when asked to synchronize to music in tapping tasks (Ayotte et al., 2002; Dalla Bella and Peretz, 2003). The vast majority of studies investigating musical production in amusia have focused on singing abilities. As expected, most of the amusic individuals (i.e., individuals with pitch perception and memory deficits) exhibit poor singing, as assessed both by peer ratings (Ayotte et al., 2002) and acoustic analyses of vocal performance (Loui et al., 2008; Dalla Bella et al., 2009; Hutchins et al., 2010b; Tremblay-Champoux et al., 2010; Williamson et al., 2012). Singing impairments are more pronounced along the pitch dimension than the time dimension (Dalla Bella et al., 2009; Tremblay-Champoux et al., 2010), reminiscent of the pattern observed for perceptual deficits. Along the pitch dimension, most amusics make interval and contour errors (Dalla Bella et al., 2009) and perform poorly on pitch-matching tasks (Hutchins et al., 2010b), whereas about half of them sing on time (Dalla Bella et al., 2009). Amusic individuals are extremely impaired when requested to sing a song without lyrics (e.g., on /la/), with, for example, half of the amusic sample of Dalla Bella et al. (2009) being unable to produce complete performances on /la/. These data thus suggest a benefit of the language dimension for the musical production.

It should be noted that not all poor singers suffer from perceptual pitch deficits: poor singing is estimated in around 10–15% of the general population (see review in Dalla Bella et al., 2012), and congenital amusia *per se* is estimated in about 4% of the general population (e.g., Kalmus and Fry, 1980), thus excluding perceptual deficits as the sole possible origin of poor singing. Poor singing without pitch perception deficits is sometimes referred to as "vocal amusia" or "oral-expressive amusia." Such a condition might arise from failures at any component of a "sensorimotor vocal loop": auditory-motor integration, motor production, short- and long-term memory (for a review, see Berkowska and Dalla Bella, 2009). Experiments with a "slider" allowing one to perform pitch matching with a non-vocal motor output supported the assumption of a variety of causes for poor pitch singing (Hutchins and Peretz, 2012a), notably poor vocal motor control and difficulties in matching a target pitch when the model has a different timbre. It is generally assumed that in congenital amusics, poor pitch singing arises from deficits in pitch perception and memory, which render it difficult to adequately monitor the acoustic feedback from the individual's own voice. In support of this claim, Dalla Bella et al. (2009) found that poor pitch singing was correlated with the ability to detect pitch changes (as tested in Hyde and Peretz, 2004; Fig. 33.2). Memory components seem also to be important in singing performance, as the limited ability of amusics to sing a song without lyrics is correlated with the score in the memory subtest of the MBEA (Dalla Bella et al., 2007; Tremblay-Champoux et al., 2010).

Most interestingly, studies of vocal performance have revealed that some amusics exhibit normal singing despite their pitch perception deficit (Ayotte et al., 2002; Dalla Bella et al., 2009; Tremblay-Champoux et al., 2010). Taken together, these observations in poor singers (pure vocal amusia) and congenital amusics suggest a dissociation between pitch perception and production. The most striking result supporting this claim was the report that congenital amusics can accurately reproduce the direction of pitch intervals whereas they cannot classify the same intervals as being rising ("up") or falling ("down") (Loui et al., 2008). However, a replication study with a larger sample of amusic participants (Williamson et al., 2012) led to a more nuanced result pattern: only two out of 14 amusic participants had better production than perception thresholds, and furthermore, amusic participants performed differently when the perceptual task used (or not) the "up/down" labels to designate the pitch intervals. Using a different approach, based on the "pitch shift reflex," which is an automatic pitch adjustment following an altered feedback of one's own voice, Hutchins and Peretz (2013) found that four out of nine amusics exhibited pitch shifts similar to that of controls after both small and large pitch changes. They labeled this subgroup "production-preserved (congenital) amusics." Pitch shifts were predicted by pitch-matching accuracy (production), but not pitch discrimination abilities (perception), which was interpreted as further evidence for a dual-route neural architecture for the processing of pitch (see below).

Further support for a dissociation between pitch perception and production has been reported for speech materials. While amusics are impaired at imitating pitch contours in statements and questions, their production abilities exceed their perceptual abilities with the same sentences (Liu et al., 2010). Two other studies point to an even clearer dissociation between amusics' pitch perception and production abilities for speech: in Nan et al. (2010), amusic speakers of a tone language (Mandarin) were impaired in distinguishing lexical tones, but not in their production; in Hutchins and Peretz (2012b), amusics were impaired only when discriminating, but not imitating, pitch variations embedded in sentences. However, recent results showed that congenital amusics who were native Mandarin speakers also showed pitch production deficits, which furthermore seemed domain-general: amusic participants were worse than control participants for the production of both songs and speech, which included pitch information (Liu et al., 2013).

Overall, pitch perception and production seem to be partly decoupled in congenital amusia, although it does not seem possible to advocate a complete dissociation. The cerebral underpinnings of this partial dissociation are not yet understood, but it has been proposed that several pathways exist for pitch encoding, resulting in an abstract code (more consciously accessible) and a vocal-motor code (Hutchins and Peretz, 2012b, 2013). Compensation through somatosensory cues has also been proposed to account for intact production abilities (Nan et al., 2010). The findings of anatomic abnormalities in auditory and frontal areas (Hyde et al., 2006, 2007; Mandell et al., 2007; Albouy et al., 2013a), as well as a reduced fiber connectivity between these areas (Loui et al., 2009), has first been taken as evidence that congenital amusia is an auditory-motor feedback disorder (Mandell et al., 2007). However, this hypothesis cannot easily account for all types of congenital amusia, notably when there are preserved pitch production abilities (Hutchins et al., 2010a). It remains a challenge for future studies to clearly establish the links between pitch perception/production abilities and abnormalities in auditory and frontal areas and their connections.

Finally, recent research has started to investigate whether poor pitch singing in congenital amusia might be improved by training. Amusics seem to benefit from singing in unison when singing without lyrics (Tremblay-Champoux et al., 2010), and from a seven-session training program, leading to improved vocal performance of familiar songs (Anderson et al., 2012).

IMPLICIT PITCH PROCESSING IN CONGENITAL AMUSIA

The partial dissociation between pitch perception and production has sometimes been taken as evidence of some implicit processing of pitch and its representations in congenital amusia (e.g., Dalla Bella et al., 2009). Similarly, preserved emotional judgments (Gosselin et al., 2009; Paquette et al., 2011 (poster presented at The Neurosciences and Music-IV Learning and Memory Conference; unpublished data)) have also been interpreted in line with this hypothesis, although musical emotions could also be based on other musically relevant information than pitch (e.g., rhythm, timbre, tempo). More generally, pitch processing has been further characterized in congenital amusia using implicit investigation methods. Such approaches are of interest because they are thought to have the potential to reveal unconscious or degraded representations (e.g., knowledge, memory) in the absence of awareness, thus contrasting with explicit knowledge, which can be consciously recollected and verbalized. Indirect investigation methods have provided evidence for the influence of spared implicit knowledge on perception and action in the presence of severe impairments in tasks requiring explicit processing. The dissociation between implicit and explicit task performance has been reported in various neurologic disorders (e.g., alexia, agraphia, aphasia, prosopagnosia), including acquired and congenital amusia. The priming paradigm is one of the implicit investigation methods and can be used to probe participants' musical knowledge (see Tillmann, 2005, for a review).

In the musical priming paradigm, a musical context (e.g., a chord sequence) is used to generate expectations about upcoming events. These music expectations arise from the regularities of the musical system to which listeners typically become sensitive via mere exposure to music of their culture (Tillmann et al., 2000). Performance (notably speed of processing) in an unrelated task (e.g., judgments on the tuning, created sensory dissonance, or used timbre) concerning the last chord of a chord sequence (i.e., the target chord) depends on the musical relatedness of this chord with its preceding context, with related targets leading to faster processing. This paradigm proved efficient in revealing implicit music knowledge in a case of acquired amusia following a stroke (Tillmann et al., 2007) and also in congenital amusia (Tillmann et al., 2012): The music priming paradigm revealed amusics' implicit sensitivity to harmonic relationships between chords belonging to a given key. Further evidence for implicit pitch representations in congenital amusia was found using a melodic priming paradigm (Omigie et al., 2012a). The probability of notes embedded in melodies (that is, their expectedness based on various dimensions, including melodic contour, tonal structures) modulated performance in an incident timbre task in amusics and controls. However, amusics were impaired relative to controls for explicit judgments of the expectedness of target notes in the same melodies. In accordance with the hypothesis of preserved implicit knowledge of musical regularities in amusia, it has recently been shown that tonal structure can, to some extent, influence amusics' performance in a short-term memory task on tone sequences (Albouy et al., 2013b). When required to compare either tonal or atonal melodies, amusics did not show a benefit of tonal structures in terms of accuracy (while controls showed increased performance for tonal melodies as compared to atonal melodies; see Schulze et al., 2012). However, both groups showed faster response times for tonal sequences than for atonal sequences, thus suggesting that amusics can process the tonal information to some extent and use it in the context of the short-term memory task. It should nevertheless be noted that not all attempts to find evidence for implicit pitch representations in congenital amusia were successful: Pfeuty and Peretz (2010) found no evidence of implicit pitch processing in a pitch–time interference task.

Additional evidence for unconscious pitch-tracking capabilities in the amusic brain comes from event-related potential (ERP) studies. Peretz et al. (2009) showed that amusic participants exhibit reliable ERPs following tones mistuned by a quarter-tone in tonal melodies, although they were not able to explicitly report the mistuning. More specifically, when controls exhibited both an early (N200) and a late (P600) ERP component following quarter-tone mistunings or out-of-key notes, amusics only exhibited the early deviance-selective response, mainly for mistuned (but not out-of-key) notes. Converging evidence for unimpaired early stages of auditory processing in passive-listening paradigms with pitch deviances was obtained in an MMN paradigm with deviances as small as an eighth of a tone (Moreau et al., 2013), with amusic participants exhibiting normal MMN despite very poor explicit detection of deviants. This study also revealed impaired long-latency deviance ERPs (P300) in congenital amusia, even when the participants could explicitly detect the deviant tones (but see Braun et al., 2008, for a different result pattern). Recently, Omigie et al. (2013) investigated amusics' brain ability to track pitch probabilities in melodies (the degree of expectedness of a tone within a melody, as in Omigie et al., 2012a) with an electrophysiologic approach. Their data revealed that both amusic and control groups presented brain responses to tones that did not respect the melodic expectations given the melodic context, and that these responses were modulated by the degree of expectedness of these tones. These results altogether argue in favor of long-term knowledge of musical structures in amusia.

Evidence for intact implicit pitch representations in congenital amusia, as revealed by priming paradigms and passive ERP protocols, together with the behavioral expressions of the deficit, raise questions about the extent to which the learning mechanisms allowing the brain to extract the regularities in the auditory environment, most notably in music and language, are preserved in congenital amusia. This internalization of rules is thought to occur involuntarily through implicit learning, also referred to as statistical learning (e.g., Tillmann et al., 2000; Saffran, 2003). In congenital amusia, statistical learning studies with new artificial musical systems have yielded mixed results. Omigie and Stewart (2011) reported intact statistical learning capabilities in amusics for both linguistic (see Peretz et al., 2012, for converging evidence for linguistic stimuli) and tone materials, albeit amusics exhibited less confidence in their responses, again suggesting a dissociation between their implicit capacities and their explicit knowledge or confidence. However, Loui and Schlaug (2012) and Peretz et al. (2012) failed to report statistical learning for various types of tone materials in amusic participants. Discrepancies in datasets might be due to the different definitions of the test phases: Omigie and Stewart (2011) used more salient violations to study the acquired artificial knowledge than did Peretz et al. (2012).

The datasets reported above point to some spared implicit processing of the pitch dimension in congenital amusia, as revealed by the musical priming paradigm, ERPs, and – in some cases– statistical learning paradigms. The studies on implicit processing of pitch, tonal structures, as well as on implicit learning suggest that congenital amusics can also have acquired partial musical knowledge about statistical regularities of the musical system. Their implicit discrimination performance might be sufficient to capture some of the regularities of associations between pitches, leading to some tonal knowledge (including the harmonic relations tested in the musical priming paradigm), even if this knowledge remains sparse and functional only at an implicit level. More generally, it has been hypothesized that congenital amusia might be linked to a disorder of awareness (e.g., Peretz et al., 2009; Omigie and Stewart, 2011; Moreau et al., 2013).

CONCLUSION

The present chapter reviewed research on congenital amusia with its deficits in music perception and production as well as the potential extensions of the deficit to the processing of other materials (in particular, speech). Research on congenital amusia investigating the underlying neural correlates provides further insights not only into the impaired neural networks of music processing for this particular disorder, but integrates also in research investigating the normal functioning of the non-amusic brain (in particular, neural correlates of auditory short-term memory or of potentially shared processes with speech processing).

The main deficit that has been studied so far for this phenomenon relates to the pitch dimension, in particular to pitch perception and pitch memory. Depending on the studies, the emphasis has been on pitch discrimination, pitch direction processing, or pitch memory. A unifying hypothesis could suggest the pitch memory deficit as the basis of the impairment as memory capacities are required in both pitch discrimination as well as pitch direction tasks (i.e., comparing tones presented in a given trial). However, future research is necessary to confirm this hypothesis and/or disentangle the contribution of the different capacities to eventually define different subtypes of congenital amusia.

While most amusic cases are impaired on the processing of the pitch dimension, some amusic cases are also impaired on the time dimension – and recent research further suggests amusic cases solely impaired on the time dimension. These observations thus contribute to the suggestion that "congenital amusia" is not a unified disorder, but might be a more general terminology to cover various forms of music perception and production deficits.

The subtypes of congenital amusia might be further revealed by self-reports, as acquired by questionnaire-based data (McDonald and Stewart, 2008; Peretz et al., 2008; Omigie et al., 2012b). For example, recent research on musical experience in congenital amusia suggests that amusic individuals might differ in their levels of music perception abilities as well as their music appreciation (and thus their music engagement in everyday life, leading to different attitudes towards music) (McDonald and Stewart, 2008; Omigie et al., 2012a). Subjective reports will help us to develop the investigation of further refinements of different subtypes of amusia, and also to indicate future directions to investigate whether congenital amusia might be related to other deficits (e.g., dyslexia) and/or have consequences for the processing of other materials (as partially addressed in research on spatial processing). Finally, the investigation approach based on music experience in everyday life further shows the need to bring this research area back to more real music (not just highly controlled and thus reduced material focusing, for example, on the pitch dimension) and emotional experience with music.

In sum, research on congenital amusia is a currently growing domain with the aim of understanding a deficit that has been known for a long time, but was missing systematic investigation. Studying the phenomenon of congenital amusia contributes not only to the understanding of the disorder itself, but also of normally functioning cognitive and neural correlates related to music processing. Furthermore, understanding the disorder itself is not restricted to the static situation in adulthood (as in most studies up to now), but requires study of its development in children, or even infancy (as started recently: Lebrun et al., 2012; Peretz et al., 2013), and the development of potential perspectives for rehabilitation. First attempts at rehabilitation have been reported recently, with a failure in children via daily music listening over 4 weeks (Mignault Goulet et al., 2012), but rather encouraging results after 7 weeks of singing workshops (Anderson et al., 2012). In this regard, the studies reporting implicit capacities of amusics related to the processing of pitch and tonal structures (e.g., Peretz et al., 2009; Omigie et al., 2012a; Tillmann et al., 2012; Albouy et al., 2013b) have also implications for the aim of developing rehabilitation programs. Indeed, as shown previously for other impaired cognitive functions (e.g., Bier et al., 2002), training perceptual and/or learning strategies might exploit the spared implicit processing resources in order to reduce the deficits in congenital amusia, or even remediate this "musical handicap," not only in children, but also in adults.

References

Albouy P, Mattout J, Bouet R et al. (2013a). Impaired pitch perception and memory in congenital amusia: the deficit starts in the auditory cortex. Brain 136: 1639–1661.

Albouy P, Schulze K, Caclin A et al. (2013b). Does tonality boost short-term memory in congenital amusia? Brain Res 1537: 224–232.

Allen G (1878). Note deafness. Mind 3 (10): 157–167.

Anderson S, Himonides E, Wise K et al. (2012). Is there potential for learning in amusia? A study of the effect of singing intervention in congenital amusia. Ann N Y Acad Sci 1252: 345–353.

Ayotte J, Peretz I, Hyde K (2002). Congenital amusia: a group study of adults afflicted with a music-specific disorder. Brain 125: 238–251.

Berkowska M, Dalla Bella S (2009). Acquired and congenital disorders of sung performance: a review. Adv Cognit Psychol 5: 69–83.

Bier N, Vanier M, Meulemans T (2002). Errorless learning: a method to help amnesic patient learn new information. J Cognit Rehabil 20: 12–18.

Bigand E, Poulin-Charronnat B (2006). Are we "experienced listeners"? A review of the musical capacities that do not depend on formal musical training. Cognition 100: 100–130.

Braun A, McArdle J, Jones J et al. (2008). Tune deafness: processing melodic errors outside of conscious awareness as reflected by components of the auditory ERP. PLoS One 3: e2349.

Brochard R, Dufour A, Despres O (2004). Effect of musical expertise on visuospatial abilities: evidence from reaction times and mental imagery. Brain Cogn 54: 103–109.

Cousineau M, McDermott JH, Peretz I (2012). The basis of musical consonance as revealed by congenital amusia. Proc Natl Acad Sci U S A 109: 19858–19863.

Cuddy LL, Balkwill LL, Peretz I et al. (2005). Musical difficulties are rare: a study of "tone deafness" among university students. Ann N Y Acad Sci 1060: 311–324.

Dalla Bella S, Peretz I (2003). Congenital amusia interferes with the ability to synchronize with music. Ann N Y Acad Sci 999: 166–169.

Dalla Bella S, Giguere JF, Peretz I (2007). Singing proficiency in the general population. J Acoust Soc Am 121: 1182–1189.

Dalla Bella S, Giguere JF, Peretz I (2009). Singing in congenital amusia. J Acoust Soc Am 126: 414–424.

Dalla Bella S, Tremblay-Champoux A, Berkowska M et al. (2012). Memory disorders and vocal performance. Ann N Y Acad Sci 1252: 338–344.

David O, Kiebel SJ, Harrison LM et al. (2006). Dynamic causal modeling of evoked responses in EEG and MEG. Neuroimage 30: 1255–1272.

Douglas KM, Bilkey DK (2007). Amusia is associated with deficits in spatial processing. Nat Neurosci 10: 915–921.

Drayna D, Manichaikul A, de Lange M et al. (2001). Genetic correlates of musical pitch recognition in humans. Science 291: 1969–1972.

Foxton JM, Dean JL, Gee R et al. (2004). Characterization of deficits in pitch perception underlying 'tone deafness'. Brain 127: 801–810.

Foxton JM, Nandy RK, Griffiths TD (2006). Rhythm deficits in 'tone deafness'. Brain Cogn 62: 24–29.

Francès R (1958). La perception de la musique, Vrin, Paris [The perception of music. Transl. W.J. Dowling (1988). Lawrence Erlbaum: Hillsdale, NJ.].

Gaab N, Gaser C, Zaehle T et al. (2003). Functional anatomy of pitch memory – an fMRI study with sparse temporal sampling. Neuroimage 19: 1417–1426.

Gosselin N, Jolicoeur P, Peretz I (2009). Impaired memory for pitch in congenital amusia. Ann N Y Acad Sci 1169: 270–272.

Griffiths TD (1999). Human complex sound analysis. Clin Sci (Lond) 96: 231–234.

Griffiths TD, Johnsrude I, Dean JL et al. (1999). A common neural substrate for the analysis of pitch and duration pattern in segmented sound? Neuroreport 10: 3825–3830.

Hannon EE, Trehub SE (2005). Tuning in to musical rhythms: infants learn more readily than adults. Proc Natl Acad Sci U S A 102: 12639–12643.

Henry M, McAuley JD (2010). On the prevalence of congenital amusia. Music Percept 27: 413–418.

Hutchins SM, Peretz I (2012a). A frog in your throat or in your ear? Searching for the causes of poor singing. J Exp Psychol Gen 141: 76–97.

Hutchins S, Peretz I (2012b). Amusics can imitate what they cannot discriminate. Brain Lang 123: 234–239.

Hutchins S, Peretz I (2013). Vocal pitch shift in congenital amusia (pitch deafness). Brain Lang 125: 106–117.

Hutchins S, Gosselin N, Peretz I (2010a). Identification of changes along a continuum of speech intonation is impaired in congenital amusia. Front Psychol 1: 236.

Hutchins S, Zarate JM, Zatorre RJ et al. (2010b). An acoustical study of vocal pitch matching in congenital amusia. J Acoust Soc Am 127: 504–512.

Hyde KL, Peretz I (2004). Brains that are out of tune but in time. Psychol Sci 15: 356–360.

Hyde KL, Zatorre RJ, Griffiths TD et al. (2006). Morphometry of the amusic brain: a two-site study. Brain 129: 2562–2570.

Hyde KL, Lerch JP, Zatorre RJ et al. (2007). Cortical thickness in congenital amusia: when less is better than more. J Neurosci 27: 13028–13032.

Hyde KL, Zatorre RJ, Peretz I (2011). Functional MRI evidence of an abnormal neural network for pitch processing in congenital amusia. Cereb Cortex 21: 292–299.

Janata P, Birk JL, Van Horn JD et al. (2002). The cortical topography of tonal structures underlying Western music. Science 298: 2167–2170.

Jiang C, Hamm JP, Lim VK et al. (2010). Processing melodic contour and speech intonation in congenital amusics with Mandarin Chinese. Neuropsychologia 48: 2630–2639.

Jiang C, Hamm JP, Lim VK et al. (2012). Impaired categorical perception of lexical tones in Mandarin-speaking congenital amusics. Mem Cognit 40: 1109–1121.

Jones JL, Zalewski C, Brewer C et al. (2009a). Widespread auditory deficits in tune deafness. Ear Hear 30: 63–72.

Jones JL, Lucker J, Zalewski C et al. (2009b). Phonological processing in adults with deficits in musical pitch recognition. J Commun Disord 42: 226–234.

Kalmus H, Fry DB (1980). On tune deafness (dysmelodia): frequency, development, genetics and musical background. Ann Hum Genet 43: 369–382.

Koelsch S, Gunter TC, Wittfoth M et al. (2005). Interaction between syntax processing in language and in music: an ERP study. J Cogn Neurosci 17: 1565–1577.

Koelsch S, Schulze K, Sammler D et al. (2009). Functional architecture of verbal and tonal working memory: an FMRI study. Hum Brain Mapp 30: 859–873.

Lebrun MA, Moreau P, McNally-Gagnon A et al. (2012). Congenital amusia in childhood: a case study. Cortex 48: 683–688.

Lidji P, Kolinsky R, Lochy A et al. (2007). Spatial associations for musical stimuli: a piano in the head? J Exp Psychol Hum Percept Perform 33: 1189–1207.

Liu F, Patel AD, Fourcin A et al. (2010). Intonation processing in congenital amusia: discrimination, identification and imitation. Brain 133: 1682–1693.

Liu F, Jiang C, Thompson WF et al. (2012a). The mechanism of speech processing in congenital amusia: evidence from Mandarin speakers. PLoS One 7: e30374.

Liu F, Xu Y, Patel AD et al. (2012b). Differential recognition of pitch patterns in discrete and gliding stimuli in congenital amusia: evidence from Mandarin speakers. Brain Cogn 79: 209–215.

Liu F, Jiang C, Pfordresher PQ et al. (2013). Individuals with congenital amusia imitate pitches more accurately in singing than in speaking: implications for music and language processing. Atten Percept Psychophys 2013 (75): 1783–1798.

Loui P, Schlaug G (2012). Impaired learning of event frequencies in tone deafness. Ann N Y Acad Sci 1252: 354–360.

Loui P, Guenther FH, Mathys C et al. (2008). Action-perception mismatch in tone-deafness. Curr Biol 18: R331–R332.

Loui P, Alsop D, Schlaug G (2009). Tone deafness: a new disconnection syndrome? J Neurosci 29: 10215–10220.

Loui P, Kroog K, Zuk J et al. (2011). Relating pitch awareness to phonemic awareness in children: implications for tone-deafness and dyslexia. Front Psychol 2: 111.

Maess B, Koelsch S, Gunter TC et al. (2001). Musical syntax is processed in Broca's area: an MEG study. Nat Neurosci 4: 540–545.

Mandell J, Schulze K, Schlaug G (2007). Congenital amusia: an auditory-motor feedback disorder? Restor Neurol Neurosci 25: 323–334.

Marin MM, Gingras B, Stewart L (2012). Perception of musical timbre in congenital amusia: categorization, discrimination and short-term memory. Neuropsychologia 50: 367–378.

McAdams S (1989). Segregation of concurrent sounds. I: Effects of frequency modulation coherence. J Acoust Soc Am 86: 2148–2159.

McDonald C, Stewart L (2008). Uses and functions of music in congenital amusia. Music Percept 25: 345–355.

Mignault Goulet G, Moreau P, Robitaille N et al. (2012). Congenital amusia persists in the developing brain after daily music listening. PLoS One 7: e36860.

Moreau P, Jolicoeur P, Peretz I (2013). Pitch discrimination without awareness in congenital amusia: evidence from event-related potentials. Brain Cogn 81: 337–344.

Näätänen R, Picton T (1987). The N1 wave of the human electric and magnetic response to sound: a review and an analysis of the component structure. Psychophysiology 24: 375–425.

Nan Y, Sun Y, Peretz I (2010). Congenital amusia in speakers of a tone language: association with lexical tone agnosia. Brain 133: 2635–2642.

Nguyen S, Tillmann B, Gosselin N et al. (2009). Tonal language processing in congenital amusia. Ann N Y Acad Sci 1169: 490–493.

Omigie D, Stewart L (2011). Preserved statistical learning of tonal and linguistic material in congenital amusia. Front Psychol 2: 109.

Omigie D, Pearce MT, Stewart L (2012a). Tracking of pitch probabilities in congenital amusia. Neuropsychologia 50: 1483–1493.

Omigie D, Müllensiefen D, Stewart L (2012b). The experience of music in congenital amusia. Music Percept 30: 1–18.

Omigie D, Pearce MT, Williamson VJ et al. (2013). Electrophysiological correlates of melodic processing in congenital amusia. Neuropsychologia 51: 1749–1762.

Paquette S, Gosselin N, Peretz I (2011). Normal use of mode and tempo in emotional judgments by amusic adults. In: The Neurosciences and Music-IV Learning and Memory Meeting, Edinburgh, UK.

Patel AD, Foxton JM, Griffiths TD (2005). Musically tone-deaf individuals have difficulty discriminating intonation contours extracted from speech. Brain Cogn 59: 310–313.

Patel AD, Wong M, Foxton J et al. (2008). Speech intonation perception deficits in musical tone deafness (congenital amusia). Music Percept 25: 357–368.

Patterson RD, Uppenkamp S, Johnsrude IS et al. (2002). The processing of temporal pitch and melody information in auditory cortex. Neuron 36: 767–776.

Peretz I (2001). Brain specialization for music. New evidence from congenital amusia. Ann N Y Acad Sci 930: 153–165.

Peretz I (2008). Musical disorders: from behavior to genes. Curr Dir Psychol Sci 17: 329–333.

Peretz I, Hyde KL (2003). What is specific to music processing? Insights from congenital amusia. Trends Cogn Sci 7: 362–367.

Peretz I, Kolinsky R (1993). Boundaries of separability between melody and rhythm in music discrimination: a neuropsychological perspective. Q J Exp Psychol 46: 301–325.

Peretz I, Zatorre RJ (2005). Brain organization for music processing. Annu Rev Psychol 56: 89–114.

Peretz I, Belleville S, Fontaine S (1997). Dissociations between music and language functions after cerebral resection: a new case of amusia without aphasia. Can J Exp Psychol 51: 354–368.

Peretz I, Ayotte J, Zatorre RJ et al. (2002). Congenital amusia: a disorder of fine-grained pitch discrimination. Neuron 33: 185–191.

Peretz I, Champod AS, Hyde K (2003). Varieties of musical disorders. The Montreal Battery of Evaluation of Amusia. Ann N Y Acad Sci 999: 58–75.

Peretz I, Brattico E, Tervaniemi M (2005). Abnormal electrical brain responses to pitch in congenital amusia. Ann Neurol 58: 478–482.

Peretz I, Cummings S, Dube MP (2007). The genetics of congenital amusia (tone deafness): a family-aggregation study. Am J Hum Genet 81: 582–588.

Peretz I, Gosselin N, Tillmann B et al. (2008). On-line identification of congenital amusia. Music Percept 25: 331–343.

Peretz I, Brattico E, Jarvenpaa M et al. (2009). The amusic brain: in tune, out of key, and unaware. Brain 132: 1277–1286.

Peretz I, Saffran J, Schon D et al. (2012). Statistical learning of speech, not music, in congenital amusia. Ann N Y Acad Sci 1252: 361–367.

Peretz I, Gosselin N, Nan Y et al. (2013). A novel tool for evaluating children's musical abilities across age and culture. Front Syst Neurosci 7: 30.

Pfeuty M, Peretz I (2010). Abnormal pitch–time interference in congenital amusia: evidence from an implicit test. Atten Percept Psychophys 72: 763–774.

Pfordresher PQ, Brown S (2007). Poor-pitch singing in the absence of tone deafness. Music Percept 25: 95–115.

Phillips-Silver J, Toiviainen P, Gosselin N et al. (2011). Born to dance but beat deaf: a new form of congenital amusia. Neuropsychologia 49: 961–969.

Piccirilli M, Sciarma T, Luzzi S (2000). Modularity of music: evidence from a case of pure amusia. J Neurol Neurosurg Psychiatry 69: 541–545.

Rusconi E, Kwan B, Giordano BL et al. (2006). Spatial representation of pitch height: the SMARC effect. Cognition 99: 113–129.

Saffran JR (2003). Musical learning and language development. Ann N Y Acad Sci 999: 397–401.

Schulze K, Gaab N, Schlaug G (2009). Perceiving pitch absolutely: comparing absolute and relative pitch possessors in a pitch memory task. BMC Neurosci 10: 106.

Schulze K, Mueller K, Koelsch S (2011a). Neural correlates of strategy use during auditory working memory in musicians and non-musicians. Eur J Neurosci 33: 189–196.

Schulze K, Zysset S, Mueller K et al. (2011b). Neuroarchitecture of verbal and tonal working memory in nonmusicians and musicians. Hum Brain Mapp 32: 771–783.

Schulze K, Dowling WJ, Tillmann B (2012). Working memory for tonal and atonal sequences during a forward and a backward recognition task. Music Percept 29: 255–267.

Sloboda JA, Wise KJ, Peretz I (2005). Quantifying tone deafness in the general population. Ann N Y Acad Sci 1060: 255–261.

Stewart L (2011). Characterizing congenital amusia. Q J Exp Psychol 64: 625–638.

Stewart L, Walsh V (2007). Music perception: sounds lost in space. Curr Biol 17: R892–R893.

Thompson WF (2007). Exploring variants of amusia: Tone deafness, rhythm impairment, and intonation insensitivity. In: The inaugural International Conference on Music Communication Science 5–7 December 2007, Sydney, Australia. Proceedings of ICoMCS, pp. 159–163.

Thompson WF, Marin MM, Stewart L (2012). Reduced sensitivity to emotional prosody in congenital amusia rekindles the musical protolanguage hypothesis. Proc Natl Acad Sci U S A 109: 19027–19032.

Tillmann B (2005). Implicit investigations of tonal knowledge in nonmusician listeners. Ann N Y Acad Sci 1060: 100–110.

Tillmann B, Bharucha JJ, Bigand E (2000). Implicit learning of tonality: a self-organizing approach. Psychol Rev 107: 885–913.

Tillmann B, Janata P, Bharucha JJ (2003). Activation of the inferior frontal cortex in musical priming. Brain Res Cogn Brain Res 16: 145–161.

Tillmann B, Koelsch S, Escoffier N et al. (2006). Cognitive priming in sung and instrumental music: activation of inferior frontal cortex. Neuroimage 31: 1771–1782.

Tillmann B, Peretz I, Bigand E et al. (2007). Harmonic priming in an amusic patient: the power of implicit tasks. Cogn Neuropsychol 24: 603–622.

Tillmann B, Schulze K, Foxton JM (2009). Congenital amusia: a short-term memory deficit for non-verbal, but not verbal sounds. Brain Cogn 71: 259–264.

Tillmann B, Jolicoeur P, Ishihara M et al. (2010). The amusic brain: lost in music, but not in space. PLoS One 5: e10173.

Tillmann B, Burnham D, Nguyen S et al. (2011a). Congenital amusia (or tone-deafness) interferes with pitch processing in tone languages. Front Psychol 2: 120.

Tillmann B, Rusconi E, Traube C et al. (2011b). Fine-grained pitch processing of music and speech in congenital amusia. J Acoust Soc Am 130: 4089–4096.

Tillmann B, Gosselin N, Bigand E et al. (2012). Priming paradigm reveals harmonic structure processing in congenital amusia. Cortex 48: 1073–1078.

Tremblay-Champoux A, Dalla Bella S, Phillips-Silver J et al. (2010). Singing proficiency in congenital amusia: imitation helps. Cogn Neuropsychol 27: 463–476.

Vos PG, Troost JM (1989). Ascending and descending melodic intervals: statistical findings and their perceptual relevance. Music Percept 6: 383–396.

Watkins KE, Vargha-Khadem F, Ashburner J et al. (2002). MRI analysis of an inherited speech and language disorder: structural brain abnormalities. Brain 125: 465–478.

Williamson VJ, Stewart L (2010). Memory for pitch in congenital amusia: beyond a fine-grained pitch discrimination problem. Memory 18: 657–669.

Williamson VJ, McDonald C, Deutsch D et al. (2010). Faster decline of pitch memory over time in congenital amusia. Adv Cognit Psychol 6: 15–22.

Williamson VJ, Cocchini G, Stewart L (2011). The relationship between pitch and space in congenital amusia. Brain Cogn 76: 70–76.

Williamson VJ, Liu F, Peryer G et al. (2012). Perception and action de-coupling in congenital amusia: sensitivity to task demands. Neuropsychologia 50: 172–180.

Wong PC, Ciocca V, Chan AH et al. (2012). Effects of culture on musical pitch perception. PLoS One 7: e33424.

Zatorre RJ, Evans AC, Meyer E (1994). Neural mechanisms underlying melodic perception and memory for pitch. J Neurosci 14: 1908–1919.

Zatorre RJ, Belin P, Penhune VB (2002). Structure and function of auditory cortex: music and speech. Trends Cogn Sci 6: 37–46.

Handbook of Clinical Neurology, Vol. 129 (3rd series)
The Human Auditory System
G.G. Celesia and G. Hickok, Editors

Chapter 34

Acquired amusia

CAMILLA N. CLARK, HANNAH L. GOLDEN, AND JASON D. WARREN*
Dementia Research Centre, UCL Institute of Neurology, University College London, Queen Square, London, United Kingdom

INTRODUCTION

So wurde eine Anschauung gewonnen über die der Perception und Production musikalischer Töne und ihrer Symbole dienenden Centren und Bahnen im Gehirn... dass sie weitgehende Analogien zu den Centren und Bahnen des Sprachvorgangs bilden würden.

[The centers and pathways in the brain subserving the perception and production of musical tones... [bear] extensive analogies... to the centers and pathways of language.]

(Knoblauch, 1888)

Amusia is impaired perception, understanding, or production of music not attributable to disease of the peripheral auditory pathways or motor system. By convention, the definition implies at least some selectivity of the deficit for music; however, the boundaries of amusia as a clinical entity are rather loosely drawn. Whereas congenital amusia (more colloquially, tone deafness) is a specific developmental perceptual disorder afflicting otherwise normal brains (see Chapter 33 in this volume), acquired amusia (henceforth here, simply amusia) results from the effects of brain disease. Interest in acquired forms of amusia has a venerable history (Steinthal, 1871; Knoblauch, 1888; Critchley and Henson, 1977) and clinical taxonomies of these amusias have been described in some detail (Wertheim, 1975; Henson, 1985; Stewart et al., 2006), based largely on observations in patients with focal brain damage, in particular stroke.

The study of amusia is challenging and a number of unresolved problems remain. Foremost among these has been our comparative ignorance of the brain mechanisms that normally process music. The field of amusia has long borrowed concepts and even clinical terminology from the better-characterized disorders of language, the aphasias. From a neurobiologic perspective, this has some justification: like speech, music is a universal and uniquely human capacity, a highly specialized auditory code that generates potentially infinite permutations from a finite set of perceptual components according to culturally and (to some extent) neurobiologically sanctioned "rules" that distinguish these specialized acoustic signals from other kinds of sounds (Alossa and Castelli, 2009). The processing of both language and music engages multiple cognitive operations requiring the coordinated activity of anatomically distributed brain mechanisms and brain substrates for processing certain attributes (for example, syntax: Patel, 2008; Fadiga et al., 2009) may be partly shared between language and music. Formal analogies can be drawn between various attributes that are putatively shared between language and music, such as vocabulary, grammar and syntax (Feuchtwanger, 1930; Lerdahl and Jackendoff, 1983; Koelsch, 2005) and both language and music establish a rich spectrum of associated knowledge and memories based on accumulated experience. On the other hand, analogies between music and language have probably been overemphasized: the clinical literature is replete with examples of patients in whom linguistic and musical abilities have become dissociated (Botez and Wertheim, 1959; Henson, 1985; Stewart et al., 2006) and superficial similarities can obscure fundamental differences (Rogalsky et al., 2011). In cognitive terms, whereas language can and often does reference things at large in the world, music is not referential in this sense; although it is an intrinsically meaningful, self-contained, abstract system. Moreover, the links between music and emotion are arguably more powerful and intimate than those between language and emotion. More fundamentally,

*Correspondence to: Dr. Jason D. Warren, Dementia Research Centre, UCL Institute of Neurology, University College London, 1st Floor, 8–11 Queen Square, London WC1N 3BG, UK. Tel: +44-(0)203-448-4773, Fax: +44-(0)203-448-3104, E-mail: jason.warren@ucl.ac.uk

speech and music differ radically in their acoustic properties, suggesting that specialized neural mechanisms might be required to encode musical and phonologic signals before they can acquire their symbolic significance. Moreover, musical training is very unevenly distributed in the general population, in contrast to linguistic competence, which is acquired (albeit with somewhat variable proficiency) by all normal human brains. There is a further complex interaction with more pervasive culture-specific influences, some of which appear to shape our perception of music from early infancy (Han et al., 2011).

These caveats also underline a number of practical difficulties in assessing amusia. There has been a dearth of standardized neuropsychologic instruments for assessing musical functions, and indeed, it is difficult to access musical capacities in musically untrained individuals. Currently used instruments tend to be vulnerable to potentially confounding cognitive effects, such as reliance on working memory or verbal labeling capacities. Amusia is more likely to be volunteered as a leading symptom in patients with at least some prior musical competence. Often amusia occurs in conjunction with other central auditory deficits or emerges with partial resolution of a generalized auditory agnosia (Motomura et al., 1986). It is particularly difficult to assess musical functions in patients with language impairment, suggesting that there may be a substantial ascertainment bias among reported cases of amusia in favor of individuals with at least relatively intact language skills.

SCOPE OF THIS CHAPTER

Several key recent developments in the field of cognitive neuroscience indicate possible solutions to these traditional problems and suggest that a further review of the topic of amusia is timely.

Firstly, sophisticated cognitive frameworks and models of music processing have been proposed that may on the one hand, allow amusia to be understood as a neuropsychologic phenomenon and on the other, make amusia an empirical test case for assessing certain aspects of the proposed organization of human central auditory processing (Peretz, 2006; Goll et al., 2010b; Koelsch, 2011a). Analogous with other domains of neuropsychology, detailed case analyses of patients with isolated loss or sparing of musical abilities are critical for establishing the relations among different components of music cognition and the extent to which these can be fractionated from one another and from non-musical competencies (Peretz and Coltheart, 2003; Patel, 2008). The elaboration of cognitive models of music processing has benefitted from analogous developments in the neurocognition of language that have seen a shift away from the methods and schemas of classical aphasiology toward cognitive information-processing approaches (Hickok and Poeppel, 2007; Rauschecker and Scott, 2009). In the realm of complex sound or auditory object processing, cognitive models have been proposed in some detail, including; initial analysis of the auditory scene, encoding of acoustic features, binding of these features into object level representations and association of these representations with data from other sensory modalities, with emotional connotations and with information stored in episodic and semantic memory (Goll et al., 2010b). It remains unclear how far the specialized signals and objects of music can be reconciled with such generic frameworks.

Secondly, the increasing availability of functional brain imaging techniques, including positron emission tomography, functional magnetic resonance imaging (fMRI) and evoked potential recordings has allowed delineation of the structure and dynamics of the neural networks that process components of music in the healthy brain (Meyer et al., 2006; Bengtsson et al., 2009; Puschmann et al., 2010). This work has shown that the brain substrates of music processing are extensive, including; temporal and parietal areas implicated in perceptual analysis of auditory signals and memories, subcortical structures implicated in reward and autonomic responses and frontal-lobe regions engaged in the evaluation of sensory signals and programming of an integrated behavioral response (Koelsch et al., 2006; Chen et al., 2008a; Peretz et al., 2009). Hypothesis-directed experiments have seen this field move beyond elementary brain mapping to discriminate among candidate neural mechanisms that support music processing (Koelsch, 2011a; Salimpoor et al., 2011). Besides illuminating the brain substrates for music processing *per se*, functional neuroimaging potentially allows an unprecedentedly direct comparison of these substrates with the neural substrates for processing speech and language (Schon et al., 2010; Rogalsky et al., 2011). The availability of techniques such as transcranial magnetic stimulation (Loui et al., 2010; Andoh and Zatorre, 2013) that can simulate the effect of regional brain lesions in healthy individuals presents fresh opportunities to link the normal and clinical literature on the neuroanatomy of music processing.

Finally, recent years have seen an explosion of interest in the neurodegenerative diseases of later life and their brain bases (Zhou et al., 2010; Rohrer et al., 2011; Warren et al., 2013). A key insight in the field of neurodegenerative disease research has followed from the finding that large-scale brain networks are specifically vulnerable to particular neurodegenerative syndromes and molecular pathologies (Zhou et al., 2010; Rohrer et al., 2011; Warren et al., 2013). No less than other complex cognitive functions, aspects of music processing appear to be vulnerable (or indeed, selectively resistant) to particular neurodegenerative pathologies (Omar et al., 2010; Hsieh et al., 2012), leading to the recognition that amusias may arise from brain network disintegration as well as focal brain damage.

In this chapter, we first consider previous taxonomies of amusia and propose a fresh framework for understanding the amusias essentially as disorders of information processing, informed by contemporary cognitive neuroscience rather than the traditional descriptive approach of bedside neurology. We then review critically the cognitive and neuroanatomic findings reported in the published literature on amusia, emphasizing well-documented cases where amusia has been a clinically relevant symptom. We do not further consider deficits of musical production here, as these tend to reflect loss of highly specialized and idiosyncratic skills: our focus is on generic cognitive and brain processes underpinning the perception and understanding of music and potentially relevant to all listeners. We contrast symptom-led case descriptions with lesion-defined case series, acknowledging that the latter are potentially valuable for establishing anatomic correlations but share the caveats of heterogeneous culprit insults (typically, strokes or temporal lobectomies) with the potential for substantial brain reorganization prior to assessment. We particularly emphasize relations with the language system, for which information-processing accounts have also been developed and a large body of neuroimaging data has been amassed; and neuropsychologic dissociations within the domain of music, as these illuminate the cognitive organization of the musical brain. For each of the key information-processing cognitive operations we propose, we review the clinical evidence for deficits targeting that operation and the extent to which this evidence can be reconciled with the picture of the musical brain emerging from cognitive neuroscience and functional neuroimaging studies. We outline a new classification of amusias based on our information-processing model together with neuropsychologic tests to characterize these disorders (Figs 34.1 and 34.2 and Table 34.1) and

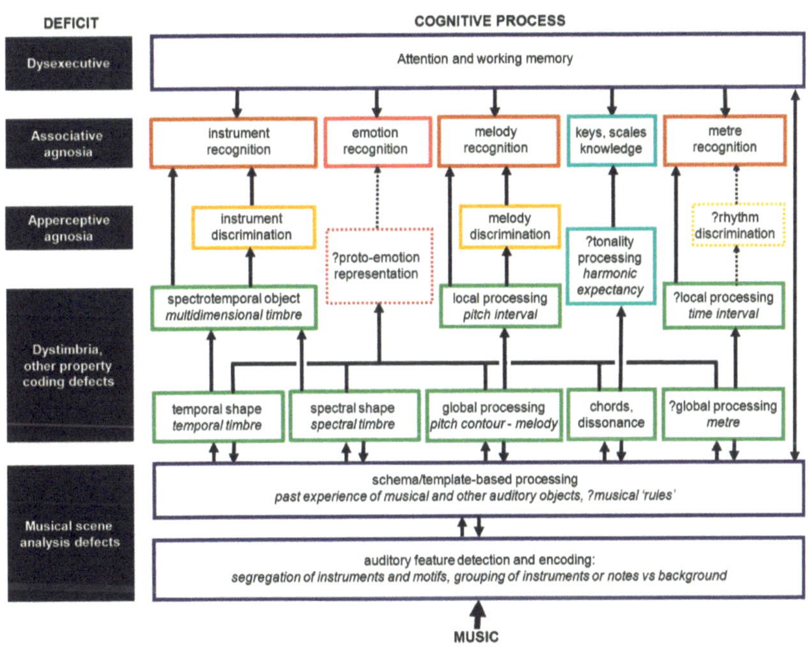

Fig. 34.1. Cognitive architecture of music processing as revealed by study of the amusias (see Table 34.1 and Fig. 34.4). General deficits corresponding to different information-processing stages (left) map on to a multicomponent modular hierarchy in which levels and modules (boxes) are cognitively separable based on the evidence of observed neuropsychologic dissociations; the primary direction of information flow through the hierarchy is indicated (arrows; in general, there are additional cognitive interactions between modules within each level, not shown here for clarity). Processes for which little direct clinical evidence is available (?) or which are inferred based on evidence in the healthy brain (dotted outlines) are also indicated. Processing of musical emotion and apperceptive processing of particular elements of music may occur partly in parallel with other dimensions of music processing, according to listening context. Brain mechanisms for processing speech and other complex sounds interact with the hierarchy according to acoustic and semantic sound characteristics. For example, extraction of temporal shape information is particularly relevant for processing speech signals and spectral shape for voice recognition: these cognitive operations may be partly shared with music before information is channeled into the separate hierarchies corresponding to the processing of phonemes and voices, respectively (not shown here; see text).

Fig. 34.2. An example of a stimulus to assess musical scene analysis. The notation indicates three simultaneous musical streams presented in the form of three-part polyphonic harmony. A highly familiar target tune (here, "Happy birthday"; notated in red) is embedded in the polyphony with two novel distractor melodies. The stimulus could be used as the basis for a test in which the task on each trial would be to identify the embedded melody or indicate whether or not a familiar tune is present.

indicate how these may relate to particular patterns of brain network damage (Fig. 34.3).

A COGNITIVE FRAMEWORK FOR UNDERSTANDING AND ASSESSING AMUSIA

The term amusia was originally coined by Steinthal (1871) in reference to deficits of musical perception and production (sensory and motor amusia) following stroke. Knoblauch (1888) synthesized early findings in cases of amusia to propose the first cognitive model of music processing, postulating an amodal integration "concept center" and hierarchic modular levels of analysis, analogous to and interacting with language-processing centers (Fig. 34.4). Knoblauch's formulation anticipated contemporary graph theoretic accounts of brain network dysfunction (Warren et al., 2013), predicting that the relative selectivity of amusic deficits produced by focal brain damage would depend on which parts of the music-processing network (e.g., hub versus connection) were disrupted. While most early attempts at classifying amusia were essentially descriptive and emphasized broad functional deficits analogous to early classification schemes for aphasia ("receptive" and "expressive," "sensory" and "motor" amusias: Critchley and Henson, 1977), there has long been interest in more specific forms of amusia that might illuminate underlying brain organization; for example, cerebral centers for particular instruments (Henschen, 1920; Kleist, 1928). Amusia has alternatively been classified according to the classic distinction made in neuropsychology between apperceptive agnosias (affecting discrimination of sensory objects, e.g., novel melodies) and associative agnosias (affecting identification of objects, e.g., familiar tunes: Eustache et al., 1990; Alossa and Castelli, 2009). In more recent work, amusia has been analyzed according to the perceptual or cognitive component of music predominantly affected (for example, disorders of pitch, rhythm, or timbre perception) (Stewart et al., 2006); the most widely used neuropsychologic instrument for assessing amusia, the Montreal Battery for Evaluation of Amusia, reflects this modular approach (Peretz et al., 2003; Table 34.1).

According to information-processing models derived from visual neuroscience (Marr, 1980), sensory objects are represented in the brain and ultimately acquire meaning as the result of a hierarchy of processing tasks that integrate particular aspects of the information embedded in the object. In the auditory domain, however, even the definition of an object is problematic (Griffiths and Warren, 2004) and this is particularly the case for musical objects. Melodies, for example, may be highly complex and extended in time while retaining their identity when transposed in pitch or transcribed for different instruments (thereby radically altering many perceptual characteristics), whereas the distinctive timbres of musical instruments convey their identities, but have many features in common with other complex sounds. One general operational, neuropsychologic definition of a musical object that can encompass these very different entities might be, "a collection of music data bound in a common perceptual representation and disambiguated from the musical scene" (Griffiths and Warren, 2004; Goll et al., 2010b). This definition is exemplified by listening to an orchestral performance in which musical coherence rests fundamentally on disambiguating the sounds of the orchestra from extraneous background noises, on binding together successive constituent notes into phrases and melodies, and perhaps also on sounding out different voices within the music. Musically untrained listeners can detect violations of certain musical properties, such as key, in keeping with internalized musical rules or templates (Cuddy et al., 1981). Musical objects once encoded will generally engage a variety of emotional, semantic, and mnestic associations. Such considerations suggest that music, like other complex sensory objects, might be amenable to a neuropsychologic analysis in terms of information-processing stages and further, that particular information-processing stages might be associated with distinctive clinical effects in the face of brain damage (Goll et al., 2010b).

To be widely applicable, information-processing accounts of music cognition should encompass the diversity of clinical amusic deficits, including the critical instances of neuropsychologic dissociation and accumulating evidence that particular cognitive functions are mediated by distributed brain networks. Computational modeling approaches suggest that these requirements can be reconciled: neuropsychologic dissociations can be simulated in artificial neural networks and do not depend on a strict partitioning or modularity of cognitive functions (Peretz and Coltheart, 2003). Cognitive modularity is computationally efficient and

Table 34.1

A framework for neuropsychologic assessment of acquired amusias as disorders of information processing

Amusic disorder	Timbre	Pitch	Timing
Anhedonia or disorder of musical emotion processing	Coding emotion from timbre Labeling of emotion from real or synthetic timbres Emotional experience of timbre ? Subjective rating of emotional responses to timbre dimensions	Coding emotion from melody Labeling emotion from melodies (e.g., mode, canonic emotions) Emotional experience of melodies Subjective rating of emotional response to melodies	Coding emotion from time cues Labeling of emotion to tempo, accent Emotional experience of timing cues ? Subjective rating of emotional response to temporal cues
Associative agnosia or disorder of musical object recognition	Recognition of instruments Naming instruments Matching instrument sounds to pictures	Recognition of tunes, chords, keys Naming tunes Matching tunes to names Detecting familiarity* Detecting errors in familiar tunes Knowledge about keys	Recognition of metres† Naming metres Matching metres to names*
Apperceptive agnosia or disorder of musical object perception	Perceptual manipulations of instrument sounds Matching of instrument timbres across different pitch ranges‡ ? Recognition of "non-canonic" instrument sounds, e.g. pizzicato, portamento	Perceptual manipulations of whole melodies Matching transformed melodies**	Perceptual manipulations of rhythmic sequences ? Matching rhythms transformed in tempo
Disorders of musical property encoding	Spectrotemporal flux perception‡‡ Discriminating real or synthetic timbres Judging direction of spectrotemporal "ripples"	Tonality (chord/scale) perception¶ Detecting key violations* Judging consonance/dissonance Judging "finished/unfinished" chord sequences	Metre perception Discriminating long-range interval structures
	Spectral feature perception‡‡ Detecting spectral envelope change Temporal feature perception ‡‡ Detecting temporal envelope change	Pitch contour (melody) perception Matching sequences of directional changes* Pitch interval perception Detecting pitch change* Judging direction of pitch change	Rhythm perception Matching sequences of interval changes Timing interval/tempo perception Detecting anisochrony Detecting internote gap, tempo changes
Disorders of musical scene analysis	Parsing superimposed instrument sounds Segregating target instrument sounds from background Grouping of notes by instrument sounds against background	Parsing superimposed pitches and pitch sequences Segregating target pitches or melody lines from background¶¶ Grouping pitches into melody lines against background	Parsing superimposed rhythms Segregating target rhythms from background Grouping notes into rhythms against background††

?Suggested test procedures.
*Tests incorporated in the Montreal Battery for Evaluation of Amusia (Peretz et al., 2003).
**Described in Foster and Zatorre (2010) and Foster et al. (2013).
†e.g., waltz, march.
††Described in Goll et al. (2012).
‡Described in Omar et al. (2010).
‡‡Described in Goll et al. (2010a, b, 2011).
¶Precise information-processing stages not well established.
¶¶See Figure 34.2.
This framework considers key amusias affecting particular information-processing stages (left) under the broad categories of pitch, timing information, and timbre, which comprise the major structural elements of music and which have been shown to be at least partly neuropsychologically separable (see text). Component processes of music cognition are shown together with test procedures to assess those processes. Underlying cognitive and neuroanatomic architectures are presented in Figures 34.1 and 34.3.

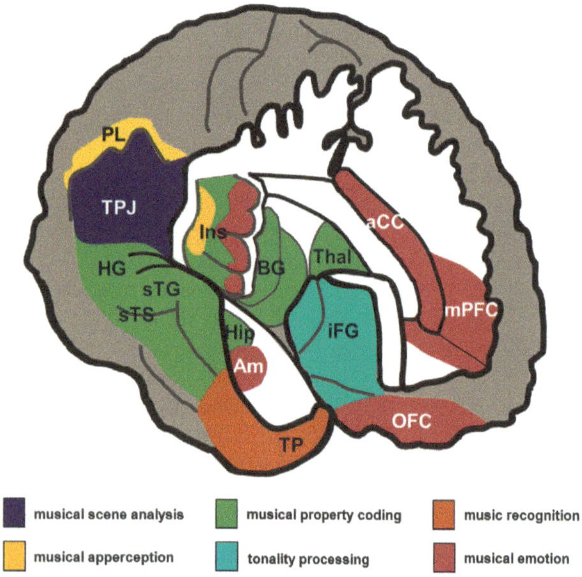

Fig. 34.3. Neuroanatomy of music processing as revealed by study of the amusias (see Table 34.1 and Fig. 34.1). This stylized view of the brain shows the right hemisphere projected forward with the overlying cortical mantle partly removed to expose deeper structures. Brain areas are color-coded according to the corresponding cognitive process coded in Figure 34.1, and indicate locations of brain damage shown to disrupt that process: dark blue, musical scene analysis; green, musical property encoding; gold, musical apperceptive processing (tentative only); red, semantic (associative) processing of music; light blue, key and tonality processing; purple, processing of emotion in music. These distributed brain correlates of amusia are consistent with functional imaging evidence in the healthy brain for separable, but overlapping brain networks with relative selectivity for processing particular dimensions of music. These networks overlap extensively with the brain substrates for processing language and other complex sounds (although the right hemisphere is emphasized here, the critical brain substrates of music processing are distributed in both cerebral hemispheres). aCC, anterior cingulate cortex; Am, amygdala; BG, basal ganglia; HG, Heschl's gyrus; Hip, hippocampus; iFG, inferior frontal gyrus; Ins, insula; mPFC, mesial prefrontal cortex; OFC, orbitofrontal cortex; PL, parietal lobe; sTG, superior temporal gyrus; sTS, superior temporal sulcus; Thal, thalamus; TP, temporal pole; TPJ, temporoparietal junction.

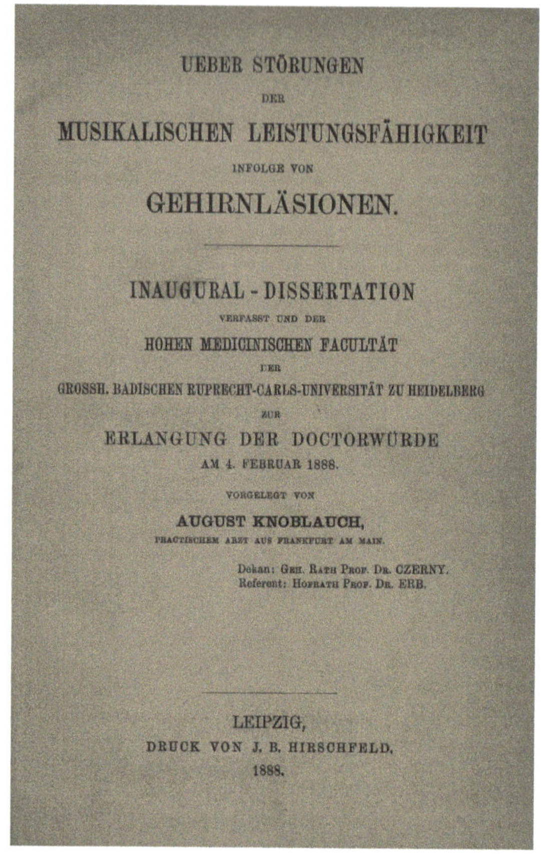
UEBER STÖRUNGEN
DER
MUSIKALISCHEN LEISTUNGSFÄHIGKEIT
INFOLGE VON
GEHIRNLÄSIONEN.

INAUGURAL - DISSERTATION
VERFASST UND DER
HOHEN MEDICINISCHEN FACULTÄT
DER
GROSSH. BADISCHEN RUPRECHT-CARLS-UNIVERSITÄT ZU HEIDELBERG
ZUR
ERLANGUNG DER DOCTORWÜRDE
AM 4. FEBRUAR 1888.
VORGELEGT VON
AUGUST KNOBLAUCH,
PRACTISCHEM ARZT AUS FRANKFURT AM MAIN.

Dekan: GEH. RATH PROF. DR. CZERNY.
Referent: HOFRATH PROF. DR. ERB.

LEIPZIG,
DRUCK VON J. B. HIRSCHFELD.
1888.

Fig. 34.4. Title page of Knoblauch's (1888) monograph "On disturbances of musical capacity after brain lesions."

therefore likely *a priori* to govern music processing to some degree (Marr, 1980; Peretz and Coltheart, 2003); while aspects of music processing may be aligned with more generic auditory cognitive operations (for example, processing of pitch intonations in melody and speech (Patel, 2008), music-specific cognitive operations have been proposed (Peretz and Coltheart, 2003; Patel, 2008). A further practical issue that arises in analyzing published reports on neuropsychologic entities such as amusia is the bias imposed by the selection criterion – a complete picture of the relations of music processing to other aspects of auditory cognition would require a balanced examination of cases with spared music processing in the face of other deficits (i.e., the documentation of double dissociations between music and other domains).

It is challenging to propose a single assessment procedure to capture the clinical diversity of amusia: the characterization of amusia requires a detailed analysis of the relevant impairment. However, as is the case for other complex neuropsychologic functions, clinicians require a practical approach to patients presenting with amusia that allows the core defect to be targeted efficiently for more detailed analysis. The Montreal Battery for Evaluation of Amusia (Peretz et al., 2003) is a normed instrument that can be used to identify deficits affecting elementary pitch and temporal processing, and provides a useful starting point for assessing patients with suspected amusia. However, certain musical deficits may not be captured with this battery, and additional screening tests may therefore be useful to detect such deficits. Following the information-processing framework outlined above, these could include a test of musical scene analysis requiring the tracking of simultaneous musical voices (Fig. 34.2), a test of musical instrument sound recognition and a test of melody recognition. Altered emotional responses to music are often best screened on taking the history of the complaint. These additional musical functions presently lack standardized assessment procedures: new cognitive tests are required, both to screen comprehensively for amusic deficits and to characterize these deficits in more detail.

In the following sections, we apply the information-processing framework at greater length to analyze key cognitive operations and propose candidate neuropsychologic tests relevant to the clinical literature on amusia (summarized in Table 34.1).

DISORDERS OF MUSICAL SCENE ANALYSIS

The perception of all auditory objects (including musical ones) requires an initial parsing of the auditory environment or "'scene" into its constituent objects (at minimum, the object of interest and the acoustic background) (Bregman, 1990). This parsing typically entails two broad subprocesses; segregation of acoustic targets from background and grouping of acoustic features arising from particular sources. Musical "scenes" may make particularly heavy demands on these processes as they typically require perceptual tracking of instruments or motifs over extended time periods and often (in polyphonic music) entail disambiguation of different musical voices. It has been proposed that the neural mechanisms of auditory scene analysis might involve a template-matching process that draws on previously stored information and top-down predictions about regularities in the acoustic environment (for example, temporal-onset synchrony and pitch similarity often signal that acoustic events are part of a single musical object).

The extent to which auditory scene analysis deficits may be selective for musical material remains unclear as reports of amusia generally do not include a specific assessment of the patient's ability to parse auditory scenes. However, symptoms compatible with a musical scene analysis defect have been documented in some cases and have been more prominent for music than for other complex sounds. One patient was able to "distinguish different instruments, but (not) perceive the whole," or in jazz pieces, the relationships between the accompaniment and the soloist (Mazzoni et al., 1993). McDonald (2006) in an analysis of his own musical deficits following a stroke, described being "struck by what seemed to be an unusual relative prominence of some of the inner parts of the musical texture" while listening to an operatic performance. Both these patients

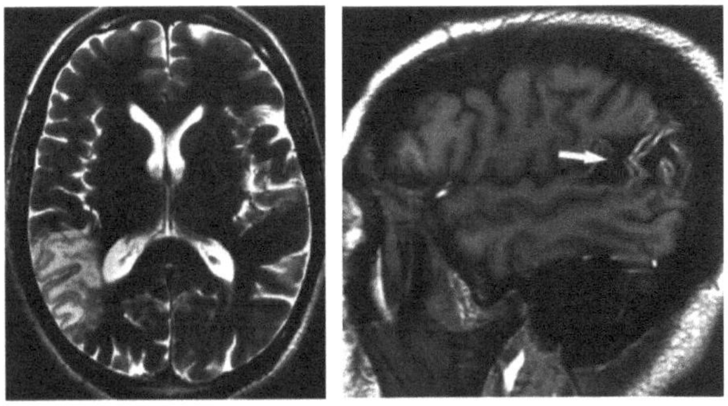

Fig. 34.5. T2-weighted axial (left) and T1-weighted sagittal (right) sections showing infarction of right supramarginal and angular gyri (arrow) in a patient with musical alexia and impaired analysis of musical scenes (the right hemisphere is shown on the left of the axial image). (Reproduced from McDonald, 2006.)

had focal lesions involving the non-dominant temporoparietal junction (Fig. 34.5). A lesion-led study of patients with Alzheimer's disease revealed deficits in the segregation and grouping of note sequences correlated specifically with regional gray-matter atrophy in the posterior superior temporal lobes and posterior cingulate cortex (Goll et al., 2012). Such clinical observations suggest candidate tests that could be used to assess musical scene analysis in future work (Fig. 34.2).

This limited clinical neuroanatomic evidence is consistent with functional neuroimaging work in the healthy brain, which has shown engagement of the temporoparietal junction, planum temporale, and inferior parietal lobe for aspects of auditory scene analysis (Zundorf et al., 2013; Gutschalk and Dykstra, 2014). A critical role for the temporoparietal junction in analysis of musical scenes is compatible with computational models according to which networks in the planum temporale parse the auditory scene via a template-matching algorithm, channeling disambiguated object representations into dorsal auditory cortical pathways that mediate auditory spatial awareness and program motor responses (Griffiths and Warren, 2002; Warren et al., 2005b; Isenberg et al., 2012).

DISORDERS OF MUSICAL PROPERTY ENCODING

Disorders of pitch perception

Pitch is often considered a building block of musical patterns; however, pitch itself is a complex percept that results from active encoding processes in auditory cortex and ascending auditory pathways (Krumhansl, 2000; Stewart et al., 2006). Pitch is not simply equivalent to physical sound frequency, as illustrated by the perceived "virtual pitch" of a harmonic series in which the fundamental frequency is missing (Krumhansl, 2000). In music, pitch values are typically combined sequentially into pitch contours or melodies, and simultaneously into harmonic complexes or chords according to culturally sanctioned "rules." For most listeners, it is relations between pitches (pitch intervals and the direction of pitch change) rather than absolute pitch values that carry musical meaning. The ability to identify absolute pitch values, absolute or perfect pitch, is possessed by a minority of (usually, musically trained) individuals and appears to be relatively resistant to the effects of, at least some forms of, brain damage (Omar et al., 2010). Although pitch value can be organized linearly by frequency (low to high), perceptually this linear ordering is strongly modulated by a cyclic cognitive equivalence corresponding to octaves (compare the similar perceptual quality of C below middle C, middle C, and C above middle C: Krumhansl, 1990). Unlike the continuous intonational variation of speech, musical pitch variation is discrete and generally organized as scales (Peretz, 2006) which (in contrast to the wide variation in the number of phonemes among languages) subdivide the octave variably into a small number of intervals (Trehub et al., 1999). Pitch relationships within scales are of fundamental importance for generating psychologic expectation and harmony (consonance–dissonance) which are in turn major sources of emotion in music (Pallesen et al., 2005; Huron, 2006) and the basis for longer-range rule-based structures or "syntax" in music (Lerdahl and Jackendoff, 1983; Koelsch, 2005; Tillmann, 2005). Pitch relationships also facilitate the establishment of distinct voices in music and (in contrast to superimposed speech sounds) their layering into a harmonious polyphonic whole (Besson and Schon, 2001). Attributes of melody and harmony together contribute to the higher-order musical organizing principle of tonality (Satoh et al., 2007). Non-musicians have implicit knowledge of tonal relationships, as shown using timed consonance–dissonance judgments (Bigand et al., 1999; Tillmann, 2005).

The processing of sequential pitch elements combined into melodies entails several specific cognitive operations. Most fundamentally, tracking of pitch intervals in melodies requires auditory short-term memory. In addition, the coding of individual pitch intervals (the gaps between successive pitch values) and the coding of pitch contour (the overall pattern of ups and downs in the melody) depend on the perceptual representation of local (interval) and global (melody contour) structure, respectively. The distinction between local and global levels of processing forms the basis of an influential hierarchic model of melody processing (Peretz, 1990) and a core component of the Montreal Battery for Evaluation of Amusia (Peretz et al., 2003). According to the Peretz model, which was motivated in part by neuropsychologic findings in groups of patients with focal brain damage (Liegeois-Chauvel et al., 1998), processing of pitch contour is primary and the contour representation is filled in to encode structural detail about individual pitch intervals in the melody. The normally integrated processing of local and global structure would be achieved via cooperative interactions between the cerebral hemispheres, the non-dominant hemisphere predominantly supplying a global framework and the dominant hemisphere a detailed intervallic readout.

Pitch-processing deficits might be classified in principle according to the aspect of pitch perception that they chiefly affect: this would imply deficits in processing pitch interval, pitch pattern, or melody and tonal structure, though in practice these levels of analysis may be difficult to separate. Impaired discrimination of pitch interval direction has been described with focal lesions involving temporoparietal cortices in the right hemisphere or bilaterally (Tanaka et al., 1987), insula (Habib et al., 1995), or subcortical structures (Motomura et al., 1986; Takahashi et al., 1992). Deficits of pitch interval processing with temporoparietal junction lesions may be at least partly attributable to deficient auditory short-term memory, as chord perception may be normal in such patients (Mackworth-Young, 1983). However, impaired pitch interval processing tends to occur as part of a more general amusia, rather than as a selective symptomatic deficit (Johannes et al., 1998; Wilson et al., 2002; Terao et al., 2006). More commonly, impaired processing of pitch contours (melodies) has been described (Mendez and Geehan, 1988; Peretz et al., 1994; Peretz and Gagnon, 1999); this rests on the assessment of unfamiliar melodies, as loss of the ability to perceive or recognize familiar tunes might reflect a deficit at any of several different processing stages. Impaired perception of pitch pattern is often accompanied by impaired processing of other kinds of elementary musical patterns such as rhythms (Tanaka et al., 1987; Johannes et al., 1998; Tramo et al., 2002), arguing that pitch processing can be affected at an early encoding stage prior to formation of whole musical objects (tunes) and that the brain mechanisms that mediate these different forms of early musical pattern encoding are at least partly shared. However, melody perception can be impaired despite retained rhythm perception following bilateral lesions of non-primary auditory cortex (Peretz et al., 1994) and there is further evidence for dissociable processing within the domain of musical pitch perception. Patients with extensive bilateral infarction of the superior temporal lobes may show retained sensitivity to musical expectation or tonality effects (for example, in deciding whether a series of chord progressions sounds finished), despite severe impairments of other aspects of music perception, including consonance judgments (Peretz et al., 2001; Satoh et al., 2007). Conversely, a patient (case GL) with bilateral temporal-lobe infarcts sparing primary auditory cortex and infarcts additionally involving right frontal opercular and left inferior parietal cortices could discriminate pitch intervals normally, but showed impaired processing of tonality and remained able to process linguistic prosody (Peretz et al., 1994). Such evidence suggests that accurate pitch encoding is dissociable from broad tuning for pitch or tonality.

Neuropsychologic findings in patients presenting with amusia have been corroborated by systematic lesion-led group studies following temporal lobectomy (Liegeois-Chauvel et al., 1998), after stroke (Ayotte et al., 2000; Schuppert et al., 2000; Sarkamo et al., 2010) and in the neurodegenerative syndrome of progressive non-fluent aphasia (Goll et al., 2011). Analysis of stroke cases has demonstrated that pitch-processing deficits are dissociable from deficits of temporal (rhythm and metre) perception (Ayotte et al., 2000). Within the domain of pitch perception, tonal memory (short-term pitch retention) is disturbed following right anterior temporal lobectomy involving secondary auditory cortex, including lateral Heschl's gyrus, but sparing primary cortex (Liegeois-Chauvel et al., 1998), while right-sided resections including Heschl's gyrus impair perception of the missing fundamental in complex tones (Zatorre, 1988). Right-sided cortectomy involving the superior temporal gyrus affects processing of both pitch interval and contour information while left-sided cortectomy has been associated with an isolated deficit of pitch interval processing (Liegeois-Chauvel et al., 1998), as predicted by the Peretz model (Peretz, 1990) and broadly consistent with findings in large series of stroke cases (Ayotte et al., 2000; Schuppert et al., 2000). Patients with progressive non-fluent aphasia and focal perisylvian atrophy have impaired pitch direction processing (Goll et al.. 2011). Right anterior temporal lobectomy has been associated with defective processing of tonal structure despite retained ability to discriminate pitch intervals, melody, and intensity (Milner, 1958; Zatorre, 1985; Warrier and Zatorre, 2004) and inclusion of Heschl's gyrus produces an additive deficit. Deficits of scale processing have been documented after stroke, but this capacity may be spared even with bilateral involvement of auditory cortex (Ayotte et al., 2000).

The cognitive relations between encoding of pitch information as embodied in music and in speech (i.e., speech intonation or prosody) are of considerable interest. One potential confound in interpreting comparisons between melody and prosody is the multidimensional nature of prosodic signals. Regional accents, emotional connotation, emphasis and intonation in speech signals generally entail pitch contour variations (Stewart et al., 2006). Though reduced comprehension of prosody has been reported in association with amusia (Mazzucchi et al., 1982; Sidtis and Feldmann, 1990; Confavreux et al., 1992), systematic parallel assessments of the musical and prosodic domains have been undertaken comparatively rarely (Piccirilli et al., 2000). One patient with an intracerebral hemorrhage involving left Heschl's gyrus and planum temporale retained comprehension of linguistic prosody despite impaired pitch pattern processing, arguing that the mechanisms that process pitch patterns in melodies and speech are dissociable (Piccirilli et al., 2000).

Though precise clinical anatomic correlation is difficult, findings in patients with pitch-processing deficits in general align with the picture emerging from functional neuroimaging and neurophysiologic studies of pitch pattern analysis in the healthy brain (Fig. 34.3). It has been proposed that pitch processing engages a music-selective computational brain network (Peretz and Kolinsky, 1993). Lateral Heschl's gyrus has been implicated in representing pitch percepts independently of physical stimulus properties, consistent with a critical role for this area as a pitch center (Kumar et al., 2011); while a distributed bilateral network beyond Heschl's gyrus, involving the planum temporale and posterior temporal lobe and extending anteriorly into planum polare and superior temporal gyrus, is engaged in processing sequential pitch patterns in melodies (Patterson et al., 2002; Warren and Griffiths, 2003) with additional engagement of inferolateral frontal areas during active listening (Overath et al., 2007). Within this network, more anterior temporal-lobe areas process tonal melodies versus atonal pitch patterns. While there is qualified neuroanatomic support for a local–global pitch-processing hierarchy in the posterior superior temporal lobes (Stewart et al., 2008), the neuroimaging evidence has suggested a profile of hemispheric lateralization that appears somewhat at odds with the neuropsychologic evidence. In line with evidence from lobectomy studies, tonal short-term working memory has been shown to be mediated by a distributed predominantly right-sided fronto-temporal network (Zatorre et al., 1994) while transcranial magnetic stimulation over the superior temporal and inferior frontal lobes reduces the accuracy of pitch matching (Loui et al., 2010). fMRI, magnetoencephalographic and evoked potential studies have implicated inferior frontal and premotor cortices, as well as anterior superior temporal gyri, in processing tonal and scale relationships (Janata et al., 2002; Koelsch, 2005). Chordal expectancies are mediated by a network including left inferior frontal lobe, posterior superior temporal sulcus, middle temporal gyrus, and supramarginal gyrus (Tillmann, 2005).

Brain regions activated during perception of pitch and pitch relationships are engaged in encoding and prediction of complex auditory sequences and overlap substantially with distributed, hierarchically organized brain networks that process linguistic syntax and other aspects of speech processing (Koelsch et al., 2013a). Anterior superior temporal cortices are engaged in processing speech streams (i.e., higher-order speech patterns), suggesting a parallel with pitch pattern analysis and the representation of higher-order structures in progressively more anterior regions of superior temporal cortex (Patterson et al., 2002); however, the left cerebral hemisphere preferentially processes speech intelligibility whereas the right hemisphere preferentially processes dynamic pitch variations in speech (Scott et al., 2000). The predilection of left-hemisphere auditory cortex for processing speech information has been proposed to reflect enhanced sensitivity to the rapid temporal transitions that are a hallmark of speech, while (following this formulation) right-hemisphere auditory cortex is more sensitive to spectral information (Zatorre and Belin, 2001). However, the true extent and mechanism for any hemispheric lateralization effect remain controversial (Belin and Zatorre, 2000; Rosen et al., 2011; McGettigan and Scott, 2012) and alternative mechanisms are also plausible (for example, preferential processing of shorter versus longer temporal integration windows: Poeppel, 2003). Other work has emphasized shared brain substrates for tonal and verbal processing. An important example is inferior frontal gyrus (Broca's area in classic aphasiology), which has been implicated in processing chord sequences (Tillmann et al., 2003; Koelsch et al., 2005) and melodies (Janata et al., 2002) as well as more complex or ambiguous syntactic hierarchies in language (Grodzinsky and Friederici, 2006) and actions (Fazio et al., 2009). Electroencephalographic data suggest that linguistic and musical syntax cognitively interact when jointly presented (Steinbeis and Koelsch, 2008) and further indicate that long-range syntactic structures are processed in music as well as speech, even by non-musicians (Koelsch et al., 2013a). Such evidence points to a generic, multimodal tuning for detection and representation of hierarchic dependencies, and might in turn facilitate programming of motor responses to complex sounds by the dorsal auditory cortical pathway (Warren et al., 2005b). On the other hand, anatomic convergence need not imply functional equivalence: in fMRI work capitalizing on a naturally occurring boundary illusion whereby verbal stimuli can be heard as spoken or sung, a musical perceptual mode was associated with enhanced activation of auditory association, parietal and inferior frontal areas, suggesting that music makes greater demands on neural resources for pitch processing and auditory-motor integration than speech (Tierney et al., 2013). Taken together, the neuroanatomic evidence provides ample substrates for the complex profiles of conjoined and dissociated deficits observed clinically, resisting any simple dichotomy between speech and musical pitch processing.

Disorders of temporal perception

Like other sounds, music necessarily unfolds over time and temporal structure is an elementary perceptual attribute of music, in addition to conveying musical meaning in its own right. Like pitch information, temporal information in music is hierarchically organized: the durations of individual notes and internote intervals generate the basic pulse or tempo and higher-order duration patterns generate rhythm and metre. By analogy with pitch processing, rhythm (which depends on relationships among successive timing intervals) could be considered a local temporal attribute and metre (which reflects long-range interval structure) a global temporal attribute (Peretz, 1990; Schuppert et al., 2000). Furthermore, as in the case of pitch, temporal structures in music often generate a sense of psychologic expectation (Patel, 2008). The rhythmic dynamics of music share certain properties with the rhythms of speech (Peretz, 2006) and indeed, geographically and historically disparate languages and musical cultures can be aligned according to those properties (Patel, 2008).

Impaired detection of temporal interval variations and rhythm violations has been described in association with strokes and other focal lesions involving the posterior superior temporal lobe and temporoparietal junction (Fries and Swihart, 1990; Fujii et al., 1990), with hippocampal sclerosis (Samson et al., 2001) and with focal degenerations involving the frontal and temporal lobes (Confavreux et al., 1992). Temporal perception is neuropsychologically dissociable from other musical attributes. Patients with damage involving right anterior temporal lobe, planum temporale or insula may show impaired perception of rhythm or the time interval despite normal perception of pitch interval (Mendez and Geehan, 1988; Fujii et al., 1990; Confavreux et al., 1992). One patient with an infarct involving right temporoparietal cortices and insula demonstrated impaired processing of short sequences of discrete notes, despite retained ability to detect equivalent frequency modulations and to process speech and environmental sounds (Griffiths et al., 1997; Fig. 34.6). These findings suggest dissociable brain mechanisms for representing continuous and discontinuous changes in the auditory stream and provide a candidate substrate for dissociated processing of speech and musical signals. Within the domain of

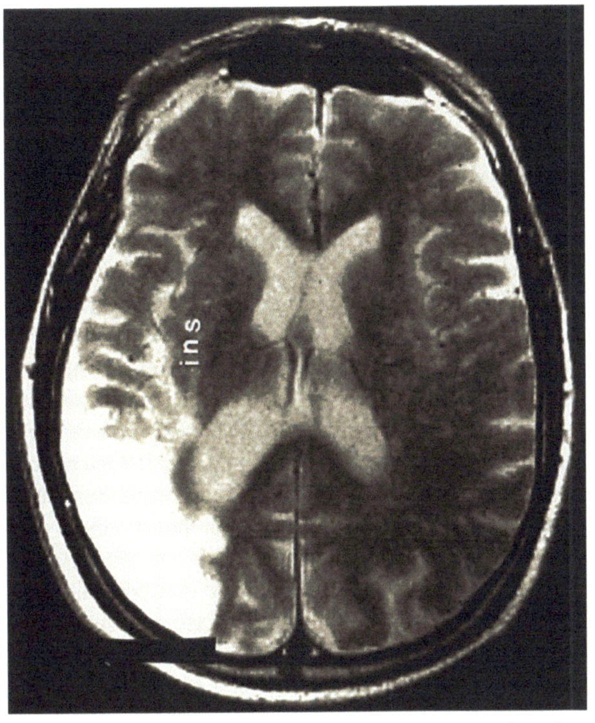

Fig. 34.6. T2-weighted axial magnetic resonance imaging brain section showing infarction between the right middle and posterior cerebral artery territories involving the temporal and parietal cortices and insula (ins) in a patient with amusia and impaired analysis of rapid note sequences (the right hemisphere is shown on the left of the image). (Case HV from Griffiths et al., 1997.)

musical time perception; dissociated deficits of time interval, rhythm and metre perception have been described (Fries and Swihart, 1990; Peretz and Gagnon, 1999; Tramo et al., 2002; Wilson et al., 2002). Together such findings support specific brain mechanisms for encoding auditory rhythms that are separable from the mechanisms that analyze more elementary information about interval and duration. However, these putative modality-specific mechanisms may be damaged alongside extramusical aspects of time sense, for example, with non-dominant parietal lesions (McDonald, 2006).

In lesion-led group studies, impaired rhythm perception has been documented following anterior temporal lobectomy involving superior temporal gyrus, particularly on the right side (Milner, 1962; Liegeois-Chauvel et al., 1998). Impaired processing of metre has been documented following right-sided and (less commonly) left-sided anterior temporal lobe resections (Liegeois-Chauvel et al., 1998; Schuppert et al., 2000). Impairments of musical rhythm perception have been found to be doubly dissociable from deficits of pitch perception (Samson et al., 2001). Within the temporal domain, doubly dissociable deficits of metre and rhythm perception have been documented and appear to be more frequent with left-hemisphere damage (Schuppert et al., 2000). Deficits may be relatively selective for the window of temporal integration, cerebellar lesions preferentially affecting perception of shorter intervals (milliseconds to seconds) and prefrontal cortical lesions, perception of longer intervals (Mangels et al., 1998); while isochrony judgments over short intervals appear more impaired following left than right temporal lobectomy (Samson et al., 2001).

Temporal structure in music is processed by distributed corticosubcortical networks in the healthy brain (Chen et al., 2008a; Grahn, 2012; Konoike et al., 2012). Temporal patterns are encoded by planum temporale and posterior superior temporal cortices (Griffiths et al., 1999; Abrams et al., 2011; Herdener et al., 2014), while engagement of cerebellum, basal ganglia, prefrontal and supplementary motor cortices is modulated by task demands (Mathiak et al., 2004; Grahn and Brett, 2007). The cortical components of this network overlap extensively with the pitch-processing hierarchy and with brain substrates for processing linguistic syntax (Griffiths et al., 1999; Patterson et al., 2002; Zatorre and Gandour, 2008), suggesting that brain mechanisms for processing rule-based auditory sequences may be at least partly independent of domain or modulated by task demands (Rogalsky et al., 2011). Anisochronous intervals that form the building blocks of more complex rhythmic patterns activate planum temporale together with Broca's area and its right-hemisphere homolog. Increasing rhythmic expertise is associated with increasing engagement of left supramarginal gyrus, also implicated in processing linguistic syntax (Limb, 2006; Chen et al., 2008b; Herdener et al., 2014). This neuroanatomic evidence is consistent with the clinical literature in according early non-primary superior temporal cortex, temporoparietal junction and their projections a primary role in the initial representation of temporal structure in music (Fig. 34.3). These auditory representations are integrated with temporal information derived from other sensory channels in the parietal lobe and anterior cortical and subcortical structures become variably engaged according to the behavioral response required (Chen et al., 2006, 2008a). Although there is substantial anatomic overlap with the brain substrates for pitch pattern analysis, the distributed network organization of temporal processing leaves substantial scope for dissociated processing of these musical dimensions: for example, processing of pitch and temporal attributes could occur over different timescales or differentially engage particular network elements (between the cerebral hemispheres or between cortical and subcortical structures).

Disorders of timbre perception

Timbre can be defined operationally as the acoustic property that distinguishes two sounds of identical pitch, duration and loudness (Bregman, 1990). It is integral to auditory object identity and links the processing of more elementary spectral and temporal musical properties to the apperceptive processing of entire auditory objects (Goll et al., 2010b). Timbre confers the individual characteristics of human voices and the voices of particular musical instruments and therefore also contributes importantly to stream binding and segregation in musical and other auditory scenes (Bregman, 1990). This property of stream segregation by timbral sequences has occasionally been exploited by composers (including Schönberg and Webern) as *Klangfarbenmelodie* (literally, sound color melody); more importantly, timbre is an independent determinant of the emotional content of music (Juslin and Laukka, 2003; Hailstone et al., 2009a). Timbre is a complex, multidimensional acoustic property including dimensions of spectral envelope (frequency structure or spectral shape, determining, for example, the relative "brightness" of woodwind voices), temporal envelope (including, for example, the distinctive onset phase or "attack" that characterizes percussion instruments) and dynamic spectrotemporal interactions (spectral flux, which helps, for example, to distinguish the sound of the oboe from brass instruments (McAdams and Cunible, 1992; Hailstone et al., 2009a)). Unlike pitch, timbre cannot therefore be ordered simply on a linear scale, and multivariate scaling techniques are required to capture the timbral relatedness of particular sounds (Samson et al., 2002).

Patients with impaired perception of timbre or dystimbria typically describe music as sounding "flat" or "mechanic" and perception of musical environmental sounds such as birdsong or human voices may be comparably blighted (Griffiths et al., 2007). Such patients generally find that music loses much of its emotional impact. Dystimbria may be selective for particular dimensions of timbre. Patients' descriptions of their symptoms often suggest disordered harmonic or spectral processing; however, temporal attributes of timbre may also contribute. In one case, a musically experienced listener had particular difficulty perceiving the timbres of keyboard and percussion instruments attributed to a selective deficit encoding rapid spectral transitions following a stroke involving the right superior temporal lobe extending to planum temporale and insula (Kohlmetz et al., 2003; Fig. 34.7). While frequently accompanied by other musical deficits (Eustache et al., 1990), dystimbria may occur in the context of preserved perception of melodies and other musical dimensions, especially with focal lesions involving the non-dominant posterior superior temporal lobe (Mazzucchi et al., 1982; Kohlmetz et al., 2003; Griffiths et al., 2007). One patient with dystrimbria following a right temporoparietal hemorrhage involving the superior temporal gurus was unable to distinguish musical instrument sounds or to derive esthetic pleasure from music (Mazzucchi et al., 1982). The timbral deficit in this case was pervasive, extending to human voices and environmental sounds; however pitch and rhythm perception were retained. Similarly, impaired discrimination of musical instrument timbres despite relatively retained competence with other aspects of music perception has been reported in a patient with semantic dementia associated with focal degeneration of the anteromedial temporal lobes (Omar et al., 2010). Though environmental sound perception is often comparably affected (Eustache et al., 1990; Johannes et al., 1998), dystimbria for musical instrument sounds may also occur despite intact perception of voices (Fujii et al., 1990; Johannes et al., 1998), environmental sounds (Tanaka et al., 1987) or both these other sound categories (Kohlmetz et al., 2003). The relations between timbre and vocal processing are of particular interest: whereas individual voices present the most straightforward analogy to musical instrument timbres, it might be argued that speech phonemes are the appropriate linguistic analogs (Koelsch, 2011a) and preserved phonemic processing has been documented in association with dystimbria (Johannes et al., 1998).

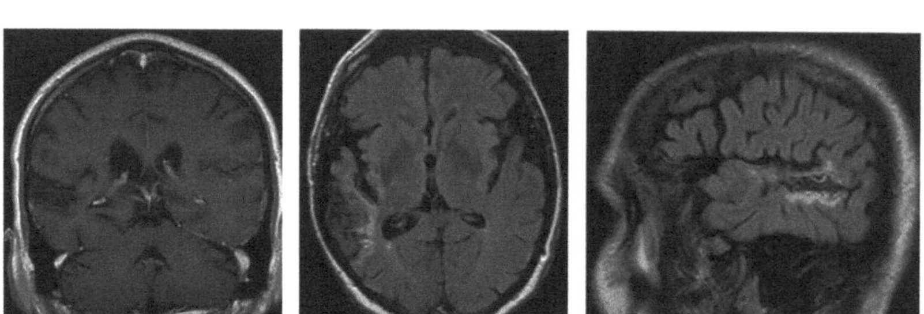

Fig. 34.7. T1-weighted coronal (left), axial (middle), and sagittal (right) magnetic resonance imaging brain sections showing infarction predominantly involving right superior temporal gyrus in a patient with selective dystimbria (the right hemisphere is shown on the left of the coronal and axial images). (Case WDK from Kohlmetz et al., 2003.)

In lesion-led studies, impaired timbre perception has been reported following anterior temporal lobectomy, particularly on the right side (Milner, 1962; Samson and Zatorre, 1994; Samson et al., 2002). The neuropsychologic findings suggest that timbral dimensions may interact cognitively with other musical parameters, particularly when timbres are embedded into musical phrases (Samson et al., 2002). In line with the temporal lobectomy evidence, impaired discrimination of spectral shape, an important dimension of timbre, has been described in cohort studies of progressive non-fluent aphasia and logopenic aphasia associated with focal degeneration of perisylvian and temporoparietal cortices (Goll et al., 2010a, 2011).

The clinical literature on dystimbria is supported by findings concerning the functional neuroanatomy of timbre processing in the healthy brain. Perception of musical instrument and complex synthetic timbres has been shown to activate a bilateral network extending from planum temporale laterally and anteriorly along superior temporal gyrus and sulcus (Menon et al., 2002; Halpern et al., 2004). Spectral shape encoding has a specific correlate in superior temporal sulcus (Warren et al., 2005a), suggesting that the representation of particular dimensions of timbre constitutes a specific neural computation that might be selectively targeted by disease (Fig. 34.3). The human voice (the prototypical musical instrument) is represented and analyzed in the right superior temporal sulcus (Belin et al., 2000). Selectivity of processing for human voices over natural environmental and synthetic sounds of comparable spectrotemporal complexity has been demonstrated in this region, although it remains uncertain to what extent such selectivity reflects specialized brain mechanisms that evolved for processing voices as uniquely behaviorally salient auditory objects or alternatively, the increased computational demands of resolving specific exemplars within a category of perceptually similar vocal sounds. Musical instruments constitute an important candidate test case for settling this issue (Bedard and Belin, 2004; Hailstone et al., 2010); however there have been few direct comparisons between voice and musical instrument processing with neuroanatomic correlation. This limited evidence has outlined a complex hierarchic organization, with category-independent responses to particular acoustic features in early non-primary auditory cortices and category-specific substrates in more anterior superior temporal cortex (Leaver and Rauschecker, 2010) and a broadly orthogonal organization for extracting specific ecologically relevant information such as the scale (acoustic size) of sound sources (von Kriegstein et al., 2007). These observations suggest potential substrates for clinically observed dissociations among patients with dystimbria and raise more fundamental questions concerning the nature of the neural mechanisms that encode the complex multidimensional property of timbre. It is likely that accurate timbre perception depends fundamentally on the conjoined encoding of spectral and temporal characteristics into "proto-object" representations and that representations corresponding to particular auditory object categories (such as instruments or voices) are preferentially activated according to behavioral context (Agus et al., 2012).

DISORDERS OF MUSICAL OBJECT PERCEPTION: APPERCEPTIVE AGNOSIA

The cognitive status of apperceptive processing for music and other complex sounds remains unclear. The concept of a specific level of cognitive processing for manipulation of whole-object sensory representations prior to the assignment of meaning to those representations was initially developed in the visual domain (Warrington and James, 1988) and informed by cases of visual deficits apparently restricted to this level of processing: apperceptive agnosias. Patients with visual apperceptive agnosia have intact (or relatively intact) elementary visual property encoding and visual object recognition, but disproportionate difficulty in identifying objects under unusual or degraded (non-canonical) viewing conditions or discriminating visually similar objects (Warrington and James, 1988). Apperceptive deficits are held specifically to disrupt stored structural representations of the three-dimensional geometric relations between features that together define a visual object and allow object constancy to be established under shifting perceptual contexts: a function assigned *par excellence* to the non-dominant parietal lobe (Warrington and Taylor, 1973). A cognitive organization of this kind has been inferred to underpin the processing of a specific highly salient category of visual objects, human faces (Bruce and Young, 1986). In the case of many sounds, however, it is not clear that an equivalent level of apperceptive analysis can be established even in principle, since the criteria for a non-canonical view of a sound are often problematic (Goll et al., 2010b). Nevertheless, certain candidates for auditory "unusual views" have been proposed, for example, phonemes spoken in a foreign accent (Hailstone et al., 2012).

From a neuropsychologic perspective, music potentially constitutes an important model system for assessing the concept of apperceptive processing in the auditory domain (Alossa and Castelli, 2009). Melodies can be regarded as musical objects in their own right and can be manipulated in various ways (for example, transposition or transcription for different instruments) that radically change their surface perceptual structure.

However, for many melodies there is no one single preeminent pitch range or timbral carrier that would define a canonical form of that melody. Alternatively, familiar melodies can be "degraded" by inserting false notes that do not violate more fundamental key relationships: this manipulation exploits the strong psychologic expectation established in normal listeners presented with a well-known tune. However, most neuropsychologic studies have tended to assess the perception of melodies as coherent auditory objects by assessing patients' ability to discriminate or match sequentially presented melodies (Lechevalier et al., 1984; Peretz, 1996; Peretz et al., 1997). The proper neuropsychologic interpretation of such studies rests on an adequate documentation of other (perceptual and semantic) levels of melody processing, in order to establish the specificity of any putative apperceptive deficit.

Limited neuropsychologic evidence from symptom-led and lesion-led studies suggests that the ability to discriminate melodies is doubly dissociable from melody or environmental sound recognition and from certain aspects of elementary musical property encoding (Eustache et al., 1990; Schuppert et al., 2000; Vignolo, 2003); such cases might be described as having apperceptive agnosia for melodies. Patients may have impaired perception of melodies despite normal facility with spoken lyrics from the same songs (Ayotte et al., 2000). This pattern of deficits seems to occur with focal lesions involving right superior temporal cortex, insula, and their subcortical projections (Fig. 34.8). There is further evidence from temporal lobectomy cases that discrimination of melodies may be disproportionately impaired following right-sided resections (Zatorre, 1985). The neuroanatomic evidence relating specifically to an apperceptive level of melody analysis in the healthy brain is (in contrast to the extensive data relating to music perception more generally) similarly limited. However, fMRI studies of perception of transformed melodies (musically equivalent to rotated views) by musicians have demonstrated a common substrate for such apperceptive processing in inferior parietal sulcus, under both pitch transposition and temporal reversal (Foster and Zatorre, 2010; Foster et al., 2013). Interestingly, this evidence would align musical apperception anatomically with the brain correlates of apperception in vision, as well as more basic aspects of auditory scene analysis, though the relationship to clinical studies remains *sub judice* awaiting a more detailed assessment of amusic patients using comparable paradigms.

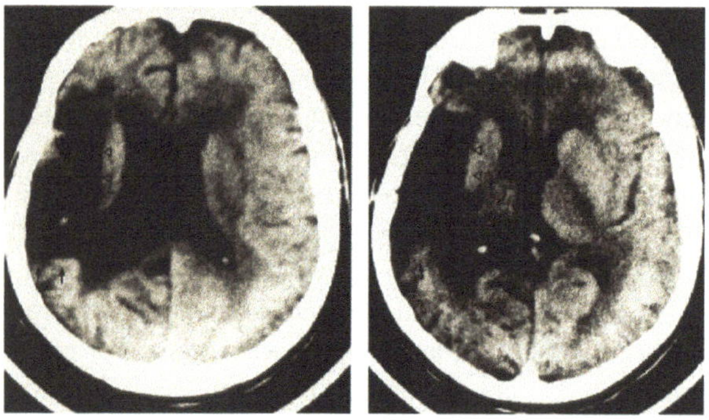

Fig. 34.8. Axial computed tomography brain sections showing infarction of right insula cortex and superior temporal lobe including auditory cortex in a patient with apperceptive agnosia for melodies (the right hemisphere is shown on the left of each image). (Case RBS16/NR from Ayotte et al., 2000.)

Apperceptive processing of musical sources such as instruments can be assessed by procedures that require matching of instrument timbres over different pitch ranges: this paradigm has revealed deficits of instrument apperception with focal degenerations of the anterior temporal lobes (Omar et al., 2010). Musical instruments can be presented in non-canonical form and indeed this has long been exploited by composers and performers for achieving particular musical effects: for example, string instruments played pizzicato, keyboard instruments played percussively, or wind instruments played portamento or at the extremes of their registers. To date, however, such effects have not been applied in neuropsychologic studies of amusic patients.

DISORDERS OF MUSICAL OBJECT RECOGNITION: ASSOCIATIVE AGNOSIAS

Disorders of melody recognition

Music, like language, is rich with meaning. There is a sense in which music is inherently meaningful, as a self-contained symbolic system bound by purely musical, but culturally sanctioned rules. In neuropsychologic studies however, music has conventionally been aligned with other complex sensory phenomena that acquire meaning through semantic associations. These extramusical associations can take the form of knowledge about musical objects (in particular, familiar melodies) or knowledge about musical sources (in particular, musical instrument sounds). While the repertoire of familiar instruments is essentially finite, the potential lexicon of tunes (like human faces or voices) is almost infinite and new exemplars can continue to become assimilated to the store throughout the lifetime (Alossa and Castelli, 2009; Peretz et al., 2009). In addition, the semantic associations of music include the notated symbolic codes that are the objects of musical reading and writing; these require special expertise and are not considered further here. The most widely described semantically based amusia – corresponding to associative agnosia for music – is impaired recognition of familiar tunes

(Eustache et al., 1990). Melody recognition is most straightforwardly assessed by asking patients to name the tune. However, this is problematic in cases with accompanying aphasia or word retrieval impairment and alternative cross-modal or within-modality matching procedures or procedures that require the patient to classify tunes by familiarity or according to a specified semantic criterion, have also been employed (Omar et al., 2010). Associative agnosia, as strictly defined, requires impaired recognition despite intact perception; however musical perceptual functions have not been assessed uniformly in the literature on melody agnosia.

Cases of associative agnosia for melodies have been described in association with rather distributed lesions involving frontal, temporal and parietal areas and insula in either cerebral hemisphere (Eustache et al., 1990; Stewart et al., 2006; Fig. 34.9) or with focal degeneration involving non-dominant perisylvian regions (Confavreux et al., 1992). Such cases show substantial neuropsychologic and neuroanatomic variation, but demonstrate that semantic processing of melodies may dissociate from semantic processing of words and environmental sounds (Eustache et al., 1990; Peretz, 1996). One patient with a right perisylvian infarct had pervasive amusia accompanied by agnosia for environmental sounds, but no evidence of verbal agnosia (Johannes et al., 1998). Limited information is available concerning the fine-grained cognitive organization of melody knowledge or the impact of brain damage on that organization. However, one detailed analysis of musically experienced patients with semantic dementia and Alzheimer's disease suggested that music may be organized analogously with other semantic knowledge domains, superordinate associated knowledge about musical pieces (genre and composer) being relatively more resilient to the effects of brain damage than more specific knowledge about the piece, and knowledge about abstract musical entities (compositions) dissociating from knowledge about physical musical sources (instruments) (Omar et al., 2010). Patients with melody agnosia may exhibit some evidence of familiarity with well-known pieces under restricted conditions, indicating that the engagement of the melody lexicon is not all or none (Peretz et al., 1994).

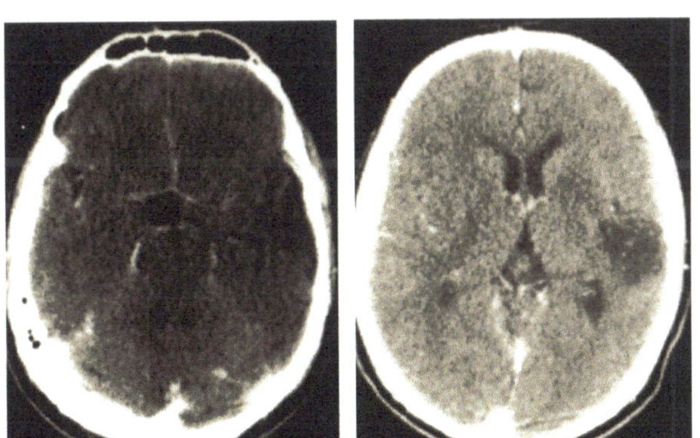

Fig. 34.9. Axial computed tomography brain sections showing infarction involving left anterior temporal lobe and temporoparietal zone in a patient with selective associative agnosia for melodies (the left hemisphere is shown on the right of each image). (Case MH from Eustache et al., 1990.)

In lesion-led group studies recognition of familiar tunes has been shown to be impaired particularly with strokes involving left perisylvian regions (Ayotte et al., 2000) but also following left-sided or right-sided temporal lobectomies (Samson and Zatorre, 1992). Cohort studies of melody recognition have been undertaken in patients with Alzheimer's disease and frontotemporal lobar degeneration with neuroanatomic correlation of regional atrophy using voxel-based morphometry: the ability to recognize familiar melodies (assessed using detection of pitch violations in familiar tunes or discrimination of familiar from unfamiliar tunes) is, in general, impaired in semantic dementia and correlates with regional gray-matter volume in right anterior temporal lobe, including the superior temporal gyrus and temporal pole and extending into mesial temporal lobe, insula, and orbitofrontal cortex (Hsieh et al., 2011; Johnson et al., 2011). Of interest, these studies further indicate that individual patients with semantic dementia may show relatively preserved melody recognition despite impairments in other semantic domains, including environmental sounds, consistent with the evidence of single case studies (Hailstone et al., 2009b; Omar et al., 2010; Weinstein et al., 2011).

On balance, while there is neuroanatomic evidence suggesting that associative agnosia for melodies results from left-sided or bi-hemispheric damage while apperceptive agnosia for melodies results from damage to the right hemisphere (Zatorre, 1985; Eustache et al., 1990; Ayotte et al., 2000), as predicted by classic models of object agnosia in other domains, neuroanatomic correlations of melody agnosia are less clear than for other forms of agnosia (Hsieh et al., 2011; Johnson et al., 2011). While it is likely that methodologic variation between studies is playing a part, there is clearly interdependence between associative and putative apperceptive processes, and anatomic attributions must remain tentative (Fig. 34.3). Relations with language processing are particularly germane to a consideration of musical associative agnosia: speech and language are the pre-eminent examples of complex semantically invested symbolic systems in the auditory modality and more specifically, melody and lyrics are closely wedded in many familiar songs. The clinical neurologic and neuropsychologic literatures contain well-documented examples of cases with dissociated competence between these two semantic domains (Eustache et al., 1990; Peretz, 1996). Additional clinical evidence that brain substrates for melody recognition overlap those for recognition of familiar faces (Hsieh et al., 2011) and voices (Hailstone et al., 2012) in the right anterior temporal lobe should be interpreted with some caution. Melodies have an abstract quality (including the potential for transfer of identity among radically different perceptual carriers under transposition and transcription) that is less clearly evident for other sensory objects, while other objects often have dense multimodal sensory associations that are not shared by melodies.

The neuroanatomic basis of melody recognition – semantic memory for music – has received much recent attention in functional neuroimaging studies of the healthy brain (Watanabe et al., 2008; Peretz et al., 2009; Groussard et al., 2010). This work has delineated a distributed bi-hemispheric network including anterior and mesial temporal, frontal and parietal-lobe areas with preponderant involvement of the superior temporal gyrus and sulcus (Plailly et al., 2007; Groussard et al., 2010; Herholz et al., 2012). This evidence suggests that the neuroanatomic substrates for semantic processing of melodies are in proximity to the temporopolar and more posterior temporal gyral substrates implicated in the semantic processing of non-musical sensory objects (Groussard et al., 2009; Lambon Ralph et al., 2009). This putative melody semantic network substantially overlaps with brain networks previously implicated in processing other semantic domains, including language (Scott et al., 2000) and shows partial convergence with the distribution of brain damage in cases of melody agnosia, while at the same time allowing potential substrates for observed dissociations. It may be possible to refine the neuroanatomic picture further using neurophysiologic techniques: in evoked potential work, N400 potentials are elicited by associative processing of both linguistic and musical material, whereas N5 potentials may be more specifically elicited by searching for intramusical meaning (Koelsch, 2011b) and may be linked to brain mechanisms that process expectancy violations in music (Koelsch, 2005, 2011b).

Disorders of musical instrument sound recognition

Knowledge about musical instruments is heavily influenced by past musical exposure and training. Partly for this reason, recognition of musical instrument sounds has not been widely studied. However, the processes involved are of considerable neuropsychologic interest for the implications they hold for models of music cognition and auditory object processing more generally. As semantically laden and highly differentiated sound sources, musical instruments have certain cognitive parallels with familiar human voices. One potentially confounding factor in comparing recognition of musical instrument sounds and voices is of course the much more restricted lexicon of familiar instrument sounds in relation to the vast store of voices accumulated by all normal listeners. Nevertheless, limited neuropsychologic evidence suggests that the cognitive mechanisms for recognition of musical instruments may be partly separable from the mechanisms mediating recognition of voices and other complex sounds. In patients with selective atrophy involving the anterior temporal lobes, ability to recognize musical instruments has been shown to dissociate from recognition of voices and environmental sounds (Hailstone et al., 2010) and may further dissociate from other dimensions of music knowledge, such as melody recognition (Omar et al., 2010). There are few well-documented studies that have systematically compared timbre perception (discrimination) and associative knowledge (instrument recognition). This is intrinsically challenging, because the perceptual and associative attributes of musical instrument sounds often interact (for example, the semantic categories of woodwind or string instruments are closely underpinned by particular sets of acoustic characteristics: Goll et al., 2010b). However, true associative agnosia for musical instrument sounds (with intact instrument sound discrimination) and sparing of environmental sound recognition has been described in the setting of bilateral superior temporal-lobe infarction (Peretz et al., 1994).

DISORDERS OF MUSICAL EMOTION PROCESSING

Disorders of musical emotion processing have not generally been included in the taxonomy of amusias. The emotional content of music is for many listeners its paramount attribute (Blood and Zatorre, 2001), though emotional responsiveness to music varies widely within the normal population (Mas-Herrero et al., 2014). Aside from its ecologic relevance, recent advances in the neuroscience of musical emotion (Gosselin et al., 2011; Salimpoor et al., 2011; Downey et al., 2013) provide a rational framework for analyzing deficits that particularly affect this aspect of music processing. The nature of musical emotions has long vexed philosophers and neuroscientists alike. Emotion in music is an important test case for more general issues in affective and social neuroscience, such as the cognitive relations between affective channels (music compared with facial and vocal expressions and with affective prosody) and the existence of dedicated multimodal brain substrates for particular emotions (Kumfor and Piguet, 2012). Here, we approach the neuropsychologic data pragmatically, in order to assess the validity of incorporating emotion into our information-processing account of amusia. From this perspective, musical emotions are complex neuropsychologic constructs built from more basic perceptual cues such as sound level, tempo, vibrato, melody, rhythm, mode, and harmony (Hailstone et al., 2009a; Eerola et al., 2013; Zatorre and Salimpoor, 2013). Normal listeners exhibit a hierarchy of emotional responses to music and these responses have dissociable physiologic (subjective experience) and cognitive (objective labeling) dimensions (Juslin and Vastfjall, 2008). The most primitive and universal emotional responses to music are associated with the consonance–dissonance dimension (Pallesen et al., 2005; McDermott et al., 2010); however, music reliably engenders powerful autonomic responses associated with a strong subjective sense of pleasure ("chills" or "shivers") that are consistent for individual listeners and elicited by particular structural features of music (Blood and Zatorre, 2001; Huron, 2006). Musical equivalents of the elementary canonical emotions described for facial expressions and vocal sounds (such as happy, sad, angry, or scary) can be labeled with reasonable consistency by normal listeners, though there is a complex interaction with cultural norms (Laukka et al., 2013). Finally, the sense of tension and resolution established via psychologic expectancies conveys a potent emotional impact across musical genres (Huron, 2006); while more complex, abstract structural features may also contribute in particular genres (such as western polyphony) (Pressnitzer et al., 2011). Whereas recognition of emotions expressed via music can be assessed using standard neuropsychologic methods (such as forced-choice labeling), assessment of emotional response is more difficult, relying on subjective ratings (such as pleasant/unpleasant: Peretz et al., 2001) or measurement of autonomic indices of arousal such as skin conductance (Johnsen et al., 2009).

Impaired music perception or comprehension tends to degrade the pleasure previously afforded by music, particularly in association with dystimbria (Mazzoni et al., 1993; McDonald, 2006); however, comprehension of musical emotions and emotional responses to music may be retained in the face of perceptual deficits (Lechevalier et al., 1984; Peretz et al., 2001). Conversely, impaired processing of musical emotions or anhedonia for music has also been reported following focal brain damage that may spare perceptual encoding of music and emotional processing of other sounds (Griffiths et al., 2004; Stewart et al., 2006; Gosselin et al., 2007; Satoh et al., 2011) and in semantic dementia despite relatively preserved recognition of melodies (Omar et al., 2010; Fig. 34.10). Reported music emotion deficits range

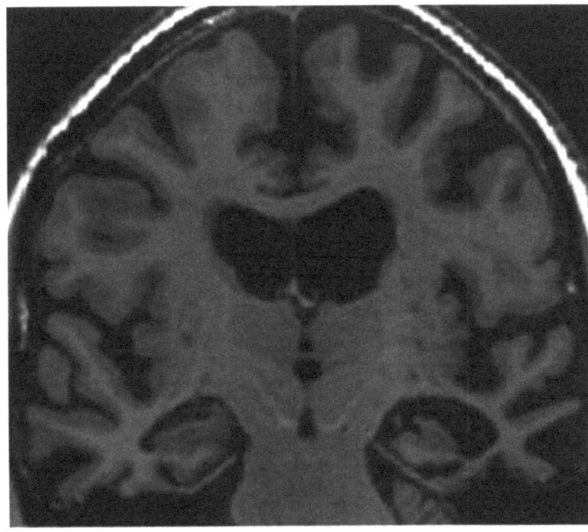

Fig. 34.10. T1-weighted coronal magnetic resonance imaging brain section showing selective atrophy of the left anterior temporal lobe in a patient with semantic dementia and impaired recognition of emotions from music (the left hemisphere is shown on the right of the image).

from impaired responses to consonance and dissonance (Peretz et al., 2001), through loss of shiver responses (Griffiths et al., 2004), impaired attribution of canonical emotions to music (Gosselin et al., 2007; Matthews et al., 2009) and deficits of more abstract affective representations (Mazzoni et al., 1993; McDonald, 2006; Satoh et al., 2011). These deficits have separable neuroanatomic correlates: impairment of more basic emotional responses to music has been associated with focal lesions involving the insula, amygdala and early auditory cortex (Peretz et al., 2001; Griffiths et al., 2004; Gosselin et al., 2007), while loss of the ability to derive pleasure from the apprehension of coherent musical structures has been described with lesions involving temporoparietal junction and parietal cortex (Mazzoni et al., 1993; McDonald, 2006; Satoh et al., 2011). Within the domain of emotion processing, recognition of emotions expressed by music may be preserved despite an impaired emotional response (Gosselin et al., 2007; Satoh et al., 2011). More "primitive" musical emotion processing such as consonance–dissonance detection or pleasurable responses to familiar music may be more resistant to the effects of certain forms of brain damage than explicit cognitive processing or labeling of emotions (Matthews et al., 2009), providing a possible neuropsychologic basis for the sometimes dramatic craving for music (musicophilia) exhibited by patients with frontotemporal lobar degeneration (Fletcher et al., 2013). However, the reverse dissociation (loss of emotional response to dissonance despite at least partial sparing of affective labeling of music) has been described in association with bilateral lesions of auditory cortex (Peretz et al., 2001).

In lesion-led work, temporal-lobe resections have been associated with various deficits of music emotion processing. Resection of mesial temporal-lobe structures, including amygdala in either hemisphere, may be associated with reduced recognition and emotional responses to negatively valenced music (Gosselin et al., 2005); while the ability to recognize and respond to happy music may be modulated by the side of the resection (Khalfa et al., 2008). Altered emotional responses to dissonance (unpleasantness) have been linked specifically to resection of parahippocampal gyrus (Khalfa et al., 2008). In one multidimensional scaling study, patients undergoing unilateral mesial temporal-lobe resections showed musical emotion similarity judgments similar to healthy controls, perhaps indicating retained implicit processing of emotion from music (Dellacherie et al., 2011). Evidence from brain-damaged individuals with various focal lesions has provided candidate brain substrates for dissociating cognitive and autonomic aspects of the emotional response to music (Johnsen et al., 2009). Damage involving right somatosensory cortex was associated with altered emotional experience of music, while damage involving ventromedial prefrontal cortex was associated with reduced autonomic responses to music. Lesion-led cohort studies of patients with frontotemporal lobar degeneration have demonstrated impaired identification of canonical emotions from music in these diseases (Omar et al., 2011; Hsieh et al., 2012); findings in Alzheimer's disease have been less consistent (Drapeau et al., 2009; Gagnon et al., 2009; Samson et al., 2009; Omar et al., 2010). Voxel-based morphometry in these cohorts has delineated a distributed brain network correlating with performance for identification of emotions from music: this network includes insula, orbitofrontal cortex, anterior cingulate and medial prefrontal cortex, anterior temporal and more posterior temporal and parietal cortices, amygdala, and the subcortical mesolimbic system.

Despite considerable theoretic interest, the processing of emotion in music and prosody has only infrequently been directly compared in clinical populations. One patient with an infarct involving the right insula and a musical perceptual deficit affecting analysis of rapid temporal changes lost subjective pleasure in music, describing it as sounding like "an out of tune child's dulcimer," yet remained able to interpret the emotional content of prosody (Griffiths et al., 1997). Another patient with bilateral selective amygdala damage due to lipoid proteinosis (Urbach–Wiethe disease) had difficulty recognizing negatively valenced emotions in music (as well as facial expressions), but preserved processing of emotional prosody (Gosselin et al., 2007). Data from a Parkinson's disease cohort provide more systematic evidence that the underlying cognitive mechanisms for processing emotion in music and prosody are at least partly dissociable (Lima et al., 2013).

It remains unclear whether there are critical brain substrates for particular musical emotions. Disproportionately impaired recognition of scary music has been described with anteromedial temporal-lobe damage (Gosselin et al., 2005, 2007), impaired recognition of sad music with anteromedial temporal-lobe resections in either cerebral hemisphere and impaired recognition of happy music in association with left-sided anterior temporal resections only (Khalfa et al., 2008). However, there is little uniformity between studies, perhaps in part attributable to methodologic differences (Dellacherie et al., 2011). Evidence that particular emotions can be comparably affected in music and other emotion modalities is similarly inconsistent, with substantial variation among emotions and study populations (Gosselin et al., 2007, 2011; Omar et al., 2011), suggesting that interactions between emotion and modality are complex. An important dimension of the neural substrates for music emotion processing is their temporal. This temporal dimension is difficult to study using conventional techniques, but may be uniquely illuminated by clinical studies, as illustrated by depth electrode work in patients with epilepsy, demonstrating sequential activation of orbitofrontal cortex, amygdala and anterior cingulate gyrus during processing of musical dissonance (Dellacherie et al., 2009).

Taken together, the clinical findings fit with recent developments in the basic neuroscience of emotion in music. Functional imaging work in the healthy brain has demonstrated very extensive neuroanatomic correlates of musical emotion processing, including engagement of salience and evaluation systems in the insula, amygdala and their limbic connections (Blood and Zatorre, 2001) and neurotransmitter (notably, dopaminergic) pathways that link cognitive and affective responses to music with mesolimbic and subcortical reward networks (Salimpoor et al., 2011) as well as prefrontal and somatosensory cortices (Koelsch et al., 2006, 2013b). A similar network has been implicated in processing affective prosody (Leitman et al., 2010). It has been suggested that extratemporal cortices may become engaged during evaluation of intrinsically ambiguous affective signals as embodied in prosody and music. Damage within this distributed network has been correlated with deficits of music emotion processing (Omar et al., 2011; Hsieh et al., 2012; Fig. 34.3). The more elemental affective attributes of dissonance and pleasantness are processed by mesial temporal structures, including amygdala, hippocampus, and parahippocampal gyrus and early auditory cortices, in addition to more widespread frontal and subcortical areas, in line with the distribution of lesions in cases with impaired response to dissonance and unpleasantness in music (Peretz et al., 2001; Koelsch et al., 2006). The coactivation of higher-order cortical areas supports clinical evidence for involvement of extratemporal cortices in defective cognitive evaluation of musical emotions (Koelsch et al., 2006, 2013b; McDonald, 2006; Johnsen et al., 2009). Loss of shiver responses to music has been linked specifically to limbic areas (amygdala and insula) implicated in mediating musical chills (Blood and Zatorre, 2001; Griffiths et al., 2004). Processing of familiarity and emotion in music has overlapping substrates in the healthy brain (Pereira et al., 2011), in keeping with findings in patient cohorts (Omar et al., 2011; Hsieh et al., 2012) and underlining the close cognitive linkage between semantic and affective processing of music (Brown et al., 2004; Lehne et al., 2014). The music emotion-processing network also overlaps mesial prefrontal and anterior temporal-lobe areas previously implicated in aspects of social cognition, notably mentalizing or theory of mind (Carrington and Bailey, 2009). Indeed, impaired processing of affective mental states represented in music has been linked to atrophy involving these prefrontal and temporal-lobe regions in patients with frontotemporal dementia and disintegration of interpersonal skills (Downey et al., 2013). Functional imaging and behavioral studies of the healthy brain are beginning to delineate the specific dynamics of musical emotional responses across neural networks and underlying changes of connectivity between cortex and subcortical structures (Koelsch et al., 2013b; Mas-Herrero et al., 2014): such work promises to corroborate electrophysiologic evidence in the damaged brain (Dellacherie et al., 2009).

CONCLUSIONS AND FUTURE DIRECTIONS

Though many challenges remain, the balance of evidence suggests that the amusias can be understood as disorders of cognitive information processing underpinned by selective dysfunction within brain networks (Figs 34.1 and 34.3). The many neuropsychologic dissociations observed attest to the modularity of music cognition (Peretz and Coltheart, 2003; Patel, 2008): the amusias target partly separable levels of an information-processing hierarchy. At the same time, however, the network organization of the musical brain substantially shares neural resources with the brain networks that support the processing of speech and other complex sounds. Contemporary cognitive neuropsychologic and functional neuroimaging evidence has made any simple dichotomy between music and language processing untenable: these complex, multicomponent modes of brain operation share comparably complex, distributed, and overlapping brain architectures. Recent interest in neurodegenerative diseases as specific disorders of brain network disintegration (Warren et al., 2013) and in the effects of these diseases on music processing (Omar et al., 2010, 2011; Hsieh et al., 2012) has opened up a new class of amusias for study and underlined the network basis of these disorders, complementing the traditional emphasis on focal brain lesions.

The information-processing model of the amusias as disorders of musical object cognition exposes several areas of difficulty that suggest directions for future work (Fig 34.1 and Table 34.1). Whereas disorders of musical property (pitch, temporal and timbral information) encoding and disorders of musical semantic processing have been relatively well defined, the status of putative disorders of musical scene analysis and musical object apperception is less clear. These levels of musical analysis have been little studied in clinical settings; however contemporary cognitive neuropsychologic models of auditory object processing (Griffiths and Warren, 2004; Goll et al., 2010b) suggest potential experimental strategies. These may include paradigms based on artificial reduced "musical scenes" (Goll et al., 2012) or stimuli that isolate musical object identity under perceptual transformations (such as transposed melodies or non-canonical instrument voices). The example of musical emotion indicates the value of extending conventional neuropsychologic models to include apparently less tractable, but essential aspects of musical experience; indeed, decoding of musical emotions may probe the interface of brain processes that are critical for abstracting complex social emotions, suggesting a biologic role for music in modeling surrogate mental states (Downey et al., 2013; Clark et al., 2014). Beyond music cognition *per se*, there is a need to analyze musical object processing in parallel with object processing in other domains, using a common cognitive framework and taking account of potential intrinsic difficulty or familiarity confounds. This may require techniques that can fractionate sound categories to isolate more elementary properties (such as the dimensions of timbre) that may be shared by different kinds of complex sounds (Labuschagne and Hanekom, 2013). Only analyses of this kind will establish the true relations between the cognitive operations that process dimensions of music and analogous dimensions of speech, voices, and other complex sounds.

This approach should be extended to functional neuroanatomy. Comparisons between brain lesion mapping and functional neuroimaging in the healthy brain are challenging; one reason for this is the extensive cerebral damage associated with many cases of amusia, precluding any very precise anatomic correlation. Nevertheless, the neuroanatomic evidence in sum suggests a coherent convergence between these approaches. Analogous with the organization of the language system, the music-processing networks that are disrupted in amusia comprise, very broadly, early auditory areas near the temporoparietal junction that are critical for parsing the musical scene and encode basic musical properties; higher-order association cortices in the anteromesial temporal lobe and its connections that represent object-level characteristics and link musical stimuli with meaning; extratemporal, dorsally directed areas that track hierarchic or syntactically organized musical information; and a partly parallel, but intimately interconnected, distributed corticolimbic network that is critical for processing musical emotion (Fig. 34.3). Functional neuroimaging paradigms can be designed to dissect specificities of music and speech-processing mechanisms while calibrating for their generally very different elementary acoustic characteristics (Tierney et al., 2013). Such paradigms may expose continuities between speech and music processing as alternative syntactically organized communication codes (Koelsch, 2011a; Koelsch et al., 2013a), while assessing the extent to which acoustic differences may influence the brain signatures of music and speech and determine the effects of strategic brain lesions. Similar considerations apply to functional neuroimaging paradigms designed to compare cognitive processes within the broad domain of music. It is important, however, to recognize certain limitations of functional neuroimaging modalities such as fMRI. These modalities may not, for example, capture dynamic interactions between brain structures; such interactions may be particularly relevant to the processing of a dynamic stimulus such as music, but may only be demonstrated using more specialized connectivity-based analysis methods or by harnessing the greater temporal resolution of electrophysiologic techniques (Meyer et al., 2006; Koelsch et al., 2013a). Macroanatomic convergence between lesions or processes does not eliminate the possibility of distinct functional mechanisms at the level of microcircuitry (for example, cortical laminae and their connections), modulated network connectivity patterns, or fine-grained areal activation profiles (Rogalsky et al., 2011).

The picture emerging from contemporary study of the amusias is more complex than Steinthal or Knoblauch could ever have envisaged. Against the need to relinquish false dichotomies must be set the promise of a unifying frame for understanding the amusias via cognitive information processing and functional neuroimaging approaches. However, any model of amusia can only be provisional and the primary role of astute clinical observation in identifying deficits, exposing counterexamples and suggesting experiments should not be overlooked. Despite the considerable sophistication of recent methods, careful characterization of patients' symptoms and deficits remains essential and the key to further progress in these diverse disorders of the musical brain.

ACKNOWLEDGMENTS

This work was undertaken at UCLH/UCL, who are supported by the Department of Health's National Institute for Health Research's Biomedical Research Centres funding scheme. The Dementia Research Centre is an Alzheimer Research UK Co-ordinating Centre. This work was supported by the Wellcome Trust, by the UK Medical Research Council and by Alzheimer's Research UK. HLG is supported by an Alzheimer's Research UK PhD Studentship. JDW is supported by a Wellcome Trust Senior Clinical Fellowship [Grant No 091673/Z/10/Z].

References

Abrams DA, Bhatara A, Ryali S et al. (2011). Decoding temporal structure in music and speech relies on shared brain resources but elicits different fine-scale spatial patterns. Cereb Cortex 21: 1507–1518.

Agus TR, Suied C, Thorpe SJ et al. (2012). Fast recognition of musical sounds based on timbre. J Acoust Soc Am 131: 4124–4133.

Alossa N, Castelli L (2009). Amusia and musical functioning. Eur Neurol 61: 269–277.

Andoh J, Zatorre RJ (2013). Mapping interhemispheric connectivity using functional MRI after transcranial magnetic stimulation on the human auditory cortex. Neuroimage 79: 162–171.

Ayotte J, Peretz I, Rousseau I et al. (2000). Patterns of music agnosia associated with middle cerebral artery infarcts. Brain 123 (Pt 9): 1926–1938.

Bedard C, Belin P (2004). A "voice inversion effect?". Brain Cogn 55: 247–249.

Belin P, Zatorre RJ (2000). 'What', 'where' and 'how' in auditory cortex. Nat Neurosci 3: 965–966.

Belin P, Zatorre RJ, Lafaille P et al. (2000). Voice-selective areas in human auditory cortex. Nature 403: 309–312.

Bengtsson SL, Ullen F, Ehrsson HH et al. (2009). Listening to rhythms activates motor and premotor cortices. Cortex 45: 62–71.

Besson M, Schon D (2001). Comparison between language and music. Ann N Y Acad Sci 930: 232–258.

Bigand E, Madurell F, Tillmann B et al. (1999). Effect of global structure and temporal organization on chord processing. J Exp Psychol Hum Percept Perform 25: 184–197.

Blood AJ, Zatorre RJ (2001). Intensely pleasurable responses to music correlate with activity in brain regions implicated in reward and emotion. Proc Natl Acad Sci U S A 98: 11818–11823.

Botez MI, Wertheim N (1959). Expressive aphasia and amusia following right frontal lesion in a right-handed man. Brain 82: 186–202.

Bregman AS (1990). Auditory Scene Analysis: the Perceptual Organization of Sound. MIT Press, Cambridge.

Brown S, Martinez MJ, Parsons LM (2004). Passive music listening spontaneously engages limbic and paralimbic systems. Neuroreport 15: 2033–2037.

Bruce V, Young A (1986). Understanding face recognition. Br J Psychol 77 (Pt 3): 305–327.

Carrington SJ, Bailey AJ (2009). Are there theory of mind regions in the brain? A review of the neuroimaging literature. Hum Brain Mapp 30: 2313–2335.

Chen JL, Zatorre RJ, Penhune VB (2006). Interactions between auditory and dorsal premotor cortex during synchronization to musical rhythms. Neuroimage 32: 1771–1781.

Chen JL, Penhune VB, Zatorre RJ (2008a). Moving on time: brain network for auditory-motor synchronization is modulated by rhythm complexity and musical training. J Cogn Neurosci 20: 226–239.

Chen JL, Penhune VB, Zatorre RJ (2008b). Listening to musical rhythms recruits motor regions of the brain. Cereb Cortex 18: 2844–2854.

Clark CN, Downey LE, Warren JD (2014). Music biology: all this useful beauty. Curr Biol 24: R234–R237.

Confavreux C, Croisile B, Garassus P et al. (1992). Progressive amusia and aprosody. Arch Neurol 49: 971–976.

Critchley M, Henson RA (1977). Music and the Brain. Heinemann, London.

Cuddy LL, Cohen AJ, Mewhort DJ (1981). Perception of structure in short melodic sequences. J Exp Psychol Hum Percept Perform 7: 869–883.

Dellacherie D, Pfeuty M, Hasboun D et al. (2009). The birth of musical emotion: a depth electrode case study in a human subject with epilepsy. Ann N Y Acad Sci 1169: 336–341.

Dellacherie D, Bigand E, Molin P et al. (2011). Multidimensional scaling of emotional responses to music in patients with temporal lobe resection. Cortex 47: 1107–1115.

Downey LE, Blezat A, Nicholas J et al. (2013). Mentalising music in frontotemporal dementia. Cortex 49: 1844–1855.

Drapeau J, Gosselin N, Gagnon L et al. (2009). Emotional recognition from face, voice, and music in dementia of the Alzheimer type. Ann N Y Acad Sci 1169: 342–345.

Eerola T, Friberg A, Bresin R (2013). Emotional expression in music: contribution, linearity, and additivity of primary musical cues. Front Psychol 4: 487.

Eustache F, Lechevalier B, Viader F et al. (1990). Identification and discrimination disorders in auditory perception: a report on two cases. Neuropsychologia 28: 257–270.

Fadiga L, Craighero L, D'Ausilio A (2009). Broca's area in language, action, and music. Ann N Y Acad Sci 1169: 448–458.

Fazio P, Cantagallo A, Craighero L et al. (2009). Encoding of human action in Broca's area. Brain 132: 1980–1988.

Feuchtwanger E (1930). Amusie. Studien zur pathologischen Psychologie der akustischen Wahrnehmung und Vorstellung undihrer Strukturgebiete, besonders in Musik und Sprache. Julius Springer, Berlin.

Fletcher PD, Downey LE, Witoonpanich P et al. (2013). The brain basis of musicophilia: evidence from frontotemporal lobar degeneration. Front Psychol 4: 347.

Foster NE, Zatorre RJ (2010). A role for the intraparietal sulcus in transforming musical pitch information. Cereb Cortex 20: 1350–1359.

Foster NE, Halpern AR, Zatorre RJ (2013). Common parietal activation in musical mental transformations across pitch and time. Neuroimage 75: 27–35.

Fries W, Swihart AA (1990). Disturbance of rhythm sense following right hemisphere damage. Neuropsychologia 28: 1317–1323.

Fujii T, Fukatsu R, Watabe S et al. (1990). Auditory sound agnosia without aphasia following a right temporal lobe lesion. Cortex 26: 263–268.

Gagnon L, Peretz I, Fulop T (2009). Musical structural determinants of emotional judgments in dementia of the Alzheimer type. Neuropsychology 23: 90–97.

Goll JC, Crutch SJ, Loo JH et al. (2010a). Non-verbal sound processing in the primary progressive aphasias. Brain 133: 272–285.

Goll JC, Crutch SJ, Warren JD (2010b). Central auditory disorders: toward a neuropsychology of auditory objects. Curr Opin Neurol 23: 617–627.

Goll JC, Kim LG, Hailstone JC et al. (2011). Auditory object cognition in dementia. Neuropsychologia 49: 2755–2765.

Goll JC, Kim LG, Ridgway GR et al. (2012). Impairments of auditory scene analysis in Alzheimer's disease. Brain 135: 190–200.

Gosselin N, Peretz I, Noulhiane M et al. (2005). Impaired recognition of scary music following unilateral temporal lobe excision. Brain 128: 628–640.

Gosselin N, Peretz I, Johnsen E et al. (2007). Amygdala damage impairs emotion recognition from music. Neuropsychologia 45: 236–244.

Gosselin N, Peretz I, Hasboun D et al. (2011). Impaired recognition of musical emotions and facial expressions following anteromedial temporal lobe excision. Cortex 47: 1116–1125.

Grahn JA (2012). Neural mechanisms of rhythm perception: current findings and future perspectives. Top Cogn Sci 4: 585–606.

Grahn JA, Brett M (2007). Rhythm and beat perception in motor areas of the brain. J Cogn Neurosci 19: 893–906.

Griffiths TD, Warren JD (2002). The planum temporale as a computational hub. Trends Neurosci 25: 348–353.

Griffiths TD, Warren JD (2004). What is an auditory object? Nat Rev Neurosci 5: 887–892.

Griffiths TD, Rees A, Witton C et al. (1997). Spatial and temporal auditory processing deficits following right hemisphere infarction. A psychophysical study. Brain 120 (Pt 5): 785–794.

Griffiths TD, Johnsrude I, Dean JL et al. (1999). A common neural substrate for the analysis of pitch and duration pattern in segmented sound? Neuroreport 10: 3825–3830.

Griffiths TD, Warren JD, Dean JL et al. (2004). "When the feeling's gone": a selective loss of musical emotion. J Neurol Neurosurg Psychiatry 75: 344–345.

Griffiths TD, Kumar S, Warren JD et al. (2007). Approaches to the cortical analysis of auditory objects. Hear Res 229: 46–53.

Grodzinsky Y, Friederici AD (2006). Neuroimaging of syntax and syntactic processing. Curr Opin Neurobiol 16: 240–246.

Groussard M, Viader F, Landeau B et al. (2009). Neural correlates underlying musical semantic memory. Ann N Y Acad Sci 1169: 278–281.

Groussard M, Rauchs G, Landeau B et al. (2010). The neural substrates of musical memory revealed by fMRI and two semantic tasks. Neuroimage 53: 1301–1309.

Gutschalk A, Dykstra A (2014). Functional imaging of auditory scene analysis. Hear Res 307: 98–110.

Habib M, Daquin G, Milandre L et al. (1995). Mutism and auditory agnosia due to bilateral insular damage – role of the insula in human communication. Neuropsychologia 33: 327–339.

Hailstone JC, Omar R, Henley SM et al. (2009a). It's not what you play, it's how you play it: timbre affects perception of emotion in music. Q J Exp Psychol 62: 2141–2155.

Hailstone JC, Omar R, Warren JD (2009b). Relatively preserved knowledge of music in semantic dementia. J Neurol Neurosurg Psychiatry 80: 808–809.

Hailstone JC, Crutch SJ, Vestergaard MD et al. (2010). Progressive associative phonagnosia: a neuropsychological analysis. Neuropsychologia 48: 1104–1114.

Hailstone JC, Ridgway GR, Bartlett JW et al. (2012). Accent processing in dementia. Neuropsychologia 50: 2233–2244.

Halpern AR, Zatorre RJ, Bouffard M et al. (2004). Behavioral and neural correlates of perceived and imagined musical timbre. Neuropsychologia 42: 1281–1292.

Han S, Sundararajan J, Bowling DL et al. (2011). Co-variation of tonality in the music and speech of different cultures. PLoS One 6: e20160.

Henschen SE (1920). Klinische und anatomische Beiträge zur Pathologie des Gehirns, Teil 5: Ueber Aphasie, Amusie und Alkalkulie. Nordiska Bokhandeln, Stockholm.

Henson RA (1985). Amusia. Vol. 1, In: JAM Fredriks (Ed.), Handbook of Clinical Neurology. Elsevier, Amsterdam.

Herdener M, Humbel T, Esposito F et al. (2014). Jazz drummers recruit language-specific areas for the processing of rhythmic structure. Cereb Cortex 24: 836–843.

Herholz SC, Halpern AR, Zatorre RJ (2012). Neuronal correlates of perception, imagery, and memory for familiar tunes. J Cogn Neurosci 24: 1382–1397.

Hickok G, Poeppel D (2007). The cortical organization of speech processing. Nat Rev Neurosci 8: 393–402.

Hsieh S, Hornberger M, Piguet O et al. (2011). Neural basis of music knowledge: evidence from the dementias. Brain 134: 2523–2534.

Hsieh S, Hornberger M, Piguet O et al. (2012). Brain correlates of musical and facial emotion recognition: evidence from the dementias. Neuropsychologia 50: 1814–1822.

Huron D (2006). Sweet Anticipation: Music and the Psychology of Expectation. MIT Press, Cambridge.

Isenberg AL, Vaden KI, Saberi K et al. (2012). Functionally distinct regions for spatial processing and sensory motor integration in the planum temporale. Hum Brain Mapp 33: 2453–2463.

Janata P, Birk JL, Van Horn JD et al. (2002). The cortical topography of tonal structures underlying Western music. Science 298: 2167–2170.

Johannes S, Jobges ME, Dengler R et al. (1998). Cortical auditory disorders: a case of non-verbal disturbances assessed with event-related brain potentials. Behav Neurol 11: 55–73.

Johnsen EL, Tranel D, Lutgendorf S et al. (2009). A neuroanatomical dissociation for emotion induced by music. Int J Psychophysiol 72: 24–33.

Johnson JK, Chang CC, Brambati SM et al. (2011). Music recognition in frontotemporal lobar degeneration and Alzheimer disease. Cogn Behav Neurol 24: 74–84.

Juslin PN, Laukka P (2003). Communication of emotions in vocal expression and music performance: different channels, same code? Psychol Bull 129: 770–814.

Juslin PN, Vastfjall D (2008). Emotional responses to music: the need to consider underlying mechanisms. Behav Brain Sci 31: 559–575, discussion 575–621.

Khalfa S, Guye M, Peretz I et al. (2008). Evidence of lateralized anteromedial temporal structures involvement in musical emotion processing. Neuropsychologia 46: 2485–2493.

Kleist K (1928). Gehirnpathologische und lokalisatorische Ergebnisse über Hörstörungen, Gerauschtaubheiten und Amusien. Mschr, Psychiat. Neurol 68: 853–871.

Knoblauch A (1888). Über störungen der musikalischen Leistungsfähigkeit infolge von Gehirnläsionen. Dtsch Arch Klin Med 331–345.

Koelsch S (2005). Neural substrates of processing syntax and semantics in music. Curr Opin Neurobiol 15: 207–212.

Koelsch S (2011a). Toward a neural basis of music perception – a review and updated model. Front Psychol 2: 110.

Koelsch S (2011b). Towards a neural basis of processing musical semantics. Phys Life Rev 8: 89–105.

Koelsch S, Fritz T, Schulze K et al. (2005). Adults and children processing music: an fMRI study. Neuroimage 25: 1068–1076.

Koelsch S, Fritz T, von Cramen DY et al. (2006). Investigating emotion with music: an fMRI study. Hum Brain Mapp 27: 239–250.

Koelsch S, Rohrmeier M, Torrecuso R et al. (2013a). Processing of hierarchical syntactic structure in music. Proc Natl Acad Sci U S A 110: 15443–15448.

Koelsch S, Skouras S, Fritz T et al. (2013b). The roles of superficial amygdala and auditory cortex in music-evoked fear and joy. Neuroimage 81: 49–60.

Kohlmetz C, Muller SV, Nager W et al. (2003). Selective loss of timbre perception for keyboard and percussion instruments following a right temporal lesion. Neurocase 9: 86–93.

Konoike N, Kotozaki Y, Miyachi S et al. (2012). Rhythm information represented in the fronto-parieto-cerebellar motor system. Neuroimage 63: 328–338.

Krumhansl C (1990). Cognitive Foundations of Musical Pitch. Oxford University Press, Oxford.

Krumhansl CL (2000). Rhythm and pitch in music cognition. Psychol Bull 126: 159–179.

Kumar S, Sedley W, Nourski KV et al. (2011). Predictive coding and pitch processing in the auditory cortex. J Cogn Neurosci 23: 3084–3094.

Kumfor F, Piguet O (2012). Disturbance of emotion processing in frontotemporal dementia: a synthesis of cognitive and neuroimaging findings. Neuropsychol Rev 22: 280–297.

Labuschagne IB, Hanekom JJ (2013). Preparation of stimuli for timbre perception studies. J Acoust Soc Am 134: 2256–2267.

Lambon Ralph MA, Pobric G, Jefferies E (2009). Conceptual knowledge is underpinned by the temporal pole bilaterally: convergent evidence from rTMS. Cereb Cortex 19: 832–838.

Laukka P, Eerola T, Thingujam NS et al. (2013). Universal and culture-specific factors in the recognition and performance of musical affect expressions. Emotion 13: 434–449.

Leaver AM, Rauschecker JP (2010). Cortical representation of natural complex sounds: effects of acoustic features and auditory object category. J Neurosci 30: 7604–7612.

Lechevalier B, Rossa Y, Eustache F et al. (1984). Case of cortical deafness sparing the music area. Rev Neurol (Paris) 140: 190–201.

Lehne M, Rohrmeier M, Koelsch S (2014). Functional neuroimaging of tension: a study with music. Soc Cogn Affect Neurosci 9 (10): 1515–1523.

Leitman DI, Wolf DH, Ragland JD et al. (2010). "It's not what you say, but how you say it": a reciprocal temporo-frontal network for affective prosody. Front Hum Neurosci 4: 19.

Lerdahl F, Jackendoff R (1983). A Generative Theory of Tonal Music. MIT Press, Cambridge, MA.

Liegeois-Chauvel C, Peretz I, Babai M et al. (1998). Contribution of different cortical areas in the temporal lobes to music processing. Brain 121 (Pt 10): 1853–1867.

Lima CF, Garrett C, Castro SL (2013). Not all sounds sound the same: Parkinson's disease affects differently emotion processing in music and in speech prosody. J Clin Exp Neuropsychol 35 (4): 373–392.

Limb CJ (2006). Structural and functional neural correlates of music perception. Anat Rec A Discov Mol Cell Evol Biol 288: 435–446.

Loui P, Hohmann A, Schlaug G (2010). Inducing disorders in pitch perception and production: a reverse-engineering approach. Proc Meet Acoust 9: 50002.

Mackworth-Young CG (1983). Sequential musical symptoms in a professional musician with presumed encephalitis. Cortex 19: 413–419.

Mangels JA, Ivry RB, Shimizu N (1998). Dissociable contributions of the prefrontal and neocerebellar cortex to time perception. Brain Res Cogn Brain Res 7: 15–39.

Marr D (1980). Visual information processing: the structure and creation of visual representations. Philos Trans R Soc Lond B Biol Sci 290: 199–218.

Mas-Herrero E, Zatorre RJ, Rodriguez-Fornells A et al. (2014). Dissociation between musical and monetary reward responses in specific musical anhedonia. Curr Biol 24: 699–704.

Mathiak K, Hertrich I, Grodd W et al. (2004). Discrimination of temporal information at the cerebellum: functional magnetic resonance imaging of nonverbal auditory memory. Neuroimage 21: 154–162.

Matthews BR, Chang CC, De May M et al. (2009). Pleasurable emotional response to music: a case of neurodegenerative generalized auditory agnosia. Neurocase 15: 248–259.

Mazzoni M, Moretti P, Pardossi L et al. (1993). A case of music imperception. J Neurol Neurosurg Psychiatry 56: 322.

Mazzucchi A, Marchini C, Budai R et al. (1982). A case of receptive amusia with prominent timbre perception defect. J Neurol Neurosurg Psychiatry 45: 644–647.

McAdams S, Cunible JC (1992). Perception of timbral analogies. Philos Trans R Soc Lond B Biol Sci 336: 383–389.

McDermott JH, Lehr AJ, Oxenham AJ (2010). Individual differences reveal the basis of consonance. Curr Biol 20: 1035–1041.

McDonald I (2006). Musical alexia with recovery: a personal account. Brain 129: 2554–2561.

McGettigan C, Scott SK (2012). Cortical asymmetries in speech perception: what's wrong, what's right and what's left? Trends Cogn Sci 16: 269–276.

Mendez MF, Geehan Jr GR (1988). Cortical auditory disorders: clinical and psychoacoustic features. J Neurol Neurosurg Psychiatry 51: 1–9.

Menon V, Levitin DJ, Smith BK et al. (2002). Neural correlates of timbre change in harmonic sounds. Neuroimage 17: 1742–1754.

Meyer M, Baumann S, Jancke L (2006). Electrical brain imaging reveals spatio-temporal dynamics of timbre perception in humans. Neuroimage 32: 1510–1523.

Milner B (1958). Psychological defects with temporal lobe excision. Res Publ Assoc Res Nerv Ment Dis 36: 244–257.

Milner B (1962). Laterality effects in audition. In: VB Mountcastle (Ed.), Interhemispheric relations and cerebral dominance. Johns Hopkins University Press, Baltimore.

Motomura N, Yamadori A, Mori E et al. (1986). Auditory agnosia. Analysis of a case with bilateral subcortical lesions. Brain 109 (Pt 3): 379–391.

Omar R, Hailstone JC, Warren JE et al. (2010). The cognitive organization of music knowledge: a clinical analysis. Brain 133: 1200–1213.

Omar R, Henley SM, Bartlett JW et al. (2011). The structural neuroanatomy of music emotion recognition: evidence from frontotemporal lobar degeneration. Neuroimage 56: 1814–1821.

Overath T, Cusack R, Kumar S et al. (2007). An information theoretic characterisation of auditory encoding. PLoS Biol 5: e288.

Pallesen KJ, Brattico E, Bailey C et al. (2005). Emotion processing of major, minor, and dissonant chords: a functional magnetic resonance imaging study. Ann N Y Acad Sci 1060: 450–453.

Patel A (2008). Music, language and the brain. Oxford University Press, Oxford.

Patterson RD, Uppenkamp S, Johnsrude IS et al. (2002). The processing of temporal pitch and melody information in auditory cortex. Neuron 36: 767–776.

Pereira CS, Teixeira J, Figueiredo P et al. (2011). Music and emotions in the brain: familiarity matters. PLoS One 6: e27241.

Peretz I (1990). Processing of local and global musical information by unilateral brain-damaged patients. Brain 113 (Pt 4): 1185–1205.

Peretz I (1996). Can we lose memory for music? A case of music agnosia in a non-musician. J Cogn Neurosci 481–496.

Peretz I (2006). The nature of music from a biological perspective. Cognition 100: 1–32.

Peretz I, Coltheart M (2003). Modularity of music processing. Nat Neurosci 6: 688–691.

Peretz I, Gagnon L (1999). Dissociation between recognition and emotion for melodies. Neurocase 5: 21–39.

Peretz I, Kolinsky R (1993). Boundaries of separability between melody and rhythm in music discrimination: a neuropsychological perspective. Q J Exp Psychol A 46: 301–325.

Peretz I, Kolinsky R, Tramo M et al. (1994). Functional dissociations following bilateral lesions of auditory cortex. Brain 117 (Pt 6): 1283–1301.

Peretz I, Belleville S, Fontaine S (1997). Dissociations between music and language functions after cerebral resection: a new case of amusia without aphasia. Can J Exp Psychol 51: 354–368.

Peretz I, Blood AJ, Penhune V et al. (2001). Cortical deafness to dissonance. Brain 124: 928–940.

Peretz I, Champod AS, Hyde K (2003). Varieties of musical disorders. The Montreal Battery of Evaluation of Amusia. Ann N Y Acad Sci 999: 58–75.

Peretz I, Gosselin N, Belin P et al. (2009). Music lexical networks: the cortical organization of music recognition. Ann N Y Acad Sci 1169: 256–265.

Piccirilli M, Sciarma T, Luzzi S (2000). Modularity of music: evidence from a case of pure amusia. J Neurol Neurosurg Psychiatry 69: 541–545.

Plailly J, Tillmann B, Royet JP (2007). The feeling of familiarity of music and odors: the same neural signature? Cereb Cortex 17: 2650–2658.

Poeppel D (2003). The analysis of speech in different temporal integration windows: cerebral lateralization as asymmetric sampling in time. Speech Communication 41: 245–255.

Pressnitzer D, Suied C, Shamma SA (2011). Auditory scene analysis: the sweet music of ambiguity. Front Hum Neurosci 5: 158.

Puschmann S, Uppenkamp S, Kollmeier B et al. (2010). Dichotic pitch activates pitch processing centre in Heschl's gyrus. Neuroimage 49: 1641–1649.

Rauschecker JP, Scott SK (2009). Maps and streams in the auditory cortex: nonhuman primates illuminate human speech processing. Nat Neurosci 12: 718–724.

Rogalsky C, Rong F, Saberi K et al. (2011). Functional anatomy of language and music perception: temporal and structural factors investigated using functional magnetic resonance imaging. J Neurosci 31: 3843–3852.

Rohrer JD, Lashley T, Schott JM et al. (2011). Clinical and neuroanatomical signatures of tissue pathology in frontotemporal lobar degeneration. Brain 134: 2565–2581.

Rosen S, Wise RJ, Chadha S et al. (2011). Hemispheric asymmetries in speech perception: sense, nonsense and modulations. PLoS One 6: e24672.

Salimpoor VN, Benovoy M, Larcher K et al. (2011). Anatomically distinct dopamine release during anticipation and experience of peak emotion to music. Nat Neurosci 14: 257–262.

Samson S, Zatorre RJ (1992). Learning and retention of melodic and verbal information after unilateral temporal lobectomy. Neuropsychologia 30: 815–826.

Samson S, Zatorre RJ (1994). Contribution of the right temporal lobe to musical timbre discrimination. Neuropsychologia 32: 231–240.

Samson S, Ehrle N, Baulac M (2001). Cerebral substrates for musical temporal processes. Ann N Y Acad Sci 930: 166–178.

Samson S, Zatorre RJ, Ramsay JO (2002). Deficits of musical timbre perception after unilateral temporal-lobe lesion revealed with multidimensional scaling. Brain 125: 511–523.

Samson S, Dellacherie D, Platel H (2009). Emotional power of music in patients with memory disorders: clinical implications of cognitive neuroscience. Ann N Y Acad Sci 1169: 245–255.

Sarkamo T, Tervaniemi M, Soinila S et al. (2010). Auditory and cognitive deficits associated with acquired amusia after stroke: a magnetoencephalography and neuropsychological follow-up study. PLoS One 5: e15157.

Satoh M, Takeda K, Kuzuhara S (2007). A case of auditory agnosia with impairment of perception and expression of music: cognitive processing of tonality. Eur Neurol 58: 70–77.

Satoh M, Nakase T, Nagata K et al. (2011). Musical anhedonia: selective loss of emotional experience in listening to music. Neurocase 17: 410–417.

Schon D, Gordon R, Campagne A et al. (2010). Similar cerebral networks in language, music and song perception. Neuroimage 51: 450–461.

Schuppert M, Munte TF, Wieringa BM et al. (2000). Receptive amusia: evidence for cross-hemispheric neural networks underlying music processing strategies. Brain 123 (Pt 3): 546–559.

Scott SK, Blank CC, Rosen S et al. (2000). Identification of a pathway for intelligible speech in the left temporal lobe. Brain 123 (Pt 12): 2400–2406.

Sidtis JJ, Feldmann E (1990). Transient ischemic attacks presenting with a loss of pitch perception. Cortex 26: 469–471.

Steinbeis N, Koelsch S (2008). Shared neural resources between music and language indicate semantic processing of musical tension-resolution patterns. Cereb Cortex 18: 1169–1178.

Steinthal H (1871). Einleitung in die Psychologie und Sprachwissenschaft. Dümmler's Verlagsbuchhandlung, Berlin.

Stewart L, von Kriegstein K, Warren JD et al. (2006). Music and the brain: disorders of musical listening. Brain 129: 2533–2553.

Stewart L, Overath T, Warren JD et al. (2008). fMRI evidence for a cortical hierarchy of pitch pattern processing. PLoS One 3: e1470.

Takahashi N, Kawamura M, Shinotou H et al. (1992). Pure word deafness due to left hemisphere damage. Cortex 28: 295–303.

Tanaka Y, Yamadori A, Mori E (1987). Pure word deafness following bilateral lesions. A psychophysical analysis. Brain 110 (Pt 2): 381–403.

Terao Y, Mizuno T, Shindoh M et al. (2006). Vocal amusia in a professional tango singer due to a right superior temporal cortex infarction. Neuropsychologia 44: 479–488.

Tierney A, Dick F, Deutsch D et al. (2013). Speech versus song: multiple pitch-sensitive areas revealed by a naturally occurring musical illusion. Cereb Cortex 23: 249–254.

Tillmann B (2005). Implicit investigations of tonal knowledge in nonmusician listeners. Ann N Y Acad Sci 1060: 100–110.

Tillmann B, Janata P, Bharucha JJ (2003). Activation of the inferior frontal cortex in musical priming. Brain Res Cogn Brain Res 16: 145–161.

Tramo MJ, Shah GD, Braida LD (2002). Functional role of auditory cortex in frequency processing and pitch perception. J Neurophysiol 87: 122–139.

Trehub SE, Schellenberg EG, Kamenetsky SB (1999). Infants' and adults' perception of scale structure. J Exp Psychol Hum Percept Perform 25: 965–975.

Vignolo LA (2003). Music agnosia and auditory agnosia. Dissociations in stroke patients. Ann N Y Acad Sci 999: 50–57.

von Kriegstein K, Smith DR, Patterson RD et al. (2007). Neural representation of auditory size in the human voice and in sounds from other resonant sources. Curr Biol 17: 1123–1128.

Warren JD, Griffiths TD (2003). Distinct mechanisms for processing spatial sequences and pitch sequences in the human auditory brain. J Neurosci 23: 5799–5804.

Warren JD, Jennings AR, Griffiths TD (2005a). Analysis of the spectral envelope of sounds by the human brain. Neuroimage 24: 1052–1057.

Warren JE, Wise RJ, Warren JD (2005b). Sounds do-able: auditory-motor transformations and the posterior temporal plane. Trends Neurosci 28: 636–643.

Warren JD, Rohrer JD, Schott JM et al. (2013). Molecular nexopathies: a new paradigm of neurodegenerative disease. Trends Neurosci 36: 561–569.

Warrier CM, Zatorre RJ (2004). Right temporal cortex is critical for utilization of melodic contextual cues in a pitch constancy task. Brain 127: 1616–1625.

Warrington EK, James M (1988). Visual apperceptive agnosia: a clinico-anatomical study of three cases. Cortex 24: 13–32.

Warrington EK, Taylor AM (1973). The contribution of the right parietal lobe to object recognition. Cortex 9: 152–164.

Watanabe T, Yagishita S, Kikyo H (2008). Memory of music: roles of right hippocampus and left inferior frontal gyrus. Neuroimage 39: 483–491.

Weinstein J, Koenig P, Gunawardena D et al. (2011). Preserved musical semantic memory in semantic dementia. Arch Neurol 68: 248–250.

Wertheim N (1975). The amusias. Vol. 4, In: PJ Vinken, GW Bruyn (Eds.), Handbook of Clinical Neurology. North-Holland Publishing Company, Amsterdam.

Wilson SJ, Pressing JL, Wales RJ (2002). Modelling rhythmic function in a musician post-stroke. Neuropsychologia 40: 1494–1505.

Zatorre RJ (1985). Discrimination and recognition of tonal melodies after unilateral cerebral excisions. Neuropsychologia 23: 31–41.

Zatorre RJ (1988). Pitch perception of complex tones and human temporal-lobe function. J Acoust Soc Am 84: 566–572.

Zatorre RJ, Belin P (2001). Spectral and temporal processing in human auditory cortex. Cereb Cortex 11: 946–953.

Zatorre RJ, Gandour JT (2008). Neural specializations for speech and pitch: moving beyond the dichotomies. Philos Trans R Soc Lond B Biol Sci 363: 1087–1104.

Zatorre RJ, Salimpoor VN (2013). From perception to pleasure: music and its neural substrates. Proc Natl Acad Sci U S A 110 (Suppl. 2): 10430–10437.

Zatorre RJ, Evans AC, Meyer E (1994). Neural mechanisms underlying melodic perception and memory for pitch. J Neurosci 14: 1908–1919.

Zhou J, Greicius MD, Gennatas ED et al. (2010). Divergent network connectivity changes in behavioural variant frontotemporal dementia and Alzheimer's disease. Brain 133: 1352–1367.

Zundorf IC, Lewald J, Karnath HO (2013). Neural correlates of sound localization in complex acoustic environments. PLoS One 8: e64259.

Handbook of Clinical Neurology, Vol. 129 (3rd series)
The Human Auditory System
G.G. Celesia and G. Hickok, Editors

Chapter 35

Hearing disorders in stroke

DORIS EVA BAMIOU*

Ear Institute, University College London and Neuro-otology Department, National Hospital for Neurology and Neurosurgery, London, UK

INTRODUCTION

Stroke is the commonest neurologic disorder, with 130 000 strokes per year and more than a million stroke survivors in the United Kingdom (MacDonald et al., 2000). It may cause physical as well as cognitive impairment, and about one in three stroke survivors are functionally dependent after 1 year (Murray and Lopez, 1996). In the majority of cases (85–90%), stroke is ischemic due to transient or permanent obstruction of a blood vessel compromising vascular supply to the brain (Deb et al., 2010). Ischemic stroke can be thrombotic, embolic, or due to systemic hypoperfusion or to venous thrombosis. Because of the presence of collateral circulation, the core area of the brain affected will undergo immediate death, while surrounding areas (penumbra) have the potential to recover, particularly when targeted with appropriate therapeutic intervention. Hemorrhagic stroke is present in 10–15% of all stroke cases and is due to rupture of a blood vessel in the brain. It may cause damage due to hypoxia, to irritant effects of the hemorrhage on the brain tissues and vasculature, or raised intracranial pressure due to bleeding (Deb et al., 2010). Both types of stroke may affect all levels of the auditory pathway and lead to hearing reception and/or perception deficits that may manifest with a variety of symptoms and with clinical presentations that start acutely before, during, or shortly after the stroke. Additionally, risk factors for stroke such as cigarette smoking, atherosclerosis and others have been associated with a more insidious onset of hearing impairment with advancing age, i.e., age-related hearing loss (Yamasoba et al., 2013).

Blood supply of the auditory system

The hearing end-organ and auditory nerve in particular are more vulnerable to acute stroke damage. This is because the vascular supply of the hearing end-organ and nerve is provided by the internal auditory artery (IAA), which is an end artery with few collaterals. This arises from the anterior inferior cerebellar artery (AICA) in the majority of cases, but may also have several anatomic variants (Wende et al., 1975). Wende et al. found that, in 158 angiographies, the IAA arises from the AICA in the majority of cases (45.4%), but other common variants included IAA origin from the superior cerebellar artery (24.4%) or the basilar artery (16%). Less commonly, the IAA arises from an anastomosis between the AICA and the superior cerebellar artery (6.7%) and it may also arise from the posterior inferior cerebellar artery (PICA: 5.4%) (Wende et al., 1975). The blood supply to the auditory nerve is also provided by the IAA, and additionally by collaterals from arteries that supply the dura mater and petrous bone within the internal auditory meatus, as well as by an anastamosing network from the AICA, PICA, and vertebral arteries at the cerebellopontine angle and the root entry zone to the brainstem level (Mazzoni, 1972). The auditory periphery, i.e., hearing end-organ and nerve, thus receives vascular supply that has a common origin to that of the low parts of the brainstem and also parts of the cerebellum. Therefore, peripheral-type hearing loss may coexist with auditory brainstemtype deficits and/or cerebellar neurologic symptoms and signs in several acute-onset poststroke hearing disorders.

*Correspondence to: Dr. Doris Eva Bamiou, MD MSc FRCP PhD, Neuro-otology Department, Box 127, National Hospital for Neurology and Neurosurgery, Queen Square, London WC1N 3BG, UK. Tel: +44-20-3448-3135, Fax: +44-20-3448-4775, E-mail: d.bamiou@ucl.ac.uk

The main blood supply of the central auditory nervous system is provided by the AICA, PICA, superior cerebellar, basilar and middle cerebral artery (reviewed in Häusler and Levine, 2000). The blood supply to the cochlear nucleus is provided by the AICA and PICA. Branches of the basilar artery and occasionally the AICA supply the superior olivary complex and lateral lemniscus, while the superior cerebellar artery supplies blood to the inferior colliculus and the basilar artery supplies the medial geniculate body. The primary and association auditory cortex and surrounding areas are supplied by the middle cerebral artery.

Hearing disorders in stroke sufferers

Auditory deficits after stroke have not been as extensively investigated as visual deficits, and unsurprisingly, recommendations for auditory function assessments in current clinical guidelines for stroke tend to be rudimentary (e.g., Royal College of Physicians, 2012) and audiologic investigations, when these do take place, appear to be rather limited. Sudden-onset hearing loss after stroke is one of the less frequent neurologic impairments (e.g., Lee, 2008), while auditory processing and perceptual deficits are less well studied and possibly underdocumented (Polster and Rose, 1998). However, studies of hearing in stroke populations indicate that hearing loss is very common (e.g., Formby et al., 1987; O'Halloran et al., 2009), and may predict long-term patient outcome (Landi et al., 2006). The present chapter will discuss the different types of hearing disorders in stroke patients classified on the basis of types of deficit and/or clinical presentation, with further discussion of the site of lesion that corresponds to such deficits and presentations. The subsections that will follow will thus discuss hearing loss after stroke, "central" deafness, with brief sections on auditory-processing deficits after stroke and poststroke onset of tinnitus and/or hyperacusis. The concluding section contains comments on current practice regarding assessment of hearing symptoms and deficits after stroke and some proposed clinical guidelines, based on this review chapter.

HEARING LOSS AFTER STROKE

Hearing loss and stroke in population studies

In the general population, a past history of stroke increases the likelihood of having hearing loss. A longitudinal study of 3526 Australian adults aged 50 years or older found that a previous history of stroke was a predictor for hearing thresholds (Table 35.1), although it did not determine deterioration of hearing loss over an 11-year period (Kiely et al., 2012). This is

Table 35.1

Hearing loss and stroke in the general population

Study	Subjects	Data collected	Design	Findings
Kiely et al., 2012	3526 participants at baseline From Australian Longitudinal Study of Ageing (ALSA): Adults >70 years + spouses aged >65 years from Adelaide metropolitan area of South Australia From Blue Mountains Hearing Study: Adults > 49 years from two postcodes in Blue Mountains, Australia	Audiometry Medical conditions Past noise exposure	Hearing tested on average two (maximum four) occasions, at 3.1–6.8 years' time interval	History of stroke independently associated with baseline levels of hearing: correlation coefficient $\beta = 2.52$, $P = 0.02$ history of stroke not predictive of change in hearing thresholds over 11 years
Gopinath et al., 2009	1394 participants with complete data at baseline From Blue Mountains Hearing Study: Adults > 49 years from two postcodes in Blue Mountains, Australia	Audiometry Subjects asked whether their hearing loss had a gradual or sudden onset Stroke diagnoses (MONICA criteria)	Hearing tested on three occasions at 5-year time intervals	OR of reporting stroke 1.55 (95% CI, 1.01–2.38) higher in subjects with vs without hearing loss marginally non-significant OR 2.04 (95% CI, 1.20–3.49) of reporting stroke with increasing hearing loss severity

OR, odds risk; CI, confidence interval; MONICA, MONItoring of trends and determinants in CArdiovascular disease.

broadly in keeping with results from a study partially overlapping the previous study Australian population of 1394 older adults (Gopinath et al., 2009), that found that, while the odds risk for reporting a previous history of stroke in those with vs those without hearing loss was raised but marginally non-significant, the odds risk of reporting stroke was significantly higher for those with moderate to severe hearing loss (Table 35.1).

Within the stroke population, hearing loss is present in the majority of stroke sufferers. Formby et al. (1987) investigated hearing impairment directly associated with stroke after excluding patients who had a previous otologic history or occupational noise exposure. They reported that 61.7% of 243 stroke patients had a pure-tone average of >25 dB at 500, 1000, and 2000 Hz for the better ear. O'Halloran et al. (2009) tested stroke patients in a stroke ward and found that 41 out of 52 patients (79%, confidence interval (CI) 67–90%) had at least mild or greater hearing impairment. A smaller study on 60 patients with past history of stroke suggested that age, the presence of lacunar stroke, multiple, bilateral ischemic focuses, and arterial hypertension increased the risk of hearing loss (Przewoźny et al., 2008).

Sudden-onset hearing loss due to ischemic stroke of the vertebrobasilar territory

Sudden hearing loss after stroke is less common than other neurologic impairments. A large prospective study of 685 consecutive cases with vertebrobasilar ischemic stroke identified sudden hearing loss in 42 and reported an overall prevalence of sudden hearing loss of 6% (Lee, 2008). Sudden hearing loss was defined as deterioration of hearing thresholds by 30 dB in at least three frequencies noted over 72 hours. AICA stroke was more likely to lead to hearing deficits than PICA stroke, with 35 AICA cases (83%) vs 5 PICA cases (12%) vs 2 isolated brainstem infarct cases (5%). Within the different infarct territory subtypes, sudden hearing loss occurred in 87.5% of cases with isolated AICA infarct, in only 3.4% of cases with isolated PICA infarct, and in 0.63% of cases with isolated brainstem infarct.

The type of hearing loss in both AICA and PICA is reported to be predominantly cochlear (Lee et al., 2002; Lee, 2008) but can also be mixed cochlear/retrocochlear, and, less frequently, retrocochlear only (Lee et al., 2002). However, it should be noted that, with the exception of tests such as auditory brainstem evoked responses and acoustic reflexes, very few studies have conducted any psychoacoustic tests, for example of localization and of temporal resolution, that may help decide if the hearing loss is peripheral or central (Ulbricht, 2003).

Lee et al. (2002) reviewed the clinical findings in 12 cases with AICA infarction who were identified over a 2-year period. Hearing loss together with tinnitus was present in 11 cases (i.e., in 91%, consistent with the 87.7% prevalence reported by Lee's later study: Lee, 2008). Six cases were thought to be cochlear, 1 retrocochlear, while in 4 it was not possible to establish the site of the lesion due to the presence of severe to profound hearing loss confounding interpretation of the absent auditory brainstem responses (ABR) and acoustic reflexes. In 7 cases there was some recovery of hearing over weeks to months. Gait and limb ataxia, thought to be of cerebellar origin, as well as vertigo and nystagmus (with absent or reduced caloric test response), attributed to involvement of the vestibular end-organ, and/or vestibular nerve, brainstem nuclei or flocculus, were present in all 12 cases. Additional clinical features included facial palsy (in 3), trigeminal sensory loss (in 6), and Horner's syndrome (in 2). The mechanism for the AICA stroke was stenosis of the lower and/or mid basilar artery in 5 patients (which could be due to atheroma or thrombus), while 2 patients had a dolichoectatic basilar artery, and 2 out of the remaining 5 patients who had normal magnetic resonance angiography (MRA) had cardiac disease. In a subsequent study, Lee (2009) reviewed 82 AICA territory cases and reported cochlear infarction in 52 (63%) with ipsilateral cochlear infarction without vestibular involvement in 3 (3%). Most of these cases (49: 60%) had a combined audiovestibular dysfunction.

Hearing loss and tinnitus may also be prodrome symptoms of AICA stroke, attributed to isolated labyrinthine infarction that may precede the AICA territory infarction by a few days (Lee et al., 2001, 2003; Kim et al., 2009), most commonly in patients who have vascular risk factors. Occasionally the hearing loss can be transient and recurrent (Park et al., 2008), possibly due to initial hypoperfusion of the IAA.

PICA stroke is less common than AICA. The clinical presentation in 5 cases with PICA infarcts (Lee, 2008) included hearing loss on the side of the stroke, with cochlear features in 3 cases and possibly combined cochlear/retrocochlear features in 2. All 5 cases had vertigo (and ipsilateral to the stroke, canal paresis on caloric testing), gait ataxia, bidirectional gaze evoked nystagmus, and severe unsteadiness on the lesion side.

Hearing loss for both AICA and PICA infarcts is mostly unilateral. However, this will depend on the affected territory, and there are case reports with bilateral hearing loss, when there is more extensive damage. Chian et al. (2013) reported the case of a patient with bilateral progressive hearing loss and mild cerebellar symptoms. Brain magnetic resonance imaging (MRI) showed infarction of the PICA territory bilaterally with inferior cerebellar involvement while MRA identified a critical basilar artery stenosis, although the right AICA could not be visualized. Local intra-arterial thrombolysis initiated 6 hours after onset of symptoms led to subsequent visualization of the right AICA and to recovery of hearing with only a mild residual hearing loss in the left ear at 6-month follow-up. On the basis of these findings, the authors suggested that a hypoperfusion injury of the cochlear nuclei in the pons (that derive their blood supply from the AICA) in addition to bilateral PICA infarction was responsible for this presentation (Chian et al., 2013). Lee et al. (2001) reported a case with bilateral hearing loss (thought to be cochlear), more prominent on the right than on the left, with additional right-sided signs of AICA infarction (left beating vestibular nystagmus and absent caloric response on the right, right facial palsy, right reduced facial sensation, right-limb ataxia). The brain MRI showed hyperintense foci in the right dorsolateral pons and cerebellar peduncle and possibly on the left cerebellar peduncle, while MRA showed stenosis of the distal vertebral artery and the middle third of the basilar artery. After anticoagulation treatment all patient symptoms improved, except the hearing loss in the right ear, which remained profound. There are also some case studies with low brainstem infarction associated with bilateral hearing loss (e.g., Shimomura et al., 1990; Yaguchi et al., 2000), in which the hearing loss is a combination of ischemia of the end hearing organ as well as of the lower pons, i.e., both peripheral and central-type hearing deficits.

Sudden hearing loss after ischemic stroke of the upper brainstem and midbrain

Because of the multiple decussations of the auditory pathway and its characteristic increasing redundancy from the periphery to the cortex, ischemic stroke above the level of the low brainstem is more likely to result in auditory-processing deficits rather than in peripheral-type hearing loss (for a review of hearing disorders in brainstem lesions, see Chapter 29), but there are exceptions. Cerrato et al. (2005) report a case with lacunar infarction of the superior cerebellar artery territory that involved the right lateral lemniscus and the inferior colliculus, with bilateral hearing loss (a sensation of bilateral "ear wadding" and hypoacusia) as one of the dominant symptoms at presentation. Hearing loss was more pronounced on the left than on the right side on audiometry, while the ABR interwave III–V interval on the left was prolonged. The authors explained the hearing deficits on the side contralateral to the stroke by the fact that acoustic fibers are mostly crossed at the level of the inferior colliculus. Lee (2008) also reported on 2 patients with upper brainstem infarction on MR, both of whom had an ipsilateral hearing loss and did not report any other symptoms, such as tinnitus or palinacusis. One of these two cases had a cochlear-type hearing loss, the other probably combined.

Sudden hearing loss after hemorrhagic lesions affecting the vertebrobasilar territory

Hemorrhage within the brainstem territory may also present with hearing loss due to involvement (by means of several different mechanisms) of cochlear nuclei, olivary complex, and/or trapezoid body. In addition to the hearing loss, such cases will have abnormalities in brainstem tests that will be consistent with the localization of the lesion. Cohen et al. (1996) reported on 4 cases of central pontine hemorrhage with documented auditory dysfunction. All cases had abnormalities of the ABR and/or acoustic reflexes and masking-level difference. Nakane et al. (2006) report a case with quadriplegia and bilateral hearing loss who had only wave I present on both sides in the auditory brainstem response test. This was due to a hematoma that extended from the lower to the upper part of dorsal pons, that appeared to damage the cochlear nuclei and trapezoid body. Goyal et al. (2010) however reported a case with central pontine hemorrhage due to capillary telangiectasia that was associated with bilateral hearing loss but normal ABR. Hearing loss had resolved 6 months later. The authors proposed that the hearing deficit was due to disruption of the ventral acoustic striae that decussates in the trapezoid body.

Hemorrhage affecting the vertebrobasilar territory may be due to rupture of vascular malformations. Cavernous hemangiomas are rare, but are frequently situated within the low brainstem, and may present with brainstem-type audiovestibular symptoms, including hearing loss (e.g., Dumas and Schmerber, 2004). Figure 35.1 presents imaging and audiologic investigations in a case with a spontaneously ruptured pontine cavernoma in a 37-year-old female patient who was admitted with a 3-day history of left-sided paresthesia, progressive weakness, and left-sided clumsiness, diplopia, frontal headaches, as well as right-ear hearing loss, tinnitus and hyperacusis, right facial palsy, and gradual drowsiness. Audiometry initially showed normal pure-tone thresholds on the left with profound hearing loss in the right ear while tympanic reflexes were only present from the left ipsilateral recording (Fig. 35.1A). Auditory brainstem evoked responses (not shown) showed absent waves on the right and normal waves on the left. There was subsequent full recovery of audiometric thresholds after resolution of the edema and the hearing loss on the right was thought to be retrocochlear. An MRI scan with MRA demonstrated a brainstem cavernoma with signal change extending medially from the posterior part of the

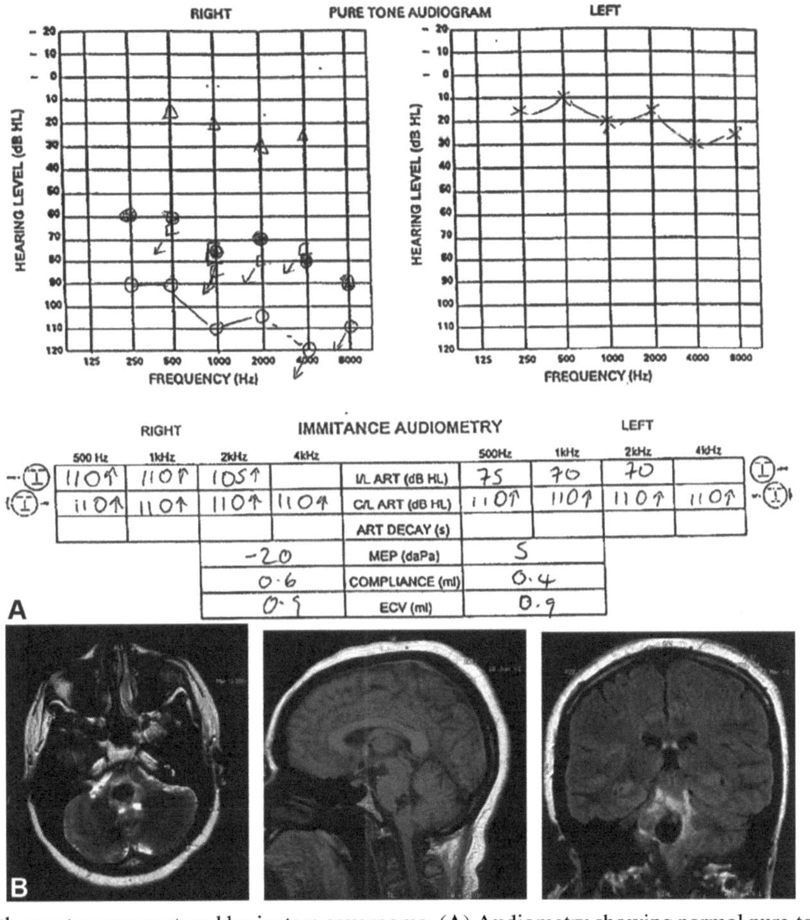

Fig. 35.1. Patient with spontaneous ruptured brainstem cavernoma. (**A**) Audiometry showing normal pure-tone thresholds on the left with profound hearing loss in the right ear and tympanic reflexes that are only present from the left ipsilateral recording. (**B**) A magnetic resonance imaging scan with magnetic resonance angiography demonstrates a brainstem cavernoma.

internal capsule into the low part of the medial geniculate body and the lateral part of each inferior brachium (Fig. 35.1B).

Aneurysms of the distal AICA are rare. Hearing loss and/or tinnitus can be presenting features, usually due to mass-type effects on the cerebellopontine angle rather than the hemorrhage (Zager et al., 2002). Hearing loss can also be a residual neurologic deficit after aneurysm surgery. Yamakawa et al. (2004) identified 56 reported cases of AICA aneurysms associated with subarachnoid hemorrhage between 1948 and 2002, including their own, who had an aneurysm within the internal auditory meatus. Following surgery of the aneurysm, hearing loss was reported in 29 (including Yamakawa's case) out of 55 cases who survived. The authors reported that, from the reviewed cases, a deeper location of the aneurysm within the meatus increases the risk for persistence of the hearing disturbance more postoperatively, regardless of the preoperative degree of hearing loss. Less common mechanisms by which hemorrhage in this territory can cause hearing loss include extension of the hemorrhage to the eighth nerve through the subpial space, as in a case of right cerebellar peduncle hemorrhage presenting with hearing loss and tinnitus (Matsuda et al., 1993).

"Central" or "cortical" deafness

Central or cortical deafness is a rare and dramatic clinical presentation. The first description of this rare disorder is attributed to Wernicke and Friedlander (1883). They reported a patient with bilateral temporal-lobe lesions but with no apparent damage to the hearing organ, who showed no awareness of sounds. Wernicke and Friedlander coined the term "cortical deafness," which has been used since as well as the term "central deafness." The term "deaf-hearing" has also been used for cases who show some response to sound, postulated to be reflexive, but with no sound awareness (Garde and Cowey, 2000).

The clinical presentation is similar to cochlear-type profound deafness, in that these patients do not show consistent response to sounds and give raised thresholds on audiometry. However, peripheral detection of sounds is either intact or partially impaired. This is indicated by present responses in lower-level tests of the sensory end-organ such as transient evoked otoacoustic emissions (e.g., Musiek et al., 2007). Auditory nerve and low brainstem tests, including acoustic reflexes and auditory brainstem evoked responses, are similarly elicited by normal or lower sensation levels than expected in view of the audiometric results (e.g., Tanaka et al., 1991; Musiek et al., 2007). These findings contrast with abnormal results in evoked potentials that arise from the thalamocortical pathway and beyond, such as middle-latency and late auditory evoked potentials, indicating that the behavioral lack of response to sounds is due to impaired perceptual awareness of the sound.

Figures 35.2 and 35.3 present imaging and audiologic test results of a case with central deafness due to stroke of bilateral auditory cortices. The patient suffered a right temporoparietal infarct in 1997 and a subsequent left posterior cerebral artery infarct in 2006, following which she presented with no response to sound and severe expressive dysphasia. On hearing tests conducted 2 years after the second stroke there was no response to sound on audiometry; however otoacoustic emissions (Fig. 35.2A), auditory brainstem evoked responses (Fig. 35.2B) and acoustic reflexes were present, indicating central deafness. The patient's axial computed tomography scan that had been conducted 3 days after the second stroke in 2006 (Fig. 35.3B) showed a cerebrospinal fluid attenuation area in the right hemisphere

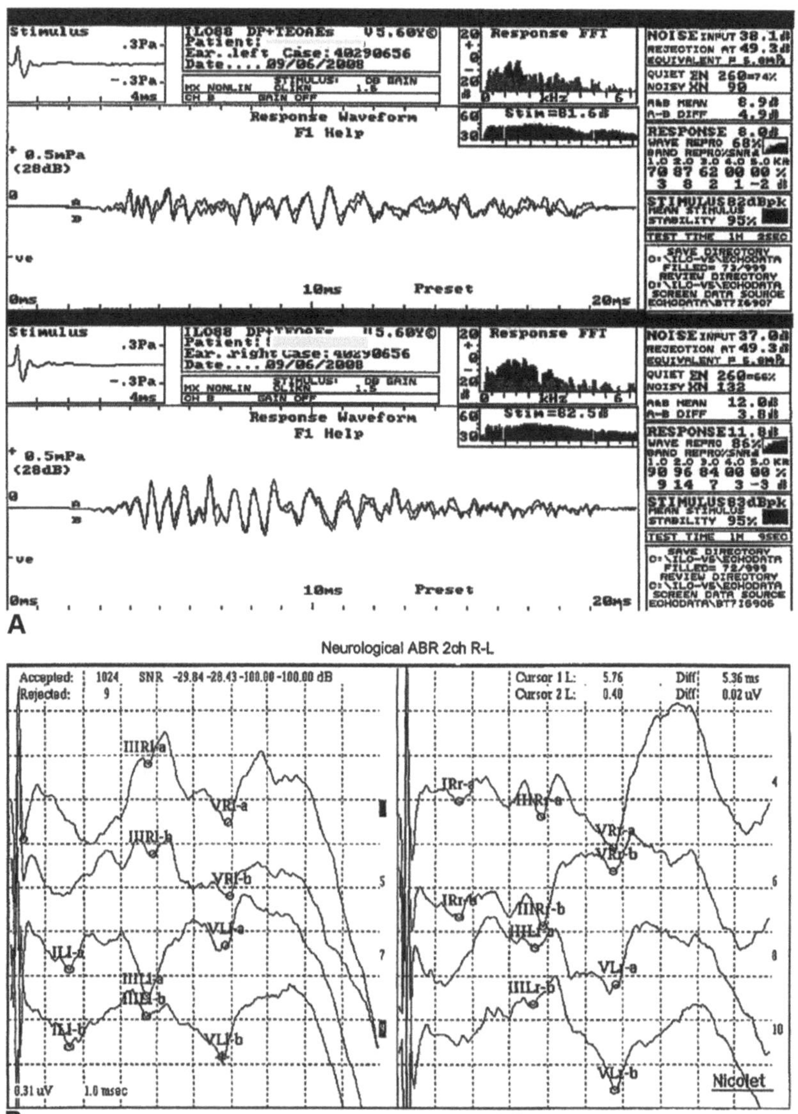

Fig. 35.2. Case of central deafness due to stroke of bilateral auditory cortices. (**A**) Normal otoacoustic emissions in right and left ear. (**B**) Normal auditory brainstem evoked responses in right and left ear.

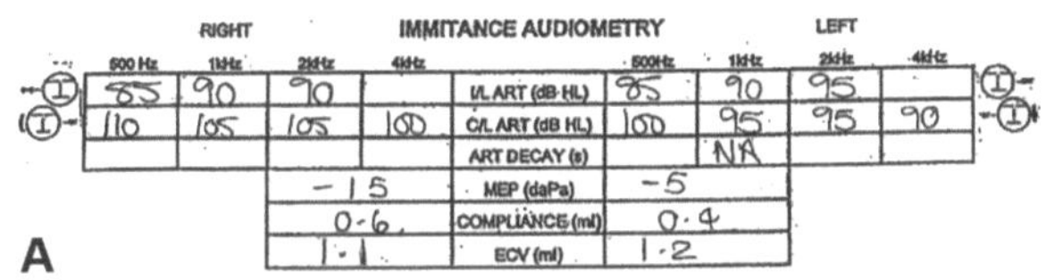

A

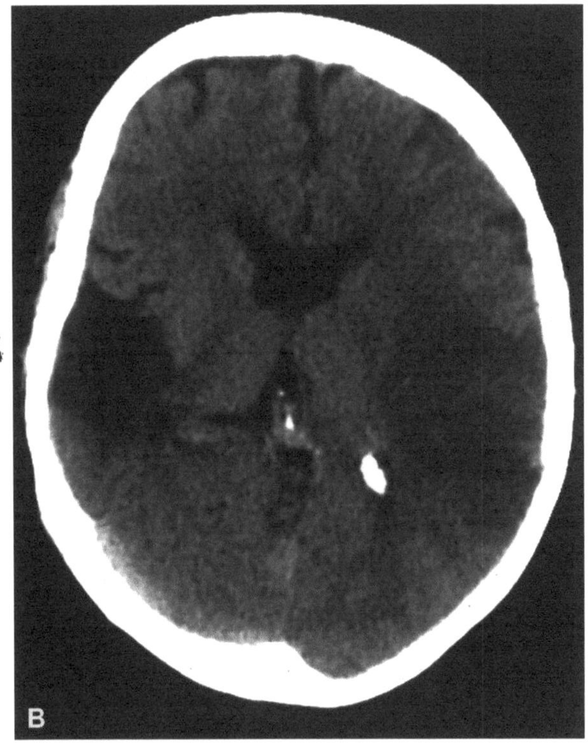

Fig. 35.3. Case of central deafness due to stroke of bilateral auditory cortices. (**A**) Normal acoustic reflexes in right and left ear. (**B**) Axial computed tomography scan after the second episode of stroke showing an old infarct on the right and a more recent infarct in the left temporoparietal region.

measuring 3.3×2.9 cm, consistent with an old infarct, and a slightly larger area of intermediate attenuation in the left hemisphere with mild mass effect.

In these cases, behavioral responses to sounds can be absent or inconsistent, while the audiometric responses when tested tend to be variable (Musiek et al., 2007). Patients may occasionally show a reaction to a sound for which they report no awareness to family members (Tanaka et al., 1991; Cavinato et al., 2012). Patients may grow tired even after short hearing test sessions (Garde and Cowey, 2000). Several cases will show no improvement (e.g., Bahls et al., 1988; Semenza et al., 2012), although both response variability (Tanaka et al., 1991) and audiometric thresholds may in some cases improve over the first few weeks to months after the stroke (Griffiths et al., 2010). There are occasional reports of patients with sudden cortical deafness who initially do not realize they are deaf, as in the case of a patient with consecutive infarcts of the right and left middle cerebral artery territory, with (premorbid) low intelligence, reported by Leussink et al. (2005), and consistent with Anton's eponymous syndrome for both blindness and deafness resulting from focal brain lesions (Anton, 1899). Finally, the clinical presentation will evolve from cortical deafness to other auditory syndromes that include agnosia (for environmental sounds, music, and/or words) and other more basic sound dimension-processing deficits (Mendez and Geehan, 1988), and it has been proposed that all these presentations might form a continuum of the auditory processing spectrum (Polster and Rose, 1998).

Anatomically, the most frequent cause for this presentation is the presence of bilateral vascular lesions that may occur simultaneously or sequentially (Musiek and Lee, 1998), that involve primary, secondary, and association auditory cortices in Heschl's gyrus and in the planum temporale (Griffiths et al., 2010). The disorder may also occur with subcortical primary lesions that do not involve the Heschl's gyri or cortical areas but result in an effective disconnection between sensory and perceptual processing (Polster and Rose, 1998). For example, Nishioka et al. (1993) reported the case of a patient with bilateral putamen hemorrhage, with resolution of hearing over 2 months. Other subcortical areas reported to be affected in similar cases include the inferior colliculus, the cerebellum, the internal capsule, the thalamic regions, and the medial geniculate body (Musiek and Lee, 1998). There are also case reports of subarachnoid hemorrhage leading to vasospasm of the middle cerebral artery with cortical deafness that resolves. For example, Tabuchi et al. (2007) reported a case of reversible cortical auditory dysfunction caused by aneurysmal

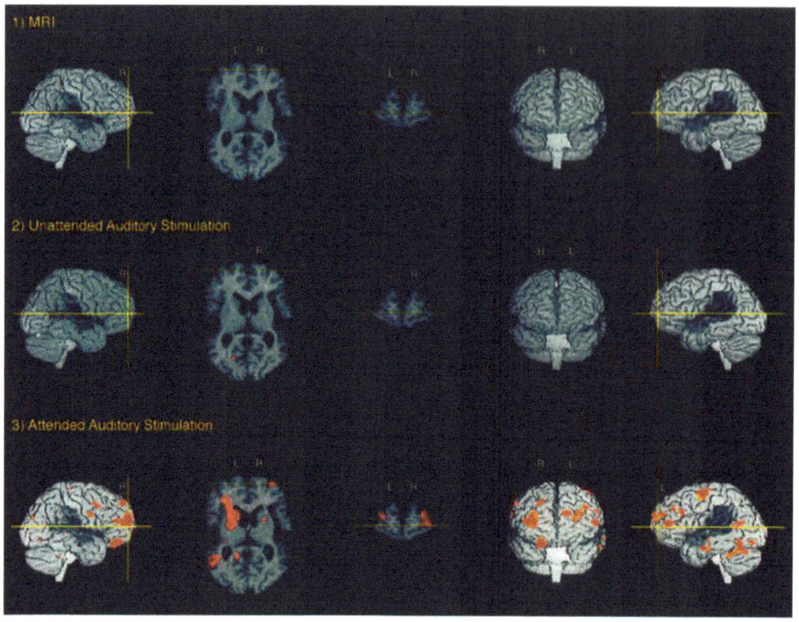

Fig. 35.4. Structural magnetic resonance imaging (MRI) and positron emission tomography (PET) activation study results in the patient reported by Engelien et al. (2000). First row: structural MRI scan. Second row: PET activation study results superimposed on to the patient's individual T1-weighted MRI for categoric comparisons of unattended stimulation versus no stimulation (rest). Third row: PET activation study results superimposed on to the patient's individual T1-weighted MRI for categoric comparisons of selectively attended versus unattended auditory stimulation. (Reproduced from Engelien et al., 2000, with permission.)

subarachnoid hemorrhage in both middle cerebral artery territories without pre-existing lesions.

There are counterarguments about the extent of neuroanatomic damage that is necessary to produce this presentation, and one must bear in mind that the central auditory nervous system is highly redundant, with several relay stations from periphery to cortex, multiple interconnections between right and left at different levels, and parallel afferent pathways to the secondary auditory cortices. The presence of extensive bilateral damage of Heschl's gyri may thus not be sufficient for cortical "deafness," and the contribution of attention-related neural substrates, that have been spared or affected by the stroke, in the clinical presentation, has been discussed by several authors. Studies of primates indicate that bilateral auditory cortex destruction will raise sound thresholds but will not completely abolish sound responsiveness (Heffner and Heffner, 1986).

Engelien et al. (2000) reported a patient with extensive bilateral damage of primary auditory cortices who showed no awareness of sound; however, when selectively attending to the auditory modality he achieved conscious perception of sound. His pure-tone audiogram tested with the standard procedure showed profound hearing loss but thresholds improved to "nearly" normal when attention-primed. Positron emission tomography conducted with the patient attending to sound demonstrated increases in cerebral blood flow in the lateral prefrontal and medial temporal cortex as well as in caudate nuclei bilaterally (Fig. 35.4).

The authors proposed that the neural substrate for selective attention in the prefrontal cortex may modulate the mediation of conscious sound awareness within the remaining auditory system, with contribution by the basal ganglia. Semenza et al. (2012) reported a case with persistent cortical deafness, on whom attentional priming had no effect. The patient had extensive cortical bilateral damage, including left and right Heschl gyri and the right frontal cortex as well as in subcortical structures. A tractography study showed extensive impairment in fractional anisotropy and diffusivity of the anterior portion of the right superior longitudinal fasciculus and the right uncinate fasciculus which connect frontal and temporal brain structures, and the authors proposed that this disconnection might have been to some extent responsible for the symptom persistence.

Information on intervention for these cases is even more limited, and it is unclear whether improvements when noted are a result of the intervention or of spontaneous resolution or a result of the natural progress of the disease. Musiek et al. (2007) implemented a speech and language rehabilitation program targeting the patient's language deficits together with use of augmentative and alternative communication devices that included a pocket talker, an FM trainer, and a word prediction program, but benefits of these communication devices were limited. Garde and Cowey (2000) attempted to teach their patient to become aware of her reflexive response to sounds; however this knowledge was not well retained between four such teaching sessions (conducted for the purposes of testing rather than rehabilitation). They also suggested that lip reading may be used as a strategy to support communication with these patients.

AUDITORY-PROCESSING DEFICITS AFTER STROKE

It is beyond the scope of this chapter to discuss severe – in terms of clinical presentation – auditory cognition disorders that may present after stroke, such as auditory neglect (see Chapter 31), auditory agnosia (see Chapter 32), or amusia (see Chapter 34). These are thought to be uncommon, but alternatively these may remain undiagnosed in the patient who does not have a language impairment, or may be thought to be unimportant in the presence of aphasia (Polster and Rose, 1998). It should be noted that such deficits, in their less severe forms, will not necessarily be reported by patients unless they are explicitly questioned. A hearing questionnaire study of patients with unilateral cerebrovascular auditory cortex lesions found that 49% of these patients reported auditory perceptual problems on the questionnaire, most commonly affecting sound localization or understanding speech in situations with simultaneous speakers (Blaettner et al., 1989), but the majority of these patients had reported intact hearing prior to the administration of the questionnaire. Our previous study of 9 patients with stroke lesions of the insula and adjacent cortex found temporal resolution and sequencing deficits (Bamiou et al., 2006). None of these patients had been referred for audiologic assessment; however, when specifically questioned, 6 out of 9 reported auditory difficulties in various domains since the stroke. We subsequently assessed 21 patients with stroke of the auditory brain documented on brain MRI in the chronic phase of stroke, excluding patients with aphasia, versus 23 age- and hearing-matched controls by means of a hearing questionnaire and auditory-processing tests (Bamiou et al., 2012). Severe and significant functional limitation (z score >3) with recognition and/or with localization of sounds was reported by 9 (43%) and with speech-in-noise recognition by 6 (28%) of the stroke patients in a range of everyday communication situations (Table 35.2). None of these patients had a previous diagnosis of auditory agnosia or neglect. These reported deficits did not correlate with hearing thresholds, but correlated strongly with some of the clinical auditory-processing tests, and with damage to the corresponding brain networks specialized for this function, as reported by others (Rey et al., 2007).

There is evidence to suggest that these deficits will persist long-term in these patients. Figure 35.5 presents auditory-processing and audiometry results in a patient tested 2 years after a left temporoparietal stroke.

Rey et al. (2007) assessed sound recognition, localization, and sound movement perception in 24 patients with

Table 35.2

Patient-reported sound recognition, sound localization, and speech in noise deficits expressed as z scores (compared to normative data) in stroke subjects. Significant deficits ($z > 3$) were identified in 9/20 subjects assessed in the chronic stage of stroke

Case	Lesion	Sound recognition z score	Sound localization z score	Speech-in-noise z score
1	Infarct, large right parietal AVM	5.13	10.50	2.8
3	Infarct, right short gyri and long gyri of the insula	3.25	6.50	1.2
5	Infarct, right putamen+insula limen	2.63	6.50	3.2
7	Infarct, left insula, striatum, frontal and temporal gyri	9.50	10.50	4.4
8	Hematoma, left posterior part striatum, posterior limb of internal capsule, lacunar infarcts in basal ganglia	8.25	5.50	2.2
10	2.8-cm diameter cavernoma hemorrhage/right brainstem and cerebellum	1.64	5.75	3.2
14	Infarct, right striatum, thalamus, posterior corona radiata, posterior limb internal capsule	5.75	5.50	3.2
16	Infarct, right insula, inferior frontal gyrus	3.67	Not computable	4
20	Infarct, right medial frontal gyrus	5.00	9.50	3.6

AVM, arteriovenous malformation.

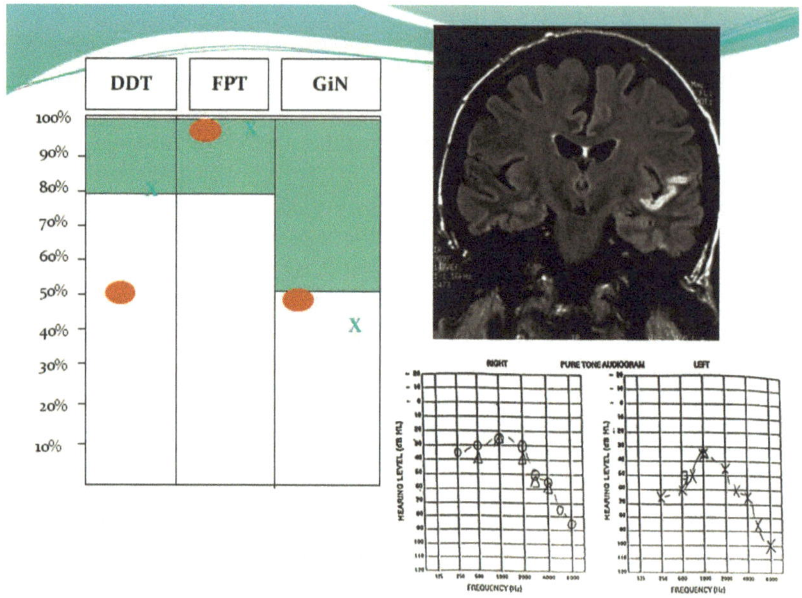

Fig. 35.5. Case study of a 58-year-old patient who suffered a left temporoparietal stroke. He had been a hearing-aid user for 3 years before the stroke and, while his hearing thresholds did not change after the stroke, he became aware of increased difficulties with listening to speech in noise. Auditory-processing test assessment 2 years after the stroke showed auditory-processing deficits consistent with the site of lesion, with a right-ear deficit on Dichotic Digits Test (DDT), which was the better-hearing ear in terms of audiometric thresholds, and abnormal Gaps-in-Noise (GiN) tests in both ears. Frequency pattern tests (FPT) were normal. Red circle, right ear, blue × left-ear test results.

focal hemispheric lesions in acute/subacute poststroke and reassessed them in the chronic stage. The authors reported an absence of a simple relation between extent of lesion, severity of behavioral test deficits, and test deficit recovery. However, patients who had deficits in both sound recognition and localization in the subacute or chronic poststroke stage tended to have lasting deficits.

The UK Royal College of Physicians (2012) makes a broad recommendation for systematic assessment for such sensory-processing deficits, that may well be justified in view of the studies discussed in the previous paragraph. At present identification of cases who require assessment relies on high clinical acumen. Everyday life communication is highly demanding and complex, and the patient reporting auditory difficulties in the chronic stage after stroke may benefit from a more detailed assessment of auditory and related functions to inform the rehabilitation plan. Language-based interventions may promote language recovery after stroke (Cicerone et al., 2011), while the role of non-linguistic-based auditory training after stroke is at present less clear. There are however reports that simply listening to music or audiotapes (Särkämö et al., 2010) poststroke may lead to broad functional improvements in communication, although auditory-training improvements may not always generalize (e.g., Slevc et al., 2011), but this is still an unexplored area of research.

OTHER AUDITORY PHENOMENA: TINNITUS, AUDITORY HALLUCINATIONS, HYPERACUSIS, AND PALINACOUSIS

Tinnitus (see Chapter 23) and music hallucinations (see Chapter 24) are "false" auditory percepts, in the sense that they are elicited without any external auditory stimuli, that are associated with abnormal "amplified" activity in brain networks that underpin sound and music perception respectively (Griffiths et al., 2010). Both these phenomena have been reported after stroke, and may present together with hyperacusis (i.e., the experience of comfortable-level sounds as being intolerably loud) and palinacousis (i.e., the experience of perseveration of the hearing sensation after the external auditory stimulus has stopped: see Chapter 25).

Tinnitus may well be the most frequent auditory symptom that is spontaneously reported by stroke patients (Häusler and Levine, 2000) that seems to be reported in association with subcortical lesions. Tinnitus with or without hearing loss is more prevalent after AICA than PICA infarction (Lee et al., 2003; Lee, 2008). There is a single case study reporting that gabapentin ameliorated poststroke bilateral tinnitus presenting a few days after a right basal ganglion hemorrhagic stroke (in the presence of an old right temporal infarct) (Chen and Yin, 2012). There are also occasional case reports of stroke in the basal

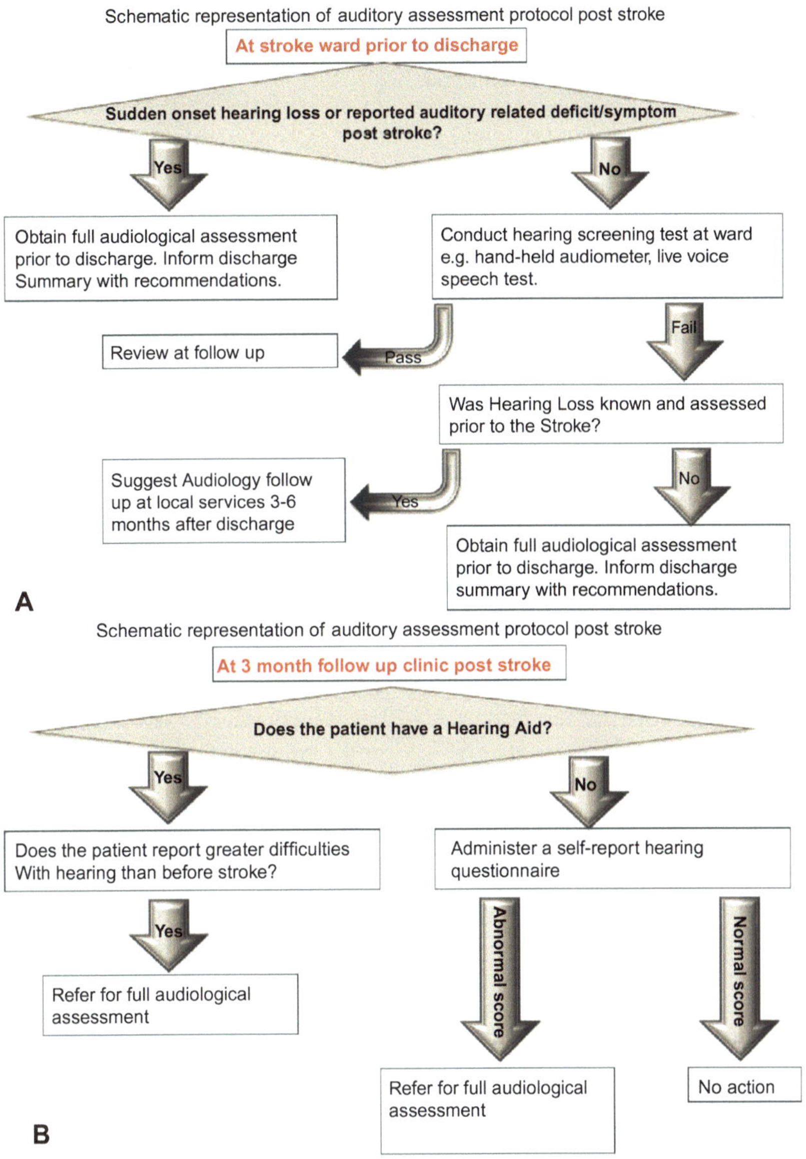

Fig. 35.6. Schematic representation of auditory assessment protocol poststroke.

ganglia territory leading to improvement of the perception of tinnitus that pre-existed the stroke. Lowry et al. (2004) reported a case of lifelong tinnitus that resolved completely following a left corona radiata vascular event, with no change in hearing thresholds. Larson and Cheung (2013) similarly reported a case with bilateral tinnitus who sustained a perioperative focal vascular injury following deep-brain stimulation for Parkinson disease. The patient reported bilateral tinnitus suppression, with complete abolition of the left-ear tinnitus and substantial reduction of the right-ear tinnitus 18 months later.

Hyperacusis (with or without tinnitus) has also been linked with stroke of subcortical brainstem and midbrain structures. Sudden-onset hyperacusis with tinnitus in the ear contralateral to the stroke were the presenting symptoms in a patient with a right dorsal pons hemorrhage, who had entirely normal hearing tests, including audiogram, acoustic reflexes, and auditory brainstem evoked responses. Tactile stimulation of the auricle exacerbated both the tinnitus and the hyperacusis, and the authors postulated that the lesion may have affected the pathway between the superior olivary nuclei and lateral lemniscus (Lee et al., 2008).

Fukutake and Hattori (1998) reported a patient with a hemorrhagic infarct of the right posterior inferior thalamus, including the medial geniculate body. The presenting feature was hyperacusis, with the sound of the television being perceived as unpleasantly loud and to some extent distorted predominantly in the right ear, while this percept sustained for another 10 minutes after the sound source had been turned off (palinacousis). Severe hyperacusis with tinnitus has also been reported as an early symptom of migrainous infarction of cerebellum and the upper pons on both sides in a 25-year-old patient with no risk factors for stroke (Lee et al., 2003). Of interest, Jankelowitz and Colebatch (2004) recorded the acoustic startle response in 32 patients with stroke (of the pons, internal capsule, and middle cerebral artery) and found an increased startle response in 8 cases in total across the three subgroups. Sound-induced flexion in a paralyzed limb was noticeable in 3 cases; however only some of the pontine lesion cases were symptomatic for the increased startle.

Auditory hallucinations may be simple (i.e., different types of noise or environmental sounds) or complex (i.e., involving music, voice, or spoken words). Both types of auditory hallucination have been described in single case studies or in case series of patients with hemorrhagic or ischemic infarction of the brainstem (Cascino and Adams, 1986; Cambier et al., 1987; Galtrey et al., 2012). These are usually bilateral, associated in the majority of cases with hearing loss with central-type features while they may coexist with other types of auditory deficit, such as localization of sound deficits. However, in a big prospective study of 521 stroke patients with no psychiatric disorders, only 4 patients reported auditory hallucinations and all 4 had an ischemic lesion of the right temporal lobe (Lampl et al., 2005). Their hallucinations consisted of hearing themselves speaking in their own voice or hearing familiar persons speaking to the patients in a well-constructed manner, and some of these events were perceived as having an unpleasant, bizarre, or even threatening content by the patients. This percept was bilateral in 3 and unilateral in 1 case only. All 4 cases had a stroke of the right hemisphere and the areas affected by the stroke included Brodmann area 41, 42, the postauditory area Brodmann 22, the middle upper region of the right temporal lobe, the right insula, and right superior temporal gyrus. All these patients had normal hearing, while 1 case had a subsequent stroke of the left temporal lobe a year later and developed central deafness. Proposed mechanisms for these illusory auditory percepts may be a disruption of the descending olivocochlear or corticofugal auditory pathways (Galtrey et al., 2012) or corticocortical connections (Lampl et al., 2005) by the stroke, leading to deafferentation, or the presence of lesion within the network for the perception/imagery of sound may lower the threshold for spontaneous activity (Calabrò et al., 2012).

CONCLUSIONS

Auditory symptoms and deficits after stroke: comments on current practice

Auditory deficits after stroke have not been as extensively investigated as visual deficits, and unsurprisingly, recommendations for auditory function assessments in current clinical guidelines for stroke tend to be rudimentary. For example, the UK Royal College of Physicians (2012) national clinical guidelines for stroke recommend that stroke patients should be assessed shortly after admission for their "ability to hear and need of hearing aids," while a broad recommendation is made that "any person who appears to have perceptual difficulties should have a formal perceptual assessment . . . followed by intervention."

It is not known to what extent recommendations for hearing assessment are adhered to for every patient in everyday clinical practice. Anecdotal experience indicates that this does not happen routinely in a systematic manner. Edwards et al. (2006) screened hearing in 53 patients with acute stroke who were admitted to an academic stroke service, and compared the screening test findings with the charted impairments. Twenty-two patients (41%) failed the bedside sound repetition hearing screening test:hearing impairment had only been charted for 3 (6%) patients before the screening, indicating that 81% of those who failed the hearing test would have remained unidentified without screening. There are also some indications that hearing impairment may well affect patient outcome. Landi et al. (2006) assessed physical functioning in 355 subjects with stroke (mean age of 77.4 years) approximately 10 months after their discharge to the community. The presence of hearing impairment, as rated by the patient and with the benefit of the patient's hearing appliance if used on the Minimum Data Set for Home Care instrument, was one out of four independent negative predictors for physical decline, with an odds risk of 1.83 (1.02–3.83). Landi et al. speculated that hearing loss may restrict the patient's participation in rehabilitation programs, and may thus lead to a lower level of physical performance.

The previous studies provide some empiric evidence to indicate that stroke patients should be systematically assessed for hearing loss that may otherwise be missed, as the presence of hearing loss may affect the patient's poststroke physical outcome. At present, there is little, if any, such information on the impact of other types of sound perception deficits poststroke on patient functioning and quality of life, while the prevalence of such deficits remains unknown, since there are even fewer studies that systematically assess such deficits in stroke sufferers than studies of hearing loss. However, when such deficits are present, they can be diverse (Polster and Rose, 1998), may go undetected unless specifically sought for (Blaettner et al., 1989), and may impact on patient communication in everyday life in the chronic stage of stroke, as reported by patients (Bamiou et al., 2012).

Assessment of auditory function after stroke: what should we do and why?

Detailed testing of every single patient may not be practical in a busy clinical setting; however screening the patient's hearing before the patient leaves the stroke ward with a short test and minimum set of questions and subsequently screening the patient's hearing needs by means of a questionnaire given at the early chronic stage poststroke, when deficits are likely to be permanent, may be a cost-effective bare-minimum assessment approach. This assessment may gather useful information for the rehabilitation team and the patient's environment for all patients and may help identify those patients who have high levels of deficits and disability in order to decide the need for additional investigation and input on a case-by-case basis (Fig. 35.6).

Subsequent detailed assessment of selected cases may include a detailed diagnostic interview and individually tailored testing of affected modules (Griffiths et al., 2010) within the audiology unit. The patient with significant auditory deficits and functional limitations may require a range of rehabilitation and remediation approaches. For example, hearing-aid amplification in a case of a patient who has both hearing loss and auditory neglect will not alleviate the neglect, since manipulation of the sound volume has no influence on the neglect (Williams and Coleman, 2008), but it may help optimize hearing on the unaffected side. In a patient with hearing loss and a language disorder, hearing aids will provide better access to the sound stimuli used for language-based interventions, which may promote language recovery after stroke (Cicerone et al., 2011), particularly when the language-related training activities have a high cognitive and linguistic load assist (Hatfield et al., 2005).

The role of non-linguistic-based auditory training after stroke is less clear. There are reports that simply listening to music or audiotapes (Särkämö et al., 2010) poststroke may lead to broad functional improvements in communication, but improvements may not generalize to untrained auditory stimuli, while temporal processing training may improve the temporal deficits, but not word deafness (Slevc et al., 2011). Other remedial strategies, such as the application of personal frequency modulation systems in stroke sufferers who report significant speech-in-noise difficulties after stroke despite the presence of reasonably preserved hearing thresholds, are unexplored, but our own pilot data indicate that such interventions may hold promise.

References

Anton G (1899). Über die Selbstwahrnehmung der Herderkrankungen des Gehirns durch den Kranken bei Rindenblindheit und Rindentaubheit. Archiv für Psychiatrie und Nervenkrankheiten (Berlin) 32: 86–127.

Bahls FH, Chatrian GE, Mesher RA et al. (1988). A case of persistent cortical deafness: clinical, neurophysiologic, and neuropathologic observations. Neurology 38: 1490–1493.

Bamiou DE, Musiek FE, Stow I et al. (2006). Auditory temporal processing deficits in patients with insular stroke. Neurology 67: 614–619.

Bamiou DE, Werring D, Cox K et al. (2012). Auditory processing deficits and disabilities in patients with stroke of the central auditory pathway. Stroke 43: 1285–1289.

Blaettner U, Scherg M, von Cramon D (1989). Diagnosis of unilateral telencephalic hearing disorders. Evaluation of a simple psychoacoustic pattern discrimination test. Brain 112: 177–195.

Calabrò RS, Baglieri A, Ferlazzo E et al. (2012). Neurofunctional assessment in a stroke patient with musical hallucinations. Neurocase 18: 514–520.

Cambier J, Decroix JP, Masson C (1987). Auditory hallucinations in lesions of the brain stem. Rev Neurol (Paris) 143: 255–262.

Cascino GD, Adams RD (1986). Brainstem auditory hallucinosis. Neurology 36: 1042–1047.

Cavinato M, Rigon J, Volpato C et al. (2012). Preservation of auditory P300-like potentials in cortical deafness. PLoS One 7 (1): e29909.

Cerrato P, Lentini A, Baima C et al. (2005). Hypogeusia and hearing loss in a patient with an inferior collicular infarction. Neurology 65 (11): 1840–1841.

Chen WH, Yin HL (2012). Successful amelioration of tinnitus in a stroke patient by low-dose gabapentin. J Stroke Cerebrovasc Dis 21: 907.

Chian CI, Chou CH, Hsueh CJ et al. (2013). Acute bilateral hearing loss as a "worsening sign" in a patient with critical basilar artery stenosis. J Clin Neurosci 20: 177–179.

Cicerone KD, Langenbahn DM, Braden C et al. (2011). Evidence-based cognitive rehabilitation: updated review of the literature from 2003 through 2008. Arch Phys Med Rehabil 92: 519–530.

Cohen M, Luxon L, Rudge P (1996). Auditory deficits and hearing loss associated with focal brainstem haemorrhage. Scand Audiol 25 (2): 133–141.

Deb P, Sharma S, Hassan KM (2010). Pathophysiologic mechanisms of acute ischemic stroke: an overview with emphasis on therapeutic significance beyond thrombolysis. Pathophysiology 17 (3): 197–218.

Dumas G, Schmerber S (2004). Cavernous haemangiomas: hearing and vestibular inaugural symptoms. Ann Otolaryngol Chir Cervicofac 121: 272–281.

Edwards DF, Hahn MG, Baum CM et al. (2006). Screening patients with stroke for rehabilitation needs: validation of the post-stroke rehabilitation guidelines. Neurorehabil Neural Repair 20: 42–48.

Engelien A, Huber W, Silbersweig D et al. (2000). The neural correlates of 'deaf-hearing' in man: conscious sensory awareness enabled by attentional modulation. Brain 123: 532–545.

Formby C, Phillips DE, Thomas RG (1987). Hearing loss amongst stroke patients. Ear Hear 8: 326–332.

Fukutake T, Hattori T (1998). Auditory illusions caused by a small lesion in the right medial geniculate body. Neurology 51: 1469–1471.

Galtrey CM, MacKinnon AD, Pereira AC (2012). Localization of bilateral auditory hallucinations and correlation to imaging in posterior circulation stroke. Neurologist 18: 418–422.

Garde MM, Cowey A (2000). "Deaf hearing": unacknowledged detection of auditory stimuli in a patient with cerebral deafness. Cortex 36: 71–80.

Gopinath B, Schneider J, Rochtchina E et al. (2009). Association between age-related hearing loss and stroke in an older population. Stroke 40: 1496–1498.

Goyal MK, Kumar G, Sahota PK (2010). Reversible sensorineural hearing loss with normal brainstem auditory evoked potentials in pontine hemorrhage due to capillary telangiectasia. Clin Neurosci 17: 1198–1201.

Griffiths TD, Bamiou DE, Warren JD (2010). Disorders of the auditory brain. In: A Palmer, A Rees (Eds.), OUP Handbook of Auditory Science, Vol. 2. The Auditory Brain, Oxford University Press, Oxford, pp. 507–540.

Hatfield B, Millet D, Coles J et al. (2005). Characterizing speech and language pathology outcomes in stroke rehabilitation. Arch Phys Med Rehabil 86 (12 Suppl 2): S61–S72.

Häusler R, Levine RA (2000). Auditory dysfunction in stroke. Acta Otolaryngol 120: 689–703.

Heffner HE, Heffner RS (1986). Hearing loss in Japanese macaques following bilateral auditory cortex lesions. J Neurophysiol 55 (2): 256–271.

Jankelowitz SK, Colebatch JG (2004). The acoustic startle reflex in ischemic stroke. Neurology 62: 114–116.

Kiely KM, Gopinath B, Mitchell P et al. (2012). Evaluating a dichotomized measure of self-reported hearing loss against gold standard audiometry: prevalence estimates and age bias in a pooled national data set. J Aging Health 24 (3): 439–458.

Kim JS, Cho KH, Lee HJ (2009). Isolated labyrinthine infarction as a harbinger of anterior inferior cerebellar artery territory infarction with normal diffusion-weighted brain MRI. Neurol Sci 278: 82–84.

Lampl Y, Lorberboym M, Gilad R et al. (2005). Auditory hallucinations in acute stroke. Behav Neurol 16: 211–216.

Landi F, Onder G, Cesari M et al. (2006). The Silvernet-HC study group. Functional decline in frail community-dwelling stroke patients. Eur J Neurol 13: 17–23.

Larson PS, Cheung SW (2013). A stroke of silence: tinnitus suppression following placement of a deep brain stimulation electrode with infarction in area LC. J Neurosurg 118: 192–194.

Lee H (2008). Sudden deafness related to posterior circulation infarction in the territory of the non-anterior inferior cerebellar artery: frequency, origin and vascular topographical pattern. Eur Neurol 59: 302–306.

Lee H (2009). Neuro-otological aspects of cerebellar stroke syndrome. J Clin Neurol 5: 65–73.

Lee H, Whitman GT, Lim JG et al. (2001). Bilateral sudden deafness as a prodrome of anterior inferior cerebellar artery infarction. Arch Neurol 58: 1287–1289.

Lee H, Sohn SI, Jung DK et al. (2002). Sudden deafness and anterior inferior cerebellar artery infarction. Stroke 33 (12): 2807–2812.

Lee H, Whitman GT, Lim JG et al. (2003). Hearing symptoms in migrainous infarction. Arch Neurol 60: 113–116.

Lee E, Sohn HY, Kwon M et al. (2008). Contralateral hyperacusis in unilateral pontine hemorrhage. Neurology 70: 2413–2415.

Leussink V, Andermann P, Reiners K et al. (2005). Sudden deafness from stroke. Neurology 64: 1817–1818.

Lowry LD, Eisenman LM, Saunders JC (2004). An absence of tinnitus. Otol Neurotol 25: 474–478.

MacDonald BK, Cockerell OC, Sander JW et al. (2000). The incidence and lifetime prevalence of neurological disorders in a prospective community-based study in the UK. Brain 123: 665–676.

Matsuda Y, Inagawa T, Amano T (1993). A case of tinnitus and hearing loss after cerebellar hemorrhage. Stroke 24 (6): 906–908.

Mazzoni A (1972). Internal auditory artery supply to the petrous bone. Ann Otol Rhinol Laryngol 81: 13–21.

Mendez MF, Geehan Jr GR (1988). Cortical auditory disorders: clinical and psychoacoustic features. J Neurol Neurosurg Psychiatry 51: 1–9.

Murray CJ, Lopez AD (1996). Evidence-based health policy – lessons from the Global Burden of Disease Study. Science 274: 740–743.

Musiek FE, Lee WW (1998). Neuroanatomical correlates to central deafness. Scand Audiol Suppl 49: 18–25.

Musiek FE, Baran JA, Shinn JB et al. (2007). Central deafness: an audiological case study. Int J Audiol 46: 433–441.

Nakane H, Sugimori H, Wakugawa Y et al. (2006). A case of hearing loss and quadriplegia after a pontine hemorrhage. J Neurol Sci 241 (1–2): 91–94.

Nishioka H, Takeda Y, Koba T et al. (1993). A case of cortical deafness with bilateral putaminal hemorrhage. Neurolog Surg (Japan) 21: 269–272.

O'Halloran R, Worrall LE, Hickson L (2009). The number of patients with communication related impairments in acute hospital stroke units. Int J Speech Langu Pathol 11: 438–449.

Park JH, Kim H, Han HJ (2008). Recurrent audiovestibular disturbance initially mimicking Ménière's disease in a patient with anterior inferior cerebellar infarction. Neurol Sci 29: 359–362.

Polster MR, Rose SB (1998). Disorders of auditory processing: evidence for modularity in audition. Cortex 34: 47–65.

Przewoźny T, Gasecki D, Narozny W et al. (2008). Risk factors of sensorineural hearing loss in patients with ischemic stroke. Otol Neurotol 29: 745–750.

Rey B, Frischknecht R, Maeder P et al. (2007). Patterns of recovery following focal hemispheric lesions: relationship between lasting deficit and damage to specialized networks. Restor Neurol Neurosci 25: 285–294.

Royal College of Physicians (2012). National clinical guidelines for stroke, 4th edn. Available online at, http://www.rcplondon.ac.uk/resources/stroke-guidelines (retrieved 22nd April 2013).

Särkämö T, Pihko E, Laitinen S et al. (2010). Music and speech listening enhance the recovery of early sensory processing after stroke. J Cogn Neurosci 22: 2716–2727.

Semenza C, Cavinato M, Rigon J (2012). Persistent cortical deafness: a voxel-based morphometry and tractography study. Neuropsychology 26: 675–683.

Shimomura T, Suzuki T, Takahashi S (1990). A case of brain stem infarction with bilateral hearing impairment and tinnitus at the onset. Rinsho Shinkeigaku 30: 1104–1108.

Slevc LR, Martin RC, Hamilton AC et al. (2011). Speech perception, rapid temporal processing, and the left hemisphere: a case study of unilateral pure word deafness. Neuropsychologia 49 (2): 216–230.

Tabuchi S, Kadowaki M, Watanabe T (2007). Reversible cortical auditory dysfunction caused by cerebral vasospasm after ruptured aneurysmal subarachnoid hemorrhage and evaluated by perfusion magnetic resonance imaging. J Neurosurg 107 (1): 161–164.

Tanaka Y, Kamo T, Yoshida M et al. (1991). 'So-called' cortical deafness. Clinical, neurophysiological and radiological observations. Brain 114: 2385–2401.

Ulbricht D (2003). AICA infarction and hearing loss: is it peripheral or central? Stroke 34: 837.

Wende S, Nakayama N, Schwedtfeger P (1975). The internal auditory artery (embryology, anatomy, angiography, pathology). J Neurol 210: 21–31.

Wernicke C, Friedlander C (1883). Ein Fall von Taubheit in Folge von doppelseitiger Läsionen des Schlafenlappens. Fortschr Med 1: 177–185 (Reprinted, "A case of deafness as a result of bilateral lesions of the temporal lobe", in G.H. Eggert (Ed.), Wernicke's Work on Aphasia. 1977, pp. 164–172).

Williams JM, Coleman AR (2008). Sound stimulus qualities and severe auditory neglect. Neurocase 14: 223–230.

Yaguchi H, Yaguchi M, Nishiwaki C et al. (2000). A case of brain stem infarction with bilateral hearing loss. No To Shinkei 52: 249–252.

Yamakawa H, Hattori T, Tanigawara T et al. (2004). Intracanalicular aneurysm at the meatal loop of the distal anterior inferior cerebellar artery: a case report and review of the literature. Surg Neurol 61: 82–88.

Yamasoba T, Lin FR, Someya S et al. (2013). Current concepts in age-related hearing loss: epidemiology and mechanistic pathways. Hear Res 303: 30–38.

Zager EL, Shaver EG, Hurst RW et al. (2002). Distal anterior inferior cerebellar artery aneurysms. Report of four cases. J Neurosurg 97: 692–696.

Handbook of Clinical Neurology, Vol. 129 (3rd series)
The Human Auditory System
G.G. Celesia and G. Hickok, Editors

Chapter 36

Hearing disorders in multiple sclerosis

MIRIAM FURST[1]* AND ROBERT A. LEVINE[2]

[1]*School of Electrical Engineering, Tel Aviv University, Tel Aviv, Israel*

[2]*Department of Ear, Nose and Throat and Head and Neck Surgery, Tel Aviv Sourasky Medical Center, Tel Aviv, Israel*

INTRODUCTION

Multiple sclerosis (MS) is a chronic inflammatory disease characterized by inflammation and focal destruction of myelin in the central nervous system (CNS) and spinal cord. As an exacerbating and remitting immune-mediated disorder of interfascicular oligodendrocyte-produced myelin, MS can impair acutely and transiently any CNS neural system, including the auditory system. However, when MS patients are not in the midst of an acute exacerbation, difficulty with hearing is a rare complaint. Yet, there are several auditory tasks that MS patients often failed to perform normally.

There are two main objective measures for detecting the involvement of MS in the auditory pathway: (1) evoked potentials (EP) and (2) magnetic resonance imaging (MRI). Auditory evoked potentials (AEP) reflect the neuroelectric activity within the auditory pathway, from the auditory nerve to the cerebral cortex, in response to an acoustic stimulus or event. The most studied AEP are the brainstem auditory evoked potential (BAEP), the auditory middle-latency response (AMLR) and the cognitive potential (P300).

BAEP, which consists of seven waves generated by one or more structures along the auditory pathway, assesses the integrity of the auditory pathway from the auditory nerve to the brainstem (see Chapter 17). AMLR, consisting of a series of waves that appear after the BAEP, reflects the activation of several subcortical structures. P300 is an endogenous, event-related potential (ERP) and its generation involves skills such as attention, auditory discrimination, memory, and semantic perspective.

In recent years, MRI has become the most reliable test for diagnosing MS (Polman et al., 2005). MS lesions that overlap the brainstem auditory pathway are rarely detected. Only when an MRI protocol is optimized for detecting and localizing MS brainstem lesions in the auditory pathway is it possible to find correlations between hearing dysfunction and MS.

In this chapter, we will first describe the MRI protocol that was specifically developed for detecting and localizing brainstem lesions, followed by a review of the auditory tasks that MS patients performed abnormally and for which correlations were found between the site of lesion and auditory performance.

PROTOCOL FOR DETECTING LESIONS IN THE BRAINSTEM AUDITORY PATHWAY

A special protocol was developed in order to detect lesions in the brainstem auditory pathway (Levine et al., 1993b; Tadmor et al., 1994; Tenny, 1994). MRI scans are obtained with T2-weighted, spin-echo pulse sequences. The data were then collected from three orthogonal planes of the brainstem: axial (perpendicular to the long axis of the brainstem), coronal (parallel to the plane of the floor of the fourth ventricle), and sagittal (perpendicular to the plane of the floor of the fourth ventricle and in the plane of the long axis of the brainstem). Contiguous 5-mm sections are obtained using T2 spin-echo sequences. A region of abnormally high signal is considered a lesion if it is found in at least two of the three orthogonal planes. A digitized anatomic model of the human brainstem was constructed from a serially sectioned adult human brain (Levine et al., 1993b). It included edges of the brainstem, auditory nuclei, and auditory fiber tracts.

A special semiautomated non-linear matching algorithm was developed in order to estimate the location of the auditory pathway on each MRI section (Tadmor et al., 1994; Tenny, 1994). Figure 36.1 is an example from a representative subject. It is a set of nine scans upon

*Correspondence to: Miriam Furst, School of Electrical Engineering, Tel Aviv University, Tel Aviv 69978, Israel. Tel: 972-3-6408050, E-mail: mira@eng.tau.ac.il

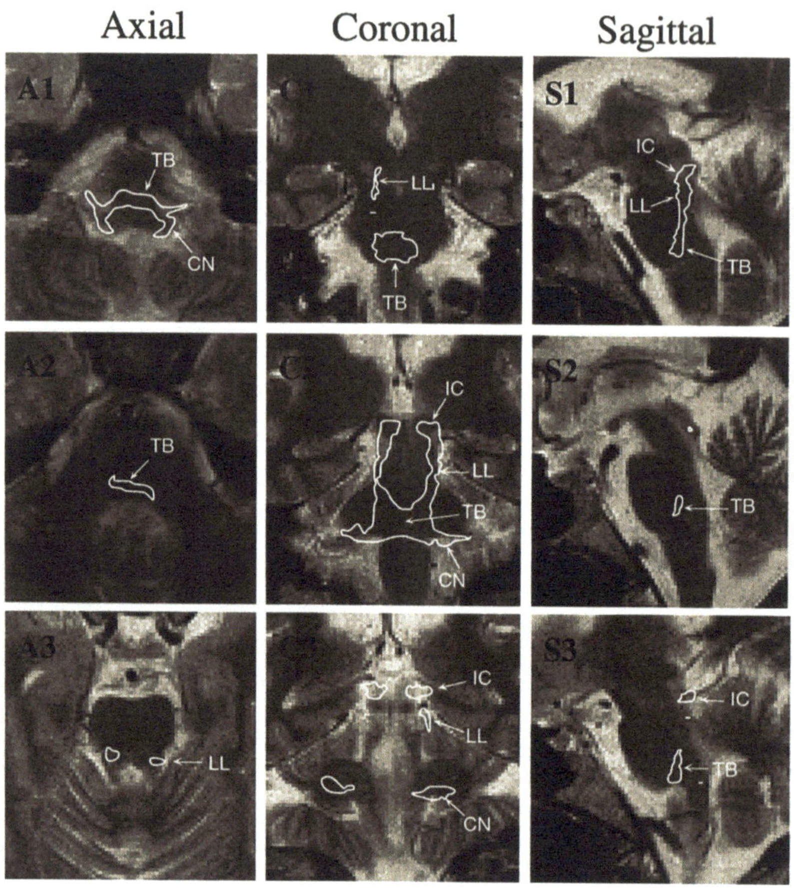

Fig. 36.1. Location of auditory pathway (white outlines) in nine sections of a typical subject magnetic resonance imaging (MRI) scan, as estimated by a semiautomated algorithm that relates the MRI scan to a computerized human brainstem anatomic atlas. First column: axial sections ordered from caudal to rostral. Middle column: coronal sections ordered from ventral to dorsal. Last column: sagittal sections ordered from left to right. CN, cochlear nucleus; TB, trapezoid body; LL, lateral lemniscus; IC, inferior colliculus. (Reproduced from Furst et al., 2000.)

which the estimate of the brainstem auditory pathway is superimposed (white outlines).

In the left column, three axial planes are shown; panels A1 and A2 represent caudal planes that include part of the trapezoid body (TB), while panel A3 represents a more rostral plane that includes part of the lateral lemnisci (LL). In the second column, coronal planes are shown from ventral (C1), where only part of the TB is included, to dorsal, where only small parts of the LL and TB are included (C3). The middle panel (C2), which is 5 mm apart from either of the other panels, includes most parts of the auditory pathway, the cochlear nucleus (CN), the TB, LL, and inferior colliculus (IC). The right column presents sagittal planes from right (S1) to left (S3) through the midline (S2).

Since the entire brainstem auditory pathway can be contained in one or two coronal planes, a schematic coronal representation of the auditory pathway (Richter et al., 1983) is used to plot the location of detected lesions, as can be seen in Figure 36.2. The detected lesions were superimposed on the schematic diagram of the brainstem auditory pathway. The schematic diagram in Figure 36.2 includes detected lesions that were found in different planes and are indicated by different patterns: vertical lines for sagittal planes, horizontal lines for axial planes, and polka dots for coronal planes.

PSYCHOPHYSICAL HEARING ASSESSMENT

In the following section we review hearing tests that are relevant to MS. It includes standard audiologic tests such as pure-tone threshold and speech intelligibility, and more specialized tests. The additional tests are binaural, since MS is a CNS disorder and the binaural tests must involve CNS processing of the outputs from each ear.

Pure-tone threshold

Pure-tone threshold are usually measured at the frequencies 0.25, 0.5 1.0, 2.0, 4.0, and 8.0 kHz at each ear. The

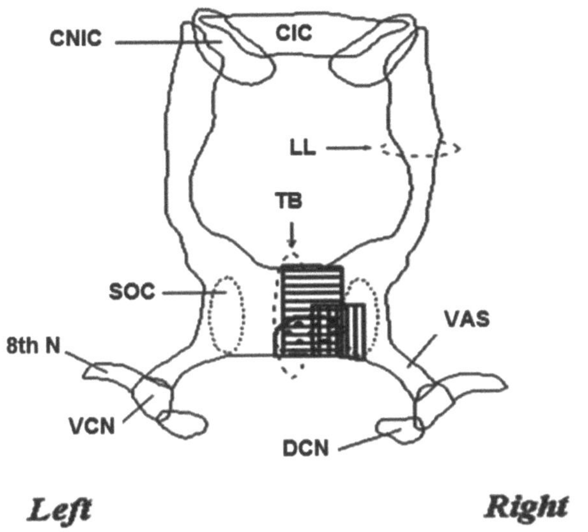

Fig. 36.2. Lesion of a patient with multiple sclerosis plotted on a coronal schematic of the brainstem auditory pathway. Regions of high signal are coded by different patterns corresponding to the magnetic resonance imaging plane on which it was detected: sagittal plane, vertical lines; axial plane, horizontal lines; coronal plane, polka dots. 8th N, auditory nerve; VCN, ventral cochlear nucleus; DCN, dorsal cochlear nucleus; VAS, ventral acoustic stria; SOC, superior olivary complex; CNIC, central nucleus of inferior colliculus; CIC, commissure of inferior colliculus; LL, lateral lemniscus; TB, trapezoid body. (Reproduced from Levine at al., 1993b.)

hearing level is determined according to the American National Standard Institute ANSI S3.6 (American National Standard Institute, 1996). In the literature there are several reports of MS-related hearing threshold deficits. These MS-related hearing losses occur when the MS involves the peripheral and brainstem auditory pathway (Jabbari et al., 1982; Daugherty et al., 1983; Shea and Brackmann, 1987; Franklin et al., 1989; Drulović et al., 1993; Bergamaschi et al., 1997; Oh et al., 2008). However, a well-controlled study of pure-tone thresholds concluded that they were not elevated in chronic MS. Moreover, even in MS patients, whose lesion overlapped the auditory pathway, the pure-tone thresholds were not significantly different from the thresholds of MS patients whose lesion did not overlap the auditory pathway (Doty et al., 2012).

Yet, there are cases where pure-tone threshold deficit is the only presenting symptom of MS, as in sudden hearing loss. Sudden unilateral hearing loss may result from etiologies affecting cochlea, eighth nerve, or more central auditory tracts. It usually appears early in the course of the disease but in most reported cases the prognosis was good, with ultimately little or no residual hearing deficit (e.g., Ozunlu et al., 1998; Oh et al., 2008; Hellmann et al., 2011).

Correlation between pure-tone threshold and BAEP in MS patients was widely studied (see review by Chiappa, 1980). The most common deficits that were observed in BAEP of MS patients are: (1) a prolonged interpeak interval between waves I and V; and (2) a reduced amplitude or absence of wave V. However, abnormality in BAEP does not necessarily indicate abnormal pure-tone threshold. There are many patients whose BAEP was abnormal, but their monaural pure-tone thresholds were all in the normal range (Jewett and Williston, 1971; Chiappa, 1980; Mustillo, 1984; Levine et al., 1993a).

Speech intelligibility

A common audiologic test is a speech reception threshold (SRT). The SRT is defined as the minimum level at which a subject can correctly identify 50% of two-syllable words. A similar test can be done at higher levels with monosyllabic, phonetically balanced words. It is referred to as a speech discrimination test and is scored by the percentage of words identified correctly. In general MS subjects perform normally in this task when the standard clinical level (70 dB above the SRT) is used. When the task is made more difficult, for example by using lower levels, then the performance of chronic MS subjects is significantly poorer than age-matched controls (Mustillo, 1984; Jerger et al., 1986; Levine et al., 1994).

Those difficulties indicate deficits in cognitive processing such as attention and auditory discrimination. They are closely related to abnormalities found in late AEP, especially latency delay of the P300 wave (Matas et al., 2010). It occurs at about 300 ms latency from the stimulus onset in young adults, with larger amplitude over the central-parietal scalp regions. The P300 is elicited by discrimination task, the oddball paradigm, which consists of a series of frequent and target stimuli, randomly administered in the proportion of 4:1 respectively. The subject's task is to evaluate the occurrence of the significant stimulus, the target one, engaging expectancy, attention, and memory during the performance.

Studies that compared MS patients with controls in such tests yielded a significant increase in P300 latencies. An example of such a difference was demonstrated by Magnano et al. (2006) and is shown in Figure 36.3. Average ERPs of two groups of subjects are shown – a group of normal subjects and a group of MS patients. The black arrow in Figure 36.3 indicates the latency of P300 in both normal subjects and MS patients. The prolonged latency of P300 in the MS patients is clearly demonstrated. Moreover, many studies indicate a significant correlation between auditory P300 latency and crucial demyelinating

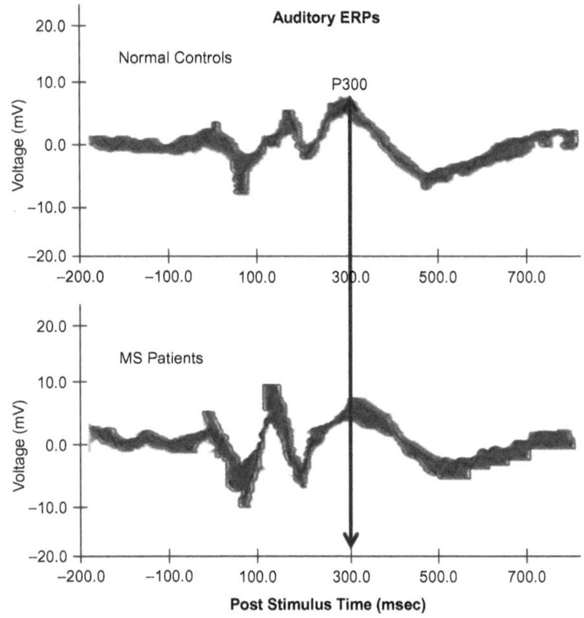

Fig. 36.3. Comparison of grand average event-related potential (ERP) responses in target detection from controls (top) and patients with multiple sclerosis (MS: bottom). The black arrow indicates the latency of P300. (Reproduced from Magnano et al., 2006.)

plaques seen in MRI on frontal horn and brainstem (Honig et al., 1992; Sailer et al., 2001; Magnano et al., 2006; Matas et al., 2010).

Binaural hearing

MS patients' abilities in binaural hearing are more vulnerable than can be detected in standard audiologic tests (pure-tone threshold and speech intelligibility), probably because MS is a CNS disorder and binaural hearing is by definition primarily a CNS function, since it integrates the sounds from both ears to make a judgment about the sound, such as where the sound is located. The auditory dysfunction in MS is probably caused by disruption of temporal processing, which is especially required when binaural integration is performed (Mustillo, 1984).

In binaural perception, both ears receive auditory stimuli and the brainstem nuclei compare the sound coming from each ear to make judgments, such as where the sound is coming from, i.e., right–left localization. For such judgments, decisions are made based upon either the differences in time parameters of the sound from the two ears (interaural time differences (ITD)) or the differences in loudness of the sound coming from the two ears (interaural level differences (ILD)). Another binaural process is involved in improving a subject's ability to detect sounds in a noisy background. Using earphones it is possible to assess separately a subject's ability to judge ITDs and ILDs. We will discuss in detail some of the most relevant tests.

Interaural discrimination tests

One way to assess binaural functioning is to determine the "just noticeable difference" (JND), which is the smallest ITD or ILD that a subject can detect reliably.

Usually the JND is estimated by using a two (or three)-interval, two-alternative, forced-choice paradigm. An example is as follows. Two successive presentations are made: (1) a reference stimulus which is diotic (both ears receive identical sounds), which is normally perceived in the midline; and (2) a test stimulus that is dichotic (the two ears receive different sounds), with either ITD or ILD, which is normally perceived to the side. In each trial, the order of the two intervals is randomly switched. The subject's task is to indicate in which direction (to the right or to the left) the two successive stimuli move relative to each other. The test interval is updated after every trial according to the subject's response. The task becomes easier when the subject's answer is wrong and more difficult if correct. The experimental session is terminated after usually six to eight "turnarounds," i.e., when the test stimulus has been changed from easy (increase ITD or ILD) to difficult (decrease ITD or ILD), or vice versa. Averaging the six to eight "turnarounds" is then defined as the JND of the parameter being tested.

Investigators have been using different types of stimulus in order to determine the time JND and level JND in MS patients. The most common stimuli are trains of clicks, noise bursts with different frequency content, or tones. The normal JND values were obtained from control normal-hearing subjects, where normal range is usually defined as the mean 3 S.D.

In Figure 36.4, four examples are shown in order to demonstrate the relationship between MS patients' performance in the time and level JND tests, their BAEPs to monaural stimulation, and the location of their lesion along the auditory pathway. In all the examples, the MS patients had lesions that involved the brainstem auditory pathway.

Figure 36.4A represent subject 572, who had a lesion in the right middle cerebellar peduncle which appeared lateral to the right LL. This patient had normal BAEPs when either ear was stimulated. The high- and low-pass level JNDs were both normal. The time JND for high-pass noise was clearly elevated, while the time JND for low-pass noise was just above the normal range.

Figure 36.4B shows subject 576, who had a pontine lesion that overlapped the left auditory pathway in the region of the rostral superior olivary complex (SOC) and TB and the caudal LL. BAEPs from right- and left-ear stimulation were similar with normal waves

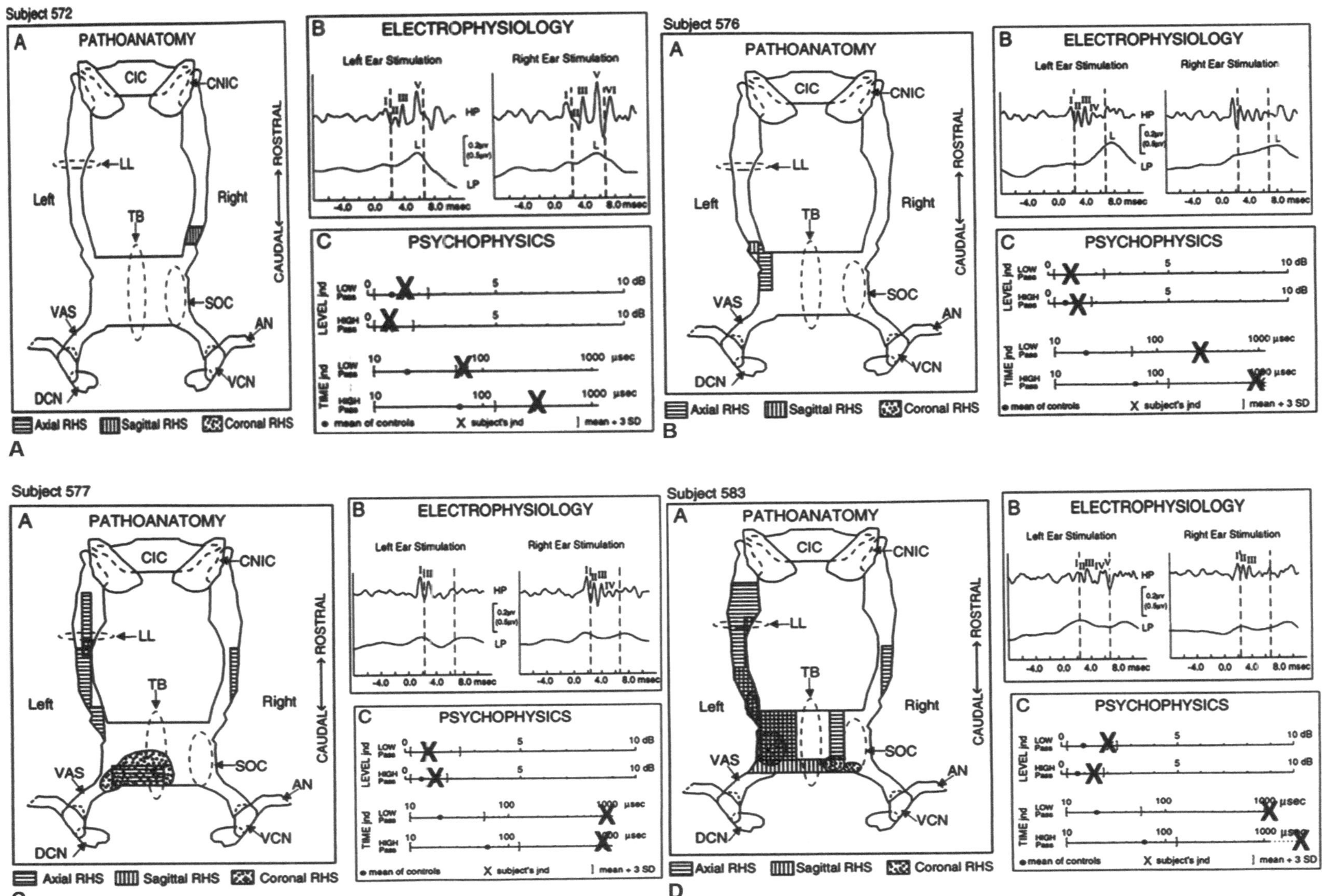

Fig. 36.4. See figure caption on next page.

I through IV, but wave V was clearly abnormal. Time JNDs were both clearly abnormal, with level JNDs both normal.

The third example (Fig. 36.4C) is from subject 577, who had a large lesion on the left involving the base of the pons. This lesion extended dorsally to involve the ventrocaudal parts of the left SOC and TB from nearly the level of the left pontomedullary junction where the ventral acoustic stria enters the TB, to just over the midline. The lesion did not extend throughout the entire region where the ventral acoustic stria meets the TB, but principally involved the ventromedial portion. A second left-sided lesion, detected in all three planes, was in the tegmentum. The BAEP of this subject were clearly abnormal. Waves I and II were normal on stimulation of the left ear, but waves III and V were absent. For right-ear stimulation only wave V was abnormal. The high- and low-pass time JNDs were highly abnormal for this subject. Both high- and low-pass level JNDs were within normal limits.

The last example (Fig. 36.4D) represents subject 583, who had a major lesion involving most of the left auditory pathway from the SOC to the rostral LL, apparently continuous in the rostral–caudal direction on both sagittal and axial scans. Two of these lesions encroached on the auditory pathway – one located in the SOC, the other in the caudal LL. Two other lesions were detected on the right: one overlapped the TB, and the other was detected lateral and dorsal to the right LL. The subject's BAEPs were abnormal. Waves I through III were normal when either ear was stimulated, but waves IV and V were absent when the right ear was stimulated. On the other hand, when the left ear was stimulated, wave V was abnormally low in amplitude. The I–V interpeak time was prolonged. The time JNDs for this subject were highly abnormal for both high- and low-pass noise bursts, while the level JNDs were both normal.

It is clear from those examples that the relationship between BAEPs, discrimination tests, and lesion location along the auditory pathway is not simple. Table 36.1 summarizes the performance of 38 MS subjects in several binaural tests and the relationship to their BAEP (Levine et al., 1994).

In studies that tested MS patients a huge variability in performance was obtained. There were subjects with performance in the normal range, others obtained significantly higher JND values, and there were those who could not perform the task at all. Among the different discrimination tests, time JND for noise bursts containing only frequencies above 4 kHz are the most commonly abnormal in MS patients (Table 36.1). Moreover, all subjects with abnormal time JND for low-pass noise bursts also had abnormal time JND for

Table 36.1

Behavioral and electrophysiologic interrelationship: percentage of abnormal performances

Psychophysics					Evoked potentials
	Interaural discrimination				
	Level difference		*Time difference*		
Masking-level difference	*Low frequency*	*High frequency*	*Low frequency*	*High frequency*	
10.5%	10.5%	7.9%	39.5%	76.3%	41.6%

Fig. 36.4—Cont'd. (**A**) Pathoanatomy, (**B**) electrophysiology, and (**C**) and psychophysics for subjects 572 (Fig. 36.4**A**), 576 (Fig. 36.4**B**), 577 (Fig. 36.4**C**), and 583 (Fig. 36.4**D**). The locations of the portions of regions of high signal (RHS) that involve the auditory pathway as determined by magnetic resonance scanning are shown at the top on a coronal projection drawing of a human brainstem. Brainstem auditory evoked potentials (BAEPs) obtained by right and left stimulations are shown with indication of waves I–V revealed by high-pass (HP) filtering of the BAEP and indication of wave L revealed from the low pass of the BAEP (LP). The subject's JNDs are shown on four horizontal scales, which are linear for the level JNDs and logarithmic for the time JNDs. The left bracket indicates the lowest possible JND score (0.25 dB for level JNDs, 10 ~LS for time JNDs), the filled circle indicates the mean score for normal controls, and the right bracket indicates 3 SD above the mean for normal controls. The subject's JND for each test is indicated by an X. If the subject's JND is within the two brackets it is considered normal. CIC, commissure of inferior colliculus; CNIC, central nucleus of inferior colliculus; LL, lateral lemniscus; TB, trapezoid body; VAS, ventral acoustic stria; SOC, superior olivary complex; AN, auditory nerve; DCN, dorsal cochlear nucleus; VCN, ventral cochlear nucleus. (Reproduced from Levine et al., 1993b.)

clicks and high-pass noise bursts (Hausler and Levine, 1980; Hausler et al., 1983; Levine et al., 1993a; Furst et al., 1995; Aharonson et al., 1998).

Most patients who performed poorly in the ITD discrimination test had abnormal BAEP in response to monaural clicks. In some cases there were no peaks after wave I. In other cases wave III was absent or its latency was prolonged (Hausler and Levine, 1980; van der Poel et al., 1988; Levine et al., 1993a). In particular, all MS subjects with abnormal evoked responses had abnormal time JND for high-pass signals, but not vice versa. On the other hand, as can be seen in Table 36.1, the time JNDs for low-pass signals and BAEPs were closely related. However, there were subjects that failed in the discrimination tests but whose BAEPs were in the normal range and vice versa.

Abnormal level JND is less frequent in MS patients; however, in studies that identified subjects with abnormal level JND, in almost all cases those subjects also had abnormal BAEPs (Furst et al., 1990; Picton et al., 1991; Levine et al., 1993a; Aharonson et al., 1998). For all other types of stimulus and for both time and level JNDs, there were MS subjects who performed abnormally, but no clear correlation was found between the psychoacoustic test and the BAEP.

Time JNDs were clearly affected by pontine lesions that involved the auditory system. Abnormal low-pass time JND was obtained by patients whose lesions were either unilateral and caudal to the TB, or bilateral and rostral to the TB (Levine et al., 1993b).

Binaural masking-level difference

Another typical binaural discrimination test is the comparison of the ability to detect signal in noise when listening with two ears rather than one. The detection of a signal in noise is improved when either the phase or level differences of the signal at the two ears are not the same as the masker. This phenomenon is called binaural masking-level difference (BMLD). Because the signal and masker do not have the same time or level relationships at the two ears, the signal and masker appear to originate from different locations in space; hence, the BMLD appears to be related to the well-known "cocktail party effect."

BMLD is typically measured with tones at low frequencies (250 or 500 Hz) and masking with gaussian white noise. The threshold for the binaural tone in the presence of continuous diotic (identical at the two earphones) noise is measured under two conditions: with the tone in phase at the two earphones or 180° out of phase at the two earphones. The degree to which the out-of-phase threshold is superior to the in-phase threshold is referred to as the BMLD. A typical normal BMLD is about 15 dB.

Studies that tested MS patients found variable numbers of patients who performed abnormally in the BMLD test. Usually in less than 40% of the patients the BMLD was smaller than in the control group (Noffsinger et al., 1972; Hannley et al., 1983; Matathias et al., 1985; Hendler et al., 1990; Levine et al., 1994). Moreover, Levine et al. (1994) found that, in MS patients, BMLD performance is highly correlated with discrimination of ITDs at high frequencies.

The correlation between BAEP and BMLD in MS patients is not very clear (Table 36.1), since there were patients whose BAEP was abnormal, with either a decrease in peak amplitude or latency increase, but who performed normally in the BMLD test. On the other hand, some MS patients who performed abnormally in the BMLD task had normal BAEP.

Lateralization experiments

Another type of experiment that tests the ability of binaural hearing is the lateralization experiment. In this experiment, subjects are asked to indicate where in their head they perceive the location of a binaural stimulus. In a matching experiment, subjects are presented with a pair of successive stimuli. The first stimulus serves as a reference, and it contains a dichotic stimulus with ITD, and the second interval includes a similar dichotic stimulus, but with ILD. The subject's task is to match the perceived position of the second stimulus to the reference by changing ILD. Among the MS patients that were tested in this experiment, some could not perform the experiment (Furst et al., 1990; Furst and Algom, 1995). Their responses were inconsistent, since they appeared to perceive the same binaural clicks in different positions on repeated trials.

A more direct procedure that tests the lateralization ability is to ask subjects to indicate where they perceive the dichotic stimulus in their head while listening to the sounds through earphones. Subjects can use a keyboard or a touch screen with equally spaced buttons, where the center button corresponds to the middle of the head. For example, in a keyboard of nine numbers, number 5 corresponds to the center of the head, numbers 1–4 correspond to the positions from the left ear toward the center, and numbers 6–9 correspond to the positions from the center toward the right ear. During the experimental session subjects are presented with dichotic stimuli with a variety of ITDs and ILDs.

A typical normal performance obtained by a normal-hearing subject while using click stimuli is presented in Figure 36.5 (Furst et al., 1995; Aharonson et al., 1998). The left panel represents the subject's performance for click stimuli with ITDs and the right panel shows the response to click stimuli with ILDs. In both panels, the y-axis represents the nine positions defined by the test procedure, and the x-axis the interaural difference. When the stimulus to the right ear was earlier than to the left ear, ITD is indicated as a positive number ranged from 0 to 1 ms. On the other hand, when the stimulus to the left ear was earlier, ITD is indicated as a negative number, ranged from −1 to 0 ms. Similarly, when the stimulus to the right ear was louder than to the left ear, ILD is indicated by a positive number ranged from 0 to 10 dB. A negative ILD is indicated when the stimulus to the left ear was louder than to the right ear.

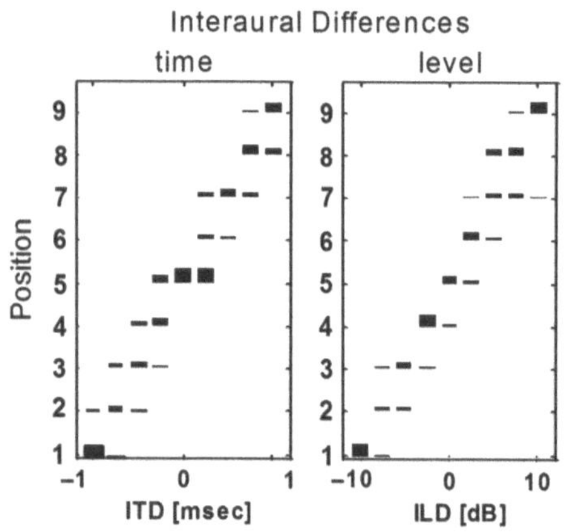

Fig. 36.5. Performance of a normal subject in the lateralization experiment for click stimuli. The histograms were obtained by counting the number of times the subject reported perceiving a certain position in response to dichotic stimuli with different interaural time differences (ITDs) and interaural level differences (ILDs). ITDs ranged from 0 to 1 ms and ILDs from 0 to 10 dB. Negative times refer to the left ear earlier. Negative levels refer to the left ear louder. (Reproduced from Aharonson et al., 1998.)

Each bin in the histograms in Figure 36.5 is equal to the relative number of times the subject indicated a certain position for a given interaural difference ITD or ILD. All the different positions were perceived as a function of ITD and ILD in about equal probability. Such a lateralization experiment can be obtained with different types of stimulus, such as click, tone bursts, and narrowband noise bursts.

Two types of abnormal performance in the lateralization test were found among MS patients: a center-oriented and a side-oriented abnormaly, as can be seen in Figure 36.6 (Furst and Algom, 1995; Furst et al., 1995, 2000; Aharonson et al., 1998). A center-oriented abnormal lateralization was obtained by those subjects who were biased toward the center positions and tended to ignore the side positions. They perceived all the dichotic stimuli in the center of their head. A side-oriented abnormal lateralization was obtained by those subjects who were biased toward the side positions and tend to ignore the center positions. They perceived all the dichotic stimuli in the sides of their head and never at the center. They generally did not confuse between right and left, but appeared to arbitrarily choose right or left when a diotic stimulus was presented. Typical performances of a center- and a side-oriented performance are depicted in Figure 36.6.

A very clear correlation was found between the lateralization test and the site of the MS lesion (Furst et al., 1995, 2000). All subjects with lesions involving the brainstem auditory pathway were found to have abnormal

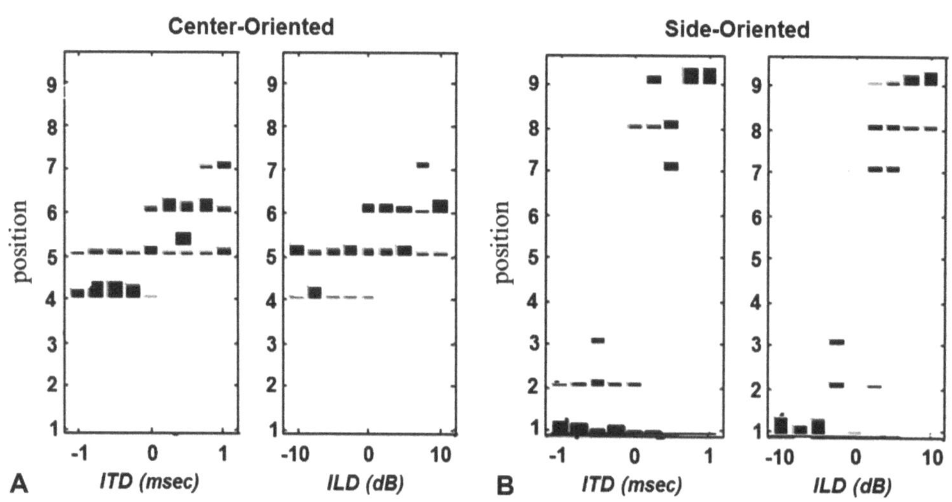

Fig. 36.6. Performances of two subjects with multiple sclerosis in the lateralization experiment. (**A**) Center-oriented abnormal performance and (**B**) side-oriented abnormal performance. The histograms were obtained by counting the number of times the subject reported perceiving a certain position in response to dichotic stimuli with different interaural time differences (ITDs) and interaural level differences (ILDs). (Reproduced from Aharonson et al., 1998.)

Table 36.2

Correlation between binaural performances and multiple sclerosis lesion site

Brainstem auditory pathway	Binaural psychoacoustic experiments				
	Lateralization		JND		
Lesion site	*ITD*	*ILD*	*ITD*	*ILD*	BMLD
None	Normal	Normal	LP: Normal HP: Normal/abnormal	Normal	Normal
Caudal to TB	Center-oriented	Normal	LP: Normal HP: Abnormal	Normal	Normal
Rostral to TB	Side-oriented	Side-oriented	LP: Normal HP: Abnormal	Normal	
Caudal + rostral of TB	Center-oriented	Center-oriented	LP: Abnormal HP: Abnormal	Normal/abnormal	Normal/abnormal

JND, just noticeable difference; BMLD, binaural masking-level difference; ITD, interaural time difference; ILD, interaural level difference; TB, trapezoid body; HP, high-pass (containing only frequencies above 4 kHz); LP, low-pass (containing only frequencies below 1 kHz).

lateralization with at least one stimulus (clicks or high-pass noise bursts), and all subjects with no detectable auditory pathway lesions had normal lateralization. Whenever a lesion was detected overlapping the auditory pathway, abnormal performance occurred. Center-oriented performance was principally associated with caudal pontine lesions and side-oriented performance with lesions rostral to the TB.

In particular, when MS lesions were restricted to the caudal pons, only the time lateralization experiment was center-oriented; level was normal. When the MS lesion was rostral to the TB then lateralization was side-oriented for both ITD and ILD. For lesions involving both the caudal pons and more rostral auditory structures, lateralization testing was a center orientation for both time and level.

Table 36.2 summarizes the correlation between MS patients' performances in the binaural psychoacoustic experiments and the site of their lesion along the brainstem auditory pathway. The MS lesions were divided into four categories: (1) no lesions overlapping the brainstem auditory pathway; (2) lesions were detected caudal to the TB; (3) lesions were detected rostral to the TB; and (4) lesions were detected both caudal and rostral to the TB. The psychoacoustic binaural experiments were divided into three categories: (1) lateralization experiments; (2) JND of interaural differences (time and level); and (3) BMLD.

It is clear from Table 36.2 that high-pass time JND was abnormal in many of the MS patients; however, it was not restricted to a specific lesion along the auditory pathway. In fact many times it was abnormal despite no detectable brainstem lesion and no other psychophysical or electrophysiologic abnormality (Table 36.1). Furthermore, whenever any other psychophysical or electrophysiologic test was abnormal, the high-pass time JND was always abnormal. There were subjects with no pontine lesions who performed abnormally in this task. On the other hand, level JND is almost always normal in MS patients; it was abnormal only when MS plaques involved large parts of the brainstem auditory pathway. The most specific binaural test is the lateralization test. Abnormal performance is obtained only when MS lesions overlap the auditory pathway. Moreover the type of abnormality indicates the site of the MS lesions.

Correlation between BAEP and binaural performances

A more direct correlation between BAEP and binaural performance was found for the binaural difference (BD) potential. BD waveform is obtained by adding the two monaural BAEPs and subtracting the corresponding binaural BAEP (Dobie and Norton, 1980; Levine, 1981; Wrege and Starr, 1981). The first positive peak (β) of the BD waveform, which occurs just after wave V peak of the monaural BAEPs, was found to be the most significant parameter for the correlation between BAEP and psychoacoustic properties. Whenever the amplitude of β was well above the noise floor, the dichotic click was perceived as a laterally fused image, namely dichotic clicks with any value of ILD (±25 dB) or dichotic clicks with ITD smaller than 1 ms. Moreover, when ITD and ILD were such that the binaural click was perceived in the same position, the latencies of β obtained by both ITD and ILD were similar, and

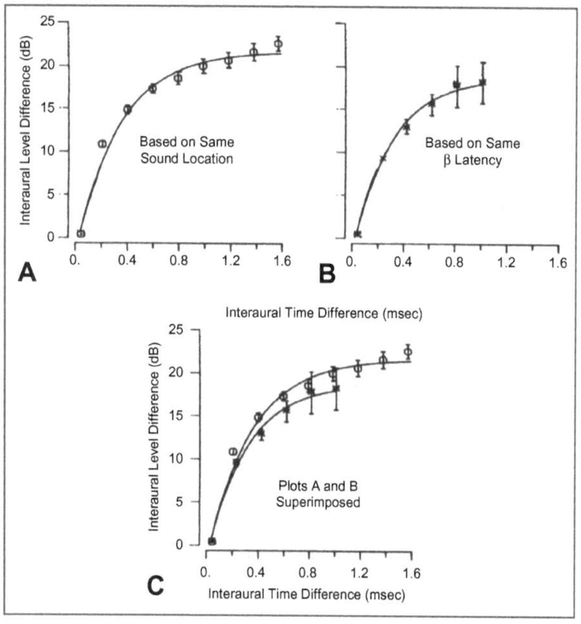

Fig. 36.7. Mean interaural time differences and interaural level differences matching curves (**A**) for psychophysical data based on sound lateralization and (**B**) for physiologic data based on latency. The error bars denote 1 SEM. The solid lines are the exponential matching curves. (**C**) The psychophysical and physiologic data are superimposed. (Reproduced from Furst et al., 1985.)

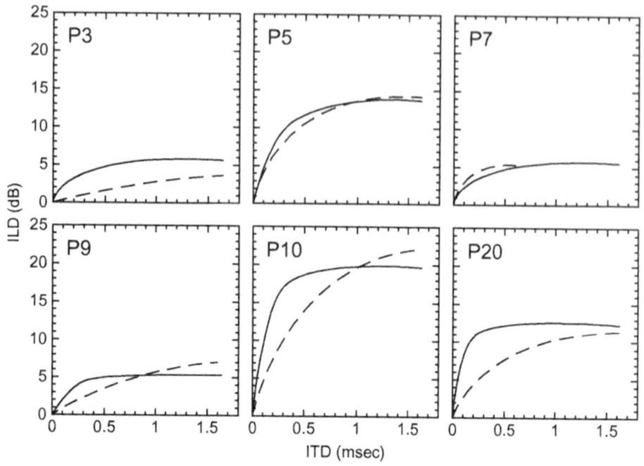

Fig. 36.8. Interaural time difference (ITD) and interaural level difference (ILD) exponential matching curves for psychophysical data based on sound lateralization (broken lines) and for physiologic data based on latency (solid lines) for 6 different patients with multiple sclerosis indicated by P3, P5, P7, P9, P10, and P20. (Reproduced from Furst et al., 1990.)

as a result the psychoacoustic matching curve for lateralization was predicted by the latency of β (Furst et al., 1985; Jones and Van der Poel, 1990). A demonstration of this phenomenon is presented in Figure 36.7. In Figure 36.7A, the mean performance of 7 normal subjects in the psychoacoustic data is shown. Each data point means that the perceived location of the sound was the same for the ITD and ILD corresponding to that point. For instance, the subjects' perceived sound location for an ITD of 0.4 ms was the same as their perceived sound location for an ILD of 15 dB. Figure 36.7B represents the physiologic matching curve that was derived from the latency of β. Each data point means that the β latency was the same for the corresponding ITD and ILD.

The solid lines in both figures (Fig. 36.7A and B) represent exponential curves that were best matched to the data points by means of minimum squared error, and expressed by the form $ILD = B(1 - e^{k \cdot ITD})$. In Figure 36.7C the two plots (Fig. 36.7A and B) are replotted together to demonstrate their similarity and to emphasize the conclusion that β latency is an adequate code for sound location.

BD waveforms were successfully obtained from many MS subjects whose monaural BAEPs were detectable (Furst et al., 1990; Furst and Algom, 1995; Pratt et al., 1998). In particular, those subjects who either performed normally in the lateralization experiment or were center-oriented in the time lateralization test had detectable BD waveforms. Subjects with center-oriented lateralization for ITD had abnormal matching curves; they reached a maximum ILD around 5 dB to match the different stimuli with ITD, whereas the average normal value of ILD that matches ITD stimuli is around 18 dB. Both normal and abnormal performances were predicted by the latency of β of the corresponding BD waveforms (Furst et al., 1990). Examples of such predictions are presented in Figure 36.8 for 6 different subjects. Each panel in Figure 36.8 represents a different MS subject, as indicated in the upper left of each panel. The solid lines were obtained by matching the β latency data of each subject to the exponential curve $ILD = B(1 - e^{kITD})$. Similarly, the broken lines were obtained from each subject's psychologic data.

THEORETIC EXPLANATIONS

Correlation between MS and audiologic tests

Classically MS has been considered to have its principal effect upon neural function by slowing conduction velocities across a focal lesion – the aftermath of acute inflammation. Of late, more and more evidence is emerging consistent with other ways that MS can chronically disrupt neural function; it has been referred to as a "neurodegenerative process." Currently, a widely accepted concept of MS is that it begins as an inflammatory process and then evolves into a neurodegenerative condition (McFarland, 2009, 2010). With the exception of high-pass time JND, most abnormal performances of MS patients can be understood by considering MS as a focal demyelinating disease.

The high-pass time JND results cannot be accounted for by focal demyelinating lesions only. Therefore it appears that high-pass time JND may be important as a marker for MS as a neurodegenerative process. Further study may prove that high-pass time JND is a valuable tool for monitoring the neurodegenerative aspect of MS.

Serial MRI scans have revealed that brains of MS patients lose volume at about 10 times the rate of healthy controls (Miller et al., 2002). Atrophy involves both cortical and deep gray matter and is more marked in cortex than white matter. Because the acute inflammation appears to be directed toward myelin, acute MS is thought to slow neural conduction through the lesion. When more severe, the process can block neural conduction altogether. Since slowing of conduction may not be uniform across the focal lesion, slowing of conduction will lead to dyssynchrony of the neural population. Hence neural processes strongly dependent upon synchrony will be more affected by an MS lesion than others.

Dyssynchrony is one theory for the difference in findings for pure-tone audiometry and speech intelligibility. Pure-tone audiometry is not dependent upon speed of neural conduction or complex neural interactions, but understanding speech, especially under adverse conditions, is dependent upon complex neural interactions. Slowing of conduction, particularly non-uniformly across a family of neurons, can be expected to disrupt such complex neural interactions.

Model for lateralization tasks

The binaural dysfunction in MS is probably caused by disruption of temporal processing. A multilevel decision-making model (Aharonson and Furst, 2001) of sound lateralization is presented in order to explain the frequent abnormal performance of MS patients whose lesions overlap the brainstem auditory pathway.

The model is presented schematically in Figure 36.9. The model assumes that each SOC receives inputs from both cochlear nuclei. In each SOC there are independent time and level comparators whose outputs then independently project to the higher center(s). The higher center(s) then uses the four possible inputs coming from the two SOCs to make a decision about the stimulus location in auditory space.

In the normal auditory system receiving sounds with little or no interaural differences (Fig. 36.9A), the outputs of the two SOCs for both time and level channels project to the higher center(s). In contrast, for dichotic stimuli with larger ITDs (1) only one time comparator (from the SOC contralateral to the leading ear) actively projects to the higher center(s) (Yin and Chan, 1990); but

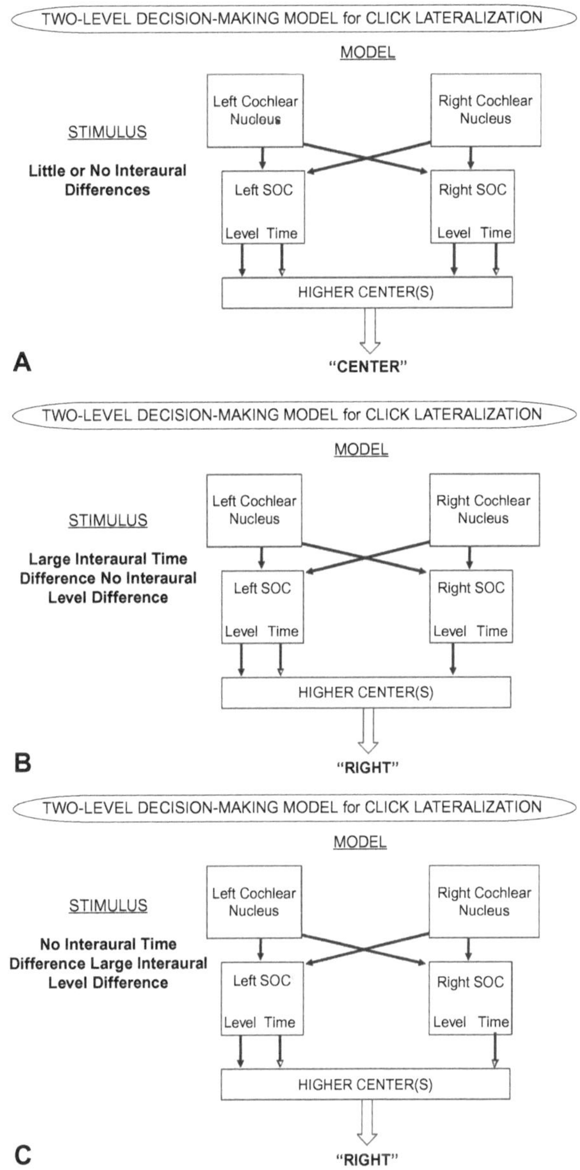

Fig. 36.9. Two-level decision-making model for sound lateralization in normal subjects. The first level is the right and left superior olivary complexes (SOC). The second level is the higher auditory center(s) rostral to the pons (inferior and superior colliculi, medial geniculate bodies, and the auditory cortex). The trapezoid body is represented by the crossing arrows connecting the cochlear nuclei to the opposite SOCs. Each lateral lemniscus connects the output of the SOC to the higher center(s) and is represented by two vertical arrows: one for time (open arrowhead) and one for level (solid arrowhead). (**A–C**) The model's active channels for three extremes of binaural stimulation. In quotation marks below the large open arrow is the correct response; it is "Right" for (**B**) and (**C**) because the right ear is receiving the earlier (**B**) or higher-level (**C**) stimulus. (Reproduced from Furst et al., 1995.)

(2) both level comparators actively project to the higher center(s), since there are no ILDs for these stimuli (Fig. 36.9B). Since latencies shorten as stimulus levels increase, for larger ILDs, only one time comparator (from the SOC contralateral to the ear receiving the more intense sound) actively projects to the higher center(s) (Fig. 36.9C). There is no level output from the level comparator of the SOC ipsilateral to the ear receiving the less intense sound.

In the case of an MS lesion involving the TB, the demyelination of fibers in the TB will critically disrupt the temporal relationships between the right and left inputs to both SOC time comparators. For any type of dichotic stimuli, time channel outputs will not project significant activity to higher center(s); however, the level comparators will function normally, since they are unaffected by slight changes in timing of neuronal discharges. Therefore lateralization for ILDs will be normal, while for dichotic stimuli with only ITDs, the higher center(s) will make a decision based upon the inputs from the SOC-level channels only. Since the level channels always indicate no interaural difference (i.e., center) for stimuli with only ITDs, this model then predicts that MS subjects with TB lesions will perceive stimuli with only ITDs as coming from the center of auditory space.

This model also accounts for the fact that MS subjects with unilateral lesions overlapping the LL perceive dichotic stimuli with little or no interaural differences as coming from the side rather than the center of auditory space. According to this model, when dichotic sounds have little or no interaural differences, the time and level channels of both SOCs actively project to the higher center(s) almost synchronously. If the inputs from the two SOCs to the higher center(s) are received nearly synchronously, this model assumes that the higher center(s) decides that the stimulus is coming from the center of auditory space (otherwise, from one side of auditory space). With a unilateral LL demyelinating lesion, the inputs of the two sides to the higher center(s) will be desynchronized; consequently, this model predicts that these stimuli will be perceived as coming from the side and not the center of auditory space.

Dichotic stimuli with large ILDs are normally projecting to the higher center(s) through only one LL for both the time and level channels. A desynchronizing lesion of one of the LLs will not affect the decision of the higher center(s), since it has no choice but to choose the one side with any kind of input from the SOC. Thus this model also accounts for correct lateralization in the case of dichotic stimuli with large ILDs.

When dichotic stimuli with large ITDs are introduced, both SOCs actively project synchronously through both level channels, while projecting through only one time channel. When the LL lesion is on the side with no actively projecting time channel, the level channel on that side will be desynchronized but the higher center(s) will decide correctly which part of auditory space the sound is coming from, since its decision will be based upon the normal time and level channels coming up the unaffected LL.

When the LL lesion is on the side with actively projecting time channel, the side of the higher center(s) with the normal LL receives a normal active level channel, but no time channel. On the side with the LL lesion, the higher center(s) receives both desynchronized time and level channels. This conflicting information is resolved by adding an additional decision process. This level acts as follows: if the higher center(s) receives any time channel input (even if desynchronized) from either side, then it ignores the inputs from both level channels and decides on the image's location only on the basis of the time channel.

Correlation between BAEP and binaural discrimination tasks

The correlation between monaural BAEP of MS subjects and their performance in the binaural discrimination tests is complex, as can be seen in Figure 36.4 and Table 36.1. From a detailed analysis, interrelationships can be found between psychophysical performance and the BAEP, which can be summarized by the following simplified model (Fig. 36.10), as was suggested by Levine et al. (1993a) based upon Jeffress' coincidence network (Jeffress, 1948).

A group of brainstem neurons (open circles, Fig. 36.10A) is innervated symmetrically by inputs from both ears through axons with cell bodies in the cochlear nuclei. Each brainstem neuron behaves as a probabilistic coincidence detector such that, when action potentials arrive nearly simultaneously from the two sides, their effects sum and greatly increase the probability of an output spike. These brainstem neurons are frequency-selective, each receiving inputs representing local regions in the two cochleas that have the same characteristic frequency (CF). High-CF neurons receive inputs from the base of the cochlea, whereas low-CF neurons receive them from the apex of the cochlea.

To a wideband transient sound, high-CF auditory nerve fibers tend to respond at the same time (group synchrony), but low-CF units tend to respond at different times depending upon their CFs (Kiang et al., 1965). This type of group synchrony from the high-CF neurons is presumably also responsible for the distinct waves in the BAEPs (Don and Eggermont, 1978) and we assume that it will be carried through to the brainstem auditory nucleus represented in Figure 36.10.

For a normal subject, the same sound delivered to both ears creates the perception of a midline-oriented

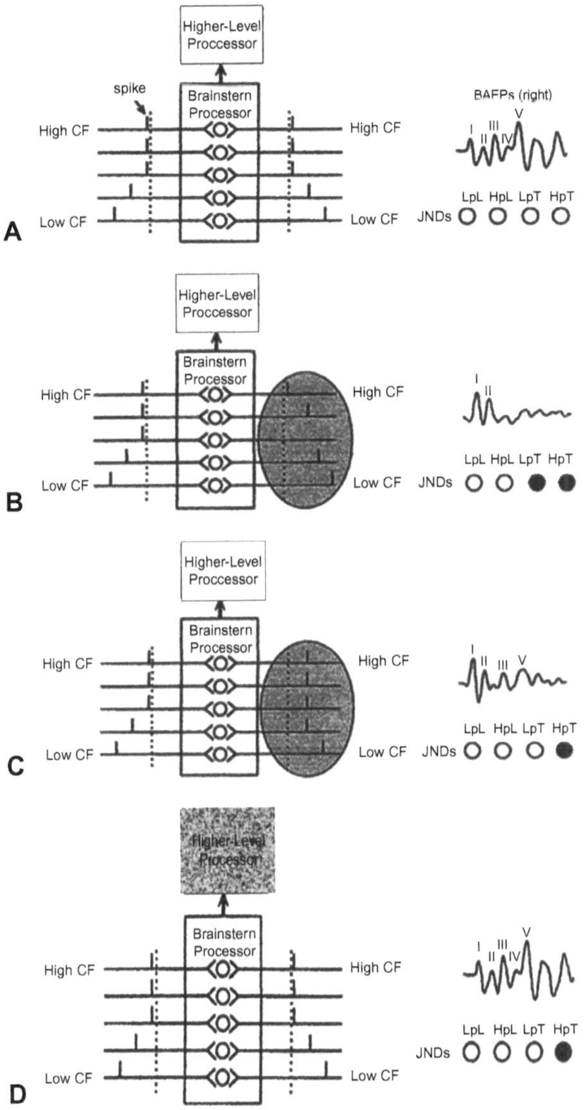

Fig. 36.10. Schematic depiction of a theoretic model to account for the brainstem auditory evoked potential (BAEP)and psychophysical results. (**A–D**) the brainstem processor (vertical rectangle) is seen as a collection of neurons (open circles) that receive symmetric inputs from corresponding regions of the cochlea through the cochlear nuclei. To the right of each model is a depiction of how the BAEPs (to clicks to the right ear) and the just noticeable differences (JNDs) (open circles, normal; filled circles, abnormal) might appear in each case. (**A**) Case of a normal subject. (**B**) Multiple sclerosis (MS) lesion, affecting the fibers from the right side. (**C**) MS lesion also alters some fiber conduction times, but uniformly, so that group synchrony of the high characteristic frequency (CF) fibers remains. (**D**) The brainstem auditory apparatus is intact, but MS lesions involve the higher-level processor (stippled area). Four types of JND values are indicated: interaural level difference of low-pass signal (LpL); interaural level difference of high-pass signal (HpL); interaural time difference of low-pass signal (LpT); interaural time difference of high-pass signal (HpT). (Reproduced from Levine et al., 1993a.)

sound source. Either amplitude or time differences in the same sound delivered to the two ears can produce the perception of a sound source off to the side of the midline and, thus, can be distinguishable from the symmetrically presented sound stimuli.

In Figure 36.10, the brainstem processor (vertical rectangle) is seen as a collection of neurons (open circles) that receive symmetric inputs from corresponding regions of the cochlea through the cochlear nuclei. The stimulus is a wideband acoustic transient presented identically at the two ears. For simplicity, only one spike per fiber is represented. The position of the spikes indicates relative times of arrival of single-unit discharges at the brainstem processor, with time of arrival increasing for spikes from nerve fibers with lower CF. Dashed vertical lines are symmetrically placed on either side of the brainstem processor to serve as reference. Each brainstem neuron is a binaural spike pattern comparator that sends its output to a higher-level processor (horizontal rectangle). To the right of each model is a depiction of how the BAEPs (to clicks to the left ear) and the JNDs (open circles, normal; filled circles, abnormal) might appear in each case.

Figure 36.10A is the case of a normal subject. In Figure 36.10B, an MS lesion, affecting the fibers from the left side (stippled oval), is seen as altering the conduction times of input neurons by varying amounts in different fibers; in one there is no spike due to conduction block. In this case many of the brainstem neurons would have a very different output than they do in Figure 36.10A. The functions of the higher-level processor that depend upon comparing the exact time of arrival of spikes to each brainstem neuron and integrating over the entire array of brainstem outputs would be disrupted, as reflected by the abnormal high- and low-pass time JNDs and center-oriented lateralization. The loss of group synchrony of the input fibers would also be reflected in the absence of recognizable waves following wave II of the BAEPs. This loss of group synchrony would also remove the type of synchrony that is essential for coherent waves in the BAEPs. Functions that require only comparing the spike counts coming from both sides within some time period might still be normal, as reflected in the normal-level JNDs and level lateralization.

In Figure 36.10C, an MS lesion also alters some fiber conduction times, but uniformly, so that group synchrony of the high-CF fibers remains. Thereby, the later BAEPs are recognizable but delayed. Only the high-pass time JND is abnormal.

In Figure 36.10D, the brainstem auditory apparatus is intact, but MS lesions involve the higher-level processor (stippled area). This situation could occur in MS subjects who performed abnormally on the high-pass time JNDs and had normal BAEPs.

If, as shown in Figure 36.10C, the MS lesion causes many fibers to have equivalent slowing of conduction, then the group synchrony would still be maintained and the BAEP waves would remain detectable, but the later BAEP waves would have a longer latency. In Figure 36.10B or C, or some combination of the two, disruption of timing from one side will disrupt the input to the brainstem neurons acting as coincidence detectors. Thus it is easy to see how the time JNDs and the BAEP amplitudes or latencies could be correlated.

High-CF neurons respond to low-frequency stimuli at higher stimulus levels, while low-CF neurons do not respond to high-frequency stimuli at any stimulus level (Kiang and Moxon, 1974). For low-pass stimuli, then, there can be cues for binaural processing not only in the response of low-CF neurons but also in the response of high-CF neurons, whereas for high-pass stimuli only high-CF neurons will respond. Furthermore, nerve fiber responses are time-locked to both the envelope and fine structure of low-frequency sounds, while they are time-locked to only the envelope of high-frequency sounds (Javel, 1980; Joris and Yin, 1992). Thus, it can be appreciated that low-pass time information, particularly at higher stimulus levels, is redundantly represented in the discharge patterns of high- as well as low-CF neurons. Consequently, low-pass time discriminations may be less vulnerable than high-pass time discriminations for MS lesions.

A third factor that is operative (whether or not a neuron is responding to the fine structure of the stimulus) may also be contributing to the robustness of the interaural discrimination for wideband low-frequency noise. This factor is related to the envelope of the correlation function and how the rate of decay of this envelope depends on the bandwidth of the effective stimulus. To understand the effects of abnormal delays, it is helpful to extend our model to include two dimensions: the CF dimension and a "characteristic-delay" dimension that spans a range of interaural delays of about 1 ms in each direction. Further, it is useful to think of the activity of these cells along the delay dimension as an estimate of the cross-correlation functions for filtered versions of the stimulus waveforms. The correlation function for a narrow-band-filtered noise waveform is an amplitude-modulated sinusoid in which the frequency of the fine structure is equal to the center frequency of the noise and the envelope is determined by the bandwidth of the filter. The activity pattern will shift when the stimulus ITD changes, and the process of discriminating time delays corresponds to distinguishing the corresponding internal patterns.

We now consider additional interaural delays imposed on the inputs to the neural model, as in Figure 36.10B or C, due to MS-induced disruption of timing. When these delays are small (<1 ms) they will be "naturally" accommodated for within the network by a simple shift of which characteristic delay neurons are active. On the other hand, large delays (relative to the characteristic delays within the model) cannot be compensated for by shifts in which cells are activated in the delay network; sensitivity to delay in this case depends on the fact that the waveform is similar to itself delayed according to its correlation function. The ability to represent time delays larger than 1 ms within the delay network depends on the envelope of the correlation function, which is determined by the filter shape, particularly by the bandwidth. Filter bandwidths for low-frequency noise are smaller than bandwidths for higher-frequency noise, so the larger delays could be better accommodated by low-frequency cells. Hence, this might explain the sensitivity to high-pass stimuli relative to low-pass stimuli in MS.

Figure 36.10D indicates that, even when the brainstem is free from lesions, psychophysical performance can be compromised by lesions at some higher level. When this occurs, the BAEPs would not need to be affected. Another possibility is that MS lesions involve the output neurons of the brainstem nuclei acting as coincidence detectors, in which case later components of the EPs could well be affected. The relative rarity of level discrimination abnormalities follows easily from the fact that spikes are not necessarily blocked by the MS lesions; only their conduction velocities are affected by demyelination. If level comparisons depend on counting spike numbers in comparable channels over some short integration time (e.g., by summed ipsilateral excitation and contralateral inhibition, such as in lateral superior olive neurons (Caird and Klinke, 1983; Yin and Chan, 1988), then the interaural time discriminator would be unaffected by small desynchronizations. However, for some lesions, conduction may actually be blocked so level discriminations can be abnormal.

An increased latency could occur from either a "group" conduction delay of the high-CF fibers or a desynchronization (or blockage) of only the high-CF fibers so that the fibers with lower CFs are generating the delayed wave (Don and Eggermont, 1978). In either case the brainstem binaural coincidence detector is receiving the low-frequency information through the lower-CF units with little change.

Consequently latency changes in wave III do not result in major low-pass time JND abnormalities, but could result in major disruption of high-pass time JND, since high-pass information is carried only in the high-CF fibers, which are either desynchronized or delayed.

A reduced amplitude (Fig. 36.10B) in a wave suggests a group desynchronization (or blockage) of both high- and low-CF fibers, Consequently, if wave III is reduced in amplitude then both high- and low-CF fibers would be supplying incorrect signals to the brainstem nucleus about both high-pass and low-pass noise. As a result, both high-pass and low-pass time JND would be abnormal.

CONCLUSION

Hearing disorders in MS are mainly manifested by poor performance on tests requiring precise neural timing. When MS plaques involve the brainstem auditory pathway, there are no silent lesions. Subjects always perform abnormally on tests highly dependent upon temporal processing.

Routine audiometry rarely detects abnormalities in MS patients even when their lesions overlap the auditory pathway, whether in pure-tone thresholds or in standard speech intelligibility tasks. Only in speech intelligibility tasks that have been modified to increase their difficulty are MS patients' performances degraded.

The sound lateralization test is highly correlated with MS plaque location. Center-oriented performance is principally associated with caudal pontine lesions and side-oriented performance with lesions rostral to the superior olivary complex. These types of abnormalities rarely affect the everyday life of MS patients.

This apparent discrepancy between lack of auditory complaints but abnormal auditory performance can be accounted for by two factors: (1) the preferential effect of MS upon neural timing; and (2) redundancy in auditory processing. For example, right–left sound localization is a binaural task. It is performed by the CNS based upon the differences in the timing and level of the sound at the two ears. Sounds closer to one ear will arrive earlier and be louder at that ear than the opposite ear. Because with earphones the timing and level properties of sounds can be controlled independently, due to the preferential effect of MS upon neural timing, often MS patients will perform abnormally for binaural timing earphone tests but normally for level tests. Since in daily living a person is receiving both level and timing information regarding the location of a sound, MS subjects with poor performance on earphone timing tests will have no complaints regarding locating sounds because of redundancy in the auditory processing of sound localization. They will rely upon the level features of sounds whose processing has been spared by MS. Consequently, despite their poor performance on tests requiring precise neural timing, these patients are functionally normal and require no auditory therapeutic intervention.

Most abnormal performances of MS patients can be understood by considering MS as a focal demyelinating disease. However, the high-pass time JND, the most frequent auditory abnormality of MS patients, cannot be accounted for by pontine focal demyelinating lesions only. Consequently high-pass time JND may be important as a marker for the neurodegenerative property of MS. Further study may prove that high-pass time JND is a valuable tool for monitoring the neurodegenerative aspect of MS.

In conclusion, hearing tests are probably not critical in establishing the diagnosis of MS. However, once the diagnosis has been established, binaural tests such as sound lateralization or interaural time discrimination of high-frequency stimuli may be useful as a means of monitoring the progression of the disease. Their advantages include being non-invasive, fast, reliable, and inexpensive.

References

Aharonson V, Furst M (2001). A model for sound lateralization. J Acoust Soc Am 109: 2840–2851.

Aharonson V, Furst M, Levine RA et al. (1998). Lateralization and binaural discrimination of patients with pontine lesions. J Acoust Soc Am 103: 2624–2633.

American National Standard Institute (1996). American National Standard Specification for Audiometers (ANSI S3.6-1996), ANSI, New York.

Bergamaschi R, Romani A, Zappoli F et al. (1997). MRI and brainstem auditory evoked potential evidence of eighth cranial nerve involvement in multiple sclerosis. Neurology 48: 270–272.

Caird D, Klinke R (1983). Processing of binaural stimuli by cat superior olivary complex neurons. Exp Brain Res 52: 385–399.

Chiappa KH (1980). Brainstem auditory evoked potentials in 200 patients with multiple sclerosis. Ann Neural 7: 135–143.

Daugherty WT, Lederman RJ, Nodar RH et al. (1983). Hearing loss in multiple sclerosis. Arch Neurol 40: 33–35.

Dobie RA, Norton S (1980). Binaural interaction in human auditory evoked potentials. Electroencephalogr Clin Neurophysiol 49: 303–313.

Don M, Eggermont JJ (1978). Analysis of the click-evoked brainstem potentials in man using high-pass noise masking. J Acoust Soc Am 63: 1084–1092.

Doty RL, Tourbier I, Davis S et al. (2012). Pure-tone auditory thresholds are not chronically elevated in multiple sclerosis. Behav Neurosci 126 (2): 314–324.

Drulović B, Ribaric-Jankes K, Kostic VS et al. (1993). Sudden hearing loss as the initial monosymptom of multiple sclerosis. Neurology 43: 2703–2705.

Franklin DJ, Coker NJ, Jenkins HA (1989). Sudden sensorineural hearing loss as a presentation of multiple sclerosis. Arch Otolaryngol 115: 41–45.

Furst M, Algom D (1995). Lateralization and discrimination of dichotic clicks: evidence from patients with brainstem lesions and normal cohorts. J Basic Clin Physiol Pharmacol 6 (2): 149–171.

Furst M, Levine RA, McGaffigan PM (1985). Click lateralization is related to the β component of the dichotic brainstem auditory evoked potentials of human subjects. J Acoust So t Am 78: 1644–1651.

Furst M, Eyal S, Korczyn AD (1990). Prediction of binaural click lateralization by brainstem auditory evoked potentials. Hear Res 49 (1–3): 347–359.

Furst M, Levine RA, Korczyn AD et al. (1995). Brainstem lesions and click lateralization in patients with multiple sclerosis. Hear Res 82 (1): 109–124.

Furst M, Aharonson V, Levine RA et al. (2000). Sound lateralization and interaural discrimination. Effects of brainstem infarcts and multiple sclerosis lesions. Hear Res 143 (1–2): 29–42.

Hannley M, Jerger JF, Rivera VM (1983). Relationships among auditory brain stem responses, masking level differences and the acoustic reflex in multiple sclerosis. Audiology 22 (1): 20–33.

Hausler R, Levine RA (1980). Brain stem auditory evoked potentials are related to interaural time discrimination in patients with multiple sclerosis. Brain Res 191 (2): 589–594.

Hausler R, Colburn S, Marr E (1983). Sound localization in subjects with impaired hearing. Spatial-discrimination and interaural-discrimination tests. Acta Otolaryngol Suppl 400: 1–62.

Hellmann MA, Steiner I, Mosberg-Galili R (2011). Sudden sensorineural hearing loss in multiple sclerosis: clinical course and possible pathogenesis. Acta Neurol Scand 124: 245–249.

Hendler T, Squires NK, Emmerich DS (1990). Psychophysical measures of central auditory dysfunction in multiple sclerosis: neurophysiological and neuroanatomical correlates. Ear Hear 11 (6): 403–416.

Honig LS, Ramsay RE, Sheremata WA (1992). Event-related potentials P300 in multiple sclerosis. Relation to magnetic resonance imaging and cognitive impairment. Arch Neurol 49: 44–50.

Jabbari B, Marsh EE, Gunderson CH (1982). The site of the lesion in acute deafness of multiple sclerosis – contribution of the brain stem auditory evoked potential test. Clin Electroencephalogr 13: 241–244.

Javel E (1980). Coding of AM tones in the chinchilla auditory nerve: implications for the pitch of complex tones. J Acoust Soc Am 68: 133–146.

Jeffress LA (1948). A place theory of sound localization. J Comp Psychol 41: 35–39.

Jerger JF, Oliver TA, Chmiel RA et al. (1986). Patterns of auditory abnormality in multiple sclerosis. Audiology 25 (4–5): 193–209.

Jewett DL, Williston JS (1971). Auditory evoked far-fields averaged from the scalp of humans. Brain 94: 681–696.

Jones SJ, Van der Poel JC (1990). Binaural interaction in the brain-stem auditory evoked potential: evidence for a delay line coincidence detection mechanism. Electroencephalogr Clin Neurophysiol 77 (3): 214–224.

Joris PX, Yin TCT (1992). Responses to amplitude-modulated tones in the auditory nerve of the cat. J Acoust Soc Am 91: 215–232.

Kiang NYS, Moxon EC (1974). Tails of tuning curves of auditory-nerve fibers. J Acoust Soc Am 55: 620–630.

Kiang NYS, Watanabe T, Thomas EC et al. (1965). Discharge patterns of single fibers in the cat's auditory nerve, MIT Press, Cambridge, MA.

Levine RA (1981). Binaural interaction in brainstem auditory evoked potentials of human subjects. Ann Neurol 9: 384–393.

Levine RA, Gardner JC, Stufflebeam SM et al. (1993a). Binaural auditory processing in multiple sclerosis subjects. Hear Res 68 (1): 59–72.

Levine RA, Gardner JC, Stufflebeam SM et al. (1993b). Effects of multiple sclerosis brainstem lesions on sound lateralization and brainstem auditory evoked potentials. Hear Res 68 (1): 73–88.

Levine RA, Gardner JC, Fullerton BC et al. (1994). Multiple sclerosis lesions of the auditory pons are not silent. Brain 117 (5): 1127–1141.

Magnano I, Aiello I, Piras MR (2006). Cognitive impairment and neurophysiological correlates in MS. J Neurol Sci 245: 117–122.

Matas CG, Matas SL, Rondina Salzano de Oliveira C et al. (2010). Auditory evoked potentials and multiple sclerosis. Arq Neuropsiquiatr 68 (4): 528–534.

Matathias O, Sohmer H, Biton V (1985). Central auditory tests and auditory nerve-brainstem evoked responses in multiple sclerosis. Acta Otolaryngol 99 (3–4): 369–376.

McFarland HF (2009). Examination of the role of magnetic resonance imaging in multiple sclerosis: a problem-orientated approach. Ann Indian Acad Neurol 12 (4): 254–263.

McFarland HF (2010). Examination of the role of MRI in multiple sclerosis: a problem orientated approach. In: R Martin, A Lutterotti (Eds.), Molecular Basis of Multiple Sclerosis, Springer, Heidelberg, pp. 287–301, 51.

Miller DH, Barkhof F, Frank JA et al. (2002). Measurement of atrophy in multiple sclerosis: pathological basis, methodological aspects and clinical relevance. Brain 125 (Pt 8): 1676–1695.

Mustillo P (1984). Auditory deficits in multiple sclerosis; a review. Audiology 23: 145–164.

Noffsinger D, Olsen WO, Carhart R et al. (1972). Auditory and vestibular aberrations in multiple sclerosis. Acta Otolaryngol 303: 1–63, Suppl.

Oh YM, Oh DH, Jeong et al. (2008). Sequential bilateral hearing loss in multiple sclerosis. Ann Otol Rhinol Laryngol 117 (3): 186–191.

Ozunlu A, Mus N, Gulhan M (1998). Multiple sclerosis: a cause of sudden hearing loss. Audiology 37: 52–58.

Picton TW, McEvoy LK, Champagne SC (1991). Human evoked potentials and the lateralization of a sound. Acta Otolaryngol Suppl 491: 139–143, discussion 144.

Polman CH, Reingold SC, Edan G et al. (2005). Diagnostic criteria for multiple sclerosis: 2005 revisions to the "McDonald Criteria". Ann Neurol 58 (6): 840–846.

Pratt H, Polyakov A, Aharonson V et al. (1998). Effects of localized pontine lesions on auditory brainstem evoked potentials and binaural processing in humans. Electroencephalogr Clin Neurophysiol 108: 511–520.

Richter EA, Norris BE, Fullerton BC et al. (1983). Is there a medial nucleus of the trapezoid body in humans. Am J Anat 168: 157–166.

Sailer M, Heinze HJ, Tendolkar I et al. (2001). Influence of cerebral lesion volume and lesion distribution on event-related potentials in multiple sclerosis. J Neurol 248: 1049–1055.

Shea III JJ, Brackmann DE (1987). Multiple sclerosis manifesting as sudden hearing loss. Arch Otolaryngol 97: 335–338.

Tadmor R, Tenny R, Furst M (1994). Algorithm for automatic matching of brainstem atlas to MR images. In: M Takahashi, Y Korogi, I Moseley (Eds.), Proceedings of the XV Symposium Neuroradiologicum, Springer-Verlag, New York, pp. 34–36.

Tenny R (1994). Algorithm for matching brainstem atlas to MR images, Faculty of Engineering, Tel Aviv University, Tel Aviv, M.Sc. Thesis.

Van der Poel JC, Jones SJ, Miller DH (1988). Sound lateralization, brainstem auditory evoked potentials and magnetic resonance imaging in multiple sclerosis. Brain 111 (Pt 6): 1453–1474.

Wrege KS, Starr A (1981). Binaural interaction in human auditory brainstem evoked potentials. Arch Neurol 38: 572–580.

Yin TCT, Chan JCK (1988). Neural mechanisms underlying interaural time sensitivity to tone and noise. In: GH Edelman, WE Gall, WH Cowan (Eds.), Auditory Function. Neurobiological Bases of Hearing, Wiley, New York, pp. 385–430.

Yin TC, Chan JC (1990). Interaural time sensitivity in medial superior olive of cat. J Neurophysiol 64: 465–488.

Handbook of Clinical Neurology, Vol. 129 (3rd series)
The Human Auditory System
G.G. Celesia and G. Hickok, Editors

Chapter 37

Hearing and music in dementia

JULENE K. JOHNSON[1]* AND MAGGIE L. CHOW[2]
[1]*Institute for Health and Aging, University of California, San Francisco, CA, USA*
[2]*School of Medicine, University of California, San Francisco, CA, USA*

INTRODUCTION

Both music and speech are complex acoustic signals that rely on a number of brain and cognitive processes to create the sensation of hearing. The transmission of auditory signals is supported by large-scale neural networks. Changes in hearing function are generally not a major focus of concern for persons with a majority of neurodegenerative diseases associated with dementia, such as Alzheimer disease (AD). However, changes in the processing of sounds may be an early, and possibly preclinical, feature of AD and other neurodegenerative diseases. For example, patients with semantic dementia often present to a physician with difficulty comprehending speech, in addition to difficulties with finding words in conversational speech. Other persons with a dementia may have difficulty following multiple auditory streams or following complex instructions. The frequency of complaints about hearing difficulties differs by the type and stage of the neurodegenerative disease.

Persons with neurodegenerative diseases, however, rarely present with significant deficits in the processing of music, compared with persons who have an auditory agnosia or amusia, an acquired impairment in the ability to process music (see Chapters 32 and 34). In contrast, there are numerous anecdotal reports suggesting that the appreciation (and perception) of music often remains preserved in persons with AD and possibly other neurodegenerative diseases.

A key question is the extent to which the core cognitive deficits associated with various neurodegenerative diseases contribute to difficulties in comprehending auditory information versus the extent to which changes in central auditory processing may contribute to auditory comprehension difficulties. In other words, because dementia, by definition, is a progressive impairment in cognition, it is sometimes difficult to separate difficulties in sound processing due to cognitive impairment versus difficulties in sound processing due to an auditory system-specific deficit.

The aim of this chapter is to review the current state of knowledge concerning hearing and music perception in persons who have a dementia as a result of a neurodegenerative disease. The fundamental questions of this chapter are: How do persons with a neurodegenerative disease process comprehend music? Do persons with different neurodegenerative diseases perceive and process music differently? How is music used as a tool to better understand both the auditory system and neurodegenerative diseases?

OVERVIEW OF DEMENTIA

Dementia is a general term that refers to a progressive decline in cognition and behavior that impacts a person's ability to function independently in everyday life (American Psychiatric Association, 2000). Dementia is a common syndrome in older adults, and there are over 50 different causes of dementia. For the purposes of this chapter, we will focus on the most common causes of dementia (see Miller and Boeve, 2009; Ames et al., 2010; Krishnamoorthy et al., 2010 for a comprehensive overview of dementia and neurodegenerative diseases).

AD is the most common cause of dementia in adults over age 65 and accounts for approximately half of all dementia diagnoses (Alzheimer's Association, 2012). Impairment in memory is the most common cognitive and often predominant deficit in AD, but deficits in other cognitive domains (language, executive function, visuospatial skills) are often present. Other neurodegenerative diseases can also cause dementia; the most common non-AD dementias include dementia with Lewy

*Correspondence to: Julene K. Johnson, Ph.D., Institute for Health and Aging, University of California, San Francisco, 3333 California St., Suite 340, San Francisco, CA 94118, USA. Tel: +1-415-476-1106, E-mail: julene.johnson@ucsf.edu

bodies (DLB), frontotemporal dementia (FTD), corticobasal degeneration, progressive supranuclear palsy, and Huntington's disease. Vascular cognitive impairment (also called vascular dementia), which is caused by vascular disease in the brain, is another common cause of dementia (Rincon and Wright, 2013). Many persons with vascular dementia also have changes in the brain associated with AD. Approximately 30–60% of persons with Parkinson disease also eventually develop cognitive dysfunction (Aarsland et al., 2003; Williams-Gray et al., 2007). There is often clinical overlap between Parkinson's disease dementia and DLB.

DLB is another common cause of dementia. It is characterized by progressive cognitive and functional decline with at least two of the following additional symptoms: (1) prominent visual hallucinations; (2) fluctuations in alertness and attention; and (3) parkinsonian motor symptoms (McKeith et al., 2005). Although visual hallucinations are a prominent feature of DLB, auditory hallucinations are rare. Corticobasal degeneration and progressive supranuclear palsy are two more rare neurodegenerative diseases that are classified as atypical parkinsonian disorders (Armstrong et al., 2013; Respondek et al., 2013). Huntington's disease is another rare neurodegenerative disease that causes involuntary motor symptoms (i.e., chorea), dementia, and psychiatric symptoms (Biglan et al., 2013).

FTD is a common cause of early age-of-onset dementia (and is caused by significant atrophy in the frontal and temporal lobes of the brain. FTD can be subdivided into two language-variant syndromes (progressive non-fluent aphasia and semantic dementia) and one behavior variant, called behavioral-variant FTD (bvFTD) (Gorno-Tempini et al., 2011; Rascovsky et al., 2011). The behavioral variant of FTD is characterized by changes in personality and social behavior (e.g., impulsivity, apathy, decreased empathy, and diminished insight). Persons with semantic dementia develop a progressive loss of the meaning (semantics) for both verbal and non-verbal concepts, which is due primarily to atrophy of the temporal lobes of the brain. Persons who have progressive non-fluent aphasia develop progressive difficulty producing language (e.g., hesitant and effortful speech). Logopenic aphasia is another variant of primary progressive aphasia characterized by slow speech and naming deficits and is thought to be caused by AD (Gorno-Tempini et al., 2008).

Mild cognitive impairment (MCI) is a term used for the preclinical stage of dementia. MCI is associated with mild declines in cognitive function, but the ability to perform daily living skills is still relatively preserved. There is a high rate of progression from MCI to dementia (Petersen, 2011).

The majority of research about hearing and music in dementia has been done with persons who have AD. For this reason, the chapter will focus primarily on AD. However, the chapter will include information about hearing function and music in non-AD dementias when it is available.

HEARING FUNCTION IN NEURODEGENERATIVE DISEASES

Hearing sounds, including music, relies on both the peripheral and central auditory systems (see Chapters 1 and 2). Both the peripheral and central auditory systems undergo change with age. The section below will provide a broad overview about hearing function in aging and neurodegenerative diseases. (For further information about hearing function in aging and dementia, see Grimes, 1995 and Weinstein, 2000.)

Hearing impairment

Hearing loss is one of the most common chronic conditions in older adults (see Chapter 21). Hearing loss in the general population increases with age, and men have higher rates of hearing impairment than women. Approximately 40% of adults over age 65 have hearing loss (>30 dB loss in at least one ear), and approximately 50% of adults over age 85 have hearing loss (Gates and Mills, 2005). Age-related hearing loss (presbycusis) first affects the high-frequency range, is progressive, and can be caused by damage to either the peripheral or central auditory system (or both). Gates and Mills (2005) note that persons with hearing loss often have difficulty understanding conversations and misperceive phonetically similar words, such as "map" and "mat." Family members or friends are often more aware of the change in hearing function than the older person with hearing loss. Hearing loss in older adults can also affect the ability to perceive and comprehend music, but few studies have explored this topic.

A number of studies have examined the relationship between hearing loss and cognitive decline in older adults. Gennis and colleagues (1991) reviewed 15 studies and concluded that hearing loss was more common in persons with dementia, but there was considerable heterogeneity in the study designs, and the underlying causes of the dementia were not always clear. Other studies suggest that the hearing loss in persons with dementia is more severe than older adults without cognitive decline (Uhlmann et al., 1989; Lin et al., 2011). Recent studies also suggest that adults with hearing loss are more likely to develop AD or other dementias over time. For example, Lin and colleagues (2013) found that hearing loss (pure-tone average >25 dB) was independently associated with both increased rate of cognitive decline and incident cognitive impairment in a large sample of community-dwelling older adults.

Several recent studies suggest that adults with Parkinson's disease may have a higher prevalence of age-related, high-frequency sensorineural hearing loss relative to healthy adults. One recent study involving 118 patients with Parkinson's disease found a higher prevalence of age-dependent high-frequency hearing loss than age- and sex-matched controls (Vitale et al., 2012). Within this cohort, 26% had mild, 63% had moderate, and 11% had severe hearing loss. The Parkinson's disease patients with hearing impairment were older, more likely to be male, and had a later age at onset than the Parkinson's disease patients with normal hearing. The authors also noted that the Parkinson's disease patients were generally unaware of their hearing loss.

It is also important to remember that the patient with dementia may not complain of hearing loss symptoms. Gates and Mills (2005) recommended an examination of hearing function in any person with cognitive decline. They also note that tinnitus is a common comorbid condition with hearing loss and should be queried in persons with cognitive decline. It is possible that hearing loss symptoms are underreported in persons with neurodegenerative diseases, as they may be regarded as secondary to concerns about memory or are confused as memory or communication symptoms. It is also possible that dementia may be overdiagnosed in persons with hearing loss.

Peripheral auditory system function

Persons with a neurodegenerative disease (with the diagnoses discussed above) are generally thought to have an intact peripheral auditory system and are only affected by typical, age-related changes. That is, most neurodegenerative diseases do not directly damage the peripheral auditory system. Autopsy studies have not detected neuropathologic changes associated with AD in the peripheral auditory system (Sinha et al., 1993). However, only a few studies have been done, and the majority of these autopsy studies focus on AD. The peripheral auditory system is just beginning to be studied in non-AD dementias.

Early studies documented normal peripheral hearing function in AD patients (Grimes et al., 1985; Eustache et al., 1995; Gates et al., 1995). For example, Kurylo and colleagues (1993) investigated a comprehensive battery of auditory function, including pure-tone thresholds, sound localization, pitch perception of complex tones, phoneme perception, timbre discrimination, and tonal memory (from the Seashore Tests of Musical Abilities). There were no differences between patients with AD and matched older adults on the majority of these tests. Another study (Strouse et al., 1995) administered immittance tests (e.g., tympanograms, acoustic reflex thresholds), pure-tone and speech audiometric testing, word recognition, and distortion product otoacoustic emissions testing to 10 mild to moderate patients with AD and matched controls. They found that the AD patients had only slightly worse low-frequency thresholds compared to the controls. Another study involving 33 patients with a diagnosis of AD, progressive nonfluent aphasia, or logopenic aphasia found only small differences on sound detection thresholds in the patient groups compared to controls (Goll et al., 2011). Mahoney and colleagues (2011) reported that 32% of patients with semantic dementia complained of tinnitus or hyperacusis, but peripheral hearing function was not examined.

A few other less common adult-onset neurodegenerative diseases occasionally involve the peripheral auditory system. For example, one study documented a 71-year-old man who developed a rapidly progressive dementia with features of DLB and also inflammation of the cartilage of the pinna (Head et al., 2006). An autopsy revealed inflammation-induced neurodegeneration caused by relapsing polychondritis, an autoimmune inflammatory disorder of cartilaginous tissues. There are also a few other rare neurodegenerative diseases or genetic mutations that can cause a dementia and sensorineural hearing loss (e.g., mitochondrial disease syndromes, adult-onset Friedreich's ataxia). A few recent studies have focused on cases of autosomal-dominant hereditary sensory and autonomic neuropathy with dementia and hearing loss (HSAN1) (Wright and Dyck, 1995). Klein and colleagues (2011, 2013; U.S. Congress, Office of Technology Assessment, 1992) described a primarily adult-onset hearing loss and neuropathy followed by memory decline or personality changes typical of FTD in the 30s and 40s with a mutation in DNA methyltransferase 1 (DNMT1). Autopsies in persons with a DNMT1 mutation in a case series from Japan revealed diffuse non-specific degeneration and glial activation especially in the frontal, temporal, parietal, and occipital cortex, hippocampus, and thalamus, but the auditory system was not mentioned (Hojo et al., 2004).

Central auditory system

Because neurodegenerative diseases primarily affect the central nervous system, the majority of studies have focused on the central auditory system. The changes to the central auditory system are thought to occur as a result of the neurodegenerative diseases, in addition to age-related changes in the central auditory system. Histopathologic and morphometric studies have found a number of neuropathologic changes in the central auditory system in different neurodegenerative diseases. The auditory system has been most extensively studied in persons with AD. Neuron loss, neuritic plaques, and neurofibrillary tangles, the classic neuropathologic lesions of AD, have been identified in auditory system brainstem and midbrain nuclei, including the olivary nucleus, inferior colliculus, and medial geniculate body (Ohm and Braak, 1989; Dugger et al., 2011). In addition, plaques and tangles are also common in primary auditory and auditory association cortex in persons with AD (Sinha et al., 1993; Chance et al., 2011). Neuropathologic studies of the central auditory system are less common in non-AD dementias. A few studies suggest that an accumulation of tau neuropathology in auditory brainstem nuclei in persons with progressive supranuclear palsy (Dugger et al., 2011) and severe neuronal loss has been documented in the auditory cortex of persons with FTD and primary progressive aphasia (Baloyannis et al., 2011). These neuropathologic changes of the central auditory system are thought to be related to the processing of sounds.

The majority of studies focus on the processing of verbal information, while studies about the processing of non-verbal information, including music and environmental sounds, are less commonly done. This section will provide a brief overview of central auditory system dysfunction in neurodegenerative diseases, and additional information about the processing of music and other non-verbal sounds will be discussed in the following two sections.

Early studies in the 1980s used dichotic listening paradigms (simultaneous presentation of different acoustic signals to the right and left ears) and other basic auditory perceptual tasks to study central auditory function in persons with AD. For example, Grimes and colleagues (1985) found that 81% of the patients with AD had abnormal scores on dichotic listening tasks. The dichotic listening scores correlated with performance on tests of cognition, volume of the temporal lobes on brain computed tomography imaging, and metabolism of the left temporal lobe using positron emission tomography imaging. That is, lower scores on dichotic listening tasks correlated with lower cognitive scores, less brain volume, and lower metabolism in the left temporal lobe. Other early studies using dichotic listening methods also documented auditory processing impairments in AD patients (Grady et al., 1989; Mohr et al., 1990).

Persons with MCI may also have central auditory function impairments. For example, Gates and colleagues (2008) examined central auditory processing using a comprehensive battery (synthetic sentence identification with ipsilateral competing message, dichotic sentence identification, and dichotic digits) in a large cohort of older adults who had mild memory impairment (but not demented) or dementia based on a cognitive screening test (Cognitive Ability Screening Instrument). The authors found that the persons with mild memory impairment performed worse than matched controls but better than the demented patients, even after controlling for age, hearing threshold, and word recognition scores. However, the underlying diagnoses of the demented participants were not reported. Another study also found that persons with MCI had an impairment on a dichotic listening task that was intermediate between healthy adults and AD patients (Idrizbegovic et al., 2011), again suggesting that deficits in central auditory processing occur before the onset of dementia and get worse after the onset of dementia symptoms.

Impairments in central auditory function may also be a risk for developing dementia. For example, one study recently found that older adults with severe central auditory dysfunction, particularly as measured by the Dichotic Sentence Identification test, were at increased risk for developing incident AD after 3 years (Gates et al., 2011). The majority of those who developed AD also had lower memory scores, as measured by the memory subscale of the Cognitive Abilities Screening Instrument. As discussed above, another recent study found that hearing loss (pure-tone average >25 dB) was independently associated with increased rate of cognitive decline and incident cognitive impairment in a large sample of community-dwelling older adults (Lin et al., 2013).

Some authors propose to frame auditory cognition deficits in terms of "auditory objects" or "auditory scene analysis," the ability to parse sound sources in the auditory environment (Goll et al., 2010b). For example, Goll and colleagues (2012a) recently assessed auditory scene analysis of both verbal and non-verbal sounds in a cohort of 21 persons with AD. The persons with AD performed much worse than matched controls on both auditory scene analysis tasks, but their performance was somewhat attenuated after accounting for visuospatial working memory. They also found that performance correlated with brain volume in several posterior cortical areas, using voxel-based morphometry of brain magnetic resonance imaging (MRI).

Central auditory processing deficits have also been studied in a few non-AD dementias. For example, Goll and colleagues (2010a) found significant difficulties in the processing of complex non-verbal sounds in 12 patients with progressive non-fluent aphasia and 8 patients with semantic dementia. The authors used several newly designed auditory tasks to assess early perceptual, apperceptive, and semantic levels of non-verbal auditory processing. While patients with progressive non-fluent aphasia were particularly impaired on the early perceptual tasks, the semantic dementia patients had particular difficulty with the processing of semantic aspects of non-verbal sounds.

In conclusion, impairments in the central auditory system are common in persons with various neurodegenerative diseases. However, the majority of studies have focused on AD, while an interest in studying central auditory function in non-AD dementias is increasing. Additional consideration of the central auditory system processing of music and non-verbal sounds can be found in the next two sections.

PROCESSING OF MUSIC IN NEURODEGENERATIVE DISEASES

As discussed in the introduction, persons with neurodegenerative diseases rarely present with significant deficits in the processing of music, compared with persons who have an auditory agnosia or amusia, an acquired impairment in the ability to process music (see Chapters 32 and 34). In contrast, there are numerous anecdotal reports suggesting that the appreciation (and perception) of music often remains preserved in persons with AD and possibly other neurodegenerative diseases. How do persons with a neurodegenerative disease process and comprehend music? Do persons with different neurodegenerative diseases perceive and process music differently? How is music used as a tool to better understand both the auditory system and neurodegenerative diseases?

Framework for examining music perception in neurodegenerative diseases

As discussed in Chapter 11, there are both perceptual and associative aspects involved in the perception of music. Basic-level perceptual hearing processes extract pitch, rhythm, timbre, and timing information, while higher-order cognitive processes help store and associate music sequences with long-term memories. All of these processes facilitate recognition of a musical sequence as meaningful or familiar.

One of the early music psychologists, Carl Seashore (1866–1949), provided a useful framework for understanding how the physical aspects of sound (frequency, intensity, duration, and waveform) are mapped on to the perceptual qualities of music (pitch, loudness, time, timbre). That is, the acoustic aspect of frequency (number of waves per second) is associated with the perception of pitch; intensity (decibels) is associated with loudness; duration (time intervals) is associated with time/meter; and wave form (tone quality from pure tone to noise) is associated with the perception of timbre. These basic physical components of musical sounds can be combined to create music intervals, scales, harmonies, melodies, and musical compositions. Figure 37.1 depicts a commonly used contemporary cognitive model of music recognition, as proposed by Peretz and Coltheart (2003). Each box represents a music-processing component, while the lines between boxes designate the path of

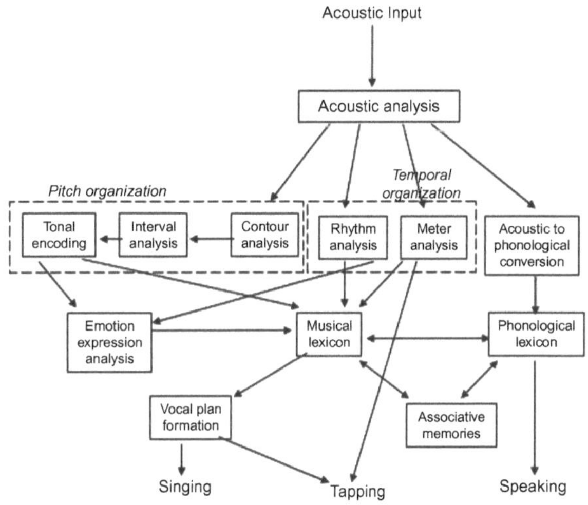

Fig. 37.1. Cognitive model of music processing. Each box represents a processing component, and arrows represent pathways of information flow or communication between processing components. A neurologic anomaly may either damage a processing component (box) or interfere with the flow of information between two boxes. All components whose domains appear to be specific to music are in green; others are in blue. There are three neurally individuated components in italics – rhythm analysis, meter analysis, and emotion expression analysis – whose specificity to music is currently unknown. They are represented here in blue, but future work may provide evidence for representing them in green. (Reproduced from Peretz and Coltheart, 2003, with permission.)

information processing between components. The acoustic signal enters from the top of the model. In addition, Stewart and colleagues (2006) identified the most important brain areas implicated in disorders of music listening. The schematic of the brain at the top provides the anatomic areas involved in music listening, while the four schematics below identify specific aspects of music processing. The authors argue that this framework can be applied to both acquired and congenital disorders of music listening. It is important to keep in mind that this framework was developed from studies of patients with focal brain lesions (e.g., stroke); however, there are no compelling reasons to think that the deficits in music processing observed in persons with dementia would not follow this framework (Fig. 37.2).

The melody is often the most conspicuous aspect of music, and many studies focus on the processing of familiar and unfamiliar melodies. For familiar melodies in western cultures, healthy adults store relatively precise knowledge about the notes and timing that comprise the melody. Recognition of a melody as familiar depends on the ability to match the mental representation/percept of the melody with a "musical lexicon," which includes all melodies known to the listener (Peretz and Coltheart, 2003). Recognition of familiar

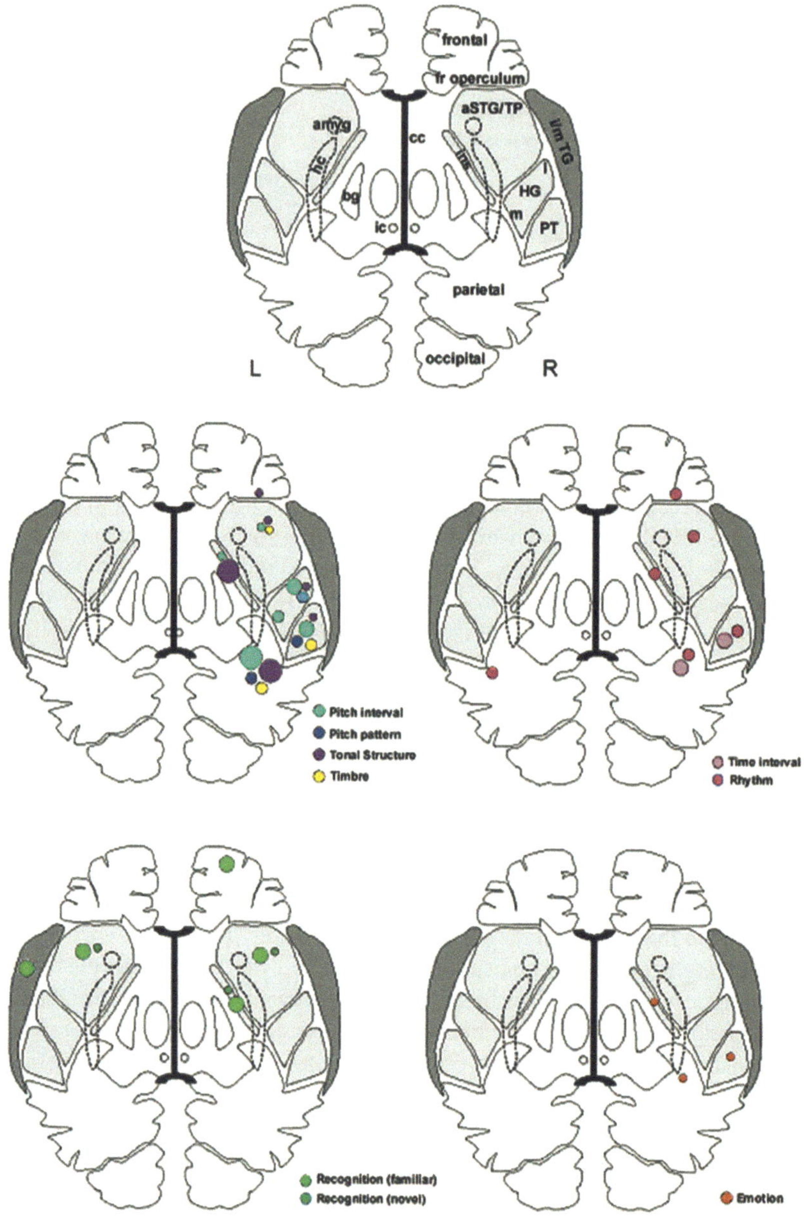

Fig. 37.2. Brain areas implicated in disorders of music listening. Critical brain substrates for musical-listening disorders across studies. Five cartoons are shown, each depicting the brain in a schematic axial section that includes all key anatomic areas involved in music listening (identified on the top cartoon); the corpus callosum (black), superior temporal plane (light gray), and middle/inferior temporal gyri (dark gray areas, in exploded view) are colored for ease of identification. Musical functions analyzed in Supplementary Table 1 (http://brain.oxfordjournals.org/content/129/10/2533.figures-only) have been grouped as follows: pitch processing (pitch interval, pitch pattern, tonal structure, timbre); temporal processing (time interval, rhythm); musical memory (familiar and novel material); and emotional response to music. Each group of functions is assigned to a separate cartoon; individual functions are identified to the right of the corresponding cartoon. Raw data from Supplementary Table 1 have been thresholded; the presence of a colored circle corresponding to a particular function in a region indicates that at least 50% of studies of the function implicate that region. The size of each circle is scaled according to the proportion of studies of the function implicating that region (see text). Meter is not represented as no brain area was implicated in 50% or more of cases. amyg, amygdala; aSTG, anterior superior temporal gyrus; bg, basal ganglia; cc, corpus callosum; r, frontal; hc, hippocampal; HG, Heschl's gyrus; ic, inferior colliculi; i, inferior; ins, insula; l, lateral; m, medial; thal, thalamus; PT, planum temporale; TG, temporal gyrus. (Reproduced from Stewart et al., 2006, with permission.)

music can be assessed using a variety of methods. Several studies asked participants to judge the familiarity of a melody to determine if the listener can discriminate familiar and novel melodies. Another method is to ask participants to listen to a melody and determine if the melody is correctly played or has an error (e.g., pitch, rhythm). The Distorted Tunes Test (Drayna et al., 2001) is one example of a pitch error detection task. Brattico and colleagues (2006) suggest that the identification of errors in melodies occurs rapidly and automatically, even with unfamiliar melodies. Several studies also ask participants to generate a title for a familiar melody, but this type of task requires different cognitive processes (e.g., verbal recall) compared with recognizing melodies with errors. The processing of unfamiliar melodies can be assessed using a number of different paradigms. For example, both the explicit and implicit memory of unfamiliar melodies can be examined. The Montreal Battery for Evaluation of Amusia (Peretz et al., 2003) includes several tasks that require comparisons of unfamiliar melodies that may differ by notes, contour, or rhythmic patterns.

The next sections will review the current literature about how music, and melodies in particular, is processed in persons who have a neurodegenerative disease. It is important to keep in mind that performance on various music tasks can be affected by the type and severity of dementia. This often makes comparisons across studies difficult. Performance on music tasks can also be affected by the amount of prior music training. The majority of case studies reviewed below focus on persons who are either professional or amateur musicians, while many of the group studies include persons with different levels of music background. Not all studies report the music background of the participants. Similar to the studies reviewed above, the majority of studies about music have been done with persons with AD. However, recent studies focus more on the processing of music in non-AD dementias.

Alzheimer disease

As mentioned in the introduction of this chapter, there are several case reports in the literature that document a relative preservation of music abilities in persons with AD. These reports primarily focus on the preserved ability to continue playing a musical instrument after the onset of dementia. Table 37.1 summarizes these case studies and provides a short summary of the findings. In several of the early case studies, Beatty and colleagues described the preserved music abilities of persons with AD who had music training (Beatty, 1988, 1999; Beatty et al., 1994, 1999; Beatty and Greiner, 1998; Cowles et al., 2003). For example, patient T was a 71-year-old amateur trombone player who continued to play in a Dixieland jazz band after he developed dementia (later confirmed to be AD by autopsy) (Beatty et al., 1994, 1997). When patient T was in the mild stages of dementia (with a Mini-Mental State Examination score of 20 out of 30), the authors made videotaped recordings of his playing during a band performance and asked raters to compare the quality of his playing at that time with recordings of him playing from approximately 30 years prior. The raters judged these two playing examples as equivalent, suggesting that there had not been any significant decline in his playing ability after the onset of dementia. Patient T was also asked to listen to 20 familiar holiday songs and generate the titles. He recalled 80% of the song titles, but his performance was slightly below a group of healthy controls who recalled 95% of the titles. In another early case report, Crystal and colleagues (Beatty and Greiner, 1998) described an 82-year-old music editor and pianist who was evaluated during the MCI stage of AD. At that time, he was able to play, from memory, 13 popular classic music compositions. However, 3 years later, he was only able to play a few bars of the same pieces. Several more recent case studies have examined both music perception and production in more detail, including comparisons with non-AD dementias (Cuddy and Duffin, 2005; Hailstone et al., 2009; Vanstone et al., 2009; Omar et al., 2010; Weinstein et al., 2011). Polk and Kertesz (1993) studied two persons with atypical presentations of AD (i.e., progressive aphasia and posterior cortical atrophy).

After the early case studies in the 1980s, the examination of music skills in persons with dementia eventually expanded. Subsequent research aimed to better understand how music is processed by persons with AD, and eventually, other types of neurodegenerative diseases. In addition, a number of group studies were conducted.

Several studies have focused on the ability of persons with AD to process both familiar and unfamiliar melodies. Persons with AD are generally able to distinguish melodies as familiar or unfamiliar (familiarity decision). This is often assessed by asking participants to listen to a melodic excerpt and decide if the excerpt is familiar or novel (new) or old. The findings with AD patients, however, are somewhat mixed. In one study, the persons with AD performed slightly lower than healthy controls (Bartlett et al., 1995) when asked to discriminate familiar and unfamiliar melodies, while other studies found no differences between AD patients and controls (Cuddy and Duffin, 2005; Hsieh et al., 2011). However, persons with AD may have difficulty with the explicit recognition of familiar melodies. One study presented eight familiar melodies and then asked 15 persons with mild to moderately severe AD to if the melodies in the next set were old

Table 37.1

Summary of case studies investigating music in persons with dementia

Study	Case	Background	Diagnosis	Main findings
Beatty et al. (1988)	GW	Music teacher and pianist	AD	Preserved ability to play piano; impaired ability to name familiar music
Crystal et al. (1989)	–	82-year-old music editor and pianist	MCI due to AD	Preserved ability to play piano; unable to name title or composers
Polk and Kertesz (1993)	CW	58-year-old guitar teacher	Primary progressive aphasia due to AD	Preserved ability to play guitar and recognize familiar melodies; unable to reproduce rhythms
Polk and Kertesz, (1993)	MA	53-year-old piano teacher	Posterior cortical atrophy due to AD	Preserved ability to name familiar tunes, discriminate melodies, identify pitch errors, and reproduce rhythmic patterns; impaired ability to play piano
Beatty et al. (1994) Beatty et al. (1997)	T	71-year-old amateur trombonist	AD (pathologically confirmed)	Preserved ability to play trombone and name familiar tunes
Johnson and Ulatowska (1996)	DN	76-year-old non-musician	AD	Preserved ability to reproduce melodies and rhythmic patterns and sing familiar songs, which declined over 2 years
Beatty et al. (1999)	ML	79-year-old amateur pianist	AD	Preserved ability to play piano; difficulty learning new song and impaired naming of familiar tunes and rhythm recognition
Miller et al. (2000)	4	49-year-old non-musician	Primary progressive aphasia	Compulsive whistling; composed songs about his bird
Miller et al. (2000)	5	78-year-old non-musician	Primary progressive aphasia	New skill in composing classic music
Miller et al. (2000)	6	71-year-old musician	Primary progressive aphasia	Composed songs that captured personalities of his acquaintances
Tzortzis et al. (2000)	MM	74-year-old pianist and composer	Primary progressive aphasia	Preserved ability to compose and play piano. Impaired naming of nonmusical stimuli, but preserved naming of musical instrument sounds
Cowles et al. (2003)	CL	80-year-old amateur violinist and pianist	AD	Learned to play new song on violin; preserved ability to play previously known tunes on violin and piano, recognize familiar songs, and reproduce rhythmic patterns
Cuddy and Duffin (2005) Vanstone et al. (2009)	EN	84-year-old amateur pianist	AD (pathologically confirmed)	Generally preserved ability to recognize familiar music
Fornazzari et al. (2006)	–	63-year-old professional pianist	AD	Preserved ability to read, interpret, and learn new musical pieces
Hailstone et al. (2009)	–	56-year-old non-musician	Semantic dementia	Correctly sang 63% of familiar melodies; increased interest in music listening and singing along
Matthews et al. (2009)	JS	30-year-old non-musician	Auditory agnosia due to neurodegenerative disease	Preserved affective response when listening to personal music; impaired performance on majority of music-processing tasks administered
Vanstone et al. (2009)	VC	83-year-old amateur pianist	AD	Preserved ability to recognize and sing familiar tunes; diminished ability to play piano
Barquero et al. (2010)	–	53-year-old music critic	Frontotemporal dementia (pathologically confirmed)	Preserved ability to recognize familiar tunes and process rhythm and music emotions;

Table 37.1

Continued

Study	Case	Background	Diagnosis	Main findings
				impaired ability to judge quality of music performance
Omar et al. (2010)	BR	56-year-old professional trumpet player	Semantic dementia	Preserved recognition of music and music symbols; impaired recognition of music emotions and recognition of musical instruments
Omar et al. (2010)	WW	67-year-old music librarian and oboist	AD	Preserved recognition of music emotions and musical instruments; impaired recognition of music and music symbols
Weinstein et al. (2011)	–	59-year-old harpsichordist	Semantic dementia	Preserved ability to play harpsichord and learn new complex pieces

AD, Alzheimer disease; MCI, mild cognitive impairment.

or new (Bartlett et al., 1995) AD patients performed worse than matched controls, suggesting that the explicit memory for familiar melodies may not remain preserved in AD. Hsieh and colleagues (2011) found that the recognition of familiar tunes correlated with the volume of the right anterior temporal lobe.

Persons with AD appear to retain the ability to recognize pitch or rhythmic errors in familiar tunes (e.g., wrong notes). For example, patient EN, who was in the severe stages of AD (Mini-Mental State Examination score of 8), was able to correctly identify familiar melodies with wrong notes (pitch errors) for 25 out of 26 tunes (Cuddy and Duffin, 2005). In another group study, Johnson and colleagues (2011) examined the ability of mildly demented patients with AD to detect pitch errors in highly familiar melodies. The AD patients performed similarly to matched controls, and, as expected, the key-violating pitch errors were easier to detect than key-preserving pitch errors for both groups. Performance on this familiar melody pitch detection task correlated with gray-matter volume in the right temporal cortex, as measured by voxel-based morphometry analysis of brain MRI. Another recent study found variability (ranging from 35% to 100% correct) in the ability of moderate to severely impaired AD patients to detect errors in familiar tunes (Vanstone and Cuddy, 2010).

Several studies have evaluated the ability of persons with AD to generate a title for familiar tune (i.e., name that tune) as another way to assess knowledge about familiar melodies. As discussed above, generating a title for a familiar melody requires a different type of cognitive process (e.g., verbal recall) compared with recognizing melodies with errors. After listening to an excerpt from a familiar melody, participants are asked to generate the title or select the correct title in a multiple-choice format. While some studies found that AD patients were able to generate a similar number of titles as controls (Hsieh et al., 2012), the majority of studies found that AD patients generated fewer song titles for familiar tunes than healthy controls (Crystal et al., 1989; Bartlett et al., 1995; Cuddy and Duffin, 2005), including several case studies of musicians with AD (Beatty et al., 1994; Cowles et al., 2003; Omar et al., 2010). However, when given a multiple-choice format, the AD patients may be able to improve their score to the level of the controls (Cowles et al., 2003; Johnson et al., 2011), suggesting that the knowledge about the tunes may be intact but the ability to access the verbal labels (titles) may be impaired.

It also appears that AD patients are generally able to discriminate two short, unfamiliar melodic excerpts that may differ in a number of melodic aspects, such as pitch or rhythm. For example, two studies used the Scale subtest from the Montreal Battery for the Evaluation of Amusia, which asks participants to listen to two unfamiliar melodies and determine if the second melody was the same as or different from the first melody. Half of the second melodies in the Scale subtest have an alteration of one pitch (note). Both studies found that mildly demented AD patients performed similarly to healthy controls, suggesting that AD patients are able to process novel melodic information that is presented in relatively brief formats (Hsieh et al., 2011; Johnson et al., 2011). Omar and colleagues (2010) examined other aspects of unfamiliar melody processing in their case report of WW, who scored below controls on tasks that required the processing of interval and timbre in unfamiliar melodies. In contrast, WW was able to discriminate unfamiliar melodies that differed by alterations in scale, contour, and rhythm.

In several other studies, the explicit short-term memory for unfamiliar melodies was examined in persons with AD. One study required participants to listen to eight unfamiliar melodies (presented twice) and then make old/new judgments about a new set of melodies (half of which were novel) (Halpern and O'Connor, 2000). Persons with AD performed similarly to controls, but worse than younger adults, on the explicit music memory task. However, both AD and age-matched controls performed near chance. Two other studies found that AD patients scored lower than controls on a task requiring explicit memory for unfamiliar melodies (Bartlett et al., 1995; Quoniam et al., 2003). Implicit memory for unfamiliar melodies may also be impaired in persons with AD. Implicit memory for unfamiliar music was examined by asking persons with AD to listen to unfamiliar melodies over several trials and rate their preference, with the idea that preference ratings increase after more exposure to the melodies ("mere exposure effect"). One study found that the preference ratings increased with the number of presentations for AD patients (Quoniam et al., 2003), while another study did not find this pattern (Halpern and O'Connor, 2000). The studies included patients with different dementia severity, and the tasks were slightly different, so it is difficult to make comparisons.

Other aspects of music processing, such as pitch discrimination, recognition of major versus minor tonality (e.g., scale, chords), timbre and meter perception, are rarely examined in detail. However, a few studies suggest that the the processing of basic music components may remain relatively intact in persons with AD. For example, several studies report intact pitch discrimination (Johnson et al., 2011; Goll et al., 2012a), meter perception (Cowles et al., 2003), tonality discrimination (Mohr et al., 1990), and timbre perception (Goll et al., 2010b).

Only a few studies have examined the auditory processing of rhythmic information in music with persons with AD. Two case studies administered a modified version of the Seashore Rhythm subtest from the Seashore Test of Musical Talent. Both ML and SL were able to complete this task, which requires participants to judge whether two rhythmic patterns are the same or different (Beatty et al., 1999; Cowles et al., 2003). The recognition of meter (waltz vs march) was intact in patient SL (Cowles et al., 2003). However, additional studies are needed to make more general conclusions about rhythm perception in persons with AD.

There are a few case reports documenting the ability of persons with AD to learn new music. One case report described an 80-year-old amateur violinist (SL) with moderately severe AD who was able to learn a new, moderately difficult violin composition over six sessions over 2 months (Cowles et al., 2003). Raters judged the quality of his violin playing as "slightly better than that of a good high school violinist." Another reported that a professional pianist in the severe stages of AD was able to learn and recall a new eight-bar composition (Fornazzari et al., 2006). However, reports like these are quite rare, and it is not yet known if the ability to learn a new musical composition is common in more than just a few exceptional cases of AD.

There has been a recent interest in studying "auditory objects" in persons with dementia. Auditory objects can be defined as a "collection of acoustic data bound in a common perceptual representation an disambiguated from the auditory scene" (p. 617) (Goll et al., 2010b) These studies often consider basic levels of musical sound processing in addition to higher levels of sound conceptualization. For example, Goll and colleagues (2011) completed a systematic study of non-verbal sound processing, including music, in 21 persons with AD (and other neurodegenerative diseases, also discussed below). They found that AD patients scored lower than controls on timbre and pitch perception; however, after adjusting for performance on a spatial working-memory task, the AD patients performed similarly to controls. As discussed below, patients with progressive non-fluent aphasia and logopenic aphasia had difficulty with timbre perception. In contrast, the AD patients had a disproportionate impairment on a task that required recognition of degraded sounds based on perceptual rather than semantic information, even after adjustment for spatial working-memory performance. On this task, participants were asked to decide whether a degraded sound was more like an animal calling or a tool being used.

Frontotemporal dementia

Although the initial studies about music and dementia focused on persons with AD, more recent studies have examined music abilities in persons with non-AD dementias, including the variants of FTD. Several case studies of persons with semantic dementia have recently been published. For example, a 56-year-old professional trumpet player and music teacher (BR) completed an extensive battery of tests to examine his music abilities after developing semantic dementia (Omar et al., 2010). On tests of music perception (from the Montreal Battery for the Evaluation of Amusia), he scored lower than controls on the recognition of melodic contour (shape of the melody), intervals, and timbre. He also had difficulty recognizing emotions in music and identifying the sounds of musical instruments. He was, however, able to play familiar pieces from memory after music cueing and continued to play his trumpet (non-professionally). As another example, a 64-year-old semiprofessional harpsichordist diagnosed with semantic dementia was able to perform technically demanding compositions and generate appropriate stylistic embellishments 5 years after the onset of semantic dementia (Weinstein et al., 2011). However, additional tests of music perception were not administered with the harpsichordist.

There have also been a few group studies involving patients with semantic dementia (who are mostly non-musicians). These studies suggest that, while several aspects of music knowledge can remain preserved in semantic dementia, they may have difficulty processing familiar melodies. When examining the recognition of familiar melodies, two studies found that patients with semantic dementia had difficulty in either detecting "wrong notes" (pitch errors) in familiar tunes or judging if a tune was familiar or novel (Hsieh et al., 2011; Johnson et al., 2011). Patients with semantic dementia also had significant difficulty generating titles for familiar melodies or selecting the famous title from a four-item multiple-choice format. However, Omar and colleagues (2010) found that their 56-year-old trumpet player with semantic dementia could match titles to several famous melodies. It is important to keep in mind, however, that the majority of case studies involve persons who have a background in music, while most of the group studies involve non-musicians with varying degrees of involvement in music.

Several other studies have documented an increased interest in music after developing semantic dementia (Miller et al., 2000; Boeve and Geda, 2001; Hailstone et al., 2009) and a change in music taste (new interest in pop music) in a person with FTD (Geroldi et al., 2000). Barquero and colleagues (2010) studied a 53-year-old music critic whose first symptom was a complaint about trouble evaluating the "quality of a musical performance." Although her neurologic and neuropsychologic examinations were normal, she had difficulty differentiating renditions of music performed by either a professional or a beginner music student. It appears that her ability to recognize various music elements (e.g., melodies, pitch errors, rhythm, and meter) was preserved at that time, although the specific tasks and results are not available. She later developed a progressive dementia syndrome and was diagnosed with FTD caused by a novel progranulin mutation. An autopsy revealed ubiquitin-positive inclusions, particularly in the frontal, temporal, and parietal cortices.

A few studies have focused on persons with progressive aphasia. For example, Miller and colleagues (2000) studied three individuals with progressive language symptoms (variants of primary progressive aphasia). One non-musician developed compulsive whistling and composed songs about a bird, and the other non-musician began composing classic music that was performed in public (despite the lack of prior training in music composition). The third patient was a skilled musician who also composed songs. Tzortzis and colleagues (2000) performed a comprehensive examination of music abilities of a professional pianist and composer with a 9-year history of progressive aphasia. They administered a series of tests that focused on receptive music skills (e.g., pitch error detection, identification of chords and musical styles) and expressive music skills (e.g., playing music from memory, reading an unfamiliar musical score). They found a consistent preservation of both receptive and expressive music skills, and the patient continued to compose music. The famous French composer, Maurice Ravel, is another example of a musician who developed a progressive aphasia and apraxia but was able to compose for a time after the onset of symptoms (Amaducci et al., 2002). The progression symptoms, however, eventually affected his ability to compose, although Ravel reported that he could "hear" the music in his head. Additional studies are needed to better understand the effect of progressive aphasia on music skills.

Parkinson's disease

Only a few studies to date have focused on the perception of music in persons with Parkinson's disease. Most studies have focused on rhythmic auditory cueing (using music) to improve gait in Parkinson's disease (e.g., Thaut et al., 2001). One study examined rhythm discrimination in 15 persons with Parkinson's disease (with Hoehn and Yahr stages 1 or 2) (Grahn and Brett, 2009). The authors administered a series of rhythm recognition tasks with and without a clear beat. The results suggested that the patients with Parkinson's disease had difficulty perceiving the beat structure, compared with matched controls. However, the patients performed similarly to controls in the non-beat conditions. Another study found that patients with Parkinson's disease had difficulty recognizing fear and anger in music (van Tricht et al., 2010). In this study, participants were asked to listen to 32 music excerpts and determine whether they expressed happiness, sadness, fear, or anger. Interestingly, the persons with Parkinson's disease were able to correctly identify happiness and sadness in the music but had more difficulty identifying anger and fear (Fig. 37.3). The authors argued that the patients did not have deficits in the processing of music, as they performed similarly to controls on two subtests from the Montreal Battery for the Evaluation of Amusia (i.e., melodies, rhythm). The difficulty identifying fear and anger in music was consistent with other studies suggesting that persons with Parkinson's disease have difficulty recognizing emotions.

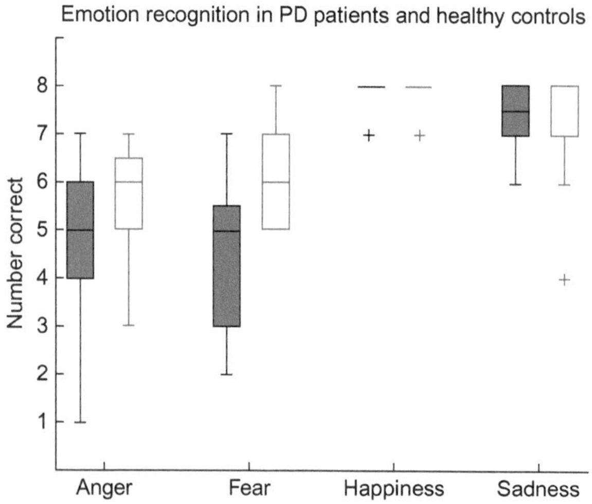

Fig. 37.3. Identification of emotions in music by persons with Parkinson's disease (PD). Box plot of the scores on the four subtests of the emotion recognition task in PD patients (black lines, gray boxes) and controls (gray lines, white boxes). On the happiness subtest all subjects obtained the maximum score, except two subjects (+). For the fear recognition subtask, none of the controls scored below 5, therefore the downward error bar is absent. (Reproduced from van Tricht et al., 2010, with permission.)

Huntington's disease

Woodie Guthrie (1912–1967) is perhaps the most famous musician to have developed Huntington's disease. Guthrie performed during the early stages of the disease, but it eventually affected his music career and ability to sing and play (Arevalo et al., 2001; Innes and Chudley, 2002). His impact on the genre of folk music resulted in his induction into the Rock and Roll Hall of Fame in 1988. The impact of Huntington's disease on the music abilities of Guthrie was never formally studied, but a few studies have examined the processing of music in persons with Huntington's disease. For example, Beste and colleagues (2011) examined the brain response (using functional MRI) to three types of auditory stimulation (music, tones, and syllables) in persons with either premanifest Huntington's disease or manifest Huntington's disease. The music condition included listening to a piano piece by French composer Charles-Valentin Alkan ("Barcarole"). Compared to controls, the persons with Huntington's disease had increased activation in the cerebellum and medial prefrontal cortex and reduced activity in the left parahippocampal gyrus and right fusiform gyrus when listening to a musical composition. In addition, the authors found that the processing of music was correlated with the severity of the movement disorder. Another study used an auditory stimulation paradigm (sequence of tones) and functional MRI methods to examine the basal ganglia-thalamic brain circuits in persons with either premanifest or manifest Huntington's disease (Saft et al., 2008). The authors found reduced activation of basal ganglia-thalamic circuits in premanifest Huntington's disease, but persons with clinically evident Huntington's disease had hyperactivation of the same structures when listening to a sequence of tones. The authors suggested that this pattern reflected functional reorganization of brain areas in an attempt to maintain auditory function.

The basal ganglia has been an area of interest in investigations of sound processing because of its role in attention-dependent temporal processing (Buhusi and Meck, 2005; Langers and Melcher, 2011). In particular, this region interacts with the supplementary motor area and the cerebellum for the temporal processing of sounds Macar et al., 2006; Ivry and Schlerf, 2008; Coull et al., 2011). Apart from these few studies, there does not appear to be any further systematic studies of music processing in persons with Huntington's disease.

Recognition of emotions in music

The link between music and emotions has been debated for centuries. An interest in how humans perceive the emotional content in music or how an emotional response to music is generated has increased in recent years (Juslin and Sloboda, 2010). There are a number of cues that convey emotions in music. For example, a listener may visually perceive the gestures of a performer that reflect emotions. Listeners might also hear musical cues, such as tempo or tonality of a piece, that convey emotional information. For the purposes of this chapter, we will focus on the recognition of emotions while listening to music in persons with dementia.

Several recent studies have examined the recognition of emotions in music in persons with both AD and non-AD dementias. Hsieh and colleagues (2011) studied the recognition of emotions in music with 11 persons with semantic dementia, 12 with AD, and 20 healthy controls. They administered 40 unfamiliar music excerpts that represented four emotions (happy, peaceful, sad, or scary). The participants were asked to listen to the excerpt and select the emotion that best represented the music. Both semantic dementia and AD patients scored lower than controls in identifying the emotions associated with the music, but the semantic dementia patients performed substantially worse than the AD patients. The authors also found that the semantic dementia patients had particular difficulty when identifying negative (sad or scary) compared to positive emotions (happy or peaceful) in the music. In addition, both AD and semantic dementia patients had more difficulty identifying emotions in music than emotions in faces.

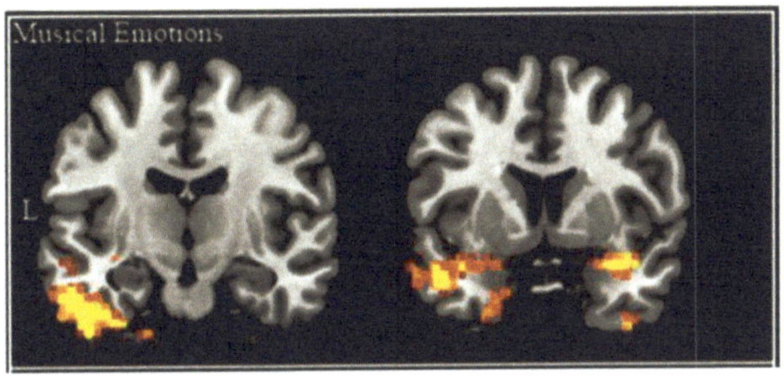

Fig. 37.4. Correlation between performance on music emotions task and brain atrophy. Voxel-based morphometry analyses showing brain regions that correlate with recognition of musical emotions (top left: Montreal Neurological Institute (MNI), $x = -46$, $y = -12$, $z = -46$; top right: MNI, $x = 40$, $y = 6$, $z = -48$). Colored voxels are significant at $P < 0.001$ uncorrected. (Reproduced from Hsieh et al., 2012, with permission.)

A similar pattern was observed in two case studies, one with AD and the other with semantic dementia (Omar et al., 2010). These initial studies suggest that persons with AD and semantic dementia have deficits in recognizing emotions in music when it is just presented as an auditory stimulus.

Hsieh and colleagues (2012) also examined the correlation between the performance on the music emotions task with the amount of atrophy on the brain MRI using voxel-based morphometry. They found that performance on this task was correlated with brain atrophy in the right anterior temporal pole, insula, and amygdala (Fig. 37.4). That is, the patients who had the most difficulty identifying the emotions in the music excerpts had the most atrophy in these areas. Interestingly, there was some overlap between the brain areas involved in perceiving emotions in music and those involved in perceiving emotions in faces. Apart from these few studies with AD and semantic dementia, the recognition of emotions in other types of neurodegenerative disorders is largely unexplored.

PROCESSING OF NON-VERBAL SOUNDS IN NEURODEGENERATIVE DISEASES

In addition to the studies involving music reviewed above, there are a number of other studies that have examined the processing of non-verbal (non-musical) sounds in persons with neurodegenerative diseases. Several authors have provided a framework for categorizing non-verbal sounds that occur in the environment. Taken broadly, there are sounds produced by both living (biologic) and non-living (non-biologic) agents. Within the living category, there are sounds produced by both humans and animals; within the non-living category, there are mechanic and environmental sources of sounds (Griffiths and Warren, 2004; Engel et al., 2009). There have been a number of studies that examine how non-verbal (non-musical) sounds are processed in neurodegenerative diseases.

In one of the early studies about non-verbal sound processing in AD, Rapcsak and colleagues (1989) used a sound–picture-matching task to evaluate the processing of non-verbal sounds in AD. The authors found that patients with mild to severe AD had considerable difficulty selecting the correct picture that corresponded to the non-verbal sound. In contrast, the AD patients had no difficulty selecting the correct picture when the examiner verbalized the names. Other studies have also documented difficulty with selecting pictures to match non-verbal sounds in persons with AD (Eustache et al., 1995; Brandt et al., 2010).

A few studies have examined the processing of environmental sounds in non-AD dementias. For example, Bozeat and colleagues (2000) used 48 sounds from six categories (domestic animals, foreign animals, human sounds, household items, vehicles, and musical instruments) to study non-verbal sound processing in 10 patients with semantic dementia. They asked the patients to match sounds to pictures, sounds to written words, and spoken words to pictures using an array with 10 within-category items. They also administered a neuropsychologic battery of general semantic knowledge. The semantic dementia patients performed below controls on all the tasks involving the environmental sounds. They also found significant correlations between the environmental sounds tasks and verbal tests of semantic knowledge, suggesting that the semantic breakdown in patients with semantic dementia extends to both verbal and non-verbal information.

In another study involving patients with two atypical parkinsonian disorders (corticobasal degeneration, progressive supranuclear palsy), AD, and FTD, Chow and colleagues (2010) examined the naming sounds generated from both manipulable (e.g., hammer) and non-manipulable objects (e.g., train). An item was considered manipulable if the action required to make the sound required manual manipulation by the hand. They found that the patients with corticobasal degeneration or progressive supranuclear palsy named significantly fewer sounds than controls or patients with AD or FTD, who typically do not have prominent motor symptoms. In addition, the corticobasal degeneration and progressive supranuclear palsy patients had a disproportionate impairment in the ability to name the sounds from manipulable objects. Interestingly, action naming and verb production are often impaired in patients with corticobasal degeneration and progressive supranuclear palsy (Bak et al., 2006; Cotelli et al., 2006). These findings suggest that the ability to physically manipulate an object is related to the ability to name an object and suggest a link between the auditory system, cognition, and the motor system. Chow and colleagues (2010) also found a correlation between performance on naming sounds of manipulable objects and gray-matter volume on MRI in the left, premotor areas and left dorsolateral prefrontal cortex.

As discussed above, Goll and colleagues (2012a) recently assessed auditory scene analysis, the ability to parse sound sources in the auditory environment, of both verbal and non-verbal sounds in a cohort of 21 persons with AD. The persons with AD performed much worse than matched controls on both auditory scene analysis tasks, but the performance was somewhat attenuated after accounting for visuospatial working memory. They also found that performance correlated with brain volume in several posterior cortical areas, using voxel-based morphometry of brain MRI. The temporal lobe has also been implicated in the central auditory dysfunction of patients with other neurodegenerative diseases. Patients with semantic dementia showed differential activation of the temporal lobe when listening to animal and tool sounds in a functional MRI paradigm (Goll et al., 2012b) (Figs 37.5 and 37.6). The paradigm involved the perceptual processing of spectrotemporally complex but meaningless sounds and for the semantic processing of environmental sound category (animal sounds versus tool sounds). Thus, anterolateral temporal cortical mechanisms may be necessary for the representation and differentiation of sound categories.

The processing of non-verbal auditory signals has also been studied using electrophysiology and other brain-imaging methods in persons with neurodegenerative diseases. For example, event-related potentials (ERPs) can be measured on the scalp (after stimulus onset) and can provide estimates of the activity of specific brain areas involved in the processing of sensory information. With auditory stimuli, electrical potentials can be measured beginning in the brainstem (as with brainstem auditory evoked potential) up to brain networks involved in both primary and secondary auditory cortices (P50, N100, N200) as well as association cortex (N200, P300). These scalp recordings are often recorded (using electroencephalograph (EEG) electrodes) when participants perform a behavioral task. The oddball task is commonly used in electrophysiologic studies of neurodegenerative diseases. The oddball task requires subjects to listen to a sequence of repeating tones (e.g., 1000 Hz) with a constant speed and press a button when they hear an oddball or deviant tone (e.g., 2000 Hz). In this oddball paradigm, the deviant tones elicit a series of electric potentials that can be measured over time using EEG methods.

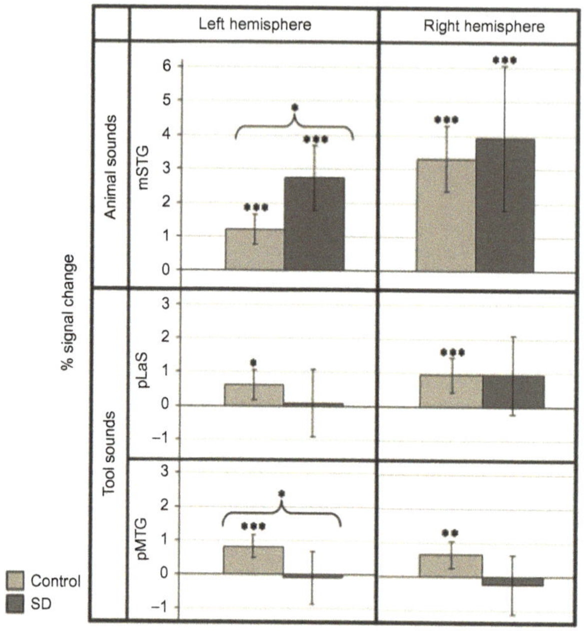

Fig. 37.5. Category-specific contrast effects sampled at previously specified foci of category-specific semantic sound processing. Bars show mean effect sizes (proportionate to percent blood oxygen level-dependent (BOLD) signal change) for the control and semantic dementia (SD) patient groups separately for the category-specific semantic contrast at prespecified foci of category-specific auditory processing (based on Lewis et al., 2005); 95% confidence intervals are also displayed. The upper panels show effects at foci previously associated with animal sound processing in the contrast assessing category-specific semantic processing favoring animal sounds, [(mful_a – mless_a) – (mful_t – mless_t)]; whilst the lower panels show effects at foci previously associated with tool sound processing in the reverse contrast assessing category-specific semantic processing favoring tool sounds, [(mful_t – mless_t) – (mful_a – mless_a)]. Asterisks above bars indicate significance levels for the control and SD groups separately; asterisks above brackets indicate significance levels for between-group comparisons. *$P < 0.05$; **$P < 0.01$; ***$P < 0.001$; mSTG, middle superior temporal gyrus; pLaS, posterior lateral sulcus; pMTG, posterior middle temporal gyrus; SD, semantic dementia. (Reproduced from Goll et al., 2012b, with permission.)

Although the various potentials have been studied in persons with dementia, the P300 brain potential is thought to be important in the study of AD, particularly in the early stages. Again, the majority of studies have been done in persons with AD. Persons with AD have long been known to exhibit prolonged peak P300 latency and decreased P300 amplitude (Polich et al., 1990; Holt et al., 1995; Polich and Pitzer, 1999) compared with healthy controls. The differences in amplitude and latency may be used to help differentiate between healthy individuals and those with preclinical or early

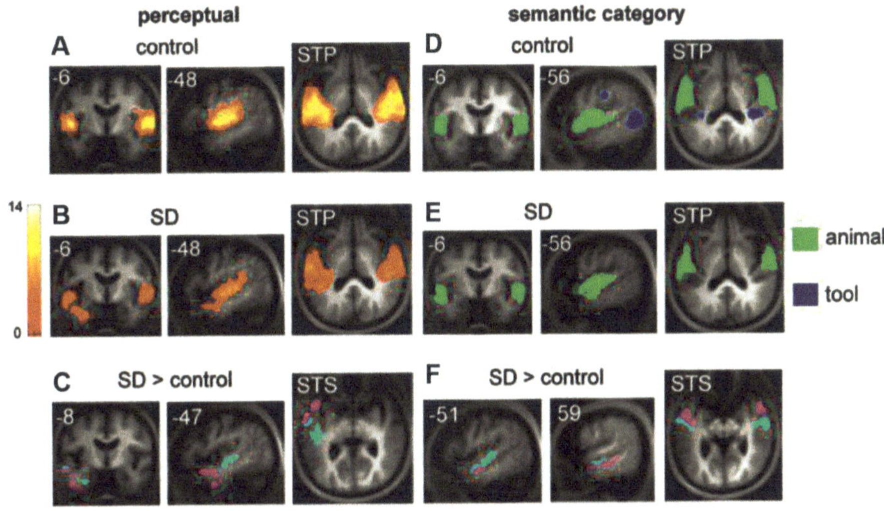

Fig. 37.6. Statistical parametric maps showing activation profiles for perceptual and semantic processing of environmental sounds in healthy controls and patients with semantic dementia. Statistical parametric maps show clusters (formed at whole-brain uncorrected height threshold $P < 0.001$) that are significant at extent threshold $P < 0.05$, family-wise error-corrected for multiple comparisons over the whole brain. Maps are rendered on a composite mean normalized structural brain image; the left hemisphere is shown on the left for all coronal and axial sections. For sagittal and coronal sections the plane is indicated using Montreal Neurological Institute (MNI) coordinates. All axial slices are tilted parallel to the superior temporal plane to show key auditory regions; the anatomic plane of view is indicated. SD, semantic dementia; STP, superior temporal plane; STS, superior temporal sulcus. Panels a and b: the color bar (left) codes voxel-wise T scores for contrast (meaningless sounds > silence). Panel c: all clusters showing a significant interaction with group (patient > control) for the contrast (all meaningless sounds > silence) are depicted in either magenta or cyan. Magenta codes voxels in which controls alone showed greater activation in the reverse contrast (silence > meaningful sounds) than the forwards (meaningless sounds > silence) contrast, indicating that the group interaction within these voxels may be driven by greater activation for controls compared to patients in the reverse contrast. However, remaining voxels, coded in cyan, are likely to be driven by greater activation for patients compared to controls in the forwards contrast. Panels d and e: green codes significant clusters in the contrast assessing the category-specific semantic processing favoring animal sounds, [(mful_a–mless_a) – (mful_t–mless_t)]; blue codes significant clusters in the contrast assessing category-specific semantic processing favoring tool sounds, [(mful_t – mless_t) – (mful_a – mless_a)]. Panel f: all clusters showing a significant interaction with group (patient > control) for the contrast assessing category-specific semantic processing favoring animal sounds are depicted in either magenta or cyan. Magenta codes voxels in which controls alone showed greater activation in the reverse contrast (category-specific semantic processing favoring tool sounds) than the forwards (category-specific semantic processing favoring animal sounds) contrast, indicating that the group interaction within these voxels may be driven by greater activation for controls compared to patients in the reverse contrast; however, remaining voxels, coded in cyan, are likely to be driven by greater activation for patients compared to controls in the forwards contrast. (Reproduced from Goll et al., 2012b, with permission.)

AD (Polich and Pitzer, 1999; Golob et al., 2002, 2007). In another study, Golob and colleagues (2007) found significantly larger P50 and N100 amplitudes (using an auditory oddball task) in amnestic MCI patients who eventually converted to AD, compared with those who remained stable. The P50 and N100 amplitudes were also larger in the MCI converters than those with mild AD, suggesting that these potentials may reflect the neural activity during the transition from MCI to AD. Figure 37.7 shows the P50 and N100 results for each of the groups studied (young adults, older adults, MCI stable, MCI converters, and mild AD). P300 latency was also increased in MCI converters, but the P300 did not differentiate the MCI converters from the mild AD patients.

ERP measures may also be helpful in differentiating dementia etiologies. For example, the P300 latency, amplitude, and topography may differ between individuals with DLB and AD (Bonanni et al., 2010). These measures may also help distinguish between patients with Parkinson's disease with and without cognitive impairment (Nojszewska et al., 2009). Although few in number, ERP methods have also been used to study non-verbal auditory processing and attention in other non-AD dementias, such as Huntington's disease (Uc et al., 2003; Beste et al., 2008), vascular cognitive impairment (Muscoso et al., 2006), and primary progressive aphasia (Onofrjet al., 1994). Thus, auditory potentials may differ according to the degree of cognitive impairment and also help distinguish dementia etiologies.

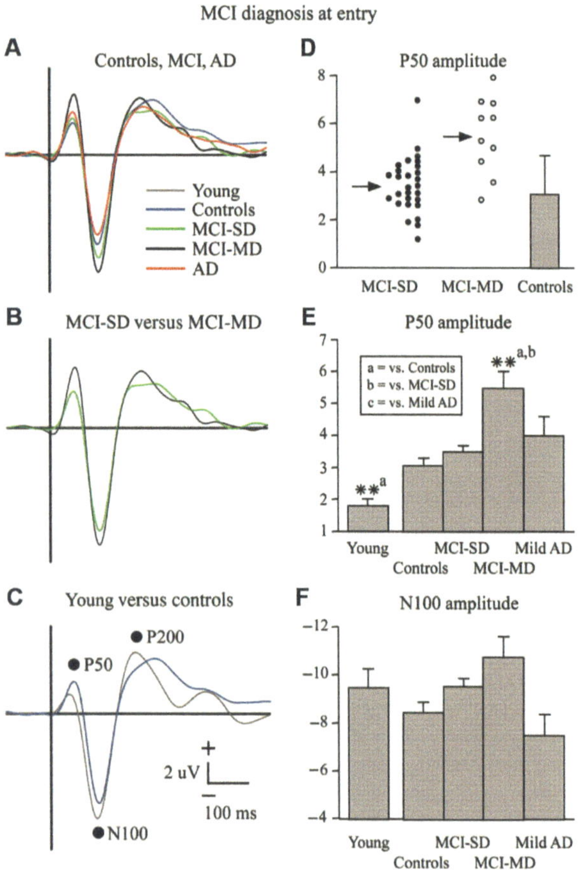

Fig. 37.7. P50 and N100 results for five subject groups. Event-related potentials to non-targets during the baseline session in all older subjects (**A**), mild cognitive impairment (MCI) subtypes (**B**), and young and older controls (**C**). (**D**) P50 amplitudes from individual MCI subjects (MCI single domain (MCI-SD) and MCI multiple domain (MCI-MD)). Mean ± 1 SD from controls are also shown for comparison. Group comparisons of P50 (**E**) and N100 (**F**) amplitudes. Note that in panel F negative potentials are plotted upwards because the N100 is negative in polarity. Vertical lines indicate stimulus onset. Asterisks show *post hoc* tests indicating significant differences between pairs of groups, shown by insert (*$P < 0.05$, **$P < 0.01$). (Reprinted from Golob et al., 2007, with permission.)

MUSIC AS THERAPY FOR DEMENTIA

Music has also been used as a therapeutic intervention for persons with dementia. The music activities range from informal (e.g., background music in assisted-living environments) to more formal approaches, such as with music therapy – defined as the use of music to accomplish specific therapeutic goals within a therapeutic relationship with a board-certified music therapist. Additional information about the use of music with dementia can be found in Clair and Memmott (2008). The early evidence for the positive impact of music on persons with dementia came out of the music therapy and nursing literature in the 1980s. These studies focused on using music activities to manage behavioral symptoms and increase alertness in persons who were in the severe stages of dementia (Norberg et al., 1986; Millard and Smith, 1989). Since these early studies, the number of studies examining the effect of music interventions on persons with dementia has increased steadily. However, it is important to keep in mind that there are concerns about the methodologic quality of many of these studies, as articulated in the most recent Cochrane Collaboration review (Vink et al., 2004), which was updated in 2011 again with the statement, "There is no substantial evidence to support nor discourage the use of music therapy in the care of older people with dementia" (Vink et al., 2011). The studies have been criticized with regard to the research design, small sample sizes, short length of interventions, and presentation of results. Keeping these limitations in mind, however, the section below will review the primary areas of research and promising leads about the effects of music interventions for persons with dementia. There is clearly a need for higher-quality studies about the therapeutic effects of music on persons with dementia, particularly because non-pharmacologic interventions, like music, can be relatively low-cost to implement and easily translated in many different settings. There is also a need to improve the tools for assessing the effect of music on persons with dementia. This area of inquiry remains of high interest, which can be inferred by the publication of several recent reviews (Wall and Duffy, 2010; McDermott et al., 2012; Raglio et al., 2012).

The majority of research about the therapeutic effects of music for persons with dementia has focused on using music to help reduce behavioral and psychologic symptoms, such as agitation, aggression, anxiety, and depression. A number of studies have concluded that music interventions, such as music listening or group music therapy sessions, were associated with reduced agitation in persons with dementia (Groene, 1993; Clark et al., 1998; Gerdner, 2000; Sung et al., 2006; Janata, 2012). It is important to note that the majority of these studies enrolled persons who met criteria for dementia, but most studies did not stipulate specific dementia diagnoses. It is likely that the effect of music on behavioral symptoms will differ depending on the dementia diagnosis. In a recent study with an improved study design, Vink and colleagues (2012) randomly assigned 77 persons with dementia (34% AD, 21% vascular dementia, 35% other or missing) to either 4 months of music therapy (twice weekly) or recreational activities (active control). Both interventions were associated with a reduction in agitation from before to after the sessions (as measured by a modified version of the Cohen-Mansfield Agitation Inventory), but there were no significant differences between the effects of music or recreational activities on agitation.

Several studies have also found a reduction in symptoms of anxiety and depression after various music interventions. For example, Guetin and colleagues (2009) randomly assigned 30 mild to moderately severe patients with AD in assisted living to either 4 months of weekly music therapy sessions or a control group. The patients with AD who completed the music therapy sessions had significantly lower scores on Hamilton Anxiety Scale scores. This effect persisted another 2 months after the music therapy sessions ended. Other studies also report a reduction in anxiety and depression after various music interventions (Ashida, 2000; Han et al., 2010; Sung et al., 2010, 2012).

Several studies have also attempted to examine the possible effect of music interventions on cognitive function in persons with dementia. For example, one study found improved spontaneous speech content and language fluency (as measured by the Western Aphasia Battery speech subscale) in a group of mild to severely impaired persons with dementia following 2 months of music therapy, compared to conversation sessions. Another study examined the effect of a 12-week Sound Training for Attention and Memory and Dementia (STAM-Dem) program on persons with mild to severe dementia compared to a usual-care group (Ceccato et al., 2012). The authors found that the group with STAM-Dem training had significantly better pre–posttest scores on attention and prose memory but not the test of global cognition (Mini-Mental State Examination). Other studies also document a lack of improvement on the Mini-Mental State Examination after various music interventions (Groene, 1993; Raglio et al., 2008; Guetin et al., 2009). This is not surprising because it is unlikely that global tests of cognition are sensitive enough to detect possible effects of music on cognition. It is difficult to know at this stage whether or not music interventions are an effective tool for managing behavioral symptoms of dementia, in particular, AD. Despite the limitations of the studies, there appear to be promising trends that suggest that music may be an effective therapeutic tool, and additional studies are needed.

Music-based movement therapy (Thaut et al., 1996) has also been used to help improve gait and balance in persons with Parkinson's disease. This type of intervention is based on the premise that the temporal structure in music helps provide meaningful auditory cues that facilitate synchronization of movement, such as gait (Madison et al., 2011). A recent meta-analysis of six randomized controlled trials involving persons with Parkinson's disease concluded that music-based movement therapy had positive effects on several motor outcomes (i.e., Berg Balance Scale, Timed Up and Go, and stride length) but not the Unified Parkinson's Disease rating motor scale (de Dreu et al., 2012). These results suggest that additional research is needed to explore the possible benefit of music-based movement therapy for persons with Parkinson's disease.

SUMMARY

This chapter provided a broad overview of hearing in persons with neurodegenerative diseases, with a particular focus on the auditory processing of music as one type of auditory stimulus. The literature suggests that the processing of music information differs by dementia etiology. Impairments in the central auditory system are common in persons with various neurodegenerative diseases. A number of case reports document a relative preservation of music abilities in persons with AD, particularly for playing a musical instrument, processing of basic aspects of music, and making judgments about familiar melodies. However, persons with AD have difficulty with the short-term memory for music excerpts. Studies with persons with semantic dementia suggest that aspects of music knowledge may remain preserved, although they may have difficulty processing familiar melodies. These studies are few and most include only musically trained individuals. Music-based movement therapy may improve some motor function for persons with Parkinson's disease, while it appears that they have difficulty with rhythm perception or recognizing some emotions in music. The processing of auditory stimuli in persons with Huntington's disease appears to engage brain circuits involved in movement. The evidence for the positive effect of music on managing behavioral symptoms in AD is encouraging, but higher-quality studies are needed. It is important to keep in mind that performance on various music tasks can be affected by the type and severity of dementia. This often makes comparisons across studies difficult. Performance on music tasks can also be affected by the amount of prior music training. The majority of case studies focus on persons who are either professional or amateur musicians, while many of the group studies include persons with different levels of music background. The studies about music processing in neurodegenerative diseases are also helping improve the understanding about how the brain processes music information.

REFERENCES

Aarsland D, Andersen K, Larsen JP et al. (2003). Prevalence and characteristics of dementia in Parkinson disease: an 8-year prospective study. Arch Neurol 60 (3): 387–392.

Alzheimer's Association (2012). 2012 Alzheimer's disease facts and figures. Alzheimers Dement 8 (2): 131–168.

Amaducci L, Grassi E, Boller F (2002). Maurice Ravel and right-hemisphere musical creativity: influence of disease on his last musical works? Eur J Neurol 9 (1): 75–82.

American Psychiatric Association (2000). Diagnostic and Statistical Manual of Mental Disorders, text revision, 4th ed. American Psychiatric Association, Washington, DC.

Ames D, Burns A, O'Brien J (2010). Dementia. CRC Press, Boca Raton, Fl.

Arevalo J, Wojcieszek J, Conneally PM (2001). Tracing Woody Guthrie and Huntington's disease. Semin Neurol 21 (2): 209–223.

Armstrong MJ, Litvan I, Lang AE et al. (2013). Criteria for the diagnosis of corticobasal degeneration. Neurology 80 (5): 496–503.

Ashida S (2000). The effect of reminiscence music therapy sessions on changes in depressive symptoms in elderly persons with dementia. J Music Ther 37 (3): 170–182.

Bak TH, Yancopoulou D, Nestor PJ et al. (2006). Clinical, imaging and pathological correlates of a hereditary deficit in verb and action processing. Brain 129 (Pt 2): 321–332.

Baloyannis SJ, Mauroudis I, Manolides SL et al. (2011). The acoustic cortex in frontotemporal dementia: a Golgi and electron microscope study. Acta Otolaryngol 131 (4): 359–361.

Barquero S, Gomez-Tortosa E, Baron M et al. (2010). Amusia as an early manifestation of frontotemporal dementia caused by a novel progranulin mutation. J Neurol 257 (3): 475–477.

Bartlett JC, Halpern AR, Dowling WJ (1995). Recognition of familiar and unfamiliar melodies in normal aging and Alzheimers-disease. Mem Cognit 23 (5): 531–546.

Beatty WW (1988). Preserved musical skill in a severely demented patient. Int J Clin Neuropsychol 10: 158–164.

Beatty WW (1999). Preserved cognitive skills in dementia: implications for geriatric medicine. J Okla State Med Assoc 92 (1): 10–12.

Beatty WW, Greiner F (1998). Preserved musical and game-playing skill in dementia. Psychol Beitr 40: 66–84.

Beatty WW, Zavadil KD, Bailly RC et al. (1988). Preserved musical skill in a severely demented patient. Int J Clin Neuropsychol 10: 158–164.

Beatty WW, Winn P, Adams RL et al. (1994). Preserved cognitive skills in dementia of the Alzheimer type. Arch Neurol 51 (10): 1040–1046.

Beatty WW, Brumback RA, Vonsattel JP (1997). Autopsy-proven Alzheimer disease in a patient with dementia who retained musical skill in life. Arch Neurol 54 (12): 1448.

Beatty WW, Rogers CL, Rogers RL et al. (1999). Piano playing in Alzheimer's disease: longitudinal study of a single case. Neurocase 5: 459–469.

Beste C, Saft C, Gunturkun O et al. (2008). Increased cognitive functioning in symptomatic Huntington's disease as revealed by behavioral and event-related potential indices of auditory sensory memory and attention. J Neurosci 28 (45): 11695–11702.

Beste C, Schuttke A, Pfleiderer B et al. (2011). Music perception and movement deterioration in Huntington's disease. PLoS Curr 3, RRN1252.

Biglan KM, Zhang Y, Long JD et al. (2013). Refining the diagnosis of Huntington disease: the PREDICT-HD study. Front Aging Neurosci 5: 12.

Boeve BF, Geda YE (2001). Polka music and semantic dementia. Neurology 57 (8): 1485.

Bonanni L, Franciotti R, Onofrj V et al. (2010). Revisiting P300 cognitive studies for dementia diagnosis: early dementia with Lewy bodies (DLB) and Alzheimer disease (AD). Neurophysiol Clin 40 (5–6): 255–265.

Bozeat S, Lambon Ralph MA, Patterson K et al. (2000). Nonverbal semantic impairment in semantic dementia. Neuropsychologia 38 (9): 1207–1215.

Brandt J, Bakker A, Maroof DA (2010). Auditory confrontation naming in Alzheimer's disease. Clin Neuropsychol 24 (8): 1326–1338.

Brattico E, Tervaniemi M, Naatanen R et al. (2006). Musical scale properties are automatically processed in the human auditory cortex. Brain Res 1117 (1): 162–174.

Buhusi CV, Meck WH (2005). What makes us tick? Functional and neural mechanisms of interval timing. Nat Rev Neurosci 6 (10): 755–765.

Ceccato E, Vigato G, Bonetto C et al. (2012). STAM protocol in dementia: a multicenter, single-blind, randomized, and controlled trial. Am J Alzheimers Dis Other Demen 27 (5): 301–310.

Chance SA, Clover L, Cousijn H et al. (2011). Microanatomical correlates of cognitive ability and decline: normal ageing, MCI, and Alzheimer's disease. Cereb Cortex 21 (8): 1870–1878.

Chow ML, Brambati SM, Gorno-Tempini ML et al. (2010). Sound naming in neurodegenerative disease. Brain Cogn 72 (3): 423–429.

Clair AA, Memmott J (2008). Therapeutic uses of music with older adults, 2nd edn. American Music Therapy Association, Silver Spring, MD.

Clark ME, Lipe AW, Bilbrey M (1998). Use of music to decrease aggressive behaviors in people with dementia. J Gerontol Nurs 24 (7): 10–17.

Cotelli M, Borroni B, Manenti R et al. (2006). Action and object naming in frontotemporal dementia, progressive supranuclear palsy, and corticobasal degeneration. Neuropsychology 20 (5): 558–565.

Coull JT, Cheng RK, Meck WH (2011). Neuroanatomical and neurochemical substrates of timing. Neuropsychopharmacology 36 (1): 3–25.

Cowles A, Beatty WW, Nixon SJ et al. (2003). Musical skill in dementia: a violinist presumed to have Alzheimer's disease learns to play a new song. Neurocase 9 (6): 493–503.

Crystal HA, Grober E, Masur D (1989). Preservation of musical memory in Alzheimer's disease. J Neurol Neurosurg Psychiatry 52 (12): 1415–1416.

Cuddy LL, Duffin J (2005). Music, memory, and Alzheimer's disease: is music recognition spared in dementia, and how can it be assessed? Med Hypotheses 64 (2): 229–235.

de Dreu MJ, van der Wilk AS, Poppe E et al. (2012). Rehabilitation, exercise therapy and music in patients with Parkinson's disease: a meta-analysis of the effects of music-based movement therapy on walking ability, balance and quality of life. Parkinsonism Relat Disord 18 (Suppl. 1): S114–S119.

Drayna D, Manichaikul A, de Lange M et al. (2001). Genetic correlates of musical pitch recognition in humans. Science 291 (5510): 1969–1972.

Dugger BN, Tu M, Murray ME et al. (2011). Disease specificity and pathologic progression of tau pathology in brainstem nuclei of Alzheimer's disease and progressive supranuclear palsy. Neurosci Lett 491 (2): 122–126.

Engel LR, Frum C, Puce A et al. (2009). Different categories of living and non-living sound-sources activate distinct cortical networks. Neuroimage 47 (4): 1778–1791.

Eustache F, Lambert J, Cassier C et al. (1995). Disorders of auditory identification in dementia of the Alzheimer type. Cortex 31 (1): 119–127.

Fornazzari L, Castle T, Nadkarni S et al. (2006). Preservation of episodic musical memory in a pianist with Alzheimer disease. Neurology 66 (4): 610–611.

Gates GA, Mills JH (2005). Presbycusis. Lancet 366 (9491): 1111–1120.

Gates GA, Karzon RK, Garcia P et al. (1995). Auditory dysfunction in aging and senile dementia of the Alzheimer's type. Arch Neurol 52 (6): 626–634.

Gates GA, Anderson ML, Feeney MP et al. (2008). Central auditory dysfunction in older persons with memory impairment or Alzheimer dementia. Arch Otolaryngol Head Neck Surg 134 (7): 771–777.

Gates GA, Anderson ML, McCurry SM et al. (2011). Central auditory dysfunction as a harbinger of Alzheimer dementia. Arch Otolaryngol Head Neck Surg 137 (4): 390–395.

Gennis V, Garry PJ, Haaland KY et al. (1991). Hearing and cognition in the elderly. New findings and a review of the literature. Arch Intern Med 151 (11): 2259–2264.

Gerdner LA (2000). Effects of individualized versus classical "relaxation" music on the frequency of agitation in elderly persons with Alzheimer's disease and related disorders. Int Psychogeriatr 12 (1): 49–65.

Geroldi C, Metitieri T, Binetti G et al. (2000). Pop music and frontotemporal dementia. Neurology 55 (12): 1935–1936.

Goll JC, Crutch SJ, Loo JH et al. (2010a). Non-verbal sound processing in the primary progressive aphasias. Brain 133 (Pt 1): 272–285.

Goll JC, Crutch SJ, Warren JD (2010b). Central auditory disorders: toward a neuropsychology of auditory objects. Curr Opin Neurol 23 (6): 617–627.

Goll JC, Kim LG, Hailstone JC et al. (2011). Auditory object cognition in dementia. Neuropsychologia 49: 2755–2765.

Goll JC, Kim LG, Ridgway GR et al. (2012a). Impairments of auditory scene analysis in Alzheimer's disease. Brain 135 (Pt 1): 190–200.

Goll JC, Ridgway GR, Crutch SJ et al. (2012b). Nonverbal sound processing in semantic dementia: a functional MRI study. Neuroimage 61 (1): 170–180.

Golob EJ, Johnson JK, Starr A (2002). Auditory event-related potentials during target detection are abnormal in mild cognitive impairment. Clin Neurophysiol 113 (1): 151–161.

Golob EJ, Irimajiri R, Starr A (2007). Auditory cortical activity in amnestic mild cognitive impairment: relationship to subtype and conversion to dementia. Brain 130 (Pt 3): 740–752.

Gorno-Tempini ML, Brambati SM, Ginex V et al. (2008). The logopenic/phonological variant of primary progressive aphasia. Neurology 71 (16): 1227–1234.

Gorno-Tempini ML, Hillis AE, Weintraub S et al. (2011). Classification of primary progressive aphasia and its variants. Neurology 76 (11): 1006–1014.

Grady CL, Grimes AM, Patronas N et al. (1989). Divided attention, as measured by dichotic speech performance, in dementia of the Alzheimer type. Arch Neurol 46 (3): 317–320.

Grahn JA, Brett M (2009). Impairment of beat-based rhythm discrimination in Parkinson's disease. Cortex 45 (1): 54–61.

Griffiths TD, Warren JD (2004). What is an auditory object? Nat Rev Neurosci 5 (11): 887–892.

Grimes A (1995). Auditory changes. In: R Lubinski (Ed.), Dementia and Communication, Singular Publishing Group, San Diego, pp. 47–69.

Grimes AM, Grady CL, Foster NL et al. (1985). Central auditory function in Alzheimer's disease. Neurology 35 (3): 352–358.

Groene RW (1993). Effectiveness of music-therapy 1/1 intervention with individuals having senile dementia of the Alzheimers type. J Music Ther 30 (3): 138–157.

Guetin S, Portet F, Picot MC et al. (2009). Effect of music therapy on anxiety and depression in patients with Alzheimer's type dementia: randomised, controlled study. Dement Geriatr Cogn Disord 28 (1): 36–46.

Hailstone JC, Omar R, Warren JD (2009). Relatively preserved knowledge of music in semantic dementia. J Neurol Neurosurg Psychiatry 80 (7): 808–809.

Halpern AR, O'Connor MG (2000). Implicit memory for music in Alzheimer's disease. Neuropsychology 14 (3): 391–397.

Han P, Kwan M, Chen D et al. (2010). A controlled naturalistic study on a weekly music therapy and activity program on disruptive and depressive behaviors in dementia. Dement Geriatr Cogn Disord 30 (6): 540–546.

Head E, Starr A, Kim RC et al. (2006). Relapsing polychondritis with features of dementia with Lewy bodies. Acta Neuropathol 112 (2): 217–225.

Hojo, Kawamata T, Tanaka C et al. (2004). Inflammatory glial activation in the brain of a patient with hereditary sensory neuropathy type 1 with deafness and dementia. Neurosci Lett 367 (3): 340–343.

Holt LE, Raine A, Pa G et al. (1995). P300 topography in Alzheimer's disease. Psychophysiology 32 (3): 257–265.

Hsieh S, Hornberger M, Piguet O et al. (2011). Neural basis of music knowledge: evidence from the dementias. Brain 134 (Pt 9): 2523–2534.

Hsieh S, Hornberger M, Piguet O et al. (2012). Brain correlates of musical and facial emotion recognition: evidence from the dementias. Neuropsychologia 50 (8): 1814–1822.

Idrizbegovic E, Hederstierna C, Dahlquist M et al. (2011). Central auditory function in early Alzheimer's disease and in mild cognitive impairment. Age Ageing 40 (2): 249–254.

Innes AM, Chudley AE (2002). Genetic landmarks through philately: Woodrow Wilson 'Woody' Guthrie and Huntington disease. Clin Genet 61 (4): 263–267.

Ivry RB, Schlerf JE (2008). Dedicated and intrinsic models of time perception. Trends Cogn Sci 12 (7): 273–280.

Janata P (2012). Effects of widespread and frequent personlized music programming on agiation and depression in assisted living facility residents with Alzheimer-type dementia. Music and Medicine 4 (1): 8–15.

Johnson JK, Ulatowska HK (1996). The nature of the tune and text in the production of songs in Alzheimer's type dementia. In: RR Pratt, R Spingte (Eds.), Music Medicine, MMB Music, Saint Louis, MO.

Johnson JK, Chang CC, Brambati SM et al. (2011). Music recognition in frontotemporal lobar degeneration and Alzheimer disease. Cogn Behav Neurol 24 (2): 74–84.

Juslin RN, Sloboda J (2010). Handbook of music and emotion. Oxford University Press, Oxford.

Klein CJ, Botuyan MV, Wu Y et al. (2011). Mutations in DNMT1 cause hereditary sensory neuropathy with dementia and hearing loss. Nat Genet 43 (6): 595–600.

Klein CJ, Bird T, Ertekin-Taner N et al. (2013). DNMT1 mutation hot spot causes varied phenotypes of HSAN1 with dementia and hearing loss. Neurology 80 (9): 824–828.

Krishnamoorthy ES, Prince MJ, Cummings JL (2010). Dementia: a Global Approach, Cambridge University Press, Cambridge.

Kurylo DD, Corkin S, Allard T et al. (1993). Auditory function in Alzheimer's disease. Neurology 43 (10): 1893–1899.

Langers DR, Melcher JR (2011). Hearing without listening: functional connectivity reveals the engagement of multiple nonauditory networks during basic sound processing. Brain Connect 1 (3): 233–244.

Lewis JW, Brefczynski JA, Phinney RE et al. (2005). Distinct cortical pathways for processing tool versus animal sounds. J Neurosci 25: 5148–5158.

Lin FR, Ferrucci L, Metter EJ et al. (2011). Hearing loss and cognition in the Baltimore Longitudinal Study of Aging. Neuropsychology 25 (6): 763–770.

Lin FR, Yaffe K, Xia J et al. (2013). Hearing loss and cognitive decline in older adults. JAMA Intern Med 173 (4): 293–299.

Macar F, Coull J, Vidal F (2006). The supplementary motor area in motor and perceptual time processing: fMRI studies. Cogn Process 7 (2): 89–94.

Madison G, Gouyon F, Ullen F et al. (2011). Modeling the tendency for music to induce movement in humans: first correlations with low-level audio descriptors across music genres. J Exp Psychol Hum Percept Perform 37 (5): 1578–1594.

Mahoney CJ, Rohrer JD, Goll JC et al. (2011). Structural neuroanatomy of tinnitus and hyperacusis in semantic dementia. J Neurol Neurosurg Psychiatry 82 (11): 1274–1278.

Matthews BR, Chang CC, De May M et al. (2009). Pleasurable emotional response to music: a case of neurodegenerative generalized auditory agnosia. Neurocase 15 (3): 248–259.

McDermott O, Crellin N, Ridder HM et al. (2012). Music therapy in dementia: a narrative synthesis systematic review. Int J Geriatr Psychiatry.

McKeith IG, Dickson DW, Lowe J et al. (2005). Diagnosis and management of dementia with Lewy bodies: third report of the DLB Consortium. Neurology 65 (12): 1863–1872.

Millard KA, Smith JM (1989). The influence of group singing therapy on the behavior of Alzheimer's disease patients. J Music Ther 26 (2): 58–70.

Miller BL, Boeve BF (2009). The Behavioral Neurology of Dementia, New York Cambridge University Press.

Miller BL, Boone K, Cummings JL et al. (2000). Functional correlates of musical and visual ability in frontotemporal dementia. Br J Psychiatry 176: 458–463.

Mohr E, Cox C, Williams J et al. (1990). Impairment of central auditory function in Alzheimer's disease. J Clin Exp Neuropsychol 12 (2): 235–246.

Muscoso EG, Costanzo E, Daniele O et al. (2006). Auditory event-related potentials in subcortical vascular cognitive impairment and in Alzheimer's disease. J Neural Transm 113 (11): 1779–1786.

Nojszewska M, Pilczuk B, Zakrzewska-Pniewska B et al. (2009). The auditory system involvement in Parkinson disease: electrophysiological and neuropsychological correlations. J Clin Neurophysiol 26 (6): 430–437.

Norberg A, Melin E, Asplund K (1986). Reactions to music, touch and object presentation in the final stage of dementia. An exploratory study. Int J Nurs Stud 23 (4): 315–323.

Ohm TG, Braak H (1989). Auditory brainstem nuclei in Alzheimer's disease. Neurosci Lett 96 (1): 60–63.

Omar R, Hailstone JC, Warren JE et al. (2010). The cognitive organization of music knowledge: a clinical analysis. Brain 133: 1200–1213.

Onofrj MFT, Malatesta G, Locatelli T (1994). Focal abnormalities of P3 ERPs unveiled in patients with cortical lesions and primary progressive aphasia by average reference recordings. Brain Topogr 6 (4): 311–322.

Peretz I, Coltheart M (2003). Modularity of music processing. Nat Neurosci 6 (7): 688–691.

Peretz I, Champod AS, Hyde K (2003). Varieties of musical disorders. The Montreal Battery of Evaluation of Amusia. Ann N Y Acad Sci 999: 58–75.

Petersen RC (2011). Clinical practice. Mild cognitive impairment. N Engl J Med 364 (23): 2227–2234.

Polich J, Pitzer A (1999). P300 and Alzheimer's disease: oddball task difficulty and modality effects. Electroencephalogr Clin Neurophysiol Suppl 50: 281–287.

Polich J, Ladish C, Bloom FE (1990). P300 assessment of early Alzheimer's disease. Electroencephalogr Clin Neurophysiol 77 (3): 179–189.

Polk M, Kertesz A (1993). Music and language in degenerative disease of the brain. Brain Cogn 22 (1): 98–117, 027.

Quoniam N, Ergis AM, Fossati P et al. (2003). Implicit and explicit emotional memory for melodies in Alzheimer's disease and depression. Ann N Y Acad Sci 999: 381–384.

Raglio A, Bellelli G, Traficante D et al. (2008). Efficacy of music therapy in the treatment of behavioral and psychiatric symptoms of dementia. Alzheimer Dis Assoc Disord 22 (2): 158–162.

Raglio A, Bellelli G, Mazzola P et al. (2012). Music, music therapy and dementia: a review of literature and the recommendations of the Italian Psychogeriatric Association. Maturitas 72 (4): 305–310.

Rapcsak SZ, Kentros M, Rubens AB (1989). Impaired recognition of meaningful sounds in Alzheimer's disease. Arch Neurol 46 (12): 1298–1300.

Rascovsky K, Hodges JR, Knopman D et al. (2011). Sensitivity of revised diagnostic criteria for the behavioural variant of frontotemporal dementia. Brain 134 (Pt 9): 2456–2477.

Respondek G, Roeber S, Kretzschmar H et al. (2013). Accuracy of the national institute for neurological disorders and stroke/society for progressive supranuclear palsy and neuroprotection and natural history in Parkinson plus syndromes criteria for the diagnosis of progressive supranuclear palsy. Mov Disord 28 (4): 504–509.

Rincon F, Wright CB (2013). Vascular cognitive impairment. Curr Opin Neurol 26 (1): 29–36.

Saft C, Schuttke A, Beste C et al. (2008). fMRI reveals altered auditory processing in manifest and premanifest Huntington's disease. Neuropsychologia 46 (5): 1279–1289. http://dx.doi.org/10.1016/j.neuropsychologia.2007.12.002.

Sinha UK, Hollen KM, Rodriguez R et al. (1993). Auditory system degeneration in Alzheimer's disease. Neurology 43 (4): 779–785.

Stewart L, von Kriegstein K, Warren JD et al. (2006). Music and the brain: disorders of musical listening. Brain 129 (Pt 10): 2533–2553.

Strouse AL, Hall 3rd JW, Burger MC (1995). Central auditory processing in Alzheimer's disease. Ear Hear 16 (2): 230–238.

Sung HC, Chang AM, Abbey J (2006). The effects of preferred music on agitation of older people with dementia in Taiwan. Int J Geriatr Psychiatry 21 (10): 999–1000.

Sung HC, Chang AM, Lee WL (2010). A preferred music listening intervention to reduce anxiety in older adults with dementia in nursing homes. J Clin Nurs 19 (7–8): 1056–1064.

Sung HC, Lee WL, Li TL et al. (2012). A group music intervention using percussion instruments with familiar music to reduce anxiety and agitation of institutionalized older adults with dementia. Int J Geriatr Psychiatry 27 (6): 621–627.

Thaut MH, McIntosh GC, Rice RR et al. (1996). Rhythmic auditory stimulation in gait training for Parkinson's disease patients. Mov Disord 11 (2): 193–200.

Thaut MH, McIntosh KW, McIntosh GC et al. (2001). Auditory rhythmicity enhances movement and speech motor control in patients with Parkinson's disease. Funct Neurol 16 (2): 163–172.

Tzortzis C, Goldblum MC, Dang M et al. (2000). Absence of amusia and preserved naming of musical instruments in an aphasic composer. Cortex 36 (2): 227–242.

U.S. Congress, Office of Technology Assessment (1992). Special care units for people with Alzheimer's and other dementias, U.S. Government Printing Office, Washington, D.C.

Uc EY, Skinner RD, Rodnitzky RL et al. (2003). The midlatency auditory evoked potential P50 is abnormal in Huntington's disease. J Neurol Sci 212 (1–2): 1–5.

Uhlmann RF, Larson EB, Rees TS et al. (1989). Relationship of hearing impairment to dementia and cognitive dysfunction in older adults. JAMA 261 (13): 1916–1919.

van Tricht MJ, Smeding HM, Speelman JD et al. (2010). Impaired emotion recognition in music in Parkinson's disease. Brain Cogn 74 (1): 58–65.

Vanstone AD, Cuddy LL (2010). Musical memory in Alzheimer disease. Neuropsychol Dev Cogn B Aging Neuropsychol Cogn 17 (1): 108–128.

Vanstone AD, Cuddy LL, Duffin JM et al. (2009). Exceptional preservation of memory for tunes and lyrics: case studies of amusia, profound deafness, and Alzheimer's disease. Ann N Y Acad Sci 1169: 291–294.

Vink AC, Birks JS, Bruinsma MS et al. (2004). Music therapy for people with dementia. Cochrane Database Syst Rev 3: CD003477.

Vink M, Bruinsma MS, Scholten RJ (2011). Music therapy for people with dementia. Cochrane Database Syst Rev 3: 1–47.

Vink AC, Zuidersma M, Boersma F et al. (2012). The effect of music therapy compared with general recreational activities in reducing agitation in people with dementia: a randomised controlled trial. Int J Geriatr Psychiatry 28: 1031–1038.

Vitale C, Marcelli V, Allocca R et al. (2012). Hearing impairment in Parkinson's disease: expanding the nonmotor phenotype. Mov Disord 27 (12): 1530–1535.

Wall M, Duffy A (2010). The effects of music therapy for older people with dementia. Br J Nurs 19 (2): 108–113.

Weinstein BE (2000). Geriatric Audiology, Thieme, New York.

Weinstein J, Koenig P, Gunawardena D et al. (2011). Preserved musical semantic memory in semantic dementia. Arch Neurol 68 (2): 248–250.

Williams-Gray CH, Foltynie T, Brayne CE et al. (2007). Evolution of cognitive dysfunction in an incident Parkinson's disease cohort. Brain 130 (Pt 7): 1787–1798.

Wright A, Dyck PJ (1995). Hereditary sensory neuropathy with sensorineural deafness and early-onset dementia. Neurology 45 (3 Pt 1): 560–562.

Handbook of Clinical Neurology, Vol. 129 (3rd series)
The Human Auditory System
G.G. Celesia and G. Hickok, Editors

Chapter 38

Future advances

GASTONE G. CELESIA[1]* AND GREGORY HICKOK[2]

[1]*Department of Neurology, Loyola University of Chicago; Chicago Council for Science and Technology, Chicago, IL, USA*

[2]*Department of Cognitive Sciences, Center for Language Science and Center for Cognitive Neuroscience, University of California, Irvine, CA, USA*

INTRODUCTION

Advances in knowledge of the function of the auditory system have been particularly astounding in the last 30 years, fueled by technical advances that have allowed us to probe the functioning brain *in vivo*. Neuroimaging technologies (e.g., positron emission tomography, magnetic resonance imaging (MRI), functional MRI (fMRI)) have increased our understanding of brain function, particularly in humans, and allowed detailed diagnosis of neurologic disorders. Technology, from cochlear implants to robot surgery, from molecular biology to neuropharmacology, has improved our ability to alleviate human suffering. But what will the future hold? Making predictions about the future is a tricky business. As Eliot points out, "among all form of mistakes, prophecy is the most gratuitous" (Eliot, 1852). Nevertheless, some educated guesses may be formulated based on the current state of the art.

NEW AND DEVELOPING TECHNOLOGIES

Technologic development is one of the biggest stories in the last 20 years of research in neuroscience and there is every reason to believe that continued development of existing methods and the invention of new technologies will drive the field in the next decade or two. Refinement of fMRI will improve the resolution of cortical mapping while diffusion MRI will allow us to push forward our knowledge of structural connectivity of the whole brain. New techniques, such as Optogenetics (Aston-Jones and Deisseroth, 2013), Brainbow, and CLARITY (Ducros et al., 2011; Chung et al., 2013) will "provide unprecedented images of neural architecture" (Livet et al., 2007; Aston-Jones and Deisseroth, 2013; Chung et al., 2013; Deisseroth, 2014). CLARITY refers to the development of see-through brains (Chung et al., 2013; Shen, 2013; Deisseroth, 2014), making tissue transparent. It offers a three-dimensional view of neural networks, which can be combined with genetically labeling neurons with multiple, distinct color information from a complex system (Chung et al., 2013) (Fig. 38.1) and the ability to probe the "structural and molecular underpinnings of physiological function and disease" (Livet et al., 2007; Ducros et al., 2011; Deisseroth, 2014). As stated by Chung et al. (2013), "CLARITY enables fine structural analysis of clinical samples, including non-sectioned human tissue from a neuropsychiatric-disease setting, establishing a path for the transmutation of human tissue into a stable, intact and accessible form suitable for probing structural and molecular underpinnings of physiological function and diseases."

Molecular fMRI is another promising technique that allows the study of neurotransmitters detectable by MRI. For example, dopaminergic signaling has been studied in rats and it demonstrated the "participation of hypothalamic circuitry in modulating dopamine response" (Lee et al., 2014).

These exciting technologies will improve our understanding of the auditory system.

NEW TECHNOLOGIES ENABLE NEW LARGE-SCALE RESEARCH PROJECTS

Similar to the human genome project of the 1990s, new methods in neuroscience have sparked several large-scale projects aimed at mapping the brain. The National Institutes of Health Brain Research through Advancing Innovative Neurotechnologies (BRAIN) initiative is aimed at getting "a dynamic picture of the brain in

*Correspondence to: Gastone G. Celesia, MD, FAAN, 3016 Heritage Oak Lane, Oakbrook, IL 60523, USA. E-mail: g.celesia@comcast.net

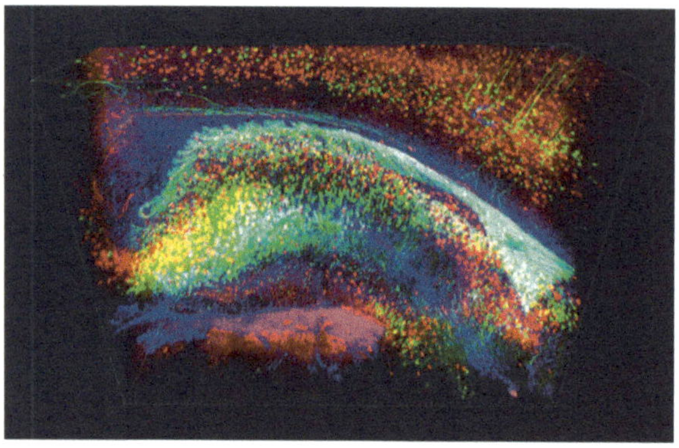

Fig. 38.1. Neurons in an intact mouse hippocampus visualized using CLARITY and fluorescent labeling. Kwanghun Chung and Karl Deisseroth, HHMI/Stanford University. (Reproduced from Shen, 2013.)

action" (Insel et al., 2013). BRAIN promises to expand the Brain Activity Map (BAM) (Alivisatos et al., 2013; Bonilha et al., 2014; Striedter et al., 2014). Another project, the Big Brain, is an ultrahigh-resolution three-dimensional model of a "human brain at nearly cellular resolution of 20 micrometers, based on the reconstruction of 7404 histological sections" (Amunts et al., 2013; Caspers et al., 2014). According to Amunts et al., it "provides a basis for addressing stereotaxic and topological positions in the brain at micrometer range" (Amunts et al., 2013).

The Human Connectome project or Connectomics (Irimia et al., 2012; Toga et al., 2012; Petrella and Doraiswamy, 2013) refers to the mapping of all neuronal connections in the live human brain. As stated by Goni et al. (2014), "Patterns of distributed brain activity are thought to underlie virtually all aspects of cognition and behavior" and, by mapping the connectivity patterns, it is hoped to achieve a better understanding of how these distributed patterns of activity give rise to human behavior. The Enhancing NeuroImaging Genetics through Meta-Analysis (ENIGMA) project is a consortium of many scientists looking at a large-scale collaborative analysis of neuroimaging and genetic data (Jahanshad et al., 2013). Its aim is to detect genetic variants that influence the brain. For example, Bakken et al. (2012) have shown that the common genetic variants in GPCPD1 gene "contribute to the proportional area of human visual cortex."

While large-scale neuroscience mapping studies are impressive and important, it is equally important to recognize that a detailed, high-resolution picture of brain structure and functional activity patterns does not, on its own, automatically reveal how the brain works, any more than the successful completion of the human genome project unraveled all the mysteries of how our genetic endowment builds a human. The key to a deeper understanding of the brain generally, and the auditory system in particular, is uncovering the nature of the computations and information-processing systems that neural circuits support, how they evolve developmentally, and how they can be modified by learning or disease. For this we need a large-scale "Human Cognome" project (Horn, 2002) aimed at mapping the information-processing systems in the mind to parallel the large-scale neuroscience mapping projects.

A combination of molecular genetics and advanced neuroimaging techniques will certainly be fruitful. Some programs, such as Connectome genetics attempts to discover how genetic factors affect brain connectivity," are presently underway (Thompson et al., 2013).

APPLICATIONS TO DISEASE

New developments in neuroscience are already having an impact on treatment. Research sparked by the Human Connectome project has been applied to Alzheimer disease and mild cognitive impairment (Reijmer et al., 2013), as well as autism spectrum disorders, in which Goch et al. (2014) report that "the network centrality of Wernicke's area is significantly" reduced, suggesting that some brain disorders may be related to defective connectivity.

Gene therapy is a particularly promising and potentially game-changing method for audition that could be facilitated by the ENIGMA project. For example, the discovery that Atoh1, a gene also known as Math1, is a "key regulator of hair cell development and induces regeneration of hair cells" (Izumikawa et al., 2005; Staecker et al., 2011) led to trials in guinea pigs and mice, showing that injection of Atoh1 into the cochlea improves hearing thresholds in the mature deaf inner ear (Fig. 38.2). A trial on humans is presently scheduled, with the hope of restoring "hearing to those that have lost it" (Thomson, 2014). Other potential strategies to repair inner-ear damage are stem-cell therapy and molecular therapy (Kesser and Lalwani, 2009; Geleoc and Holt, 2014).

Other technologic advances have greatly improved digital hearing aids and recently nanotechnology has demonstrated the possibility of extracting energy from the biologic battery of the inner ear (Mercier et al., 2013). These authors have shown that the endochochlear potential can generate energy that can be utilized by a microchip (Fig. 38.3). It may be possible in the future to utilize this inner-ear "battery" for therapy of hearing disorders and specifically to energize "cochlear implants and stapes prosthesis" (Mercier et al., 2013).

This is an exciting period for neuroscience, and the next decades will certainly result in stunning improvements in our understanding and treatment of hearing disorders.

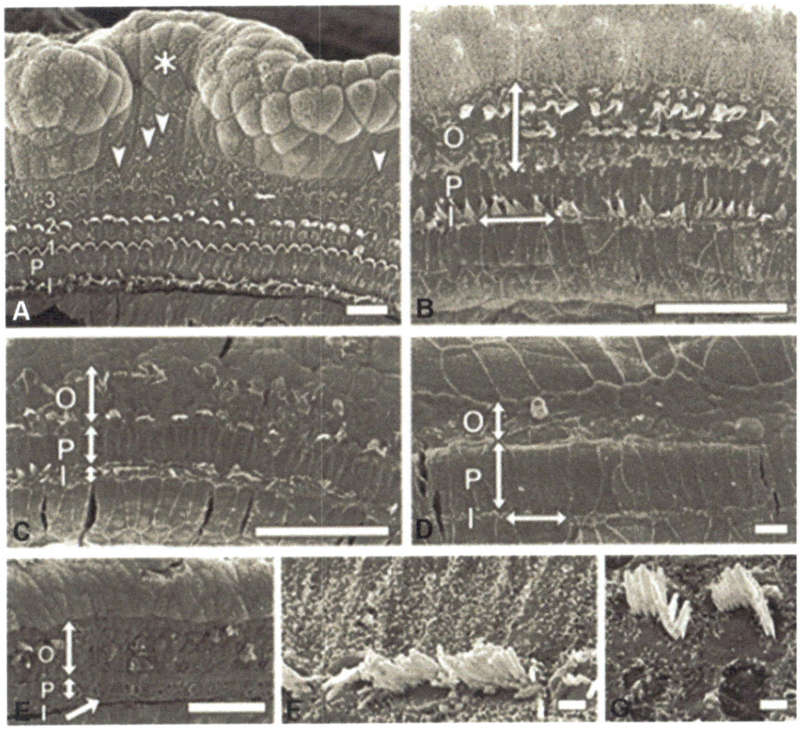

Fig. 38.2. Hair cells reappear in deaf ears treated with Ad.Atoh1. Scanning electron micrograph view of deafened cochlea 2 months after Atoh1 inoculation (**A–C**), contralateral cochlea (**D**), Ad.empty-inoculated ear (**E**), and higher magnification of inner hair cell (IHC) (**F**) and outer hair cell (OHC) (**G**) 2 months after Atoh1 inoculation. (**A**) The site of inoculation in the second cochlear turn (asterisk) is shown along with numerous stereocilia bundles at the normal sites of IHCs (I) and OHCs (rows 1–3). Pillar cells (P) are present between IHCs and OHCs. Ectopic bundles (arrowheads) are seen lateral to the third row of OHCs. (**B**) In some Atoh1-treated ears the morphology of IHCs (I) and OHCs (O) is less well differentiated. (**C**) In other Atoh1-treated ears, air cell reappearance is incomplete and third-row OHCs are missing. (**D, E**) Second cochlear turns of right (**D**, contralateral to **A**) or Ad. empty-inoculated left cochlea (**E**), showing complete absence of hair cells. (**F, G**) Stereocilia bundle organization is relatively normal in IHCs (**F**) and OHCs (**G**) but supporting cells between neighboring air cells are narrow and not well defined. Scale bar, 25 μm in (**A**); 50 μm in (**B**), (**C**), and (**E**);10 μm in (**D**), and 5 μm in (**F**) and (**G**). (Reproduced from Izumikawa et al., 2005.)

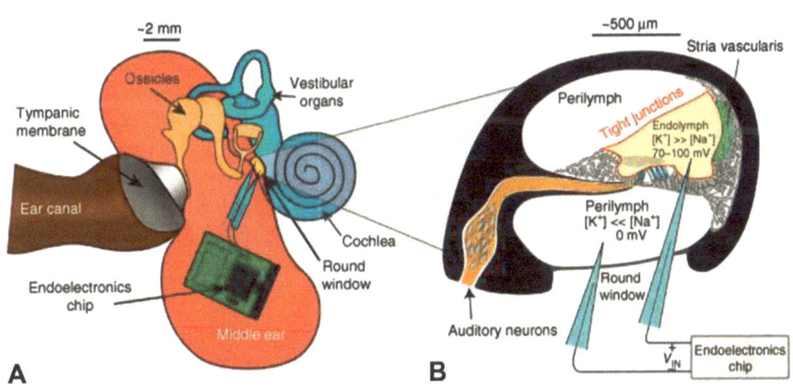

Fig. 38.3. Anatomy and physiology of inner ear. (**A**) Schematic of a mammalian ear, including the external, middle, and inner ear, which contains the cochlea and vestibular end organs. The endoelectronic chip is illustrated in one possible location, although the experiments were done with the chip located outside the middle-ear cavity. (**B**) Cross-section of a typical cochlear half-turn, showing the endolymphatic space (yellow) bordered by tight junctions (red), the stria vascularis (green), and hair cells (blue), which are contacted by primary auditory neurons (orange). (Reproduced from Mercier et al., 2013.)

References

Alivisatos A, Chun M, Church GM et al. (2013). The Brain Activity Map. Science 339: 1284.

Amunts K, Lepage C, Borgeat L et al. (2013). BigBrain: an Ultrahigh-Resolution 3D Human Brain Model. Science 340: 1472–1475.

Aston-Jones G, Deisseroth K (2013). Recent advances in optogenetics and pharmacogenetics. Brain Res 1511: 1–5.

Bakken TE, Roddey JC, Djurovic S et al. (2012). Association of common genetic variants in GPCPD1 with scaling of visual cortical surface area in humans. Proc Natl Acad Sci U S A 109 (10): 3985–3990.

Bonilha L, Nesland T, Rorden C et al. (2014). Asymmetry of the structural brain connectome in healthy older adults. Front Psychiatry 4: 186.

Caspers J, Zilles K, Amunts K et al. (2014). Functional characterization and differential coactivation patterns of two cytoarchitectonic visual areas on the human posterior fusiform gyrus. Hum Brain Mapp 35 (6): 2754–2767.

Chung K, Wallace J, Kim SY et al. (2013). Structural and molecular interrogation of intact biological systems. Nature 497 (7449): 332–337.

Deisseroth K (2014). Circuit dynamics of adaptive and maladaptive behaviour. Nature 505 (7483): 309–317.

Ducros M, van 't Hoff M, Evrard A et al. (2011). Efficient large core fiber-based detection for multi-channel two-photon fluorescence microscopy and spectral unmixing. J Neurosci Methods 198 (2): 172–180.

Eliot G (1852). Middlemarch, Collins' Clear-Type Press, London, p. 98.

Geleoc GS, Holt JR (2014). Sound strategies for hearing restoration. Science 344 (6184): 596.

Goch CJ, Stieltjes B, Henze R, Hering J, Poustka L, Meinzer HP, Maier-Hein KH (2014). Quantification of changes in language-related brain areas in autism spectrum disorders using large-scale network analysis. Int J Comput Assist Radiol Surg 9 (3): 357–365.

Goni J, van den Heuvel MP, Avena-Koenigsberger A et al. (2014). Resting-brain functional connectivity predicted by analytic measures of network communication. Proc Natl Acad Sci U S A 111 (2): 833–838.

Horn RE (2002). Beginning to conceptualize the human cognome project, Manuscript, Stanford University. Available online at, http://www.stanford.edu/~rhorn/a/topic/cognom/artclCncptlzHumnCognome.pdf.

Insel T, Landis SC, Collins FS (2013). The NIH BRAIN Initiative. Science 340: 687–688.

Irimia A, Chambers MC, Torgerson CM et al. (2012). Circular representation of human cortical networks for subject and population-level connectomic visualization. Neuroimage 60: 1340–1351.

Izumikawa M, Minoda R, Kawamoto K et al. (2005). Auditory hair cell replacement and hearing improvement by Atoh1 gene therapy in deaf mammals. Nat Med 11 (3): 271–276.

Jahanshad N, Kochunov PV, Sprooten E et al. (2013). Multisite genetic analysis of diffusion images and voxelwise heritability analysis: a pilot project of the ENIGMA-DTI working group. Neuroimage 81: 455–469.

Kesser BW, Lalwani AK (2009). Gene therapy and stem cell transplantation: strategies for hearing restoration. Adv Otorhinolaryngol 66: 64–86.

Lee T, Cai LX, Lelyveld VS et al. (2014). Molecular-level functional magnetic resonance imaging of dopaminergic signaling. Science 344 (6183): 533–535.

Livet J, Weissman TA, Kang H et al. (2007). Transgenic strategies for combinatorial expression of fluorescent proteins in the nervous system. Nature 450 (7166): 56–62.

Mercier PP, Lysaght AC, Bandyopadhyay S et al. (2013). Energy extraction from the biologic battery in the inner ear. Nat Biotechnol 30 (12): 1240–1243.

Petrella JR, Doraiswamy PM (2013). From the bridges of Konigsberg to the fields of Alzheimer: connecting the dots. Neurology 80 (15): 1360–1362.

Reijmer YD, Leemans A, Caeyenberghs K et al. (2013). Disruption of cerebral networks and cognitive impairment in Alzheimer disease. Neurology 80 (15): 1370–1377.

Shen H (2013). See-through brains clarify connections. Technique to make tissue transparent offers three-dimensional view of neural networks. Nature News 496: 7444.

Staecker H, Praetorius M, Brough DE (2011). Development of gene therapy for inner ear disease: using bilateral vestibular hypofunction as a vehicle for translational research. Hear Res 276 (1–2): 44–51.

Striedter GF, Belgard TG, Chen CC et al. (2014). NSF workshop report: discovering general principles of nervous system organization by comparing brain maps across species. Brain Behav Evol 83 (1): 1–8.

Thompson PM, Ge T, Glahn DC et al. (2013). Genetics of the connectome. Neuroimage 80: 475–488.

Thomson H (2014). Deaf people get gene tweak to restore natural hearing. In: NewScientist, Reed Business Information, New York, pp. 8–9.

Toga AW, Clark KA, Thompson PM et al. (2012). Mapping the human connectome. Neurosurgery 71 (1): 1–5.

Index

NB: Page numbers in *italics* refer to figures and tables.

A

Absolute threshold, outer/middle ear 3–4
Acamprosate 420
Acetylcholine
associative behavioral memory 135–136, *135*, *136*
representational plasticity (RP) 134
Acoustic contour training 551
Acoustic reflex (AR) tests 314–315, *314*
Acoustic stimulus parameters 125–126
Acoustic trauma
acute *410*, 417–418
auditory nerve damage 501
Acquired amusia 607–632
cognitive framework, assessment 610–613, *610*, *611*, *612*
vs congenital amusias 589
future trends *609*, *610*, *611*, *612*, 625–626
musical emotion processing *611*, 622–625, *623*
musical object perception *611*, 619–620, *620*
musical object recognition *611*, 620–622
melody *612*, 620–622, *621*
musical instruments 622
musical property encoding *611*, 613–619
pitch perception *612*, 613–616
temporal perception 616–617, *617*
timbre perception 618–619, *618*
musical scene analysis *611*, 613, *613*, 625
overview 607–608
scope 608–610, *609*, *610*, *611*, *612*
Acquired auditory synesthesias *390*, 399–403
auditory-gustatory 403
auditory-tactile 403
auditory-visual *390*, 399
lesional 399–401, *400*
neurophysiologic studies 400
non-lesional 401
mechanisms 401–403
Acquisitions, signal-to-noise ratio (SNR) 262–263
Activation likelihood estimation (ALE) 188
Acute acoustic trauma *410*, 417–418
Acute experiments 228
Adaptation designs 267–268, *268*
Adaptive auditory analyzer 122
Adaptive Test of Temporal Resolution (ATTR) 320
Adaptive tracking algorithm 57
Affective agnosia, auditory 581–582
Age factors
auditory development 68–69
auditory neuropathy (AN) 501
Age-related hearing loss (ARHL) 357–373
central auditory system 369–370
auditory cortex 369
calcium homeostasis 369–370
cortical disconnection, theory of 369
consequences 371
definitions/terminology 358
diagnosis/clinical manifestation *360*, *361*, 370
epidemiology 358–360, *359*
general aspects 357
peripheral auditory system 364–369, *364*
endocochlear potential *360*, 366–367, *368*
fibrocytes/fibroblast growth factor *362*, 367–368
gender/hormones 368–369
genetics 365
mitochondria 365–366
oxidative stress 364–365, *364*, *365*
rehabilitation 370–371
Schuknecht's classification 360–364
cochlear conductive presbycusis *360*, *361*, 363–364
indeterminant presbycusis 362
mixed presbycusis 362–363
neural presbycusis 360–361, *361*
sensory presbycusis 360, *360*
strial presbycusis 361, *361*, *362*
spiral ligament, atrophy 361, *363*
Agnosia *see* Auditory agnosia
Air conduction audiometry test *484*
Alloacusis 563
Alzheimer disease (AD) *see* Dementia
American Academy of Audiology 539, 545–546
American National Standards Institute (ANSI) 650–651
American Speech-Language-Hearing Association 538, 540, 545–546
Amisulpride 445
Amplification *see* Hearing aids
Amplitude modulation detection 503
Amusia 581
see also Acquired amusia; Congenital amusias
Amygdala, anatomy *36*, 39
Anhedonia *see* Musical emotion processing
Anterior auditory field (AAF) 14, *16*, 17–18, 19, 20, *74*, 110
Anterior ectosylvian sulcus (AES), 19, *74*
Anterior inferior cerebellar artery (AICA) 510, 516, 530, *530*
stroke 633, 634, 635–636, 642–643
Anterior optic pathways 399–400
Anterior parietal cortex, anatomy *29*, 38
Anterior temporal lobe (ATL) 152, 153
Anteroventral cochlear nucleus (AVCN) 102, 103
sound localization 8–10, *8*, *9*
Anticonvulsant drugs 446
Antidepressant drugs 446
Antipsychotic drugs 445–446
Aphasia 157, 459–460, 577
Apparent diffusion coefficient (ADC) 278
Apperceptive auditory agnosia 574–575
verbal (word deafness) 578–579, *579*
see also Musical object perception
Arousal, representational plasticity (RP) 132
Art, synesthesias 404–405
Associative auditory agnosia 574–575
verbal (word deafness) 579–580
see also Musical object perception
Asymmetric sampling in time (AST) theory 94–95, *94*
Attack dimension 191, 192
Attention deficit hyperactivity disorder (ADHD) 544–545
Attractor network models 449–450
Attractor state 450
Audiometry 293, 511–512, *512*, 527–528
Auditory agnosia 573–588
affective 581–582
amusia 581
definitions 573–576, *575*, *576*
environmental sounds 580–581
general 576
overview 573
phonagnosia 580, *580*
treatment *579*, 582–583, *583*

Auditory agnosia *(Continued)*
verbal (word deafness) 574, 576–580, *577*, *578*
apperceptive 578–579, *579*
associative 579–580
Auditory brainstem responses (ABRs) *484*
see also Auditory neuropathy (AN)
Auditory coding
development 59–60
sensorineural hearing loss (SNHL) 484–486
sound, encoding behavioral importance 130, *131*
temporal *see* Temporal coding
Auditory comprehension 140–141
Auditory cortex
age-related hearing loss (ARHL) 369
imaging *see* Functional magnetic resonance imaging (fMRI)
loudness *see* Loudness
music *see* Music perception
primary *see* Primary auditory cortex (PAC)
studies *see* Electrocorticography (ECoG) research
temporal coding *see* Temporal coding
Auditory cortex, anatomy 15–20, *16*, 27–54, *458*
definitions 28, *28*
frequency-band strips 17, 18
hierarchical tiers 225
human vs non-human 41–44
location 41–42
regions/areas 42–44, *42*
organization, principles of 28–41
regions, principle 1 *28*, 29–30, *29*, *30*
areas, principle 2 30–31, *31*
tonotopy, principle 3 17–18, *17*, 31–32, *31*
thalamic inputs, principle 4 32–33, *32*
connection features, principle 5 33–35, *34*
topography, principle 6 *32*, 35–41
single neuron responses 18–19
sound localization 19, 65–66, 110–111
spatial representation *109*, 110–111
stimulus complexity 19–20
Auditory development 55–72
age factors 68–69
behavioral testing/psychoacoustics 56–58, *58*
coding, features 59–60
frequency/intensity 59
loudness 60
sound detection 59
ear 56
masking/segregation 60–64
backward masking 63–64
co-modulation masking release (CMR) 61–62
energetic masking 61
grouping/segregation 60–61
informational masking 62–63
Auditory development *(Continued)*
overview 55–56
spatial/binaural hearing 64–68
binaural unmasking 66–67, *66*
cues 64, *64*
precedence effect 67–68, *67*
sound localization 64–66, *65*
Auditory discrimination tests 543, *545*
Auditory edges 250, *250*
Auditory evoked potentials (AEPs) 326
Auditory feedback control 162, 165–167, *166*
Auditory Figure Ground subtest (SCAN:3A/3C) 323–324
Auditory hallucinations (AHs) 433–456
brainstem-related hearing disorders 521–525, *522*, *525*
classification 434, *435*
conceptualization/demarcation 434
definition 433–434
etiology 437–439, *438*
pathology, absence of 438–439, *439*
metatheoretic considerations 447–450
network science 449–450, *449*, *451*
philosophy of science 447–449
non-verbal, stereotyped, repetitive *410*, 411
overview 433
pathophysiology 439–445, *440*
functional auditory network 440–443, *441*, *442*
processes 444–445
structural auditory network 443–444
phenomenologic characteristics 434–437, *435*
non-verbal auditory hallucinations 436
perceived location 434–437
verbal auditory hallucinations 435–436, *435*
psychotic illness 461–463, *463*
stroke 642–644
treatment 445–447
pharmacotherapy 445–446
non-pharmacologic methods 446–447
Auditory memory 460
Auditory neglect 557–571
alloacusis 563
behavioral deficits 558–565
clinical presentation 557–558
defined 567
extinction 558–559, *558*
dichotic speech tests 559–561, *560*
temporal order 561
neural basis 565–567
functional imaging 565–567, *566*
lesion sites 565, *565*
models of neglect 565
spatial hearing deficits 564–565
non-spatial auditory deficits 562, 569
open questions 568–569
sound lateralization 563–564, *563*
target detection tasks 561–562, *562*
treatment 568
workups 567–568
Auditory neuropathy (AN) 495–508
auditory processing 503–504
brainstem-related hearing disorders *514*, 516, *517*
clinical expression 501–502
diagnosis 496–500
audiology/psychoacoustics 499–500
electrophysiologic tests 293, 496–499, *497*
neuroimaging 499
etiology 500
future trends 505
hearing disorders, associated 502–503
auditory dyssynchrony 496
sound detection thresholds 502
speech perception 502–503, *502*, *503*
overview 495–496, *496*
pathologies 500–501
auditory nerve 500–501, *500*
treatment 504–505
cochlear implants 505
electric stimulation 420
hearing aids 504–505
signal clarity 504
see also under Sensorineural hearing loss (SNHL)
Auditory neuropathy spectrum disorder (ANSD) 337
Auditory objects 670
Auditory pathways 1–26
auditory cortex *see* Auditory cortex, anatomy
brainstem, dorsal auditory stream 11–12
dorsal cochlear nucleus (DCN) 8, *8*, *9*, 11, 13
posteroventral cochlear nucleus (PVCN) 8, *8*, *9*, 11–12
ventral nucleus of lateral lemniscus (VNLL) *10*, 11, 12, *21*
brainstem, ventral auditory stream 8–11
anteroventral cochlear nucleus (AVCN) 8–10, *8*, *9*
lateral superior olive (LSO) 8–9, 10–11, 12, *21*
medial superior olive (MSO) 9–11, *9*, 12, 13, *21*
outputs 10–11, *10*
central auditory nervous system (CANS) 7–8, *8*
centrifugal system 21–22, *21*
cochlea 4–7
anatomy, overall 4, *5*
anatomy, relative to function 4–6, *5*, *6*
output 6–7, *7*
inferior colliculus (IC) 12–13, *12*, 14–15
central (ICC) 12–13, *12*
external/dorsal cortex nucleus (DCN) 13
medial geniculate body (MGB) *10*, 13–15, *21*
anatomy/inputs 14, *14*
medial/dorsal nuclei 15
ventral nucleus 14–15

Auditory pathways *(Continued)*
outer/middle ear 3–4
absolute threshold 3–4
overview 3
Auditory perception
perceived loudness, defined 75
perceptual learning 122
perceptual release theory 442–443
timescales 85–86, *86*
Auditory scene 111–112, 187
analysis (ASA) 60–61, 670
Auditory steady-state responses (aSSR) 251
Auditory task classes *124*, 129, 130
Auditory-motor integration *151*, 153–154, *154*, 155, *155*, *156*
Aura model, auditory hallucinations (AHs) *440*
Autoimmune inner-ear disease *410*, 415
Autophony (voice echo) 410–411, *410*, 460–461, 463
Average evoked potential (AEP) waveforms *237*
Ayahuasca 404

B

Backward masking 63–64
Bastian, Henry Charlton 161
Bayesian random-effect modeling 280
Behavioral matters
habituation/dishabituation 56–57
sound 130, *131*
stimulus generalization gradient 120
techniques 420
testing 56–58, *58*
central auditory 542–547, *545*
training 57–58
Benzodiazepines 418–420
Big Brain model 689–690
Bilingualism 214, *215*
Binaural hearing
discrimination tasks *653*, 660–663, *661*
interaction tests *314*, 316–319, *317*, 543
masking level differences (BMLD) 66–67, 655
multiple sclerosis (MS) 652
unmasking 66–67, *66*, 489
Bioelectric signal averaging techniques 227
Bipolar disorders *438*
Blood oxygenation level-dependent (BOLD)
contrast 259, 260, 261, 262, 265, 270–271
response 264, *264*, 267, 271, 272, 273–274
Blood supply, auditory system 633–634
Blowing tinnitus 410–411, *410*
Bone conduction audiometry test *484*
Borderline personality disorder *438*
Brain Activity Map (BAM) 689–690
Brain regions
comparisons 272, *273*
pitch perception 616
speech processing 178–179, *178*
speech production 163, *163*
Brain Research through Advancing Innovative Neurotechnologies (BRAIN) 689–690
Brainstem
auditory hallucinosis model *440*
auditory pathway, lesions 649–650, *650*, *651*
dorsal auditory stream 10–11
music and language 213, *213*
ventral auditory stream 8–11
Brainstem auditory evoked potentials (BAEPs) 293, 397, 400, 401, 515–516, *515*
intraoperative monitoring 300–301, *301*, *302*
multiple sclerosis (MS) *653*, 657–658, *658*, 660–663, *661*
see also under Electrophysiologic tests
Brainstem strokes *526*, 529–531
hearing loss 636
internal auditory artery, occlusion *511*, *526*, 529, *530*
lateral inferior pontine syndrome *526*, 530–531, *530*
lateral superior pontine syndrome *526*, *530*, 531
sound lateralization 513, *513*
Brainstem-related hearing disorders 509–536
anatomy 509–510, *510*, *511*
auditory hallucinations (AHs) 521–525, *522*, *525*
clinical disorders 525–531, *526*, *527*
evaluation 511–521, *512*
brainstem auditory evoked potentials (BAEPs) 515–516, *515*
neuroimaging 514–515, *514*, *515*
psychoacoustic tests 512–513, *512*, *514*
isolated lesions
auditory nerve *514*, 516, *517*
cochlear nuclei, unilateral lesions 516, *517*
inferior colliculus (IC), bilateral lesion 520–521
inferior colliculus (IC), unilateral lesions 518–519, *519*
lateral lemniscus, unilateral lesions 516–518, *517*
pontomedullary region, bilateral lesion 521
superior olivary complex, unilateral lesions 516, *517*
trapezoid body, unilateral lesions 518
management 531
multiple sclerosis (MS), hearing abnormalities *526*, 531
prevalence 510–511, *512*
British Society of Audiology 538

C

Calcium homeostasis 369–370
Canadian Guidelines on Auditory Processing in Children and Adults (CISGSLPA) 538–539
Canadian Inter-organizational Steering Group for Speech-Language Pathology and Audiology (CISGSLPA) 538–539
Cannabis 404
Carbamazepine 411–412, 420, 427
Carboplatin *483*
Cardiac gating 264–265, *265*
Categoric designs 265–266, *266*
Caudal-rostral axis *34*, 35
Caudolateral area (CL) 110–111
Caudomedial area (CM) 110–111
Central auditory nervous system (CANS)
age-related hearing loss 369–370
dementia 669–671
see also Central auditory processing disorder (CAPD)
Central auditory processing disorder (CAPD) 313–332, 537–556
case study 551–552, *552*, *553*
central auditory tests *314*, 315–325
binaural interaction procedures *314*, 316–319
listening *see* Dichotic listening, test
speech *see* Monaural low-redundancy speech tests
temporal *see* Temporal processing tests
cost-effectiveness 326
definitions/conceptualizations 537–542
behavioral central auditory tests 542–547, *545*
electrophysiologic tests 547–548, *547*
language/learning/communication 540–542, *541*
diagnosis 542–548
differential diagnosis 327, *327*
intervention 548–551, *549*
central resources training 548–550
direct remediation 550–551
environmental modifications 548, *549*
overview 313, 537, *538*
peripheral tests 313–315, *314*
acoustic reflex (AR) 314–315, *314*
otoacoustic emissions (OAEs) 314, *314*
pure-tone thresholds/speech recognition 313–314, *314*
psychophysical tests 326
Central deafness, stroke 637–641, *638*, *639*, *640*
Central inferior colliculus (ICC) 12–13, *12*
Central nervous system (CNS) 401
tinnitus *410*, 415, 417
see also Central auditory nervous system (CANS); Central auditory processing disorder (CAPD)
Central resources training 548–550
Centripetal model, auditory hallucinations (AHs) *440*
Cerebellopontine angle tumors *410*, 417
see also Vestibular schwannomas
Cerebral cortex, new model 141

Change response 249–251
Characteristic frequency (CF) 123
Charcot–Marie–Tooth (CMT) disease 497, *498*, 505
Charles Bonnet syndrome *437*
Children
  auditory neuropathy (AN) *498*, 501
  deaf 68–69
  see also Congenital hearing loss
  language 216
  music 207–208, 209, 212, 589–590
Cholinergic implantation of specific memory 139–140
Chromatic scale 194–195, 197
Chronic acoustic trauma *410*, 415
Chronic experiments 228–232, *230*, *231*, *232*
Chronic progressive symmetric hearing loss *410*, 415
Circle of fifths *195*
Cisplatin *483*
Citalopram 445–446
CLARITY (Clear, Lipid-exchanged, Acrylamide-hybridized Rigid, Imaging/immunostaining compatible, Tissue hYdrogel) 689
Clicking *410*, 411–412
  click trains 236, *236*
Clogged ear 460
Clomipramine 446
Closed-field conditions 99–100
Clozapine 445–446
Coarse intermittent sounds coincident with jaw/head movements 410, *410*
Cochlea, anatomy *see under* Auditory pathways
Cochlear battery 360
Cochlear development 56
Cochlear disorders
  conductive presbycusis *360*, *361*, 363–364
  nuclei, unilateral lesions 516, *517*
  see also Sensorineural hearing loss (SNHL)
Cochlear implants (CI) 61, 87
  auditory neuropathy (AN) 505
  cognition/learning 350–351
  congenital hearing loss 342, 347–348, *348*
  sensorineural hearing loss (SNHL) 491
Cochlear microphonics (CMs) 290–291, *291*, 495
Cochlear summating potential 291–292, *292*
Cochrane Collaboration review, dementia 682
Cocktail-party effect 63
Coding *see* Auditory coding
Cognitive ability
  age and 370
  congenital hearing loss 349–352
Cognitive Ability Screening Instrument 670
Cognitive behavioral therapy 420, 446
Cognitive conjunctions 266–267
Cognitive model of music processing 610
Cognitive subtraction 265
Communication 207–208, 211, 540–542, *541*
Co-modulation masking release (CMR) 61–62
Competing Sentences test 513, 551
Complex musical objects 192–199
Compound action potential (CAP), eighth-nerve 289
Computer-based auditory skills training 551
Concept center 610
Conduction aphasia 169–170
Conductive hearing loss *410*, 416
Conductive presbycusis, cochlear *360*, *361*, 363–364
Congenital amusias 589–605
  implicit pitch processing 600–601
  neural correlates 593–596
    anatomic 593–595, *594*
    functional investigations 595–596
  overview 589–591, *590*
  pitch dimension, deficit 591–593
    memory for pitch *590*, 592–593, *593*
    pitch perception 591–592, *591*
  pitch/music production disorders 598–599
  processing deficits 596–598
    spatial processing 598
    speech processing 596–597, *597*, *598*
    temporal processing 596
    tests 596
Congenital hearing loss 333–356
  audition/speech 344–349
    cochlear implants 347–348, *348*
    literacy development 349
    mild bilateral/unilateral conditions 346–347, *347*
    mild to severe conditions 344–346, *345*
  cognition/learning 349–352
    cochlear implants 350–351
    theory of mind (ToM)/social development 351–352
  current context 342–343
    cochlear implants 342
    expectations 342–343
    neonatal screening 342
  effects on child/family 338–342
    auditory plasticity 338–339, *338*, *339*
    child family experience, impact 340–342, *341*
    language learning, sensitive period 339–340
  neurocognitive development 343–344, *343*
    neurodevelopmental rehabilitation 344
  overview 335
  prevalence/epidemiology 335–338, *336*
    additional disabilities 337–338, *337*
    onset 335–336, *336*
    progressive 336
    severity 337
Congenital hearing loss *(Continued)*
  types *336*, 337
Conjunction analysis 266–267
Connectomics 690
Consonance 193–194, 207–208
  -dissonance judgments 613–614
Contour *191*, 193–197, *195*
Contralateral hemifield spatial tuning 108
Coping-with-voices protocols 446–447
Core-belt-parabelt hierarchic model 34–35, *34*, 225
Co-registration 270
Corollary discharge model *440*
Cortical cooling 239–240, *240*
Cortical deafness 637–641, *638*, *639*, *640*
Cortical disconnection theory 369
Cortical oscillations 91
Cortical potentials *496*, 497–498, *497*, *497*, *498*, 499, *499*
Cost-effectiveness, diagnostic tests 326
Critical bands (CBs) 76, *77*, 79–80
Cross-activation model 397
  /hyperbinding model 398–399

D

Data analysis 232–234, *233*
Data quality 263–265
Deafferentiation model *440*
Deaf-hearing 637
Deafness
  central 637–641, *638*, *639*, *640*
  children 68–69
  word *see* Verbal auditory agnosia; *see also* Age-related hearing loss (ARHL); Congenital hearing loss; Sensorineural hearing loss (SNHL)
Decreased sound tolerance 375–388
  definitions 375–377
  diagnosis 378–380, *379*
  mechanisms 380–383, *382*
  neurophysiologic model 383, *383*
  overview 375
  prevalence/epidemiology 380
  problems 377–378, *378*
  treatment 383–385
    tinnitus retraining therapy (TRT) 383, 384–385
Dementia 667–688
  hearing function 668–671
    central auditory system 669–671
    impairment 668–669
    non-verbal sounds, processing 679–681, *680*, *681*
    peripheral auditory system 669
  music, processing 671–679
    Alzheimer disease (AD) 673–676, *674*
    emotion, recognition 678–679, *679*
    frontotemporal dementia 676–677
    Huntington's disease 678
    Parkinson's disease 677, *678*
    perception, framework for examining 671–673, *671*, *672*
  musical therapy 682–683
  overview 667–668

Dementia with Lewy bodies (DLB) 667–668, 669
Depressive disorder *438*
Desensitization 384
Development *see* Auditory development
Developmental auditory synesthesias 390–399, *390*, *391*, *392*, *393*
with auditory concurrent *393*, 394
with auditory stimuli as inducer 390–393, *391*, *392*
auditory-olfactory/auditory-gustatory 393–394
auditory-visual 390–393
bidirectional 394
drugs, effect of 404
neural models 397, 398–399, *398*, *399*
neurobiologic investigations 394
functional studies 395, *395*, *396*
neurophysiologic studies 395–397, *396*
structural studies 394–395
prevalence/genetics 394
Diagnostic and Statistical Manual of Mental Disorders (American Psychiatric Association) 438–439
Dichotic auditory target detection tasks 561–562, *562*
Dichotic CVs (consonant-vowels) test 315
Dichotic Digits (DD) test 315, 370, 513, 544, 551
Dichotic listening 99–100, 550, 551
test *314*, 315–316
findings 316, *316*
methodology 315–316
Dichotic Rhyme test 551
Dichotic Sentences test 315, 370
Dichotic speech tests *541*, 543, *544*, *545*
Dichotic Words test 315
Diffusion imaging tractography 277–288
applications 280–282, *281*, *282*
auditory pathways, reconstructions 282–285, *283*, *284*, *285*, *286*
deterministic vs probabilistic 282
diffusion tensor imaging (DTI) 279–280, *279*
future directions 285
magnetic resonance imaging (MRI) 278, *278*
non-tensorial models *279*, 280
overview 277
Diffusion orientation distribution function (dODF) 280
Diffusion spectrum imaging 280
Diffusion tensor imaging (DTI) 177, 279–280, *279*
Digital signal processing hearing aids 504–505
Diotic presentation 99–100
Diplacusis (diplacousis) *see* Decreased sound tolerance
Directions Into Velocities of Articulators (DIVA) model 163–169, *163*, *164*, *166*, *167*, 170–171, *170*, 172
Disabilities 337–338, *337*
Disconnection syndrome 577–578
Discretization 89–90
Discrimination
auditory tests 543, *545*
frequency/intensity 59
suppression 107
tasks 196
binaural *653*, 660–663, *661*
interaural (ITD) tests 652–655, *653*, *654*
temporal processing tests 320–321
Disinhibited feedback model 397
Disjunctivism 448
Dissociative disorder *438*
Dissonance 193–194, 207–208
Distance localization 104–105
Distorted Tunes Test 671–673
Distortion product otoacoustic emissions (DPOAES) 495
Dopamine (DA) 136–137
Dorsal cochlear nucleus (DCN)
auditory pathways 13
sound localization 104
stimulus analysis 8, *8*, *9*, 11, 13
Dorsal cochlear nucleus (DCN) hypothesis 420–426
implications 424–426, *425*, *425*
non-lateralized tinnitus 424
tinnitus pathway 426
Dorsal nucleus
of lateral lemniscus (DNLL) *10*, 11, 12, *12*, *21*
of medial geniculate body (MGB) 15
Down syndrome 337
Dual-stream model *see under* Speech perception
Duplex theory 100–101
Duration Pattern Test (DPT) 319–320
Dynamic attending theory 209
Dyslexia 209
Dyssynchrony 496

E

Ear
development 56
'effect' 551–552
mind's 169–170
Early Hearing Detection and Intervention programs 345–346
Early-right anterior negativity (ERAN) 198
Earworms 436–437, *437*
Echo
heard/unheard 68
precedence effect (PE) 67–68, *67*, 107
threshold 107
voice 410–411, *410*, 460–461, 463
Echolalia 463–464
Eighth-nerve compound action potential (CAP) 289, 290–293, *292*, *293*, 294, 295, 296–297
Eighth-nerve disorders *526*
vascular compression *410*, *411*, 413–414, 416
Einstein, A. 278
Electric stimulation 226–227
auditory neuropathy (AN) 420
brainstem 296
focal 238
mapping 226, 239–240, *239*
tract tracing 238
transcranial magnetic stimulation (TMS) 426
Electrocochleography (ECochG) 498–499
see also under Electrophysiologic tests
Electroconvulsive therapy (ECT) 446, 447
Electrocorticography (ECoG) research 223–244, 246
acute experiments 228
chronic experiments 228–232, *230*, *231*, *232*
data analysis 232–234, *233*
functional connectivity 238, *238*
functional lesioning 239–240, *239*, *240*
overview 225–226
spectrotemporal processing 234–238, *235*, *236*, *237*
subjects 227–228, *229*
validity 240–241
Electrode fabrication 227
Electroencephalography (EEG) 246
musicogenic epilepsy 472–474, *472*
Electromagnetic recording 245–256
overview 245–246
techniques 246
see also Magnetoencephalography (MEG)
Electrophysical recording 226
Electrophysiologic responses, reduced *541*
Electrophysiologic tests 289–312
brainstem auditory evoked potentials (BAEPs) 289, 294–302, *294*
frequency-following responses 301–302, *302*
generators 294–296, *295*
interpretation *297*, 298–301
recording techniques *292*, 296–298, *297*
brainstem disorders *512*
electrocochleography (ECochG) 289–293
clinical applications 292–293
components 290–292
middle-latency auditory evoked potentials (MLAEPs) 289, *290*, 302–304, *302*, *303*, *304*, 305
clinical utility 304
components/generators 302–304
long-latency auditory evoked potentials (LLAEPs) 289, *290*, 304–307, *304*
clinical/research applications 307
mismatch negativity (MMN) 305–306, *306*, 307

Electrophysiologic tests *(Continued)*
P1-N1-P2-N2 complex 304–305
P3/P300 component *292*, *293*, 306–307, *307*
overview 289, *290*, *291*
Emotion *see* Musical emotion processing
Endocochlear potential *360*, 366–367, *368*
Enhancing Neuroimaging Genetics through Meta-Analysis (ENIGMA) project 690
Environmental modifications 548, *549*
Environmental sounds 580–581
Epilepsy
non-lesional auditory-visual synesthesia 401
research 227, 228–232
see also Musicogenic epilepsy
Equal-temperament tuning 190
European First Episode Schizophrenia Trial (EUFEST) 445
Event-related band power (ERBP) 233, 236, *237*
Event-related potential (ERP) 600–601
Evoked potentials, auditory 370
Experiential responses 458
Experimental design 265–268
Expertise, musical 199–202, *200*, *201*, 213–214, *214*
Exploding-head syndrome 410, *410*
Extended speech signal 87
External auditory hallucinations (AHs) 434–435
Extinction *see under* Auditory neglect
Extraoperative diagnostic studies 299–300, *299*, *300*, *302*

F

Factorial designs 266–267, *266*
False discovery rate (FDR) 272
Family, effects of childhood hearing loss 338–342
Family-centered practice 341
Family-wise error rate (FWE) 272
Far-field conditions 99–100
Far-field potentials 289
Fear, representational plasticity (RP) 132
FEAT (fMRI Expert Analysis Tool) 473
Feedforward control *164*, 167–169
Fiber orientation distribution function (fODF) 280
Fibroblast growth factor 367–368
Fibrocytes *362*, 367–368
Field strength 261–262
Fixed lateralization 106
Flashbacks 139, 458
Flutter percept 235–236
Fluttering 410, *410*
FM-listening systems 504
Focal cooling 226
Focal electric stimulation 238
Food and Drug Administration (FDA) 348
Forced-choice methods 57
Forward transduction 6
Fractional anisotropy (FA) 279–280, 281–282
Free-field conditions 99–100
Frequency
discrimination 59
Pattern, Test, (FPT) 319–320, 513, 544, 551
perception 487–488, *487*
poor selectivity 488
resolution 4, 503
Frequency-band strips 17, 18
Frequency-following responses 301–302, *302*
recording technique 193–194
Front/back localization 103–104
Frontotemporal dementia (FTD) 667–668, 669–670, 676–677, 679–680
Functional auditory network 440–443, *441*, *442*
Functional connectivity 238, *238*
Functional lesion 226
Functional magnetic resonance imaging (fMRI) 257–276
analysis 268–272
spatial preprocessing 269–270, *269*
statistical *269*, 270–272
applications 257–258
data quality 263–265
cardiac gating 264–265, *265*
scanner acoustic noise 263–264, *264*
experimental design 265–268
adaptation designs 267–268, *268*
categoric designs 265–266, *266*
factorial designs 266–267, *266*
parametric designs *266*, 267
key principles 258–261, *258*
definitions 258–259, *258*
scanning parameters 259–261, *260*, *260*
musicogenic epilepsy 472–474, *473*, *474*
overview 257, *258*
signal-to-noise ratio (SNR) 261–263
acquisitions, number of 262–263
field strength 261–262
radiofrequency coils 262
receiver bandwidth *258*, 263
voxel size 262
Function-based tractography approach 177
Furosemide *483*

G

GABA (gamma-aminobutyric acid) 425, 449–450
GABAergic inhibition 450
Gamma-band oscillations 137, *138*
assessment/treatment 139
Gap detection
tasks 503
threshold (GDT) 320, *541*
Gaps-in-Noise (GIN) measure 320–321
Gedankenlautwerden 435–436, *435*
Gender
age-related hearing loss (ARHL) 368–369
language organization 157
Gene therapy 690
General linear model (GLM) 270–272
Generalization gradients 129
Genetics 365, 394
GingerALE application 188
Goal-directed movement 162–163
Gradient order DIVA (GODIVA) model 163–164, 169
Guthrie, Woodie 678

H

Habit vs memory 127
Hallucinations *see* Auditory hallucinations (AHs); *see also* Musical hallucinations
Hallucinogens 403–404
Haloperidol 445
Harmony *189*, *195*, 197–198
harmonic expectancy violation 198
harmonic intervals 193–194
harmonic sequences 198
Hashish 404
Head trauma *410*, 419
Head-orienting response 65
Head-related transfer functions (HRTFs) 100, 104
Heard echo 68
Hearing, measurement 292–293
Hearing aids 490–491, 504–505
Hearing in Noise Test (HINT) 324
Hearing loss
chronic progressive symmetric *410*, 415
cochlear/auditory nerve disorders *see* Sensorineural hearing loss (SNHL)
stroke *see under* Stroke
sudden, idiopathic *410*, 417
tinnitus 420–421
see also Age-related hearing loss (ARHL); Congenital hearing loss; Deafness; Sensorineural hearing loss (SNHL)
Hearing-motion synesthesia 394
Heisenberg's uncertainty principle 95–96
Hemodynamic imaging *see* Functional magnetic resonance imaging (fMRI)
Hemorrhagic lesions, stroke 636–637, *637*
Hereditary hearing loss *410*, 415
sensorineural (SNHL) 481
Hereditary sensory and autonomic neuropathy with dementia and hearing loss (HSAN1) 669
Herpes zoster oticus (Ramsay Hunt syndrome) *410*, 417
Hierarchical State Feedback Control (HSFC) model 163–164, 169–171, *170*
Hindrance-modulated orientation anisotropy (HMOA) 281–282
Histological Studies on the Localization of Cerebral Function (Campbell) 117
History, synesthesias 404–405
Horizontal localization 100–103

Hormones, age-related hearing loss (ARHL) 368–369
Hughlings-Jackson's Darwinian thesis 442–443
Human Connectome project 690
Huntington's disease 678
Hybrid depth electrodes (HDEs) 227, 229, 230, 234
Hybrid non-tensorial diffusion approach 280
Hyperacusis
  stroke 642–644
  tinnitus 420
  see also Decreased sound tolerance
Hyperacusis Questionnaire (Khalfa's) 379, *379*
Hyperbinding model 398–399
Hypoxia 482

I

Idiopathic conditions
  non-lesional auditory-visual synesthesia 401, *402*
  sudden hearing loss *410*, 417
  tinnitus *410*, 419
Illusion 457–458
Implicit learning 211
Indeterminant presbycusis 362
Individual vs group level analysis *269*, 271–272
Induced auditory synesthesia *390*
Inducer-concurrent couplings 389, *391*, *392*, *393*
Infarction *530*
Inferior colliculus (IC) 245–246
  auditory pathways 12–13, *12*, 14–15
  bilateral lesion 520–521
  unilateral lesions 518–519, *519*
Informational masking 62–63
Information-processing models 610–612
Inner hair cells (IHCs) 382
  ribbon synapses 501
  sensorineural hearing loss (SNHL) 479–480, *480*
Inner-speech model, auditory hallucinations (AHs) *440*
Intensity discrimination 59
Intensity theory 100–101
Interaural level differences (ILDs) 64
  binaural hearing 652, *657*
  binaural unmasking 489
  sound localization 101, 102–103, 105–106
  spatial hearing 488–489
Interaural pitch difference (IPD) 377, 380
Interaural time differences (ITDs) 64
  binaural hearing 652, *657*
  binaural unmasking 489
  brainstem lesions 510–511, 513, 516–518
  sound localization 100–102, *102*, 103, 105–106, 107, 111–112
  spatial hearing 488–489
  tests 652–655, *653*, *654*
    dichotic speech 559–560
Internal auditory artery (IAA) *511*, *526*, 529, *530*, 633
Internal auditory hallucinations (AHs) 434–435
Internal synesthesia 401
International Classification of Functioning, Disability and Health (ICF) (WHO) 343–344, 538–539
Interval *191*, 193–197, *195*
  harmonic 193–194
  melodic 194, *195*
Intraoperative BAEP monitoring 300–301, *301*, *302*
Intraoperative neurophysiologic monitoring (IONM) 293, 294, 297–298, 300
Invasive recordings
  see Electrocorticography (ECoG) research
Ipsilateral Competing Message (ICM) test 323–324
Ischemic stroke *see* Stroke
Isochronous sequence 198–199
Isofrequency lines (frequency-band strips) 17, 18
Isolated auditory objects 188–192
Isolated lesions *see under* Brainstem-related hearing disorders

J

Jaw proprioception 424
Jeffress ITD-versus-place model 102
Jervell and Lange-Nielsen syndrome 481
Just noticeable difference (JND) 652, *657*

K

Kant, I. 448
Key *195*, 197–198
Klotho 365
Knowles electronics manikin for auditory research (KEMAR) 100

L

Lamotrigine 445–446
Landmarks hypothesis (Stevens) 90
Language
  disturbance 459–460
  see also Music and language; Speech perception; Speech processing; Speech production
Lateral inferior pontine syndrome *526*, 530–531, *530*
Lateral lemniscus
  dorsal nucleus (DNLL) *10*, 11, 12, *12*, *21*
  ventral nucleus (VNLL) *10*, 11, 12, *21*
Lateral lemniscus, unilateral lesions 516–518, *517*
Lateral nucleus of trapezoid body (LNTB) 102
Lateral superior olive (LSO) 103, 105
  sound localization 8–9, 10–11, 12, *21*
Lateral superior pontine syndrome *526*, *530*, 531
Lateralization discharge model *440*
  see also Sound lateralization
Learning
  central auditory processing disorder (CAPD) 540–542, *541*
  congenital hearing loss 349–352
  music and language 212
  /plasticity, models 211–212
  rule-based 211–212, *211*
  sensitive period 339–340
Learning and memory 117–148
  contemporary approaches
    see Representational plasticity (RP)
  definitions 118
  general implications 140–141
  overview 117–118
  perceptual learning 122
  research 118–119
  treatment implications 138–140
Left frontal cortex 178–179, *178*
Left parietal cortex 179, *181*
Left temporal cortex 179, *180*, *181*
Limbic mediation model 399
Listening
  in-noise 318–319, 504, 543
  complaints 540
  systems 504
  tactics 504
Listening-in-noise 504
Listening in Spatialized Noise - Sentence (LISN-S) test 318–319, 543
Literacy development 349
LMNA gene 365
Local cooling of the cortex 226
Local field potentials (LFPs) 119, 226, 229
Localization
  of objects 245
  see also Sound localization
Long-latency auditory evoked potentials (LLAEPs) 397
  see also under Electrophysiologic tests
Loudness 73–84
  animal physiology *75*, 76–80, *78*
  auditory development 60
  discomfort levels (LDLs) 375, 378–380, 385
  exposure to 418–419
  human cortical studies *78*, 80–82, *81*
  perception 486, *487*
  psychophysics 75–76, *75*, *77*
  sensory processing, overview 73–75, *74*
Low-pass filtered speech tests (LPFS) *314*, 323, 513
LSD (lysergic acid diethylamide) 404

M

McGurk effect 208
Magnetic resonance imaging (MRI), diffusion 278, *278*

Magnetoencephalography (MEG) 246–252, *247*
auditory steady-state responses (aSSR) 251
neuronal oscillations 251–252, *252*
transient evoked responses 248–249
transition responses 249–251, *250*
Major key 198
Maskers 60
Masking 60–64
backward 63–64
co-modulation masking release (CMR) 61–62
energetic 61
informational 62–63
temporal *314*, 321–322, *321*
Masking level differences (MLDs) 317, *317*, 503–504, 512–513
binaural (BMLD) 66–67, 655
Mean diffusivity (MD) 279–280, 281
Mean length of utterance (MLU) 345–346
Medial geniculate body (MGB) 225, 458–459
anatomy *see under* Auditory pathways
auditory cortex 28, 32–33, 40, 41–42
connections *32*, 33
dorsal nucleus 15
medial nucleus 15
ventral nucleus 14–15
Medial geniculate complex (MGC) *32*
dorsal division (MGd), inputs *32*, 33
magnocellular division (MGm), inputs *32*, 33
ventral division (MGv), projection 33
Medial nucleus
of medial geniculate body (MGB) 15
of trapezoid body (MNTB) 8–9, 10, 12, *21*, 102, 103
Medial superior olive (MSO) 102
sound localization 9–11, *9*, 12, 13, *21*
Medial temporal-lobe epilepsy 475
Medical Research Council (UK), Institute of Hearing Research 264
Medication-related tinnitus *410*, 415–416
Melody *189*, *191*, 193–197, *195*
familiar 197
melodic intervals 194, *195*
melodic intonation therapy 209
recognition *612*, 620–622, *621*
song and 197
Memory
auditory 460
vs habit 127
for pitch *590*, 592–593, *593*
role of 442–443
strength substrate 131, *132*, 139
see also Learning and memory
Ménière's disease 293, 414–415, 416
formes frustes 416
Mescaline 404
Meter *191*, *193*, *195*, 198–199
Midbrain stroke 636
Middle ear, auditory pathways 3–4
Middle-latency auditory evoked potentials (MLAEPs) *see under* Electrophysiologic tests
Middle-latency response (MLR) 547
Migraine 401
Mild cognitive impairment (MCI) 668, 670
Mind's ear 169–170
Minimum audible angle (MAA) 64–65, 67–68, 105–106
thresholds 65, *65*
Minimum audible movement angle (MAMA) 105–106
Minor key 198
Mismatch field/negativity (MMF/MMN) 199–202, 249–251, 397, 566–567, 595, 600–601
long-latency auditory evoked potentials (LLAEPs) 305–306, *306*, 307
paradigm 196, 198
Misophonia *see* Decreased sound tolerance
Mitochondria, age-related hearing loss (ARHL) 365–366
Mixed presbycusis 362–363
Modulations
rates 87
sensitivity to *86*, 88–89, *89*
Monaural conditions
sound localization 103
spectral cues 103
Monaural low-redundancy speech tests *314*, 322–325, *541*, 543, *545*
classification 322–323
Montreal Battery for Evaluation of Amusia (MBEA) 581, 671–673
acquired amusia 612–613, 614
congenital amusias 589–590, 592, 595, 596
Montreal Battery for Evaluation of Musical Abilities (MBEMA) 589–590
Motion aftereffects 106
Motion perception 105–106
Motivation 211
Motivational states/tasks 126
Motor system 208–209
Motor-auditory integration *151*, 153–154, *154*, 155, *155*, *156*
Multidimensional scaling 191
Multiple rule tasks 126–130, *128*, *129*
Multiple sclerosis (MS) 649–666
brainstem lesions 513, *513*, *526*, 531
auditory pathway 649–650
overview 649
psychophysical assessment 650–658
binaural hearing 652
binaural masking level differences (BMLD) 655
brainstem auditory evoked potentials (BAEPs) *653*, 657–658, *658*, 660–663, *661*
interaural discrimination (ITD) tests 652–655, *653*, *654*
lateralization experiments 655–657, *656*, *657*
pure-tone threshold 650–651
Multiple sclerosis (MS) *(Continued)*
speech intelligibility 651–652, *652*
theoretic explanations 658–663
binaural discrimination tasks *653*, 660–663, *661*
lateralization tasks, model 659–660, *659*
tests, correlation 658–659
Multiple-Activity Scale for Hyperacusis (MASH) 379, *379*
Multiple-compartment modeling 280
Multivoxel pattern analysis (MVPA) 196
Music and language 207–222
clinical implications 216, *216*
language, ability/impairment 215–216, *215*
learning/plasticity, models 211–212
rule-based learning 211–212, *211*
neural plasticity 212–216
auditory brainstem 213, *213*
selective enhancement 213–215
origins 207–208
communication 207–208, 211
structural 207–208
overview 207
temporal processing 208–211
oscillatory rhythms 210–211, *211*
rhythm, as temporal map 209–210
Music perception 187–205
complex musical objects 192–199
harmonic sequences 198
harmony/key *189*, *195*, 197–198
interval/contour/melody *189*, *191*, 193–197, *195*
rhythm/meter *189*, *191*, *193*, *195*, 198–199
dementia 671–673, *671*, *672*
disorders *see* Acquired amusia; *see also* Congenital amusias
expertise/plasticity 199–202, *200*, *201*
future directions *191*, *193*, 202
isolated auditory objects 188–192
pitch chroma 188–190
timbre *189*, 190–192, *193*
overview 187–188
auditory scene 187
literature 187–188
neuroimaging 188, *189*, *191*, *193*
Musical emotion processing 202, 209–210
acquired amusia *611*, 622–625, *623*
dementia 678–679, *679*
Musical hallucinations 436–437, *437*, 446
palinacousis 463, 466
pathophysiology 444–445
Musical instruments, recognition 622
Musical object perception *611*, 619–620, *620*
recognition *611*, 620–622
Musical priming paradigm 600
Musical processing, dementia 671–679
Musical production disorder 598–599
Musical property encoding *611*, 613–619
Musical scene analysis *611*, 613, *613*, 625

Musical sound analysis testing 581
Musical therapy, dementia 682–683
Musical training 551
Musicogenic epilepsy 469–478
clinical aspects 470
diagnostic evaluation, seizures 470–471, *471*
epilepsy/music, relevance 474–476
functional imaging 472–474, *472*, *473*, *474*
illustrative case 471–472, *472*
seizures/reflex epilepsies 469–470
Myofascial disorders 428

N

National Institutes of Health (NIH) 689–690
Near-field conditions 99–100
Near-field potentials 289
Neck proprioception 423–424
Neglect *see* Auditory neglect
Neonatal screening 342
Network science 449–450, *449*, *451*
Neural circuitry of language model (Wernicke) 153
Neural excitability *86*, 92
Neural oscillations *see* Oscillatory mechanisms
Neural pitch salience 193–194
Neural plasticity
learning models 211–212
music and language 212–216
musical expertise 199–202, *200*, *201*
Neural presbycusis 360–361, *361*
Neural synchrony 137, *138*
Neurocognitive development 343–344, *343*
Neurodegenerative diseases *see* Dementia
Neurodevelopmental rehabilitation 344
Neurofibromatosis type 2 (NF2) 529, *529*
Neuron responses 18–19
Neuropathy *see* Auditory neuropathy (AN)
N-methyl-D-aspartic-acid (NMDA) receptors 449–450
Noise
listening in 318–319, 504, 543
pink noise therapy 384
sensorineural hearing loss (SNHL) 481–482
see also under Functional magnetic resonance imaging (fMRI)
Non-verbal sounds, processing 679–681, *680*, *681*
Norepinephrine (NA) 137
Notched noise technique 485

O

Observer-based psychophysical method (OBPM) 57
Occipital cortex, anatomy *36*, 37–38
Octave equivalence 188
Octotoxic drugs 482, *483*
Olanzapine 445
Open-field conditions 99–100
Optokinetic stimulation 568
Oral-expressive amusia 598–599
Ordering/sequencing tests *314*, 319
Organ of Corti 4, *5*, *6*
Orientation distribution functions (ODFs) 280
Oscillatory mechanisms *86*, 91–92
dysfunctional 95
gamma-band 137, *138*
magnetoencephalography (MEG) 251–252, *252*
rhythms 210–211, *211*
Otoacoustic emissions (OAEs) 56, 289
age-related hearing loss (ARHL) 370
sensorineural hearing loss (SNHL) 484, *484*
tests 314, *314*
tinnitus *410*, 416
Otoferlin (OTOF) 495, 501, 503
Outer ear, auditory pathways 3–4
Outer hair cells (OHCs) 380–382, *382*, 479
Oxidative stress 364–365, *364*, *365*

P

P1-N1-P2-N2 complex 304–305
P3/P300 component *292*, *293*, 296, 307, *307*
Palilalia 464
Palinacousis 457–468
anatomy 458–460, *459*
associated phenomena 466
musical hallucinations 466
palinopsia 466
auditory memory 460
clinical characteristics 460–461
content 460
latency 461
lesion location 461, *462*
quality 460–461
sound lateralization 461
defining characteristics 457–458
illusion 457–458
perseveration 457
differential diagnosis 461–464
auditory hallucinations (AHs) 461–463, *463*
echolalia 463–464
palilalia 464
postictal psychosis (PIP) 463
tinnitus 464
etiology 466
historical perspective 458, *458*
overview 457
pathophysiology 464–466
ictal *462*, 464, *465*
postictal 465–466
stroke 642–644
Palinopsia 466
Panoramic location coding 108
Paradoxical extinction 560–561
Parallel processing 92–95, *93*, *94*
Parametric designs *266*, 267
Parkinson's disease 669, 677, *678*
Pathways *see* Auditory pathways
Pattern recognition models 488
Pediatric Speech Intelligibility with Ipsilateral Competing Message (PSI-ICM) test 323–324
Pediatric Speech Intelligibility (PSI) test *314*, 324
Pendred syndrome 481
Perception *see* Auditory perception
Peretz model 614, 615
Perilymphatic fistula *410*, 416
Peripheral auditory nervous system
aging 364–369, *364*
dementia 669
tests 313–315, *314*
Perseveration 457
Persistent angular structure MRI 280
Phantom sensations 82
Phase locking 484–485
Phasic alerting 568
Philosophy of science 447–449
Phonagnosia 580, *580*
PING (pyramidal interneuron gamma network) model 92–93
Ping-pong interaction 91–92
Pink noise therapy 384
Pitch
chroma 188–190
classes 187, 194–195, *195*
constancy 196
dimension, deficit 591–593
errors 677
memory *590*, 592–593, *593*
onset response 249–251
production disorder 598–599
shift reflex 599
Pitch perception 487–488, *487*
acquired amusia *612*, 613–616
congenital amusias 591–592, *591*
'pitch percept' 235–236
Pitch processing 614
deficits 614–615
implicit 600–601
Place-frequency mapping 56
Plotting results 272
Polyacusis *see* Decreased sound tolerance
Pontomedullary region, bilateral lesion 521
Posterior auditory field (PAF) *16*, 17–18, 19, 20
sound localization 110
Posterior inferior cerebellar artery (PICA) 633, 634, 635–636, 642–643
Posterior parietal cortex, anatomy *36*, 37
Posteroventral cochlear nucleus (PVCN) 8, *8*, *9*, 11–12
Postictal psychosis (PIP) 463
Postinfectious tinnitus *410*, 419
Post-traumatic stress disorder (PTSD) 139
Precedence effect (PE) 67–68, *67*, 107
Prefrontal and cingulate cortex, anatomy *36*, 37
Presbycusis *410*, 415
see also Age-related hearing loss (ARHL)
Prestin 4–6
Primary auditory cortex (PAC) *16*, 88
anatomy 459, *459*
see also Learning and memory

Principal diffusion direction (PDD) 280–281
Prism adaptation 568
Production-preserved (congenital) amusics 599
Progressive non-fluent aphasia 615
Pseudohallucination 434–435
Psychedelics 403–404
Psychical responses 226–227
Psychoacoustics 56–58, *58*, 499–500
  tests 512–513, *512*, *513*, *514*
    pattern discrimination (PPDT) 561–562
Psychometric functions, measurement 57
Psychophysics 75–76, *75*, *77*
  tests 326
Psychosis (PIP), postictal 463
Psychotherapy 446
Psychotic disorders *438*, 461–463, *463*
PSYRATS questionnaire *451*
Pulsatile (cardiac synchronous) tinnitus *410*, *411*, 412–414, *413*
Pure insertion 265
Pure-tone audiogram 512, *512*
Pure-tone threshold, assessment 313–314, *314*, 650–651

Q

Q-ball imaging 280
Quasi-hallucination 434–435
Quetiapine 445

R

Radial diffusivity (RD) 279–280, 281
Radiofrequency coils 262
Ramsay Hunt syndrome *410*, 417
Random Gap Detection Test (RGDT) 320
Rate place coding 484–485, *485*
Rate-level function (RLF) *78*
Rayleigh, Lord 100–101
Receiver bandwidth *258*, 263
Receptive field
  analysis 120
  shift 117
Re-entrant feedback model 397
Reflex epilepsies 469–470
Region of interest (ROI) analysis 272
Rehabilitation
  age-related hearing loss (ARHL) 370–371
  neurodevelopmental 344
Reinforcement outcome prediction 118–119
Renormalization 127
Reorientation 269
Reperception model *440*
Representational plasticity (RP) 117, 119–122
  acoustic stimulus parameters 125–126
  across species 120–122
  auditory cortical dynamics, specificity 120, *121*
  auditory task classes *124*, 129, 130
Representational plasticity (RP) *(Continued)*
  characteristics *124*, *125*, *126*
  findings, challenges to 131–134
    cortical lesions 133–134
    fear/increased arousal 132
    tuning shifts 132–133
  forms of 122–123, *123*
  functions, primary auditory cortex (PAC) 130–131
  individual vs group analysis 127–129, *130*
  mechanisms 134–138
    circuit/synaptic processes 137–138
    gamma-band oscillations 137, *138*
    neuromodulators 134–137
  motivational states/tasks 126
  multiple rule tasks/learning strategy 126–130, *128*, *129*
  representational gain, reversal/loss 127
  specific memory traces (SMTs) 122, 123–126
  treatment implications 138–140
  two discipline synthesis 119–120
Resolution, temporal processing tests 320–321
Responses to sound, ventral nucleus 15
Resting-state networks (RSNs) 273–274
Reverse transduction 4–6
Rhythm *189*, *191*, *193*, *195*, 198–199
  oscillatory 210–211, *211*
  perception 208
  as temporal map 209–210
Right-ear advantage (REA) 315
Root-mean-square (RMS) error 64, 65, *65*, *67*, 68
Rule-based learning 211–212, *211*

S

Salicylic acid *483*
Salvia divinorum 404
SBUTT (sudden, brief, unilateral, tapering tinnitus) 409–410, *410*, 414, 426
SCAN:3A (Auditory Processing Disorders in Adolescents and Adults) screening test 323–324
SCAN:3C (Auditory Processing Disorders in Children) screening test 323–324
Scanner acoustic noise 263–264, *264*
Schuknecht's classification 360–364
Seashore Tests of Musical Ability 669
  Seashore Rhythm subtest 676
Segregation 60–64
  grouping 60–61
Seizures
  diagnostic evaluation 470–471, *471*
  palinacousis *462*, 464, *465*
  reflex epilepsies 469–470
Semantic specificity 208
Sensitive periods 212
  auditory development 338–339
  language learning 339–340
Sensorineural hearing loss (SNHL) 479–494
  auditory coding 484–486
    auditory nerve damage 486
    cochlear damage 485–486
  brainstem lesions 514, 515, 520–521, 527–528
  defined 358
  diagnosis 483–484, *484*
    auditory nerve dysfunction 484
    cochlear damage 484
    otoacoustic emissions 484, *484*
    threshold measurement 483–484
  etiology 481–483, *482*
    aging 482–483, *483*
    hereditary causes 481
    hypoxia 482
    noise 481–482
    octotoxic drugs 482, *483*
  perceptual consequences 486–490
    cochlear damage 486–490, *487*
  structures affected by 479–481
    auditory nerve 480–481
    inner hair cells 479–480, *480*
    outer hair cells 479
    stria vascularis 480, *481*
  treatment 490–491
    cochlear implants 491
    hearing aids 490–491
Sensory gating 307
Sensory hearing loss, defined 358
Sensory presbycusis 360, *360*
Sensory processing 73–75, *74*
Sequencing tests *314*, 319
Serotonin (5HT) 137, 381
Sex differences
  age-related hearing loss (ARHL) 368–369
  language organization 157
Short-latency auditory evoked potentials *see* Brainstem auditory evoked potentials (BAEPs)
Sign language 338–339
Signal clarity, auditory neuropathy (AN) 504
Signal-to-noise ratio (SNR) 63, 261–263
Singing, poor 589
Single-sided deafness 69
Slice timing 269–270
Social development 351–352
Somatic modulation (SM) 418–419, 421, 423, 426–427
Somatic tinnitus *410*, 418–419, *418*
Somatosensory feedback control 167, *167*
Somatosensory pulsatile tinnitus syndrome *410*, *411*, 412, *413*
Somatosensory system 421, *422*, *422*, 423, *423*
Sound
  encoding behavioral importance 130, *131*
  -field amplification systems 504
  identification 245
  pressure level (SPL) 75

Sound *(Continued)*
recognition *641*
tolerance, decreased *see* Decreased sound tolerance
Sound detection 59
thresholds 502
Sound lateralization
auditory neglect 563–564, *563*
brainstem strokes 513, *513*
multiple sclerosis (MS) 513, *513*, 655–657
palinacousis 461
tasks, model 659–660, *659*
tests *314*, 317–318
Sound localization 99–116
auditory cortex 19
auditory development 65–66, *65*
auditory neuropathy (AN) 504
brainstem 8–11
central representation 107–111
auditory cortex 108–110
specialized cortical areas 110–111
superior colliculus (SC) 107–108
defined 488–489
distance localization 104–105
dominance localization 67–68, 107
horizontal localization 100–103
duplex theory 100–101
interaural level differences (ILDs) 101, 102–103, 105–106
interaural time differences (ITDs) 100–102, *102*, 103, 105–106, 107, 111–112
monaural conditions 103
measuring 64–65
motion perception 105–106
overview 99
precedence effect (PE) 67–68, *67*, 107
spatial hearing mechanisms 111–112
stroke *641*
terms/techniques 99–100
tests *314*, 317–318
vertical and front/back 103–104
Sound Training for Attention and Memory and Dementia (STAR-Dem) 683
Sparse scanning method 264, *264*
Spatial hearing 99, 488–489, *489*
Spatial listening, test *314*, 318–319
Spatial normalization 270
Spatial preprocessing 269–270, *269*
Spatial processing 598
Spatial release from masking (SRM) 66, *66*
Spatial smoothing 270
Specific memory traces (SMTs) 122–130
Spectogram, speech sounds 86–87, *86*
Spectral centroid dimension 192
defined 191
Spectral flux 191
Spectrotemporal features 88–89, 190–192
Spectrotemporal processing 234–238, *235*, *236*, *237*
Spectrotemporal receptive fields (STRFs) 234
Speech
audition and 344–349
error, correction 169–170
intelligibility 651–652, *652*
reception threshold (SRT) 318, 651
recognition test 313–314, *314*
sounds, temporal structure 85, 86–87, *86*
Speech perception 149–160
auditory neuropathy (AN) 502–503, *502*, *503*
dual-stream model 149–156, *150*
auditory-motor integration *151*, 153–154, *154*, 155, *155*, *156*
bilateral organization/parallel computation 149–150, *151*
clinical correlates 156–157
computational asymmetries 150–151
dorsal stream 153–156
lexical-semantic access 152
phonological processing/superior temporal sulcus (STS) 151–152
ventral stream 149–153
hallucinations 442
overview 149
sensitive period 339–340
sensorineural hearing loss (SNHL) 489–490, *490*
sex differences 157
Speech Perception in Noise (SPIN) test 324
Speech processing 177–186
brain regions 178–179, *178*
left frontal cortex 178–179, *178*
left parietal cortex 179, *181*
left temporal cortex 179, *180*, *181*
central auditory processing disorder (CAPD) 540–542, *541*
congenital amusias 596–597, *597*, *598*
language pathways 179–184
dorsal 180–182, *181*, *182*, *183*
ventral 182–184
overview 177–178
Speech production 161–176
brain regions 163, *163*
future directions 171–172
goal-directed movements, planning 162–163, *163*
hallucinations 442
neurocomputational models 163–171
Directions Into Velocities of Articulators (DIVA) model 163–169, *163*, *164*, *166*, *167*, 170–171, *170*, 172
gradient order DIVA (GODIVA) model 169
Hierarchical State Feedback Control (HSFC) model 163–164, 169–171, *170*
overview 161–162
Speech-in-noise 540, *641*
perception 211, *211*
tests *314*, 323–325
Speech-to-noise ratio 491
Spherical deconvolution 280
Spiral ligament, atrophy 361, *363*
Spontaneous otoacoustic emissions (SPOAE) 382–383, 384
S-segment 10
Staggered Spondaic Word test 315
Statistical analysis *269*, 270–272
Statistical inference 272
Stevens, K.N. 90
Stimulation
optokinetic 568
see also Electric stimulation
Stimulus-specific adaptation (SSA) 13
Streaming, auditory development 61
Stress, tinnitus and 418–419
Stria vascularis 480, *481*
Strial presbycusis 361, *361*, *362*
Striatum, anatomy 38
Stroke 633–648
assessment protocol *643*, 645
brainstem *see* Brainstem strokes
current practice 644–645
hearing loss 634–641
central/cortical deafness 637–641, *638*, *639*, *640*
hemorrhagic lesions 636–637, *637*
population studies 634–635, *634*
upper brainstem/midbrain 636
vertebrobasilar area 635–637, *637*
overview 633–634
blood supply, auditory system 633–634
hearing disorders 634
processing deficits 641–642, *641*, *642*
tinnitus/hallucinations/hyperacusis/ palinacousis 642–644
Structural auditory network 443–444
Sudden, brief, unilateral, tapering tinnitus (SBUTT) 409–410, *410*, 414, 426
Sudden idiopathic hearing loss *410*, 417
Sulpiride 445–446
Superconducting quantum interference devices (SQUIDS) 246
Superior cerebellar artery (SCA) 510, 531
Superior colliculus (SC) 107–108
Superior olivary complex, unilateral lesions 516, *517*
Superior semicircular canal dehiscence (SSCD) *410*, 416–417
Superior temporal cortex, anatomy 36–37
Superior temporal sulcus (STS) 151–152
The Synesthesia List 394
Synesthesias 389–408
history/art 404–405
terminology 389
types 389–404
acquired *see* Acquired auditory synesthesias
affectivity 390, *390*
automaticity 389–390
consistency 389
developmental *see* Developmental auditory synesthesias
dynamicity 390
induced 403–404
drug-induced 403–404
mechanisms 404
sensory deprivation 403

Synthetic Sentence Identification with Ipsilateral Competing Message (SSI-ICM) test *314*, 323–324

T

T1 recovery 259
T2 decay 259
Tapping tasks 598
Target detection tasks 561–562, *562*
Technical advances 689–692
  applications 690–691
  new/developing technologies 689, *690*
  research projects, large-scale 689–690
Telomeres 365
Temporal coding 85–98
  continuous sound stream 89–95
    discretization 89–90
    dysfunctional oscillatory sampling 95
    multiple timescales, analysis *86*, 90–91
    neural excitability *86*, 92
    neural oscillations *86*, 91–92
    parallel processing 92–95, *93*, *94*
  overview 85–87
    auditory perception, timescales 85–86, *86*
    speech sounds, temporal structure 85, 86–87, *86*
  poor 488
  sensorineural hearing loss (SNHL) 484–485, 486
  spectrotemporal features 88–89
    modulations, sensitivity to *86*, 88–89, *89*
    temporal modulations, sensitivity 88
Temporal dimension 187
Temporal modulations 235–236
  sensitivity 88
Temporal patterning tests *541*
Temporal perception *612*, 616–617, *617*
Temporal processing 617
  age and 370
  congenital amusias 596
  improving 582
  music and language 208–211
Temporal processing, tests *314*, 319–322, 543, *545*
  integration *314*, 321
  masking *314*, 321–322, *321*
  ordering/sequencing *314*, 319
  resolution/discrimination *314*, 320–321
Temporal structure, regularity/variability 209–210
Temporal-lobe association areas 459
Temporal-lobe epilepsy 226–227
Theory of mind (ToM) 351–352
Thresholds
  absolute 3–4
  -based testing 592
  gap detection (GDT) 320, *541*
  measurement 483–484
  minimum audible angle (MAA) 65, *65*
Thresholds *(Continued)*
  phenomenon 415, 426–427, *427*, 483–484
  pure-tone tests 313–314, *314*
  sound detection 502
  tinnitus 424–425
Timbral dimension 187
Timbre perception 618–619, *618*
Time-compressed speech tests *314*, 325
Time-Compressed Speech with and without Reverberation test 551
Timescales 85–86, *86*, 90–91, 92
Tinnitus 82, 377–378, *378*, 409–432
  characteristics *410*
    non-specific quality 414–419, *414*
    specific quality 410
  etiology 409–419
  hyperacusis 420
  neurology 420–427
    dorsal cochlear nucleus (DCN) hypothesis 420–426
    threshold phenomenon 415, 426–427, *427*
    typewriter tinnitus *411*, 427
  vs palinacousis 464
  stroke 642–644
  treatment 419–420, 446
    retraining therapy (TRT) 383, 384–385
Tonal dimension 187
Tone
  audiogram 370
  duration 126–127
  onset to error (TOTE) strategy 126–127, *129*
Tonotopic maps 17–18, *17*, 31–32, *31*, 123
Tractometry 281–282
Training 550–551, *550*
  behavioral 57–58
  central resources 548–550
Transcortical sensory aphasia 157
Transcranial magnetic stimulation (TMS) 426
Transformation between sounds and actions 245
Transient evoked responses 248–249
Transition responses 249–251, *250*
Trapezoid body (TB)
  lateral nucleus (LNTB) 102
  medial nucleus (MNTB) 8–9, 10, 12, *21*, 102, 103
Trapezoid body, unilateral lesions 518
Travelling wave 75–76
Triggering factors, tinnitus 415
Tufted neurons 14
Tuning shifts 132–133
Turkeltaub non-additive algorithm 188
Typewriter tinnitus *410*, 411–412, *411*

U

Uncertainty principle, Heisenberg's 95–96
Unheard echo 68
Unilateral tonal/noise tinnitus 414
Universal newborn hearing screening (UNHS) programs 335–336, 337, 342
Usher syndrome 481

V

Ventral nucleus
  of lateral lemniscus (VNLL) *10*, 11, 12, *21*
  of medial geniculate body (MGB) 14–15
Ventral posterior field (VP) *16*
Verbal auditory agnosia (word deafness) 156–157, 574, 576–580, *577*, *578*
  apperceptive 578–579, *579*
  associative 579–580
Veridical perceptions 447, 448
Vertebrobasilar ischemic stroke 635–637, *637*
Vertical localization 103–104
Vestibular schwannomas *514*, 525–529, *526*
  diagnostic testing 527–528
  neurofibromatosis type 2 (NF2) 529, *529*
  neuroimaging 528, *528*
  physiopathology 528, *528*
  symptomatology 525–527, *527*
  treatment 529
Vestibular symptoms *410*, 416
Virtual auditory space 100
Visual reinforcement audiometry 56–57
Vocal amusia 598–599
Vocal performance studies 599
Vocoded speech 87
Voice echo (autophony) 410–411, *410*, 460–461, 463
Voice onset time (VOT) 236–238, *237*
Voxel comparisons, multiple *258*, 272
Voxel size 262
Voxel-based-morphometry (VBM) 593–595

W

Waterfall illusion 106
Waveform
  components *297*, 298
  speech sounds 86–87, *86*
Wernicke–Lichtheim model 156
Wernicke's aphasia 157
Wernicke-type aphasia 459–460
White-matter pathways *see* Diffusion imaging tractography; *see also* Speech processing
Williams syndrome 381
Within-subject realignment 269
Word deafness *see* Verbal auditory agnosia
World Health Organization (WHO) 343–344, 358–359, 538–539
Wrong notes (pitch errors) 677

Z

Ziprasidone 445

FSC
www.fsc.org
MIX
Papier | Fördert
gute Waldnutzung
FSC® C083411

Kolibri 360 GmbH
Ferdinand-Jühlke-Straße 7
99095 Erfurt, Deutschland
produktsicherheit@kolibri360.de